JON HICKSON

Magneto-stratigraphy			Chronostratigraphy			(1st order)	(2nd order)	(3rd order)	Inferred eustatic curves
Magnetic anomalies	Polarity	Polarity Chronozones	System	Series	Stages	Sequence	Supercycle sets	Supercycles / Cycles	Time (yr × 10⁶)

Time (yr × 10⁶)

Magnetic anomalies / Polarity Chronozones:
29 C29 / 30 C30 / 31 C31 / 32 C32 / 33 C33 / 34 C34
(CRETACEOUS MAGNETIC QUIET ZONE)
M0 CM0 / M1 CM1 / M2 CM2 / M3 CM3 / M4 CM4 / M5 CM5 / M6 / M7 CM7 / M8 CM8 / M9 CM9 / M10 CM10 / M10N CM10N / M11 CM11 / M12 CM12 / M13 CM13 / M14 CM14 / M15 CM15 / M16 CM16 / M17 CM17

Chronostratigraphy:

System	Series	Stages	Sub	Sequence	Supercycle sets	Supercycles	Cycles
		PALEOCENE — DANIAN		TEJAS (T)	TEJAS A (TA) TA-1		1.2 / 1.1
CRETACEOUS	UPPER	MAASTRICHTIAN (UPPER / LOWER)	SENONIAN			UZA-4	4.5 / 4.4 / 4.3 / 4.2 / 4.1
		CAMPANIAN (UPPER / LOWER)		UPPER ZUNI (UZ)	UPPER ZUNI A (UZA)	UZA-3	3.5 / 3.4 / 3.3 / 3.2 / 3.1
		SANTONIAN (UPPER / LOWER)					
		CONIACIAN					2.7 / 2.6
		TURONIAN (UPPER / MIDDLE / LOWER)				UZA-2	2.5 / 2.4 / 2.3 / 2.2 / 2.1
		CENOMANIAN (UPPER / MIDDLE / LOWER)					
		ALBIAN (UPPER / MIDDLE / LOWER)				UZA-1	1.5 / 1.4 / 1.3 / 1.2 / 1.1
	LOWER	APTIAN (UPPER / LOWER)		LOWER ZUNI (LZ)	LOWER ZUNI B (LZB)	LZB-4	4.2 / 4.1
		BARREMIAN (UPPER / LOWER)				LZB-3	3.5 / 3.4 / 3.3 / 3.2 / 3.1
		HAUTERIVIAN (UPPER / LOWER)	NEOCOMIAN			LZB-2	2.5 / 2.4 / 2.3 / 2.2 / 2.1
		VALANGINIAN (UPPER / LOWER)					1.6 / 1.5
JURASSIC	BERRIASIAN	RYAZANIAN (UPPER / LOWER) / PORTLANDIAN					1.4

Stage boundary ages (Ma): 66.5, 74, 84, 88, 89, 92, 96, 108, 113, 116.5, 121, 128, 131

Inferred eustatic curves — scale: 250 200 150 100 50 0M
LONG TERM / SHORT TERM

Time (yr × 10⁶): 65, 70, 75, 80, 85, 90, 95, 100, 105, 110, 115, 120, 125, 130

PRINCIPLES OF
SEDIMENTARY DEPOSITS

PRINCIPLES OF SEDIMENTARY DEPOSITS

STRATIGRAPHY AND SEDIMENTOLOGY

Gerald M. Friedman

BROOKLYN COLLEGE, GRADUATE SCHOOL OF THE CITY UNIVERSITY OF NEW YORK
AND NORTHEASTERN SCIENCE FOUNDATION

John E. Sanders

HOFSTRA UNIVERSITY AND NORTHEASTERN SCIENCE FOUNDATION

David C. Kopaska-Merkel

BROOKLYN COLLEGE AND NORTHEASTERN SCIENCE FOUNDATION
AND GEOLOGICAL SURVEY OF ALABAMA

Macmillan Publishing Company
NEW YORK

Maxwell Macmillan Canada
TORONTO

Maxwell Macmillan International
NEW YORK   OXFORD   SINGAPORE   SYDNEY

Copyright © 1992 by Macmillan Publishing Company, a division of Macmillan, Inc.

PRINTED IN THE UNITED STATES OF AMERICA

Macmillan Publishing Company
866 Third Avenue, New York, New York 10022

Macmillan Publishing Company is part
of the Maxwell Communication Group of Companies.

Maxwell Macmillan Canada, Inc.
1200 Eglinton Avenue East
Suite 200
Don Mills, Ontario M3C 3N1

LIBRARY OF CONGRESS CATALOGING-IN-PUBLICATION DATA

Friedman, Gerald M.
    Principles of sedimentary deposits : stratigraphy and
sedimentology / Gerald M. Friedman, John E. Sanders, David C.
Kopaska-Merkel.
        p. cm.
    Based on Principles of sedimentology / Gerald M. Friedman and John
E. Sanders. 1978.
    Includes index.
    ISBN 0-02-339359-9
    1. Sedimentology. 2. Geology, Stratigraphic. I. Sanders, John
Essington. II. Kopaska-Merkel, David C. III. Friedman,
Gerald M. Principles of sedimentology. IV. Title.
QE471.F73 1992
551.3—dc20   91-27866
                                                                    CIP

Printing: 1 2 3 4 5 6 7 8   Year: 2 3 4 5 6 7 8 9 0 1

Contents

PART III   From Layers to Sequence- and Seismic Stratigraphy   149

PART IV   Process Sedimentology and Environments of Deposition   225

CHAPTER 9   *Deep-Water Settings*   319

CHAPTER 10   *Shelf Seas and Epeiric Seas*   349

CHAPTER 11   *Beaches and Barriers*   403

CHAPTER 12   *Marginal Flats: Peritidal Environments*   443

CHAPTER 13   *The Mouths of Rivers: Estuaries, Deltas, and Fans at the Sea Shore*   475

CHAPTER 14   *Nonmarine Environments: Deserts, Rivers, Lakes, and Glaciers*   515

PART V   Large-Scale Patterns of Sedimentary Deposits   567

CHAPTER 15   *Extraterrestrial Forcing Functions*   569

CHAPTER 16   *Principles of Stratigraphy*   595

CHAPTER 17   *Basin Analysis*   627

PREFACE

We have based this book on *Principles of Sedimentology* (Friedman and Sanders, 1978), which served students and professional geologists since 1978. We present here an updated book.

The decade of the 1980s was a dynamic one for the study of sedimentary deposits. Numerous new concepts on seismic- and sequence stratigraphy, plate tectonics, basin analysis, facies architecture, extraterrestrial forcing functions relating to sedimentology, and eustatic- and tectonic controls of regional- and global sequences brought about profound changes. In addition, break-throughs in computer- and seismic technology, and in satellite-image processing, extended the impacts of the foregoing on the study of sedimentary deposits. For geoscience education in the 1990s, these new developments demand not just a new edition, but a book that encompasses a global view of sedimentary deposits.

The 1978 book defined sedimentology in the widest possible sense as the geology of sedimentary deposits. Stratigraphy was included as a branch of sedimentology. However, in the 1980s stratigraphy expanded through sequence- and seismic stratigraphy, and the emerging discipline of basin analysis gained much importance. Accordingly, as reflected by the new title, the scope of this book has been broadened and the emphasis altered.

An important innovation in this book is our chapter on extraterrestrial forcing functions, a new approach to the study of terrestrial sedimentary deposits. The following example illustrates why we feel it necessary to discuss extraterrestrial forcing functions. In 1978–1979, coinciding with the publication date of the previous book, a symposium on extinctions in the fossil record was held. One of the speakers was physics Nobel Laureate Louis Alvarez, who presented to an unbelieving- and hostile audience of geologists his hypothesis that at the end of the Cretaceous Period an asteroid impact had caused mass extinctions. Little more than ten years later, most geologists agree that an asteroid or comet did indeed strike the Earth at the end of the Cretaceous, with catastrophic effects. Hence, extraterrestrial factors have become recognized as important processes affecting the formation of sedimentary deposits. In the 1978 book only a brief section in a complement to the main text explained the effects of the Moon and Sun on the mechanics of tides. By contrast, in this book we devote an entire chapter to extraterrestrial forcing functions. Within it are such topics as planetary orbits, Milankovitch climatic factors, oceanic tides, and the Sun's orbit.

In addition to the greatly enhanced attention to global- and extraterrestrial effects on sedimentary deposits, the 1978 book's chapter on sedimentary tectonics has been vastly expanded and revised to cover basin analysis. The current emphasis in the science on basins as entities requires that the student be introduced to this material. Our classification of basins is founded upon the tenets of plate tectonics, which have invigorated ideas of how basins form, evolve, and become filled with sediment.

Another feature of the present book is the organization of specialized subjects and of some descriptive examples into Boxes. These are concise summaries of material germane to the theme of the book, but out of its main stream. The Boxes are independent of the narrative, but enhance it by providing examples of phenomena described in the main part of the text or by exploring in detail topics only mentioned in passing in the general text. The purpose of this organizational structure is twofold. It permits increased flexibility for the instructor in planning lessons and in assigning readings. The Boxes also serve to emphasize some topics, such as plate-tectonic concepts (Box 17.1), that might be lost to view if buried within the chapters. These topics are important to many aspects of sedimentary deposits, and are Boxed so that they cannot be overlooked.

We retain with minor modification the 1978 classification of sediments and of sedimentary rocks. In this classification, we distinguish four kinds of sediments or of rocks: intrabasinal, extrabasinal, carbonaceous, and pyroclastic. Our reasons for adopting this classification are explained in detail in Chapter 4. In brief, the first two categories account for most sediments and sedimentary rocks, namely, those that form from the waters of the basin (intrabasinal; e.g., carbonates, evaporites) and those that are brought into the basin from outside (extrabasinal; e.g., sandstones, claystones). The other two categories are somewhat different, but are employed because they represent sediments and sedimentary rocks that are neither consistently intrabasinal nor extrabasinal. Carbonaceous rocks are composed of organic plant material; for example, coal. Pyroclastic rocks consist of particles ejected from volcanoes and not redeposited by ordinary sedimentary processes (these rocks are known as tephra).

One of our areas of emphasis in this book is the relationship between current knowledge and its historical development. The history of stratigraphy and sedimentology fascinates students, but geologists in industry commonly find it less rewarding. The latter's concern often is with a tool or technique handy for immediate use, and how useful tools were developed does not really matter. To accommodate the latter point of view, some recent (1990, 1991) stratigraphy- and basin-analysis books cite few pre-1970 references. Perusal of the reference section of some new books on stratigraphy or basin analysis would seem to indicate that these disciplines sprang from the forehead of Jove, fully formed, at a time when the 1978 book, weighing 2.5 kilograms and including 4,500 references, was going to press. To ignore the rich tradition of the subject does the student a great disservice. Science deals with an everchanging pattern of emphasis. If the contributions of today are not worth citing in twenty years' time, then what should students learn now, if, two decades hence, their knowledge is presumed no longer to matter? In fact, much of what we can learn from Henry Clifton Sorby (1826–1908), or even Nicolaus Steno (1638–1686), is as valid today as ever. The real basis for a learning program is to put the subject into context, which involves an appreciation for and understanding of its history. Therefore, in this book our emphasis includes the historic background of key concepts in the study of sedimentary deposits.

About thirty years ago, one of us was involved in a Ph.D. examination in stratigraphy. Two hours were spent discussing sedimentation patterns in geosynclines. A colleague tried to steer questioning into other areas of stratigraphy, but the examining faculty contended that only geosynclines really mattered. Plate tectonics may have pushed aside geosynclines, yet it is shocking that the term geosyncline is not even listed in the subject index of current textbooks of stratigraphy and basin analysis. Can one effectively teach a subject out of context?

Interestingly, some reviewers of the manuscript of the 1978 book objected to the inclusion of vertical tectonics.

We quote: "there is no such thing as vertical tectonics, all tectonic movements are laterally driven—take out the offending section." In compliance with the reviewers' requests, but under duress, this subject was removed. Today vertical tectonics is a hot subject in basin analysis, and is included.

This book is intended to be used as a text in undergraduate- and graduate courses of stratigraphy and sedimentology. Our objective is to provide a comprehensive text that can fill the needs of any instructor in these fields, regardless of the exact design of the course. We believe the book you now hold meets this objective. In writing it, we have aimed primarily at undergraduate juniors and seniors. Our style of writing in the early chapters is consistent with lower-level undergraduate courses, but we believe that, as the student advances, he or she progressively should become accustomed to a more-professional style. Therefore, the later chapters are written at a professional level. To those who simply scan the book, this lack of uniformity may seem perplexing, but it is deliberate and was carried out with the students in mind. Beginning graduate students will also find this text useful. Chapters 5 to 17 are central to the professional geologist employed in oil- and gas exploration or -development and in the study of ground water or economic geology in either industry or government.

In our treatment of references we have departed from tradition. Instead of interrupting discussion with literature citations, we cite authors' names and dates at the ends of appropriate sections. (References cited in figure captions are not duplicated in the source lists.) The bibliography at the end of the book contains a full alphabetical list of authors. (However, in most cases, citations appearing in our previous book are repeated here only if they appear in figure captions.) We adopted this format at the suggestion of many undergraduates and after consultation with many colleagues who teach undergraduates. To encourage search of the literature by students, we have prepared a list of suggested readings at the end of each chapter.

Some recent stratigraphy texts have republished in full the *North American Stratigraphic Code* (1983). Initially we were tempted to do so as well, to make this code readily available. However, inclusion of this 34-page document would increase the length and hence the cost of the book. Because the *Code* was published in the *Bulletin of the American Association of Petroleum Geologists*, a publication widely circulated, we decided that it was not necessary to include it in our book. In Chapter 16 (Principles of Stratigraphy) the *Code* has been included in the list of suggested readings.

The 1978 book contained about 200 pages of references, a total of 4,500. By comparison, the reference section of this book is small. When we wrote our previous book, we knew that readers would find its comprehensive reference list indispensible. The writing of the 1978 book required laborious searches for worldwide sources in library shelves of several towns and universities; we desired to pass on the fruits of this labor to our readers. Today no such effort is necessary; searches can be carried out with GeoRef of the American Geological Institute (AGI). Online telephone computer hook-up is especially helpful; only keywords need to be provided, and a comprehensive set of current references will be printed out.

Some terms used in stratigraphy and sedimentology have been misused so many times that one is commonly not certain of the intent when one encounters them in print. In order to ensure that our usage is clear, we shall define what we mean by *unconformity, facies, sequence,* and "*system.*"

Unconformity. The proper use of this term is as a particular stratigraphic relationship, not as a material thing. We avoid using "unconformities" and use instead "surfaces of unconformity," following the usage in the AGI glossary.

Sequences. We adhere to the L. L. Sloss formal definition (1963; large bodies of strata on cratons that are set off by surfaces of unconformity of regional extent), and tag it by using an initial capital S, as in Sequence. We use sequence with a small s for small-scale units of various kinds as in Bouma sequence, facies sequences, etc. In between are several ranks of things variously referred to as "sequences." We use parasequence for the small ones; parasequence sets for the next-larger category; then we propose a new term, mesosequence, for what many "sequence stratigraphers" designate as "sequence," following the Exxon group. For the next-higher category we use the term megasequence. (See Chapters 5 and 6.)

System. This word, when used as part of a formal name or technical term, should be restricted to its long-established stratigraphic sense of the strata deposited during a geologic period: Cambrian System and so forth. We do not use the neologism "systems tract," because Teichert (1958; as also reflected in the AGI glossary) expressed the very same concept by his long-prior use of "facies tract." Accordingly, we use "high-stand facies tract," "low-stand facies tract," and so forth.

We deplore the sloppy terminology that renders much so-called technical writing very difficult to read. We have made every effort to be precise and consistent in our own usage. A sampling of the way we've dealt with "problem" words follows. "Laminae" are things; "lamination" is an attribute or process. "Laminations" is not a word. "Velocity" is a vector quantity in physics; "celerity" and "speed" are appropriate words to use when writing about how fast fluids move. Those things that compose sediments are "particles," not "grains." Thus, "fine-grained" should be a term of the past; we use "fine-textured." "Geometry" is a branch of mathematics; objects possess "shapes." Cross strata are just that and written without a hyphen. We hope that the reader will find our pickiness an aid to comprehension.

ACKNOWLEDGMENTS

Sixty reviewers of various drafts of chapters and six reviewers of the entire manuscript helped in the preparation of the 1978 book, *Principles of Sedimentology*, on which this book is based. These reviewers included not only stratigraphers and sedimentologists, but also biologists, log analysts, geophysicists, geochemists, paleontologists, and structural geologists. These reviewers, together with many donors of illustrations, were acknowledged in the previous book.

For reviewing this book our thanks are extended to Mark Graber, Raymond V. Ingersoll, Robert S. Merkel, Laurence L. Sloss, and Daniel A. Textoris, who read several chapters. David M. Rubin read part of Chapter 14. Gordon S. Fraser, William E. Galloway, Anthony R. Prave, and Richard A. Paull read the entire manuscript. We are most grateful for the constructive criticism of these reviewers. Graduate students of the Department of Geology of Brooklyn College and the Graduate School and University Center of the City University of New York read chapters of the manuscript and wrote critical evaluations of their contents. We express our thanks to Allan P. Bennison for interesting suggestions relating to deltaic sedimentation in the mid-continent of the United States. Victoria Spain, Hofstra staff of reference librarians, helped us track down historical information on our most difficult case, Usiglio. Robert A. McConnin, senior editor of Macmillan Publishing Company, provided encouragement while we wrote the text. In the preparation of the line drawings we acknowledge the understanding cooperation of Ronald K. Levan, Dora Rizzuto, and Anna Yip. Margaret Comaskey pulled it all together during the production stage.

The following institutions and colleagues provided illustrations: Academic Press; K. L. Acker; D. B. Alsop; American Association for the Advancement of Science; American Association of Petroleum Geologists; American Geological Institute; Annual Reviews, Inc.; J. H. Barwis; Desiree Beaudry; W. C. Beckmann; W. M. Blom; J. C. Boothroyd; B. D. Bornhold; D. J. Bottjer; P.-A. Bourque; M. J. Burgis; Cambridge University Press; Canadian Society of Petroleum Geologists; D. J. Cant; H. S. Chafetz; J. A. Cherry; J. M. Coleman; P. A. Dickey; M. L. Droser; A. J. Eardley; Elsevier Scientific Publishers; W. H. Fertl; R. V. Fisher; R. A. Freeze; R. W. Frey; W. E. Galloway; Geological Association of Canada; Geological Society of America; R. N. Ginsburg; Pierre Giresse; Government of Canada; R. B. Halley; W. K. Hamblin; B. U. Haq; J. C. Harms; M. J. Haworth; M. O. Hayes; B. C. Heezen; H. D. Holland; Ian Hutcheon; International Association of Sedimentologists; M. T. Jervey; C. G. St. C. Kendall; D. J. J. Kinsman; R. A. Kirby; W. C. Krumbein; R. P. Langford; A. S. Laughton; Mark W. Longman; Bernard Luskin; A. Matter; McGraw-Hill, Inc.; R. S. Merkel; A. D. Miall; R. M. Mitchum; G. F. Moore; P. Morris; National Association of Geology Teachers; C. S. Nelson; Norwegian University Press; Dag Nummedal; P. E. Olsen; R. S. Patterson; PennWell Publishing Co.; H. W. Posamentier; Prentice-Hall, Inc.; G. Reiger; J. M. Rine; M. J. Risk; A. C. Roberts; K. Rützler; R. Sarmiento; Martine Savard; J. W. Schmoker; W. C. Schwab; SEPM (Society for Sedimentary Geology); Shell Oil Co.; R. E. Sheridan; Springer-Verlag, Inc.; W. L. Stokes; M. Sturm; Marie Tharp; S. Thompson; Erik Thomsen; D. J. Timko; M. E. Torresan; University of South Carolina; P. R. Vail; J. C. Van Wagoner; John Warren; R. A. Wheatcroft; John Wiley & Sons; Wiley-Interscience, Inc.; B. H. Wilkinson; and Lee Wilson.

Gratefully we acknowledge Steve Buttner, Louise Koenig, and Baiying Guo for their help with the mechanical phase of manuscript preparation. Steve Buttner also assisted with editorial work. The Geological Survey of Alabama, and its head, Ernest A. Mancini, graciously permitted DCKM to spend time consulting with the other two authors during the late stages of book preparation. DCKM thanks Sheila and Morgan Kopaska-Merkel, who bore the roles of textbook widow and textbook orphan with remarkable equanimity. DCKM also thanks Gretchen Luepke for her understanding patience in regard to her own deadlines. JES thanks Barbara W. Sanders for keeping the home ship afloat while JES spent countless weeks in Troy on book business. Finally we express our thanks to Sue Friedman for her help in doing the many essential things behind the scene: she assisted in ways too numerous to list.

PART I

INTRODUCTION

Before you read the text of this book, turn first to the Preface. If you have not read it, we suggest that you do so now.

This book is a comprehensive summary of the geology of sedimentary deposits. We include the bases for interpreting sedimentary deposits in terms of processes, environments of deposition, vertical successions of strata, and regional- and global effects on sedimentation. Our approach to the vast domain of stratigraphy and sedimentology differs from that followed in several well-known books entitled *Sedimentary Rocks* or *Sedimentary Petrology*. In most of these books, which have been widely used for teaching undergraduate courses dealing with sedimentary deposits, many chapter headings are the names of sedimentary rocks, such as sandstones, shales, limestones, evaporites, coal, and zeolites. Although we discuss sedimentary rocks, we prefer a chapter organization that emphasizes the dynamic aspects of stratigraphy and sedimentation. This approach was pioneered in the 1978 Friedman-Sanders *Principles of Sedimentology*, and it has gained much favor with the authors of recent sedimentology and stratigraphy texts.

Our general progression is to begin with sedimentary particles, on scales ranging down to that of the electron microscope. From particles we progress to sedimentary processes, sedimentary products, sedimentary sequences, and large-scale patterns of sedimentary deposits, including how all of these topics interrelate with tectonic effects, sea-level fluctuations, and extraterrestrial forcing functions. We have organized our material into five major, but unequal, parts.

Part I introduces our subject. Its one chapter presents some of the large themes of stratigraphy and sedimentology in the context of their history, their impact, and the effects of climate, atmosphere, and extraterrestrial factors.

In Part II we summarize sedimentary particles and their aggregates: sediments and sedimentary rocks. Chapter 2 describes the kinds of particles that compose sedimentary deposits, Chapter 3 is devoted to the conversion of sediments to sedimentary rocks through the process of diagenesis, and Chapter 4 concerns the subject of classifying and naming sedimentary rocks, as well as some of their characteristics.

Part III, consisting of two chapters, presents the advances made in sequence- and seismic stratigraphy, the importance of eustatic sea-level changes, and global stratigraphic cyclicity. Chapter 5 begins with the properties of individual layers and progresses to the kinds of things one can expect to see at exposures that display layers in sectional view. Chapter 6 summarizes the large-scale stratigraphic relationships that are collectively known as sequence stratigraphy.

Part IV consists of eight chapters and is concerned with process sedimentology and facies architecture. Chapter 7 begins with physical-, biological, and chemical processes. Chapter 8 relates the circulation in the atmosphere, in modern oceans, and in basin waters. Chapters 9 to 14 analyze sedimentary environments. In Chapter 9 deep-water settings are discussed. Chapter 10 concentrates on the numerous aspects of shallow-water marine sediments, Chapters 11 and 12 present transitional environments, where land and sea interface (shoreface-, beach-, barrier-, and peritidal environments), Chapter 13 takes up transitional environments where rivers enter the sea (estuaries, deltas, and fans at the sea shore), and Chapter 14 deals with nonmarine environments: deserts, rivers, lakes, and glaciers.

Part V, entitled "Large-Scale Patterns of Sedimentary Deposits," includes three chapters. Chapter 15 concerns extraterrestrial forces that affect the Earth's sedimentary deposits, including planetary orbits, oceanic tides, Milankovitch cycles, and the Sun's orbit. Chapter 16 takes on the task of analyzing sedimentary strata on the surface and in the subsurface through field study and subsurface borings, including instrumental logs, and relates the principles of classical stratigraphy. Chapter 17 concerns the fundamentals of basin analysis, which is the study of how basins form, how they evolve, and especially how they are filled with sediment. The current approach to basin study is based upon the concepts of plate tectonics.

CHAPTER 1

A Global View of Sedimentary Deposits: Is It Big Enough?

In modern study of sedimentary deposits, a global viewpoint is becoming increasingly important. This new emphasis on taking a planetary perspective on geologic subjects lies behind the new discipline of "Earth-system science."

At least five factors have coalesced to support this new-type world view of sedimentary deposits. These are (1) the new global understanding of tectonic processes made possible by plate tectonics; (2) the space-age capability of being able to look at the world as a unit via images from orbiting satellites and of being able to make computer models using global-scale information; (3) results from the satellite probes to other planets in the solar system which have established a renewed awareness of the importance of interactions between processes indigenous to a planet and factors that are external to that planet (with reference to the Earth, the external factors are known as extraterrestrial factors); (4) the worldwide search for petroleum and resultant acquisition and detailed analysis of multichannel continuous seismic-reflection profiles from virtually all of the world's sedimentary basins, which have provided powerful support for the idea that sea-level changes exert a fundamental influence on the deposition of sediments; and (5) the collection and analysis of an extensive worldwide suite of cores from holes drilled through the sediments underlying the modern deep-sea floors and accompanying development of the discipline of paleoceanography. A final factor in any global view of the modern world is the ever-increasing impact of human activities on Planet Earth's lithosphere, hydrosphere, and atmosphere.

In many respects, this new emphasis on a global view of modern processes is an attempt to bring studies of modern-day Earth processes (known by the general term of actualism) to the same scale as some important geological syntheses made during the late nineteenth and the first half of the twentieth century. For example, Edouard Suess' (1831–1914) three-volume compilation of the ge-

ologic record entitled *Das Antlitz der Erde* (1888), which was translated into English under the title *The Face of the Earth*, presented a global view. Somewhat later, A. W. Grabau's (1870–1946) books *Pulsation Theory* (1936) and *The Rhythm of the Ages* (1940) were attempts at a synthesis of the stratigraphic record based on the recognition of worldwide changes in sea level.

In this chapter we show how some of these factors interact and affect the study of sedimentary deposits. We begin by summarizing some of the important geological ideas that have been expressed about the connection between sedimentary deposits and changes of sea level. We move from there to the relationship between sedimentary deposits and climate, showing how the comparison between modern- and ancient glacial deposits contributed to the geologic proofs of the former extent of the Pleistocene continental glaciers and their connection to global climate and changes of sea level. We conclude with a look at some of the major crosscurrents within modern studies of global climate, a topic intricately involved with the composition of the Earth's atmosphere and the impacts on the atmosphere of human activities, global-scale natural processes, and extraterrestrial forcing functions. A key challenge facing students of sedimentary deposits is to examine the stratigraphic record to obtain from it as much as possible in the way of data for comparing the effects of past climates with concepts based on interpretations of the current situation.

Some Early Geologic Ideas About the Connection Between Sedimentary Deposits and Changes of Sea Level

In the eighteenth century, during the early stages of the development of what we now refer to as geology, various students of natural history (whom we would now call

3

geologists) occupied themselves with concocting "theories of the Earth." This popular pastime generally was not based on any clear understanding of sedimentary deposits. Rather, its chief requirement seems to have been a vivid imagination and a facile pen. The "rules" of the "game" were to begin with some initial condition (a ball of fire torn loose from the Sun, or a sphere having a molten interior and a cold crust that was entirely bathed in seawater, for example) from which the author would trace the course of events that led to the conditions observable today. Depending on how firmly an author was locked into the chronology of Earth history developed by biblical scholars using the *begat* method (lists of all the names in the Old Testament: A begat B, B begat C, etc.) and on how many events he felt obliged to include, his "theory" called for slow or fast action.

The "Universal Ocean" and the Worldwide Emergence Associated with Its Shrinkage

An exceptional example of this approach was written in 1715–1716 by the French naturalist Benoit de Maillet (1656–1738) under the title of *Telliamed* (the author's name spelled backward). De Maillet's manuscript (reconstructed and translated into English by Carozzi, 1968) contained many original ideas about evolution of celestial bodies and presented the results of his numerous studies of the ocean and of marine sediments. Based on the emergence from the sea of a rock noticed by his grandfather, de Maillet decided that such recent lowering of sea level was part of a more-general process. Accordingly, he undertook studies of modern nearshore areas, including observations of the shallow sea floor using a diving apparatus.

He concluded that what he named *"Primitive"* mountains had been formed by the action of marine currents when the sea was much higher than today. He supposed, in fact, that the sea had been high enough to submerge even the highest peaks and thus was a *"Universal Ocean."* As the level of the universal ocean dropped, the primitive mountains were uncovered and, along their shores, wave action deposited coastal sediments. He introduced the concept that slow natural changes could achieve large results by working during a long period of time. He even expressed the idea that, in the shallow seas marginal to the primitive mountains, marine organisms flourished and eventually evolved into terrestrial life (Carozzi, 1968, 1969).

De Maillet deserves much more credit in the founding of geology than he has received heretofore. Many of the naturalists of the eighteenth century in France and elsewhere clearly borrowed liberally from de Maillet's writings. Some of this "borrowing" may have resulted from de Maillet's insistence that his manuscript not be published until ten years after his death. Nevertheless, de Maillet's work was widely read in France, both in manuscript form and as a book after its formal publication in 1748. (His work formed a part of the "forbidden literature" of France, where the all-powerful king's official censors decreed what books could and could not be published and circulated.) *Telliamed* was translated into English and editions appeared both in London (1750) and in Baltimore (1797).

Nearly all early naturalists who wrote about the universal ocean relied on it as a mechanism for establishing a definite chronological order within the geologic record and as a means of correlating sedimentary deposits in widely scattered localities. These early writers expressed many ideas about the relationship between their universal ocean and the geologic record. We illustrate the spectrum of these ideas by comparing the views of John Whitehurst (1713–1788) with those of Abraham Gottlob Werner (1750–1817).

Both Whitehurst and Werner visualized a universal ocean in which islands would eventually form. But they differed drastically on the kinds of processes they thought had taken place in the universal ocean to form the primitive islands.

Whitehurst supposed that the raw materials of the primitive islands consisted of a kind of primordial chaos in the form of ". . . an universal pulp, the solids would equally subside from every part of its surface, and consequently become equally covered with water" (Whitehurst, 1778, p. 26) and that the lunar tides would move these solids about, eventually forming islands (as modern tidal currents form sand banks). Once these islands (A B C of Figure 1-1) had formed, they would exert an incumbent weight on the "stratum of subterranean fire" (F F of Figure 1-1). The resulting "expansive force" of F F would cause the low areas (D D of Figure 1-1) to "ascend," and this upward movement would raise the surface of the sea. "Consequently the island ABC, became more or less deluged, as the bottom of the sea was more or less elevated; and this effect must have been more or less universal, as the fire prevailed more or less universally, either in the same *stratum*, or in the central part of the earth" (Whitehurst, 1778, p. 87). In modern terms, Whitehurst can be said to have visualized that the chief actions of the universal ocean involved only physical processes: the settling of the particulate chaotic "pulp" and the shifting of these particles by tidal currents to form primitive islands. Whitehurst further supposed that these initial particles would become cemented so that the primitive islands would be underlain by solid sedimentary rock.

By contrast, Werner argued that the first work of the universal ocean had been chemical precipitation and that such precipitation was the mechanism for manufactur-

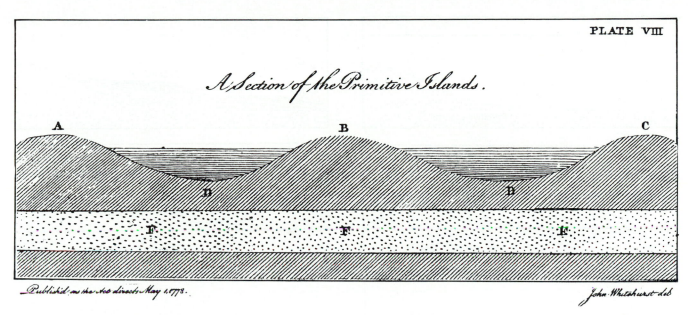

PLATE VIII

A Section of the Primitive Islands.

A   B   C

D   D

F   F   F

Published as the Act directs May 1.1778.   *John Whitehurst del.*

FIGURE 1-1.   Schematic profile through outer part of Earth and remnants of the "Universal Ocean" showing three "Primitive" islands (A B C), according to John Whitehurst (1778). The subsurface layer (F F F) consists of "universal fire." Unequal loading of F caused by thicker material under A B C causes the top of F to push up on D D, thereby elevating the bottom of the "Universal Ocean" (closely spaced horizontal lines), diminishing the volume of the ocean basins, and forcing the level of the sea to rise and flood the "Primitive" islands. (J. Whitehurst, 1778, pl. VIII; plates follow p. 199.)

ing rocks displaying crystalline textures. In particular, Werner argued that the universal ocean had precipitated a worldwide layer of granite and that the upper surface of this granite layer was not smooth, but contained high areas destined to become primitive islands as the level of the universal ocean dropped (Figure 1-2, A). Werner classified the granitic rocks composing these islands as "primitive." Notice the contrasts between Whitehurst and Werner. Whitehurst's small "primitive" islands were underlain by sediments that had been moved physically by lunar tidal currents and cemented into sedimentary rock; he asserted that these small islands had formed before the great Biblical flood (their age was "antediluvian"). He argued that they differed totally from the much-larger modern-day mountains, to which he assigned a "postdiluvial" age. Werner's interpretation of the relationship between certain modern mountains and the universal ocean was just the opposite of Whitehurst's. Whitehurst claimed no connection between the primitive islands and modern mountains; Werner asserted that certain modern mountains were nothing more than former primitive islands—that they *are* these primitive islands minus the water.

According to Werner, solid sediment particles and their movement and deposition in strata by physical processes could not take place until after the level of the universal ocean had dropped to the point where the primitive islands had appeared. (See Figure 1-2, B.) He

visualized that the oldest stratified rocks displayed two diagnostic properties: (1) They possessed an initial dip (having been deposited on the steep subaqueous slopes surrounding the primitive islands) and (2) they were mixed rocks, partly mechanical (debris eroded from the primitive islands) and partly chemical (crystals that had been precipitated out of the waters of the universal ocean and that formed the cement between the mechanically deposited particles). Werner classified these as the "Transition" rocks; he thought their typical representative was the tough, hard variety known to the German miners as "grauwacke."

With yet-further drop in sea level, sediment eroded from the ever-enlarging islands, both from their cores of granitic primitive rocks and from the dipping Transition rocks on their flanks, formed the Secondary territory underlain by layers that encircled the central high regions and were confined to the "foothills" of the higher mountains.

In its last low stage before dropping to its modern-day level, the universal ocean deposited the loose alluvial (Tertiary) formations.

Werner's position on the origin of crystalline rocks solely as a product of precipitation at low temperature from the universal ocean that took place only at its highest levels led him to two controversial conclusions about granite and basalt: (1) All granite is entirely older than all the stratified rocks and (2) basalt likewise was precip-

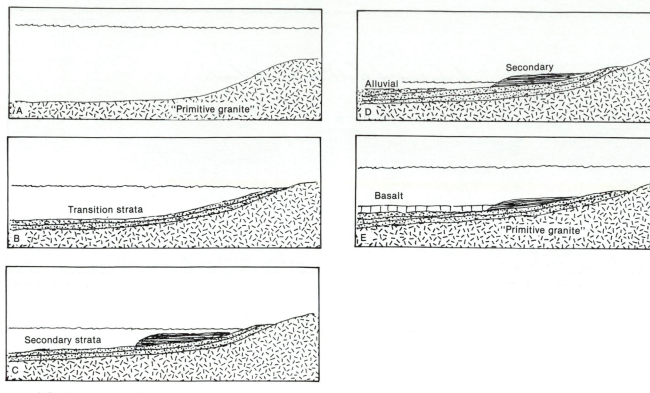

FIGURE 1-2.   Schematic profile sections illustrating Abraham Gottlob Werner's views on the geologic work of the "Universal Ocean." (Suggested by fig. 2.13 in R. H. Dott, Jr., and R. L. Batten, 1981, p. 32.)

A. "Universal Ocean" precipitates granite having irregular surface.

B. Ocean level drops, exposing islands composed of "Primitive" granite; erosion of granite forms Transition strata composed partly of sediment deposited mechanically as initially dipping layers on flanks of islands and partly of additional chemical material precipitated from the "Universal Ocean" (the supposed chemical component forming the cement of the Transition rocks).

C. With further lowering, Secondary strata, horizontal and consisting entirely of mechanical deposits, overlie the Transition strata.

D. At a still-lower level the "Universal Ocean" deposits alluvial (Tertiary) strata, which are confined to the lowlands.

E. Layer of basalt, containing prismatic joints that Werner took to be crystal faces and from which he inferred that the basalt was a chemical deposit, overlies alluvial strata. In order to precipitate this basalt, the "Universal Ocean" had not merely to return, but to be deep enough to submerge all the islands and thus to shut off any supply of mechanical sediment, thus fulfilling the supposed requirements for precipitation (no lands exposed).

itated out of the universal ocean. In the Rhine Valley, Germany, sheets of basalt overlie some of the alluvial formations. Werner interpreted each sheet of basalt as the product of a universal ocean that reversed its general shrinkage and had deepened once again to the point of crystallization, a condition he thought required all land areas to be submerged.

Naturalists who purported to explain all rocks as products of the universal ocean became known as Neptunists. And by far the most-influential leader among Neptunists was Abraham Gottlob Werner. What is more, Werner's firmly indoctrinated students dominated geology from about 1790 to 1830.

Many of the authors of books with titles like *Theory of the Earth* called upon the good old universal ocean

to perform various functions that showed they did not understand de Maillet's profound insights into the geologic activities taking place along today's coasts. Most authors discussed at great length how this great hypothetical body of water shrank progressively from its maximum extent to its present level, and in so doing created both the entire geologic record and the modern world. (Where the water went to was an awkward problem, typically skipped over lightly by supposing that cracks opened in the crust and that the water drained into the interior.) As the water level dropped, high spots on the irregular sea floor became islands.

We cite this early emphasis on the universal ocean as an example of the importance geologists have always placed on the need to know what goes on in the sea as a basis for understanding the geologic record. The modern Dutch geologist Ph. H. Kuenen (1902–1964) captured this point in his phrase "no geology without marine geology."

SOURCES: Dott and Batten, 1988; Geikie, 1905; Sanders, 1978; Whitehurst, 1778.

Sea-Level Changes and Associated Systematic Patterns of Sedimentary Strata

The French chemist G. F. Rouelle (1703–1770) was interested in mineralogy and included among his lectures in chemistry at the Jardin du Roi in Paris references to a "theory of the Earth" that included a universal ocean and its great shrinkage. Historians do not know for certain whether James Hutton (1726–1797), a young student from Edinburgh who spent nearly two years (1748–1749) in Paris studying medicine as the best means of learning about chemistry, attended Rouelle's lectures. But historians do know that 16 years later a promising young French student, Antoine Lavoisier (1743–1794), was sufficiently inspired by Professor Rouelle's material in mineralogy and geology to divert from chemistry for a while and to begin his scientific career as a geologist.

ANTOINE LAVOISIER: MODERN MARINE DEPOSITS AND THEIR BEHAVIOR DURING CHANGES OF SEA LEVEL

Antoine-Laurent Lavoisier is usually thought of not as a geologist but as a world-famous French chemist. However, before he devoted himself to chemistry, he worked as a geologist, assisting one of France's best geologists, Jean Guettard (1713–1785), in making a mineralogic atlas of France on the quadrangle maps then becoming available. Lavoisier measured the stratigraphic sections. He used a barometer to establish the altitudes of key parts of the landscape. He then tied his generalized columnar section for each quadrangle into these control points. By this means, he determined the thicknesses of the horizontal strata.

In 1789, the year of the fall of the Bastille (taken as the initial event in the chain of events constituting the French Revolution), Lavoisier presented to the French Royal Academy of Sciences a memoir devoted to the patterns of sediments formed along a modern shore and how these would shift if the sea advanced or retreated (Figure 1-3). He described how, at today's sea level, coarse nearshore sediments (in his categories, the littoral sediments) give way in an offshore direction to fine-textured (or pelagic) sediments. Lavoisier wrote that if the sea were to advance, the pelagic sediments would shift landward and overlie older littoral sediments. Similarly, if the

sea were to retreat, littoral sediments that are not eroded away would overlie older pelagic sediments.

Lavoisier presented three measured sections from parts of the Paris basin (Figure 1-4) and interpreted them in light of his views on changes of sea level and the associated systematic patterns of sediment. He stated that the chalk (later assigned a geologic age of Cretaceous) and the sediments overlying it had been deposited by a sea which had both advanced and retreated, and had not,

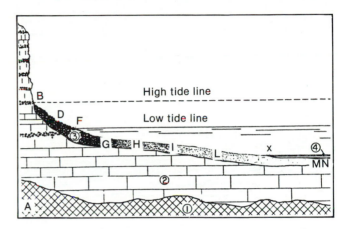

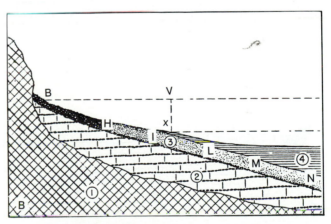

FIGURE 1-3. Schematic profiles at right angles to shore along the coast of France to illustrate Lavoisier's ideas about the shore-parallel zones in nearshore marine deposits and how these shift as sea level changes. Circled numbers indicate (1) old rocks on which the chalk (2) was deposited; (3) littoral beds [the name Lavoisier gave to the sequence of deposits (BDFG) formed along the French coast, ranging from wave-rounded beach gravel (B) composed of chert (black) from the upper part of the chalk, to coarse sand (H), fine sand (I), and "an impalpable dust of siliceous earth (L)," each forming belts parallel to shore]. Farther offshore are the pelagic beds, the deposits of the open sea "composed of pure calcareous matter resulting from the accumulation of shell-bearing organisms without any admixture of foreign particles" (4, seaward of X); and (5) littoral deposits composed of debris eroded from the old rocks.

A. Conditions at existing sea level.

B. Conditions after sea level has risen far enough to overlap all of the upper, cherty part of the chalk and has started to erode a coastal cliff in a high part of the old rocks not covered by the chalk.

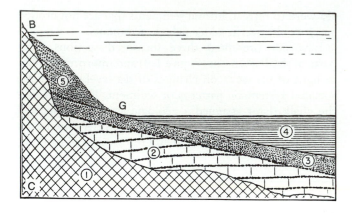

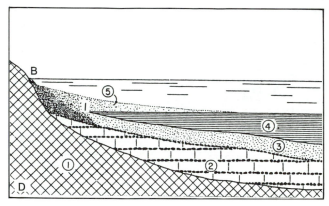

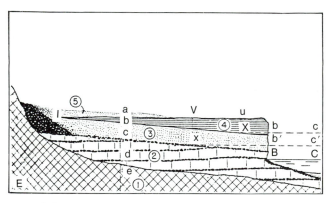

FIGURE 1-3. (*Continued*)

C. After still-further rise in sea level, the sea has eroded a thick body of debris from the old rocks and has deposited the coarse parts as littoral beds (5) and has built up a thick succession of pelagic beds (4), which now overlie the earlier-formed littoral beds composed of chert gravel from the upper part of the chalk.

D. As the sea is lowered, it spreads the littoral beds composed of debris from the old rocks (5) over the pelagic beds (4).

E. After a great drop in level and a subsequent advance, the sea cuts a cliff in the upper chalk and overlying marine deposits formed during the earlier advance-retreat cycle. A vertical sequence along the line to the left of the circled No. 1 consists of 5 units, from top downward: (a) littoral beds derived from the old rocks (circled 5); (b) pelagic beds formed during the maximum post-chalk advance (circled 4); (c) littoral beds (circled 3) formed of chert debris eroded from the upper chalk; (d) the chert-bearing upper chalk (circled 2); and (e) the old rocks (circled 1). (Redrafted by A. V. Carozzi, 1965, from the original figures in Antoine Lavoisier's 1789 memoir.)

as he had been taught and as most naturalists of the time believed, simply retreated. According to Lavoisier, in one of its advancing phases, the sea had deposited the pelagic chalk. Subsequently, sea level had dropped and the chalk had been eroded. Then, during another advancing phase, the sea deposited a coarse littoral bed. At its maximum stand, it deposited a pelagic clay. Finally, during a retreat phase, it deposited another coarse littoral bed.

Lavoisier clearly expressed in 1789 ideas that are generally thought to have been first formulated more than a century later by Johannes Walther (1860–1937) in 1893. These are widely known as Walther's Law of Facies. We discuss this concept in Part III of this book. If true justice were to be done to Lavoisier's contribution, this principle should be renamed Lavoisier's Law of Facies.

Finally, Lavoisier wrote that advances and retreats of the sea involved long time periods. Moreover, his field observations led him to suspect that "most certainly the so-called older rocks consist also of an association of littoral beds and do not represent the primitive earth" (translation by Carozzi, 1965, p. 83).

SOURCE: Middleton, 1973.

Modern Views: Spectrum of Sea-Level Changes and Related Cyclic Deposits

We now move from Lavoisier to modern-day concepts of sea-level changes and associated cyclic deposits. In doing so, we need to jump in time more than two centuries, but in the realm of ideas, our leap needs to span a much-smaller gap. This is a tribute to Lavoisier's remarkable insights. The main point we emphasize here is that the level of the sea undergoes many changes on time scales ranging from those associated with surface waves (a few seconds) and with astronomic tides (half a day to monthly and more), to those having periods of millions of years. One can visualize these changes as if they were waves having various periods (or the inverse of periods, frequencies).

A useful way to analyze waves is to sort them according to their periods. Such an approach is known as compiling a wave spectrum, or applying spectral analysis. The fundamental principle of spectral analysis is that each kind of wave displays a peak frequency that is characteristic of its origin. Table 1-1 classifies orders of changes of sea level according to period.

At this point, we examine the relationship between climate and sediments by considering how the Pleistocene climate shifted back and forth between the Earth's two predominant climate modes: *ice house* and *greenhouse*. To these, we propose to add a third: the *boiler room*. These climate shifts are intimately linked to changes of sea level.

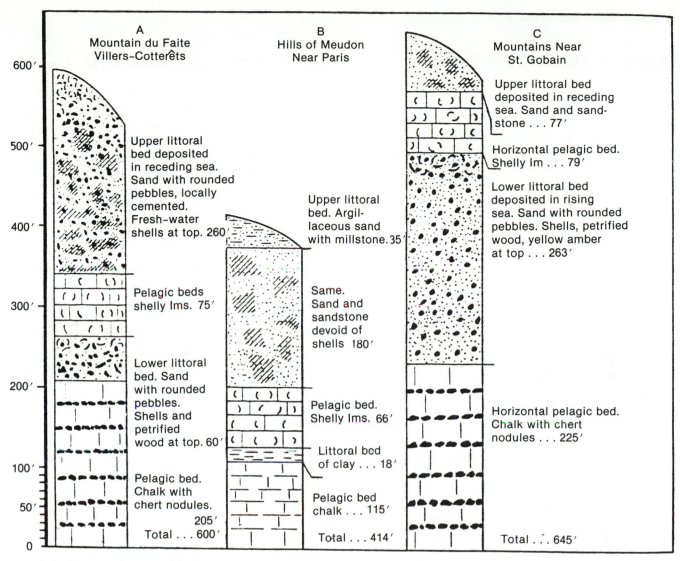

FIGURE 1-4. Three stratigraphic columns measured by Lavoisier in the Paris basin, each showing chalk (pelagic bed) at the base overlain by a lower littoral bed deposited at the edge of an advancing sea, a middle pelagic bed, and an upper littoral bed deposited in a receding sea. (Redrafted by A. V. Carozzi, 1965, fig. 7, p. 82, from an original in Antoine Lavoisier's 1789 memoir.)

TABLE 1-1. Orders of sea-level changes from infrequent (first order) to frequent (fifth order)

Order	Period (yr)	Inferred cause
1	2 to 3 $\times 10^8$	Supercontinents form, break up
2	1 to 8 $\times 10^7$	Changing volumes of ocean basins as product of sea-floor spreading and/or changes in the geoid
3	1 to 10 $\times 10^6$	Mid-Ocean Ridge changes? Growth and decay of continental glaciers
4	$\approx 10^5$	Climatic changes
5	$\approx 10^4$	Climatic changes

SOURCE: Duration data from Vail, Mitchum, and Thompson, 1977.

9

Sedimentary Deposits and Climate: The Pleistocene Ice Ages

A second global-scale association exists between climate and sediments. In many following chapters we emphasize the relationships in warm climates. Here, we single out the development of ideas about the connections between glaciers, distinctive sediments, and features sculpted by glaciers on bedrock.

In a far-reaching analysis of ancient environment based on the physical properties of the sedimentary deposits, Louis Agassiz (1807–1873) inferred in 1840 that during the past tens of thousands of years North America had been covered by continental ice sheets. Agassiz grew up in Switzerland and there had become familiar with such features of the Alpine glaciers as erratic boulders, striae on bedrock, and various kinds of moraines. His initial investigations were devoted to mapping the evidence that the former extent of the Alpine glaciers had been greater than that of the modern Alpine glaciers. After he had demonstrated the transition from features made by the modern Alpine glaciers to those made by formerly larger Alpine glaciers, he was prepared to take on the task of studying the northern parts of Europe and North America in localities far removed from the Alps. His research, and the work of many who came after him, has resulted in a comprehensive compilation of evidence on all aspects of the Pleistocene ice ages: former extent of the glaciers, relationship of glaciers to sea level, and relationship of sea level and climate.

Extent of the Former Continental Ice Sheets

In numerous localities in northern Europe and in Great Britain, Agassiz found deposits and markings on the bedrock comparable to those that he had used to reconstruct the formerly great extent of the Alpine glaciers. Later, after he had moved to Cambridge, Massachusetts, he found that in nearly every place he visited in eastern North America he could recognize comparable erratic boulders, striated bedrock, and moraines. Despite the fact that in these parts of North America no glaciers exist today, Agassiz was confident that glaciers had been present in North America but had vanished. He thus used sedimentary deposits to reconstruct the extent of the former glaciers (Figure 1-5).

After maps such as that in Figure 1-5 had been compiled, glaciated areas were determined. The areas of modern glaciers aggregate 14.9 million km². The areas occupied by Pleistocene ice at its maximum extent (regardless of age) total 44.4 million km². The difference is 29.5 million km², and Pleistocene glaciers covered nearly three times the area currently under glacial ice.

Relationship Between Glaciers and Sea Level

A discussion of the relationships between glaciers and sea level depends on several factors, one of which is the water-equivalent volume of glaciers (91.7% of the volume of ice). Other factors include (1) the amount of water locked up in the maximum Pleistocene glaciers compared with the water locked up in modern glaciers, (2) the configuration of the shallow parts of the ocean basins, and (3) the changes in the lithosphere that result from changing the positions of the ice- or water loads. For this discussion, we include only the water-equivalent volumes.

The total volume of ice is arrived at by multiplying the areas of the glaciers by some number representing thickness. Based on evidence from the modern ice caps in Greenland and Antarctica, estimates of the thicknesses of the continental glaciers are placed at 1.5 to 2 km. Accordingly, the water-equivalent volume of modern glaciers is about 24 million km³ and that of the Pleistocene-maximum glaciers about 71.3 million km³. The difference is 47.3 million km³, which is thought to translate into a difference in sea level of 132 m.

SOURCE: Flint, 1971, p. 73–85.

Relationship Between Sea Level and Climate: Ice House, Greenhouse, and Boiler Room

Given the data from the volumes of ice and amounts of lowering of sea level, one cannot doubt that for continental glaciers to be widespread, the Earth's climate must be in an ice-house mode and sea level must be low. By contrast, when large continental glaciers are absent, the Earth's climate will be in its greenhouse mode and sea level will be high. Because climate clearly controls the volume of glacial ice, climate and sea level are directly linked.

Although sea level and climate are thus clearly linked to these two contrasting climate modes, impressive geologic evidence from the outer parts of the modern continental shelves supports the view that the times of warmest climate coincide not with the highest sea levels, but with the lowest. Short episodes of extremely warm climate are not closely coupled to high sea level, but are just the opposite—out of phase with high sea level. These extremely warm intervals may be required to terminate the ice-house mode; they precede the greenhouse mode. We suggest that these short, extremely warm episodes be informally designated as the Earth's boiler-room climate modes.

SOURCES: Fischer, 1986; Sanders and Friedman, 1969.

What Caused the Ice Ages? Effects of Extraterrestrial Forcing (Milankovitch Cycles)?

Various mechanisms have been proposed to explain the Earth's climatic oscillations from glacial- to nonglacial modes. Some of these have included variable amounts of volcanic dust in the Earth's atmosphere, cyclic variation in the proportion of carbon dioxide in the Earth's atmosphere, variable solar activity, and variations in the Earth's orbital parameters. In this chapter, we mention variations in carbon dioxide and in Earth-orbital parameters. We discuss variable solar activity in Chapter 15.

In 1864, the self-educated Scottish man of science James Croll (1821–1890) proposed the hypothesis that the variations in the Earth's orbital elements through time were sufficient causes of the climatic changes during the Pleistocene Period in which glacial ages alternated with nonglacial ages. This idea languished until the twentieth century, when Milutin Milankovitch (1895–1945) used the variations in the astronomic elements of the Earth's orbit to calculate a Northern-Hemisphere climatic curve (Fig. 1-6).

Milankovitch's curve indicated that the ice ages displayed a definite periodicity. Milankovitch claimed that these astronomic factors could introduce significant cyclic variation into the Earth's climate even with no changes in solar activity, with no changes in the Earth's albedo, and with no changes in the composition of the Earth's atmosphere. Thus Milankovitch sidestepped three factors that have been cited as contributing to climatic variation. Of these, the factor of possible solar variation clearly is extraterrestrial, whereas albedo and composition of the Earth's atmosphere probably would not be considered extraterrestrial factors. In view of the ionospheric interactions between molecules of gases (e.g., ozone) and solar radiation, the atmosphere does not fit cleanly into either category, terrestrial or extraterrestrial.

Milankovitch did not deny the possible contributions of these three factors; he was simply trying to establish the point that the variations in the Earth's orbital elements alone could trigger significant climate changes.

These Milankovitch climate cycles, as they were later called, contained two particular features that most geologists specializing in the Pleistocene rejected. (1) The Milankovitch scheme specified that the ice ages should have alternated between the Northern Hemisphere and the Southern Hemisphere. Evidence that the ice ages were synchronous in both hemispheres led to a general rejection of the Milankovitch approach. (2) The Milankovitch climate curve indicated a warm peak about 10,000 years ago, when the sea was low. Because the idea was widely accepted that sea level and climate moved together in lockstep fashion, geologists argued that this Milankovitch warm peak should have coincided with a high sea level, not a low one. This mismatch between warm climate and low sea level served as an additional reason for rejecting the Milankovitch method. This seemingly contradictory peak may represent a boiler-room episode.

Why, then, despite these two seemingly fatal flaws, has the Milankovitch concept emulated the biblical Lazarus by coming back from the dead? The answer lies in the results of geochronologic investigations made on deep-sea sediments. Evidence suggests that the periods of Quaternary climatic oscillations matched the astronomic periods calculated by Milankovitch. In other words, spectral analysis of climate showed coincidences with the spectra of orbital-element variations. This spectral coincidence has convinced many doubters that the Milankovitch scheme must be correct. No revised explanations of the rejected Milankovitch predictions have been proposed. Rather, the two sticking points have simply been swept aside. Despite them, the current Milankovitch vogue flourishes. The Milankovitch revival is based solely on the geochronologic results that connect the spectrum of climatic changes with the so-called Milankovitch periods.

Despite the Milankovitch revival based on spectral coincidence, several important problems remain. Use of computers to generate graphs of the Milankovitch equations of the variable orbital elements has resulted in the pronouncement that the computer-generated Milankovitch cycles become unstable after a few million years. Based on these results, astronomers warned geologists that in pre-Quaternary sediments, Milankovitch-type cycles should not exist. Evidence from the geologic record demonstrates that some fundamental flaw exists in the Milankovitch equations or in the computer programs: strata several hundred million years old display climatic cycles having the so-called Milankovitch periods. As we shall see in Chapter 15 and in a following section of this chapter, the variable orbital factors emphasized by Milankovitch affect the circumstances under which the Earth interacts with solar radiation. These include contrasts between the seasons (summer versus winter), alternating coincidences between Northern-Hemisphere summers and the orbital position of minimum Earth-Sun separation (perihelion) and of maximum Earth-Sun separation (aphelion), and alternating times of uniform Earth-Sun separation (virtually circular orbit) and of nonuniform Earth-Sun separation (orbit having maximum ellipticity). The question remains whether these changes are sufficient mechanisms for the inferred repetitive, oscillatory changes in the Earth's cli-

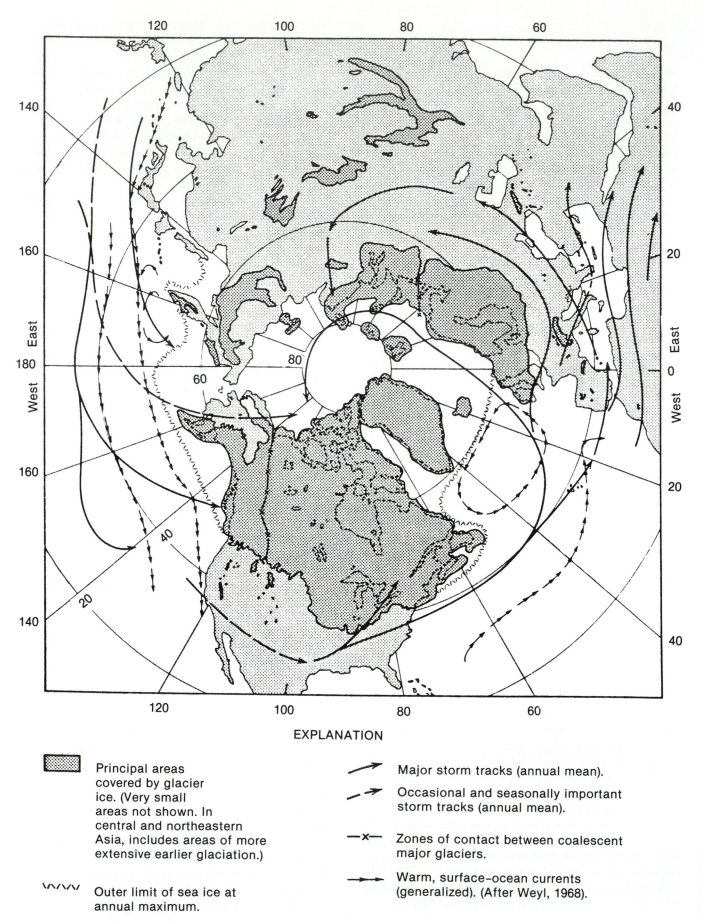

EXPLANATION

Principal areas covered by glacier ice. (Very small areas not shown. In central and northeastern Asia, includes areas of more extensive earlier glaciation.)

Major storm tracks (annual mean).

Occasional and seasonally important storm tracks (annual mean).

Zones of contact between coalescent major glaciers.

Outer limit of sea ice at annual maximum.

Warm, surface-ocean currents (generalized). (After Weyl, 1968).

FIGURE 1-5. Map of Northern Hemisphere showing extent of ice sheets during the Quaternary Period. (R. F. Flint, 1971; A, fig. 4-8, p. 74; B, fig. 4-9, p. 75.)

A. Extent of present-day glaciers.

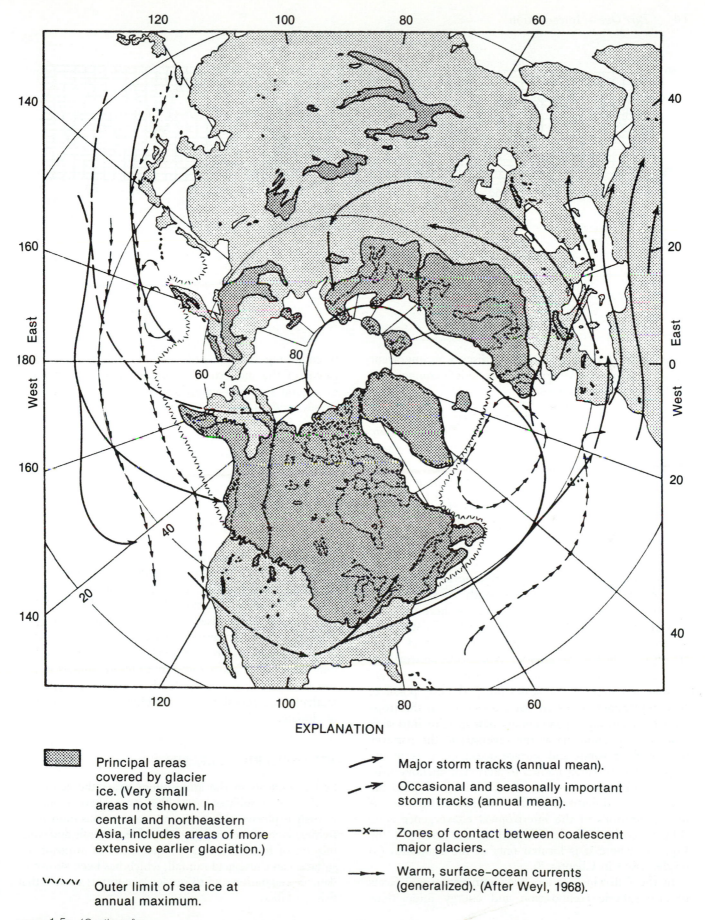

EXPLANATION

Principal areas covered by glacier ice. (Very small areas not shown. In central and northeastern Asia, includes areas of more extensive earlier glaciation.)

Outer limit of sea ice at annual maximum.

Major storm tracks (annual mean).

Occasional and seasonally important storm tracks (annual mean).

—x— Zones of contact between coalescent major glaciers.

Warm, surface-ocean currents (generalized). (After Weyl, 1968).

FIGURE 1-5. (Continued)
B. Extent of glaciers during a Pleistocene glacial maximum (without regard to age).

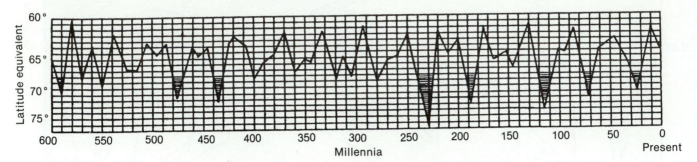

FIGURE 1-6.   Milankovitch climate curve computed for the last 600,000 years with summer insolation at latitude 65°N expressed in terms of latitude equivalent. Thus the curve at latitude 65° would be "normal"; the curve at any latitude higher than 65° represents times when the climate would be warmer than "normal"; and the curve at any latitude lower than 65° represents times when the climate would be colder than "normal." Large pointed troughs that have been filled in are inferred to coincide with particular European glacial episodes. (M. Milankovitch, 1941, fig. 48, p. 415.)

mate. In the following section we discuss an important climatic factor that Milankovitch excluded: composition of the Earth's atmosphere.

SOURCES: Berger, 1974; Broecker and Denton, 1990; Hays, Imbrie, and Shackleton, 1976; Olsen, 1986.

The Earth's Atmosphere: Greenhouse Effect, Human Activities, Global-Scale Natural Processes, and Extraterrestrial Factors

Many developments in the atmospheric sciences are of fundamental importance to the study of sedimentary deposits. Thanks to satellites, the atmosphere can now be viewed in a global context. Moreover, modern computers have been used to create mathematical models of atmospheric circulation. Worldwide networks of measuring stations have been established to record atmospheric conditions on scales not previously possible. New understanding of air-sea interactions in the tropics is contributing fundamentally new insights into such large-scale phenomena as the reversals in the patterns of prevailing winds and surface-ocean currents in the tropical Pacific (long known to Peruvian anchovy fishermen and named by them the El Niño) and into the monsoons and hurricanes that are related to the annual migrations of the intertropical convergence zone (ITCZ) both to the north and to the south of the Equator, where it is located only at the equinoxes (to be discussed in Chapter 8).

In the following sections we summarize some facets of atmospheric composition and energy interactions (highlighting sulfates, radiocarbon, and the *greenhouse gases*); of the modern-day increase in carbon dioxide; of natural variation in carbon dioxide and its possible causes; and of possible effects of adding heat to the Earth's atmosphere. We close by mentioning the atmospheric transport of sediment in two contrasting kinds of suspensions, low-altitude suspensions and high-altitude suspensions, as an example of another important global-scale process affecting climate.

Composition and Energy Interactions

Many new results from research into atmospheric physics and atmospheric chemistry are of great value to the study of sedimentary deposits. Atmospheric sulfates affect the acidity of rainfall. Reactions high above the Earth between atmospheric gases and the electromagnetic radiation from the Sun and from beyond the solar system (cosmic radiation) produce radioactive isotopes of carbon (carbon-14) and ozone. In addition, studies of the so-called greenhouse gases (such as carbon dioxide and methane) are revealing new insights into the dynamics of climate.

ATMOSPHERIC SULFATES

Sulfate radicals in the atmosphere affect the acidity of rainfall. The sulfates enter the atmosphere from extremely explosive volcanoes and from various human activities, such as the smelting of sulfide minerals and combustion of fossil fuels. Large quantities of atmospheric sulfates can cause acid rainfall, which has been shown to damage vegetation and to render lake waters so acid that fish are killed.

RADIOCARBON

In the upper atmosphere, incoming energy reacts with isotopes of nitrogen and carbon to form a radioactive isotope of carbon, ^{14}C or radiocarbon. Two reactions are as follows:

$$^{14}N - 1 \text{ electron} \rightarrow {}^{14}C \text{ (radioactive)} \qquad \text{(Eq. 1-1)}$$

By radioactive decay:

$$^{14}C + 1 \text{ electron} \rightarrow {}^{14}N \qquad \text{(Eq. 1-2)}$$

Although the founder of the radiocarbon-dating method, W. F. Libby (1908–1980), supposed that the natural production of radiocarbon is constant, subsequent research has demonstrated that the amount of radiocarbon produced varies in response to variations in the Sun's output (Figure 1-7). These variations in radiocarbon pro-

duction have been proved by an extensive series of measurements made on meticulously chosen samples of wood from the annual growth increments of trees whose ages have been established by the methods of dendrochronology (tree-ring analysis).

sources: Damon, Leman, and Long, 1978; Eddy, 1978; H. E. Suess, 1980a, 1981.

GREENHOUSE GASES

The reactions between incoming energy waves and the gases in the atmosphere vary according to the sizes of the gas molecules and the wavelengths of the energy waves. Figure 1-8 shows a spectrum of incoming radiation based on wavelengths. The incoming energy warms the surface of the Earth. The temperature of the Earth's

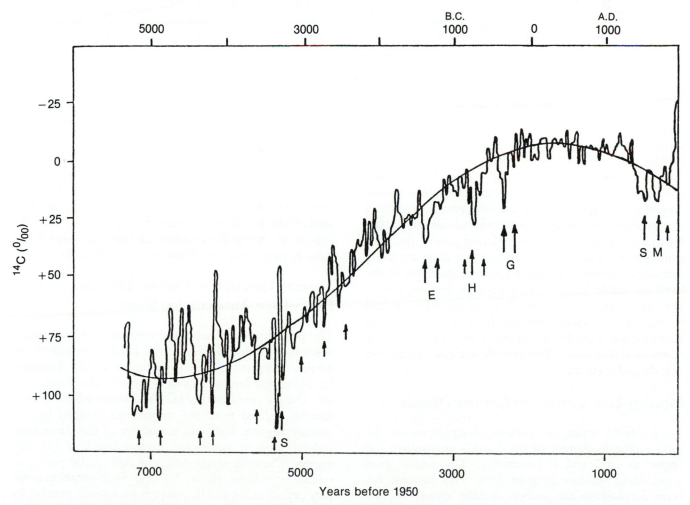

FIGURE 1-7. Chart showing radiocarbon deviations expressed as parts per mil (parts per thousand) from 5300 B.C. to 1950, derived from dating of samples from tree rings of known calendar age. (P. E. Damuth, 1977, in T. Landscheidt, 1987, fig. 25-6, p. 429.)

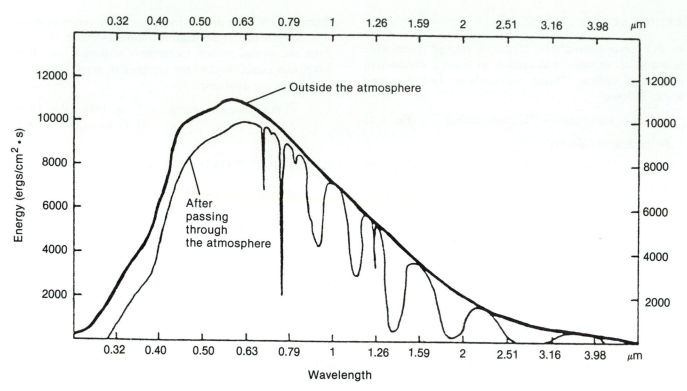

FIGURE 1-8.   Spectrum of radiation received by the Earth outside the atmosphere (upper curve) compared with that after the Sun's rays have passed through the atmosphere, shown on a graph of intensity of energy versus wavelength of radiation. (H. H. Lamb, 1972, fig. 2-11, p. 42.)

surface averages 285.5 K and ranges between 225.3 K for the polar zone 80 to 90° South latitude to 298.7 K for the equatorial zone 0 to 10° North latitude. According to Wien's Law relating the absolute temperature of a radiating body and the wavelength of emitted energy waves, the energy waves radiated outward by the Earth are concentrated in the infrared part of the electromagnetic spectrum. The greenhouse gases are those that allow incoming radiation (peak of spectrum in the band of visible light) to pass through but do not allow the backscattered infrared waves radiated outward by the Earth to escape upward. Instead, the gases trap heat by causing the Earth's radiated infrared waves to be re-radiated downward. Two greenhouse gases are carbon dioxide and methane.

Modern-Day Increase in Carbon Dioxide

A systematic series of modern observations of the carbon-dioxide content of the Earth's atmosphere was begun in 1957 and is continuing. The results show a persistent increase (Figure 1-9). This increase has been ascribed to the carbon dioxide created by the combustion of fossil fuels (coal, petroleum). According to this interpretation, the natural content of carbon dioxide in the Earth's atmosphere is considered to be constant. As the proportion of carbon dioxide in the Earth's atmosphere increases, so does the amount of re-radiated heat (greenhouse effect). The result is a carbon-dioxide heat increment that is added to the heat situation in the base climate, a general term for the sum total of all other factors that affect the heat content of the atmosphere apart from carbon dioxide and other trace gases having the same effects as carbon dioxide.

Natural Variation in Carbon Dioxide: Cores from Antarctic Ice Sheet

The presumption of supposed constancy in the natural amount of carbon dioxide in the Earth's atmosphere is invalid. Deep cores have been drilled in the Antarctic Ice Sheet, the longest one at Vostok, in the Soviet sector. The ice contains tiny bubbles, within which are gas samples derived from the ambient air trapped by the crystallizing ice. Laboratory analyses of the cores have yielded values of atmospheric carbon dioxide at different depths. These lie in the range of 200 to 300 parts per million by volume (Figure 1-10, A). Such variation is totally preindustrial and therefore is exclusively natural. In addition, from the same core, measurements have been made of the stable isotopes of hydrogen (deuterium) and of oxygen (^{18}O; Figure 1-10, B), which provide the basis

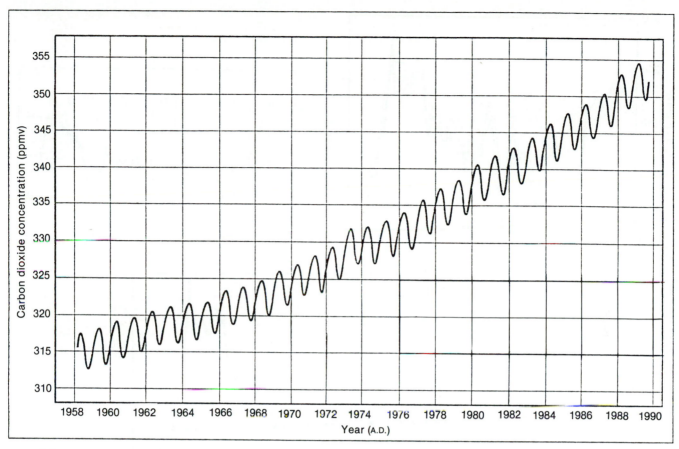

FIGURE 1-9.   Variation in carbon dioxide in the Earth's atmosphere, 1958 to 1989, at Mauna Loa Climate Observatory, Hawaii, shown by monthly averages that display the seasonal variation resulting from the uptake of carbon dioxide by deciduous plants as well as the general increase from 315 parts per million by volume in 1958 to about 353 parts per million by volume in 1989. (R. M. White, 1990, fig. on p. 39.)

for a temperature curve (Figure 1-10, C). Granted the validity of the relationships that have been developed between depth in the ice and age, then the Vostok ice core extends a record of atmospheric composition and temperature variations back 160,000 years. (See Figure 1-10, C.)

The results from the Vostok core clearly mean that any hypothesis for explaining the variations in the Earth's climate must include a mechanism by which the carbon-dioxide content of the Earth's atmosphere can oscillate cyclically at the same periods as the Milankovitch cycles. Contending candidate mechanisms for such oscillation include natural activities of the biosphere, changes in the circulation of the deep water in the ocean basins, and volcanic activity (discussed further in the following section). Chapter 15 discusses these in connection with a global view of the Earth's climate.

SOURCES: Broecker and Denton, 1990; Ekdahl and Keeling, 1973; H. D. Holland, 1978; Oeschger and Langway, 1989; J. Tyndall, 1861.

Atmospheric Suspensions

Particles from volcanic eruptions (and other kinds of particles) may be transported in suspension in the atmosphere around the globe. Deposits of such suspensions provide other important clues to ancient climates. The basic principles that govern the suspension of particles in a turbulent fluid are discussed in Chapter 7. However, we need to anticipate that discussion here. Without getting into the details, we simply note that the same principles apply whether the fluid is water or air. Because of air's low density and viscosity, one might suppose that *suspensions in air* (**eolian suspensions**) would always consist of small particles and that the particles would all be about the same size. As we shall see, this is true for some eolian suspensions, but not for others. Because of the structure of the Earth's atmosphere, suspended particles can travel at two distinct levels, forming *low-altitude suspensions* and *high-altitude suspensions* (Figure 1-11).

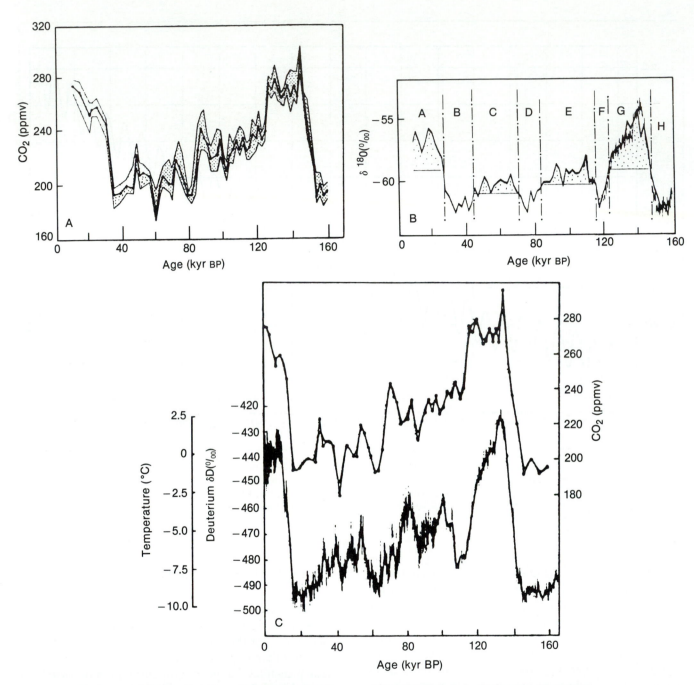

FIGURE 1-10.   Variations of carbon dioxide, δ oxygen-18, and δ deuterium with depth in Vostok ice core, Antarctica.

(A and C, from J. M. Barnola, D. Raynaud, Y. S. Korotkevich, and C. Lorius, 1987, fig. 1, p. 408 and fig. 2, p. 410; B, from C. Lorius, J. Jouzel, C. Ritz, L. Merlivat, N. I. Barkov, Y. S. Korotkevich, and V. M. Kotlyakov, 1985, fig. 1, p. 592.)

A. Carbon dioxide (parts per million by volume) recovered from bubbles in the ice. Notice that the range of values is from about 180 to 300 parts per million. Such variation is not in any way related to industrial-age combustion of carbon-bearing fuels, a factor held to be responsible for the upward trend of the curve of carbon dioxide in the modern atmosphere (shown in Figure 1-9). The present-day climate is thought to lie within the range of climates during the Pleistocene Epoch, yet the present-day proportion of carbon dioxide in the atmosphere (now about 370 parts per million by volume) lies well outside the range shown from these ice-core measurements. The significance of these relationships is yet to be determined.

B. δ¹⁸O; estimates of age of ice shown at top. Kyr = one thousand years.

C. δ deuterium (lower curve, with temperature scale) and carbon dioxide (upper curve) plotted against inferred age. B.P. = before present.

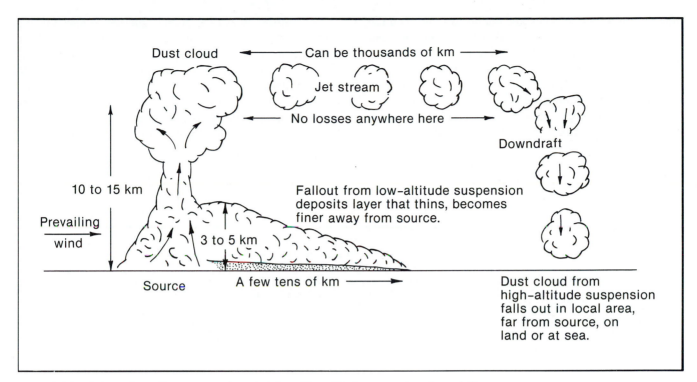

FIGURE 1-11.   Schematic profile contrasts characteristics of low-altitude suspensions (lower left) and high-altitude suspensions (top and right). Further explanation in text. (J. E. Sanders, in G. M. Friedman and J. E. Sanders, 1978, fig. 4-19, p. 103; based on L. Moldvay, 1961.)

LOW-ALTITUDE SUSPENSIONS

Low-altitude suspensions are *suspensions of particles that are confined to the lower 2 to 5 km of the atmosphere.* Within this zone are (a) suspensions blown upward by strong surface winds and (b) most of the clouds of material exploded out of volcanoes. In low-altitude suspensions, a clear relationship exists between the parent area of the suspended particles and the final deposit. In the deposits of low-altitude suspensions, the thicknesses of deposited layers and sizes of the particles within them become smaller downwind from the parent area.

HIGH-ALTITUDE TRANSPORT OF SEDIMENT BY JET STREAMS

By contrast, **high-altitude suspensions** are *eolian suspensions that travel in the Earth's high-altitude jet streams.* Particles are transferred upward from the Earth's surface into the jet streams only under especially energetic circumstances. These circumstances include explosions of nuclear weapons; violent volcanic explosions; and dry, twister-type windstorms. Once aloft, the particles tend to remain as distinct dust clouds. They are held aloft by such strong winds that generally no particles fall out from them until the suspension has reached some point of downdraft. There the suspended particles return to Earth.

High-altitude suspensions commonly contain admixtures of sand with the usual silt. No obvious connection exists between the deposit from a high-altitude eolian suspension and its parent area. In fact, the location of the parent area may be a complete enigma. The parent areas are known for a few examples. Volcanic material exploded from Hekla Volcano, Iceland, on 29 March 1947 reached an altitude of 9 km. The volcanic material formed a cloud that moved southward 880 km and then northeastward nearly 3000 km before landing in southwestern Finland 51 hours later. No particles fell out between Iceland and Finland. The mean speed of this high-altitude suspension was 75 km/h.

Some high-altitude suspensions remain aloft for years. Before coming to rest they may circle the Earth many times. High-altitude suspensions are capable of affecting the Earth's radiation balance and thus the Earth's climate. High-altitude suspensions are also the ones that leave erratic geologic records. We mention this point because, in evaluating the geologic record to check on ideas about volcanoes and climate, geologists have concentrated exclusively on the effects of low-altitude suspensions. By concentrating on products of low-altitude suspensions, they may have been looking in the wrong places. If any connection exists between the volume of volcanic material deposited by low-altitude suspensions on the one hand, and climate on the other, it is indirect, and functions via some connection between the volume of deposits of low-altitude suspensions and that of high-altitude suspensions. No known basis exists for making any connection between the quantities of deposits from these two kinds of suspensions. Therefore, in view of the foregoing, we think that the critical evaluation of the geologic record of volcanic activity with respect to ice-age climate has yet to be made.

An asteroid impact has been suggested to have formed dust clouds of high-altitude suspensions that lasted long enough to change the Earth's climate and create a darkness that caused mass extinctions at the end of the Cretaceous.

SOURCES: L. W. Alvarez, W. Alvarez, Asaro, and Michel, 1980; Moldvay, 1961; Thorarinsson, 1954.

Suggestions for Further Reading

BARNOLA, J. M.; RAYNAUD, D.; KOROTKEVICH, Y. S.; and LORIUS, C., 1987, Vostok ice core provides 160,000-year record of atmospheric CO_2: Nature, v. 329, p. 408–414.

CAROZZI, A. V., 1965, Lavoisier's fundamental contribution to stratigraphy: Ohio Journal of Science, v. 65, no. 2, p. 71–85. (Presents a translation into English from Lavoisier's original 1789 memoir of the French Royal Academy of Sciences, entitled: Observations generales sur les couches horizontales, qui ont été déposées par la mer, et sur les consequences qu'on peut tirer de leurs dispositions, relativement à l'anciennete du globe terrestre, including redrafted figures.)

CAROZZI, A. V., 1969, de Maillet's Telliamed (1748): an ultra-Neptunian theory of the Earth, p. 80–99 *in* Schneer, C. J., ed., Toward a history of geology: Cambridge, MA, Massachusetts Institute of Technology Press, 469 p.

CHEN, C-T. A., and DRAKE, E. T., 1986, Carbon dioxide (*sic*) increase in the atmosphere and oceans and possible effects on climate, p. 201–235 *in* Wetherill, G. W., Albee, A. L., and Stehli, F. G., eds., Annual Review of Earth (*sic*) and Planetary Sciences, v. 14, 1986: Palo Alto, CA, Annual Reviews Inc., 593 p.

DOTT, R. H., JR., 1988, Perspectives: something old, something new, something borrowed, something blue—a hindsight and foresight of sedimentary geology: Journal of Sedimentary Petrology, v. 58, p. 358–354.

DUVEEN, D. I., 1956, Lavoisier: Scientific American, v. 194, no. 5, p. 84–94.

LORIUS, C.; JOUZEL, J.; RITZ, D.; MERLIVAT, L.; BARKOV, N. I.; KOROTKEVICH, Y. S.; and KOTLYAKOV, V. M., 1985, A 150,000-year climatic record from Antarctic ice: Nature, v. 316, p. 591–596.

PLASS, G. N., 1959, Carbon dioxide and climate: Scientific American, v. 201, p. 41–47.

RAPPAPORT, RHODA, 1967, Lavoisier's geologic activities, 1763–1792: Isis, v. 58, p. 375–384.

SCHUTZ, L., 1980, Long range (*sic*) transport of desert dust with special emphasis on the Sahara: New York Academy of Sciences Annals, v. 338, p. 515–532.

SUESS, H. E., 1980, Radiocarbon geophysics: Endeavour (new series), v. 4, p. 113–117.

PART II

PARTICLES, SEDIMENTS, AND SEDIMENTARY ROCKS

Up to this point we have been focusing on global patterns of sedimentary strata. Now we must shift gears and look at small-scale features. We shall then work our way up to larger and larger elements until we arrive once again at a global perspective. This approach is necessary because the large-scale patterns and features cannot be understood unless their component parts have been mastered.

In this part we begin with the ultimate building blocks of sedimentary rocks: the particles. In Chapter 2 we first discuss the attributes of individual particles, including their sizes and the classification of particles by size, their shapes, and how they are classified according to composition and origin. We do this because it is instructive to approach sediments and sedimentary rocks with an understanding of their component parts. Only with a thorough grounding in the nature of sedimentary particles is one ready to tackle the behavior of the huge masses of particles that constitute bodies of sediment.

After classifying particles on the basis of mode of origin (extrabasinal, carbonaceous, pyroclastic, and intrabasinal), and also by mineral composition and biological influence, we consider properties of particle aggregates. These include bulk density, porosity, permeability, and magnetic properties. Throughout much of the remainder of this book we shall deal with the bulk properties of sediments (e.g., color, mineralogic composition, texture, and fabric); how sediments and depositional environments interact by temperature and pressure; and how sedimentary layers combine to form patterns characteristic of different environmental settings and different formative processes. Ultimately, however, it all begins with particles.

Chapter 3 moves on to the all-important subject of **diagenesis**, *the sum of physical, inorganic chemical, and biochemical changes in a sedimentary deposit after its initial accumulation and excluding metamorphism.* A key diagenetic change is the conversion of sediments into sedimentary rocks. Chapter 3 therefore is the link between sand and sandstone, mud and shale, carbonate sediment and limestone. We begin by discussing water, Planet Earth's ubiquitous liquid solvent, the physical- and chemical characteristics of water, and its effects on the common sedimentary minerals, especially silica and calcium carbonate.

We then take up diagenetic processes that operate in the near-surface realm. These include processes operating above and below the water table, in arid- and humid climatic settings, on the sea floor, and in the shallow subsurface. Then we consider diagenetic processes and products that characterize the deeper subsurface realm. As with particles, the subject of Chapter 2, we shall in a sense leave diagenesis behind in later parts of this book. However, in another way it will always be with us, for we shall again and again hark back either to diagenetic processes or to their products. A basic comprehension of the ways in which sedimentary rocks form from sediments is essential background material for their further study.

In the final chapter of this part we classify sedimentary rocks. Fundamentally, if we are to discuss sedimentary rocks intelligently, then we must agree on names for them. Even beyond the simple need to acquire a common vocabulary, classifications reflect ideas about the origins and nature of what we talk about. Therefore Chapter 4 is far more than a simple list of words and their definitions. In this chapter we consider the differences between descriptive- and genetic classifications, and then describe some of the important classification schemes that have been devised for sedimentary rocks. We shall see that classification schemes depend intimately on ideas about what are the essential formative processes that create sedimentary rocks.

At the end of Part II the reader should be armed with a thorough knowledge of what sediments are made of, how they become rocks, and how and why we classify them as we do. This should provide a firm foundation for subsequent discussions of stratigraphy, process sedimentology, and environments of deposition in Parts III and IV.

CHAPTER 2

Sediments: Names, Particles, Bulk Properties

In this chapter we start our systematic discussion of sedimentary deposits by looking first at sedimentary particles. We begin with properties of individual particles and move on to properties of aggregates of particles. Under individual particles, we present systematic definitions of their sizes and shapes. Next follows a classification of sediments based mostly on particle sizes. Then we return to particles again and classify them by source, and within each of the categories, by mineral composition. We close with some important bulk properties: color, fabric, bulk density, porosity, permeability, acoustic impedance, electrical properties, magnetic properties, and radioactivity.

Individual Particles

Sizes

The sizes of sedimentary particles enter into so many important aspects of sedimentology that a standard scale of sizes has been established. We discuss grade scales, summarize the names of particles based on size, and present ways of naming particle aggregates based on their ranges of particle sizes, a concept known as sorting.

SEDIMENT GRADE SCALES

As a basis for standardizing analyses of sediments, a specific grade scale has been established (Figure 2-1). Familiarity with the important boundaries that define the major categories of the easily visible groups is so important that students should do whatever is necessary to acquire the skill of recognizing these categories. Study of a standard reference set of sieved sand grades is a good way to master the skill of recognizing the standard subdivisions of sand.

NAMES OF PARTICLES BY SIZE CLASS

As shown in Figure 2-1, the major classes of sediment particles defined by size, from largest to smallest are *boulders*, *cobbles*, *pebbles*, and the various grades of *sand*, *silt*, and *clay*. The sizes of some common reference objects (volleyball, tennis ball, head of a wooden match, and lower limit of visibility with the unaided eye) have been added to Figure 2-1.

The size classes defined in Figure 2-1 form an octave scale: a geometric scale in which the adjacent orders within the scale differ by a factor of 2. This means that going toward the smaller size classes, the ratio is one half, whereas going in the direction of the larger sizes, the ratio is 2. The bounding sizes of each class can be expressed not only by their dimensions in millimeters but also by the negative logarithm of this dimension to the base 2. Such a scale having $\log_2$ for the class boundaries is known as the ϕ scale. Hence, in the ϕ scale, 8 mm, which is 2^3 mm, is written as -3ϕ; 4 mm (2^2 mm), becomes -2ϕ, and so on, as shown in the figure. The mathematical expression of the ϕ scale is

$$\phi = -\log_2 d \qquad \text{(Eq. 2-1)}$$

where d is the particle diameter in millimeters. ϕ values are dimensionless numbers, and because they are based on millimeters, the ϕ scale is metric.

The principal advantages of the ϕ scale are that (1) the distribution of particle sizes can be plotted with ease on arithmetic graph paper, and calculation of statistical parameters such as mean, median, standard deviation, and skewness is simplified; (2) limiting particle diameters for each size class become whole numbers instead of fractions of millimeters or micrometer values employing two or three digits (See Figure 2-1.); and (3) because of the use of the negative log, increasing ϕ values

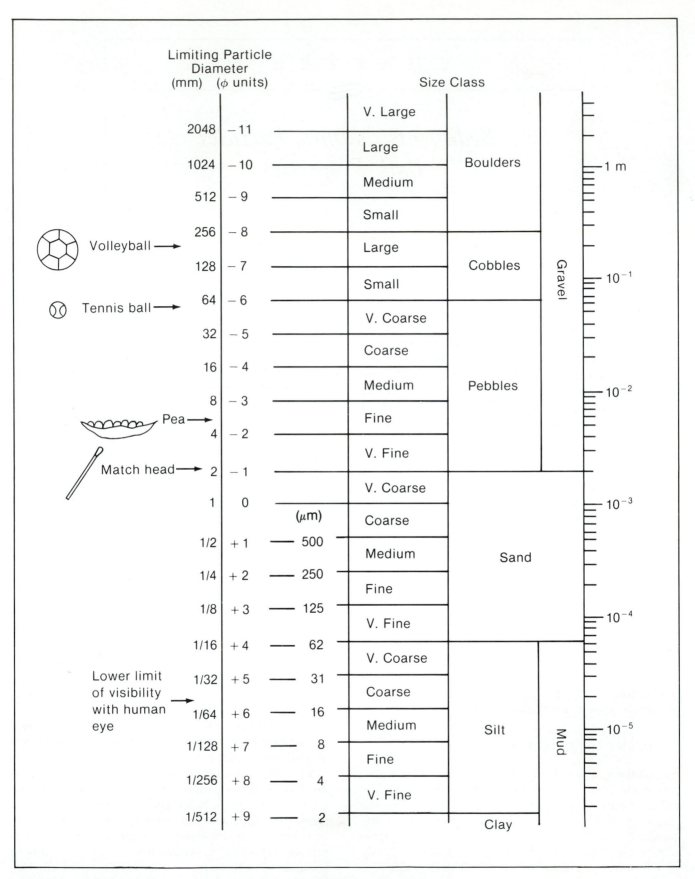

FIGURE 2-1.   Standard size classes for sediment particles. Text explains φ units. (Compiled from W. C. Krumbein, 1936; W. F. Tanner, 1969; J. A. Udden, 1914; and C. K. Wentworth, 1922). This scale follows the recommendations of the Intersociety Grainsize Study Committee of the Society of Economic Paleontologists and Mineralogists, under Tanner (1969). Several published versions that do not follow this committee's recommendations differ from the one shown here.

24

correspond to decreasing particle sizes. This usage agrees with the geological practice of plotting larger sizes to the left, and smaller sizes to the right on graphs. Modern sedimentologists use ϕ units instead of micrometers and millimeters; ϕ units can be derived either from Eq. 2-1 or from a nomograph (Figure 2-2).

Shapes

The concept of shape or form can be resolved into two geometric components: (1) sphericity and (2) roundness. **Sphericity** (equidimensionality) is defined as *the degree to which a particle approximates a spheroid (sphere or cube)*. In other words, sphericity expresses how nearly equal the three mutually perpendicular internal axes are.

Roundness refers to *the sharpness or degree of curvature of corners or edges of a particle*. For example, a cube is an example of a spheroid having sharp edges and corners. Its roundness is thus zero. By contrast, a sphere possesses perfect roundness. Box 2.1 discusses how these two attributes are measured and expressed numerically, and includes some examples of attempts to use shape in geologic studies, as well as some examples of varied shapes secreted by organisms.

Names of Particle Aggregates Based Mainly on Sizes and Sites of Origin of Particles

Sediments can be classified and named in a purely descriptive, objective, and precise manner. Such a classification is a descriptive classification. Contrasting with a descriptive classification is a genetic classification, which conveys what is known, inferred, or believed about the origin of a given sediment or sedimentary rock. Some classifications combine both objectives and are descriptive-genetic. In the following discussion we begin with a descriptive classification of sediments based on size (and in some cases on shape). We conclude with an example of a descriptive-genetic classification that is based on sites of origins of the particles.

Sediments Classified by Particle Sizes

The individual particles of a sediment may vary in size from large boulders down to submicrometer size. Figure 2-1 defines the standard classes of sediments based on size as *clay* and *silt* (or *mud*), *sand, pebbles, cobbles,* and *boulders*.

These classes are sometimes referred to by the Latin terms proposed by A. W. Grabau (1904). These are

rudite (from *rudus*, rubble), for gravel-size materials; *arenite* (*arena*, sand), for sand; and *lutite* (*lutum*, mud), for silt-clay mixtures.

Mixtures of sizes may be expressed in diagrams that contain 3 or 4 end members. Mixtures with 3 end members can be expressed on triangular diagrams. Those with 4 end members require tetrahedral diagrams (Figure 2-3). In a depositional setting, these may be mixed in almost any proportion. In Figure 2-4 textural class limits are shown for a triangular diagram with clay, silt, and sand as end members. Silt and clay are commonly grouped as mud. This is not altogether appropriate, for in origin, mineral composition, and response to physical forces silt and clay differ.

Figure 2-5 (on page 41) is a textural triangle having as end members mud, sand, and gravel as well as sediments composed of varying mixtures of these. Table 2-1 shows some common names used for sedimentary aggregates.

For carbonate sediments, the end members are *lime mud, lime sand,* and *lime gravel,* and mixtures may once again be of any proportion, as indicated in Figures 2-4 and 2-5.

This classification is equally valid for **tephra**, *a general term used for all particles exploded from volcanoes*. In the classification of tephra by size, three categories are used. These are not the same as those used for other sediments. The tephra size categories are (1) particles larger than 64 mm, (2) particles between 64 and 2 mm, and (3) particles smaller than 2 mm. (See Table 2-1.)

Tephra larger than 64 mm are further divided into two groups based on shape. These are **bombs**, *ejected as liquid lava that solidified during flight*, and **blocks**, *ejected from a volcano as solids*. As a result of their solidification during flight, bombs display twisted forms and concentric structure. By contrast, blocks consist of angular fragments of volcanic rock.

Particles in the intermediate size range are named **lapilli**, defined as *pyroclastic particles 2 to 64 mm in diameter,* and **ash**, *pyroclastic particles in the fine size range*. Ash may be divided into coarse ash (62 μm to 2 mm in diameter) and fine ash ($\leq$ 62 μm). Figure 2-6 (on page 41) is a textural triangle of end members for plotting tephra.

SOURCES: Kittleman, 1979; Lajoie, 1979; Schmincke, 1988.

Concept of Sorting

A fundamental property of a deposit that implies important information about the conditions of transport and deposition is the extent to which the coarser particles (commonly sand size or larger), which are collectively

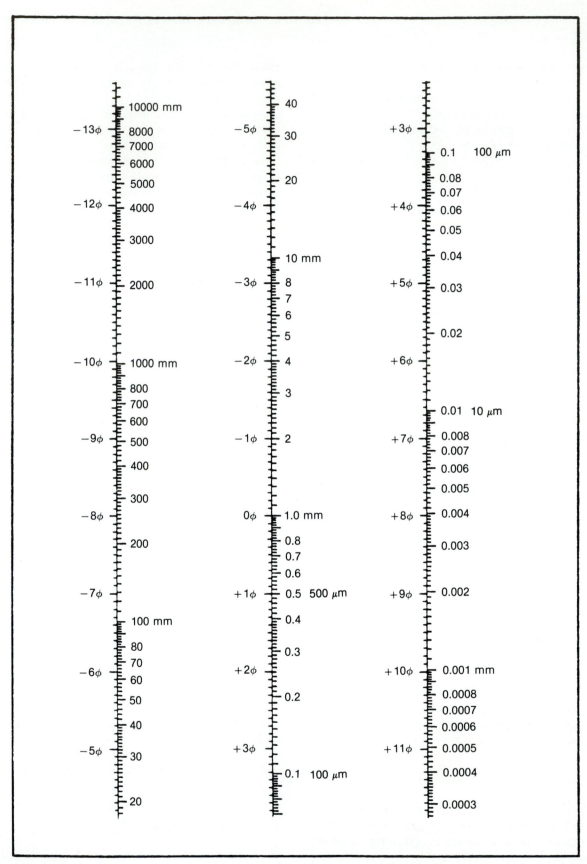

FIGURE 2-2. Nomograph for converting values of ϕ scale to millimeters and vice versa. (Compiled from various sources.)

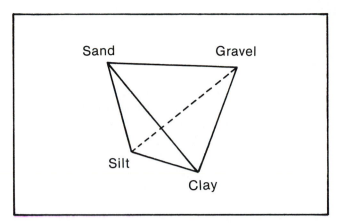

FIGURE 2-3. Tetrahedron having as corners clay, silt, sand, and gravel. (W. C. Krumbein and L. L. Sloss, 1963, fig. 5-6, p. 158.)

TABLE 2-1. Names of aggregates of sediments and of tephra based on limiting diameters

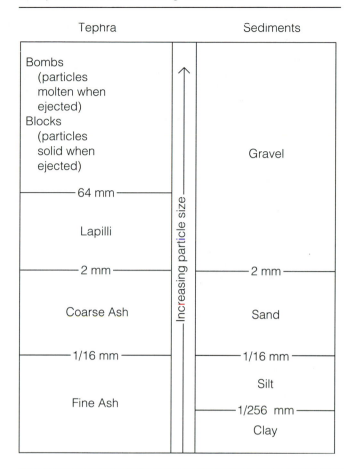

Tephra	Sediments
Bombs (particles molten when ejected) Blocks (particles solid when ejected)	Gravel
——— 64 mm ———	
Lapilli	
——— 2 mm ———	——— 2 mm ———
Coarse Ash	Sand
——— 1/16 mm ———	——— 1/16 mm ———
	Silt
Fine Ash	——— 1/256 mm ———
	Clay

(arrow labeled "Increasing particle size" between columns)

SOURCE: C. R. Longwell, R. F. Flint, and J. E. Sanders, 1969, part of table C-3, p. 625, with modification based on the Intersociety Grainsize Study Committee, W. F. Tanner, 1969.

designated as the *framework*, have become separated from the finer particles (commonly silt- and clay-size, and sometimes referred to as mud) that are known as the *matrix*. The term **sorting** refers to *the selection, during transport, of particles according to their sizes, specific gravities, and shapes*. Deposits that contain only a small range of particle sizes are referred to loosely as well sorted (this term also can be defined numerically). Box 2.1 explains how numerical data on the size distribution of particles form the basis for expressing sorting numerically. Here we want to introduce a few general concepts and definitions.

In many depositional settings, sediments spanning a wide range of particle sizes have not been sorted. A general term for such a sediment is a **diamicton**, defined as *any nonsorted- or poorly sorted terrigenous sediment that consists of sand and/or larger particles in a muddy matrix*.

Where framework particles are abundant enough to be in contact with one another, the term **particle-supported fabric** is used (Figure 2-7, A). *Where the matrix is so abundant that the coarser particles are not in contact with one another*, the fabric is described as **matrix supported** (Figure 2-7, B). In such a fabric, coarse particles are suspended or "floating." Depending on the dominant size class in the matrix, a sediment can be further specified as being *clay supported, silt supported,* or *mud supported*. Rarely, particles coarser than silt size can form the matrix of very coarse sediments.

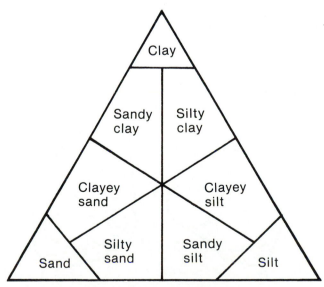

FIGURE 2-4. Triangular diagram having sand, silt, and clay as end members; areas within triangle enclosed by lines show names for sediments consisting of varying mixtures of the end members. (W. C. Krumbein and L. L. Sloss, 1963, fig. 5-7, p. 159.)

BOX 2.1

Shapes- and Sizes of Sedimentary Particles

This section discusses how the attributes of shape, sphericity and roundness, are measured, presents examples of how measurements of shape and roundness have been used in geologic studies of sedimentary deposits, follows with some examples of the varied shapes secreted by organisms, and closes with a discussion of numerical methods of particle-size analysis.

Measurement of Sphericity

A standard method of referring to the *axes* of a sedimentary particle has been established. These *axes* of a particle are defined by visualizing a rectangle that encloses the particle projection in such a way that the longest dimension of the particle (defined as axis *a*) is parallel to the length of the rectangle. The intermediate axis (defined as axis *b*) of the particle is then parallel to the width of the rectangle. The shortest axis (defined as axis *c*) is determined at right angles to the other two axes.

Sphericity can be specified numerically by means of two ratios between these axes: b/a and c/b (Box 2.1 Figure 1).

Using the value of 0.67 as a dividing line, the field of the diagram can be subdivided into four regions and thus can be used to define four categories of objects based on shape: (1) spheroids (both b/a and $c/b > 0.67$); (2) rods ($b/a < 0.67$; $c/b > 0.67$); (3) blades (both b/a and $c/b < 0.67$); and (4) disks ($b/a > 0.67$; $c/b < 0.67$).

The properties displayed in processes of transport and deposition of these categories of particles that are based on shape differ. Rods resemble spheres in tending to roll. Disks and blades do not roll.

Measurement of Roundness

Roundness refers to the sharpness or degree of curvature of corners or edges of a particle (Box 2.1 Figure 2). Two sets of reference objects have been sketched in Box 2.1 Figure 3 to show how equal sphericity can be accompanied by contrasting roundness. Roundness and sphericity of particles are largely independent; for example, wind is an extremely effective rounding agent but hardly affects sphericity.

Box 2.1 Figure 3 shows outlines of sand particles that range from ovals to circles (expressions of sphericity in two dimensions) with varying angularity, ranging from very angular (at left) to well rounded (at right).

The shapes of terrigenous particles reflect their origins, internal lattices, structures, and histories.

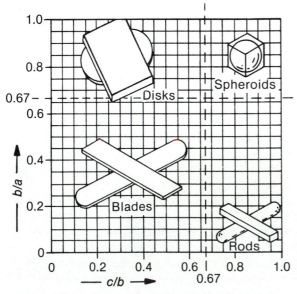

BOX 2.1 FIGURE 1.   Graph of sphericity with angular- and rounded rods, blades, and disks shown for comparison. Text defines ratios. (After W. C. Krumbein and L. L. Sloss, 1951, fig. 4-7.)

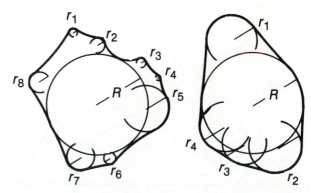

BOX 2.1 FIGURE 2.   Roundness as illustrated by idealized cross sections through two sedimentary particles showing definitions of radii of individual corners (or edges) as r_1, r_2, etc., and of maximum inscribed circle of radius R. (W. C. Krumbein, 1940, fig. 11, p. 670).

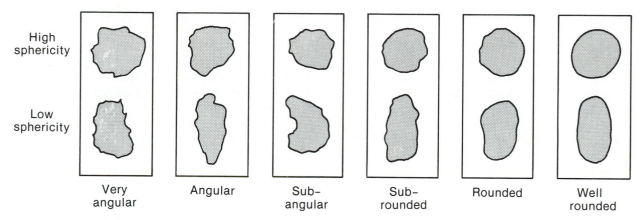

BOX 2.1 FIGURE 3.   Six roundness classes defined by sections through particles having high- and low sphericities. (After M. C. Powers, 1953, fig. 1, p. 118.)

Geologic Significance

Shapes of particles depend on (1) initial form, including shapes that may be inherited from previous cycles of erosion, transport, and deposition; (2) composition (whether a particle consists of one or several minerals or of rock fragments); (3) hardness, brittleness, or toughness; (4) inherited partings, such as crystal boundaries, fractures, bedding, schistosity, or cleavage; (5) size; (6) agent of transport; (7) rigors of transport, including distance and energy of the agent of transport; and (8) random effects.

Weathering of bedrock releases terrigenous particles having multitudes of shapes that reflect their conditions in the parent rock. Beautiful shapes of euhedral quartz crystals (Box 2.1 Figure 4) that have grown in a limestone host have fooled trusting souls into thinking that they owned dia-

monds. True diamonds may accumulate in gravel, fully preserving their euhedral form. Loosely consolidated sandstone of Pennsylvanian age on the north flanks of the Wichita-Amarillo arch in Oklahoma and Texas consists of very coarse euhedral- to subhedral quartz crystals that were liberated from volcanic rocks as the arch was uplifted and stripped. Derivation of these crystals from volcanic rocks can be inferred from their shapes, for euhedral quartz derived from volcanic rocks is distinctive. Other shapes that are diagnostic of origin include the glass shards of tephra (Box 2.1 Figure 5)

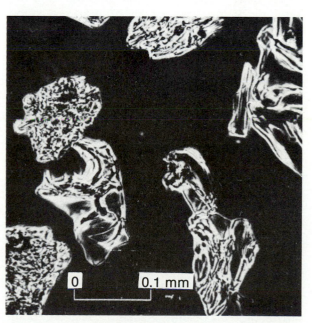

BOX 2.1 FIGURE 5.   Tephra, consisting of glass shards having irregular shapes, exploded from Mount Mazama (ancestral peak at Crater Lake), Oregon. Photomicrograph of specimen from Creston Bog, Washington. (R. B. Taylor, U.S. Geological Survey.)

BOX 2.1 FIGURE 4.   Euhedral quartz crystals released from limestone (*Herkimer diamonds*), Little Falls, New York. (Authors.)

and the spindle-shaped bombs thrown out by volcanoes.

During transport, as particles collide with one another or with bedrock, their edges and corners are worn off. Thus angular particles tend to become rounded. The assumption has been made that by examining a round pebble or sand grain, one could learn something about how far it may have traveled. In fact, the obsession of many sedimentologists with this aspect of the problem of shape has led to intensive mathematical analysis and quantification of shape in particles of terrigenous sediment. Some successes have been achieved, but in the final analysis this quantification has not led to the hoped-for insight into particle histories. Because the factors affecting the ultimate shapes of particles are so numerous, no definitive quantitative relationship between particle shape and particle history has yet been achieved.

Particle shape is important for another reason; shape determines how a particle is transported. Shape influences mode of transport in water, for example, for it partially determines whether a particle will roll or will be carried in suspension; hence shape partly controls the behavior of a particle in falling through the fluid. During settling through water, a rod-shaped particle falls faster than a disk-shaped particle having the same volume and density. Hence a water current carrying both kinds of particles in suspension will drop the rods before it drops the disks. Heavy minerals are sorted not only by their specific gravities, as emphasized in Chapter 7, but also by their shapes. For example, one might expect that, because they are six or seven times denser than most common accompanying minerals, all gold particles would settle rapidly to the bottom of gravel. However, only compact gold particles settle to the bottom. By contrast, very small gold particles that become scaly and flat are readily suspended. In rivers of moderate grade, these tiny flakes can be carried hundreds of kilometers, and may finally settle in thin streaks on river bars. In the zone of back-and-forth movement of waves on beaches, disk-shaped gravel particles tend to be carried up the beach by the swash. Spherical- and rod-shaped gravel particles tend to roll down the beach with the backwash. Thus this selective movement of particles based on shape is capable of separating gravel into two discrete assemblages, one dominated by disks and the other by spheres plus rods.

Despite previous failures in relating quantitative expressions of shape to geological processes, careful analysis of particle shape can contribute impor-

tant insights in the study of sedimentary deposits. We need to consider quantitative methods that offer the best approach for new understanding using carefully designed field experiments. Such experiments attempt to study one variable at a time, keeping other variables constant. One such experiment might include the release of carefully shaped and measured blocks of various lithologies, one lithology at a time, in a stream at a given location. Another way in which the evaluation of shape may be used to enhance its utility for determining the environments of transport or of deposition is to use other criteria to constrain possible interpretations of particle shape. For example, facies relationships and fabric criteria were used to identify Precambrian sediments of Svalbard (Norway) as glacial. Particle shape, fabric, and surface characteristics were then used to assign these sediments to glacially related subenvironments.

Distinctly shaped sedimentary particles deposited by one process may preserve these shapes even after they have been eroded and redeposited by another process. Sand-size particles that were transported by running water do not commonly preserve shape characteristics imposed upon them by the water because they tend to retain shape characteristics imposed by their parent deposits or during previous sedimentary cycles. Thus shapes may yield information about provenance or about transporting- or depositing agents, but not necessarily about the agents responsible for the particles' most-recent episode of transport and -deposition.

Shapes of Skeletal Materials

The shapes of biocrystals and of tests secreted by organisms range from simple to exceedingly complex. For example, coccoliths secrete disks (Box 2.1 Figure 6), fusulinids are spindle shaped, crinoid columnals are button shaped, calcispheres are spherical, and the spines of some sea urchins range from cones to irregular clublike objects (Box 2.1 Figure 7). Some tests that occur as particles are beautifully ornate (Box 2.1 Figure 8); others are simple. Some are bilaterally symmetrical, others radially symmetrical, and still others asymmetrical.

The Quantitative Analysis of Particle-Size Distribution

The movements of air and water commonly separate particles by their sizes; thus many sedimentary deposits consist of pure sand, silt, or clay. However, where sediments from contrast-

BOX 2.1 FIGURE 6.   Greatly enlarged views of calcareous biocrystals (coccoliths in Lower Cretaceous nannoplankton) recovered from Munk Marl Bed in core E-1, Danish North Sea, at a depth of 2.5 km as seen in scanning-electron micrographs. (Erik Thomsen, 1989, fig. 2, p. 716.)
   A. Plates exclusively of *Micrantholithus obtusus.*
   B. Plates exclusively of *Axopodorhabdus dietzmannii.* Scale bars in both indicate 100 μm.

ing parent deposits converge, mixtures of sizes are common. Such mixtures of sizes, known as populations, are defined in three categories that are collective terms for certain groups of sizes: gravel, consisting of particles that individually may be boulders, cobbles, or pebbles; sand, which may be very coarse, coarse, medium, fine, or very fine; and mud, which may consist of clay and various size classes of silt. (See Fig. 2-1.) Even these three categories are subject to mixing, and the name of the less-common size class reverts to an adjective—thus muddy gravel, muddy- or silty sand, and pebbly silt. The proportions of each size class can be visually estimated in the field or in hand sample and can be measured in the laboratory and expressed as percentages.

It should be understood that these terms refer strictly to size and make no reference to composition. Although sands commonly consist of the mineral quartz, the beaches of Bermuda are composed of sand consisting of skeletal particles. In places on the island of Hawaii the particles of the beach sand consist mostly of the mineral olivine, and the sands are dark green. Near Rome, Italy, beaches consist of volcanic rock fragments. Along Bristol Bay, Alaska, the beaches in places are formed of black sands, chiefly magnetite.

BOX 2.1 FIGURE 7.   Spines of echinoid. At left, heap of individual spines separated from skeleton. At right, upper surface of skeleton with spines still attached. South shore, Sicily. (Authors.)

BOX 2.1 FIGURE 8.   Tests of radiolaria; sketches made by viewing through a microscope. (Ernst Haeckel, 1925, pl. 4, p. 68; scale not given.)

Beaches of south India, near Madras, are rich in monazite. The beaches of Lake Champlain, in New York, are enriched in red garnet sand. Beaches in populous areas commonly contain significant admixtures of plastic particles. Among the commonest constituents of most beach sands are calcium-carbonate skeletons. Thus sand is a size term only. As a size term, clay does not equate with clay minerals. In fact, some clays contain more quartz than clay minerals.

The distribution of sizes in sediments relates to (1) the availability of different sizes of particles in the parent material, (2) processes operating during sediment transport and -deposition, particularly the competency of flow, and (3) diagenetic processes that operate after deposition to change the sizes of some particles and to destroy other particles. Certain sediments have never been sorted by size and inherit a wide range of particle sizes from a single origin. These include the deposits of pyroclastic hot-particle flows and of debris flows composed of pyroclastic debris (lahars) that evolve from pyroclastic hot-particle flows. As discussed in Chapter 7, certain kinds of particle-transport mechanisms, such as pyroclastic hot-particle flows and debris flows, possess a very limited particle-sorting ability and therefore deposit sediments that contain particles having a wide range of sizes. However, the range of particle sizes in most sedimentary deposits is restricted. To quantify the significance of the distribution of sizes in terms of processes, we introduce the statistical concept of the normal distribution.

The Normal Distribution

The *normal distribution* (also known as the Gaussian distribution) is a powerful function that is applied to populations; its applications in geology are numerous. The normal distribution is used in sedimentology, especially in the study of the particle sizes of sediments. The word normal is a statistical term and does not mean the opposite of abnormal nor the direction perpendicular to some plane.

The normal curve is a special kind of frequency-distribution curve. A **frequency-distribution curve** is *a plot of the frequency with which some variable occurs within arbitrarily defined subclasses of a population.* For example, in sedimentology, a size-frequency curve is made by plotting the proportion, in percent, of various particle sizes whose class limits are expressed by the units on the ϕ scale. Conventionally, frequency is plotted along the vertical axis (ordinate) and the ϕ sizes along

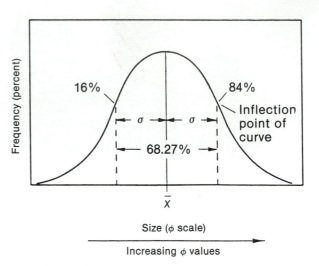

BOX 2.1 FIGURE 9.   A normal frequency-distribution curve. Text explains symbols.

the horizontal axis (abscissa). If the particle sizes are distributed normally, the central part of the population contains the bulk of the particle sizes and the coarser- and finer parts are equally distributed on each side of this central part. When the frequencies of these sizes are plotted, a bell-shaped curve results (Box 2.1 Figure 9). By definition, all normal distributions plot as symmetrical, bell-shaped curves. Normal distributions must also satisfy other requirements. The two points of inflection on the flanks of the curve are positioned such that the area under the curve between these points is equal to 68.27% (rounded to 68%) of the area under the entire curve, which is equal to 100%. (See Box 2.1 Figure 9.) Half of this middle 68%, or about 34%, lies on each side of the center of the distribution, which is the 50% point. The percentage borders of the middle 68% lie at values of 16 and 84%, respectively. This distance from the middle of the distribution is the *standard deviation* (σ). The peak of the normal curve occurs exactly in the middle of the distribution and is known as the *mean* ($\overline{X}$). As a first approximation, a normal curve is defined in terms of the numerical values of the mean ($\overline{X}$) and of the standard deviation (σ).

Curves of normal distributions having the same means and enclosing the same total areas may display different shapes (Box 2.1 Figure 10). This difference of shape dictates that the number of ϕ classes on the abscissa between the points of inflection will not be the same for each curve. (See Box 2.1 Figure 10.) If a curve is steep, the number of ϕ classes between the two points is small (Curve A in Box 2.1 Figure 10); if a curve is flat

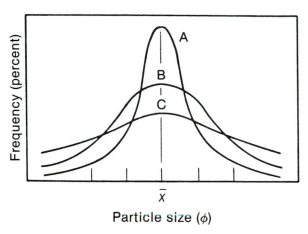

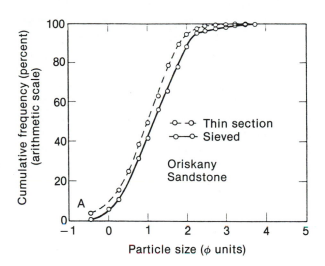

BOX 2.1 FIGURE 10. Normal size-frequency-distribution curves of various shapes. Further explanation in text.

and wide, the number of ϕ units between these points is large (Curve C in Box 2.1 Figure 10). This spread of the distribution about the mean defines the concept of sorting. A frequency curve of a well-sorted sediment is sharp peaked and narrow; this means that the sample includes only a few size classes. In contrast, the frequency curve of a poorly sorted sediment is low and wide; the sample includes many size classes. The specifics of sorting and the frequency distribution of sediments in terms of standard deviation are discussed farther along.

The Cumulative-Frequency Curve

In simple frequency curves, the only kind of curve we have examined thus far, each point plotted represents the frequency of that size class only. However, in the study of populations of particle sizes, another kind of curve, the *cumulative-frequency curve,* is particularly useful. The cumulative-frequency curve differs from the simple-frequency (incremental-frequency) curve in that each point represents not merely the frequency of each class but the sum of all percentages of the preceding size classes. A cumulative-frequency curve starts from 0% at the left and climbs toward 100% on the right as each frequency class is added. Various kinds of vertical scales can be used; the same ϕ scale is used on the horizontal axis. Two common vertical scales are arithmetic and probability.

When cumulative-frequency curves of normal distributions are plotted using an arithmetic vertical scale, S-shaped- or ogive curves result (Box 2.1 Figure 11, A). No further comments are needed about the arithmetic scale.

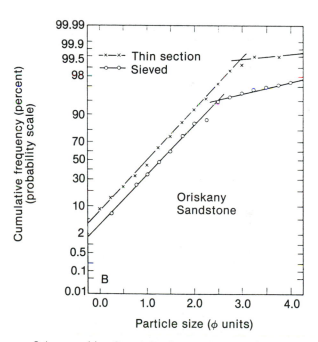

BOX 2.1 FIGURE 11. Cumulative-frequency curves having arithmetic (A) and probability (B) scales for ordinates. (After G. M. Friedman, 1958, fig. 1, p. 404.)

A probability scale is constructed so that the points in the central part of the graph, between 30 and 70%, are close together and the points at the ends of the graph, especially less than 10 and more than 90%, are widely spaced. On probability paper, normal distributions plot as straight lines. (See Box 2.1 Figure 11, B.)

To determine if a distribution approximates normality, simply plot the data graphically on probability paper. (Many statistical software packages contain routines to construct automatically all of the simple kinds of graphs, as well as to calculate the various statistical parameters, that are discussed in this section.) Probability plots of cumulative curves are more useful than ogive curves with arithmetic scales because (1) they test for normality of a distribution, (2) interpolation for statistical measures is more accurate, (3) the slope of the line is a function of the standard deviation of the distribution (a steep slope means a low value for the standard deviation and a gentle slope means a high value), and (4) separate subpopulations, if they exist, may be identifiable as individual straight-line segments. It is also possible to estimate graphically the relative proportions of different subpopulations.

Probability Plots

When cumulative-frequency distributions of the particle sizes of sediments are plotted on probability graphs, one of the most-striking observations is that most of them do not come out as one continuous straight line. In fact, commonly two or three or even more straight-line segments are present, each having a different slope and separated by a sharp break between the segments (Box 2.1 Figure 12). These segmented curves indicate the inappropriateness of considering the entire sample as one normally distributed population. The slope of each straight-line segment and the positions of the breaks between segments are functions of the mechanisms of deposition. In sands with three straight-line segments, the central- and larger segment may represent a subpopulation that moved in the current by a jumping motion (saltation; Chapter 7), the segment at the fine end of the distribution commonly resulted from deposition of particles carried in suspension, and that at the coarse end from particles that rolled or slid. (See Box 2.1 Figure 12.)

The line segments that appear on cumulative-probability plots have been interpreted as truncated normal distributions, but this may not be appropriate. One can generate breaks lacking reality in cumulative distributions, by assuming, for example, that the total distribution as well as those of segments are normal if they are not. Other theoretical distributions may more closely approximate the actual size distribution of particles in a sample than do normal distributions. Thus equations based on these distributions may better predict or describe observed size-frequency distributions. These potential problems are not unique to the segmentation analysis of cumulative size-frequency distributions of

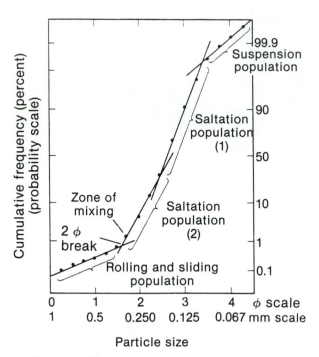

BOX 2.1 FIGURE 12.   Cumulative-frequency curve of beach sand, drawn on probability paper, consisting of four straight-line segments. (After G. S. Visher, 1969, fig. 4, p. 1979.)

sedimentary particles, but are inherent to the interpretation of any cumulative-frequency distribution. The lesson to be learned is that the results of numerical- or statistical analyses are not *necessarily* any more valid than are qualitative interpretations of nonnumeric data.

A kind of distribution that is commonly applied to geological data is the *log-normal distribution*. In a log-normal distribution the logarithms of the data points are normally distributed, but the actual data points form a distribution in which the values of many observations are small and those of fewer and fewer observations are successively larger. The values of most observations are smaller than the mean. This kind of distribution is exemplified by the results of a lottery: most people win nothing, a much-smaller number of people win moderate amounts of money, and only a few people win very large amounts of money. A geological example of the log-normal distribution is the size-frequency distribution of discovered oil fields.

Size-frequency distributions of replicate samples taken from a single site and a single sedimentation unit commonly show considerable variation in statistical parameters (e.g., mean or standard deviation) of the several segments defined on cumulative curves. Therefore, meaningful interpretation of cumulative size-frequency curves can only be made when within-sample variation has been determined. In fact, the necessity of collecting replicate samples to determine the scale of local variation is a general principle of quantitative analysis.

The pattern of cumulative curves of particle sizes on probability plots is a function of the processes that formed the sedimentary deposits as well as of the availability of particles in the various size classes in the parent materials. A glacier picks up any sediment in its path and also creates huge quantities of new sediment by ripping loose blocks from the bedrock and by grinding its load, both old sediments and new blocks of bedrock, into smaller particles. This till deposited directly by the flowing ice of a glacier includes a wide range of particle sizes: everything the glacier carried (Box 2.1 Figure 13). The very gentle slope of the lines of the distributions indicates that tills are extremely poorly sorted.

The wind can strip the fines from sediment that is otherwise poorly sorted and thus improve the sorting of what it leaves behind. The fine-textured particles that the wind removes may accumulate as a separate deposit. In Pleistocene times such fine-textured windblown sediments derived from glacial sediments formed widespread loess deposits. The cumulative-frequency curves of loess resemble those of modern dust falls. Although

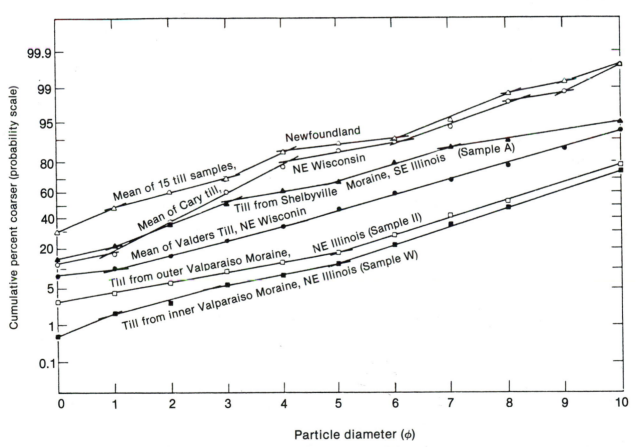

BOX 2.1 FIGURE 13.   Cumulative-frequency curves of various tills. Finest-textured till (from Valparaiso Moraine) consists largely of ground-up debris from limestones and shales (95%) with but minor crystalline rocks (3%). Coarsest till (from Newfoundland) consists chiefly of ground-up debris from crystalline rocks. (Data for Illinois samples from W. C. Krumbein, 1933, table 1, p. 388; for northeast Wisconsin samples, from R. C. Murray, 1953, fig. 2, p. 144; and for Newfoundland samples, from R. M. Slatt, 1972, fig. 4, p. 289.)

one might suppose that eolian suspensions would always be well sorted, particle-size analyses of windblown suspended sediments disclose that sorting is poor in the fine-textured sediments deposited by the wind.

Tills may be washed by waves, and the result would be a layer of boulders, perhaps one boulder thick, surrounded and overlain by well-sorted sand. Such well-sorted sand may be a beach sand (See Box 2.1 Figure 12.); the steep slope of the lines of the distribution indicates good sorting. Wind and water acting on sediment, such as till, serve as a mechanism of selection and removal of particles; such selection and removal constitute the process of sorting. The cumulative-frequency curves of sediments reflect this sorting, but also reflect the distribution of the initial particle sizes of the parent material which the sorting process affected. Box 2.1 Table 1 presents a classification of sands into sorting classes based on standard deviations. This table shows the ranges of standard deviations for sands of various origins.

Graphic Measures

The statistical measures of a frequency distribution can be determined graphically. This method uses the cumulative-frequency curve to determine the ϕ values of various percentile points. Such a graphic approach yields values that can be inserted into formulas designed to approximate the values computed by using the equations of moments. The method of moments is a more-rigorous way of calculating the statistical measures of frequency distributions which is discussed on page 41.

In the graphic approach, ϕ values from the curve are read for various percentiles. Many investigators have employed the 25th and 75th percentiles, which define the so-called *quartile deviations*. The quartile deviations represent the central part of the curve and hence involve the bulk of the particles. Other workers rely on the 16th and 84th percentiles, which represent sizes lying one standard deviation on either side of the mean in a normal distribution (though most sediment samples are not normally distributed). These values may be combined with those of the 5th and 95th percentiles, which are convenient numbers for describing the ends (or tails) of the distribution.

The **median** is *the particle size in the exact middle of the population;* half of the particles (by weight) are finer and half coarser. The median can be read directly from the cumulative curve; it is the ϕ intercept of the point where the cumulative-frequency curve crosses the 50% mark on the cumulative-frequency scale.

BOX 2.1 TABLE 1.   Classification of sands into sorting classes based on standard deviations

Ranges of values of standard deviation (ϕ units)	Sorting class	Environments of sands
<0.35	Very well sorted	Coastal- and lake dunes; many beaches (foreshore); common on shallow marine shelf.
0.35 to 0.50	Well sorted	Most beaches (foreshore); shallow marine shelf; many inland dunes.
0.50 to 0.80	Moderately well sorted	Most inland dunes; most rivers; most lagoons; distal marine shelf.
0.80 to 1.40	Moderately sorted	Many glacio-fluvial settings; many rivers; some lagoons; some distal marine shelf.
1.40 to 2.00	Poorly sorted	Many glacio-fluvial settings.
2.00 to 2.60	Very poorly sorted	Many glacio-fluvial settings.
>2.60	Extremely poorly sorted	Some glacio-fluvial settings.

SOURCE: G. M. Friedman, 1962, table 4, p. 750; table 5, p. 752.

NOTE: This table shows the ranges of standard deviations for medium- and fine sands of various origins.

Another size factor that can be read from a graph is the **mode**, *the most-frequently occurring particle size.* The mode is the peak of an incremental frequency curve (Box 2.1 Figure 14). The mode cannot be easily or accurately determined from the cumulative-frequency curve.

In a normal distribution the ϕ values of mode, median, and mean coincide; in other distributions, they do not coincide. (See Box 2.1 Figure 14.)

The mode is an important statistical parameter, especially in sediments containing several subpopulations, each of which may possess its own mode. The presence of several modes in a sand suggests that the particles have been derived from several parent deposits. In such multipopulation- or polymodal sands, the ϕ values and magnitudes of the modes give information on mixing of sediments. In sand bodies, modes can be traced in a downcurrent direction. Changes in the modes reflect the history of the sand. Box 2.1 Figure 15 is an example of analysis of the modes in sands from the continental shelf off Long Island, New York. All sand samples shown contain modes

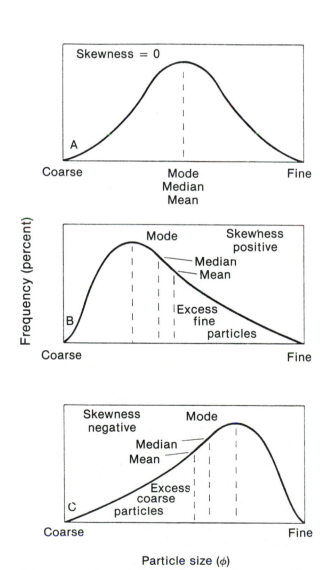

BOX 2.1 FIGURE 14. Frequency-distribution curves having various skewness values. (G. M. Friedman and J. E. Sanders, 1978, fig. 3-18, p. 75.)

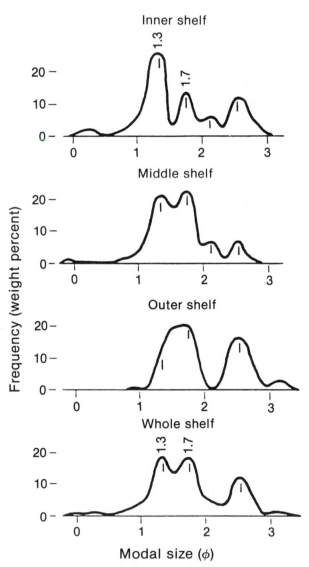

BOX 2.1 FIGURE 15. Frequency distributions of sands from continental shelf off Long Island, New York, showing various modes. Explanation in text. (T. F. McKinney and G. M. Friedman, 1970, fig. 17, p. 233.)

falling in the medium-sand (1ϕ to 2ϕ)- and in the fine-sand ranges (2ϕ to 3ϕ). From the inner-across the middle- to the outer shelf, the proportion of the coarser mode of the medium-sand size (1.30ϕ) decreases whereas that of the finer mode in this size range (1.7ϕ) increases. The increase in the finer medium-sand mode across the shelf at the expense of the coarser medium-sand mode can be related to the geologic history of the shelf. When the shelf was emergent during the last ice age, streams deposited sand there. The shift in position of the modes implies a downstream decrease in the availability of coarser medium-sand sizes. In polymodal sediment samples, the mean and median are not informative about the shape of the size-frequency distribution or about the genesis of the sand body, because both of these parameters are only meaningful if the distribution is approximately normal.

Once the various percentile values have been read from the probability plots, the approximate statistical parameters are determined using the following formulas (the subscripts are the percentile values read from a probability plot of the cumulative-frequency curve):

GRAPHIC MEAN

$$M_z = \frac{\phi_{16} + \phi_{50} + \phi_{84}}{3}$$

INCLUSIVE GRAPHIC STANDARD DEVIATION

$$\sigma_I = \left(\frac{\phi_{84} - \phi_{16}}{4}\right) + \left(\frac{\phi_{95} - \phi_5}{6.6}\right)$$

SIMPLE SORTING MEASURE

$$So_s = \frac{1}{2(\phi_{95} - \phi_5)}$$

INCLUSIVE GRAPHIC SKEWNESS

$$Sk_I = \left(\frac{\phi_{84} + \phi_{16} - 2\phi_{50}}{2(\phi_{84} - \phi_{16})}\right) + \left(\frac{\phi_{95} + \phi_5 - 2\phi_{50}}{2(\phi_{95} - \phi_5)}\right)$$

SIMPLE SKEWNESS MEASURE

$$\alpha_s = (\phi_{95} + \phi_5) - 2\phi_{50}$$

GRAPHIC KURTOSIS

$$K_G = \frac{\phi_{95} - \phi_5}{2.44(\phi_{75} - \phi_{25})}$$

The skewness measures provide information on the symmetry of the frequency curve. A positive value for skewness indicates a trailing off of the curve to the right of the mean and an excess of fine particles. (See Box 2.1 Figure 14, B.) By contrast, a negative number means a trailing off of the curve to the left of the mean and an excess of coarse particles. (See Box 2.1 Figure 14, C.) Kurtosis measures the peakedness or broadness of the curve.

Differences in Size Distributions Among Sediments of Various Origins

The distribution of particles in a sand will be influenced greatly by the conditions that determine what happens to the fine particles forming the suspended load. Sands deposited by rivers almost invariably contain fine particles from the suspended load. Sands deposited on the parts of beaches where the breaking waves continuously wash thin sheets of water back and forth invariably lack admixtures of fine-textured sediments. Water motion is so vigorous that the fine particles are always kept moving and do not come to rest with the sands. The positions of the breaks between the straight-line segments on the probability plots are predictable and are functions of the processes that deposited the sediment.

These relationships between the contrasting one-way flow of water in rivers and the back-and-forth flow of water on parts of beaches and their characteristic size distributions, so clearly indicated by the cumulative-probability plots, likewise become apparent from applications of the older assumption that only one statistical size function covers all size classes of a sample. In the following paragraphs we compare the probability-plot approach with the one-function approach. In Chapter 7 we consider the mechanics of deposition of particles at greater length. Probability plots of river sands typically show two straight-line segments with a break commonly occurring at a cumulative frequency of less than 90% and falling within the range of 2.5ϕ to 3.5ϕ (Box 2.1 Figure 16, A). Thus the segment representing the fine fraction usually constitutes 10% or more of the total distribution. This fraction has resulted from the trapping of suspended particles among coarser particles during movement of the current and from deposition as the current waned.

On the frequency curve using the one-function approach, the presence of these abundant fine particles is shown by the stretching out or trailing off of the curve to the right of the mean. (See Box

2.1 Figure 14, B.) Such a frequency distribution is positively skewed. Despite many exceptions, the frequency distributions of most river sands display positive numerical values for skewness. This results in large part from the presence of the fine-textured fraction. (See Box 2.1 Figure 14, B.)

Probability plots of one variety of beach sand, the kind deposited by the back-and-forth water motions mentioned previously, indicate that the segment representing the suspended particles usually constitutes less than 1%. (See Box 2.1 Figure 16, B.) In addition, the central parts of the graphs tend to show two separate straight-line segments, each representing a different saltation population.

The one-function approach shows that frequency distributions of such beach sands tend to be symmetrical (See Box 2.1 Figure 14, A.), negatively skewed (See Box 2.1 Figure 14, C.), or slightly positively skewed. These frequency distributions are not stretched out to the right of the mean, as were the distribution curves of the river sands.

As mentioned previously, the slope of the cumulative curve on a probability plot reflects the degree of sorting. This is true for the general trend of the entire sample even if several subpopulations are present, and also for each straight-line segment. Hence the standard deviation must be a process-sensitive parameter, even if we apply the one-function approach. Comparison of Box 2.1 Figure 16, A, and Box 2.1 Figure 16, B, shows how slopes of each straight-line segment compare with the general slope of the entire curve (the line that would be formed by connecting the lowermost point at the left with the uppermost point at the right). The slopes of the segments representing the saltation populations are generally similar for both river- and beach sands. The slope of the entire distribution for the beach sand in Box 2.1 Figure 16, B, is parallel to the slope of the entire distribution for the stream sand in Box 2.1 Figure 16, A. However, in many other examples beach sands are better sorted than river sands and therefore the slopes of curves for beach sands tend to be steeper than those for river sands. The same relationship emerges by a computation of the standard deviation for beach- and river sands using the one-function approach. Because of the effects of the back-and-forth flow of thin sheets of water from breaking waves, beach sands subjected to such oscillating flow are generally well sorted; the values for their standard deviations are low. River sands tend to be less well sorted than such beach sands and the values for

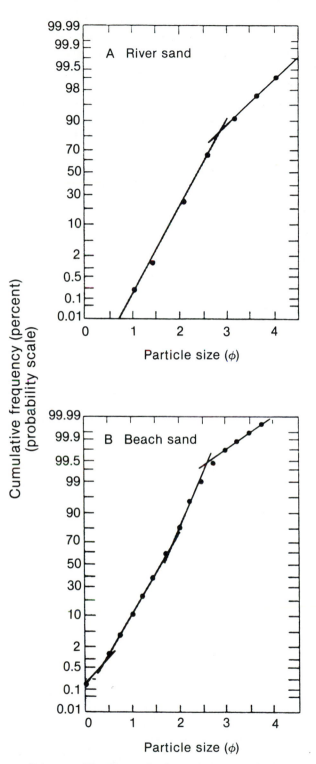

BOX 2.1 FIGURE 16.   Cumulative-frequency curves. A, river sand (Arkansas River); B, beach sand (Bald Point, Florida). (After G. S. Visher, 1965b, fig. 10, p. 126, and 1969b, fig. 7, p. 1082.)

the standard deviations of river sands are higher. As noted previously, the skewness is sensitive to geologic processes. Therefore a scatter plot of the two process-sensitive factors, standard deviation (sorting) and skewness, is likely to be informative. On such a plot, despite some overlap, the beach sands (of the kind we have been discussing) occupy one area and the river sands another (Box 2.1 Figure 17). Whereas the spread of values for sorting and skewness indicates that the frequency distribution for river sands varies considerably, the spread of values for beach sands is much less. Bivariate scatter plots of sorting versus mean particle size and of sorting versus kurtosis have been effective in differentiating eolian sediments formed in different environments. A bivariate scatter plot, as shown in Box 2.1 Figure 17, just like the straight-line segments of probability plots, reveals how a study of sizes of particles relates to processes forming sedimentary deposits.

Application of the parameters analogous to mean, sorting, skewness, and so forth, to size-frequency distributions other than the normal distribution, such as the hyperbolic distribution, has been successful in some cases in distinguishing sedimentary deposits formed in different environments. These different distributions and

parameters can be used just as the more-familiar parameters of the normal distribution are (e.g., with bivariate scatter plots) to attempt discrimination between sedimentary deposits formed in different environments. In the future, alternative distributions will be employed more and more in the evaluation of sediment size-frequency distributions as geologists become more sophisticated. This trend owes a great deal to the development of algorithms for rapid calculation of parameters of alternative distributions using modern computers, especially personal computers.

An important aspect of particle-size distributions is their accurate measurement. Ultimately, all numerical interpretations, and indeed all interpretations of depositional energy, transport mechanisms, and environmental conditions, depend upon accurate data collection. Large particles present a particular problem for study of particle-size distributions. Individual large particles may be measured readily, but the standard method of generating size-frequency data for coarse sediments is by sieving. This method will not work for particles with diameters greater than about 1/10th the diameter of a sieve. Yet many gravels include significant numbers of particles far too large for sieving. It is next to impossible to estimate the abundance of such particles in a sediment accurately for two reasons. First, a very large sediment sample is required to encompass a statistically valid number of large particles. For example, if the concentration of particles larger than this book is $1/m^3$, and one wishes to measure 100 large particles to describe the size distribution of the coarse part of the size-frequency distribution accurately, then one must sift through 100 m^3 of sediment and remove and measure all the large particles. Thus very large volumes of sediment must be processed. The second reason for the difficulty of getting accurate size-frequency data for large particles is the fact that a concentration of $1/m^3$ is high for particles the size of this book or larger. Thus it is often impossible to collect a sufficient number of larger particles to get a statistically valid sample, especially if the large particles encompass more than a few ϕ classes.

Although quantitative studies of size and sorting and various statistical parameters have become a part of sedimentological practice, one need not depend entirely on them. Observations with the hand lens can go a long way in relating size fractions to geologic processes. For example, if silt and clay are present in a sand, then that sand probably was not deposited in the zone of back-and-forth water motions on a beach where the fine

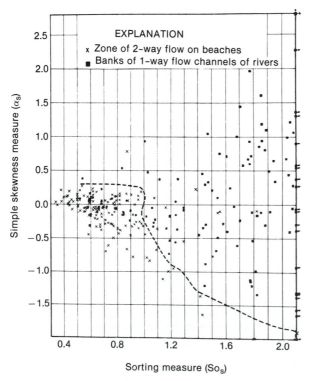

BOX 2.1 FIGURE 17.   Scatter plot, simple skewness measure versus simple sorting measure, for beach- and river sands. (G. M. Friedman, 1967, fig. 18, p. 342.)

fractions remain suspended and are deposited somewhere else.

The Method of Moments

Heretofore we have discussed graphic estimation of statistical parameters in some detail. Nowadays, of course, most such parameters are calculated using available statistical software and computers. These calculations do not employ the graphic estimation methods but rely on a more-accurate method known as the method of moments. Graphic measures were used in the days of paper-and-pencil calculations when the more-complex calculations required for moment measures took prohibitively long times to complete. The term *moment* was introduced into statistics by analogy with mechanics. In mechanics, the moment of a force about a point of rotation, such as a fulcrum, is computed by multiplying the magnitude of that force by the distance from the force to the point of rotation. In statistical moments the force of mechanics is replaced by a frequency function (such as the percentage of the distribution within a given class interval). The concept of the distance to a point is the same in both kinds of moments. In statistical moments the point of rotation of mechanics is replaced by an arbitrary point, commonly either the origin of the curve or its mean. The distance is determined by the spacing of the size class from the arbitrary point.

The statistical moment is the moment of a given size class with respect to the arbitrary point (i.e., to the origin of the curve). The **statistical moment of the distribution** is *the moment per unit frequency.* It is determined by finding the moments of each size class (frequency percentage in each size class times the distance of each from the origin), adding them up, and dividing the sum by 100.

The first moment is the mean:

$$\overline{X} = \frac{\sum f m_\phi}{100} \qquad \text{(Eq. 2.1-1)}$$

where *f* is frequency in percent for each size class and m_ϕ is the midpoint of each ϕ class. In the first moment the distance term (m_ϕ) appears in the first power. Successively higher moments are defined by raising the distance term to progressively higher powers. In the higher moments the distance term is represented with respect to the mean, so the higher moments are moments about the mean, whereas the first moment is the moment with respect to an arbitrary point. The second moment is the square of the standard deviation. The third moment is the skewness, and the fourth moment is the kurtosis.

SOURCES: Friedman, 1961, 1979; Friedman and Sanders, 1978.

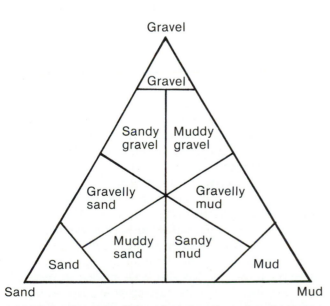

FIGURE 2-5.   Triangular diagram having mud, sand, and gravel as end members, with marked-off areas within the triangle to indicate names for sediment mixtures. (Authors.)

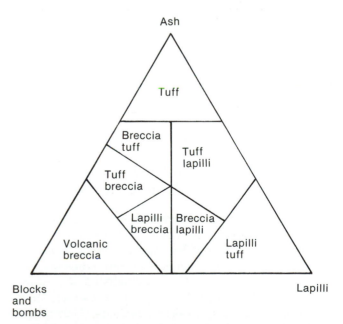

FIGURE 2-6.   Triangular diagram for naming tephra having as end members ash, lapilli, and bombs or blocks, with marked-off areas within the triangle shown for names of mixtures. (Authors.)

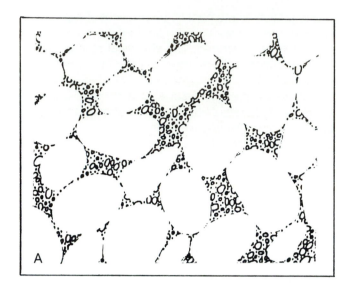

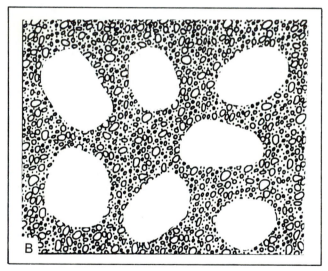

FIGURE 2-7. Fabric in mixtures of coarse- and fine sediment.
(Authors.)
  A. Particle (or framework) support; large particles touch one
another.
  B. Matrix support; large particles do not touch one another.

Sediments Classified by Site of Origin with Respect to Water of Basin of Deposition

Depending on where they come from with respect to the water of the basin in which they are deposited, sedimentary particles can be divided into four broad groups: (1) **extrabasinal**, that is, *solids derived from outside the water of the basin of deposition;* (2) **carbonaceous**, *derived from plant debris* (excluding debris of plants that secrete mineralized skeletons, such as calcareous algae); (3) **pyroclastic** or **tephra**, *derived from exploding volcanoes;* and (4) **intrabasinal**, *solids that are derived from within the water of the*

basin of deposition (Table 2-2). The definitions of groups (2) and (3) are not equivalent to the basinal groups, but are set apart because (a) their compositions are distinctive, and (b) they cannot be unambiguously classified as either extrabasinal or intrabasinal because they can come from either. Some compositional overlap also creeps into the basinal varieties. Recycling of lithified carbonaceous- or volcanogenic sedimentary strata is a source of extrabasinal particles. For example, extrabasinal particles include redeposited volcanic rocks eroded from older pyroclastic deposits, kerogen or coal particles eroded from "oil shales" or coal seams, and carbonate particles recycled from older (intrabasinal) carbonate rocks. However, autochthonous coal seams are true carbonaceous rocks.

The extrabasinal group consists of **terrigenous particles**, *solids that are breakdown products of older deposits.* The carbonaceous group includes solid hydrocarbons and plant matter that has been lithified as coal. Minor amounts of animal-derived organic matter may be incorporated in carbonaceous strata. The pyroclastic group includes whatever may be exploded out of volcanoes. Newly formed lava liquid and associated crystals exploded out of volcanoes are collectively designated as tephra. As noted above, we assign particles eroded from older volcanic rocks to the terrigenous group. Pyroclastic material may accumulate in the same basin in which the volcanoes are located, or the particles may travel long distances by wind or water ultimately to accumulate in a remote basin. Whether from subaerial- or submarine volcanoes, tephra are not derived from the water of the basin of deposition, and thus we classify them as their own distinctive group.

The intrabasinal group consists of materials that grew as solids in the water of the depositional basin as (1) **biocrystals** (*minerals secreted by organisms*) or (2) crystals that were precipitated chemically. The precipitates include evaporites and authigenic sediments, both defined in following sections. Our discussion of particles treats each of these kinds in turn.

Most calcium-carbonate particles are secreted as biocrystals by organisms within the basin of deposition; a few are of inorganic chemical origin. Intrabasinal particles can be plentiful even in the absence of any supply of extrabasinal terrigenous sediment.

An important point about intrabasinal sediments relates to the sediment/water interface. Materials that form from the water, either above the sediment or as nonattached particles upon or within it, constitute the evaporitic- or authigenic classes. To complicate matters, the term authigenic has also been applied to materials, such as mineral cements, that grow in place as crystals within sediment or rock. These are diagenetic materials, not subject to becoming sedimentary particles.

TABLE 2-2.   Names and origins of four major groups of sediments

SEDIMENTS								
Pyroclastic	Extrabasinal				Intrabasinal			Carbonaceous
Volcanic	Terrigenous				Biochemical	Evaporites	Authigenic	
PARTICLES						CRYSTALS		
New rocks	Derived from pre-existing rocks				Skeletal remains Ooids Other non-skeletal particles	Transported crystals	In situ	
Volcanic explosions	Weathering products (solids)	Survival products						
Subaqueous eruption products		Volcanic	Non–volcanic					
		Igneous	Meta-morphic	Sedi-mentary				

NOTE: Further explanation in text.

Extrabasinal Particles

Terrigenous Particles

As mentioned, all particles eroded as solids from the land constitute terrigenous particles. The word terrigenous comes from Latin, in which *terra* means land, signifying derivation from the rocks of preexisting land. Other adjectives that have been applied to terrigenous sediments include *detrital, fragmental, siliciclastic,* or *clastic.* The term clastic derives from the Greek language and means broken.

Most terrigenous particles ultimately come from the bedrock. They become detached from their parent bedrock by (1) weathering, (2) wind and water erosion, (3) catastrophic mass-wasting events, or (4) glacial activity. Thus terrigenous particles either escaped or survived chemical weathering, or they consist of secondary alteration products, chiefly clay minerals and iron oxides, that are the solid products formed during chemical weathering (ions in solution are the other products; they enter streams and are delivered to the water of depositional basins). Chemical weathering tends to destroy rock fragments and feldspars. Accordingly, we can make use of the presence or absence of these "doomed" particles to define the concept of compositional maturity. Total compositional maturity (100% maturity) represents a case in which *a sedimentary deposit derived from bedrock consisting of a feldspathic parent rock (such as a granite) lacks rock fragments and contains no feldspar.* Compositional immaturity refers to *a sedimentary deposit containing rock fragments plus feldspar (and other easily destroyed minerals).* The materials mentioned can be used to define compositional maturity even if their parent area has not been determined or remains unknown.

We discuss terrigenous sediment under the following headings: (1) *rock fragments,* (2) *quartz,* (3) *feldspars,* (4) *heavy minerals,* and (5) *layer-lattice silicates.*

ROCK FRAGMENTS

Particles having recognizable characteristics of their parent rock are designated as **rock fragments**. The qualification of retaining characteristics that are recognizable is a critical part of this definition. In a very important sense, all terrigenous sediments consists of fragments of former bedrock. However, if a piece of parent bedrock has been broken into its individual mineral constituents, the all-important textural characteristics, which are used to identify kinds of rocks, can no longer be recognized. For this reason, particles consisting of individual minerals are not classified as rock fragments.

Factors Affecting Survival.   Many factors affect the survival or nonsurvival of rock fragments. These factors can be classified as (1) characteristics of the parent rock, (2) weathering and transport, and (3) postdepositional weathering and stresses created during cementation.

Table 2-3 summarizes some common kinds of parent rocks by arranging them in three columns according to the average sizes of their individual constituents. The initial sizes of rock fragments are determined by the block-making pattern of intersecting joints, -faults, and -bedding-surface partings. Before the blocks are moved, further breakdown may take place along these partings.

The persistence of rock fragments depends significantly on six factors. These factors are (1) the ratio between the original sizes of the mineral constituents in the parent rock and the sizes of the sedimentary particles, (2) the degree of interlocking among the constituents of the parent rock, (3) the spacing of partings in the parent rock, (4) the chemical stability of rock fragments, (5) the kinds and intensities of weathering and transport, and (6) postdepositional weathering and stresses of cementation. In weathering and in many kinds of transport, three general tendencies operate widely: (1) the sizes of rock fragments tend to be reduced, (2) individual fragments

containing more than one mineral species tend to break apart to form particles consisting of one mineral species, and (3) the feldspars and other unstable minerals tend to disappear from the sand sizes and be transformed into clay minerals whose particles lie mainly in the clay-size range. Rock fragments composed of more than one mineral species tend to be chemically less stable than particles composed of only one mineral species. This results in part from the fact that the most-unstable minerals tend to be destroyed in the process of converting multi-mineral rock fragments to one-mineral particles.

Chemical weathering of feldspars tends to convert blocks of granite into masses of individual resistant minerals, usually consisting of quartz and the heavy accessory minerals, including both transparent- and opaque varieties. Abrasion during transport may significantly reduce the sizes of some rock fragments and may even be responsible for total destruction of others. Abrasion during transport also rounds off sharp corners of rock fragments.

Even after they have survived weathering from their parent-area bedrock and abrasion during transport, and have been deposited, rock fragments are subject to further changes, including both further chemical decompo-

TABLE 2-3.   Some common parent rocks arranged in groups by sizes of constituents

Coarse-textured rocks (average particle size > 1 mm)	Medium-textured rocks (average particle size < 1 mm > 0.1 mm)	Aphanitic (= fine-textured) rocks (average particle size < 0.1 mm)
Igneous Rocks		
Granite	Microgranite	Rhyolite
Diorite	Andesite	Felsite
Gabbro	Dolerite	Basalt
Various coarse-textured porphyries and porphyritic rocks	Various medium-textured porphyries and porphyritic rocks	Various fine-textured porphyries and porphyritic rocks
Volcanic breccias		
Metamorphic Rocks		
Gneisses	Medium-textured schists; phyllites	Slates
Coarse-textured schists	Medium-textured quartzites	
Coarse-textured quartzites	Medium-textured marbles	
Coarse-textured marbles		
Amphibolites		Mylonites
		Hornfels
Sedimentary Rocks		
(> 2 mm)	(< 2 mm > 0.06 mm)	(< 0.06 mm)
Conglomerates	Sandstones	Shales
Sedimentary breccias		Siltstones, claystones
Coarse-textured limestones	Medium-textured limestones	Fine-textured limestones
Coarse-textured dolostones	Medium-textured dolostones	Fine-textured dolostones
		Cherts

sition and physical breakage. For example, as a result of postdepositional chemical weathering, granitic rock fragments in Quaternary deposits that are older than about 50,000 years crumble easily. The ease of crumbling becomes more pronounced in the rock fragments enclosed within still-older Quaternary glacial deposits. Rock fragments in surface gravels commonly break easily along discrete planes. Such breakage has been ascribed to the fatigue effects associated with stresses caused by the crystallization of cement. In Holocene sediments that have been affected by glaciation, the degree of postdepositional weathering of particles has been used, with some assumptions about rates, to date exposure surfaces and to estimate the duration of exposure.

As a result of weathering, transport, and postdepositional processes just described, the coarse-textured varieties of bedrock are destined to disappear from the sedimentary record. Instead of remaining as rock fragments, the coarse-textured varieties of bedrock attain immortality as so many individual mineral particles; some are of sand size, but a great many more are smaller.

SOURCES: Füchtbauer, 1988; MacClintock, 1940; A. J. Moss, 1972b; Pettijohn, Potter, and Siever, 1987.

Use in Provenance Studies. Rock fragments that can be recognized by simple inspection in the field are easiest to use and provide the most-direct provenance information. All one needs to do is to match the pieces directly with the corresponding bedrock. (Box 2.2 gives examples of provenance studies.) Unfortunately, as mentioned, most fragments of still-identifiable bedrock tend to be destroyed.

Other rock fragments are not readily identified by field study, but can be employed to establish provenance by examining them with a binocular microscope (Figure 2-8) or, better still, by cutting thin sections and studying them with a polarizing microscope.

In thin-section study, many rock fragments are easy to identify (Figure 2-9). However, certain kinds of rock fragments pose difficulties. The fine-textured siliceous rocks such as cherts, quartzose siltstones, metamorphosed cherts, and fine-textured quartzites, and mixed quartz-feldspar volcanic rocks present many similarities (Figure 2-10). These rocks offer real challenges even to experienced observers. In addition, rock fragments that consist of one mineral species only, such as quartz, could be counted as rock fragments because they are polycrystalline or as quartz because they contain only one mineral.

Because of the tectonic significance attached to volcanic activity, rock fragments of volcanic provenance generally are given special consideration. We classify volcanic debris that has been eroded as a special variety

FIGURE 2-8.   Rock fragments in terrigenous sediment, Hartsdale, New York, viewed through binocular stereomicroscope. (B. M. Shaub.)

of terrigenous rock. Accordingly, despite their common connection as volcanic, we treat such debris separately from the pyroclastic materials, which we do not assign to the terrigenous category because pyroclastic materials became sediments as a result of explosions and not as a result of erosion.

Rock fragments that were eroded from older carbonate rocks (Figure 2-11) are usually treated as a special group because they consist of calcite or of dolomite and not of rock-making silicate minerals. Because of their compositions they are commonly classified with "limestones." They are also more susceptible to weathering than are terrigenous sediments containing abundant quartz. (See next section.) However, their origins are the same as those of other terrigenous particles.

QUARTZ

In nearly all igneous- and most metamorphic rocks except quartzites, in which it may form 100% of the rock, the proportion of quartz ranges from traces to about 40% by volume. By contrast, in typical terrigenous sediments, quartz is the dominant mineral (Figure 2-12). Quartz makes up about 65% of the average sandstone and about 30% of the average shale. This contrasting abundance is a result of chemical weathering. During

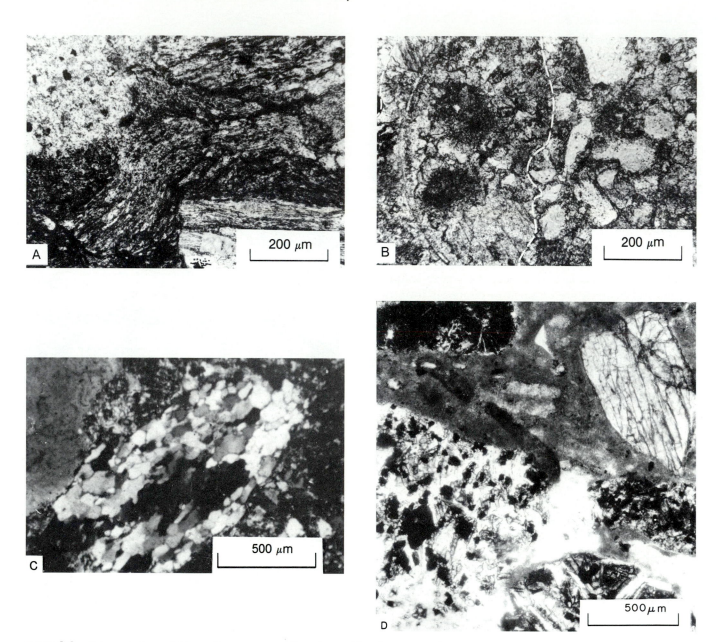

FIGURE 2-9.   Enlarged views of thin sections as seen with petrographic microscope; plane-
polarized light except as otherwise noted.

A. Phyllite fragments (dark) having bent foliation. Skaneateles Formation (Middle Devonian),
Catskill Front, New York State. (Authors and J. A. Smith.)

B. Rock fragment of limestone (within dashed lines) containing pelecypod shell (crescent
shaped) and various calcite crystals. Recrystallization has obscured outline of limestone fragment.
Small quartz particles at right. Skaneateles Formation (Middle Devonian), Catskill Front, New York
State. (Authors and J. A. Smith.)

C. Fragment of metamorphic quartzite viewed in cross-polarized light. Hamilton Group (Middle
Devonian), Catskill Front, New York State. (J. H. Way, Jr., and authors.)

D. Volcanic debris. At bottom are fragments of basalt containing opaque matrix (black)
and feldspar laths (white); light-colored particle at upper right showing prominent cleavage
is augite. Medium-gray material between fragments is cryptocrystalline carbonate cement.
Holocene, Fuerteventura, Canary Islands. (G. M. Friedman, 1968, fig. 2, p. 13; sample collected by
Gerd Tietz.)

FIGURE 2-10.   Fine-textured fragments (chert; black with tiny specks of white) in sandstone, viewed in cross-polarized light. Chert composes about 50% of the particles; most of the remainder are quartz but a few are feldspar. Core from oil well, through Morrow Formation (Pennsylvanian), western Oklahoma. (Authors.)

FIGURE 2-11.   Exposure of conglomerate of Late Cretaceous age composed of fragments eroded from carbonate parent rocks of Late Paleozoic age. Near Price, Utah. (Authors.)

FIGURE 2-12.   Quartz in sediments and sedimentary rocks.
A. Modern sediment from sea floor, continental shelf off Virginia, 0.25- to 0.125-mm fraction viewed through binocular microscope. (B. M. Shaub.)
B. Sandstone containing quartz particles of two sizes, photomicrograph of thin section in plane-polarized light. Core from test boring, Inmar Formation (Jurassic), Negev Desert, Israel. (Authors.)

BOX 2.2

Provenance Studies

Some rock fragments are so distinctive that they can be identified by inspection in the field. For example, an unusual ridge-making conglomerate, the Green Pond Conglomerate (Lower Silurian), containing pebbles of white quartz and red jasper in a maroon-colored matrix, forms a narrow outcrop belt about 45 km long in northwestern New Jersey. Boulders and cobbles of Green Pond Conglomerate are abundant in the mid-Quaternary glacial sediments (pre-Wisconsinan) being eroded in the coastal cliffs of southeastern Staten Island (S.I.). These boulders, as well as other distinctive materials of known provenance, such as anthracite coal, prove that the glacier flowed across Staten Island from the northwest to the southeast (Box 2.2 Figure 1).

Two especially intriguing examples having economic significance include (1) gold in California,

where the bedrock source was found, and (2) diamonds in the Great Lakes region, where the bedrock source has not yet been found.

The successful search for the *Mother Lode* (the name applied to the gold's parent deposit) in California was stimulated in 1848 when placer gold was discovered in the diggings in modern stream gravels for a water mill at Sutter's Fort, near Coloma. A year later, when the gold fever was in full swing and the hordes of forty-niners had arrived, panning and digging took place in all modern streams. Digging upstream resulted in the discovery of richer ancient placers in uplifted and eroded stream gravels of Tertiary age, which are exposed in the sides of modern valleys. Eventually the parent deposit of the gold was located in the bedrock of the foothills of the Sierra Nevada, in a belt 1.6 km wide and nearly 200 km long (Box 2.2 Figure 2).

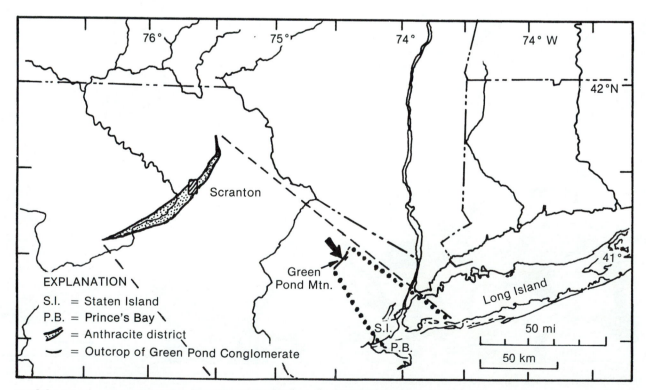

BOX 2.2 FIGURE 1.   Map of part of northeastern United States showing locations of parent deposits of rock fragments found in Quaternary glacial sediments in New York City. Further explanation in text. (J. E. Sanders in G. M. Friedman and J. E. Sanders, 1978, fig. 2-1, p. 27.)

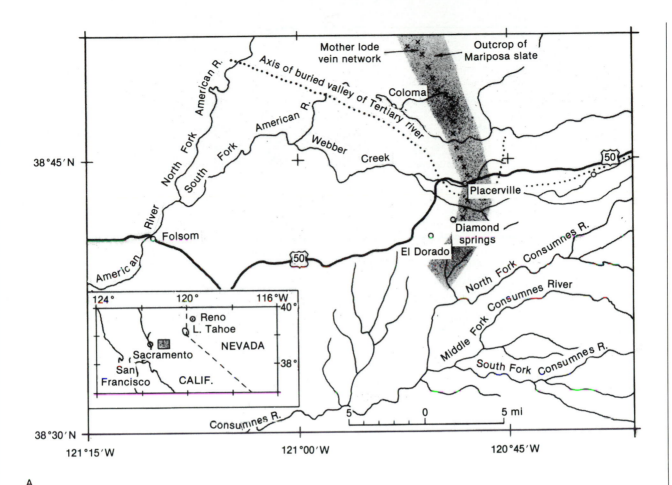

A

BOX 2.2 FIGURE 2.   Areas east of Sacramento, California, where gold was discovered in 1848 at Sutter's Mill (near Coloma). News of this discovery triggered the great California gold rush of 1849. (J. E. Sanders, in G. M. Friedman and J. E. Sanders, fig. 2-11, A, p. 40.)

The diamonds found in the Quaternary glacial sediments of the north-central part of the United States come from localities lying in the general vicinity of Lake Michigan. Most are situated in Wisconsin, but others are in Michigan, Indiana, and southwestern Ohio. The parent deposit was long thought to be the bedrock in some part of Canada, possibly near Hudson's Bay (Box 2.2 Figure 3). However, kimberlite pipes have been located and mapped by the U. S. Geological Survey along the Michigan-Wisconsin state line. It is too early to tell if the mother lode has at last been found.

When heavy minerals have been determined from a sample network of regional extent, the distribution of certain species may form a distinct areal pattern. An area having closely similar suites of heavy minerals is a *heavy-mineral province*. Such provinces depend on provenance differences, but the provinces can be defined even if knowledge of provenance is not complete. For example, among modern sands in the Gulf of Mexico, two heavy-mineral provinces can be identified (Box 2.2 Figure 4). The heavy minerals in the western Gulf province, derived from the Mississippi River and other rivers to the west of it, consist of a hornblende-pyroxene assemblage containing substantial amounts of epidote and garnet. By contrast, the heavy minerals of the eastern Gulf province, derived from the Appalachians and vicinity, contain a kyanite-staurolite assemblage. Modern computerized multivariate data analysis permits more-sophisticated interpretation of heavy-mineral assemblages and identification of and correction for complicating factors such as mixing of heavy-mineral assemblages from different provinces and sorting of heavy minerals by currents before final deposition.

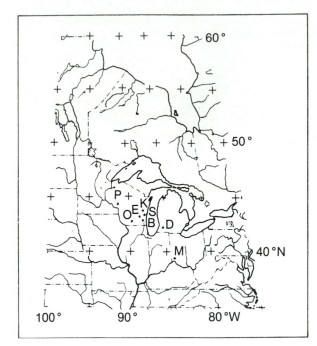

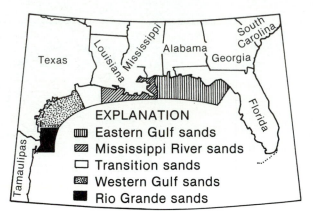

BOX 2.2 FIGURE 4.   Map showing the distribution of the four chief kinds of modern sands along the shores of the Gulf of Mexico, based on their contents of heavy minerals. (J. E. Sanders, 1981, fig. 19.15, A, p. 491, based on A. Goldstein, Jr., 1942, fig. 1, p. 78.)

BOX 2.2 FIGURE 3.   Locations of diamonds found in Pleistocene glacial sediments. Bedrock parent area(s) of these diamonds not known. Localities of finds abbreviated as follows: P, Plum Creek; K, Kohlsville; O, Oregon; E, Eagle; S, Saukville; B, Burlington; D, Dowagiac; M, Milton. (Localities from W. H. Hobbs, 1899, p. 382, replotted by J. E. Sanders on map having sinusoidal projection. Base map from Marshall Kay. J. E. Sanders in G. M. Friedman and J. E. Sanders, 1978, fig. 2-11, B, p. 40.)

chemical weathering, the feldspars, which predominate among most igneous- and many metamorphic rocks, are altered to clay minerals and the quartz accumulates in the weathered residue.

Another contrast between quartz in igneous- and metamorphic parent rocks and quartz in terrigenous sediments is in constituent sizes. The typical sizes of quartz constituents in such parent rocks range from 0.5 to 1.0 mm. In terrigenous sediments, large quantities of quartz particles are smaller than 0.06 mm. The origin of these small quartz particles is not surely known. Many of the small particles may be products of subglacial grinding.

Except for a few colored varieties that can be recognized with a hand lens and thus may be useful locally for provenance studies during field study, most particles of terrigenous quartz yield their provenance information only to those who are willing to have thin sections cut and then to study the sections with a polarizing microscope. Features that can be seen in thin sections include (1) kind of extinction, (2) inclusions of other minerals, (3) fluid inclusions, (4) shape, (5) whether the particles are monocrystalline or consist of polycrystalline aggregates, (6) kind of boundaries between adjoining individuals within polycrystalline aggregates, and (7) deformation lamellae. The relationships of these features to kinds of parent rock are summarized in Table 2-4. Photomicrographs of a few kinds of quartz are shown in Figure 2-13.

Well-rounded quartz particles suggest, but do not necessarily prove, a parentage from an older sedimentary deposit. Quartz particles having circular outlines have been reported from soils, from volcanic rocks, and from inclusions within feldspars of metamorphic rocks. (See Figure 2-13, F.)

FELDSPARS

Although feldspars form the dominant group of rock-making silicate minerals and predominate in the ultimate parent rocks, feldspars almost never are the dominant minerals in sedimentary deposits. Feldspars form only 10 to 15% of modern terrigenous sediments.

In a given thin section of an igneous- or metamorphic parent rock, the compositions of all of the plagioclase crystals are presumed to be the same (neglecting zoning). But in a thin section of a sedimentary deposit, each plagioclase particle bears no necessary compositional similarity to other plagioclase particles.

Many feldspar particles do not display diagnostic twinning or cleavage. Hence staining techniques should be applied routinely to enable such feldspars to be distinguished easily from quartz and to separate the potassium feldspars from the plagioclase feldspars.

Survival. Intense chemical weathering can destroy feldspars. This statement implies that generally, feldspar is doomed not to survive in sediments and does so only where the rate of mechanical detaching of pieces of the bedrock exceeds the rate of chemical weathering. Accordingly, in order to understand the survival of feldspars, we need to examine how topographic relief and climate interact with respect to these two principal kinds of weathering.

Several possibilities involving climate and relief can be visualized in which, although many feldspars will be destroyed, a few will survive: (1) If relief is high, no matter what the climate, steep cliffs will undoubtedly be present where fresh bedrock is exposed and where rapid detachment of nonweathered rock fragments can take place. Even if interfluves are blanketed by a thick regolith within which feldspars are decomposing, steep cliffs exposing relatively unaltered bedrock are likely to be present along the walls of stream gorges. (2) If the climate is dry, either in a hot desert- or a frigid arctic zone, chemical weathering is negligible, plant cover is absent, and fresh bedrock is exposed even without high relief. (3) Glacial erosion of feldspathic bedrock rips loose and grinds up large quantities of feldspars. This can happen in valley glaciers and piedmont glaciers in mountainous areas in nearly any latitude; continental glaciers affect all kinds of landscapes.

Use in Provenance Studies. The value of feldspars in provenance studies could be great, but it depends on how much specific information is available concerning feldspars from potential parent deposits.

High-calcium plagioclase (An > 50) is generally derived from mafic igneous rocks. Many kinds of twinning are present in feldspars, and, to a certain extent, twinning suggests parent rocks (Figure 2-14). Certain kinds of twinned feldspars are easily recognized and hence are helpful in provenance studies. Table 2-5 summarizes the common kinds of feldspars and their twinning habits. As the table shows, almost no variety of twinned feldspar is restricted to any single kind of rock. Feldspars from volcanic rocks may contain bleblike inclusions that originate from volcanic glass (Figure 2-15), as in volcanic quartz.

In the 1960s geochemists found that feldspars can be eroded from their parent bedrock and transported into a body of terrigenous sediment without losing their radiogenic argon during weathering. Accordingly, K-Ar dating techniques can be employed to determine the age of crystallization (or of last recrystallization) of the terrigenous feldspars. Although such age determinations do not necessarily determine directly the provenance of the feldspars, such ages may enable certain otherwise-prospective parent rocks to be rejected.

In-situ postdepositional alteration of feldspars to clay minerals or to other minerals, for example, the albitization of plagioclase, destroys their provenance information. This is an extremely common process and one of the major factors limiting the use of feldspars for provenance studies. It also is a major factor in creating secondary porosity and in reducing permeability by growth of authigenic clays, subjects discussed at greater length in following sections of this chapter and in Chapter 3.

SOURCES: Abdel-Monem and Kulp, 1968; Blatt, 1982; Füchtbauer, 1988; Krynine, 1935; Suttner, 1989; van der Plas, 1966.

HEAVY MINERALS

Igneous rocks and some metamorphic rocks include a group of minerals that generally are resistant to chemical weathering and that are denser than quartz [specific gravity (sp gr) = 2.65] and feldspars (sp gr = 2.56 to 2.76). The specific gravities of these dense minerals range upward from 2.9; they are thus known collectively as *heavy minerals*.

On the average, the heavy-mineral content of most sands is less than 1 or 2% by weight. However, within a given sand, and among different sands, the proportion of heavy minerals to light minerals varies. As the overall size of the sediment particles decreases, the proportion of heavy minerals usually increases. In some sands the heavy minerals have been concentrated by various mechanical processes to form layers, a millimeter to several tens of centimeters thick, in which they may constitute 50% or more of the total.

Kinds of Heavy Minerals. Heavy minerals include many kinds of opaque- and transparent species. The opaque group consists of oxides, sulfides, and other so-called ore minerals. Typically, the opaque minerals predominate. The transparent group consists of rock-making silicates (Table 2-6). Because they transmit polarized light, they can be examined routinely in mounts of sand-size particles on a glass slide for study with a petrographic microscope (Figure 2-16). Some particles of rock-making silicate minerals have been so thoroughly altered or coated that they are no longer transparent.

TABLE 2-4.   Petrographic characteristics of terrigenous quartz (and some quartzose rock fragments) derived from various parent rocks

| Parent rock | Kind of extinction | | Inclusions | | |
	Nonundulatory estimated percentage	Undulatory estimated percentage	Mineral	Fluid	Shape
Granites	10 to 20	80 to 90	Zircon Monazite Rutile Apatite Tourmaline Vermicular chlorite	Common	Irregular
Quartz veins		×		Particularly abundant	Irregular
Gneisses		Nearly 100		Common	Much is irregular, some spherical quartz inclusions in feldspars
Schists		Nearly 100	Staurolite Kyanite Sillimanite	Rare	Irregular
Slates, phyllites, fine-textured schists	×				
Quartzites		100			Irregular
Sandstones	40 to 45 (proportion increases up to 80 as amount of quartz in rock increases)	55 to 60	Depends on ultimate parent material		Rounded

Use in Provenance Studies.   Heavy minerals can be employed in provenance studies in a general way to recognize broad categories of possible parent rocks, or more specifically, to pinpoint the provenance of the particles.

A few species of heavy minerals are diagnostic of particular kinds of parent rock. Mere identification of such species suffices to determine provenance. For example, olivine, platinum, diamond, and chromite occur only in

| Internal structure and particle size | | | | | | |
| Monocrystalline | | Polycrystalline aggregates | | | | |
Estimated percentage	Particle size (mm)	Estimated percentage	Particle size (mm)	Boundaries in polycrystalline aggregates	Deformation lamellae	Remarks
~ 50	0.5	50 Number of individuals in aggregates is small (2–5)	1.0	Irregular		
20 to 25	0.2	75 to 80 Large (> 5) numbers of elongated individuals in aggregates: mixture of two distinct sizes	1.0	Linear, shapes formed by outlines of crystals Linear or irregularly interlocking, elongate shapes predominate	×	Quartz from metamorphic parent rocks is characterized by its inclusions and by the elongate shapes and bimodal size distributions of polycrystalline aggregates.
40	0.25	60	0.55		×	
					×	
90 and more		10 and less				Abraded overgrowths of former cement that grew in optical continuity with rounded particles are diagnostic indicators of derivation from an older sandstone

mafic igneous rocks. Sillimanite, kyanite, andalusite, glaucophane, wollastonite, garnet, epidote, and cordierite are confined to various metamorphic rocks. Cassiterite comes from the margins of granitic plutons. (See Table 2-6.)

Many other heavy minerals, including tourmaline, zircon, magnetite, fluorite, hypersthene, topaz, rutile, monazite, sphene, beryl, hornblende, ilmenite, and biotite, occur in both igneous- and metamorphic rocks. More

TABLE 2-4.   *(Continued)*

| Parent rock | Kind of extinction | | Inclusions | | |
	Nonundulatory estimated percentage	Undulatory estimated percentage	Mineral	Fluid	Shape
Cherts	Nearly 100		Rare		Rounded
Volcanic rocks	Nearly 100		Glass inclusions may be present	Absent	Ranges from euhedral bipyramid to spheres to much-embayed crystals

SOURCE: Mostly after H. Blatt, 1967, and H. Blatt, G. V. Middleton, and R. C. Murray, 1972.

NOTE: X = present but no estimate of proportions available.

identification of heavy-mineral species within this group does not suffice to determine provenance. The only way these minerals can be used in provenance studies is to match some characteristic of the microscopic morphology or chemical composition of a particle with a corresponding characteristic in the minerals from a known parent deposit. These techniques work because the trace-element compositions or other characteristics of the heavy minerals in many parent deposits are distinctive. Box 2.2 gives examples of the use of heavy minerals in provenance studies.

Other Uses of Heavy Minerals.   Heavy minerals have been studied extensively by oil companies because useful information can be obtained from small samples, such as those brought to the surface during drilling of exploratory borings. In the oil fields of southern California, heavy minerals have proved to be a valuable means for distinguishing one Tertiary sand from another in single holes and in matching sands from one hole to another. Such uses are possible even where the provenance of the particles is not known.

Purple zircons (var. hyacinth) are derived only from ancient Precambrian rocks. In the Lake Superior region of the United States, purple zircons are indigenous only to granites of early pre-Huronian age (>2,600 m.y.). Hyacinth zircons have likewise been reported from the Lewisian granites and gneisses of the Scottish Northwest Highlands, from the gneisses of the New Jersey Highlands, and from other granites of early Precambrian age. The purple color results from long periods of radioactive bombardment with alpha particles; hence purple zircons are not found in parent deposits that are younger than about 2500 m.y.

TABLE 2-4. *(Continued)*

| Internal structure and particle size | | | | | | |
| Monocrystalline | | Polycrystalline aggregates | | | | |
Estimated percentage	Particle size (mm)	Estimated percentage	Particle size (mm)	Boundaries in polycrystalline aggregates	Deformation lamellae	Remarks
		Nearly 100	Average size of individual crystals from 3 to 5 µm	Linear or irregular; may contain radial fabrics indicating tiny cavity fillings		Unless stains or other special techniques are used to identify feldspars, it may be difficult or even impossible to distinguish chert from fine-textured siliceous– or felsic volcanic rocks.
100	Up to 1 mm or larger					Study of individual minerals usually is not a sufficient basis for distinguishing volcanic particles of terrigenous origin from those of pyroclastic origin.

An important event in the geologic history of an area is the date of uncovering of a pluton or of a basement complex. Commonly, dates of unroofing can be established by determining the age of the oldest sands containing heavy minerals traceable to the newly exposed parent deposits. A classic study in which heavy minerals in sands have been examined for the purpose of determining the date of unroofing of a pluton was carried out in southwestern England (Box 2.3).

Chemical Stability of Heavy Minerals. Studies of the chemical stability of heavy minerals have been conducted in soil profiles and in various sandstones. Investigations of both kinds have shown that certain species, such as garnet, hornblende, epidote, and apatite, can be dissolved. The effect is to remove these minerals and to concentrate the chemically stable species, such as zircon, rutile, monazite, and tourmaline. Removal of minerals that have been previously deposited in a body of sand is

designated as intrastratal dissolution (Chapter 3). Naturally, both weathering and intrastratal dissolution can reduce the variety of heavy minerals which become a part of the geologic record and thus impede provenance determination of heavy minerals in ancient sedimentary rocks.

SOURCES: Basu and Molinaroli, 1989; Darby, 1984; Darby and Tsang, 1987; Force, 1976; Füchtbauer, 1988; Milner, 1962; A. C. Morton, 1985.

LAYER-LATTICE SILICATES

The rock-making minerals we have discussed thus far tend to form nonflaky particles. By contrast, another large group of minerals found in terrigenous sediments tends to form flakes of various sizes. X-ray studies have shown that these flakes are expressions of crystal lattices built of sheets.

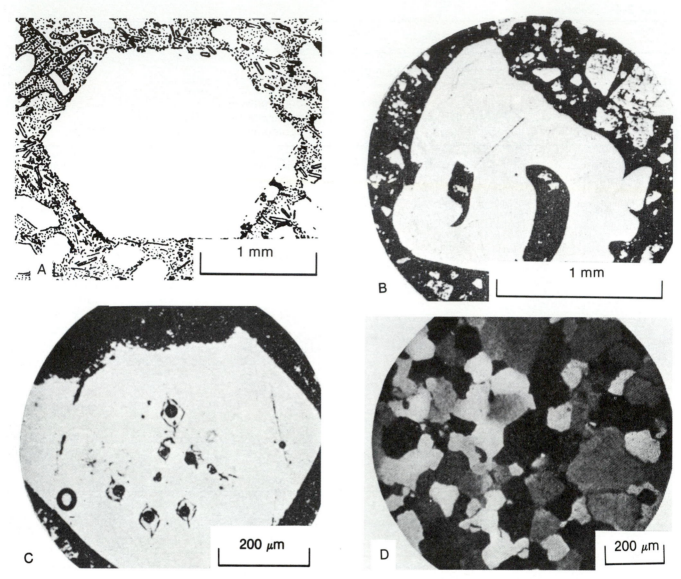

FIGURE 2-13.   Enlarged views of various kinds of quartz in volcanic- and metamorphic rocks; photomicrographs and sketches of thin sections.

A. Phenocrysts of high quartz (white) in microcrystalline groundmass, sketch. Riebeckite-acmite microgranite, Mynydd Mawr, Carnarvonshire, Wales. (F. H. Hatch, A. K. Wells, and M. K. Wells, 1949, fig. 91, p. 218; reproduced by permission of George Allen & Unwin Ltd, London.)

B. Rounded and embayed quartz phenocryst in volcanic rock, viewed in plane-polarized light. Quartz porphyry, Scharfenstein, Münsterthal, Black Forest, Germany. (H. Rosenbusch and J. P. Iddings, 1905, pl. III, fig. 3.)

C. Quartz containing bleblike inclusions of glass, seen in plane-polarized light. Quartz porphyry, Dosenheim, Germany. (H. Rosenbusch and J. P. Iddings, 1905, pl. VI, fig. 3.)

D. Mosaic of quartz crystals from metamorphic rock showing linear boundaries (polygonized quartz), viewed in cross-polarized light. Specimen and locality data not given. (E. H. Weinschenk, 1916, pl. V, fig. 1.)

Layer-lattice silicates consist of stacked sheets in various combinations of two fundamental units: (1) tetrahedral layers containing linked silicon–oxygen tetrahedra, and (2) octahedral layers in which hydroxyl ions (OH) occur in two planes, one above and one below a plane of magnesium or aluminum ions (Figure 2-17). The metallic ions are octahedrally coordinated with the hydroxyl ions.

The octahedral layer consists of two important structural varieties: (1) the brucite structure, in which all available cation spaces in the octahedral layer are occupied (trioctahedral arrangement), and (2) the gibbsite

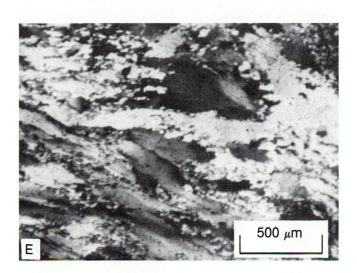

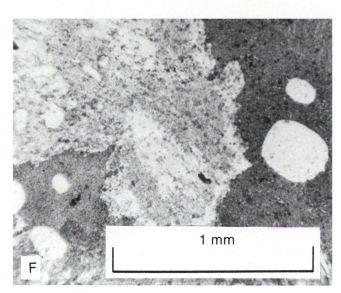

FIGURE 2-13. (*Continued*)

E. Elongated quartz crystals of two distinct sizes showing strain shadows from intensely sheared metamorphic rock: sutured borders; viewed in cross-polarized light. Pre-Triassic, Branford, Connecticut. (Authors.)

F. Circular quartz inclusions, lacking strain shadows, within feldspar crystals of metamorphic rock, a coarse-textured granitic gneiss; cross-polarized light. Pre-Triassic, Branford, Connecticut. (Authors.)

FIGURE 2-14. Enlarged views of feldspar in sedimentary rocks; photomicrographs of thin sections in cross-polarized light.

A. Grid twinning of microcline. Nubian sandstone (Shehoret Formation, Cambrian), Negev Desert, Israel. (T. Weissbrod.)

B. Polysynthetic twinning of sodic plagioclase. Upper Triassic part of New Haven Arkose, Branford, Connecticut. (Authors.)

TABLE 2-5.   Kinds of feldspars, their chief habits of twinning, and their geologic occurrences in plutonic-, volcanic-, and metamorphic rocks

Kind of feldspar		Twinning	Geologic occurrence
Potassium feldspars	Orthoclase	Carlsbad Baveno Manebach	Igneous rocks
	Microcline	Cross-hatch (or grid) pattern (combined albite and pericline laws)	Plutonic- and metamorphic rocks; rare in volcanics
	Sanidine		Volcanic rocks; high-temperature metamorphic rocks
Sodium–calcium feldspars (plagioclase)	Albite	A-twins (lamellar; includes albite, pericline, acline laws)	Any igneous- or metamorphic rock of appropriate composition
	Intermediate varieties		
	Anorthite	C-twins (other; includes simple twins and their modifications-Manebach, Baveno, parallel, complex)	
	Plagioclase crystals may be zoned; composition of feldspars in various zones may change progressively from being more calcic in the center to less calcic on the margins or may oscillate from zone to zone		

SOURCE: Modified from E. D. Pittman, 1970.

NOTE: All the kinds of feldspars shown in this table can occur as particles in which twinning is not present.

Intergrowths of Na- and K- feldspars are known as perthite (or microperthite), mesoperthite, and antiperthite, with the names used as follows:

K-spar >> Na-spar, perthite
K-spar = Na-spar, mesoperthite
K-spar << Na-spar, antiperthite

Glide twins result from intense deformation of feldspars.

FIGURE 2-15.   Feldspar phenocrysts in pyroclastic rock. Dark, bleblike inclusions are volcanic glass. Small semicircular embayments at upper left of large feldspar phenocryst have resulted from dissolution. Photomicrograph of thin section in plane-polarized light; Cretaceous volcanics near Lucea, Jamaica. (Authors.)

structure, in which only two thirds of the available cation spaces in the octahedral layer are occupied (dioctahedral arrangement).

The linkages of tetrahedral- to octahedral layers are made between the otherwise unshared oxygens at the apices of the tetrahedra, which take the positions of two thirds of the hydroxyls in one plane of the octahedral layers. Other hydroxyls are located in the centers of the hexagons defined by the linked tetrahedra.

The combination of tetrahedral- and octahedral layers forms a basis for the major subdivision of layer-lattice silicates into three main groups: Group I, one tetrahedral layer with one octahedral layer (a two-layer, 1-1 arrangement); Group II, two tetrahedral layers, one above and one below the one octahedral layer (a three-layer or 2-1 arrangement); Group III, sets of three-layer arrangements, as in Group II, alternating with an octahedral

TABLE 2-6.   Some heavy minerals, their densities, and parent rocks

Minerals	Specific gravity	Igneous		Hydrothermal (veins, pegmatites)	Metamorphic		Sedimentary
		Mafic	Felsic		High-rank	Low-rank	
Anatase	3.82–3.97	×					
Augite	3.23–3.52	×					
Brookite	4.14–0.06	×					
Cassiterite	6.99		×	×			
Chromite	4.5–4.8	×					
Hypersthene	3.42–3.84	×					
Ilmenite	4.72–0.04	×	×				
Leucoxene	2.09–2.16	×	×				
Magnetite	5.175	×	×		×		
Olivine (Mg)	3.275–4.32 (Fe)	×					
Rutile	4.23	×					
Serpentine		×					
Apatite	2.9–3.1		×				
Biotite	2.7–3.4		×	×	×		
Hornblende	3.02–3.27		×	×	× Blue-green variety		
Monazite	4.6–5.4		×	×			
Sphene	3.45–3.55		×				
Tourmaline	4.7		× Small pink euhedra	× Typically blue (indicolite)		× Small pale-brown euhedra, carbonaceous inclusions	
Zircon	4.6–4.7		×				
Fluorite	3.18			×			
Garnet	3.7–4.3			×	×		
Topaz	3.49–3.57			×			
Andalusite	3.13–3.16				×		
Epidote	3.35–3.5				×	×	
Kyanite	3.53–3.67				×		
Sillimanite	3.23–3.27				×		
Staurolite	3.65–3.83				×		
Zoisite	3.533				×		
Barite	4.50			×			×
Galena	2.58						

SOURCE: Densities from W. L. Roberts, G. R. Rapp, Jr., and Julius Webers, 1974; table after F. J. Pettijohn, 1975, Table 13-1, p. 487.

layer (a four-layer or 2-1-1 arrangement). (See Figure 2-17, B.) Because in each of these groups the thickness of the combined layers varies, the largest spacings observed in their resultant X-ray diffraction patterns vary. The largest spacing is 0.7 nm in Group I, 1.0 nm in Group II, and 1.4 nm in Group III (Table 2-7).

A second major basis for classifying layer-lattice silicates is by the cation content of the octahedral layer.

Thus Groups I and II can be further subdivided into dioctahedral- (gibbsite structure) and trioctahedral (brucite structure) minerals.

A third basis for classification is the nature of the stacking of the 1-1 and 2-1 layer groups.

A final item applies to the 2-1 lattices; some are expandable and others are not. An expandable lattice is one that can incorporate layers of water and thus can increase

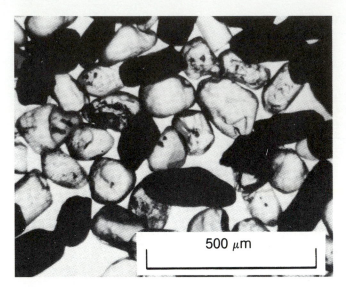

500 μm

FIGURE 2-16.   Opaque heavy minerals, including rutile and leucoxene, not distinguishable in this view (black), and transparent heavy minerals (light), all zircons. Some zircons contain many inclusions; others lack inclusions. Concentrate from Pleistocene sand attained in commercial operation by various separation techniques, viewed through binocular stereomicroscope. North Stradbroke Island, Queensland, Australia. (Minsands Exploration Pty., Ltd.)

the dimensions of the lattice. The most-expandable lattice is found in montmorillonite, a variety of smectite and a common constituent of altered volcanic-ash layers.

Layer-lattice silicates are alteration products of rock-making silicate minerals, chiefly the feldspars and ferromagnesian silicates (Figure 2-18).

SOURCES: Bailey, 1980; Bennett, Bryant, and Hulbert, 1991; Grim, 1968; Heling, 1988; O'Brien and Slatt, 1990; Pevear and Mumpton, 1989; Weaver, 1967.

Carbonaceous Sediments

Solid particles of carbonaceous organic matter include two kinds: (1) solid carbonaceous materials reworked from older formations and (2) modern plant detritus.

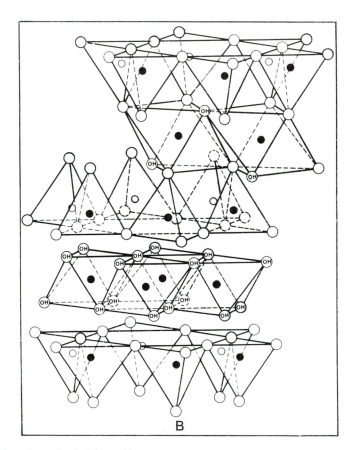

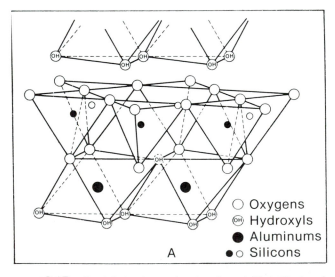

- ○ Oxygens
- ⊙ Hydroxyls
- ● Aluminums
- ●○ Silicons

FIGURE 2-17.   Crystal structures of various layer-lattice silicate minerals; schematic sketches with ions (circles) in expanded positions. In the minerals, the ions can be considered as spheres that touch one another.

A. A Group I mineral, kaolinite, with a 1-1 arrangement of layers. Tetrahedral layer in upper part, with apices of tetrahedra pointing downward. The Si^{4+} ions are shown in black- or open circles to indicate their positions within different rows of tetrahedra. Octahedral layer below, with Al^{3+} ions surrounded by O^{2-} and OH^- ions.

B. Example of a Group III layer-lattice silicate mineral (chlorite), showing the tetrahedral-octahedral-tetrahedral, or 2-1 arrangement at top and the in-between octahedral layer, below, together forming a four-layer 2-1-1 pattern. At bottom is tetrahedral layer of the top of the next underlying unit. (R. E. Grim, 1968, A, fig. 4-14, p. 58, based on J. W. Gruner, 1932; B, fig. 4-17, p. 100, based on R. C. McMurchy, 1934; used with permission of the McGraw-Hill Book Co.)

BOX 2.3

Minerals Through Time: Dates of Exposure or of Burial of Distinctive Parent Rocks; Unroofing Studies

The distributions of terrigenous particles within a succession of strata may contain important clues to the geologic history of a region, from which the dates of the availability of a given parent area can be determined. This availability may be related to the uncovering of ancient basement rocks long buried by younger strata, to the unroofing of a distinctive pluton, or to the burial and later uncovering of distinctive sedimentary strata. Examples of each follow.

As a first approximation, the positions of particles in the sediments filling a depositional basin are in-

versely related to their positions in an elevated parent area. That is, particles derived from formations that were at the top of an elevated area tend to be found at the bottom of the filling strata. As streams cut deeper into the parent rocks, they erode particles that appear higher up in the filling strata.

The potential value of a radiometric date from the minerals of a pluton is enhanced as a direct function of the geologic evidence on the date of intrusion. In many cases, the lower (younger) limit on the age of a pluton can be established only by determining the age of the oldest post-pluton

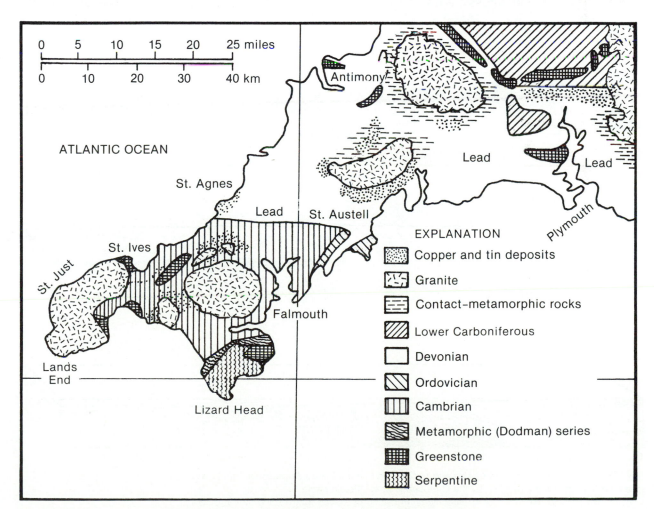

BOX 2.3 FIGURE 1.   Geologic sketch map, southwestern England, showing plutons of Dartmoor Granite that were intruded late in the Paleozoic Era and elevated and eroded during the Mesozoic Era, finally becoming unroofed early in the Cretaceous Period. (J. E. Sanders, 1981, fig. 6.30, p. 171.)

sands containing heavy minerals traceable to the newly exposed parent pluton. A classic study in which heavy minerals in sands have been examined for the purpose of determining the date of unroofing of a pluton was carried out in southwestern England and involved distinctive detritus (certain kinds of zircons, octahedrite, tabular anatase, monazite, brookite, and manganiferous garnets) from the Dartmoor Granite (Box 2.3 Figure 1). This granite intrudes folded-and metamorphosed marine sedimentary strata of Devonian- and Carboniferous ages of the Variscan fold belt that trends E–W across southern England. The folded rocks of the Variscan belt are overlain unconformably by nonmarine red sandstones of Permian- and Triassic ages (New Red Sandstone).

The first indication that erosion had cut close to this granite is found in Permian sands. Likewise, sands of Triassic- and Jurassic ages contain minerals from the contact aureole, but not from the granite itself. Finally, in early Cretaceous time, the granite became exposed and the heavy minerals of Dartmoor provenance flooded eastward. All sands younger than Cretaceous contain Dartmoor minerals, indicating that once it had become unroofed, this granite continued to supply sediments to the surrounding areas. Thus, in southern England, heavy minerals of Dartmoor provenance are reliable indicators of sands of post-early Cretaceous age.

The relationship of the basin-filling strata in the Newark Basin of the northeastern United States to the coastal-plain strata that overlap them illustrates how mineral data contribute to a contrasting kind of geologic problem: the possible former updip extent of the eroded edge of a group of strata. Considerations of the relationship of rivers in the Appalachian region led to the hypothesis that the coastal-plain strata formerly

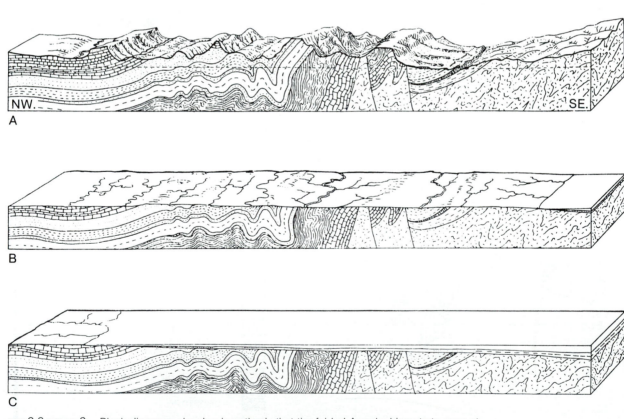

BOX 2.3 FIGURE 2.   Block diagrams showing hypothesis that the folded Appalachian strata were at one time unconformably overlain by the Cretaceous coastal-plain strata.
 A. Medial Jurassic elevation of Appalachians and tilting of Newark basin-filling strata to NW.
 B. Erosion creates Fall-Zone peneplain.
 C. Late Cretaceous sea shown as having submerged the entire width of the Appalachian chain.
[D. W. Johnson, 1931 (reprinted 1967), figs. 1–3, p. 15.]

extended updip far enough to bury the entire Appalachian fold belt (Box 2.3 Figure 2). If this concept of the former updip extent of the coastal-plain strata is correct, then no minerals derived from the presumably buried rocks would have been available to the rivers feeding sediment into the nearshore regions where the coastal-plain strata (of Late Cretaceous- and Early- and Middle Cenozoic ages) were accumulating. Studies of the minerals in the sands in the coastal-plain strata indicate no contributions from the distinctive Newark strata, but continued contributions from the Precambrian rocks of the elevated central belt of the Appalachians. Therefore the mineral data support a former updip extension great enough to have buried the outcrop areas of the tilted Newark strata but not great enough to have buried the central parts of the Appalachian chain.

TABLE 2-7.   Classification of some layer-lattice silicates occurring in sedimentary deposits

Arrangement	Two-layer (1-1) 0.7 nm		Three-layer (2-1) 1.0 nm				Four-layer (2-1-1) 1.4 nm
Number of layers; largest spacing on X-ray diffraction patterns							
Kind of lattice	Nonexpandable		Expandable		Nonexpandable		Nonexpandable
Cation content of octahedral layer	Dioctahedral	Triocta-hedral	Dioctahedral	Triocta-hedral	Dioctahedral	Triocta-hedral	
Mineral name	**Kaolinite**	No minerals common in sediments	**Smectite**	No minerals common in sediments	**Muscovite (sericite)**	**Biotite**	**Chlorite**

Dioctahedral; most layers are nonexpandable but a few are expandable

Illite[a, b]

Glauconite[a, c]

Mixed-layer clay minerals

Mixed-layer clay minerals

SOURCE: Modified from C. M. Warshaw and Rustum Roy, 1961, p. 1464.

NOTES: Clay minerals are shown in boldface type.

[a] Illite and glauconite consist chiefly of mica layers but do include some expandable layers. Despite this mixing of layers these two minerals are not considered to be of mixed-layer type.

[b] Illite designates a fine-textured mineral having almost the same structure as muscovite but containing slightly less potassium than muscovite.

[c] Glauconite contains larger amounts of Fe^{3+} than of Al^{3+} in the octahedral layer.

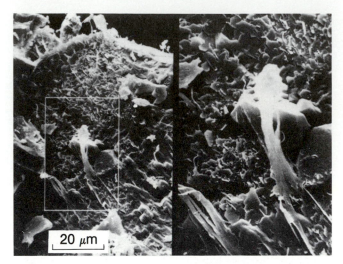

FIGURE 2-18.   Greatly enlarged view of authigenic illite fibers growing on altered plagioclase particle as seen in scanning-electron micrograph. (Ian Hutcheon, 1983, fig. 4, C, p. 7; photo by A. E. Oldershaw. Permission granted by Geological Association of Canada.)

Solid Carbonaceous Materials Reworked from Older Formations

Solid carbonaceous organic matter reworked from older formations includes detrital bituminous- and anthracite coal, amber, solid pieces of wax, and kerogen, which is a collective name applied to a great variety of insoluble solid hydrocarbons having long-chain polymer structures. Only a few chemical formulas have been determined for kerogens. A specimen from the Green River Formation (Eocene) of Colorado yielded $C_{215}H_{330}O_{12}N_5S$. Kerogen is the most-abundant and widespread form of organic matter in the world; its average abundance in sedimentary rocks is 0.3% by weight. Using the estimate of 10^{18} tons for the world's total quantity of sedimentary deposits, the world's total amount of kerogen is computed at 3×10^{15} tons. This is about 600 times as much as the world's known reserves of coal, which have been estimated at 5×10^{12} tons. The so-called oil shales are more correctly designated as kerogen rocks. These "shales" contain no oil; their organic matter is in the form of kerogen. If its composition is appropriate, kerogen can be converted to petroleum by heating in the laboratory to temperatures of 375°C.

Kerogens form by low-temperature geothermal alteration of organic matter, much of which comes from plants. Kerogens are extremely stable. They will not dissolve in organic solvents or acids and do not oxidize at ordinary temperatures. Particles of kerogen recycled from older formations have been identified in the sediment

that is coming out from beneath the Antarctic ice sheet and is being deposited on the floor of the Ross Sea.

SOURCES: J. M. Hunt, 1979; Sackett, Poag, and Edie, 1974.

Modern Plant Detritus

In regions having abundant rainfall and warm- to tropical temperatures, modern plants form a nearly continuous cover. In humid-temperate seasonal climate zones, tremendous quantities of leaves are dropped from the trees each fall. Much of the plant litter becomes soil humus, but many fallen leaves can become parts of the deposits of ponds, lakes, swamps, rivers, and even the sea.

Other plant detritus includes twigs, tree trunks, and microscopic seeds, pollen, and spores. Plant material that collects in swamps forms peat.

Much other organic matter, derived from the soft parts of organisms, is not solid, but occurs as liquid hydrocarbons and other organic compounds. Because the affinity of this organic matter for oxygen is so strong, and thus the organic matter is so easily oxidized, its preservation depends on its rate of supply and burial compared with the abundance of oxygen. The importance of these relationships to the color of sediments is discussed in Chapter 8.

Pyroclastic Particles

Particles derived explosively from volcanoes are **tephra**, or pyroclastic debris. The word pyroclastic is derived from two roots: *pyro*, which refers to fire, and *clastic*, which signifies breaking; whereas tephra comes from a Greek word meaning ash. Pyroclastic debris is easily recognizable and, because of its volcanic origin, merits separate recognition.

Silicic, rhyolitic magmas are more viscous and contain more volatiles and hence produce more tephra (erupt more explosively) than do mafic basaltic magmas. Silicic magmas at 900°C can be nearly 10 trillion times as viscous as motor oil at room temperature. Some pyroclastic particles may be truly igneous in the sense that when they became particles and were propelled out of the volcano, they were molten. However, once launched, they follow the same laws that govern the movement of particles of other kinds of sediment.

Pyroclastic particles can include rock fragments, single crystals, or bits of volcanic glass. These particles have been grouped together under the headings lithic (rock fragments), crystal and crystal fragments, and vitric (glassy) and pumice, respectively. Vitric particles are usually the most abundant. Pyroclastic particles range in size from a few micrometers to more than 1 meter, a range of more than a millionfold.

Rock fragments can consist of previously solidified volcanic rocks or of any kind of rock through which the volcanic gases and the lava passed on their way to the surface. The crystals must have already grown within the magma. The glass particles represent blebs of molten lava that solidified so rapidly that their atoms did not form crystal lattices. Many glass particles in tephra are shards of the thin walls of gas bubbles that formed in the magma as it cooled and expanded during its rise to the surface in a volcano (Figure 2-19). These particles are called bubble-wall shards. When viewed under the microscope, they display a bewildering variety of shapes and surface textures. Other glassy particles are fragments of droplets, and larger glassy fragments may be bits of pumice, a frothy rock filled with voids that is commonly light enough to float on water.

Certain features characterize mineral particles from volcanic sources: doubly terminated prisms of high quartz, rounded- and embayed quartz crystals, oscillatory zoning in plagioclase, and bleblike inclusions within quartz or feldspar. The rock fragments usually consist of only a few minerals and are often coated with a thin,

glassy film that bears the impressions of tiny bubbles. Tephra (See Box 2.1 Figure 5.) generally is so distinctive that it is readily recognized. The use of tephra in provenance studies requires that petrographic study be supplemented by determinations of the index of refraction of glass particles and analysis of trace elements.

Pyroclastic debris can travel in five ways: (1) as *turbulent hot-particle flows that are ejected from volcanic vents and move along the ground* (**nuées ardentes** or "ash flows"); (2) as debris flows (lahars) or other kinds of fluid flows that evolve from turbulent hot-particle flows with cooling and with addition of solid material or water; (3) as low-altitude suspensions in the atmosphere; (4) as high-altitude suspensions in the stratosphere (as high as 80 km for the smaller fragments); and (5) in the case of some larger particles, as ballistic projectiles that are launched from vents and travel only short distances. Large pumice particles tend to accumulate very near the vents because their low densities make them very poor projectiles. The processes of pyroclastic-particle movement are explained in Chapter 7.

Tephra have been spread widely from some spectacular volcanic explosions. As mentioned in Chapter 1, tephra from the 1947 eruption of Mt. Hekla landed 3800 km away in Helsinki. Tephra in the rock record have been recognized as much as 2000 km from their parent volcanoes. If the tephra can be traced to their parent volcano, and if they were discharged at a known date, then they can serve as valuable time markers. An example of the use of tephra layers as time markers is the analysis of cores of sediment from the floor of the eastern Mediterranean Sea. Many cores contain two layers of tephra from the volcano Santorini (formerly Thera). The older layer is estimated to be about 25,000 years old, whereas the younger layer has been dated at about 1,450 B.C. This date approximately coincides with the abrupt demise of the Minoan civilization on the island of Crete. The larger particles from the explosions remained near the vent (Figure 2-20), but the finer material was carried up into the atmosphere and spread southeastward for hundreds of kilometers (Figure 2-21). The record of this eruption has even been preserved as a distinct layer in ice cores from the Greenland ice cap. Proximity to the source may be indicated by coarse particle size, good sorting, and greater tephra-layer thickness.

Because tephra are deposited quickly, even if their dates of eruption are not known, an individual layer of tephra can serve as an excellent relative time marker. Tephra eruptions may last as long as 6 months, but the vast majority of the tephra is usually deposited in just a few days.

SOURCES: Bitschene and Schmincke, 1990; Kittleman, 1979; Lajoie, 1979; Oeschger and Langway, 1989; Schmid, 1981; U.S. Department of Energy, 1988.

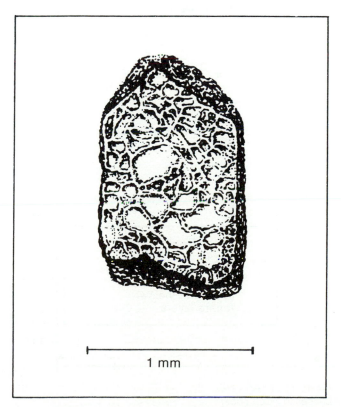

1 mm

FIGURE 2-19.   Bubble-wall texture in tephra from Recent pyroclastic flow resulting from catastrophic eruption of Mount Mazama (ancestor of Crater Lake), Oregon. (R. V. Fisher, 1963, fig. 4, A, p. 226.)

FIGURE 2-20.   Coarse tephra close to vent of violently explosive eruption from Santorini Volcano, Grecian Archipelago. (Authors.)

Intrabasinal Particles

Intrabasinal particles include various solids that grew biochemically or chemically in the waters of the depositional basin. These include carbonate biocrystals and other carbonate particles, silica biocrystals, particles composed of evaporite minerals, and certain authigenic minerals, such as glauconite, that grew at the water/sediment interface.

Carbonate Biocrystals and Other Carbonate Particles

Most skeletal materials secreted by the metabolic activities of living organisms are composed of calcium

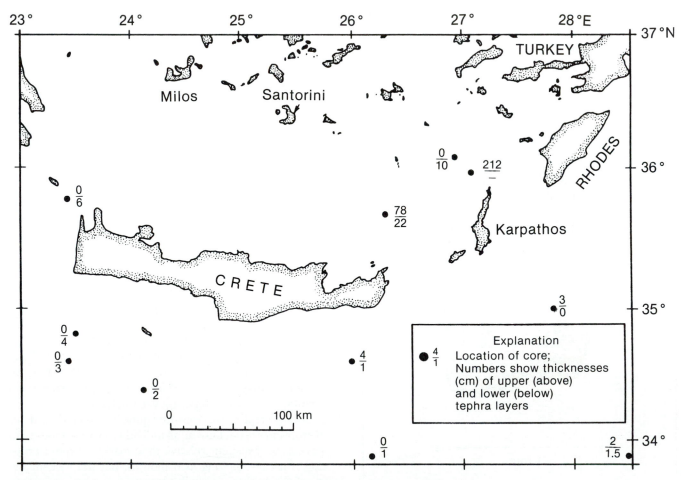

FIGURE 2-21.   Sketch map of eastern Mediterranean showing locations of bottom-sediment cores containing layers of tephra erupted from Santorini Volcano. Core west of Rhodes in which upper tephra layer is 212 cm thick did not penetrate deep enough to encounter lower tephra layer. Further explanation in text. (Modified from D. Ninkovich and B. C. Heezen, 1965, fig. 162, p. 425, and fig. 169, p. 439.)

carbonate. The minerals secreted by organisms form **bio-crystals**, that is, *solids having the lattice properties of minerals but distinctive shapes that are not crystal faces*. In the modern marine environment, widely varying groups of organisms produce skeletal debris. In the past, some groups of organisms that are now important sediment producers were insignificant, and some once-important groups are now minor or no longer exist at all (Figure 2-22).

MINERAL COMPOSITION OF CALCIUM-CARBONATE SKELETAL DEBRIS

In modern carbonate sediments the two principal minerals are calcite and aragonite. This fact was established in 1859 by Henry Clifton Sorby (1826–1908), who also showed that calcite tends to predominate in sand-size skeletal debris, whereas aragonite predominates in lime mud.

Although the chemical formulas of both aragonite and calcite are the same, $CaCO_3$, these minerals differ in crystallographic arrangement, in density, in hardness along certain crystallographic directions, in their con-

tents of trace elements, and in their solubilities in various fluids (discussed further in Chapter 3).

Aragonite crystallizes in the orthorhombic crystal system and consists of essentially pure calcium carbonate. Calcite crystallizes in the rhombohedral class of the hexagonal system and contains various amounts of magnesium and other elements. Calcite ($CaCO_3$) and magnesite ($MgCO_3$) form an isomorphous series within which two varieties of calcite exist. One has been termed ordinary- or low-magnesian calcite, and the other high-magnesian calcite. The distinction between the two forms of calcite is arbitrary and varies from one researcher to another. We shall define high-magnesian calcite to contain more than 10 mol % $MgCO_3$ in solid solution; it may contain more than 30 mol % $MgCO_3$. Low-magnesian calcite has less than 8 mol % $MgCO_3$. A calcite containing 8 to 10 mol % $MgCO_3$ may be considered either high- or low-magnesian calcite. The composition of the mineral calcite, which represents both kinds, can be given by the formula $(Ca_{1-x}Mg_x)CO_3$, where $0 < x < \sim 0.30$. Modern shallow-water skeletal carbonate particles consist mostly of high-magnesian calcite and aragonite.

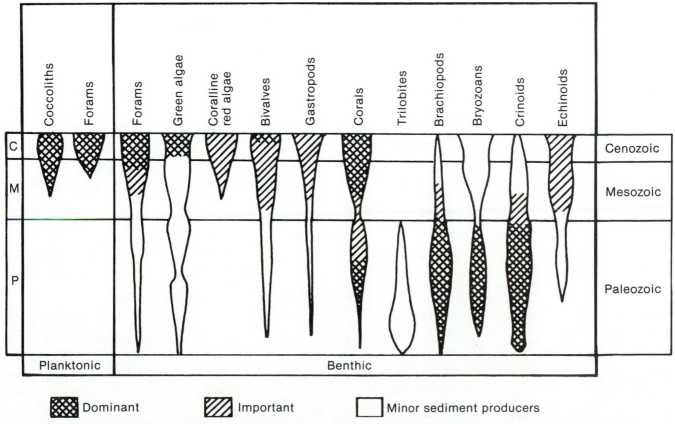

FIGURE 2-22. Diversity, abundance, and relative importance of various taxonomic groups of marine organisms as producers of calcium-carbonate sediment. P = Paleozoic, M = Mesozoic, C = Cenozoic. (B. H. Wilkinson, 1979, fig. 1, p. 526.)

Henry Clifton Sorby distinguished aragonite and calcite by careful measurements of their specific gravities (calcite = 2.71; aragonite = 2.93). He also found that aragonite will scratch a cleavage plane of Iceland spar in the line of the shorter diagonal of the rhomb face, but that calcite will not.

In terms of their specific compositions, most tests or skeletal fragments of plants and animals consist either of aragonite or of high-magnesian calcite (uncommonly of low-magnesian calcite). Corals, most mollusks, and green algae synthesize skeletons of aragonite. Red algae, some calcareous sponges, echinoids, crinoids, and bottom-dwelling calcareous foraminifers secrete high-magnesian calcite. Oysters, *Pecten* (a pelecypod), barnacles, and surface-drifting foraminifers are among the organisms that build skeletons of low-magnesian calcite.

In high-magnesian calcite, the mole percentage of $MgCO_3$ varies greatly, depending on the taxonomic affiliation of the organism (Table 2-8). Even in skeletons composed of aragonite or low-magnesian calcite, the concentrations of Mg^{2+} and other minor elements such as Sr^{2+} and Na^+ vary systematically among taxonomic groups of organisms.

A few uncommon minerals have been found in the calcareous skeletons of some organisms. These minerals include unstable calcium-rich protodolomite in the teeth of modern echinoids; brucite $[Mg(OH)_2]$ in the red alga *Goniolithon;* magnetite in the denticles of chitons; and goethite and opal in the denticles of gastropods. For some extinct groups of organisms, original shell composition is not known with certainty.

SOURCE: Lowenstam, 1981.

SAND- AND GRAVEL-SIZE CALCIUM-CARBONATE SKELETAL DEBRIS

Much carbonate skeletal debris includes whole skeletons of calcium-carbonate-secreting organisms, such as

TABLE 2-8.   Proportion of $MgCO_3$ and kind of calcite in calcareous skeletons of various marine organisms

Name of organism	Location	Proportion of $MgCO_3$ (mol %)	Kind of calcite
Foraminifers			
Elphidium spp.	Point Barrow, Alaska	0.3	Low-magnesian
Homotrema rubrum	Bermuda	13.7	High-magnesian
Peneroplis spp.	Florida Keys	13.5	High-magnesian
Orbitolites spp.	Palau	15.9	High-magnesian
Echinoids			
Encope spp.	Honshu, Japan	10.6	High-magnesian
Echinometra lucanter	Bermuda	14.7	High-magnesian
Eucidaris tribuloides	Florida Keys	11.4	High-magnesian
Crinoids			
Hypalocrinus naresianus	Philippines	10.2	High-magnesian
Crinometra concinna	Cuba	11.7	High-magnesian
Pelecypods			
Ostra spp.	Bermuda	1.3	Low-magnesian
Pecten spp.	Palau	2.0	Low-magnesian
Barnacles			
Mitella polymerus	Pacific Grove, California	2.1	Low-magnesian
Scapellum regium	British Columbia	0.2	Low-magnesian
Red algae			
Lithothamnium spp.	La Jolla, California	14.1	High-magnesian

SOURCE: After K. E. Chave, 1954, table 1, pp. 269–273.

FIGURE 2-23. Benthic foraminiferan *Homotrema rubrum* encrusting coral fragment. Islas Los Roques, off coast of Venezuela, South America. (Authors.)

FIGURE 2-24. View of bedding surface of limestone composed chiefly of skeletal debris of bryozoans (branched, twiglike colonies containing rows of tiny pits) and of stem plates of crinoids (button-shaped objects). Lowville Limestone (Middle Ordovician), near Kingston, Ontario, Canada. (Authors.)

foraminifers and mollusks, as well as broken pieces of the hard parts secreted by these and other organisms (Figure 2-23).

Many skeletal sands accumulate only after the skeleton-secreting organisms have died. Individual organisms, such as mollusks and brachiopods, and colonial organisms, such as encrusting bryozoans or isolated coral heads, supply such nonreef skeletal sands. Boring organisms are significant factors both in the death of some shelled organisms, such as mollusks, and in the subsequent breaking apart of the skeletal material. In order to harvest the soft parts inside a mollusk shell, many boring predators, such as gastropods, can drill holes right through a mollusk's shell. The first hole drilled through a shell by such an enemy results in the death of the shell's inhabitant. After the mollusk has died, however, its shell may be riddled by other borers, such as sponges, algae, or fungi. This intense boring ultimately breaks down the mollusk shell into a heap of small calcium-carbonate particles whose origin becomes difficult to determine.

Some skeletal materials become skeletal sands without such violent biologic disintegration. Parts of some initially solid skeletons are easily detachable and at death simply fall away from the rest of the skeleton without much outside assistance. For example, spines of sea urchins (a variety of echinoid) are important constituents of some modern skeletal sands and some ancient limestones composed of lithified skeletal sands. (See Box 2.1 Figure 7.)

The plates of other members of the phylum Echinodermata (crinoids, blastoids, cystoids) are individual crystals of calcite. As recognized by Sorby in the middle of the nineteenth century, these biocrystals are of sand size initially. For example, sand-size skeletal debris from Paleozoic crinoids constitutes voluminous deposits, now limestones. Fragments of bryozoan colonies are signifi-

cant constituents of many Paleozoic limestones (Figure 2-24). Other complete tests are of sand size and therefore no breakage is required for them to become constituents of skeletal sands. These include many small foraminifers and some larval mollusks.

Much skeletal debris becomes skeletal particles as a result of the life processes of the organisms; no dying is involved. Box 2.4 summarizes some ways in which this happens.

Two other kinds of carbonate particles include lime mud (Box 2.5) and sand-size debris not composed of identifiable skeletal debris (Box 2.6).

SOURCES: Füchtbauer and Richter, 1988; Scoffin, 1987; Tucker and V. P. Wright, 1990.

Siliceous Skeletal Materials

The major contributors of siliceous skeletal remains are diatoms (Figure 2-25, page 82) and radiolarians. (See Box 2.1 Figure 8.) Contributions from dinoflagellates and sponges are minor. Radiolarians and diatoms live either among the free-floating organisms in the near-surface water of the sea or at the bottom. In addition, some

BOX 2.4

Organisms That Create Skeletal Debris Without Dying

Ordinarily, one supposes that organisms have to die to contribute their skeletal material to sediments. Although death is involved in the origin of much fossil skeletal material, numerous noteworthy exceptions are known. These include various single-celled marine floating organisms, arthropods, sharks, lime-secreting tropical-marine plants, and reefs.

Among single-celled organisms, two members of the "organic soup" that floats in the surface waters of the oceans, a group of plants known as coccoliths and of animals named Foraminifera, illustrate our point. Coccoliths shed plates (Box 2.4 Figure 1) during the lifetime of the organism. Many foraminifers keep making tests throughout their lives. These one-celled organisms reproduce by cell division. Hence, in order to reproduce, the soft parts must be outside their solid tests. Early in the cycle of reproduction the organisms discard their old tests. Thereafter, the naked cell divides and each of the new cells secretes a new solid test. The discarded tests sink toward the sea floor. If they reach bottom without being dissolved en route, they become significant constituents of many marine sediments. Eventually an individual

foraminifer does die, of course, and its skeleton likewise sinks toward the sea floor and may become part of the bottom sediments. By and large, the livelier the foraminifers are, the more skeletal material they generate.

Like the foraminifers, most marine arthropods (such as ostracodes) periodically discard their skeletons. They do not do so during reproduction, but, because their growth proceeds in spurts by a process known as *moulting*. In order to grow, the animal must break out of its tight-fitting former external skeleton, named an *exoskeleton*, enlarge rapidly in an exposed condition, and then quickly secrete a new, larger exoskeleton. Marine-arthropod skeletons commonly consist of a complex organic substance known as *chitin*, which may include calcium carbonate. A notable exception is the extinct Paleozoic trilobites, whose exoskeletons appear to have consisted entirely or almost entirely of calcium carbonate. In some Lower Paleozoic strata, carbonate debris from trilobite exoskeletons is abundant enough to form skeletal carbonate sands or even skeletal carbonate "gravels." It is often possible to determine whether a particular trilobite exoskeleton was shed by a living trilobite during moulting or was being worn by a trilobite that died. Exoskeletal parts shed by a moulting trilobite are left in characteristic positions, are split along certain seams known as sutures, and are commonly considerably thinner than exoskeletons that enclosed dead individuals of the same species. Among some trilobite species, each individual may have produced several dozen exoskeletons.

Sharks' teeth are a third kind of fossil skeletal material that is deposited in abundance without requiring the death of numerous individual sharks. Fossil sharks' teeth attracted the attention of early European naturalists. One of Nicolaus Steno's (1631–1687) important scientific contributions was his demonstration that certain fossil curiosities were sharks' teeth (Box 2.4 Figures 2 and 3).

Sharks possess the capability of continuously replacing the teeth that they regularly lose. This means that during its lifetime each individual shark can make hundreds or even thousands of skeletal contributions to the bottom sediments. Sharks' teeth are phosphatic; they consist of apatite. Obviously, the teeth are only a small part

BOX 2.4 FIGURE 1.   Greatly enlarged view of skeleton of coccolith (*Coccolithus* cf. *C. barnesae*). Isfiya Chalk (Upper Cretaceous), Mount Carmel, Israel, as seen in scanning-electron micrograph. (A. Bein.)

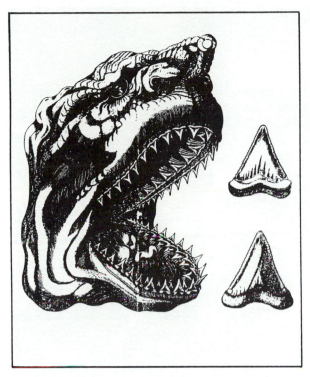

BOX 2.4 FIGURE 2. Head of a shark and its teeth, sketched by Michelle Mercat and first published by Nicolaus Steno (1631–1687) illustrating that shark's teeth and fossil "tongue stones" (*Glossopetrae*) are identical. Steno and G. G. Leibnitz (1646–1716), from whose study *Protagea* (1749) the head and teeth have been reproduced (pl. VII), argued in detail that the *Glossopetrae* were fossil teeth of very large sharks. (History of geology collection, G. M. Friedman.)

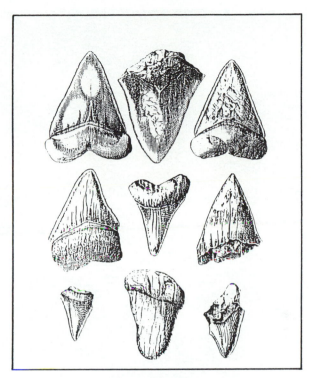

BOX 2.4 FIGURE 3. *Glossopetrae* or fossil "tongue stones" identified by Steno as fossil teeth of sharks (Nicolaus Steno, 1667, pl. 6.) (History of geology collection, G. M. Friedman.)

of a shark, hence are not comparable to the series of complete exoskeletons secreted by individual foraminifers and trilobites. However, because shark skeletons are cartilaginous rather than bony, they are almost never preserved in the fossil record. Hence sharks' teeth are their only commonly preserved skeletal remains. In most strata, sharks' teeth are not widespread, but some Tertiary deposits contain millions of them. Because of their high resistance to abrasion, sharks' teeth can be recycled from older formations into modern sediments. For example, along the banks of Aquia Creek in northern Virginia, sharks' teeth are being eroded from a Tertiary formation and are being concentrated in the modern beach sediments.

Minor non-fatal contributions of skeletal material to carbonate sands are made by *Halimeda* (parts of lime-secreting tropical-marine plants that are broken off by storms that do not kill the plants). In some lagoons on Pacific coral islands, *Halimeda*

plates form a major proportion of the recognizable skeletal debris.

Other organisms that contribute to skeletal sands without having to die to do so include organic reefs and other organisms that live attached to the bottom. These organisms can create skeletal debris without dying, but while this is happening they may be, so to speak, nearly "nibbled to death" by fish and other predators. Many kinds of fish surround modern organic reefs and several species, such as the parrot fish (Box 2.4 Figure 4) and trigger fish, graze on the reef. These predatory fish eat the whole thing; they bite off bits of the reef colony, digest the edible organic matter, and then excrete the indigestible carbonate particles. Other predators on reefs include boring sponges.

Major storms are an important producer of reef-derived sediment, as well as one of the most-important agents of sediment redistribution on and near reefs. Storms break off portions of some organisms (such as corals) and completely destroy others. Hurricanes may convert up to 100% of the living corals on large parts of a reef into gravel- and sand-size detritus.

BOX 2.4 FIGURE 4.   Fish surrounding modern reef, Gulf of Aqaba, Red Sea. (D. Popper.)

Although the total volume of skeletal material of modern- and ancient reef-building organisms that is preserved intact is small, reefbuilders sup- ply copious amounts of skeletal sand that is dis- tributed around the reef. A reef is analogous to a factory that produces a steady stream of goods.

diatoms inhabit fresh waters. The diatoms that live on the bottom may form sticky organic mats. Diatoms also may be important contributors to organic mats composed of combinations of several groups of mat-forming organisms, including algae and bacteria.

Siliceous organic remains accumulate in the deeper parts of the oceans, as well as locally in enclosed shallow arms of the sea, and in lakes. Although, in general, siliceous skeletal particles are not abundant on the floor of the ocean, locally they have been found to compose up to 40% of the deep-sea sediments. The abundance of siliceous skeletal particles is related to three factors: (1) the rate of production of siliceous tests, (2) the dilution of these tests by terrigenous- and carbonate particles, and (3) the dissolution of siliceous tests in the water column and especially on the floor of the ocean.

Particles Composed of Evaporite Minerals

In places where evaporite minerals are precipitated, such as in modern playas, the chemically precipitated crystals may later be transported physically as particles. Physical transport may be by wind or water. The wind may pick up gypsum crystals and pile them up as particles in dunes. The dune sands of White Sands National Monument, New Mexico, consist of precipitated

BOX 2.5

Lime Mud

The silt- and clay-size components of carbonate sediments are collectively designated as lime mud. Most lime mud consists of tiny needles and platelets of carbonate crystals that are too small to resolve with binocular- or petrographic microscopes. Accordingly, interpretations of the particles of lime mud have been based on the notion that they shared a common origin, namely inorganic precipitation from saline water. Studies with the scanning-electron microscope have shown that, among the supposedly uniform constituents of lime muds, great variety exists. We review three mechanisms for forming lime mud: (1) inorganic precipitation, (2) biochemical secretion, and (3) breakup of sand-size (or larger) aragonitic skeletal debris.

Inorganic Precipitation of Aragonite

Inorganic precipitation may be an important source of lime mud. According to a recent estimate by Shinn and others (1989), the amount of lime mud precipitated inorganically on the Great Bahama Bank may be much greater than that secreted by algae. The tiny needles of aragonite precipitated inorganically are dispersed in the water and form *drifting "clouds" of milky-white water* called **whitings**. Sedimentation rates from whitings are high, yet the whitings appear to persist for months. This implies that, despite the losses from the fallout to the bottom, the components of the whitings are being renewed. A plausible mechanism for such renewal is rapid precipitation in the water column of aragonite needles (and possibly small particles of high-magnesian calcite). The tiny individual aragonite crystals precipitated by whitings are indistinguishable from aragonite biocrystal needles secreted by codiacean green algae (described below).

SOURCE: Shinn, Steinen, Lidz, and Swart, 1989.

BOX 2.5 FIGURE 1.   Colony of the green alga *Penicillus*, colloquially called the shaving-brush alga. Continuous filaments of this alga are organized into holdfast, stem, and head. In living specimens, holdfast (at bottom) is embedded in sediment. Florida Bay. (K. W. Stockman, R. N. Ginsburg, and E. A. Shinn, 1967, part of Fig. 3, p. 635; courtesy R. N. Ginsburg.)

Biochemical Secretion of Biocrystals

Within their cells, certain algae secrete solid skeletal components that consist of tiny biocrystals of aragonite. Much lime mud forms by the accumulation of such tiny biocrystals, which are released onto the bottom when the algae die.

Codiacean or green algae, especially the alga *Penicillus* sp. (Box 2.5 Figure 1), which resembles a shaving brush and hence is popularly known as the shaving-brush alga, produce fine biocrystals (needles) of aragonite (< 15 μm) within the sheaths of their filaments (Box 2.5 Figure 2). When the organism dies, the filaments disintegrate and release the needles to the sea, where they accumulate as lime mud. The modern-day rate of production of aragonite needles by *Penicillus* sp. alone can account for all the mud

BOX 2.5 FIGURE 2.   Electron micrograph of aragonite needles secreted as biocrystals within the cells of the stems of the green alga *Penicillus*. Florida Bay. (K. W. Stockman, R. N. Ginsburg, and E. A. Shinn, 1967, fig. 2b, p. 635, courtesy R. N. Ginsburg.)

in the inner Florida Reef tract and one third of the mud in northeastern Florida Bay. Two more-or-less-equally abundant green algae (*Udotea* and *Rhipocephalus*) are likewise active contributors of lime mud in this area. Codiacean algae are known in rocks as old as Ordovician, and they may have contributed aragonite needles to ancient lime muds.

Coralline red algae secrete skeletons composed of high-magnesian calcite. High-magnesian calcite, probably derived from the breakdown of red-algal skeletons, is abundant in carbonate muds in Florida Bay and on the Great Bahama Bank west of Andros Island. Red algae and serpulid worms living on leaves of the marine grass *Thalassia* may produce as much or more lime mud than does *Penicillus*.

Comminution of Sand-Size Aragonitic Skeletal Debris

Some lime mud results from the physical breakage of sand-size- and larger aragonitic skeletal material. Such mud can originate by (1) decomposition of organic matter, such as conchiolin, which binds the small hard parts together; (2) weakening

or comminution of the skeletal material by boring organisms such as fungi, algae, or sponges; (3) feeding activities of predatory organisms, and (4) physical breakage by abrasion in agitated water.

A modern organism whose activities create noteworthy quantities of lime mud is the boring sponge *Cliona*. These sponges create mud-size particles by breaking down skeletal material from reefs and from individual or other colonial organisms. Boring sponges convert the solid masses of calcium carbonate secreted by the reefs into an ever-increasing supply of small-size skeletal debris, which accumulates in the vicinity of the reefs. Bivalves, foraminifers, sea urchins, sipunculans (wormlike organisms), polychaete worms, algae, fungi, and bacteria also bore holes of various sizes into reefs and into calcium-carbonate skeletal debris. (See Figure 2-40.) The products of all this boring are akin to the sawdust on the floor of a carpentry shop, except that in the sea, the "sawdust" from boring is composed of tiny fragments of calcium carbonate. (See Chapter 7 for further discussion.)

The lack of fossils in ancient fine-textured limestones is frequently cited as evidence in favor of an origin by inorganic precipitation rather than by accumulation of skeletal debris. However, as noted previously, when they die, *Penicillus* and other fragile green algae disintegrate completely. Thus even in modern sediments, the aragonite biocrystal needles they contribute do not resemble recognizable skeletal debris.

Electron microscopy has revealed that many fine-textured, apparently nonfossiliferous limestones of deep-sea origin consist almost entirely of coccoliths, the shells of calcareous free-floating algae known as *Coccolithaceae*. (See Box 2.1 Figure 6.) The individuals of some kinds of coccoliths secrete an intricately organized structure composed of calcite crystals between 1/4 and 1 μm in diameter. When the individual plates are together they form spherical- to oval discs about 2 to 20 μm broad in their planes of flattening. (See Box 2.4 Figure 1.)

gypsum crystals that became particles. The same is true of many dunes marginal to playas in West Texas. Miocene gypsum deposits of Sicily contain distinctive strata of the kinds made only by currents that drive sand-size sediment along the bottom. Therefore, these gypsum deposits must have originated by the lithification of

gypsum particles. Ooids of halite occur in the Dead Sea. If these halite ooids can be interpreted in the same way as ooids composed of calcium-carbonate minerals, then in the Dead Sea, precipitation of halite must have been taking place on the surfaces of particles that were being shifted on the bottom.

BOX 2.6

Nonskeletal Carbonate Sands

The evidence that numerous and various sand-size carbonate particles are abundant among the modern carbonate sediments of the Bahamas was assembled by L. V. Illing (1954). Many of these consist of microscopic needles of aragonite (Box 2.5) that compose lime mud. Lime mud can be converted into sand-size particles by desiccation and subsequent breakage of the mud chips; by the activities of shrimps and crabs, who make tiny mud balls to line their burrows in sand-size sediment; by excretion from the digestive tracts of deposit-feeding organisms; and by the activities of microorganisms that bore into and otherwise modify skeletal debris. The process of recrystallization is also capable of converting sand-size objects of various origins into cryptocrystalline particles.

Based on their appearances under the petrographic microscope, these particles can be grouped into two classes: (a) those lacking concentric coatings and (b) those possessing concentric coatings. The noncoated group includes *intraclasts, pellets*, and *peloids*. Within the coated group are *ooids*, *pseudooids*, and *pisolites*. *Clumps of some or all of the various sand-size particles* can be *loosely aggregated to form composite individual particles* known as **grapestones**.

Intraclasts

The term **intraclast** refers to *sand-size or larger particles, texturally analogous to rock fragments, broken from consolidated- or hardened materials in one locality and redeposited at another locality within the basin of deposition. Intra* means within, in this context within the basin of deposition, and *clast* denotes broken (Box 2.6 Figure 1). Many intraclasts are recycled fragments of coherent sediment. Intraclasts are of various sizes and shapes. Many are angular and their diameters may exceed 2 mm.

Sand-size intraclasts that have become rounded are referred to as peloids if their origin is uncertain (more details in a following section). Peloid is a general term that can be applied on the basis of shape when the origin has not been determined.

Pellets

Spherical- or ellipsoidal sand-size particles of calcium carbonate are called **pellets**. Internally, pellets are commonly homogeneous (Box 2.6 Figure 2), although some pellets contain parallel longitudinal tubules. Although they are of sand size, most pellets may be compared to tiny mudballs, and usually consist of aragonite. Most pellets are formed by deposit-feeding organisms that eat the mud. These organisms digest organic matter from the mud and excrete the undigested lime mud in the form of fecal pellets. Evidence for the fecal origin of pellets came from observations in the Bahamas where geologists using face masks and snorkels have watched sea cucumbers excreting long strings of pellets. In aquaria, gastropods from the Bahamas have obliged geologists by excreting fecal pellets. Burrows of the marine crustacean *Callianassa* are surrounded by small piles of uniformly shaped fecal pellets.

In modern carbonate sediments, pellets are probably the most-common single kind of particle. Because a few individual organisms can excrete thousands of pellets, the abundance of pellets is easy to understand. Although pellets are common in ancient limestones, they seem to be less common than one would expect from observations of modern carbonate sediments. The explanation seems to be that the shapes of the pellets may not be preserved. Pellets are discrete sand-size particles of lime mud, and during changes after deposition, their boundaries commonly become obscured and erased (they may become squashed, or simply fused together). Thereafter the limestone appears as if it had been lithified from homogeneous lime mud.

Peloids

A useful term, **peloid**, has been introduced for *particles that resemble pellets but for which no particular origin is implied*. Not all pelletlike particles are of fecal origin; some are lime-mud aggregates that originated when lime mud dried out on exposure to the atmosphere. When lime mud is so exposed, desiccation cracks (the so-called mud cracks) form (Box 2.6 Figure 3), and small chips of dried-out mud spall off. (See Box 2.6 Figure 3, B.) These particles may be larger than sand-size pellets, but they are very fragile, and are quickly comminuted to rounded, sand-size, pelletlike particles. (See Box 2.6 Figure 3, C.) Many peloids are sand-size, rounded intraclasts. Fecal pellets and pellet-shaped mud aggregates commonly cannot

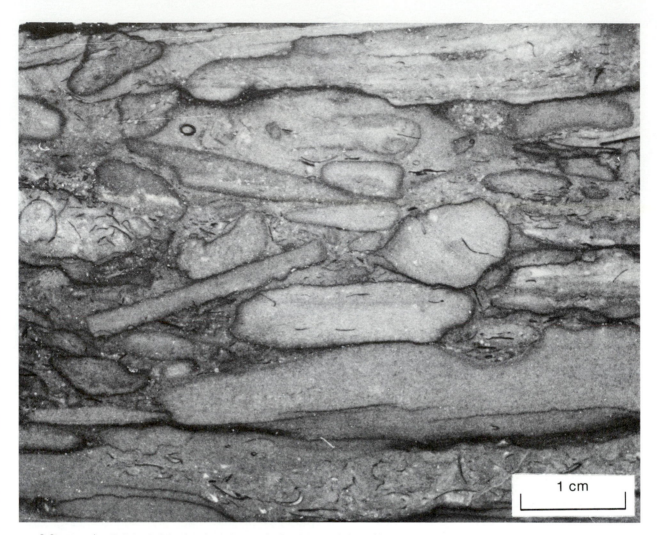

BOX 2.6 FIGURE 1.   Polished slab of carbonate conglomerate, consisting of intraclasts, cut perpendicular to layers. Each intraclast is an aggregate of lime mud; slablike shapes of cross sections result from breaking along stratification planes. (See Box 2.6 Figure 3.) Tribes Hill Formation (Lower Ordovician), Mohawk Valley, New York. (M. Braun and G. M. Friedman, 1969, fig. 7, p. 119.)

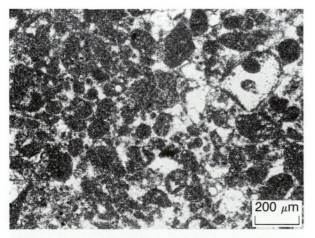

BOX 2.6 FIGURE 2.   View of limestone composed largely of pellets but including a few skeletal fragments; photomicrograph in plane-polarized light of thin section, Zohar Formation (Jurassic) recovered from test boring, coastal plain, Israel. (Authors.)

be distinguished. Peloids may also form by recrystallization of ooids.

Coated Particles

Coated particles are composite. In them, a central nucleus, such as another calcareous particle (skeletal particle, a foraminifer, an intraclast, pellet, or peloid) or a noncalcareous fragment (such as a particle of quartz), is enclosed by a coating of one or more layers of calcium carbonate that may be transparent and display concentric- or radial fabric or may be opaque and lack apparent fabric. The coating layers may result from precipitation of calcium carbonate or from the activities of microorganisms. Some coated particles can be recognized in hand specimen, but most can be identified only by petrographic examination of thin sections.

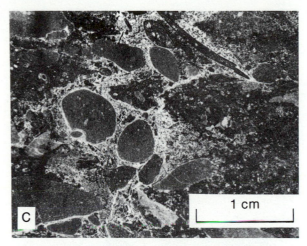

BOX 2.6 FIGURE 3. (*Continued*)
C. View of polished and etched surface of limestone containing sand-size and larger aggregates of lime mud. Rounded particles in center resemble fecal pellets. Tribes Hill Formation (lower Ordovician), Mohawk Valley, New York. Larger particles of same kind of lime-mud aggregates from same locality are shown in Box 2.6 Figure 1. Photograph printed directly from acetate peel. (M. Braun and authors.)

BOX 2.6 FIGURE 3. Views of sequence of events beginning with desiccation cracks and ending with sand-size aggregates of lime mud that resemble fecal pellets.

A. Desiccation cracks in carbonate-evaporite-terrigenous mud, sea-marginal flat, Gulf of Aqaba, Sinai Peninsula. Small dead fish were left by falling tide. Small shovel and single footprint (lower center) give scale. (Authors.)

B. Lime mud, recently covered by high water of storm tide, now exposed to the Sun. Drying has created chips, which have broken along vertical desiccation cracks and then have curled up and spalled off parallel to stratification. These dried-out chips may be washed into the sea, where they may accumulate as the particles of an intraformational conglomerate (See Box 2.6 Figure 1.) or be reduced to sand size. Crane Key, Florida. (Authors.)

Sand-size coated particles are ooids (two or more layers of coating material) and pseudooids (only one layer of coating material). Larger coated particles include various kinds of pisolites.

Ooids

The sand-size coated particles known as ooids derive their name from a Greek word which means egg or egglike because under the microscope these particles resemble the roe of fish (Box 2.6 Figure 4, A). Ooids usually are spherical or elliptical.

The rims of most modern marine ooids consist of aragonite; a few are composed of

high-magnesian calcite. In most modern aragonitic ooids the long axes of the individual aragonite crystals are tangential to the rims. This is known as *concentric* fabric. In some ooids the fabric is *radial*: the long axes of the aragonite crystals are normal to the rims and therefore diverge away from the center of the ooid. Modern ooids having radial fabric (mostly composed of high-magnesian calcite) occur in hypersaline waters, such as in the Great Salt Lake of Utah; in Baffin Bay, off Laguna Madre, Texas; or in thick accumulations of algal mats as found in sea-marginal ponds of the modern Red Sea (Box 2.6 Figure 5) and along the west coast of Australia. In addition to higher salinity, locations of formation of radial ooids are characterized by less-energetic water movement. It appears, therefore, that low-energy conditions favor growth of radial ooid fabric whereas high-energy conditions favor growth of tangential (concentric) ooid fabric. However, this applies only to aragonitic ooids; modern marine high-magnesian calcite ooids are exclusively radial.

On the basis of their internal fabrics, most ancient marine ooids may be classified into two groups: (1) Group one consists of ooids having either (a) concentric aragonitic fabric or (b) coatings consisting of crystalline mosaics of calcite (or dolomite), commonly lacking any vestige of tangential or radial structure. These mosaics may consist of minute euhedral rhombs only a few micrometers across or of anhedral

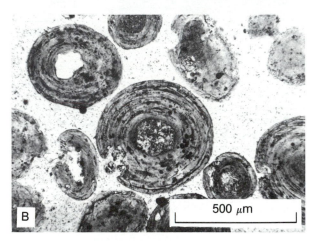

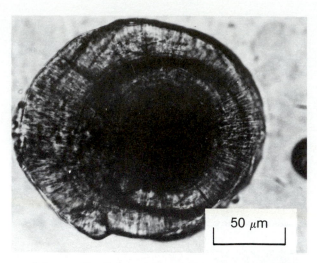

BOX 2.6 FIGURE 5.   Enlarged view of thin section through ooid having radial fabric; photomicrograph in plane-polarized light. Algal mat from sea-marginal hypersaline pond, Gulf of Aqaba, Red Sea. (G. M. Friedman and others, 1973, fig. 11, p. 550.)

BOX 2.6 FIGURE 4.   Views of ooids.

A. Enlarged view of exteriors of ooids, modern sediments, South Cat Cay, Bahamas. Seen through binocular microscope. (Authors.)

B. View of thin section cut through modern ooids, Ras Sudar, Gulf of Suez, Red Sea; photomicrograph in plane-polarized light. Concentric shells surround nuclei consisting of quartz particles (white) or of fine-textured carbonate materials. (B. Buchbinder.)

spar crystals nearly as large as the ooids themselves. The structure of these ooids has been substantially altered by diagenesis. Many of these ooids contain high levels of the trace element Sr^{2+}; others do not. (2) Group two consists of calcite ooids in which the radial fabric is well developed. The levels of Sr^{2+} in group-two ooids are lower than those in group-one ooids. Rocks of a given age are characterized by an abundance of either group-one ooids or group-two ooids, but not by abundant occurrences of both kinds of ooids.

Where ooids of groups one and two occur in rocks of the same age (e.g., the Jurassic of the U.S. Gulf Coast), commonly each kind is found in rocks interpreted as having been deposited in contrasting environments. A few cases have been reported (for example, in the Cambrian of New-

foundland) of bimineralic ooids. In these ooids the fabrics of group one and of group two alternate in concentric zones. Such ooids have been interpreted as having formed under fluctuating water conditions.

Aragonitic ooids contain elevated levels of Sr^{2+}. They commonly exhibit tangential fabric, but are readily recrystallized to the more-stable polymorph, calcite. The result of this transformation typically is a crystalline mosaic that displays few or no remnants of the primary group-one fabric. High-magnesian calcite ooids contain low levels of Sr^{2+} and are characterized by radial fabric. They are mineralogically resistant to recrystallization, but are instead converted to low-magnesian calcite by a process that preserves their primary group-two fabric. (See Chapter 3 for a discussion of constraints and modes of carbonate-mineral transformations.) Bimineralic ooids may have been precipitated from waters intermediate in composition between those favoring the precipitation of calcite and those favoring precipitation of aragonite. If this interpretation is correct, then changes in ambient conditions could have resulted in alternating precipitation of aragonite and of high-magnesian calcite.

The meaning of these observations is not yet entirely clear. Current interpretations suggest that throughout geologic time the chemical environment in seawater has alternated between conditions favoring inorganic precipitation of aragonite (aragonite seas), including the present day, and conditions favoring inorganic precipitation of calcite (calcite seas). Just what these conditions are is a matter of hot debate, but suggested controls on the mineral composition of ooids include (1) changes in the concentration of $CO_3{}^{2-}$ in seawater and of CO_2 in the atmosphere and (2) changes in the Mg/Ca ratio in seawater, both controlled by sea-level changes (degree of flooding of continents), rates of submarine weathering, and sea-floor hydrothermal activity.

The ooids in algal mats may be precipitated by the algae themselves, yet the origin of most ooids is still problematic. Most modern ooids occur in or close to the intertidal position in a zone in which the waves break and pound. Many geologists consider that in this turbulent, shallow-water environment, ooids may form inorganically as the cooler water from adjoining deep zones spreads across the shoals, gives up some of its CO_2, and becomes warmer.

The geologic literature is replete with papers describing examples in which ooids and skeletal particles from modern marine environments recrystallized to peloids. Transitional stages reveal the direction of change (Box 2.6 Figure 6). This process is called *micritization,* and is commonly accomplished by the actions of boring and encrusting microorganisms. Thus peloids found in rocks may have originated in diverse ways. Therefore the origin of ancient peloids must be interpreted with caution.

SOURCES: Füchtbauer and Richter, 1988; Kopaska-Merkel, 1987b; Sandberg, 1983; Scoffin, 1987; Tucker and Wright, 1990; Wilkinson, 1982.

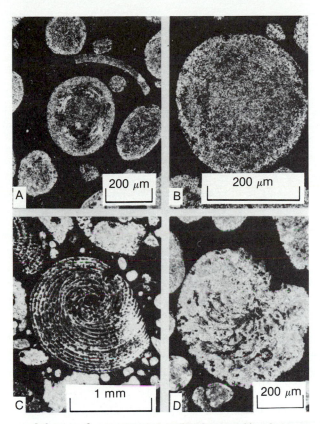

BOX 2.6 FIGURE 6.   Enlarged view showing transitional stages in recrystallization of ooid and foraminifer; photomicrographs in cross-polarized light.

A. Partly recrystallized ooid.

B. Completely recrystallized ooid in which only indistinct remnants of concentric shells remain. Compare with internal structure of pellet (Box 2.6 Figure 2).

C. Peneroplid foraminifer partly recrystallized at lower right.

D. Almost completely recrystallized peneroplid foraminifer. Only indistinct organic structure remains in fine-textured carbonate; particle begins to resemble pellet. Modern sediments, Great Bahama Bank. (E. G. Purdy, 1963a, pl. 5, facing p. 348.)

Pisolites

Spherical- or elliptical coated particles that exceed 2 mm in diameter are known as **pisolites**. The division between pisolites and ooids is one of size; ooids are smaller than 2 mm. Despite this seemingly arbitrary size differentiation, the origins of ooids and pisolites are not the same. Pisolites that are merely oversized ooids are not common, either in modern carbonate sedimentary environments or in ancient limestones. We shall discuss four common kinds of pisolites: *oncolites, rhodoliths, vadose pisolites,* and *cave pisolites.*

Oncolites.   Oncolites (also known as algal pisolites) very closely resemble vadose pisolites (to be described). Their origins are, however, quite different. Oncolites consist of encrustations on various particles (Box 2.6 Figure 7). When particles, commonly skeletal, roll about intermittently on the sedimentation surface, microorganisms, especially cyanobacteria (blue-green algae or cyanophytes), but also green algae, bacteria, and diatoms, attach themselves to and repeatedly coat these particles with concentric laminae by trapping and binding sediment. The microorganisms coat the exposed portions of the oncolites during periods of quiescence, and may be partially abraded from the surfaces of the oncolites during the intervening periods of rolling. Thus oncolites commonly are composed of irregular- and incomplete laminae (Box 2.6 Figure 8). They may also contain the skeletons of small encrusting

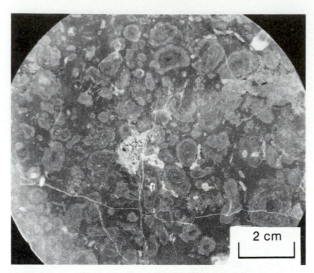

BOX 2.6 FIGURE 7.   View of oncolites on polished surface cut parallel to bedding. Evidence of the effects of algae (the criterion of oncolites) consists of cell filaments (not visible here, but evidence that specialists can recognize in thin sections). In the absence of algal evidence, oncolites cannot be distinguished from vadose pisolites. Test boring through limestone of Inmar Formation (Jurassic), Negev Desert, Israel. M. Goldberg and G. M. Friedman, 1974, pl. 4, fig. 3.)

organisms, such as the red encrusting foraminifer *Homotrema* and coralline red algae, which colonized the surface of the oncolite when it was temporarily resting on the sea bottom. Oncolites from the Red Sea consist of high-magnesian calcite. Oncolites may also form in lakes.

Vadose Pisolites.   The most-common pisolites are known as caliche- or vadose pisolites (or vadolites) (Box 2.6 Figure 9). Vadose pisolites form in the weathering zone as caliche. Vadose pisolites are not truly particles because they form

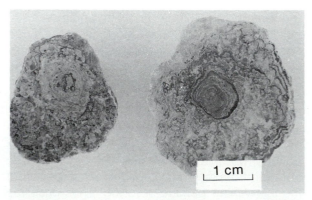

BOX 2.6 FIGURE 8.   View of oncolites sampled from Llunelly Formation (Lower Carboniferous), Blaen Onneu, South Wales, United Kingdom. (Authors.)

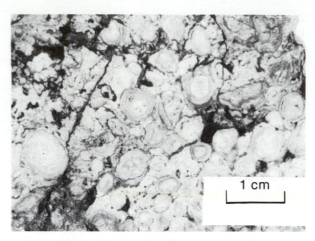

BOX 2.6 FIGURE 9.   Vadose pisolites seen in cross section on polished face of dolomitic carbonate rock. Compare with Box 2.6 Figures 7 and 8. Carlsbad Formation (Permian), New Mexico-Texas border. (Authors.)

chemically *in situ* by evaporation in a semiarid climate. Vadose pisolites may be reworked into sedimentary particles that become transported to a basin of deposition. Particles of vadose pisolite have been concentrated along surfaces of unconformity. In the Negev Desert of Israel, vadose pisolites were deposited in hollows on a karst surface by rivers of Early Jurassic age, which drained an ancient landscape underlain by carbonate rocks. Vadose pisolites generally consist of low-magnesian calcite.

Vadose pisolites are difficult to distinguish from oncolites and yet, for interpreting the sedimentary environment, this distinction is very important. An oncolite in a limestone indicates that the rock was formed in the saline water of a marine environment or in a saline lake. In contrast, vadose pisolite forms in the weathering zone in a semiarid climate. Although their origins and significance differ drastically, oncolites and vadose pisolites are so similar that even specialists have commonly experienced difficulty in distinguishing oncolites from vadose pisolites.

Cave Pisolites.   Cave pearls, or cave pisolites, are pisolites that are true particles; they form in a manner analogous to ooids, but in pools of water in caves. When the cave water is agitated, it rolls the cave pisolites and causes them to be concentrically coated by low-magnesian calcite. In at least some cave pisolites, filamentous microorganisms have been calcified. The nuclei of cave pisolites include calcareous particles, dendritic calcite crystals, or bits of fossil bones.

Grapestones (Lumps)

Particles that have been designated as grape-stones or lumps are composed of skeletal particles or nonskeletal particles (such as ooids or pellets) that have become cemented together (Box 2.6 Figure 10). The term grapestone comes from the appearance under the microscope of these particles; they resemble bunches of grapes. Grapestones form in areas where short periods of bottom agitation are followed by prolonged periods of bottom stability. During the stable times, cement is precipitated. Cyanobacteria may aid in precipitating the cement. (See Box 2.6 Figure 10.) The stability of the bottom is necessary to keep the particles together while the cement is being precipitated.

Grapestones consist of a mixture of aragonite and high-magnesian calcite. The framework particles composing the grapestones evidently become cemented with tiny crystals of aragonite. In addition, microbial borings become filled with high-magnesian calcite. The ultimate result is a complex mosaic of original particles (commonly consisting of aragonite and/or high-magnesian calcite), interparticle aragonite cement, and pore-filling high-magnesian calcite.

Although grapestones or lumps are common in the modern carbonate sediments of the Bahamas, such particles are not common in other modern environments where carbonate sediments are being deposited. Moreover, grapestones have not been unequivocally reported from ancient limestones.

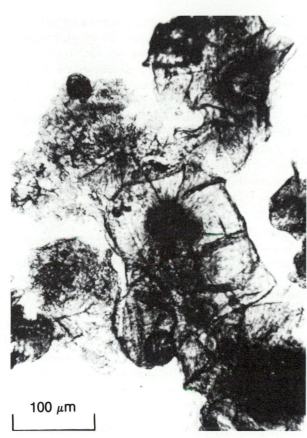

BOX 2.6 FIGURE 10. Enlarged view of thin section cut through grapestone clump, algal mats in modern sediments, sea-marginal hypersaline pond, Gulf of Aqaba, Red Sea; photomicrograph in plane-polarized light. Particles are mostly radial ooids. (G. M. Friedman and others, 1973, fig. 13, p. 551.)

100 μm

Evaporite minerals that began as crystals but became particles are not common. Nevertheless they do occur sporadically in the rock record. They are significant in that they indicate evaporite crystals did not crystallize where they are now found, but in some other place, and afterward were moved by currents in the same way as many other solid particles.

SOURCE: Füchtbauer and Valeton, 1988.

Glauconite Particles

Particles composed of glauconite are common in some marine strata. Much confusion has arisen because the term glauconite has been used in one way by geologists examining samples from bore holes or modern sediments based on what they see looking with a binocular stereomicroscope, and in a different way by mineralogists using X-ray diffraction. Geologists who work in the field or who examine sedimentary particles or rocks under the binocular microscope rely on external appearance; they tend to identify as "glauconite" any sand-size, green, earthy peloids in marine sediments. The shapes of such "glauconite" peloids vary; they may be ovoid, spherical (Figure 2-26), or botryoidal. Glauconite may fill the chambers of foraminifers. After the calcite of the glauconite-filled tests has been dissolved, as commonly happens, the internal shapes of the foraminifers are preserved by the glauconite as internal molds. Similarly, glauconite molds may preserve the internal shapes of dissolved-away siliceous radiolarian tests. Commonly, however, the shapes of these green peloids defy precise description. Hence, despite their variability of shape, they are monotonously described as small, round, green peloids devoid of internal structure. In such usage, "glauconite" is largely a morphologic- and color term.

FIGURE 2-25.   Greatly enlarged view of siliceous skeletons of three kinds of diatoms as seen in scanning-electron micrograph. Large circular test in center is *Cyclotella comta*; small circular tests in foreground and background are *Cyclotella glomerata*; hollow tubular skeletons at left and top are *Syndera ulna*. Pitted background is 0.45-μm membrane filter. From fresh-water samples, Lake George, New York. (S. L. Williams.)

Mineralogists use *glauconite* as a mineral name rather than as a morphologic term. X-ray diffraction reveals that "small, round, green peloids" consist of four groups of minerals: (1) peloids displaying the three basal diffraction peaks of a micaceous 1.0 nm lattice (Figure 2-27), the structural characteristics of *ordered glauconite*; (2) peloids which likewise display the lattice properties of the mineral glauconite but in which the heights of the peaks on the X-ray diffractogram are subdued, their bases are broad, and their sides asymmetric (*disordered glauconite*) (Figure 2-28); (3) peloids which consist of interlayered clay minerals; and (4) peloids which consist of a mixture of clay minerals, usually illite with montmorillonite and illite with chlorite.

Glauconite forms as an authigenic sediment in the marine environment by replacement of preexisting particles. Commonly, glauconite first infills pores and fractures within particles (Figure 2-29) and then replaces the preexisting material.

Both ordered- and disordered glauconite are layer-lattice silicates containing potassium and ferric iron. The chemical composition of the unit cell of an ordered glauconite is exemplified by samples from the Cambrian Bonne Terre Formation of Missouri. The formula of such glauconite may be written as:

$$K_{1.58}(Ca)_{0.10}(Al_{0.70}Fe_{2.11}{}^{3+}Fe_{0.49}{}^{2+}Mg_{0.82})(Si_{7.22}Al_{0.78}) -$$
$$O_{20}(OH)_4$$

As the composition of the Bonne Terre glauconite shows, ordered glauconites contain a potassium-atom equivalent

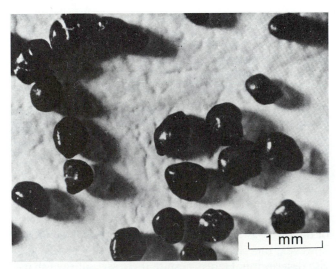

FIGURE 2-26.   View of spherical- and ovoidal pellets of glauconite as seen using binocular stereomicroscope. Dark green color shows here as black. Lisbon Formation (Eocene), Choctaw County, Alabama. (D. M. Triplehorn, 1966, fig. 1, p. 249.)

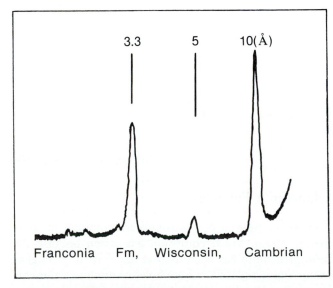

FIGURE 2-27.   X-ray diffractogram of ordered glauconite. (J. F. Burst, 1958, part of fig. 1, p. 486.)

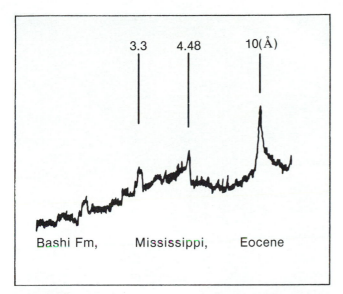

FIGURE 2-28.   X-ray diffractogram of disordered glauconite. (J. F. Burst, 1958, part of fig. 1, p. 486.)

FIGURE 2-29.   Greatly enlarged views of glauconite replacing smectite (glauconitic smectite) as seen in scanning-electron micrographs. (B. D. Bornhold and P. Giresse, 1985, fig. 7, D, E, p. 657.)
   A. Rosettes of glauconitic smectite filling void within shale particle. Scale = 10 μm.
   B. Mature boxwork crystal cluster of glauconitic smectite in shale particle. Scale = 5 μm.

greater than 1.4 per unit cell. Because this amount of potassium is small, the potassium supplies less-than-usual binding power to the lattice and disordering begins. When the potassium-atom equivalent falls below approximately 1.4 per unit cell, the ordered stacking becomes obscured. On the X-ray record, disordering is revealed by a broadening and loss of symmetry of the peaks and by the development of a third diffraction peak, indicative of disordering, between the two basal peaks.

Glauconite can be found in minor amounts in many marine sediments. In a few places, such as along the coasts of New Jersey or southeast England, glauconite is abundant in the modern marine sediments because it is being eroded offshore from older glauconite-rich strata. In this case, glauconite is classified as extrabasinal particles.

SOURCES: Logvinenko, 1982; Van Houten and Purucker, 1985.

Characteristics of Particle Surfaces

The surface characteristics of particles result from various processes that range from strictly mechanical to biophysical. The features found on particle surfaces include well-defined, localized linear striae, crescentic depressions, various pits, and overall characteristics such as polish, matte surface ("frosting"), and thin coatings. These range in size from the easily visible striae on boulders that have been abraded by glaciers (Figure 2-30) to features that can be seen only at magnifications of thousands

of times in the images made using a scanning-electron microscope (Figure 2-31).

An overall matte finish or "frosted" appearance of some quartz particles contrasts with the clear- or polished surfaces of others. Such characteristics are easily seen when sands are examined by binocular stereomicroscope. The matte finish is an expression of many minute irregularities on the surfaces of the particles. These can be tiny pits caused by impact of other particles during transport by the wind or by other agents (Figure 2-31), depressions etched during dissolution or peripheral replacement of the quartz by calcite, or bumps created by the growth of tiny crystals of quartz or other minerals on particle surfaces. Definitive examples of

FIGURE 2-30.   Striated and polished boulder, eroded from tillite of Precambrian age and incorporated in a tillite of Permian age, Halletts Cove, South Australia. (Authors.)

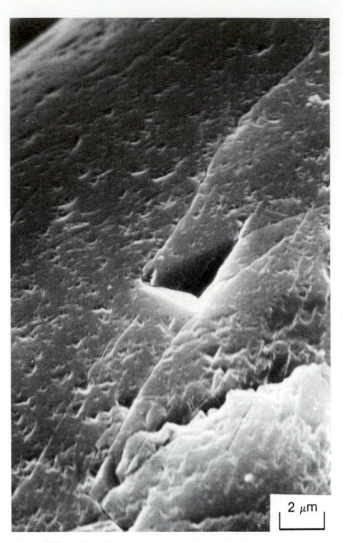

FIGURE 2-31.   Greatly enlarged view of segment of surface of quartz particle showing V-shaped depressions resulting from impacts; continental shelf off Argentina; scanning-electron micrograph. (D. H. Krinsley and I. J. Smalley, 1972, fig. 5, p. 290.)

frosting that came about from sand blasting are provided by the appearance of glass bottles left on a beach. Figure 2-32 shows how the twine holding a glass sphere has protected parts of the surface of the glass from being frosted during the time when the rest of the sphere was being sand blasted by wind as the sphere lay on a beach.

Although many attempts have been made to use the surface characteristics of particles as the "indices of refraction" of particular environments, the geologic history can introduce complications that defy simple-minded applications of the results. For example, particle-surface characteristics acquired in one environmental setting can be changed if the particle is moved into a different environment. Abrasion by river transport can destroy glacial striae on boulders or change the matte finish of desert sands into polished surfaces. However, movement of a particle from one environment characterized by much abrasion into another characterized by little or no abrasion may permit the surface characteristics acquired in the first environment to be preserved as relicts on particles collected from the second environment. Such relict features may deceive the unwary.

Some Properties of Particle Aggregates

The next level of our study deals with some properties of aggregates of sedimentary particles. Included here are color, bulk density, fabric, effects of particle-size distribution, porosity, permeability, acoustic impedance, electrical properties, magnetic properties, and radioactivity.

Color

During the visual examination of a sedimentary deposit, one of the first characteristics noticed is the color. Although color is not a constant property, it does reflect the presence of certain minerals and these, in turn, may have formed under well-defined environmental conditions. In this section we review some of the major factors affecting the colors of sedimentary deposits.

Color is caused by minerals, in the form of framework minerals having distinctive colors, as the rock matrix, or as a micrometer-size coating on the surfaces of framework particles. *The color of framework minerals* is known as the **intrinsic color**. Some examples include reddish-brown Pleistocene sands in southern Connecticut,

FIGURE 2-32.   Glass sphere (Japanese fisherman's float) found on Bering Sea coast of Alaska showing effects of abrasion. Twine has been moved to show clear glass areas of the sphere formerly beneath the twine that the twine protected from being abraded. (Authors.)

southeastern New York, and northern New Jersey whose color comes from sand-size rock fragments from the red-brown Newark Group (Upper Triassic-Lower Jurassic); the black sands in many localities, some derived from volcanic rocks and others from intense weathering of igneous rocks containing magnetite or ilmenite as accessory minerals; the red beach sands of southern Long Island, New York, and of Lake Champlain, New York, formed where the waves have concentrated garnet; and the green sands of the Upper Cretaceous and Cenozoic formations underlying the coastal plain of New Jersey and elsewhere that contain abundant glauconite.

Where color results from a powderlike coating on the framework particles, the term **pigment** is used. Most sedimentary pigments are iron minerals: black, from microscopic pyrite or other sulfides; reddish brown, from hematite; and yellowish or yellowish brown from limonite.

Some dark-colored sediments contain more than the usual amount (say 1 or 2%) of organic matter, usually as plant debris. In many cases, it is not possible to determine by visual inspection how much of the dark color has resulted from the organic matter directly and how much from the pyrite that nearly always forms where organic matter is abundant.

The hematite and limonite iron-oxide pigments are products of oxidizing conditions, in contrast to the reducing conditions where sulfides form. Subaerial exposure and weathering of dark-colored sediments pigmented by pyrite usually results in a change in color from dark gray or black to brownish or yellowish brown. In some cases, oxidation of sulfides yields elemental sulfur (colored yellow) or whitish sulfate minerals. Further discussion of these color-related chemical changes and their environmental significance is found in Chapter 7.

Bulk Density

The **bulk density** of a sedimentary deposit, *the weight per unit volume*, is an attribute affecting the gravitational attraction of the deposit and the speeds of passage of seismic waves through it. Bulk density is controlled by the densities of the mineral particles and the amount of void space among the particles (a function of the porosity, packing, and quantity and mineral composition of cement).

The densities of common sedimentary minerals range from the 2.16 g/cm^3 of halite to the 5.18 g/cm^3 of magnetite. Densities in the range of 2.6 to 2.9 are typical of the so-called light minerals, which include quartz (2.65), feldspars (2.57 for orthoclase; 2.62 to 2.76 for the plagioclase groups), calcite (2.71), and dolomite (2.87). As noted in a previous section, minerals whose densities exceed 2.9 g/cm^3 are known as heavy minerals. (See Table 2-6.)

The bulk density of a turbid water-sediment mixture may be about 1.05. The bulk densities of freshly deposited fine sediments (muds) of Lake Maracaibo, Venezuela, lie in the range of 1.23 to 1.45. In one area the density of mud in the top meter of the sediment is 1.30, but at a depth of 5 m, has increased to 1.45. After further settling (compaction), the sediment density becomes 2.0 or slightly more. Figure 2-33 shows the inverse relationship between porosity and bulk density. Table 2-9 gives examples of typical bulk densities for common kinds of sediments and sedimentary rocks.

Fabric

The term **fabric** refers to *the arrangement of the constituents (particles or crystals) of a deposit in terms of (1) orientation of nonspherical individuals, (2) packing, and (3) overall sorting.*

ORIENTATION OF NONSPHERICAL INDIVIDUALS

Orientation of nonspherical individuals refers to the positions in space of axes or planes. Examples include the long axes of rod-shaped particles, long axes or planes

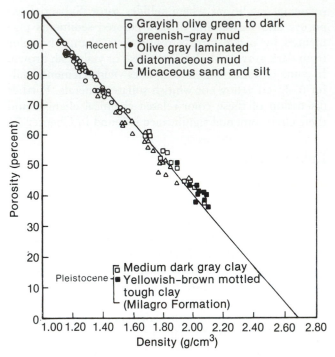

FIGURE 2-33. Graph of porosity vs. bulk density for sediments of Recent and Pleistocene ages cored from Lake Maracaibo, Venezuela. (R. Sarmiento and R. A. Kirby, 1962, fig. 20, p. 720.)

of flattening of blades, and planes of flattening of disks. (See Box 2.1 Figure 1.)

Many sedimentary fabrics are created during deposition. Some fabrics are completely **random** or **isotropic**; in them, *the individuals are oriented in such a way that the properties of the sediment are essentially uniform in all directions*. Other fabrics display a **preferred orientation**; in them, *long axes or planes display a distinct parallelism* (Figure 2-34). In some cases, long axes become aligned parallel to the direction of flow of the depositing current. In other cases, the current may have rolled particles along the bottom and thus oriented the long axes at right angles to the current. Disk-shaped micas tend to settle so that their planes of flattening are parallel to the depositional surface. Hence, in the geologic record, micas tend to be parallel to stratification. In some gravels, *the planes of flattening of the blade-shaped and disk-shaped particles are positioned in a shinglelike, overlapping fabric*, or **imbricate arrangement**. The planes of flattening of imbricated particles systematically dip upcurrent (Figure 2-35).

PACKING

Another important fabric element is the **packing** or *spatial density of the constituent individual particles in a sedimentary deposit*. Packing exerts a large influence on

porosity and permeability, as is explained further in following sections.

Using spheres as examples, one can distinguish two packing arrangements: open and close (Figure 2-36).

Sedimentary fabrics may be strongly controlled by (1) depositional processes, (2) the shapes and other characteristics of the particles, (3) postdepositional processes such as burrowing, slumping, or injection of sedimentary dikes (sediment disturbed by upward escape of fluids), or (4) by some combination of these.

OVERALL SORTING

Fabrics in fine sedimentary aggregates need to be investigated by X-radiography, by thin-section petrography, and by scanning-electron microscopy (SEM). Initially, many flaky clay minerals may display a wide-open "house-of-cards" kind of packing that is very porous (Figure 2-37). Clumps, or floccules, of clay, dispersed (nonclumped) clays, and the churning of the sediment by burrowing organisms all result in distinct microfabrics of the clayey seabottom (soupy, soft, or firm). SEM fabric analysis of fine sediments and -sedimentary rocks can yield insights into depositional conditions. For example, flocculated clays are most common in soupy- or soft bottoms and the flocculated fabric may be preserved in ancient shales in which no other preserved evidence exists for the original soft bottom (Figure 2-38). Other kinds of microfabric features within fine sedimentary aggregates are visible in thin sections examined under a polarizing microscope. A great deal more work needs to be done before the nature and origin of microfabrics in fine-textured sedimentary aggregates are well understood.

Collective Effects of Particle-Size Distribution

Two important effects of particle size are its control on the bonding among individuals in particle aggregates and on the behavior of fluids in the pores, particularly as this relates to the pore pressure- and capillarity of water.

In an aggregate of coarse-silt-size- or coarser particles, the only significant factor keeping the particles together is gravity acting on the individual particles at their points of contact. Such aggregates are described as being *cohesionless*; in the engineering literature, they are referred to as "friction soils." Aggregates of cohesionless particles can sustain heavy loads, present well-drained surfaces, and form what are known as angle-of-repose slopes. Furthermore, as is explained in Chapter 7, they respond to shearing by dilating, and under certain conditions, the body of particles can lose its internal frictional forces and flow much as if it were a fluid.

TABLE 2-9.   Bulk densities of some sediments and sedimentary rocks

Kind of sediment	Bulk density (saturated with water) (g/cm^3)	Kind of sedimentary rock
Ocean-bottom sediment, Atlantic Ocean	1.73–1.83	
Globigerina ooze, Pacific Ocean	2.81	
Rock salt (dry)	2.1	
Soil, Recent, Princeton, NJ	1.78	
Silt, Recent, Rosebud Co., MT.	1.81–2.15	
Sand, Modern, Beaver R., OK	2.02–2.15	
Gravel, Modern Beaver R., OK	2.13	
Clay, in Cohansey Sand (Miocene), Crossley, NJ	2.03–2.05	
	2.11	Clay Frontier Fm. (U. Cret.), Afton quad., WY
	2.40	Woodside Shale (Triassic), Afton quad., WY.
	2.86	Shale, Batesville, AR
	2.60	Shale, Carboniferous, England
	2.46	Cherokee Shale (Penn.), Fulton, MO
	1.98	Bearpaw Shale (U.Cret.), Rosebud Co., MT.
	2.59	Shale depth 4000 ft, N. OK
	2.45	Limestone (Carboniferous), Buxton, England
	2.64	Limestone (Devonian), Dundee, MI
	1.96–2.40	Chalk (Cret.), England
	2.39	Portland Limestone, England
	2.40	Chert, England
	2.62	Preuss Sandstone (Jur.), Afton quad., WY
	2.44	Ephraim Conglomerate (L. Cret) Afton quad., WY
	2.66	Thaynes Limestone (Triassic) Afton quad., WY
	2.68	Bighorn Dolostone (Ord.), Afton quad., WY
	2.71	Jefferson Dolostone (Dev), Afton quad., WY
	2.45	Dolomitic sandstone, England
	2.53	Medina Sandstone (Sil.) Niagara, NY
	2.43	Woodbine Sandstone (U. Cret.) Shreveport, LA
	2.45	Big Injun Sandstone (Miss.), Mannington, WV

SOURCE: H. C. Spicer, 1942, table 2-6, p. 19–26.

By contrast, aggregates containing chiefly fine-silt- and clay-size particles are subject to cohesive forces. The electrostatic forces caused by nonbalanced surface charges of the crystal lattices exceed the frictional forces caused by gravity and the particles begin to "stick together." In the engineering literature, aggregates containing small particles chiefly of clay- and silt sizes are referred to as "cohesive soils." In addition, certain clay minerals tend to attract water molecules to their surfaces. Moreover, the sizes of the openings among the individual particles of fine-textured aggregates are so small that the surface-tension effects of water become significant and this causes capillary phenomena and negative pore pressures (a condition under which some fluids tend to

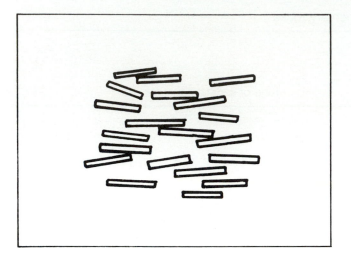

FIGURE 2-34.   Parallelism of platy particles (schematic).

be pulled into the pores) and enables slopes underlain by such aggregates to be vertical. Aggregates consisting chiefly of clay- and silt-size particles cannot sustain heavy loads and respond to shearing by collective internal deformation. However, fine-textured aggregates that have been consolidated by cementation and/or compaction can sustain heavy loads.

Porosity

Porosity is *the ratio, in percent, of the volume of void space to the total volume of the deposit.* Porosity can be expressed mathematically as:

$$\text{porosity} = \frac{\text{total volume} - \text{volume of solids}}{\text{total volume}} \times 100$$

(Eq. 2-2)

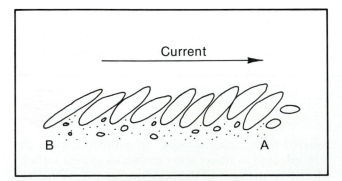

FIGURE 2-35.   Schematic profile, in vertical plane parallel to current, showing shinglelike, overlapping (imbricate) arrangement of platy particles. Deposit grew upstream by first depositing particle at A, and then by adding other particles in succession toward B. (Authors.)

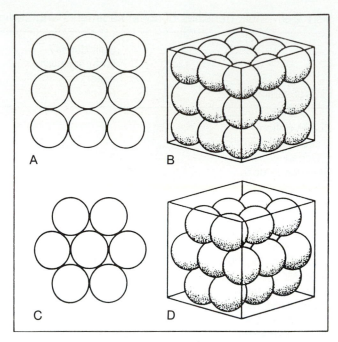

FIGURE 2-36.   Open packing and close packing of spheres. A and B, open (cubic) packing in plan view (A), and in perspective (drawn inside cube to emphasize the point that spheres packed this way are not stable by themselves). C and D, close (hexagonal) packing in plan view (C), and in perspective (D). (J. E. Sanders, 1981; A and B, fig. 4.2, p. 94; C and D, fig. 4.4, p. 95.)

The pore space in a sedimentary deposit can be filled with interstitial fluids such as air, fresh water, formation water, ore-forming fluids, CO_2, helium, H_2S, and hydrocarbons (oil and natural gas). Most often, fluids of more than one kind occupy the pore space in a sedimentary

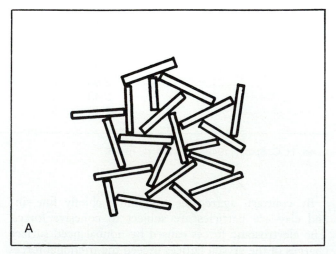

FIGURE 2-37.   House-of-cards (or cardhouse) packing of clay platelets.

A. Sketch of platy particles in random orientation. (R. H. Meade, 1966, fig. 5, p. 1091.)

FIGURE 2-38.   Equidimensional undisturbed clay floccules, muddy sand from Shelikof Strait, Alaska; Scanning-electron micrograph. (M. E. Torresan and W. C. Schwab, 1987, p. 414, fig. 9, A. Courtesy of Michael Torresan and U.S. Geological Survey.)

FIGURE 2-37.   (*Continued*)
B. Greatly enlarged view showing open packing and random orientation (cardhouse packing) of crystals of the clay mineral halloysite. Lou Veigne, Belgium: electron micrograph (P. Buurman and L. van der Plas, 1968, fig. 5, p. 347, courtesy L. van der Plas; photo, T. F. D. L., Wageningen, The Netherlands.)

aggregate. Understanding of porosity is a prerequisite to any analysis of the behavior of interstitial fluids.

Porosity refers to all openings within a deposit, and may be classified as **primary**, which is *porosity that existed when a sediment was deposited*, and **secondary**, which is *porosity that was created some time after deposition.* For the purposes of this discussion, we shall concentrate on primary porosity. We discuss the important subject of secondary porosity in Chapter 3.

Primary porosity may be divided into **interparticle porosity**, that is, *the pore spaces among the particles of a sediment or a rock*, and **intraparticle porosity**, which refers to *the pore spaces within sedimentary particles.*

Interparticle porosity is a function of (1) the initial void spaces, as controlled by the sizes, shapes, and fabric of the particles; and (2) any postdepositional changes, such as the mechanical infiltration from above of interstitial fines or precipitation of mineral cement in the pores.

Sizes and shapes of particles are closely interrelated and are thus not easily isolated in connection with porosity. If one keeps shape constant, by considering only spheres, for example, then porosity depends only on the packing and ranges from a maximum of 48% for open-packed spheres to a minimum of 26% for close-packed spheres (Figure 2-39). This relationship is valid for all aggregates of one-size spheres no matter what their diameters.

In natural sediments, however, not all particles are spheres. The increase of porosity with decreasing particle sizes shown in Figure 2-39 is a reflection of the change from predominantly spheroidal quartz and feldspar of the frameworks of sands to the predominantly flaky micas and clay minerals, which form the bulk of fine aggregates. Some of these aggregates are deposited in an initial open condition having much porosity. (See Figure 2-37.)

Intraparticle porosity, like the interparticle porosity just described, is directly related to the characteristics of the particles, but unlike interparticle porosity, intraparticle porosity is not related to the fabrics of particle aggregates. The intraparticle porosity of calcium-carbonate particles is either intrinsic to the mode of formation of the particle, as in the stereom of echinoderms, or was formed by some other process, such as boring by microorganisms (Figure 2-40; see also Box 2.1 Figure 6, and Box 2.6 Figure 7, C).

The destruction of porosity by cementation and its creation by dissolution are the subjects of Chapter 3.

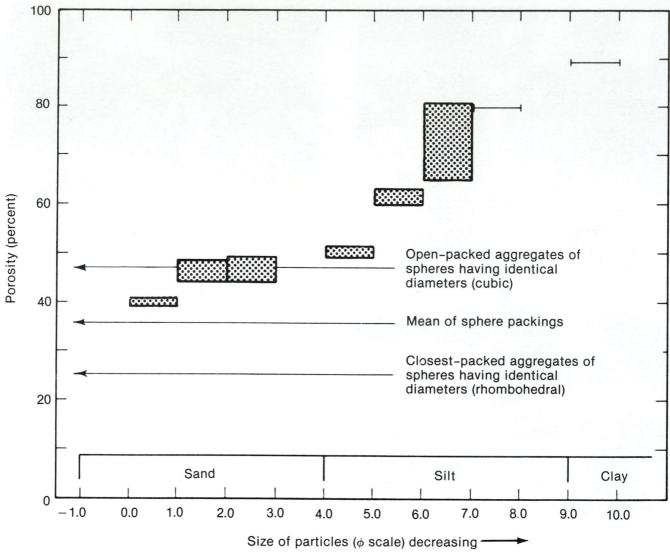

FIGURE 2-39.   Variation of porosity with particle size in modern (noncompacted) sediments. (Compiled from various sources.)

Permeability

The storage of fluids in a deposit is a function of the total porosity, with the exception that some pore spaces are too small to admit the molecules composing fluids. For example, the diameters of large oil molecules can be 30 nm or more, which is within the range of the sizes of the narrow connections between interparticle pores. By contrast with mere storage, however, the flow of fluids is controlled by the effective porosity, or degree of connection among the pores, and by the **permeability**, which is defined as *the capacity of a porous material to transmit fluids*. Permeability depends on the factors of the sizes, shapes, and degree of connection among pore spaces and on the viscosity of the fluid and the pressure driving it.

Henry Darcy (1803–1851), a French engineer, was the first person to analyze the flow of fluids through porous media. He studied the flow of ground water through sand. He derived the following flow equation:

$$Q = \frac{K \mu A (P_2 - P_1)}{L} \qquad \text{(Eq. 2-3)}$$

where

Q = quantity of fluid transmitted in a unit time (cm^3/s)

K = coefficient of permeability (related to the sizes, shapes, and interconnection of pores)

A = cross-sectional area of connected pores at right angles to direction of flow (cm^2)

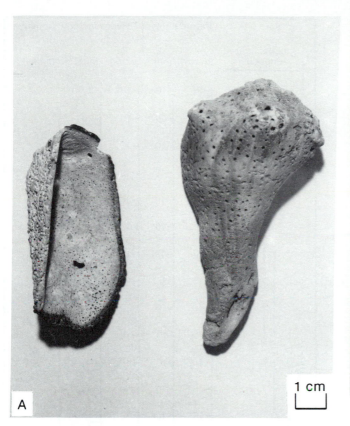

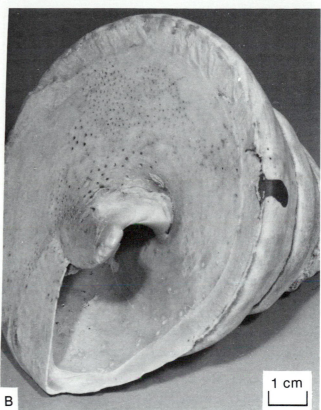

FIGURE 2-40.   Modern mollusk shells riddled with tiny bores.
  A. Side views of two specimens from St. Augustine, Florida.
  B. End view of specimen from island in Strait of Malacca, southeastern Asia. (V. Buffaline, 1978, and authors.)

L = length along which pressure differential is measured (cm)

$P_2 - P_1$ = pressure difference (atmospheres) between two points separated by distance L

μ = viscosity of fluid (centipoises), a factor influenced by temperature.

The standard unit of permeability is a **darcy**, *that permeability which allows 1 cubic centimeter per second of a fluid having a viscosity of 1 centipoise to pass through a pore cross-sectional area of 1 square centimeter under a pressure gradient (measured at right angles to the face of flow) of 1 atmosphere per centimeter.* Because the permeability values of most sedimentary rocks are much less than one darcy (Figure 2-41), a unit *1/1000 of a darcy*, the **millidarcy**, is generally used.

Permeability and porosity are related in complex ways. One of the chief factors affecting both is the packing. For example, both permeability and porosity decrease as packing changes from open (cubic) to close (hexagonal; see Figure 2-36).

If we keep packing constant, then a change of size among identically shaped particles does not change the porosity but it does affect permeability. Recall that the porosity of identically packed spheres is constant, no matter what the sizes of the spheres may be. But the permeability of an aggregate of large spheres exceeds that of an aggregate composed of small spheres. This is because the openings among the large spheres are larger than among the small spheres. In general, although other factors affect permeability, as particle size decreases, porosity increases (a result of changes in shape and in packing), but permeability decreases (the sizes of the connections between adjacent interparticle pores become smaller).

Permeability also varies with sorting. In well-sorted coarse sand, in which the interparticle pores are large and have not been blocked by any admixed fines, permeability may amount to thousands of millidarcies (Figure 2-42). Now, if fines are added to the coarse sand, the sizes of the pores are reduced and therefore, so are the porosity and the permeability. In this example, mean pore size has decreased and this reduces the permeability.

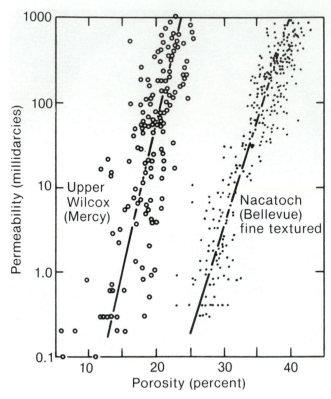

FIGURE 2-41. Measured values of porosity and permeability of two petroleum-reservoir sandstones plotted against each other. (G. E. Archie, 1950, fig. 1, p. 945.)

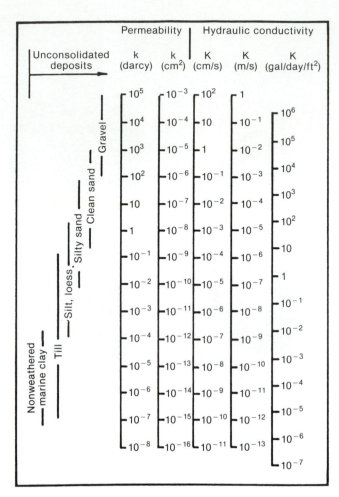

FIGURE 2-42. Typical ranges of permeability values of selected sediments ranging in size from clay to gravel shown opposite 5 scales that express permeability (small *k*) or hydraulic conductivity (large *K*) in different units. (R. A. Freeze and J. A. Cherry, 1979, table 2.2, p. 29. Reprinted by permission of Prentice-Hall, Inc., Englewood Cliffs, New Jersey.)

Both porosity and permeability have been measured on many samples of petroleum-reservoir rocks. A certain porosity-permeability trend may characterize a given geologic formation or hydrocarbon field. (See Figure 2-41.)

Finally, permeability is affected by any heterogeneity within a deposit. Permeability is at a maximum parallel to the direction of any preferred orientation within the fabric. *Any feature that impedes or prevents the throughflow of fluids* is a **permeability baffle** or **barrier**. An example is a layer of clay or of anhydrite that prevents fluids from flowing upward from one sandstone to another. Instead of flowing across such a permeability barrier, the fluid stops or flows along the contact at the base of the barrier. A **flow unit** is *a three-dimensional body of rock that is separated by permeability barriers from other bodies of rock.* Moving subsurface fluids tend to remain within flow units. If the configuration of flow units permits, then fluids may be trapped against permeability barriers.

In natural rock formations, one of the most-important factors affecting permeability is the number of fluids. For example, three fluids may be present: oil, gas, and saltwater. Darcy's simple equation refers to one fluid only. Where more than one fluid is present, individual relative permeability values may be established, depending on interactions with the other fluids as well as with the rock material. A significant factor in relative permeability is *wettability*. *A fluid that clings to the rock surface and forms a thin film having a low angle of contact with the rock (measured within the fluid)* is a **wetting fluid**. If the **fluid** is **nonwetting**, it *forms beads, as does water on a freshly waxed automobile; its internal contact angle is high* (Figure 2-43). A nonwetting fluid maintains a minimum surface area of contact with the rock surface. The relative permeability values of nonwetting fluids are higher than those of wetting fluids; hence nonwetting fluids are more mobile than wetting fluids.

In sandstones, water is usually wetting and oil, nonwetting. In many limestones, the reverse is true. Wettability is controlled by the electrochemical attraction between the fluid and the surfaces of the rock. Accordingly, wettability is affected by the roughness or

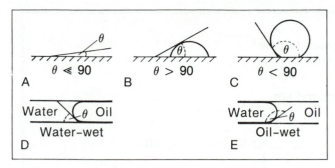

FIGURE 2-43. Wetting of solid surfaces by liquids shown in schematic sketches. (S. K. Ghosh and G. M. Friedman, 1989, fig. 1-2, p. 50; A, C, and D after A. I. Levorsen, 1967, figs. 10-5 and 10-6, p. 446 and 447.)

A. Contact angle is nearly zero; the liquid completely wets the surface. For most solids, water is a good wetting agent.

B. Contact angle of intermediate value; wettability of the liquid is also intermediate.

C. Contact angle > 90°; liquid is nonwetting. Mercury is an example of a nonwetting agent for most solids.

D. Water wet, as in sandstones

E. Oil wet, as in limestones

smoothness of the rock surface and by the presence or absence of organic films on the surface.

Porosity and permeability are part of the subject of **petrophysics**, which deals with *the physics of rocks, their pore systems, and their contained fluids.*

SOURCE: Freeze and Cherry, 1979; Ghosh and G.M. Friedman, 1989.

Acoustic Impedance

Acoustic impedance is defined as *the product of the speed of sound waves times the bulk density.* If the speed of sound is expressed in centimeters per second and the bulk density in grams per cubic centimeter, the dimensions of acoustic impedance are grams per square centimeter per second. If the speed of sound is expressed in kilometers per second (as is commonly the case), then the acoustic impedance is 10^5 g/cm^{2}s. Acoustic impedance and porosity of modern sediments of Lake Maracaibo, Venezuela, are related as shown in Figure 2-44.

Acoustic impedance is an important factor in determining how sound waves behave as they pass through solid bodies. The speed of sound in a solid is a function of both the density and rigidity of the solid. The bulk density of a well-cemented quartz sandstone is the same as that of a poorly cemented, friable quartz sandstone having the same porosity as the well-cemented sandstone. However, sound waves pass much more readily through the well-cemented sandstone because the firm contacts between the quartz particles transmit the sound as readily as do the particles themselves. The weak- or

discontinuous contacts between the particles of a poorly cemented sandstone impede the passage of sound waves.

Because of the differing minerals among their contained particles and because of the differing degrees of firmness of the contacts between particles, sedimentary aggregates transmit sound waves at different celerities. Thus, at the boundary between two different sediments or sedimentary rocks, an acoustic-impedance contrast may exist. Because they can reflect sound waves, *surfaces having acoustic-impedance contrasts* are known as **reflection interfaces**. This is the basis for the technique of recording seismic-reflection profiles, which are used as the basis for seismic stratigraphy (Chapter 6) as well as in study of the structural configuration of subsurface strata.

Electrical Properties

All sedimentary aggregates can transmit, or conduct, electricity at least to some degree. However, some sedimentary aggregates are better conductors than others. This variation is the basis for two electric well-logging tools that record the self-potential or spontaneous potential and electrical resistivity of the formations penetrated by the bore hole (discussed in Chapter 16). **Electrical resistivity** is *the inverse of conductivity.* The electrical properties of sedimentary aggregates are determined both by the contained fluids and by the particles and other rock materials. The self- or spontaneous electrical potential of most sandstones, in which quartz particles predominate, is low. The spontaneous potential of shales, in which clay minerals usually predominate, is high. The conductivity of most sulfide minerals is high. Of these, pyrite is the most common. The conductivity of a pyrite-cemented sandstone is much higher than would be expected from the abundance of quartz.

The electrical properties of sedimentary aggregates owe as much to the fluids they contain as they do to the composition and other characteristics of the particles. Salt water is a good conductor whereas oil and gas are poor conductors.

Magnetic Properties

The magnetic properties of aggregates of sedimentary particles derive from their content of magnetic minerals, of which the most important are magnetite and pyrrhotite. In order to understand the magnetic properties of sediments it will be helpful to begin by reviewing the behavior of magnetic minerals during the cooling of lava flows.

As a lava flow cools and hardens, its magnetic minerals, which typically crystallize early, are widely separated within the lava. Because they are able to move, they

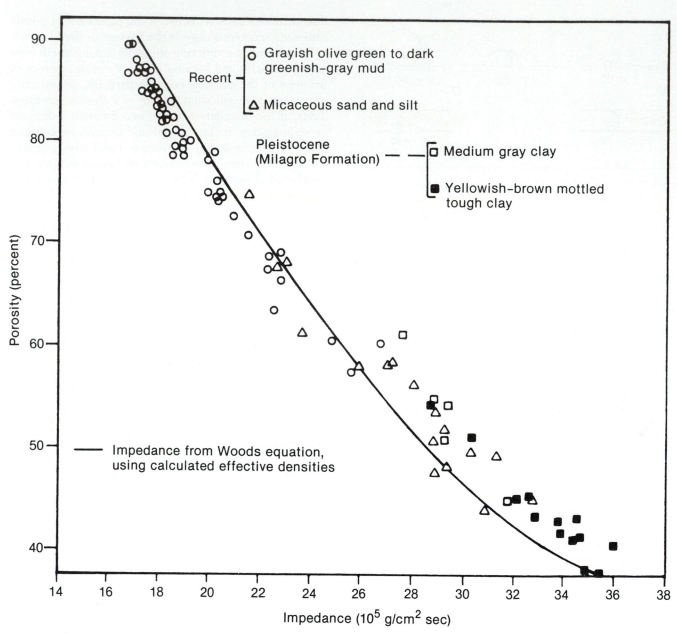

FIGURE 2-44.   Graph of porosity vs. acoustic impedance for sediments of Recent and Pleistocene ages cored from Lake Maracaibo, Venezuela. (After R. Sarmiento and R. A. Kirby, 1962, fig. 23, p. 723.)

commonly align themselves with the Earth's magnetic field. Just one process is involved: the response to the Earth's magnetic field by individual magnetic minerals that are free to move within the liquid magma/lava. By contrast, in sediments the alignment of magnetic particles can be affected by several processes. These include (1) the direct influence of the Earth's magnetic field on particles < 10 μm in size (or as small as 5 μm for fine sediments, which compares with the behavior of magnetic particles in a cooling igneous rock); (2) the alignment during deposition of flake-shaped detrital micas that

imparts not only a preferred orientation by shape, but incidentally establishes a magnetic-susceptibility anisotropy because the magnetic properties are determined by the properties of the crystal lattice that likewise determine particle shapes; (3) burrowing organisms that churn the sediment disrupt the depositional fabric; moreover, the burrowers simultaneously introduce enough water into the sediment to increase pore space and this increase may enable small particles to rotate; and (4) small sedimentary particles may settle into open burrows, where the primary magnetic orientations they assume can be modified by postdepositional effects. Processes 1, 2, and 4 lead to *primary* magnetic fabric, whereas process 3 produces *secondary* magnetic fabric.

In many sedimentary aggregates the energy of the transporting medium (hydro- or aerodynamic- and gravitational effects) on larger magnetic particles (> 10 μm) easily overwhelms the influence on them of the Earth's magnetic field. Only very tiny particles settling through still water or small particles that are free to rotate while being deposited in highly fluid muds acquire a preferential magnetic fabric governed by the Earth's magnetic field.

Any primary preferential alignment of magnetic particles can be destroyed by burrowing organisms. Thus the magnetic fabric of sediments that have been intensely bioturbated is essentially random.

The orientation imparted to magnetic particles at the time of origin of the rock, and which remains in the rock unless destroyed by intense heating, is called **remanent magnetism.** This orientation is a record of the direction of the Earth's magnetic field at the time and place when the rock formed. Therefore, if the rocks can be restored to their original locations and orientations, the orientation of the Earth's magnetic field at that time and place can be discovered. The term remanent magnetism derives from a misspelling of remnant when the term was defined; the misspelling was perpetuated and is now regarded as correct in this usage.

Radioactivity

Because the rock-making minerals and related substances contain radioactive isotopes, all sedimentary aggregates and sedimentary rocks are in some measure radioactive. The commonest radioactive isotopes are uranium-238 and uranium-235, thorium-232, radium-226 or radium-223 or radium-224, rubidium-87, potassium-40, and carbon-14.

The uranium isotopes are present in certain heavy minerals. Thus the uranium in these minerals is subject to being concentrated mechanically in placers, as for example, are gold particles.

Most of the natural radioactive isotopes in sedimentary deposits are concentrated in the clay minerals that contain potassium, and thus potassium-40, the radioactive isotope that decays to argon-40. Other faintly radioactive minerals common in sedimentary deposits include glauconite (sometimes considered a clay mineral), hornblende, and feldspars, which contain not only argon-40 and potassium-40, but rubidium-87, which decays to strontium-87.

The natural radioactivity of sedimentary deposits forms the basis for gamma-ray well logs, in which the logging tool measures the radioactivity of the sedimentary strata near the tool. (Further details are in Chapter 16.)

Certain igneous rocks, notably granites, contain uranium-bearing zircons. In the uranium-lead decay chain is a gas, radon, which is also radioactive. Radon released from the granite (or other uranium-containing deposits, such as some "black" shales) percolates through the regolith and may enter the basements of houses. Because many modern houses have been tightly sealed to make them energy efficient, radon entering from the basement tends to accumulate indoors, and may reach levels that have been declared hazardous.

Suggestions for Further Reading

BASAN, P. B., ed., 1978, Trace fossil (*sic*) concepts: Tulsa, OK, Society of Economic Paleontologists and Mineralogists Short Course 5, 181 p.

BATES, N. B., and BRAND, UWE, 1990, Secular variation of calcium-carbonate mineralogy (*sic*); an evaluation of ooid and micrite chemistries: Geologische Rundschau, v. 79, p. 27–46.

CHOQUETTE, P. W., and PRAY, L. C., 1970, Geological nomenclature and classification of porosity in sedimentary carbonates: American Association of Petroleum Geologists Bulletin, v. 54, p. 207–250.

FRASER, H. J., 1935, Experimental study of porosity and permeability of clastic sediments: Journal of Geology, v. 43, p. 910–1010.

FRIEDMAN, G. M., 1979, Differences in size distributions of populations of particles among sands of various origins: Sedimentology, v. 26, p. 3–23.

FRIEDMAN, G. M., and JOHNSON, K. G., 1982, Exercises in sedimentology: New York, John Wiley and Sons, 208 p.

ILLING, L. V., 1954, Bahamian calcareous sands: American Association of Petroleum Geologists Bulletin, v. 38, p. 1–95.

KITTLEMAN, L. R., 1979, Tephra: Scientific American, v. 251, p. 160–177.

KRINSLEY, D. H., and SMALLEY, I. J., 1972, Sand: American Scientist, v. 60, p. 286–291.

LAJOIE, JEAN, 1979, Facies models 15. Volcaniclastic rocks: Geoscience Canada, v. 6, p. 129–139.

LOWENSTAM, H. A., 1981, Minerals formed by organisms: Science, v. 211, p. 1126–1131.

SHINN, E. A., Steinen, R. P., Lidz, B. H., and Swart, P. K., 1989, Whitings, a sedimentologic dilemma: Journal of Sedimentary Petrology, v. 59, p. 147–161.

SUTTNER, L. J., 1989, Recent advances in study of the detrital mineralogy (*sic*) of sand and sandstone: implications for teaching: Journal of Geological Education, v. 37, p. 235–240.

TANNER, W. F., 1969, The particle size (*sic*) scale: Journal of Sedimentary Petrology, v. 39, p. 809–812.

VAN HOUTEN, F. P., and Purucker, M. E., 1985, On the origin of glauconite (*sic*) and chamosite granules: Geo-Marine Letters, v. 5, p. 47–49.

WARDLAW, N. C., 1976, Pore geometry (*sic*) of carbonate rocks as revealed by pore casts and capillary pressure: American Association of Petroleum Geologists Bulletin, v. 60, p. 245–257.

Burial Diagenesis and Other Ways of Turning Sediments into Sedimentary Rocks

This chapter, devoted to the geologic relationships involved in converting sediments into sedimentary rocks, forms the all-important link in the chain of geologic thought established by James Hutton. In Chapter 2 we reviewed some important points about sediments. In Chapter 4 we explore sedimentary rocks. Our general objective in this chapter is to focus on the kinds of changes that take place in the general process of **lithification**, *the conversion of loose sediments to masses of solid rock*. To do this we must first get acquainted with mineral reactions, such as precipitation, dissolution, and replacement, and be familiar with some of the major chemical relationships between fluids and minerals. We must then review the various natural fluids, including those at the Earth's surface and those found exclusively at depth. We shall need to understand the conditions of temperature and pressure that prevail in subsurface environments and how the common minerals of sedimentary deposits react in these subsurface environments. Do all rock-making minerals behave the same? Or are some dissolved? How can we recognize that a mineral has been dissolved? Which minerals are precipitated?

We shall be emphasizing the textural characteristics that enable us to infer the circumstances under which carbonate minerals and quartz, chiefly, grew. For example, was it into an empty space? Or by replacing a pre-existing mineral? The ultimate goal of textural study of minerals is to establish a sequence of events. Finally, we summarize briefly some distinct geologic effects of deep burial: maturation of kerogen to petroleum, formation of fluid inclusions, origin of certain kinds of dolomite, and some special effects in sands and sandstones. In short, we shall be exploring the vast field of **diagenesis**, *the sum of physical-, inorganic-chemical-, and biochemical changes in a sedimentary deposit after its initial accumulation and excluding metamorphism*.

As we shall see, depth of burial is a great geologic equalizer. Many kinds of sediment, fashioned in the host

of local depositional environments that we shall be exploring in Chapters 9 through 14, enter at the top of the subsurface realm. But, because of the great lateral uniformity of subsurface conditions controlled by depth of burial, such as temperature, pressure, and the composition of fluids, distinctive diagenetic products, keyed to the Earth's horizon, can crosscut this "Tower-of-Babel" array of the products of local depositional environments.

We begin with some fundamental properties of waters and minerals.

Waters and Minerals

Minerals that form cements in sedimentary rocks have been precipitated from various natural waters, at the Earth's surface, within the soil zone, in the shallow subsurface, or in the deep subsurface. A firm basis for understanding mineral cements requires familiarity with the compositions of these waters and with the factors that control how minerals behave in them. A rich field of research is flourishing in what is known as *low-temperature geochemistry*, by which is meant the study of these natural waters and how minerals react with them. We review a few general relationships about natural waters, emphasizing some of the major factors that control the solubility of minerals in water solutions. Included are salinity and concentration by evaporation, the factor known as pH, dissolved gases, the factor of Eh, and stability relationships of calcium carbonate and silica.

Salinity

An important property of any water is the **salinity**, or *the total amount of dissolved solids*. Salinity is expressed by various methods. Two of these are parts per thousand (abbreviated ‰, sometimes expressed as parts per mil, not to be confused with parts per million), and weight (in grams, milligrams, or micrograms) of dissolved solids per kilogram of water. The concentrations of individual constituents, such as chloride ions (Cl^-) or sodium chloride (NaCl), are expressed in parts per million (ppm).

Three widespread kinds of water containing large amounts of dissolved solids are seawater (salinity about 35‰, 35 g/kg); waters of saline lakes, such as the Dead Sea (salinity of surface water about 200‰); and formation waters, the highly saline waters found in pores of formations in deep subsurface environments (salinity in the range of 75,000 to 300,000 ppm, which is about twice to nearly 10 times that of seawater).

CONCENTRATION BY EVAPORATION

From the earliest days, people have been harvesting salt from the sea by the simple process of allowing the water to evaporate. As the water evaporates, the salinity of the remaining water progressively increases. Ultimately, salt crystallizes. Because brines form an environment generally hostile to simple organisms, the reactions that take place in brines are as inorganic as one can expect to find in nature. Even on salt flats, however, some hardy organisms flourish.

An elaborate series of experiments to measure the relationships between the salinity of sea water and *salts deposited as a result of evaporation* (that is, **evaporite minerals**), was carried out along the shore of the Mediterranean Sea, at Cette, France, by the Italian chemist J. Usiglio. For his experiments Usiglio collected two samples of 5 l each, from two stations, one 3 km out from the shore, and the other 5 km out. He collected the samples at night, two from each station. One sample came from the surface, and the second from 1 m below the surface.

For his evaporation experiments Usiglio prepared a hothouse at a controlled temperature of 40°C. He placed the water samples in porcelain vats and let them evaporate. From time to time he collected samples for analysis. He carried out the analytical work after allowing the samples to cool to 21°C. He measured the density and proportion of the original water remaining and determined the kinds and amounts of solids precipitated. Figure 3-1 shows his results in graphic form. The sequence of solids precipitated is Fe_2O_3, $CaCO_3$, $CaSO_4 \cdot 2H_2O$, NaCl.

Usiglio's experiments seemed to have settled the questions associated with the origin of evaporite minerals from brines. However, as will be seen in Chapters 10, 12, and 14, the interpretation of some aspects of the origin of evaporites is still under debate. Additional data pertinent to the debate come from shallow pools marginal to the Red Sea in an area where evaporation is exceptionally intense and in which bedded gypsum accumulates. We review some of the measurements from the waters of these pools that allow us to confirm Usiglio's results on the interrelationship between progressive increase in the concentration of brines and the precipitation of evaporite minerals.

The shallow pools marginal to the Red Sea lose more than 2.5 m of water by evaporation each year. This loss of water is balanced by influx of water from the Red Sea. During the spring the waters in the pool become saltier (Figure 3-2). In June, at a salinity of about 120‰ (120×10^3 mg/l) and prior to the imminent precipitation of gypsum, the sulfate concentration has built to its highest level. But between the months of June and August, when total salinity reaches 330‰ (330×10^3 mg/l), gypsum is precipitated and the concentration of sulfate drops.

Experimental work on both natural- and artificial seawater has shown that at a concentration of about 3.35 times that of seawater (at 30°C) or at a salinity of about 124‰ (124×10^3 mg/l), gypsum is precipitated. Although experience in sedimentology commonly teaches that, because of conflicting variables, laboratory experiments are not necessarily duplicated in nature, the natural precipitation of gypsum as a chemical reaction in the Red Sea pools confirms the laboratory results.

During fall and winter, the salinity in the pools drops to a level of near 220‰ (220×10^3 mg/l). As the salinity drops, the sulfate concentration increases. This indicates that after the initial precipitation of gypsum, the sulfate levels build up again.

SOURCES: G. M. Friedman and Krumbein, 1985; A. C. Kendall, 1979; Nissenbaum, 1980.

pH: Definition and Reactions

A fundamental property of water is expressed as the **pH**, which is *the activity in a solution of the hydrogen ions*, a factor that is expressed as acidity or alkalinity. Because the activity of a solute is not generally the same as its concentration, activity may be defined as *ideal* or *effective concentration*. By definition, the activity of a pure substance such as water is taken as equal to 1. Concentrations and activities may differ. The **activity** may be calculated as *the product of the concentration and the activity coefficient*. The activity coefficient may be thought of as the ratio of effective concentration to actual concentration. The activity coefficients of some major ions in seawater are

$$Cl^- = 0.63 \qquad Mg^{2+} = 0.25$$
$$SO_4^{2-} = 0.068 \qquad CO_3^{2-} = 0.021$$

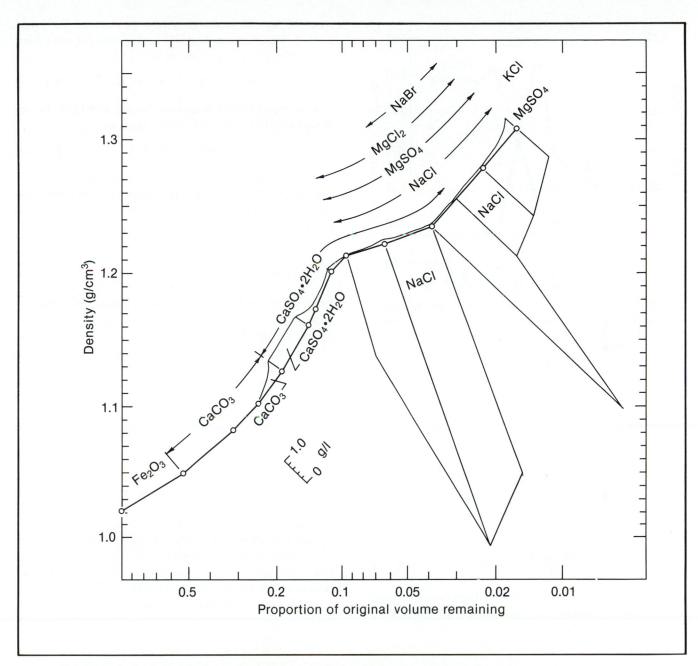

FIGURE 3-1.   Results of Usiglio's experiments on the evaporation of seawater plotted on a graph of density (increasing upward) vs. proportion of original volume of water remaining (decreasing toward the right, $\log_{10}$ scale). The density of the water increased from the 1.025 of the initial sea-water to 1.21 at the point where only 1/10 of the original volume remained and halite (NaCl) began to precipitate. Sequence of solids precipitated shown by chemical formulas above the curve. Amounts of each precipitated at the various sampling points (circles) are shown by lengths of lines normal to the curve (in grams per liter; bar scale on lower left of figure). (Data based on Usiglio, 1849, and taken from A. W. Grabau, 1920, table vii, p. 54.)

The **pH** of a solution is formally defined as *the negative logarithm to the base 10 of the hydrogen-ion activity.* Thus if the activity of hydrogen ions (expressed as the ion symbol in square brackets) is 0.0001, this is written $[H^+] = 0.0001$. The number 0.0001 can be expressed as 10^{-4}. The logarithm to the base 10 of 10^{-4} is -4; the negative logarithm of -4 is 4. Thus the pH of the hydrogen-ion activity of 0.0001 is equal to 4.

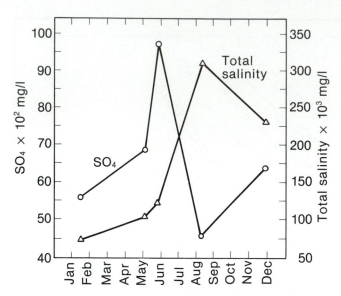

FIGURE 3-2.   Seasonal changes for 1970–1971 in salinity (triangles; scale at right) and concentration of sulfate (circles; scale at left) in water of sea-marginal pool subjected to intense evaporation along shores of Red Sea. Further explanation in text. (G. M. Friedman, A. Sneh, and R. W. Owen, 1985.)

Pure water dissociates slightly into hydrogen ions and hydroxyl ions:

$$H_2O \rightarrow H^+ + OH^- \qquad (Eq.\ 3\text{-}1)$$

The law of mass action stipulates that the rate of a chemical reaction is directly proportional to the active masses of the reacting substances or to the molar concentrations of the reacting substances. This law can be written as a chemical equation as

$$\frac{[H^+][OH^-]}{[H_2O]} = K_w \qquad (Eq.\ 3\text{-}2)$$

As mentioned, the activity of pure liquid water is taken to be unity. Therefore the denominator of Eq. 3-2 be-

comes 1 and K_w reduces to the product of $[H^+]$ and $[OH^-]$. Experiments at 25°C and 1 atm pressure have shown that the value of K_w is 10^{-14}. Thus

$$[H^+][OH^-] = K_w = 10^{-14} \qquad (Eq.\ 3\text{-}3)$$

A neutral solution is defined as one in which the activities of $[H^+]$ and $[OH^-]$ are equal, or

$$[H^+] = [OH^-] \qquad (Eq.\ 3\text{-}4)$$

In order for Eq. 3-3 to be satisfied, the values of these two ions are 10^{-7}, of which the negative logarithm to the base 10 (the pH) is 7 (Figure 3-3). The pH of the neutral point changes slightly with temperature and pressure.

Dissolved Gases

Many gases are dissolved in natural waters. Of particular importance to the subject of mineral cements are oxygen, carbon dioxide, hydrogen sulfide, and methane. It would take us far afield into several branches of chemistry to include a thorough exposition of the interactions of these gases and natural waters. The following short summary is intended to give the reader a general appreciation of the importance of these gases.

OXYGEN

The oxygen dissolved in surface waters comes from two sources: (1) the atmosphere, which consists of about 20% oxygen, and (2) photosynthesis by aquatic plants. Water in contact with the atmosphere acquires dissolved oxygen whose saturation value varies with the salinity, temperature, and pressure; the saturation value of oxygen decreases as the salinity increases. Waters having salinities exceeding 300‰ contain no dissolved oxygen (Figure 3-4).

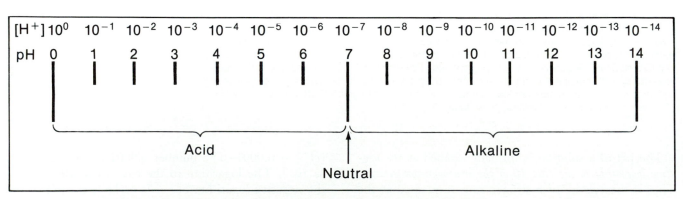

FIGURE 3-3.   Graphic representation of the relationship between $[H^+]$ and pH in solutions. (G. M. Friedman and J. E. Sanders, 1978, fig. 5-18, p. 132.)

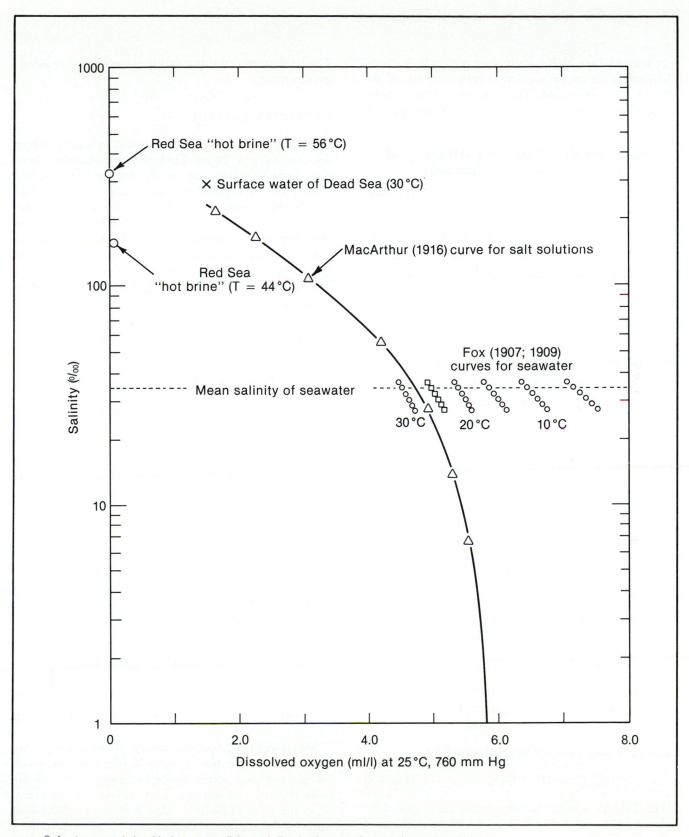

FIGURE 3-4. Inverse relationship between salinity and dissolved oxygen in natural waters and in salt solutions. Notice that the amount of dissolved oxygen diminishes notably in waters that are more saline than seawater (dashed horizontal line). Closely spaced sets of small open circles (at right) show saturation values of oxygen (from normal dry atmosphere) in seawater at temperatures shown. (Original data in C. J. J. Fox, 1907, and 1909; values taken from H. U. Sverdrup, M. W. Johnson, and R. H. Fleming, 1942, table 38, p. 188.) Curve through open triangles, solubility of oxygen in salt solutions of varying concentrations at 25°C and 760 mm Hg pressure. (Original data in G. G. MacArthur, 1916; values taken from G. M. Friedman, 1975b, table 1, p. 394.)

In waters where high salinity does not exclude plants, it is possible for the proportion of oxygen to exceed the value of 100% saturation. This can come about because during photosynthesis the plants release free oxygen according to the equation

$$\text{sunlight} + 6\ CO_2 + 6\ H_2O \rightarrow \underset{\text{(glucose)}}{C_6H_{12}O_6} + 6\ O_2$$

$$(\text{Eq. 3-5})$$

A liter of well-aerated surface fresh water and most seawater contains about 8 ml of dissolved oxygen, written 8 ml/l. Some of the dissolved oxygen is used in respiration by gill-bearing animals, such as fish and various invertebrates. Much of the dissolved oxygen is removed from the water by reacting with first-cycle protoplasmic organic matter.

CARBON DIOXIDE

The carbon dioxide dissolved in water usually comes from the respiration of organisms, from the oxidation of first-cycle organic matter, or from the oxidation of fossil fuels. Carbon dioxide is the chief gaseous waste product of oxygen-breathing organisms. Even plants, which during photosynthesis take in CO_2 and give off O_2 (Eq. 3-5), generate CO_2 as a waste product of their nonphotosynthetic metabolism. Unlike other gases that dissolve in water without being combined chemically, carbon dioxide reacts with the water to form carbonic acid. In the following equations for dissolved CO_2 in water, (aq) designates a dissolved species, (l) stands for liquid, and if a solid phase is involved, (c) stands for crystalline solid:

$$CO_2(aq) + H_2O(l) \rightarrow \underset{\text{(carbonic acid)}}{H_2CO_3(aq)} \qquad (\text{Eq. 3-6})$$

The carbonic acid ionizes as follows:

$$\underset{\text{(carbonic acid)}}{H_2CO_3(aq)} \rightarrow H^+ + \underset{\substack{\text{(bicarbonate} \\ \text{ion)}}}{HCO_3^-} \qquad (\text{Eq. 3-7})$$

The bicarbonate ion itself can be ionized into

$$HCO_3^- \rightarrow H^+ + CO_3^{2-} \qquad (\text{Eq. 3-8})$$

The carbonic acid can become part of the system of bicarbonate ions and carbonate ions, which enter the solid carbonate-mineral phases:

$$CO_2\ (\text{dissolved}) \rightarrow \underset{\substack{\text{(carbonic} \\ \text{acid)}}}{(H_2CO_3)} \rightarrow \underset{\substack{\text{(bicarbon-} \\ \text{ate ions)}}}{(HCO_3^-)} \rightarrow \underset{\substack{\text{(carbonate} \\ \text{ions)}}}{(CO_3^{2-})}$$

$$(\text{Eq. 3-9})$$

The implications of this reaction are discussed in subsequent sections.

HYDROGEN SULFIDE

Hydrogen sulfide, a lethal gas, appears in waters lacking dissolved oxygen. In any kind of water, abundant first-cycle organic matter can totally deplete the dissolved oxygen. Or the high salinity may exclude dissolved oxygen. In the absence of dissolved oxygen, anaerobic bacteria may be the sole surviving organisms. Among the most-important geologically active bacteria are sulfate reducers, especially *Desulfovibrio*. These bacteria obtain oxygen by breaking down the sulfate radicals dissolved in most waters. A lethal by-product from this reaction is hydrogen sulfide; other important by-products include calcium carbonate, native sulfur, and pyrite. In the following equations we show the bacterial breakdown of calcium sulfate, typically gypsum (Eq. 3-10).

$$CaSO_4 \rightarrow Ca^{2+} + SO_4^{2-} \ (\text{gypsum dissolves})$$

Bacteria attack the dissolved sulfate radical:

$$SO_4^{2-} + 2\ \underset{\substack{\text{(organic} \\ \text{matter)}}}{CH_2O} \rightarrow \underset{\substack{\text{(hydrogen} \\ \text{sulfide)}}}{H_2S} + 2\ HCO_3^-$$

Calcium dissolved in the water reacts with the bicarbonate ions:

$$Ca^{2+} + 2\ HCO_3^- \rightarrow CaCO_3 + H_2O + CO_2$$

The sum of these reactions is

$$CaSO_4 + 2\ \underset{\substack{\text{(organic} \\ \text{matter)}}}{CH_2O} \rightarrow$$

$$\underset{\text{(calcite)}}{CaCO_3} + H_2O + CO_2 + \underset{\substack{\text{(hydrogen} \\ \text{sulfide)}}}{H_2S}$$

$$(\text{Eq. 3-10})$$

Sulfate-reducing bacteria can also alter gypsum or anhydrite to native sulfur, deposits of which are known in modern salt lakes and in the caprocks atop many salt diapirs. Reduction of sulfates creates sulfides and these may be in the form of finely divided iron monosulfides that impart a black color to the sediment. Iron monosulfides eventually change into pyrite.

METHANE

The hydrocarbon gas methane (CH_4) can form naturally in at least two ways: (1) by the bacterial reduction of carbon dioxide at low temperatures and (2) by the thermal cracking of solid hydrocarbons (kerogens) at higher temperatures (on the order of 75°C or more). Whereas some bacteria create methane, other bacteria destroy it. The *methane* found in sediments that have not been deeply buried doubtless is *of bacterial origin* (**bacteriogenic methane**). An example is so-called marsh gas.

If methane migrates from the interstitial water in sediments into an overlying body of water that contains dissolved oxygen, then the methane will be oxidized to form carbon dioxide and water:

$$3\ CH_4\ +\ 6\ O_2 \rightarrow 3\ CO_2\ +\ 6\ H_2O \quad \text{(Eq. 3-11)}$$
(methane)

The carbon in the carbon dioxide derived by the oxidation of methane is easily recognized by the high negative values of the ratio of its stable isotope with respect to the accepted standard Pee Dee Belemnite (expressed as PDB) ($\delta^{13}C$ PDB greater than -25).

SOURCES: Beauchamp and others, 1989; Friedman, 1988a, 1991; Schoell, 1988.

Eh (Redox Potential)

Another kind of process capable of modifying sediments in their environment involves oxidation or reduction. As an example, consider the surface waters of the Dead Sea, where sulfate ions are precipitated as gypsum. In shallow waters, where the supply of oxygen is ample, this sulfate is preserved. As shown in Eq. 3-10, below the wave-influenced zone of oxidation, sulfate-reducing bacteria break down the sulfate and the dissolved sulfur is largely in the form of H_2S or HS^-. This ability of an environment to reduce sulfur or to oxidize it, or the ability to cause any other changes in oxidation or reduction, is expressed by the Eh or redox potential, terms that are used interchangeably with oxidation potential or oxidation–reduction potential. Although oxidation and reduction are commonly biological, as the example from the Dead Sea shows, many kinds of oxidation- or reduction reactions are purely inorganic chemical reactions.

Although the term oxidation suggests combination of an element with oxygen and the term reduction the removal of oxygen, as in the reactions

$$2\ Fe\ +\ O_2 \rightarrow 2\ FeO \text{ (oxidation)} \quad \text{(Eq. 3-12)}$$

$$2\ FeO \rightarrow 2\ Fe\ +\ O_2 \text{ (reduction)} \quad \text{(Eq. 3-13)}$$

the addition of hydrogen, even without removal of oxygen, as in

$$S\ +\ H_2 \rightarrow H_2S \quad \text{(Eq. 3-14)}$$

is considered reduction. Because hydrogen is electropositive and oxygen electronegative, oxidation increases the proportion of electronegative products and reduction increases the proportion of electropositive constituents. A species that loses electrons is said to be oxidized, one that gains electrons, reduced. Thus in the reaction

$$Fe^{2+} \rightarrow Fe^{3+}\ +\ e^- \quad \text{(Eq. 3-15)}$$

iron has lost an electron (oxidation), and in

$$Fe^{3+}\ +\ e^- \rightarrow Fe^{2+} \quad \text{(Eq. 3-16)}$$

iron has gained an electron (reduction). (In Eqs. 3-15 and 3-16 the symbol e^- represents the electron, or negative charge.) In these reactions the valence of iron has changed. These two reactions involving iron are examples of oxidation and reduction that can take place without the presence of oxygen or hydrogen. Accordingly, **oxidation** can be defined as *chemical reactions in which the participating ions lose orbital electrons and thus their valence numbers increase*, and **reduction**, as *chemical reactions in which the participating ions gain orbital electrons and thus their valence numbers decrease*.

An important process in sedimentary environments involving oxidation and reduction is the change in iron during the transformation of magnetite to hematite or vice versa. Magnetite (Fe_3O_4 or $Fe_2O_3 \cdot FeO$; two ions Fe^{3+} and one Fe^{2+}) changes to hematite (Fe_2O_3 with both ions Fe^{3+}) as follows:

$$2\ Fe_3O_4\ +\ H_2O \leftrightharpoons 3\ Fe_2O_3\ +\ 2\ H^+\ +\ 2\ e^-$$

$$\text{(Eq. 3-17)}$$

This is a reversible reaction. If electrons are removed, the reaction goes to the right; the iron in the form of Fe^{2+} in Fe_3O_4 oxidizes and changes to Fe^{3+}. If electrons are added, the reaction goes to the left, and some of the Fe^{3+} ions are reduced to Fe^{2+}.

Eh, or **redox potential**, is *an expression of the relative intensity of oxidation or reduction in solution or of the electron concentration in a solution*. Because oxidation and reduction are electrical properties, measurements are made with an electrolytic cell. The numbers used to represent Eh, which record relative intensity, have been chosen arbitrarily with reference to the reaction

$$2\ H^+\ +\ 2\ e^- \leftrightharpoons H_2 \quad \text{(Eq. 3-18)}$$

At 25°C and 1 atm pressure, the Eh of Eq. 3-18 is the zero reference. Therefore, in measurement of Eh, a hydrogen electrode is employed as a reference. Thus Eh can be defined as *the equilibrium potential of an oxidation–reduction reaction relative to the potential of a standard hydrogen electrode*.

SOURCES: Barnaby and Rimstidt, 1989; Brookings, 1988.

Eh AND pH IN SEDIMENTARY ENVIRONMENTS

In sedimentary environments Eh and pH are commonly interdependent. Eh designates the concentration of electrons in solutions, and pH that of protons (hydrogen ions). Because electrons neutralize protons, many reactions depend on both Eh and pH. High values of Eh, representing a low electron content, are generally accompanied by low values of pH (high proton content; Figure 3-5).

The strongest oxidizing agent in sedimentary environments is the oxygen of the atmosphere. Stronger agents would oxidize water, hence liberate oxygen, and water would be unstable.

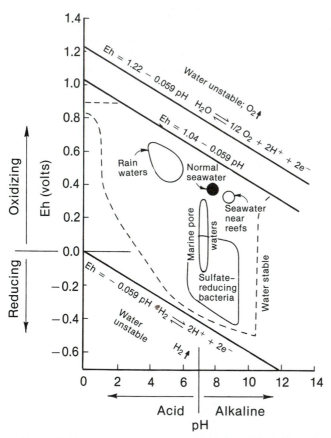

FIGURE 3-5. Characteristics of various waters plotted on a graph of Eh vs. pH. Dashed line encloses limit of measurements made to date in sedimentary environments. The reactions shown on the lines limiting the field of water stability are half-equations or redox couples; the oxidized form of any couple having a higher potential than that of the upper line theoretically will decompose water and form oxygen gas. The reduced form of any couple having a lower potential than that of the lower line theoretically will decompose water and generate hydrogen gas. (Data mostly from L. G. M. Baas Becking, I. R. Kaplan, and D. Moore, 1960; seawater and pore waters from G. M. Friedman, 1968, and G. M. Friedman, B. P. Fabricand, E. S. Imbimbo, M. E. Brey, and J. E. Sanders, 1968.)

Because the boundaries of the stability of water are known, we can plot these in a diagram with Eh as ordinate and pH as abscissa. (See Figure 3-5.) These boundaries give the limits of Eh in waters of sedimentary environments. The upper boundary for pH in sedimentary environments is about 11; however, such a high pH is not common and results from photosynthesis or from bacterial reactions, as in Eq. 3-10, or from hydrolysis of Na_2CO_3 or $NaHCO_3$ in evaporite lakes. The lower boundary of pH is 1 to 2; lower pH's mean strong acids, such as sulfuric acid. In fresh-water swamps and bogs pH values may be as low as 4 or even slightly lower; pH in these waters is controlled chiefly by weak acids, including carbonic- and organic acids. On this diagram we can indicate the Eh–pH values that have been empirically determined for waters of various sedimentary environments. (See Fig. 3-5.)

In contrast to field measurement of pH, which can be done precisely, field measurements of Eh are imprecise and at best are qualitative to semiquantitative. Although we are discussing pH–Eh relationships under the heading of chemical reactions, we should recall that for the most part Eh is driven by biological processes, such as photosynthesis, respiration, bacterial reactions involved in the sulfur cycle, and the decomposition of organic matter.

SOURCES: Baas Becking, Kaplan, and Moore, 1960;
Langmuir, 1971.

Stability Relationships of Calcium Carbonate and Silica

Calcite and related carbonate minerals and quartz plus its closely associated varieties of silica constitute an overwhelming majority of mineral cements. Although we shall be discussing them individually farther along, at this point we need to review how they respond to the fundamental environmental factors of salinity, temperature, pressure, and pH.

EFFECT OF SALINITY ON CALCIUM CARBONATE

The effects of salinity are best illustrated by considering the contrasting ways in which the two commonest forms of calcium carbonate, aragonite and calcite, behave in seawater and in fresh water. Near-surface seawater is nearly everywhere supersaturated with respect to $CaCO_3$. For the most part, shallow continental fresh waters are not saturated with calcium carbonate and so readily dissolve it.

The chemical statement of the dissolution of $CaCO_3$ is written as [as before, (c) for crystalline; (l) for liquid phases]

$$CaCO_3(c) + H_2O(l) \rightarrow Ca^{2+} + HCO_3^- + OH^-$$
$$\text{(calcite)}$$

$$\text{(Eq. 3-19)}$$

The reaction governing the dissolution of $CaCO_3$ may be simplified as

$$CaCO_3(c) + H_2O(l) + CO_2(aq) \rightarrow Ca^{2+} + 2\,HCO_3^-$$
$$\Updownarrow$$
$$CO_2(gas)$$

$$\text{(Eq. 3-20)}$$

Of the two mineral forms of calcium carbonate, aragonite dissolves more readily in fresh water than does low-magnesian calcite. Dissolution of the more-soluble aragonite and precipitation of the less-soluble calcite is intimately involved in the process of making sedimentary particles into strata of rock. This is one of the most-important reactions at the surface of the Earth.

Because of its supersaturation with respect to calcium carbonate, one might expect that near-surface seawater would be precipitating carbonate minerals. Several factors are at work to inhibit such precipitation. Calcium-carbonate surfaces that might serve as sites for the nucleation of carbonate-mineral growth adsorb organic molecules from solution in such a way as to block nucleation. Once these organic compounds have been removed, the reaction equilibrates and calcium carbonate is precipitated. But the organic compounds are removed only at high values of pH. Precipitation of calcium carbonate from seawater is also inhibited by magnesium, by phosphate adsorption, and to a lesser degree, by the sulfate ions in solution. Despite considerable study and discussion by geologists, the nature of this inhibiting effect is not yet understood.

The CO_2 to make carbonic acid in fresh waters comes mostly from the bacterial decay of dead plants and other organic matter and from respiration of living plants. Organic acids, such as lactic acid, derived from rotting vegetation, and sulfuric acid, produced by the weathering of sulfide minerals, especially pyrite, accelerate dissolution of calcium carbonate. Such dissolution increases the concentration of calcium carbonate in the waters; ultimately, calcium carbonate may be precipitated as in stalactites or stalagmites in caves or as mineral cements (discussed in following sections).

Late in the nineteenth century the Royal Society of London sponsored an expedition to Funafuti Atoll, in the southwest Pacific, to drill a hole to test Charles Darwin's (1809–1882) ideas about the origin of atolls as a result of subsidence (discussed further in Chapter 10). A surprising result was that many of the samples recovered were best described as "crystal mush"; they consisted of carbonate skeletal debris that had not been cemented.

From this finding, the idea became widespread that marine carbonate skeletal debris could remain in contact with seawater indefinitely without being cemented. As we shall see, numerous examples have been found in which seawater has precipitated cement consisting of calcium-carbonate mineral(s).

EFFECT OF TEMPERATURE ON CALCITE AND SILICA

The effect of temperature on the solubility of calcite and silica exemplifies the contrasting ways in which these minerals are affected by a given variable. As shown in Figure 3-6, the solubility of calcite decreases with temperature, whereas that of silica increases. *The decrease of solubility with an increase in temperature*, as exhibited by calcite, has been named **retrograde solubility**. *The increase of solubility with increase in temperature*, as in silica, is **normal solubility**.

EFFECT OF PRESSURE ON CALCITE AND SILICA

In general, the solubility of a solid in water increases as pressure increases. This relationship is known as the Riecke principle. The effects of this phenomenon are

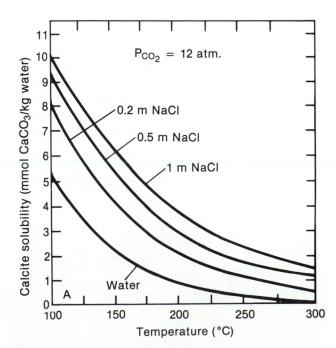

FIGURE 3-6.   Contrasting relationships of solubility of calcite and forms of silica in waters as a function of temperature with calcite and forms of silica.

A. Solubility of calcite in water and in three saline solutions at a partial pressure of carbon dioxide of 12 atm.

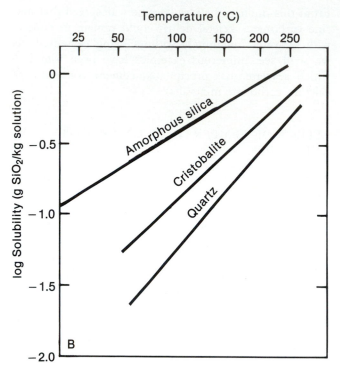

Temperature (°C)

FIGURE 3-6.   (*Continued*)
  B. Comparison of the solubilities of amorphous silica, cristo-balite, and quartz in water in the temperature range of 25°C to 250°C. (W. S. Fyfe, N. J. Price, and A. B. Thompson, 1978; A, fig. 4-13, p. 71; B, fig. 4-16, p. 75.)

commonly termed pressure dissolution. This kind of dissolution commonly yields a kind of zigzag or sutured contact where one solid has interpenetrated another. An example is **stylolites**, which consist of *sutured surfaces along which nonsoluble material has accumulated to build a residue* (Figure 3-7). Stylolites are common in carbonate rocks and sandstones and have even been found in granites.

SOURCE: Finkel and Wilkinson, 1990.

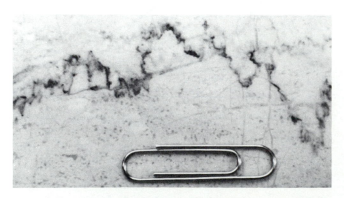

FIGURE 3-7.   Stylolite seam viewed on polished slab of limestone cut normal to stratification. (J. E. Sanders, 1981, fig. 10.24, p. 255.)

EFFECT OF pH ON CALCITE AND SILICA

Calcium carbonate and silica respond to pH in opposite senses. At low levels of pH, calcium carbonate dissolves and silica is stable. At high levels of pH, calcium carbonate is stable and silica dissolves (Figure 3-8). Thus the opposite relationships of calcite and silica with pH match those of calcite and silica with temperature.

During photosynthesis, aquatic plants withdraw CO_2 from the surrounding water and thus raise the pH. In reducing sulfate, anaerobic bacteria may raise the pH to values of 9.5 or higher. Although such increases in pH are induced biologically, the resulting reactions, such as the precipitation of calcium carbonate or dissolution of silica, probably should be classified as being *biologically mediated* inorganic chemical reactions.

The precipitation of silica as chert is another example of the control by pH of a natural reaction. In the waters of the Coorong, a lagoon in South Australia, at pH values greater than 10, caused by algal photosynthesis, quartz particles are dissolved. When the pH drops, the silica

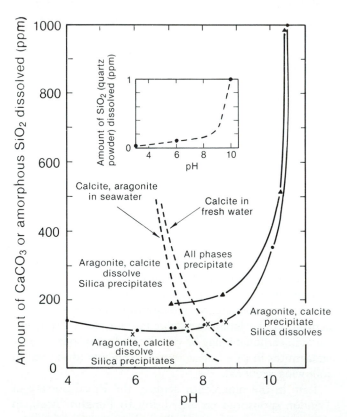

FIGURE 3-8.   Effect of pH at approximately 25°C on the solubilities of calcium carbonate, quartz, and amorphous silica. (Data points from G. B. Alexander, W. M. Heston, and R. K. Iler, 1954, dots; K. B. Krauskopf, 1956, crosses; and G. Okamoto, O. Takeshi, and G. Katsumi, 1957, triangles. Dashed lines for aragonite and calcite from C. W. Correns, 1950; for quartz, from A. Heydemann, 1966.) (G. M. Friedman; A. J. Amiel; and N. Schneidermann, 1974, fig. 18, p. 823; G. M. Friedman, 1975a, fig. 18, p. 391.)

is precipitated in the form of crystallites of cristobalite. At some later time this cristobalite will probably become chert. (See Chapter 4.)

The contrasting behavior of calcite and silica with respect to pH offers an explanation for a common petrographic observation: quartz replaces calcite or vice versa.

Early Diagenesis: Cements Precipitated in Near-Surface Environments

The Vadose- and Phreatic Zones

The zone above the ground-water table is the **vadose zone** (Figure 3-9). Most of the time, the pore space between the particles in this zone is filled with air, except near particle-to-particle boundaries, where surface tension maintains a film of water between the particles. This film of water is called a meniscus. In the vadose zone the pores generally remain open. Where undersaturated meteoric waters pass through the sediment and rock in the vadose zone, calcium carbonate commonly is dissolved.

However, supersaturated waters may pass through the vadose zone, possibly the same waters that earlier dissolved calcium carbonate elsewhere in the vadose zone and thus became saturated with respect to calcium carbonate. CO_2 outgassing could be the trigger that renders these waters supersaturated and initiates precipitation of carbonate cement in the vadose zone. Uptake of CO_2 by plants can exert the same effect. In the presence of supersaturated waters, *a partial cement consisting of equant, anhedral crystals grows in the thin films of water that persist near particle-to-particle boundaries*. Such partial cement is **meniscus cement** (Figure 3-10). Meniscus cement may grow until the crystals reach the surface of the water film. In such a case, the body of cement acquires a distinctive configuration that defines the original curving surface of the film of water and results in pores having rounded outlines.

In arid climatic zones rainfall is sparse, but when it rains, it pours! In brief desert storms the waters are undersaturated with respect to calcium carbonate. These fast-moving undersaturated waters pass rapidly through carbonate sediments; their effect is to dissolve calcium

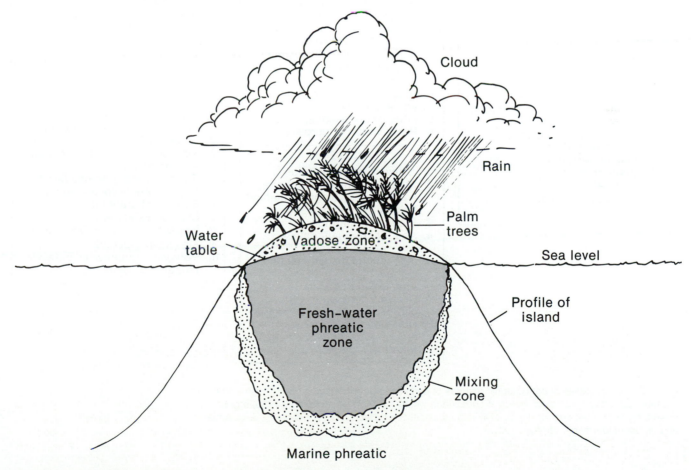

FIGURE 3-9.   Schematic profile through a steep-sided island, such as an emergent reef, showing the distribution of the subsurface waters into vadose-, fresh-water phreatic-, mixing-, and marine phreatic zones. (Authors.)

Fresh-water vadose environment

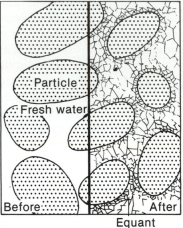

Meniscus water film

Zone of Dissolution

1. Dissolution by undersaturated meteoric water
2. Production of CO_2 in soil zone aiding solution
3. Extensive dissolution
4. Preferential removal of aragonite if present
5. Formation of vugs in limestone

Zone of Precipitation

1. Meniscus or pendant distribution of water
2. CO_2 loss or evaporation
3. Minor cementation
4. Meniscus cements
5. Pendant cements
6. Equant calcite
7. Preservation of most porosity

A

Fresh-water phreatic environment

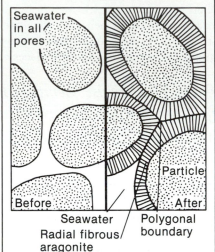

Equant
calcite cement
Crystal boundaries are generally
more irregular than shown

Zone of Dissolution

1. Dissolution by undersaturated meteoric water
2. Development of moldic- and/or vuggy porosity

Stagnant Zone

1. Little or no water movement
2. Water saturated with $CaCO_3$
3. Little cement
4. Stabilization of Mg-calcite and aragonite
5. Little or no leaching
6. Preservation of porosity

Active Zone

1. Active water circulation
2. Some leaching of aragonite; leaching may be accompanied by calcite replacement
3. Rapid cementation
4. Abundant equant calcite cement
5. Isopachous bladed calcite cement
6. Interlocking crystals
7. Crystals coarsen toward center of pores (drusy mosaic)
8. Complete replacement of aragonite by equant calcite
9. Relatively low porosity

B

Marine phreatic environment

Seawater in all pores

Particle

Before   After

Seawater   Polygonal
Radial fibrous/   boundary
aragonite

Stagnant Zone

1. Little or no water circulation through sediment
2. Bacterial (?) control on cementation
3. Water saturated with $CaCO_3$
4. Little cement except in skeletal micropores
5. No leaching
6. No alteration of particles

Active Zone

1. Water forced through sediments by waves, tides, or currents
2. All pores filled with seawater
3. No leaching in shallow marine environments
4. Random aragonite needles
5. Isopachous fibrous aragonite
6. Botryoidal aragonite
7. Cryptocrystalline calcite
8. Isopachous fibrous Mg-calcite
9. Mg-calcite peloids
10. Polygonal boundaries between isopachous cements

C

FIGURE 3-10. Sketches of petrographic textures of cements precipitated in vadose-, fresh-water phreatic-, and marine phreatic zones of subsurface waters. (After M. W. Longman, 1980, fig. 2, p. 464; fig. 6, p. 468; and fig. 13, p. 474; with modifications based on G. M. Friedman; A. J. Amiel; and N. Schneidermann, 1974.)

carbonate. Secondary pores may develop as aragonite of ooids or of skeletal biocrystals is leached to form molds. On a large scale, such waters may dissolve masses of solid carbonate rock to form caves.

The zone below the ground-water table is the **phreatic zone.** (See Figure 3-9.) If impermeable strata prevent the waters in the phreatic zone from circulating, these waters attain saturation and start to precipitate calcite. Phreatic calcite cement crystals are equant, commonly rhombic, and may surround some particles with **isopachous coatings** (defined as *mineral coatings, on a particle, whose thickness is uniform*; Figure 3-11). The remainder of the pore space typically fills with a drusy mosaic in which crystals coarsen toward the center of each pore. (See Figure 3-10, B and Figure 3-11.) In general, in the phreatic zone cement is both more ubiquitous and more uniformly distributed than in the vadose zone. Ultimately, original pores may become plugged with drusy calcite mosaic and the limestone may become so tight that it will no longer transmit fluids. If, however, the phreatic waters are able to circulate, then they may remain undersaturated with respect to calcium carbonate and dissolve carbonate and, like the waters of the vadose zone, create further pore space. Because undersaturated phreatic waters tend to become saturated by dissolving carbonate material, the

boundary between vadose- and phreatic zones tends to be marked by a system of caverns, formed when undersaturated waters entered the phreatic zone and dissolved calcium carbonate to become saturated. Deeper into the phreatic zone, cementation is common.

SOURCE: Goter and Friedman, 1987.

Caliche and Silcrete: Evaporitic Crusts and -Cements

In arid and semiarid regions, where evaporation exceeds rainfall, the predominant direction of movement of soil moisture is upward. Vadose waters carrying CO_2 dissolve calcium compounds from the soil particles. By capillary attraction, these interstitial waters are drawn toward the surface, where they evaporate. As evaporation increases the salinity of these interstitial waters, they may precipitate calcite and various other minerals as surface- or near-surface crusts or as cements between the soil particles. *Surface- or near-surface crusts composed of calcite* are known as **caliche, nari,** or **calcrete.** Such *crusts composed of silica* are **silcretes.**

Calcretes are characterized by a host of peculiar features that help to identify them, including **tepee structures,** *inverted-V-shaped ridges formed where expanding sheets of calcrete meet and are forced upward by the pressure of crystallization.* Commonly, tepees outline large polygons that are readily visible from the air (Figure 3-12).

SOURCES: Füchtbauer and Valeton, 1988; S. E. Phillips and Self, 1987; Tucker and V. P. Wright, 1990; Warren, 1983.

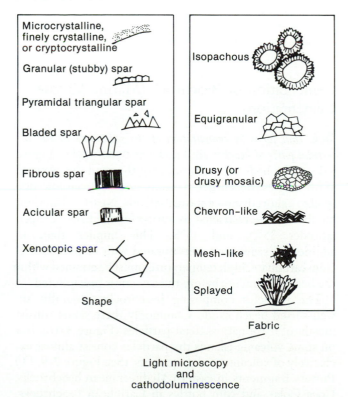

Shape

Fabric

Light microscopy and cathodoluminescence

FIGURE 3-11.   Classification of cement textures. (After M. Savard and P.-A. Bourque, 1989, table 1, p. 793, with modifications by present authors.)

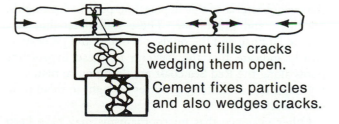

A   Thermal contraction at night forms cracks

Sediment fills cracks wedging them open.

Cement fixes particles and also wedges cracks.

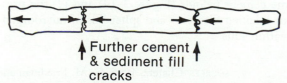

B Thermal expansion during the day causes the crust to arch up and overthrust

↑Further cement↑ & sediment fill cracks

FIGURE 3-12.   Schematic profiles showing formation of tepee structures. (C. G. St. C. Kendall and J. K. Warren, 1987, fig. 3, p. 1012.)

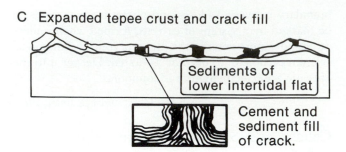

C Expanded tepee crust and crack fill

Sediments of
lower intertidal flat

Cement and
sediment fill
of crack.

FIGURE 3-12.   *(Continued)*

Biochemical Precipitation in Microbial Mats

Cyanobacteria are important factors in precipitating carbonate cement. A particularly instructive example in which a self-evident relationship exists between cyanobacterial mats and carbonates is found in the mats that form continuous carpets at the bottoms of hypersaline pools along the Red Sea coast. Commonly, the sticky mucilaginous sheaths of such mats trap samples of whatever sediment particles are present in the overlying water. In this area the only available extraneous sediment consists of intrabasinal carbonate biocrystalline skeletal particles from the Red Sea or bits of extrabasinal terrigenous debris from the Precambrian bedrock. However, among the solids associated with the cyanobacteria, only a few consist of these extraneous particles.

In sectional view, the cyanobacterial deposit contains alternating millimeter-thin laminae of crustlike calcium carbonate and soft organic matter (Figure 3-13). The carbonate minerals in these paper-thin carbonate laminae include both high-magnesian calcite (predominant) and aragonite. Other carbonate particles associated with the cyanobacteria of this hypersaline pool include ooids, oncolites, and grapestone. These millimeter-thin carbonate laminae and the ooids, oncolites, and grapestone clearly indicate that the bacteria in these hypersaline pools along the Red Sea coast are doing more than simply trapping adventitious particles brought to them from elsewhere.

Other evidence that microorganisms may have been factors in precipitating calcium carbonate includes (1) the presence of bacterial cells in the high-magnesian calcite of some cement peloids (Figure 3-14) and (2) the presence of calcified filaments and spheroids of microorganisms within soils.

SOURCES: Chafetz, 1986; G. M. Friedman and Krumbein, 1985; G. M. Friedman, Sneh, and Owen, 1985.

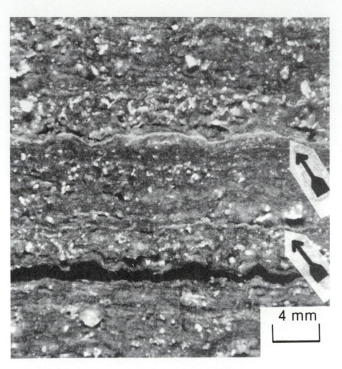

4 mm

FIGURE 3-13.   Vertical section through core that penetrated cyanobacterial mats composed of alternating laminae of soft microbial cells and millimeter-thin laminae composed of high-magnesian calcite and aragonite (two of which are indicated by arrows). Scattered through the soft microbial material are various sediment particles (ooids, oncolites, and grapestones). Sea-marginal hypersaline pool, Gulf of Aqaba (Elat), Red Sea. (G. M. Friedman; A. J. Amiel; M. Braun; and D. S. Miller, 1973, fig. 9, p. 550.)

Cementation of Beachrock (Marine Vadose Cementation)

Rocks that form by cementation of the sediments in the intertidal parts of beaches are known as **beachrocks** (Figure 3-15). The beds of beachrocks are thin and dip seaward at low angles. Beachrock typically extends from low-tide level to a short distance above high tide. Nearly all modern beachrocks are restricted to warm climatic belts between latitudes 35°N and 35°S. This implies that, in addition to seawater supersaturated with respect to calcium carbonate, high temperatures must be attained within the beach sediments before the cement is precipitated.

The particles composing beachrocks generally are well-sorted beach sands. Commonly these sands consist mostly of carbonate skeletal particles (Figure 3-16), but on some volcanic islands the particles consist almost exclusively of volcanic rock fragments. (See Figure 2-9, D.) Pottery fragments in eastern Mediterranean beachrocks, Coca Cola- and rum bottles in Caribbean beachrocks, debris from World War II in Pacific beachrocks, and even-younger materials in Hawaiian beachrocks affirm

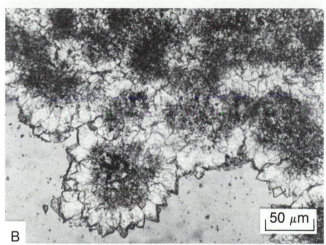

FIGURE 3-15.   Modern beachrock exposed at low tide (center of view), Mediterranean coast of Libya; ruins of Roman temple in foreground. (Authors.)

FIGURE 3-14.   Peloids resulting from bacterially induced precipitation of carbonate.

A. Much-magnified view of marine peloid in a cavity within a coral head, as seen in scanning-electron micrograph. A fossil bacterial clump is surrounded by a rim of euhedral crystals of high-magnesian calcite (corroded during sample preparation). Specimen from Holocene-Pleistocene reef tract, southern Florida. (H. S. Chafetz, 1986, fig. 2F, p. 814.)

B. Silt-size nonmarine peloids in travertine, enlarged view of thin section seen in a photomicrograph. These peloids are composed of bacterial remains embedded in single crystals of calcite that are surrounded by rims of calcite rhombohedra. These nonmarine travertine peloids closely resemble marine peloids (such as that illustrated in A), which likewise contain bacterial remains. (H. S. Chafetz, 1986, fig. 3D, p. 815.)

In May, 1919, with the help of the staff of the Marine Laboratory at Tortugas (Florida), a large cask was filled with shelf detritus, dug up at a point offshore from the laboratory. The sides and ends of the cask were perforated, to permit of the easy flow of seawater (sic) through the mass, and the whole was buried at the proper depth in the beach. It was hoped that, after a year or two, the cask might be opened, with the object of seeing whether its contents had become

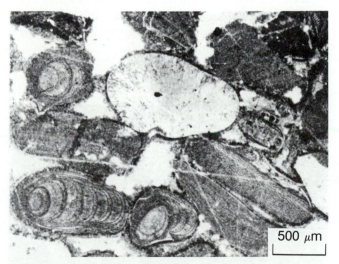

FIGURE 3-16.   Enlarged view of beachrock composed of particles consisting of two kinds of carbonate biocrystals: coralline algae (semiopaque, with fine cellular pattern) and mollusk (transparent, above center); petrographic thin section in plane-polarized light. Cement consists of (1) finely crystalline high-magnesian calcite (dark, opaque thin rim adjacent to particles) and (2) aragonite (tiny, partly transparent, thin, discontinuous layer projecting from cryptocrystalline rims into interparticle voids). Nahariya, Mediterranean coast, Israel. (B. Buchbinder, 1977.)

that cementation in these beachrocks is occurring at the present time.

The rate at which loose particles can be cemented into beachrock is very rapid. R. A. Daly (1871–1957) attempted an *in-situ* experiment to determine the cementation rate. The experiment was not a success; however, a rapid rate was proved, as Daly explained vividly:

in any degree cemented together. Unfortunately, the hurricane of September, 1919, tore up the cask and carried it several hundred meters along the key. There left exposed to the air, this material cannot be used as a test. The experiment should be duplicated. On the other hand, this hurricane itself improved upon an artificial experiment. Like that of 1910, it threw up on the beach a large quantity of shelf sand, which, a year afterwards, was found by Dr. [A. G.] Mayor to have been hardened into beach-rock (sic).

The size-frequency distribution of the particles in most beachrocks is usually the same as that of nearby beach sand. However, not all beachrocks consist of sand-size particles. A few beach gravels consisting of limestone fragments have become cemented to form coarse beachrock. One of the first geological accounts of coarse beachrock is that of Charles Darwin (1809–1882) in his book *Geological Observations on the Volcanic Islands and Parts of South America Visited during the Voyage of HMS Beagle*, in which he described his observations of 1832 in the Cape Verde chain. Under the heading of "Recent Conglomerate" he noted:

I found fragments of brick, bolts of iron, pebbles, and large fragments of basalt, united by a scanty base of impure calcareous matter into a firm conglomerate. To show how exceedingly firm this recent conglomerate is, I may mention, that I endeavored with a heavy geological hammer to knock out a thick bolt of iron, which was embedded a little above lowwater mark, but was quite unable to succeed.

The chief mineral cements in beachrocks are aragonite and high-magnesian calcite, either singly or in combination. Of the two, aragonite, which is commonly fibrous, is more abundant. High-magnesian calcite is cryptocrystalline or, less commonly, bladed.

The origin of the mineral cements in beachrocks has been a subject of great interest for many years. *Cyanobacteria* and algae are almost certainly involved, not only in stabilizing the beach sediment but also in raising the pH by photosynthesis to a level at which calcium carbonate is precipitated. Levels of pH high enough to precipitate calcium carbonate have actually been measured in modern beachrock. Biologically controlled $CaCO_3$ precipitation (principally of aragonite) by pH changes has been demonstrated in laboratory experiments using a variety of bacterial strains. Although the pH in the waters of beachrocks rises to 9.5 during the day, calcium carbonate is precipitated at night after the waters of high pH have drained into the sediment. Thus organically mediated cementation in beachrock probably occurs extracellularly (as it does in laboratory experiments involving calcium-carbonate precipitation by bacteria), as a result of fluctuations in the concentration of CO_2 and HCO_3^- that are controlled organically. Films of nonliving organic matter coating mineral particles may also be involved in chemical- or biochemical processes that lead to cementation.

In some beachrocks, high-magnesian calcite or aragonite have replaced quartz, chert, and feldspar particles (Figures 3-17 and 3-18). At a pH level of 9 or 10, silica is dissolved and calcite or aragonite may be precipitated. (See Figure 3-8.)

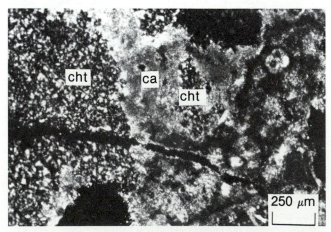

FIGURE 3-17. Enlarged view of modern beachrock in which a calcium-carbonate mineral (aragonite or high-magnesian calcite, ca) has replaced a rock fragment composed of chert (cht) as seen in petrographic thin section in crossed-polarized light. Specimen from Sinai Peninsula, Red Sea. (M. H. Helsinger, 1973, and authors.)

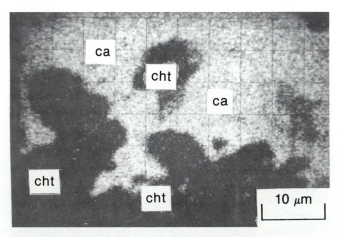

FIGURE 3-18. Very highly magnified image showing chert rock fragment (dark, labeled cht) that has been partially replaced by calcium-carbonate mineral (light color, labeled ca). Backscattered image from electron-microprobe in which white dots mark locations of calcium atoms. (M. H. Helsinger, 1973, and authors.)

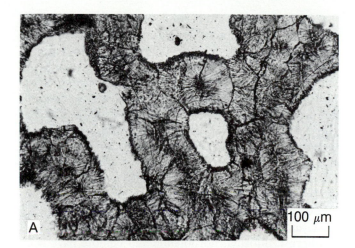

FIGURE 3-19.   Enlarged views of corals, showing progressive cementation; petrographic thin sections in plane-polarized light of specimens from same locality, Sinai Peninsula, Red Sea. (G. M. Friedman; A. J. Amiel; and N. Schneidermann, 1974, figs. 2, 3, and 5, p. 817 and 818.)

A. Specimen of living coral lacks cement; white areas are void spaces.

B. Sample from 10 cm below surface of reef displays rim cement of aragonite crystals (remaining void spaces appear as white).

Cementation of Reefs (Marine Phreatic Cementation)

Although by secreting a rigid calcium-carbonate framework, reef-building organisms form solid carbonate rock directly, nevertheless, within their rigid frameworks they leave much initial pore space. As sediment infiltrates from above, hole-dwelling organisms construct their skeletons, and cement is precipitated, these pore spaces within the reef become progressively filled. Cementation begins in the dead parts of the reef only a short distance beneath the living part. In thin section under the petrographic microscope this cement is one of three kinds: (1) fibrous, composed of aragonite or of high-magnesian cal-

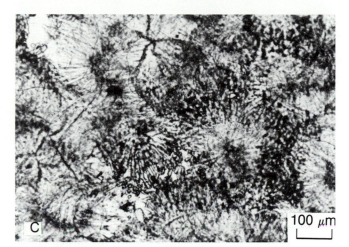

FIGURE 3-19.   (*Continued*)

C. Sample from 70 cm below surface of reef in which aragonite cement has virtually obliterated all void space.

cite (Figures 3-19, A and 3-20); (2) microcrystalline ("micritic" or "micrite" cement), usually high-magnesian calcite (See Figure 3-16.); and (3) minute spherical masses composed of crystals of high-magnesian calcite.

As with beachrock, a possible clue to the origin of cement in reefs may be contained in the evidence that high-magnesian calcite or aragonite have replaced quartz

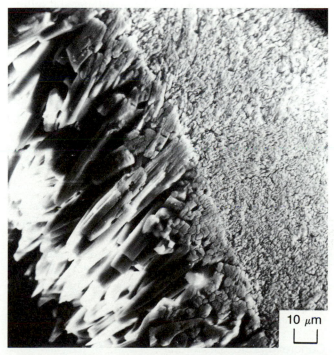

FIGURE 3-20.   Very highly magnified view of part of coral biocrystal (right) to which aragonite crystals forming a rim cement have been attached (left); scanning-electron micrograph of specimen from Sinai Peninsula, Red Sea. (G. Gvirtzman and G. M. Friedman, 1977.)

particles. A high level of pH (9 to 10), probably resulting from photosynthesis, has been measured in reefs of the Red Sea. Here, also, etched, corroded, and partially replaced quartz particles have been found. In reefs, photosynthesis and respiration of the biomass cause a shift in the bicarbonate buffer system of seawater with the uptake of CO_2. In this process, microlevels of pH 10 and even 10.5 may be reached and maintained in thin jell-like or monomolecular layers. At such high pH levels, silica dissolves and calcite or aragonite may be precipitated. (See Figure 3-8.) Such loss of carbon dioxide would promote the precipitation of calcite.

Of interest is the relationship that in Red Sea reefs the amount of fibrous aragonite cement increases progressively downward with distance from the live corals at the surface of the reef. Within less than 60 cm of the surface, pore space has been almost entirely eliminated. (See Figure 3-19.) This implies that the longer the time of exposure to the interstitial fluids, presumably seawater, the greater the amount of cement. Such a relationship is diametrically opposite to the older observation that some carbonate sediments appear to remain exposed to seawater indefinitely without becoming cemented.

More-recent studies of the interiors of coral reefs have complicated this concept that reefs are progressively cemented by the seawater within them. The pH of pore waters in the interiors of many reefs are low. The waters are anoxic, rich in dissolved organic matter and inorganic nutrients, and high in methane, sulfate, and sulfide. Waters having such properties are considered to be incapable of precipitating $CaCO_3$ cement. What is more, net Ca^{2+} flux from some reefs suggests that within the reefs calcium carbonate may be dissolving.

Diurnal pH fluctuations associated with daytime photosynthesis and nighttime respiration in the outer parts of reefs and in cyanobacterial mats suggest that the chemical composition of the pore water within reefs may vary cyclically. During some parts of the chemical cycle, $CaCO_3$ may be precipitated. During other parts of the cycle, just the reverse may take place.

SOURCES: Buczynski and Chafetz, 1990; Epstein and G. M. Friedman, 1982; G. M. Friedman and Foner, 1982; Guo and G. M. Friedman, 1990; MacIntyre, 1985; Sansone and others, 1988; Schroeder and Purser, 1986; Scoffin, 1987; Tucker and V. P. Wright, 1990; K. R. Walker and others, 1990; Zhong and Mucci, 1989.

Concretions

Concretions are *subspherical-, oblate-, or irregular solids formed by chemical precipitation from aqueous solution around a nucleus: an object, buried in sediment, such as a leaf, bone, pebble, or an entire organism, such as a fish.*

Concretions provide a good way to check on possible effects of intrastratal dissolution of framework particles, such as the heavy minerals of sands, by comparing heavy minerals inside concretions with those outside the concretions. The calcite cement forming the concretions probably was precipitated very soon after the sediment had been deposited. Once any early formed calcite has encased them, heavy-mineral species that typically are dissolved in alkaline solutions will be preserved. Outside the concretions, alkaline solutions will destroy many heavy minerals, but, because alkaline solutions do not dissolve calcite (See Figure 3-8.), those inside the concretion are protected.

Within various concretions in Miocene sandstones of southern California, hornblende and epidote have been found. Outside the concretions, these two minerals have been practically obliterated. The abundance of apatite in the sandstone outside the concretions is further proof that alkaline intrastratal solutions dissolved the hornblende and epidote but left untouched the calcite concretions and their internal contents. Like calcite, apatite dissolves readily in acid solutions.

SOURCES: Gautier, 1982; H. Irwin, 1980; Mozley, 1989; Pearson, 1985; Raiswell, 1988; Savrda and Bottjer, 1988; Scoffin, 1987; Weeks, 1953, 1957.

Hardgrounds

Areas of the sea floor where the sediments have become cemented are called **hardgrounds**. Ancient hardgrounds in the rock record can be identified by criteria indicating that the cement was precipitated before the layer of sediment was buried by other sediment. These criteria include borings; attached organisms, such as oysters or crinoid holdfasts; evidence that shells lying on the former sea floor were eroded; and thin crusts of manganese oxides, for example, which blacken the surface. In the future, hardgrounds may also be recognized by the human artifacts that have been cemented along with the natural particles.

Hardgrounds are geologically significant for two reasons: (1) They constitute evidence of slow- or zero deposition in the marine environment, and (2) they are relatively impermeable and hence form permeability barriers, or seals, to the vertical movement of hydrocarbons or of pore waters. Modern examples of hardgrounds include the top of Atlantis Seamount in the Atlantic Ocean, parts of the Persian Gulf, and steep escarpments in the Bahamas.

On Atlantis Seamount, part of the mid-Atlantic ridge, at a depth of 300 m, a cryptocrystalline cement composed of high-magnesian calcite not only has bound together tests of planktonic- and benthic organisms, most of them gastropods and foraminifers, but also has replaced some of the aragonite biocrystals. The cement crystallized between 9,000 and 12,000 years ago. Since that time, no sediment has accumulated on top of the seamount.

In large areas on the floor of the Persian Gulf where the depth of water is only 2 to 3 m and where new sediments are accumulating only very slowly, modern marine carbonate sediments have been cemented by fibrous aragonite. The cement binds not only the carbonate particles, but also artifacts, including pottery, glass, and iron bolts thrown overboard from ships off the end of a jetty that is only a few decades old.

The precipitation of the fibrous aragonite is thought to have been controlled by two factors, both of which can influence the flow of water, and possibly the diffusion of ions, through sediment: (1) rate of sedimentation and (2) particle size.

The cement is most abundant at the water/sediment interface and decreases downward. None is present at subbottom depths ranging between 10 and 15 cm. As more and more cement plugs the pores, less and less interstitial water can circulate. This provides an automatic limit to the depth of cementation. Therefore, a cemented layer thicker than 15 cm can form only where the rate of addition of new sediment does not exceed the rate at which cement is precipitated. If the rate of sedimentation increases, the amount of sediment that becomes cemented will be reduced. If sediment is deposited fast enough, the effect is to prevent any cement from being precipitated.

SOURCES: Scoffin, 1987; Tucker and Wright, 1990.

Subsurface Environments and Diagenesis; Subsurface-Precipitated Cements

As strata are buried to progressively greater depths, the materials composing them are subjected to higher temperatures, to greater pressures, and to brines of varying salinities and other chemical conditions. Two processes, discussed in following sections, effect the consolidation (or lithification) of sediments. These processes are cementation and compaction. Compaction reduces bulk volume, and cementation bonds particles together by filling the gaps between them with new material. The resultant rock is commonly both less voluminous and more rigid than its precursor sediment.

Geothermal Gradient

The increase of temperature downward in the Earth is known as the **geothermal gradient**. The value of this gradient depends on the amount of heat that is moving upward

from the interior of the Earth to the surface and on the thermal conductivity values of the strata. In holes drilled for petroleum in the United States the average geothermal gradient has been found to be about 3.6°C/100 m (2°F/100 ft). For example, at the bottom of a typical oil well in Oklahoma that is 5000 ft (1524 m) deep, the temperature is 160°F (71°C). This temperature results from the mean annual air temperature of 60°F (15.5°C) plus 100°F (55.5°C) increase on a gradient of 2°F/100 ft (3.6°C/100 m). Geothermal- and paleogeothermal gradients vary for different basins (Figure 3-21).

Subsurface Pressures

COMPACTION

With progressively deeper burial, sediments are subjected to ever-greater load from the weight of the overlying strata. This load induces several effects, notably a mechanical volume strain, a reduction in volume in re-

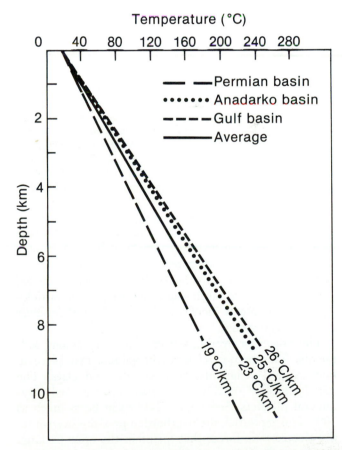

FIGURE 3-21. Graph of temperature versus depth, on which are plotted the inferred average paleogeothermal gradients for three major North American basins and their average. (G. M. Friedman, 1987a, fig. 7, p. 83.)

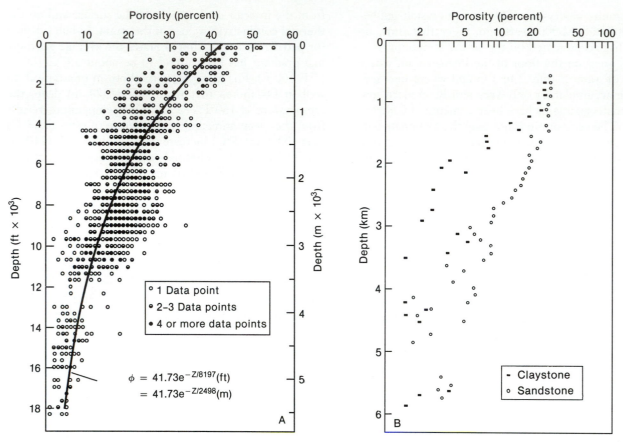

FIGURE 3-22.   Graphs of porosity vs. depth.

A. Graph of porosity vs. depth for specimens of carbonate sediments and -rocks measured in 15 wells in south Florida. Curve based on least-squares exponential fit to all data. (J. W. Schmoker and R. B. Halley, 1982; P. A. Scholle and R. B. Halley, 1985, fig. 3, p. 314.)

B. Graph of porosity vs. depth for sandstone (circles) and claystone (rectangles). (Szalay and Koncz, 1980.)

sponse to stresses imposed by the external load (from burial or from deformation). As the volume decreases, so does the porosity (Figure 3-22). Simultaneously, the density increases (Figure 3-23). The particles are forced into denser-packed fabrics, and the pressure at particle-to-particle contacts increases. The particles themselves may undergo strain.

The effects of overburden pressure vary among sediments. The greatest changes in volume take place in fine-textured sediments, such as silts and clays. The initial porosity of tiny (<2 μm) open-packed clay-mineral platelets (See Figure 2-37.) can be as much as 70%. During burial, the overburden pressure overcomes the electrostatic forces that maintain the open packing. Compaction of the fine-textured sediments depends on the rate at which water can be expelled. Finally, the mechanical readjustment of the platelets is such that they

become preferentially oriented parallel to one another. (See Figure 2-34.) Interestingly, clay-mineral platelets attain such parallel orientation at a very early stage of compaction, at pressures near 1 kg/cm^2.

The porosity of both clays and sands diminishes as depth increases, but the relationships are not the same. (See Figure 3-22, B.) The greatest divergence is between porosity values of about 20% and 2 to 5% and in the depth range of 1 to 3 km. The claystone points reach 5% porosity at a depth of about 2 km. The porosity of the sandstones reaches the 5% value at a depth of 3 km.

Insights into the behavior under load of oolitic limestones come from laboratory experiments in which modern ooids were subjected to elevated pressures and -temperatures over periods of several days. At the end of the experiments, the artificially compacted ooids displayed many of the textural features common in oolitic

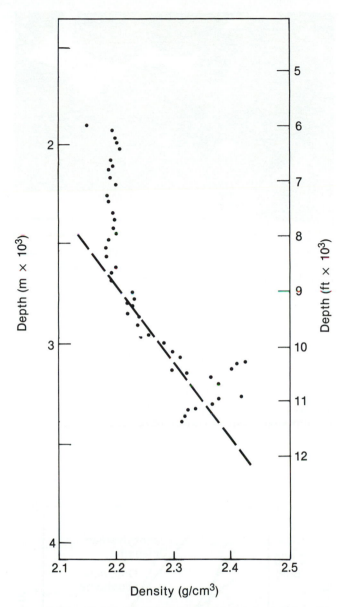

FIGURE 3-23.   Graph of rock density vs. depth, based on specimens measured from various borings. (After D. P. Helander, 1983, fig. 8-52, p. 197; as modified on the basis of W. H. Fertl and D. J. Timko, 1970.)

These experimental results suggest that increased pressure and -temperature accompanying burial of carbonate sediments bring about many changes. Particles are compacted mechanically. Particles are rearranged, deformed, and broken. Particles display the effects of pressure dissolution (adjacent particles interpenetrate each other along sutured and concavo-convex contacts). Aragonite recrystallizes to calcite.

The factors that influence compaction in sands are mostly the shapes and sorting of particles in the sand and the depth of burial. During compaction, quartz particles respond by shifting into more-dense packing arrangements; hence porosity decreases. Angular, low-sphericity, and poorly sorted sands are more compressible than rounded, high-sphericity, and well-sorted sands. However, sands with high proportions of unstable rock fragments or of shale clasts compact more quickly than quartzose sands because the rock fragments and shale clasts deform readily.

The thickness of sedimentary sequences may be reduced where soluble strata, especially limestone or rock salt, in contact with water are dissolved as a result of enhanced solubility through increased pressure.

SOURCES: Bhattacharyya and G. M. Friedman, 1979, 1983b.

Subsurface Waters

Below the water table, fluids occupy all the pores in sedimentary strata. Exceptionally, oil and gas may be present. But, apart from such hydrocarbons, water is universally present. This pore water contains ions in solution in varying concentrations. Except for the shallow subsurface setting, where water may be derived from rainfall, rivers, lakes, and related sources of fresh-water runoff, the bulk of the subsurface water comes from seawater. Burial changes the chemical composition of this seawater. With progressive burial, the salinity of *subsurface waters trapped in the pores of sedimentary strata that have been out of contact with the atmosphere for an appreciable part of a geologic period*, known as **formation waters**, increases (Figure 3-25). The salinity values of formation waters in sands and sandstones, which may become ten times greater than that of seawater, exceed those of shales, which may be only half that of the contiguous coarser layers. The processes causing this increase in salinity with depth are not well understood. Processes mentioned in the geologic literature include the following: (1) Subsurface strata behave as semipermeable membranes, and (2) during compaction from muds to shales, the latter extrude salt from their larger pores; thus salt is lost from shales and goes to the formation waters of the adjacent sands and sandstones. Formation waters

limestones (Figure 3-24). Their packing was reorganized. Some ooids deformed plastically (Figure 3-24, A). Others fractured. The laminae of some buckled (Figure 3-24, B). Sutured- and concavo-convex- and linear contacts between ooids, showing truncation of ooid laminae, formed as well. In addition, aragonitic ooids recrystallized to anhedral calcite mosaics. This recrystallization, during which the gross concentric laminae were preserved, was found to be a function of temperature.

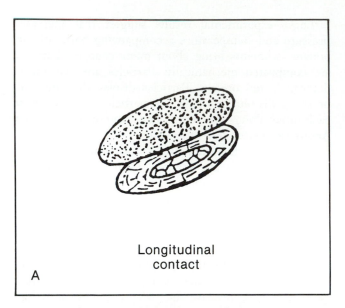

Longitudinal
contact

A

B   100 µm

FIGURE 3-24.   Views showing compaction of ooids.
A. Sketch of plastically deformed ooids with longitudinal contact. Concavo-convex contacts between ooids are also common,
and they may be sutured, as in Figure 3-24, C. (A. Bhattacharyya
and G. M. Friedman, 1984, fig. 1, C)
B. Photomicrograph of buckled ooid. (Authors.)

can be usefully visualized as a kind of "universal ocean."
As the salinity increases in these subsurface brines, oxygen
becomes excluded. (See Figure 3-4.) When oxygen disappears, sulfate-reducing bacteria become exceedingly active
and degrade the available $CaSO_4$ and precipitate calcite
(Eq. 3-10). Although formation waters are highly concentrated brines, their contents of dissolved sulfate are not as
high as one might expect. The bacterial destruction of sulfate is responsible. This biochemical reaction involving loss
of sulfate and corresponding precipitation of calcite may
provide a key answer to the perennial question of where calcite cement comes from.

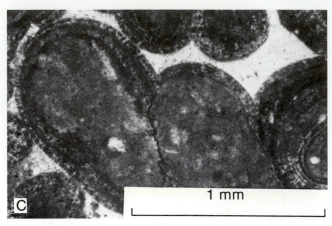

C   1 mm

FIGURE 3-24.   (*Continued*)
C. Photomicrograph showing concavo-convex sutured contact
between ooids. This kind of contact results from pressure dissolution. Middle Jurassic (Dogger), Paris Basin, France. (R. Cussey
and G. M. Friedman, 1977, fig. 6.)

In the deep subsurface, minerals can be preserved
indefinitely that readily dissolve when they come into
contact with the near-surface zone of fresh water. For example, halite is a widespread constituent of deeply buried
marine strata, but is almost never seen at the surface.
Instead of the halite that characterizes the subsurface

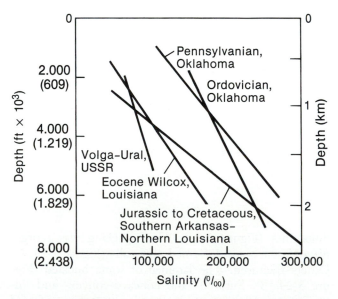

FIGURE 3-25.   Graph of salinity versus depth shows that the concentration of dissolved solids in formation waters increases with
depth, based on samples collected from sandstones penetrated
in various boreholes. (J. M. Hunt, 1979, fig. 6-3, p. 194, after P. A.
Dickey, 1969.)

FIGURE 3-26.   View of collapse breccia in dissolution pit. Angular clasts of dolostone and limestone formed when evaporites underlying continuous strata of carbonate rocks were dissolved and the strata collapsed. Lower Ordovician, Comstock, New York. (S. J. Mazzullo, 1978.)

realm, one finds only collapse breccias (Figure 3-26) at the surface. The example of halite, present down below but absent at the surface, is only one of many kinds of contrasts between exposed strata and deeply buried strata.

SOURCE: Dickey, 1969.

Changes of Organic Matter with Depth

Because certain organic compounds are extremely sensitive to temperature, we can use them as organic geothermometers. For example, crude oils contain porphyrins. In the laboratory, porphyrins have been found to break down at 200°C. From this finding we infer that crude oils containing porphyrins have never been subjected to temperatures in excess of 200°C.

As "raw" organic matter is buried to progressively greater depths it undergoes changes, first to kerogen, then to crude oil, and finally (in extreme cases) to natural gas. The evolutionary series is controlled by both time and temperature. In other words, increasing temperatures may cause the maturation of organic matter, or long-term application of moderately high temperatures can exert the same effect. In the Pliocene and Miocene shales of the Los Angeles basin, California, the composition and structure of the solvent-soluble hydrocarbons extracted from samples of the shales do not resemble petroleum at depths less than 8000 ft (2425 m). Thus these shallow hydrocarbons are said to be immature. The change from immature- to petroleum hydrocarbons in these extracts from the shales takes place at depths

corresponding to a geothermal temperature of 115 to 143°C (239 to 289°F). Other subsurface changes in organic matter with depth include an increase in the specific gravity of oil and a decrease in its naphthene/paraffin ratio.

SOURCE: J. M. Hunt, 1979.

Subsurface Diagenesis

The effects of deep burial on sediment and on rock have been ignored by many, but are now being studied more intensively. Many things remain to be learned. For example, the influence of pressure and temperature is known for conditions of metamorphism and in the shallow subsurface zones but is less well known for the zones in between. As we have seen, with increasing depth of burial, the temperature increases (See Figure 3-21.), porosity decreases (See Figure 3-22.), and density increases (See Figure 3-23.); even the salinity of the formation waters changes with depth. (See Figure 3-25.) Not long ago many people believed that most cementation took place at relatively shallow depths. Now it is quite apparent that cementation in the deep subsurface is important.

LITHIFICATION BY CALCITE

Porosity loss in carbonates is progressive to depths up to 8 km or more. The mechanisms of porosity loss are compaction, pressure dissolution, and cementation. Compaction effects seem to occur mostly early in the burial of a carbonate sediment. However, some mechanical compaction occurs at great depth, especially in sediments that were deeply buried rapidly, so that overburden pressure became great before much cement could be precipitated. Low cementation rates can exert the same effect. Mechanical collapse of diagenetic fabrics that formed in the subsurface has been observed in strata (in deep cores from bore holes) that, since their burial, have never been uplifted to near the surface. In the deep subsurface, pressure dissolution and cementation are both important processes.

Pressure dissolution in limestones is responsible for massive redistribution of calcium carbonate in the subsurface. An abundant supply of water as well as a high degree of permeability are necessary to dissolve the carbonate and to redistribute the solute. Permeability studies have shown that permeability across stylolites is very small, but that along stylolites permeability may be substantial. Thus, in the subsurface, fluids in limestones may use stylolites as conduits. This inference is supported by the common occurrence of euhedral authigenic dolomite crystals that postdate stylolite formation and that are found only or primarily along and near stylolites. Authigenic crystals of other minerals, such as barite, are also found along stylolites and provide additional evidence for fluid flow along stylolitic surfaces.

Little is known about the transfer of material in solution in the subsurface or about the reprecipitation of dissolved carbonate as a late cement. Reduction in thickness of limestone strata of from 30 to 50% as a result of pressure dissolution may be common. Examples of extensive dissolution are given by stylolites that are parallel to the bedding. (See Figure 3-7.) Pressure dissolution may provide some of the solute for late-stage precipitation of sparry cement that occupies a large volume (up to about 50%) in some ancient limestones.

In the deep subsurface, most fluids are in approximate equilibrium with $CaCO_3$. Therefore their potential for precipitating or dissolving $CaCO_3$ is small. Nevertheless, as just mentioned, significant evidence exists that, in the subsurface, $CaCO_3$ has been extensively dissolved.

SOURCES: Bathurst, 1975, 1980, 1983; Ricken, 1986.

LITHIFICATION BY SILICA

Silica cement consists of several mineral phases, the most common of which is quartz. Typically, the quartz cement tends to grow outward from each particle as if particle and cement were a single crystal. In other words, the crystallographic arrangement of the quartz particle being cemented governs the crystallographic orientation of the quartz forming the cement. Such *quartz cement added in crystallographic continuity with a quartz particle* is called an **authigenic overgrowth** (Figure 3-27). Authigenic overgrowths characteristically develop around particles that are single crystals. Most quartz particles are single crystals, whereas most carbonate particles are polycrystalline; hence carbonate cement, both in limestones and in sandstones, generally makes a mosaic of crystals that occupies the pores between particles. (See Figure 3-10.)

Because of the crystallographic continuity between quartz particle and quartz overgrowth, it may be difficult to find the boundary between them. In thin sections viewed under the petrographic microscope the boundary between particle and overgrowth can be recognized if solid- or liquid inclusions (vacuoles), iron oxides, or clay minerals rim the particle as a dusty border. (See Figure 3-27.) Another way to separate particle and overgrowth by eye in microscopic view is to study the density of inclusions. The density of inclusions in the quartz particle may differ considerably from that in the quartz of the overgrowth.

The stage of early burial is commonly, but not necessarily, characterized by precipitation of an incomplete quartz cement. Much void space remains among the particles. As burial progresses and pressure dissolution becomes effective, quartz particles begin to interpene-

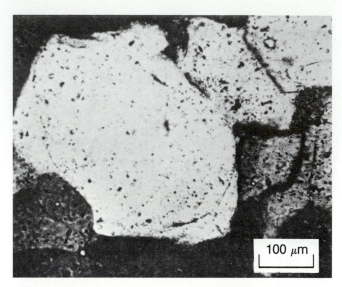

FIGURE 3-27.   Enlarged view of quartz-cemented sandstone showing large round quartz particle (outline marked by "dusty" border) to which has been added, in crystallographic continuity, a quartz overgrowth, as seen in a photomicrograph of a petrographic thin section. Concavo-convex contact between two quartz particles is at lower left. Simpson Sandstone (Lower or Middle Ordovician), Anadarko Basin, Oklahoma. (Authors.)

trate and form an interlocked granular framework whose particles display concavo-convex- and sutured contacts. With further burial these contacts become deeply interpenetrated, even microstylolitic. Ultimately, particles become completely intergrown and welded. They form a fabric that is a precursor to the fabric of metamorphic quartzites. Such complete interpenetration results from the association of elevated temperatures and -pressures, and the presence of interstitial waters of high ionic concentration that are found at considerable depth of burial or where structural deformation is in progress. Although such a sequence of progressive interpenetration leading to metamorphic quartzites is common, the quartz of some deeply buried sandstones and even quartzites does not display the diagnostic features resulting from pressure dissolution. These rocks may have been cemented early in their diagenetic history.

In contrast to the direct particle-to-particle contacts among detrital quartz particles, which commonly are concavo-convex or sutured, such concavo-convex or sutured contacts between the quartz of cements that grew outward from particles are rare. This rarity leads to the interpretation that the quartz cement was precipitated after the particles themselves had adjusted to pressure in the absence of cement.

An inverse relationship exists between the amount of quartz cement and the abundance of detrital clay matrix. Evidently, the presence of clay minerals retards the growth of quartz cement. One likely explanation for this relationship is the low porosity and -permeability of sandstones rich in clay matrix. In addition, the presence of clay-mineral crystals on the surfaces of quartz particles reduces the surface area of "clean" quartz on which quartz overgrowths can nucleate. In some sandstones devoid of clay it can be shown that lack of porosity can result from the presence of an impervious cement rather than from another petrographic variable, such as the nature of the particle-to-particle contacts.

The solubility of quartz at ordinary temperatures in near-neutral solutions is about 6 to 10 ppm, that of chalcedony and opal about 32 to 34 ppm, and that of freshly precipitated amorphous silica 120 to 140 ppm. During a rise in temperature from 25 to 120°C, the solubility of quartz increases by a factor of 10. (See Figure 3-6, B.) Increasing salinity of the pore waters partially offsets the temperature effect.

The configuration of the contacts between particles is related to (1) original packing, and (2) the amount of pressure dissolution. The number of contacts per particle and kind of contact, whether tangential, long or straight, concavo-convex or sutured, depends on pressure increase resulting from weight of overburden or from structural deformation; the effects of structural deformation are more pronounced than those of weight of overburden. Tangential contacts are the result of original packing. Long- or straight contacts result from the interplay of three factors: (1) original packing, (2) pressure, and (3) precipitated cement. Concavo-convex- and sutured contacts (Figure 3-28) are generally the result of pressure. Strained quartz particles show many more concavo-convex and sutured contacts than do nonstrained quartz particles. This behavior probably results from the greater solubility of strained crystals. Carbonate particles in sandstones or in limestones may likewise display concavo-convex- and sutured contacts that are the result of pressure dissolution. (See previous discussion of experimental compaction of ooids.)

Reduction in pore space and in thickness of sand- and sandstone strata results from (1) mechanical compaction caused by rotation, fracturing, or plastic deformation of particles; (2) dissolution at contacts between particles; and (3) a combination of these two factors. Reduction of pore space, but not of thickness, may result from precipitation of cement. Cementation, especially if early, may prevent mechanical compaction or dissolution at particle contacts. Both compaction and cementation are accelerated by high temperatures, moving water solutions, and high pressures equivalent to great depths of burial. As a rule, porosity decreases with increasing depth of burial. (See Figure 3-22.)

Although dissolution of the tests of siliceous organisms, such as the spicules of sponges or the biocrystals of radiolarians, has been shown to be a major factor in

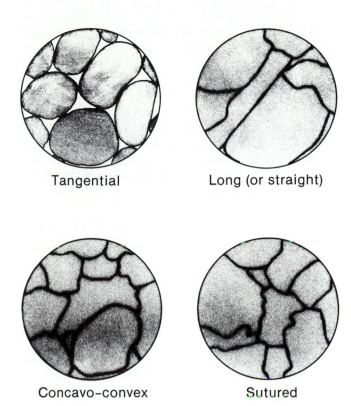

Tangential   Long (or straight)

Concavo-convex   Sutured

FIGURE 3-28.   Kinds of contacts between particles in sandstones; schematic enlarged views of petrographic thin sections. A, Tangential; B, long or straight; C, concavo-convex; D, sutured. (After J. M. Taylor, 1950, pl. 1 and 2, p. 711 and 712.)

the supply of silica to form chert, such organisms are thought to be only minor contributors to the supply of silica that cements sandstones and shales. Considerations of mass balance indicate the inadequacy of the organic source. The major source of silica is thought to be that dissolved from dissolution of particles which interpenetrate, especially where particles have been "welded" to each other. In some sandstones, the amount of silica cement precipitated balances the amount dissolved during particle interpenetration. In such sandstones effective pressures and point contacts of quartz particles increase the solubility of the quartz. Thus, at these contacts, especially where clay films are present between the particles, silica dissolves. As silica is liberated, the pore waters become supersaturated with respect to silica and quartz is reprecipitated as overgrowths. Increase of pH above 9.5 (See Figure 3-8.) and increase of temperature (See Figure 3-6, B.) increase the solubility of silica. In some fresh-water aquifers, waters having pH above 9.5 have been measured. Such high pH values, as with increased temperatures, may lead to a condition of supersaturation with respect to silica in pore waters and thus to the ultimate precipitation of quartz. This happens if subsurface waters are able to transport solutions rich in silica from a site where silica is being dissolved to another site where it is precipitated.

An additional source for silica is mud and shale. When these fine sediments are compacted they release silica-supersaturated waters. In sandstones the amount of silica cement has been found to increase toward a contact with a shale. This is considered to be support for the idea that waters expressed from muds can transport dissolved silica into sands. Likewise, mineral transformations in which clay minerals, feldspars, and other silicates react with meteoric waters that trickle into the subsurface may provide dissolved silica that can be reprecipitated as quartz cement in sandstones. A final source is dissolution of volcanic glass.

Another mode of silica precipitation has been proposed to explain the common occurrence of silicified fossils and precipitation of silica in the immediate vicinity of fossils in carbonate sediments. It is well known that at shallow depths of burial, bacterial reduction of sulfate and oxidation of organic matter are dominant processes in many sediments. (See Eq. 3-10.) These processes can liberate carbon dioxide and can alter the salinity and pH of the pore waters. Thus pore waters can become undersaturated for $CaCO_3$ and (possibly) supersaturated for silica. A criterion for assigning a burial origin to cement is any fabric relationship indicating that cement postdates effects of compaction, such as fractures and stylolites.

SOURCES: Füchtbauer, 1988; Füchtbauer and Valeton, 1988; Pittman, 1981.

Suggestions for Further Reading

BEAUCHAMP, B.; KROUSE, H. R.; HARRISON, J. C.; NASSICHUK, W. W.; and ELIUK, L. S., 1989, Cretaceous cold-seep communities and methane-derived carbonates in the Canadian Arctic: Science, v. 244, p. 53–56.

BHATTACHARYYA, AJIT, and FRIEDMAN, G. M., 1984, Experimental compaction of ooids under deep-burial diagenetic temperatures and pressures: Journal of Sedimentary Petrology, v. 54, p. 362–372.

BROOKINGS, D. G., 1988, Eh–pH diagrams for geochemistry: Berlin-Heidelberg-New York, Springer-Verlag, 176 p.

FINKEL, E. A., and WILKINSON, B. H., 1990, Stylolitization as a source of cement in Mississippian Salem Limestone, west-central Indiana: Geological Society of America Bulletin, v. 74, p. 174–186.

FRIEDMAN, G. M., 1975, The making and unmaking of limestones or the downs and ups of porosity: Journal of Sedimentary Petrology, v. 45, p. 379–398.

VOLLBRECHT, R., 1990, Marine (*sic*) and meteoric diagenesis of submarine Pleistocene carbonates from the Bermuda carbonate platform: Carbonates and Evaporites, v. 5, p. 13–95.

CHAPTER 4

Classifying and Naming Sedimentary Rocks

A classification begins when people assign names to things. When rocks such as sandstone, shale, and limestone were given these distinctive names to mark certain aspects of their appearances, the classification of sedimentary rocks was born. In this kind of classification, in which popular names are given in a very loose manner, definitions lack precision. As sedimentary rocks attracted the attention of geologists, another kind of classification became necessary, in which the most-suitable grouping of rocks became the most-important objective. The naming of the rocks became less important and was governed by the grouping. A variety of approaches can be employed in classifying and naming rocks, as we shall discuss in the following pages.

To offset the danger that a chapter on classifying and naming rocks may degrade into a catalog, we shall discuss and explain some of the properties of rocks as well as their genesis.

Descriptive- and Genetic Classifications

As with sediments, sedimentary rocks can be classified and named in a purely descriptive, objective, and precise manner; a classification that serves this purpose is a descriptive classification. Contrasting with a descriptive classification is a genetic classification, which conveys what is known, inferred, or believed about the origin of a given rock. Some classifications combine features of both and thus are descriptive-genetic.

Descriptive Classifications

In terms of communication, even descriptive classifications are not as straightforward as one might suppose. Consider, for example, the simple term sandstone. To some geologists, this term means a sedimentary rock consisting of particles of a certain size (62 μm to 2 mm), as we have seen in Chapter 2. No particular composition of particles is implied. To these geologists, a carbonate rock having only ooids as particles would be an example of a sandstone. If a geologist shows you such a sample and says, "Here, look at this interesting sandstone," you may look at it, note its softness by scratching it with a knife, and watch it effervesce by dropping some acid on it, and feel compelled to reply, "But this rock is a limestone." You have used composition in naming this rock. Some geologists use both composition and texture and employ the term "lime sand"; but because the rock is consolidated it should more correctly be termed "lime sandstone." Thus one variable, such as particle size or composition, is not enough to classify a sedimentary rock satisfactorily. Both particle size and -composition must be employed. A. W. Grabau (1870–1946) understood this and left a lasting legacy in sedimentology by introducing terms that employed both particle size and -composition. Let us take our carbonate rock having only ooids as particles and see how Grabau would have classified it. For rock names, Grabau abandoned the common sedimentary terms gravel, sand, and clay, and for particle size introduced the Latin terms *rudite* (from *rudus*, rubble), *arenite* (*arena*, sand), and *lutite* (*lutum*, mud). (See also Chapter 2.) The ending -ite refers to a rock. Therefore, to Grabau, because the particles in our rock consisting of ooids, measure between 0.5 and 1 mm in diameter, the rock would be an arenite. Grabau would express composition with appropriate prefixes, such as *calc* (a,i)-, *argil*-, and *silic* (i,a)-. Accordingly, in the Grabau scheme, our oolitic rock would be a *calcarenite*. By contrast, *a sandstone consisting of quartz particles* would be a **silicarenite**. The terms *calcilutite*, *calcarenite*, and *calcirudite* are commonly employed today, as are the terms *lutite*, *arenite*, and *rudite*, and the term *arenaceous.* In fact, one of the modern limestone classifications, which we shall examine later, uses one of these terms. Yet for some reason not known to us, the prefix *silic-* has fallen into disuse, and the term *argillutite* (there can be no *argilrudite* or *argilarenite*) is only rarely used for shales. But, back to our oolitic rock. The term *calcarenite* expresses the particle size and composition of the rock. If we add the adjective oolitic, we have conveyed quite specifically the essential characteristics of the rock.

Genetic Classifications

This kind of classification is based entirely on the origin of a rock. Grabau introduced many terms to relate the rock to its agent of deposition. These terms include such artificial names as biosilicipulveryte or hydrogranulyte, which we do not wish to describe here. As a genetic classification, Grabau's system is complete and logical. However, geologists have not adopted it, not only because of the unfamiliar mixtures of Latin terms (unfamiliar names, including terms that have been called "Greek barbarisms," have been accepted by geologists before), but mostly for very practical reasons. In Chapters 9 to 14 we shall appreciate the difficulty of pinning down specifically the depositional agent responsible for the formation of a given rock. Commonly, more than one kind of depositional process may have been involved; or from a study of the rock it may not be possible to infer which of several different depositional agents was operative. Moreover, a geologist must be able to classify and name a sample of rock before its origin is determined. Use of a strictly genetic classification "puts the cart before the horse" for most geologic purposes. Thus, although several of Grabau's descriptive terms are still popular, as stated in the previous paragraph, Grabau's genetic classification is not currently in use. The best classification is based on purely descriptive parameters, but into which genetic interpretations, where they can be reasonably inferred, can be judiciously blended.

SOURCES: Friedman and Sanders, 1978; Füchtbauer, 1959, 1988; Grabau, 1913; Holmes, 1928; Rodgers, 1950; Sabine, 1974.

Classifying Sedimentary Rocks

We find it convenient to divide sedimentary rocks into four major kinds, based on the groups of sediments discussed in Chapter 2. These are (1) intrabasinal, (2) extrabasinal, (3) pyroclastic, and (4) carbonaceous (Table 4-1).

TABLE 4-1. Classifying and naming major kinds of sedimentary rocks

Group	Kind of rock
1. Intrabasinal rocks	Carbonate rocks: limestones and dolostones
	Evaporites
	Authigenic rocks
2. Extrabasinal rocks	Terrigenous rocks
3. Pyroclastic rocks	Pyroclastic rocks
4. Carbonaceous rocks	Peat and coal

We shall discuss the classification and naming of sedimentary rocks beginning with intrabasinal rocks. The reason is the availability of an end-member classification system specific for one kind of intrabasinal rocks, namely limestone, which lends itself to extension to extrabasinal rocks, specifically sandstone. Our classification of sandstone parallels that for limestone.

Classifying and Naming Intrabasinal Rocks

Intrabasinal rocks, that is, *sedimentary rocks formed by the lithification of intrabasinal sediments* (defined in Chapter 2), consist of carbonate rocks (limestones and dolostones), authigenic rocks, and evaporite rocks. Carbonate rocks, which are widespread and compose an estimated 20% of the total mass of sedimentary rocks underlying the land area of the Earth, are defined as those containing more than 50% carbonate minerals. Two broad compositional divisions are (1) limestones and (2) dolostones. We discuss them separately.

Classifying and Naming Limestones

We discuss two different classifications of limestones. Both are descriptive, but they possess genetic overtones.

Except for reefs, which framework builders construct; for consolidated lime mud; and for limestones formed by the bacterial breakdown of calcium sulfate, the basic constituents of most limestones are recognizable sand-size particles and the spaces between these particles. The interparticle spaces may be occupied by one or more of (1) *microcrystalline- or cryptocrystalline particles of calcium carbonate commonly smaller than 4 μm,* known as **micrite,** that may be lithified lime mud, which implies that only incomplete sorting took place in the environment of deposition; (2) optically clear spar (commonly calcite) in crystals that are easily visible in thin sections of normal thickness, and (3) void space, which may be filled with fluids, such as water or petroleum (oil or gas). Any of the foregoing kinds of limestones may be so altered that the only appropriate way to describe them is as recrystallized limestones.

We characterize briefly these chief components of typical limestones and then show how they have been employed as a basis for two widely used methods for classifying and naming such limestones.

The sand-size particles in most limestones are those we discussed in Chapter 2: skeletal particles, ooids, intraclasts, grapestone, and pellets or peloids. The sizes of the individual mineral constituents of micrite are usually 1 to 4 μm. In hand sample, micrite is dull and opaque; in a thin slice viewed under the petrographic microscope it is subtranslucent with a faint brownish cast. Sparry cement is the clear, crystalline component, having vitreous luster, that occupies some or all of the voids among the sand-size particles; crystals of sparry calcite cement measure tens or hundreds of micrometers in length.

The following two classifications, which were proposed nearly simultaneously by R. J. Dunham and by R. L. Folk, show how the same three constituents of most limestones: sand-size particles of calcium carbonate, micrite, and spar, can be employed in methods of classifying and naming limestones. We begin with the Dunham classification.

DUNHAM CLASSIFICATION

Dunham recognized two first-order categories among limestones: (1) those in which the original particle components were bound together (named collectively as boundstones) and (2) those in which the original components were not bound together (no collective name proposed for this group) (Figure 4-1).

Boundstones are *limestones showing evidence that particles being deposited were bound by organisms or that they consist of frameworks constructed by organisms.* This group includes reefs, stromatolites, and travertine. A. F. Embry and J. E. Klovan have divided Dunham's boundstone group into subgroups according to the nature of the binding. Thus *limestones bound together with microbial laminae* are **bindstones.** *Limestones that originated in place* (autochthonous limestones) *as frame-built reefs* are **framestones.** *Limestones that consist predominantly of sediment trapped by baffling organisms* are **bafflestones.** This expansion of the Dunham classification is congruent with ecological classifications of dominant reef organisms into guilds (constructors, binders, bafflers). These subcategories were designed for use in describing reefs and are commonly used only in studies of reefal limestones.

If one has determined that a given limestone falls within the second group (unnamed group in which binding was not a factor), then one must determine if the sand-size (or larger) constituents are in contact (particle-supported fabric) or whether these constituents do not touch one another (matrix-supported fabric; see Figure 4-1).

Particle-supported rocks are subdivided into two further classes: (1) *particle-supported limestones devoid of lithified lime mud,* known as **grainstones,** and (2) *particle-supported limestones containing some lithified lime-mud matrix, but not enough to keep the sand-size particles from touching one another,* known as **packstones.**

Matrix-supported limestones are also subdivided into two classes: (1) *mud-supported limestones containing at most 10% sand-size- or larger particles,* known as **mudstones,**

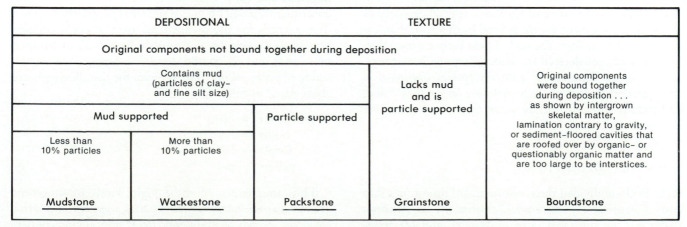

CLASSIFICATION ACCORDING TO DEPOSITIONAL TEXTURE

DEPOSITIONAL			TEXTURE	
Original components not bound together during deposition				Original components were bound together during deposition . . . as shown by intergrown skeletal matter, lamination contrary to gravity, or sediment-floored cavities that are roofed over by organic- or questionably organic matter and are too large to be interstices.
Contains mud (particles of clay- and fine silt size)		Lacks mud and is particle supported		
Mud supported		Particle supported		
Less than 10% particles	More than 10% particles			
Mudstone	Wackestone	Packstone	Grainstone	Boundstone

FIGURE 4-1.   Classifying and naming limestones according to depositional texture: the scheme of Dunham.

and (2) *mud-supported limestones containing at least 10% sand-size- or larger particles*, known as **wackestones**. Notice that in the Dunham classification the term micrite is not used. Instead, the fine- size carbonate rock is called *mudstone* (better, *lime mudstone*). (See Figure 4-1.)

In employing the Dunham classification to name limestones, one selects the appropriate class from the four listed and then determines the kind of sand-size particles present. The rock name consists of three parts in the following order: (a) a single word or words for the particles, (b) the term *lime*, and (c) the name of one of these four classes. Examples are fusulinid lime packstone, ostracode lime mudstone, or coated-particle lime packstone. When all the rocks being described are limestones, it is common practice to eliminate the word "lime" from the Dunham-classification examples just mentioned.

A distinctive feature of the Dunham classification is the fundamental division of the mainstream group limestones in which binding was not involved into *mud-supported* and *particle-supported* categories. According to Dunham, the names assigned to limestone samples are tentative; yet the names are short, meaningful English nouns that apply only to the texture of the limestone. The term wackestone has been criticized by some; the other terms have been readily accepted.

The rationale of this classification is that it allows one to map gradients in rate of production of sand-size particles relative to rate of accumulation of lime mud. In calm waters, lime mud, if present, settles on the bottom and remains there. Hence limestones derived by lithification of an original lime mud deserve to be contrasted with those derived by lithification of a carbonate sediment devoid of lime mud, regardless of the number and size of the sand-size particles. As mentioned, this relationship between sand-size particles and lime mud generally (but not always) distinguishes an original sediment deposited in calm water from a sediment deposited in agitated water. This distinction is fundamental, and in the Dunham classification is incorporated in the class name.

According to Embry and Klovan, a special category of limestones consists of debris derived from a reef, that was transported away from the reef. These form a group they have named reef-derived (or allochthonous) limestones. Examples are **floatstones**, *matrix-supported limestones in which more than 10% of the particles are larger than 2 mm*; and **rudstones**, *limestones in which the coarse particles are clast supported*.

FOLK CLASSIFICATION

R. L. Folk subdivided the constituents of limestones into two categories: (1) *allochemical constituents* or *allochems* (particles, meaning sand-size or larger), and (2) *ortho-*

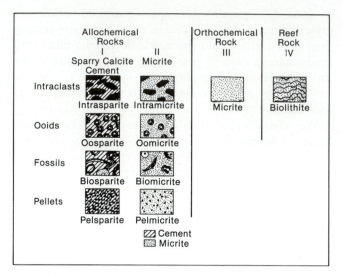

FIGURE 4-2.   Classifying and naming limestones according to the scheme of Folk.

chemical constituents (collectively designated by the word *orthochems* and consisting of micrite, presumably lithified original lime mud; and cement, sparry calcite; Figure 4-2). Folk's classification, which we have modified slightly, lists four kinds of allochems (particles): (1) intraclasts (including grapestone), (2) ooids (including some other kinds of coated particles), (3) skeletal debris (fossils or fossil fragments), and (4) pellets (including peloids).

The interstices among the allochems are filled with the orthochemical constituents (micrite or sparry calcite cement). Combinations of allochems and orthochems provide a basis for recognizing eight kinds of limestones. In addition, *micrite* may lack sand-size particles and hence stand alone as a ninth kind of limestone. Finally, a tenth kind is *reef rock* (**biolithite**). (See Figure 4-2.)

Folk assigned names to eight of these ten kinds of limestone by using composite words consisting of two parts: (a) an initial abbreviated expression for the allochems, and (b) a word designating the orthochemical constituents, based on one of the two groups listed above. These words for the orthochems are (1) *micrite* (microcrystalline carbonate, presumably lime mud or lithified lime mud), and (2) **sparite** (*limestones having a cement consisting of sparry calcite*). The prefixes for the allochems are abbreviated as follows:

$$
\begin{aligned}
\text{intraclasts} &= \text{intra} \\
\text{ooids} &= \text{oo} \\
\text{fossils} &= \text{bio} \\
\text{pellets} &= \text{pel}
\end{aligned}
$$

Thus the names of these eight kinds of limestone are *intrasparite, intramicrite, oosparite, oomicrite, biosparite, biomicrite, pelsparite,* and *pelmicrite*. If several kinds of particles are important constituents of a limestone, their abbreviations are strung together in order of increasing abundance. For example, a limestone with orthochemical constituents falling in the micrite category and having allochems composed of 10% intraclasts, 20% skeletal debris, and 70% peloids would be an *intrabiopelmicrite*. If an oolitic limestone includes both micrite and spar among its orthochemical constituents with micrite as the lesser component, then the rock name is *oomicsparite*. In a comparable three-component mixture in which sparite is the minor orthochemical constituent the rock name becomes *oosparmicrite*.

The Folk classification allows for particles larger than sand size, which are, however, not particularly common in limestones. For particles of pebble size, such as the intraclasts of flat-pebble conglomerates, the shells of lag concentrates, or the pellets of organisms having large anal diameter, Folk continued the tradition of one of Grabau's terms, *rudite*. A limestone composed of large pellets, if accumulated in a micrite, is known as *pelmicrudite*; flat pebbles in a micrite would be *intramicrudite*. If sparry cement occurs between these same pebble-size particles, the limestones would be designated *pelsparrudite* and *intrasparudite*.

The distinctive aspect of Folk's classification is its use of names for eight major kinds of limestones based on what lies between the particles (spar or micrite). When this classification was first introduced, many geologists rebelled against it because they felt that the proposed new names were too strange. Others feel that many of the distinctions made in the Folk classification are difficult to recognize in the field. However, once one has mastered Folk's classification, it proves to be both practical and logical.

LIMITATIONS IN CLASSIFYING AND NAMING LIMESTONES

As we have seen in the preceding sections, modern classifications of limestones are based on the relative concentrations of micrite and sparry cement. If these classifications were purely descriptive, this philosophy of using micrite and sparry cement would not raise any difficulties. However, these classifications imply genetic relationships. They are based on the idea that, in calm waters, tiny particles of lime mud are (a) available and (b) able to settle on the bottom and remain there. By contrast, in agitated waters, particles of micrometer size, such as those in lime mud, even if available, are thought to remain in suspension and thus not be deposited. Although this concept is satisfactory from the point of view of dynamics in the depositional environment, lime mud does not necessarily stay put after it has been buried. Three other problems involve (1) transfer of lime mud; (2) precipitation of cement, not as sparry calcite, but in cryptocrystalline form that appears the same as micrite; and (3) recrystallization. We examine these in Box 4.1.

Neither Dunham nor Folk established categories for an important group of limestones that form as a byproduct of the bacterial destruction of gypsum (Eq. 3-10).

The categories in both the Dunham and the Folk classifications refer only to those limestones that are purely intrabasinal. However, in the real world, particles of intrabasinal origin commonly have been mixed in all proportions with particles of extrabasinal origin. In Figure 4-3 we present a method for naming such mixed rocks. We illustrate this approach using mixtures of sand-, silt-, and clay-size quartz particles (the most-common mineral in extrabasinal rocks) and carbonate. Mixtures of quartz and carbonate can be plotted on a triangular diagram whose end members are carbonate minerals, clay- and silt-size quartz, and sand-size quartz. This classification is congruent with the division of rocks into intrabasinal- and extrabasinal kinds, and it also uses the standard method of modifying rock names by listing minor constituents in order from least- to most common. The names are simple and descriptive, but could easily be further modified for more-precise usage.

Rocks containing less than 10% quartz are *carbonate rocks* and can be subdivided further using the Dunham or Folk classifications. For carbonate—quartz mixtures containing between 10 and 60% quartz, the general word *limestone* (or *dolostone*) is used, preceded by one or more modifiers indicating the relative proportions of clay- plus silt- or sand-size quartz. Thus a rock consisting of 70% calcite and 30% quartz silt would be a *silty limestone*. A rock with 65% calcite, 23% quartz sand, and 12% quartz silt would be a *silty sandy limestone*. The lower limit of rocks whose primary name refers to the carbonate content is placed at 40% carbonate, rather than 50% carbonate, because these rocks tend to have formed by addition of quartz to an environmental setting in which carbonate material was regularly and actively accumulating.

Rocks containing from 60 to 90% quartz are classified as *calcareous siltstones* or *calcareous sandstones* (modified as appropriate with the words sandy or silty inserted after the word calcareous). Thus a rock containing 68% quartz sand, 18% calcite, and 14% quartz silt would be a *silty calcareous sandstone*. Rocks containing more than 90% quartz are extrabasinal rocks and are classified as discussed in a following section.

SOURCES: Cuffey, 1985; Dunham, 1962; Embry and Klovan, 1971; Folk, 1959.

BOX 4.1

Microcrystalline Carbonate (cf. Micrite) That Is Not Original Lime Mud (Micrite)

As is evident from the emphasis placed on it by both Dunham and Folk, a well-sorted and cleanly washed residue of sand-size carbonate particles contrasts significantly with a poorly sorted mixture in which sand-size- and finer particles did not become separated in the depositional environment. Here we consider three cases in which a microcrystalline-carbonate component can be present along with sand-size carbonate particles and the correct interpretation does not imply anything about conditions within the depositional environment. The cases are (1) the trickling down of lime mud into the pore spaces of an underlying well-sorted, cleanly washed lime sand, (2) precipitation of a cement that is not sparry calcite but rather cryptocrystalline aragonite or high-magnesian calcite that mimics micrite, and (3) recrystallization.

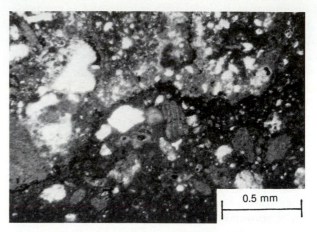

BOX 4.1 FIGURE 1.   Modern reefrock from the Red Sea in which particles have been cemented by cryptocrystalline carbonate that resembles micrite. Photomicrograph of thin section, plane-polarized light. (G. M. Friedman, 1985b, fig. 1, p. 119.)

Trickling Down of Lime Mud

Lime mud from an overlying low-energy deposit may trickle down into empty pores of an underlying high-energy deposit. Waters carrying fine particles of cryptocrystalline carbonate in suspension may flow through the pores of a well-sorted carbonate sediment composed chiefly of sand-size carbonate particles and, in the process, drop off some of the fine particles, thus transferring them from one layer to another. Given that examples of such transfers have been found, anyone who bases an interpretation entirely on particle/micrite ratios and does not consider transfer of lime mud from a low-energy- to a high-energy deposit may be deceived.

Precipitation of Cryptocrystalline Cement

An even-more-complicated problem than the physical transfer of mechanically deposited lime mud is the precipitation of cryptocrystalline cement; that is, under the microscope, its crystals are too small to be recognized and separately distinguished. This cement looks just like micrite (Box 4.1 Figure 1). Such a cement has been found to have been precipitated within millimeters to centimeters of the surfaces of such high-energy de-

posits as reefs (Box 4.1 Figure 1) and beachrocks. (See Figure 2-9, D.) Because such a cement can be precipitated within the pore spaces of sediments in even the highest of high-energy environments, noncritical use of the particle/micrite ratio as a reliable clue to depositional environment could lead to error.

In summary, the philosophy of using the particle/micrite ratio as a criterion for interpreting the depositional environment of a limestone is based on the common observation that, in agitated waters, pores among the particles remain open and later, during diagenesis, are filled with a mineral cement that is always sparry calcite. In many cases this idea is correct. Therefore the classifications of limestones discussed in this chapter are logical and probably practical. However, one should be aware that interpretations based on noncritical acceptance of the concepts behind these classifications are not infallible, and, as in all geological interpretations, caution is necessary.

Recrystallization

The carbonate minerals composing limestones are very susceptible to being recrystallized. After these minerals have recrystallized it may be difficult or impossible to decide, even with a petrographic microscope, what kind of original carbonate sediment was ancestral to the limestone being studied. Because the former particles take on new shapes as crystals, recrystallization is commonly referred to by the name **neomorphism** (*neo* = new; *morph* = shape). We can define neomorphism as *a transformation between one mineral and itself or a polymorph* (another mineral having the same chemical composition but different crystal structure and physical-chemical properties), *which results in the growth of new crystals that are larger or simply different in shape from the previous particles or crystals*. Generally, in neomorphism crystals of calcite replace original particles that were composed either of aragonite or of high-magnesian calcite.

Neomorphism in limestones may proceed in at least two ways: (a) by **micrite enlargement**, *a process by which particles of lime mud that measure only a few micrometers in diameter enlarge to sizes measuring tens or even hundreds of micrometers in diameter*, and (b) by a process in which crystals of calcite, generally of the order of tens to hundreds of micrometers in size, may replace former micrite, sparite, and particles, disrespecting any previously existing boundaries and con-

tinuing until the entire limestone consists of these replacing crystals.

Micrite enlargement creates neomorphic products resembling cement in crystal size. Depending on the crystal size of the new product, the result is known as microspar or pseudospar. **Microspar** is *a mosaic of neomorphic crystals having diameters ranging from 4 to 50 μm*; **pseudospar** is *a mosaic of neomorphic crystals larger than those of microspar* (Box 4.1 Figure 2).

Distinguishing Between Neomorphic Products and Cements in Limestones, and Inferring Depositional Fabric

In microspar and pseudospar the distribution of the sizes of individual crystals is usually random (Box 4.1 Figure 3). Microspar or pseudospar may exist in isolated patches and apparently "floating" in the midst of micrite. By contrast, as we have seen in Chapter 3, true cement is mostly distributed between particles and the sizes of individual crystals of the cement tend to increase away from the wall of the original void. (See Figure 3-20.) Even a sample of limestone that has undergone extensive micrite enlargement may display relic patches of nonreplaced micrite, either locally or throughout. These patches reveal that the limestone originally was a micrite rather than

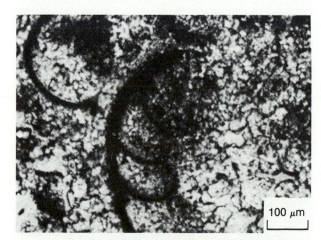

BOX 4.1 FIGURE 2. Limestone showing neomorphic crystals of calcite that have replaced the former fabric of the rock. Crystals < 50 μm are microspar; those ≥ 50 μm are pseudospar. Ghosts of foraminifers are preserved as dark lines of carbonate containing opaque inclusions of unknown composition. Pleistocene limestone from borehole, Mediterranean coast of Israel; photomicrograph in plane-polarized light of a thin slice. (L. Ordan and authors.)

BOX 4.1 FIGURE 3. Limestone showing neomorphic crystals of calcite that have replaced the former fabric of the rock, including skeletal fragments. Dark parallel lines in large crystal in upper central part of photograph are probably organic matter defining possible ghost structure of a mollusk shell. Pleistocene limestone from borehole, Mediterranean coast of Israel; photomicrograph in plane-polarized light of thin slice.(L. Ordan and authors.)

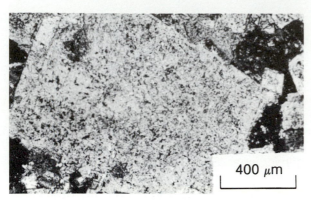

BOX 4.1 FIGURE 4.   Dolostone showing crystals of two sizes: (1) a large crystal or porphyrotope and (2) small crystals enclosing it. Black areas are unfilled pores. Madison Limestone (Mississippian), Lander, Wyoming; photomicrograph in cross-polarized light of a thin slice. (G. M. Friedman, 1965b, fig. 4, p. 650.)

a sparite. These features are helpful in naming, classifying, and understanding neomorphically altered limestones.

Crystals of calcite (or dolomite or evaporites) on the order of tens or hundreds of micrometers in diameter commonly replace any or all of the components of a calcium-carbonate sediment or of a limestone: particles, cement, and/or micrite. These new crystals cut across existing boundaries. Such replacing crystals may be **porphyrotopes**, which are *large crystals surrounded by smaller particles or crystals* (Box 4.1 Figure 4); or

they may be **poikilotopes**, which are *large crystals among smaller ones that enclose original particles or newly formed crystals* (Box 4.1 Figure 5). Where porphyrotopes or poikilotopes are abundant, especially where they coalesce and merge, the original texture may be entirely obliterated. Fortunately, the original boundaries of particles may be preserved as **ghosts**: *recognizable outlines of the original particles*. Within porphyrotopes or poikilotopes, the original pattern of ornamentation of fossils may be preserved. Ghosts preserve original boundaries because organic matter is concentrated along such boundaries. Where proper conditions prevailed, pyrite may have formed along the boundaries of original particles (Box 4.1 Figure 6).

Because of the ready susceptibility of carbonate minerals to neomorphic changes, even experienced sedimentary petrographers may not always be able to recognize the original particles, cement, and micrite. Neomorphism in many limestones advances to a stage where all crystals of calcite abut and mutually interfere with one another. In such mutual interference, individual crystals cannot develop their own **crystal faces**, that is, *the outward planar growth surfaces of crystals that reflect their internal structure or lattice. Crystals having boundaries that are not crystal faces* are said to be **anhedral**. *An arrangement of anhedral crystals* is known as a **xenotopic fabric** (Box 4.1 Figure 7). In *a xenotopic fabric the boundaries between crystals that mutually interfered with one*

BOX 4.1 FIGURE 5.   Poikilotopic gypsum (white areas with closely parallel light- and dark striae inclined from upper left to lower right) enclosing particles of quartz and feldspar. Holocene, Great Salt Plains near Jet, Oklahoma; photomicrograph in cross-polarized light of a thin slice. (G. M. Friedman, 1965b, fig. 9, p. 652.)

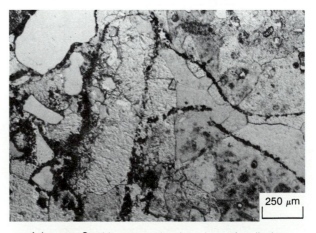

BOX 4.1 FIGURE 6.   Limestone showing ghost of mollusk fragment (approx. center) whose original boundary is preserved by pyrite (black). Pleistocene limestone from borehole, Mediterranean coast of Israel; photomicrograph in plane-polarized light of a thin slice. (L. Ordan and authors.)

BOX 4.1 FIGURE 7. Dolostone showing xenotopic fabric in which all preexisting depositional texture has been obliterated. Text defines xenotopic fabric. Cool Creek Formation (Ordovician), near Mill Creek, Oklahoma; photomicrograph in plane-polarized light of a thin slice. (G. M. Friedman, 1965b, fig. 1, p. 648.)

another during growth are known as **compromise boundaries**. To reemphasize, these boundaries are not crystal faces. In limestones displaying a xenotopic fabric, all evidence of former particles, micrite, or cement may have been erased, unless ghosts are preserved that transect the crystals of the xenotopic fabric. Yet even in less-severely altered limestones, experienced petrographers may be unable to recognize the depositional texture or distinguish cement from pseudospar.

Classifying Recrystallized Limestones

Only careful study of thin sections with a petrographic microscope (or using modern techniques, such as cathodoluminescence- or fluorescence microscopy) helps in unraveling the effects of neomorphism and thus in naming and classifying a recrystallized limestone on the basis of the precursory calcium-carbonate sediment. The white-card technique, invented by R. L. Folk, in which a white piece of paper is placed beneath a thin section while it is being examined with a petrographic microscope in reflected light, may also be extremely revealing when applied to recrystallized limestones. Commonly, even such careful work is inconclusive, and one must be satisfied by naming the rock simply as a *recrystallized limestone* or as a *neomorphically altered limestone*. If the depositional fabric of a recrystallized limestone can be inferred, and if cement and neomorphic crystals can be distinguished, then the rock can be classified using either the Dunham or the Folk classifications but with the word recrystallized added before the other name. Alternatively, such rocks may be classified simply as recrystallized- or crystalline limestones. If *most crystals in a neomorphic limestone are smaller than 50 μm,* the rock is named a **microsparite**. *If most crystals in a neomorphic limestone are larger than 50 μm,* the rock is called a **pseudosparite**.

SOURCES: Folk, 1965; G. M. Friedman, 1968, 1975a, 1985b; G. M. Friedman, Amiel, and Schneidermann, 1974.

Classifying and Naming Dolostones (Carbonate Rocks Consisting of the Mineral Dolomite)

Compositionally, dolostones are of two kinds: (1) **calcareous dolostones**, *carbonate rocks containing 50 to 90% of the mineral dolomite*, and (2) **dolostones**, *carbonate rocks containing 90% or more of the mineral dolomite*. Carbonate rocks containing several percent of dolomite are common, but are classified as limestones under either the Dunham or Folk systems (sometimes with the modifier dolomitic preceding the rock name).

Dolostones may or may not preserve the original textures of their antecedent calcium-carbonate sediment. Where dolostones preserve such textures, the outlines of original calcareous particles, now dolomite, such as skeletal fragments, ooids, or pellets, may be readily ap-

parent or may be obscured but still recognizable as ghosts. A *dolostone in which the shapes of the original particles are preserved* is the product of **dolomitization** that is said to be **mimetic-** or **selective**. *Dolostone in which all preexisting textures have been obliterated and no identifiable particles are present* is a product of **nonmimetic-** or **nonselective dolomitization**. (See Box 4.1 Figure 7.)

In Box 4.1, under the heading of recrystallization in limestones, we introduced the terms *anhedral* and *xenotopic*. The opposite of anhedral is *euhedral*, which implies that a crystal is wholly bounded by crystal faces. *A fabric in which the predominant constituents are euhedral crystals* is known as **idiotopic** (Figure 4-4). Idiotopic dolostone is one of the most-important carbonate reservoirs for oil and gas because it is abundant, porous, and permeable. Because of their sparkling, sugary appearance, *idiotopic dolostones* are commonly known as **sucrosic dolostones**.

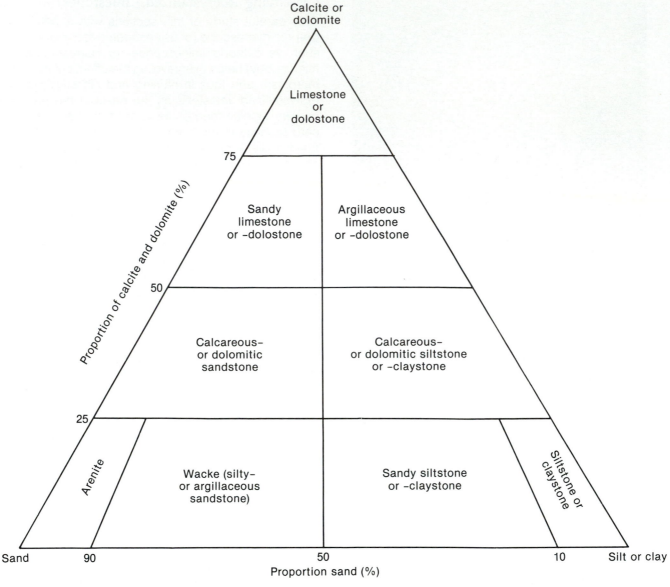

FIGURE 4-3.   Classifying and naming sedimentary rocks composed of mixtures of carbonate particles and clay-, silt-, and sand-size quartz. Carbonates are intrabasinal, whereas sand, clay, and silt are extrabasinal. (Authors.)

Crystals partly bounded by crystal faces are subhedral. *A fabric in which most of the constituent crystals are subhedral* is hypidiotopic (Figure 4-5).

Another dolomite classification, proposed by D. F. Sibley and J. M. Gregg, is based on crystal form and rock fabric, but in addition takes into account the factor of crystal size. In the Sibley-Gregg scheme, fabrics are separated into two groups: (1) *unimodal* and (2) *polymodal*. Then dolostones are further subdivided using crystal boundaries (*planar* versus *nonplanar*). Planar fabrics are further divided into those composed of dominantly eu-

hedral crystals as contrasted with those composed predominantly of subhedral crystals. Thus six classes are defined: *planar-e (euhedral), planar-s (subhedral),* and *nonplanar fabrics,* for both unimodal- and polymodal crystal sizes. In addition, adjectival modifiers are used to identify the kinds of particles, matrix, cement, and porosity, where these characteristics are determinable. This classification focuses attention on the distinction between planar- and nonplanar crystal boundaries, which is a result of temperatures and saturation states during crystal growth. It also forces the petrographer to take note

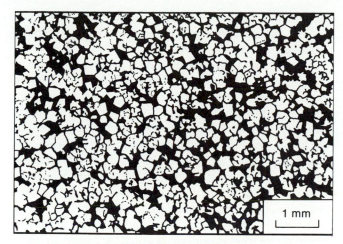

FIGURE 4-4.   Dolostone with fabric largely idiotopic and consisting of noninterpenetrating euhedral crystals. Dark areas between crystals are open pores. Lower Permian, Tunisia; sketch of a thin slice viewed under polarizing microscope. (G. M. Friedman, 1965b, fig. 3, p. 649.)

Dolostones may also be classified genetically into three major groups: (1) syngenetic dolostone, (2) diagenetic dolostone, and (3) epigenetic dolostone. Syngenetic dolostone here means *dolostone that has formed penecontemporaneously in its environment of deposition.* This kind of dolostone contrasts with diagenetic dolostone, which is *dolostone formed by replacement of calcium-carbonate sediments or of limestones during or following consolidation,* such as within beds of carbonate sediments or of limestones. In borderline cases the distinction between syngenetic and diagenetic becomes difficult or impossible to make. Epigenetic dolostone here means *dolostone that has formed by localized replacement of limestone along postdepositional structural elements, such as faults and fractures.* Many, but not all, epigenetic dolostones are genetically associated with metallic ore deposits, notably of lead- and zinc minerals.

SOURCES: Bathurst, 1975; Dunbar and Rodgers, 1957; G. M. Friedman, 1965; G. M. Friedman and Sanders, 1967; Sibley and Gregg, 1987.

of polymodal crystal-size distributions, which probably result from multiple crystal-nucleation events. Possible problems with the classification include the fact that the unimodal-polymodal distinction appears to be a continuum and the fact that the nonplanar-planar distinction is founded on a genetic interpretation.

Classifying and Naming Authigenic Rocks

Authigenic rocks are chemically and mineralogically diverse; their only common feature is their authigenic origin. The most-important kinds of authigenic rocks are (1) chert, (2) phosphate rock, and (3) sedimentary iron ores. Some evaporites formed by precipitation in the interstitial waters of a preexisting sediment and thus qualify as being authigenic. However, we treat all evaporites together in a following category.

CHERT

Chert is *a tough, brittle siliceous rock exhibiting a splintery-to conchoidal fracture and a vitreous luster.* The silica to form most cherts is thought to come from the tests of siliceous organisms.

The tests of siliceous organisms, such as those of radiolarians and diatoms, as well as the spicules of sponges consist of opal. Likewise, siliceous shells in cherts of Tertiary age are mostly opaline. In contrast, nearly all silica in Paleozoic cherts occurs as quartz and chalcedony. Using a length of 35 μm as a boundary (as seen in thin sections), the quartz of chert can be subdivided into (1) *microquartz* (quartz crystals < 35 μm) and (2) *megaquartz* (quartz crystals > 35 μm) (Figure 4-6).

Other silica minerals present in chert include quartzine and lutecine, which are fibrous varieties of quartz, and cristobalite (opal-CT). Quartzine and lutecine may form at the expense of sulfate minerals in evaporite deposits. These silica minerals may be the only testimony to the former presence of such sulfates, now

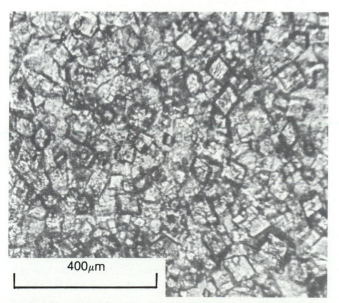

FIGURE 4-5.   Dolostone showing hypidiotopic fabric. Madison Limestone (Mississippian), Dubois, Wyoming; photomicrograph in plane-polarized light of thin slice. (G. M. Friedman, 1965b, fig. 11, p. 654.)

FIGURE 4-6.   Chert showing crystals of megaquartz (bulk of photograph) filling cavity lined by a zone of fibrous chalcedony. Long axes of chalcedony fibers stand normal to wall of cavity. Fort Ann Formation (Lower Ordovician), New York State; photomicrograph in cross-polarized light of thin slice. (S. J. Mazzullo.)

replaced and vanished. These silica minerals may therefore be useful indicators of ancient environments.

In thin slice under the petrographic microscope, chert may be cryptocrystalline, fine crystalline, medium crystalline, coarse crystalline (See Figure 4-6.), or fibrous. The fibrous variety is chalcedony, a polymorph of quartz, which displays a distinctive pattern of radial fibers; the others are quartz. Porcellanite consists of cristobalite or opal; cristobalite is a metastable polymorph of quartz; opal is a hydrous silica which contains between 4 and 9% water. This water is physically adsorbed; most of it is driven off at 100°C. The cristobalite of porcellanite is highly disordered and is known as opal-CT.

The silica of chert may be (1) an alteration product of volcanic rock, such as smectite and volcanic glass or (2) a precipitate derived from the dissolution of tests of siliceous organisms. The most-probable source of silica in the cherts of the central Pacific Ocean is dissolution of radiolarian tests. The silica in most oceanic cherts probably is of biogenic origin. Biogenic opal is dissolved and reprecipitated as finely crystalline cristobalite or opal-CT, which inverts to quartz. The end product is a classic dense, vitreous chert.

Cherts are forming by maturation of opal-CT to quartz and by replacement of carbonate sediments at the present day in Cenozoic- and Mesozoic deep-sea sediments.

In chert, impurities generally constitute less than 10%. Such impurities commonly include both replaced- and nonreplaced fossils, disseminated detrital quartz, rhombs of carbonate minerals, pyrite, hematite, organic carbon, and clay minerals.

Nonmarine cherts have been precipitated in hypersaline lakes in the African Rift Valley. The saline water leaches and dehydrates a sodium-silicate mineral precursor. Some unusual volcanoes in the rift valley spew tephra composed of pure Na_2CO_3.

Chert occurs (1) in nodules and (2) in strata (bedded or ribbon cherts). Common synonyms for chert or varieties of chert include *jasper*, *flint*, *novaculite*, and *porcellanite*. Jasper is yellow-, red-, brown-, or green-colored chert. Flint is gray to black cryptocrystalline chert that exhibits conchoidal fracture and whose color derives from included organic matter. Novaculite is a chert that in hand specimen is milk-white in color. Porcellanite is a low-density and dull-lustered siliceous rock having the texture and appearance of nonglazed porcelain.

The origin of chert has been under debate since the beginning of the twentieth century. A factor in prolonging this debate is the effect of diagenesis. Processes that operate during diagenesis generally have obliterated most preexisting textural features and faunal remains.

Although the source of silica for many cherts evidently is the dissolved siliceous tests of marine organisms, many chert nodules are inferred to have grown by the simultaneous action of (a) dissolution of carbonate and (b) precipitation of silica in thin films of water at the boundaries of the carbonate being dissolved. This is confirmed by two facts: (1) Within chert nodules, carbonate particles that have not been silicified have been corroded whereas similar particles near but outside nodules have not been corroded; (2) the fine structure of former carbonate particles is often preserved within chert nodules in the form of micrometer-size inclusions whose arrangement preserves ghosts of former particles.

During chertification, pore waters are supersaturated with respect to opal-CT or quartz but need not be undersaturated with respect to calcium carbonate, because the replacement process is thought to be controlled by the force of silica crystallization. Thus nodular cherts form primarily by replacement of carbonate rocks or -sediment. Bedded cherts, by contrast, are inferred to form by recrystallization of siliceous oozes and not by replacement of carbonates. Siliceous oozes consist primarily of the siliceous opaline tests of marine organisms and underlie much of the ocean floor where terrigenous input is low and where the sea floor is below the carbonate-compensation depth (the depth at which calcium carbonate in the water column dissolves completely). In addition, hydrothermal exhalations from the basaltic ocean crust may be enriched in silica, and contribute to the formation of bedded cherts that are relatively close to sources of hydrothermal fluids. When hot mafic lava and seawater react on the deep-sea floor, silica is released into the water. Such a reaction is the likely source of the silica for the chert that accompanies pillowed basalts.

Although most cherts are probably the end product of the reaction

$$\text{biogenic opal} \rightarrow \text{cristobalite or opal-CT} \rightarrow \text{quartz}$$

the possibility of direct crystallization of quartz should not be excluded.

SOURCES: Calvert, 1974; Flörke, Jones, and Segnit, 1975; Jones and Segnit, 1971; Kastner, 1981; Kastner and Keene, 1975; Knauth, 1979; Laschet, 1984; Riech and von Rad, 1979; Tada and Iijima, 1983; L. A. Williams and Crear, 1985; L. A. Williams, Parks, and Crear, 1985.

PHOSPHATE ROCK

Sedimentary phosphate deposits usually consist of a carbonate fluorapatite $[Ca_5(PO_4)_3(F, CO_3)]$; a variety of apatite known as francolite is the basic raw material for many phosphorous-containing compounds on which modern technological developments depend. When the apatite-like phase cannot be identified, the name *collophane* is commonly applied. Such phosphates, or phosphorites (the two terms are synonymous), are here referred to as phosphate rock. Within the essential mineral of phosphate rock, other elements can substitute for the Ca, P (or PO_4), and F. Some of the elements that participate in these substitutions are potentially of economic interest. Many trace elements thus become concentrated in higher-than-normal values. Perhaps at some time in the future, uranium, rare earths, and other trace elements may be extracted from phosphate rocks.

Phosphate rock occurs on the modern sea bottom and in ancient deposits. Phosphate-rich sediments are present on the modern continental shelves of Africa, southeastern United States, southern California, northern South America, Australia, and New Zealand. Phosphate rock forms nodules on the modern sea bottom, extending from shallow- to great water depths. In some nodules the concentration of phosphate reaches 96%, but in others abundant impurities are included, such as organic matter, siliceous tests of organisms, calcareous shells, some of which are partly phosphatized, sharks' teeth, and various terrigenous particles.

Ancient deposits of phosphate rock form extensive phosphogenic provinces. These provinces include (1) the Upper Precambrian province of central and southeast Asia, (2) the Cambrian province of central and southeast Asia, extending into northern Australia, (3) the Permian province of North America, specifically the Phosphoria Formation, which extends through Wyoming, Utah, Colorado, Idaho, Montana, and Nevada, (4) the Jurassic-Lower Cretaceous province of eastern Europe, (5) the

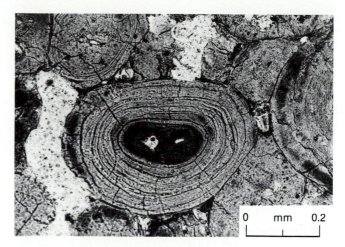

FIGURE 4-7.   Phosphate rock showing ooids composed of collophane. Phosphoria Formation (Permian), Idaho; photomicrograph in plane-polarized light of thin slice. (P. J. Cook.)

Upper Cretaceous-Eocene province of the Middle East and North Africa (Morocco, Egypt, Israel), extending into West Africa and the northern part of South America, and (6) the Miocene province of the southeastern United States. Each of these provinces extends for several thousand square kilometers.

In ancient deposits, phosphate rock may occur as bone fragments, sand-size pellets of possible fecal origin (100 to 400 μm in diameter), ooids (Figure 4-7), pisoids, phosphatized tests of calcareous shells, and large pellets (2 to 5 mm), known as **coprolites**, defined as *remains of animal feces, mostly of vertebrates, commonly 1 to 15 mm in length, whose color is light to dark brown or black, and shape usually elongate to ovoid*. Some phosphate rock displays cross strata. The average phosphate rock of the Phosphoria Formation contains 80% apatite; impurities include quartz, illite, organic matter, dolomite, calcite, and iron oxides.

The chemical reaction that results in the precipitation of phosphorite is still not known. Phosphorites are thought to form where deep phosphate-rich waters upwell adjacent to a shallow shelf. However, upwelling by itself may not be enough to cause precipitation. In places, phosphorites have replaced carbonate sediments or -rocks (See Figure 4-7.) and are composed of skeletal particles, ooids, and pellets. In this reaction apatite replaced $CaCO_3$, but the geochemical setting, the timing, and the depositional and diagenetic conditions are not known.

SOURCES: Altschuler, Jaffe, and Cuttitta, 1956; Bentor, 1980; P. J. Cook and Shergold, 1979; Germann, Bock, and Schröter, 1984; Kolodny, 1981; Notholt and Jarvis, 1989, 1990; Notholt, Sheldon, and Davidson, 1989; Riggs, 1986; Sheldon, 1987.

SEDIMENTARY IRON ORES

Iron-rich sedimentary rocks have been defined as sedimentary iron ores. In this usage the term is a general rock name rather than a value term based on economics. However, valuable iron-rich sedimentary rocks, true ores in the fullest sense of that term, have influenced modern history. England achieved greatness through exploitation of her sedimentary iron ores, while at the same time mining her rich coal deposits. During the nineteenth- and early part of the twentieth centuries Germany and France were at each other's throats over possession of the vital Lorraine deposits, the famous oolitic minette ores of Jurassic age. Discovery of the vast Lake Superior iron ore deposits in 1844 ushered in the industrial age for the United States. In Europe and in North America, where sedimentary iron ores and coal met, large industrial centers arose, particularly in England, France, Germany, and the United States.

Although all sedimentary rocks contain readily detectable amounts of iron, the term iron-rich deposits designates those consisting predominantly of iron minerals, notably oxides, hydroxides, carbonates, silicates, and sulfides. The oxides are hematite and magnetite; the hydroxide mineral is goethite (including limonite); the carbonate mineral is siderite; the silicate minerals are chamosite, greenalite, and glauconite; and the sulfide minerals are pyrite and pyrrhotite.

One of the remarkable features of sedimentary iron ores is that the mineral compositions in those of Precambrian age contrast with those of younger (Phanerozoic) ages. In Precambrian deposits hematite typically is interbedded with chert. By contrast, in Phanerozoic deposits hematite replaces ooids and fossil fragments or forms an earthy matrix. This change in the distribution of hematite with time is probably related to the biologically driven buildup of oxygen in the Earth's atmosphere during the late Precambrian. As a result, oxidized iron in the form of hematite is thought to have been precipitated on a massive scale.

Goethite is found mostly as a replacement of ooids. Siderite forms sparry cement or microspar or replaces chamosite or the various kinds of calcareous particles, notably ooids, as well as matrix or cement of limestones. Siderite also forms layers or may be restricted to discontinuous concretions. Such layers and concretions, which are in fact siderite rock, are commonly referred to as clay ironstones. Chamosite, the most-common Phanerozoic iron-silicate mineral, occurs mostly as chamosite mudstone or as a constituent of ooids and less commonly as scattered flakes in the matrices of oolitic deposits. Greenalite is almost entirely restricted to Precambrian rocks; in these it forms small dark-green granules. Because this discussion is limited to authigenic rocks, we exclude such detrital deposits as magnetite- or ilmenite sands or -sandstones.

Sedimentary iron ores are classified into three broad groups: (1) *bog-iron deposits*, (2) *ironstones*, and (3) *iron formations*. The basis of the classification combines mineral composition, occurrence, and geologic age, as will be apparent in the ensuing discussion. In addition, sedimentary iron ores may be named according to their dominant iron mineral; e.g., siderite rock, hematite rock, or magnetite rock.

Swamps and lakes of glaciated areas are the sites of bog-iron deposits, composed of goethite. The particles of such deposits may be oolitic and are cemented to form lenses or layers. In some lakes such deposits contain high concentrations of manganese (Chapter 2). In the waters of swamps and lakes, carbon dioxide, a product of plant decay, is abundant. In the vadose zone (zone above the ground-water table) CO_2 dissolved in running water combines with iron to form ferrous bicarbonate $Fe(HCO_3)_2$ (Figure 4-8). If the freely moving ground water in the phreatic zone contains oxygen, then the waters above the ground-water table, rich in iron bicarbonate, and the waters below the ground-water table, rich in oxygen, interact. The ferrous bicarbonate is oxidized and limonite forms along the ground-water table. (See Figure 4-8.) Bacterial activity may also be involved in the formation of these deposits.

Ironstones are ferruginous rocks that are (1) of Phanerozoic age and (2) essentially devoid of chert. In these two respects ironstones differ from iron formations. Ironstones compose the deposits that have served as the bulk of European iron resources. These deposits stretch over areas of many square kilometers. Ironstones are interbedded with limestones, shales, sandstones, and conglomerates. In the United States, the Silurian Clinton iron ores extend discontinuously from New York to Alabama. One kind of Clinton ferruginous rock consists of ooids and fragments of bryozoa, crinoids, trilobites, and various other skeletal debris partly coated with or replaced by iron oxide. Another consists entirely of ooids of hematite in a calcite matrix. A third is composed of small, flattened concretions of hematite containing fragments of fossils replaced by hematite. A fourth is a pebbly conglomerate or sandstone in which all particles have been coated with hematite and the rock has been cemented with hematite and calcite. The Middle Jurassic minette ores of Alsace-Lorraine, the former bone of contention between France and Germany, are oolitic limonite-chamosite-siderite rocks, just as is one kind of sedimentary iron ore of the same age in England, known as the Northampton sand ironstones.

Both the American- and the European rocks are interpreted as products of nearshore, restricted, shallow-marine environments. The source of the iron for these

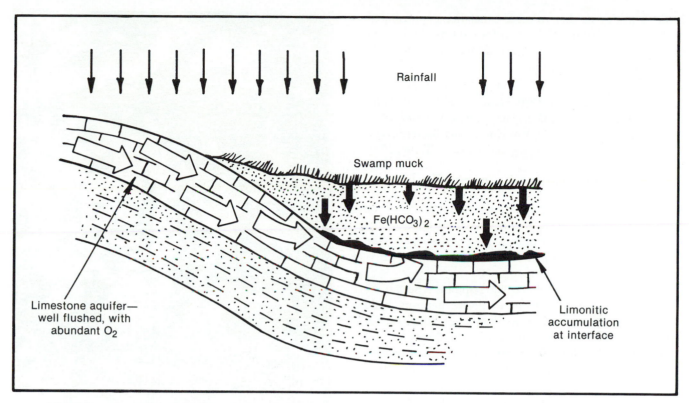

FIGURE 4-8.   Schematic profile-section illustrating deposition of iron oxides as a result of the interaction of soil waters from overlying marsh with underlying well-oxygenated water of limestone aquifer. (Modified from H. Borchert, 1960.)

deposits is thought to have been deeply weathered continental rocks. In all igneous rocks of continental land masses, small amounts of iron are found in disseminated ilmenite, -hematite, -magnetite, and -pyrite; ferrous iron is present in ferromagnesian silicates, such as olivine, pyroxenes, amphiboles, and biotite. Deep weathering releases this iron from the rocks and various iron salts form. After the iron has thus been released from the bedrock, streams carry away these iron salts in suspension or in solution to the sea, where ferrous iron is oxidized and deposited. Such precipitation also involves replacement of carbonate particles and matrix or cement under as-yet-unknown conditions. Alternative- or additional sources for the iron may be submarine volcanoes or upwelling currents carrying iron derived from deeper parts of the ocean floor.

Iron formations are of Precambrian age. Typically, these ferruginous rocks are prominently interbedded with quartz or chert. Such well-developed bedding has led to the name *banded-iron formation* for these deposits (Figure 4-9). Hematite, magnetite, and less commonly, siderite are the constituent minerals of iron formations. The iron mineral of certain iron formations is pyrite. Some of these deposits extend through hundreds of kilometers. The classical examples of banded-iron ores are those of the Labrador-Quebec-Ontario-Wisconsin-Michigan-Minnesota belt, especially those near the southwestern tip of Lake Superior. They are well known from a long history of geologic research. Yet despite this research, the origin of these deposits is controversial. They are little better understood today than they were 50 years ago.

FIGURE 4-9.   Exposure of banded-iron formation showing interbedded magnetite (dark gray to black) and quartz (light gray to white). Precambrian, Algoma District, Ontario, Canada. (Authors.)

Little doubt exists that volcanic exhalations contributed to the origin of some of the iron formations. The origin of others is in doubt. The subject of their origin has been a fertile one for speculation.

SOURCES: Button, 1976; Dimroth, 1975; G. M. Friedman, 1959; G. A. Gross, 1980; Guilbert and C. F. Park, Jr., 1986; Gygi, 1981; D. B. James, 1966; Maynard, 1983; Van Houten, 1990; Van Houten and Bhattacharyya, 1982; T. P. Young and W. E. G. Taylor, 1989.

Classifying and Naming Evaporite Rocks

Evaporites form by precipitation from brines whose salinity values have been greatly increased by evaporation, as explained in Chapter 3. Evaporation may take place in closed- or in open systems, in numerous environmental settings.

Although almost 40 different precipitate-type minerals have been recorded from evaporite deposits, only about 20 are present in more than trace amounts. Of these, only two kinds are common, (1) sulfates and (2) halides, both of which form extensive deposits of sedimentary rock. Evaporite rocks are named from their constituent minerals. *Evaporite rocks composed of calcium sulfate* are gypsum rock or anhydrite rock. *Evaporite rocks composed of sodium chloride* are halite rock (rock salt). Sometimes the word "rock" is dropped and these rocks are simply termed gypsum, anhydrite, and halite; either usage is acceptable.

Two kinds of sulfate minerals are common in sedimentary rocks: (a) gypsum ($CaSO_4 \cdot 2H_2O$) and (b) anhydrite ($CaSO_4$). Although anhydrite forms in the modern sea-marginal flats of Abu Dhabi in the Persian Gulf, for all practical purposes, gypsum, as the hydrated sulfate, occurs at the Earth's surface, and anhydrite, the anhydrous sulfate, in the subsurface. The change from one kind of sulfate to the other depends on the geothermal gradient as well as on the salinity of the subsurface brine. Anhydrite generally will replace gypsum at depths of 300 to 700 m (Figure 4-10). Depending on the availability of water, gypsum from anhydrite will appear at varying depths, normally between 100 and 300 m.

Both anhydrite and gypsum rocks may be nodular, massive, or laminated (Figure 4-11). Gypsum rock may display graded layers, cross strata, scour-and-fill structures, ripples, and other features resulting from the interaction between a current and sediments. Gypsum rock consisting of selenite crystals may grow in rosette- or radial patterns, with some crystals 2 to $2\frac{1}{2}$ m in length. In the subsurface, anhydrite is usually compact and devoid of pore space. Such anhydrite may act as a caprock for reservoirs of oil and gas.

Among the halides, halite rock is the most common. It forms successions up to 1000 m thick. Halite rock is commonly laminated; dark, inclusion-rich layers al-

1 cm

FIGURE 4-10.   Acid-etched sample of core from borehole showing laths of anhydrite pseudomorphs after gypsum in a limestone matrix. The anhydrite displays the habit of gypsum, proving the original sulfate mineral was gypsum; in the subsurface the gypsum changed to anhydrite. Salina Group (Silurian), Michigan. (R. D. Nurmi.)

ternate with light, inclusion-poor layers. Inclusions consist predominantly of anhydrite. Although they are much less abundant than anhydrite, other impurities include dolomite, quartz, calcite, and various clay minerals. The clay mineral talc is of authigenic origin in evaporites and is especially common in halite rock. Freshly broken samples of halite rock commonly emit a strong odor of H_2S, suggesting that much sulfate has been reduced according to Eq. 3-10. Nonrecrystallized and nondeformed halite rock may show various sedimentary structures, including

400 μm

FIGURE 4-11. Anhydrite rock showing anhydrite nodules with stellate arrangement of crystals; small anhydrite laths are interspersed between the nodules. From Miocene strata encountered in borehole below floor of western Mediterranean Sea (Site 124, Deep Sea Drilling Project); photomicrograph in plane-polarized light of thin slice. (G. M. Friedman, 1972b, fig. 5, p. 698.)

those of stromatolites and nodules (Figure 4-12). Such halite rock has formed at the expense of organic matter, carbonate, and anhydrite. Anhydrite and dolomite are commonly interlaminated with the halite.

Halite rock commonly recrystallizes and moves as a solid; this movement is referred to as salt flowage. Such recrystallized halite is composed of centimeter-size crystals having a tightly interpenetrating framework.

SOURCES: Müller, 1988; Schreiber and Friedman, 1976; Shearman, 1966; Warren, 1989.

Classifying and Naming Extrabasinal Rocks

As mentioned, extrabasinal rocks consist mainly of one group. These are the *terrigenous* rocks.

Classifying and Naming Terrigenous Rocks

As we have seen in Chapter 2, where we discussed terrigenous particles, a fundamental basis for classification is size. As with particles, so it is with rocks formed by lithification of the particles. Therefore, on the basis of increasing particle size, we recognize *three main kinds of terrigenous rocks*: (1) *shales*, (2) *sandstones*, and (3) *conglomerates* and *sedimentary breccias*. Earlier in this chapter, in our discussion of the Grabau system of classification, we introduced the terms *lutite* for shale, *arenite* for sandstone, and *rudite* for conglomerates. Some geologists recognize a fourth kind of terrigenous rock, *siltstone*. For it, the term *siltite* has been introduced. (The addition of *ite* from the Greek to the ordinary English word silt is a true example of a Greek barbarism.) We shall discuss how to classify and name each of the three main kinds of rocks separately. We discuss siltstones along with shales.

SHALES

Fine-textured terrigenous rocks compose about two thirds of the sedimentary rock record, yet because their particles are so small, they are poorly known and incompletely understood. Naming and classifying shales is complicated. Various bases for naming these rocks have been proposed, including (1) particle size, (2) proportion of clay minerals, and (3) **fissility**, *the property of splitting easily into thin layers parallel to the bedding.*

Probably the most-satisfactory definition of shale is according to particle size. In terms of size, the name shale is the lithified equivalent of mud. We have defined mud as a sediment consisting of clay- and silt-size particles. Likewise, **shales** are *terrigenous rocks composed of clay- and silt-size particles.* Surprisingly, despite the technique of breaking down a shale into its constituent particles using immersion of a sample in a bath of an ultrasonic transducer, few reliable data exist on the size-frequency distribution in shales. This is true partly because clay minerals are commonly deposited in clumps called floccules, which are typically of silt size. Simple hand-specimen inspection indicates that many rocks named shales by geologists are really siltstones; these rocks consist of particles of silt- rather than of clay size (Chapter 2). Also, clay-mineral crystals may grow after deposition, and so become welded together. Despite the use of the term shale for fine-textured sedimentary rocks spanning both the clay- and silt-size ranges, it is important to distinguish the dominantly *clay shales* from the dominantly *silt shales*. This is so because clay- and silt-size particles differ mineralogically (dominantly clay minerals for clay-sizes versus dominantly quartz for silt-size particles), respond differently to shear stress (particles of clay minerals are cohesive), indicate different environmental energy

FIGURE 4-12.   Samples of core from borehole showing stromatolites preserved in anhydrite and halite. Light gray projecting layers consist of anhydrite; dark gray recessed layers consist of halite. Salina Group (Silurian), Michigan. (R. D. Nurmi.)

levels, and originate in different ways. Most silt-size particles are composed of quartz and feldspar fragments of originally larger particles in igneous- and metamorphic rocks, whereas clay-size particles are dominantly clays, which are chemical precipitates or alteration products. Thus clays enter the fine end of the shale particle-size distribution by crystal growth and flocculation, whereas detrital quartz and feldspar particles enter the coarse end by comminution. Some geologists like to make a distinction between shales and siltstones; others are satisfied to lump under the term shale all fine-textured terrigenous rocks.

The proportion of clay minerals in shales varies widely. The commonly accepted notion that shales are composed almost entirely of clay minerals is not correct. In most shales, quartz may compose between one quarter and one half of the total; in some shales, quartz together with feldspar exceeds 50% of the rock. Increasing proportions of quartz and feldspar generally parallel increasing particle size; thus coarse shales generally contain greater proportions of quartz and feldspar, whereas fine shales contain less of these and a greater proportion of clay minerals. *Fine-textured terrigenous sedimentary rocks having abundant quartz (and commonly also feldspar) particles of silt size and that as a result, closely resemble sandstones,*

are commonly referred to as **siltstones**. Generally, siltstones lack fissility. By contrast, fine-textured shales contain more clay minerals, which are generally (but not necessarily) oriented parallel to bedding, imparting fissility to the shale. Thus fine shales differ from sandstones and from siltstones not only in particle size but in bulk physical properties as well.

Fissility is a characteristic of the depositional fabric that is maintained only when burrowing organisms are absent. Therefore the presence or absence of fissility in shales is of considerable environmental significance (i.e., burrowers excluded as a result of environmental restriction). A fissile shale was not burrowed before it was lithified. According to the definition used by some geologists, a shale must display fissility. More generally, however, the term shale has been applied to all very fine-textured sedimentary rocks (rocks with particles in the clay- to silt-size range) that contain substantial amounts of clay minerals, whether these rocks are fissile or not. On surface exposures, where weathering has taken place, it is easy to distinguish fissile from nonfissile fine-textured rocks. In subsurface cores or samples fissility is less obvious. The terms *mudrock, mudstone, claystone,* and *argillite* have been employed for shales lacking fissility; argillites are more-indurated shales than mudstones and claystones. Recall that in the Dunham classification discussed earlier in this chapter, the term mudstone refers to fine-textured limestone (in which case the rock should be called lime mudstone to avoid confusion with fine-textured siliciclastic rocks).

A dark, thinly laminated carbonaceous shale, exceptionally rich in organic matter (5% or more of total organic carbon) and sulfides (FeS$_2$), is known as **black shale**. It forms by partial anaerobic decay of organic matter in a reducing setting in which water circulation is restricted and deposition slow. Black shale is commonly radioactive as a result of uranium enrichment, and can be very useful for correlation.

SOURCES: Bates and Jackson, 1987; Byers, 1974; Chamley, 1989; Dunbar and Rodgers, 1957; Füchtbauer and Leggewie, 1984; Heling, 1988; Pettijohn, 1960, 1975; Potter, Maynard, and Pryor, 1980; Spears, 1980; C. E. Weaver, 1980.

SANDSTONES

Sandstones are *terrigenous sedimentary rocks in which sand-size particles predominate*. (See Figure 2-1.) As with limestones, sandstones can be classified on the basis of (1) particles, (2) matrix, and (3) cement. Particles are mostly those derived from the lands, hence are terrigenous, but pyroclastic and even calcium-carbonate particles may be abundant. In Chapter 2 we discussed in some detail the kinds of particles that compose sands and sandstones. The essential particles of sandstones are (1) quartz, (2) feldspar, and (3) rock fragments. As we have seen in Chapter 2, many other kinds of particles occur in sands and sandstones, but those others are not common and are not considered essential in naming and classifying sandstones. Quartz occurs in several varieties, and feldspars and rock fragments are of various kinds. The matrix is the mud: physically deposited material that consists mostly of clay minerals and quartz. Cement is the chemically precipitated filling of original void spaces.

The distinction between framework particles and matrix or cement is our basis for classifying sandstones into major groups. Some sandstones lack any matrix; only mineral cement occupies the spaces among the framework particles. This cement is equivalent to the spar

of limestones. As with the micrite of limestones, the argillaceous matrix of sandstones is considered to imply (1) availability of mud and (2) a low-energy regime in the depositional environment. (Note that the difficulty of eroding cohesive clay particles allows extrabasinal mud to be deposited and preserved in an environment which episodically experiences relatively high energy levels; this is not true of lime muds.) Also, as with micrite in limestones, an argillaceous matrix in a sandstone may filter postdepositionally into the pores among sand-size particles, or during diagenesis clay minerals may be precipitated in pore spaces. Clays may be mixed with sand-size particles by bioturbation or by mass wasting (e.g., slumping) after deposition of fine- and coarse particles in separate layers. Even with the scanning-electron microscope it may be difficult or impossible to distinguish a primary argillaceous matrix that resulted from the conditions of sorting in the environment of deposition from diagenetically formed clay minerals which are, in fact, cement. Very often, perhaps, the argillaceous matrix is derived from *in-situ* degradation of unstable sand-size rock fragments or from crushing of sand-size clay clasts. Unfortunately, we do not yet know the relative abundance in sandstones of each of (1) mechanically deposited clay-mineral matrix, (2) fine-textured matrix filtered postdepositionally into pores, (3) finely crystalline clay-mineral cement, and (4) other sources of clay-size material in sandstones.

Depending on the presence or absence of argillaceous matter, we recognize two end-member groups of sandstones: (1) argillaceous sandstones (sandstones containing about 15% or more of clay-size material), and (2) ordinary sandstones, which contain less than about 15% clay-size material. The value of 15% is arbitrary; in a qualitative study, the obvious presence of a fair amount of clay suffices to classify a sample as argillaceous.

Based on the proportions of quartz, feldspar, and rock fragments, we can divide each of these two first-order groups (argillaceous sandstones and ordinary sandstones) into three second-order groups. Thus, as indicated in Figure 4-13, we can recognize six sandstone end members: (1) *quartz sandstones,* (2) *feldspar sandstones,* (3) *rock-fragment sandstones,* (4) *argillaceous quartz sandstones,* (5) *argillaceous feldspar sandstones,* and (6) *argillaceous rock-fragment sandstones.*

End-member sand-size particles	Idealized end-member rock names	
	< 15% matrix	> 15% matrix
Quartz	Quartz sandstone	Argillaceous quartz sandstone
Feldspar	Feldspar sandstone	Argillaceous feldspar sandstone
Rock fragments	Rock-fragment sandstone	Argillaceous rock-fragment sandstone

FIGURE 4-13. Classifying and naming sandstones.

These end-member names would be used only in those examples where all the sand-size material consists exclusively of quartz, feldspar, or rock fragments. Although sandstones consisting exclusively of quartz are not unusual, few are known in which the sand-size debris consists entirely of anything else. The sand-size debris of most sandstones includes chiefly quartz, but with the quartz may be variable amounts of feldspar or rock fragments or feldspar and rock fragments. These mixtures are named by incorporating the names of all kinds of sand-size debris present, listed in increasing order of abundance. Thus if a sandstone contains 70% quartz, 20% feldspar, and 10% rock fragments, we would name it a rock fragment-feldspar-quartz sandstone. The presence of the two modifiers preceding the word quartz indicates that the rock is not a pure quartz sandstone. If sufficient matrix is present for the adjective argillaceous to be used, then argillaceous always precedes the names based on kinds of sand-size particles. Thus a rock containing 20% of fines as matrix and the proportions of sand-size debris given above, would be named an argillaceous rock fragment-feldspar-quartz sandstone.

Although classifying and naming sandstones are descriptive procedures, just as with limestones, genetic implications lurk beneath the surface. It has been generally considered that the presence or absence of clay-size particles is a function only of the level of energy in the original depositional environment. However, as we have pointed out, other factors, such as the availability of mud in the depositional environment, postdepositional filtering of clay minerals into a sand, and diagenetic formation of clay-mineral cements within a sandstone, may confound the simple use of the argillaceous content for interpreting the energy levels at the site of deposition. We considered, but rejected, the idea of using the term *muddy* instead of argillaceous. This is because the term muddy carries the implication that the matrix was *deposited with* the sand-size particles, and *in a fine-textured form*. As we have just explained, this may often not be the case.

The kinds of feldspars or rock fragments reflect (1) the parent terrain or provenance from which the particles in a sample of sandstone have been derived and (2) their preservation during weathering, transportation, and diagenesis.

The pioneer sedimentary petrographer Paul D. Krynine (1902–1964) provided the modern impetus in naming and classifying sandstones. Krynine's (1948) original article is a classic in geology; it combines astute observation in the field and of samples with a deep insight and appreciation for the fundamental principles of geology. Krynine's contribution is felt to this day. In his classification of sandstones, Krynine introduced triangular diagrams for plotting his three fundamental end members: *orthoquartzite*, *graywacke*, and *arkose* (Figure 4-14).

The use of the last two of these terms has since led to misunderstanding and confusion. The earliest description of graywacke dates from 1789, and the term arkose was introduced in 1823. The history of both terms had been checkered even before Krynine included them in his classification. According to Krynine, a **graywacke** is *a sandstone composed of angular quartz (and chert) particles and abundant metamorphic rock fragments, with little or no cement and feldspar, and containing more than 12 to 17% micas and chlorite (either in the clay matrix or as metamorphic rock fragments).* Krynine felt that the mica and chlorite between the particles were indeed a mechanically deposited matrix and represented poor sorting; we now know that this can no longer be assumed. When deeply buried and subjected to appropriate geothermal gradients and migrating solutions, framework particles may be dissolved and new clay minerals, such as chlorite and illite, may form. This diagenetic neoformation of clay-mineral cements within sandstones may generate a rock composition fitting the term graywacke. Yet Krynine and many others use this term to refer to a rock having a mechanically deposited matrix and poor sorting. To avoid the term graywacke, some authors substituted the term *wacke* (Figure 4-15). In his classification, Krynine used the term **arkose** for *a sandstone with more than 30% feldspar.*

In our scheme of naming sandstones, a rock defined by Krynine as a *graywacke* would be named an *argillaceous rock fragment-quartz sandstone* if it contained no feldspar, and an argillaceous rock fragment-feldspar-quartz sandstone if it contained feldspar in greater abundance than rock fragments, or an *argillaceous feldspar-rock fragment-quartz sandstone* if the rock fragments were more abundant than the feldspar. The arkose of Krynine would

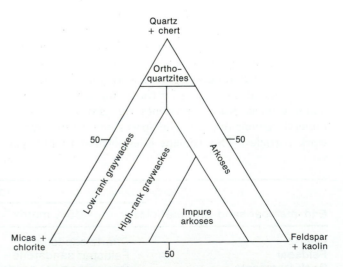

FIGURE 4-14.   Triangular diagram; Krynine's classification of sandstones. (After P. D. Krynine, 1948, fig. 4, p. 137, reprinted with permission of the University of Chicago Press.)

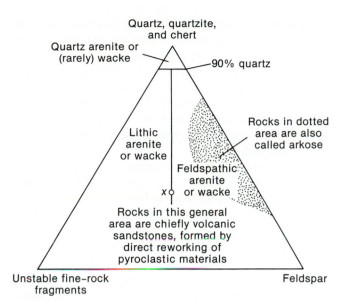

FIGURE 4-15. Triangular diagram; C. M. Gilbert's classification of sandstones. (American Geological Institute, 1989, Data Sheet 32.2. From a diagram supplied by C. M. Gilbert. Reprinted from the AGI Data Sheets, 3rd edition, by permission of the American Geological Institute.)

be named by us a *feldspar-quartz sandstone* if the quartz is more abundant than the feldspar, but a *quartz-feldspar sandstone* if feldspar abundance exceeds quartz abundance.

Since Krynine proposed his original classification of sandstones, approximately 50 new classifications, based mostly on Krynine, have been proposed in more than ten countries and in seven languages. Most of these classifications are variants on his theme. Because we prefer simple words in naming and classifying sandstones and use visual estimates for compositions, we have adopted in this chapter the simple scheme of Figure 4-13 rather than the elaborate classifications based on Krynine or variants thereof. Our classification of sandstones parallels those for limestones, which we discussed earlier in this chapter. No agreement currently exists among geologists about which sandstone classification is preferred, and no signs suggest that such an agreement is in the offing.

SOURCES: Chang, 1967; Folk, 1956; Füchtbauer, 1988; C. M. Gilbert, 1954; G. deV. Klein, 1963; McBride, 1962; Sanders, 1978; F. L. Schwab, 1975.

CONGLOMERATES AND SEDIMENTARY BRECCIAS

Coarse terrigenous rocks are formed by the lithification of gravel. The degree of rounding of the particles defines two first-order categories. **Conglomerate** is defined as *a coarse terrigenous sedimentary rock formed by the lithification of rounded gravel.* **Sedimentary breccia** is a *coarse terrige-*

nous sedimentary rock formed by the lithification of angular gravel-size or larger particles. In conglomerates and sedimentary breccias more than 30% of the large particles exceed 2 mm (-1ϕ) in diameter. The particles may consist of pebbles, cobbles, boulders, or mixtures of these sizes.

Conglomerates and sedimentary breccias may be named and classified by (1) the proportion of gravel-size particles, (2) the kind of matrix, and (3) the kinds of gravel-size particles.

According to the proportion of gravel-size particles and the kinds of matrix, (1) a sample containing 80% pebbles, cobbles, and/or boulders is termed a conglomerate proper, and (2) one having between 30 and 80% of such coarse particles is (a) a sandy or arenaceous conglomerate or (b) a shaly or argillaceous conglomerate. The matrix between the coarse particles in a conglomerate may also be calcareous or sideritic. As we have seen in this definition, a rock is named a conglomerate with only about 30% coarse particles; that is, the bulk volume of the sandy- or shaly material may exceed that of the gravel-size particles. The reason for lowering the limit for gravel-size particles below 50% in defining the rock as conglomerate is the significance attached to the presence of such coarse particles. The proportion of gravel is a function of (1) the highest current speed at the time of deposition and (2) the availability of such coarse particles. Moreover, experience has shown that in outcrop or in hand samples, geologists will name a rock a conglomerate with the proportion as low as, and even lower than, 30% gravel-size particles. Pebbles just stand out like plums in a pudding, and accordingly the rock is named conglomerate.

Based on the variety of pebbles, cobbles, and/or boulders composing them, we can classify conglomerates into two kinds: (1) *conglomerates consisting only of pebbles, cobbles, and/or boulders of a single kind of rock* (such as one or various varieties of chert and quartzite or of other rock), or **oligomictic conglomerates**; and (2) *conglomerates containing pebbles, cobbles, and/or boulders of many kinds of rock*, known as **polymictic conglomerates**. The term *mictic* derives from the Greek word *miktos*, meaning mixed; *oligo-* comes from *oligos*, meaning few, and *poly-* from *polys* meaning many, both terms also Greek.

An example of an oligomictic conglomerate is the conglomerate of the Cardium Formation (middle Upper Cretaceous) of Alberta, Canada (Figure 4-16). As with many oligomictic conglomerates, the Cardium Conglomerate contains pebbles that are resistant to wear and decomposition: quartzite, chert, siliceous shale, and radiolarian chert; some quartzite pebbles have been brecciated and healed with vein quartz. The pebbles are markedly worn and well rounded. Their surfaces are smooth and polished, and many exhibit indentations or pits. The colors of individual chert pebbles include white,

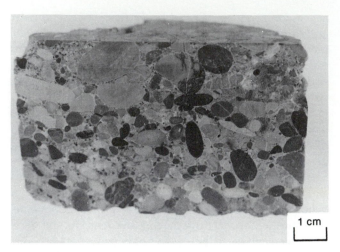

FIGURE 4-16. Core from borehole showing well-rounded chert pebbles in oligomictic conglomerate. Cardium Formation (Cretaceous), Pembina, Alberta, Canada. (Authors.)

gray, green, blue, and black. Pebbles generally make up an estimated 65 to 85% of the bulk composition of this conglomerate. Ellipsoidal pebbles are locally oriented with their axes of elongation parallel to the bedding. The finer particles between the pebbles consist of quartz and chert of sand- and silt size as well as of small pebbles of a size that falls close to the upper boundary of sand size.

Some geologists claim that oligomictic conglomerates are derived from a deeply weathered terrain in which, by long transport, all but the most-resistant pebbles have been eliminated. Among the kinds of rock most prone to wear and tear and early elimination are limestone clasts, which do not survive extensive abrasion. Yet numerous examples of oligomictic conglomerates consisting of limestone clasts exist in the rock record. Most of these conglomerates accumulate on the downthrown side of an intermittently active fault where the upthrown side contains limestones. Examples are conglomerates (fanglomerates) with limestones and boulders along the fault scarps near the margin of the rift valley of Israel; the limestone fragments were derived from local Cretaceous rocks. (See Figure 14-7.) Another example, in the Book Cliffs near Price, Utah, consists of a Cretaceous limestone conglomerate in which the carbonate-rock debris accumulated in a subsiding area adjacent to a rising tectonic arch. (See Figure 2-11.) It is clear that the diversity of rock types in the bedrock of the parent area, as well as the extent of weathering and particle transport, exerts a powerful influence on the composition of particles in conglomerates.

Rocks composed of limestones of pebble size that are intrabasinal, such as the *intramicrudite* or *intrasparrudite* that we noted earlier in this chapter (for example,

the flat-pebble conglomerates of tidal environments of Chapter 12), are considered in this chapter under limestones. Strictly speaking, such rocks are conglomerates. Indeed, they are oligomictic conglomerates, but they are also limestones. As we have remarked earlier in this chapter, rocks may be classified by composition, such as limestone, or by texture, such as conglomerate. In the classification followed in this chapter, we have used chiefly composition.

An example of a polymictic conglomerate is the Upper Devonian Twilight Park Conglomerate of the Catskill Mountains in New York State (Figure 4-17). The pebbles are composed of vein quartz and of various sedimentary- and metamorphic rock fragments. The fragments of sedimentary rocks include fossiliferous limestones and dolostones which bear striking similarity to older limestones and dolostones exposed nearby, fine- to medium-textured sandstones, various kinds of shale including siltstones, and metamorphic rock fragments, such as phyllites, quartz-muscovite schists, and quartz-chlorite schists. The pebbles are mostly well sorted and -rounded. This conglomerate is probably a product of braided streams. Streams of high energy must have been at work to transport pebbles ranging in diameter from 3 to 10 cm. This product of a complex stream system developed on a slope of steep gradient, most probably in association with a series of coalescing fans that spread westward from the high, tectonically active parent terrain to the east. To the west, these conglomerates interfinger with fluvial- and deltaic sediments, which we shall discuss in Chapters 14 and 13.

The term **diamictite** has been proposed for *a nonsorted, noncalcareous, terrigenous sedimentary rock composed of sand-*

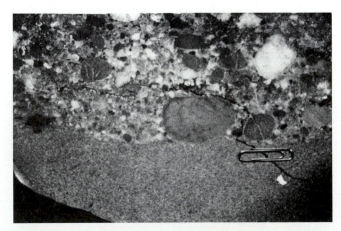

FIGURE 4-17. Slab of polymictic conglomerate overlying sandstone showing rounded pebbles of vein quartz and of various sedimentary- and metamorphic rock fragments. Upper Devonian, Hunter, New York. Paperclip = 3 cm. (Specimen from Rensselaer Center of Applied Geology.)

FIGURE 4-18.   Exposure of diamictite, a lithified former till, showing pebbles and boulders, mostly gneisses, embedded in sandy- and argillaceous matrix. Large boulder in center of photograph is an erratic of gneiss. Adelaide System (Precambrian), Sturt Gorge near Adelaide, South Australia. (Authors.)

size and/or larger particles dispersed through a fine-textured matrix. Many, if not most, diamictites are conglomerates (Figure 4-18). Diamictites originate by landslides, earth flows, mudflows, solifluction, ice rafting, subaqueous slumping and sliding, and/or by glaciers. A non-lithified diamictite is a diamicton.

SOURCES: Dunbar and Rodgers, 1957; Flint, Sanders, and Rodgers, 1960a, b; Füchtbauer, 1988; Hubert, 1960; Pettijohn, 1975.

Classifying and Naming Pyroclastic Rocks

Pyroclastic rocks are lithified tephra. Their particles originate as explosive igneous material but are deposited as sediments. Sedimentary petrographers have neglected the study of these rocks, in part because the particles are of igneous origin and in part because most early classical studies in sedimentology took place in areas removed from active volcanism. Furthermore, because of their chemical instability, pyroclastic rocks are subject to rapid diagenetic alteration, which makes those of Precambrian-, Paleozoic-, and even Mesozoic ages difficult to identify.

Volcanoes and their products, both on land and in the sea, are the major source of particles for pyroclastic rocks, especially in areas of tectonic activity and in island arcs. Particles originating by explosive ejection from volcanic vents spread out over great areas. (See Figure 2-21.) On the continents, pyroclastic rocks far exceed the volume of extrusive igneous rocks. With increasing distance

from their parent volcanoes, tephras from low-altitude suspensions become thinner and finer. (See Figure 1-10.) Also, tephras are chemically and texturally distinctive, so ancient tephras may be used for correlation. Mapping of particle size and thickness of tephra deposits assists in locating the parent volcano.

Figure 4-19 shows how sizes of tephra particles can be used to classify pyroclastic rocks. The terms bombs, blocks, lapilli, and ash are defined in Chapter 2.

Pyroclastic rocks composed mostly of blocks are **volcanic breccia**; pyroclastic rocks composed mostly of lapilli are **lapilli tuffs**; and *pyroclastic rocks composed of ash* are **tuffs** in the strict sense of the word.

Classifying a pyroclastic rock on the basis of composition, that is, by the kinds of particles composing it, is usually difficult. The individual constituents of many pyroclastic rocks consist of tiny crystals or pieces of glass. Of these, glass predominates. Where glass is absent and the constituents are large enough to be recognized in a thin slice under the petrographic microscope, further subdivision becomes possible. Nonglass constituents include volcanic rock fragments or crystals of various minerals, as shown in Figure 2-9.

The three end members of pyroclastic rocks are (1) volcanic rock fragments, (2) crystals, and (3) glass. In a thin slice under the petrographic microscope an operator estimates (or point counts) the relative proportions of these three end members. If by size a rock is a tuff and rock fragments exceed glass, this tuff may be described as a *glass, rock-fragment tuff.* End members are listed in order of increasing abundance. By this approach one can recognize *crystal-, rock-fragment-,* and *glass tuffs.* Adjectival modifiers may be added.

Because of the chemical reactivity and instability of their constituents, pyroclastic rocks are highly susceptible to diagenetic alteration. Glass alters to clay min-

Limiting Particle Diameter mm φ units	Standard Size Classes of Sediments	Size Classes of Tephra Particles	Size Classes of Pyroclastic Rocks
	Cobbles	Blocks / Bombs	Agglomerate / Volcanic breccia
64 ——— −6	Pebbles	Lapilli	Lapilli tuffs
2 ——— −1	Sand	Coarse ash	Tuffs
1/16 ——— +4	Silt	Fine ash	

FIGURE 4-19.   Size classes and terms for tephra particles and pyroclastic rocks. (Modified from R. V. Fisher, 1966b.)

erals, especially to smectite, zeolites, chalcedony, opal, quartz, or to a microcrystalline material, which in a thin slice under the petrographic microscope resembles chert. Glass is commonly found only in mid-Tertiary and younger rocks. Even zeolites are absent from Lower Paleozoic rocks. Such old zeolites probably become altered to feldspars. An alteration product of tephra is **bentonite**, *a plastic clay or shale composed for the most part of the clay mineral montmorillonite (a variety of smectite) that formed by the alteration of tephra.*

One kind of tuff that carries a special name is **ignimbrite** or **welded tuff**, *a nonsorted pyroclastic rock deposited from a* nuée ardente *while the particles were still in a plastic condition.* Ignimbrites may occupy large areas and attain considerable thicknesses. Ignimbrites are of economic interest as well, because their low permeability values have aroused interest in the potential storage of hazardous wastes within subsurface ignimbrites. At Yucca Mountain, Nevada, a succession of tuffs up to 3300 m thick formed between 8 and 16 million years ago in Middle- to Late Miocene time. The U.S. Department of Energy is studying the possibility of storing nuclear waste in a welded tuff near the middle of the volcanic pile. If the site is approved it would become the first geological repository for spent nuclear fuel and high-level waste in the United States.

SOURCES: R. V. Fisher, 1961, 1966a; Lajoie, 1979; Schmincke, 1988; U.S. Department of Energy, 1988.

Classifying and Naming Carbonaceous Rocks

Carbonaceous rocks, primarily coal, have been classified in two ways: (1) by rank (content of carbon and caloric value) and (2) by petrographic characteristics.

In order of increasing rank, the classes of coal are (1) lignite, (2) subbituminous coal, (3) bituminous coal, and (4) anthracite. *The progression of carbonaceous material through the continuous series of lignite (or its precursor, peat) through bituminous coal to anthracite* is known as **coalification**. During coalification the color of the products darkens and their luster increases. Also during this progression, their relative contents of carbon and caloric value gradually rise, and their contents of moisture, volatile matter, and oxygen decrease (Figure 4-20). Except in anthracite (in which hydrogen decreases), the proportion of hydrogen remains roughly constant through this series. Coalification is accomplished in two stages: (1) biochemical and (2) geochemical. In the biochemical stage, bacteria convert plant material to peat. During the geochemical stage, both chemical- and physical changes result from the action of temperature and pressure. The increase in temperature is derived chiefly

from the geothermal gradient, but could also be related to plutons and possibly to heat associated with deformation. The pressure results from deep burial and from deformation. With application of still-greater heat and pressure, anthracite can be converted to graphite.

In essence, the rank of coal is a product of the temperature history. Older coals are more likely to have been more deeply buried, and hence exposed to greater geothermal temperatures than younger coals. Also, the more time a coal spends at high temperature, the higher its rank becomes. Thus the ranks of Carboniferous coals are generally higher than those of Tertiary coals. Where the geothermal gradient is low, coals may be subbituminous; where the gradient is high, in contrast, the coal may be bituminous at a comparable depth. Thermal metamorphism near igneous intrusives likewise leads to increasing rank.

In tightly folded (and previously deeply buried) strata of Pennsylvanian age in Pennsylvania, anthracite is present. By contrast, in gently folded strata of the same age, bituminous coal occurs. Tertiary coals in Alaska that have been folded are of high rank, whereas coals of the same age in non-folded strata in the Dakotas are still lignite. The ranks of coals from more-intensely folded strata are not necessarily higher than are those of coals of the same age from nonfolded strata. Therefore, the older idea that deformation was a large factor in coalification seems to need revision. Deformation may be of less importance than depth of burial and resultant exposure to geothermal heat.

Using petrographic characteristics, one can recognize four kinds of coal constituents (macerals): (1) **vitrain** (or **anthraxylon**), *a bright, glossy, vitreous or glassy-looking material having conchoidal fracture,* (2) **durain**, *a dull material, lacking luster and having an earthy appearance,* (3) **clarain**, *a coal having a smooth surface when broken at right angles to the bedding plane, with broken faces having a pronounced gloss or shine and silky luster,* and (4) **fusain** (or *mineral charcoal* or *mother of coal*), *carbonized wood resembling charcoal;* fusain is the constituent of coal that soils the hands when touched. These four constituents of coal can be recognized in hand samples, but are more readily studied by examination of polished surfaces under the microscope using reflected light with oil-immersion objectives. Although we have listed them as individual entities, these four kinds of coal commonly are intimately commingled.

Coals have been recorded in strata ranging in age from Precambrian to Holocene. However, only after the Late Silurian Period, when land plants had become established, was it possible for plant material to accumulate on a scale to form large deposits of coal. In the history of the Earth, two particularly rich coal-forming periods are known: (1) the Carboniferous (principally Pennsylvanian) and Permian periods; and (2) the Late Cretaceous and Early Tertiary Periods.

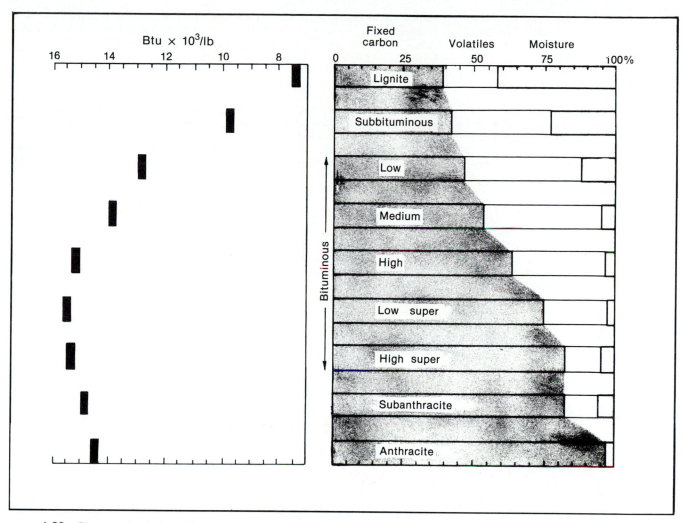

FIGURE 4-20.   Diagram showing coalification. During changes from lignite (top) to anthracite (bottom), contents of fixed carbon and caloric value (measured in Btu values) generally increase, whereas contents of moisture and volatiles generally decrease. (Modified from M. R. Campbell, 1930, fig. 1, p. 681, and fig. 4, p. 688.)

Most coals form from accumulated plant debris in sea-marginal swamps or in closed fluvial basins. In position with respect to the shoreline, most sea-marginal coals are analogous to anhydrite rock accumulating in seamarginal flats. The difference between anhydrite and coal as a product in analogous depositional environments is one of climate. Whereas anhydrite forms under arid conditions, coal forms under humid conditions. The warmer and wetter the climate, the more luxuriant is the flora and the greater the accumulation of plants that constitute the raw materials of possible future coals. Coals are associated with cyclical sequences of strata that resulted from the growth and submergence of deltas, as discussed in Chapter 13. Times of widespread accumulation of plant matter that later became converted to coal coincided with times of widespread emergence of the continental masses.

SOURCES: M. R. Campbell, 1929; Dutcher, Hacquebard, Schopf, and Simon, 1974; Murchison and Westoll, 1968; Stach, Mackowsky, M. Teichmuller, G. H. Taylor, Chandra, and R. Teichmuller, 1982; M. Wolff, 1988.

Suggestions for Further Reading

BATHURST, R. G. C., 1975, Carbonate sediments and their diagenesis, 2nd ed.: Amsterdam, Elsevier Publishing Company, 658 p. (Chapter 12, Neomorphic processes in diagenesis, p. 475–516.)

CHAMLEY, HERVE, 1989, Clay sedimentology: New York-Heidelberg-Berlin, Springer-Verlag, 623 p.

DEAN, W. E., and SCHREIBER, B. C., eds., 1978, Notes for a short course on marine evap-

orites: Tulsa, OK, Society of Economic Paleontologists and Mineralogists Short Course No. 4, 188 p.

DUNHAM, R. J., 1962, Classification of carbonate rocks according to depositional texture, pp. 108–121, *in* Ham, W.E., ed., Classification of carbonate rocks: Tulsa, OK, American Association of Petroleum Geologists Memoir 1, 279 p.

FISHER, R. V., and SCHMINKE, H.-U., 1984, Pyroclastic rocks: New York, Springer-Verlag, 528 p.

FOLK, R. L., 1956, The role of texture and composition in sandstone classification: Journal of Sedimentary Petrology, v. 26, p. 166–171.

FOLK, R. L., 1959, Practical petrographic classification of limestones: American Association of Petroleum Geologists Bulletin, v. 43, p. 1–38.

FRIEDMAN, G. M., 1965, Terminology of crystallization textures and fabrics in sedimentary rocks: Journal of Sedimentary Petrology, v. 35, p. 643–655.

FRIEDMAN, G. M., 1985, The problem of submarine cement in classifying reefrock: an experience in frustration, p. 117–121 *in* Schneidermann, N.; and Harris, P. M., eds., Carbonate cements: Tulsa, OK, Society of Economic Paleontologists and Mineralogists Special Publication 36, 379 p.

FRIEDMAN, G. M., and SANDERS, J. E., 1967, Origin and occurrence of dolostones, p. 267–348 *in* Chilingar, G. V., Bissell, H. J., and Fairbridge, R. W., eds., Carbonate rocks, origin, occurrence and classification: Amsterdam, Elsevier Publishing Company, 471 p.

GROSS, G. A., 1980, A classification of iron formations based on depositional environments: Canadian Mineralogist, v. 18, p. 215–227.

JAMES, H. L., 1954, Sedimentary facies of iron-formation (*sic*): Economic Geology, v. 49, p. 235–293.

LAJOIE, JEAN, 1979, Facies models 15. Volcaniclastic rocks: Geoscience Canada, v. 6, p. 129–139.

POTTER, P. E., MAYNARD, J. B., and PRYOR, W. A., 1980, Sedimentology of shale: New York, Springer-Verlag, 270 p.

SANDERS, J. E., and FRIEDMAN, G. M., 1967, Origin and occurrence of limestones, p. 169–265 *in* Chilingar, G. V., Bissell, H. J., and Fairbridge, R. W., eds., Carbonate rocks, origin, occurrence and classification: Amsterdam, Elsevier Publishing Company, 471 p.

SHELDON, R. P., 1981, Ancient marine phosphorites: Annual Reviews in Earth and Planetary Sciences, v. 9, p. 251–284.

SIBLEY, D. F., and GREGG, J. M., 1987, Classification of dolomite rock textures: Journal of Sedimentary Petrology, v. 57, p. 967–975.

PART III

FROM LAYERS TO SEQUENCE- AND SEISMIC STRATIGRAPHY

Now that we have considered individual particles and how they combine to form sediments, as well as how sediments are transformed to sedimentary rocks, it is time to continue our progression from the very small to the very large. We turn now to layers. Individual layers and sets of layers at various scales, dealt with in this part, are the subject matter of the science of stratigraphy.

The Birth of Stratigraphy: Steno's Four Principles

Niels Steensen (1638–1687) was a great Danish genius who excelled in whatever he undertook. Steensen eventually moved to Italy and changed his name to Nicolaus Steno. He established four important principles about strata (layers). In doing so, he founded the science of stratigraphy, even though a century or more passed before the study of strata was revived and stratigraphy came into its own.

After he had studied the strata exposed in Tuscany (Italy), Steno wrote his monumental treatise (1669) entitled *De solido intra solidum naturaliter contento dissertationis prodromus*. In this book he expressed four fundamental principles that govern the deposition of strata. He used his principles to synthesize a chronology for Tuscany.

STENO PRINCIPLE NO. 1

The materials composing rock strata were deposited initially as sediments. Strata may contain fossils, and these are the remains of once-living organisms that were buried with the sediments. (See Box 2.4, Figure 3.) Some fossils are the remains of land-dwelling organisms; others, of marine organisms. Strata containing remains of marine organisms had been deposited by the sea. Those containing remains of land-dwelling organisms had been laid down by streams.

STENO PRINCIPLE NO. 2

The boundaries of strata display characteristic features. The surfaces on which loose sedimentary strata had been deposited were solid. The lowermost stratum, therefore, had become firm before the sediments composing the next stratum had been deposited on top of it. The bottoms of strata conform to and thus bury any relief features on the underlying surface, but the tops of strata tend to be horizontal.

STENO PRINCIPLE NO. 3 (NOW KNOWN AS THE PRINCIPLE OF SUPERPOSITION)

The lower strata are older and each stratum had been fully in place before the next-younger stratum accumulated on top of it.

STENO PRINCIPLE NO. 4

Strata were originally continuous throughout their extent. In Steno's words, a stratum must have been terminated laterally against an older solid body, or else extended over the whole Earth.

Steno's principles have stood the test of time, and remain the cornerstone of the modern study of sedimentary strata. At this point we must divide stratigraphy into 2 parts. We discuss some material here. However, the subjects of correlation of strata, definition of units, and related subjects are deferred to Chapter 16. This is because a solid grounding in depositional processes, their products, and facies analysis (Part IV) is a prerequisite for the study of correlation of strata.

Chapter 5 begins with a discussion of layers: their composition, configuration, internal features, features of their surfaces, and criteria for their recognition.

The chapter continues with a discussion of *layers* and *time*. We consider rates of sediment accumulation, layers as records of the passage of time, and truncated layers as indicators of gaps in the stratigraphic record. These gaps are important because they form one of the bases for defining the boundaries of groups of layers.

At the end of Chapter 5 we consider successions of layers. We emphasize two contrasting kinds of *cyclic-* or *patterned sequences*: (1) *successions resulting from changes within depositional settings such as lateral redistribution of sediment* (autocyclic sequences) and *those resulting from large-scale external changes of the depositional setting, such as changes in climate, sea level, subsidence, or uplift* (allocyclic sequences). These two kinds of cyclic sequences are closely connected in *parasequences*, the next-larger unit in the hierarchy of stratigraphic units. *Parasequences* are genetically related combinations of allocyclic layers and of autocyclic layers that resulted from accumulation of sediments along a slope in such a way that the slope migrated.

Chapter 6 presents the all-important geologic concept of *base level* and how changes of base level and the sediments at the margins of basins interact. Our discussion follows J. L. Rich's recognition of the three critical environments of shelf, slope, and basin floor, which he called unda-, clino-, and fondo environments. This large-scale tripartite subdivision of the depositional settings at the margins of basins forms the basis for classifying sedimentary layers broadly into undathems (sediments deposited in the unda- or shelf environment), clinothems (sediments deposited in the clino- or slope environment), and fondothems (sediments deposited in the fondo- or basin-floor environment).

Next we introduce the discipline of *seismic stratigraphy*, in which continuous seismic-profile records are interpreted in terms of patterns of layers. These patterns of layers have resulted from the fundamental differences in undathems, clinothems, and fondothems, whose locations shift in response to fluctuating base level (sea level, in the ocean basins).

The changes in successions of strata that result from the responses of the sediments in these three critical environments to changes of sea level form the basis for *sequence stratigraphy*. This new branch of stratigraphy combines study of continuous seismic data with an understanding of the effects of base-level changes (especially sea-level changes) for predicting the patterns of layers deposited on passive-divergent lithospheric-plate margins.

In the last part of Chapter 6 we return to the subject treated at the end of Chapter 5. Here we consider the entire hierarchy of sequence-stratigraphic units: *parasequence sets*, *facies tracts*, and *mesosequences*. We also relate these successions to changes of base level.

Included within Chapter 6 are boxes reviewing the history of the use of sound in the oceans, that culminated in the invention and refinement of continuous seismic-reflection profilers. The acquisition of continuous seismic data across many of the world's basins starting in the 1960s established a firm foundation for the systematic classification of groups of layers too large to study conveniently in outcrop.

At the end of Part III the reader should be familiar with layers, groups of layers, and the importance of base level in controlling patterns of deposition of layers. Also in Part III we shall have spent a lot of time thinking about sequence-stratigraphic units. These are not, however, much like the formal units of classical stratigraphy, which are described in Chapter 16. In Part IV we shall take up the subjects of depositional processes and depositional environments (process sedimentology and facies analysis).

CHAPTER 5

Layers and Their Associated Primary Sedimentary Structures: A First Look

Layers and the primary sedimentary structures associated with them are diagnostic features of sedimentary deposits. Both are closely connected with sedimentary processes and with environments of deposition. In addition, certain aspects of layers are greatly affected by conditions of burial and of diagenesis.

Several important points about layers can be determined by beginners even before they have acquired the background needed for interpreting the environmental significance of the layers. For example, study of *primary sedimentary structures* is an important basis for determining the order of layers—the way up—that must be established before one even begins to look for patterns among layers. Other primary sedimentary structures can be used to infer the direction of transport of the currents that deposited the sediment. The study of the products of paleocurrents is important, even if one has not yet begun the complex task of reconstructing the ancient depositional environment. Finally, layers are related to time. We introduce the layers–time connection in this chapter and expand on it in Chapters 15 and 16.

Our objective in this chapter is to become familiar with individual layers and their associated primary sedimentary structures. After we have explored these and looked at two examples of how they can be used, we work our way upward in the hierarchy to large groups of layers.

Individual Layers

The general collective term for *all layers that were deposited with the oldest at the bottom* is **strata**. The singular is stratum. Thus layered deposits are described as being *stratified* (Figure 5-1).

An important corollary of the origin of strata is that the material composing a stratum was once at the Earth's solid surface (Steno Principle No. 2). The bottom of

each new layer buried an older part of the Earth's solid surface, and the top of each new layer became, for a short while at least, a part of the Earth's solid surface. This relationship of burial means that the configuration of each layer contains information by which the details of this burying can be reconstructed. The relationship of burial also suggests a logical way to proceed in studying each stratum: Determine the composition and overall shape, paying particular attention to its thickness, lateral extent, and dip with respect to the dips of adjacent strata. Then look into the interior of the layer. Next, notice special features of the top of the layer and of the bottom. Finally, study the relationships of any internal features to the top- and bottom surfaces. After one has attended to these things, one is ready to study groups, or successions, of strata. As explained in Chapter 16, beyond the study of successions of strata at a given locality is the broad topic of their large-scale lateral (or regional) relationships.

Constituents

A necessary prelude to a treatment of the constituents or the materials composing strata is a brief discussion of two postdepositional factors that affect strata. A starting point is the comparisons and contrasts between the layers in modern sediments and the beds in ancient bedrock. Sediments that have not been deeply buried have not experienced the changes that are associated with burial diagenesis. Before such diagenesis, the most-obvious aspects of the layering are likely to be changes in particle sizes. After such diagenesis, some beds may be distinctly set off from others by prominent surfaces along which the rock splits or parts. These are known as *bedding-plane partings*. Some changes may have resulted from processes of subsurface diagenesis, such as dissolution and cementation, which accentuate

FIGURE 5-1.   Contrasting expressions of strata.
. A. Well-defined strata of sandstones in wall-like remnant not yet eroded, Canyonlands National
Park, Utah. In general, most of these strata are uniform and persist unchanged across the field of
view (no scale shown, but possibly as much as 100 m from left to right). (U.S. Department of the
Interior, National Park Service Photo by M. Woodbridge Williams, not dated.)

FIGURE 5-1.   (*Continued*)
    B. Vaguely defined strata in boulder conglomerate, roadcut on
Route 128 south of Boston, Massachusetts, dip about 15° to the
right. (J. E. Sanders, 1981, fig. C.20, p. 555.)

differences between layers; these processes produce *diagenetic bedding*. Other postdepositional changes can affect the distinctness of expression of bedding. For example, in core samples or new exposures of fresh rock the bedding may be very obscure. By contrast, after the same layers have been weathered, the bedding may become very distinct.

In general, each bed consists of material having consistent properties. The beds are set off from one another along surfaces where these properties change. The properties of these materials include texture, fabric, and mineral composition. Thus the bedding may be a function of the ways in which particles are distributed in a sediment (Figure 5-2, A) or in an ancient sedimentary rock (Figure 5-2, B). Textural changes commonly define the layering. For example, a layer of pebbles may lie next to a layer of sand (Figure 5-3, A). Layers may be expressed

FIGURE 5-2.   Basis for stratification.

A. Strata in modern beach sediment defined by change of color and composition of particles from nearly white, light-mineral sand composed predominantly of quartz (below) to a dark red–black layer, about 7 cm thick, of heavy-mineral sand composed predominantly of garnet–ilmenite–magnetite (above). Corner of trench dug at Cupsogue Beach, Westhampton, Long Island, New York, 17 April 1977. (J. E. Sanders, 1981, fig. 8.9, p. 203.)

B. Thick layer of uniform sand cemented into a sandstone bed approximately 15 m thick. (Dinosaur National Monument, Utah; U.S. Department of the Interior, National Park Service photo by Dick Frear, not dated; J. E. Sanders, 1981, fig. 8.11 (a), p. 205.)

The task of describing sedimentary rocks can be as challenging and detailed as an individual observer cares to make it. The things one records from field work are influenced by the scope and purpose of the study, by the variety displayed by the rocks, by the experience and skill of the observer, and by the specifications laid down by the observer's boss. Routine description should begin with features visible in a hand specimen. These include the rock name, color, particle size, sorting of particles, shapes of particles, composition of particles (including fossils), kind of cement or matrix, and any special features. Larger-scale sedimentary structures should be recorded.

Configuration

Under this heading we emphasize thickness and lateral extent of individual layers. In a following section, we expand the coverage to include the relationships of the boundaries (*contacts*) of individual strata and groups of related strata to other strata.

THICKNESS

The thicknesses of layers are functions of the sizes of the particles, the availability of particles, the kinds of processes responsible for the deposition of the particles, and how long these processes operated. Obviously, the limiting minimum thickness of a layer is the diameter of its largest particles. Hence the thicknesses of layers consisting chiefly of pebbles will be in the range of centimeters or tens of centimeters (Figure 5-4, A). By contrast, the thickness of a layer consisting chiefly of silt can be as little as a fraction of a millimeter (Figure 5-4, B). The maximum thickness of a layer depends largely on the environment of deposition and on the supply of a given kind of sediment.

The thicknesses of strata vary through a wide range, from many tens of meters (See Figure 5-2, B.) to a fraction of a millimeter. (See Figure 5-4, B.) Many strata measure a few centimeters in thickness. Within this broad range, the variation of thickness seems to be continuous. Therefore measurements of the thicknesses of strata are amenable to statistical analysis using the same strategy that has been applied to the study of the sizes of particles. (See Box 2.1.) However, only a few systematic statistical studies have been made of the thicknesses of strata (the results of four such studies are shown in Figure 5-5), and we do not yet know if any natural thickness classes exist that could be the basis for a set of standard thickness-class limits.

The thicknesses of strata may remain virtually constant within a given stratigraphic unit or may change system-

by contrasting composition; beds of sandstone and shale may alternate with those of limestone (Figure 5-3, B).

In the field study of strata, the most information nearly always comes from strata of sandstones and limestones. These strata contrast decidedly with finer-textured strata above and below. The strata of sandstones and limestones typically form ledges. Being thus set off, the ledge-forming strata typically display two bedding surfaces, an upper and a lower, and a sectional view that exposes the features within the bed. The following paragraphs summarize some things to study when one examines ledge-making strata.

FIGURE 5-3.   (*Continued*)
B. Alternating layers having contrasting chemical compositions. Thick white layers consist of limestone, darker-colored layers, of sandstones and shales. Prominent white ledge near top is 1 m thick. Ardon Formation (Lower Jurassic), Nahal Aradon, Makhtesh Ramon, Negev Desert, Israel. (M. Goldberg and G. M. Friedman, 1974, pl. 2.)

FIGURE 5-3.   Basis for stratification.
A. Thick layers in coarse sediment (top); thinner layers in fine sediment (bottom). Pleistocene sediments, near Rensselaer, New York. Diameter of largest particle (just above center at left) is 10 cm. (Authors.)

atically through the unit. In an **upward-thickening succession**, *the thicknesses of strata increase progressively toward the top* (Figure 5-6, A). By contrast, an **upward-thinning succession** is one in which *the thicknesses of strata decrease progressively toward the top* (Figure 5-6, B).

Layers composed of peat, clay, carbonate sediment, and evaporites are subject to great thickness modifications as a result of diagenesis. Freshly deposited peat and clay consist largely of water (up to 90%). With time and under the weight of overlying sediments (or of glacial ice, in some cases), the water is squeezed out. A layer of freshly deposited clay 10 m thick having a porosity of 90% may be compressed into a shale only 2 m thick.

Dramatic evidence of the extent of compaction of some shales is given by complexly deformed sandstone dikes. Presumably, the sand was originally tabular from having been injected into an early-formed planar joint. As compaction took place, the tabular dike was crumpled. In the example shown in Figure 5-7, the compaction after emplacement of the sandstone dike reduced the thickness of the shale to about 50% of its value at the time the dike was emplaced (the shale may already have been greatly compacted before the dike formed).

During diagenesis, carbonate sediments are prone to dissolution, either selectively or completely (Chapter 3). Tangible evidence of total dissolution is given by stylolites that are parallel to bedding. (See Figure 3-7.) Stylolite surfaces are known to be impermeable barriers to the movement of fluids perpendicular to the plane of the stylolites, but function as channels for the movement of fluids along the stylolites. Because stylolites are commonly subparallel to bedding, their presence tends to restrict movement of fluids perpendicular to bedding and to promote fluid movement parallel to bedding.

As mentioned previously, strata of halite and of other easily soluble evaporites can be dissolved completely. This happens when the strata are uplifted from positions of deep, salty formation waters and the evaporites are brought into contact with near-surface fresh ground

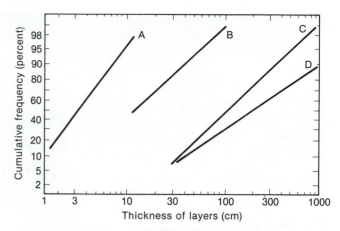

FIGURE 5-5. Thicknesses of strata, cumulative-frequency curves with cumulative frequency shown on a probability scale. (Probability scale explained in Box 2.1.) Curve A, sandstones, and Curve B, shales, Upper Carboniferous, Cantabrian Mountains, Spain (M. H. Nederloff, 1959.) Curve C, sandstones, and Curve D, shales, Upper Carboniferous, Joggins, Nova Scotia (J. H. Way, Jr., 1968).

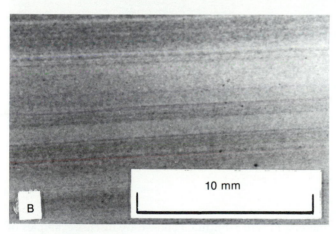

FIGURE 5-4. Contrasting thicknesses of strata.
 A. Thick layers containing abundant boulders. Modern sediments, Wheeler campground, near Ojai, California. (Authors.)
 B. Rock containing thin layers composed of abundant fine sand and silt. Middle Ordovician, Route 44, west of Mid-Hudson Bridge, Highland, New York. Enlarged view through microscope of thin section. (B. Caplan.)

addition, the thicknesses of some strata are changed greatly during folding. Shales tend to be thinned on the limbs and thickened in the crestal parts of some folds.

FIGURE 5-6. Changes of thickness in successions.
 A. Upward-thickening succession, mid-Ordovician strata, south side of U.S. 44-N.Y. 55, west of Mid-Hudson bridge, Highland, New York. (J. Obradovich.)

water. Layers of salt at depth can be greatly thinned as a result of diapiric upward flow of salt plugs, as is common in the U.S. Gulf Coast.

In all these examples, diagenetic reactions greatly reduce the thicknesses of some strata and remove parts or all of other strata. Other mechanisms for eliminating strata are erosion at the land surface and subaqueous slumping, which remove strata from one place and pile them up at another locality downslope. In

FIGURE 5-6.   (*Continued*)
   B. Upward-thinning succession, western wall of Val Lasties, Sella Group, Dolomites, eastern
Alps. (Agi Nadi, in J. E. Sanders, 1981, fig. 3(b), p. 4.)

LATERAL EXTENT

The lateral extents of layers are determined by the geologic conditions under which the sediment was spread out and deposited. Recall from the introduction to Part III, Steno Principle No. 4, which expresses the concept of the original continuity of strata and also the idea that the strata ended against older solid bodies or else would be of worldwide extent.

Where both the top- and the bottom of a given stratum are plane surfaces that are parallel to each other, and this condition persists laterally across an exposure, the shape of the bed is said to be *tabular*. The original attitudes of many plane parallel strata are horizontal. (See Figures 5-1, A and 5-3, A.) We discuss exceptions in a following section.

The lateral persistence of some tabular beds is noteworthy. We confine ourselves here to examples of strata whose thickness is uniform on the scale of a typical exposure (usually not more than a few tens, rarely hundreds of meters; Figure 5-8). In Chapter 6 we discuss examples of tabular strata whose extent is regional. Some layers end abruptly where they are cut off by the margins of the trough or channel in which they were deposited (Figure 5-9). Numerous examples are channels related to ancient streams. The long axis of a channel parallels the azimuth of the ancient current but does not in itself indicate the direction of flow along this azimuth.

In the next section we present the first of many discussions of the relationships of strata under the heading of cross strata.

Features Within Strata

Between its bounding surfaces, the material forming a stratum can be arranged in diverse ways. The sizes of the constituent particles may be essentially uniform throughout or may vary systematically across- or along the bed. The material may lack internal layers or possess various kinds of layers that are plane and parallel to the bounding

FIGURE 5-7. Sandstone dike, injected as a tabular body, intricately folded because, after it had been emplaced, the thickness of the enclosing shale was reduced by about 50% during compaction. Karoo System (Upper Carboniferous to Lower Jurassic), Coffee Bay, Transkei, South Africa. (J. F. Truswell, 1972, fig. 9, p. 582.)

FIGURE 5-8. Laterally persistent layers of limestone (Pennsylvanian) exposed along the steep sides of the valley in which the San Juan River flows, southeastern Utah. (Authors.)

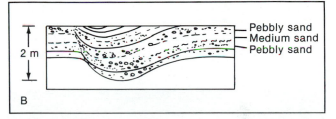

FIGURE 5-9. Strata that end at the edges of a channel. View toward shore along the axis of a channel that was cut and filled in the crest of a beach during a storm. The long axis of the channel is normal to the shore.

A. Photograph; Machete at left, about 70 cm long, gives scale.

B. Sketch from photograph Cupsogue Beach, Westhampton, Long Island, New York, April 1970. (Authors.)

surfaces, that are inclined to the bounding surfaces, or that display the effects of deformation. In addition, sediment that originally was arranged into distinct internal layers may have been so completely rearranged by the effects of burrowing organisms that scarcely any vestiges of the layers can be seen. Collectively, *all features formed during deposition or shortly thereafter* are designated as **primary sedimentary structures**. Included are the beds themselves, features on the bedding surfaces, and features within the beds.

In the following paragraphs we describe primary sedimentary structures and comment on their use in determining original top direction and direction of flow of ancient currents.

PARTICLE SIZES

If the sizes of the particles are uniform throughout the bed, then the bed appears as the coarse conglomerate of Figure 5-1, B or the thick sandstone shown in Figure 5-2, B. In contrast is a **graded layer**, defined as *a layer in which the particle sizes change according to a systematic gradient in a vertical- and/or lateral direction.*

Among graded layers, two contrasting arrangements are known: (1) **normal grading**, which refers to *the con-*

dition within a layer in which the particles become systematically finer upward (Figure 5-10) and (2) **inverse grading**, which describes *the condition within a layer in which the particles become systematically coarser upward.* Each of these arrangements is "normal" in the sense that the end product is the usual result of the operation of distinctive- and contrasting depositional processes. The term "normal" for the upward-fining arrangement was proposed when that kind was thought to be the only kind and could be reliably employed as a criterion for determining original top direction in deformed strata.

INTERNAL LAYERS

The interior of a bed may display various layers that are thinner than the main bed. Some of these *sublayers*

are essentially parallel to the bounding surfaces of the main bed. The following discussion applies to them. Later we describe internal layers that are not parallel to the bounding surfaces of the main bed.

Laminae. "Thin" layers found in medium- and fine-textured sands (Figure 5-11, A), in silts (Figure 5-11, B), in clay-size sediments (Figure 5-11, C), and in carbonates and evaporites (Figure 5-10, D) are designated *laminae* (singular, lamina).

No agreement has yet been reached upon the limiting thickness and general definition of laminae with respect to the thicker layers. Two contrasting proposals are shown in Table 5-1. In the proposal of McKee and Weir (1954), which was amended slightly by Ingram (1954), **beds** are defined as being *strata that are thicker than*

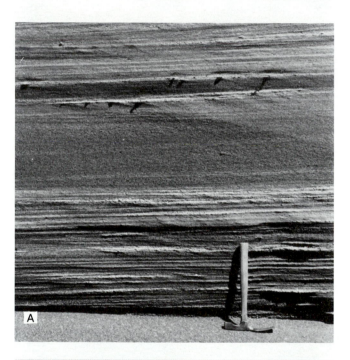

FIGURE 5-11.   Plane, parallel laminae in various sedimentary materials.

A. Thinly laminated Pleistocene sand, Union Pacific Railway ballast pit, between King Hill and Glenns Ferry, Elmore County, Idaho. (H. E. Malde, U.S. Geological Survey, 02 September 1962, in J. E. Sanders, 1981, fig. 8.13, p. 206.)

B. Laminated silt deposited on the living-room floor of the residence of the paleontologist A. F. Foerste (1862–1936) in Dayton, Ohio, during the flood of March 1913. Specimen given to Charles Schuchert (1858–1942), Yale Peabody Museum. (C. O. Dunbar and John Rodgers, 1957, fig. 17, p. 33.)

FIGURE 5-10.   Graded layer, containing pebbles at the base and fine sand at the top. Pliocene filling of Ventura basin, exposed along Santa Paula Creek, California. (Authors.)

FIGURE **5-11.** (*Continued*)
 C. Thickly laminated Pleistocene very fine sand and silt (light color) and clay (gray), Columbia River valley, Ferry County, Washington. (F. O. Jones, U.S. Geological Survey, in J. E. Sanders, 1981, fig. 13.19, p. 321.)

FIGURE **5-11.** (*Continued*)
 D. Thinly laminated calcite and gypsum, Castile Formation (Permian), west Texas. (J. E. Sanders, 1981, fig. 8.11(b), p. 205.)

1 cm, and **laminae** as *layers that are thinner than 1 cm.* **Thin laminae** are *layers thinner than 0.3 cm.*

In the proposal of C. V. Campbell (1967) a bed is considered to be the stratum that reveals the principal layering of the deposit, and laminae are defined as the smallest megascopic strata within a bed. In the Campbell scheme some laminae could be thicker than some beds.

In general, both proposals agree in their use of beds for "thick" layers and laminae for "thin" layers. McKee and Weir use an arbitrary boundary of 1 cm to make a nonoverlapping subdivision between beds and laminae. In Campbell's proposal layers as thick as 20 cm or more could be assigned to laminae, among which he established 5 thickness classes. The thicknesses of the three thickest laminae classes overlap with those of the three thinnest classes of beds. (See Table 5-1.)

Cross Strata.   Some packets or sets of layers, defined by upper- and lower bounding surfaces that were deposited as essentially horizontal planes, enclose other layers that never were horizontal, but that make various angles with these bounding surfaces. *Sets of strata that are not parallel to the main layers* are **cross strata** (Figure 5-12).

Cross strata may be thick or thin and may be parts of groups known as *cosets* that are small (Figure 5-13, A) or large (Figure 5-13, B). In general, as one might expect, small cross strata form parts of thin cosets, whereas thick cross strata are parts of thick cosets. Because the range of thicknesses of cross strata spans part of the range of thicknesses of plane, parallel strata discussed in a previous paragraph, the same two contrasting approaches shown in Table 5-1 have been applied to cross strata. In the scheme of McKee and Weir, cross strata are arbitrarily classified into nonoverlapping thickness categories with a class boundary at a thickness of 1 cm. McKee and Weir advocate use of cross strata as a general term, subdivisible into cross beds (thicker than 1 cm) and cross laminae (thinner than 1 cm). (See Figure 5-13, A.)

TABLE 5-1.   Thickness classes of strata comparing two kinds of usage

Thickness
(cm)

Strata (Stratum)							
	Beds (Bed)	Very thick	100	Very thick			
		Thick	30	Thick			BEDS
		Medium	10	Medium	Very thick		
		Thin	3	Thin	Thick		
		Very thin	1	Very thin	Medium		LAMINAE
	Laminae (Lamina)	*1*	0.3	Thin			
		Thin	0.1	Very thin			

¹ No special name has been suggested for laminae having this thickness; the unmodified adjective "laminated" has been used for describing laminae in the thickness class from 0.3 to 1.0 cm.

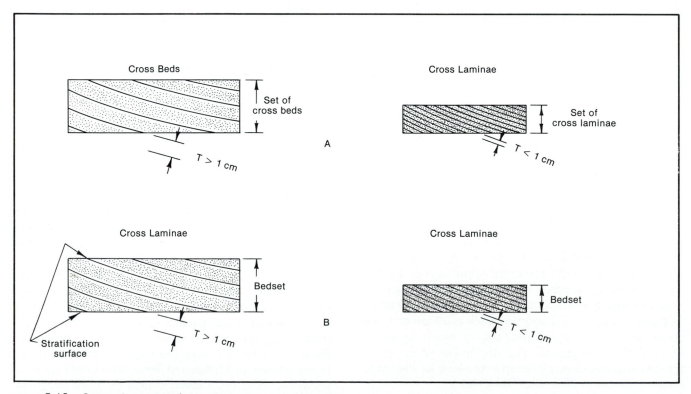

FIGURE 5-12.   Contrasting usages for naming varieties of cross strata.
   A. Usage based on thickness of inclined layers with arbitrary boundary at 1 cm. (E. D. McKee and G. W. Weir, 1953.)
   B. Usage based on concept that all diagonal layers within a bedset, regardless of thickness, should be named cross laminae. Further explanation in text. (C. V. Campbell, 1967.)

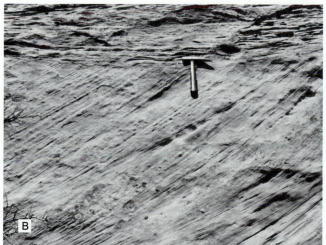

FIGURE 5-13.   Cross strata of contrasting scales.

A. Small-scale cross strata in Pleistocene sand, view eastward, side of stream valley eroded into coastal cliffs, north shore of Long Island. The dip of these cross strata to the south proves that the water which deposited the sand flowed from left (north) to right (south). (J. E. Sanders, 1981, fig. 4(b), p. 5.)

B. Large-scale cross strata, Navajo Sandstone (Jurassic), Utah, of the kind made by migration (from right to left) of an eolian dune. (Authors.)

In the contrasting usage proposed by C. V. Campbell, the term cross beds does not appear. Instead, all layers that are not parallel to the bounding stratification surfaces are named cross laminae. Collectively, *the intervals defined by the bounding stratification surfaces* are defined as **bedsets**. Any subdivisions within such bedsets are termed *laminae*. (This follows from Campbell's definition of laminae as subdivisions within beds.) In this case, because they are not parallel to the bounding surfaces, they are named *cross laminae*. Thus in Campbell's scheme a bedset could include cross laminae (or even cross strata) but could not include cross beds.

Although Campbell's scheme offers a useful, consistent approach to cross strata, it flies in the face of the deeply entrenched geologic usage of cross bed and cross bedding. In view of the demonstrated tenacity which geologists have shown for adhering to their ingrained language, one can predict at best sporadic progress for Campbell's proposal to eliminate *cross bed*.

Further discussion of cross strata and the scale of lateral persistence of strata requires us to introduce the concept of bed forms and to discuss how cross strata form by the interactions among currents, new sediment, and bed forms. We begin these topics in a following section devoted to features on tops of beds. The close connection between bed forms (features of the surfaces of layers) and cross strata (found within beds) illustrates an important point about the origin of some beds. This point is that, as new sediment is added, a structure that forms on the surface of the bed becomes buried and thus enters the geologic record as a feature within a bed.

Cross strata dip in the direction of the slope on which the sediment was deposited. Where this slope was planar and formed the downcurrent side of a linear ridge that was transverse to the current, the strike of the cross strata is perpendicular to the ancient current and the dip of the cross strata is in the direction of the ancient current. Not all cross strata were formed by the migration of such linear flow-transverse ridges. Some cross strata were deposited on slopes that formed the sides of troughs. As a result, dip directions of such cross strata scatter through an azimuth of 180°. Only the long axes of these troughs were parallel to current direction.

Cross strata can be used to determine original tops where they are concave up, truncated at their tops, and tangential at their bases. Not useful for top determinations are planar cross strata that are truncated at both top and bottom and curved cross strata that are not truncated at all (Figure 5-14).

Deformed Internal Strata.   The strata within a bed, whether plane, parallel laminae; cross strata; or some

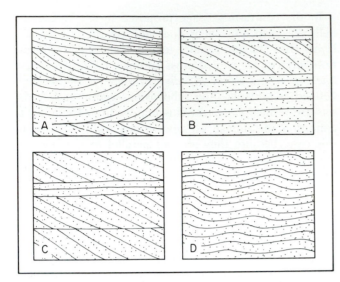

FIGURE 5-14.   Schematic sections of four kinds of cross strata. (J. E. Sanders in G. M. Friedman and J. E. Sanders, 1978, fig. 13-12, p. 414.)

A. Concave-up cross strata truncated along upper sides and becoming tangential at lower contacts.

B. Concave-down cross strata that are truncated along their lower contacts and tend to be tangential along their upper contacts. Concave-down cross strata are not abundant, but enough of them do exist to trap the unwary and to complicate any simple-minded application of rules for using cross strata of the kind illustrated in A to determine original top direction.

C. Planar cross strata truncated along both upper- and lower bounding surfaces.

D. Climbing-ripple cross strata, not truncated.

FIGURE 5-15.   Beds containing internal strata that have been deformed.

A. Large-scale folds in coarse Casper Sandstone (Pennsylvanian), southwest of Laramie, Wyoming. (R. W. Fairbridge, July 1957, in J. E. Sanders, 1981, fig. 8.18, p. 208.)

B. Small-scale anticlines that grew diagonally upward across a bed of fine sandstone while new sediment was being added and the bed was growing upward. Middle Ordovician, west side of Hudson River, Highland, New York (opposite Poughkeepsie). (J. E. Sanders, 1981, fig. 8.17, p. 208.)

mixture of both, may display complex folds indicating that the strata have been deformed (Figure 5-15). The time of origin of the deformation with respect to the time of deposition of the enclosing nondeformed beds is not always easy to determine. In some cases, however, the arrangement of the deformed layers proves that deformation was proceeding while new sediment accumulated (Figure 5-15, B). **Convoluted laminae** are examples of *primary sedimentary structures originating from the simultaneous deposition and deformation of sediment ripples.* As a result, when the sediment forming the bed containing them stopped being deposited, the deformed laminae within were already in place.

BURROW-MOTTLED SEDIMENT

In many environments of deposition, *organisms that live in burrows or other shelters in the sediment* (collectively named the **infauna**) as well as many *organisms that live on the sediment* (the **epifauna**) rework sediment as part of their daily activities. Laboratory experiments in aquaria have demonstrated that the effects of invertebrate sediment-moving activities can completely rework

what was initially a well-stratified sediment. The term **bioturbation** has been proposed as a general term for *a collective name for all the physical stirring and churning of sediments done by organisms in sediments* (Figure 5-16). Bioturbation is so rapid and comprehensive that well-preserved strata imply conditions of swift deposition or an environment not suitable for habitation by bottom-working organisms. Bioturbation contrasts with boring, a process by which some organisms can drill holes into solid rock.

SOURCES: C. V. Campbell, 1967; Collinson and Thompson, 1982; Kuenen and Migliorini, 1950; McKee and Weir, 1953; D. G. Moore and Scruton, 1957; Rubin, 1987; Way, 1968.

Features of Sediment Surfaces

Many features that are important for the interpretation of sedimentary deposits are found on the surfaces of the

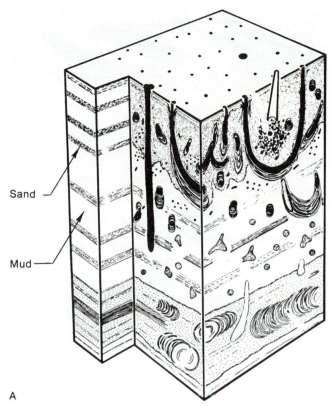

FIGURE 5-16.   Effects of burrowing organisms in sediment.
   A. Block diagram of modern sediment consisting of interlayered sand (stippled) and mud (white) (nonburrowed layers shown at left) that have been burrowed by various organisms (at right, where almost no strata are visible). (H.-E. Reineck, W. F. Guttmann, and G. Hertweck, 1967, fig. 10, p. 252.)
   B. Relief peel made from vertical face of box core that penetrated the sediments deposited on a lobe of a flood-tidal delta. Protruding rough parts of peel are coarse brown sand; recessed smooth parts are fine gray sand. Burrowing organisms have nearly obliterated all traces of former stratification. Moriches tidal delta, Westhampton, New York. (D. M. Caldwell, 1971 ms., fig. 15, p. 64.)

sediment. In modern deposits such features are visible on the tops of bodies of subaerially exposed sediments, on sediment bodies that may be subaerially exposed intermittently (as on intertidal flats when the tide is out, on river floodplains after flood waters have drained away, on lake bottoms exposed by loss of water from the lake), or on underwater photographs of sediment/water interfaces. Many kinds of features are known: some are made by reactions between the sediment and a flowing current; others are made during loss of water; still others are made by organisms.

BED FORMS IN COHESIONLESS SEDIMENT

As is explained in Chapter 7, one of the results of shearing along an interface between a moving fluid and a body of cohesionless sediment particles is that the interface is shaped into *a regular, systematic, repeated pattern of relief features* known as **bed forms**. Our chief emphasis at this point is on the connection between the shapes of the bed forms and the kinds of strata deposited. Later we classify bed forms and discuss their relationships to flow conditions.

Among bed forms, scales and patterns span a wide range. Important attributes are size, overall plan view, profiles of individual features, and internal structures.

Sizes.   In size, bed forms range from lines of individual particles, to small-scale ripples having crest-to-crest spacings (wavelengths) of only 2 to 3 cm (Figure 5-17) and heights of less than 1 cm, to enormous ridges (dunes) having wavelengths of tens or even hundreds of meters (Figure 5-18) and heights of tens of meters.

The dimensions of bed forms are controlled by several variables that are not altogether understood. Clearly, particle size exerts a lower limit: a bed form can be no

FIGURE 5-17.   Active ripples, nearly symmetrical in transverse profile, in shallow water (3 to 5 cm deep) at margins of a modern lake. Wavelengths of ripples are about 1.5 cm; their crests are pointed, and their troughs rounded and concave up. Wavelength of water waves is almost 30 cm. In coarser sediment at water's edge (above leaves at right center), where waves are breaking, no ripples are present. (Authors.)

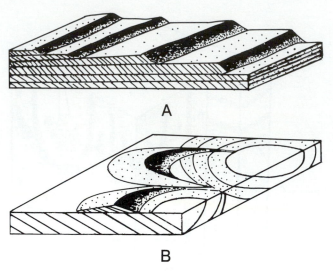

FIGURE 5-19.   Block diagrams of a few kinds of bed forms showing relationship between plan views and associated cross strata formed by the addition of new sediment. Current from left to right in A through C; direction of current for D not known.
  A. Straight-crested bed forms yield planar cross strata.
  B. Cuspate bed forms yield trough-shaped cross strata that formed on curved downcurrent sides that are concave downcurrent.

smaller than the average particle size. The dimensions of bed-form ridges usually are several tens of times larger than particle diameter. Depth of water limits the sizes of bed forms created by water currents. Finally, the sizes of waves on the water surface place limits on the sizes of certain bed forms.

Plan Views.   The plan views of bed forms can be described by determining the continuity and orientations of the crests (Figure 5-19). Some ripple crests are nearly straight and persist laterally. The degree of scatter of dip azimuths of such ripples is small. Other crestlines are slightly sinuous. Still others consist of a series of short, curved segments (a pattern referred to as *cuspate*). Still-other bed forms lack linear crestlines and exist as isolated individuals or *hummocks*. As we shall see in a following

section, the ground-plan shapes of bed forms affect the kinds of strata formed.

The uniformity of orientation of the crests of bed forms can be specified numerically by measuring the azimuths of the dips of one side (the steeper side of asymmetric bed forms). An expression of the scatter of values about a mean direction is the *percent vector magnitude*. The average direction of dips of the steeper sides of asymmetric bed forms closely approximates (within about 10°) the direction of current that created the bed forms.

Profiles.   In the study of bed forms, the profiles of main interest are those along a line at right angles to the trends of the crests. Major subdivision is based on symmetry: the profiles of bed forms are either symmetrical (Figure 5-20) or asymmetrical. (See Figure 5-18.)

Symmetrical bed forms are products of oscillatory shearing that is equal in both directions, as under certain shoaling waves. Such bed forms are typified by ripples having pointed crests and rounded, concave-up troughs. (See Figure 5-20.) Other symmetrical varieties include ripples having equally rounded crests and -troughs (See Figure 5-20.); narrow pointed ridges and nearly flat troughs (See Figure 5-20.); and broad, convex-up crests separated by narrow, groovelike troughs. (See Figure 5-20.)

Asymmetrical bed forms are products of unidirectional shearing, as along the bed of a stream, or of oscillatory shearing in which the intensity of shearing in each direc-

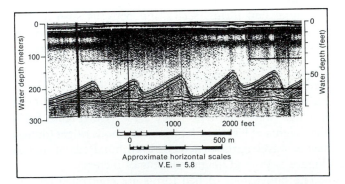

FIGURE 5-18.   Continuous seismic-reflection profile through large bed forms, composed of gravel, on floor of Minas Basin, Nova Scotia. (Huntec, Ltd., courtesy Roger Hutchins.)

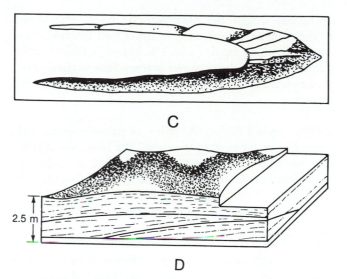

C

D

FIGURE 5-19. *(Continued)*
C. U-shaped or linguoid (convex side downcurrent) bed forms yield cross strata (not rendered here) that dip outward through an arc of 180°.
D. Hummocky bed forms yield cross strata that intersect underlying former bedding surfaces at low angles and that dip in all directions. (Suggested by fig. 6.1, p. 60 in J. D. Collinson and D. R. Thompson, 1982; with hummocks added from J. C. Harms, J. B. Southard, and R. G. Walker, 1982, fig. 3-15, p. 3-31.)

2.5 m

tion is not equal. Such conditions prevail under parts of the fields of shoaling waves and also where shoaling waves and various unidirectional currents act simultaneously.

The profiles of asymmetric bed forms range from angular to smoothly curved. Angular ripples are formed by unidirectional shearing; the planar downstream sides typically are inclined at the angle of repose of the sand and form a sharp angle at the crest with the otherwise curved, convex-up, asymmetric part of the profile that is inclined in the upcurrent direction at a low angle. Smoothly curved asymmetric ripples are steeper on the side toward which the predominant shearing is directed. In the ripples formed beneath most fields of shoaling waves, the landward sides are steeper.

Many attributes of ripples have been measured with a view toward determining whether the ripples originated by a unidirectional current or by an oscillating current. One useful pair of measurements has been used to define the ripple index (RI):

$$\text{Ripple index} = \text{ripple wavelength} / \text{ripple height}$$

$$(\text{Eq. 5-1})$$

If RI > 15, the ripples were formed by a unidirectional current. If RI < 4, then the origin was by an equally

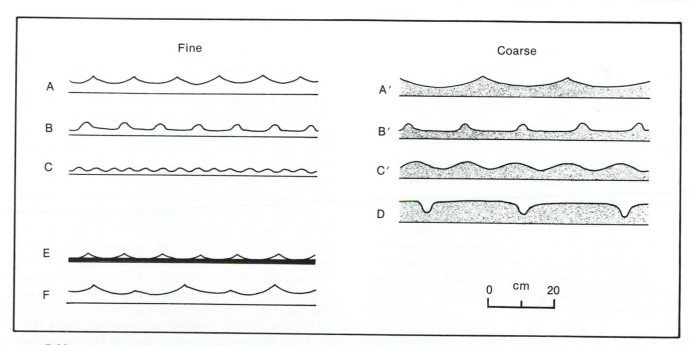

FIGURE 5-20.   Profiles at right angles to various wave-generated ripples.
A–F. Symmetrical ripples showing contrast in sizes governed by coarseness of bottom sediment (small ripples in fine sand, left; larger ripples in coarse sand, right). A and A', pointed crests with broadly circular, concave-up troughs; B and B', narrow, peaked crests with nearly flat troughs; C and C', rounded peaks and -troughs of equal curvature; D (coarse sand only), broad, flat crests with narrow, circular troughs; E (fine sand only), incomplete ripples of type A, having normal pointed crests but with underlying nonsandy material exposed in troughs; F (fine sand only), mixture of two sets of ripples of type A; small peaks appear in center of broadly circular, concave-up troughs. (Based on R. A. Davis, Jr., 1965a, fig. 2, p. 858.)

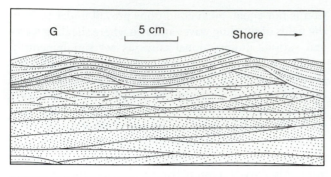

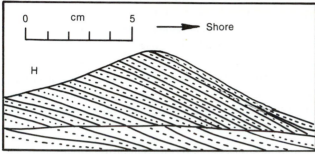

FIGURE 5-20. *(Continued)*
G, H. Sketches drawn from photographs of relief peels prepared from slices cut perpendicular to bottom at right angles to shore in wave-rippled sediment collected using a box corer from locations off German coast of Baltic Sea, where water depths are less than 1.6 m. G. Slightly asymmetrical ripples in which most cross laminae dip toward shore (below). A few cross laminae dip offshore (middle). Wavy laminae (top) drape over underlying ripples. Solid lines, boundaries of sets of laminae or cross laminae; dotted lines, laminae within sets. (R. S. Newton, 1968, fig. 2, p. 282.) H. Asymmetrical ripple with all cross laminae dipping toward shore. (R. S. Newton, 1968, pl. 2, fig. B, p. 280.)

oscillating current (shoaling waves). An RI from 4 to 15 is ambiguous with respect to origin.

The long axes of many ripples are perpendicular to the direction of the current. Among asymmetrical ripples, the steeper sides are downcurrent. The shapes of symmetrical ripples having pointed crests and broadly rounded, concave-up troughs indicate original top direction. The shapes alone of asymmetrical ripples do not show top direction. If combined with internal cross strata that are concave up, truncated at their tops, and tangential at their bases, however, ripples can be used for indicating tops.

Particle-Size Distribution.   The particle-size distribution among bed forms is not exactly uniform across the profiles. That is to say, some kind of sorting action has operated so that coarser particles are segregated from finer particles. This subject is complex and has not been exhaustively investigated. From what has been found, we know that in some bed forms (for example, ripples formed by water currents and most kinds of dunes blown

up by the winds), the finer particles have accumulated in the troughs. By contrast, in other bed forms (for example, climbing ripples where sediment is falling out of suspension), just the opposite is true: finer particles have accumulated at the crests and coarser particles are in the troughs.

Fabric.   In bed forms, the fabric of particles is not uniform. For example, in ripple crests, the long axes of elongate particles tend to be at right angles to the trends of the crests and thus parallel to the current direction. By contrast, long axes of particles in the troughs tend to be oriented perpendicular to the current.

Internal Structures.   The internal structures of bed forms result from the addition of new sediment without along-bottom migration, from migration along the bottom without addition of new sediment, and from combinations of addition of new sediment and migration along the bottom (Figure 5-21).

The sizes of the migrating bed forms dictate the upper limit of the scale of cross strata. Migrating small-scale ripples create ripple-scale cross laminae (Figure 5-22). By contrast, migrating large ripples create medium-scale cross strata (Figure 5-23).

SOLE MARKINGS

The bottoms of beds are referred to as the **soles**; *any relief features on the bottoms of beds* are **sole markings**. Typically, sole markings are found on the bases of beds of well-cemented sand-size- or coarser sediment (initially cohesionless particles) that overlie mudstones or shales (initially cohesive sediment). The sole markings become visible only when the underlying layer has been eroded. In order to visualize the origins of these features, it is necessary to consider several possibilities, some of which took place on the surface of the fine-textured cohesive layer before the sand-size cohesionless sediment was deposited, some of which took place while the sand-size sediment was being transported, and some of which took place after the sand-size sediment had been deposited. Sole markings show which side of a bed is the original bottom; many also indicate the direction of the paleocurrent that deposited the sand-size sediment.

Consider first an example of dinosaur tracks (Figure 5-24) impressed into mud (later to become a mudstone) and then covered by sand (later to become a sandstone). The original feature made in the sediment was the concave-up imprint of the dinosaur's three-toed foot (Figure 5-24, A). (Of course, the ultimate original is the dinosaur's foot.) After being filled with sand that was later cemented, the depression in the mud is preserved as a positive-relief sole marking on the base of a

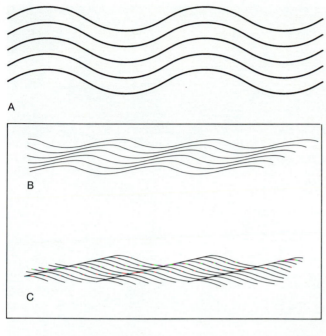

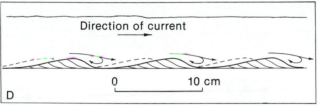

FIGURE 5-21.   Internal structure of selected linear bed forms shown by sketches in vertical plane parallel to current direction.

A. Rolling ripples. Ripple shapes buried by addition of new sediment without ripple migration.

B. Climbing ripples in which sediment upbuilding was nearly equal to downcurrent shifting. Sediment was deposited on both the upcurrent- and the downcurrent sides of the ripples.

C. Product of upbuilding at a slower rate than downcurrent shifting; upcurrent sides of ripples eroded slightly.

D. Product of downcurrent ripple migration where no new sediment was added. Erosion of upcurrent sides provides sediment for additions to downcurrent sides. The resulting internal structure is truncated ripple cross laminae. Arrows show flow within the water of the main current (above) and of eddies (close to ripples). (Based on various sources.)

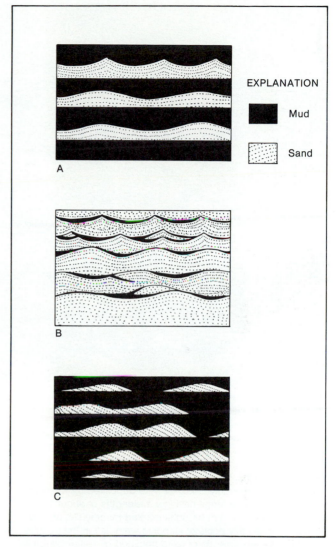

FIGURE 5-22.   Sketches of internal structure of small-scale ripples alternating with layers of mud under contrasting conditions of sediment abundance.

A. Wavy beds in rippled sand; continuous layers of both sand and mud. Irregular thickness of sand layers results from relief on the ripples.

B. Rippled sand in complete layers, but with draping mud as minor constituent.

C. Lenticular layers where sand supply was not sufficient to enable complete layers to be deposited, but where mud was abundant enough to form complete layers. (Modified from H.-E. Reineck and F. Wunderlich, 1968, fig. 1, p. 99.)

sandstone bed (Figures 5-24, B and 5-24, C). In this case, the sole marking on the bottom of the sandstone is a counterpart, or mold, of the original footprint in the mud. If one took some modeling clay and pressed it against the positive-relief sole marking, and then removed the clay, the surface of the clay would be a duplicate, or cast, of the original footprint.

In many cases the current that transported the sand interacted with the cohesive-sediment substrate and eroded down to a level where the cohesive sediment was firm enough to preserve *features made by the moving current*, collectively designated as **current marks**. An extensive list

of current marks has been compiled; we here consider only flutes, mud wisps ("flame structures"), longitudinal mud ridges, and a few tool marks.

Flutes.   Where turbulent eddies from a current impinge upon a cohesive substrate they may scour it into distinctive curved features known as **flutes**, defined as *current-scoured depressions in cohesive bottom sediment and having*

FIGURE 5-23.   Relationship between sizes of bed forms and sizes of cross strata shown by vertical face dug parallel to direction of current (above) and on relief peel made from box core with vertical side parallel to current direction.

A. (*above*) View toward north of two sets of cross strata of contrasting sizes. The climbing sets of thicker planar cross strata (below; compare with Figure 5-20, C) were deposited by westward migration of two bed forms having wavelengths of about 1 m and amplitudes of 15 cm (troughs marked by gray film of mud left behind when tiny pools of standing water left in the troughs vanished). As this episode of sand transport by water neared its close, the current continued and the water level dropped. During this dropping-water stage the top surface of the larger set of planar cross strata at right was modified by westward migration of ripples, which resulted in deposition of a thin veneer of ripple-laminated fine sand. (B. Caplan.)

B. (*right*) View in same orientation of relief peel made from the vertical face of a box corer collected two weeks after the photo in A was made. The surface of the small-scale ripples (wavelengths about 15 cm; heights, 1.5 to 2 cm) is at a depth of 9 cm; it has been covered by horizontally stratified windblown sand.

deeper, pointed upcurrent ends and wider, shallower, and flared downcurrent ends. Flutes are nearly always known only from the preservation of their *counterparts*, or *molds*, on the soles of overlying sandstones (Figure 5-25). Proof that flutes formed as the sand was being transported is available from the characteristics of the flute-filling sand. Some flute fillings consist of cross-laminated sand in which the downcurrent inclination shows that the flute was filled from the steeper side (Figure 5-26). Some flute fillings contain coarser particles than are found at

the bottom of the sandstone outside the flutes. This indicates that the flutes trapped a few of the in-transit coarser particles, most of which were deposited elsewhere.

Mud Wisps ("Flame Structures").   Some sand-carrying currents pulled up ripple-like ridges of mud, some of

FIGURE 5-24.   Dinosaur footprint as impression in ancient mud (now mudstone) and as counterpart on base of a sandstone bed.

A. Impression of dinosaur footprint at top of mudstone bed, East Berlin Formation (Lower Jurassic), Middlefield, Connecticut. (Authors.)

B. Counterpart of a different dinosaur footprint on base of a sandstone bed, Newark Supergroup (Lower Jurassic), formation and locality not known. Specimen mounted on wall outside office of Department of Geological Sciences, 5th Floor, Schermerhorn Extension, Columbia University, New York, New York. (Authors.)

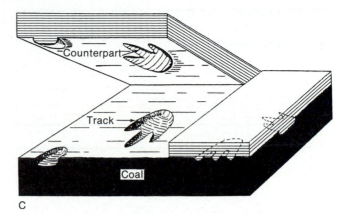

C

FIGURE 5-24.   (*Continued*)

C. Sketch of the footprint and counterpart on the base of the overlying bed of a giant Cretaceous dinosaur, Wyoming. The footprint sketched, about 1 m long and 0.3 m deep, was made in swampy material that became coal. It was covered by dark-colored mud that became a black shale. (R. R. Shrock, 1948, fig. 133, p. 178.)

ridges that are separated by pointed furrows (Figure 5-28). When the original bottom relief is reconstructed from its sole-mark counterpart, the result is a series of linear, sharp-crested mud ridges separated by slightly wider, rounded, concave-up troughs. The long axes of both ridges and troughs trend parallel to the inferred current direction. Accordingly, these have been named *longitudinal mud ridges*.

Flutes, mud wisps, and longitudinal mud ridges are examples of the collective interaction between the turbulent eddies within a sediment-carrying current and its cohesive substrate. Other possible interactions between the transported sediment and the bottom include the contact between individual particles being transported and the bottom. An individual particle becomes the tool; how it behaves determines the kind of mark. Collectively, they are *tool marks*, next described.

which are transverse to the current and others, parallel to it. In the lee of some transverse ridges, cross-laminated sand was deposited (Figure 5-27), proving that the microrelief of the wisp existed when the sand was deposited. Because of their similarity to the profile of a candle flame, these *mud wisps* have been named "flame structures."

Some mud wisps formed as ripple-like ridges perpendicular to the current. Others are parts of ridges that were parallel to the current. The wisps are reliable top indicators: the mud is on the former bottom and the sand on the former top.

Longitudinal Mud Ridges.   The soles of some sandstone beds are covered with narrow, linear, rounded, parallel

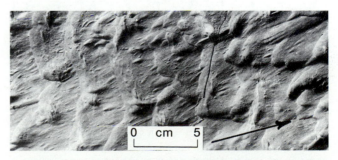

FIGURE 5-25.   Counterparts of flutes, current from lower left to upper right. Base of sandstone, Krosno beds (Oligocene), Wernejowka, Carpathians, Poland. (S. Dzulynski and J. E. Sanders, 1962, pl. IV, B.)

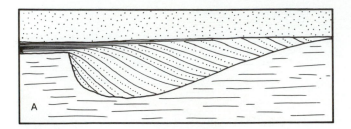

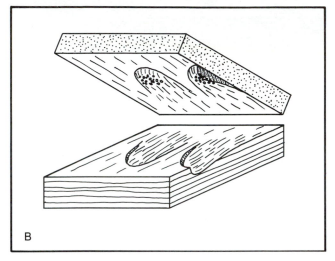

FIGURE 5-26.   Flute fillings, schematic sketches.

A. View in plane through median axis of flute filling parallel to current direction shows that sand filled the flute from the steeper, upcurrent end.

B. Blocks showing original flutes in cohesive mud with the overlying sandstone lifted up to show flute fillings containing coarser particles (black dots) than are present in other parts of the covering sandstone bed.

Tool Marks.   On the scale of a few meters, the length of many exposures of bed soles, tool marks can be subdivided into *continuous tool marks* and *discontinuous tool marks.* An example of *a continuous tool mark* is a **groove**, *where the tool moved across the bottom along a straight course,*

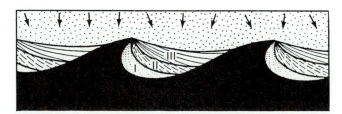

FIGURE 5-27.   Mud wisps with lee-side cross strata. Schematic profile of pointed mud wisps (black) and three sets of lee-side cross laminae (numbered I, II, and III from oldest to youngest) in vertical plane parallel to current (from left to right). Cross laminae of set I were slightly deformed prior to deposition of set II. Set II was in turn deformed (and the older set I, further deformed) prior to deposition of set III. Length of profile, about 10 cm. (J. E. Sanders, 1965, fig. 1, p. 198.)

FIGURE 5-28.   Photo: counterparts of longitudinal mud ridges exposed on base of Ordovician graywacke, west side of U.S. 9W, 5 km northwest of Coxsackie, Greene County, New York, the same locality as the specimens photographed in Plate 59 of P. E. Potter and F. J. Pettijohn, 1964. Lens cover = 6 cm (Authors.)

leaving behind a linear depression that extends for a few- to many meters. After the groove has been filled, it is usually displayed as a **groove mold,** *a sole mark that is a counterpart of a groove* (Figure 5-29).

Discontinuous tool marks result from brief, intermittent contact between tool and bottom. Some *tools graze the bottom as part of a curved trajectory that takes them up*

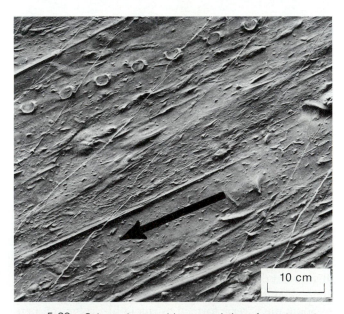

FIGURE 5-29.   Sole-mark assemblage consisting of counterparts of many kinds of tool marks. The nine circular features in a line at upper left were made by the impact and turning over, while in contact with the bottom, of a fish vertebra. Arrow shows direction of current. Thin parallel lines crossing the specimen from upper right to lower left are joints and tiny faults. Krosno beds (Oligocene), Rudawka Rymonowska, Carpathians, Poland. (S. Dzulynski and J. E. Sanders, 1962, p. XIV, B.)

into the current again (**brush marks**) or *strike the bottom and remain within it* (**prod marks**). Some *tools roll along the bottom like tiny wheels*, leaving **roll marks**, whereas other *circular tools skip along the bottom, cutting into it at each point of contact as if they were tiny cookie cutters* (**skip marks**). In general, tool marks are characterized by their linearity; their long axes are parallel to the direction of the current.

SOURCE: Dzulynski and Sanders, 1962.

As a final category we discuss what may happen along the interface between a cohesionless sand and a cohesive mud after the sand has been deposited. Included are load-flow structures and various invertebrate tunnels.

Load-Flow Structures. For several reasons, newly deposited sand tends to founder downward, especially if the underlying sediment yields easily to load, or, in other words, is in a *hydroplastic condition*. The sand/mud interface becomes deformed. In sectional view, the parts of the sand that move downward are usually bulbous and broadly curved. By contrast, the parts of the mud that project relatively upward may narrow to sharp points which resemble the flame structures previously discussed (Figure 5-30). In some examples, the entire layer of sandstone has been separated into these downward-yielding masses. The terms "ball-and-pillow structure" and "pseudo-nodules" have been used.

Tunnels Made by Invertebrates. After a layer of sand has been deposited, its base may be influenced by invertebrates that tunnel along a sand/mud interface. The top halves of the small tunnels appear as concave depressions on the sole of the sandstone bed. In many cases it may not be possible to prove conclusively whether a given feature made by organisms was already present in

the mud when the sand was deposited or was made later, after the sand had been deposited.

MISCELLANEOUS FEATURES

We include under this miscellaneous heading two other features found on surfaces of beds: shrinkage cracks and raindrop-impact pits.

Shrinkage Cracks. What geologists refer to as **mud cracks** are the familiar fractures, *linked polygons that form in cohesive sediment as it shrinks during drying* (Figure 5-31). In plan view, mud cracks form five- or six-sided polygons. In sectional view, mud cracks narrow downward and the edges of the polygons may curl up. With continued drying, the parts of the polygons nearest a

FIGURE 5-31. Shrinkage cracks in modern mud and mudcracks in ancient sedimentary rock.
A. Cracks formed by the drying out of mud. The set of large cracks involves a thicker layer than the set of small cracks. (United Nations Food and Agriculture Organization photo by H. Null, in J. E. Sanders, 1981, fig. 8.19(c), p. 209.)
B. Mudcracks on bedding surface of Manlius Limestone (Devonian), Bossardsville, Pennsylvania. (R. W. Fairbridge, in J. E. Sanders, 1981, fig. 8.19(d), p. 209.)

FIGURE 5-30. View of load-flow structures in Recent silts, Lake Roosevelt, exposed at mouth of Harker Canyon, Washington. (C. A. Kaye (1916–1983), U.S. Geological Survey, photo taken in mid-1950s.)

crack may become detached from the underlying sediment. This results in the formation of flat mud chips. These platy particles, or pieces of them, may be redeposited as mud-chip conglomerates. Although mud cracks are easily destroyed by the action of water, they form so readily that they are common features seen in ancient rocks.

Shrinkage cracks are common in subaerial environments where wet sediments become exposed to the atmosphere and the rays of the sun and thus dry out. Shrinkage also accompanies loss of water during the aging of a gel or of a flocculated colloidal suspension. As a result, cracks may form under water. Shrinkage cracks formed in sediment by the aging of a gel have been named *syneresis cracks*.

Raindrop-Impact Pits. Raindrop-impact pits form when raindrops strike wet, cohesive sediment. (See Figure 14-15, A, B.)

Criteria for Recognizing Strata

Strata result from differences. If all sediments were absolutely uniform in all respects, then many of the strata that are so evident would not be visible. The degree to which strata are apparent depends on the extent of the contrasts between one stratum and the next. The surfaces along which the changes have taken place typically are the preferred surfaces along which sedimentary rocks part or split. *Strata-bounding surfaces along which the rock parts* are known as **bedding-surface partings** (or **bedding-plane partings**; in some European literature the term **stratification joints** is used). Prominent sets of parallel bedding-surface partings are a striking aspect of many exposed strata. Some prominent sets of partings that are parallel to one another are not parallel to bedding. Indeed, the most-prominent sets of parallel partings in many deformed siltstones and shales constitute slaty cleavage, which is of tectonic origin (Figure 5-32). During deformation, the sheet silicates within the body of material become aligned parallel to a direction that is related to the deformation. The orientation of the slaty cleavage and its resulting parting are parallel to the aligned sheet silicates. Such cleavage follows directions that range all the way from perpendicular to parallel with the stratification.

The differences in stratal characteristics are expressed in various ways other than the bedding-surface partings. These include particle size, color, resistance to weathering, mineral composition, and layer-to-layer changes in degree of cementation or of compaction, in abundance of concretions or of chert, and in abundance of fossils.

FIGURE 5-32.   Closely spaced, nearly horizontal partings along slaty cleavage cutting across overturned strata (defined by light- and dark-colored layers) that dip steeply to the left. St. Francis Group (Ordovician), Moe River, Compton County, Quebec. (K. C. Bell, Canada Geological Survey, in P. E. Potter and F. J. Pettijohn, 1964, pl. 6A and J. E. Sanders, 1981, fig. C.13, p. 554.)

The stratification of uniform-looking sand-size sediments can be seen better by making *relief peels*. Such peels can be prepared in the field from carefully smoothed surfaces cut perpendicular to the stratification of unconsolidated sediments or in the laboratory on suitable faces cut from cylindrical cores or box cores. (See Figures 5-16, B, and 5-23, B.) Another laboratory technique for enhancing the visibility of strata of sand- to clay-size materials is X-radiography of oriented thin slabs of the sediment or of the rock (Figure 5-33).

Strata and Time

Strata and time are connected in many ways. Most of these are topics that we treat in Chapter 16. But here we need to take up three points: (1) rates of sediment accumulation, (2) strata as direct time indicators, and (3) truncated strata as indicators of gaps in the record.

Rates of Sediment Accumulation

Two rates of sediment accumulation have been the focus of most studies: (1) the short-term or instantaneous rate of addition of new sediment and (2) the long-term or average rate of accumulation of strata. If short-term rates are multiplied by the long time required to reach a standard for geological comparisons, say meters of sediment thickness per million years, then large numbers

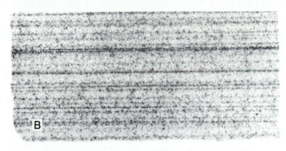

FIGURE 5-33. View of smoothed face of sandstone (A) compared with X-radiograph (B). The laminae, so prominent on the positive print from the X-ray negative (B), are scarcely noticeable on the smoothed face. (W. K. Hamblin, 1962, parts of figs. 1 and 2, pp. 204 and 205.)

result. For example, from a compilation of hundreds of measured rates, using an average time base of one day, the computed number of meters per million years is 10^5 (100 km). In contrast, geologically determined rates, using a time base of 10 million years, average out to 10^{-2} m, or 1 cm, per million years. This is a difference of seven orders of magnitude.

The contrast in these two kinds of rates of accumulation suggests that in a succession of sedimentary rocks consisting of even, tabular beds, with no immediately apparent signs of significant gaps in the record, the time represented by the materials composing the beds is trivial compared with the time represented by the planes between the beds. Put in another way, this implies that the stratigraphic record accumulates episodically; short-term, rapid events of sediment addition are separated by much-longer times when no new sediment is added; indeed, previously deposited sediment may be removed.

Another expression of this contrast in rates between rapid active sediment addition and low average rates of long-term accumulation is the existence of laminae in marine strata deposited in environments where burrowing organisms were abundant. Given the ubiquitous effects of bioturbation in such environments, slow upward accumulation of sediment would enable organisms to destroy any initial laminae. The presence of laminae

implies some combination of circumstances that would get around the effects of bioturbation.

Because they represent times, however brief, during which deposition was interrupted, the *boundaries between strata resulting from times when sedimentation was interrupted* are called **surfaces of discontinuity.** *The time represented by a discontinuity* is known as a **hiatus.** Although along surfaces of discontinuity little or no new sediment may be added, a few geologic processes that may affect the sediment surface leave clues which may be recognizable in the rock record. Shells lying on a depositionally neutral sea floor may be bored and dissolved; indeed, if the sediment is composed of calcium carbonate, it, too, may be dissolved. Equivalent places on the modern sea floor are the hardgrounds described in Chapter 3. Other parts of the sea floor or of a saline lake may become sites where authigenic minerals, such as phosphorite, glauconite, or manganese oxides accumulate. (Authigenic sediments were discussed in Chapters 3 and 4.)

In nonmarine strata a soil may form and be recognizable because of *in-situ* plant roots, by chemical gradients resulting from the effects of near-surface oxidation and downward-percolating waters, or by minerals that were chemically precipitated near or at the surface by the upward movement and evaporation of soil moisture. Examples include caliche, calcrete, and silcrete.

SOURCE: Sadler, 1981.

Strata As Time Indicators

Some strata, mostly in certain kinds of lakes, accumulate on a seasonal basis so that the deposit of each year can be identified. *The sediment layer(s) deposited in a year* constitute a **varve.** In terms of time, varved sediments are very useful. They can be counted like tree rings to yield a chronology based on years. Varved deposits are typical of lakes found near the margins of melting glaciers. (See Figure 5-11, C.) Accordingly, the counting of varves and matching of varved sequences in the sediments from one lake basin to another was employed to establish a chronology of retreat of the last Scandinavian ice sheet and of the last North American continental glacier. This topic is discussed further in Chapter 14.

The seasonal reversal of trade winds in tropical regions is a regularly repeating phenomenon that may leave its impact on sediments. Tropical varves resulting from these changes have not been described, but that may be related more to ignorance of the sedimentary consequences of reversing winds than to their effects on sediments.

Other strata, such as layers of tephra, are deposited essentially instantaneously. If the date of the responsible volcanic explosion can be determined, then a tephra layer serves as a datum surface of known date.

SOURCES: Antevs, 1922, 1925; DeGeer, 1912; Sauramo, 1923; Schove, 1987.

Truncated Strata As Indicators of Gaps in the Record

In the foregoing discussion of the discontinuities inherent even in what appears to be an uninterrupted stratigraphic record, we dealt only with tabular beds having plane, parallel boundaries that persist essentially unchanged across one's field of view and thus display no obvious indications that anything is missing. Numerous other gaps in the stratigraphic record are obvious because layers are not continuous, but end in some fashion. In discussing these indications of missing parts of the record, it is helpful to separate the features associated with the lower contacts of strata from those associated with upper contacts.

RELATIONSHIPS AT LOWER CONTACTS

Where horizontal strata terminate against a sloping surface underlain by older deposits (or gently dipping strata end against a surface whose inclination is in the same direction but is greater than the dip angle of the strata) and the upslope extent of upper strata is greater than that of the lower layers, the relationship is defined as **onlap** (Figure 5-34, A). Similarly, *initially dipping strata at the base of a depositional embankment that grows laterally may end against a surface in such a way that the younger layers extend farther along the surface than the older layers* (Figure 5-34, B; see also Figure 5-26, A). Such a relationship is defined as **downlap**. *Onlap and downlap are collectively designated as* **baselap**.

RELATIONSHIPS AT UPPER CONTACTS

As a result of erosion, layers may be removed along an irregular contact, giving *a relationship of unconformity* (Figure 5-34, C). *If the lower strata were deformed so that they are no longer parallel to the layers within the covering strata,* then the resulting relationship is one of **discordance**, or of **angular unconformity** (Figure 5-35).

Toplap is *a stratigraphic relationship in which the updip ends of dipping strata are truncated at a depositional surface.* In some cases it may not be possible to distinguish a toplap relationship from one of angular unconformity. Two useful clues for identifying toplap are (1) the preservation of the depositional embankment to which the inclined strata are parallel (upper right corner of Figure 5-34, D) and (2) the presence of updip segments that are convex up and whose curvature tends to form an asymptotic relationship with the horizontal surface (upper left corner of Figure 5-34, D).

Among subaerial dunes, sand that formerly projected well above the surroundings may be removed down to the water table. The result is that the large-scale cross strata formed by the forward migration of the dune are truncated and the surface of truncation overlain by a horizontal layer of sand (Figure 5-36). Two contrasting time relationships are represented by these dune strata. Clearly, the inclined strata become progressively younger toward the right. The time planes are inclined and coincide with the cross strata. By contrast, the horizontal

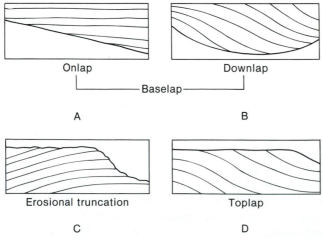

FIGURE 5-34.   Ending of strata at lower- and upper contacts. Text explains terms. (R. M. Mitchum, Jr.; P. R. Vail; and S. Thompson, III, 1977, fig. 2, p. 58.)

FIGURE 5-35.   Relationship of angular unconformity between Siluro-Devonian carbonates (dipping about 45° to the left) and Ordovician siltstones and graywackes (at right, appearing nearly vertical but also dipping to the left though much more steeply than the carbonates). Cuts on north side of exit ramp from new location of N.Y. Route 23 to Jefferson Heights, Catskill, New York. (Authors.)

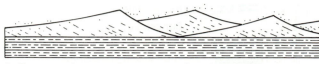

1. Dune sand accumulates on previous level substratum.

2. Sand accumulation continues. water table rises in sand.

3. Wind action removes sand to water table.

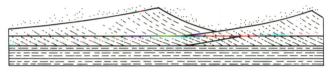

4. Second dune field accumulates on water-table surface.

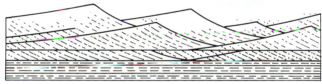

5. Water table rises to new position in dune field.

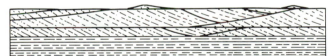

6. Wind action removes sand to second water table.

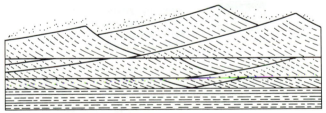

7. Third dune field accumulates, etc.

FIGURE 5-36.   Large-scale planar cross strata from an eolian dune that have been truncated along horizontal surfaces controlled by the ground-water table. (W. L. Stokes, 1968, fig. 1, p. 512.)

surface that truncates the inclined strata probably represents a time surface related to the water table and erosion of any dry sand above it. Therefore, one or more horizontal time planes are parallel to the horizontal strata.

Certain surfaces of discontinuity are key elements in the study of successions of strata, discussed in the next section.

Successions of Strata

When one moves from the study of an individual bed to the study of several beds, two contrasting possibilities are available. The first is to examine adjacent beds to determine if any patterns exist in the order and characteristics of the beds. Many kinds of patterns have been described; they are important characteristics of depositional environments and -processes. The second is to look for some basis for organizing many layers into large-scale units that are needed in order to carry out certain kinds of broad-scale regional analysis. The term "sequence" has been applied to the strata related to both kinds of objectives. In order to sort out the usages and show the basis for each we first summarize two cases: (1) *autocyclic successions* of strata and (2) *allocyclic successions*. After that, we review several sequence definitions.

Patterned Successions (Autocyclic Strata) Deposited Along a Sloping Surface That Shifts Laterally

In numerous depositional settings, the details of which are presented in succeeding chapters, autocyclic successions are deposited when *materials within the depositional setting are redistributed without any change in the total setting, in the total energy, or in the material input.* An example is the sediment deposited along a sloping surface. For various reasons, the sediments deposited on the higher parts of the slope differ from those deposited on the lower parts. As new sediment is added, the location of the depositional slope shifts laterally. As it shifts, the slope leaves behind a succession of strata having a fixed pattern. The positions of the sediments within this pattern are determined by the locations that the various materials occupy along the slope. The height of the slope determines the thickness of the autocyclic succession (Figure 5-37).

As indicated in Chapter 1, Antoine Lavoisier described how nearshore marine strata are differentiated into depth-defined belts paralleling the shore. He also discussed the effects on these belts of changes of water level, a topic we discuss in a following section.

In the central parts of the United States, coals of Pennsylvanian age are exposed within groups of strata that display repeating patterns, or cycles. This cyclicity was first described by J. A. Udden (1859–1932) in 1912, from his studies of the Peoria quadrangle, Illinois (Figure 5-38). In 1931 the Illinois Geological Survey published *Bulletin 60*, a book devoted to cyclic Pennsylvanian strata from many states. The name *cyclothem* was later proposed for these patterned successions (Wanless and Weller, 1932, p. 1003). Some parts of cyclothems are autocyclic suc-

(allocyclic) depositional product of the submergence of the body of prograded sediment.

Parasequences may be as thin as a few centimeters, but more commonly range in thickness from tens of centimeters to several tens or even hundreds of meters or more.

The Lower Devonian strata that crop out on the escarpment at John Boyd Thacher State Park, southwest of Albany, New York, illustrate the characteristics of marine parasequences in carbonate rocks (Figure 5-39). Two of the several parasequences are constituted as follows: a skeletal grainstone is overlain by patch reefs and these, in turn, are overlain by interreef grainstone (interval between the upper PS (parasequence surface) and the scale mark for 20 m on Figure 5-39). An underlying parasequence consists of skeletal grainstone that grades up into stromatolites (algal-laminated mudstone). Each of these two parasequences consists of strata that were formed when a depositional slope prograded seaward. The surfaces bounding the parasequences (labeled PS in Figure 5-39) are inferred to have resulted from rapid submergence. A set of several repeating parasequences, as shown in Figure 39, is known as a **parasequence set**, defined as *"a relatively conformable succession of genetically related parasequences bounded by surfaces (called parasequence*

set boundaries) of erosion, non-deposition, or their correlative conformities."

Concepts identical to those just set forth, and developed independently of the definition of parasequence in seismic stratigraphy were formulated under the name of punctuated aggradational cycles (PACs). What have been named PACs are thin (1-5 m) upward-shoaling cycles whose boundaries are defined by the depositional products of episodes of rapid submergence.

Another example of a parasequence comes from the Holocene deposits of the Rhône delta that was being built into the Mediterranean Sea as the water level rose as a consequence of the rapid melting of the Pleistocene ice sheets (Figure 5-40). The profile-section based on data from a line of closely spaced borings shows two surfaces formed during rapid submergence (marked by closely spaced letter M's). After each of these episodes of rapid submergence, sea level continued to rise, but at a much-slower rate. During such times of slow submergence, a lobe of the delta prograded seaward, depositing a sheet of sand that overlies a silty clay. The strata between the two lines of M's represent a parasequence.

According to the scheme of Exxon and credited to Van Wagoner (1985), the next higher-ranking unit is the *sequence*, defined as "a relatively conformable succession of genetically related parasequence sets (or parasequences) bounded by surfaces (called sequence boundaries) of erosion or their correlative conformities." To avoid confusion with other usages of the term sequence already discussed in this section, we propose the term *mesosequence* for the "sequence" of Exxon. **Mesosequences**

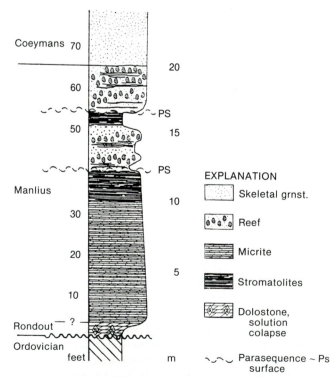

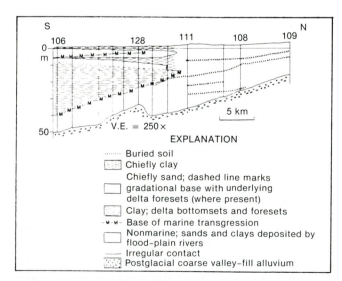

FIGURE 5-39.   Columnar section showing parasequences of Lower Devonian limestones at Indian Ladder Trail, John Boyd Thacher State Park, New York. (G. M. Friedman, 1990, fig. 3, p. 16.)

FIGURE 5-40.   Subsurface stratigraphic section through Rhône delta in Mediterranean based on interpretation of 12 borings, five of which are located along the transect located on Figure 13-29. Further discussion in text. (Data from E. Oomkens, 1970, fig. 10, p. 221; reinterpreted as shown here by J. E. Sanders.)

are *units that are bounded by surfaces of unconformity (in basin-marginal areas that were eroded during times of emergence) or by the basinward extensions of such surfaces that can be followed as continuous lines on seismic records* (Chapter 6). Mesosequences consist of many parasequences, which are commonly grouped into parasequence sets having distinctive patterns of facies and that are bounded by surfaces associated with episodes of submergence of greater amplitude than those associated with parasequence boundaries. In turn, parasequence sets can be grouped into facies tracts that constitute the mesosequences (Chapter 6).

In Chapter 6 we examine how study of seismic profiles has formed the basis for the new discipline of seismic stratigraphy.

SOURCES: Goodwin and Anderson, 1985; Goodwin, Anderson, and Goodman, 1986; Klüpfel, 1917; Van Wagoner, 1985; Van Wagoner and others, 1988, 1990.

Suggestions for Further Reading

CAMPBELL, C. V., 1967, Lamina, lamina set, bed, and bedset: Sedimentology, v. 8, p. 7–26.

GOODWIN, P. W., and ANDERSON, E. J., 1985, Punctuated aggradational cycles: a general hypothesis of episodic stratigraphic accumulation: Journal of Geology, v. 93, p. 515–533.

HARMS, J. C., 1975a, Stratification produced by migrating bed forms, Chapter 3, p. 45–61 *in* Harms, J. C., Southard, J. B., Spearing, D. R., and Walker, R. G., eds., Depositional environments as interpreted from primary sedimentary structures and stratification sequences: Tulsa, OK, Society of Economic Paleontologists and Mineralogists Short Course 2 Lecture Notes, 161 p.

HARMS, J. C., 1975b, Stratification and sequence in prograding shoreline deposits, Chapter 5, p. 81–102 *in* Harms, J. C., Southard, J. B., Spearing, D. R., and Walker, R. G., eds., Depositional environments as interpreted from primary sedimentary structures and stratification sequences: Tulsa, OK, Society of Economic Paleontologists and Mineralogists Short Course 2 Lecture Notes, 161 p.

KLEIN, G. DEV., and WILLARD, D. A., 1989, Origin of the Pennsylvanian coal-bearing cyclothems of North America: Geology, v. 17, p. 152–155.

KUENEN, PH. H., and MIGLIORINI, C. I., 1950, Turbidity currents as a cause of graded bedding (*sic*): Journal of Geology, v. 58, p. 91–127.

MCKEE, E. D., and WEIR, G. W., 1953, Terminology for stratification and cross-stratification (*sic*) *in* sedimentary rocks: Geological Society of America Bulletin, v. 64, p. 381–389.

NELSON, C. M., 1985, Facies in stratigraphy: from "terrains" to "terranes": Journal of Geological Education, v. 33, p. 175–187.

SADLER, P. M., 1981, Sediment accumulation (*sic*) rates and the completeness of stratigraphic sections: Journal of Geology, v. 89, p. 569–584.

VAN WAGONER, J. C., MITCHUM, R. M., CAMPION, K. M., and RAHMANIAN, V. D., 1990, Siliciclastic sequence stratigraphy in well logs, cores, and outcrops: concepts for high-resolution correlation of time and facies: Tulsa, OK, American Association of Petroleum Geologists Methods in Exploration Series 7, 55 p.

VAN WAGONER, J. C.; POSAMENTIER, H. W.; MITCHUM, R. M.; VAIL, P. R.; SARG, J. F.; LOUTIT, T. S.; and HARDENBOL, J., 1988, An overview of sequence stratigraphy and key definitions, p. 39–45 *in* Wilgus, C. W.; Hastings, B. S.; Posamentier, H. W.; Van Wagoner, J.; Ross, C. A.; and Kendall, C. G. St. C., eds., Sea level (*sic*) changes: an integrated approach: Tulsa, OK, Society of Economic Paleontologists and Mineralogists Special Publication 42, 407 p.

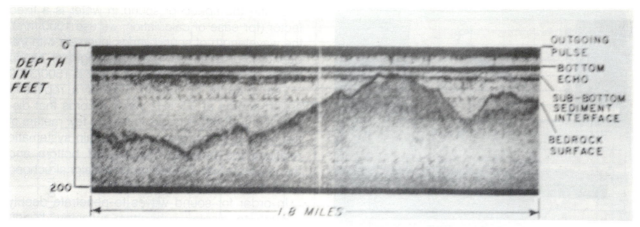

BOX 6.2 FIGURE 5.   Segment of a record from a continuous seismic-reflection profiler in Long Island Sound, New York, showing irregular surface of bedrock buried beneath sediment, displaying horizontal reflecting interfaces. (Walter Beckman, in J. B. Hersey, 1965, fig. 4, p. 55.)

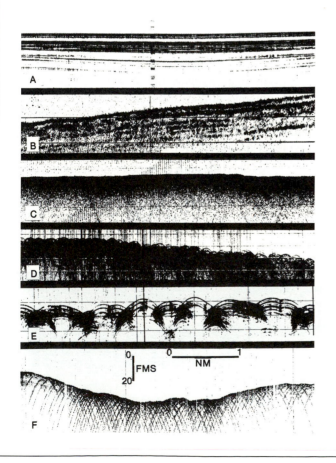

BOX 6.2 FIGURE 6.   Examples of traces of sound waves reflected from contrasting kinds of deep-sea floor in the South Atlantic off northeastern South America with 3.5-kHz sound source and PDR. A, B, and C are traces of sound reflected from smooth bottoms; D, E, and F, from rather rough bottoms. (Lamont-Doherty Geological Observatory of Columbia University, courtesy J. E. Damuth.)

A. Sea floor underlain by fine sediments lacking interbeds of sand or silt. The topmost line is the trace of sound reflected from the bottom; dark lines beneath it and generally parallel to it are traces of sound reflected from layers beneath the water/sediment interface.

B. Trace of reflection from sea floor underlain by sediments whose upper parts contain high concentrations of bedded silt and -sand interstratified with fine-textured deep-sea sediment.

C. Trace of echo from sea floor underlain by sediments having high concentrations of silt and sand, as in B, but with no distinct traces of sound reflected from subbottom interfaces. Bottom traces such as this have been referred to as "prolonged echoes."

D. Trace of echo from sea floor having numerous small-scale bed forms.

E. Trace of echo from sea floor consisting of large bed forms.

F. Trace of echo from sea floor consisting of sediment that has been eroded by deep currents.

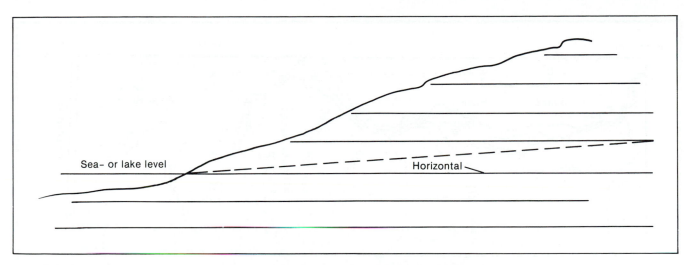

FIGURE 6-1. Base level for stream and lake or sea. For streams, base level is a surface (dashed line in this view) that is nearly parallel to the Earth's horizon extending from the level of the lake or the sea beneath the strata of sediment or sedimentary rock. Because base level is gently inclined, even when streams have reached base level, the water can still flow. Lake level is a temporary base level. The ultimate base level is sea level. (Authors.)

to grasp the fundamental points. In the next section we examine several examples of the importance of base level in controlling the sediments when the sediments reflect large changes in water level. Thereafter, we move on to effects of sea-level changes.

Changes of Base Level: Examples from Lakes

We examine three cases: (1) Lake Mead, Arizona and Nevada, (2) Great Salt Lake-Lake Bonneville, Utah, and (3) the lake in the Newark basin (Late Triassic and Early Jurassic age) in New Jersey-Pennsylvania in which the Lockatong Formation was deposited.

The sediments that began to accumulate in Lake Mead (Figure 6-2) during the 6 years, 6 months, and 6 days (1935–1941) while the water was backing up behind Hoover Dam have been studied in considerable detail. Box 6.2 discusses the density currents that were noticed passing through the discharge pipes and their relationship to floods in the upstream reaches of the Colorado River. From the point of view of the former river bed, the rising lake level was a submergence. What had been a river bed became a lake bottom. The water deepened progressively until the level of 1170 ft was attained. (See Figure 5-39.) At that level the water stabilized, and a delta lobe began to prograde out into the lake. The schematic profile-section shows a thin basal layer of sand that formed in the old river bed as lake level rose. This represents sediments deposited during submergence. The delta that

prograded into the lake after the water level had stabilized constitutes sediments deposited at a high stand. (See Figure 5-39.)

The sediments of Great Salt Lake, Utah, in the Bonneville basin (Figure 6-3), dramatize the contrasts associated with large changes of water level. In this basin, an extensive record of changes has been revealed by samples recovered from the deep

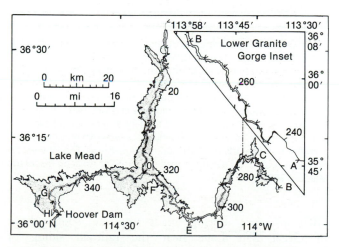

FIGURE 6-2. Index map of Lake Mead, Nevada and Arizona. Numbers show distance along the Colorado River at 20-mile intervals (5-mile marks shown by inverted V's), measured from the U.S. Geological Survey's concrete gauge well opposite the mouth of the Paria River, near Lees Ferry, Arizona (lat. 36°52′N; long. 110°35′W). (G. M. Friedman and J. E. Sanders, 1979, fig. C-4, p. 513; original in H. R. Gould, 1951, fig. 1, p. 35.)

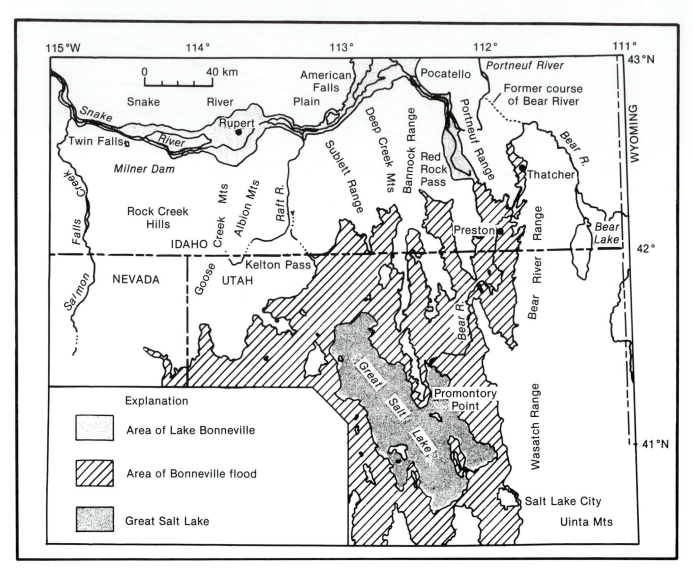

FIGURE 6-3.   Pleistocene Lake Bonneville and its drainage route to the north via Red Rock Pass and into the Snake River (and from there to the Pacific Ocean via the Columbia River), northern Utah and -Nevada and southern Idaho. About 18,000 years ago, changes in drainage caused the lake level to be lowered by about 30 m, to the level of the Bonneville shoreline, which was controlled by a natural bedrock dam at Red Rock Pass (altitude 1550 m, 5085 ft). An estimated 1583 km^3 of water drained from the lake at a rate that has been inferred to have ranged between 424,800 and 50,976 m^3/s. Flow at the higher rate would have drained Lake Bonneville in 6 weeks; at the lower rate, in 1 year. The amount of water involved equals about half the yearly discharge from the Amazon, the modern world's largest river. The inferred upper rate of Bonneville outflow is slightly more than twice the maximum rate of Amazon discharge (203,000 m^3/s). (After H. E. Malde, 1968, fig. 1, p. 3.)

(307 m) core boring made on the south shore of Great Salt Lake near the railroad siding of Burmester, Utah (Figure 6-4). Within the top 100 m, at least 15 levels displaying ancient soils and 9 others indicating dryness, but without a soil, are interlayered with sediments among which 17 levels are thought to have been deposited in the middle of a large, deep lake. It is important to realize that lake-level changes not only can be observed directly (Figure 6-5) but also can be inferred with confidence from studying the sedimentary record. In the Bonneville basin, the difference between no water and "deep" water has been as great as 300 m.

Studies of the cyclic Lockatong Formation (an ancient lake deposit of Late Triassic age, Newark basin, central

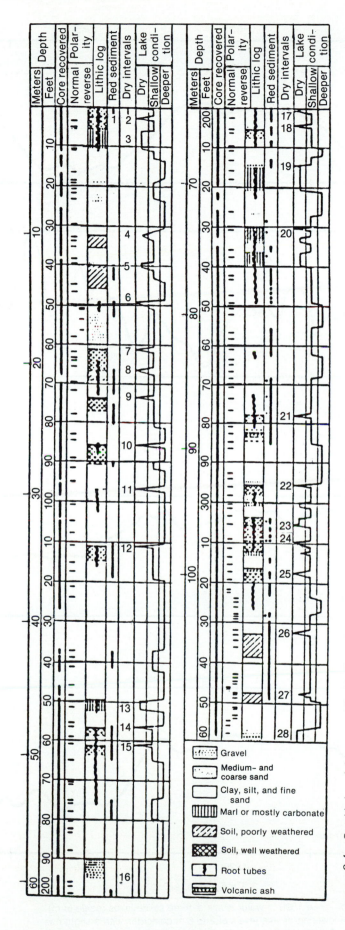

FIGURE 6-4. Graphic log of deep core drilled in Great Salt Lake, Utah, showing kinds of sediments encountered, magnetic polarity, and inferred shifts from dry conditions to deep-lake conditions. (A. J. Eardley, R. T. Shuey, S. Gvosdetsky, and others, 1973, fig. 1, p. 212.)

193

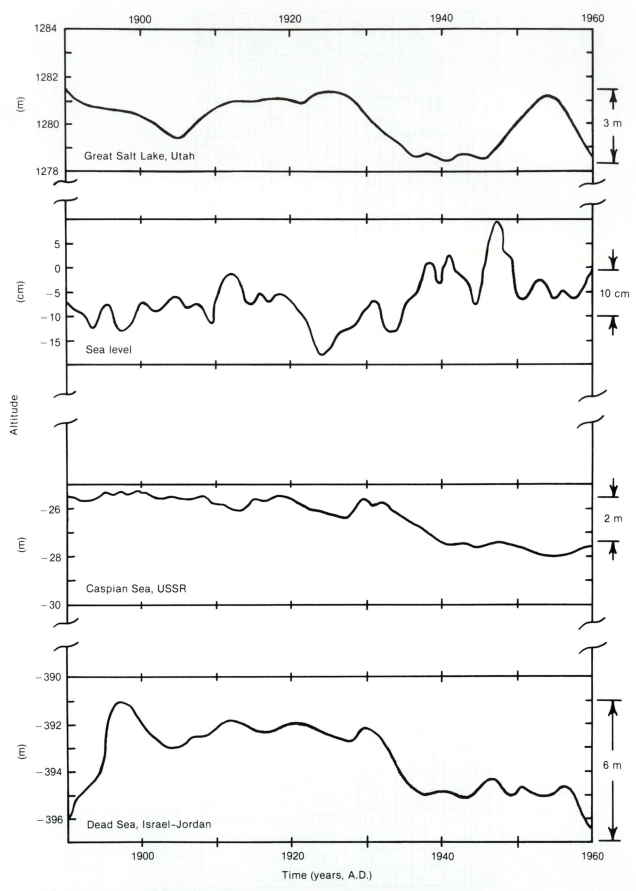

FIGURE 6-5.   Graphs showing variation since 1890 in water levels of three saline lakes (scale in m) and of the sea (scale in cm). General lowering of lake levels of 2 to 6 m coincides with a slight rise of world sea level of 10 cm. (Redrawn from D. Neev and K. O. Emery, 1967, fig. 19, p. 36.)

New Jersey) indicate that the water levels fluctuated greatly and on various time scales (Figure 6-6).

After this brief look at the effects on sediments of changes of water level in lakes, we turn to the larger subject of sea level, the ultimate base level.

SOURCES: Eardley, Shuey, Gvosdetsky, and others, 1973; H. R. Gould, 1951, 1960a; Malde, 1968; Neev and Emery, 1967; Olsen, 1986.

Changes of Sea Level

Many factors affect the position of the sea relative to the land. Three that we discuss here are changing volumes of continental glaciers, changing volumes of the ocean basins, and local crustal movements. In discussing sea level it is important to keep in mind the distinction between *worldwide changes of sea level*, or eustatic changes of sea level, and local changes. As far as sediments are

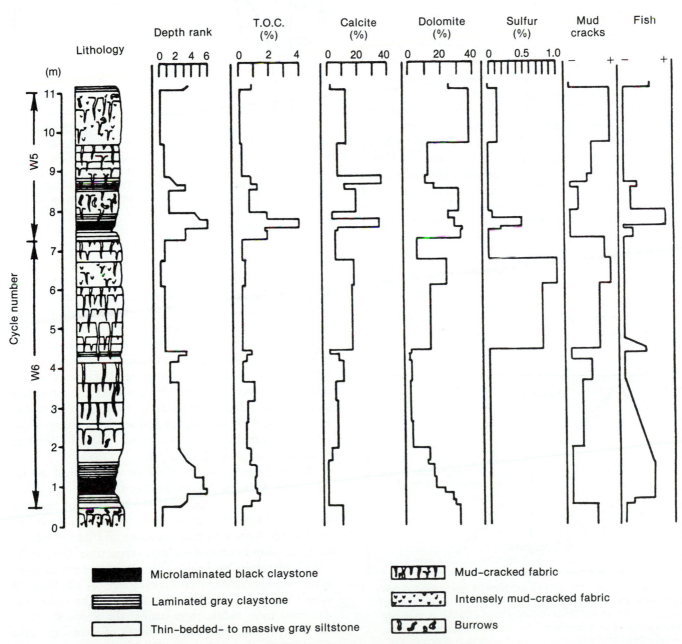

FIGURE 6-6. Two parasequences, known as Van Houten cycles, of sedimentary rocks from Lockatong Formation (Upper Triassic), Eureka quarry, Tradesville, Pennsylvania. A Van Houten cycle begins with a submergence, continues with deposits made in a deep-water phase, and ends with a retreat of the waters and creation of soils or deposition of nonlake sediments. (P. E. Olsen, 1986, fig. 2, p. 844.)

concerned, the important point is whether *the sea is rising relative to the land* (a **submergence**), for whatever combination of reasons, or whether *the sea is falling relative to the land* (**emergence**; Figure 6-7). In addition to changes in the vertical sense, *the sea may be moving inland* (a **retrogression** or a **marine transgression**) or *abundant sediment may be pushing the land outward* (**progradation** or **depositional regression**). The sedimentary record is usually unambiguous about which of these sea-level situations prevailed.

Sea-Level Curves

During the Pleistocene Epoch, each time that continental glaciers spread over the northern lands, world sea level dropped, a eustatic change. After the glaciers had melted, sea level rose again, another eustatic change. R. A. Daly (1871–1957) described this reciprocal relationship between water in the ocean basins and continental glaciers on the lands as being one of "robbery and restitution." In a general way, as mentioned in Chapter 1, a clear connection exists between climate and sea level. That is, times of glacial maxima match times of low sea level (the Earth's ice-house modes), and interglacial times of glacial minima match times of high sea level (the Earth's greenhouse modes). But we caution that the relationship among "cold" climates, "warm" climates, and sea level is not as straightforward as one might suppose. For example, the times of the warmest Pleistocene climates, indicated by the maximum distance from the Equator to which tropical shallow-water carbonates spread, do not coincide with times of highest sea levels.

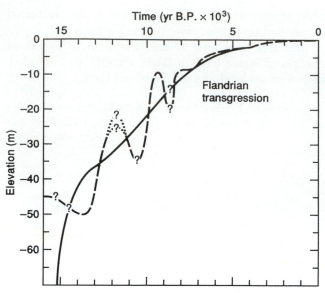

FIGURE 6-8.   Sea-level curve for last 15,000 yr according to the Shepard interpretation. Dashed curve illustrates "the apparent fluctuation of the rising sea level that appears to have occurred along the Texas coast" (quotation from F. P. Shepard, 1960, p. 342). Solid curve represents approximate mean of dates compiled by J. R. Curray. It implies a mean rise of 9 mm/yr for about 10,000 yr. (After F. P. Shepard, 1960, fig. 3, p. 342; and J. R. Curray, 1965, fig. 2, p. 725.)

Just the opposite! The warmest times thus indicated, which seem to have been required to start the meltdown phases at the end of a glacial age, match times when sea levels were lowest (the Earth's boiler-room modes). Clearly, volumes of glacial ice and levels of the sea are reciprocally related and both are responsive to climate. But it is a gross oversimplification to equate highest sea levels with warmest climates and lowest sea levels with coldest climates. However climate and sea level are ultimately shown to be related, the evidence from the Pleistocene indicates that climatically caused sea-level changes can take place rapidly, on the order of 100 m in 6500 yr (1 m/65 yr, or about 1.5 cm/yr).

One of the by-products of the development of the radiocarbon dating technique has been a large-scale effort by many scientists to calibrate the behavior of sea level during the Holocene submergence (known widely in Europe as the Flandrian transgression); in other words, to try to develop a curve of sea level versus time. One of the surprising results has been a near-total polarization of research workers into two contrasting schools: the Shepard school and the Fairbridge school. Using the same fundamental tool, radiocarbon dating, the Shepard school has interpreted their results in terms of a smooth curve (Figure 6-8). By contrast, the Fairbridge school presents a curve showing many ups and downs (Figure 6-9).

Neither of these schools has placed much emphasis on the stratigraphic record; indeed, they both have sought

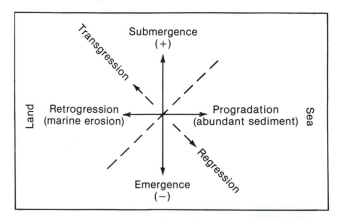

FIGURE 6-7.   Combination of circumstances governing the position of the sea relative to the land, shown in schematic diagram in plane at right angles to shore. Either land or water or both can move in the vertical direction, positive if up, negative if down. Even without any relative change in the vertical direction, the shoreline can be shifted landward by marine erosion (retrogression or transgression; negative direction) or seaward by prograding sediment (depositional regression; positive). (Suggested by figure in H. Valentin, 1959.)

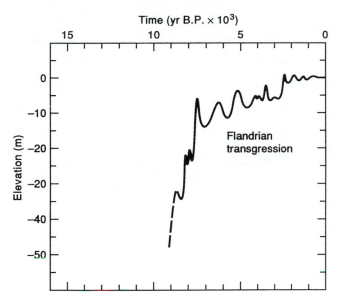

FIGURE 6-9. Sea-level curve for last 15,000 yr according to the Fairbridge interpretation. The oscillations shown imply rates of sea-level rise exceeding 24 mm/yr during periods of about 500 yr. (R. W. Fairbridge, 1961, fig. 14, p. 156.)

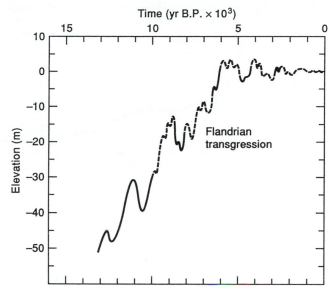

FIGURE 6-10. Sea-level curve of Flandrian transgression, starting about 9500 yr B.P., on French Normandy coast. The thousands of control points have been omitted. (After Mireille Ters, 1987, fig. 12-2, p. 208–209.)

to acquire data points along coasts that they regarded as "stable" in contrast to coasts that have actively subsided. A synthesis of the subsurface stratigraphy obtained from many borings along the French Normandy coast has yielded evidence in support of a Holocene sea-level curve very close to the Fairbridge curve and in contrast to the Shepard curve (Figure 6-10).

After nearly three decades of deadlock, the dilemma of the Holocene transgression will doubtless be resolved by study of the stratigraphic relationships in places where deposition accompanied submergence. Although in such places, points showing absolute water levels at any given instant may not be established, the effects of changing base level on the sediments should make it possible to decide whether the sea rose in a single, progressive submergence (Shepard view) or as a series of small rises that alternated with drops (Fairbridge view). If it were possible to collect a sufficiently detailed continuous seismic-reflection profile off the Mississippi delta, south Louisiana, the record should show how sea level behaved.

Major eustatic changes of sea level result from the interaction between volume of seawater and volume of the ocean basins, both of which are products of global tectonics, particularly of sea-floor spreading. When plates spread actively, the relief on mid-ocean ridges is greatest. This is because younger, hotter, and therefore less-dense, rocks are more abundant. The rising ridge reduces the volume of the ocean basins and forces water to spread over the continental platforms. When plate spreading becomes dormant, the ridges subside. This increases the volume of the ocean basins and causes water to retreat from the continental platforms.

Realization of the mobility of the deep-ocean floors has established the basis for understanding the worldwide changes of sea level that stratigraphers have been recognizing from study of the stratigraphic record on the continents. But this subject has been given a whole new impetus by the relationships between sea levels and the sedimentary record that have grown out of careful study of continuous seismic-reflection profiles.

SOURCES: Curray, 1965; Fairbridge, 1987; Kopaska-Merkel, 1989; Milliman and Emery, 1968; Sanders and G. M. Friedman, 1969; Shepard, 1963; Terse, 1984.

From here, our path leads us to the sediment terraces that form at the margins of basins, and then to the effects on these terrace sediments of submergence, emergence, and progradation. That will bring us to the fundamentals of seismic stratigraphy and hierarchies of sequences. We close this chapter with the results of attempts to use seismic stratigraphy to construct a global record of sea-level changes.

The Depositional Terraces at the Margins of a Basin: Three Critical Environments

At the margins of any marine basin accumulating sediments, a depositional terrace forms. A typical example of such a depositional terrace that has accumulated along the western margin of the Atlantic Ocean basin is shown in a profile drawn from New Jersey to Bermuda (Figure 6-11). Going from the shore seaward, the surface of this

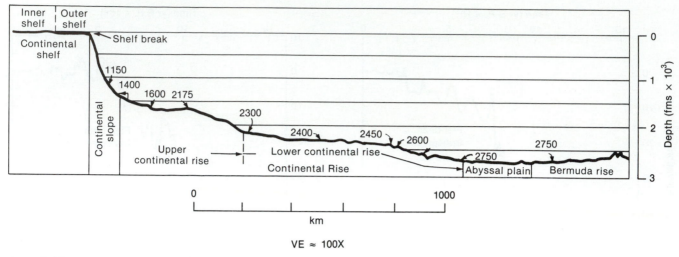

FIGURE 6-11.   Bathymetric profile, perpendicular to depth contours, off northeastern United States, shows flat continental shelf, steep continental slope, and the gradually sloping continental rise that ends seaward at an abyssal plain. (B. C. Heezen and others, 1959; from B. C. Heezen, 1965, fig. 5, p. 236.)

terrace can be divided into shelf, slope, continental rise, and abyssal plain.

In 1951 John L. Rich (1884–1956) wrote a profound paper on how such a depositional terrace affects sediment deposition. His insights were intended to apply to any basin of any size (Figure 6-12) that contains a body of standing water in which noticeable wave action is present. Examples include oceans, inland seas, and large lakes. Quoting from Rich (1951):

> The shelf is that part of the bottom of the water body which is above wave base and is therefore subject to wave action.
>
> The slope is that inclined part extending down from wave base to the deeper, generally relatively flat bottom of the water body. The angle of inclination of the

slope may range from the angle of subaqueous repose to one that approaches closely to zero.

In these definitions the depth of wave base is the critical factor, because it separates two strongly contrasted environments, the shelf and the slope. In the one the water is continually being stirred by waves and tidal currents; in the other it is generally quiet. The boundary between slope and bottom is generally gradational and not sharply marked. A slope may extend down thousands of feet, as does the continental slope of eastern United States, or it may extend downward only a few feet from a shallow wave base to the bottom of a shallow lake; and it may be relatively steep or almost flat.

The terms shelf, slope, and bottom have always been useful and continue to be so. However, when Rich

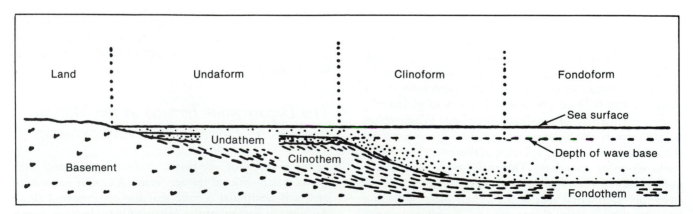

FIGURE 6-12.   Schematic profile at right angles to shore from a land area into the water filling a bathymetric basin, showing the three major surfaces of the depositional terrace and sediments associated with each. (Relabeled from J. L. Rich, 1951, fig. 1, p. 3.)

submitted the manuscript of his paper to the Geological Society of America for publication in the *Bulletin*, it was accepted on the condition that he follow the pioneers in geology and coin Greek terms for these environments. The Geological Society's editor felt that simple English terms would give rise to confusion and make geologists think that Rich's concepts apply only to the continental shelf and continental slope of the modern ocean and not to all marine basins and also to lakes. In response to the editor's insistence, Rich chose three Greek words as general ones for the three critical environments. These are (1) **unda** environment for *the shallow water overlying the shelf*, (2) **clino** environment for *the deeper water overlying the slope*, and (3) **fondo** environment for *the deepest water covering the bottom of the basin.*

As general designations for the surfaces involved, Rich joined the English geomorphological term *form* with the Greek words. Thus he proposed (1) *undaform* for any surface underlying an unda environment, such as the shallow water overlying a continental shelf; (2) *clinoform* for any surface underlying a clino environment, as in the case of the continental slope; and (3) *fondoform* for any surface underlying a fondo environment and forming a generally flat, deep part of the water body. Rich's names for these surfaces drew the disdain of the linguistic purists. By combining Greek- and English words, he had perpetrated a Greek barbarism.

Rich's terms are now universally known in the Earth-Sciences community, especially among geophysicists working with continuous seismic-reflection profiles. But in the early 1950s his new words were scorned. For example, at a meeting of the Geological Society of London in 1952, in the discussion following the presentation of a paper by Ph. H. Kuenen (1902–1976), O. T. Jones (1878–1967), Woodwardian Professor of Geology at Cambridge University and past president of the Geological Society of London, stood up and remarked that he "hoped that the barbaric nomenclature devised by J. L. Rich for what corresponded essentially to shelf, slope, and bottom environments would not find a place in British geological journals." Professor Jones has lost out; Rich's terms have gained acceptance. This, despite Rich's initial hesitation about proposing new names. According to Rich, the **undaform** is

the more or less flat topographic surface that exists above wave base in an aqueous environment where bottom sediments are moved or stirred by waves and currents, particularly during storms. The **clinoform** is *the sloping surface extending from wave base down to the generally flat floor of the water body. The generally flat floor of the water body is called* the **fondoform.** The rock units formed in each of these environmental settings are called *undathem, clinothem,* and *fondothem.*

SOURCES: J. L. Rich, 1950, 1951.

Unda Environment, Undaform, and Undathem

Rich elaborated on these features as follows:

The most significant process of the unda environment is the repeated agitation of the water by waves and currents. Wave action frequently throws the finer material into suspension. Currents then readily move it, and ultimately...most of it [is]...carried out to settle permanently in deep water.

This gradual elimination of the finer particles causes a concentration of the coarser elements of the sediment load on the floor of the undaform.

The cumulative effect of all...processes acting on [an undaform] must be to eliminate the finer material gradually by working it out to deeper water; to leave behind a concentrate of the coarser sediment on the floor of the undaform; and, because of repeated reworking, to produce in this coarse material an irregular lenticularity of bedding, cross-bedding (*sic*) on a moderate scale, and ripple marks....

Exceptions to the prevailing coarseness of [undaform sediments occur]...near the mouths of large rivers bringing in great quantities of sand, clay, and silt. Under such conditions the time available for reworking by waves on the undaform may not be great enough to permit the winnowing out and removal of all fine material....

[On] a sinking undaform, subsiding slowly enough so that the sediment supplied always keeps the surface of the undaform within the range of wave action, great thicknesses of sediment could accumulate, all having [undathem characteristics].

Conditions in the unda environment favor development of a luxuriant bottom-dwelling fauna of the larger shellfish, corals, and brachiopods. The unda zone lies entirely in depths through which light penetrates, and, except in the immediate vicinity of the mouth of a large muddy river, the water is clear except during severe storms because the bottom is composed mainly of coarser sediments left behind as the finer material is put into suspension by the waves and carried away into deeper water.

Clino Environment, Clinoform, and Clinothem

In Rich's words

...The most distinctive features of the clino environment are the inclination of the surface...and the freedom from wave-caused disturbances of the water.

The universal factor affecting all sediments deposited on subaqueous slopes is the tangential component of the Earth's gravity. By the action of this gravity component (See Figure 7-1.), individual sediment particles and even entire bodies of sediment may be moved as a result of being pulled from the front, as it were, instead of being pushed along from the back, as for example, are the bed-load sediments in a river. Processes of subaqueous gravity displacement of sediment can be divided into two large groups: (1) turbidity currents, which involve the suspended-load sediment, and (2) various other processes, which are concerned with bed-load sediments. These other processes include liquefied cohesionless-particle flows, subaqueous rockfalls, and subaqueous slumps and subaqueous debris flows, discussed in Chapter 9.

Despite all these conceptual developments, the internal structure of the continental terrace, as known from a few scattered borings, was inferred to consist of a series of blanketlike formations, thickening seaward as a result of having been deposited on a surface that was subsiding by greater amounts with increasing distance from the hinge line (Figure 6-13).

Fondo Environment, Fondoform, and Fondothem

According to Rich:

> In the fondo environment the water is generally quiet, though locally and/or temporarily it may be moved by general currents and by waning density currents descending from the clinoform. The latter could affect only relatively small areas adjacent to the foot of the clinoform.

> Deposition on the fondoform must be mainly from suspension, but bottom-dwelling organisms undoubtedly contribute a minor part. The sediment settling from suspension consists mainly of fine terrigenous material and of the remains of pelagic organisms living mostly in the surface water.

> Remains of the larger bottom-dwelling organisms and of the larger forms of pelagic life, such as the shells of cephalopods or the teeth of sharks, are scattered through [a fondothem] like plums in a pudding, entirely without sorting.

> Deposition in the fondo environment must ordinarily be slow, and a long time would generally be represented by only a small thickness of sediment...

giving what has been referred to subsequently as a *condensed section*.

Rich's analysis of the three critical environments was published just prior to the great explosion of information about the sea floor made possible by the precision depth recorder and its wide use in oceanographic investigations. (See Box 6.2.) For example, three important features not known to Rich are (1) abyssal plains (See Box 6.2 Figure 4; such plains can be considered as a special kind of fondoform.), (2) continental rises, and (3) deep-sea fans (these last two are important features of the clino environment).

Because features comparable to those referred to as "abyssal" and "deep-sea" have been found in basins where water depths were not oceanic, slight modifications have been proposed in the terms for general usage. Thus the general fondoform equivalent of the abyssal

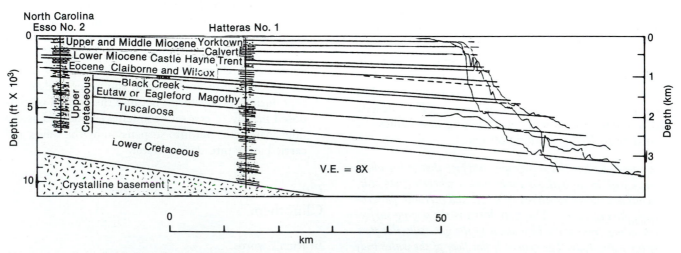

FIGURE 6-13.   Inferred internal structure of upper part of the continental terrace off North Carolina, based on the samples from two exploratory borings and the shape of the continental slope as determined from traces on Precision Depth Recorder (PDR). (B. C. Heezen and H. W. Menard, 1965, fig. 10, p. 242; based on B. C. Heezen and others, 1959.)

plain is now the *basin-floor plain*. Similarly, the general-basin analog of a deep-sea fan is a *basin-floor fan*.

SOURCES: J. L. Rich, 1950, 1951.

Base-Level Controls on Deposition of Extrabasinal Sediments at Basin Margins

Now that we have reviewed the general concept of base-level control on sedimentation and Rich's analysis of the three general settings at the margins of a basin, we have dealt with two of the three major points of background needed for understanding the basis for seismic stratigraphy. The final point is the effect of changes of base level on the sediments at the margins of a basin. We emphasize extrabasinal sediments and changes of sea level at the margins of a marine basin. In Chapter 10 we discuss some modifications needed to include intrabasinal sediments. Despite this emphasis, we remind the reader that most of the major principles involved apply as well to changes of lake level at the margins of nonmarine basins.

We need to extend Rich's analysis by adding two important aspects that he did not emphasize: (1) outbuilding (progradation) of coastal sediments across an undaform, and (2) transport of sediment from undaform to fondoform via subaqueous canyons. In this section, we outline the effects on basin-marginal depositional terraces of changes in relative sea level (submergence and emergence) and of progradation. Each of these conditions yields *characteristic arrays of genetically related sediments* that we shall refer to as facies tracts.

A high-standing sea submerges the coast; as a result, extrabasinal sediments tend to be kept near the coastal zone. The environments lying farther seaward, the unda- and clino environments, will tend not to receive extrabasinal sediment. In other words, they experience a condition of sediment starvation. By contrast, during the opposite condition, a low stand of the sea, the coast emerges and large quantities of extrabasinal sediment are spread outward and downward. We shall trace the general possibilities, starting with a low stand. We follow the progression of a cycle that runs from low stand to high stand and back to low stand again.

At the start of a low stand, one of two contrasting situations may be established as a function of how far out onto the former undaform the shoreline has shifted: (1) the shoreline may shift all the way to the edge of the former undaform, or (2) the shoreline may shift only part way across the former undaform.

If the shoreline shifts all the way to the outer edge of the undaform, then the former undaform will be eroded, streams will incise their valleys and will deliver vast quantities of sediment to heads of submarine canyons. From

there gravity can propel the sediment down the canyons. The sediment does not stop on the clinoform, but continues downward and outward. Most likely, it will be deposited at the lower ends of the canyons as fans that prograde basinward across the adjacent fondoform. This pattern of sediments has been named the *low-stand facies tract* (Figure 6-14, A).

If the condition of emergence persists, this early phase of deposition of basin-floor fans on the fondoform may be followed by a second, later phase in which extrabasinal sediment is so abundant that its delivery into the deep part of the basin is not confined to submarine canyons but is added to the entire clinoform. This phase is characterized by deposition of a **low-stand wedge**, *a wedge-shaped (in cross section) body of sediment that builds out onto the fondoform adjacent to the base of the undaform.* A fully developed low-stand wedge builds both upward against the adjacent clinoform and outward across the adjacent fondoform. In so doing, it engulfs any earlier low-stand fans (Figure 6-14, B).

If the shoreline shifts only part way across the former undaform, then much of the former undaform will not be significantly eroded and the extrabasinal sediment will be deposited all along the outer edge of the undaform; an undathem will prograde outward over the clinothem. Although, with time, the undaform/clinoform boundary may shift outward considerably, the clinothem/fondothem facies boundaries will shift outward (basinward) by only small amounts or not at all. The resulting pattern of sediments has been named the *shelf-margin facies tract* (Figure 6-15).

Now, let us examine what happens when the cycle reverses and sea level begins to rise. If the shoreline has been at the edge of the former undaform, then during the initial stages of the new cycle of submergence, conditions on the clinoform and fondoform may not change very much, and the low-stand facies tract will continue to accumulate. However, as the rate of submergence increases and the exposed undathem is flooded, the sediment supply to the offshore region may be cut off abruptly. As the shoreline transgresses rapidly landward across the undaform, a *transgressive facies tract* is formed (Figure 6-16). Typically, a transgressive facies tract consists of a very thin, discontinuous sheet of sand (Chapter 11). When the point of maximum submergence has been reached, the transgressive tract stops being deposited. The general lack of extrabasinal sediment to the clinoform and fondoform results in no deposition there or in deposition of strata that are much thinner than normal. Such thin deposits on the fondoform are known as *condensed-section fondothems*. At the time of maximum submergence, these attain their most-landward extent.

Where extrabasinal sediment is abundant, the coast may start to prograde. *A body of coastal sediment prograded*

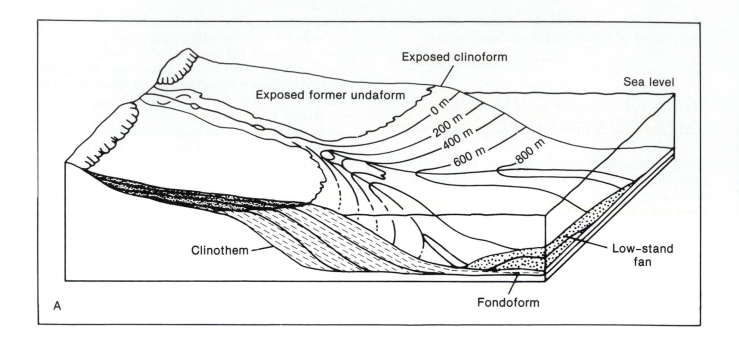

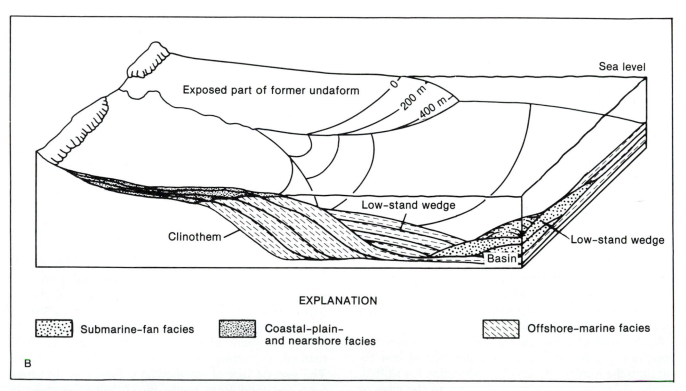

FIGURE 6-14.   Relationships associated with origin of low-stand facies tract; schematic block diagrams of a portion of a basin-marginal terrace. (After H. W. Posamentier, M. T. Jervey, and P. R. Vail, 1988, figs. 2 and 3, p. 112.)

A. Prograding during submergence first builds a basin-marginal terrace composed of thin undathem (light stipple) and thick clinothem (dashed pattern). As a result of emergence, a river has eroded a valley across the subaerially exposed undaform and upper part of clinoform, depositing a low-stand fan.

B. Progradation of the submerged clinoform results in deposition of the low-stand wedge.

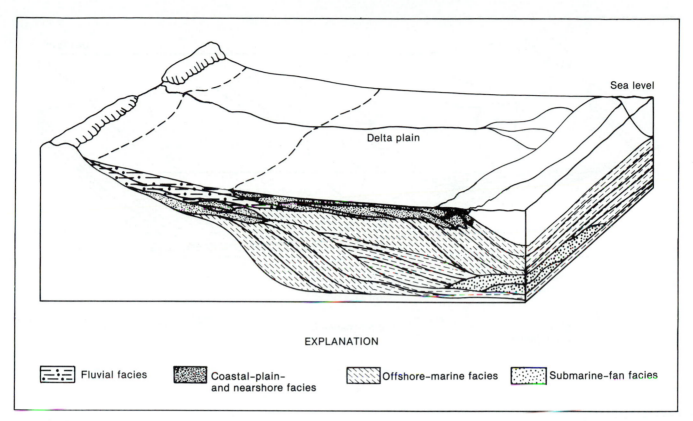

FIGURE 6-15. Geologic relationships associated with deposition of a shelf-margin facies tract; schematic block diagram through a segment of a basin-marginal terrace. Despite continuous submergence (emergence, according to the Exxon view), a delta has built a lobe to the seaward edge of the undathem. Because of submergence, the undathem continues to thicken while the clinoform builds seaward, giving the sigmoid or complex sigmoid-oblique pattern shown in Figure 16-18, C and D (H. W. Posamentier, M. T. Jervey, and P. R. Vail, 1988, fig. 6, p. 114.)

EXPLANATION

Fluvial facies   Coastal-plain- and nearshore facies   Offshore-marine facies   Submarine-fan facies

basinward during a high stand of the sea constitutes a **high-stand facies tract** (Figure 6-17). The sediments constituting a high-stand facies tract overlie those deposited during the preceding transgression.

In many cases, intrabasinal sediment is so abundant that a high-stand facies tract continues to prograde even if submergence continues. The relationships at the boundary between the coastal sediments and marine sediments can be used to infer whether a high-stand facies tract was being prograded at a stable sea level or during submergence. At stable sea level, the boundary between coastal- and marine sediments forms a horizontal surface; nonmarine deposits are thin. If a high-stand facies tract was being prograded during submergence, then the boundary between coastal- and marine sediments forms a surface that rises seaward; nonmarine deposits thicken correspondingly.

When the cycle reverses and sea level starts to drop, coastal prograding ceases and rivers begin to incise the coastal deposits. The shoreline moves outward across the undaform. With further emergence, the cycle proceeds to its low-stand situation.

Introduction to Seismic Stratigraphy

We begin with fundamental geometric relationships among the dark lines, which are traces of the sound waves reflected upward from subsurface layers having large acoustic-impedance contrasts. (Chapter 2 defines acoustic impedance.) We start with some simple principles that are the same as those we developed in Chapter 5 for individual layers. However, with the new continuous seismic-reflection profiles, two important points must be kept in mind. (1) Although the dark traces on the seismic-profile records display many of the configurations that one can see in typical exposures (or even in some hand specimens), the scale of the interfaces from which sound was reflected is determined by the wavelengths of the sound waves used in the survey. (For example, the wavelength of sound waves having a frequency of 100 Hz is 15 m.) (2) The continuity of the line marking the returned sound from a particular interface enables one to trace a given acoustic-impedance-contrast interface from its termination beneath the undaform, across the clinoform, and down to the fondoform.

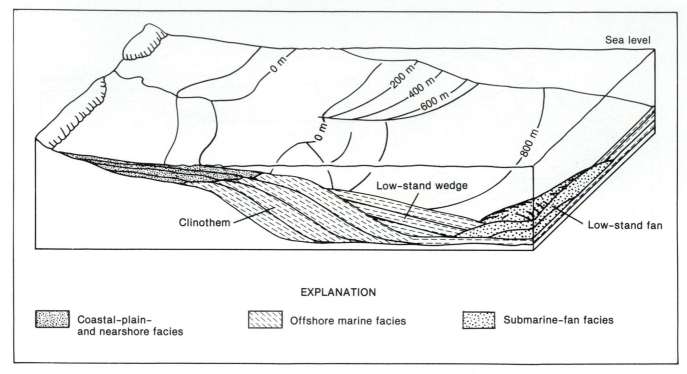

FIGURE 6-16.   Relationships responsible for depositing a transgressive facies tract; schematic block of a segment of a basin-marginal terrace. The formerly exposed undaform is rapidly submerged. Fluvial strata begin to accumulate in the coastal region, but the rising sea tends to trap sediments there and thus the formerly active low-stand wedge becomes essentially inactive and will not be buried unless a delta lobe progrades to the outer margin of the undaform. (H. W. Posamentier, M. T. Jervey, and P. R. Vail, 1988, fig. 4, p. 113.)

Reading the Lines and What Is Between the Lines

Seismic stratigraphy focuses on the continuity of and relationships among the lines on the record. It is based on the proposition that these lines can be considered as the edge views of planes and that these planes can be viewed as bedding planes. In other words, the analysis consists of studying individual prominent lines and then of finding out how the less-prominent lines in between are disposed with respect to the prominent lines. (The use of the word lines in this discussion is not to be confused with the term seismic line for the map location of the survey traverse.)

The first step in seismic stratigraphy is to select a prominent line on the reflection record and see how far it extends. Can it be traced completely across the record? Or does it end somewhere? We discuss these two categories further, in turn.

Continuous Reflection Traces

The common possibilities among lines on the records are shown in Figure 6-18. The simple sketches have been

paired with examples of a continuous seismic-reflection profile showing the sketched relationships.

The simplest case is that of the *parallel group* (Figure 6-18, A). These imply that uniformity prevails. Next comes the *subparallel group*, with slight variations in thickness (Figure 6-18, B). The groups marked parallel and subparallel typify undathems and fondothems.

The *sigmoidal group* (Figure 6-18, C) results from upbuilding and outbuilding of the outer edge of a basin-margin sediment terrace. The individual strata, composed mostly of clinothems, extend from undaform, across the clinoform, and onto the fondoform. They imply simultaneous upbuilding and outbuilding, with the greatest thickness in the clinothems.

The group named *complex sigmoid-oblique* (Figure 6-18, D) differs from the sigmoid type in that the preserved boundary between undaform and clinoform is a sharp angle and the upbuilding and outbuilding takes place in the absence of any significant change in the angles of these two segments of the bottom. As a result, the location of the undaform-clinoform slope break migrates upward and to the right on the figure.

What has been named a *migrating-wave pattern* (Figure 6-18, E) is a large-scale analog of the climbing-ripple

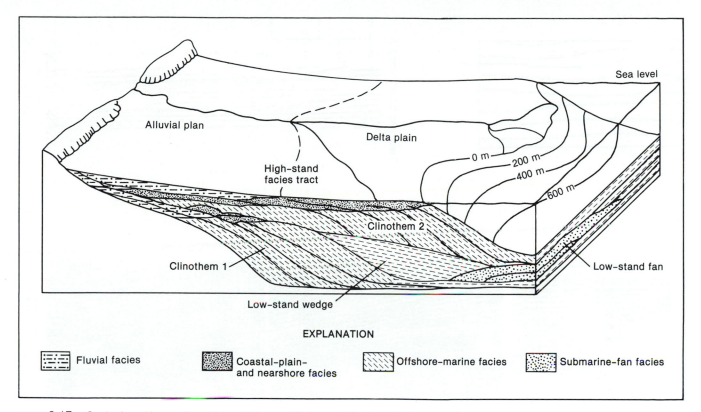

FIGURE 6-17. Geologic setting under which a high-stand facies tract is deposited; schematic block diagram of a segment of a basin-marginal terrace. A delta lobe has prograded across the undathem, thus enabling a copious supply of sediment to be delivered to the formerly starved undaform. In the Exxon figure from which this has been redrawn, the relationships shown here are indicated as taking place during a gradual emergence. As indicated in Figure 6-31, the pattern of deposits at the undathem-clinothem boundary prove just the contrary: the prograding was accompanied by gradual submergence. (H. W. Posamentier, M. T. Jervey, and P. R. Vail, 1988, fig. 5, p. 113.)

FIGURE 6-18. Concordant reflector patterns, not terminated, in sketches and views of seismic-profile records. Where two panels of seismic-profile records are shown, they are the same, but in each case, interpretive lines have been added to the lower panel. (R. M. Mitchum, Jr.; P. R. Vail; and S. Thompson, III, 1977, fig. 2, p. 58; and R. M. Mitchum, Jr.; P. R. Vail; and J. B. Sangree, 1977, various figures.)

A. Even, parallel reflection traces in sketch at left and on a segment of a seismic profile (right). (R. M. Mitchum, Jr.; P. R. Vail; and J. B. Sangree, 1977; sketch, fig. 4, p. 123; segment of seismic-profile record, fig. 5a, p. 124.)

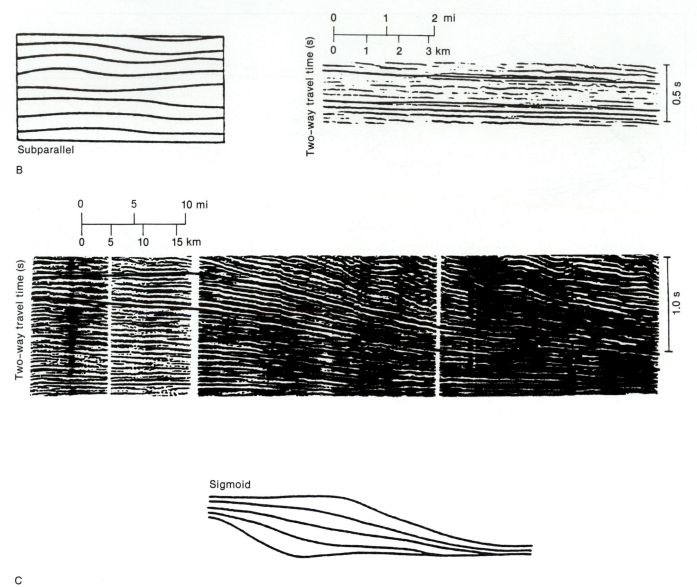

Subparallel

B

Sigmoid

C

FIGURE 6-18.   (*Continued*)
   B. Subparallel reflection traces. (*Ibid*.; sketch, fig. 4, p. 123; segment of seismic-profile record, fig. 5b, p. 124.)
   C. Sigmoid pattern. (*Ibid*.; sketch, fig. 6, p. 125; segment of seismic-profile record, fig. 7a, p. 126.)

pattern shown in Figure 5-14, D. In the segment of the seismic-profile record shown in Figure 6-18, E, the wavelengths of the migrating waves are nearly 6.4 km (4 miles).

The *divergent* category (Figure 6-18, F) is what one would expect where the bottom (possibly the basinward side of an undaform) is subsiding unequally along a hinge line located off the figure to the left while more-or-less uniform sedimentation continues. Thus the right side has subsided more than the left, and this makes the layers thicker and tilts them differentially as they form. As

a result, the dip of the lower layers is greater than that of the upper layers. The interpretation of the entire continental terrace off the eastern United States made prior to modern multichannel continuous seismic-reflection profiles can be characterized as being a large-scale example of the divergent-strata pattern. (See Figure 6-13.)

Four other concordant, curved groups of reflector traces have been recognized (Figure 6-19), but we do not include any examples of seismic records to match the drawings, as we did in Figure 6-18. In the relationship named *concordance* (Figure 6-19, A), the strata

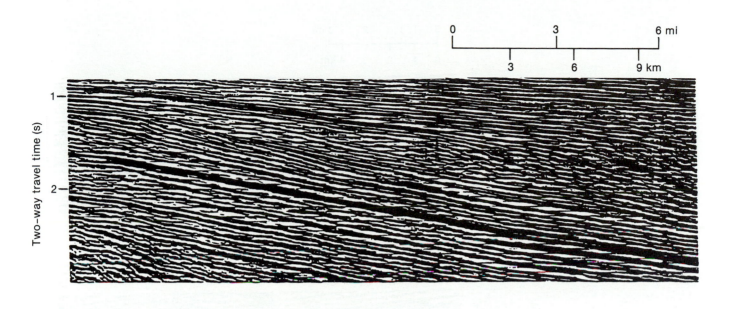

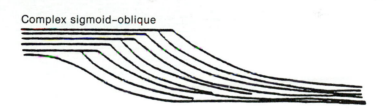

D

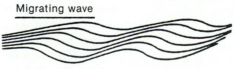

E

FIGURE 6-18. (*Continued*)
   D. Complex sigmoid-oblique reflection traces. (*Ibid.*; sketch, fig. 6, p. 125; segment of seismic-profile record, fig. 7d, p. 126.)
   E. Migrating-wave pattern of reflection traces. (*Ibid.*; sketch, fig. 13, p. 132; segment of seismic-profile record, fig. 14, p. 132.)

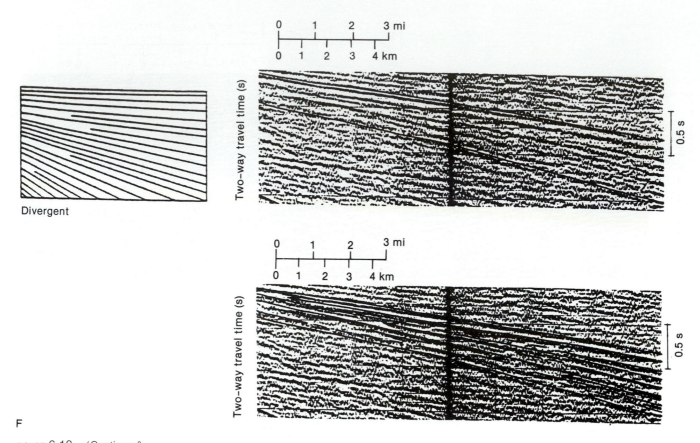

F

FIGURE 6-18. (*Continued*)
　F. Divergent pattern. (*Ibid.*; sketch, fig. 4, p. 123; segment of seismic-profile record, fig. 5c, p. 124.)

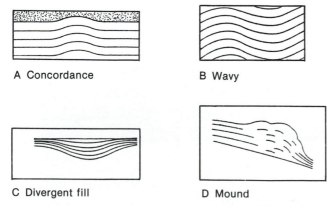

FIGURE 6-19.　Curved concordant reflection-trace patterns.
　A. Concordance of traces over a convex-up feature. (R. M. Mitchum, Jr.; P. R. Vail; and S. Thompson, III, 1977, fig. 2, A, 3, p. 58.)
　B. Wavy pattern. (R. M. Mitchum, Jr.; P. R. Vail; and J. B. Sangree, 1977, fig. 11, p. 130.)
　C. Divergent-fill pattern of reflection traces confined to a trough. (*Ibid.*, fig. 15, p. 133.)
　D. Local mound on inclined reflection traces; possibly from a subaqueous slump structure. (*Ibid.*, fig. 9, p. 128.)

are uniformly convex upward around some initial curved high spot on the bottom. In the *wavy group* (Figure 6-19, B), the strata may have been gently folded or are parts of large-scale bed forms that are being uniformly buried by additions of new sediment (as in the ripples sketched in Figure 5-21, A). The pattern of the reflector traces in the category named *divergent fill* suggests a gentle synclinal trough that could have been subsiding while sediment accumulated (Figure 6-19, C). The horizontal upper surface of the topmost layer indicates that subsidence had ceased. The clinothem with the mound having only short, discontinuous reflector traces is inferred to indicate a body of sediment that slumped down the clinoform, moving toward the right (Figure 6-19, D).

A special kind of generally concordant, more-or-less parallel, but slightly curved reflection pattern is the *hummocky variety* (Figure 6-20).

Continuous lines have been interpreted as energy returns from widespread interfaces that are time surfaces. If this is true, then the causative acoustic-impedance contrast probably has resulted from diagenesis (Chapter 3).

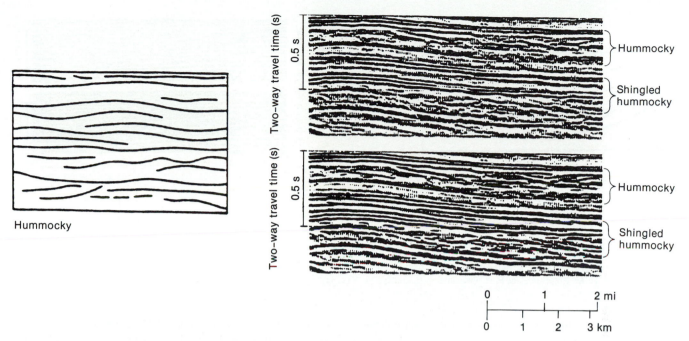

FIGURE 6-20.   Reflection traces in hummocky pattern. (R. M. Mitchum, Jr.; P. R. Vail; and J. B. Sangree, 1977, sketch, fig. 11, p. 130; segment of seismic-profile record, fig. 8b, p. 127. Upper- and lower panels are the same segments, but interpretive lines have been drawn on the lower panel.)

SITUATIONS IN WHICH REFLECTION TRACES END

If the trace of a reflecting interface ends on the record, one should check to see if it matches one of the kinds of terminations as sketched for individual layers in Figure 5-34. Is it at the base or the top of the associated set of lines? If at the base, then does it display the *onlap* pattern (Figure 6-21, A)? Or *downlap* (Figure 6-21, B)?

Top-discordant patterns are numerous. The most-important kind to identify is that resulting from erosional truncation of the underlying layers (Figure 6-22).

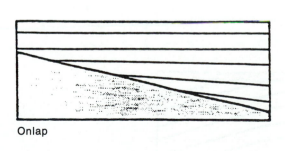

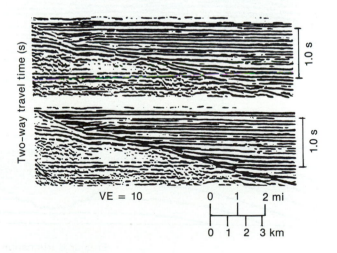

A

FIGURE 6-21.   Sketch and segments of continuous seismic-profile record illustrating contrasting base-discordant reflector traces. Profile-record segments are same views repeated; interpretive lines have been drawn on the lower panel.

A. Onlap. (Sketch, R. M. Mitchum, Jr.; P. R. Vail; and S. Thompson, III, 1977; mirror image of fig. 2, B, 1, reversed to match the record segment, which is in R. M. Mitchum, Jr.; P. R. Vail; and J. B. Sangree, 1977, fig. 3b, p. 120.)

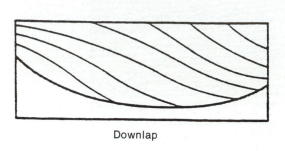

Downlap

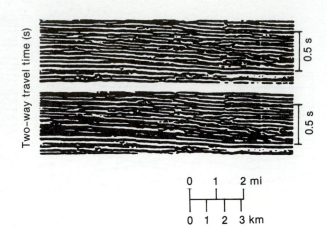

B

FIGURE 6-21. (*Continued*)
   B. Downlap. (Sketch, R. M. Mitchum, Jr.; P. R. Vail; and S. Thompson, III, 1977, fig. 2, B, 2; record segment, R. M. Mitchum, Jr.; P. R. Vail; and J. B. Sangree, 1977, fig. 3c, p. 120.)

Such erosional surfaces serve as first-order boundaries for recognizing units.

Another top-discordant pattern is *toplap* (Figure 6-23, A), which shows how the layers within the clinothem deposited by the basinward progradation of a clinoform have been truncated along a horizontal surface. Such a surface is thought to be the product of a current-swept bottom, rather than of subaerial erosion resulting from a drop in sea level. The pattern named *oblique tangential* (Figure 6-23, B) resembles toplap.

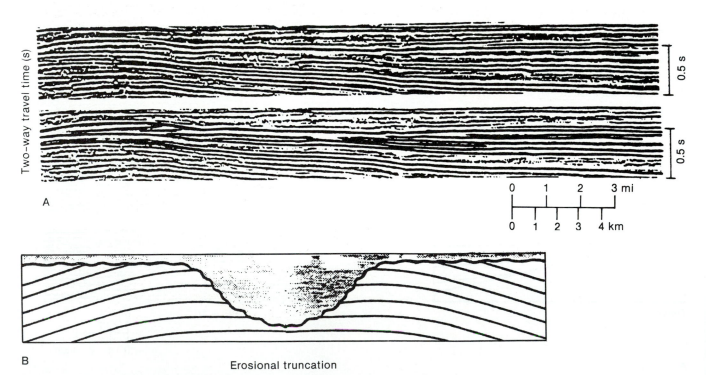

Erosional truncation

FIGURE 6-22.   Seismic-reflection traces of the effects of erosional truncation, which yields top-discordant relationship.
   A. Segment of seismic-profile record. (R. M. Mitchum, Jr.; P. R. Vail; and J. B. Sangree, 1977, fig. 2a, p. 119.)
   B. Sketch showing anticlinal fold with a small valley located at the anticlinal crest. (R. M. Mitchum, Jr.; P. R. Vail; and S. Thompson, III, 1977, fig. 2, A, 1, with right half formed by repeating the original sketch as a mirror image.)

Under certain conditions a clinoform may prograde basinward in such a way as to leave planar strata showing both base-discordant and top-discordant relationships (Figure 6-24). In other words, these inclined reflector traces display both downlap and toplap.

MISCELLANEOUS SPECIAL FEATURES

As a final category, we show what we refer to as miscellaneous special features. In one of these the patterns of the reflector traces are a series of curved line segments having diverse orientations, named a *chaotic pattern* (Figure 6-25, A). In total contrast is a pattern typical of some shales in which, within an interval defined by other lines, few or no reflection-trace lines are present. This pattern has been named *reflection free* (Figure 6-25, B).

TROUGH-FILLING PATTERNS

The strata filling well-defined troughs or channels display many of the patterns previously discussed, but on a smaller scale. In particular, notice in Figure 6-26 that mostly onlap relationships characterize the basal reflector traces and that the tops are usually horizontal and continuous, but may display toplap.

NONCARBONATE MOUND PATTERNS

In a few cases, sediments have not been spread out in flat-topped, horizontal layers but form mounds. Some are built by purely physical processes, and others by organisms. We defer those built by organisms until Chapter 10. Here we include only those built by physical processes.

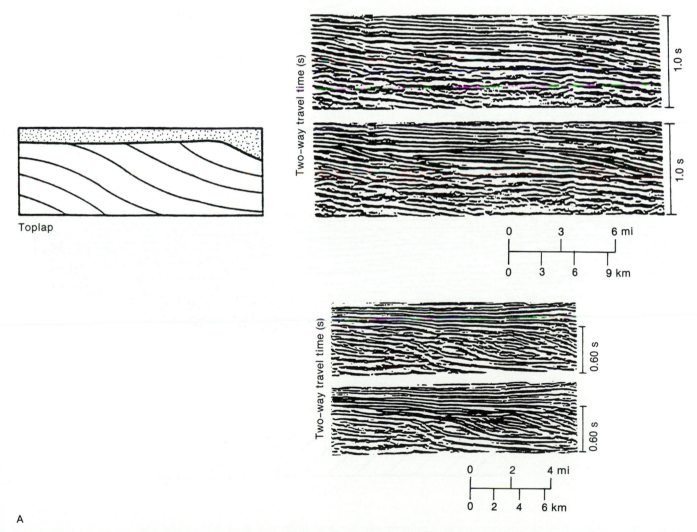

Toplap

A

FIGURE 6-23.   Sketches and seismic-profile record segments of top-discordant reflector traces.
A. Toplap. (Sketch, R. M. Mitchum, Jr.; P. R. Vail; and S. Thompson, III, 1977, fig. 2, A, 2; record segments, R. M. Mitchum, Jr.; P. R. Vail; and J. B. Sangree, fig. 2, c and d, with c in mirror image to match direction of inclination on the sketch.)

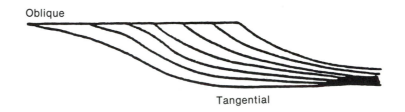

B

FIGURE 6-23.   (*Continued*)
B. Clinothem displaying top truncated along a horizontal surface and basal parts curving around to become tangential to fondothem. (R. M. Mitchum, Jr.; P. R. Vail; and J. B. Sangree, 1977; sketch, fig. 6, p. 125; record segment, fig. 7b, p. 126.)

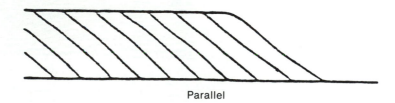

A

FIGURE 6-24.   Reflector-trace terminations at both upper- and lower contacts; sketches and seismic-profile record segments.
   A. Pattern named parallel; planar clinothem layers truncated at both top and bottom. (R. M. Mitchum, Jr.; P. R. Vail; and J. B. Sangree, 1977; sketch, fig. 6, p. 125; record segment, fig. 7c, p. 177.)

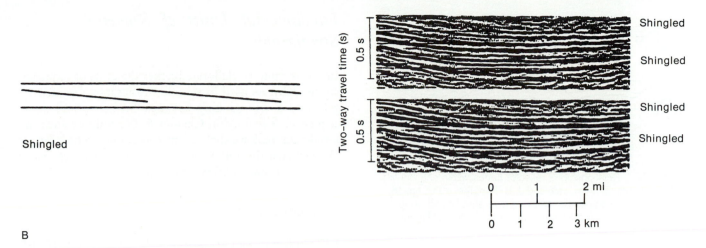

FIGURE 6-24.   (*Continued*)
   B. Shingled pattern. (*Ibid*.; sketch, fig. 6e, p. 125, record segment, fig. 8a, p. 127.)

In Chapter 5 we described small-scale examples of bed forms. Under certain conditions analogous features on a much-larger scale become visible on continuous seismic-reflection profiles, as shown in the migrating-wave pattern of Figure 6-18, E. Other features that build positive-relief forms are fans, volcanic cones, and slumps (Figure 6-27).

Fans are important features that may blur an otherwise clear-cut separation between clinoform and fondoform. A fan that progrades out onto a basin-floor fondoform

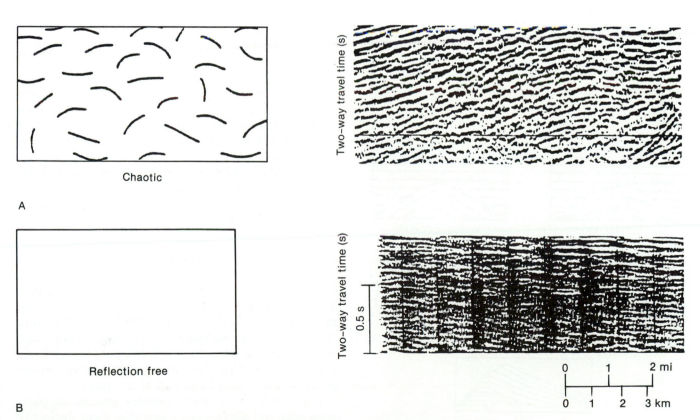

FIGURE 6-25.   Miscellaneous special kinds of reflector-trace patterns, sketches, and seismic-profile record segments.
   A. Chaotic pattern; internal reflectors in short, curved segments having irregular orientations.
(R. M. Mitchum, Jr.; P. R. Vail; and J. B. Sangree, 1977; sketch, fig. 9b, p. 128; record segment, fig. 10b, p. 129.)
   B. Reflection-free pattern. (*Ibid*.; sketch, fig. 9c, p. 128; record segment, fig. 10c, p. 129.)

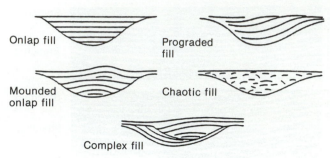

FIGURE 6-26.   Relationship of reflector traces within channel fills. (R. M. Mitchum, Jr.; P. R. Vail; and J. B. Sangree, 1977, fig. 15, p. 133, slightly modified by removing divergent fill and using mirror image of prograded fill to match the orientation of Figure 5-9.)

becomes a local clinoform. The internal structure of fans may be somewhat simple (as in Figure 6-27, A), with convex-up top and baselap relationships at the bottom. Fans may resemble volcanic mounds (Figure 6-27, B). Or the fan mounds may be complex and built of several fan lobes, thus becoming compound (Figure 6-27, C). The fan in Figure 6-27, A had been covered by fondothem strata; that of 6-27, C by clinothem strata.

The example of Figure 6-27, E is an isolated sediment mound built by bottom-following currents carrying fine sediment in suspension, as explained in Chapter 9. The pattern of reflector traces in this sediment mound should be compared with that of Figure 6-18, E.

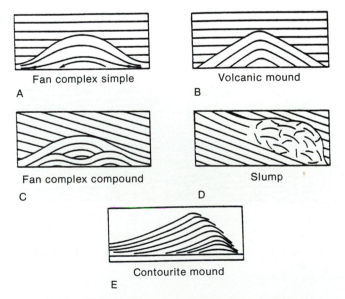

FIGURE 6-27.   Sketches of noncarbonate mounds and their covering layers. Further explanation in text. (R. M. Mitchum, Jr.; P. R. Vail; and J. B. Sangree, 1977, part of fig. 13, p. 132, rearranged.)

Fundamental Units of Sequence Stratigraphy

In this section we elaborate further the units of sequence stratigraphy. As used in seismic stratigraphy, "sequence" refers to a unit ranking between Sequence and parasequence, as defined in Chapter 5. For units of such intermediate rank we shall use mesosequence (a new term). We describe the following units: *parasequence sets*, *facies* ("systems") *tracts*, and *mesosequences*.

Parasequence Sets

Parasequences can be grouped to form parasequence sets. According to the conditions inferred from the patterns of seismic reflectors within the units that were prograded at the edge of the undaform, three kinds of parasequence sets can be recognized (Figure 6-28). The depositional patterns involved resulted from varying interactions between outbuilding of the undaform-clinoform boundary and submergence. Where the parasequence sets continue to prograde basinward during submergence, the coastal sediments build outward over marine sediments, and the nonmarine strata thicken. If each unit that progrades basinward extends farther toward the basin than the preceding unit, then the pattern is named a *progradational parasequence set*.

The opposite situation exists where each younger unit that progrades basinward does not extend as far basinward as the preceding unit. As a result, the marine strata thicken and extend generally farther toward the basin margin with time. This forms the *retrogradational parasequence set*.

Finally, the units that prograde may be arranged one on top of the other so that with time, their basinward extents do not change. This forms what has been named an *aggradational parasequence set*.

Facies ("Systems") Tracts

Parasequence sets are combined into units that the Exxon lexicon designates as "systems tracts," but for which we shall substitute Teichert's prior (1958) term *facies tract* (Figure 6-29). As explained in a previous section, such tracts consist of deposits that formed during differing parts of a cycle of base-level change that goes from emergence to submergence and back to emergence.

Each such facies tract includes groups of parasequences and parasequence sets that form distinctive vertical- and lateral patterns of units. These units are bounded by seismic interfaces that formed during times of maximum submergence. During each major cycle of emergence-submergence-emergence, these facies tracts

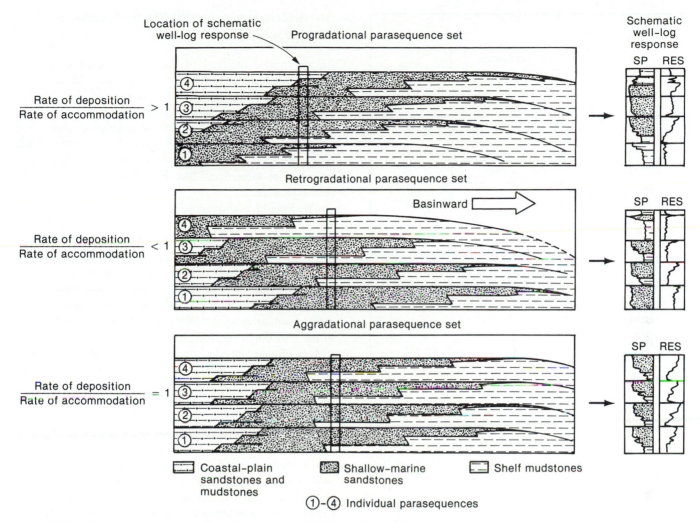

FIGURE 6-28. Diagrammatic representation of the three kinds of parasequence sets with schematic appearance of logs from bore holes through each kind of set. (H. W. Posamentier, R. W. Mitchum, P. R. Vail, J. F. Sarg, T. S. Loutit, and Jan Hardenbol, 1988, fig. 1, p. 40.)

succeed each other in a predictable fashion and occupy predictable positions relative to the mesosequence boundaries and to one another.

Mesosequences

The next-higher-ranking unit that can be identified from continuous seismic profiles is known as a *mesosequence* (our term for the "sequence" of the Exxon language). Mesosequences are bounded by surfaces of unconformity (in basin-marginal areas that were eroded during times of emergence) or by the basinward extensions of such surfaces that can be followed as continuous lines on seismic records into parts of the basin that may have been eroded by subaqueous currents or to areas where the lines on the seismic profile are plane and parallel and thus presumably indicate strata that are concordant

and "conformable." The key to seismic stratigraphy is the continuity of the reflecting interfaces from areas that were eroded into the basin where deposition was not interrupted by erosion (Figure 6-30).

In addition to the evidence of erosion associated with mesosequence boundaries, the seismic expression of a mesosequence may include (1) a basinward shift in the position of the configuration formed by coastal onlap, (2) onlap by overlying strata, and (3) evidence that streams were rejuvenated during the period of emergence. These attributes of mesosequences are not displayed by smaller-scale emergence-submergence cycles.

The concept of sea-level change at mesosequence boundaries requires further comment. Commonly associated with parasequence boundaries are diagnostic facies transitions which suggest that the water was deep-

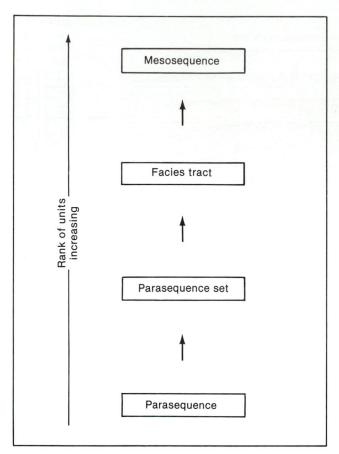

FIGURE 6-29.   Hierarchy of mesosequence-stratigraphic units. Parasequences (and their synonymous PACs) are defined in Chapter 5. (Authors.)

An Alternative Classification of Genetic Stratigraphic Units

An alternative method of defining mesosequences has been developed by William E. Galloway, based upon D. E. Frazier's (1931–1976) concepts of depositional episodes and genetic stratigraphic sequences. This classification scheme uses genetic stratigraphic units that are defined on the basis of products associated with times of maximum submergence. Frazier called these units "facies sequences" but they clearly are the same as parasequences.

The Frazier-Galloway scheme is based upon units that are synonymous with parasequences, as is the Exxon scheme. However, the two are fundamentally incompatible because the Frazier-Galloway mesosequences are bounded by features formed during maximum submergence (*downlap surfaces*). The Exxon mesosequences are bounded by features formed during maximum emergence (undaform subaerially exposed, with related effects in the clinoform-fondoform realm).

According to the concepts of sea-level change as outlined in this chapter, the *numbers* of major downlap surfaces and major subaerial-exposure surfaces should be the same, because they are both controlled by the cycles of sea-level change. However, the Galloway mesosequences and the Exxon mesosequences will be almost exactly out of phase. The explanation for this situation follows from the contrasting bases for defining the mesosequences. The Exxon mesosequence boundaries correspond to the falling inflection point on the eustatic sea-level curve. By contrast, the Galloway mesosequence boundaries come just after the rising inflection point on the eustatic sea-level curve. Exxon-group mesosequence boundaries correspond to times of maximum emergence, whereas Galloway mesosequence boundaries correspond to times of maximum submergence (hence of maximum transgression), and, in the fondothem, of onlap by condensed sequences.

Some strengths of the Galloway system are the following:

1. The boundaries of genetic sequences are homologous to parasequence boundaries; both result from the effects of maximum submergence. This presents the Exxon scheme with a conceptual difficulty in which the bounding surfaces of all lower-level units are marine-flooding surfaces, whereas those of higher-level units are not.

2. Many depositional basins are filled through a repetitive alternation of strata that were deposited rapidly and thus caused large-scale downlap (high-stand- and lowstand facies tracts) and intervals of nondeposition or very slow deposition (transgressive facies tracts). Thus it is argued that major depositional units that are deposited relatively

ening while they were being deposited. For example, in the parasequences within the Lower Devonian limestones (See Figure 5-40.), limestones displaying algal laminae are overlain by skeletal grainstones. In contrast, the inferred water-depth pattern at mesosequence boundaries may be one of an upward transition from a deeper-water facies to a relatively shallow-water facies of a lowstand facies tract. In the associated fondothem, the relationships at a mesosequence boundary may lack evidence that water depth changed. But when this surface is followed onto the adjacent undathem, evidence of the effects of subaerial exposure may be present to identify the sequence boundary itself.

Mesosequences consist of many parasequences, which are commonly grouped into parasequence sets having distinctive patterns of facies and bounded by surfaces associated with major episodes of submergence. In turn, parasequence sets can be grouped into facies tracts that constitute the mesosequences.

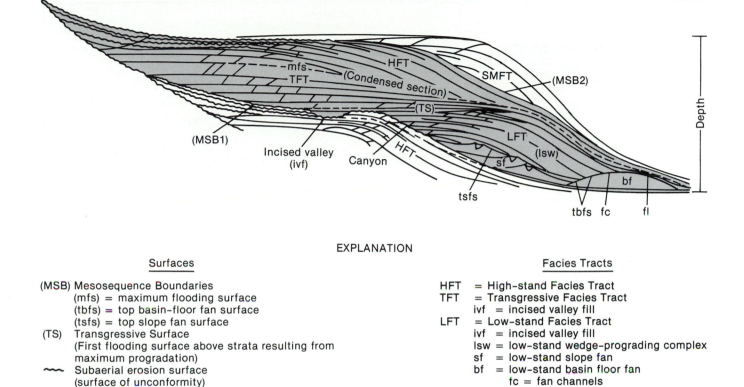

EXPLANATION

Surfaces

(MSB) Mesosequence Boundaries
     (mfs) = maximum flooding surface
     (tbfs) = top basin-floor fan surface
     (tsfs) = top slope fan surface
(TS) Transgressive Surface
     (First flooding surface above strata resulting from
     maximum progradation)
〜〜 Subaerial erosion surface
     (surface of unconformity)
— — Condensed section
     (coincides with mfs)

Facies Tracts

HFT = High-stand Facies Tract
TFT = Transgressive Facies Tract
     ivf = incised valley fill
LFT = Low-stand Facies Tract
     ivf = incised valley fill
     lsw = low-stand wedge-prograding complex
     sf = low-stand slope fan
     bf = low-stand basin floor fan
        fc = fan channels
        fl = fan lobes
SMFT = Shelf-Margin Facies Tract

FIGURE 6-30. Schematic profile section through basin-marginal terrace (land to left, basin to right) showing how the strata have accumulated as a result of shifting base level and progradation to build a mosaic of undathems, clinothems, and fondothems. Shown are one complete mesosequence (shaded) and parts of two others. Further explanation in text. (After P. R. Vail, 1987, p. 4, no figure number.)

continuously under homogeneous sea-level- and tectonic regimes are separated by what we call downlap surfaces, and not by major surfaces of unconformity that resulted from subaerial erosion.

3. The boundaries of the Exxon-group mesosequences often correspond to very subtle surfaces that formed as a result of subaerial erosion that affected only restricted areas. Throughout most of the basin, these surfaces may be difficult to recognize. In contrast, on seismic-reflection profiles, downlap relationships are nearly always very obvious. This is especially true in basins dominated by extrabasinal sediments.

4. Surfaces displaying downlap are not only readily recognizable on seismic profiles, but are commonly characterized by the effects of erosion by submarine currents, hardgrounds, lag deposits with glauconite- or manganese-nodule deposits, or condensed sedimentary veneers. Thus these surfaces may be as recognizable in cores and in outcrop as Exxon-group mesosequence boundaries (Figure 6-31). It has been suggested that the fossiliferous condensed section which usually overlies the Galloway-type mesosequence boundary might facilitate paleontological dating and correlation between nonmarine- and marine strata. However, condensed sections are not particularly useful for stratigraphic paleontology. Still, Galloway-type mesosequence boundaries do form widespread surfaces that extend across facies boundaries and facilitate *physical* correlation between nonmarine- and marine strata.

Although we have chosen to present the Exxon scheme of classifying mesosequences as the method of choice in this book, we do not mean to imply that the Galloway classification is less well designed or less useful. Rather, the Exxon scheme has been more thoroughly explored theoretically, and is both more widely understood and more widely used. However, we present the Galloway classification as an alternative because it offers an equally valid way of analyzing the stratigraphic record. Future stratigraphic studies will determine which of these two methods is more useful in a practical sense.

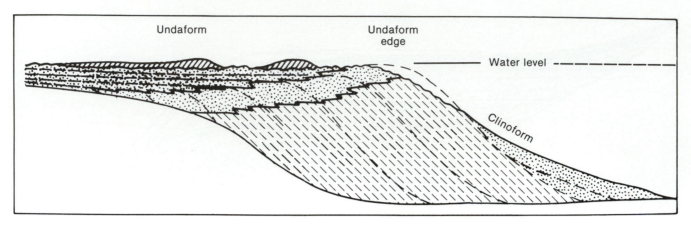

FIGURE 6-31.   Schematic profile through basin-marginal terrace (land to left, basin to right), showing results of prograding during submergence. As submergence takes place, base level rises, and this enables the undaform to be built upward, depositing an undathem (at left). Simultaneously, the clinoform builds seaward and the zigzag boundary between the undathem-edge deposits (stippled) and the clinothem rises in a basinward direction. This relationship is a diagnostic indicator that prograding and submergence were taking place simultaneously. (After W. E. Galloway, 1989, fig. 3, p. 129.)

Attempts to Infer Sea-Level History from Seismic Profiles

To a large extent, patterns of sediment deposition are determined by the relative position of the sea against the land. In this chapter we outlined the indications on seismic profiles of changes in relative sea level. Recall that extrabasinal sediments are mostly terrigenous and result from erosion of the land. By contrast, intrabasinal sediments are derived from within the waters of the depositional basin; in them, carbonates predominate. Each of these two major categories of sedimentary materials react differently to sea-level changes. In this chapter, we include only the characteristics displayed by extrabasinal sediments, which have been studied more intensively by sequence stratigraphers (Box 6.3).

Our analysis of the relationship between extrabasinal sediments and sea-level changes involves four simplifying assumptions:

1. That the basin-marginal terrace subsides by tilting downward toward the basin along a hinge line that is landward of the strandline so that the rate of subsidence increases basinward.

2. That the frequency of eustatic sea-level change exceeds that of rate of subsidence (Figure 6-32) and that, as a first approximation, we can assume a zero rate of change of subsidence.

3. That the maximum rate of eustatic sea-level fall is not much greater than the rate of subsidence (which is of opposite sign). The result of this assumption is that, except during periods of maximal rate of eustatic sea-level fall, a condition of submergence prevails. (This simplifies the interpretation of the resultant stratigraphic pattern.)

4. That variations in the rate of sediment supply or of sediment compaction are small relative to variations in the rate of eustatic sea-level change. This is also a reasonable assumption.

If one or more of these assumptions is not satisfied, then the stratigraphic interpretation becomes complicated.

Mesosequences are interpreted as the products deposited between two points on the eustatic sea-level curve where a change takes place between an emergence and a following submergence. Thus a mesosequence begins to form when eustatic sea level is falling at approximately its greatest rate. Mesosequence boundaries can be followed as continuous lines on seismic records. A given mesosequence ends when the condition of submergence is supplanted by rapid emergence. Conditions of emergence have been selected for defining mesosequence boundaries because most of the prominent surfaces of unconformity are eroded at times of rapid emergence.

The end papers of this book show sea-level curves based on the synthesis of seismic stratigraphy by Exxon Research Laboratories.

SOURCES: Cross, 1990; Galloway, 1990; Galloway and Hobday, 1983; James and Leckie, 1988; Posamentier and others, 1988; Vail, 1987; Van Wagoner, 1985; Van Wagoner and others, 1988, 1990; Visher, 1990.

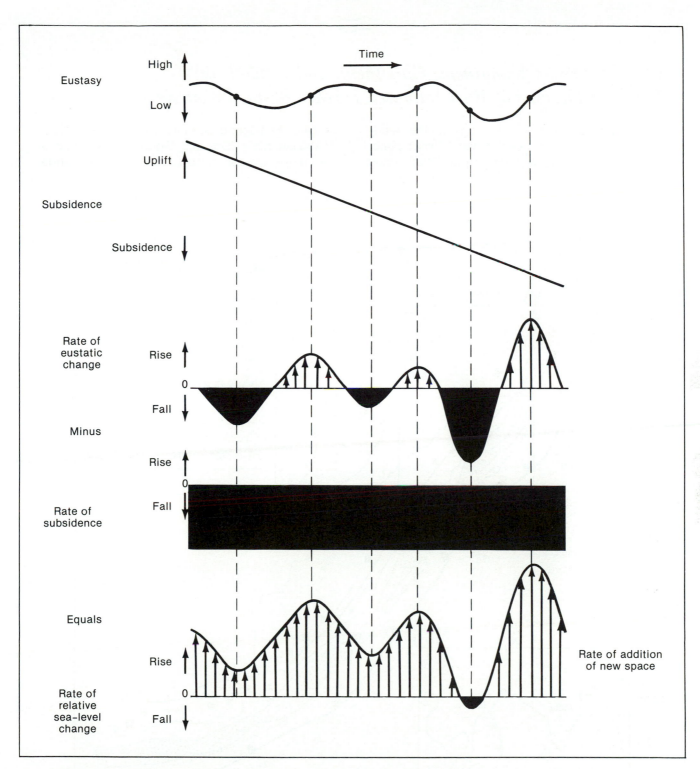

FIGURE 6-32.   Relative sea level as a function of eustasy and subsidence. (H. W. Posamentier, M. T. Jervey, and P. R. Vail, 1988, fig. 10, p. 116.)

BOX 6.3

Case History of Undathem, Clinothem, and Fondothem: Cenozoic Deposits of West Sumatra Forearc Basin, Indonesia

The island of Sumatra is part of the leading edge of the Southeast Asian (China) plate, which in this area is overriding Upper Cretaceous to lower Paleogene oceanic crust of the Indian-Australian plate (Box 6.3 Figure 1). An unusually high rate of subduction into the Sunda

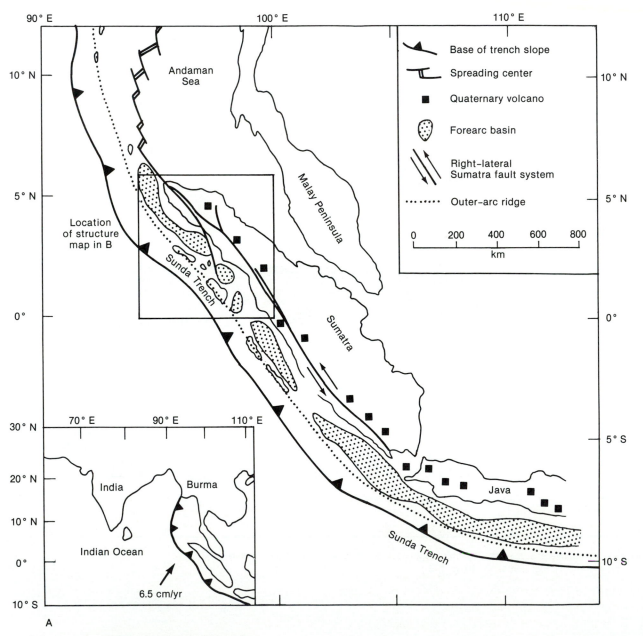

BOX 6.3 FIGURE 1.   Western Sunda arc and vicinity.
A. Regional map showing location of local index map.

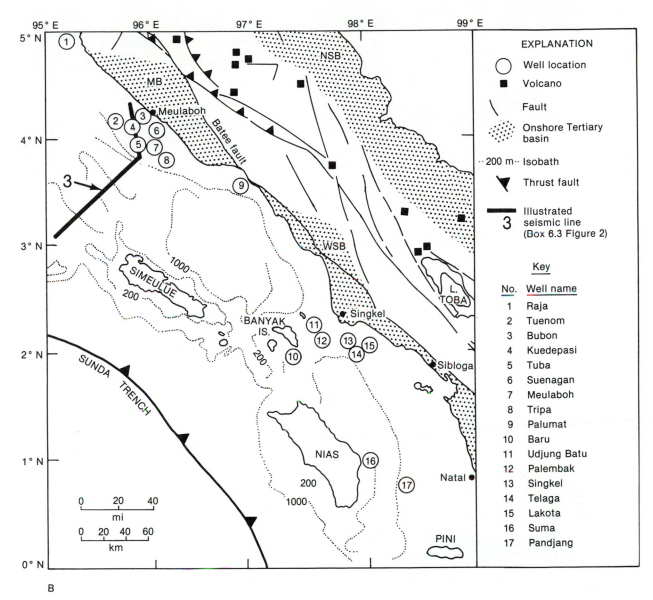

EXPLANATION

○ Well location

■ Volcano

＼ Fault

Onshore Tertiary basin

··200 m·· Isobath

▼ Thrust fault

▬▬ Illustrated seismic line (Box 6.3 Figure 2)

3

Key

No.	Well name
1	Raja
2	Tuenom
3	Bubon
4	Kuedepasi
5	Tuba
6	Suenagan
7	Meulaboh
8	Tripa
9	Palumat
10	Baru
11	Udjung Batu
12	Palembak
13	Singkel
14	Telaga
15	Lakota
16	Suma
17	Pandjang

B

BOX 6.3 FIGURE 1. (*Continued*)
B. Structure map, northern Sumatra, showing location of seismic-reflection profiles of Box 6.1 Figures 2 and 3. (Desiree Beaudry and G. F. Moore, 1985; A, fig. 1, p. 743; B, fig. 2, p. 744.)

Trench of about 7 cm/yr makes this a very active plate boundary.

Between the Sunda Trench and the island of Sumatra (believed to be underlain by Lower Paleozoic continental crust) lies a southeast-northwest-trending forearc basin. To the northeast is a basin-marginal terrace that has been subsiding nearly continuously throughout the Neogene (Box 6.3 Figure 2). To the southwest lies a deeper basin into which a sedimentary wedge of Miocene and younger age has prograded.

Note in the Miocene (labeled 3), the distribution of undathem, clinothem, and fondothem. Similarly, in another seismic-reflection profile (Box 6.3 Figure 3), seismic reflections show features labeled topset, foreset, and bottomset, which show the distribution of the products (undathem, clinothem, and fondothem) of the three critical environments: undaform, clinoform, and fondoform.

SOURCE: Beaudry and Moore, 1985.

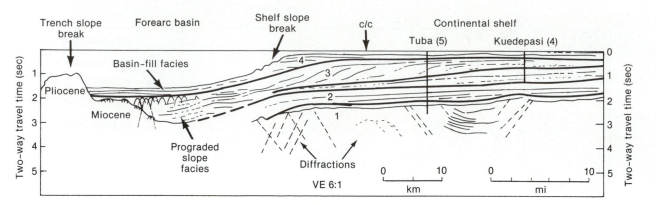

BOX 6.3 FIGURE 2.   Line drawing of multichannel seismic-reflection profile across the basin-marginal terrace adjoining the forearc basin between the island of Sumatra and the Sunda Trench (off diagram to left). Heavy black lines delineate four mesosequences labeled 1, 2, 3, and 4. Note the prominent clinothem in mesosequence 3, resulting from prograding with only slight submergence. By contrast, in mesosequences 2 and 4, the undaform built upward, and a prominent undathem accumulated. (Desiree Beaudry and G. F. Moore, 1985, fig. 3, p. 745.)

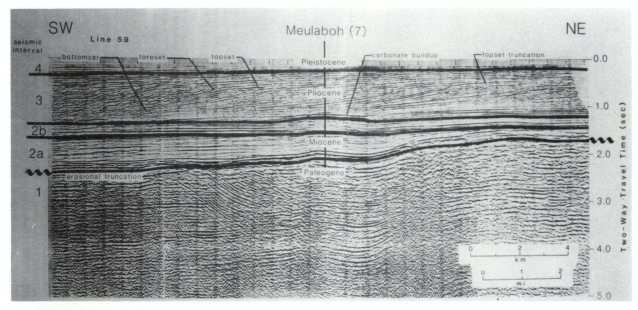

BOX 6.3 FIGURE 3.   Segment of a multichannel continuous seismic-reflection profile of the Sumatra forearc basin. Note the distribution of undathem, clinothem, and fondothem (labeled topset, foreset, and bottomset) in the Pliocene. (Desiree Beaudry and G. F. Moore, 1985, fig. 6, p. 748.)

Suggestions for Further Reading

ANSTEY, N. A., 1982, Simple seismics: Boston, Massachusetts, International Human Resources Development Corp., 168 p.

CHRISTIE-BLICK, N., and others, 1988, Chronology of fluctuating sea levels since the Triassic, comments and replies: Science, v. 241, p. 596–602.

CROSS, T. A., and LESSINGER, M. A., 1988, Seismic stratigraphy: Annual Reviews of Earth and Planetary Sciences, v. 16, p. 319–354.

FAIRBRIDGE, R. W., 1987, The spectra of sea level in a Holocene time frame, p. 127–142, *in* Rampino, M. R., Sanders, J. E., Newman, W. S., and Konigsson, L. K., eds., Climate: History, Periodicity, and Predictability: New York, Van Nostrand Reinhold, 588 p.

FRAZIER, D. E., 1974, Depositional episodes: their relationship to the Quaternary stratigraphic framework in the northwestern portion of the Gulf Basin: Texas University at Austin, Bureau of Economic Geology Circular 74–1, 28 p.

GALLOWAY, W. E., 1989, Genetic stratigraphic sequences in basin analysis I: Architecture and genesis of flooding-surface bounded depositional units: American Association of Petroleum Geologists Bulletin, v. 73, p. 125–142.

HAQ, B. U.; HARDENBOL, JAN; and VAIL, P. R., 1987, Chronology of fluctuating sea levels since the Triassic: Science, v. 235, p. 1156–1167.

NUMMEDAL, DAG; PILKEY, O. H.; and HOWARD, J. D., eds., 1987, Sea-Level Fluctuation and Coastal Evolution: Tulsa, OK, Society of Economic Paleontologists and Mineralogists Special Publication 41, 267 p.

SLOSS, L. L., 1988, Forty years of sequence stratigraphy: Geological Society of America Bulletin, v. 100, p. 1661–1665.

VAIL, P. R., 1987, Seismic stratigraphy interpretation procedure, p. 1–10 *in* Bally, A. W., ed., Atlas of seismic stratigraphy, v. 1, Tulsa, OK, The American Association of Petroleum Geologists Studies in Geology 27, 125 p.

VAN WAGONER, J. C., MITCHUM, R. M., CAMPION, K. M., and RAHMANIAN, V. D., 1990, Siliciclastic sequence stratigraphy in well logs, cores, and outcrops: Concepts for high-resolution correlation of time and facies, *in* Methods in Exploration, v. 7: Tulsa, OK, American Association of Petroleum Geologists Methods in Exploration 7, 55 p.

WHEELER, H. E., 1958, Time stratigraphy: American Association of Petroleum Geologists Bulletin, v. 42, p. 1047–1063.

WILGUS, C. K., HASTINGS, B. S., KENDALL, C. G. St. C., POSAMENTIER, H. W., ROSS, C. A., and VAN WAGONER, J. C., eds., 1988, Sea-Level Changes: An Integrated Approach: Tulsa, OK, Society of Economic Paleontologists and Mineralogists Special Publication 42, 407 p.

PART IV

PROCESS SEDIMENTOLOGY AND ENVIRONMENTS OF DEPOSITION

Now that we have discussed the constituents of sedimentary deposits, have reviewed what can happen as they are converted into sedimentary rocks, have summarized the major kinds of sedimentary rocks, and have taken our first look at sedimentary strata, we are prepared to examine the subjects of Part IV, the largest in the book. Here we deal with depositional processes, depositional environments, and their products in the geologic record.

We begin with two chapters devoted to processes: Chapter 7 (covering physical-, biological-, and chemical processes) and Chapter 8 (circulation in the atmosphere, in modern oceans, and in basin waters). In the following six chapters we systematically describe many of the Earth's important environments of deposition. We start in deep-water basins and procced upward and outward toward basin margins and, finally, to nonmarine environments. Most extrabasinal particles begin in nonmarine environments and are transported from there to the sea. Moreover, the average person is likely to be more familiar with nonmarine environments than with those under the waters of a basin. Despite the obvious logic in the progression from nonmarine to marine, we have chosen just the reverse order for one overriding reason: to emphasize the importance of basins.

Our objective is to show how geologists start with features in the geologic record and work backward to infer depositional processes and depositional environments. In doing so, geologists employ scientific reasoning that is specific to geology and foreign to other fields of science. In physics and chemistry, for example, known processes are followed in carefully controlled experiments to see what unknown products will result. After the products have formed they are directly observable and can be related exactly to the formative processes. The reasoning is forward from process to product. By contrast, in many geological investigations the rock is the known product and the formative processes are the unknowns. Instead of doing the experiment and seeing the result, a geologist starts with the result and asks, in effect: "What was the experiment?" (Figure IV-l).

How does a geologist proceed to unravel the results of those natural "experiments" that may have taken place millions, or even hundreds of millions, of years ago?

A geologist must be familiar with modern sediments and with laboratory experiments with sediments where close relationships between processes and products have been established. In the case of sand, as an example, Figures 5-13, B, IV-2 and 7-23 show various kinds of cross strata that laboratory experiments and observations in modern sediments have established as being diagnostic of certain kinds of flow. Finding these

Chemistry, physics, engineering

$$\text{Processes} \xrightarrow{\text{yield}} \text{Product (controlled experiment)}$$

Geology, sedimentology

$$\text{Product} \xrightarrow[\text{in terms of}]{\text{interpreted}} \text{Processes}$$

FIGURE IV-I.   In chemistry, physics, and engineering, controlled experiments, that is, processes, yield a product. In geology, especially sedimentology, the processes that shaped the product are interpreted from a study of the product. The product may be rock strata, or samples from it, such as hand specimens or cuttings from boreholes, or indirect indicators of it, such as electric logs from boreholes or continuous seismic-reflection profile sections displaying subsurface relationships. The processes that resulted in the product are inferred in terms of depositional- and diagenetic events. (Authors.)

kinds of cross strata in sedimentary rocks enables a geologist to infer that these particular processes of flow operated to form the sediments that are now rocks. This matching of observed features in a rock with experience from modern sediments and laboratory experiments is reasoning by analogy, the geologist's stock in trade.

Because analogy does not constitute proof, a scientific discipline that operates by analogy is beset by many inherent problems. For example, many products cannot be produced in the laboratory, hence information on processes cannot be backed by laboratory experience. Or processes in modern sediments cannot be observed because they operated beneath 5000 m of water, where continuous observations are out of the question. Conversely, we may know more about the processes than about the products.

Finally, various kinds of formative processes may yield essentially the same products. Thus the product itself might not convey information

FIGURE IV-2.   Steeply dipping cross strata formed by advance of sand from right to left. Modern dune, Lybia (R.A. Medeiros, H. Schaller, and G.M. Friedman, 1971, fig.4, p. 20).

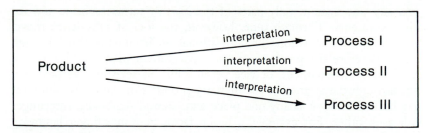

FIGURE IV-3.   A given geologic product, such as rock strata, samples from it (hand specimens or cuttings from boreholes), or indirect indicators of it (electric logs from boreholes or continuous seismic-reflection profiles) are susceptible to interpretation by two, three, or more processes. Thus a sandstone may be interpreted as of eolian origin, or as a product of deposition in submarine sand ridges, or even by a third possible formative process. Commonly the geologist arrives at multiple possible solutions, that is, multiple origins, rather than a unique interpretation for a sedimentary deposit. (Authors.)

pointing to a single specific process, but may be susceptible of interpretation by two, three, or more processes (Figure IV-3).

In dealing with the geologic record, geologists are confronted, not with the products of processes that operated in isolation, but with products of associated processes that operated collectively in what we know as depositional environments. We can define a depositional environment as a natural geographic entity in which sediments accumulate. In attempting to infer the history of a sedimentary deposit, a geologist needs to do more than simply infer the physical- or chemical processes that operated. The ultimate goal is reconstruction of the pattern of ancient depositional environments. This is not easy, because what appears to be a single sedimentary product may have formed by contrasting processes operating in different depositional environments. (See Figure IV-2.) For example, large-scale planar cross strata (See Figure 5-13, B.) are known from modern eolian sand dunes and from linear submarine sand ridges. (See Figure 10-13.)

An experience of a geologist who had studied the characteristics of supposed ancient river deposits in a Midcontinent area of the United States is a case in point. The geologist had used all the latest approaches in the physical analysis of the strata: he had analyzed the bed forms, measured directions of dip of cross strata, and finally had worked out a three-dimensional interpretation. On a field trip, a paleontologist discovered fossils that proved the strata were marine. The revised interpretation in light of the marine fossils was that the supposedly fluvial deposits had been deposited under intertidal conditions in a marginal-marine environment. Some of the physical characteristics of sediments deposited in intertidal channels are similar to those deposited in fluvial channels. Although some minor differences exist that can enable one to distinguish the products of these two environments without the fossils, such differences are not always readily apparent.

The aspect of a sedimentary deposit that reflects the depositional environment is known as the **facies**, which is defined as *the lithologic- and paleontologic characteristics of a sedimentary deposit, imparted by the depositional environment*. The term facies, initially used by Nicolaus Steno in 1669 to express a fundamentally important concept in geology, proved to be so useful that multiple definitions proliferated. (See the third edition of *AGI Glossary*, R. L. Bates and J. A. Jackson, 1987.) The foregoing definition will be used herein.

At any one instant, only parts of the Earth's surface subside and receive sediments. Through time, however, the loci of subsidence move. Accordingly, the positions of depositional environments shift through time. Therefore the products of one environment may overlie those of an adjacent environment, or vice versa. Thus, as seen in a vertical section through the strata, the sedimentary products of these environments may be interbedded at any one place and, viewed regionally, interfinger with each other. For example, a beach facies may interfinger landward with a coastal-dune facies and this, in turn, may interfinger landward with fluvial- and other nonmarine facies. In the seaward direction, a beach facies may interfinger with various shallow-marine facies, which ultimately may interfinger with deep-sea facies.

The modern usage of facies with its specific emphasis on facies changes stresses this interfingering. In recent years, the characteristics of depositional environments and their products have been idealized as "facies models." Such models assist in relating what has been observed in the geologic record to modern environments and -processes.

A given facies is deposited only within the area occupied by its specific depositional environment. In other words, at any one time, geographic relationships determine the horizontal dimensions of each facies. This being true, how is it possible for a facies to occupy eventually a wider area than that occupied by its depositional environment? The boundaries of the environments shift with time. As these boundaries migrate, the characteristic deposits of the environment become even more widely distributed. In this way, many facies are spread laterally with time.

Facies change is an important concept in economic geology. Porous- and permeable sandstones of barrier-island origin may interfinger seaward with offshore shales and landward with lagoonal shales. Such interfingering in the subsurface results in the isolation of porous- and permeable sandstones within less-porous and less-permeable shales. Petroleum hydrocarbons that migrate into the porous- and permeable sandstones may be trapped to form an oil reservoir. Modern oil exploration for so-called **stratigraphic traps**, that is, *traps formed as a result of permeability barriers created by lateral facies changes*, relies on the prediction of such lateral facies changes. The geologist tries to predict the existence of a porous- and permeable potential reservoir sandstone enclosed in a tight-seal facies.

The importance of the concept of facies, including lateral facies changes and depositional environments, requires that we discuss each separate environment whose products may be represented in the rock record. The purpose of such a discussion is to study modern depositional environments to develop characteristics for recognizing the deposits of these environments in the rock record. Many geologists employ loosely the term *criteria* for features found in the rock record that resulted from the action of a given depositional environment. We shall use the term *criterion* sparingly and only where we can establish the environment without fail; by definition, a criterion is foolproof. However, as we have pointed out, various formative processes may yield essentially the same product. Because of this element of uncertainty, we prefer to speak of *characteristics* of the products of depositional environments rather than of criteria.

The widely employed cliche "the present is the key to the past," phrased in the nineteenth century by Sir Archibald Geikie (1835–1924), was based on James Hutton's revolutionary eighteenth-century approach to the geologic record. Hutton reversed the procedure followed by his contemporaries: instead of assuming some given initial condition(s) and proceeding toward the present, he worked backward from the present into the past. Although Hutton's method has formed the foundation of geology, one needs to stick to the philosophical concept but avoid being trapped in too many specifics. For example, the conditions (emergent continents and waning continental glaciers) prevailing at the present time, known stratigraphically as the Holocene, differ significantly from the conditions that prevailed during much of the past as inferred from the rock record. Through long periods of geologic time, the continents were submerged beneath shallow seas. Associated with the waxing and waning continental ice sheets were rapid changes in sea level (drops during the waxing and rises during the waning). During the past 12,000 years, a rapid submergence and accompanying transgression (named the Flandrian in northwestern Europe) have provided us with an excellent modern analogue for rapid transgression. Less well known are examples of the products of rapid progradation or the products of periods of standstill. Interestingly, many petroleum reservoir sandstones are parts of successions that were deposited when the coastline was prograded. Since sea level reached its present position only about 3,500 years ago, the thickness of strata deposited in many places is not comparable with those of more-ancient strata in the geologic record. For example, a reef deposited at the current high stand (during the last 3,500 years) may be only one meter thick. By contrast, the thickness of an ancient reef in the rock record may be 400 meters. Hence attempts to compare the present with the past typically face a scale problem. Perhaps the old cliche "the present is the key to the past" should be replaced by a two-part motto. Such a motto would state that (1) "the present is the key to geological processes" and (2) "geological processes are the key to the past." In modern environments, processes can be observed and measured. Applications of the results of such observations and measurements to the geologic record have resulted in major advances in the understanding of facies.

SOURCES: Albritton, 1963; R. L. Bates and J. A. Jackson, 1987; Dunbar and Rodgers, 1957; W. L. Fisher and McGowen, 1969; Miall, 1982; A. B. Shaw, 1964.

CHAPTER 7

Sedimentary Processes: Physical, Biological, and Chemical

In this chapter we examine three major categories of processes that operate in many depositional environments. These are (1) physical processes (mostly interactions between currents and sediment particles), (2) biological processes (emphasizing secretion of skeletons, organism/sediment interactions, bacterial reactions that change minerals, and weathering), and (3) chemical processes (chiefly those that give rise to evaporites, zeolites, and feldspars).

Physical Processes

Under physical processes we examine some fundamentals of the ways in which sediments are transported and deposited. The purpose is to be able to connect processes and products. The objective of this section is to prepare a foundation on which the reader can build the skill of interpreting the physical conditions under which sediments were deposited. The chief characteristics of a sediment that are used in such interpretations are textures, kinds of sedimentary structures, such as bedding, and sequences and associations of sedimentary structures.

Specifically, this section deals with the concept of shearing and how shearing affects bodies of particles, not only when gravity is the sole agent responsible for the shearing, but also when currents of air or of water are responsible. Such a discussion involves the dynamics of fluids.

Concept of Shearing

The mechanics of materials is a subject that deals with the application of forces to materials and the ways in which the materials respond to these applied forces. The key concept related to the movement of bodies of particles involves **shearing**, *a deforming- or potentially de-*

forming condition that sets up within a body the tendency for parts to slide over other parts along a series of parallel planes. These planes are named *shear planes;* along them, the tendency for slipping-type movement is maximized. The term tendency is used because the body being subjected to shearing may be strong enough to withstand the forces applied and thus not move noticeably. An internal condition of *shearing stresses* exists whether or not the external applied force is great enough to cause noticeable movement. The intensity of a stress is expressed in terms of the amount of force acting within a unit area.

Bodies of sediment particles move when the intensity of the shearing to which they have been subjected overcomes their internal resistance to shearing.

RESISTANCE TO SHEARING: BODIES OF COHESIVE- AND COHESIONLESS PARTICLES

A body of particles is able to resist shearing if the particles are able to remain in contact with one another. Whether they can do this or not depends on (1) gravity, (2) conditions within any interstitial fluids present, and (3) proportional effect of electrostatic forces resulting from uneven distribution of charges over the surfaces of the particles.

The effect of gravity is to cause particles to be attracted toward the center of the Earth. This tends to hold the particles in place on horizontal surfaces. However, as is explained in a following section, on a slope, a gravity component exists that tends to pull the particles down the slope.

The effect of interstitial fluids varies with the position of the top surface, or free surface of the fluid, for example, the water table. Above this level is capillary water; below it, the pores contain an interconnected body of water whose *hydrostatic pressure* increases regularly downward. The effect of capillarity is to cause films of water to

231

remain above the free-surface level. The weight of the water above this level exerts what has been designated by soils engineers as *negative pore pressure*. The effect of negative pore pressure is to allow sand, for example, to stand in a vertical slope. By contrast, sand that is totally dry or totally beneath the free surface of the water will make slopes no steeper than the angle of repose of the particles. The effect of negative pore pressure is to increase resistance to shearing.

The effects of electrostatic forces vary with particle size and become noticeable in particles smaller than coarse silt (5ϕ, 0.03 mm). Where electrostatic charges of attraction loom proportionately large with respect to the masses of the particles, the effect is dramatic. The particles behave as tiny magnets; the electrostatic forces are strong enough to hold the particles together and thus to increase their shearing strength. In engineering usage, bodies of fine particles held together by electrostatic forces are known as "cohesive soils."

In contrast are aggregates of particles larger than fine sand (2ϕ, 0.25 mm). These *aggregates of sand-size- or coarser particles in which the only significant factor keeping the particles together is gravity acting on the individual particles at points of contact* consist of **cohesionless particles**. These are the "friction soils" of engineering usage. Bodies of cohesionless particles are not bound together by electrostatic forces, hence are readily eroded and dispersed.

Particles of very fine sand and coarse silt lie in the intermediate size range. Such particles may be cohesive or cohesionless. The conditions that govern the variation in their behavior are not known.

PATTERNS OF SHEARING

The patterns of shearing that may be applied to particles are diverse. The top surface of the body of particles may be inclined or horizontal; on an inclined surface, the direction of shearing may be up or down the slope (Figure 7-1).

Shearing down a slope can be applied by the tangential component of gravity, which acts along and down a slope (See Figure 7-1, A.), or by a moving fluid, which itself is being impelled down a slope by the tangential component of gravity. (See Figure 7-1, B.) In addition, shearing may be applied to sediments by a fluid in a direction that is up a local slope (See Figure 7-1, C.), parallel to a slope, or along a horizontal surface. (See Figure 7-1, D.)

Gravity Shearing: Dynamic Dilatancy (Bagnold Effect)

One of the most far-reaching concepts connected with the physical transport of a body of sediment particles concerns the collective response to shearing. The key point, which has emerged from studies of the shearing

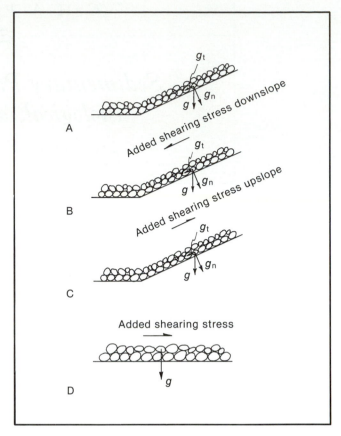

FIGURE 7-1. Various combinations of shearing stresses affecting bodies of cohesionless sediment.

A. Particles on a slope; only downslope force is g_t.

B. Added shearing force acts down the slope; adds to effect of g_t.

C. Added shearing force acts up the slope; opposes the effect of g_t.

D. Added shearing force applied to particles on horizontal surface. Not shown is added shearing force acting along the slope.

of aggregates of particles, is that almost nothing important about their behavior en masse can be predicted from analysis of how an individual particle behaves. Another significant point is that gravity shearing can transport particles even where no fluids are present. This conclusion is supported by evidence that gravity has displaced bodies of regolith on the Moon, which today possesses neither an atmosphere nor a hydrosphere and presumably never did.

Gravity shearing results from the component of gravity that acts along and down a slope (Figure 7-2, A). As long as the friction force, created by g_n and the roughness of the particles, exceeds g_t, the particles do not move. As long as the particles maintain their point-to-point contacts, the body of particles behaves as a solid. It resists shearing and will propagate shear waves (S waves of seismology).

If the tangential component of gravity exceeds the frictional forces within a body of particles, then the particles

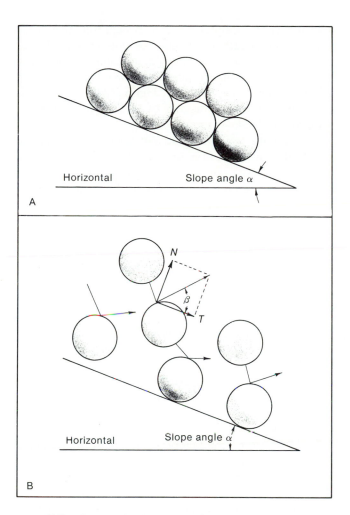

FIGURE 7-2.   Groups of spheres on a slope.

A. This group is held in position because $g_n > g_t$.

B. If a shearing force is applied in a direction that is downward parallel to a slope having an angle α with the horizontal, the spheres are driven apart and move readily. In their shifting parallel to the slope (direction T), the spheres collide with other spheres and thus are dispersed away from the slope (direction N). Angle β defines the mean angle of travel of spheres resulting from collisions. Further explanation in text. (Based on R. A. Bagnold, 1954.)

begin to move. As they do so, the volume they occupy increases. Technically, this is known as *dilatancy*, and the density of the body of particles at the point of flow is the *critical density*.

Why should shearing cause dilatancy? The answer is that the particles moving along the slope collide with one another and thus are displaced, bounced, or tipped upward, as it were, away from the slope. (See Figure 7-2, B.) *The dilatation of a body of cohesionless particles being sheared* has been named **dynamic dilatation**.

Dynamic dilatation has also been named the *Bagnold effect*, after its discoverer, R. A. Bagnold. The magnitude of the Bagnold effect is a function of the intensity of shearing and of the viscosity of the ambient fluid.

SHEARING AND DISPLACEMENT OF PARTICLES AWAY FROM SURFACE OF SHEARING

The shearing resistance of particles can be resolved into two components, T, the component parallel to the plane of shearing, and N, the component normal to the plane of shearing. These two are related as

$$\frac{N}{T} = \tan \beta \qquad \text{(Eq. 7-1)}$$

where β is mean angle of encounters (can be considered as a dynamic analog of the coefficient of friction).

The spacing of the particles is based on the geometric relationships among spheres. In linear form, the concentration λ is

$$\lambda = \frac{\text{diameter of sphere}}{\text{mean radial separation distance}} \qquad \text{(Eq. 7-2)}$$

The volume concentration C reaches its maximum value C^* when all the spheres are in contact. The value of C equals $(1.00 - \text{porosity})$. The linear- and volume parameters are related as

$$\lambda = \frac{1}{(C^*/C)^{1/3} - 1} \qquad \text{(Eq. 7-3)}$$

When the spheres are in contact, λ is infinite. General shearing becomes possible only when λ is 22 or less. For values of λ between 22 and 14, a particle-water mixture behaves as a granular paste. At values of λ less than 14, the dispersion of particles behaves as a Newtonian fluid.

The particle shearing stresses, expressed as T, define two ranges as a function of the rate of shearing. In the *inertial range*, T is a function of the square of the rate of shearing:

$$T_{\text{inertial}} = 0.013\sigma(\lambda D)^2 \left(\frac{dU}{dy}\right)^2 \qquad \text{(Eq. 7-4)}$$

where

$\sigma =$ density of particles

$D =$ diameter of particles

$y =$ direction normal to plane of shearing

$dU/dy =$ rate of shearing; change of speed U with distance y from fluid/sediment interface.

In the *viscous range*, T is a function of the first power of the rate of shearing:

$$T_{\text{viscous}} = 2.2\lambda^{3/2}\eta \left(\frac{dU}{dy}\right) \qquad \text{(Eq. 7-5)}$$

where

$\eta =$ viscosity of fluid (poises).

SOURCE: Bagnold, 1956.

LIQUEFACTION AND FLUIDIZATION

Displacement by shearing is not the only way to destroy the point-to-point contacts within a body of particles. Flow of fluids through the particles can separate them; in engineering usage this is designated *excess pore pressure*. The particles can also be displaced by the passage through them of seismic waves or by volcanic explosions. By whatever mechanism it is done, the effect is the same—the body of particles dilates, ceases to behave as a solid, and starts to behave as a liquid. *A body of cohesionless particles, dilated past its critical density, that behaves as a liquid* is a body of liquefied particles. Liquefaction designates *any process that causes a body of particles to change its behavior from that of a solid to that of a liquid. The special case of liquefaction created by upward flow of fluids within the pores of the body of particles* is the condition of fluidization.

Bodies of liquefied particles flow down the slightest slopes. Such bodies of particles spread out and come to rest in the lowest places available to them and there build layers of sediments having horizontal upper surfaces (Chapter 5).

SOURCES: J. R. L. Allen, 1982; G. M. Friedman and Sanders, 1978; Hsü, 1989.

Subaerial Gravity-Displaced Sediments

Subaerial gravity-displaced sediments include everything from water-soaked sod that has crept slowly a short distance downslope, to dry bodies of rock debris that traveled far and at amazing speeds, to bodies of tephra that were literally red hot. Here we concentrate on some major kinds of liquefied coarse-particle flows.

The liquefied coarse-particle flows of interest here share two characteristics: (1) some traveled at speeds of more than 150 km/h (41.7 m/s) and (2) their destructive effects can be catastrophic.

ROCK AVALANCHES

A typical rock avalanche is illustrated by the body of rock that was shaken loose from Shattered Peak, Alaska, by the seismic waves associated with the great Prince William Sound earthquake of 24 March 1964 (Figure 7-3). One part of the rock debris traveled along an azimuth of about 290° (direction of WNW). Part of this body of material turned northward and flowed along the axis of a valley glacier tributary to Sherman Glacier. But another part continued on a course of 290°, crossed Spur Ridge, where the local relief is 130 m, and came to rest on the north side of Sherman Glacier. On the west side of Spur Ridge, a stand of trees was left intact. The rock avalanche literally flew over these trees.

In order to emphasize the conclusion that rock avalanches flow rather than slide or ride on air cushions, as many geologists have inferred, a change of name has been suggested. The proposed new word is *sturzstroms*, a German term used by Albert Heim (1849–1937) in his report on the famous disaster of 11 September 1881 at Elm, Switzerland.

We next discuss two relatively uncommon kinds of gravity-displacement flow mechanisms, followed by one that is much more common. We discuss the exotica first because, although they do not occur often, their deposits are distinctive and their effects can be catastrophic.

SOURCES: Fraser, 1989; G. M. Friedman and Sanders, 1978; Graf, 1984; A. Heim, 1882, 1932; Hsü, 1989; Nilsen, 1982; Savage, 1984.

BASE-SURGE DEPOSITS

A base surge is *an expanding, dense basal cloud of gas and/or particles that travels outward from an intense explosion*. The explosion may be a nuclear blast, a volcanic outburst, or a meteor impact. The Earth's commonest base-surge deposits are those associated with maar-type volcanoes, where explosions in the crater liquefy a body of particles and set them in motion.

In a base surge, the kinetic energy is in the particles. The body of particles travels outward along the ground and beneath what may be a passive segment of the atmosphere, or, on the Moon, in a vacuum (Figure 7-4, A). If the surge travels beneath an atmosphere, a shearing couple is established between the body of fast-moving particles and the overlying, possibly passive, air. In this shearing couple, the relative direction of drag above is toward the explosive source, and below it is away from the explosive source. Although, in fact, the outward-moving particles below may be dragging the air above, the relative sense of shear is the same as if the air were moving toward the explosion and dragging the particles inward toward the source. Because heat from the explosion causes air to rise in a chimneylike column, no doubt surface winds do indeed blow radially inward toward the explosion to replace the rising air. Such inward-blowing surface winds increase the drag at the interface between the air and the outward-moving particles. (See Figure 7-4, B.)

Along any interface where a body of cohesionless particles forms a shearing couple with a fluid that generally lacks such particles, the resistance to shearing consists of two major components. These are (1) the drag created between the individual particles and the fluid (*particle resistance*), and (2) the drag created by the configuration of the top surface of the body of particles (*form-drag resistance*). Under certain conditions the rate of shear-

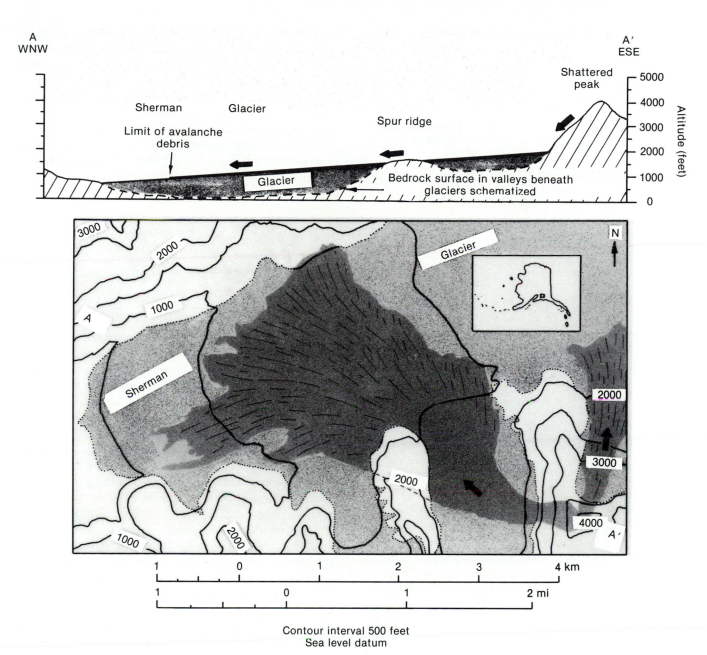

FIGURE 7-3.   Map and section of Sherman avalanche, southern Alaska. Short lines in area of avalanche debris (shaded) on map indicate large grooves in the surface of the deposit. (Map modified from R. L. Shreve, 1966, fig. 2, p. 1640; profile by J. E. Sanders, 1981, fig. 11.15b, p. 267.)

ing can increase faster than the particle resistance. (The particle resistance is a function of the number of particles in motion, and therefore increases, albeit commonly more slowly than the rate of shear, as the rate of shear increases.) When this happens, additional resistance to shearing appears by an increase in the surface roughness. The roughness results from the appearance of bed forms (Chapter 5). The longitudinal profile of a typical bed form (See Figure 5-21, D.) is asymmetric, with a gentler slope on the upshear side, the side toward the di-

rection from which the upper shearing couple acts. (See Figure 7-4, B.) Some asymmetric bed forms migrate in the downshear direction. They do so because sediment is eroded from their gentler upshear sides and is deposited as cross strata on the steeper downshear sides (Figure 7-5). Bed forms are discussed at greater length in subsequent sections.

SOURCES: G. M. Friedman and Sanders, 1978; Sigurdsson and others, 1980.

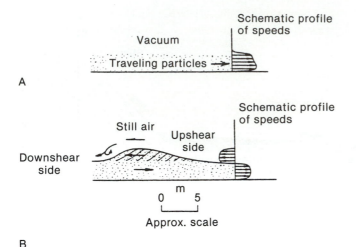

FIGURE 7-4. Two possible configurations of shearing at top of base surge being propelled outward from a volcanic source at left.

A. Top of base surge is plane.

B. Top of base-surge sediments shaped into asymmetric bed form that migrates toward left, in downshear direction relative to the air. These same arrangements could exist if the fluid were the only active agent. Further explanation in text. (Authors.)

NUÉES ARDENTES

Another kind of deposit created by violently explosive volcanoes is intermediate between sedimentology and igneous petrology. These are the materials deposited by **nuées ardentes**, *incandescent glowing clouds of tephra exploded from a volcano.* Many of the tephra in a *nuée ardente* consist of tiny clots of magma or of hot solid particles. At its time of emplacement, the temperature within a *nuée ardente* may range from 550 to 950°C. As a result, the deposits made by a *nuée ardente* commonly contain charcoal and display effects of flowage and welding that occurred as the hot particles came into contact with one another and were subjected to the weight of overlying materials while they finished cooling.

FIGURE 7-5. Sectional view of cross strata in base-surge deposits discharged from volcano of Pleistocene age near Rome, Italy, and dipping toward the volcanic source, which is inferred to have been located to right of this view. (Courtesy W. Alvarez.)

The great mobility and high speeds of flow displayed by *nuées ardentes* have been ascribed to fluidization. According to some geologists, this condition was created by gas escaping from the particles, as has been observed in active submarine lava flows in the Hawaiian Island chain. The quantitative effects of this inferred gas jetting from individual particles have not been established. If the mechanism responsible for cold rock avalanches can operate with *nuées ardentes*, then explanations for the mobility of *nuées ardentes* need not involve gas jetting and other kinds of unusual fluidization mechanisms.

SOURCES: G. M. Friedman and Sanders, 1978; Pettijohn, Potter, and Siever, 1972; Schmincke, 1988.

DEBRIS FLOWS: VOLCANIC AND NONVOLCANIC

A **debris flow** is *a flowing muddy mixture of water and fine particles that supports and transports abundant coarser particles.* A debris flow differs from a turbidity current (Chapter 9), which is a turbulent suspension of sediment in water; in a debris flow, particles are not suspended, and turbulence is suppressed. *A debris flow composed largely of volcanic particles* is a **lahar**. Debris flows and lahars differ from the kinds of flows previously discussed in that abundant water creates a viscous mud. Ordinary debris flows are common in semiarid climates, where plant cover is sparse and where torrential rains can mix with exposed regolith. The water to create lahars can come from the melting of snow or ice that may have accumulated on a volcanic cone between eruptions or from great rains that are created by the updrafts of hot gases during an eruption or created by subsequent storms.

Lahars generated on the steep slopes of volcanic cones are potential hazards in valleys that extend away from the cones. Volcanic cones in the Cascade Range of the northwestern United States have generated lahars that flowed as much as 80 km from the cone.

Debris flows and lahars generally deposit poorly sorted mixtures that lack internal layer structures. Some show normal grading (Figure 7-6); others are inversely graded; and still others show no internal arrangements but consist of chaotic mixtures of debris of many sizes. (See Figure 5-4, A.)

An additional kind of liquefied coarse-particle flow is represented by loose sand that flows down subaerial slopes (Figure 7-7). This kind of liquefied coarse-particle flow may be a result of liquefaction caused by a sudden shock.

SOURCES: Fraser, 1989; G. M. Friedman and Sanders, 1978; Hsü, 1989; Leg 123 Shipboard Scientific Party, 1988; Miall, 1978; Savage, 1984.

FIGURE 7-6.   Sectional view of lahar resting on silty sand deposited by a stream. Age of lahar is 4800 radiocarbon years. Mud Mountain, east of Tacoma, Washington. (D. R. Crandell, U.S. Geological Survey.)

Concepts of Fluid Mechanics

In this section we summarize the aspects of fluid mechanics that are important for understanding how fluids transport sediments physically. Fluid mechanics is basically an engineering subject, but its geologic applications are numerous. Although engineering research has been extensive and a quantitative basis has been established for designing aircraft, ships, systems of pipes, and so forth, nevertheless, many of the problems of greatest concern for geologists are of least concern to engineers. For example, in a flume experiment where a flow of water transports sand, engineers go to elaborate lengths to establish

FIGURE 7-7.   Thin lobes of dry sand that flowed down steep bank of St. Mary's River, northeast of Chester, Florida. (Authors, 03 May 1970.)

a steady, uniform flow, so that while it prevails, they can make various measurements. Flume operators work hard to prevent secondary flows and to blot out wave action, both of which are important factors in natural streams. Moreover, a geologist might find much more value in the conditions many engineers have tended to neglect, namely the decay of the flow and the deposition of its sediment load. Some flume experiments that have been the basis for new geologic insights are of no engineering interest; virtually no measurements of any aspects of the flow were made. In what follows, we shall try to distill from all possible sources the concepts that we regard as important for understanding the origin of those features of sediments that are created by currents.

The most-fruitful geologic approach may be to make critical measurements in the field where natural currents operate. In that way, the size limitations of the engineers' flumes are overcome and direct relationships between the current and the sediments can be established. Here we discuss the properties of fluids and boundary conditions.

PROPERTIES OF FLUIDS

Fluids are characterized by their viscosities, densities, negligible resistance to shearing (except in fluids of extremely high viscosity), and ability, under appropriate conditions, to flow turbulently. Fluids do not transmit shear waves (**S** waves of seismology). The two most-common fluids of geologic importance are air and water. In both, the viscosities and densities vary with temperature. Moreover, in water these two properties vary with amount of dissolved materials (salinity; Table 7-1; see also Figures 3-1, 8-34, and 8-35).

A fluid's resistance to shear varies according to the rate of shear. At low shear rates, resistance varies with the first power of the rate of shear, and at high shear rates, with the square of the rate of shear. At low rates of shear the fluid displays a condition known as **laminar flow**, which is defined as a *condition of flow in which parts of the fluid slide over one another along surfaces that conform to the shape of the boundary of the fluid and the directions of flow at every point do not change with time* (Figure 7-8).

At higher rates of shear, complex flow paths appear; typical features are curved and spiral flow paths known as *vortices* and *eddies*. *Fluid flow characterized by vortices and eddies* is **turbulent flow**. (See Figure 7-8, C.) In a turbulent flow, the eddies moving away from the shear surface represent an upward force that is capable of supporting sediment. However, the law of continuity requires that flows away from the shear surface must be equaled by flows toward the shear surface. This inevitably leads to complex patterns that are known as *secondary flows* or *secondary circulation*. Of particular geologic interest are the cylindrical "flow tubes" whose axes lie along the boundary of the flow and are parallel to the direction of the

TABLE 7-1.   Density and dynamic viscosity of water as function of salinity and temperature at atmospheric pressure.

Salinity (parts per thousand)	Temperature (°C)													
	0		5		10		15		20		25		30	
	ρ^a	μ^b	ρ	μ	ρ	μ	ρ	μ	ρ	μ	ρ	μ	ρ	μ
0	0.999	17.9	0.999	15.2	0.999	13.1	0.998	11.4	0.997	10.1	0.996	8.9	0.995	8.0
10	1.008	18.2	1.008	15.5	1.007	13.4	1.007	11.7	1.006	10.3	1.004	9.1	1.003	8.2
20	1.017	18.5	1.016	15.8	1.015	13.6	1.014	11.9	1.013	10.5	1.012	9.3	1.011	8.4
30	1.025	18.8	1.024	16.0	1.024	13.8	1.023	12.1	1.021	10.7	1.020	9.5	1.019	8.6
35	1.028	18.9	1.028	16.1	1.027	13.9	1.026	12.2	1.025	10.9	1.024	9.6	1.023	8.7

SOURCE: After N. E. Dorsey, 1940.

[a] ρ = density (grams per cubic centimeter).
[b] μ = dynamic viscosity (poises $\cdot 10^{-3}$).

main flow. Around each flow tube the threads of flow are helicoidal (Figure 7-9). The diameters of these flow tubes range from a centimeter or so in thin flows of water to hundreds of meters in the atmosphere or in the ocean. An extreme case of a single flow tube in the atmosphere is the high-altitude jet stream. Flow tubes leave sediment in streaks or ridges that are parallel to the main direction of flow.

SOURCES: Bagnold, 1968; G. M. Friedman and Sanders, 1978.

BOUNDARY CONDITIONS

The flow of a fluid past a boundary is affected by the *surface roughness* and by the composition of the boundary. Particularly striking contrasts appear depending on whether the fluid flows over a fixed surface, such as a ledge of bedrock; over a substrate underlain by cohesive sediment; over a surface underlain by movable (cohesionless) particles (Figure 7-10); or the flow is wind blowing over water (Figure 7-11).

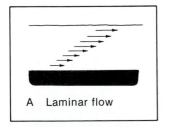

A   Laminar flow

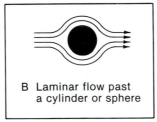

B   Laminar flow past a cylinder or sphere

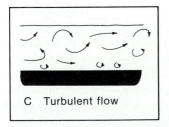

C   Turbulent flow

FIGURE 7-8.   Within a body of fluid undergoing laminar flow (A and B), the paths of flow are simple lines parallel to the boundary of the flow. If the boundary is plane (A), the lines are straight. If the boundary is curved (B), the lines curve. Within a body of fluid flowing turbulently (C), the paths of flow are complex curves. (Authors.)

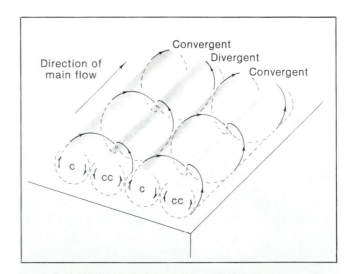

FIGURE 7-9.   Schematic block diagram through base of a body of fluid undergoing turbulent flow in which boundary layer has developed a secondary-flow pattern including helicoidal lines of flow (arrows). Small curved arrows at "ends" of "cylinders" show sense of rotation in each "cylinder." Further explanation in text. (Modified from S. Dzulynski, 1965, fig. 3, p. 197.)

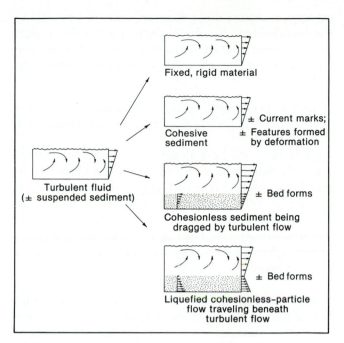

FIGURE 7-10. Ranges of boundary conditions at base of turbulent flow of water that may or may not be transporting a load of suspended sediment. Schematic rectangular segments of longitudinal profiles through current, flowing from left to right, also include at right idealized plots using $\log_{10}$ vertical scale (not shown) of speed versus height. Further explanation in text. (After J. E. Sanders, 1965, fig. 3, p. 216.)

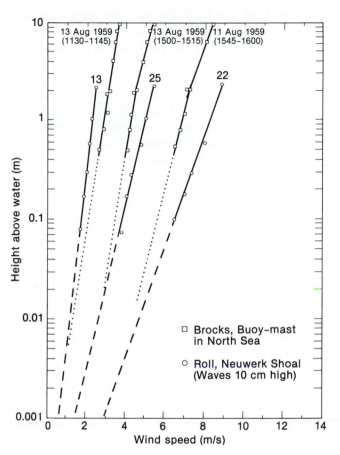

FIGURE 7-11. Plot of wind speed versus height above the sea surface, using $\log_{10}$ of height. The convergence of the lines through points measured by Roll (nos. 13, 25, 22) indicates that the surface roughness was the same in each. The lack of convergence of lines drawn through points measured in the North Sea by Brocks suggests that surface roughness was not the same for all series of his measurements. (After H. U. Roll, 1945, fig. 5; K. Brocks in E. L. Deacon and E. K. Webb, 1962, fig. 5, p. 56.)

Fluids and Particles

In this section we examine how a moving fluid transports particles. The subject is further divided into the ways in which the fluid transfers energy to the particles and an analysis of the two major methods of particle transport, suspension and traction.

METHODS BY WHICH FLUIDS ENERGIZE PARTICLES

In order to move a particle that is locked in a bed among other particles and hence not free to roll easily, a fluid must transfer enough energy to overcome the downward force exerted by gravity on the net weight of the particles. In the case of water, this net weight is known as the *immersed weight*. A fluid can energize a particle directly and acting alone in one of four ways: (1) fluid *lift force*, (2) changes in fluid pressure and surges related to wave action, (3) fluid impact, and (4) support from flow within turbulent eddies.

Fluid "Lift Force." The flow of a fluid above a stationary particle creates a fluid lift force which is related to the airfoil principle that enables airplanes to fly. Careful observation of sand in wind tunnels has shown that the

lift force of the air can raise quartz particles of sand size. Three comments need to be made: (1) Not only are sand particles lifted by a strong-enough air current, but at the same time they are subjected to a rotational couple so that they spin vigorously. (2) The lift force can be shown to stop operating by the time the clearance beneath the lifted particle is about half the diameter of the particle. (3) Despite the effects of item (2), lifted particles can be observed to rise to heights equal to several particle diameters. This is interpreted to mean that the lift force accelerates the particles and that once they are moving, the inertia of the particles keeps them going for a short distance after the effectiveness of the lift force stops and before the drag created by viscosity becomes totally effective. Once the effect of viscosity overtakes the inertia, the particle stops rising. If nothing else happens, fluid impact will push the particle in the direction of local flow, and gravity will pull it back to the bed.

The effects of the returning particles on other particles involve special topics that we must defer until we have finished our list of energizing effects on particles by fluid.

The best- and simplest way to appreciate how the fluid lift force works is to watch the wind blow dry leaves along a paved road or -sidewalk. Some leaves simply skid along the surface. Other leaves, especially any circular ones, may be tipped on their sides and roll along as tiny wheels. All these are being dragged but not appreciably lifted. Now and then, however, one or more leaves may make small excursions away from the pavement. After a brief interval of transport at a short distance above the ground along a direction parallel to the wind, the leaf returns to the ground, where it may stop, skid along the surface, or roll downwind. Later, it may make another jump. If you really want to make a full-scale experiment, measure the speed of the wind at different heights above the ground at two points about 1 m apart, and record the sizes of the leaves and how high and how far leaves of a given size leap. Both the heights of the leaps and the distances of downwind advance should work out to be direct functions of the speed of the wind and difference in the wind speeds at the two heights.

SOURCES: Bagnold, 1956; G. M. Friedman and Sanders, 1978.

Changes in Fluid Pressure and Surges Related to Wave Action.   Within a turbulent flow, short-term pressure fluctuations are numerous. These pressure fluctuations can affect sediments in water; they take place not only in the fluid above the water/sediment interface but also in the interstitial water within the sediments. An abrupt decrease in the fluid pressure in the water above the water/sediment interface may be a factor in dislodging sediment particles. An upward impulse would result from a momentary lower pressure in the fluid above the water/sediment interface compared with the pressure within the interstitial water.

Pressure fluctuations beneath a turbulent current are thought to be responsible for causing large particles on the bed of the flow to oscillate up and down *in situ*. Oscillations at 10 to 15 times per second through distances of several millimeters have been recorded. Such oscillations abrade large particles without moving them downcurrent.

When water-surface waves pass a given point, the changes in height of the water surface create pressure fluctuations in the bottom water. In addition, a wave advancing through water that is shallower than half its wavelength causes the bottom water to move in short, turbulent surges. As the depth decreases, these surges intensify. During each part of the wave cycle a verti-cal surge is created that affects not only the water above the particles but also the interstitial water for a short distance down into the bottom sediment. The upward surge is strong enough to lift some particles above the bed.

SOURCES: Fraser, 1989; G. M. Friedman and Sanders, 1978; Schumm and Stevens, 1973.

Fluid Impact.   We have already mentioned fluid impact, a subject that perhaps should not be listed as a separate factor, because it is involved in all the other three mentioned. What is meant by fluid impact is the full-face collision between the moving water and the particle. Those who have ever been in the forefront of a crowd that firefighters dispersed by using high-pressure streams of water know what fluid impact is all about.

Support from Flow Within Turbulent Eddies.   The upward flow of fluid within turbulent eddies is a source of energy for moving particles. As long as the upward speeds of flow within the eddies exceed the settling speeds of the particles, the particles will remain within the flow.

Settling of Spheres in Water and in Air.   One of the key factors that determines whether a sediment particle will or will not be transported in turbulent suspension is the relationship between the upward components of flow within the fluid and the terminal speed of fall, or *settling speed*, of the particle. The methods of analyzing the settling speed are based on the simplifying assumption that all the particles are spheres of a single mineral, say quartz, so that the densities of all are the same. The two chief controls on settling speed are (1) density and (2) size (as expressed by diameter of the spheres). Density affects settling according to the first power. Spheres smaller than 0.18 mm settle according to the squares of their diameters, whereas spheres larger than 0.18 mm settle according to the square roots of their diameters (Figure 7-12).

Spheres of minerals denser than quartz settle faster than quartz spheres of the same diameters. Therefore, on a graph such as that of Figure 7-12, the settling curves for minerals denser than quartz would plot above that of quartz. Where settling speeds alone control depositional processes, quartz particles of any given size are accompanied by smaller heavy-mineral particles. A general term that describes *all particles that were deposited together by any process operating in air or in water* is sedimentation equivalence. *All particles that settle through still water at the same speeds* are hydraulically equivalent; correspondingly, *all particles that settle through still air at the same speeds* are aerodynamically equivalent.

In some analyses of modern sediments, settling time has been used directly as a single variable that combines

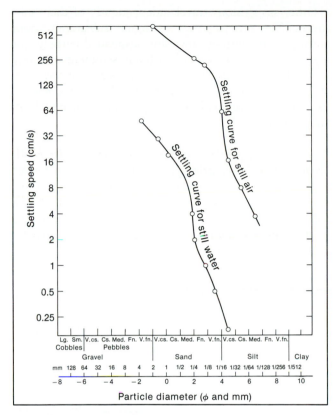

FIGURE 7-12.   Graph of settling speed versus particle diameter for settling of quartz spheres in water (left) and in air (right). In this graph, and in all other graphs in this book showing particle diameters on the horizontal axis (abscissa), particle size will be shown as decreasing toward the right, increasing toward the left. (Replotted by J. E. Sanders; data from still water by R. J. Gibbs, M. D. Matthews, and D. A. Link, 1971, fig. 1, p. 11, and table 2, p. 12; data from still air from L. Moldvay, 1961.)

the effects of size, shape, and density into a single function. By analogy with the $-\log_2 \phi$ scale for particle diameters (See Eq. 2-1.), a ψ scale has been devised for expressing settling speeds:

$$\psi = \log_2 c \qquad \text{(Eq. 7-6)}$$

where

$$c = \text{settling speed (cm/s)}$$

An approximate rule of thumb worth remembering is that the average speed of upward flow within turbulent eddies of water is about 1/12 the mean forward speed of the current. The mean speed typically prevails at a depth that is 0.4 of the average depth down from the surface toward the bottom of a flow (such as a river) (or 0.6 up from the bottom).

SOURCES: Bagnold, 1941; Blatt, Middleton, and Murray, 1980; G. M. Friedman, 1961; Hallermeier, 1981; Hand, 1967; Pettijohn, Potter, and Siever, 1987.

Transport Mechanisms

Currents transport sediments physically by two contrasting mechanisms that are the bases for defining two kinds of loads: (1) **suspended load**, a sediment load *dispersed within a current of air or water* (mainly silt-size and clay-size particles with varying proportions and sizes of sand) *and transported within the main body of the flow because the effect of the Earth's component of gravitational acceleration on the particles is overcome by upward components of turbulent flow*, and (2) **bed load**, the name for *the sediment load that moves in response to the several processes taking place as a result of shearing at the boundary of the flow*. Although the task of collecting samples of sediments being transported in these two loads may be difficult, and although sediments at times may pass imperceptibly back and forth between the two, nevertheless the mechanical principles controlling each load are not the same. In fact, the physical principles involved are quite distinct. The difference between the two kinds of loads is demonstrated by a wind that blows dry leaves. Recall that the jumping leaves traveled along parallel to the wind and now and then made short excursions upward, forward, and back to the pavement. Leaves that hug the pavement or slide, bounce, and roll along it are being transported in the bed load. In contrast are the leaves that are caught by swirling eddies; such leaves undergo twisting spiral motions and may travel high off the ground and be blown long distances before they return to the ground again. The high-flying, swirling leaves are traveling in the suspended load. In both their paths and their positions the leaves traveling in the suspended load differ from the leaves being transported in the bed load.

TRANSPORT IN SUSPENSION

The basic driving force for transport of the suspended load is turbulent flow, which is a fluid's unique way of responding internally to shearing stresses. Suspended-sediment transport operates in water and in the atmosphere. We shall discuss each separately, emphasizing the requirements for suspending fine sand.

Suspensions in Water. *A suspension of particles in water* is an **aqueous suspension**. Within a current of water and sediment, turbulence can be maintained up to the point where the density of the water/sediment mixture becomes ca. 2.0 g/cm³. Addition of more sediment inhibits turbulence, and additional movement, if any, is as a thick paste or a debris flow. The transformation of a turbulent aqueous suspension to a debris flow (a *mudflow* if fine sediment predominates) is a common process in regions having semiarid climates where shallow flows of water start up during flash floods. In large rivers and the sea, the loads available do not test the water's capacity for suspending sediment. However, subaqueous debris flows do occur when fluidized mixtures of particles and water do not mix with, and hence do not become diluted by, the overlying water.

A few aqueous suspensions actively interchange fine sand, very fine sand, and coarse silt with their substrates. The limit for fine sand is provided by the settling speed of quartz in water, 4 cm/s. To maintain fine sand in suspension, the average forward speed of a current of water must be about 50 cm/s. *A turbulent suspension that is exchanging sediment with its substrate and within which the quantity and particle-size distribution of the suspended load decrease upward* is a **graded suspension**. In a graded suspension both the quantity and particle sizes of the suspended load decrease upward through the flow.

Other suspensions travel independently of their substrates. They may pass over a sediment substrate without picking up or putting down any suspended particles, or they may not be in contact with any sediment substrate at all. These *suspensions that are independent of their substrates and within which the quantity and particle sizes of the load are more or less uniform throughout* are named **uniform suspensions**.

Atmospheric Suspensions at Low Altitudes and High Altitudes. The basic principles that govern *a suspension of particles in air,* an **eolian suspension,** are the same as those that govern aqueous suspensions. Both consist chiefly of fine particles. In the fine sediments deposited by the wind, sorting is poor. The population of particles spans as many as 10 ϕ classes (Figure 7-13).

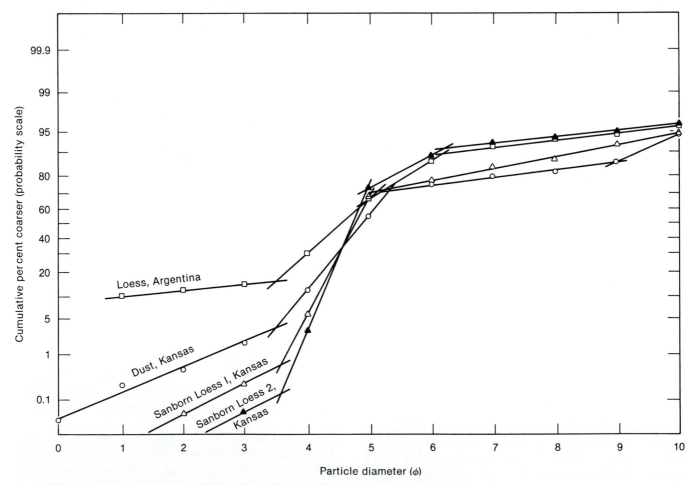

FIGURE 7-13. Cumulative-frequency curves of three samples of loess compared with a sample from a modern dust fall, which in 2 h in September 1939 deposited a layer about 1.5 mm thick on a freshly cleaned surface of a piece of furniture in a third-floor room, 7 m above the ground, in the Lakeway Hotel, Meade, Kansas. All curves consist of two populations that are poorly sorted ($< 3.5\phi$ and $> 5\phi$), separated by a population of very fine sand and coarse silt whose sorting is somewhat better. Analyses by sieves (sand sizes) and pipette (particles smaller than 4ϕ). (Data for Argentine loess from M. E. Teruggi, 1957, table 2, p. 325, sample No. 60; dust and two samples of Sanborn Loess from Kansas, from A. Swineford and J. C. Frye, 1945, table 1, p. 250, samples 1 and 2.)

Because of the structure of the Earth's atmosphere, suspended particles can travel at two distinct levels, forming *low-altitude suspensions* and *high-altitude suspensions*. (See Figure 1-11.) Low-altitude suspensions are confined to the lower 2 to 5 km of the atmosphere. These suspensions are blown upward by strong surface winds and include many tephra clouds generated by nearly all volcanic explosions.

High-altitude suspensions travel in the Earth's high-altitude jet streams. Particles reach the height of the jet streams only if they are activated under especially energetic circumstances. Some of these include nuclear-weapons tests in the atmosphere and extraordinary volcanic explosions. Low- and high-altitude suspensions were described in Chapter 1.

SOURCES: G. M. Friedman and Sanders, 1978; Moldvay, 1961; Passega, 1957.

TRANSPORT BY TRACTION

The term **traction** is *a collective name for the processes by which sediment is moved as bed load.* Traction has been found to be a function of the inertia of the particles; many of its effects result from collisions. The principles of traction are easiest to understand by studying eolian transport of sand. After we describe windblown sand, we shall examine sand in water currents.

Eolian Traction. Observations in wind tunnels or in nature where the wind blows across dry sand have shown that the shearing drag of the wind activates *a sheet of bed-load sediment, usually thin (≤ 10 cm but controlled by wind speed), that is pushed along at the base of a current of air in the direction of the wind.* Within this sheet of sand, named a **traction carpet,** particles move vigorously. Many bounce and jump along. The force of impact of collisions between particles moving downward at the ends of their trajectories and particles at rest on the ground pushes the particles on the ground forward. *The bouncing motion of particles in a traction carpet moved by a fluid* is **saltation.** *The forward movement of particles within a traction carpet and resulting from collisions with saltating particles* is **surface creep.** Measurements in wind tunnels and in the field indicate that in tractional transport by the wind about 75% of the particles move by saltation and about 25% by surface creep.

But what happens after saltation has begun and many particles are affected? To answer this question, we need to refer again to wind-tunnel experiments. As we mentioned previously, the speed of a fluid, including the wind, increases away from the surface of shear, which for the wind is usually upward from the ground. Because the increase is parabolic on an arithmetic scale, a plot of wind speed versus height on a $\log_{10}$ scale is a straight line (Figure 7-14, curves A, B, C). When a traction carpet has de-

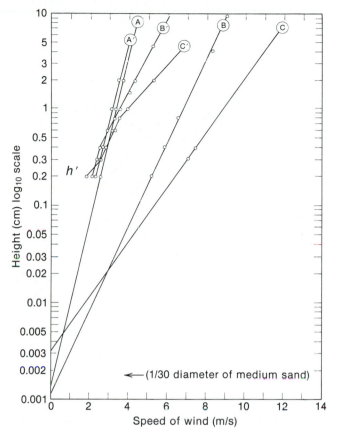

FIGURE 7-14. Graph of wind speed vs. log height for three pairs of winds. Each pair possesses the same speed gradient and applies about the same shearing force; hence their two curves are parallel. One wind of each pair (A, B, C) blows over a fixed sand surface. The corresponding other wind (A', B', C') blows over a traction carpet of saltating sand. At height of 10 cm, the speeds of winds A and A' are similar (4.5 and 4.3 m/s, respectively). Compare this with differences between the other two pairs at this height (B, 9 m/s versus B', 6 m/s; and C, 12.3 m/s versus C', extrapolated to 8.3 m/s). These differences result from the drag exerted on the wind by the sand at the top of the traction carpet. (Data from wind-tunnel measurements by R. A. Bagnold, 1941, fig. 17, p. 58.)

veloped fully, new relationships are established between wind speed and height. Now the straight lines intersect at a new point, h', which is governed by the height of the top of the traction carpet. (See Figure 7-14, curves A', B', and C'.) This change upward in the location of the point of intersection of the wind-speed lines means that the zone of maximum drag has shifted to a position related to the top of the traction carpet. For practical purposes, within the traction carpet, we may therefore consider that the wind speed is zero. No longer does the wind lift single particles. Rather, when each particle reaches the top of the traction carpet (the maximum heights of the saltation trajectory), the wind blows the particle forward and gravity pulls it back to the ground. When the particle strikes the ground it may (1) bounce

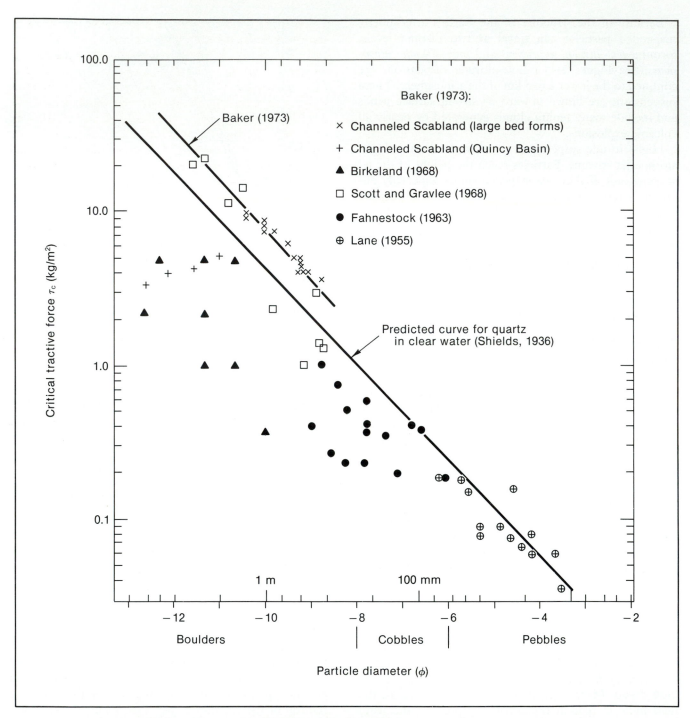

FIGURE 7-15.   Relationship between critical tractive force in flowing water and particle size for gravel. Birkeland's data (▲) from Pleistocene breakout flood, eastern California, flow depths and bottom slopes estimated; Scott and Gravlee data (□), from dam failure on Rubicon River, California, in 1964, flow depths and bottom slopes estimated; Fahnestock's data (●) from measurements in modern shallow stream; Lane's data (⊕) from engineering studies of stable channel bottoms of modern rivers; Channeled Scabland data from V. R. Baker (1973) (×, large bed forms; +, Quincy Basin), with inferred slope of water surface taken as energy slope. (Replotted and converted to metric units by J. E. Sanders, from V. R. Baker, 1973, fig. 13, p. 24.)

upward again (as a basketball being dribbled down the court), (2) strike other particles with sufficient force to spatter them upward, or (3) strike (an)other particle(s) at rest with sufficient impact to drive the particle(s) forward as part of the surface creep.

Within the traction carpet, the particles move almost entirely as a result of their own momentum and inertia. Only at the top of the traction carpet are the particles activated by the fluid impact and hence exert a shearing drag on the moving air above. Thus, in contrast to the behavior of isolated single particles, which "feel" the wind everywhere, the individual particles within the body of a traction carpet "feel" the wind only at the tops of their saltation trajectories. Elsewhere, although they are out of direct contact with the drag of the wind, the particles continue to move. And, by collisions, they cause other particles to move as well.

SOURCES: Bagnold, 1941; Blatt, 1982; Blatt, Middleton, and Murray, 1980; G. M. Friedman and Sanders, 1978.

Traction in Water. Granted that the remarkable kind of inertia-and-impact traction just described operates where quartz is driven by the wind, the question arises whether a comparable mechanism can operate in currents of a much-more-viscous fluid such as water. The fundamental point to determine is whether particles being driven along at the same speed as the water and moving essentially parallel with the bottom make excursions away from the bottom that are independent of other motions of the fluid. In short, is there any evidence that particles can punch their way through the fluid, so to speak, in a direction that is away from the surface of shearing? The answer is yes. The conclusive evidence can be seen where swift-moving thin sheets of water transport pebbles. Many such pebbles not only travel upward through the water but are projected out of the water and up into the air as well. Clearly, such pebbles are in no way supported by fluid turbulence nor have they been propelled by athletically inclined fish. Additional proof that sheared particles move upward through a liquid without receiving upward impulses from turbulent eddies is given by experiments with viscous fluids and low-density particles in which the only fluid flow is laminar. In such experiments the particles move away from the surface of shearing. In this case, impulses from turbulent eddies cannot be involved. The conditions of the experiment specifically excluded turbulence.

SOURCES: R. S. Anderson and Hallet, 1986; Bagnold, 1954, 1955, 1956; Francis, 1973; G. M. Friedman and Sanders, 1978; P. R. Owen, 1964.

Various attempts have been made to establish the flow conditions at the start of general movement of the bed load. One concept is known as the critical tractive force.

Other expressions employ the critical shearing stress or critical mean speed of a flow of a given depth. The relationship between critical tractive force and particle diameter for gravel sizes is shown in Figure 7-15. The shearing stress is found in several ways. One is by multiplying an empirical friction factor by the density of the fluid times the square of the mean speed of flow. Another (the Du Buys equation for boundary shear) is the product of the density of the fluid, the mean depth, and the energy slope. The relationship between mean speed of flow and gravel-size particles is shown in Figure 7-16.

Another relationship for showing how properties of a flow and particle sizes are related makes use of a dimensionless shearing parameter, the Shields parameter (Figure 7-17). The relationships between particle sizes for coarse sediment and rate of sediment transport are shown in Figure 7-18.

HJULSTRÖM EFFECT

As one might expect, it takes a swifter current to move larger particles than small particles. The most-movable sediment is fine sand. Contrary to intuition, particles smaller than fine sand are difficult to erode. Because of cohesion, such fine particles tend to stick together. To erode these fine cohesive sediments the current must be as swift as one capable of transporting pebbles (Figure

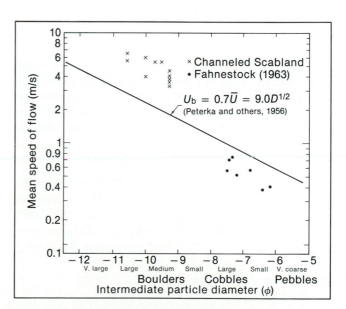

FIGURE 7-16. Relationship between mean speed of flow and particle size (based on intermediate diameter). Channeled Scabland data ($\times$) from uniform reaches where large-scale bed forms are present, depths 60 m and more. Fahnestock's data ($\bullet$) are measurements in a modern shallow river. Line is based on relationship established by Peterka and others (1956) from studies of rip-rap sizes in stilling basins. (Replotted and converted to metric units by J. E. Sanders, from V. R. Baker, 1973, fig. 14, p. 25.)

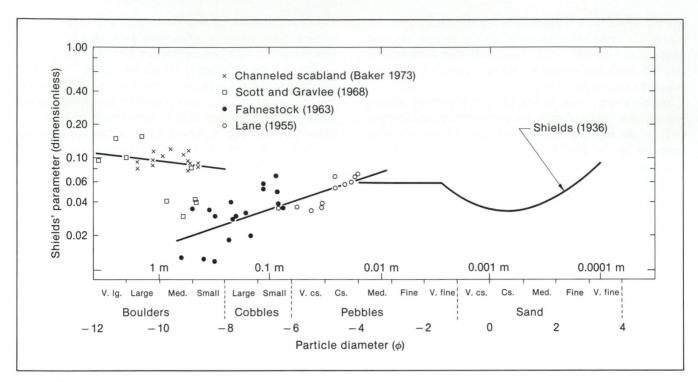

FIGURE 7-17.   Summary diagram relating Shields' (1936) parameter θ (the critical shear stress divided by the immersed weights of the particles) to particle size (intermediate particle diameter). Channeled Scabland data ($\times$) from V. R. Baker (1973) taken from reaches containing large-scale bed forms. Other data: Scott and Gravlee ($\square$); Fahnestock ($\bullet$); Lane ($\circ$). The reasons for the divergence between the two groups of coarse particles are not known but are thought to be related to depth of flow. The Channeled Scabland flows were deeper than 60 m, the others possibly only a few meters deep. (Replotted and converted to metric units by J. E. Sanders from V. R. Baker, 1973, fig. 16, p. 27.)

7-19). A graph showing the relationship between speed of current and sizes of particles is known as a *Hjulström diagram. The relationship whereby fine sediments resist erosion and require a faster current than that needed to move larger particles* has been referred to as the **Hjulström effect** after the Swedish geographer Filip Hjulström, who published his now-famous diagram in 1935.

SOURCES: Bedient and Haber, 1988; Blatt, 1982; Blatt, Middleton, and Murray, 1980; G. M. Friedman and Sanders, 1978; Graf, 1984; Hjulström, 1935, 1939; Mehta, Hayter, Parker, Krone, and Teeter, 1989; Parchure and Mehta, 1985.

SYSTEMATIC SERIES OF BED FORMS; FLOW REGIMES

In flowing over cohesionless sediment, a current can create a variety of features ranging from groups of linear streaks of slightly coarser sediment that are parallel to the current to various bed forms. With increasing current speed, a systematic sequence of bed-form patterns arises. These are of great importance because the kind of bed forms and the way they migrate determine the kinds of strata that form in sediment being transported by traction.

Bed forms can be organized into three categories: (1) *ripples*, which are small-scale features not found in sediments having median diameters larger than 0.6 mm (median diameter or d_{50} = 1.2 or 1.3 ϕ); (2) *dunes*, of several kinds and sizes, which are not restricted by particle size, which are out of phase with the shape of the water surface, and some of which are limited by the depth of flow; and (3) *in-phase waves*.

Ripples.   When clear water shears over fine sand at a speed of about 20 cm/s, the sand particles are set in motion. Almost as soon as movement of the topmost part of the sand becomes general, the surface of the sand becomes rippled. The wave lengths of such *ripples* range from 10 to 30 cm and their heights from 0.6 to 3 cm. H. C. Sorby (1826–1908) described ripple formation in a shallow stream that flowed through his estate. After an initial episode of rapid growth, the general configuration of the rippled sediment becomes stabilized and all the

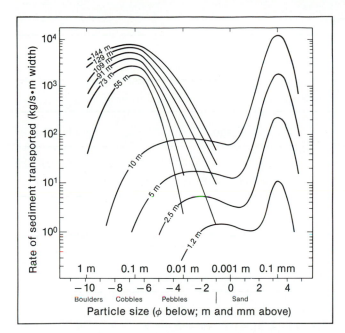

FIGURE 7-18.   Graph of rate of sediment transported vs. sizes of cohesionless particles moved by currents having depths indicated. Except for overlap in center of graph, deeper flows traveling at the same speed transport more sediment than shallower flows. (Replotted by J. E. Sanders from V. R. Baker, 1973, fig. 20, p. 31, and fig. 21, p. 32.)

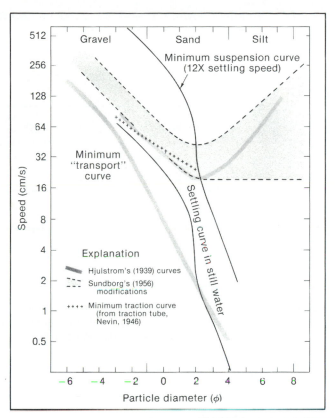

FIGURE 7-19.   Hjulström diagram, a graph of speed of current vs. particle sizes, showing relationships among erosion, transport, deposition, and settling speed. Truncation of minimum traction curve by minimum suspension curve at a point defined by $+2\phi$ and 25 cm/s means that at forward current speeds > 25 cm/s, particles having diameters < $+2\phi$ will always travel in suspension. Settling curve in still water from Figure 7-12.

ripple trains migrate slowly downcurrent. They migrate because sediment is eroded from their gently dipping upcurrent sides and is deposited on their steeply inclined downcurrent sides. (See Figure 5-21, D.) If the depth of the water exceeds several times the heights of the ripples, then the surface of the water is not noticeably influenced by the ripples.

The kind of observations made on ripples by Sorby were duplicated by G. K. Gilbert (1843–1918), who carried out a series of systematic investigations in five different experimental flumes ranging in size from a small indoor apparatus about 4 m long, 30 to 50 cm high, and 20 cm wide, to a large outdoor installation about 45 m long, 30 to 50 cm high, and 30 cm wide. In these flumes, the water was recirculated, but the sediment was not. In the large flumes, Gilbert was able to extend his experiments to include features larger than ripples. Since Gilbert's day many investigators have carried out studies of water/sediment interaction in flumes. As a result, much information is now available on the relationships between currents and many aspects of sediment behavior.

Dunes.   Features that we shall refer to as *flow-transverse subaqueous dunes* (in accordance with the newly established standard terminology of SEPM, the Society for Sedimentary Geology) originate in one of two ways: (1) from rippled sediment at speeds of flow of about 50 cm/s; and (2) from a plane bed of sediment coarser than 0.6 mm. We shall refer to eight classes of such dunes (Table 7-2). All are characterized by downshear slopes that are near the angle of repose (about 30°). The small dunes (types A and E) can form in shallow flows (depth < 0.5 m), the medium-size and larger subaqueous dunes do not form in flows that are shallower than about 4.7 m. (See Figure 5-18.) Over the crests of flow-transverse subaqueous dunes the flow separates. Between the point of separation at the crest of the dune and the point of reattachment on the back of the next dune downcurrent is a *separation eddy*. The water in the bottom part of the separation eddy travels in the opposite direction from the main current at speeds of from one-third to one-half that of the main flow. If the sediment is fine sand and the backflow speed at the bottom of the separation eddy is great enough, then the backflow may create small-scale ripples whose cross laminae dip in a direction opposite to that of the main current (Figure 7-20). Moreover, large dunes create irregularities on the

TABLE 7-2.   Kinds of flow-transverse subaqueous bed forms

Dunes (lower-flow regime)

	Dimensions (m)						
	Small		Medium		Large		Very large	
	L	H	L	H	L	H	L	H
Shape	0.6 to 5	0.075 to 0.4	5 to 10	0.4 to 0.75	10 to 100	0.75 to 5	> 100	> 5
2-D[1]	Type A		Type B		Type C		Type D	
3-D[2]	Type E		Type F		Type G		Type H	

Transition Forms

| Type I | Transition state between lower-flow-regime dunes and upper-flow-regime in-phase waves; dips of downcurrent sides are low—about 10° or less; thus their downcurrent migration creates low-dipping cross strata. |

In-Phase Waves (upper-flow regime)

| Type J | Antidunes |
| Type K | Standing waves |

SOURCE: After G. H. Ashley, chm., 1990, Table 6, p. 169.

[1] 2-D shapes are straight crested with planar downcurrent slopes; downcurrent migration yields planar cross strata. (See Figure 5-19, A.)

[2] 3-D shapes are cuspate; downcurrent slopes are concave surfaces opening downcurrent; migration yields trough cross strata. (See Figure 5-19, B.)

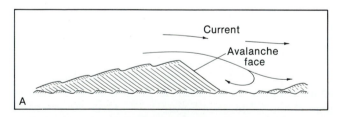
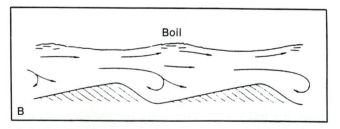

FIGURE 7-20.   Water flowing over dunes composed of fine sand (A) and coarse sand (B), seen in schematic profiles in vertical plane parallel to current. Over crests of dunes, the surface of the water is depressed. Over troughs between dunes, the surface of the water is heaped up into small, irregular "mounds" named boils.

A. Dunes composed of fine sand, with ripples, including those on upcurrent slope and backflow ripples in the troughs. Ripples on upcurrent slope advance to crest of dune and there disappear when the sand composing them avalanches down the downcurrent slope of the dune. Ripples in the trough migrate in a direction that is opposite to that of the main current. These backflow ripples are preserved beneath the dune as sand from the avalanche face, which shifts downcurrent, buries them. Although the ripples sketched here are confined to the trough, in some streams, the ripples migrate up the lower parts of the avalanche faces of the dunes. The combination of small-scale cross strata of the backflow ripples with large-scale cross strata of dunes is unique to fine sand transported by flowing water. (Modified from J. R. Boersma, 1967, fig. 12, p. 231.)

B. Dune in coarse sand. Small-scale ripples, as in Figure 7-20, A, do not form in coarse sand. (Modified from D. R. Simons, E. V. Richardson, and C. F. Nordin, Jr., 1965, fig. 3, p. 36.)

water surface. Above the crests of the dunes, the water surface is drawn down. Above the troughs of the dunes, the water surface is piled up. Another way of expressing this relationship is to state that the waves on the water surface are not in phase with the dunes.

At greater rates of shearing, transition bed forms of type-I appear. Type-I features lack associated small-scale ripples and their downstream sides are inclined at angles of about 10° or less; this angle contrasts with the 30° of all classes of flow-transverse subaqueous dunes. The water flowing downstream maintains continuous contact with all parts of the profiles of type-I bed forms; the flow does not separate (Figure 7-21).

Plane Bed. At still-greater speeds of flow, the washed-out type-I features disappear and the sediment/water interface becomes flat. Sediment continues to travel downcurrent rapidly, but now it does so as a series of flat sheets.

In-Phase Sediment Waves. An important limit on the behavior of a fluid is provided by the **Froude number**, *F*, which is defined as *a dimensionless number that is proportional to the ratio of the inertial- to gravity forces within a fluid; it is equal to the average speed of a flow divided by the square root of the product of the gravitational acceleration and the depth, D.* In symbols, this becomes

$$F = \frac{\overline{U}}{(gD)^{1/2}} \qquad \text{(Eq. 7-7)}$$

The denominator, $(gD)^{1/2}$, is also the formula for long water-surface waves having wavelengths greater than 20 times the depth of water. At $F = 1$, $\overline{U} = (gD)^{1/2}$; the celerity of the long waves equals the average downcurrent speed of the water. At Froude numbers larger than 0.84, rounded sediment waves appear that are in phase with the water surface (Figure 7-22). The *in-phase sediment waves that migrate upcurrent* are called **antidunes** (type-J bed forms).

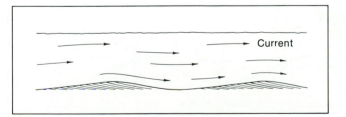

FIGURE 7-21. Water flowing over type-I bed forms, seen in schematic profile in vertical plane parallel to current. The surface of the water lacks appreciable relief and no separation eddy is present. (Based on D. R. Simons, E. V. Richardson, and C. F. Nordin, Jr., 1965.)

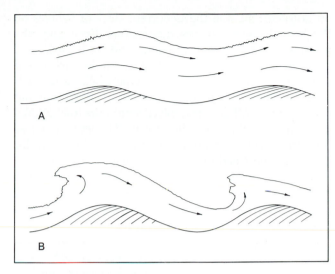

FIGURE 7-22.   In-phase waves seen in schematic profiles in vertical plane parallel to current direction.
  A. When water over the troughs is always deeper than the heights of the crests of the antidunes, standing-wave antidunes form. The surface of the water above the upstream sides of the antidunes is billowy, but no breaking takes place.
  B. When the water in the troughs becomes shallower than the heights of the antidunes, water waves break, and crash over in an upcurrent direction. Great turbulence is created when water waves over antidunes break. (Based on D. R. Simons, E. V. Richardson, and C. F. Nordin, Jr., 1965; and J. C. Harms and R. K. Fahnestock, 1965.)

Flow Regimes. The concept of *flow regimes* is based on the observed set of systematic relationships between the flowing water and its tractionally transported load of cohesionless particles. A **flow regime** is defined as *the aggregate of relationships prevailing among the stream of water, the shape of the water/sediment interface, the mode of sediment transport, the process of dissipation of energy within the current, and the phase relationships between the bed forms and the surface of the water.* The two flow regimes, lower and upper, are separated by a transitional stage.

In the *lower-flow regime,* the water/sediment interface is shaped into ripples alone, ripples plus dunes, or dunes alone. Sediment transport is minor and intermittent. Transport takes place by the motion of a traction carpet up the backs of the ripples or of the dunes and then by gravity avalanching of particles down the steep downcurrent slopes of these features. Some reversal in the direction of transport is possible in the troughs of dunes—on backflow ripples. The current's energy is dissipated by the roughness of and inertial drag by the particles but chiefly by means of *form drag* exerted by the bed forms and by backflow within the separation eddies. The undulations on the surface of the water are not in phase with the bed forms. The sediment is sorted in a downcurrent direction. The material left in the bed forms is slightly coarser and better sorted than the sediment that passes on downcurrent. On the upcurrent sides of pebbles on the bed, the water scours depressions. Eventually, the pebbles roll into these depressions and are buried.

In the transitional stage, great variability exists, but the characteristic features are the type-I bed forms. Sediment movement is tending to become continuous but some storage still does take place within the low-dipping cross strata of the type-I bed forms. However, as noted previously, sediment does not avalanche down the lee slopes of type-I bed forms. Energy within the current is dissipated by means of the inertial drag and roughness of the moving particles. The flow does not separate. The water surface tends to flatten and its shape tends not to be related to the low, long bed forms.

In the *upper-flow regime*, the water/sediment interface is plane or shaped into various ephemeral undulations. Sediment transport is large and continuous. Pebbles travel downcurrent at about half the mean speed of the current. Energy within the flow is dissipated by means of the roughness of and inertial drag by the particles, by the breaking of antidune waves, and by the creation and dissipation of sediment undulations. On the water surface, undulations are in phase with bed forms. No downcurrent sorting occurs; the size distribution of the sediment that travels on downstream is the same as that of the sediment composing the bed. These relationships are summarized graphically in Figures 7-23 and 7-24.

SOURCES: Simons, Richardson, and Albertson, 1964;
Simons, Richardson, and Nordin, 1965.

Selective Transport of Sediment

Flowing water transports sediments selectively with the result that sediments are sorted. We discuss three exam-

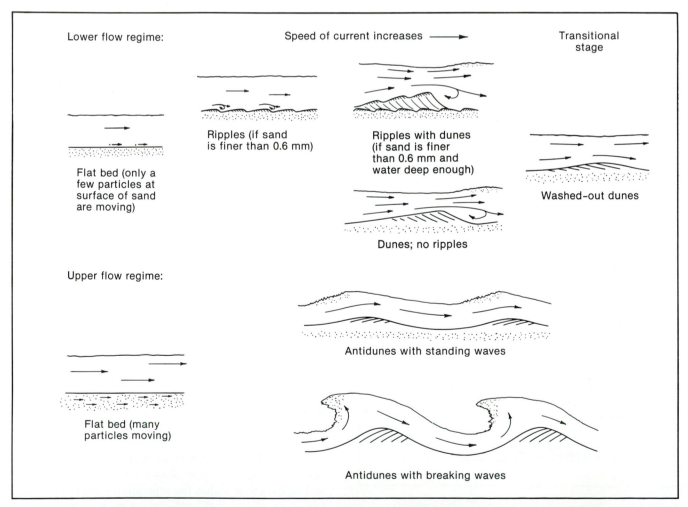

FIGURE 7-23.   Relationships of water and cohesionless sediment as related to flow regimes; schematic profiles of water and sediment in vertical plane parallel to current. Speed of current increases progressively from upper left to lower right. Further explanation in text. (Based on D. R. Simons, E. V. Richardson, and C. F. Nordin, Jr., 1965; and J. C. Harms and R. K. Fahnestock, 1965.)

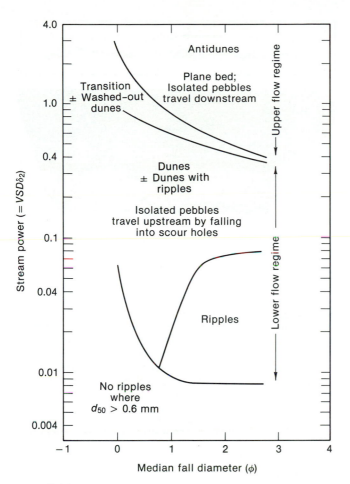

FIGURE 7-24.   Graph of stream power (expressed as a product of *V*, mean speed of flow; *S*, slope of water surface; D, depth; and δ_2, specific weight of water) vs. particle size showing boundaries within which various bed conditions prevail. Median diameter is abbreviated as d_{50}. (Replotted from D. B. Simons, E. V. Richardson, and C. F. Nordin, Jr., 1965, fig. 21, p. 52.)

ples: (1) lag concentrates, (2) concentration of fine sand, and (3) downcurrent decrease in particle sizes.

LAG CONCENTRATES

Lag concentrates form when the sediment in a current consists of many sizes, only some of which the current can move. Typical examples are numerous where till consisting of a wide range of particle sizes is reworked by water. The water takes away the sand and finer particles and leaves behind a residue of coarser particles, cobbles and boulders (if present in the till).

CONCENTRATION OF FINE SAND

Simple shearing experiments with various sand mixtures clearly demonstrate several important contrasts between fine sand and coarser particles. For example, shearing of a mixture of coarse sand and pebbles brings the pebbles

to the surface. The result is inverse grading. By contrast comparable shearing of a mixture of pebbles and fine sand creates the opposite result: the pebbles sink to the bottom. The result is normal grading. If the original mixture includes roughly equal parts of fine sand, coarse sand, and pebbles, a water current will segregate the fine sand. If the current is swift enough, ripples appear in the fine sand. In the coarser component, larger bed forms may appear. If so, then the result will be that coarse sediment in small dunes will migrate over rippled fine sand.

DOWNCURRENT DECREASES IN PARTICLE SIZE

A downcurrent decrease in particle size has been observed in many localities. Such changes can result from selective transport or progressive abrasion and breakage. In the gravels deposited by the catastrophic floods of Pleistocene age in eastern Washington, two key localities (Soap Lake constriction and Sentinal Gap) served as reference points and two kinds of debris, granite and basalt, constitute tracer materials. As a result, a diagram of particle diameter versus distance transported could be compiled (Figure 7-25). Downstream from Soap Lake constriction the median particle size decreases 50% every 6 km (1ϕ unit per 6 km). Downstream from Sentinal Gap the change is a 50% decrease every 1 km downstream (1ϕ unit/km).

SOURCES: V. R. Baker, 1973; C. W. Ellis, 1962.

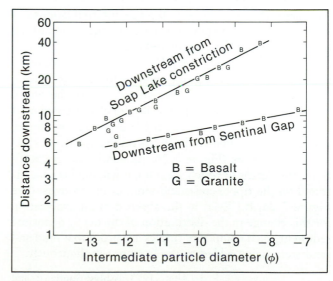

FIGURE 7-25.   Graph of decrease of particle size with distance downcurrent caused by catastrophic flood in eastern Washington as a result of the release of waters from ice-dammed Lake Missoula. Further explanation in text. (Replotted by J. E. Sanders from V. R. Baker, 1973, figs. 17 and 18, p. 28.)

Current/Substrate Interactions

In the sense used here, reworking excludes any deforming effects. If deformation takes place, the current is categorized as *depositing/deforming*. No insurmountable barrier separates these two categories of currents. In fact, while the shearing forces are still being applied by the current, the newly deposited sediment may be cohesionless at some times and cohesive shortly afterward. If the cohesionless sediment undergoes nothing but simple traction, then the current is defined as *depositing/reworking*. If the cohesive sediment is deformed, then the current is defined as depositing/deforming. Thus, as the circumstances within the bottom sediment change, the same current can shift from one category to the other and back again.

EFFECTS OF DEPOSITING/REWORKING CURRENTS

A **depositing/reworking current** is *a current that has slowed to the point where it has begun to deposit sediment, yet is still moving swiftly enough to be creating tractional effects in the particles it has deposited.* The main difference between an eroding/reworking current and a depositing/reworking current is that in the latter, sand-size sediment formerly carried in suspension falls out upon the bottom and builds the bed upward while the deposited sand is being tractionally shifted.

A depositing/reworking current may create plane laminae but its diagnostic product is superimposed ripples of a kind H. C. Sorby named "ripple drift with deposition from above," later generally referred to simply as *climbing ripples.* (See Figure 5-21, B.) The internal laminae of ripple-drifted sediment result from the interaction of the forward speed of the current and the ripples and the rate of upbuilding of the bottom. While the ripples are moving, the sediment falling out upon them becomes sorted. At times, very fine sand may spread across the entire rippled surface. At other times silt and clay become concentrated in the troughs or on the downcurrent slopes, whereas the very fine sand collects on the upcurrent sides and on the crests (Figure 7-26, A). Under other conditions, the details of which are not known, mud is concentrated on the ripple crests and sand, in the troughs. (See Figure 7-26, B.) Even in the absence of an explanation for this change in the distribution of the newly deposited sediment, such composite ripples are proof that mud and sand were being added to the bottom simultaneously.

SOURCES: J. R. L. Allen, 1973; Ashley, Southard, and Boothroyd, 1982; Friedman and Sanders, 1978; R. E. Hunter, 1977; Jopling and R. G. Walker, 1968; E. D. McKee, 1938, 1939; Sanders, 1963a; Sorby, 1859, 1908; Środoń, 1974.

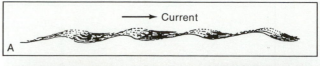

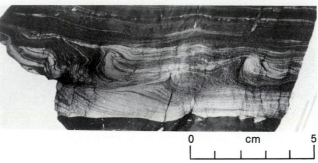

FIGURE 7-26.   Composite climbing ripples that grew upward by simultaneous additions of very fine sand and of mud.

A. Sand (stippled) concentrated on ripple crests, mud (black) in ripple troughs, schematic profiles in vertical plane perpendicular to the current. Persistence of this pattern through a thickness of more than a few centimeters can yield a large-scale "false bedding" having an inclination determined by the angle of climb of the ripples. (J. E. Sanders in G. M. Friedman, J. E. Sanders, and I. P. Martini, 1982, fig. 20, p. I-57, based on features exposed in cuts on U.S. Route 44-N.Y. Route 55, Highland, New York, west of Mid-Hudson Bridge.)

B. Changes during growth from very fine sand (white) on crests to mud (black) on crests; photograph of a large thin section prepared at Yale University by P. A. Scholle in 1963 from specimen of Martinsburg Formation (Middle Ordovician), cuts along N.Y. Route 17, 2 mi (3.2 km) NW of Goshen, New York. (J. E. Sanders, 1965, pl. IIIA, p. 204.)

EFFECTS OF DEPOSITING/DEFORMING CURRENTS: CONVOLUTED LAMINAE

As mentioned, a depositing/deforming current is one in which the shearing drag of the current deforms the newly deposited sediment. Recall that cohesionless sediment reacts to shearing by dilating and forming ripples. But if initially cohesionless sediment becomes cohesive, then its reaction to shearing is to be deformed. An extremely common result is for a rhythmic pattern of ripple-scale features to form and then to be deformed. Where the shearing affects a layer of cohesive mud over which fine sand is being transported and eventually starts to be deposited, then the result may be flame structures and lee-side cross-laminated fine sand. (See Figure 5-27.) Where newly deposited fine sand is rippled, the shearing may deform the ripple laminae, initially by simply causing the downcurrent-dipping laminae to become oversteepened (Figure 7-27). In other situations the deformed ripple laminae tend to be drawn upward into small anticlinal "folds" and in the adjacent depressions (the "synclines") cross laminae accumulate. When sand is deposited as cross laminae, it is cohesionless. But soon thereafter it may become co-

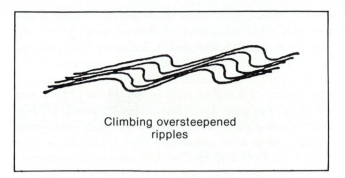

FIGURE 7-27.   Climbing oversteepened ripples; schematic profile in vertical plane parallel to current direction. (J. E. Sanders in G. M. Friedman, J. E. Sanders, and I. P. Martini, 1982, fig. 18, p. I-56.)

hesive, whereupon it responds to shearing by deforming. This is demonstrated by deformation of some cross laminae that formed early. A complicated succession of events may follow, consisting of alternate episodes of deposition of new cross laminae (cohesionless) and their subsequent deformation (sediment now cohesive). The earliest sets record the cumulative result of all episodes of deformation. The fact that the youngest set may not be deformed (Figure 7-28) at all proves that the deformation was taking place while the current was flowing, and that when the current finally stopped, the newly deposited sediment contained complexly deformed internal laminae. Because sediment was being added to the bottom while the deformation was in progress, any attempt to calculate "total layer-parallel shortening" of the convolute "anticlines" yields fictitious results. The lay-

ers were not deposited as horizontal sheets and afterward deformed (the assumption made in the "shortening" attempts). On the contrary, while the layers were being rippled and deformed they were accreting new sediment. The amount of mass in the upper laminae, therefore, may be very much greater than that in lower laminae. Figure 7-29 illustrates the appearance of convolute "anticlines" in outcrop.

SOURCES: J. R. L. Allen, 1982; Ashley and others, 1990; Blatt, 1982; Blatt, Middleton, and Murray, 1980; Davis, 1983; Fraser, 1989; Fredsoe, 1982; Graf, 1984; Kennedy, 1980; Nakato, 1990; Reading, 1986; Sanders, 1960, 1965; Van Rijn, 1984a, b, c; R. G. Walker, 1984.

Biological Processes

In sediments, many reactions are biochemical; that is, organisms drive these reactions. We begin with these biological processes and biochemical reactions. In a following section we shall concern ourselves with chemical processes. Some of the products of biological processes, such as skeletal debris and pellets, have already been discussed in Chapter 2. Some important chemical- and biochemical processes that affect the transformation of sediments into sedimentary rocks appear in Chapter 3.

This difference between biochemical- and chemical reactions may not seem of any consequence to beginners. But as soon as the subject of calcium carbonate in sediments is raised, one encounters conflicting

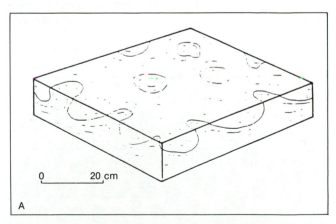

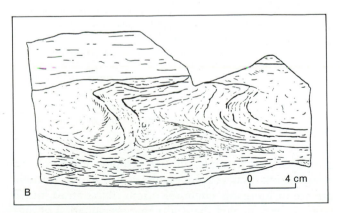

FIGURE 7-28.   Convoluted laminae sketched in three dimensions, and from a face perpendicular to axis of current.

A. Block diagram showing irregularly shaped "anticlines" and domes.

B. Mushroom-like "anticline" that grew upward, deforming, even overturning cross-laminated sand on its flanks. This feature is inferred to have been drawn upward by the interaction between two oppositely directed flow "tubes" whose long axes were parallel to the current. Compare Figure 7-9. At the time it was deposited, the cross-laminated sand was cohesionless, but it must have become cohesive almost immediately upon being deposited and thereafter responded to current-imposed shearing by deforming. (H. G. Davies, 1965; A, fig. 3, p. 312; B, fig. 2, p. 311.)

FIGURE 7-29. Small "domes" aligned along the crests of ripple-like ridges formed by tops of convolute "anticlines" in Devonian strata of Percé, Quebec, Canada.

A. General view of inclined bedding surfaces. (Length of sledgehammer is about 1 m.)

B. Close view of two "domes" nearest the end of the wooden handle of the sledgehammer in A. The quaquaversal dips of the laminae within the "domes" suggest that these features may be analogous to the larger-scale bed forms known as hummocks. (Authors.)

(2) subsequent destruction of these skeletons by predators and other destroyers of skeletal material to form various kinds of skeletal debris, including lime sand and -mud; (3) trapping and baffling by organisms of lime mud and its accumulation as laminae, beds, or mound-like deposits; (4) pelletization by organisms of lime mud into sand-size particles discussed in Chapter 2; (5) burrowing and stirring, hence mixing and even homogenizing of sediment, by animals in their search for food or shelter; and (6) activities of microorganisms that drive various chemical reactions, which promote precipitation of diverse minerals, including calcite, native sulfur, and pyrite.

SOURCE: Shinn and others, 1989.

Secretion of Calcium-Carbonate Skeletons by Organisms

Some living organisms deposit calcium carbonate as external skeletons, commonly shells, to protect and support their soft parts, or as internal skeletons to support their soft parts.

The precipitated skeletal calcium carbonate serves most organisms in ways other than as mere protection from predators; for example, it enables corals and coralline algae to inhabit an ecologic niche where motion of the turbulent surface waters is at a maximum. The patterns of the skeletal framework of corals are complex at several levels. One level is the shape of the colony as seen with the unaided eye (Figure 7-30). The great strength of coral skeletons is achieved by densely packing together individual *corallites*, which in turn possess great strength as shown by views through a microscope of the internal plates (Figure 7-31). Such strength en-

positions. For example, in the geologic literature are many claims that certain examples of calcium carbonate are of inorganic chemical origin. By contrast, other workers have denied that inorganic precipitation of marine calcium carbonate is significant. Evidently most calcium carbonate in sediments is secreted biochemically by organisms, but a substantial contribution from inorganic precipitation (*whitings*) has been confirmed.

In this section we take up the following kinds of biological processes: (1) metabolic activities of organisms that lead to the secretion of calcium-carbonate skeletons;

FIGURE 7-30. Skeleton of colony of the modern coral *Favia favus* from Gulf of Aqaba, Red Sea. Massive colony consists of individual corallites, each one of which was attached to the dark hollows having radiating plates called septa. Flat (lower) side of colony was cemented to sea floor. (Authors.)

Septa   Theca

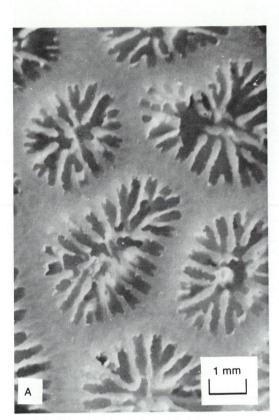

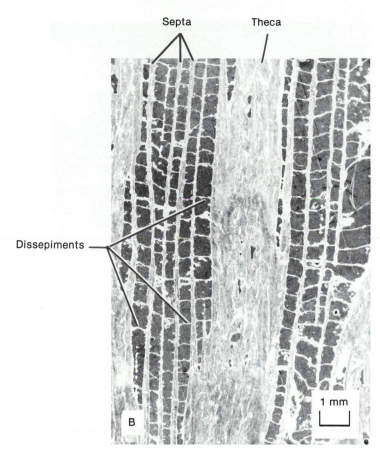

Dissepiments

FIGURE 7-31. Views of a modern coral colony in transverse- and longitudinal sections.
A. View through microscope looking down on top of individual corallites of the coral *Favia*, showing skeletons of two complete and several incomplete corallites. Massive partitions (*thecae*) join radiating septa secreted by individual corallites to form sturdy, but porous, stony mass.
B. View through a microscope of thin slice of coral colony cut at right angles to A. Seen in side view are two kinds of vertical partitions (septa and *thecae*), and one kind of horizontal partition (dissepiment), which reinforces septa and *thecae* to provide added strength. (Photo by G. Kuslansky, in G. Gvirtzman and G. M. Friedman, 1977, fig. 4B, p. 363.)

ables corals to build a rigid structure that may form hazards to navigation (Figure 7-32) and successfully defies the mechanical forces imposed upon it by the constant battering of the breaking waves, yet allows for exposure of maximum surface area to serve the physiological needs of the organisms.

In shallow seas, especially in tropical seas, many kinds of organisms secrete skeletons of calcium carbonate. Organisms that live attached to the bottom are referred to as *sessile*. Some of these, such as corals (Figure 7-33) and coralline algae, are veritable biochemical "factories" that catalyze a large-scale transfer of dissolved calcium- and bicarbonate ions from seawater into their solid calcium-carbonate biocrystalline skeletons. The ultimate thickness of strata of such skeletons may total thousands of meters. How marine organisms transfer ions from solution to lattices of solid crystalline calcium carbonate is not fully understood.

As we have seen in Chapter 2, the skeletons of organisms may consist of aragonite or of calcite, and the calcite may be of the high-magnesian- or low-magnesian variety. Chemical analyses of skeletal aragonites indicate that in these aragonites the proportions with respect to calcium of both magnesium and strontium may vary. Thus, among the aragonites we can recognize both high- and low-magnesian (Figure 7-34) as well as high- and low-strontium varieties (Figure 7-35). Different groups of organisms secrete *biocrystals* displaying these varieties.

How an organism secretes a biocrystal composed of a particular polymorph or chemical variety of a mineral is not well understood. The variables involved include (1) growth rate, (2) the presence of certain organic- and inorganic compounds, and (3) temperature. For instance, some invertebrates secrete calcium-carbonate biocrystals in association with an organic matrix. One suggestion is that the chemical structure of this organic matrix may

FIGURE 7-32.　Coral reef that proved to be a hazard to navigation. Liner Hey Daroma stranded on outer edge of reef, Gulf of Aqaba, Red Sea. Dark area in lower right is deep water; white area in lower left is surf from breaking waves. Patchy gray areas in top of photograph are areas of reef devoid of live calcifying organisms (not covered by water when this photo was taken). (Authors.)

determine which polymorph of calcium carbonate the organism precipitates. Further discussion of the importance of at least one organic compound, a long-chain or complex sugar, is contained in a subsequent paragraph. Temperature is important; in organisms which deposit both aragonite and calcite, the proportion of aragonite may be higher in the biocrystals precipitated by those animals that live in warmer waters.

In green- and red algae, which are important contributors to the total volume of calcium carbonate precipitated both in modern- and in ancient seas, calcification may be a relatively simple reaction. The waters in which these organisms live are supersaturated with respect to calcium carbonate. During photosynthesis the algae withdraw CO_2 and this reduces the capacity of the water to

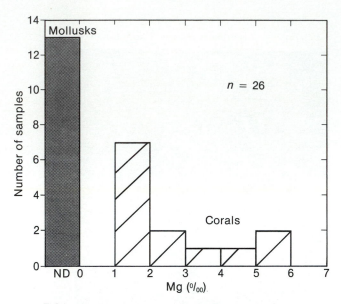

FIGURE 7-34.　Bar graphs showing magnesium contents of skeletal aragonite of specimens collected from Gulf of Aqaba, Red Sea. Corals secrete high-magnesian aragonite, mollusks, low-magnesian aragonite. ND below mollusk bar indicates level of magnesium too low to be detected by analytical technique employed. (G. M. Friedman, 1968, fig. 24, p. 913.)

hold Ca^{2+} and CO_3^{2-} and causes calcium carbonate to be precipitated. Even where photosynthesis is not involved, skeletal carbonate can be precipitated. Nevertheless, in modern oceans, photosynthesis may be among the most-important factors involved in the precipitation of calcium carbonate.

Photosynthesis is significant even in secretion of calcium-carbonate biocrystals by corals. Among the common modern corals, the *Scleractinia* or stony *hexacorals*,

FIGURE 7-33.　Underwater view seaward at edge of reef showing grove of coral colonies, Gulf of Aqaba, Red Sea. Darker patches at upper right and upper left are reflections from underside of water surface of water surface of coral colonies. (Photo by W. Halpert, in G. M. Friedman, 1968, fig. 18, p. 908.)

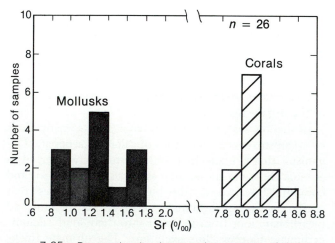

FIGURE 7-35.　Bar graphs showing strontium contents of skeletal aragonite of the same specimens shown in Figure 7-34. Corals secrete high-strontium aragonite; mollusks, low-strontium aragonite. Scale for strontium concentration is not continuous between mollusks and corals. (G. M. Friedman, 1968, fig. 23, p. 913.)

we can distinguish two kinds: (1) reef-building, or *hermatypic*, and (2) nonreef, or *ahermatypic*. Most hermatypic corals live symbiotically with unicellular dinoflagellate algae known as *zooxanthellae*. These algae are contained within cells in the endodermal epithelium of the corals. Algal photosynthesis is thought to influence the precipitation of calcium-carbonate biocrystals by zooxanthellate corals as follows. The algae remove CO_2 from the aqueous medium within the coral and cause the equilibrium reaction

$$Ca^{2+} + 2\ HCO_3^- \leftrightharpoons CaCO_3 + H_2O + CO_2 \quad (Eq.\ 7\text{-}8)$$

to go to the right. Other mechanisms may also be involved. *Zooxanthellae* actually move organic-carbon products of photosynthesis to coral tissues; impedance of this migration hinders calcification.

Photosynthesis by the *zooxanthellae* raises the concentration of intracellular oxygen. This in turn increases the rate and efficiency of the coral's metabolism. As metabolism increases, so does the rate with which soluble waste products are generated in the cells of the corals. For their photosynthesis and reproduction, *zooxanthellae* require those very substances, such as CO_2, nitrates, sulfates, and ammonia, that corals must eliminate. Indeed, *zooxanthellae* have been shown to carry on photosynthesis at such high rates that they not only use all the inorganic phosphate produced by the host coral, but in addition, they may even absorb phosphate from the surrounding seawater. This rapid *zooxanthellae*-driven waste removal from the host cells of the corals probably results in increased metabolic efficiency. In addition to their direct function in influencing the precipitation of calcium carbonate, *zooxanthellae* can be regarded as the combined intracellular lungs and -kidneys of the corals. In these two capacities they release still-more energy for precipitation. Other kinds of plants, such as filamentous green algae and cyanobacteria (blue-green algae), live in the skeletons of many corals. The total plant biomass in a live hermatypic coral head may exceed the animal biomass, although a recent estimate put the plant biomass at no more than 3% of the total polyp biomass. Still, metabolically at least, corals can be considered part plant. Interestingly, *zooxanthellae* often grow many times faster than their coral hosts, and excess symbionts are either expelled or consumed by the corals. No quantitative study of these phenomena has yet been made.

We note in passing that some zooxanthellate corals are not reef builders. Therefore, the common assumption that reef building and possession of symbiotic *zooxanthellae* invariably go together is not correct.

In the absence of *zooxanthellae* or in darkness corals are forced to rely on diffusion alone to get rid of their soluble inorganic metabolic waste products. Diffusion is a slow process, and coral metabolism is far less efficient without the zooxanthellate contribution. Nevertheless, in deep- and cold water some non-zooxanthellate corals build reefs. Hundreds of deep-water reefs have been found in the eastern Atlantic as far north as Norway. However, these deep-water reefs never become as large as some of the shallow-water reefs built by zooxanthellate corals.

Corals secrete tiny bundles of minute fibers, each of which is a single crystal of aragonite (Figure 7-36). The source of Ca^{2+} and some HCO_3^- ions for the secretion of calcium carbonate is seawater that contains these ions. Corals are able to absorb Ca^{2+} and HCO_3^- ions directly from seawater; they evidently derive little calcium from their food. In experiments, corals have been shown to secrete biocrystals even if they lack particulate food. Still other experiments prove that corals derive some HCO_3^- ions from their food. The proof consisted of finding, in the coral's skeleton, atoms of radiocarbon (^{14}C), which had come from mouse tissue that was fed to the polyps. The relative contributions of carbonate from seawater and from metabolism are still not known.

In absorbing ions from seawater, corals are not able to discriminate calcium from strontium. Hence corals deposit both elements in their skeletons in the proportions at which these elements are present in seawater. By contrast, most organisms that secrete biocrystals from calcium derived through the food chain, such as mollusks, are able to discriminate between these two elements. Because corals do not fractionate calcium and strontium, coral biocrystals are composed of high-strontium aragonite. Mollusks discriminate against strontium substitution for calcium in aragonite and thus synthesize low-strontium aragonite. (See Figure 7-35.)

After corals have absorbed Ca^{2+} from seawater, they transfer the ions to the sites of calcification by a series of processes that are poorly understood. Here Ca^{2+} is adsorbed on a complex organic matrix, a long-chain- or complex sugar, upon which the biocrystals are secreted. Certain organic molecules evidently form a structural template for calcium-carbonate precipitation, whereas other organic molecules actually cause crystals to nucleate and grow. At calcification sites Ca^{2+} combines with HCO_3^-; a possible series of reactions follows:

$$Ca^{2+} + 2\ HCO_3^- \rightarrow CaCO_3 + H_2CO_3 \quad (Eq.\ 7\text{-}9)$$

Actually, H_2CO_3 is mostly dissolved CO_2. Removal of CO_2 by algal photosynthesis is catalyzed by the enzyme *carbonic anhydrase*, as follows:

$$H_2CO_3 \rightarrow H^+ + HCO_3^- \quad (Eq.\ 7\text{-}10)$$

and

$$H_2CO_3 \rightarrow CO_2 + H_2 \quad (Eq.\ 7\text{-}11)$$

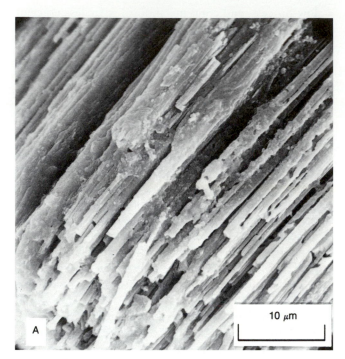

FIGURE 7-36.   Much-enlarged view of aragonite fibers, the fundamental unit of coral biocrystals, in faviid coral; scanning-electron micrographs.
  A. Longitudinal view.
  B. Transverse view exposing tips of fibers. (G. Gvirtzman and G. M. Friedman, 1977, fig. 3A, p. 362.)

Small changes in the speed of reaction of Eqs. 7-10 and 7-11 may accelerate or decrease the speed of reaction of Eq. 7-9. In Eqs. 7-10 and 7-11 algal photosynthesis fixes CO_2 and HCO_3^-, thus increasing the reaction rate of the system, provided other reactions do not limit the rate. Where *zooxanthellae* are absent, as in most ahermatypic corals, this reaction still takes place but carbonic anhydrase must act alone. Hence non-zooxanthellate corals and organisms secreting calcium carbonate in darkness do so at much slower rates than do the organisms that are aided by photosynthesis. Organisms that secrete calcium carbonate and that live symbiotically with *zooxanthellae* generally secrete biocrystals much faster and grow to larger sizes than do organisms lacking *zooxanthellae*, such as those living in deep water below the photic zone. The development of modern corals as important reef-building organisms during the Cenozoic Era may have been related to the evolution of this symbiotic relationship between some hermatypic corals and *zooxanthellae*.

The general principle of taking ions out of seawater to make calcium carbonate, as just discussed, applies to other organisms that secrete calcium carbonate, whether they do it in symbiosis with algae or not. Yet how other organisms, such as mollusks, secrete biocrystals is still imperfectly understood. The calcium-carbonate biocrystals of mollusk shells are enclosed in an organic matrix

that consists largely of amino acids and a long-chain- or complex sugar. The organic matrix consists of an insoluble portion, called *conchiolin*, and a soluble portion, consisting of acidic macromolecules rich in the amino acid *aspartic acid*. Conchiolin forms a porous framework in which the pores are filled by the acidic soluble organic material. This organic matrix (1) forms before the biocrystals and (2) probably controls biocrystal growth. The suggestion has been made that the atoms of the insoluble organic matrix may be arranged in one plane parallel to one of the planes of the calcium-carbonate crystals. The carbonate of the biocrystal would then grow on the matrix by a process of crystal overgrowth, the carbonate being a crystalline extension of the organic matrix (this is called *epitaxy*). The insoluble matrix probably controls the direction of growth of calcium carbonate and the shapes of biocrystals. By contrast, the soluble organic material may control (1) nucleation, (2) the polymorph to be precipitated (whether aragonite or calcite), and (3) the actual process of mineral growth. The mantle of the mollusk can vary the kind of organic matrix that it produces and thus can determine the kind of shell structure formed.

The amino acid aspartic acid constitutes as much as 30% or more of the soluble fraction of the organic matrix of the mollusk shell. Aspartic acid is also abundant in the organic matrix of gorgonian spicules ($> 50\%$) and of larval sea-urchin spicules. Macromolecules rich in aspartic acid preferentially bind Ca^{2+} and are involved in the precipitation of calcium-carbonate biocrystals by many, if not all, calcifying invertebrate groups.

If we examine the precipitation of calcium-carbonate biocrystals by mollusks on a larger scale, we see that carbonate moves to the mantle in the blood, which in mollusks is saturated with respect to calcium carbonate. The mollusk precipitates biocrystals in a fluid phase between the mantle and the surface of the shell. New shell material forms on a substrate of old shell; this older shell serves as a crystal template to direct the precipitation of calcium carbonate from a solution that the mantle tissues maintain. The mantle shows greater permeability to transferring ions on the side facing the shell than on the opposite side. Indeed, the mantle is quite remarkable in being more permeable to calcium than to sodium and potassium. Movement of calcium across the mantle proceeds by diffusion rather than by active transport. Temperature, salinity, nutrient levels, and age of the shell influence calcification. As with corals, the enzyme carbonic anhydrase is involved in the precipitation of calcium-carbonate biocrystals.

Phosphates act as crystal poisons and inhibit the growth of calcium carbonate. Phosphates, if present, attach themselves to surfaces of calcium salts and thereby interrupt the orderly crystal lattice and so disrupt and stop crystal growth. The shells of some animals, such as the modern brachiopod *Lingula*, consist of calcium phosphates. The mechanisms by which calcium phosphate becomes part of biocrystals are not known, but as with calcium-carbonate biocrystals, they do involve the presence of an organic matrix. The evolutionary sequence of biocrystals of most groups of shell-bearing invertebrates probably started with phosphatic compounds and continued with calcium carbonate. For example, the shells of most Cambrian brachiopods are phosphatic. From the Ordovician Period onward, the numbers and diversity of brachiopods having calcareous shells became predominant. As discussed elsewhere in this chapter, organic material can also inhibit the growth of calcium carbonate. Mg^{2+} also poisons the calcium-carbonate lattice. However, such poisoning only slows the rate at which calcium carbonate is precipitated; it does not prevent precipitation.

As we have read in earlier chapters, some organisms secrete siliceous tests, but the processes involved are not at present understood. Calcium carbonate and silica are the two materials most commonly precipitated as biocrystals. However, within the biocrystals of various modern-day organisms, some 60 minerals have been identified. In the future, no doubt, more will be discovered.

SOURCES: Coyle and Evans, 1987; Crenshaw, 1989; Darwin, 1896; Fagerstrom, 1987; G. M. Friedman and

Sanders, 1978; Kingsley, 1989; Leadbetter and Riding, 1986; Lowenstam and Weiner, 1989; Pirazzoli and others, 1987; Schuhmacher and Zibrowins, 1985; Scoffin, 1987; Watabe, 1989; Wilkerson and others, 1988; A. Williams, 1989.

Degradation of Calcium-Carbonate Skeletons into Skeletal Debris: Lime Sand and -Mud

Almost all calcium carbonate in the ocean has been secreted as biocrystals within the cells of organisms. Calcium carbonate functions as a skeleton and protective armor for these organisms. The soft parts of some organisms are potential food for other kinds of organisms. Predators have to crush the protective calcium-carbonate armor to get at the edible soft parts. Thus, in large part, predators are responsible for creating sedimentary debris out of skeletal materials. The shells of different groups of organisms differ in their resistance to destruction; the skeletons of different coral species even differ in crystal size and ultrastructure and therefore in solubility and mechanical strength. Ultimately, however, despite their protective armors, most animals in the sea are consumed by predators. The true "law of the sea" is eat or be eaten.

As we noted at the outset, most calcium carbonate in modern sediments occurs as skeletal debris. Even large modern coral reefs typically contain more than 90% of sand-size skeletal debris produced by degradation of the reef. As we read in Chapter 2, biting and boring convert the solid colonies of calcium-carbonate skeletons secreted by the reef organisms into an ever-increasing supply of skeletal particles that accumulate in the vicinity of the reefs. On reefs, the soft parts of the corals are draped in large part on the outside of the complex supporting framework. In order to get the edible matter, a fish has to bite off small chips of reef. Observations in modern reefs indicate that reefs are attacked by predatory fish of two kinds: (1) those that dart at the reef, nip off a chip, and withdraw; and (2) those that browse on the reef and remove mouthfuls of skeletal material. Both kinds of fish harvest the reef for food; they nibble off the calcium carbonate to get at the soft organic material attached to it. Marks of teeth may be recognized where the fish removed plant- or animal tissue from the calcium carbonate. With their crushing jaws, the fish separate the organic material and get rid of the indigestible calcium-carbonate particles by egesting them or, on occasion, by regurgitating them. As long ago as the middle of the nineteenth century, Charles Darwin demonstrated how important fish are in creating skeletal sand. He sliced open their intestinal tracts and found their guts full of carbonate skeletal particles derived from the reef. The numerous, beautiful schools of brilliantly colored grazing fish, such as parrot fish (See Box 2.4 Figure 4.), have been called the "cows" of the reef; perhaps "goats" would be a better analogy.

Fish are not the only organisms that attack reefs. During the late nineteenth century, in his classical studies of the reefs fringing the Sinai Peninsula, Johannes Walther observed that scavenging crustaceans create significant amounts of skeletal sand from reefs. Sponges of the genus *Cliona* bore into reefs. The boring activity of the sponge, combined with wave surge, leads to the formation of coarse fore-reef rubble in the deep water along the reef front. This organism dissolves about 2% of the reef's substrate, removing the rest as chips that are about 60–80 μm across. Huge volumes of sand-size- and finer detritus may be generated by the erosive activities of these organisms. In their destruction of reef or other calcium-carbonate rock, populations of *Cliona lampa* are capable of yielding 6 to 7 kg of lime mud per square meter of surface every 100 days (Figure 7-37). *Cliona caribbaea* removes at least 8 kg/m²/yr of substrate and occupies, on average, 5% of the surface in areas where it is abundant. This rate of erosion equals or exceeds the Holocene erosion rate of the subtidal pavement off Grand Cayman Island, a known habitat of this species.

In many organisms living in or out of reefs, edible organic matter may be distributed through the calcium-carbonate biocrystals or be protected within the biocrystalline skeleton. Predators can get the edible matter only by swallowing and crushing the calcium carbonate together with the edible bits and then digesting the organic matter and egesting the solid calcium carbonate.

In still-other examples, shells of calcium carbonate protect the soft parts and predators have to crack open the shells. Crabs and lobsters are able to open a pelecy-

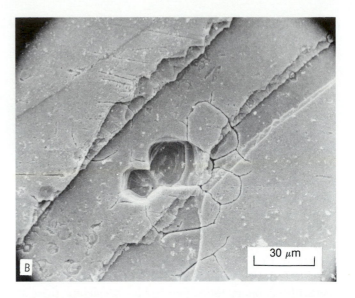

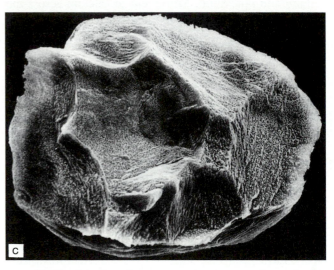

FIGURE 7-37.   *(Continued)*
   B. Much-enlarged view of a calcareous surface from which two chips have been removed by a boring sponge and in which four are partially etched; scanning-electron micrograph. (K. Rützler and G. Rieger, 1973, p. 149, pl. 2, fig. 3.)
   C. Much-enlarged view of a typical chip eroded by a boring sponge, showing microborer infestation; scanning-electron micrograph. (K. L. Acker and M. J. Risk, 1985, fig. 8, p. 709.)

FIGURE 7-37.   Destruction of calcium-carbonate particles by boring *Cliona* sponges.
   A. *Cliona caribbaea* attacking the coral *Siderastrea radians*. Dark area on left has been destroyed by sponge. Scale in centimeters. (L. Acker and M. J. Risk, 1985, fig. 6, p. 708.)

pod shell with a nutcracker-like claw, crushing the shell like a walnut. To make sure that it finds all edible matter, the crab or lobster cuts the broken segments of the shell into innumerable sand-size fragments.

Some sponges, algae, fungi, worms, echinoids, foraminifers, and pelecypods find sheltered living quarters by boring into solid material (Figure 7-38). The clam *Tridacna* excavates into coral reefs. The boring sponge *Cliona celata* is able to penetrate the shells of five-year-old oysters. Other organisms, such as some gastropods,

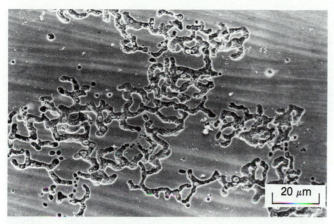

FIGURE 7-38.   Greatly enlarged view of a small sector of a mollusk shell from Bass Basin, southeastern Australia, that has been bored by an unidentified microorganism; scanning-electron micrograph. (W. M. Blom and D. B. Alsop, 1988, fig. 6J, p. 276.)

bore through the shells of other mollusks for purposes of predation. (See Figure 2-40.) Borings, especially microborings such as those made by cyanobacteria, commonly form centripetally, and the outer parts of shells are more intensely bored than are the interiors. However, bioerosion is a rapid process; a shell may become totally infested with boring organisms in as little as a few months. Although the precise mechanisms by which many of these organisms accomplish their boring is not known or only partially understood, such borings may substantially weaken the material that has been bored (Figure 7-39; see also Figure 7-37.) and render it more susceptible to chemical- and physical destruction than it would otherwise be. Boring organisms may locally strengthen skeletal material: the boring pelecypod *Lithophaga obesa* cements the walls of its borings with calcium carbonate that is more dense and therefore stronger than that of its coral host. In general, however, bored material is significantly weaker than nonbored material. Mechanical forces of wave surge finally disintegrate the bored skeletal material into sand- or mud-size particles. In agitated waters, physical- and biological processes of particle-size reduction coincide.

Microorganisms also attack large calcium-carbonate particles and thereby create large quantities of fine skeletal debris. As much as 350 g/m²/yr of calcium carbonate may be destroyed by microboring organisms in shallow-marine low-latitude settings. This may be a substantial fraction of the total sediment influx. Fungi, algae, and cyanobacteria bore into and thus cause the total disintegration of skeletal material. As one passes from tropical- to temperate latitudes, the abundance of boring cyanobacteria decreases whereas that of boring red- and green algae increases. Mollusk shells are more prone to attack than other shells, hence their skeletal parts are selectively removed. A possible reason for this selective destruction is that fungi may feed preferentially on the

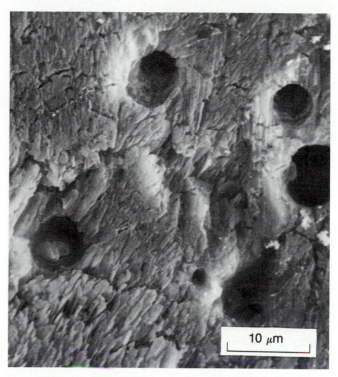

FIGURE 7-39.   Much-enlarged view showing that removal of aragonite from a cluster of cylindrical bores has weakened this modern faviid coral skeleton; scanning-electron micrograph. (G. Gvirtzman and G. M. Friedman, 1977, fig. 22, p. 378.)

organic matrices of the mollusk shells, which to them may be unusually tasty.

Even after it has been buried to a depth of as much as several meters, skeletal material can be broken down further. This results from the activities of burrowing organisms. In order to obtain food and shelter, some burrowers rework the sediments as much as several meters below the water/sediment interface.

Thus we see that biological processes are responsible for the secretion of calcium-carbonate biocrystals but that the resulting skeletal debris is subjected to physical breaking. These skeletons may be reef- or nonreef, colonies or solitary individuals. However, we must here repeat from Chapter 2 that lime mud also forms by at least two other processes: (1) the accumulation of tiny skeletal components secreted within the cells of certain algae and other lightly calcified marine organism, such as serpulid worm tubes, and (2) by inorganic precipitation in whitings. Therefore, lime mud is not solely the product of the destruction of skeletal material.

SOURCES: Barthel, 1981; Ells and Milliman, 1985; Farrow and Fyfe, 1988; Hixon and Brostoff, 1982; Nelsen and Ginsburg, 1986; Pestana, 1985; Scoffin, 1987, 1988; Scoffin and others, 1980; Scott and Risk, 1988; Shinn and others, 1989; Tudhope and Risk, 1985; Venec-Peyre, 1987; Young and Nelson, 1988.

Trapping and Baffling of Sedimentary Particles by Organisms

Other important biological factors in the accumulation of sediment are the trapping and baffling carried out by various organisms. Trapping, usually in shallow water, is best displayed by filamentous cyanobacteria. These organisms secrete mats or sheaths containing sticky organic matter, known as *mucilage*. Fine extraneous particles, usually lime mud that is washed across the mucilaginous mats, stick to the mats, in about the same way that dust or lint clings to exposed masking tape. After they have trapped and bound a sheet or lamina of extraneous particles, the cyanobacteria grow another layer of mat through and over the trapped- and bound particles. By repeating this trapping and binding of extraneous particles from suspension and this growing of mats over the bound or "glued" particles, these organisms produce a deposit characterized by millimeter-thin laminae (Figure 7-40). *A laminated, lithified deposit consisting of alternating layers of material formed by microorganisms, usually cyanobacteria, and of extraneous particles that have been cemented* is a **stromatolite**. Such laminated structures are common in ancient rocks (Figure 7-41).

After the organic matter of the cyanobacteria has decomposed, commonly the only indication of their former presence is the finely laminated calcium-carbonate sediment or limestone. As we read in Chapter 2, cyanobacteria also precipitate millimeter-thin laminae of solid calcium carbonate. In the present discussion, however, we are concerned only with their trapping- or baf-

FIGURE **7-41.** Stromatolite in a dolostone rock, Tribes Hill Formation (Lower Ordovician), Fort Hunter, New York. (M. Braun and G. M. Friedman, 1969, fig. 3, p. 117; G. M. Friedman, 1972a, fig. 5, p. 21.)

fling action on extraneous particles. These extraneous particles can consist of anything, but typically, fine carbonate particles predominate.

Stromatolites form many shapes, including domed cabbage heads (Figure 7-42). Stromatolites date well back into the Precambrian (ca. 3.5 billion years ago), were common throughout the Phanerozoic, and are known from modern environments where carbonate sediments are being deposited. In many parts of the world, stromatolites dominate thick successions of carbonate rocks.

Another kind of trapping is illustrated by *turtle grass* (*Thalassia*), which traps lime mud in the modern mud mounds of south Florida. In places in these modern

FIGURE **7-40.** Dug-out sample of laminated algal deposit, Abu Dhabi, Persian Gulf. Horizontal laminae, consisting of mucilaginous algal mats (dark layers) and trapped lime mud (light layers, more numerous at the base), are parallel to modern surface (at top). (Authors.)

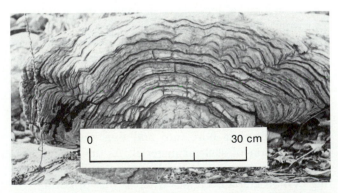

FIGURE **7-42.** Sectional view of algal stromatolite showing domed laminae known as cabbage-head structures, Hoyt Limestone (Upper Cambrian), Saratoga, New York. (R. W. Owen and G. M. Friedman.)

FIGURE 7-43.   Thicket of turtle grass on shallow sea floor, Great Bahama Bank. (Authors.)

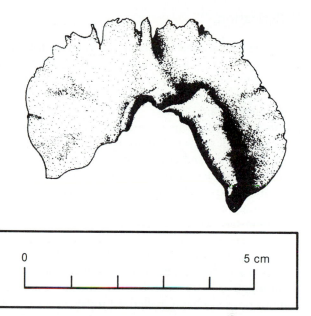

FIGURE 7-44.   Sketch of leaflike habit of modern brown alga. *Ivanovia*, a phylloid Pennsylvanian alga, is presumed to have resembled this leaflike shape. (L. C. Pray and J. L. Wray, 1963, fig. 8, p. 217.)

mounds, turtle grass is so abundant that it forms carpets that act as baffles (Figure 7-43). By reducing the speeds of currents, such baffles promote the deposition of lime mud. The intertwined roots and rhizomes create a protective skin that prevents erosion of the underlying sediment in the same way that terrestrial grasses stabilize the sand in dunes, or marsh grasses trap silt and mud on an intertidal marsh. Turtle grass may increase depositional rate by more than 60% as compared to similar but nonvegetated seabottom. If turtle grass meadows are killed by storms or by polluted water, then the trapped fine sediment is rapidly eroded.

In the same mud mounds, mangroves with their roots act as baffles by impeding the movement of suspended lime mud and, once the mud has been deposited, by preventing its subsequent removal. The lime mud accumulates to form elongate mud banks that stand 1.5 to 3 m above the floor of the bay.

Phylloid algae (phylloid means leaflike), such as *Ivanovia*, which colonized the so-called mud mounds of certain Pennsylvanian and Permian strata in the southwestern and central United States, probably are functional analogs of the modern turtle grass and mangroves. Phylloid algae were several centimeters tall, may have possessed appendages attached to the sea floor, and consisted of a single or of a limited number of broad "leaves" (Figure 7-44). Some phylloid algae (e.g., *Ivanovia*) were codiacean green algae and probably grew as upright plants, as do modern codiaceans. Others of these Pennsylvanian algae were rhodophytes (red algae) and, like many living rhodophytes, their growth habits were recumbent or encrusting. Despite their very different taxonomic affinities, the shapes of codiacean- and rhodophyte phylloid algae are similar. In ancient marine communities their ecological functions may have been the same as those of their modern counterparts. Phylloid algae were so abundant that they formed meadows,

carpets, and groves on the shallow sea floor. With their broad "leaves," these algae served as effective bafflers and trapped pelleted lime mud. At times, other attached organisms, such as crinoids, performed a similar function. Recumbent phylloid algae (rhodophytes) may have been sediment binders rather than -bafflers. To some extent the "leaves" of phylloid algae themselves formed deposits of coarse sediment, and phylloid algae therefore also functioned as the calcified alga *Halimeda* does in modern tropical seas.

At the water/sediment interface in estuaries and perennial streams in temperate and tropical regions, a mesh of intertwined plants prevents the movement of particles. Thus the physical laws governing the transport of particles, discussed in previous parts of this chapter, may be offset by effective baffling action of plants. Such offsetting illustrates the complexity or difficulty of interpreting the geologic record. In trying to compare the particle-size distributions of modern- and ancient sediments on the basis of physical sedimentation alone, one must be certain that no extraneous factors, such as baffling plants, influenced the distribution. In very ancient sediments, no plants existed in the depositional environment, which complicates comparison of these ancient rocks with modern sedimentary environments.

SOURCES: Almasi and others, 1987; Awramik, 1984; Coyle and Evans, 1987; Dill and others, 1986; Fagerstrom, 1987; Hine and others, 1987; Pentecost and Riding, 1986; Scoffin, 1987.

Pelletization

As we have explained in Chapter 2, deposit-feeding organisms pelletize soft lime mud. They eat the mud, pass it through their guts, pack it together, and excrete it as sand-size particles. The excreted pellets commonly contain much organic matter. The lengths and widths of pellets range from tens of micrometers (silt to very fine sand) for small snails to hundreds of micrometers (coarse sand) for worms and crustaceans.

After mud-size sediment has been squeezed and bound into sand-size particles, its physical behavior ceases to be that of mud and becomes that of sand. Once a body of lime mud has become largely pelletized, it is more accurately termed a sand than a mud. *Pelletization* greatly enhances permeability, and this change substantially increases the rates of many diagenetic processes.

Although we can observe that modern muds of all compositions are being pelletized today, nevertheless, in modern sediments and in ancient rocks, indurated pellets are found chiefly in carbonates and rarely in shales. As a mud composed of clay minerals is compacted on its way to changing into shale (Chapter 3), former pellets are destroyed. However, recent research has shown that clay fabrics, as revealed by X-radiography and SEM analysis, are materially and permanently affected by bioturbation. Shales that have been subjected to different degrees of bioturbation can be distinguished. Thus in the future it may be possible to apply similar techniques to identify formerly pelleted shales. Unlike mud composed of clay minerals, lime mud may be cemented by calcium carbonate after it has been pelleted and before it has been compacted. Even if the rock is subjected later to loads sufficient to cause compaction, the cement protects the pellets from deformation.

Burrowing

Many organisms, including worms, mollusks, crustaceans, and insects, burrow into sediment, finding food or shelter. Burrowing organisms are so numerous and their activities are so extensive that the effects of burrowing are important biological processes that affect sediments deposited in many geologic environments, both continental and marine. Burrowing accelerates weathering on land and destroys preexisting sedimentary structures.

A classical study of burrowers on land was published by Charles Darwin in 1881 (entitled *The formation of vegetable mould through the action of worms*). Darwin estimated that an acre of average land contains about 53,000 earthworms and that annually these worms collectively pass about 10 tons of soil through their bodies. They continually turn over the soil and spread fresh soil on the sur-

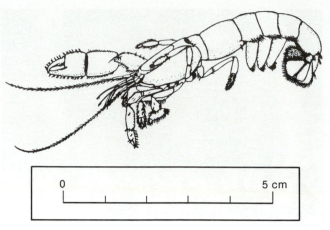

FIGURE 7-45.   Burrowing shrimp, *Callianassa*, from intertidal flats, Mugu Lagoon, California; note single appendage modified as a large digging organ. (M. R. Schmidt, in G. E. MacGinitie, 1934; Perkins, 1971, frontispiece.)

face at an average rate of 2.5 cm in 5 yr. Worms promote weathering not only by turning over the soil and hence transporting it to the surface where it is altered chemically, but also because by chewing and grinding down particles during ingestion they increase the surface area.

On intertidal flats and on the floors of shallow shelf seas, especially in tropical- and subtropical climates, the shrimp *Callianassa* (Figure 7-45) is the most-active burrower. It burrows to depths of 1 m or more. It lines the walls of its living quarters with tiny mud balls. Thus a longitudinal slice through the dwelling part of its burrow resembles a longitudinal slice through a cob of corn. The callianassid shrimp transfers particles upward that accumulate as conical mounds standing 6 to 50 cm above the bottom (Figure 7-46). At the top of the mound is

FIGURE 7-46   *Callianassa* mounds and pellets.
A. Conical mounds of sediment on exposed intertidal flat, pierced by exhalant holes of *Callianassa* burrows, Abu Dhabi, Persian Gulf. Lengths of black segments on painted rod are 10 cm. (Authors.)

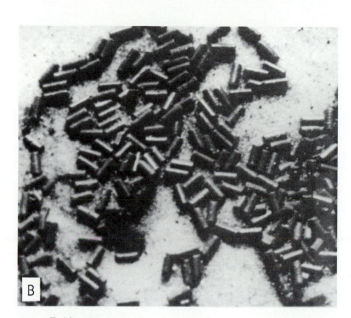

FIGURE 7-46.   (*Continued*)
B. Fecal pellets of the ghost shrimp *Callianassa major* on a Georgia beach. Bar scale = 1 cm. (R. W. Frey and R. A. Wheatcroft, 1989, fig. 3, p. 263.)

Distinct
(boundaries sharply defined)

Indistinct
(boundaries poorly defined)

FIGURE 7-47.   Schematic sketch of distinct- and indistinct mottled structures. (After D. G. Moore and P. C. Scruton, 1957, fig. 2, p. 2726.)

a small crater rimming the burrow's exhalant opening, from which the sediment issues when the shrimp pumps it up from below. Because freshly pumped-up sediment emerges from a reducing environment below, its color is dark gray. Placed 20 to 50 cm away from the mounds are funnel-shaped entrances, inhalant holes, that lead down into complex passages having diameters as large as 3cm. These shrimps build mounds in modern carbonate- and terrigenous sediments of the Gulf of Mexico, the Caribbean, the Red Sea, and other areas. Shrimp burrows and shrimp fecal pellets are known from rocks as old as Late Jurassic. Although the burrows found in ancient rocks have been named *Ophiomorpha*, the discovery of fossil remains in some of these burrows has shown that the burrowers were callianassid shrimps.

On the non-carbonate intertidal flats of northwestern Europe, the worm *Arenicola* makes burrows and mounds that are analogous to those of *Callianassa*. Other worms and some pelecypods are likewise active burrowers. *Fillings of burrows that result in discontinuous lumps, tubes, pods, and pockets of sediment randomly enclosed in a matrix of sediment having contrasting texture* are designated as **mottled structures** (Figure 7-47). Sediments containing extensive mottled structures are described as *mottled*. They are products of bioturbation (Chapter 5). Extensive burrowing homogenizes the sediment and destroys preexisting sedimentary structures. (See Figure 7-46, A.) Many ancient homogeneous rocks, particularly limestones, may be products, not of some uniform physical process, but instead, of the busy activity of burrowing organisms.

Burrows, as well as tracks and trails in sediments and rocks, are known as trace fossils, *lebensspuren* (German: trace of the presence of life), or *ichnofossils*. A valuable property of ichnofossils is that in most cases they cannot be transported away from their sites of origin. Because they are part of the sedimentary fabric, they may be destroyed as the sediment is reworked. Ordinarily they are not moved. However, some ichnofossils are more likely to be preserved in the sediments from some environments than in the sediments from others. This introduces a bias into the trace-fossil record. For example, waves usually destroy traces formed on the sediment surface in an intertidal zone. By contrast, the chances are good that deep burrows formed in the sediments deposited in low-energy settings will be preserved.

To facilitate comparison between bioturbated strata, systematic estimates of the intensity of bioturbation can be made. To this end, many classifications of bioturbation textures have been proposed (Figure 7-48). Different environmental settings, particularly those with varied rates of deposition, are characterized by different kinds and numbers of trace fossils. Accordingly, the characterization of burrowing intensity in ancient rocks may assist in environmental interpretation. In addition, because certain kinds of trace fossils are restricted to particular environments, assemblages of trace fossils that are called *ichnofacies* may also yield helpful clues in environmental interpretation. We return to this subject in Part V.

Burrows may provide information on the depositional history of strata. Where the bottom builds upward slowly, organisms can rework the sediment completely and homogenize it. By contrast, where the bottom builds up rapidly, burrowing organisms keep pace by moving upward in the newly deposited sediment. The time available may not be sufficient for extensive bioturbation. Where individual layers accrete upward very rapidly, their upper parts may be bioturbated and their lower parts nonburrowed (Figure 7-49). In some strata,

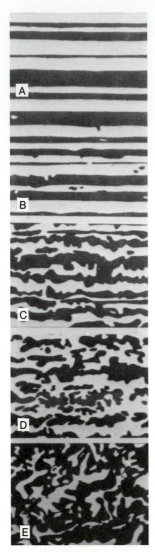

FIGURE 7-48.   Schematic diagrams of ichnofabric indices 1 through 5. Ichnofabric index 1 = no bioturbation; successively higher index numbers correspond to greater degrees of bioturbation. Index 6, total- or near-total homogenization, is not illustrated. (M.L. Droser and D.J. Bottjeg 1986, fig. 1, p. 558.)

   A. Ichnofabric index 1.
   B. Ichnofabric index 2.
   C. Ichnofabric index 3.
   D. Ichnofabric index 4.
   E. Ichnofabric index 5.

FIGURE 7-49.   Bioturbated light greenish gray nannofossil chalk, the upper part of an abyssal turbidite, overlain by dark greenish gray claystone. The chalk was deposited so rapidly that the ubiquitous abyssal burrowing organisms were not able to homogenize its lower part. However, abundant and varied burrows in the upper part of the nannofossil chalk, filled with darker claystone, attest to partial bioturbation of the top of the chalk during deposition of the basal portion of the overlying pelagic clay. The claystone is massive, showing no vestige of lamination, a result of intense bioturbation of this slowly deposited unit. Middle Miocene, Ocean Drilling Program Leg 123, Site 765, section 765C-11R-4. (The interval shown alongside the meter stick from 124 to 135 cm is referenced to the top of the section at 450.5 m below the sea floor in 6000 m of water). (Leg 123 shipboard scientific party, 1988, fig. 3B, p. 204.)

large numbers of chimneylike burrows of callianassid shrimps have been horizontally truncated. In contrast, others have not been truncated and extend above- and below the surfaces of truncation. The latter were formed after the truncation (erosive) event.

SOURCES: Aller, 1982; Cuomo and Rhoads, 1987; Darwin, 1881; Ekdale and others, 1984; Frey and Pemberton, 1983; McCall and Tevesz, 1982; O'Brien, 1987.

Effects of Plants

The biochemical effects on sedimentation of macroscopic plants are limited in the marine realm, but can be striking in continental settings. Soils are formed by the action of organisms, including macroscopic plants.

Growing plant roots mechanically break apart rocks along preexisting fractures. In addition, plant roots form avenues along which meteoric water can enter the soil. If chemical conditions are such that calcium carbonate can be precipitated, rhizoliths, cylindrical structures of well-cemented sediment (Figure 7-50), are formed. This cementation is promoted both by the introduction of meteoric water into the sediment and by the growth of algae, fungi, and bacteria in the hospitable environment created by the plant roots. Cementation may proceed much faster in the immediate vicinity of roots than in the surrounding sediment. This introduces heterogeneity into the physical- and chemical properties of the soil.

Decay of plant material generates organic acids that are capable of dissolving minerals through a process known as *chelation*. Chelating agents effectively remove ions such as Mg^{2+}, Fe^{2+}, and Ca^{2+} from solution by surrounding and bonding to them to form complexes that remain in solution. The organic acids "sequester" these cations and keep them from being "seen" by the water. Thus additional cations may enter solution and more mineral matter dissolves.

SOURCES: Almasi and others, 1987; G. M. Friedman and Sanders, 1978; Hine and others, 1987; Leadbetter and Riding, 1986.

Effects of Microorganisms

Actual fossil remains of microorganisms, such as bacteria, have been recognized in Precambrian rocks more than 3 billion years old. Microorganisms have been credited as effective geologic agents in a variety of chemical processes. Their influence is still not clearly understood and generally it is not possible to distinguish reactions in which microorganisms are involved from those which are purely chemical. What is more, we usually find the products of reactions in which microorganisms were important rather than finding the microorganisms themselves. In many chemical reactions at ordinary temperatures, bacteria function as catalysts. Most of these reactions, if purely inorganic, would require not only the passage of considerable time but also the addition of considerable energy in the form of heat, pressure, or ultraviolet light. Bacteria can catalyze chemical reactions at burial depths as much as 140 m below the sea floor.

Weathering of rocks is in part caused or accelerated by bacteria. Other simple organisms, such as fungi, algae, and lichens, also participate in weathering processes. Soils contain dense populations of bacteria. Minerals, such as certain feldspars and micas, decompose at least twice as rapidly in sediment in which bacteria are active as in sterile sediment that lacks bacteria. Fungi and **heterotrophic bacteria**, defined as *bacteria that obtain their*

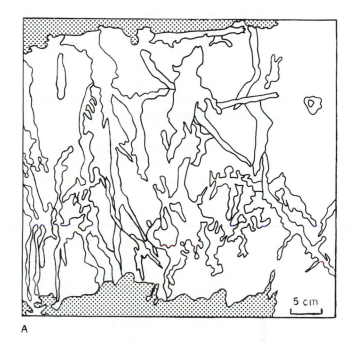

A

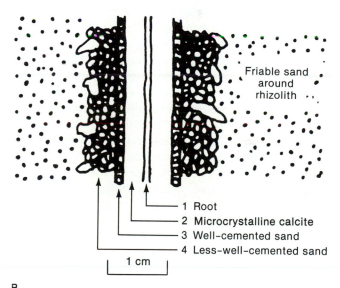

B

FIGURE 7-50. Rhizoliths, calcified tubular structures that form in soils around plant roots. (After B. Jones and K.-C. Ng, 1988, figs. 2 and 3, p. 458.)

A. Sketch of rhizoliths exposed in cross-sectional view showing downward branching patterns and variable angles of rhizoliths.

B. Rhizolith showing the four constituent zones (1–4), schematic diagram. Zone 1, which is a tube occupied by a root, is not always present. Where zone 1 is absent, zone 2 forms the entire core of the rhizolith.

energy through the oxidation of organic matter, produce simple organic acids (oxalic acid, acetic acid, lactic acid, and keto-gluconic acid). During weathering, these acids decompose feldspars and other silicate minerals. **Autotrophic bacteria**, defined as *bacteria that obtain their* *energy through oxidation of inorganic matter*, produce inorganic acids (H_2SO_4, HNO_3) that dissolve limestone and some other geologic materials. By introducing CO_2 and various organic acids, all of which promote the breakdown of solid rock to masses of particles, bacteria acidize the soil. Boring algae degrade only limestones, but the kind of acid involved and the exact mechanism of degradation are not known.

Among the most-important geologically active bacteria are sulfate reducers, especially *Desulfovibrio*. These bacteria are involved in the formation of H_2S, calcium carbonate, native sulfur, and pyrite, and in the reduction of the sulfate ion in formation waters (Chapter 3).

Bacteria, which may induce or facilitate the precipitation of calcite, have been implicated in the formation of marine calcitic peloidal cement. Also, magnesian-calcite- and aragonite cements grow in the tiny borings made in calcium-carbonate shells by bacteria, algae, and fungi. This cement growth is probably encouraged by the decay of organic material within these tiny holes, creating a microenvironment favorable for precipitating calcium carbonate. In the waters of the modern Dead Sea, calcite forms by bacterial action at the expense of gypsum. In this process H_2S is given off. In the early 1890s, the doughty pioneer-geologist M. Blankenhorn reported that gypsum, although precipitated continuously from the surface waters of the Dead Sea, is deposited and preserved only in shallow waters where the supply of oxygen is ample. Actually, because bacterial reduction of the sulfate cannot keep pace with the precipitation of gypsum, some amounts of gypsum are preserved in the Dead Sea at all water depths. Below the wave-influenced zone of oxidation, sulfate-reducing bacteria (which flourish in the absence of oxygen) break down calcium sulfate, and as a result, calcite forms. The mechanism for this biochemical reaction depends on the ability of the bacteria to extract oxygen from the sulfate in gypsum. With this oxygen they are able to oxidize organic matter and thus to produce energy. When the HCO_3^- produced in this bacterial oxidation of organic matter combines with calcium from the decomposed calcium sulfate, calcite forms, as is shown in Eq. 3-10. Certain nonfossiliferous carbonate sediments in modern seas and some similar ancient limestones have been attributed to this biochemical reaction. Examples include modern deep-water carbonate sediments of the Red Sea, deep-water limestones of Permian strata in Texas, Devonian shelf carbonates of western Canada, limestone caprock atop salt plugs in the Gulf Coast area of the United States, and calcite cement that forms at the oil/water contact of hydrocarbon reservoirs.

In the absence of solid calcium sulfate, this reaction can proceed from dissolved calcium- and sulfate ions in seawater. In the laboratory, sulfate reduction by *Desul-*

fovibrio is usually studied by using soluble compounds, and the bacteria attack the sulfate ion directly in solution. Although, as we shall see in the section on chemical processes, chemical precipitation of $CaCO_3$ in the ocean, where Ca^{2+} and HCO_3^- are buffered, requires a high pH, bacteria living on the degradation of organic compounds, such as proteins, sugars, or their acids and salts, precipitate $CaCO_3$ at pH values as low as 7.0.

No one who has visited The Netherlands or the city of Venice at the height of summer can fail to be impressed with the stench of H_2S emanating from the canals where the sulfate-reducing bacteria attack the inorganic sulfates present in the sewage. In south Florida, the Bahamas, or the algal flats of Abu Dhabi in the Persian Gulf H_2S rises when pits are sunk into the sediment or when interstitial water is pumped to the surface from the shallow-water environments where carbonate sediments are accumulating. When salt-bearing strata or deep-water limestones as old as Paleozoic are freshly broken, H_2S is emitted. H_2S can evolve on the sea floor from gypsum and erupt violently. Such H_2S has traveled in the air as far as 65 km inland from offshore exhalations near the coastal town of Swakopumund, Walvis Bay, in southwestern Africa. There, every few years, a stretch of sea bottom 320 km long and 40 km wide generates H_2S. In 1951 the exhalation lasted several months. The atmosphere resembled "a London fog, metal work turned black, public clocks were blotted out by deposit, thousands of fish [were] strewn on beach (*sic*), [and] sharks came into surf (*sic*) gasping on the evening tide" (newspaper report quoted by K. R. Butlin, 1953). The bottom sediments of this area yielded pure cultures of sulfate-reducing bacteria.

Reduction of gypsum or of anhydrite by sulfate-reducing bacteria forms native sulfur. Examples of such sulfur have been found in modern salt lakes and playas or in caprocks atop salt plugs.

In lakes as well as in shallow- and deep marine environments, for example, in the Black Sea, bacteria reduce sulfates to sulfides; the sulfides occur as black, finely disseminated iron monosulfides and impart a black color to the sediment. Below the water/sediment interface, monosulfide changes later to pyrite. Bacterial iron reduction in anoxic marine sediments can also result in formation of siderite, which is important to paleomagnetic studies. Bacteria also reduce manganese and nitrates, and generate methane gas.

Some evidence exists that bacteria not only dissolve quartz and silicate minerals but transfer and precipitate amorphous silica.

Thus we see that a variety of sedimentary material results from bacterial activity. We emphasize that most of this material is involved in the sulfur cycle.

Even so-called "inorganic" precipitation of calcium carbonate in the marine realm, as in the formation of aragonite-, calcite-, and magnesian-calcite cement crystals in marine sediments, is affected by organic materials. Organic matter rich in aspartic acid, which is the most-abundant amino acid in skeletal protein, preferentially binds to calcite mineral surfaces. This process may be akin to the way in which the molluscan mantle, which is organic, controls precipitation of its calcium-carbonate biocrystals. However, adsorption of organic matter to crystal surfaces commonly impedes crystal growth by occupying bonding sites for Ca^{2+} and CO_3^{2-} ions. Interestingly, the concentration of amino acids in sediments is essentially reduced to zero at a depth of a few meters below the sea floor. This implies that aspartic-acid poisoning of crystal surfaces would not be effective at greater depths in the sediment. In sum, the inorganic chemical reactions described in the next section may be greatly affected by biologic materials and processes.

SOURCES: Aston, 1980; Butlin, 1953; Chafetz, 1986; Cohen and others, 1984; Ellwood and others, 1988; G. M. Friedman and Sanders, 1978; Skyring and others, 1983.

Chemical Processes

In this section we discuss various kinds of chemical processes that take place in solutions within sediments, especially processes that influence pH and Eh. In addition, we shall study weathering as a chemical process, certain reactions of unknown or uncertain mechanism, and chemical processes leading to the formation of certain minerals, such as gypsum, zeolites, and feldspars. Some aspects of pH and Eh in sedimentary environments were described in Chapter 3, to which the reader should refer if a review of these subjects is needed.

Movement in Solution

Chemical changes take place in solutions. Examples of these solutions include waters in lakes and rivers, runoff on the ground, waters in the sea, and formation waters in the pores of sediments. If one examines carefully the effect of purely inorganic chemical reactions in these solutions, one is forced to conclude that the importance of such reactions may be minor. Some natural waters are full of organisms. Simple organisms participate in most chemical changes that occur in natural waters. Changes in pH in waters, which in turn influence dissolution and precipitation, are largely regulated by biological processes. As we read in Chapter 3, physicochemical changes become more important after sediments have been deposited because of the reactions that can take place between the particles and contained waters. Thus parts of the rock record may be erased by dissolution. Examples of dissolution can be found along stylolites in limestones (See Figure 3-7.) or in sediment turned into rock where, as cements are precipitated, dissolution commonly also proceeds. Even the formation of cements, through the control of pH, may be directly or indirectly influenced by simple organisms. And in deeply buried strata, formation waters are impoverished in the sulfate ion because of the activities of sulfate-reducing bacteria. These bacteria attack and reduce sulfate ions in solution and, as shown in Eq. 3-10, cause calcite cement to form at oil/water contacts. Here we are treading on ground that is in a state of flux at the research level. Yet, wherever we turn, it seems that biochemical processes, considered here as biological processes, are more important than purely inorganic reactions. Purely inorganic chemical reactions can take place only where simple organisms are totally absent. At the surface of the Earth, environments devoid of such organisms are not common. Therefore in this chapter we have discussed biological processes before inorganic chemical processes. Purely inorganic processes include the formation of evaporite minerals and the precipitation of zeolites and feldspars. Even here we may predict that, as research progresses, challenges will come. Sooner or later, some of these reactions may be discovered to result in part or in full from biological processes. As an example, the formation of manganese nodules in the depths of the sea has generally been considered to be an inorganic chemical reaction. However, this concept has now been challenged by those who claim that nodules form by bacterial action or are stromatolitic in origin.

The pH of river waters ranges from generally weakly alkaline to weakly acid, whereas the pH of surface seawater is weakly alkaline and tends to be nearly constant at a value of about 8.3.

To explain the function of pH in creating changes within sediments, we shall discuss the effect of pH on the dissolution and precipitation of calcium carbonate. Although the distribution of calcium carbonate results principally from biological processes, calcium carbonate is sensitive to changes of pH. To understand the effect of pH on the dissolution and precipitation of calcium carbonate, we have to say a few words about the carbonate-seawater system. Near-surface tropical and subtropical seawaters are supersaturated with respect to calcium carbonate, yet at a pH of 8.3 inorganic precipitation does not occur. The calcium carbonate-seawater system is generally out of equilibrium. Experimental evidence suggests that carbonate particles adsorb an amorphous organic coating that protects them from dissolution in seawater. This adsorption occurs at a rate faster than carbonate can be dissolved. In an experimental solution of HCl having a pH as low as 6.0, this same organic coating protects even the finest suspended carbonate particles from being dissolved. As we shall see in the next section, in waters deeper than 5,000 m, where dissolution does occur, if they are sufficiently protected with such a coating, $CaCO_3$ particles persist.

SOURCES: Aller, 1982; Aston, 1980; Bathurst, 1983; Berner, 1989; Decima and others, 1988; Demaison and Moore, 1980; G. M. Friedman and Sanders, 1978; Garber and others, 1987; Hunt, 1979; Morse and Mackenzie, 1990; Olausson, 1980; Sansone and others, 1988; Scoffin, 1987; Tribble and others, 1990.

DISSOLUTION OF CALCIUM CARBONATE, A FUNCTION OF pH

As already discussed in Chapter 3, when the pH of waters decreases, calcium carbonate dissolves. In warm, shallow seas at a pH of about 8.3, almost no inorganic processes dissolve aragonite- and calcite particles. However, since the classical studies of the Challenger expedition in the last century, it has been known that the proportion of calcium-carbonate particles in sea-floor sediments decreases as depth of water increases (Table 7-3). At depths between 4000 and 6000 m, this decrease is particularly rapid. Although the reasons for this decrease have been debated, the evidence suggests that calcium carbonate dissolves because the CO_2 concentration increases with depth. The control on CO_2 appears to be in part biological; it results from biological oxidation of organic-carbon compounds. Also, the water masses at greater depth were derived from the polar region; their temperature is lower and the water contains more dissolved CO_2. Increased concentration of CO_2 is in turn reflected by lower pH,

which leads to calcium-carbonate dissolution. However, increase of pressure with depth may also be involved; such increase affects the dissociation of carbonic acid. (See Eqs. 3-7 and 3-8.) The depth at which calcium-carbonate concentration in the sediment decreases most rapidly is known as the **carbonate-compensation depth**, defined as *the depth at which the rate of dissolution of solid calcium carbonate equals its rate of supply*.

The progressive decrease in calcium carbonate and also the change in mineral facies with increasing depth can be illustrated by considering the sediments deposited on a sloping sea floor in a tropical region where depth ranges from 1000 m to 8000 m (Figure 7-51). At depths between 1000 and 2500 m, the tests of pteropods, which consist of aragonite, predominate in fine-textured sediments; the proportion of calcium carbonate may reach 90%. At about 2700 m aragonitic pteropod tests begin to disappear. At 3600 m, they are practically absent. In some seas, such as the Red Sea, pteropods disappear from the bottom sediments at water depths of less than 1000 m. Their former presence is indicated by internal molds of pteropods in calcite-rich sediment. The original aragonitic shell has been dissolved (Figure 7-52); transitional stages indicating progressive dissolution range from unaffected to corroded and partially dissolved tests. As the tests of pteropods disappear, those of *Globigerina*, which consist of low-magnesian calcite, become dominant. At the pH range prevailing below 4000 m, calcite is less soluble than aragonite. At a depth below 4000 m, and especially below 6000 m, so much calcium carbonate has been dissolved that only insoluble siliceous material collects as brown clay on the floors of the deep ocean. Although it is likely that in the geologic past the carbonate-compensation depths differed from those in modern oceans, evidence that tests have been partially dissolved at the sediment/water interface has been used for indicating deposition below an ancient carbonate-compensation depth. From this inferred water-depth indicator, subsidence rates of ancient ocean bottoms have been computed. These criteria have been supplemented by lithologic- and faunal evidence that does not depend on dissolution depths.

Most shallow continental fresh waters are not saturated with calcium carbonate and so readily dissolve it. As we saw in Chapter 3, aragonite is more prone to dissolution than low-magnesian calcite, a behavior which parallels that in the depth of the ocean, where aragonitic tests of pteropods dissolve more readily than calcitic tests of *Globigerina*. Carbonic acid (See Eqs. 3-6 and 3-7.) is the most-common geologic solvent of calcium carbonate. The CO_2 of carbonic acid comes mostly from the bacterial decay of dead plants and other organic matter and from respiration of living plants. Organic acids, such as lactic acid derived from rotting vegetation, and

TABLE 7-3. Decrease of calcium carbonate in bottom sediments with increasing depth of water

Depth of water (m)	Proportion of $CaCO_3$ in bottom sediments (weight percent; mean)
1–1000	86.0
1000–2000	66.9
2000–3000	70.9
3000–4000	69.6
4000–5000	46.7
5000–6000	17.4
6000–7000	0.9
> 7000	0.0

SOURCE: Data from *Challenger* expedition, J. Murray and R. Irvine, 1891.

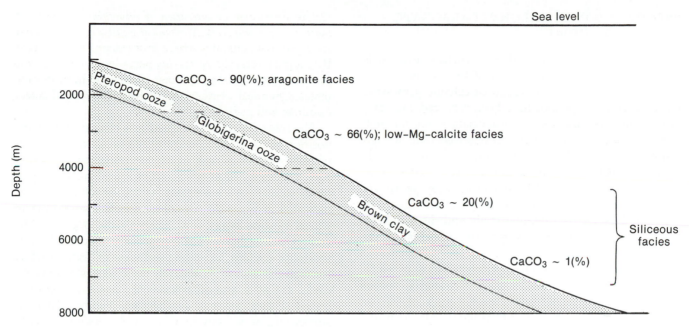

FIGURE 7-51. Generalized profile (schematic) of a portion of a deep-sea basin in the tropics showing progressive disappearance of calcium carbonate in sediments with increasing depth of water. Metastable aragonite of pteropod shells disappears at a shallower depth than does the more-stable low-magnesian calcite of *Globigerina* tests. Resistant siliceous material persists at greatest depths.

sulfuric acid produced by the weathering of sulfide minerals, especially pyrite, accelerate dissolution of calcium carbonate. Such dissolution increases the concentration of calcium carbonate in the waters; ultimately, calcium carbonate may be precipitated as in stalactites or stalagmites in caves or, as discussed in Chapter 3, as cement between particles.

$CaCO_3$ dissolution may be written

$$CaCO_3(c) \rightarrow Ca^{2+} + CO_3^{2-} \qquad \text{(Eq. 7-12)}$$

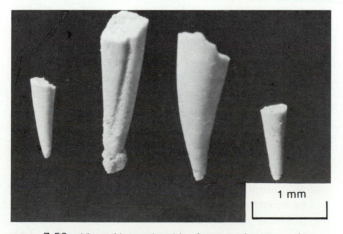

FIGURE 7-52. View of internal molds of pteropod tests, modern sediments underlying floor of Red Sea at depth of about 1500 m. (Photo by L. V. Rickard, in G. M. Friedman, 1965, pl. 1, p. 1192.)

The carbonate ions react with H^+ to produce bicarbonate ions:

$$CO_3^{2-} + H^+ \rightarrow HCO_3^- \qquad \text{(Eq. 7-13)}$$

This reaction upsets the equilibrium of Eq. 7-12 and more calcium carbonate goes into solution. As a consequence of this same reaction, H^+ is used up and more dissolved CO_2 reacts with water to produce more carbonic acid, which again drives the reaction. (See Eqs. 3-6 and 3-7.) As more carbonic acid forms, the equilibrium between CO_2 partial pressure of air and water is upset and CO_2 diffuses from air into the water; this leads to further dissolution of $CaCO_3$.

The reaction governing the dissolution of $CaCO_3$ may be simplified as shown in Eq. 3-20.

Chapter 3 demonstrates the importance of this reversible reaction in the dissolution of calcium carbonate, specifically that of the more-soluble aragonite and the precipitation of the less-soluble calcite. This reaction is one of the most important at the surface of the Earth because it involves the process of making sedimentary particles into strata of rock.

SOURCES: Aller, 1982; Bathurst, 1983; Boggs, 1987; G. M. Friedman and Sanders, 1978; Morse and Mackenzie, 1990; Scoffin, 1987; Vacher and others, 1990; Walter and Burton, 1990.

PRECIPITATION OF CALCIUM CARBONATE, A FUNCTION OF pH

As mentioned above, although near-surface seawater is supersaturated with respect to $CaCO_3$, organic compounds prevent the precipitation of calcium carbonate. Once organic compounds have been removed, the reaction equilibrates and calcium carbonate is precipitated. But the organic compounds are removed only at high values of pH, so pH controls the precipitation of calcium carbonate. Calcium-carbonate precipitation from seawater is also inhibited by magnesium, by phosphate adsorption, and to a lesser degree by the sulfate ions in solution. Despite considerable study and discussion by geologists, the nature of this inhibiting effect is not yet understood.

Biological processes, such as photosynthesis, in which CO_2 is withdrawn, raise the pH; in bacterial reduction of sulfate the pH may reach values of 9.5 or higher. Although increase in pH is induced biologically, precipitation of calcium carbonate is probably an inorganic chemical reaction. Our information on the pH requirements of the precipitation of calcite is in part based on observations of the effects of the replacement of quartz by calcite. Whereas calcite is insoluble in alkaline solutions having pH greater than 9, in such alkaline solutions silica dissolves. (See Figure 3-8.) Thus from petrographic examination of thin sections and from experimental work on the solubility of amorphous silica, quartz, calcite, and organic compounds, we gain information on the requirements for the precipitation of calcite in sediments. A pH level of 9 to 10, resulting from biological processes, appears to be a necessary condition for the simultaneous dissolution of silica and precipitation of calcium carbonate in marine waters. Such pH levels have been measured in Red Sea reefs. In Chapter 3 we dealt with the formation of calcite as a cement.

The precipitation of silica as *chert* is another example of the control by pH of a natural reaction. In the waters of the Coorong, a lagoon in South Australia (See Box 12.1 Figures 1, 2, and 3.), at pH values greater than 10, caused by algal photosynthesis, quartz particles are dissolved. When the pH drops, the silica is precipitated in the form of crystallites of cristobalite. At some later time this cristobalite will probably become chert.

SOURCES: Bathurst, 1983; Boggs, 1987; G. M. Friedman and Sanders, 1978; Morse and Mackenzie, 1990; Scoffin, 1987; Vacher and others, 1990.

Weathering

Weathering is *a complex of subaerial processes that result in (1) the breakdown of solid bedrock into loose particles and (2) the destruction or alteration of minerals, including formation of new minerals. Mechanical weathering* is confined mostly to cold climates where *frost wedging* is important. By contrast, *chemical weathering* pervades all climates, but proceeds faster in warm, moist climates than in dry climates. Chemical weathering involves *solutions* and *solutes*, *hydration* and *hydrolysis*.

The role of chemical weathering is related to the composition of waters, temperature, and presence or absence of plant matter as well as of microorganisms, such as bacteria. Rocks most prone to chemical weathering in moist climates are limestones, which dissolve readily, as indicated in Eq. 3-20. Dissolution in limestones occurs within existing pores. This enlarges the pores and increases permeability. The progressive increase in permeability enables increasing volumes of water to flow through the deposit. Ultimately these waters dissolve out caves, *sink holes*, and other kinds of *karst*. Rocks consisting of silicate minerals weather less readily than limestones but, as we saw in Chapter 2, *regolith* forms by chemical weathering of silicate rocks. Without chemical weathering, the Earth's surface would resemble the lunar surface, and life in its present form could not exist. Agriculture would not be possible. Many valuable mineral resources, such as bauxite, some iron ores, clays, and other materials with which modern civilization has been built are created by chemical weathering.

Natural waters dissociate slightly into hydrogen and hydroxyl ions, as expressed in Eq. 3-1. These ions react with minerals, at whose surfaces atoms and ions possessing unsatisfied valencies are located. Hydroxyl ions become bonded to the exposed cations and hydrogen ions to exposed oxygens and other anions, and thus hydration results. An example is the reaction in which feldspar weathers (Figure 7-53). In this reaction, oxygen atoms from the feldspar lattice combine with hydrogen ions from the water to form hydroxyls and potassium is released. Aluminum, which was originally present in tetrahedral coordination with oxygen, assumes its preferred octahedral coordination. As a result of these processes, feldspar weathers into clay minerals and metallic ions are removed in solution. Weathering reactions continue as repeated flushing by percolating waters removes the soluble constituents. Thus dissolution is essential to chemical weathering. The minerals break down; their soluble constituent ions and atoms are carried away in solution, and the relatively insoluble constituents remain behind as clays or hydroxides of iron, magnesium, and aluminum.

The extent of chemical weathering and kind of residual products formed depend mostly on the degree of leaching of the parent rock, as shown in Table 7-4. This table shows that, among the elements of silicate rocks, aluminum compounds are the least-soluble weathering

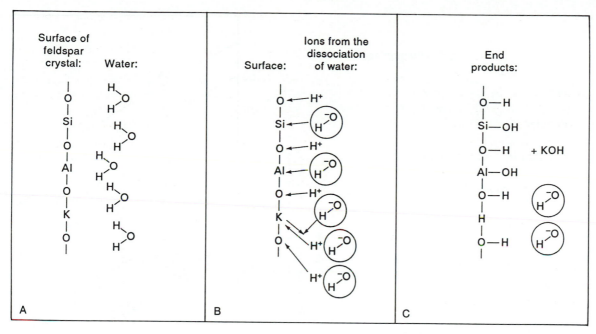

FIGURE 7-53.   Decomposition of feldspar lattice by water; schematic. Further explanation in text. (Authors.)

products; hence under conditions of chemical weathering, bauxite forms. Likewise, under oxidizing conditions, Fe^{3+} salts concentrate as a lag deposit in *laterite*, but under reducing conditions they may dissolve. TiO_2 as rutile becomes part of the resistant heavy-mineral residue and becomes concentrated in placers. However, if released from the parent mineral, such as ilmenite or titaniferous magnetite, as $Ti(OH)_4$ at a pH of less than 5, titanium can become mobile. Ions of Si^{4+} do not exist in solution but $H_3SiO_4^-$ and $H_2SiO_4^{2-}$ are readily leached in highly alkaline solutions (pH > 9.5); such alkaline solutions are not common in nature. At lower pH the solubility of

H_4SiO_4 is low. At the mobile end, Ca^{2+}, Mg^{2+}, Na^+, and K^+ are leached readily; however, K^+ and Mg^{2+} may be captured by illite, smectite, or other clay minerals and thus lose their mobility.

Despite the relative ease with which ions of K^+ and Na^+ are mobilized during weathering, muscovite, which is rich in K^+, and sodium-rich plagioclase may be remarkably resistant to chemical weathering. Thus another variable, crystal structure, may influence the rate of decay of minerals. The stability sequence of various minerals in chemical weathering, known as Goldich's stability sequence (Figure 7-54), parallels the sequence of crystallization in a silicate melt, known as Bowen's reaction sequence. Minerals that crystallize at high temperatures in a magma, and hence are the first to precipitate, are also those that are most prone to decay in chemical weathering. The minerals that precipitate last, such as muscovite and quartz, offer greatest resistance to weathering. Thus the minerals that are stable at high temperatures are not stable at the Earth's surface. Despite considerable debate, the precise mechanism by which crystal structure is involved in determining the rate of decay in Goldich's stability sequence is still not completely understood. One of the variables that control resistance to weathering is the kind of chemical bond in the mineral.

Apart from the effect of temperature, which influences rate of reaction, other aspects of climate are important in chemical weathering, especially because they influence the availability of water and distribution of organisms

TABLE 7-4. Mobilities of some common cations in chemical weathering

Readily leached	Relatively immobile
Ca^{2+}, Na^+	H_4SiO_4, solubility low.
K^+ and Mg^{2+} readily leached, but may be captured by illite, montmorillonite, or other clay minerals.	Ti^{4+}, immobile in rutile (TiO_2); some mobility, if released as $Ti(OH)_4$.
	Fe^{3+} concentrates in laterite under oxidizing conditions.
	Al^{3+} concentrates in bauxite

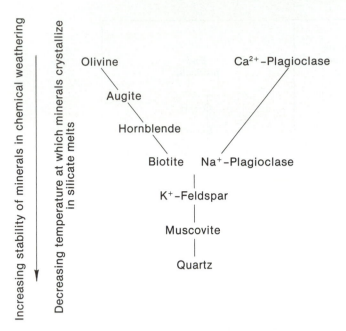

FIGURE 7-54.   The stability sequence in chemical weathering of common rock-making minerals. (S. S. Goldich, 1938, table 18, p. 56.)

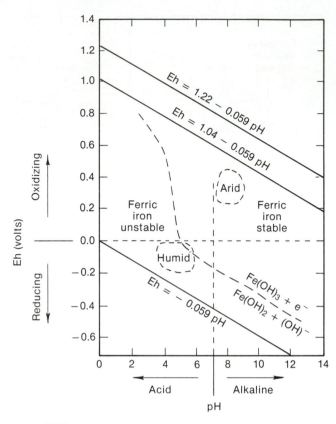

FIGURE 7-55.   Graph of Eh versus pH for water solutions showing stability field for ferric iron and characteristics of ground waters in arid- and humid terrains. In the field where ferric iron is not stable, ferrous iron is stable. (Data from I. G. M. Baas Becking, I. R. Kaplan, and D. Moore, 1960; F. C. Loughnan, 1969.)

Chemical weathering proceeds through hydrolysis at the surfaces of minerals (Eq. 3-1), after which hydrogen ions replace cations. Although in arid regions, water penetrates the rocks in flash floods, during the long dry spells, the ground water moves up to the surface and evaporates. Thus soluble constituents that form strong bases, such as Na^+, K^+, Ca^{2+}, and Mg^{2+}, are retained and the environment becomes alkaline (pH 7.5 to 9.5); oxidation destroys plant debris. Under such conditions of high pH and oxidation, ferric oxides are stable (Figure 7-55) and hematite and limonite color the sediments red, brown, or yellow. In humid climates, where vegetation is abundant, shallow ground waters tend to be acid (3.5 to 5.5) and reducing. Under these conditions soluble constituents are easily removed and the residue is progressively enriched in minerals containing abundant aluminum hydroxides. These include kaolinite and the bauxite minerals gibbsite and diaspore, and boehmite. Ferric iron is unstable (See Figure 7-55.); hence the sediments tend to be colored various hues of gray or green, but never red, brown, or yellow.

Chemical weathering not only creates new minerals such as clay minerals and iron hydroxides or -oxides, but also makes available, by freeing from the parent rock, resistant heavy minerals that may be concentrated as residues to placers, such as those containing gold, diamonds, or rutile (a source of titanium).

SOURCES: Boggs, 1987; G. M. Friedman and Sanders, 1978; Tandon and Friend, 1989.

Reactions of Unknown or Uncertain Mechanism

In the modern ocean and in the rock record are found many chemical products, such as phosphorites and manganese oxides, which provide for us a record of geologic events even though we do not understand the kind of chemical reactions that caused these materials to form. Phosphorites and manganese salts commonly accumulate where other sediments are absent or are deposited very slowly. In the geologic record, accumulations of phosphorites and manganese salts accompany surfaces where the sedimentation of other material has been interrupted.

In the modern ocean, both phosphorite and manganese salts occur as nodules and coverings over extensive areas. Phosphorites are thought to form where phosphate-rich waters from depth upwell adjacent to a shallow shelf. However, upwelling by itself may not be enough to cause precipitation of phosphorites. $^{234}U/^{238}U$ dating indicates that the age of some "modern" phosphorites exceeds the maximum limit of the dating technique, which is 800,000 yr. Hence in some areas of modern upwelling, phosphorites evidently are not now forming. The phosphorites are ancient and not related to modern conditions. Accordingly, the chemical reactions that result in the precipitation of phosphorites are still not known.

Manganese nodules occur both in the sea and in some lakes (Figure 7-56). Nodules on the sea floor may be of multiple origins; some may form by bacterial processes and others by interaction of submarine volcanism and seawater. Manganese from volcanic sources probably oxidizes from Mn^{2+} to Mn^{4+} and is precipitated. In this reaction, seawater must contain abundant oxygen.

Manganese-iron nodules were known in lakes before they were discovered in the ocean. In fact, such nodules have been mined from Swedish lakes since the middle of the nineteenth century. In 1860 these Swedish sources yielded approximately 10,000 to 12,000 tons. Removal of this quantity of nodules did not exhaust the supply. Instead, more nodules grew during the following year. The supply was fully replenished in cycles of 30 to 50 years. Under oxidizing conditions in lake waters, where the rate of sedimentation is low, manganese- and iron oxides precipitate on nuclei, such as clay, plant-spore capsules, bark, or detrital particles. As pore waters enriched in manganese and iron become oxidized, manganese- and iron compounds are precipitated as nodules. Currents on the floor of the lake keep the nodules exposed and supply manganese and iron in oxygen-enriched waters.

SOURCE: G. M. Friedman and Sanders, 1978.

Inorganic Chemical Processes Leading to the Formation of Minerals

Earlier in this chapter, we stated that purely inorganic chemical reactions that form minerals in sediments and

FIGURE 7-56.   Enlarged view of manganese nodules formed in fresh water, Lake George, New York, photomicrograph of petrographic thin section seen in plane-polarized light. (M. Schoettle and G. M. Friedman, 1971, fig. 2, p. 103.)

in sedimentary rocks are rare. Yet, although uncommon, where simple organisms are absent, such reactions do take place. Very few organisms can live in brines and in sediments that have been buried. Although some simple organisms, such as cyanobacteria and sulfate-reducing bacteria, may persist in such places, these organisms do not affect the chemical processes that we shall discuss. We discuss the formation of evaporites, zeolites, and feldspars.

EVAPORITES

Evaporite minerals, such as gypsum, anhydrite, and halite, which form from brines, are examples of products of inorganic natural chemical reactions. Brines form in an environment generally hostile to simple organisms and inorganic chemical reactions proceed. As early as 1849, with his classical experiments, Usiglio evaporated Mediterranean seawater in a vessel and determined the kinds of evaporite minerals that formed. His experiments seemed to have settled the origin of evaporite minerals from brines. Although, as will be seen in Chapters 8, 12, and 14, the interpretation of the origin of deposits of evaporites is still under debate, data from shallow pools marginal to the Red Sea confirm the results of Usiglio's experiments, which demonstrated the interrelationship between progressive increase in the concentration of brines and the precipitation of evaporite minerals. (See Chapter 3.)

SOURCES: G.M. Friedman and Krumbein, 1985; Friedman, Sneh, and Owen, 1985; Warren, 1989.

ZEOLITES

Until recently, zeolites were known chiefly as well-formed crystals in vugs and cavities in basalts and other mafic volcanic rocks. The older literature contains only a few references to scattered occurrences of zeolites in sedimentary rocks. However, zeolites have been found to be among the most-abundant and widespread authigenic silicate minerals. In fact, zeolites are so abundant that each year between 100,000 and 200,000 tons are mined from sedimentary deposits for use as fillers in the paper industry, as soil conditioners, as dietary supplements in animal husbandry, as ion exchangers in pollution abatement, in pozzolanic cements and in concretes, and as acid-resistant adsorbants in the drying of gases. Zeolites are hydrated aluminosilicate minerals, containing calcium, sodium, and potassium. In composition zeolites resemble feldspars, but feldspars are not hydrated.

The most-common zeolites are clinoptilolite, phillipsite, laumontite, and analcite. Zeolites have been recognized in sedimentary rocks from mid-Paleozoic to

Holocene, but not from rocks of Precambrian and early Paleozoic ages. Based on origin, two kinds of zeolites can be recognized: (1) those related to tephra and (2) those formed in saline, alkaline lakes devoid of tephra material.

Zeolites of pyroclastic origin commonly form by the reaction of volcanic glass with various kinds of waters, such as (1) waters of saline, alkaline lakes, (2) seawater, (3) fresh water, and (4) saline interstitial solutions. Commonly, the end product of this reaction is potassium feldspar. Burial of tuffs results in distinct zones, probably caused by increasing temperature, which successively downward are (1) fresh glass, (2) clinoptilolite, (3) analcite, (4) laumontite, and (5) feldspar. In this sequence, one kind of zeolite changes to another, and ultimately a stable feldspar forms. An example of a geologic unit containing abundant zeolites derived from pyroclastic debris is the analcite-rich part of the Popo Agie (pronounced popozhuh) Member of the Chugwater Formation (Triassic) in west-central Wyoming. Another example includes the zeolites, especially phillipsite, of pyroclastic origin, that are widely distributed in the Pacific Ocean. They form at the expense of volcanic glass at the ocean bottom.

Zeolites that crystallize in saline, alkaline lakes devoid of pyroclastic or tephra material derive their calcium locally from the lakes. A modern example is the sediment of Lake Natron, Tanzania. An ancient example is the Lockatong Formation (Lower Jurassic) of the Newark Group in central New Jersey.

SOURCES: Blatt, Middleton, and Murray, 1980; Helmold and Van de Kamp, 1984.

FELDSPARS

Authigenic feldspars, occurring as microcline ($KAlSi_3O_8$) or as albite ($NaAlSi_3O_8$) in sedimentary rocks, were discovered more than 100 years ago. Although they are only accessory constituents in most sedimentary deposits, in upper Precambrian and Cambro-Lower Ordovician rocks they are commonly dominant. In shales of these ages, authigenic potassium feldspars may be so abundant that the concentration of K_2O is raised to near-commercial levels. The concentration of K_2O in the insoluble residues of Cambrian and of Lower Ordovician carbonate rocks may be comparable to that of feldspar-rich shales. These levels of concentration of K_2O are about three times those of the average shale.

Volcanic glass alters successively to various zeolites and finally to feldspar. The composition of zeolites is extremely variable; hence the kind of end-product feldspar, whether microcline or albite, can be variable and may de-

pend on the composition of the zeolite precursor. Potassium feldspar is the end product of the alteration of many zeolites. The tephra at the margin of a Quaternary lake in Tanzania change mostly to phillipsite and clinoptilolite, and those in the center of the lake to potassium feldspar and phillipsite. Because zeolites have not been found in rocks of Precambrian and Early Paleozoic ages, and because feldspar is the stable end product, we think it probable that zeolites older than Early Paleozoic have changed to feldspar. Some of the Cambro-Ordovician rocks that are rich in K_2O are known to contain pyroclastic material. The limestones that are likewise enriched in K_2O formed as shoals near the edge of a deep ocean in which volcanoes were active. Interestingly, in some Cambro-Lower Ordovician carbonate rocks, thin laminae of dolomite containing only a little feldspar alternate with thin laminae in which numerous feldspar crystals are embedded in a dolomitic matrix (Figure 7-57). The feldspar crystals in these carbonate rocks may have replaced wind-transported tephra that accumulated among microbial mats on the hypersaline shoals at the margin of the deep ocean in which the volcanoes were active. Intercalations of shaly stringers rich in feldspar and the high feldspar content in the carbonate rocks possessing 30 to 50% insoluble residue may have resulted from abundant eruptions and wind transport of debris from ancient deep-sea volcanoes.

SOURCES: Blatt, 1982; G. M. Friedman, 1990; G. M. Friedman and Sanders, 1978.

FIGURE 7-57.   Enlarged view of specimen of dolostone, Tribes Hill Formation (Lower Ordovician) Mohawk Valley, New York showing abundant feldspar crystals, many of which are euhedral, as a horizontal layer across the middle, with feldspar-poor layers that consist of dolomite above and below; photomicrograph of thin section in plane-polarized light. (M. Braun and G. M. Friedman, 1969, fig. 4, p. 117.)

Suggestions for Further Reading

AITKEN, J. D., 1967, Classification and environmental significance of cryptalgal limestones and dolomites, with illustrations from the Cambrian and Ordovician of southwestern Alberta: Journal of Sedimentary Petrology, v. 37, p. 1163–1178.

ALLEN, J.R.L., 1970, Physical processes of sedimentation: London, Allen and Unwin, 248 p.

ALLER, R. C., 1982, The effects of macrobenthos on chemical properties of marine sediments and overlying water, p. 53–102 *in* McCall, P. L., and Tevesz, M. J. S., eds., Animal–sediment relations (*sic*): The biogenic alteration of sediments: New York, Plenum Press.

ANDERSON, R. S.; and HALLET, BERNARD, 1986, Sediment transport by wind: toward a general model: Geological Society of America Bulletin, v. 97, p. 523–535.

BAGNOLD, R. A., 1941, The physics of blown sand and desert dunes: London, Methuen and Company, 165 p. (Reprinted 1954.)

BAKER, V. R., 1973, Paleohydrology and sedimentology of Lake Missoula flooding (*sic*) in eastern Washington: Geological Society of America Special Paper 144, 69 p.

BATHURST, R. G. C., 1975, Some chemical considerations, Chapter 6 *in* Carbonate sediments and their diagenesis, 2nd ed.: New York, Elsevier Publishing Co., 658 p.

BLATT, H., 1982, Sedimentary petrography: San Francisco, W. H. Freeman and Company, 564 p.

CHAFETZ, H. S., 1986, Marine peloids: a product of bacterially induced precipitation of calcite: Journal of Sedimentary Petrology, v. 56, p. 812–817.

CLOKE, P. L., 1966, The geochemical application of Eh-pH diagrams: Journal of Geological Education, v. 14, p. 140–148.

FAGERSTROM, J. A., 1987, The evolution of reef communities: New York, John Wiley & Sons, 600 p.

HSÜ, K. J., 1975, Catastrophic debris streams (sturzstroms) generated by rock falls: Geological Society of America Bulletin, v. 86, p. 129–140.

MIDDLETON, G.V., and SOUTHARD, J.B., 1984, Mechanics of sediment movement: Eastern Section, Society of Economic Paleontologists and Mineralogists Short Course Notes 3, 2nd ed., 401 p.

SANDERS, J. E., 1965, Primary sedimentary structures formed by turbidity currents and related resedimentation mechanisms, p. 192–219 *in* Middleton, G. V., ed., Primary sedimentary structures and their hydrodynamic interpretation: Tulsa, OK, Society of Economic Paleontologists and Mineralogists Special Publication 12, 165 p.

SHINN, E. A.; STEINEN, R. P.; LIDZ, B. H.; and SWART, P. K., 1989, Whitings, a sedimentologic dilemma: Journal of Sedimentary Petrology, v. 59, p. 147–161.

CHAPTER 8

Circulation in the Atmosphere, in Modern Oceans, and in Basin Waters

In this chapter we examine how the circulation of the atmosphere and the oceans affects the condition of waters in various basins and how the condition of the waters, in turn, affects sediments. We begin with the large-scale circulation of the Earth's atmosphere.

Global Atmospheric Circulation

Several interpretations have been proposed of the large-scale circulation of the Earth's atmosphere. Interest in this subject is at an all-time high and several general-circulation computer models have been constructed. The general features of this global atmospheric circulation control many aspects of the origin and distribution of sediments in the modern world. Moreover, the principles operating today presumably can be extended into the past for at least as long as the composition of the Earth's atmosphere approximated its modern-day composition. Of particular importance to the study of sediments are (1) the zone of great rainfall within the rising air of the Intertropical Convergence Zone (ITCZ) and its flanking equatorial trade-wind belts; (2) the great zones of aridity within the belts of descending air lying about 30° north and south of the Equator; (3) the midlatitude belts where the polar air masses interact with the tropical air masses; and (4) the cold polar air masses.

In our brief portrayal of the global atmospheric circulation here, we shall emphasize the connection between movements of the atmosphere and their effects on the surface waters of the oceans. That is, we shall focus on the major climatic zones, locations of deserts, and sites of great rainfall. The physical aspects of the wind blowing sediment particles were discussed in Chapter 7.

The atmospheric circulation is driven by solar radiation, conditioned by the Earth's rotation, modulated by the back-and-forth seasonal migration across the Earth's surface of the *subsolar point*, and finally forced into an accommodation with the configuration of the oceans and continents. The **subsolar point** is *the point on the Earth's surface where the Sun's rays are at normal incidence and contain the most energy per unit area.* A further variable is the composition of the atmosphere, a factor that can change as a result of variable volcanic activity, of circulation of the oceans, or of human activities. As with so many situations in geology, these factors all interact and are so complex that attempts to understand even small facets of them have challenged the intellects of our brightest and best investigators. In order to deal rationally with this maze of interlocking processes, we shall take them up one at a time and proceed by the method of successive approximations.

Solar Radiation: Uneven Distribution

Solar radiation is most intense at the *subsolar point* (Figure 8-1). Solar radiation is least intense at the poles, where the Sun's rays are oblique to the Earth's surface and must traverse the greatest thickness of atmosphere. This results in a persistent excess of heat in the equatorial parts of the Earth and a persistent deficit of heat at the poles. If no counterbalancing tendency existed, then with time, the equatorial regions would become progressively hotter and the poles, colder. The fact that this is not happening proves that a large-scale atmospheric circulation redistributes the solar heat and maintains a generally uniform temperature regime.

Before we proceed, we shall outline the proofs that the tropical regions have not been getting hotter with time. The most-convincing evidence comes from reef-building corals. Among the modern order of corals, the *Scleractinia*, two major groups are recognized: (1) the hermatypic, tropical reef-building corals, and (2) the ahermatypic, or nonreef-building corals. Several factors limit

279

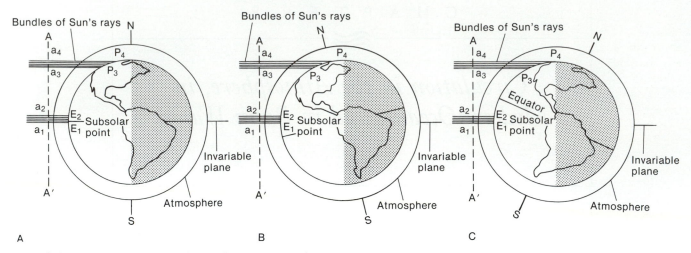

FIGURE 8-1.   Subsolar point viewed from a distant space station.
   A. Spring and fall equinox.
   B. Summer solstice.
   C. Winter solstice.

the growth of hermatypic corals. Of interest here is temperature. Modern reef-building corals do not thrive where temperature drops below 20°C nor rises above 30°C. A map of the distribution of hermatypic corals in the modern world (Figure 8-2) shows clearly their confinement to the tropical regions. Fossils of hermatypic corals are largely restricted to rocks that are inferred to have been deposited in ancient tropical regions, ind-

icating that the ancient temperature distributions were generally similar to the modern one.

The Earth's Rotation and the Coriolis Effect

The Earth rotates on its polar axis at a rate of once per day. The motion is from west to east, but the sense of rotation differs depending on one's vantage point. From

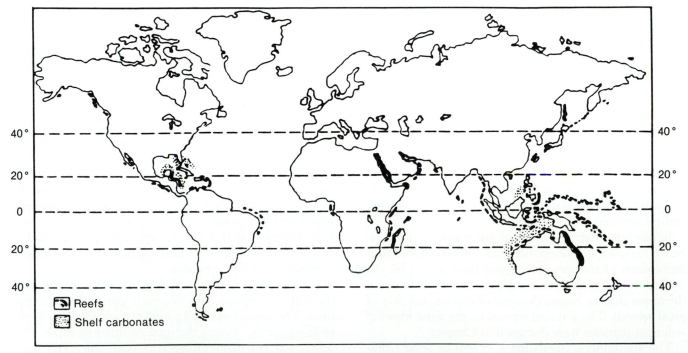

FIGURE 8-2.   Distribution of modern hermatypic corals. (C. S. Nelson, 1988, fig. 1, p. 4; based on J. L. Wilson, 1975, p. 2.)

a point above the North Pole, the Earth is seen to be rotating counterclockwise (Figure 8-3, A). Yet from a point above the South Pole, the motion is clockwise (Figure 8-3, B). This rotation affects all winds and currents of water. In the Northern Hemisphere, where the sense of rotation is counterclockwise, currents are deflected toward their right (Figure 8-4). That is, a wind blowing from the north toward the south will be experienced at the Earth's surface as a wind moving from northeast to southwest. Because winds are named by the directions

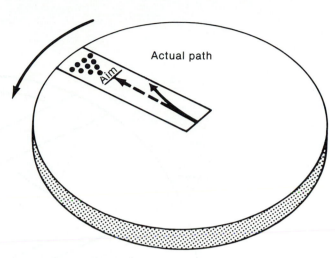

FIGURE 8-4.   Coriolis effect illustrated by what happens to a bowling ball when the alley is oriented like a spoke on a large turntable. Because of the counterclockwise rotation of the turntable, a ball aimed at the headpin will roll into the right-hand gutter. This illustrates the right-deflection rule of the Coriolis effect that applies to winds and currents in the Northern Hemisphere. If the turntable were to rotate clockwise, then the ball would be deflected to its left and roll into the left-hand gutter. This would illustrate the left-deflection rule of the Coriolis effect that prevails in the Southern Hemisphere. (J. E. Sanders, 1981, fig. 15.21, p. 375.)

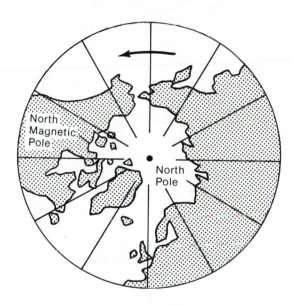

from which they blow, the would-be north wind becomes in fact a northeast wind.

Air–Sea Interactions

The relationships between winds and ocean currents are complex. At first glance, one's intuition holds that a wind should move surface water forward, straight in the direction in which the wind is blowing, as the wind blows a sailboat on a downwind reach. That this is true in a general way is indicated by the large-scale agreement between a map showing the directions of the world's major wind belts (Figure 8-5) and a map showing the directions of ocean surface currents (Figure 8-6).

In contrast are the results from observations of wind directions and directions of drift of icebergs in the Norwegian Sea. Here the icebergs were not driven in the direction of the wind, but rather moved at right angles to the wind. This curious relationship has been explained as a general result of the Coriolis effect and is included in a phenomenon known to oceanographers as the Ekman spiral. We shall refer to the surface-water effects of this as the Ekman relationship. According to the *Ekman relationship*, in the Northern Hemisphere, winds and water movements are related as shown in Figure 8-7. A systematic relationship exists among wind direction, direction of surface current, and changes of level

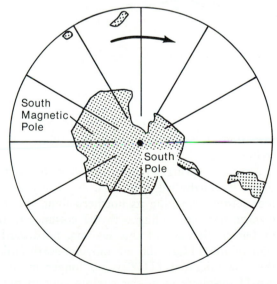

FIGURE 8-3.   Rotation of the Earth, from west to east, appears not to be the same as seen from two different vantage points.

A. From above the North Pole, the sense of rotation is counterclockwise.

B. From above the South Pole, the sense of rotation is clockwise. (J. E. Sanders, 1981, fig. 15.22a, b, p. 375.)

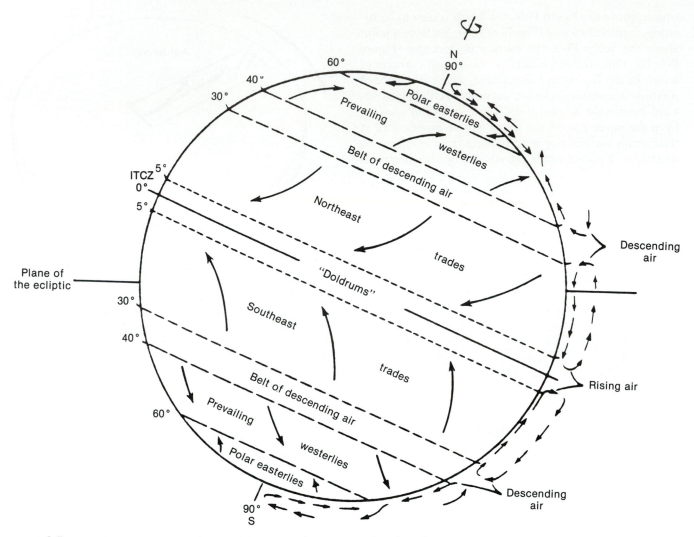

FIGURE 8-5.   Wind belts of the world. Sketched at the time of an equinox when the subsolar point is above the Earth's Equator. (After C. R. Longwell, R. F. Flint, and J. E. Sanders, 1968, fig. 2-10, p. 36.)

that give rise to **downwelling**, *vertical downward motion of water resulting from a tendency of the surface to pile up in a given region*, and **upwelling**, *the rising of sub-surface water toward the surface as a result of wind-driven lowering of the surface*. Along some equatorial coasts in the Northern Hemisphere, such as that of the Arabian Peninsula off Yemen and Oman, which trends northeast-southwest (Figure 8-8), the seasonal wind shifts cause a corresponding seasonal change of water movement from downwelling to upwelling. These wind shifts in the Arabian Sea have been known to Arab sailors for centuries. They even gave it a name: *monsoon*, which means time or season. At the spring equinox the ITCZ lies above the Equator and the wind belts are symmetrical, as depicted on all textbook diagrams: north of the ITCZ is the belt of northeast trades, and south of it, the corresponding

belt of south-east trades. During the prevailing northeast winds, the coastal water mass is subjected to downwelling.

After the spring equinox, however, the ITCZ continues its northward migration, eventually, at the summer solstice, reaching its northern limit, the Tropic of Cancer, here coinciding with the entrance to the Gulf of Oman. Because of this seasonal northward migration of the ITCZ, the air moving northward toward the now-migrated ITCZ is no longer in the Southern Hemisphere as at the equinox, but in the Northern Hemisphere. Accordingly, it no longer follows the left-deflection rule, which governs the southeast trades, but now must follow the right-deflection Coriolis effect. This means that it must become a southwest wind. Thus, for a few weeks near the summer solstice, southwest

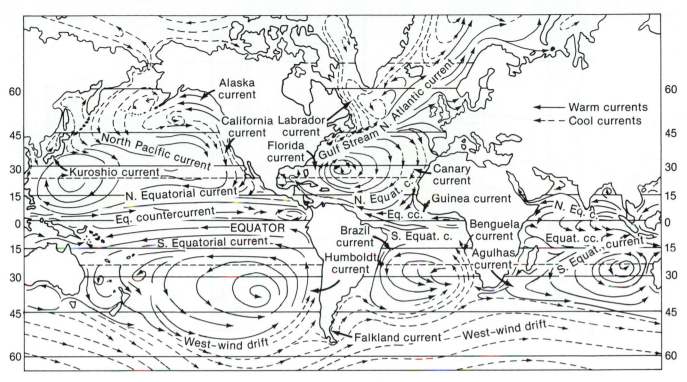

FIGURE 8-6.   Surface ocean currents of the world.

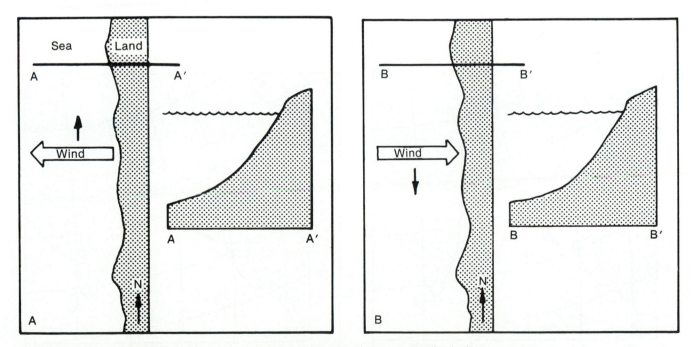

FIGURE 8-7.   Ekman effect illustrated by winds oriented at various directions to a shoreline in the Northern Hemisphere that trends north-south. In each frame a schematic map of the coast is shown at left and a profile at a right angle to the shore at right. (After R. H. Fleming and R. Revelle, 1938, fig. 18, p. 122.)

   A. East wind, blowing offshore, generates surface current to north.

   B. West wind, blowing directly onshore, generates surface current to the south.

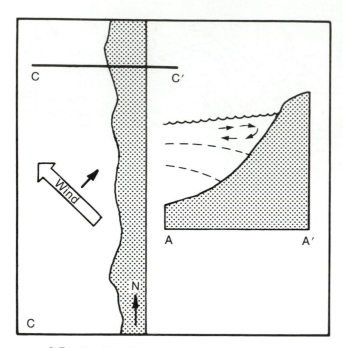
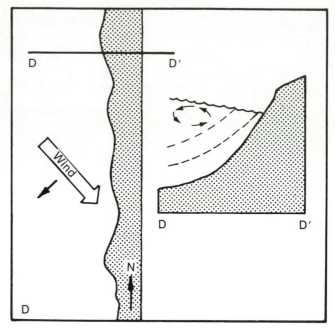

FIGURE 8-7. (*Continued*)

C. Southeast wind, blowing diagonally offshore, creates surface current toward the northeast, which drives water against shore, raising the level. This imbalance creates downwelling and a subsurface current that flows offshore. Dashed lines are isopycnals, lines representing depths having equal pressure.

D. Northwest wind, blowing diagonally onshore, creates surface current toward the southwest, away from shore. This tends to lower the water level, to establish a condition of upwelling, and to create a subsurface current flowing toward shore.

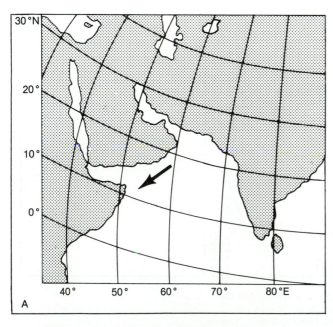
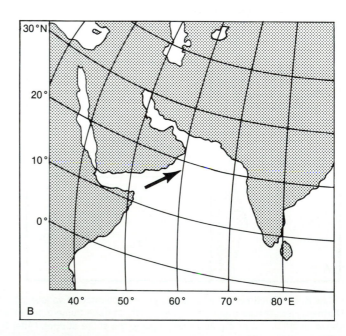

FIGURE 8-8.   Map of Arabian Sea with seasonal winds.

A. Northeast trade winds prevail for most of the year; coastal water downwells.

B. Southwest winds of summer in a belt that widens northward as the ITCZ migrates toward the Tropic of Cancer. Coastal water upwells. (Authors.)

winds enable sailors to go from as far south as Somalia all the way to India on a downwind reach for the entire trip. While the southwest winds are blowing, the coastal waters are subjected to upwelling. After the summer solstice, the north-south width of the belt of these southwest winds progressively diminishes. It is replaced by the more-typical belt of northeast winds (and corresponding coastal downwelling) that characterizes Northern Hemisphere tropical air moving southward toward the ITCZ. This pattern of northeast winds and coastal downwelling does not change even when the ITCZ continues moving southward during its winter excursion into the Southern Hemisphere.

A belt of analogous seasonal wind shifts exists off the coast of Sumatra, a northwest-southeast-trending coast that lies along the northeast margin of the Indian Ocean, but in the Southern Hemisphere (Figure 8-9). During much of the year, when the ITCZ is above the Equator or on its northward excursion, this coast is situated within the belt of southeast trades. In the coastal waters, downwelling accompanies the southeast winds. Soon after the fall equinox, however, the ITCZ migrates south past this area. As a result, a seasonally widening belt of air flowing southward toward the ITCZ develops in the equatorial parts of the Southern Hemisphere. Within this belt the south-flowing air is left-deflected and thus becomes a northwest wind. For a short time near the winter solstice, therefore, it is possible to sail on a downwind reach in front of a northwest wind all the way from Padang to the North West Cape of Western Australia. In the coastal water, upwelling coincides with the times of northwest winds.

At this point it will be helpful to look at the Nile River in Egypt as an example of how an important

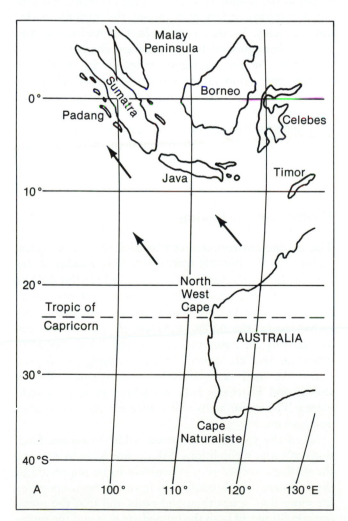
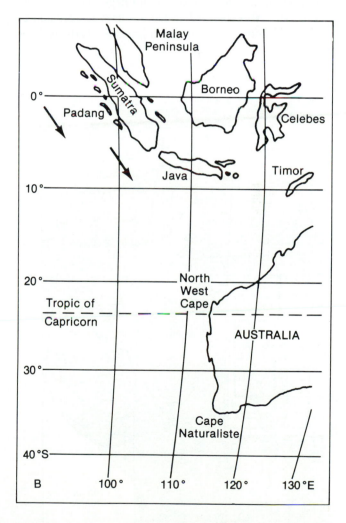

FIGURE 8-9.   Map of northeastern Indian Ocean with seasonal winds.

A. Southeast trade winds result in coastal-water downwelling.

B. As ITCZ migrates southward toward the Tropic of Capricorn, an ever-widening belt of northwest winds is established, and with them, coastal-water upwelling. (Authors.)

tant global climate system operates, knowledge that was made possible only by a twentieth-century, space-age view of the world's general atmospheric circulation. Nineteenth-century geographers and explorers thought that the explanation of the Nile's annual flood was to be found in variations in the level of Lake Victoria, which is the source of the *Blue Nile* (Figure 8-10).

During the nineteenth century the American meteorologist William Ferrel (1817–1891) argued that the seasonal flood of the Nile resulted from the annual northward migration of the intertropical convergence zone (ITCZ) into the drainage network of the White Nile, especially the Abyssinian highlands. According to Ferrel, the year-round supply of water to the Nile comes chiefly via the Blue Nile, whose source lies south of the Equator. Accordingly, the Blue Nile collects rainfall from the ITCZ during all months. When the tropical zone of torrential rainfall moves into the drainage network of the White Nile, an expanded collecting area is available to catch the abundant rain. Such a migration neatly explains

the exact start of the annual flood and its more-or-less regular termination. If this seasonal effect is the complete explanation, then the amount of the Nile flood should be more or less the same year after year. Observations of the heights of this precious flood have been made by the Egyptians for thousands of years (Figure 8-11). They prove that irregular variation, not equality, is the rule. The observed variation in flood heights requires us to bring in variable amounts of rainfall. The causes of such variations are not fully understood, but they seem to be connected to certain large-scale factors in the global atmosphere.

Modern Oceans

Systematic scientific study of the modern oceans, which began in the middle of the nineteenth century, has provided extensive information about the various marine environments. Viewed as a whole, the marine environment is distinguished by the chemical characteristics of its waters, by the enormous range of its depths, by a distinctive temperature structure that is related to global climatic zones, by numerous dynamic physical processes, and last, but not least, by remarkable groups of organisms having life styles that are based on adjustments to various environmental subdivisions of the sea. Marine environments are affected by adjacent land masses.

Environmental Realms

The marine environment can be divided into two great realms: (1) the **benthic realm**, which is *a collective designation for all the bottom of the sea*, and (2) the **pelagic realm**, that is, *all the ocean water lying seaward of low-tide level*. The pelagic realm consists of two major provinces: (a) the **neritic province**, *the coastal-ocean water overlying the continental shelves*, and (b) the **oceanic province**, *the "blue-water" regions overlying the deep-sea basins*. We shall examine various subdivisions of these realms and provinces later on when we discuss sediments. The major units will suffice for our broad-scale descriptions.

All of the specific environments just listed are functions of today's conditions. We do not have to examine the geologic record very extensively to be impressed by the fact that today's conditions have not been operating for more than a few thousand years and that the previous conditions, of both the immediate past and the more-remote past, differed enormously from today's. As we set forth on our journeys of exploring the modern oceans, we need to keep constantly in mind our two chief objectives: (1) understanding how the modern environments

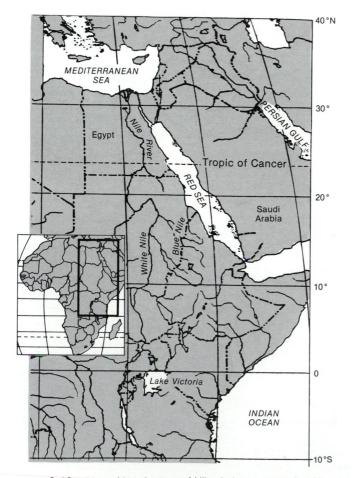

FIGURE 8-10.   Map of headwaters of Nile drainage network, with inset map of Africa.

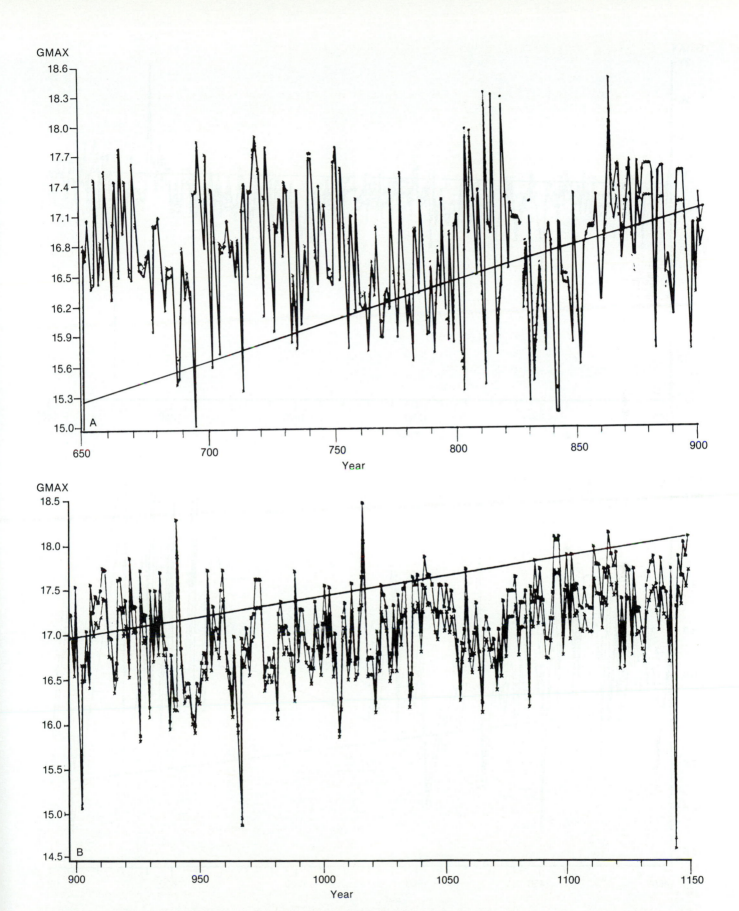

FIGURE 8-11. Heights of the Nile floods vs. time. (F. A. Hassan and B. R. Stucki, 1987, fig. 1-1, p. 39–41.)

A. A.D. 650 to 900.
B. A.D. 900 to 1150.

287

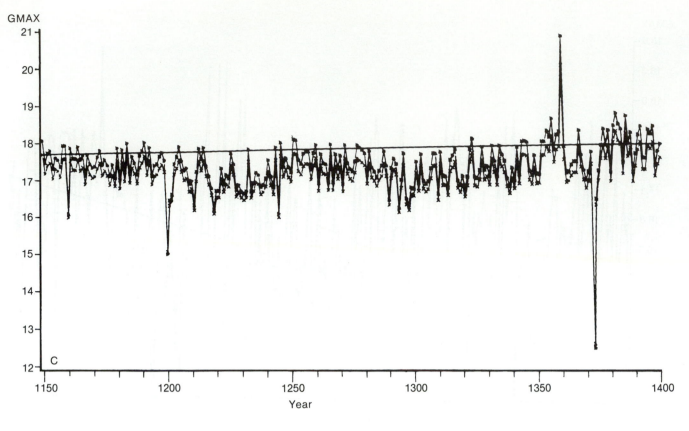

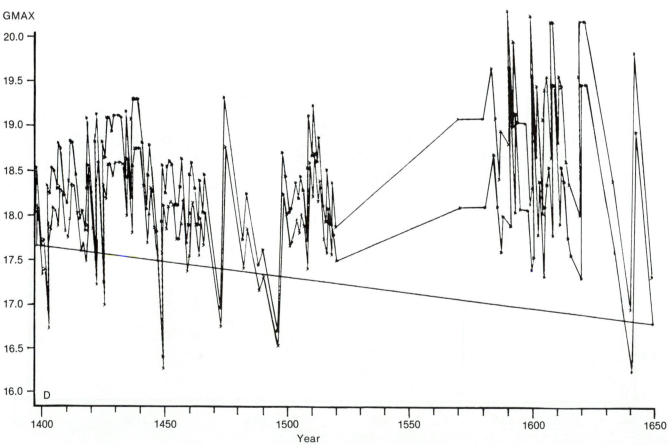

FIGURE 8-11. (*Continued*)
  C. A.D. 1150 to 1400.
  D. A.D. 1400 to 1650.

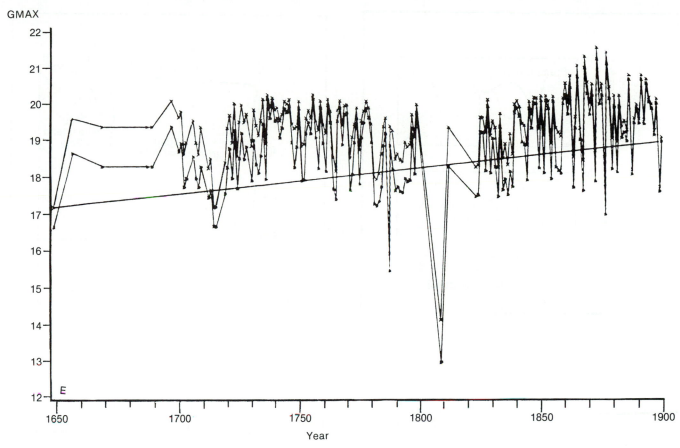

GMAX

FIGURE 8-11. (*Continued*)
E. A.D. 1650 to 1900.

operate, and (2) establishing a sound basis for interpreting the geologic record. We begin with the chemical characteristics of seawater.

Chemical Characteristics of Seawater

The composition of the sea-salt residue recovered by evaporating away many samples of seawater from all over the world is remarkably constant. Although many elements have been identified in such sea salts, only six ions form 99% of the bulk of sea salts (Table 8-1). The two chief species are Na^+ and Cl^-. Hence the chemical composition of seawater usually is expressed by a measurement of one or the other of these ions. An expression of the chemical composition of water is its *salinity*. Commonly salinity is expressed as parts per thousand (abbreviated ‰). The salinity of average seawater is 35‰.

Although the composition of sea salt is thought to remain essentially constant throughout the oceans, the salinity of seawater varies from place to place. Abundant rainfall and large entering rivers reduce the salinity. Evaporation of water, which is especially vigorous in arid climates, increases salinity (Figure 8-12).

Other important chemical characteristics include pH (Chapter 3), amount of dissolved gases (notably oxygen and carbon dioxide), and Eh (Chapter 3); the Eh is chiefly a function of the dissolved gases. Seawater is a *buffered solution*, which means that its pH changes but little as the amount of carbon dioxide, for example, changes. Normal seawater is slightly alkaline. (See Figure 3-5.) The precipitation of calcium carbonate, the chief material

TABLE 8-1. Six predominant ions in sea salts

Ion	Proportion (percent)
Cl^-	55.1
Na^+	30.6
SO_4^{2-}	7.7
Mg^{2+}	3.7
Ca^{2+}	1.2
K^+	1.1
Total	99.4

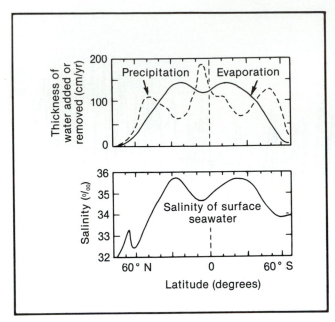

FIGURE 8-12.   Relationships among latitude, evaporation-precipitation, and salinity of surface seawater. (After G. Wüst, W. Brogmus, and E. Noodt, 1954; also M. G. Gross, 1972, fig. 7–12, p. 184.)

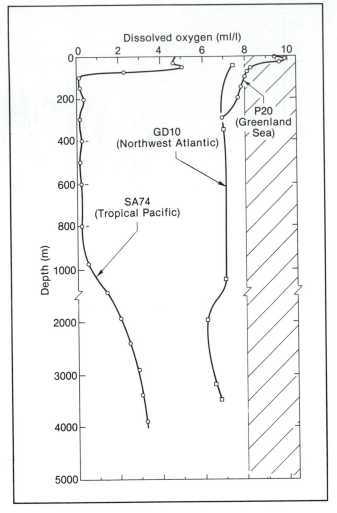

FIGURE 8-13.   Variations in content of dissolved oxygen with depth in various oceans. (Note change in scale at 1000 m.)

SA74 curve from Albatross Station 74, eastern tropical Pacific, 11°39′N, 114°15′W, 22 September 1947 (L. Bruneau, N. G. Jerlov, and F. F. Koczy, 1953.)

In cold waters off coast of east Greenland (P20 curve from Polarbjorn Station 20, 14 August 1932, A. Jakhelln, 1936), oxygen content in upper 100 m exceeds 100%-saturation value of 8 ml/l; in upper 30 m, the value exceeds 9 ml/l.

GD10 curve from Bodthaab Station 10 (56°56′N, 51°17′W, 03 June 1928 (Bull. Hydrographique for 1929). (Replotted from F. A. Richards, 1957, fig. 1, p. 188, and fig. 2, p. 191.)

composing the skeletons of marine organisms, is more efficient in alkaline solutions than in neutral solutions. (See Figure 3-8.)

Although the near-surface waters may contain 5 to about 10 ml/l of oxygen, in many parts of the ocean the abundance of oxygen decreases with depth to a low value, known as the *oxygen minimum* (Figure 8-13, curve SA74). Water masses having low values of dissolved oxygen have been affected by the oxidation of organic matter, which was created in the overlying surface waters or sank downward with a water mass that was previously located at the surface. If such moving masses of water leave the surface carrying a load of suspended first-cycle organic matter, then with time, this organic matter will be oxidized. Thus, with distance from its point of sinking, the oxygen content of such a moving mass of water will progressively decline. In the polar regions, where low temperatures impede biological activity and where large masses of cold water sink to the bottom without carrying much organic matter with them, the large amounts of oxygen that are present descend with and stay with the water. (See Figure 8-13, curve GD10.) Because such oxygen-rich waters are effectively isolated from the near-surface zones of biological activity, they tend to conserve their oxygen.

Where oxygen-minimum water comes into contact with the bottom, the sediments may display a zone of maximum values of organic matter. An example of such a relationship has been reported from the Gulf of Mexico (Figure 8-14). The oxygen-minimum zones (OMZs) in regions of coastal upwelling are often characterized by phosphorite deposits, organic-rich fine-textured sediments, and early diagenetic chert. Some OMZs in regions of coastal upwelling lack these features, but exhibit characteristic lithofacies and biofacies that might permit their recognition in the rock record. The sediments of some OMZs differ from adjacent parts of the sea bottom in benthic-organism diversity, bioturbation intensity, sediment-particle size, and in abundance of glauconite, carbonate, organic carbon, and depth-indicating foraminifers.

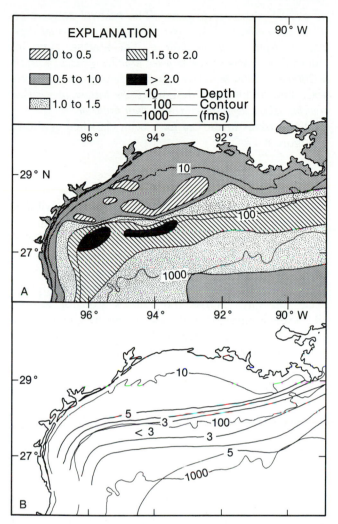

FIGURE 8-14.   Coincidence between higher-than-usual values of organic matter in bottom sediments and lower-than-usual proportions of dissolved oxygen in water next to bottom.

A. Organic matter (based on 1.8 times measured content of organic carbon) in bottom sediment. (Original, P. D. Trask, 1953, fig. 3, p. 158; taken from F. A. Richards, 1957, fig. 10, p. 216.)

B. Oxygen content (ml/l) of waters next to bottom. (F. A. Richards, 1957, fig. 11, p. 217.)

The value for dissolved oxygen ranges from about 5 ml/l for the surface ocean waters of tropical and temperate regions to about 7.5 ml/l for surface ocean waters in the polar regions. The solubility of oxygen in water varies with temperature, pressure, and salinity. (See Figure 3-4.)

Depth Zones

The depths of the oceans range all the way to 11,000 m, the maximum depth measured in the Marianas trench, western Pacific Ocean. In a general way, the depths of the sea are closely related to the two major subdivisions of the Earth's crust: (1) continental masses and (2) oceanic blocks. In general, the surfaces of continental masses stand high and those of the oceanic blocks lie low. Wherever this relationship prevails it results in shallow water (depths to a few hundred meters or so) over continental masses, and deep water (average about 4000 m) over the oceanic blocks.

One of the principal effects of depth on the properties of seawater relates to the penetration of solar energy. Sunlight does not penetrate very deeply into seawater. Hence plants, which require sunlight for photosynthesis, can grow only in the surficial parts of the sea. The amounts of sunlight provide a basis for recognizing three life zones in the sea (Figure 8-15). The **photic zone** is *the topmost, lighted layer of seawater, up to 80 m thick, in which light is sufficient to support plant growth.* The **disphotic zone** is *the transitional, fading-light layer of seawater, having its lower boundary at a depth of 600 m.* The **aphotic zone** is *the lightless bulk of the ocean lying deeper than 600 m in which plants cannot grow and in which all life depends on downward movement of food from above or upon upward percolation of minerals from the Earth below.*

The plants that live in the sea display contrasting life styles related to depth zones of the sea. Many plants, including diatoms (See Figure 2-25.), dinoflagellates, coccolithophorids (Figure 8-16, and see Box 2.4 Figure 1.), a few cyanobacteria, and a few green algae, float freely in the waters of the photic zone. These are the phytoplankton. Other plants attach themselves to the bottom. These include kelp, various green algae (See Box 2.5 Figure 1.), and the turtle grass *Thalassia.* (See Figure 7-43.) Two important boundaries on a continental shelf, both related to light penetration, are the outermost points where large attached algae are present and the lower limit of abundant reef-coral growth.

Other depth zones are based on the general characteristics of the bottom, on the properties of the water, or

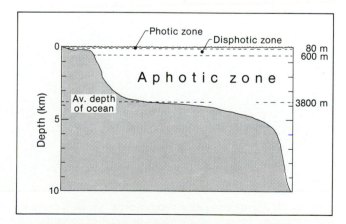

FIGURE 8-15.   Relationships of three major depth zones in the ocean based on light penetration compared with schematic profile of ocean depths. Text defines terms. (After J. W. Hedgpeth, 1957a, fig. 3, p. 22.)

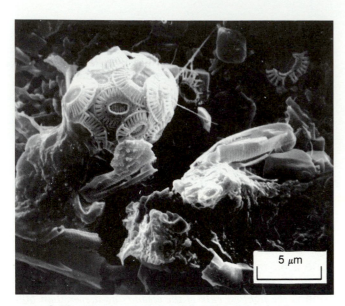

FIGURE 8-16.   Tiny suspended particles filtered from surface waters of the western Atlantic Ocean at 34°05'N, 77°00'W, 30 December 1971. The prominent particle in upper left consists of bound-together coccospheres (calcareous nannoplankton); coccospheres on right have been bound to unidentified particle, probably of organic matter. Scanning-electron micrograph. (J. W. Pierce.)

on great hydrostatic pressure. The **bathyal zone** refers to *the environment of the upper parts of the continental slopes down to depths of about 2000 m*. The **abyssal zone** includes those parts of *the deep-sea floor lying between depths of 2000 and 6000 m, where water temperatures never exceed 4°C*. The **hadal zone** designates *the bottom in sea-floor trenches where depths are greater than 6000 m*.

Depth zones in the sea have been classified as shown in Figure 8-17.

Temperature Belts; Water Masses

As is well known, solar energy is not distributed uniformly over the Earth's surface. Maximum solar heat is received where the Sun's rays are normal to the surface. Minimum solar heat is received where the Sun's rays strike the surface at low angles. The zone of normal-incident solar rays migrates back and forth across the Earth's Equator seasonally. (See Figure 8-1.) It lies at the Tropic of Cancer in the Northern Hemisphere summer and at the Tropic of Capricorn in the Southern Hemisphere summer (Figure 8-18). Because the regions between these two Tropics always receive sunlight at high angles, the temperature of the ocean's surface water is always warm (> 26°C) and varies but little seasonally (26 to

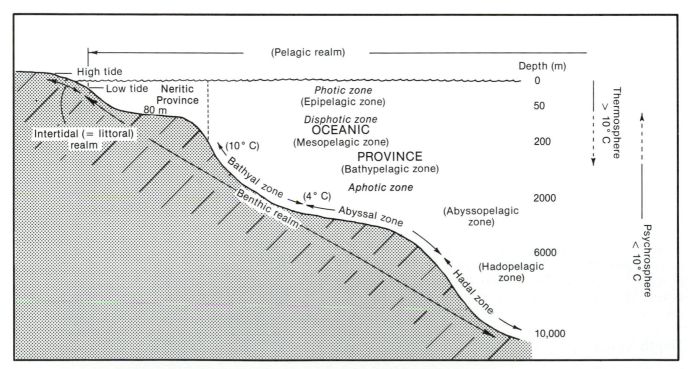

FIGURE 8-17.   Schematic profile through ocean showing ecologic subdivisions of bottom and of water. Important boundaries are based on depth of penetration of sunlight, depth where water temperature becomes 10°C (100 to 700 m), and depth below which water temperature is always 4°C or colder (about 2000 m in Atlantic Ocean; 1000 to 1500 m in Pacific and Indian Oceans). (Based on J. W. Hedgpeth, 1957a, fig. 1, p. 18, and A. F. Bruun, 1957, p. 641–645 and fig. 1, p. 642.)

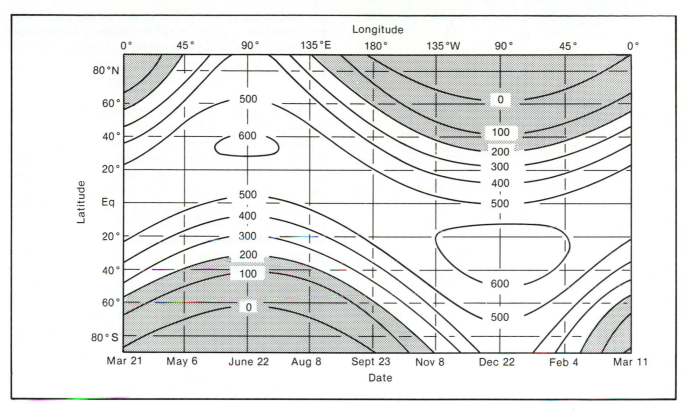

FIGURE 8-18.   Contours of daily insolation in calories per square centimeter in different areas of latitude and longitude on different days. (After B. Haurwitz, 1941; also R. W. Holmes, 1957, fig. 2, p. 112.)

28°C, Figure 8-19). Rainfall is abundant and this tends to lower the salinity of the surface waters. (See Figure 8-12.) The surface temperature of seawater in the low-latitude zones is regarded as one of the great oceanic constants that has changed little since the salinity of seawater reached its present level.

At the other extreme is the surface water of the polar zones. These areas receive no sunlight during winter. Even though they receive continuous insolation during the summer, the temperature of the surface water remains at or close to the freezing temperature of seawater (−1 to −2°C) and ice persists for most of the year.

In between these two extremes lie other zones where the characteristics of the surface waters vary considerably from season to season. These zones coincide closely with the Earth's major wind belts and resulting surface currents. In subpolar areas, the sea ice migrates seasonally. In the winter, pack ice is present, but the water temperature may be 5°C. During the summer, insolation is great enough to melt the ice and to raise the temperature of the surface seawater to about 10°C. Great bursts of diatom growth take place in the summer.

The *temperate* regions feature great winter storms having gale-force winds and much precipitation. Surface sea-water is driven eastward (in the areas of the West Wind Drift of the Southern Hemisphere, the surface water flows completely around the world). In the Northern Hemisphere, the northern temperature boundary of this region is about +5°C in the winter and +10°C in summer; the temperature of the southern boundary ranges from 15°C in winter to about 23°C in summer. The strong winds mix the surface waters. This brings up the nutrient elements from below and promotes great plankton productivity. As a result, all marine organisms are abundant.

The *subtropical* parts of the oceans coincide with the great gyres created by the surface winds. Beneath the high-pressure wind cells of the midlatitudes, the seas are calm and the winds are weak (these are the so-called "*doldrums*"). Cloud cover is scanty. The abundant sunshine creates an excess of evaporation over precipitation. As a result, the salinity of the surface seawater is elevated. (See Figure 8-12.) The upper parts of the water column are little mixed; a strong pycnocline is characteristic. Temperature boundaries vary between 15 and 23°C in the winter and between 23 and 25°C in summer.

The *tropical* parts of the oceans coincide with the great trade-wind belts, which drive the great west-flowing equatorial currents. The nearly constant trade winds cre-

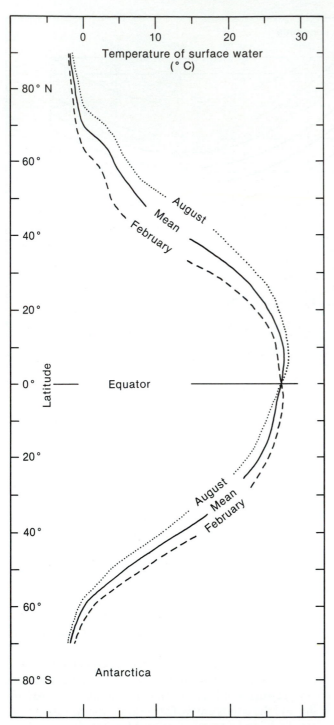

FIGURE 8-19.   Graph showing temperature of surface water of ocean plotted against latitude. Water masses having temperatures of 4°C or colder extend continuously from the surface of the sea in both polar regions to abyssal depths at all latitudes. Details of crossing of curves at Equator schematic. (Replotted by J. E. Sanders from R. H. Fleming, 1957, fig. 1, p. 89, which was based on G. Wüst, W. Brogmus, and E. Noodt, 1954.)

ate moderate-size waves. The surface waters of the tropical areas that originate in the neighboring high-salinity subtropical regions likewise display high salinity. As these waters reach the equatorial zone of great rainfall, their salinity decreases. Temperatures range from about 23 to 28°C.

The temperature ranges for various climate zones are shown graphically in Figure 8-20.

These variations in temperature and salinity affect marine organisms in diverse ways. Some organisms always remain within a particular kind of water mass; their tolerances for changes in temperature or salinity are small. In cold waters the number of species tends to be low compared to the vast numbers of individuals in each species and the individual organisms tend to be large. In some species of Foraminifera the cold-water forms coil in a right-handed fashion, whereas the warm-water forms of the same species coil in a left-handed style. Where salinity varies greatly, either up or down from that of normal seawater, the marine organisms tend to be dwarfed.

In the next section we discuss how these climate zones of the sea influence the movements of seawater.

Dynamic Physical Processes

The dynamic physical processes in the ocean include surface waves, tides, internal waves (including internal tides), surface- and subsurface water currents, and various kinds of gravity-powered flows of sediment along the bottom. We discuss the first four here and the last in Chapter 9.

WAVES

We begin with some fundamental definitions of waves, then follow with discussions of the origin of waves and

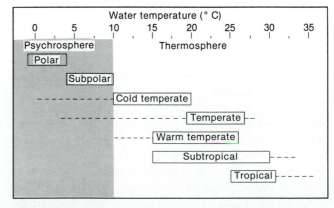

FIGURE 8-20.   Temperature ranges for various geographic climate zones. (Data from J. W. Hedgpeth, 1957; modified from M. G. Gross, 1972, fig. 14-17, p. 410.)

wave dynamics in deep water. We take up the important topics of the zones of reaction between the waves and the bottom in Chapters 10 and 11.

Definitions.   Waves are described by the following five characteristics: (1) **wavelength (L)**, *the horizontal distance between successive crests*; (2) **wave height (H)**, *the vertical distance between the highest point of a wave crest and a wave trough* (Figure 8-21); (3) **wave celerity (C)**, *the speed of advance of a wave*; (4) **wave period (T)**, *the time required for one wavelength to pass a fixed point*; and (5) **wave steepness (H/L)**, *the ratio of wave height to wavelength*. The steepest crestal angle that a water wave can maintain is 120° (Figure 8-21, B). When this angle is reached, the steepness $H/L = 1/7 = 0.143$.

Kinds and Origins.   The waves on the surface of the sea are of three chief kinds: (1) sea waves, (2) swells, and (3) tsunami. The tidal bulges, discussed in a following section of this chapter, are also waves.

What have been named **sea waves** are *steep, irregular waves that are being actively blown by the wind*. The periods of sea waves range up to about 9s. The sizes of wind-generated waves are governed by three factors: (1) speed of the wind; (2) duration of the wind; and (3) *length of open water across which the wind from a given direction can blow*, a distance known as **fetch**. Most lakes are so small that the limiting factor is the length or width of the lake. It has often been written that in the open sea, "fetch is unlimited." This statement is not correct. In the sea, the expanse of open water may be more than 15,000 km; for most purposes, it may be considered to be "unlimited" and therefore, the available open water does not restrict fetch. But in most parts of the world, wind direction is related to size of and position within a cyclonic storm. Because the definition of fetch involves the wind direction, if any factor limits the distance that wind from a given direction can blow, then that factor likewise restricts fetch. As far as winds generated by a cyclonic storm are concerned, the maximum possible fetch is determined by a distance that equals no more than one-quarter of the circumference of the storm. Because the circumference of a storm rarely exceeds 1000 km, the upper limit of open-sea fetch with respect to cyclonic storms is about 250 km (Figure 8-22). In an ocean, the "unlimited" factor is propagation distance. The importance of this factor is illustrated by the second category of waves.

In contrast to the irregular, choppy sea waves are **swells**: *long, low, regular waves on the sea surface, originally generated as sea waves by the winds associated with a cyclonic storm, but which have propagated outward beyond the boundary of the storm, and by the coalescence of sea waves, have become more regular than sea waves.*

Swells on the open sea are of such large scale that it is helpful to visualize some smaller-scale analogies. If one drops a pebble into a pond when the surface of the water is mirror smooth, one sees a series of smooth, regular ripples that move away from the point of impact as a series of concentric circles. The larger ripples travel faster, so after a few seconds, the ripples in the concentric pattern have become organized by size: smaller in the center grading outward to largest at the periphery. On a pond, wind waves that enter stands of reeds along the shore are almost instantly converted to regular waves analogous to swells.

Swells generated by large storms radiate across the open sea in a concentric pattern like that of the ripples from a dropped pebble. Swells from winter storms in the Antarctic Ocean south of New Zealand have been traced all the way to the beaches of Alaska. In the open sea, the periods of long swells that have radiated far outward from storm centers may be from 14 to 20 or 22 s; those of normal swells are 6 to 14 s. Waves having these periods are much larger than any wind-generated waves, either in the sea or on lakes. Systematic relationships exist between the speed and duration of the wind

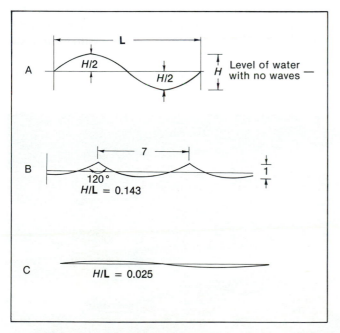

FIGURE 8-21.   Sketch profiles, in vertical plane parallel to direction of propagation, of various waves on the surface of a body of water.

A. Idealized sketch of a symmetrical wave whose profile is that of the graph of the sine of an angle. Much vertical exaggeration.

B. Symmetrical waves having critical limiting steepness. Waves would break before they could become any steeper.

C. Symmetrical wave (true scale) having the same shape as waves that are taken as the boundary between steep and low. (J. E. Sanders in G. M. Friedman and J. E. Sanders, 1978, fig. A-1, p. 465.)

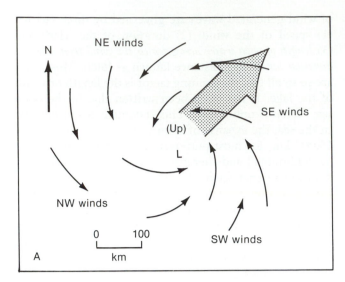

FIGURE 8-22.   Views of cyclonic centers in atmosphere of Northern Hemisphere. In sketches, large arrow shows direction of motion of cyclonic cell; small arrows, the direction of surface winds in various parts of the cell.

A. Low-pressure storm cell with counterclockwise air circulation. Note that in SE quadrant, forward speed of storm is in same direction as the circumferential SW winds. Thus what one experiences on the ground is a wind speed equal to the forward speed of the storm added to the speed of the circumferential wind. In NW quadrant, the wind on the ground is equal to the speed of the circumferential wind minus the forward speed of the storm. The difference in the ground speeds of the winds in these two quadrants, therefore, is twice the forward speed of the storm. (J. E. Sanders in G. M. Friedman and J. E. Sanders, 1978, fig. A-2, A, p. 465.)

and the heights, periods, and steepnesses of waves generated (Figure 8-23). Moreover, the steepnesses of waves that move outward from a storm change progressively with distance traveled (Figure 8-24). Wave steepness is a factor that also influences wave behavior close to shore, as is discussed in Chapter 11.

As anyone who visits a lake soon realizes, the surface of the water may be glassy smooth at some times and ruffled by whitecaps at others. The whitecaps and related waves persist only as long as the wind is blowing. When the wind dies down, the waves soon disappear. By contrast, the sea surface is nearly always wavy. Even if the local wind is calm, the shore continues to be pounded by the ever-present swells. As mentioned, the swells are waves generated by winds from distant storm centers which may be so far away that one is not aware of them.

Our final category of wave is named **tsunami**. These are also known as seismic sea waves. They are *exceptionally long-period (12- to 15-min) waves on the sea surface caused by displacement of the sea floor*. Such displacements result from several causes. These include gravity-powered mass movements of bottom sediments that were triggered by earthquakes or from collapse of the tops of volcanic cones. Tsunami is a Japanese word meaning "harbor waves." In Japanese, this word is spelled the same in both singular and plural forms. However, the term "tsunamis" appears widely in the scientific literature. Another popular error about tsunami is to refer to them as "tidal waves." The wavelengths of tsunami are so long

FIGURE 8-22.   (*Continued*)
   B. Storm at sea in North Atlantic Ocean as photographed looking obliquely downward from Apollo 7 spacecraft. The width of the field of view is about 200 km. (NASA, in J. E. Sanders, 1981, fig. 15.3, p. 361.)

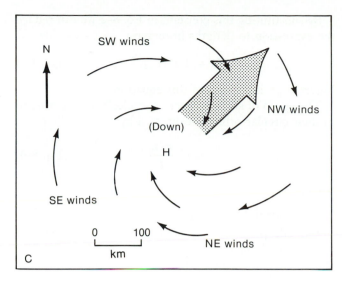

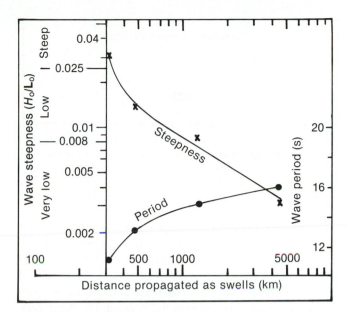

FIGURE 8-22. (*Continued*)

C. High-pressure cell with clockwise circumferential winds. Maximum ground-experienced winds are in NW quadrant (SW circumferential winds added to forward speed of cell); minimum winds in SE quadrant (NE circumferential winds subtracted from forward speed of cell). (J. E. Sanders in G. M. Friedman and J. E. Sanders, 1978, fig. A-2, B, p. 465.)

FIGURE 8-24.   Graph of propagation distance versus properties of swells starting at a point 320 km away from the storm center, a distance the waves would travel in 12 h. Scale for steepness curve is at left; that for period at right. (J. E. Sanders in G. M. Friedman and J. E. Sanders, 1978, fig. A-4, p. 466; based on data in J. M. Caldwell, 1966.)

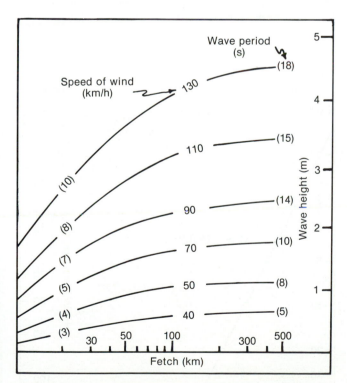

FIGURE 8-23.   Graph of fetch versus wave height on which are plotted curves showing properties of waves (periods in parentheses) for winds of six different speeds (shown by numbers breaking the curves). (Replotted by J. E. Sanders from C. A. M. King, 1959, fig. 3-8, p. 78; based on data of Darbyshire, 1952, as originally graphed by Silvester, 1955.)

(hundreds of kilometers) and their heights in the open sea are so small that they cannot be detected by eye.

Wave Dynamics. According to the ways in which the water moves as the waves pass, we can classify waves on the water surface into two groups: (1) *waves that cause masses of water to oscillate but not to undergo appreciable net displacement in the direction of wave advance* (**waves of oscillation** or **oscillatory waves**) and (2) *waves in which the masses of water beneath the moving crests are displaced in the direction of wave advance* (**waves of translation** or **translatory waves**). We discuss only waves of oscillation here. Waves of translation are discussed in Chapter 11 in the section on beaches.

The profiles of an idealized train of waves traveling across the surface of water that is deeper than half the wavelength can be described as a series of symmetrical crests and troughs resembling the graph of the sine of an angle (Figure 8-25).

As each crest passes, the water surface is elevated and all the water "particles" immediately below the surface down to the depth where wave influence is no longer significant are moved forward in the direction of wave advance (Figure 8-25, points A and A'). This forward motion is happening simultaneously beneath all the wave crests. After a wave crest has passed, the water surface is gradually lowered and the wave-moved water "particles" are displaced downward (Figure 8-25, point B). As each

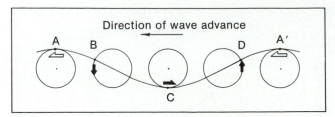

FIGURE 8-25.   Sketch profile, in vertical plane parallel to direction of propagation, of idealized symmetrical wave of oscillation
in deep water, showing direction of wave travel and movement
of water "particles" in their circular orbits. (J. E. Sanders in G. M.
Friedman and J. E. Sanders, 1978, fig. A-5, p. 466.)

wave trough passes, the water surface attains its lowest elevation and the underlying water "particles" move
in a direction opposite to that of wave advance (Figure
8-25, point C). After a trough has passed, the height of
the water surface begins to increase and the water "particles" move upward (Figure 8-25, point D). After each
complete wave form has gone by, the water "particles"
have described a circular orbit lying in a vertical plane
whose edge parallels the direction of wave advance.

The various motions are expressed mathematically as
follows.

The celerity of the wave is

$$\mathbf{C} = \left(\frac{g\mathbf{L}_0}{2\pi} \tanh \frac{2\pi y}{\mathbf{L}_0} \right)^{1/2} \qquad \text{(Eq. 8-1)}$$

where

$\mathbf{C}$ = speed of wave advance (m/s)

g = acceleration of gravity

$\pi \approx 3.14$

$\tanh$ = hyperbolic tangent (term related to
exponential functions that is defined by
$\tanh v = (e^v - e^{-v})/(e^v + e^{-v})$,
where v = a variable and e = base of
natural logarithms, i.e., Napieran base)

y = depth of water (m)

$\mathbf{L}_0$ = deep-water wavelength (m)

The subscript zero will be used to indicate relationships for deep-water waves.

Where the water depth denoted by y is greater than
$\mathbf{L}_0/2$, the term $2\pi y/\mathbf{L}_0$ converts to $2\pi/\mathbf{L}_0/2/\mathbf{L}_0 = \pi$.
The hyperbolic tangent of π is about equal to 1; hence,
for waves in deep water we can simplify Eq. 8-1 to

$$\mathbf{C} = \left(\frac{g\mathbf{L}_0}{2\pi} \right)^{1/2} \qquad \text{(Eq. 8-2)}$$

The term $(g/2\pi)^{1/2}$ is a constant computed as (length
in meters)

$$\left(\frac{9.80}{6.28} \right)^{1/2} = (1.56)^{1/2} = 1.25$$

By substituting this constant in Eq. 8-2 we can reduce
the expression to (lengths in meters)

$$\mathbf{C} = 1.25(\mathbf{L}_0)^{1/2} \qquad \text{(Eq. 8-3)}$$

Figure 8-26 is a graph of this equation.

The speed of surface water "particles" moving around
in their circular orbits is expressed by

$$u = \frac{\pi H_0}{T} \qquad \text{(Eq. 8-4)}$$

where

u = speed of water "particles" (m/s)

H_0 = deep-water wave height (m) = d_s (diameter
of circular orbits of surface water "particles";
the subscript s refers to surface conditions)

T = wave period (s)

$\pi \approx 3.14$

We can use these equations to calculate some of the
important properties of waves. For example, the celerity
of typical ocean waves having wavelengths of 64 m is
calculated as

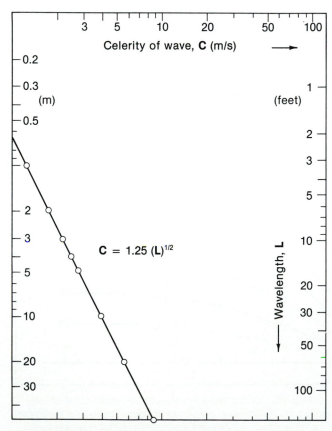

FIGURE 8-26.   Graph of relationship between celerity of deep-
water waves of oscillation and wavelength. (J. E. Sanders in
G. M. Friedman and J. E. Sanders, 1978, fig. A-6, p. 467.)

$$\mathbf{C} = 1.25(64)^{1/2}(\text{m/s})$$

$$= 10 \text{ m/s}$$

The wave period is written mathematically as

$$T = \mathbf{L}/\mathbf{C} \qquad \text{(Eq. 8-5)}$$

where

$\mathbf{L}$ = wavelength (m)

The period of waves having wavelengths of 64 m is $64/10 = 6.4$ s.

In waves having profiles that are pure sine curves, wave period and wavelength are related as

$$\mathbf{L_0} = \frac{g}{2\pi}T^2 \qquad \text{(Eq. 8-6)}$$

The $(g/2\pi)$ part of Eq. 8-6 is a constant (using meters for length units) equal to 9.80/6.28 or 1.56. Therefore, Eq. 8-6 becomes

$$\mathbf{L_0} = 1.56T^2 \qquad \text{(Eq. 8-7)}$$

Figure 8-27 is a graph of wave period versus wavelength for deep-water waves of oscillation.

Let us assume that the steepness of the waves we have been discussing is 0.025, a value commonly used as the boundary between steep- and low waves (Figure 8-28). This means that deep-water wave height, H_0, is $64 \times 0.025 = 1.6$ m. Using Eq. 8-4, we can calculate the speed of motion of the surface-water "particles" in their circular orbits as

$$u = \frac{\pi 1.6}{6.4} = 0.785 \text{ m/s} \qquad \text{(Eq. 8-8)}$$

The diameters of the circular orbits generated by the passing waves diminish downward according to an exponential decline function (Figure 8-29). At the surface the diameters of the orbits, d_s, are equal to wave height, H_0. At a depth y beneath the still-water surface the diameters of the orbits, d_y, are given by the equation

$$d_y = d_s e^{-2\pi y/\mathbf{L_0}} \qquad \text{(Eq. 8-9)}$$

where

d_y = diameter of orbit at depth y (m)

d_s = diameter of orbit at water surface (deep-water wave height, H_0) (m)

e = base of natural logarithms

y = depth beneath still-water surface (m)

At a depth equal to 1/9 the deep-water wavelength, the diameters are equal to half their value at the surface. At a depth of half the deep-water wavelength, the diameters of the orbits are 1/23 of their value at the surface—so small that they can be neglected. We can show this by solving Eq. 8-9 with the dimensions of the waves we have

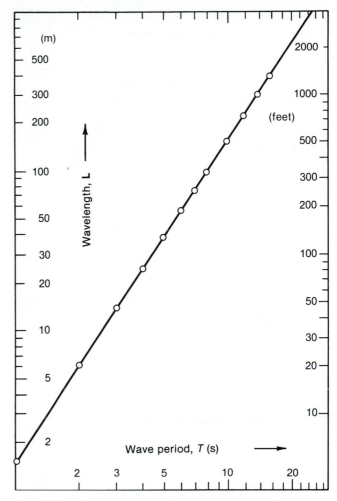

FIGURE 8-27.   Graph of wavelength versus wave period for deep-water waves of oscillation. (J. E. Sanders in G. M. Friedman and J. E. Sanders, 1978, fig. A-7, p. 468.)

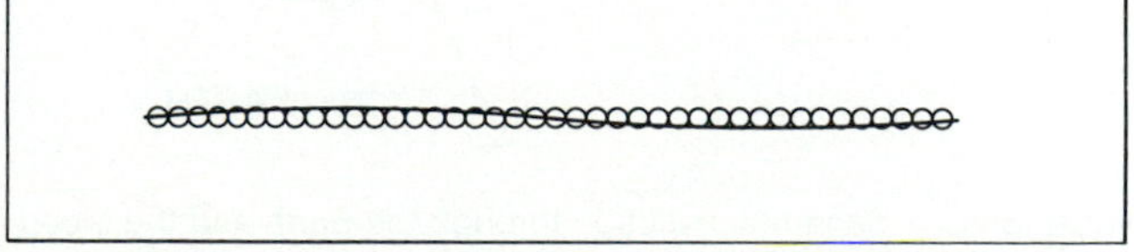

been studying ($d_s = H_0 = 1.6$ m; $\mathbf{L_0} = 64$ m), where $y = 32$ m:

$$d_y = 1.6e^{-2\pi \cdot 32/64} = 1.6e^{-3.14} = \frac{1.6}{23} = 0.069 \text{ m}$$

SOURCES: Clifton and Dingler, 1984; Friedman and Sanders, 1978; Inman and Bagnold, 1963; C. A. M. King, 1972; Komar, 1976; Tricker, 1964.

FIGURE 8-28.   Sketch profile, in vertical plane parallel to direction of propagation, of wave in Figure 8-21, C, enlarged to show circular orbits followed by water "particles" at the surface. (J. E. Sanders in G. M. Friedman and J. E. Sanders, 1978, fig. A-8, p. 468.)

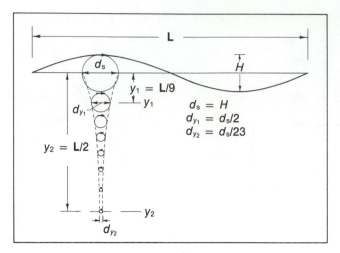

FIGURE 8-29. Schematic profile, in vertical plane parallel to direction of propagation of waves of oscillation in deep water, showing rapid decline downward in diameters of circular orbits described by water "particles" as a result of passage of waves on the water surface. Further explanation and definitions in text. (J. E. Sanders in G. M. Friedman and J. E. Sanders, 1978, fig. A-9, p. 468.)

OCEANIC TIDES

Nearly everyone who has visited the seashore is aware of the daily changes of water level known as the tides. Moreover, this awareness may extend to the notion that tidal action is somehow connected to the Moon and/or the Sun. But beyond that, a certain vagueness is likely to prevail. In order to keep all readers on the same level, the following paragraphs review the principles of the tides, with emphasis on the factors that create cyclic variability.

In analyzing the ocean tides, the first principle that one needs to understand is a fundamental precept of celestial mechanics. In any pair (or more) of orbiting bodies, the focus of the elliptical orbits is *not* the center of any of the orbiting bodies. This is true even if one body is very much larger than the others. The true focus is *the common center of mass* (**barycenter**) *of all the bodies.* We can illustrate by considering the relationships associated with the Earth–Moon barycenter.

Orbit of the Earth and the Moon Around the Earth–Moon Barycenter. As for the Earth and the Moon, their common center of mass, or barycenter, lies within the Earth in a zone that ranges in depth from the surface between 1436.5 km at lunar apogee and 2047.5 km at lunar perigee (the mean value is 1707 km; Figure 8-30).

To a first approximation, tidal rise and fall of the sea can be visualized as resulting from the interaction of a rapidly rotating oblate spheroid, the solid Earth, and a slower-moving prolate spheroid of the Earth's oceans that results from and moves around with the orbiting

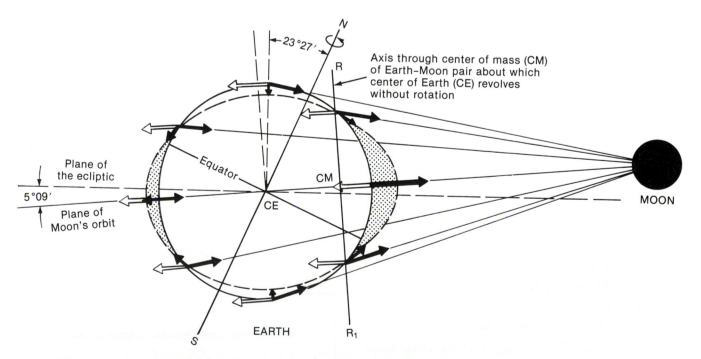

FIGURE 8-30. Schematic section through the Earth and the Moon drawn perpendicular to the plane of the Earth's orbit around the Sun (plane of the ecliptic), showing how the Moon's gravitational attraction and the centrifugal force of the orbital motion of the Earth–Moon pair around their common center of mass (the Earth–Moon barycenter, CM on diagram) forms two water bulges, the one on the side of the Moon being larger than the one on the opposite side. (C. R. Longwell, R. F. Flint, and J. E. Sanders, 1969, fig. 14-5, p. 321).

Earth–Moon pair. This concept forms what is known as the equilibrium theory of tides (Figure 8-31). Another way of describing the prolate spheroid is to visualize the ocean as consisting of two "bulges," one on the side of the Earth that is toward the Moon, and the other on the opposite side. These two "bulges" result from the ocean water's combined interactions with the Moon's gravitational attraction, which varies cyclically as the longitude of lunar perigee shifts, and the centrifugal force of the Earth–Moon pair's orbit around the Earth–Moon barycenter (not to be confused with the centrifugal force associated with the Earth's daily rotation on its polar axis, which results in the oblate shape of the solid Earth).

Lunar Variables. The three chief variables that can be noticed easily on a monthly time scale are phase, Earth–Moon separation distance, and declination.

The phases of the Moon result from the ever-changing alignments of the Sun, the Earth, and the Moon. Twice during each orbit, the Moon experiences what are known as syzygy phases; that is, *times when the centers of the Sun, the Earth, and the Moon lie along a straight line.* At the syzygy of New Moon, the Moon lies between the Earth and the Sun; at the syzygy of Full Moon, the Earth lies between the Sun and the Moon. During these syzygy alignments the tidal amplitudes are greater than normal, yielding what are known as *spring tides* (Figure 8-32, A and C).

When the Moon is at a right angle to a line from the Earth to the Sun, the Moon is in a quarter phase. At quarter phases the gravitational pull of the moon and

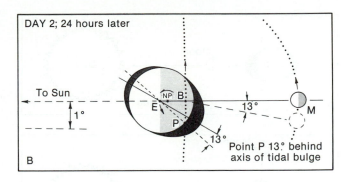

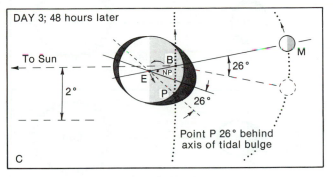

FIGURE 8-31.   (*Continued*)
B. After 24 h, the Moon has advanced ca. 13° in its orbit, pulling the long axis of the water bulges with it. At point P, high tide is 51 min away.
C. After 48 h, the Moon and tidal bulge have advanced ca. 26° from position on day 1; at point P, high tide is 2 × 51 min away. (J. E. Sanders in G. M. Friedman and J. E. Sanders, 1978, fig. E-1, p. 538.)

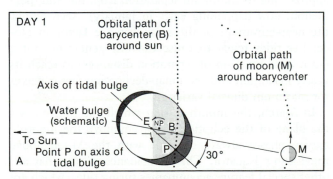

FIGURE 8-31.   Equilibrium concept of tides shown in schematic sketch of Earth–Moon pair in orbit around the Sun (out of figure at left), as seen looking down from vertically above the plane of the ecliptic (lunar declination omitted). The long axis of the prolate water spheroid of the Earth's oceans (shown by the dark bulges) does not point directly toward the Moon; this lack of alignment is illustrated here by assigning an arbitrary displacement of 30°. (B = Earth–Moon barycenter; NP = Earth's North Pole of rotation, with direction of rotation shown by curved arrow having only one barb.) In its orbit around the Sun, the Earth–Moon barycenter advances about 1° per day.
A. High tide on day 1 at point P, which lies along the axis of the tidal bulge on the side of the Full Moon.

the Sun are at right angles. This yields tidal amplitudes that are less than normal, or *neap tides.* (See Figure 8-32, B and D.)

The Moon's orbit is distinctly elliptical. As a result, the *Earth–Moon separation distance* changes during each lunar orbit. The point of closest approach of the Moon to the Earth is known as *perigee*; the point of maximum separation between the Moon and the Earth is named *apogee.* Because the Moon's gravitational pull on the Earth and the rate of rotation of the Earth–Moon pair about their common barycenter are greatest at the perigee position, the maximum tidal range also happens at perigee.

The height of the Moon above the horizon is known as *lunar declination.* This declination changes regularly because the Moon's orbit lies in a plane that makes an angle of 5°09' with the plane of the Earth's orbit. (See Figure 8-30.) Therefore, in each lunar orbit the Moon will reach its maximum north declination and maximum south declination once, and will pass above the Earth's Equator (point of zero declination) twice, once during each passage from the points of maximum declination.

Lunar declination affects the relative amplitudes of tides in localities having two high tides and two low tides

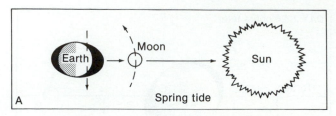

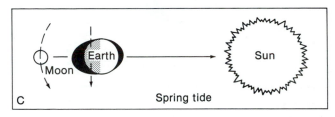

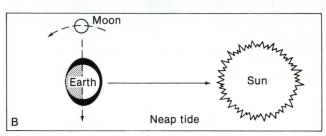

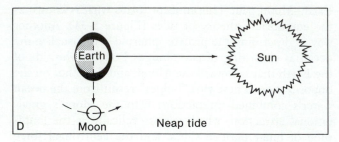

FIGURE 8-32.   *(Continued)*

D. Last quarter phase of the Moon; neap tides. The center of the Earth lies along the orbital path. (Authors, corrected from J. E. Sanders, 1981, fig. 15.11, p. 367.)

FIGURE 8-32.   Spring tides, neap tides, and phases of the Moon, as seen looking down from above; distances schematic; angles of plane of the ecliptic with invariable plane and of Moon's orbit plane with plane of the ecliptic omitted. Dashed lines show orbital paths (straight for Earth–Moon barycenter; curved for Moon), with arrows indicating direction of motion. Long solid arrows show direction of maximum gravitational pull on the Earth from the Moon and the Sun.

A. New Moon; Earth, Moon, and Sun aligned, but with Moon between the Sun and the Earth. The technical term in astronomic language for an Earth–Moon–Sun alignment is a syzygy phase. The large tidal ranges that coincide with syzygy phases are known as spring tides. Notice that because of the Moon's position toward the Sun, the center of the Earth has shifted off the orbital path of the Earth–Moon barycenter in a direction away from the Sun.

B. First quarter phase of the Moon; Sun, Earth, and Moon form 90° arrangement. The small tidal ranges that coincide with the Moon's quarter phases are known as neap tides. The center of the Earth lies along the orbital path of the Earth–Moon barycenter.

C. Full Moon; Moon, Earth, and Sun in syzygy, but with Moon on side of the Earth opposite the Sun. As in A, spring tides. Because of the Moon's position away from the Sun, the center of the Earth has shifted off the orbital path of the Earth–Moon barycenter in a direction toward the Sun.

each day. The technical term for such daily changes in tidal amplitude is *diurnal inequality*. When the Moon is at its north- or south positions, diurnal inequality is greatest. When the Moon is above the Earth's Equator, the diurnal inequality of tidal heights disappears.

The effects of these three factors on the predicted tidal heights at Fire Island Inlet, New York, are shown in Figure 8-33. Clearly evident are the two spring-tidal bulges associated with the syzygy phases, but because of the coincidence of perigee with Full Moon, the bulge associated with Full Moon exceeds that of New Moon. Also notice the diurnal inequality associated with the times of maximum declination (N for north on 3 September and S for south on 19 September).

The relationships between phase and declination shift on a regular basis monthly, but these two factors always coincide in the same way with the key positions of the Earth's declination with respect to the Sun. This creates an annual (seasonal) phase-declination lunar cycle.

For example, every December, the month of the winter solstice and thus of minimum solar declination as seen in the Northern Hemisphere, the monthly syzygy phase coincides with maximum south lunar declination. December is also a time when the Earth–Moon pair is approaching its minimum separation from the Sun (perihelion, now happening early in January). Accordingly, the near-minimum of the solar-distance factor in December tides tends to cause the tide-generating force (an inverse function of separation distances) to reach its yearly maximum and the lunar-declination factor makes for maximum diurnal variation.

In March, the month during which the Sun crosses the plane of the ecliptic (spring equinox), the monthly syzygy phases always coincide with the Moon above the Earth's Equator. This coincidence means that another special feature accompanies spring tides, which are themselves special because of their greater-than-normal amplitudes. This added feature is that these higher-than-normal spring-tidal amplitudes will prevail for 5 or 6 successive tides. Stated in another way, the diurnal inequality vanishes just at the time of maximum tidal amplitudes.

In June, the month of the summer solstice when the Sun reaches its maximum height in the northern sky, the monthly syzygy phases and maximum declination coincide. Because the Earth–Moon pair now attains its maxi-

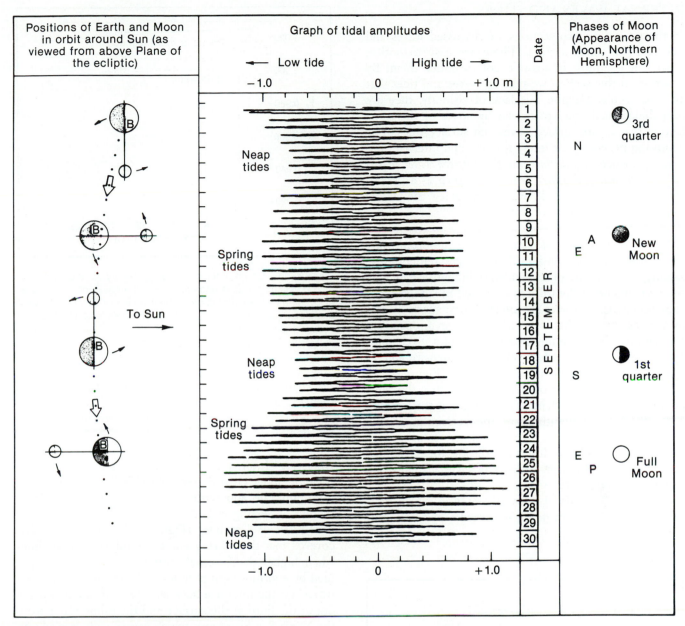

FIGURE 8-33.   Orbit of Earth–Moon pair (left), predicted tidal heights for Fire Island Inlet for September 1988 (middle), and selected astronomic data (phases of the Moon, lunar declination, and dates of apogee and perigee; right), showing how tidal heights and diurnal variations are controlled by the relationships among lunar phase, lunar declination, and times of apogee and perigee. Further explanation in text. (Redrawn by J. E. Sanders using September 1988 tidal data from J. E. Sanders in G. M. Friedman and J. E. Sanders, 1978, fig. E-2, p. 539.)

mum separation distance from the Sun (aphelion) early in July, the solar-distance factor in June tides approaches its maximum value and this causes the Sun's tide-generating force to approach its minimum values. June syzygy tides resemble those in December in that diurnal inequality is always at maximum values.

In September, the month of the autumnal equinox, during which the Sun again crosses the plane of the eclip-

tic, the monthly syzygy phases coincide with the Moon above the Earth's Equator. This repeats the March pattern: syzygy-phase spring-tidal amplitudes accompanied by zero diurnal inequality. (On Figure 8-33, the syzygy phases are New Moon on the 11th and Full Moon on the 25th; notice how both coincide with the Moon's position above the Earth's Equator and thus zero diurnal inequality.)

INTERNAL WAVES AND -TIDES

In the preceding two sections we discussed waves and tides on the water surface. These are easy to notice; indeed, it is almost impossible not to notice them. By contrast, in this section we take up waves and tides that are not visible. Despite this lack of visibility, they are nonetheless real and are driven by some of the same factors that create surface waves and surface tides. The condition necessary for internal waves and internal tides is the presence of *a distinct interface between water masses having contrasting densities*, which is known as a **pycno-cline**. Before we get to the waves, we review the factors that control the density of water.

Factors Affecting Density of Water.   Three important variables affecting the density of water are temperature, salinity, and quantity of suspended sediment. We include the first two here; the third is discussed in Chapter 9.

Sedimentary processes are affected particularly by two properties of water that vary with changes in temperature and salinity: (1) the point where water of a given temperature and salinity attains its maximum density, and (2) the temperature at which freezing begins (Figure 8-34). A remarkable property of fresh water is that the temperature of maximum density is not right at the freezing point (0°C), but is slightly above the freezing point, at a temperature of +3.98°C for water of zero salinity (Figure 8-35). This relationship is responsible for the annual pattern of circulation in fresh-water lakes in temperate climate zones (Chapter 14).

A zone in a stratified body of water in which density varies rapidly with depth is named a pycnocline. If

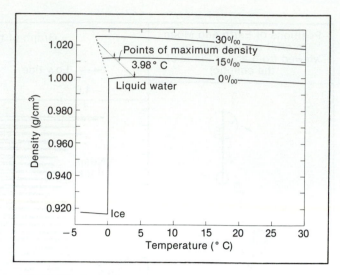

FIGURE 8-35.   Graph of temperature vs. density for liquid water having various salinities and for ice. Small arrows and line of dots at upper left mark temperatures where waters of various salinities attain their maximum densities. (Based on data in M. G. Gross, 1972, fig. 6-8, p. 150.)

the change of density results from a change in salinity, then the term *halocline* is appropriate. *A marked change of density with depth that is caused by temperature* is a **thermocline**.

Discovery.   Internal tides were discovered by the Swedish oceanographer Otto Pettersson's (1848–1941) investigations into the apparently erratic appearances of large schools of herring in Gullmarfjord on the coast of Sweden north of Goteborg (Figure 8-36). Pettersson discovered that the herring entered Gullmarfjord at times when higher-than-normal spring tides were accompanied by enormous (amplitudes to 15 m) bulges (internal waves) on the interface between the fresh water flowing out of the fjord at the surface and the saline water from the North Sea-Skagerrak, which occasionally flowed into the fjord at a lower level. Pettersson devised an oil-filled copper float that he could trim so that it would sink through the surficial layer of fresh water but would not sink through the underlying salt water. A calibrated rod 5 m long extended upward from the float. By noting the length of the rod protruding from the water, Pettersson was able to monitor on a daily basis the depth of the fresh water/salt water interface within the fjord opposite his marine station at Borno. The depth to this interface ranged from about 5 m, the usual situation, to zero, when the float appeared at the surface. This meant that a great influx of salt water had entered the fjord and had lifted the fresh water, causing it all to flow outward. He found that the level of the pycnocline varied as a function of the lunar tide. Accordingly, from astronomic data con-

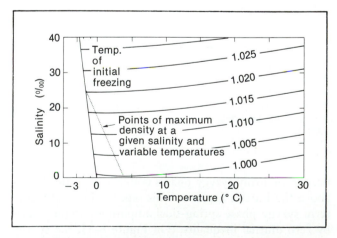

FIGURE 8-34.   Graph of salinity vs. temperature showing how density (shown by labeled lines inside graph with units of density expressed as g/cm³) is related to maximum density and to temperature of initial freezing. (Based on data in M. G. Gross, 1972, fig. 7-15, p. 188.)

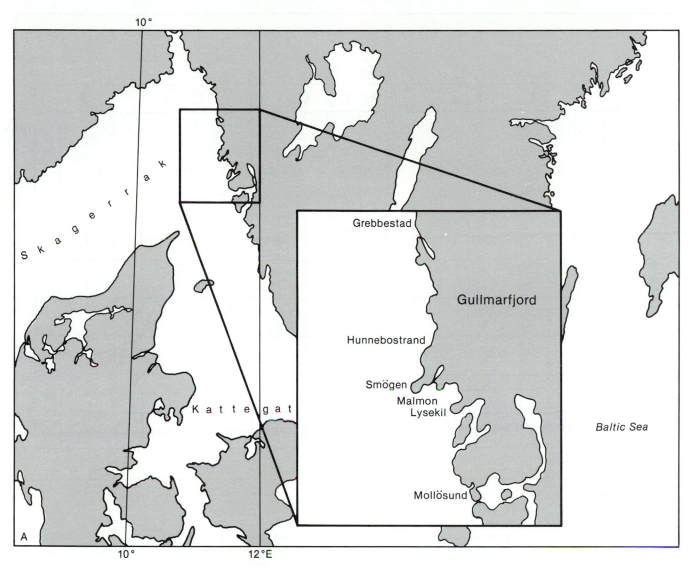

FIGURE 8-36. The "Moon waves" of Gullmarfjord, Sweden.
A. Location map with inset map of Gullmarfjord. Index map shows coasts of Norway, Sweden, and Denmark facing the North Sea gulfs known as Skaggerrak and Kattegat.

cerning variations in the tide-generating force, he was able to predict when the schools of herring would enter Gullmarfjord (Figure 8-36, B).

Pettersson referred to these perturbations on the pycnocline as the "Moon waves of Gullmarfjord." His results have been virtually ignored by oceanographers.

SOURCES: Apel, 1981; Fu and Holt, 1984; Komar, 1976; O. Pettersson, 1912, 1930; Wunsch, 1978.

Circulation and Stagnation of Waters in Basins

The water in almost any basin experiences changes of level, of density, or of both, and these cause currents to flow. The currents, in turn, may cause the basin waters to circulate or to become stagnant. Of particular importance to sediments are the patterns of circulation that lead to the development of anoxic conditions on the one hand, and to deposition of evaporites on the other.

General Principles

The significance of the circulation of waters from the deep sea into basins marginal to a deep-sea basin was discovered and emphasized in the nineteenth century by the English scientist W. B. Carpenter (1813–1885).

After the first reliable measurements of water temperatures at various depths in the sea had been made, the observed changes with depth were at first interpreted simply as functions of depth. By contrast, the tempera-

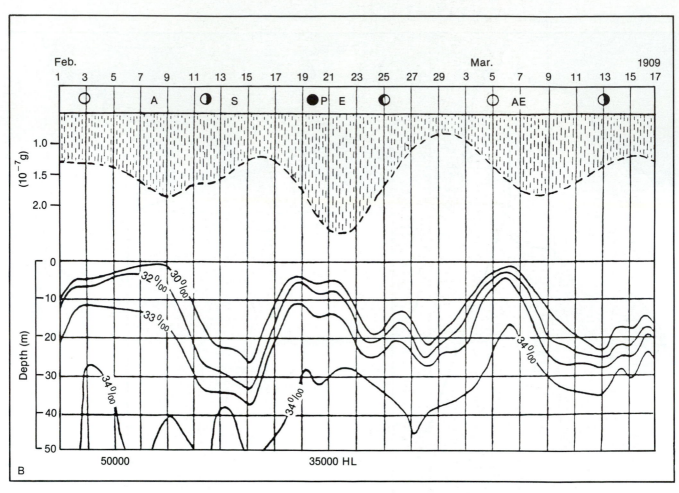

FIGURE 8-36. (*Continued*)

B. Graph of relationships in February and the first half of March, 1909. At top are shown the astronomic data of Moon phase, declination (N for maximum north; E, Moon over the Earth's Equator; S, for maximum south), and Earth–Moon separation distance (A, apogee; P, perigee). The dashed-line curve indicates the calculated tide-producing force. The bottom curves show the depth of the pycnocline, as indicated by lines of equal salinity. The numbers at the bottom margin represent the number of hectaliters of herring caught in the Gullmarfjord. The fishermen fished daily, but most of the time their nets were empty. (B, after O. Pettersson, 1930, fig. 2, p. 270.)

ture in the water of the Mediterranean Sea was found to be nearly the same from top to bottom and to be equal to that of the surface water entering from the Atlantic at the Strait of Gibraltar (Figure 8-37). Carpenter correctly analyzed the implications of the temperature/depth profiles in the Mediterranean and explained that two contrasting kinds of water circulation existed. The Mediterranean is an example of a semi-isolated basin in which surface water of normal salinity flows in at the surface, becomes denser by evaporation, and sinks. The denser water flows out of the basin as a deeper countercurrent beneath the entering surface water (Figure 8-38). By contrast, Carpenter cited the Black Sea and the Baltic Sea as examples of semi-isolated basins in which a strong outflow of fresh water occupies nearly the entire cross section of the narrow entrance to the basin. The outflowing fresh

water prevents enough salt water from entering these two basins to ventilate the bottom waters. As a result, the dissolved oxygen in the deeper waters of these two basins is consumed by oxidizing first-cycle organic matter much faster than it is replenished from contact with the atmosphere. Carpenter not only recognized the importance of these two kinds of circulation patterns for modern seas, but also pointed out their paleogeographic significance for interpreting ancient sedimentary strata. For some reason not known to us, Carpenter's work was not mentioned in the Challenger reports. Therefore his remarkable papers on water circulation into and out of marginal basins have not been read by later workers who assumed that the Challenger scientists had included all previous work. As a result, Carpenter's fundamental contributions have been generally ignored.

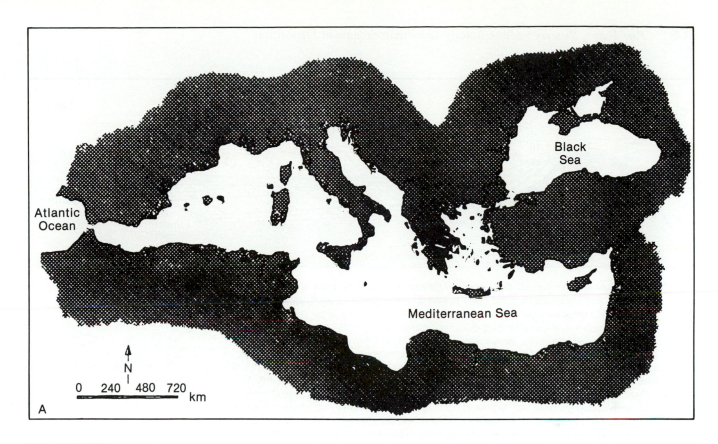

A

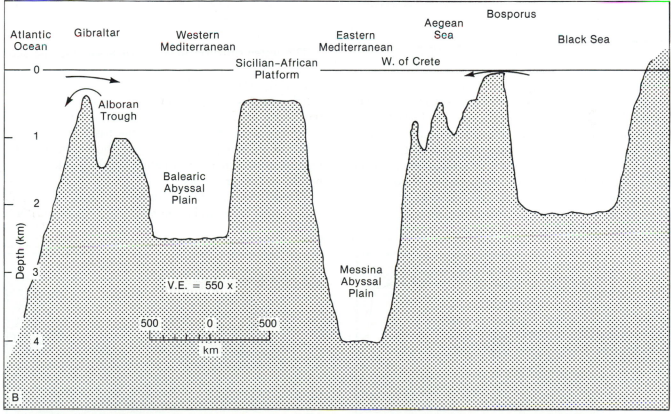

B

FIGURE 8-37.   Atlantic Ocean, Mediterranean Sea, and Black Sea, and the generalized flow patterns as recognized by W. B. Carpenter.

    A. Map of Mediterranean and Black seas. (After A. Brambati, 1972, fig. 2, p. 716.)

    B. Profile from Black Sea to Atlantic Ocean through Bosporus, Aegean Sea, and the Mediterranean. Arrows show general direction of water flow. (Authors.)

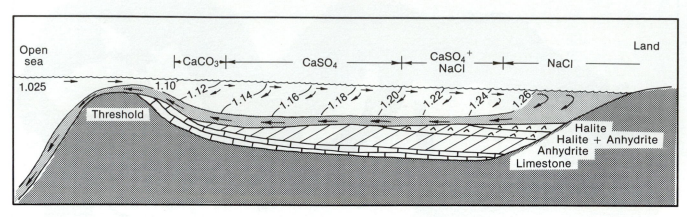

FIGURE 8-38.   Schematic longitudinal profile-section through semi-isolated basin that is located in a hot, arid climate and separated from the open sea by a narrow portal whose threshold depth, although shallow, is still great enough to permit two-way flow of water. Lines show inferred density (g/cm³); arrows, directions of currents. Evaporitic products shown on floor of basin are derived from the relationships between brine densities and precipitates in Usiglio's experiments (Figure 3-1), and on the supposition that evaporite materials will accumulate on the bottom directly beneath zones where corresponding densities exist in the surface waters. The deposit marked "limestone" does not refer to rocks formed by cementation of carbonate sediments, but to a chemical precipitate. "Anhydrite" has been used for the calcium-sulfate material; this usage reflects the fact that petroleum geologists deal with data from deeply buried formations where anhydrite is the stable form and gypsum is not present. In the kind of evaporite basin shown, the calcium sulfate would be gypsum initially. However, any evaporitic gypsum would become a part of the stratigraphic record only if it escaped destruction by sulfate-reducing bacteria. (After P. C. Scruton, 1953, fig. 4, p. 2505.)

The morphologic setting for these patterns of water circulation is a basin that is separated from the open sea or another basin by a narrow strait, known as the portal, in which the depth is much less than that inside or outside the basin (Figure 8-39). The shallow entrance to a semi-isolated basin is sometimes termed a "sill." This usage is based on analogy with a raised sill in a doorway or a window. However, in geology the term sill has been preempted for concordant tabular plutons. For the shallow zone at the entrance to a basin, the word *threshold* seems a preferable alternative and we adopt it.

The water situation in marginal basins may be subjected to long-term changes connected with climate and eustatic sea level. In moist climates, abundant fresh water will flow from the surrounding lands into and, eventually, out of the basin. In hot, dry climates, evaporation will lower the level of the water in the basin and seawater will flow in to equalize the level. Short-term modifying effects on general flow patterns include tides, winds, floods, and other factors.

The geologic importance of the circulation pattern of basin waters began to be emphasized by modern workers

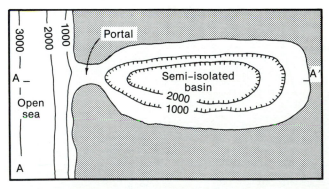

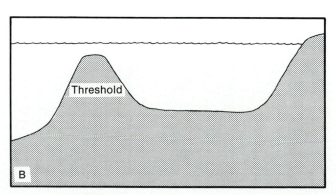

FIGURE 8-39.   Semi-isolated, landlocked basin having narrow connection to open sea, morphologic setting. (Authors.)
    A. Schematic map; depth contours in meters.
    B. Schematic profile with large vertical exaggeration. No attempt has been made to show the characteristics of water masses. These vary according to local conditions.

during the 1930s with the publication of papers summarizing hydrographic research on Norwegian fjords, the *Snellius* reports on the hydrography and sediments of the basins in the Indonesian region, and the book entitled Recent Marine Sediments (1939).

SOURCES: Byers, 1977; W. B. Carpenter, 1871, 1874, 1882; Jenkyns, 1986.

Basins Having Ventilated Waters

In semi-isolated basins where fresh water flows out but where the thickness of this outflow is less than the depth at the threshold, a two-way flow is established in the portal (Figure 8-40). At the surface, fresh water flows out of the basin; at depth, salt water flows in.

Examples of semi-isolated basins having ventilated waters are most of the fjords on the coast of British Columbia (locally referred to as inlets) and many Norwegian fjords. In these fjords the color of the fine-textured bottom sediments is light gray or greenish gray.

Basins Having Anoxic Waters

In semi-isolated basins where the thickness of the outflowing fresh water equals the depth of water at the threshold, only one-way flow prevails through the portal (Figure 8-41). The fresh water may not block off all salt water, but may allow such a small amount to pass that its overall ventilating effect is negligible. In such basins the level of the main pycnocline lies deeper than the threshold. The outflowing fresh water prevents oxygen-bearing salt water from the open sea from entering the basin or allows only small amounts to enter. As a result, the bottom water within the basin is fully cut off from any possibility of renewing its supply of dissolved oxygen or receives only a small influx. The anoxic conditions thus established are analogous to those in a stratified lake during the summer (See Figure 14-48.) or to those in a permanently stratified tropical lake, such as the Dead

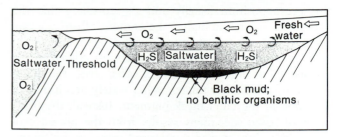

FIGURE 8-41.   Semi-isolated basin with anoxic bottom water. Outflowing fresh water occupies full depth of shallow threshold. Schematic longitudinal profile with open sea at left. Slope of water surface vastly exaggerated. (After K. M. Strøm, 1939, fig. 1, p. 358.)

Sea. (See Figure 14-53.) From the surface waters, where organisms are abundant, first-cycle organic matter can "rain" down through the pycnocline onto the basin floor.

It is a fundamental principle of organic chemistry that all organic matter that is capable of being oxidized will tend to be oxidized. But this can happen only where a supply of dissolved oxygen exists. Therefore, below the pycnocline, the falling first-cycle organic matter will tend to consume any available dissolved (free) oxygen. If this supply of oxygen becomes exhausted, the first-cycle organic matter ceases to be oxidized and begins to accumulate. Anaerobic bacteria proliferate. Their metabolism generates hydrogen sulfide (Eq. 3-10). Modern examples of semi-isolated basins having anoxic conditions include parts of the Baltic Sea; the Black Sea; Kaoe Bay, on the coast of Halmahera, Indonesia; a few Norwegian fjords; and the Cariaco basin, Venezuela.

The characteristic deposit of an anoxic basin having stagnant bottom water is black silt and -clay. The fine-textured black sediments of the Black Sea contain 25 to 35% by weight organic carbon. In the fine sediments of Norwegian fjords having stagnant bottom waters, amounts of organic carbon as high as 23.4% have been recorded. Organic-carbon contents in fine-textured fjord sediments in excess of 12% are common. By contrast, the proportion of organic carbon in fine-textured sediments deposited under oxidizing bottom waters ranges between 1 and about 2.5%.

In the stratigraphic record of the bedrock underlying continents, black shales are common. These dark-colored rocks have attracted considerable study. Therefore, the properties of some of them are known in elegant detail. In the study of a black shale, the usual matters to be resolved include (1) nature of the black pigment, (2) quantity of organic carbon in the sediments, and (3) environment of deposition, including the inferred water depth and conditions of circulation of the waters in the depositional basin.

According to a common presupposition, the color of black shales results from "organic matter." Adherents to

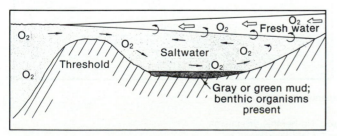

FIGURE 8-40.   Semi-isolated basin with all water ventilated. Thickness of outflowing fresh water is less than depth of threshold. Schematic longitudinal profile with open sea at left. Slope of surface water vastly exaggerated. (After K. M. Strøm, 1939, fig. 1, p. 358.)

this presupposition usually add the corollary that the darker the color, the higher the content of organic carbon. However, a few chemical analyses have shown that the intensity of black color is not always related to the abundance of organic carbon. In black shale, the quantity of organic matter ranges rather widely, and it may or may not be the chief black pigment. Instead, the black color of some specimens results from the presence of thoroughly dispersed, fine-textured iron sulfides (pyrite or hydrotroilite). Furthermore, some fine-textured strata that contain abundant organic carbon in the form of kerogen are not black at all. An example is the so-called "oil shales" of the Green River Formation (Eocene) of western Colorado, in which the strata having the highest contents of organic carbon are brownish, not black.

The black color of some rocks results from various manganese oxides. In the pigments of most of the widespread black shales, however, manganese oxides are not major contributors.

The environment of deposition of a black shale needs to be inferred from its stratigraphic relationships. Black shales range from local to regional in extent, and can be deposited in nonmarine-, transitional-, and marine environments whose geographic locations range from well within the interior of continental masses to the deep-sea floor. Two common kinds include (1) the so-called "roofing slates" that form somewhat localized lenses overlying coal seams; and (2) widespread thin units, such as the Chattanooga Shale and related formations (Upper Devonian-Lower Mississippian), which were deposited in the central United States and Canada in an area amounting to perhaps 500,000 km^2. The black shales of the coal-bearing sequences evidently were deposited in swamps or marshes. The environment of deposition

of widespread black shales such as the Chattanooga has been the subject of much controversy. Points of view have ranged from organic-rich soils to deep-water marine sediments in a basin having stratified water masses and stagnant bottom waters.

Whatever the environment, the key relationships for the sediments depend on the position of the zero-Eh interface. In basins with stagnant bottom waters this interface lies within the water mass and thus bottom-dwelling organisms are excluded (Figure 8-42, A). In some cases, however, the zero-Eh interface lies within the sediments. In this circumstance the bottom waters can be oxidizing and thus can support a normal population of bottom-dwelling organisms. However, within the interstitial waters in the sediments, below the zero-Eh interface, reducing conditions prevail and bacterial reactions color the sediments black (Figure 8-42, B). Study of the fossil remains in a black shale or other black sedimentary deposit should permit one to infer where the zero-Eh interface was located. This information can be valuable in an attempt to work out the rest of the environmental reconstruction but in itself does not point to any distinctive environmental setting.

SOURCES: Byers, 1977; Jenkyns, 1986; Ross and Degens, 1974; Schlanger and Jenkyns, 1976; Thunnell, D. F. Williams, and Belyea, 1984; van Hinte, Cita, and van der Weijden, 1987.

Basins Whose Waters Are Ventilated Occasionally

If the supply of fresh water to a basin having anoxic conditions were to decrease sufficiently (as in winter or during a drought), then the quantity of outward flow

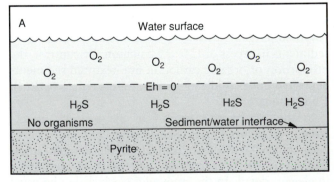
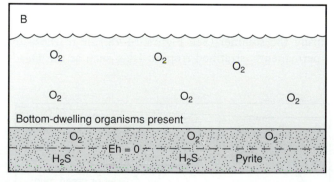

FIGURE 8-42.   Schematic profile through body of water and bottom sediments showing variable relationships that result from different positions of the Eh = zero interface.

A. Zero-Eh interface lies within the body of water. Bottom water is anoxic. No benthic organisms live on or in the bottom and pyrite forms in sediment. Water beneath zero-Eh interface forms an anoxic environment.

B. Zero-Eh interface lies within sediments. Bottom water is oxygenated. Benthic organisms can live on and in the bottom. Upper parts of sediment contain oxygenated interstitial waters and are colored brownish; lower levels of sediment lack oxygen and are gray to black and contain pyrite. (Modified from W. C. Krumbein and R. M. Garrels, 1952, fig. 5, p. 20.)

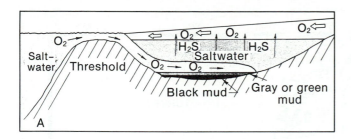

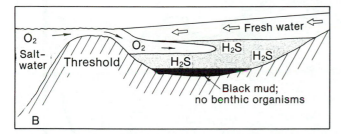

FIGURE 8-43.   Semi-isolated basin subjected to occasional inflow of oxygenated water from adjacent open sea. Schematic longitudinal profiles with open sea at left; slope of water surface vastly exaggerated. (After K. M. Strøm, 1939, fig. 1, p. 358.)

A. Inflowing water from open sea, denser than all the water in the basin, flows to bottom of basin and displaces former bottom water upward. As long as this freshly supplied water contains oxygen, the bottom deposits are gray- or green muds.

B. Inflowing water from open sea, denser than the upper part of the water in the basin but less dense than the bottom water, forms an interflow and ventilates only part of the basin. Black mud continues to accumulate beneath anoxic bottom water that is not affected by the influx of oxygenated seawater.

could become less than that requiring the whole cross section of the portal. If this happens, then large amounts of oxygen-bearing salt water from the open sea can flow into the basin (Figure 8-43). Or, as found by Otto Pettersson at Gullmarfjord, during times of greater-than-normal-tide-producing conditions, internal tides may enable salt water to enter semi-isolated basins.

For whatever reason, when large amounts of seawater enter the basin, several things can take place. If the entering seawater is denser than all the water in the basin, the seawater will flow along the bottom and displace the basin water upward (Figure 8-43, A). If enough seawater can flow in, it may displace upward all of the former bottom water. The old, stagnant bottom water, full of H_2S, rises to the surface. The level of water in the basin may be raised considerably, and much or all of the surface layer of fresh water discharged rapidly out through the portal. The H_2S causes mass mortality of the organisms that normally live in the ventilated water above the pycnocline. Eventually the H_2S in the displaced water is oxidized, the water level resumes its former position, and organisms recolonize the aerated surface water.

If the entering seawater is not denser than all the water in the basin, then the incoming water will spread out

at some intermediate depth as an **interflow,** *a density current that flows along a pycnocline.* In this case the newly arrived seawater will displace only part of the water upward. Only the upper part of the water column becomes ventilated (Figure 8-43, B). If the volume of fresh water increases to its former proportions, then the inflow of salt water from the open ocean will stop once again and a new cycle of stagnation will commence. The only geologic indication of a short period of ventilation may be a large number of dead organisms. A longer period of ventilation could result in the deposition on the basin floor of a layer of gray sediment.

At various times during the Quaternary Period, episodic influxes of aerated salt water from the Mediterranean are inferred to have ventilated the Black Sea (Figure 8-44). The most-recent product of ventilation has been radiocarbon dated at 9000 yr B.P. The times of ventilation are thought to have caused the locations of anoxic conditions to shift from the Black Sea to the previously ventilated eastern basin of the Mediterranean. The mechanism responsible for this change is thought to have been a large influx into the Black Sea of well-aerated salt water from the eastern Mediterranean through the portal at the Dardanelles when the rising sea reached a level 40 m below its present level. The entering salt water displaced a vast amount of fresh water out of the Black Sea. The displaced fresh water is inferred to have spread out across the eastern basin of the Mediterranean and not to have mixed immediately with the subjacent salt water. Until it eventually dissipated, this upper layer of fresh water kept oxygen from reaching the lower layer of salt water, which thus became stagnant. Stagnation may have been facilitated because the water in the main Mediterranean basin was not so vigorously renewed, as it is now, as a consequence of less-than-present influx of aerated Atlantic surface water across the Sicilian Platform during a lower sea level.

During these times when anoxic conditions prevailed in the Eastern Mediterranean, dark-colored, organic-rich, fine-textured sediments known as *sapropels* were deposited. Core RC9-181, which is 10 m long and extends downward to sediment thought to be about 400,000 years old, contains 12 layers of sapropel (Figure 8-45).

Closed Basins on the Sea Floor

The kinds of circulation patterns just described prevail not only in landlocked basins but operate as well in many closed basins on the sea floor. Although what are now closed marine basins on the sea floor may have been bordered by land at times in their past histories, a highly modified kind of water circulation operates even if land is nowhere in sight today. Modern examples are known from the Moluccas, Indonesian archipelago; from

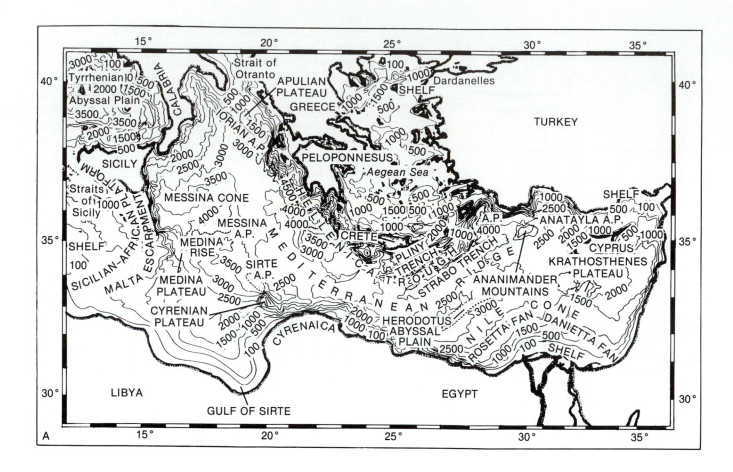

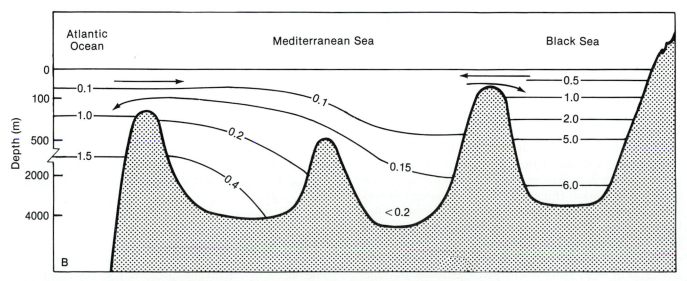

FIGURE 8-44.   Black Sea and connection via Dardanelles and Aegean Sea to eastern Mediterranean deep basin.

A. Bathymetric map of eastern Mediterranean. Contours in corrected meters. (After W. F. B. Ryan, D. J. Stanley, J. B. Hersey, D. A. Fahlquist, and T. D. Allan, 1970, fig. 5, p. 400.)

B. Schematic profile from Atlantic Ocean to Black Sea showing content of phosphate phosphorus in the water (lines in mg atoms per cubic meter of water). Arrows show general direction of water flow into and out of the Mediterranean. (After A. C. Redfield, B. H. Ketchum, and F. A. Richards, 1963, fig. 13, p. 64.)

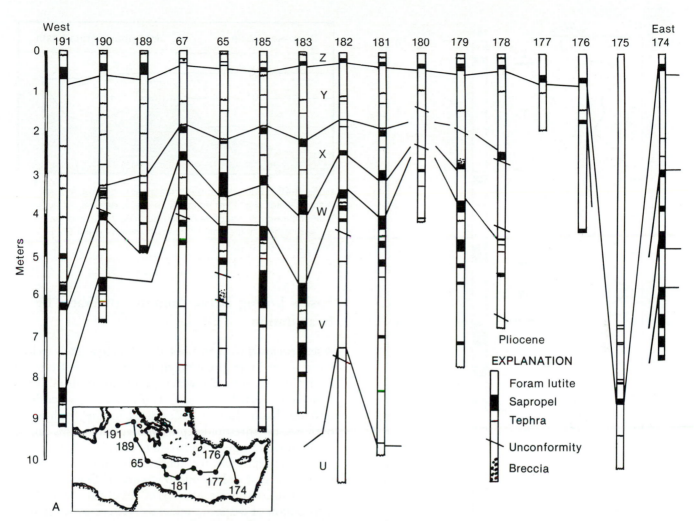

FIGURE 8-45.   Sapropels in Quaternary sediments, eastern Mediterranean.
A. Cores collected by R. C. Conrad, Lamont-Doherty Geological Observatory of Columbia University, along a traverse extending from off the Nile delta to the southwestern tip of Italy (locations on inset map). Correlations shown by boundaries of zonal scheme using a reverse-alphabet letter designation (youngest is Z, then Y, X, and so forth). Sapropels are shown in black. "Foram lutite" is a fine-textured carbonate containing tests of Foraminifera. (Modified from W. B. F. Ryan, 1972, fig. 8, p. 165 by reversing the orientation of the line of cores.)

off southern California; and from the Cariaco basin, off northern Venezuela.

The characteristics of the waters in the bathymetric basins on the sea floor are governed by two chief factors. These are (1) the temperature-depth relationships in the water of the adjacent open sea and (2) the depths of the thresholds of the basin (Figure 8-46).

Between the land and the deep Pacific basin off southern California is an extensive region, known as a continental borderland, that is characterized by shallow banks or islands and deeper bathymetric basins (Figure 8-47). The relationship between water temperature and depth in the open Pacific is so well known that it is possible to infer the depth of the threshold of a bathymetric basin simply by measuring a temperature-depth profile of the water filling it (Figure 8-48). The basins are inferred to

have formed as a result of crustal movements that began during the Miocene Epoch and are still in progress. Two of the basins, the Los Angeles basin and the Ventura basin, are now on land. Formerly they were marginal bathymetric basins, but became filled with thick bodies of sediment of Miocene-, Pliocene-, and Pleistocene ages. Many of the sands in these basins are prolific producers of oil.

The Cariaco basin, off Venezuela, is defined by the 100-m depth contour; the depth at the threshold is slightly less than 100 m (Figure 8-49). At the present time the bottom waters of the Cariaco basin are anoxic.

SOURCES: Hulsemann and Emery, 1961; F. A. Richards, 1975.

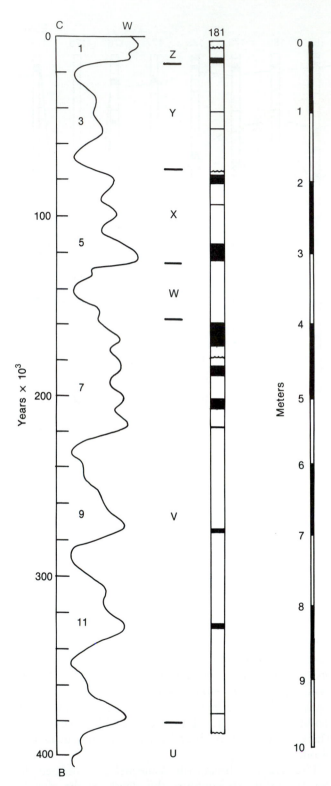

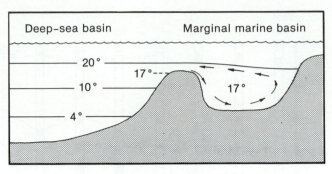

FIGURE 8-46.   Bathymetric basin marginal to deep-sea basin fills with water of uniform temperature that is determined by the temperature-depth structure of the deep-sea water and the depth of the threshold of the basin. Schematic profile. (After Ph. H. Kuenen, 1950, fig. 25, p. 47.)

Basins Losing Large Amounts of Water by Evaporation

A semienclosed basin in a hot, dry climate is likely to lose more water by evaporation than flows into it from the land. Consequently, the water level tends to drop. If the basin is connected to the sea, then surface seawater flows into the basin. Just as with basins having fresh-water outflows, the threshold depths of evaporite basins can be great enough to permit a two-way flow, in at the surface and out at depth. Or, the depth at the threshold may be so shallow that only a one-way current flows—inward. These two contrasting patterns of circulation control the distribution of evaporite minerals. Where two-way flow takes place through the portal, the most-saline evaporite minerals are precipitated at the margins of the basin. Where one-way flow prevails, the most-saline evaporites are precipitated in the center of the basin.

PORTALS WITH TWO-WAY FLOW: MOST-SALINE EVAPORITES AT MARGINS OF BASIN

A modern example of a basin losing large amounts of water by evaporation and having two-way flow through its portal is the Red Sea.

Modern Red Sea.   The Red Sea (Figure 8-50) is an example of a modern basin undergoing evaporitic water losses and having two-way flow through its portal. However, the Red Sea is an example of a basin whose circulation is complicated by seasonal shifts in the prevailing winds. The Red Sea is 1800 km long, about 270 km wide, and is connected to the Indian Ocean by a portal about 140 km long; the minimum threshold depth of 100 m is located 140 km inside the portal. No rivers drain into the Red Sea; net loss by evaporation is about 3.5 m/yr. Despite

FIGURE 8-45.   *(Continued)*
   B. Core RC-181 showing sapropels (black), thickness scale in meters, age by reverse-alphabet and by numbered zones, and an inferred climatic curve (C = cool; W = warm) opposite inferred time scale. (Modified from W. B. F. Ryan, 1972, by changing scale of graphic log of RC-181 from fig. 8, p. 165, and adding it and the thickness scale to the time scale and inferred climatic curve of fig. 9, p. 166.)

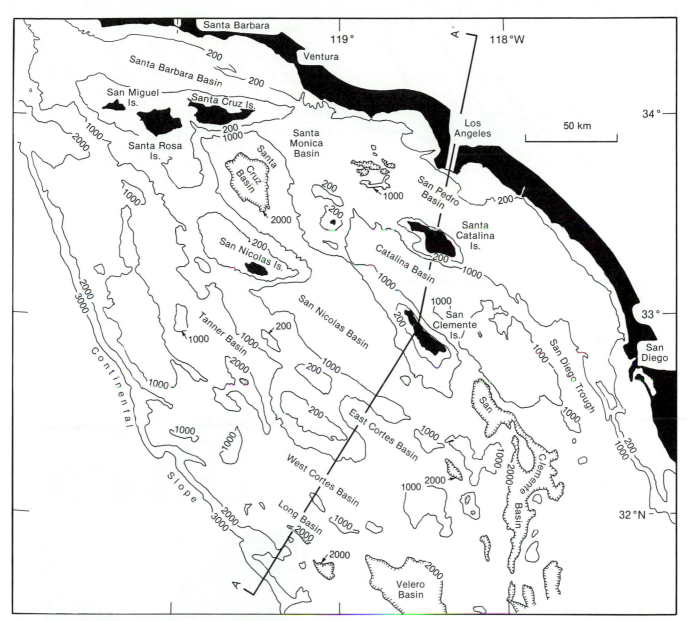

FIGURE 8-47. Map of bathymetric basins of continental borderland off southern California.
Line A-A′ locates profile-section of Figure 8-48, A. (After K. O. Emery, 1960b, fig. 69, p. 81.)

this loss, which is three times that of the Mediterranean, the Red Sea is not an evaporite basin. Its maximum salinity, at the northwest end, is about 40 to 41‰.

The interchange of water between the Red Sea and the Indian Ocean is not a simple function of evaporative lowering of the basin water and surface inflow through the portal. Strong winds, which change with the seasons, and the tides affect the surface flow of water. The winter winds blow from the SSE and promote an inward surface flow from the Indian Ocean. The summer winds blow from the NNW; they force the surface water to flow out of the Red Sea. These reversing winds and the tides

complicate attempts to determine a salt budget. Accordingly, it is not known with certainty whether the Red Sea receives salt from the Indian Ocean or exports salt to it.

If the configuration of the portal has not been altered by tectonic movements since Pliocene time, then during Quaternary glacial low sea levels, the Red Sea would have become an isolated saline lake comparable to the modern Dead Sea. Cores of glacial-age sediment from the floor of the Red Sea include calcite-rich muds and layers that have been cemented by aragonite. The calcite-rich muds are inferred to be the products of the bacterial destruction of gypsum, as takes place in the Dead Sea today.

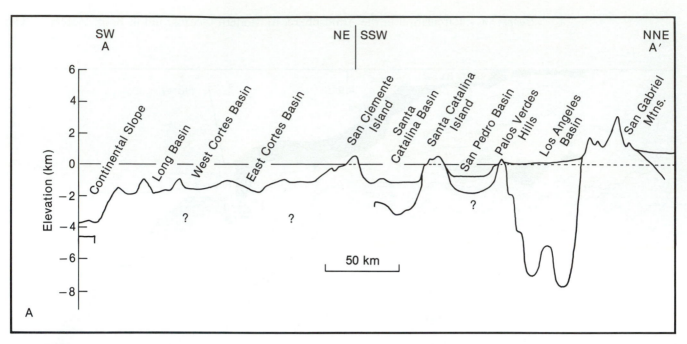

FIGURE 8-48. Profile- and depth relationships of basins in continental borderland off southern California.

A. Profile and partial section along line A-A' of Figure 8-47. The Los Angeles basin, which has been filled with sediment, is no longer a bathymetric basin.

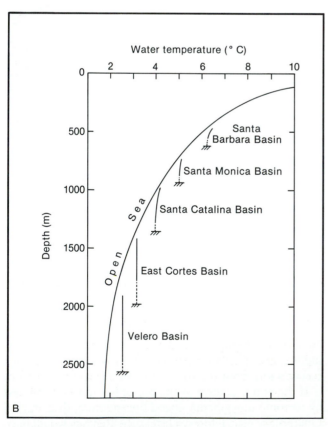

FIGURE 8-48. (*Continued*)

B. Graph of temperature vs. depth showing relationship between properties of water in the open sea and those of several bathymetric basins. Temperatures of waters in the bathymetric basins are a direct function of the basin-threshold depths. (After K. O. Emery, 1960b: A, fig. 69, p. 81; B, fig. 96, p. 109.)

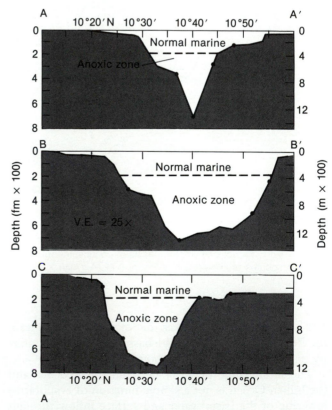

FIGURE 8-49. Cariaco basin off Venezuela.

A. Profiles along lines shown in B on which the boundary between normal marine water and the anoxic zone is indicated. (After F. A. Richards, 1975, fig. 2, p. 13.)

316

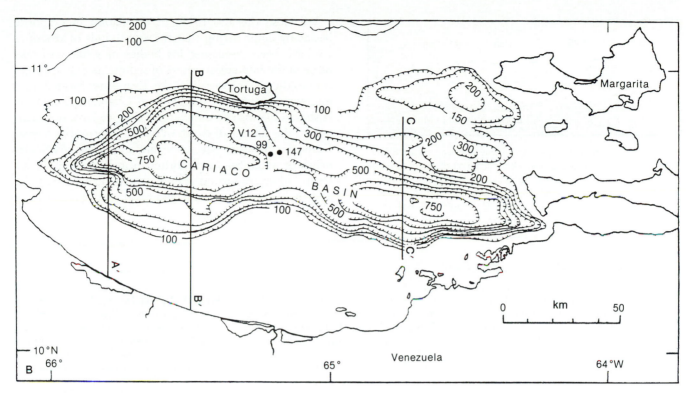

FIGURE 8-49. *(Continued)*
B. Bathymetric sketch map showing depth in meters. A shelf area where water depths are about 100 m separates the Cariaco basin from the Caribbean Sea. Black dots mark locations of two cores of bottom sediments; V12-99, Lamont-Doherty Geological Observatory, Columbia University; 147, JOIDES program. (After Edgar and others, 1973, fig. 1, p. 170.)

The aragonite cement may have resulted from chemical precipitation from extremely saline and hot (> 35°C) bottom water.

PORTALS WITH ONE-WAY FLOW: MOST-SALINE EVAPORITES AT CENTER OF BASIN

If the depth of the threshold is shallow and the inflow of water to an evaporitic basin very powerful, then the only flow through the portal will be inward. This situation is the reverse of that shown in Figure 8-41, where a surface outflow of fresh water held out denser salt water. In evaporite basins having only one-way flow, the densest brine accumulates in the deepest part of the basin (Figure 8-51).

No modern examples of a basin having this pattern of circulation are known, but ancient examples may be reasonably inferred from the patterns of strata composed of evaporite minerals. In such basins, the halite, for example, is at the center of the basin, not at the margins. Precipitation may result from the mixing of brines.

SOURCES: O. B. Raup, 1970; Schmalz, 1990; Warren, 1989.

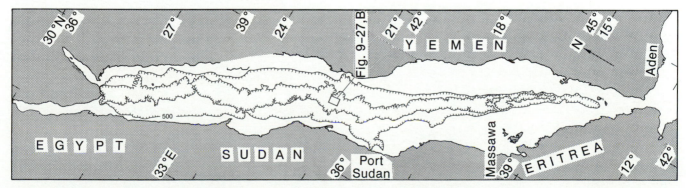

FIGURE 8-50.   Bathymetric map of Red Sea. Depth contours in meters; contour interval, 500 m. (After Y. Herman and P. E. Rosenberg, 1969, fig. 1, p. 449.)

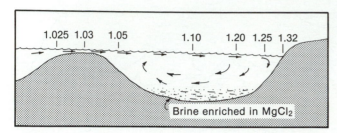

FIGURE 8-51.   Evaporite basin having only one-way inward flow through portal, schematic profile with inclination of water surface much exaggerated. Densest brine, enriched in $MgCl_2$, accumulates at center of basin. Numbers at top are approximate densities of surface waters (g/cm^3). (After O. B. Raup, 1970, fig. 7B, p. 2254.)

We close this discussion of circulation in evaporite basins by emphasizing that the same kinds of reactions we have been discussing for bodies of water also take place in the interstitial waters of sediments that are adjacent to the sea or to a saline lake. Evaporites formed from interstitial waters are discussed in Chapters 12 and 14. It should likewise be emphasized that we have not mentioned the subject of depth of water in evaporite basins. Evaporite deposits per se do not imply any particular depth of water. Some evaporites form slightly above sea level. Other evaporite mineral crystals become particles and are moved by currents and by gravity. These can be deposited at any depth.

Suggestions for Further Reading

BENSON, R. H., 1972, Ostracodes as indicators of threshold depth in the Mediterranean during the Pliocene, p. 63–75 *in* Stanley, D. J., ed., assisted by Kelling, Gilbert; and Weiler, Yehezkial, The Mediterranean Sea: a natural sedimentation laboratory: Stroudsburg, PA, Dowden, Hutchinson, and Ross, 765 p.

BYERS, C. W., 1977, Biofacies patterns in euxinic basins: A general model, p. 5–17 *in* Cook, H. E., and Enos, P. E., eds., Deep-water carbonate environments: Tulsa, OK, Society of Economic Paleontologists and Mineralogists Special Publication 25, 336 p.

EKDALE, A. A., and MASON, T. R., 1988, Characteristic trace-fossil associations in oxygen-poor sedimentary environments: Geology, v. 16, p. 720–723.

HAY, W. W., 1988, Paleoceanography: A review for the GSA centennial: Geological Society of America Bulletin, v. 100, p. 1934–1956.

HEDGPETH, J. W., 1957, Classification of marine environments, p. 17–27 *in* Hedgpeth, J. W., ed., Treatise on marine ecology and paleoecology: Geological Society of America Memoir 67, v. 1, Ecology, 1296 p.

HULSEMANN, JOBST, and EMERY, K. O., 1961, Stratification in recent sediments of Santa Barbara Basin controlled by organisms and water character: Journal of Geology, v. 69, p. 279–290.

JOHNSON, D. A., 1982, Abyssal teleconnections: Interactive dynamics of the deep ocean (*sic*) circulation: Palaeogeography, Palaeoclimatology, and Palaeoecology, v. 38, p. 93–128.

KELLER, G., and BARRON, J. A., 1983, Paleoceanographic implications of Miocene deep sea (*sic*) hiatuses: Geological Society of America Bulletin, v. 94, p. 590–613.

PARRISH, J. T., 1982, Upwelling and petroleum source rocks, with reference to the Paleozoic: American Association of Petroleum Geologists Bulletin, v. 66, p. 750–774.

PARRISH, J. T., ZIEGLER, A. M., and SCOTESE, C. R., 1982, Rainfall patterns and distribution of coals and evaporites in the Mesozoic and Cenozoic: Palaeogeography, Palaeoclimatology, and Palaeoecology, v. 40, p. 67–101.

REA, D. K., LEINEN, M., and JANECEK, T. R., 1985, Geologic approach to the long-term history of atmospheric circulation: Science, v. 227, p. 721–725.

ROSS, D. A., and DEGENS, E. T., 1974, Recent sediments of Black Sea, p. 183–199 *in* Degens, E. T., and Ross, D. A., eds., 1974, The Black Sea—geology, chemistry, and biology: Tulsa, OK, American Association of Petroleum Geologists Memoir 20, p. 633

SCHLANGER, S. O., and JENKYNS, H. C., 1976, Cretaceous ocean anoxic events, causes and consequences: Geologie en Mijnbouw, v. 55, p. 179–184.

THUNNELL, R. C., WILLIAMS, D. F., and BELYEA, P. R., 1984, Anoxic events in the Mediterranean Sea in relation to the evolution of late Neogene climates: Marine Geology, v. 59, p. 105–134.

VAN HINTE, J. E., CITA, M. B., and VAN DER WEIJDEN, C. H., eds., 1987, Extant (*sic*) and ancient anoxic basin (*sic*) conditions in the eastern Mediterranean: Marine Geology, v. 75 (special issue), 281 p.

WRIGHT, E. K., 1987, Stratification and paleocirculation of the Late Cretaceous Western Interior seaway of North America: Geological Society of America Bulletin, v. 99, p. 480–490.

CHAPTER 9

Deep-Water Settings

Death of the Quiet Ocean: Discovery of Gravity-Powered Bottom-Sediment Flows

Late in the nineteenth and during the first half of the twentieth century, geologists who studied sedimentary strata that had been deposited on the floors of ancient seas and lakes based their interpretations on what they thought was a single, universally valid generalization. According to this generalization, coarse sediment is confined to nearshore zones, and in the deeper water, only the finest sediments accumulate. The dying out of wave action downward had been proved by experiment, confirmed by the observations of a few divers, and predicted by mathematical analysis. The large-scale distribution of sediments in the world's oceans had been determined by the scientists of the *Challenger* expedition (work at sea, 1872–1876). The geologists who had analyzed the sediment samples prepared a comprehensive monograph, which was based not only on specimens from the *Challenger*'s stations, but also on study of nearly every sample that anyone had ever collected previously. These sediment samples greatly strengthened the prevailing generalization. Based on their studies of the grab samples, the *Challenger* geologists, led by Sir John Murray (1841–1914), stated positively that nothing but the finest particles are present in the deep-sea basins and that sands and gravels are confined to continental shelves and nearshore areas. Accordingly, the effect of the *Challenger* reports was to give geologists renewed confidence in their interpretation that ancient marine sandstones were nearshore deposits. Moreover, if the sandstone in question contained ripple marks and plant debris, then its nearshore interpretation was considered to be beyond any doubt. Writers about the deep-sea floor waxed poetic. They described the abyss as a quiet place subjected only to an "eternal snowfall" of tiny skeletons discarded by the microorganisms that live in the near-surface waters.

The only hint of activity related to deep-water sediments to disturb the poets came from a few geologists who reported evidence for subaqueous slumping, which they named *intraformational disturbances*. By this they meant contorted strata enclosed by noncontorted strata. These geologists inferred, correctly, most would now agree, that the contortions had resulted from occasional downslope shifting of still-soft sediments.

Another much-favored interpretation during this period of time held that not only marine sandstone but indeed nearly all the ancient marine strata now exposed on land had been deposited in "shallow" water. Those who adopted this "universal" shallow-water viewpoint were not eager to discuss downslope movements of sediment on the sea floor. Quite the contrary: such movement implied that the ancient sea floor sloped perceptibly, and a sloping sea floor always leads from "shallow" into "deeper" water. It is probably no coincidence that most of the papers on the subject of subaqueous disturbances were published in Europe. In contrast to most American geologists, many European geologists accepted the interpretation that numerous ancient marine strata now exposed on land had been deposited in "deep" water.

The "shallow-water" ideas just reviewed were so firmly entrenched that they provided the basis for what was considered on other grounds to be an "outrageous" interpretation of the first-found deep-sea sands. These sands had been discovered in 1909 by scientists of the German South Polar Expedition. E. Philippi, the geologist who examined and described the samples, concluded that these sands had been deposited on a beach. After deposition, he argued, enormous subsidence had taken place, dropping the former beach to the deep-sea floor. Philippi's inference, though soundly based from the point of view of the prevailing interpretation of sands, created a conflict with a second fundamental geologic precept of the day, namely, that continents and oceans are permanent, ancient features of the Earth's crust. Because the tectonic concepts were thought to be of greater significance than the sedimentologic interpretations, it was easier to gloss over Philippi's work than to attempt to resolve the mighty dilemma that it posed.

But even before the German expedition of 1909 some geologists had discovered, and realized the significance of, evidence that the deep-sea floor is not the tranquil place that others supposed it to be. In 1897 the geophysicist John Milne (1835–1900) wrote at length about evidence of dynamic activities on the deep-sea floor. Milne was aware of the association between earthquakes and broken submarine telegraph cables. What is more, Milne inferred that submarine cables had broken because they had been struck by undersea avalanches triggered by earthquakes.

The first detailed bathymetric survey of a continental slope was made during the mid-1930s, using radio methods for navigation and the best-available echo sounder to determine depths. The results, published by A. C. Veatch (1878–1938) and P. A. Smith (1900–1980) in 1939, focused attention on submarine canyons and stimulated a vigorous debate over the origin of these canyons.

The history of this controversy, beginning with the hypothesis for the origin of submarine canyons, that during the Quaternary Ice Ages sea level had been greatly lowered and the canyons had been cut subaerially by the headward erosion of rivers, and continuing through R. A. Daly's (1871–1957) (1936) inference that density currents could have been responsible for the cutting of submarine canyons, to Ph. H. Kuenen's (1902–1976) experiments in the late 1930s and 1940s, was outlined in Chapter 6 because these events spurred the development of seismic stratigraphy. But what happened to ideas about density currents after Kuenen presented the results of his experiments to the European geological community in 1948?

As mentioned in Chapter 6, the American John L. Rich (1884–1956) attended the meeting in Britain at which Kuenen presented his paper in 1948. Rich was impelled by Kuenen's ideas to look for evidence of turbidity-current deposition in ancient rocks. Ultimately he divided basins into three morphologic parts as described in Chapter 6: an upper region of low slope, called the *unda* environment; a middle region of steeper slope, called the *clino* environment; and a lower region of low slope, called the *fondo* environment. However, other Americans had to wait until Kuenen came to the United States in 1950 as an American Association of Petroleum Geologists Distinguished Lecturer. Kuenen showed a movie he had made of his experimental turbidity currents, and the response demonstrated that a moving picture was worth any number of printed words. The effects of Kuenen's visit were like skyrockets, exploding into whole new avenues of thought.

Bruce Heezen (1924–1977) attended Kuenen's lectures at Columbia University in New York in 1950 and realized that here was an explanation for the breaking of submarine cables off the Grand Banks in 1929 (Figure

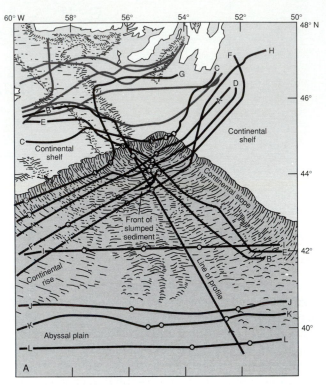

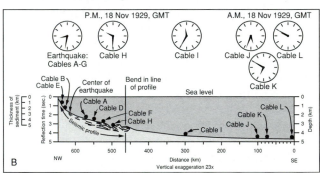

FIGURE 9-1.   Grand Banks cable breaks, off Newfoundland, Canada.

A. Bathymetric map of sea floor showing locations of submarine telegraph cables (lettered A through L) broken during and shortly after the earthquake of 18 November 1929.

B. Profile showing locations of cables, a continuous seismic profile, and times when cables broke. (Seismic profile after B. C. Heezen and C. L. Drake, 1964a, fig. 1, p. 222.) (C. R. Longwell, R. F. Flint, and J. E. Sanders, 1969, fig. 15-16, p. 363.)

9-1). Heezen inferred that the cables had been broken by gravity-powered bottom-sediment flows triggered by a major earthquake, the Acadian-Newfoundland quake of 18 November 1929. He went on (with others) to develop the concept of the geostrophic- or contour currents, which are important factors in reworking sediments on the continental rises.

The paleontologist M. L. Natland also attended one of Kuenen's lectures and found therein the key to the problem of petroleum-reservoir sandstones which he had been studying. Natland had observed that the reservoir sandstones were intercalated with shales that contained what appeared to be deep-water Foraminifera. He therefore suggested that the sandstones themselves must have been deposited in deep water. None of his colleagues accepted this explanation, enamored as they were with the idea that the deep-water environment is characterized by eternal tranquility, and Natland was discredited. Kuenen provided the evidence that Natland needed to show that sands can indeed be deposited in deep water.

From our perspective it is somewhat surprising that, although evidence of high-energy events in the deep sea was reported before 1900, and sands were recovered in cores from the sea floor as early as 1909 by Philippi and again in 1936 [core descriptions were published by M. N. Bramlette (1896–1977) and W. H. Bradley (1899–1979) in 1940], the implications of these facts were not widely appreciated until 1950.

The realization by sedimentologists that density currents are important agents of erosion and deposition in deep-water basins solved another puzzle too. Since early in the nineteenth century, the subject of primary sedimentary structures had been investigated by many geologists, for diverse purposes. Primary sedimentary structures formed one of Sir Charles Lyell's (1797–1875) chief arguments when he established the view that ancient bedrock strata can be interpreted by comparison with modern sediments. A few geologists, most notably H. C. Sorby (1826–1908), showed that many sandstones contain evidence of the activities of ancient currents. Sorby measured and listed the directions of dip of cross strata from hundreds of localities throughout Great Britain. Field geologists mapping in structurally complex areas had developed the use of primary sedimentary structures for inferring the former top direction of the strata. R. R. Shrock's monograph on this aspect of primary sedimentary structures, which was published in 1948, stimulated widespread interest in all aspects of the study of primary sedimentary structures. The origins of many of the features discussed in Shrock's book were well known. One particular exception was "graded bedding" (or graded deposits, as we refer to them). Graded deposits were widely recognized but their origin had not been satisfactorily determined.

Shrock's book was followed in 1950 by publication of a classic paper of Kuenen and Migliorini in which turbidity currents were shown to be a cause of "graded bedding". A. H. Bouma, a student of Kuenen, applied himself to the study of graded deposits and of turbidity-current deposits. Bouma developed a vertical-sequence model (published in 1962) for such deposits that forms the basis for all modern studies of turbidite sequences (Figure 9-2).

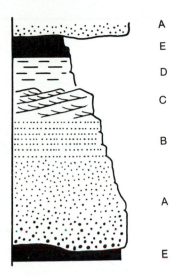

FIGURE 9-2. Vertical sequence in sediments deposited by gravity-powered bottom flows. This sequence consists ideally of five divisions, labeled A, B, C, D, and E. The sequence is named the Bouma sequence after A. H. Bouma (1962).

A. Either a graded sandstone in which the particle size decreases systematically upward or a massive sandstone; the base of the original sand of this "high-speed" depositional layer is sharp; it divides the sand from the underlying "low-speed" shaley layer of the preceding sequence. Division A typically is a product of liquefied cohesionless-particle flow.

B. Parallel-laminated sandstone that represents conditions of upper-flow regime, hence is likewise a "high-speed" structure. (See Chapter 7, especially Figure 7-24.)

C. Ripple cross-laminated fine- or very fine sandstone that represents the lower-flow regime, hence is a "low-speed" structure. (See Chapter 7, Figure 7-24.)

D. Faint parallel laminae of mudstone.

E. Shaley layer at top of sequence. At the contact between E and the overlying sandstone A of the next sequence, abundant sole marks may be present. (See Figures 5-23, 5-24, and 5-27.) The fine-textured fallout from the tail of a turbidity current may be difficult or impossible to distinguish from pelagic sediments. Further discussion in text. (After A. H. Bouma, 1962; R. G. Walker, 1976, fig. 1, p. 26.)

The work of Bouma led, in the 1960s and 1970s, to attempts to explain, not just the vertical sequences of strata that might be deposited by certain processes, but the three-dimensional relationships of strata that accumulate under the influence of multiple processes, all affecting the same part of the basin floor. In the case of turbidites, the early work culminated in a seminal study by two Italian geologists, Mutti and Ricci-Lucchi (1972), of the depositional patterns that characterize modern submarine fans. Mutti and Ricci-Lucchi applied what they knew about sedimentary processes and about deposits inferred to have been laid down in submarine fans to explain the distribution of submarine-fan sediments in terms of sediments deposited within channels, sediments deposited between channels, the formation of lobes of sediment as channels shift their positions over

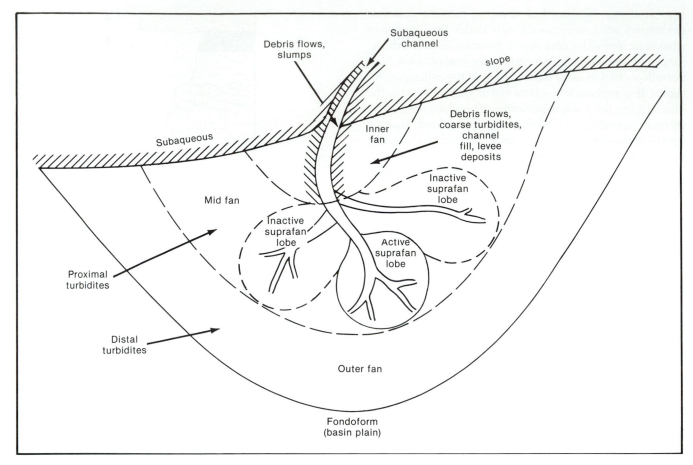

FIGURE 9-3.   Idealized submarine fan, showing upper-, mid-, and lower fans. Further discussion in text. (Based upon various sources.)

time, and the less-frequent shifting of the locus of deposition from one lobe to another (Figure 9-3). Because the processes that form submarine-fan deposits are comparable to those that form fluviodeltaic deposits (described in Chapter 14) fans from these two contrasting settings are much alike.

J. E. Sanders, another Kuenen student, observed evidence for longitudinal transport by turbidity currents in an elongate basin, refuting the widely held belief that bottom currents could only flow across basins from their sides and not along them from their ends.

The total impact of all the developments that followed from the early work of Daly, Kuenen, and others was truly revolutionary. As a result, a more-balanced view now prevails concerning the ranges of depths of ancient marine waters. Although few criteria for using sediments to make estimates of paleowater depths are foolproof, realistic quantitative estimates of the depth of ancient seas are available now. The former rigid insistence, chiefly by many American geologists, that all parts of ancient seas were "shallow" has disappeared.

Before discussing the various sediment-distribution- and sediment-deposition processes in deep-water basins,

we need to consider the major morphologic divisions of deep-water basins as they relate to sediments. This is the subject of the next section.

SOURCES: Bouma, 1979, 1991; Bouma, Normark, and Barnes, 1985; Bramlette and Bradley, 1940; Brunner and Normark, 1985; Daly, 1936; Edwards, 1991; Embley, 1980; Heezen, Hollister, and Ruddiman, 1966; Hiscott, Pickering, and Beeden, 1986; Kennett, 1982; Kessler, 1990; Kirwan, Doyle, Bowles, and Brooks, 1986; Kuenen, 1937; Kuenen and Migliorini, 1950; Kuenen and Sanders, 1956; Marjanac, 1990; Mutti, 1979, 1985; Mutti and Johns, 1978; Mutti and Ricci-Lucchi, 1972; Natland and Kuenen, 1951; Pickering, 1989; Prothero, 1990; Ricci-Lucchi, 1978; Rich, 1951a; Stow, 1985.

Major Morphologic Subdivisions of Deep-Water Basins as Related to Sediments

In this section we examine three major morphologic subdivisions: (1) the basin margins, (2) various depositional settings on the basin slope and rise, and (3) various parts of the deep-sea floor (including abyssal plains and deep-sea trenches). These three major morphologic subdivisions of basins are conveniently subdivided into seven different settings, described below.

Basin Margins

The margins of the ocean basins have been named in various ways by geologists whose chief interests lay with the geology of continents and with the fundamental geophysical contrasts between continental masses and ocean basins. One indication of their concern for continents is their name for the common boundary between a continental mass and an ocean basin: the continental margin. This usage was consistent with the kind of crust underlying the area; for example, sialic- or continental crust is thought to extend outward to the places where the water depth is about 2000 m. Therefore every marginal feature shallower than this bears the prefix "continental." From the point of view of sediments, however, a basin is a basin. From this point of view, ocean basins differ from other kinds of basins mainly in being larger. Morphologically, and regardless of the kind of crust lying below, the basin slope begins at the outer edge of the shelf, also known as the **bathyal zone**, *the sea floor in the depth range of 200 to 2000 m*. Accordingly, in our discussion we shall emphasize the morphologic boundaries that affect movements of sediments. We shall use many of the well-established names for the features involved.

Throughout the following discussion we shall endeavor to keep firmly in mind the fact that most of the sedimentary processes that take place in the ocean basins are equally characteristic of other basins, and that the kinds of sedimentary deposits found in ocean basins, are, for the most part, also found in other water-filled basins. Therefore, except where noted, the features discussed are meant to apply to all water-filled basins, even if the examples are oceanic. (After all, the ocean basins collectively cover nearly 3/4 of the Earth's surface; they are the largest of the Earth's basins; and they have received the most-intensive study by sedimentologists.)

In the discussion that follows, a distinction must sometimes be made between the major ocean basins of the world and all other basins, including the peripheral marine basins, such as the Gulf of Mexico, which may in some respects more closely resemble nonmarine basins than they do the major ocean basins. Specifically, peripheral basins may be shallower than the major ocean basins, and they may be underlain by transitional- or continental crust rather than by oceanic crust. Major cratonic basins, such as the Michigan Basin of North America, may be comparable in area, water depths, and other characteristics to large peripheral marine basins like the Gulf of

Mexico. At some times during their histories, many major cratonic basins have been connected to the sea (for example, the Michigan Basin during much of the Devonian and Silurian). These basins, in their sedimentation processes and depositional features, closely resemble marine peripheral basins, but are underlain by continental crust and are more readily isolated from marine waters. When we refer only to the major ocean basins, we speak of oceanic basins.

The basin-margin areas we shall discuss include basin slopes (and -escarpments), subaqueous canyons and -fans (and subaqueous cones), subaqueous aprons, basin rises, and deep-sea trenches (Figures 9-4, 9-5, and 9-6). The deep-sea trenches are found only in marine basins.

Basin Slope (or -Escarpment)

The term **continental slope** designates *the relatively steep (commonly 3 to 6°) slope that lies seaward of the continental shelf*. (See Figure 9-4.) Similarly, the term *basin slope* applies to the same feature in any water-filled basin. In many areas, the corresponding shelf-edge slope is much steeper and the term *escarpment* is used. For example, the slopes of the Blake, Florida, and Campeche escarpments are 15° or more; locally, the Bahamas escarpment is vertical and even overhanging.

In Chapter 6 we introduced the term clinoform, applicable to continental slopes (including the *bathyal* zone), escarpments, and indeed to any subaqueous slope.

The lower limits of the continental slope along eastern North America range between depths of 500 and 5000 m; elsewhere, in deep-sea trenches, it may locally exceed 10 km.

Continental slopes are dissected by submarine canyons. The two other major morphologic features of continental slopes are the interchannel areas and small basins that form on the slope by a variety of mechanisms. Smaller counterparts of all of these features (except deep-sea trenches) are known from the basin slopes of large lakes or of peripheral marine basins.

Subaqueous Canyons and -Fans (and Subaqueous Cones)

Indenting the continental slope in nearly all parts of the world where terrigenous sediments underlie the continental terrace (and also indenting the slopes of large nonmarine basins) are numerous **submarine canyons** (See Box 6.1 Figure 1.), *sinuous V-shaped valleys that may be hundreds of kilometers long with axes sloping seaward as much as 80 m/km*. The walls of the canyons are steep and contain outcrops of bedrock underlying the continental terrace. Sand-size sediment covers the floors of some canyons; the surface of the sand is ripple marked. The axes of these ripples trend across the canyon, and

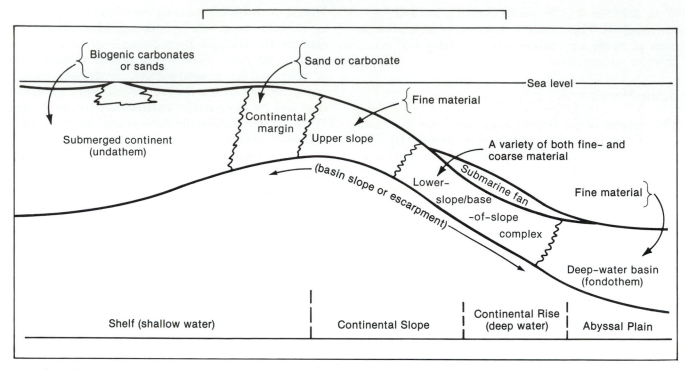

FIGURE 9-4.   Schematic profile of basin margin from shelf to basin. Typical sediment types for each part of the profile are indicated. Note location of submarine fan. (After B. D. Keith and G. M. Friedman, 1977.)

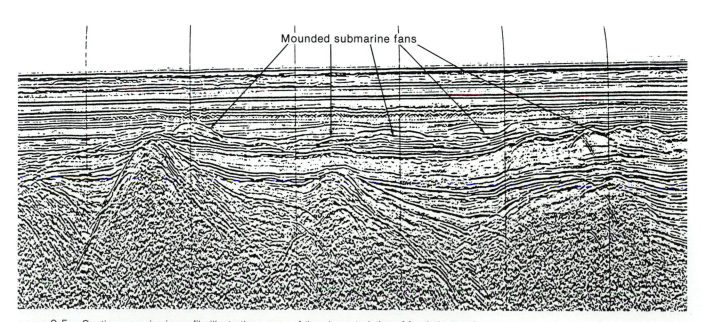

FIGURE 9-5.   Continuous seismic profile illustrating some of the characteristics of fondothems. A mounded seismic facies overlies a basement displaying high relief. These consist primarily of turbidites and other gravity-displaced deposits. Submarine fans form near the edges of fondothems where they interfinger with clinothems. The submarine-fan deposits are overlain by more-distal fondothem strata, which present an even, parallel aspect. The parallel seismic reflections of these basin-plain strata result from an intercalation of distal turbidites and pelagic sediments. (Courtesy of Merlin Geophysical.)

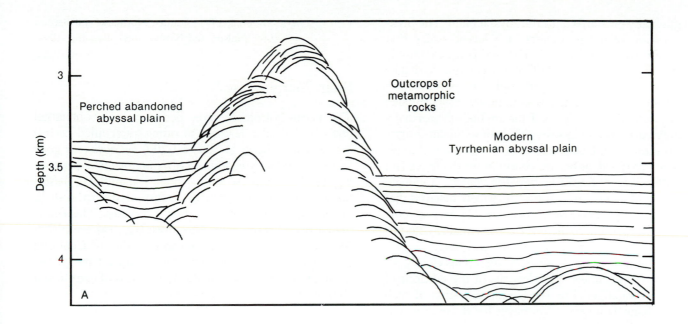

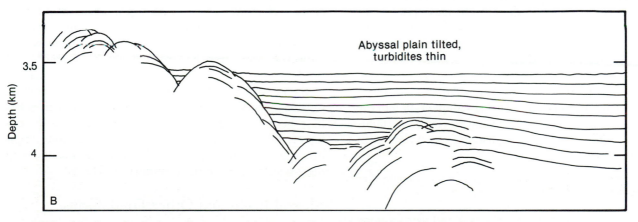

FIGURE 9-6.   Sketches of seismic profiles across ridges and adjacent abyssal plains. Flat-topped gravity-displaced sediments are *ponded* in depressions, known as *sediment ponds*. The rugged ridges lack sediment cover. In places, the sediments have been deformed. Central Tyrrhenian Sea (Western Mediterranean). (After B. C. Heezen, C. Gray, A. G. Segre, and E. F. K. Zarudski, 1971, fig. 3, p. 329.)

their steep sides are in the downcanyon direction. Reversing currents having tidal periods have been measured in several canyons. Erosion is an important process in the upper parts of many submarine canyons, and was the dominant factor in the formation of all submarine canyons. Because of the present high stand of sea level, many submarine canyons do not appear to be experiencing much erosion nowadays.

The lower ends of many canyons end in sediment fans called **deep-sea fans** (or **deep-sea cones** if they are particularly large), *subaqueous fans that are (morphologically) analogous to fans built on land in desert basins* (subaerial fans; Chapter 14). The volumes of sediment composing some subaqueous fans are 10,000 times the volumes of the associated canyons. This relationship indicates that the canyons are serving as conduits for sediment. Such sediment moves into their upper ends and is discharged at their lower ends. Typical fans are thickest and coarsest textured at their landward ends, where they receive sediment from subaqueous canyons, and thin and fine gradually outwards onto the abyssal plain. The proximal portion of a fan consists of a leveed valley that feeds sediment to the mid fan. The mid fan is commonly channeled, and is composed of lobes analogous to deltaic lobes. (See Figure 9-3.) Usually only one is active at a given time. The outer fan is commonly nonchanneled and generally lacks fan lobes. Fans are built primarily by turbidity currents and by other kinds of sediment-gravity flows (Figure 9-5). Coarse-textured debris flows, slumps, and other coarse deposits are common on the inner fan. **Turbidites**, *deposits formed by turbidity currents*, dominate the mid- and outer-fan regions. It is widely believed that most deep-sea fans consist of turbidites. However, some fans have been found to include deposits made by slumping or by other gravity-displacement processes.

Remarkably sinuous channels on some fans closely resemble fluvial channels. The deposits of the portions of fans that were affected by the lateral migration of such sinuous channels might be mistaken for those of meandering streams and vice versa.

Some fans are fed by remobilized prodeltaic sediments. Because delta lobes switch, the primary locations of the source of the fan sediment changes episodically. Therefore fans fed by deltas may show somewhat different stratigraphic sequences than fans fed by subaqueous canyons not related to deltas.

Deposition in submarine fans and -cones is greatest when sea level is low or beginning to rise (lowstand facies tract). In general, lateral sedimentation in the fondo environment is greater during sea-level lowstands and is less during sea-level highstands, as explained in Chapter 6.

Subaqueous Aprons

In contrast to subaqueous canyon-fan- and canyon-cone complexes are **subaqueous aprons**, *coalesced subaqueous fans*. Subaqueous fans are fed by single submarine canyons, whereas multiple close-together submarine canyons generate a subaqueous apron. The term subaqueous apron thus describes the situation in which sediment is pouring off the continental shelf and onto the continental slope along a more-or-less continuous region that may extend hundreds of kilometers parallel to the shelf edge. The processes that deposit sediments in subaqueous aprons are essentially the same as those that operate in subaqueous fans and -cones. However, because subaqueous aprons are greatly elongated parallel to the continental margin, the three-dimensional shapes of their sedimentary deposits differ greatly from those of the deposits of subaqueous fans and -cones. Also, the bodies of sediment that constitute subaqueous aprons are generally at least an order of magnitude larger than those of subaqueous fans and -cones.

SOURCES: Bouma, 1991; Bouma, Normark, and Barnes, 1985; Nelson and Nilsen, 1984; Reading, 1991; Stow, 1986.

Basin-Margin Rises

The ocean-bottom features designated as **continental rises** (and similar features in other water-filled basins) are *the gently sloping* (1:100 to 1:700) *top surfaces of great bodies of sediment lying seaward of the continental slopes*. (See Figure 9-4.) The sediment bodies underlying basin-margin rises consist of debris transported off the shelf, mainly by sediment-gravity-flow processes. The continental rise offshore from New York City, eastern North America, consists of a tremendous wedge of Cenozoic sediment that prograded out over thinner latest Jurassic and Cretaceous sediments. The widths of continental rises are hundreds of kilometers.

Deep-Sea Trenches

A **deep-sea trench** is *a narrow, elongated, steep-sided, commonly rock-walled depression that generally is deeper than the adjacent sea floor by 2000 m or more*. Many deep-sea trenches are situated along the margins of ocean basins. However, some trenches, notably in the western- and southwestern Pacific Ocean, lie far from the margins of the ocean basin. In many modern deep-sea trenches, practically no sediment has accumulated. Other trenches have been filled partially or even completely with sediment. All deep-sea trenches are located at convergent lithospheric-plate boundaries. No features analogous to deep-sea trenches occur in nonmarine basins.

Abyssal Plains and Other Fondoforms

An **abyssal plain** is *the flat or imperceptibly inclined upper surface (gradient less than 1:1000) of a body of gravity-displaced sediment lying on the deep-sea floor*. (See Figure 9-4.) This surface may be draped by pelagic sediment. Commonly, abyssal-plain sequences consist of intercalated gravity-displaced-deposits (mainly turbidites) and pelagic sediments. Two important aspects of abyssal plains need to be emphasized. (1) The notion that the water depths must be abyssal restricts the term to features found on the modern deep-sea floor or to ancient examples where abyssal depths can be reasonably inferred (commonly this is difficult). (2) Flat surfaces and bodies of gravity-displaced sediments exactly analogous to modern abyssal plains have been found in the modern seas at shallower depths (for example, the flat surfaces of small bodies of sediment "ponded" in various closed depressions high up on the mid-Atlantic Ridge or in basins of a continental borderland; Figure 9-6).

Moreover, features analogous to oceanic abyssal plains have been found in lakes (as in Lac Léman, Switzerland and France). We prefer not to stretch the term abyssal by using abyssal plain to refer to the flat tops of these other bodies of gravity-displaced sediments that are not covered by abyssal depths of water or are not even located in the oceans. These surfaces are *basin plains* or fondoforms.

Other Parts of a Basin Floor

The mid-ocean ridges are *submarine volcanic mountain chains formed by the igneous activity along the divergent boundary where lithospheric plates are separating.* These are not sites of major sediment production or accumulation, except in small closed depressions, and they are not considered further here. However, they are the sites of sea-floor hydrothermal processes creating ore deposits.

Apart from the remarkably smooth basin plains, the basin floor consists of rugged rocky features that have been but thinly veneered by sediments or are devoid of sediment cover. Among these irregular, rough rocky masses are various elongate depressions, many of which contain flat-topped bodies of sediment (the so-called *sediment ponds*). (See Figure 9-6.)

Now that we have described the different parts of basins, it is time to consider the kinds of sedimentation processes and of sedimentary deposits that pertain to basins.

SOURCES: Bouma, Berryhill, and Knebel, 1982; Brunner and Normark, 1985; Damuth and Embley, 1981; Dingler and Anima, 1989; Fraser, 1989; Friedman, 1988; Gardner and Kidd, 1987; Goodwin and Prior, 1989; Gradstein, Ludden, and others, 1990; Grotzinger, 1986; Leg 123 Shipboard Scientific Party, 1988; Ludvigsen, 1989; Melvin, 1986; Mullins, Heath, van Buren, and Newton, 1984; Mutti, 1979; Nilsen, 1980; Normark, 1989; Piper and Normark, 1983; Reading, 1986, 1991; Rona and others, 1990; Twichell, Grimes, Jones, and Able, 1985; Van Wagoner, Mitchum, Campion, and Rahmanian, 1990.

Lateral- and Vertical Sedimentation

A fundamental division can be made between two kinds of sedimentary processes that typify large basins (especially marine basins). Lateral sedimentation refers to *deposition of sediment along an inclined surface that migrates in one direction when more sediment is added.* Lateral sedimentation characterizes basin slopes, or clinoforms, where the dominant processes are those of gravity-displacement. *Gravity-displacement processes* trans-

fer sediment from clinoforms to fondoforms. *Contour currents,* discussed in the next section under processes of slope-influenced sedimentation, represent another kind of lateral-sedimentation process, in which sediments are moved on the bottom parallel to the slope and thus remain on the clinoform rather than being transferred to the neighboring fondoform. Vertical sedimentation is *the upward accumulation without lateral extension of a sloping depositional surface that results from widespread addition of sediment from above.* Most vertically sedimented material on the deep-sea floor consists of particles generated in the water column, such as the tests of floating microorganisms, and particles that enter the water from the air, such as volcanic ash. These particles settle down through the water column under the influence of gravity, and come to rest on undaforms, clinoforms, and fondoforms alike. After initial deposition by vertical sedimentation, sedimentary materials may be remobilized by lateral-sedimentation processes. These materials may either be transferred along a clinoform by contour currents or may be moved downslope onto a fondoform by gravity-displacement processes.

The sedimentary deposits found beneath *fondoforms,* called *fondothems,* thus are made up of laterally sedimented gravity-displaced deposits, such as turbidites, slumps, and debris flows, intercalated with vertically sedimented fine-textured pelagic sediments. The thicknesses of gravity-displaced deposits and of pelagic deposits in a given set of strata may be comparable. However, gravity-displaced deposits were formed by brief pulses of activity separated by long periods of quiescence. A given gravity-displaced stratum may have taken no more than a few hours to form (See Figure 9-11, B). By contrast, pelagic sediments are forming all the time, at a slow but steady rate. This is the gentle rain which nineteenth-century oceanographers and sedimentologists thought was the only kind of sedimentation on fondoforms. A stratum of pelagic sediment may have taken millennia to build up, particle by particle, continuing to increase in thickness until the next turbidity flow or other lateral-sedimentation event came along. This contrast between the time scale of laterally sedimented- versus vertically sedimented deposits will be taken up again in Chapter 16.

Processes of Slope-Influenced Sedimentation and Their Products

The sediments deposited under the influence of subaqueous slopes include those underlying the continental slopes; those underlying the flat floors and mantling the steep walls of submarine canyons; those underlying abyssal fans and cones, continental rises, abyssal plains, and sediment ponds; and those underlying the equivalent features in nonmarine basins. Two general terms that were introduced in Chapter 6 are useful for designating such sediments; these are *clinothems* and *fondothems,* collective terms for sediments underlying *clinoforms* and *fondoforms,* respectively.

Viewed broadly, most slope-influenced sediments fall into two categories: (1) **gravity-displaced deposits,** which are *sediments that travel along the bottom down a subaqueous slope and, if they go beyond the bottom of the slope, spread themselves out on the nearest available fondoform;* and (2) *fine-textured, well-stratified, well-sorted, tractionally reworked sediments deposited by contour currents* (**contourites**). **Contour currents** are *currents that flow along subaqueous contours.* Contourites are found on the lower parts of slopes; they build clinothems basinward. Other sediment-transport mechanisms that can emplace or move sediments on slopes are varied, but involve much less sediment than do the two categories just mentioned. Organic buildups (reefs or bioherms) may form in slope- or even basin-floor settings. In the modern world, these deep buildups are commonly dominated by azooxanthellate corals and/or sponges (Chapter 7); some ancient slope- and basinal buildups were composed primarily of bryozoans and crinoids. Suspended fine-textured sediment may be transported into deep water by wind; in near-surface currents; and as **nepheloid layers,** *near-bottom masses of turbid water containing fine-textured sediment that has been suspended by waves or currents or by being injected into the sea or a lake by an inflowing stream of fresh water.* In addition, sediments that fall from above may accumulate on any slope. These are, in the oceans, the pelagic sediments described in a following section.

Gravity Displacement

As we have seen, gravity-displaced deposits are all sediments transported by all the various processes that are energized by the tangential component of gravity acting along a slope. In effect, all gravity-displaced sediments are pulled from the front, rather than being pushed along from the back (as are the bed-load sediments of a stream, for example).

The characteristics of gravity-displaced deposits are determined by the kind of sediment available at the top of and along the slope and by the process(es) involved in moving it down the slope. The available sediment may consist wholly of terrigenous debris (including second-cycle carbonates), wholly of first-cycle shallow-water carbonates, wholly of pelagic sediments, or of any mixture of these three. We shall discuss five kinds of clinothems that are deposited by five kinds of processes: (1) subaqueous taluses, (2) products of subaqueous slumps,

(3) channel deposits of subaqueous canyons and -fans, (4) interchannel deposits of subaqueous fans and -cones, and (5) contourites; and two kinds of slope-influenced fondothems: (6) liquefied cohesionless-particle flows, and (7) turbidites.

Triggering mechanisms for gravity displacement of sediments are all processes that permit the downslope pull of gravity to exceed the shear strength of the material. These include earthquake shocks, influx of sediment-laden river water, oversteepening by erosion or deposition, postdepositional tilting, excess pore pressures created by rapid sedimentation, artesian fluid flow or interstitial gas generation in organic-rich sediments, and wave loading, in which water is forced into sediment by the impacts of waves.

Subaqueous Taluses and Other Accumulations of Rock Fragments

Individual large blocks can be broken from a reef or from a coastal cliff and can fall into deep water, there to accumulate as *subaqueous taluses* or other accumulations of rock fragments. The angle of repose of a pile of large rock fragments is about 45°; therefore, if a reef is providing the debris and the supply is plentiful enough, such a steep depositional slope may prograde and the reef may grow seaward over its **reef talus,** *coarse forereef sediments, consisting of sand- and cobble-size fragments of skeletal debris derived from a reef and deposited in strata having an appreciable initial dip* (Figures 9-7 and 9-8).

If the blocks are broken from a cliff at the shore, they will consist of whatever rocks are exposed in the cliff. Because carbonate banks as well as reefs tend to form steep submarine escarpments, the chances are good that carbonate rocks from them will accumulate as taluses in

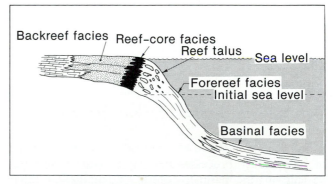

FIGURE 9-7.   Idealized profile and section across shelf-edge reef, as seen in many parts of the rock record, showing massive reef core (black), backreef (stippled), and forereef facies (showing steep initial dip, slightly exaggerated, toward basinal facies). (Authors.)

FIGURE 9-8.   Slump rubble of skeletal limestone in strata of Early Jurassic age that had advanced across bedded succession of forereef of Triassic Steinplatte reef. The slope descending from the reef scarp down which the rubble moved, locally exceeded 30°. Note wavy slide surface between talus and bedded forereef beds. Kammerkohr Alm, northcentral Austria. (R. E. Garrison and A. G. Fischer, 1969, fig. 17, p. 40.)

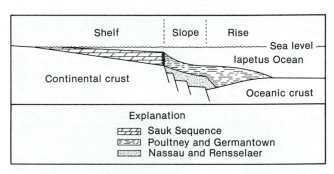

FIGURE 9-10.   Schematic profile and section, Cambrian and Early Ordovician basin margin and abyssal plain, eastern New York State. Carbonates floored the former shelf (Sauk Sequence) and terrigenous deposits floored the former slope and rise (Poultney and Germantown formations). The paleoslope was probably an active hingeline between the continent to the west and the ocean to the east. Slump rubble of shallow-water limestone moved down this steep slope to oceanic depth by slides, slumps, turbidity currents, and mudflows, to come to rest at the deep-water basin margin, where dark mud accumulated. (After L. V. Rickard and D. W. Fisher, 1973, fig. 2A, p. 587.)

deep water. In order for other kinds of materials to form similar deposits of comparable size, they would have to be exposed on a steep scarp that extended from above water level. Such a scarp could be created by faulting or by glacial erosion.

The rocks from a cliff may be undermined by the bio-erosion activities of organisms or by the effects of storms or ordinary gravity effects (Figure 9-9). The pounding of waves against coastal cliffs compresses the air within the joint systems. The result may be a sudden, explosive springing loose of large blocks of rock, which would continue to move underwater as rock falls.

Widespread examples of inferred ancient subaqueous accumulations of coarse rock fragments, not reef taluses,

are in the Lower Paleozoic strata of the Appalachians, northeastern North America (See Figure 17-11.) and in the Upper Paleozoic strata of the Ouachita chain, southeastern Oklahoma. In these examples, blocks of many kinds of shallow-water carbonate rocks are enclosed in a terrigenous dark gray or black shale that is devoid of calcium carbonate (Figures 9-10 and 9-11). One of the most-famous examples is found in the Cow Head Group, in which boulders of shallow-water carbonates containing Cambrian trilobites are enclosed within deep-water strata containing Ordovician fossils.

Products of Subaqueous Slumps and -Debris Flows

Sediments that have come to rest layer by layer on a subaqueous slope may afterwards move en masse down a curving slump surface (a fault). At its upslope end, such a surface cuts the strata at a high angle. At its lower end, the surface becomes parallel with the layers, but the overlying displaced strata dip steeply down toward it. The body of sediment thus moved is said to have slumped. During slumping, the sediment may behave in two ways: (1) *the strata within the displaced mass remain intact and may be much deformed* (coherent slump or slide) (Figure 9-12) or (2) *the entire body of sediment may become so thoroughly mixed and churned that nearly all traces of stratification are obliterated* (incoherent slump). An *incoherent slump* may consist of *a fine-textured matrix carrying large blocks as a debris flow* (Figure 9-13). A term applied to *a debris-flow deposit containing large blocks* is olistostrome.

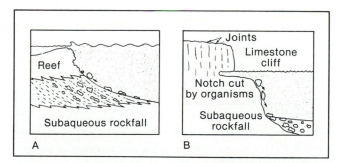

FIGURE 9-9.   Subaqueous rockfalls related to two kinds of rocky shores. (A after R. H. Dott, Jr., 1963, fig. 7, p. 110; B, authors.)

A. Blocks fall from steep front of active reef.

B. Blocks fall from limestone cliff, having broken along joints after being undercut by notch near sea level that was excavated by organisms.

FIGURE 9-11.

A. Ancient slump deposits. Slump rubble of skeletal- and pel-letal limestone resembling breccia displaced from shallow sea behind shelf edge into deep-water, dark-colored shale. This lime-stone rubble is known as brecciola. West Castleton Formation (Lower Cambrian), campus of Rensselaer Polytechnic Institute, Troy, New York. (G. M. Friedman, 1972, fig. 8, p. 28.)

B. Graded layer; the deposit of a coherent slump. The overly-ing fine pelagic strata took far longer to form than did the slump, which may have accumulated in a few hours. West Castleton For-mation (Lower Cambrian), about 16 km south of Troy, New York. (Authors.)

Where the surface of displacement is steep and the strata cut by it are nearly horizontal, then the strata and the surface of displacement may form an angle close to 90°. (See Figure 9-15.) This angle will be maintained

FIGURE 9-12.   Intraformational folds among skeletal talus de-posits of forereef formed as a result of downslope slumping. Bone Springs Formation (Permian), Crew's Canyon, near Pine Springs, Texas. (Authors.)

as long as the stratification persists in the slumped sed-iment. As the slumped sediment moves downslope, the displaced material may be transported across a surface of displacement whose dip progressively decreases in the downslope direction. Accordingly, the strata within the slumped mass will dip back toward the slope, always

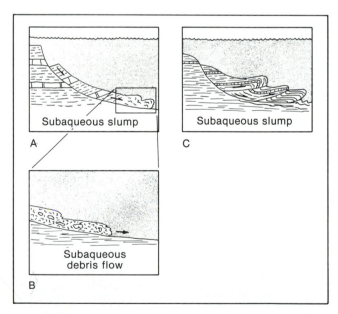

FIGURE 9-13.   Contrasting effects at toe of a subaqueous slump: debris flow and thrusting.

A. Toe of subaqueous slump becomes incoherent, forming sub-aqueous debris flow.

B. Enlarged view of toe of A, showing closer view of subaque-ous debris flow.

C. Subaqueous slump, whose lower end did not form a debris flow but instead was stacked up in a series of thrust sheets. (After R. H. Dott, Jr., 1963, fig. 7, p. 110.)

maintaining the same angle that was established back upslope where the surface of displacement was steep and the strata horizontal.

Typically, the downslope edge of a slumped mass of sediments mixes with water to the extent that the stratification is destroyed and the material becomes a chaotic mixture of water and particles of whatever sizes are avail-

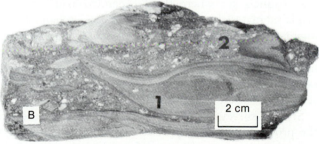

FIGURE 9-14. (*Continued*)

C. Intensely deformed part of a fold that formed at the toe of a slumped mass: 1, overturned graded layer; 2, several layers bent together until their free ends touch, as in B1, forming a nappe-like refolded isoclinal fold. (J. E. Sanders, 1968, A, pl. 2, fig. 2; B, pl. 2, fig. 3, following p. 280.)

FIGURE 9-14. Photographs of material that slumped on the bottom of an ancient lake, Portland Formation (Jurassic), Branford, Connecticut.

A and B. Poorly sorted deposits of subaqueous debris flows. In A, 1 designates large quartz fragment; 2, remnant of bedding; and 3, detached remnant of formerly continuous bed of fine sand, not bent. In B, numbers mark intensely crumpled remnants of formerly continuous layers: 1, several layers bent back until their free ends touch; 2, a graded layer pinched together and arched upward, former top at left.

able (Figure 9-14). After such mixing has taken place, the process of movement is that of a subaqueous debris flow. (See Figure 9-13, A and B.)

If a subaqueous slump does not become an incoherent subaqueous debris flow, its lower end may push out over nondisturbed sediments or over portions of itself, thus duplicating the thickness as in a bedding thrust. (See Figure 9-13, C.) Along the snout of a slump in which such duplication or thrusting has taken place, the strata may be overturned and complexly folded. (See Figure 9-14.) These exhibit, in miniature, the structures characteristic of foreland fold-and-thrust belts. The upslope edge of a slumped mass may become stretched, and may exhibit extensional faulting and boudinage.

Numerous continuous seismic-reflection profiles made perpendicular to the continental slope have shown disturbed strata that probably resulted from submarine slumping (Figure 9-15). Identical features, except for size, underlie the sloping bottoms of nonmarine basins. Slumps and debris flows are particularly common on large subaqueous fans and subaqueous cones.

Submarine slumps can involve tremendous quantities of sediment. A slump off the northwest African continental margin consists of 600 km³ of sediment that became a debris flow that covered an area of 30,000 km² of the sea floor. This deposit is far larger than the largest known turbidite.

Many examples of ancient slumped strata have been reported. A few examples from carbonate rocks include the forereef deposits of the Steinplatte reef (Triassic), Austria (See Figure 9-8.); the Permian of west Texas (See

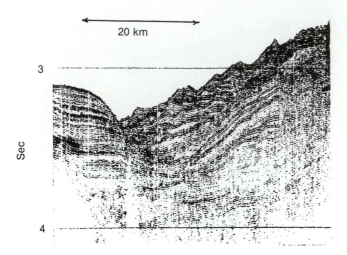

FIGURE 9-15.   Seismic-reflection profile showing disturbed strata that probably resulted from submarine slumping at the base of the continental slope off Baranof Island, Alaska. (Lamont-Doherty Geological Observatory of Columbia University, R. W. Embley, 1975 ms., fig. V-59, caption on p. 208.)

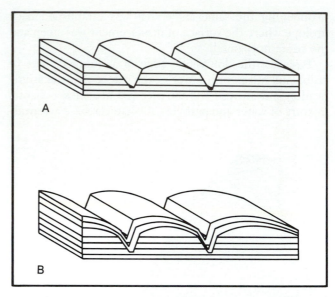

FIGURE 9-16.   Block diagrams of submarine canyons and intercanyon areas on upper continental slope off Cape Hatteras, southeastern United States.
   A. Canyons are incised into planar strata.
   B. Convex-upward strata are deposited in intercanyon areas and form the present canyon walls. (After P. A. Rona, 1970, fig. 7, p. 150.)

Figure 9-12.); and the Cambrian limestones in eastern New York. (See Figure 9-11.) Debris flows that probably originated as slumps are common in the Cambro-Ordovician Cow Head Group of western Newfoundland. Examples from ancient terrigenous strata include the Carboniferous of New South Wales, Australia; the Silurian of western Wales, Great Britain; and the Upper Cambrian of northwest Tasmania.

Channel Deposits of Subaqueous Canyons and Fan-Valleys

Subaqueous canyons serve as avenues of sediment transport between a shelf and a basin plain. These canyons are incised into sedimentary strata or into other kinds of bedrock. The sediments underlying the floor of a subaqueous canyon may fill the canyon to varying depths. Two kinds of channels are distinguished: erosional channels and depositional channels. The former typify the upper reaches of subaqueous canyons, in which little or no sediment is deposited and in which the underlying material is eroded. By contrast, sediment is deposited within depositional channels. In addition, depositional channels carry sediment from the shelf to the slope, sediment which may then be deposited outside of the depositional channels, as described below. Off Cape Hatteras, southeastern United States, however, processes of sediment removal clear the canyon of sediment over the sites of original incision, whereas in the intercanyon areas, accumulation of strata builds up the walls of the canyons (Figure 9-16). Where fill is deposited within the canyon, this fill may be thick enough to display a comparatively flat surface, with a central channel flanked by

natural levees. Between the natural levees and the wall of the canyon may be flat areas that are analogous to the flood plains of meandering rivers on land. This analogy of morphology extends to processes as well, but with different end products as a result of the contrast in environments. Various kinds of coarse sediments are confined to the channels, whereas finer materials spread from the channels to the adjacent areas. In effect the channels are the confined pathways for the "bed-load" material displaced by gravity, whereas the broader areas next to the channels are where the "suspended-load" gravity-displaced material accumulates. The coarser channel sediments are laminated and cross laminated; some contain large rounded siltstone clasts (shale pebbles). These are interstratified with plant debris. Bioturbation structures occur here and there, but are more abundant in the finer sediments outside the channels.

Seismic surveys of submarine canyons have revealed the products of a variety of kinds of sediment-fill- and canyon-evolution processes. These include large-scale slumps with rotation of coherent blocks; channel-fill sequences consisting of chaotic sediments transported by debris flows and other gravity-displacement processes; and the products of channel widening with formation of reentrants and isolation of channel-side knolls (Figure 9-17).

Some sediment flows pass completely through the canyons and cross the subaqueous fans and stop on the basin plains. Others drop some or all of their loads in

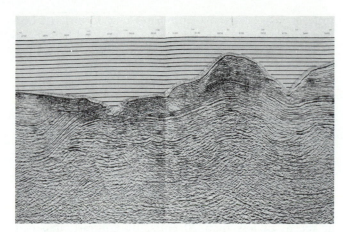

FIGURE 9-17.   Segment of continuous seismic-reflection profile across submarine canyons. (Courtesy Merlin Geophysical.)

the canyons or on the fans. The sediment flows may be in the form of pure suspended loads (*low-density turbidity currents*); mixed bed loads and suspended loads (*high-density turbidity currents* ± an accompanying sand flow); or pure bed loads (*liquefied coarse-particle flows*). Lateral shifting of channels creates complex internal relationships. However, it is not known if such channel migration creates a definite pattern of parasequences, as in a point-bar succession, for example.

Interchannel Deposits of Subaqueous Fans and -Cones

When they are freshly deposited, the interchannel sediments on subaqueous fans and -cones are distinctly stratified. However, soon afterward, burrowing organisms may destroy these strata. Graded layers are typical; these are thought to have been deposited by the parts of turbidity currents (described in a subsequent section of this chapter) that were higher than the natural levees and thus spilled over them. The interchannel deposits constitute the bulk of subaqueous fans and -cones. Minor amounts of pelagic sediment (to be discussed in a following section) are present (Figure 9-18).

Contour Currents and Contourites

The existence of deep, bottom-following currents (*contour currents*) of dense cold water flowing along the continental rises, predicted by the physical theory of ocean circulation, has been demonstrated by direct measurement, by analysis of photographs and of side-scan sonar records of the sea floor, and by study of continental-rise sediments (Figure 9-19).

Where contour currents are active, the sea floor shows the effects of sediment deposition. On a large scale, the topographic expression may be smooth, but on a smaller

FIGURE 9-18.   Lower part of carbonate turbidite (light color) overlying hemipelagic clayey siliceous ooze (darker color). Sand-size massive basal portion of turbidite (radiolarian-foraminiferal ooze; 122.5–112 on scale) grades up into finer, vaguely laminated interval (nannofossil radiolarian-foraminiferal ooze; 107–112), and finally into clay- and silt-size upper massive portion (clayey nannofossil ooze; 107 and above). Scale in centimeters. Ocean Drilling Program Leg 123, Site 765 (6000 m water depth, Argo Abyssal Plain, northeast of North West Shelf of Australia), Hole B, core 2, section 3. Base of turbidite 10.5 m below sea floor. Late Pleistocene or Holocene in age.

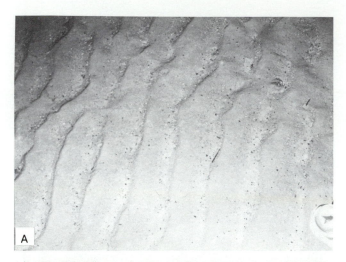

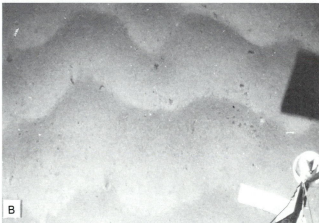

FIGURE 9-19.   Photographs of bottom-sediment surface of the Blake Escarpment, western Atlantic Ocean. The visible sedimentary structures are the result of reworking by contour-following bottom currents.

A. Strongly scoured bottom ripples with sand in troughs; strong current to south (note direction of compass arrow). Water depth 2442 m.

B. Large linguoid ripples in mud show current direction to the south (top of photograph; note compass arrow). Water depth 3430 m.

C. Hard-packed rippled bottom; current direction to south to southwest (note compass arrow). Water depth 4000 m. (Lamont-Doherty Geological Observatory of Columbia University, E. D. Schneider, 1970 ms, fig. 18, p. 80–81.)

scale, it may be wavy. Contourites may form large bed forms with wavelengths as great as 1 km and heights of several meters in sediment drifts in 2000 to more than 4000 m of water. Contourites are generally cross laminated and laminated throughout, commonly with heavy-mineral- or shell layers, and may be either normally- or inversely graded.

Most of the material deposited as contourites probably came from some upslope source; the sediment was displaced downslope by gravity and then came under the influence of the contour currents. The flow of contour currents is much slower than that of most gravity-displacement processes: contour currents typically flow at speeds of less than 20 cm/s. By contrast, liquefied cohesionless-particle flows (described in the next section) have been clocked at celerities of greater than 300 cm/s.

Contourites (Figure 9-20) have been accumulating in the western Atlantic since Miocene time. Presumably this date marks the time when the Atlantic Ocean was wide enough and deep enough to enable cold polar water to flow southward beneath the Gulf Stream from the Arctic Ocean toward the Equator. However, the Miocene marks a time when deep-water sedimentation in many of the world's oceans was accelerated (lowstand facies tracts were forming), and therefore the onset of contourite sedimentation in the western Atlantic may be related to global sea-level change. This, in turn, may be related to the opening of the Atlantic itself. Contourites in the eastern Atlantic off Spain and Portugal are as old as

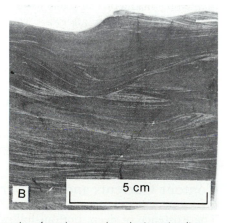

FIGURE 9-20.   Photographs of modern- and ancient contourites.

A. Core of modern contourite raised from Caicos Outer Ridge, Bahamas, western Atlantic Ocean. This sediment is a well-sorted, medium-textured skeletal sand; note horizontal laminae. (Lamont-Doherty Geological Observatory of Columbia University, E. D. Schneider.)

B. Polished slab of inferred contourite of Cambrian age sampled near campus of Rensselaer Polytechnic Institute, Troy, New York. This inferred contourite is a current-ripple cross-laminated pelletal limestone; quartz silt accentuates the laminae. (B. D. Keith.)

Eocene, but a surface of unconformity separates Eocene from uppermost Miocene and younger contourites in this area.

Examples of inferred ancient contourites that are older than sediments in the modern ocean include parts of the lower Cenozoic of the Swiss Alps, parts of the Middle Ordovician Martinsburg Formation in the Appalachians, and most of the Cambro-Ordovician Taconic sequence of the Northern Appalachians. (See Figure 9-20, B.)

Liquefied Cohesionless-Particle Flows

The features we refer to as **liquefied cohesionless-particle flows** are *bodies of cohesionless sediment particles that have become dilated past their critical densities* (Chapter 7). Some geologists classify liquefied cohesionless-particle flows into two kinds. All such flows derive their motive power from the force of gravity, coupled with the dilation of the bodies of particles. However, dispersive pressure causing *dilation* may result from upward flow of pore fluids through the sediment (fluidized cohesionless-particle flows) or by particle collisions (grain flows). In both cases, a shock, such as an earthquake, may trigger the dilating process. The subject of the independent existence of liquefied cohesionless-particle flows is not one on which all geologists agree. Moreover, those who do consider these flows as important factors in deep-water sedimentation have not yet agreed on a single name for them. Terms proposed have included inertial flows, grain flows, mass flows, rheologic bed stage, fluidized cohesionless-particle flows, and the one used here, liquefied cohesionless-particle flows. Because such flows commonly accompany turbidity currents, some investigators have supposed that the liquefied cohesionless-particle flows are part and parcel of turbidity currents and therefore do not merit a special name.

The fundamental argument in favor of the separate recognition of liquefied cohesionless-particle flows is that even when they are moving along beneath a turbulent current, as, for example, in a river, *the quantity of sediment discharged by the cohesionless-particle flow becomes independent of the turbulent current*. In fact, when a cohesionless-particle flow is present, the usual conditions of drag at the base of the turbulent current are reversed. Instead of the turbulent current's pushing the carpet of sand, as in a traction carpet, and thus being enormously dragged by the cohesionless particles, a liquefied cohesionless-particle flow gives the overlying turbulent current a free ride, so to speak. Indeed, if anything, the cohesionless-particle flow drags the overlying turbulent flow along with it.

The concept of the independence of liquefied cohesionless-particle flows is further based on (1) the direct observation of sands flowing down submarine canyons powered only by the influence of the tangential component of gravity; (2) flows of sand along the floors of Norwegian fjords and elsewhere; (3) injection of coarse tongues or sheets as sill-like intrusives between the strata of channel-marginal sediments; and (4) the occurrence of coarse, well-sorted gravel in the channels of the fills of some submarine canyons. Liquefied cohesionless-particle flows may evolve into turbidity currents as they flow, and possibly, vice versa.

Sands flowing down the steep slopes of San Lucas Submarine Canyon, off Baja California, have been observed and photographed by divers (Figure 9-21). Motion pictures of these flows reveal two important aspects. First, the flows develop from traction carpets energized by wave action. The seaward surge of bottom water beneath wave troughs starts the sand moving as a traction carpet downslope. (See Figure 7-1, B.) Once the sand has been energized by wave action, the slope-parallel component of gravity keeps the sand flowing downslope. When the next wave crest arrives, it creates a landward surge of bottom water that temporarily reverses the flow of some of the sand. (See Figure 7-1, C.) Despite such reversal, however, much sand escapes from the zone of wave action and continues to flow down the canyon.

Second, the point where the speed of the sand flows reaches a maximum is at the margins of the flow, next to the rock wall of the canyon. This location of the locus of maximum celerity is readily explained by the concept that the basic mechanism of flow is the Bagnold effect (Chapter 7). The sand moves faster next to the rock wall because the individual particles make larger rebounds

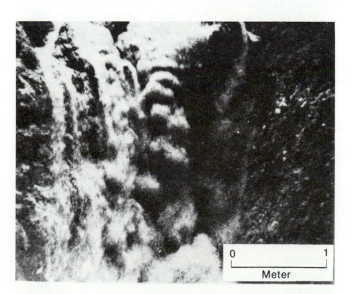

FIGURE 9-21. Underwater photograph of cohesionless-particle flow of sand, which is cascading like a waterfall over a steep submarine slope off San Lucas Bay, Baja California, Mexico, at depth of about 10 m. (William Bunton, U. S. Navy Underwater Systems Laboratory, San Diego, California, courtesy R. F. Dill.)

when they encounter a solid surface than when they strike other particles of about their own size. Where the particles rebound farther, they travel faster.

Other sand flows down submarine canyons take place off southern California, where the heads of the canyons intercept the belt of nearshore sediment being shifted by wave action. The discharges into the canyons are well-established factors in the budgets of nearshore sand.

Some flows of liquefied cohesionless particles may begin their careers in the channel of a river during a flood stage. For example, whenever the Congo River reaches its flood stage, the sand bar at its mouth disappears temporarily and the submarine cables that have been laid parallel to the coast all break. Coarse sediment from the river is transported to the Congo abyssal fan. Another possible site of liquefied cohesionless-particle flow formation is at a coastline characterized by eolian sand dunes. Storms may undercut dunes, which slump and then almost immediately liquefy. Water-saturated basal parts of dunes may also liquefy in response to shocks, with resultant flowage. This may be an important mechanism for the destruction of eolian sand seas and the offshore transport of large quantities of sand during coastal submergence.

Flows of sand along the floors of Norwegian fjords have broken telephone cables that had been laid on the bottom. On 2 May 1930, in Orkdalsfjord, Norway, a sand flow from the shore was triggered at an extremely low tide. Possibly the seepage of ground water started the flow. Two telephone cables on the bottom were broken, one at a distance of 20 km from the origin of the flow. The average speed of the flow has been computed at 10 km/h (Figure 9-22).

Another episode of breaking of submarine cables, which has achieved worldwide fame, took place in the previous year, on 18 November 1929, off the Grand Banks, south of Newfoundland, as a result of the Acadian-Newfoundland ("Grand Banks") earthquake, whose initial shocks were recorded at 2032 GMT. At the time the earthquake was recorded, seven cables, all lying near the epicenter, broke instantaneously. (See Figure 9-1.) Later, five other cables were snapped, each lying progressively farther south of the epicenter and in progressively deeper water. The last cable to break, which lay 625 km distant from the epicenter on an abyssal plain where the depth of water is about 4800 m, ceased operating at 0950 GMT, 19 November 1929, 13 h and 17 min after the initial shock of the earthquake. Broken segments of cable were transported far from the places where the cables had been laid and were buried in coarse sediment.

The cables that broke in sequence after the earthquake parted where they crossed large submarine canyons. When the frayed ends of the cables were brought up

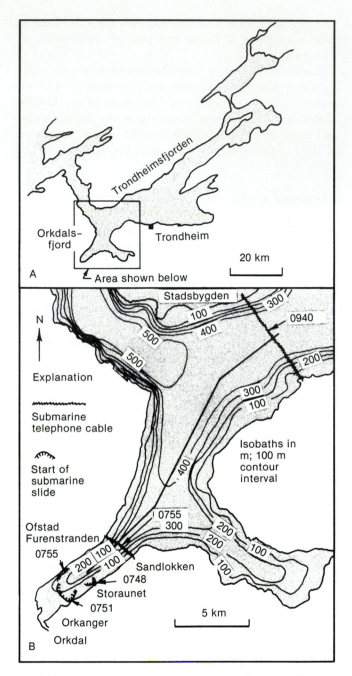

FIGURE 9-22.   Submarine slides and cable breaks, Orkdalsfjord, Norway, 02 May 1930.

A. Index map of Trondheimsfjorden, showing tributary, Orkdalsfjord.

B. Bathymetric map of Orkdalsfjord showing places and times (24-h notation) where submarine slides began on morning of 02 May 1930 (after an unusually low tide) and locations of submarine telephone cables broken by flow of displaced sediment along bottom.

to be repaired, the steel strands were seen to have enclosed small pebbles. Evidently a catastrophic flow of coarse sediment, including pebbles, traveled along the axes of the canyons and the impact of the moving sediment broke these cables.

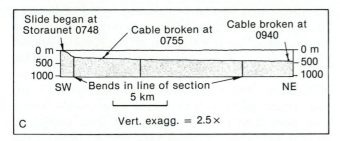

FIGURE 9-22. (*Continued*)

C. Profile from Storaunet to Stadsbygden-Varpenesset submarine cable along line shown in B. Assuming that sediment starting at Storaunet at 0748 broke both cables, then the speed of flow along the bottom was 26 km/h from Storaunet to the Furenstraden-Sandlokken cable and averaged 10 km/h from Storaunet to the Stadsbygden-Varpenesset cable. (After A. Andresen and L. Bjerrum, 1967; A, fig. 2A, p. 231; B, fig. 7, p. 235; C, profile drawn by J. E. Sanders from Andresen and Bjerrum's fig. 2A.)

If it is assumed that the displaced sediment began from the steep (1:50) continental slope at the epicenter, then the locations of the cables and the times of their breakage can be used to plot a graph of the speed of the displaced sediment. (See Figure 9-1.) From the presumed place of origin to the first broken cable, the calculated speed is 22.5 m/s. Between the last two breaks, where the bottom slope is 1:2000, the computed speed is 6 m/s.

In the early 1950s, when these and other cable breaks were being restudied, the only mechanism generally thought to be capable of transporting sediment rapidly along the deep-sea floor was a turbidity current. Therefore, the broken cables were considered to have timed the flow of a turbidity current. (However, it was suggested by K. C. Terzaghi that "soil" liquefaction might have been involved.)

The finding in 1968 of coarse, well-sorted gravel in the axis of the eastern of the two large submarine canyons involved suggests that the active process may not have been a turbidity current but a catastrophic subsea avalanche—a liquefied flow of cohesionless sediment (Figure 9-23). Whatever the mechanism was, we know that it caused coarse sediment to flow down two submarine canyons with extraordinary speed and that the moving sediment packed a tremendous wallop.

Sedimentary sills have been found extending laterally from the coarse filling of a channel and penetrating between strata adjacent to the channel. How such penetration by sheets of cohesionless sediment takes place if the sediment traveled in suspension is a complete mystery. By contrast, it is not unusual behavior for a thick and fast-moving liquefied cohesionless-particle flow.

Coarse, well-sorted gravels have been found on the floors of some submarine canyons and in the subbottom

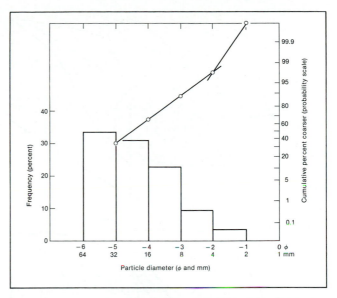

FIGURE 9-23. Cumulative-frequency graph (on probability scale) and histogram (using arithmetic scale) of coarse, well-sorted gravel collected by J. E. Sanders in August 1968 from axis of submarine canyon off Grand Banks, at lat. 43°32'N, long. 55°12'W, in water 4000 m deep. If such sediment was emplaced as an aftermath of the earthquake of 18 November 1929, then it probably traveled as a cohesionless-particle flow and not as a turbidity current. (J. E. Sanders, unpublished.)

sediment underlying the floors of other canyons. The origin of these narrow tongues and ribbons of coarse sediment is not surely known. The thickness of these ribbons of coarse sediment was less than the heights of natural levees which flank the channels. Sediment that spilled over the levees, undoubtedly the suspended load of turbidity currents, generally consists of particles of silt- and clay size, but locally includes sand-size particles (Figure 9-24).

The sedimentary structures produced by all kinds of liquefied cohesionless-particle flows closely resemble one another. Particularly characteristic are (1) inverse grading, (2) downslope coarsening, and (3) rapid lateral change in particle size and -fabric. Fluidized cohesionless-particle flows may be distinguishable from grain flows if they contain fluid-escape structures: *sand volcanoes, fluid-escape pipes,* and *dish structures.* Grain flows are not expected to contain these features. However, not enough is known about the distribution and frequency of pure fluidized cohesionless-particle flows and grain flows as opposed to hybrid liquefied cohesionless-particle flows, nor about the detailed behavior of particles in these flows as they come to rest, to settle this question.

SOURCES: Lowe, 1979; Narden, Hein, Gorsline, and
Edwards, 1979; Stow, 1986.

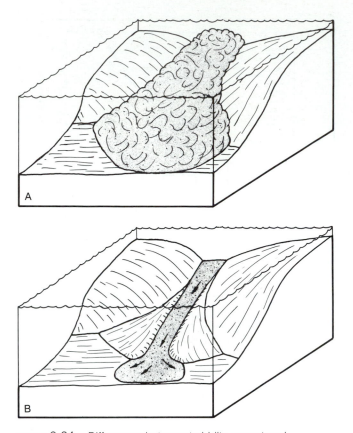

FIGURE 9-24.   Differences between turbidity current and cohesionless-particle flow, schematic sketches of lower end of submarine canyon.

A. Suspended sediment of turbidity current spreads over wide area, filling canyon from wall to wall.

B. Tongue of bed-load cohesionless-particle flow is confined by natural levees at sides of channel on basin-floor fan at mouth of canyon. (Authors.)

Turbidity Currents and Turbidites

The historical background on the study of *turbidity currents* and other subaqueous gravity-displacement processes, as well as the definition, classification, and mechanics of turbidity currents were discussed in Chapters 6 and 7 and in previous sections of this chapter.

Turbidity currents can occur in any basin where the slope of the bottom is great enough to permit density currents to flow. This critical slope may be less than 1°. The following discussion focuses on (1) the relationships between the current and its deposits and the bottoms across which it flows; (2) the relationship of the deposits of turbidity currents, by definition currents that transport gravity-displaced sediment in turbulent suspension, to the deposits of liquefied cohesionless-particle flows; (3) aspects of the origin of marine turbidity currents; (4) turbidity currents and abyssal plains; and (5) turbidity currents and organic matter.

Turbidity-current flows are initiated by shocks (e.g., from earthquakes) or by excess pore pressures. A turbid-

ity current may flow at a maximum speed of between 20 and 25 m/s. Turbidites may transport astonishing amounts of sediment: a turbidite on the Hatteras Abyssal Plain off eastern North America is 500 km long, 200 km wide, and is estimated to consist of more than 100 km^3 of sediment. This is equivalent to more than 700 million box cars filled with sediment, which would form a train more than 11 million km long (long enough to reach to the Moon and back more than 15 times). Even this tremendous turbidite is minuscule compared to some large slumps and subaqueous debris flows.

A turbidity current flows because of its suspended load—no suspended load, no current. When the load has been deposited, the current ceases to flow and that particular depositional episode is finished. Therefore, as the speed of the current begins to decrease and sediment starts being deposited, two things happen: (a) the current reworks the sediment it has just deposited, and (b) the current keeps slowing down as the load drops. Therefore, any features made by a current that depend on the speed of the current will be systematically deposited and always in the same order. The entire list of possible structures may not be deposited by each current. After all, not all turbidity currents are identical in size nor in sediment load. And because of different bottom slopes, all turbidity currents do not travel with the same speed nor cover identical distances. The critical point to emphasize is that in a waning current, the structures that form only when sediment is being transported at high speeds (if they form at all) will always appear first. Hence these "high-speed" depositional structures will always underlie the genetically related "low-speed" structures. (See Figures 5-10 and 9-2.) Obviously, if great eddies or secondary flow systems are present within the current, this simple picture of a gradual decrease in speed at a single point may not take place. Instead, the speed may shift back and forth within a wide range. However, despite such variations, the inevitable overall progression is from fast to slow. The thickness of a turbidity-current deposit tends to be proportional to particle size, at least in well-sorted sediments.

Dilute, fine-textured turbidity currents are called nepheloid flows or *nepheloid layers*. Deep-water nepheloid layers may be continually fed by fine sediments transported basinward from shallow water, and may be semipermanent features of deep basins. Some nepheloid layers may not be dilute turbidity currents, but rather consist of turbid water containing suspended fine sediment stirred up by other kinds of bottom currents.

In addition to doing the two things just explained before it begins to deposit any sediment, a turbidity current is capable of interacting with the substrate over which it flows. Ordinarily, such a substrate will be fine textured—either a silty clay deposited by the diluted tail end of a previous turbidity current or a pelagic ooze. When they pass over bottoms underlain by fine sediment, many currents mark and sculpt the bottom in various ways. (See Figures 5-25, 5-26, 5-28, and 5-29.)

After the current has left its marks on the bottom and has covered them with its cohesionless sediment (silt size and coarser, commonly sand), we may have to wait for a long time before we can get a good look at what happened. The best evidence is preserved only after the cohesionless particles have been cemented. After that important event, the bottom surface of the now-cemented sediment composed of cohesive particles, typically a sandstone, may display all kinds of *sole marks* (Chapter 5). A general name for these features, which is popular in Europe and which was invented because some of the features resemble ancient writings, is "hieroglyphs."

A turbidity current is an *autosuspension*; the tangential component of gravity acting on the excess density of the fluid and suspended-sediment system pulls the mass downslope. The forward speed of the fluid creates turbulent eddies and the upward motion of the threads of flow within these eddies overcomes the immersed weight of the particles, thus suspending them.

The coarser particles in a liquefied cohesionless-particle flow also must be kept dispersed. By whatever means these cohesionless-particle flows do it, the evidence from the sediments is clear that they can travel faster than an accompanying turbidity current. When the two move together, the cohesionless-particle flow acts as a frictionless basal carpet for the turbidity current. The cohesionless-particle flow pushes out ahead of the turbidity current. Within the fast-traveling basal flow, the coarsest particles move to the front. After they have come to rest, the finer particles catch up and many outdistance them. Proof that this differential movement has occurred is shown by horizontal- as well as vertical grading and by the deposition, in any small depressions in the bottom, of coarser sediment than is found elsewhere nearby. (See Figure 5-26, B.) Where both a liquefied cohesionless-particle flow and a turbidity current traveled together and both deposited their loads in close succession, the material dropped by the cohesionless-particle flow underlies the turbidite. Typical turbidite sequences are shown in Figures 5-10 and 9-2. Vertical sequences in sediments deposited by gravity-powered bottom flows consist ideally of five divisions, labeled A, B, C, D, and E (described in the caption to Figure 9-2). The sequence is named the Bouma sequence after A. H. Bouma (1962). Some refer to this sequence as simply the "turbidite sequence," with subdivisions labeled T_a through T_e, but we prefer the established usage that honors its discoverer.

Sequences of turbidites commonly consist of monotonously interbedded alternating and laterally persistent layers of sandstones and shales. (See Figure 5-6.)

Not all divisions of the Bouma sequence need always be present; sequences may consist of any combination of the five divisions. The characteristics of gravity-powered bottom flows include (1) sharp base with sole marks, (2) divisions of Bouma sequence (See Figure 9-2.), (3) graded layer or massive sandstone, and (4) monotonously interbedded alternating and laterally persistent sandstones and shales (See Figure 5-6.). However, items (3) and (4) may also occur in other associations.

Complete turbidite sequences, including all five Bouma divisions, are common on mid-fan regions. Outer-fan regions are characterized by turbidite sequences lacking the A division. The inner fan is characterized by coarse deposits that lack the upper (fine) divisions. Products of slumps and debris flows are common on inner parts of a fan. Basin-plain turbidites are thin and fine textured; commonly they consist only of the D and E Bouma divisions.

Prior to the 1980s, the main interest in turbidity currents and turbidites centered on the effects of a current traveling down one slope and coming to rest on a flat basin-floor plain. A new interest has developed in the ability of turbidity currents to transport sediment considerable distances up a second slope, such as a local bathymetric high. How far upslope such transport takes place is a function of the thickness of the current.

In addition to transporting sediment up a second slope, a turbidity current may move upslope, stop, and then reverse track and flow back down this second slope, just as the swash on a beach moves up the beach face, stops, and flows back down again as backwash. Such *turbidity currents that have been backreflected after traveling upslope* have been named **contained turbidity currents** and *the sediments deposited by contained turbidity currents* are known as **contained turbidites**. The Ordovician Cloridorme Formation of the Gaspé Peninsula, Quebec, Canada, contains thick-bedded calcareous wacke (given the name of TCW beds) that has been interpreted as contained turbidites. These are characterized by interlayered cosets of cross laminae showing opposite directions of current flow that are capped by thick layers of mudstone/siltstone that were deposited after the turbidity current stopped sloshing back and forth between the oppositely dipping slopes.

Turbidity currents and organic matter are related in several significant ways. Many marine examples are known in which turbidites include land-plant debris, for example, twigs and leaves that were still green when collected. This plant material may be the food on which the abyssal animals live. The clay fractions of turbidites contain up to 2% or so of total organic matter. Turbidites likewise contain kerogen of unknown provenance. Kerogen is so resistant to most reagents that some or all of it could be recycled.

Turbidites have attracted the attention of oil-company geologists, especially after the evidence became conclusive that the prolific reservoirs of Pliocene age in the oil fields of the Los Angeles Basin and Ventura Basin were turbidite deposits (Figure 9-25). The discovery of oil in turbidite sand prompted the idea that turbidity currents could transport potential reservoir sands into a deep-water basin where the organic matter, presumed to be falling through the water from the plankton, was accumulating in the fine-textured supposed source rocks of the center of the basin. The various kinds of organic matter in turbidites raise the possibility that the organic matter may be brought into the deep water with the turbidites. If so, then turbidites could be their own source rocks.

Inferred examples of ancient marine turbidites and liquefied coarse-particle flows (commonly these two have not been distinguished, but both are identified simply as "turbidites") have been reported from every fold chain. The general term "flysch" has been applied to sequences containing many such layers. (Flysch is discussed further in Chapter 17.) Exposed examples range in age from Precambrian to Pliocene. In the Alpine chains the flysch consists largely of turbidites (plus contourites as well). The Martinsburg Formation (Ordovician) and comparable rocks in the Appalachians of eastern North America include interbedded turbidites and contourites. Other examples include the Pennsylvanian of the Ouachita-

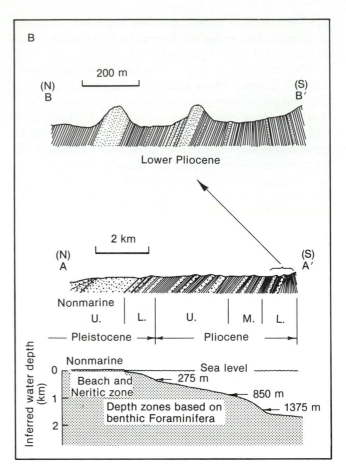

FIGURE 9-25.   (*Continued*)
B. Profile-section, along line A-A′ of A, showing Pliocene and Pleistocene strata. These strata consist of coarse sands and conglomerates interstratified with foraminiferal shales. Geologists formerly interpreted such coarse deposits as of shallow-water origin. M. L. Natland, however, compared the benthic Foraminifera of these deposits to those from modern bottom samples of the adjacent Pacific Ocean. This faunal evidence, as indicated in the water-depth diagram of this figure, enabled him to determine depth zones. He discovered that during most of its late Tertiary history, water depth in the basin had been hundreds of meters. The inferred depth of water during accumulation of the strata is shown. The coarse deposits are inferred to have formed as products of turbidity currents and submarine slides. (After M. L. Natland and Ph. H. Kuenen, 1951, figs. 1, 2, and 4, p. 77, 78, 80.)

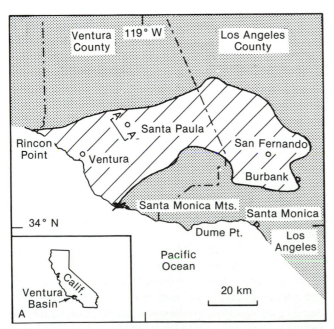

FIGURE 9-25.   Ventura Basin, California; an example of a deep basin of a continental borderland.
A. Index map of Ventura Basin. Line A-A′ is cross section of strata shown in B.

Marathon belt of southeastern United States and numerous formations of Cretaceous and younger ages in the California Coast Ranges. Most of these examples consist largely of noncarbonate terrigenous sediments. Most turbidites are deposited in deep-sea fans or -cones, for these large sediment piles form wherever large quantities of sediment enter the deep-sea environment.

SOURCES: Edwards, 1991; Elmore, Pilkey, Cleary, and Curran, 1979; Enos, 1969a, b; Kessler, 1990; Komar, 1985; Laval, Cremer, Beghin, and Ravenne, 1988;

Marjanac, 1990; Muck and Underwood, 1990; Pickering and Hiscott, 1985; Pickering, Hiscott, and Hein, 1989; Pilkey, Walker, and Cleary, 1989; Reading, 1991.

Examples of Gravity-Displaced Deposits

Examples of gravity-displaced carbonate sediments in the deep sea include those of the western Atlantic Ocean surrounding the island of Bermuda, known as the Bermuda apron, and the Tongue of the Ocean contiguous to the Great Bahama Bank. An example of a mixed carbonate-noncarbonate gravity-displaced deposit floors the Gulf of Aqaba, one of the northern branches of the Red Sea. (See Figure 14-11.)

The Bermuda Apron designates the steep submarine slopes that surround Bermuda and lead into the surrounding deep water from the shallow-water platform on which the islands are located. At depths of 4500 to 5000 m the sediments from the Apron are essentially identical to those near shore. However, the Apron sediments show well-developed graded layers attributed to turbidity currents. The lower part of each graded layer consists of sand-size particles containing typical nearshore skeletal debris. The tops of the graded layers consist of mud-size particles that are composed of calcite. The coarse particles in the lower part of each graded layer, although likewise containing calcite, consist to a large part of aragonite and high-magnesian calcite. The depths of 4500 to 5000 m are at or below the compensation depth for aragonite and below that for high-magnesian calcite, but evidently the coarse sizes of the particles and their rapid downslope transport prevented them from being dissolved. Among the fine particles of the tops of the graded layers, only calcite has not been dissolved.

The Tongue of the Ocean, a narrow channel more than 2000 m deep, is surrounded on three sides by the Great Bahama Bank. The very steep rocky walls of the channel are composed of Cretaceous and Tertiary limestone bedrock. The Tongue of the Ocean is underlain for the most part by carbonate sediments with well-developed graded layers that contain displaced shallow-water organisms. In this respect, it is similar to many other deep-water areas that are adjacent to steep slopes and contiguous shallow-water zones. The nearshore debris was originally deposited along the marginal escarpment of the Bahamas and was later displaced downslope by gravity, probably by turbidity currents.

The Gulf of Aqaba is a narrow, steep-sided tectonic valley. (See Figure 14-11.) The Gulf forms part of the Great Rift Valley that extends from Turkey to the Red Sea and beyond through East Africa. This trench is so narrow and steep-sided that Leopold von Buch (1774–1853) described it as "a crevasse in the earth's crust." The submarine slopes of the gulf are fault surfaces; they are virtual precipices with a normal gradient of 60° to 70°. The shore cliffs are equally precipitous. The gulf sediments are graded and the grading has been attributed to turbidity currents. The content of carbonate ranges from 20 to 75%. The remainder was derived from the Precambrian crystalline rocks of the Arabo-Nubian Shield, which form the shores of the Gulf.

During Early Paleozoic times in eastern North America, the submarine slope declined precipitously into the deep proto-Atlantic Ocean. (See Figure 9-10.) Slump rubble of shallow-water skeletal and pelletal limestone, displaced from the shelf edge or from the shallow sea landward of the shelf edge, is overlain and underlain by normal bedded sequences of shale (Figure 9-11). This shale could mean that the bottom lay below the compensation depth of calcium carbonate. Or it could be a terrigenous contourite brought from far away.

The shape of the rubble fragments ranges from irregular to tabular and from angular to rounded. The fragments show considerable variation in size, up to 60 cm thick and 240 cm long. *Deposits composed of rubble of carbonate rocks, usually angular, interstratified with dark-colored marine shale* are known as **brecciolas**. (See Figure 9-11.) The brecciolas have been interpreted as products of turbidity currents, gravity slides, and debris flows. Such brecciolas formed along hundreds of kilometers of the original eastern edge of the carbonate shelf. They mark the former margin of the basin during the Cambrian and Ordovician periods. Slides, slumps, turbidity currents, mudflows, and sand falls moved down the steep unstable slope beyond the shelf edge. Brecciolas and other gravity-displaced deposits can originate at the shelf margin or in deeper water on the slope.

In the Jurassic Period a hinge line formed a precipitous slope near the present-day Mediterranean coast of Israel between shelf edge and deep sea. Carbonate sediments composed chiefly of skeletal debris, but with subordinate intraclasts of pellets, were displaced downslope by turbidity currents to make graded layers 1 to 7 cm thick (now skeletal-pellet limestones). Overlying and underlying the graded layers are black- and green shales containing abundant coccoliths.

Basinal limestones are generally dark gray or black and are laminated on a millimeter scale. The dark laminae are caused by concentrations of organic matter, iron sulfides, or argillaceous material. Some laminae show micrograding; the bases of such small-scale graded layers may show microscopic cut-and-fill structures. In these limestones, planktonic organisms may be abundant. Depending on the geologic ages of the strata, the plankton may have included coccoliths, pteropods, globigerinid Foraminifera, radiolarians, or diatoms, as well as various other planktonic forms. In deposits of Early Paleozoic age the plankton included graptolites. Because the sea bottom on which these deposits accumulated was probably anoxic, bottom-scavenging organisms were absent. The sediments were not bioturbated and the original laminae were preserved, serving as one of the diagnostic characteristics of basinal limestones. The fine particles composing the basinal limestone represent in part the results of vertical sedimentation of the tests of planktonic organisms and in part the products of lateral downslope gravity displacement. Fondothems, like clinothems, may exhibit characteristic seismic signatures. In the case of fondothems, the intercalation of turbidites (and cohesionless-particle flows) with pelagic sediments produces an even pattern of parallel seismic reflections (See Figure 9-5.) that has been called a *railroad-track pattern.*

SOURCES: Alvarez, Colacicchi, and Montanari, 1985; Blewett, 1991; Brunner and Normark, 1985; Chough and Hesse, 1985; Damuth and Embley, 1981; Dingler and Anima, 1989; Edwards, 1991; Faugeres, Stow, and Gauthier, 1984; Fraser, 1989; Friedman, 1988; Gardner and Kidd, 1987; Garfunkel, 1984; Gradstein, Ludden, and others, 1990; Hiscott and James, 1985; Hiscott, Pickering, and Beeden, 1986; Kasper, Larne, and Meeks, 1987; Kessler, 1990; Kirwan, Doyle, Bowles, and Brooks, 1986; Larne, 1985; Leckie and Krystinik, 1989; Leg 123 Shipboard Scientific Party, 1988; Lowe, 1988; Marjanac, 1990; Melvin, 1986; Middleton and Neal, 1989; Normark, 1989; Pickering, Hiscott, and Hein, 1989; Piper and Coleman, 1982; Piper and Normark, 1983; Reading, 1986, 1991; Schwab and Lee, 1988; Stow, 1985; K. C. Terzaghi and Peck, 1948; Underwood, 1986.

Pelagic Deposits, Modern and Ancient

The pelagic realm of the sea contributes its own special kind of intrabasinal deposits. These **pelagic deposits** are *open-sea deposits containing less than 20% of terrigenous sediments or of tephra having diameters greater than 10 μm (fine silt).* Pelagic deposits include accumulations of (1) the skeletal remains of microorganisms that live in seawater, (2) cherts derived from the dissolution and reprecipitation of siliceous tests of microorganisms, (3) the ultrafine insoluble products that settled from above, (4) alteration products of clay minerals and (5) chemical precipitates (or their secondary alteration products) from hot saline bottom waters or from hypersaline surface waters. Many pelagic deposits contain small perfect crystals of authigenic dolomite (usually < 10μm), almost always in very low concentrations. Pelagic deposits are not found in lake basins, although some large lakes do contain some vertically sedimented materials, especially chemical precipitates.

The chief kinds of pelagic sediments are oozes; brown clays; diagenetic products, such as chert and pelagic lime-

stone; iron-rich variegated sediments; and black mud. Evaporites may also form in the deep sea.

OOZES

The most-widespread pelagic deposits are **oozes**, *pelagic sediments containing at least 30% skeletal remains of microorganisms.* Two varieties of oozes are calcareous and siliceous. On the modern deep-sea floor, calcareous oozes consist of the tests of Foraminifera (particularly those of the genus *Globigerina*) (Figure 9-26), of pteropods (planktonic gastropods), and of coccolithophorids. (See Box 2.4 Figure 1 and Figure 8-16.) Thus we can recognize foraminiferal (*Globigerina*) ooze, pteropod ooze, and coccolith ooze. In *Globigerina* ooze >30% of the sediment consists of tests of planktonic Foraminifera, including chiefly *Globigerina*, but also many other genera of foraminifers. In *Globigerina* ooze whole tests of Foraminifera are set in a fine-textured matrix of coccoliths and broken foraminiferal tests. Although a few genera of planktonic Foraminifera can live at depths greater than 500 m, most live within the upper 100 m of the ocean. *Globigerina* ooze is the most widespread of pelagic deposits; it carpets nearly half of the deep-sea floor. Such ooze is the dominant deep-sea sediment in the Atlantic and in much of the Indian and Pacific oceans. Pteropods build tests of aragonite (Chapter 7); they are distributed worldwide in the upper 100 m of the ocean. Pteropod ooze occurs in the tropical and subtropical parts of all oceans. Examples are known from the Red Sea, from the Atlantic Ocean, and from the Pacific and

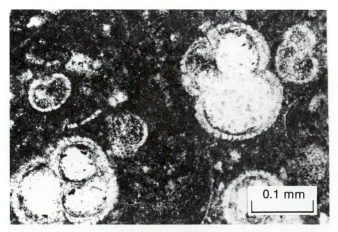

FIGURE 9-26.   Calcareous ooze of Pleistocene age consisting of tests of *Globigerina* that have been cemented by micritic calcite; photomicrograph of thin section in plane-polarized light. Core from 96.8 m below the sea floor, Deep-Sea Drilling Project, Site 229, southern Red Sea, where water depth is 852 m. (J. A. Schrank and G. M. Friedman; samples courtesy of S. Ali and P. R. Supko.)

Indian oceans. Pteropod ooze is more abundant in water less than 2500 m deep; it is rare where the water is deeper.

Coccolith ooze is the dominant pelagic sediment in the Mediterranean and Black seas. In the Pacific, Atlantic, and Indian oceans the skeletal remains of coccoliths (which are composed of calcite) constitute up to 50% of the carbonate fraction of the calcareous oozes.

The depth distribution of the various particles in calcareous oozes is a function of particle size and of the relative solubility of the calcium-carbonate mineral that the organism secreted. (See Figure 7-51.) The most-resistant skeletons are the comparatively large calcite *Globigerina*. Next come the tiny calcite parts of coccolithophorids. Least resistant are the comparatively large but very thin and soluble aragonitic skeletons of pteropods.

Siliceous oozes consist of more than 30% tests composed of silica. The two chief builders of siliceous skeletons are the planktonic unicellular diatoms, which are plants (See Figure 2-25.), and radiolaria, which are animallike protists. (See Box 2.1 Figure 8.)

Diatoms are the dominant microorganisms in cold polar waters and in the floating ice of the ice shelf that surrounds the Antarctic continent. Hence diatom ooze forms an almost continuous belt on the sea floor around Antarctica and another belt that crosses the North Pacific.

Radiolarians live at all depths in all parts of the ocean, but in the bottom sediments, tests of radiolarians generally are overshadowed by those of the organisms that secrete tests of calcium carbonate. Radiolarian ooze forms a belt in the deep parts of the tropical Pacific and is present in parts of the tropical Indian Ocean as well.

Nannofossil (*coccolith*) ooze becomes *chalk* when lithified. Diatom ooze becomes *diatomite*, also known as *diatomaceous earth*. Radiolarian ooze becomes *radiolarite* or *radiolarian chert*.

Because various physical- and chemical factors restrict the distribution of both benthic- and planktonic organisms, their distribution in sediments can be both a biostratigraphic- and an environmental indicator. The distribution of biogenic particles in oozes and other marine sediments is complicated when sediment-gravity flows transport the skeletons of shelf- and slope-dwelling organisms into deeper water.

BROWN CLAY

As used by the early workers who first described deep-sea sediments, the "red clay" was supposed to be a pelagic sediment and thus to differ from variously colored "muds," which were classified as terrigenous sediments. This concept was based solely on examination of the tiny particles with a polarizing optical microscope.

Thanks to the sophisticated techniques for studying the submicroscopic particles of deep-sea sediments using X-ray diffractometers, scanning-electron microscopes, and electron probes that have recently become available, the relationships among deep-sea sediments are better understood. The brown clay (originally "red clay") is still regarded as a pelagic sediment. But many of its constituents, such as clay minerals and fine-textured quartz and feldspars, are of terrigenous origin. Some of these particles fall out of suspension in seawater; some of the quartz and feldspars were transported out to sea by high-altitude winds. Brown clay contains less than 30% skeletal debris of microorganisms. Brown clay floors the deep sea where the water is deep and the calcareous skeletal debris has been dissolved, and where siliceous organisms are not abundant.

DIAGENETIC PRODUCTS: CHERT AND PELAGIC LIMESTONE

As the kinds of oozes just described become buried, they undergo various diagenetic alterations that affect both the mineral phases and the fabrics (Chapter 3). Skeletal particles dissolve completely. The material composing them goes into solution and is later precipitated. Material dissolved from siliceous organisms creates cherts. Dissolution of calcareous skeletal material and its later precipitation creates micritic pelagic limestone in carbonate sediments below the floor of the Red Sea. (See Figure 9-26.)

IRON-RICH VARIEGATED SEDIMENTS

On the floor of the Red Sea (Figure 9-27) are several small enclosed basins where water temperatures are about 56°C (the temperature of normal Red Sea bottom water is 22°C) and salinity is about 225‰, compared to an average value of 41‰ in surface Red Sea water. The largest area enclosing the bodies of hot brine measures 12 km long and 5 km wide. The oxygen content of the hot brine is only 0.1 ml/l or less, which contrasts with the value of 1 to 2 ml/l of oxygen dissolved in the normally oxygenated water at depths of less than 1980 m.

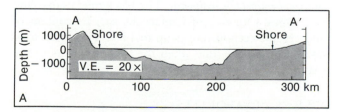

FIGURE 9-27.   Location of "hot-brine" pools in Red Sea.
   A. Bathymetric profile across Red Sea in area of "hot-brine" pools. (After Y. Herman and P. E. Rosenberg, 1969, fig. 1, p. 449.)

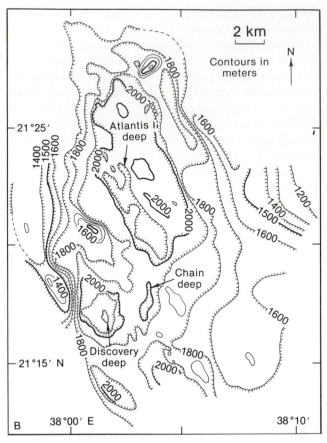

FIGURE 9-27.   (*Continued*)
B. Detailed bathymetric chart of closed depressions filled with hot brines. Deeps were named after the research vessels from which the hot-brine areas were located. (After T.-L. Ku, 1969, fig. 1, p. 513.)

The sediments beneath these hot, oxygen-poor brines consist of variously colored (variegated) amorphous iron oxides and goethite.

BLACK MUD

Although, strictly considered, it is not a true pelagic deposit, the widespread black mud underlying much of the Atlantic Ocean is discussed here. Such transitional materials between ultrafine pelagic deposits and the coarse terrigenous sediments are sometimes designated as *hemipelagic sediments*. The black mud underlying the Atlantic Ocean is of Cretaceous age. Its content of total organic carbon ranges up to 14%. This black mud was deposited at a time when the bottom waters in the Atlantic Basin lacked oxygen.

DEEP-SEA EVAPORITES

In a hot, dry climate, evaporites may be precipitated and accumulate in semienclosed basins. No modern counterparts of deep-sea evaporites are yet known, and it is

likely that, as in the modern Dead Sea, early formed evaporite minerals, such as gypsum, may be degraded by sulfate-reducing bacteria. (See Eq. 3-10.) Any deep-sea evaporites might include halite. But they could include gypsum or anhydrite only where sulfate-reducing bacteria are not present.

DISTRIBUTION OF PELAGIC SEDIMENTS

Various factors affect the distribution of pelagic sediments. Obviously, special conditions are required for the black muds, variegated iron-rich sediments, and evaporites. Leaving those aside, we can concentrate on the factors that influence the distribution of microorganisms in the water, the circumstances that accompany their transfer from the water to the bottom, and what happens to them on the bottom.

In addition to the effect of light, in the surface waters of the modern oceans the distribution of planktonic organisms is controlled largely by temperature. For example, as we remarked earlier, diatoms flourish in the polar seas and various kinds of Foraminifera are restricted to certain water masses.

But does knowledge of the distribution of planktonic organisms in the surface waters enable one to predict the distribution of their skeletal debris on the deep-sea floor? After all, we know that the rates of settling of the tiny particles are very slow (a particle of fine-sand size requires about 10 days to settle 4000 m; for very fine clay, the time exceeds 100 yr), that they have a long way to fall, and that both surface- and subsurface currents are flowing. Despite all the possible complications, a map of the distribution of the accumulated tests of planktonic organisms in the present-day bottom sediments almost exactly matches a map of the distribution of the living forms in the near-surface waters. This astonishing coincidence can mean only one thing: the tiny particles are not settling as isolated individuals. Instead, they probably make the long trip downward clustered together in fecal pellets, which settle relatively rapidly.

However it happens, the fact that the pattern of skeletal debris of pelagic microorganisms in the bottom sediments so closely reflects the patterns of these same organisms living in the surface waters is very significant. We can map the boundaries between various kinds of microorganisms in ancient pelagic sediments and compare these boundaries with today's. Deviations from the present pattern have presumably been caused by changes in climate (but evolutionary changes in environmental tolerances of marine pelagic microorganisms are known to occur). The abundance of the skeletal debris of microorganisms in pelagic sediments depends on three factors: (1) the productivity of the surface waters, (2) the rate at which carbonate particles are dissolved, and

(3) the production of other particles in the water (Chapter 7).

Production of tests of planktonic organisms is highest near the mouths of the world's major rivers and in areas of upwelling, such as those peripheral to the major gyres of oceanic circulation, the Gulf Stream, for example. Productivity is lowest within central oceanic gyres, such as the Sargasso Sea. Near areas of equatorial upwelling the production of all planktonic organisms, both siliceous and calcareous, rises.

Ancient examples of pelagic sediment include many of the materials encountered in the holes made by the Glomar Challenger for the Deep-Sea Drilling Project (the Joint Oceanographic Institutions' Deep Earth Sampling project, hereafter abbreviated JOIDES) and those found by its successor program, the Ocean Drilling Program (ODP) and the JOIDES Resolution. Examples exposed on land include the Tertiary chalks and marls on the island of Barbados, West Indies; the Cretaceous-Eocene succession overlying the pillowed basalts at the top of the Troodos ultramafic complex on Cyprus; the red nodular limestones (Ammonitico Rosso), Knollenkalk of Jurassic age in the Alpine Mediterranean region; lower Mesozoic brown clays and ferromanganese nodules of Borneo, Timor, and Rotti in Indonesia; and the Devonian-Carboniferous limestones of the Carnic Alps, Austria.

One example of Jurassic pelagic (and deep-water) strata from the Austrian Alps overlies a coral-reef complex of Late Triassic age. Just above the reefs are Jurassic limestones containing a mixed fauna of fossil benthic- and pelagic organisms. Still higher are coccolith oozes and radiolarian cherts, both comparable to modern oceanic pelagic sediments. A progressive deepening of the basin took place—from sea level to depths inferred to have been as great as 3000 to 4000 m. What was part of the continent during the late Triassic Period became part of the ocean floor during the Jurassic.

GLACIAL-MARINE SEDIMENTS

Near the world's two great modern ice sheets, on Antarctica and Greenland, and to a lesser extent wherever glaciers are found, icebergs carrying a frozen-in load of terrigenous sediment float out to sea, melt, and drop their ice-rafted rocky freight on the sea floor. Glacial meltwater streams deliver poorly sorted sediments to the nearshore marine environment. Meltwater, milky with fine, glacially ground-up rock particles, flows into the sea. Coarse sediments are reworked by waves and currents in shallow water, and are redeposited by sediment-gravity flows onto the outer shelf-, slope-, and basin areas. In these and related ways, **glacial-marine sediments** are deposited. Such sediments are defined as *marine terrigenous sediments derived from a glacier and including nonsorted particles (dropstones) of nearly any size.*

Glacial-marine sediments form a continuous belt around the Antarctic continent and make smaller areas in the Norwegian Sea and North Pacific. During the Pleistocene glacial ages, the boundaries of these areas shifted toward the Equator. During Pleistocene interglacial ages, the outer limits of glacial-marine strata migrated poleward. Older examples of glacial-marine strata are found in the Upper Paleozoic bedrock of Australia.

TEPHRA

Among sea-floor sediments, tephra are widely distributed. Tephra include individual glass shards scattered through other sediment and layers composed exclusively or almost exclusively of volcanic debris. Most tephra probably settled through the water column (vertical sedimentation) but some may have been gravity-displaced along the bottom. Tephra entering the sea from the atmosphere may take years to settle. Locally in the water off Indonesia in the 1920s, Dutch geologists found that fine tephra from the explosive eruption of Tambora in 1815 was still settling. Two layers of tephra from Santorini (Thera) are widespread in the deep-sea sediments of the eastern Mediterranean. (See Figure 2-21.)

DEEP-SEA VOLCANISM

Volcanoes and volcanic rocks are much more abundant on the sea floor than on the continents. Because of the problem of accessibility, however, our knowledge of volcanic activity is based largely on what happens during subaerial eruptions. When a volcanic eruption takes place beneath seawater, the environment is not the same as when the eruption occurs on land. In particular, as depth increases, so does the hydrostatic pressure. The rate of increase of pressure is one atmosphere for every 33 ft (9.9 m). The depth of the critical point for water (217 atm) is 2160 m.

In theory, the hydrostatic pressure of the deep sea can prevent the escape of water vapor from the lava. In fact, the reverse might take place; seawater could enter the lava and react with the silicate minerals. Calcium might not enter plagioclase, but go into the seawater. As a result, the plagioclase feldspar might become albite instead of labradorite, for example.

Densities of volcanic rock and sizes of vesicles have been found to decrease systematically in volcanic rocks dredged up off Kilauea, Hawaii. At depths of less than 800 m, vesicles having mean diameters > 0.5 mm occupied 10% of a rock whose density measured 2.8 g/cm^3. At depths over 800 m, the vesicles were rare, their diameters averaged < 0.1 mm, and the density of the rock measured 3.0 g/cm^3.

Commonly associated with ancient pillowed basalts, interpreted as products of submarine eruptions (depths not always known or specified), are **radiolarian cherts**, *bedded cherts containing recognizable remains of radiolarians.* Some geologists have suggested that the silica to make the cherts came somehow from the volcanic activity. If the silica of volcanic origin did enter the seawater, its abundance probably triggered a great bloom of siliceous organisms. Later, most of the siliceous skeletons were dissolved and reprecipitated as chert. Alternatively, the silica may have precipitated directly from the water.

MÉLANGES

In some mountain chains, certain fine-textured marine strata contain complex slices and large blocks (up to several kilometers in length) of strata unlike the rocks enclosing them. Both slices and blocks and their enclosing matrix show evidence of having been intensely sheared and broken. The term **mélange** designates such *a heterogeneous mixture of angular, poorly sorted, and exotic tectonic fragments with pervasively sheared fine-textured matrix of marine origin.*

Mélanges are somewhat comparable to the carbonate clast, black-shale diamictites mentioned previously. The difference is that mélanges include tectonic slices, whereas the carbonate-clast, black-shale diamictons are accumulations of sediments. Where this distinction cannot be made, the origin of the deposit will remain in doubt.

Mélanges are like epeiric seas; the only known examples come from the geologic record. Mélanges are features of convergent plate margins, like the western margin of north America, and are thought to consist of continent-derived- and ocean-floor sediments that were mixed together in a subduction zone. If any mélanges exist beneath the floors of the present-day oceans, they have yet to be recognized.

SOURCES: Cook, Hine, and Mullins, 1983; Droxler, Morse, and Kornicker, 1988; Fraser, 1989; Gradstein, Ludden, and others, 1990; Kidder, 1985; Lumsden, 1988; Reading, 1986; Thiede and Suess, 1983; Warme, Douglas, and Winterer, 1981.

Marine Nondepositional Environments; Authigenic Sediments and Condensed Sections

On parts of the deep-sea floor, currents prevent particles from accumulating to form any of the kinds of pelagic- or terrigenous sediments we have previously described. Such areas are described as nondepositional environments. This term is something of a misnomer, for certain chemical- or biochemical reactions may add materials directly to the bottom. However, the rate of addition is slow. Examples include manganese oxides and phosphorite. Along with these authigenic accumulations may be found such large fossil debris as shark's teeth, fish scales and vertebrae, and the resistant earbones of whales. Strata that accumulate under these conditions are known as *condensed sections* (Chapter 6).

Nodules and crusts of ferromanganese oxides underlie large areas of the central Pacific and eastern Indian Ocean.

SOURCES: Fraser, 1989; Gradstein, Ludden, and others, 1990; Kidder, 1985; Lumsden, 1988; Reading, 1986.

Biogeography of Open-Marine Organisms

The subject of **biogeography** deals with *the study of the distribution of organisms in the modern world and the factors that influence this distribution.* Knowledge of the distribution of organisms depends on an understanding of the life styles of these organisms and of the ways in which organisms can move from place to place. *Aquatic organisms that float freely in the water and that are incapable of swimming or are capable of swimming only short distances* collectively constitute the **plankton**. *Organisms which live on or in the bottom* are the **benthos**. *Epifauna live on the bottom* and **infauna** *live below the bottom.* Some benthic organisms are able to move around on the bottom (*vagrant benthos*). Others attach themselves to the bottom and remain in one spot throughout their adult lives. Such attached organisms (*sessile benthos*) may live as isolated individuals or cluster together to build various kinds of colonies. *Organisms capable of swimming long distances* are collectively designated as the **nekton** (nektonic organisms). In deep-water basins, benthic organisms contribute little sediment. The biogeography of benthic organisms is treated in Chapter 10. We consider in the following sections the biogeography of the plankton and the nekton, both of which can make substantial contributions to the sediments of deep basins.

From the point of view of travel, nektonic organisms can go just about everywhere. If they can survive in the cold deep water of the open sea, then their ranges may be virtually worldwide. Planktonic organisms can be distributed widely by surface currents. Organisms of the vagrant benthos can crawl short distances. Even the organisms composing the sessile benthos can get around, at least during their embryonic stages. Most marine invertebrates reproduce by discharging eggs and sperm directly into the water or by making egg cases or otherwise attaching the eggs to the bottom. The larval stages of nearly all marine invertebrates involve a planktonic stage. Therefore, even though the adults may spend their entire lives stuck in one place, the youths are not thus confined.

Because of the ways they influence sediments, four special groups of marine organisms merit further discussion. These are (1) planktonic organisms, (2) fish and other vertebrates, (3) reef-building organisms, and (4) various other invertebrates. As mentioned, groups 1 and 2 are treated here; groups 3 and 4 are discussed in Chapter 10.

Planktonic Organisms

Apart from their vital role at the base of the marine food chain, planktonic organisms are widely represented in marine bottom sediments. Many planktonic organisms secrete hard parts composed of calcium carbonate or of silica (Table 9-1).

Among the most-prolific contributors of skeletons to the bottom sediments are the Foraminifera (or forams). These one-celled creatures reproduce by cell division. In order to do so, the soft parts must be outside the skeleton. Hence every time an individual foram reproduces, it discards the old skeleton, divides, and each of the two new individuals promptly secretes a new skeleton. Thus no dying had to take place to start a foram shell on its way to the bottom. Quite the contrary, the livelier the forams are, the more skeletons they secrete and drop toward the bottom. (See Box 2.4.)

Fish and Other Vertebrates

Fish and other free-swimming aquatic vertebrates (turtles, whales, porpoises) contribute distinctive hard parts to bottom sediments. The scales of fish, the earbones of whales, and the teeth of sharks are resistant to dissolution and thus appear in bottom sediments. In addition, burrow-dwelling fish have been implicated in widespread bioerosion of firm substrates, and in the tropics, coral-eating fish (e.g., parrot fish) are responsible for considerable destruction of reef framework and production of silt- and sand-size sediment particles.

SOURCES: Brunner and Normark, 1985; Droxler, Morse, and Kornicker, 1988; Fraser, 1989; Mullins, Thompson, McDougall, and Vercoutere, 1985; Reading, 1986; Sanders, 1981; Twichell, Grimes, Jones, and Able, 1985; Whittington, 1985.

Characteristics of Deep-Water Deposits

Ancient marine basin-margin deposits (clinothems) include a wide range of features, from reef talus to submarine canyons, most of which contain coarse debris. A common size range of sediment deposited on clinothems outside submarine canyons and inside the channels on sea-floor fans and cones is silt to fine sand. Much sediment is even finer than silt. Evidence of gravity displacement is common. Deposits from channels of submarine canyons and sea-floor fans (or -cones) usually are coarser than the adjacent interchannel sediments. Although these channels spread out sands as they shift, no definite sequence has been described as a product resulting from the lateral shifting of these channels. Contourites are characterized by silt size, and by numerous thin interstratified layers of fine sand that is horizontally laminated or ripple cross stratified on a small scale. Slump structures may be present; strata that slumped and were deformed may be closely associated with a debris flow and may be overlain by a turbidite.

Marine deep-sea deposits (fondothems) consist of interbeds of vertically sedimented pelagic deposits and laterally transported turbidites or products of coarse-particle flows. If bathymetric indications can be inferred from the fossils, a systematic difference will exist. The nonmarine- or shallow marine organisms will be in the coarse layers and planktonic- or deep-water benthic organisms in the fine layers. If turbidity currents were active, many of the fine-textured strata may be turbidites; carbonate may show the effects of progressive dissolution with depth. If the water depth was sufficient to place the bottom below the carbonate-compensation depth, then most calcareous skeletal debris of microorganisms will be dissolved. In that case, carbonates will be found only in the coarser layers that were gravity displaced from shallow water.

TABLE 9-1. Major groups of planktonic organisms that secrete hard parts

| Organism (P = plants) | Material forming skeleton | |
	Silica	Calcium carbonate (mineral phase)
Diatoms (P)	×	
Radiolarians	×	
Silicoflagellates	×	
Foraminifers		× (calcite)
Coccolithophorids (P)		× (calcite)
Calcispheres (P)		× (calcite)
Pteropods		× (aragonite)
larval mollusks		× (calc./arag.)
larval arthropods		× (calc./chitin)

Evidence of transport by turbidity currents may be found in evaporites, in cherts, in pelagic oozes, as well as in the coarser terrigenous sands and carbonate sediments consisting of displaced shallow-water organisms.

Bedding in fine marine fondothems may have been obliterated by burrowing organisms. Interstratified coarse- and fine sediments form beds of uniform thickness, some of which may be traceable for many kilometers or even hundreds of kilometers.

In the next three chapters we consider undathems, their processes and sediments, with particular emphasis on marine deposits.

Suggestions for Further Reading

BOUMA, A. H., 1979, Continental slopes, p. 1–15 *in* Doyle, L. J., and Pilkey, O. H., eds., Geology of continental slopes: Tulsa, OK, Society of Economic Paleontologists and Mineralogists Special Publication 27, 374 p.

BOUMA, A. H., NORMARK, W. R., and BARNES, N. E., eds., 1985, Submarine fans and turbidite systems. New York-Berlin-Heidelberg-Tokyo, Springer-Verlag, 351 p.

FISCHER, A. G., and ARTHUR, M. A., 1977, Secular variations in the pelagic realm, p. 19–50 *in* Cook, H. E.; and Enos, Paul, eds., Deep-Water carbonate environments; Tulsa, OK, Society of Economic Paleontologists and Mineralogists Special Publication 25, 336 p.

FRIEDMAN, G. M., 1965, Occurrence and stability relationships of aragonite, high-magnesian calcite, and low-magnesian calcite under deep-sea conditions: Geological Society of America Bulletin, v. 76, p. 1191–1196.

GROSS, T. F., and NOWELL, A. R. M., 1990, Turbulent suspension of sediment in the deep sea, p. 167–181 *in* Charnock, H.; Edmond, J. M.; McCave, I. N.; Rice, A. L.; and Wilson, T. R. S., eds., The deep sea (*sic*) bed: its physics, chemistry and biology: London, UK, The Royal Society, 194 p.

KEITH, B. D., and FRIEDMAN, G. M., 1977, A slope-fan-basin-plain model, Taconic Sequence, New York and Vermont: Journal of Sedimentary Petrology, v. 47, p. 1220–1241.

KOMAR, P. D., 1985, The hydraulic interpretation of turbidites from their grain sizes and sedimentary structures: Sedimentology, v. 32, p. 395–407.

NARDIN, T. R., HEIN, F. J., GORSLINE, D. S., and EDWARDS, B. D., 1979, A review of mass movement (*sic*) processes, sediment (*sic*) and acoustic characteristics, and contrasts in slope (*sic*) and base-of-slope systems versus canyon-fan-basin floor (*sic*) systems, p. 61–73 *in* Doyle, L. J., and Pilkey, O. H., eds., geology of continental slopes: Tulsa, OK, Society of Economic Paleontologists and Mineralogists Special Publication 27, 374 p.

PICKERING, K. T., and HISCOTT, R. N., 1985, Contained (reflected) turbidity currents (*sic*) from the Middle Ordovician Cloridorme Formation, Quebec, Canada: an alternative to the antidune hypothesis: Sedimentology, v. 32, p. 373–394.

STOW, D. A. V., 1986, Deep clastic seas. Chapter 12, p. 399–444 *in* Reading, H. G., ed., Sedimentary environments and facies (*sic*), 2nd ed: Oxford, UK, Blackwell Scientific Publishers, 615 p.

CHAPTER 10

Shelf Seas and Epeiric Seas

In this chapter, we focus on the factors that determine the kinds of sediments deposited on the floors of shelf seas and epeiric seas. As we shall see, many variables are involved. These include such things as the breadth and depth of the shallow sea, kind of junction with the open ocean, air–sea interactions, the astronomic tides, biofacies relationships, changes of climate and of sea level (which may or may not be closely coupled to changes of climate), relationships between extrabasinal- and intrabasinal sediments, and tectonic activities.

We begin with some general definitions and a review of the physiographic/geologic setting of these shallow seas. Then we survey the environmental- and tectonic factors. We conclude with examples of sediments from modern shallow seas and of ancient rocks that have been interpreted as products of shallow seas.

Definitions and General Physiographic Settings

Shallow-water seas are of two kinds: (1) **shelf seas**, defined as *submerged basin-rimming shelves that are not very wide*, and (2) **epeiric seas**, *broad shallow seas that develop during times of high sea level, when the cratons are flooded*.

Shelf Seas

A shelf sea is characterized by the following: (1) it borders a continental landmass on one side and a deep-sea basin on the other, (2) its waters are subjected to astronomic tides, and (3) the properties of its water column may be extremely variable and respond to various air–sea interactions.

Shelf seas, which cover the peripheral regions of continental blocks, are also known as *neritic zones, pericontinental seas, and coastal oceans* (Figure 10-1, A, right). Modern

shelf seas are synonymous with **continental shelves**, *the submerged margins of continental masses rimming the modern ocean basins*. The mean width of the continental shelves surrounding present-day continental masses is 75 km; the range is from less than 0.5 to 1300 km. Viewed regionally, the surface of the continental shelf is extremely flat; its mean slope is $0°7'$ (1 in 500). However, local relief may amount to tens of meters. There are elongate- and irregular ridges as well as open- and closed depressions.

Most shallow-water areas adjoin deep water and the shallow water ends at a pronounced change in slope. The depth to *the pronounced change in slope at the outer edge of the continental shelf* (**shelf break** or **shelf margin**) varies from 130 m to > 350 m. Most wide continental shelves coincide with passive continental margins. (Similar breaks in slope are found at the margins of many platforms; they are known as *platform margins*.)

On their landward sides shelves are bounded by another steeper segment, the **shoreface**, *the relatively steep concave-up slope extending from the outermost line of breakers to the more-nearly horizontal shelf surface* (Figure 10-2). The shelf, then, can be characterized as the generally flat area between two steeper slopes, the shoreface on the landward side and the shelf break on the seaward side.

Because of their intermediate positions between lands and the deep-sea basins, the continental shelves represent zones across which sediments from the lands must pass in order to reach the deep-sea basins. As mentioned in Chapter 6, two contrasting conditions are those associated with high stands of the sea (shelves submerged, as now) and those associated with low stands of the sea (shelves emerged, as during a glacial age). During high stands the shoreface and shelf break are distinctly separated, and only a few mechanisms for across-shelf transport of coarse sediment can operate. During low stands the shoreface may coincide with the shelf break, and

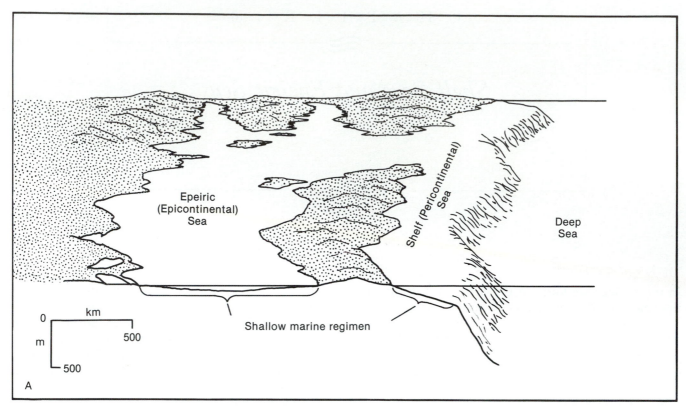

FIGURE 10-1.   Schematic views of shelf- and epeiric seas.

A. Shelf- (pericontinental) and epeiric (epicontinental) seas. Shelf seas cover margins of modern continental blocks. Epeiric seas, which in the geologic past spread across large parts of the interiors of continental masses, are not present in the modern world. Text explains distinction between shelf- and epeiric seas. (After P. H. Heckel, 1972, fig. 1, p. 227.)

B. (*below*) Carbonate ramp.

C. (*below*) Rimmed shelf. A wide shelf lagoon separates the shelf-marginal reef tract from the mainland.

D. (*below*) Tidal sea covering a carbonate platform (surrounded on all sides by deep water).

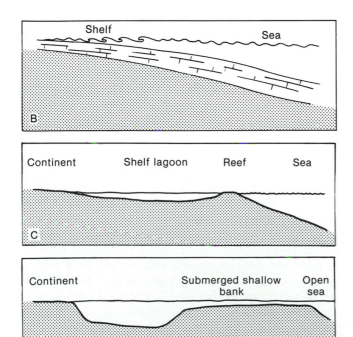

several pervasive mechanisms (streams, the wind, glaciers) can transport coarse sediment across the shelf areas.

Depending on their kinds of sediments, modern, highstand shelves can be subdivided into two first-order categories: (1) **allochthonous shelves,** *shelves that receive large amounts of extrabasinal sediment from the adjacent land,* and (2) **autochthonous shelves,** *shelves on which little or no extrabasinal sediment is entering from the adjacent land. The bottom is composed of intrabasinal sediment or of extrabasinal sediment that was spread out at a lower sea level and was reworked during the Holocene submergence and/or is being reworked by the sea now.*

Allochthonous shelves include those near modern deltas, near localities where fans are active at the shore, and off deserts where the wind is blowing sand into the water.

Autochthonous shelves on which intrabasinal carbonate sediments predominate are known as **carbonate shelves.** Because we have elected to integrate the relationships that are common to extrabasinal- and intrabasinal sediments, we shall use *carbonate shelf* as a general term in the same

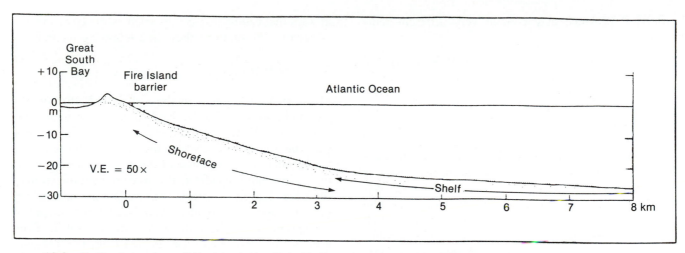

FIGURE 10-2.   Profile of shoreface off Fire Island, New York. (Profile and conversion to metric units by J. E. Sanders from bathymetric chart, contoured by N. Kumar in 1970, and appearing in J. E. Sanders and N. Kumar, 1975a, fig. 3, p. 67.)

sense that "carbonate platform" is employed by some specialists in carbonates (for example James Lee Wilson, 1975). Those who use "carbonate platform" as a general term in the way we use *carbonate shelf* apply "isolated platforms" to the features that we call *platforms* (defined in an upcoming paragraph).

Among carbonate shelves, contrasting conditions at their seaward edges form the basis for recognizing three major kinds: (1) carbonate ramps, (2) rimmed shelves, and (3) carbonate platforms.

A **carbonate ramp** is *a carbonate shelf that slopes gently seaward so that the bottom passes from shallow water into deep water without any sharp break in slope or rimming reef complex.* (See Figure 10-1, B.) We can emphasize the key elements of this definition of carbonate ramp by listing four characteristics: (1) the gentle seaward slope (usually 1° or less); (2) the shallow zone agitated by everyday wind-generated waves passes downslope without break into deep water; (3) reefs are not present; and (4) in view of (3), no reef talus or reef-derived breccias are present. Examples of modern carbonate ramps are the shelves off the west coast of Florida; on the Campeche Bank off the Yucatan Peninsula, Mexico; and on the Sahul shelf off northwestern Australia.

A **rimmed shelf** is *a carbonate shelf having a reef at its outer margin landward of which is a wide shelf lagoon,* a wide (> 20 km) shallow body of water that occupies the position of a shelf sea, but is sheltered by the outlying reef tract. (See Figure 10-1, C.) A rimmed shelf displays the following: (1) the rimming reef coincides with the shelf margin; (2) as a consequence of (1), reef rocks and skeletal grainstones are present along the shelf margin; (3) the slopes on the seaward sides of the reefs are steep (three degrees to tens of degrees); and (4) also as a consequence of (1), reef-talus breccias and turbidites containing reef-derived skeletal debris are present in the

deep water adjacent to the rimming reef. Some modern examples of shelf lagoons landward of rimming reefs include the bodies of water lying between the Great Barrier Reef and northeastern Australia; the Gulf of Batabano, Cuba; and the waters between the shelf-edge reefs and the mainland of Belize (formerly British Honduras).

A **carbonate platform** is *a large, mostly flat shoal that is covered by a shallow tidal sea and surrounded by steep slopes leading to the deep sea on all sides; the only associated land consists of low-lying islands.* (See Figure 10-1, D.) As mentioned, the sharp break in slope at the outer edge of a platform is known as the *platform margin.* An example is the Bahamas Platform. Recall that some specialists in the study of carbonates employ the term "carbonate platform" for carbonate shelf; they use "isolated platforms" for the features we designate as carbonate platforms.

In a few places in the modern world, at tectonically active margins, the outer part of a continental block is not a comparatively smooth region (as on a continental shelf), but rather consists of one or more closed depressions (basins) and intervening banks or shoals. The best-known example is off southern California, where a distinctive underwater basin-and-bank morphology exists. (See Figure 8-47.) To emphasize its contrasts with a typical passive-margin continental terrace, this region off southern California has been named a *continental borderland.*

Because of the high relief of the adjacent California coast ranges, great quantities of extrabasinal sediments are being shed to the continental-borderland province. Gravity transport along the bottom distributes this sediment according to the pattern of low areas between the coast and various basins. Because they may be isolated by the intervening ridges and thus cut off from long-distance bottom transport of coarse extrabasinal sediment, neighboring basins within a continental-borderland province may contain very different sediments.

SOURCES: Crevello, James Lee Wilson, Sarg, and J. F. Read, 1989; J. F. Read, 1985; Scoffin,1987; Swift, 1976; Tucker and V. P. Wright, 1990.

Epeiric Seas

The shallow epeiric (or epicontinental) seas of the geologic past spread across the continental masses, either as great embayments (See Figure 10-1, A left.) or as vast seaways that may have submerged all or nearly all of the area underlain by the continental block. During much of geologic time, epeiric seas having negligible morphologic relief were widespread. In such seas, normal seawater containing typical supplies of nutrients circulated freely. Locally, however, the water became hypersaline.

Past geologic thought held that the action of astronomic tides would be confined to the marginal parts (known as pericontinental seas) and that toward the interior of the craton, tidal effects would disappear (as does the modern tide in crossing Florida Bay). As a result, epeiric seas were thought not to have been subject to astronomic-tidal fluctuations.

According to current thinking, this non-tidal view has been modified. New evidence has shown that along the shoaling parts of ancient epeiric seas, intertidal flats were very extensive and that the deposits of epeiric seas display characteristics of **peritidal sediments**, defined as *sediments that are influenced by short-term variations of sea level caused by the tides* (Chapter 12). In ancient epeiric seas, astronomic tides would be damped only if the sea floor displayed great morphologic roughness.

SOURCES: Bouma, Berryhill, Brenner, and Knebel, 1982; Fraser, 1989; H. D. Johnson and Baldwin, 1986; Kennett, 1982; G. de V. Klein, 1982; Nio and Nelson, 1982; Sellwood, 1986; Swift and Niedoroda, 1985; R. G. Walker, 1984.

Environmental Factors

The basic environmental factors are related to the interactions between the water and the bottom. These are the result of various air–sea interactions as well as the tides. Marine invertebrates, distributed according to many complex factors, exert numerous profound influences on shallow-water marine sediments. Changes of sea level and of climate create first-order changes in the environmental factors that leave major marks on the bottom sediments. The amounts of extrabasinal- and intrabasinal sediment and tectonic activities may not strictly qualify as "environmental," but they affect the sediments

in ways that can be just as important as the environmental factors. We examine these in a following section.

Air–Sea Interactions

Much of what happens in these shallow bodies of water results from distinctive kinds of air–sea interactions. One important air–sea interaction is the mixing of oxygen from the atmosphere into the surface water (Chapter 8). Another is the response of the water to air temperature, which not only affects the mid-level stratification of the water but also controls the conditions under which marine evaporites may form. Wind-generated waves and changes in the height of the water surface resulting from wind action and from changes in barometric pressure are contrasting kinds of air–sea interactions that regulate motion of the water overlying the bottom sediments (*bottom-boundary layer*). In comparable fashion, perturbations of any midwater stratification interfaces (pycnoclines) can also cause motion in the bottom boundary layer. We emphasize here the temperature effects and the principles of wave action by which perturbations of the water surface and of a pycnocline initiate motions in the bottom-boundary layer.

TEMPERATURE EFFECTS

Air- and water temperature are complexly linked. Viewed regionally and on a long-term basis, air temperature controls water temperature. But locally and on a short-term basis, water temperature influences air temperature. Temperature affects both the properties of the water masses overlying a shelf and the kinds of organisms that are present. At this point, we illustrate two temperature effects: (1) the midwater stratification of temperate regions and (2) marine evaporites. We discuss the temperature effect on organisms in a following section on biofacies.

Midwater Stratification. The temperature, salinity, and density structure of the seawater overlying most shelves in the temperate-climatic zones (chiefly areas lying poleward of the Tropics of Cancer and of Capricorn) vary seasonally (Figure 10-3).

During the winter the water column is well mixed and attains its minimum temperature and maximum salinity. At other times of the year, warmer- and less-saline surface water overlies the bottom water along one or more well-defined pycnoclines.

Marine Evaporites. In addition to establishing midwater stratification, hot, dry air masses may be responsible for a permanent condition in which evaporation exceeds precipitation. (See Figure 8-12.) The result can

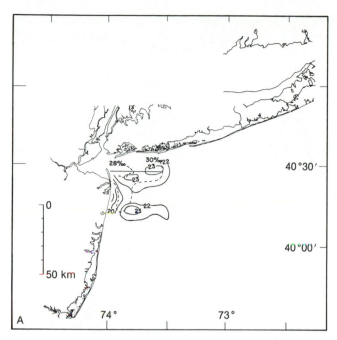

FIGURE 10-3. Distribution of water temperature and salinity at surface and depth in New York Bight.

A. Map of New York Bight showing the distribution of surface-water temperature and -salinity in July 1948. Approximately east-west-trending shoreline is that of New York State; north-south shoreline is that of New Jersey.

be hypersaline water, as in Hamelin Pool Basin, Shark Bay, Western Australia. In this pool, salinity reaches 60 to 65‰ (Figure 10-4). This is regarded as high but is still much less than that required to precipitate evaporites. (See Figure 3-1.)

Shark Bay is an elongate, complex embayment that trends NW-SE and faces the Indian Ocean at coordinates approximately lat.25°S and long.113°E. The climate is semiarid; the mean annual rainfall is 20 to 22 cm, and most rain falls during the winter (May-July) or in summer tropical storms (December-March). Annual evaporation is 200 to 220 cm, 10 times greater than the annual rainfall.

Water temperatures range from 22°C (winter) to 24°C (summer) in the adjacent Indian Ocean to a low of 17 or 18°C (winter) and a high of 25 to 27°C (summer) in Hamelin Pool Basin at the southeast extremities of the Bay. Salinity of Indian Ocean water is usually 36 to 37‰.

In Shark Bay are two other water masses: (1) meta-haline water (salinity 45 to 50‰) in the middle reaches of the two forks of the bay; and (2) hypersaline water (salinity 60 to 65‰) in Hamelin Pool Basin. Where the salinity measurements were made, no horizontal stratification of water masses was found. Instead, the boundaries between water masses are vertical. If any dense, hypersaline water flows seaward along the bottom, it does so elsewhere.

Given the situation in Shark Bay, the question arises as to whether the water in a basin having a restricted

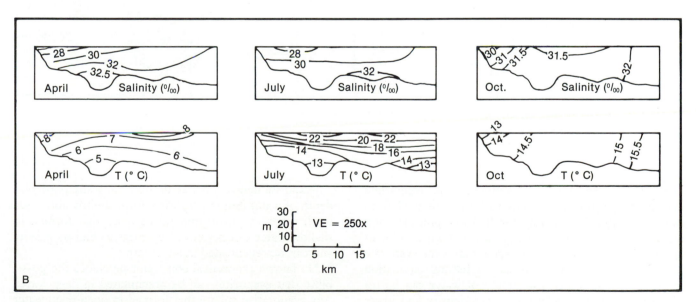

FIGURE 10-3. *(Continued)*

B. Distribution of temperature and salinity with depth in New York Bight in April, July, and October 1948 at same site as A, where contours are shown for temperature and salinity. (Modified from B. H. Ketchum, A. C. Redfield, and J. C. Ayers, 1951, figs. 5, 6, 7, 9, p. 11, 12, 13 15; and M. G. Gross, 1972d, figs. 12-15, 12-16, 12-17, 12-18, 12-21, p. 345, 346, 348.)

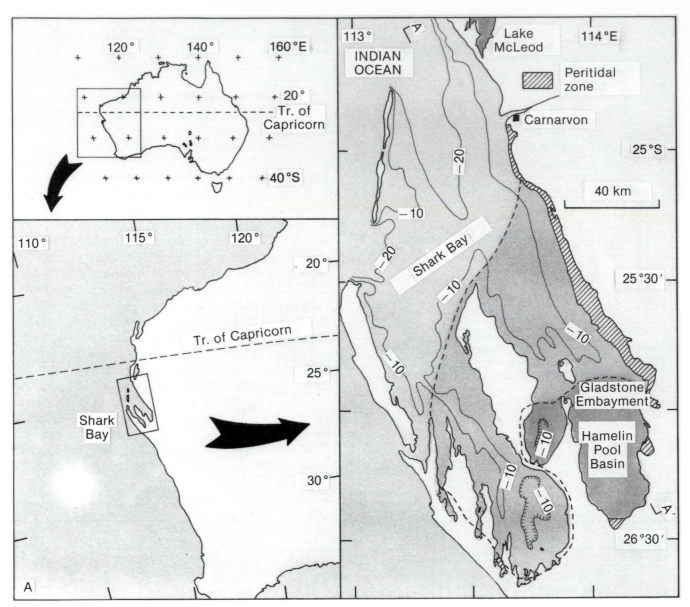

FIGURE 10-4.   Increase of salinity landward in waters of Shark Bay, Western Australia, as a result of evaporation.
A. Location maps.

connection to the open sea could become hypersaline enough for evaporite minerals to be precipitated from the open water. Presumably, the hydrographic situation would have to resemble that shown in Figure 8-38. In an open system, no body of water may ever evaporate to the point of dryness. Instead, a natural circulation is established in which the supply of water can be replenished. Water of relatively lower density and lower salinity flows toward the place(s) where evaporation is in progress. Such circulation can take place in a body of open water or within the pores of sediments at the margins of the water body.

Evaporite deposits per se do not imply any particular depth of water. Some evaporites form slightly above sea level; other evaporite mineral particles that formed in shallow water can be moved by currents and by gravity and can be redeposited at any depth.

No known theoretical constraint precludes the possibility that evaporites can be precipitated in deep water. Any evaporative sulfate that formed in open deep water may not enter the geologic record as a sulfate mineral. Rapid diagenetic changes can convert sulfates to carbonates (Chapter 3). Thus it is not yet clear whether evaporites deposited in deep basins formed only when those

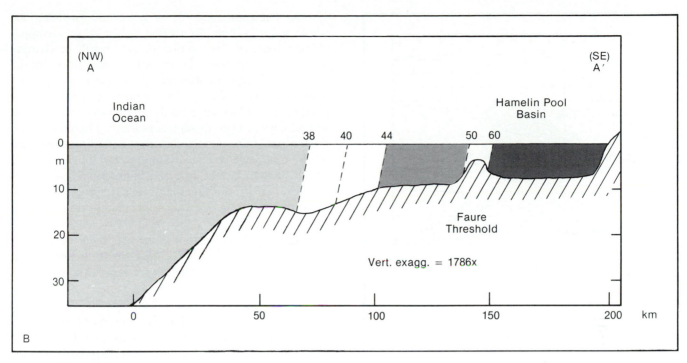

FIGURE 10-4. (*Continued*)
B. Bottom profile and hydrographic section along A-A' of A. Nearly vertical dashed lines are values of equal salinity (indicated by numbers, parts per thousand). Further explanation in text. (Redrawn from B. W. Logan and D. E. Cebulski, 1970, A, fig. 2, p. 3; B, fig. 7, p. 14.)

basins desiccated to dryness or near dryness, or whether some of the evaporites formed from deep- but very saline waters.

Commonly in the rock record, such as, for instance, in Silurian strata of the Michigan Basin, a sharp lateral contact exists between evaporites and reefs. Cursory examination may suggest that the reef-building organisms lived at the time when evaporite minerals were being precipitated from brines (Figure 10-5). However, reefs and evaporites are mutually exclusive. Reef-building organisms do not survive under the saline conditions in which evaporite minerals are precipitated. Organisms prefer not to be pickled in brine! Yet in the rock record, reefs and evaporites commonly are in lateral contact. The sequence of events resulting in such lateral contact is: (1) organisms build reefs under conditions of normal marine salinity; (2) the sea recedes, or the reefs emerge, or the water salinity increases dramatically; and (3) in a tropical-arid setting, marine-, sea-marginal-, or continental evaporites migrate across the subaerially exposed reefs. An interval of emergence commonly separates reef construction from precipitation of evaporites.

Where evaporites overlie a karst surface dissolved from carbonates, one can infer that the climate changed from humid (to provide the rainfall for dissolving the carbonates) to arid (to create the evaporites). Many pinnacle reefs that are surrounded by evaporites may have grown on the pedestals of an older karst landscape.

A common site of marine-evaporite formation is in juvenile ocean basins that open after rifting. These narrow basins commonly are poorly connected to the open ocean. The resultant restricted circulation may permit deposition of great thicknesses (up to several kilometers) of subaqueous evaporites. The lakes of the East African Rift are thought by some to represent a modern example of a nascent ocean basin. In these lakes, evaporation of waters having distinctive compositions gives rise to many unusual chemical species.

SOURCES: Bellanca and Neri, 1986; Catacosinos and Daniels, 1991; Dean and Schreiber, 1978; Decima, McKenzie, and Schreiber, 1988; Friedman and Kopaska-Merkel, 1991; Handford, Loucks, and Davies, 1982.

MOTIONS OF WATER

In Chapter 8 we discussed the relationship between winds and coastal waters with particular reference to the seasonal reversals of prevailing winds in tropical regions that accompany the annual migration of the subsolar point and the ITCZ. Here we emphasize another

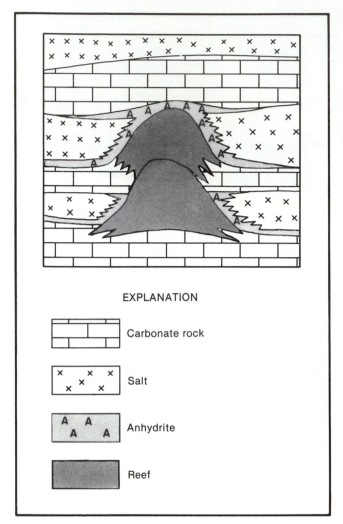

EXPLANATION

☐ Carbonate rock

☐ Salt

☐ Anhydrite

☐ Reef

FIGURE 10-5.   Generalized subsurface section of Silurian deposits in northern Michigan Basin showing pinnacle reefs and contiguous strata. Reefs, which are about 200 m thick and 1.5 km across, are shown as having sharp contacts with the enveloping anhydrite. The significance of the reef-anhydrite contact is a much-debated subject that is explained further in the text. (Modified from K. J. Mesolella, J. D. Robinson, L. M. McCormick, and A. R. Ormiston, 1974, fig. 6, p. 40.)

distinctive feature of the water overlying marine shelves, namely that air–sea interactions which perturb the water surface or a midwater stratification interface cause the water in the bottom-boundary layer to apply shearing stresses to the bottom. The basic principles are those of wave mechanics.

In the zone of shoaling waves, the water in the bottom boundary layer may move in three ways: (1) in short oscillations associated with the passage of the shoaling waves having periods of a few seconds up to about 14 s or so (wind-generated sea waves and swells); (2) in longer oscillations associated with the passage of shoaling waves having periods between 14 s (long-period swells) and

15 min (tsunami); and (3) as a current involving the whole water column, generated either as an accompanying feature of wave action or as a result of changes of level brought about by the astronomic tides, by the winds, by changes in atmospheric pressure, or by other disturbing agents. Many of these currents are in fact related to "waves" having periods ranging from six hours to several days or even to several months. Therefore, what the bottom boundary experiences as a unidirectional flow on the time scale of tens of minutes and longer time intervals, may in fact be half of a bidirectional oscillation associated with a wave on the water surface whose period is so long that one does not immediately think of it as being a wave.

In Chapter 8 we examined the origins of deep-water waves. Here we analyze waves that are not in deep water. We discuss wave base, shoaling transformations, and the effects of large waves.

Wave Base.   A water depth of $L_0/2$ ($y/L_0 = 0.5$, where L_0 is wavelength and y is depth of water) has been widely regarded by geologists as *wave base*. Seaward of wave base, the passage of waves on the water surface does not induce perceptible movement of the bottom-boundary layer. Landward of wave base, however, waves moving across the water surface do begin to induce movement of the bottom-boundary layer and thus to cause it to apply shearing stresses to the bottom sediments. Just as a spectrum of wavelengths and wave periods exists among the water-surface waves, so a corresponding spectrum of wave bases must be recognized. Remember that wave base is not a specific number for all waves, but rather is a depth defined by a ratio whose value changes with wavelength, hence also with wave period (Figure 10-6).

Once the waves have begun to react with the bottom, they undergo various changes known as *shoaling transformations*. These changes affect the celerities and shapes of the waves, with consequences explained in Box 10.1. Eventually, shoaling waves disappear by breaking.

Shoaling waves are responsible for the distinctive dynamic aspects of the shoreface. Between wave base and transformation base, the orbital motions created by the waves are still essentially circular. (See Box 10.1 Figure 1.) This means that the water in the bottom-boundary layer is not significantly shearing against the bottom. From transformation base toward shore, however, the orbits of the shoaling shallow-water waves become more-and-more elliptical and the intensity of the shearing applied to the bottom by the bottom-boundary layer increases. (See Box 10.1 Figure 1.)

The depth of transformation base for everyday wind-generated waves and swells serves as a boundary dividing the shoreface into inner- and outer parts. In the inner part of the shoreface, the water in the bottom-boundary

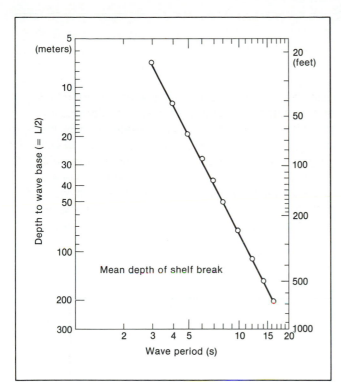

FIGURE 10-6.   Graph of depth to wave base vs. wave period. At the mean depth of the shelf break in many parts of the world, swells having periods of greater than about 8 s cause motions in water within the bottom-boundary layer that begin to affect the sediments. (J. E. Sanders in G. M. Friedman and J. E. Sanders, 1978, fig. A-10, p. 469.)

layer oscillates back and forth at a right angle to shore every few seconds.

SOURCES: Friedman and Sanders, 1978; Komar, 1976;
Leeder, 1982.

Wave-Generated Oscillations in Bottom-Boundary Layer Overlying the Inner Shoreface. Within the bottom-boundary layer, the oscillatory-elliptical motion is part of a repeated cycle having four distinct and contrasting parts: (1) upward lift, (2) shearing toward shore, (3) downward push, and (4) shearing away from shore. As each shoaling wave passes, this cycle is repeated, always in the same sequence. As a crest of a shoaling shallow-water wave nears, the water particles in the bottom-boundary layer move upward. Under each shoaling shallow-water wave crest, the bottom water applies a shearing force directed landward. After the crest of the shoaling wave has passed, the bottom water surges downward briefly. Under each shallow-water wave trough, the bottom water applies a shearing force directed away from shore.

SOURCE: Kolp, 1958.

If the shearing in the bottom boundary layer against cohesionless bottom sediments is sufficiently intense,

ripples form (Figure 10-7). During any instant in time at any point on the inner shoreface, the moving water in the bottom-boundary layer applies only one-directional shearing; asymmetrical ripples may form. The profiles of these ripples are similar to those of current ripples in a stream bed; their steep sides dip in the direction of shear. However, as the four-part cycle proceeds, the direction of shearing will reverse. When that happens, then new asymmetric ripples appear, but their steep sides are on the gently sloping sides of the previous set of ripples. This repeated reversing of the direction of the shearing applied to the bottom by water within the bottom-boundary layer and corresponding appearance of first one set of asymmetric ripples and then a second set having exactly opposite asymmetry creates a resultant set of ripples possessing sharp, symmetrical peaks that point upward. The crests of these *wave-generated ripples* (also known as *symmetrical ripples* or *oscillation ripples*) are separated by troughs whose profiles are symmetrical, broadly rounded, and concave up. (See Figures 5-20, A and A′, E, and F.)

In general, within the inner shoreface, a given set of shoaling shallow-water waves tends to create two belts of ripples: (1) an outer belt of symmetrical ripples, and (2) an inner belt of asymmetrical ripples having their steep sides toward shore.

Depending on what the waves are doing, these trains of ripples on the inner shoreface may shift toward shore or offshore. As the ripples migrate first one way and then the other, they acquire a distinctive internal stratification. (See Figure 5-20, G.)

Not all wave-generated ripples are shaped like those just described. Other kinds of ripple profiles are symmetrical with different curvature of crests and troughs. (See Figure 5-20, B, C, C′, and D.) Still-other wave-generated ripples are asymmetrical. (See Figure 5-20, H.)

The profiles of asymmetrical wave-built ripples are indistinguishable from those created by unidirectional currents. (Compare Figure 5-20, H with Figures 5-21, B and 5-21, A and C.) In some cases, however, the internal cross strata of these two kinds of ripples are not the same, and hence it is possible to distinguish wave-built asymmetrical ripples. (See Figure 5-20, G.) In other cases the internal cross laminae of wave-built asymmetrical ripples are the same as those in asymmetrical ripples created by unidirectional currents. (Compare Figures 5-20, H and 5-21, B.) The symmetry or degree of asymmetry of wave-generated ripples is a function of the relative intensities of the onshore-directed- and offshore-directed shearing motions established by shoaling waves in the bottom-boundary layer.

A widely repeated statement found in the scientific literature on shoaling waves asserts that the elliptical orbits associated with shoaling shallow-water waves

BOX 10.1

Shoaling Transformations

In theory, shoaling transformations begin when waves have passed into water shallower than wave base. However, for practical purposes, the waves advance well landward of wave base before shoaling transformations become significant. For some distance landward of wave base, the waves can still be treated mathematically as deep-water waves. The orbits of water "particles" are still circles (Box 10.1 Figure 1).

In the mathematical analysis of the effect of shoaling transformations on wave celerity, it is important to realize that in deep water and in the outer part of the zone of shoaling waves, wavelength exerts the predominant control. By contrast, in the inner part of the zone of shoaling waves, and especially in connection with breaking, wave celerity is controlled by two other factors: (1) wave height or H_0 and (2) water depth. A useful way to

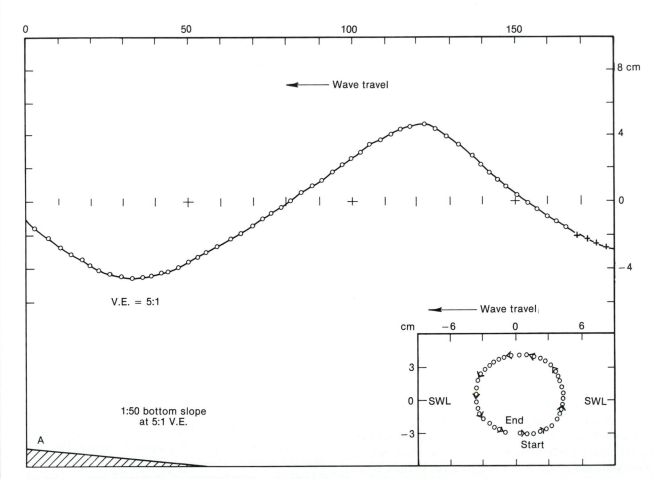

BOX 10.1 FIGURE 1.   Shoaling transformations shown by experimental waves in a large wave tank (0.3 x 1 x 20 m). Profiles of waves, and corresponding orbits of surface-water "particles" were determined by taking motion pictures, through glass side of tank, of neutrally buoyant particles (formed by a mixture of carbon tetrachloride, xylene, and zinc oxide) added to the water, with a clock and a calibrated grid in the field of view, and by projecting the developed movie film one frame at a time and plotting the results.

A. Shoaling waves; H =9 cm, L = 158 cm; period = 1.06 s; water, 36 cm deep over bottom sloping 1:50. H/L = 0.057; y/L = 0.228. Orbit of surface-water particles (lower right) is a nearly perfect circle.

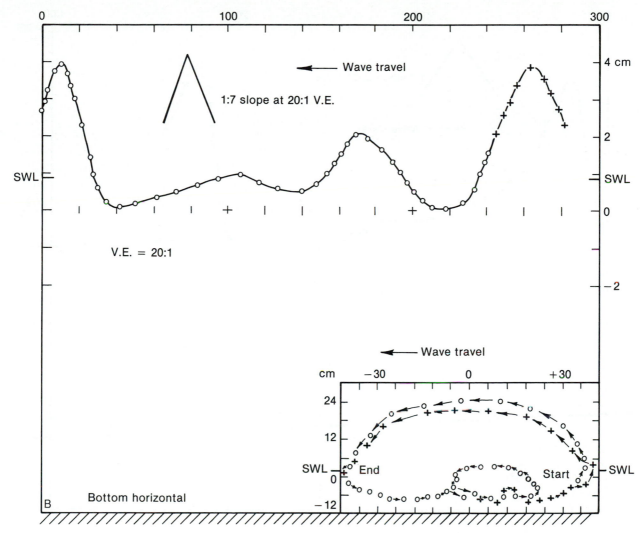

BOX 10.1 FIGURE 1. *(Continued)*
B. Much-transformed waves; H = 3.8 cm; L = 254 cm; period = 2.67 s; water, 8.9 cm deep, bottom horizontal; H/L = 0.015; y/L = 0.035. Secondary wave in flattened trough generated a second, smaller elliptical path at the base of the main ellipse. Such double orbits may be responsible for creating multiple sets of symmetrical ripples, as in Figure 5-20, F. Small circles denote positions at equal time intervals on the first wave's orbit.

deal with these changes is to examine the relationships between two ratios: (1) depth/wavelength = y/L_0 and (2) wave height/depth = H_0/y (named *relative wave height*). These are related via wave steepness (H_0/L_0) as:

$$\frac{\text{steepness}}{\text{relative wave height}} = \frac{y}{L_0} \quad \text{(Box 10.1 Eq. 1)}$$

In the wave shown in Box 10.1 Figure 1,A, the relative wave height is 0.25. Off southern Califor-

nia, studies on the bottom beneath shoaling waves have shown that for swells with L_0 = 300 m, H_0 = 1.5 m, H_0/L_0 = 0.005, period T = 12 s, and C = 22 m/s. The results predicted by deep-water theory begin to deviate seriously from observations where relative wave height, H/y, is 0.04. This is at a depth of 37.5 m, y/L_0 = 0.125, or 1/8 of the deep-water wavelength. We shall use the term **transformation base** for *the depth of 1/8 of the deep-water wavelength*, L_0. At depths of less than $L_0/8$, the deep-water formula is no longer appropriate. *Waves traveling in water depths of*

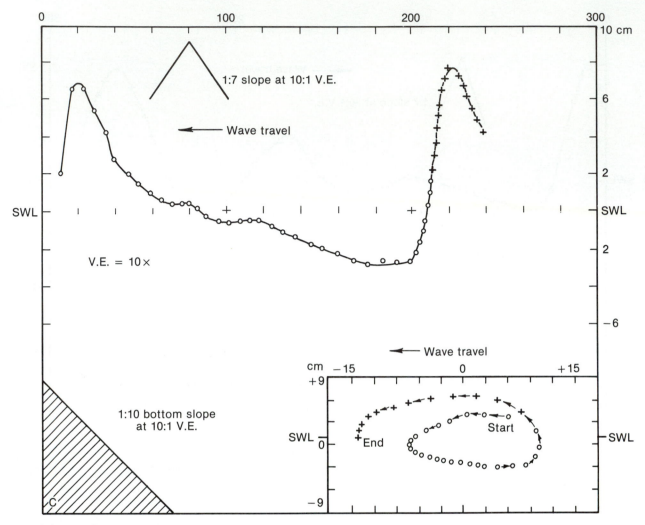

BOX 10.1 FIGURE 1. (*Continued*)
   C. Profile of wave just prior to breaking. Height of breakers, 11.3 cm; *H/L* of waves outside breaker zone = 0.0206; period of breaking waves, 1.51 s; depth of water at breakers = 9.94 cm. Bottom slope = 1:10. Movement of water "particles" is an elliptical orbit with mass transport landward. (Data of J. R. Morison and R. C. Crooke, 1953, figs. 4, p. 10; 10, p. 11; and 12, p. 12, replotted on metric grid by J. E. Sanders in G. M. Friedman and J. E. Sanders, 1978, fig. A-11, p. 470–471.)

less than $L_0/8$ are referred to herein as **shallow-water waves**.

When the depth becomes very small with respect to wavelength, shallow-water waves may undergo another transformation and become **very-shallow-water waves**, defined as *waves traveling through water that is shallower than 1/20 of their deep-water wavelengths*. In other words, the critical depth (y) at which waves become very-shallow-water waves is the ratio $L_0/20$ or $y/L_0 = 0.05$. Waves having periods of 6.4 s and wavelengths of 64 m become very-shallow-water

waves at a depth of 64/20 = 3.2 m. With very-shallow-water waves, the term tanh $(2\pi y/L_0)$ in Eq. 8-1 becomes about equal to $2\pi y/L_0$. Thus we can dispense with the hyperbolic tangent, and Eq. 8-1 becomes

$$C = \left(\frac{gL_0}{2\pi} \frac{2\pi y}{L_0}\right)^{1/2}$$

which clears to

$$C = (gy)^{1/2} \qquad \text{(Box 10.1 Eq. 2)}$$

where

C = celerity of wave (m/s)

g = acceleration resulting from gravity (9.80 m/s^2)

y = depth of water (m)

Because g is a constant whose square root is 3.13, we can write Box 10.1 Eq. 2 as

$$C = 3.13(y)^{1/2} \qquad \text{(Box 10.1 Eq. 3)}$$

(all units of length in meters).

Box 10.1 Eqs. 2 and 3 state that the celerity of a very-shallow-water wave is proportional to the square root of the water depth. At the depth (3.2 m) where they become very-shallow-water waves, the celerity of waves having wavelengths of 64 m and periods of 6.4 s is $C = 3.13 \times (3.2)^{1/2} = 5.6$ m/s. This is only a little more than half their celerity as deep-water waves (10 m/s). Box 10.1 Figure 2

is a graph of wave celerity versus depth for very-shallow-water waves.

In the zone between $L_0/8$ and $L_0/20$, the mathematical analysis of waves is very complex.

SOURCES: D. O. Cook and Gorsline, 1972; Inman, 1963a; Komar, 1976b.

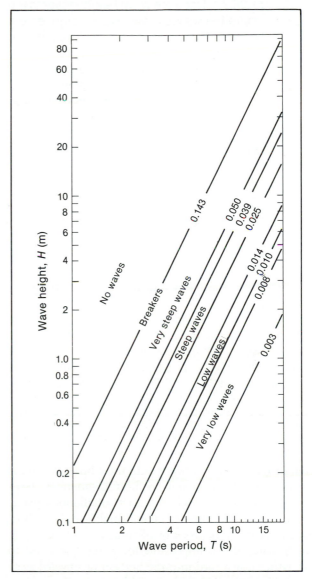

BOX 10.1 FIGURE 3. Graph of wave height vs. wave period (both on log$_{10}$ scales) for water waves of indicated steepnesses. In the open ocean, wind-generated sea waves in a storm center start out steep ($H_0/L_0 > 0.025$), but after becoming swells, they are changed into low waves ($H_0/L_0 < 0.014$). As swells approach shore, they steepen again and eventually break. Waves having steepness > 0.039 break before they become very-shallow-water waves. Further explanation in text. (J. E. Sanders in G. M. Friedman and J. E. Sanders, 1978, fig. A-13, p. 473.)

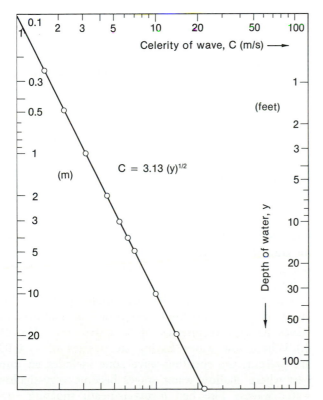

BOX 10.1 FIGURE 2. Graph of celerity (log$_{10}$ scale, origin at left) vs. water depth (log$_{10}$ scale, origin at top) for very-shallow-water waves. Celerity of very-shallow-water waves is proportional to the square root of water depth (Box 10.1 Eqs. 2 and 3). (J. E. Sanders in G. M. Friedman and J. E. Sanders, 1978, fig. A-12, p. 472.)

The landward limit of the zone of shoaling waves is the **breaker zone**, *the line where shoaling waves first collapse*. Waves collapse, or break, because the water in their crests has been heaped up beyond the limiting crestal-peak angle of $120°(H_0/L_0 = 0.143)$. (See Figure 8-21, B.) According to some formulations, this happens at a water depth of 1.28 times the deep-water wave height ($1.28H_0$). Because wave height governs the depth where the waves break, wave height enters the definition of wave steepness and wavelength governs the depth ($y/L_0 = 0.05$) of transformation to very-shallow-water waves. It is possible to calculate a limiting wave steepness of waves that will break before they become very shallow-water waves:

$$\text{breaker depth} = 1.28H_0 = 0.05L_0$$

(Box 10.1 Eq. 4)

$$\text{limiting steepness} = H_0/L_0 = 0.05/1.28 = 0.039$$

(Box 10.1 Eq. 5)

Box 10.1 Figure 3 is a graph of wave steepness versus wave periods showing the relationships between this steepness limit and the steepness of breakers ($H_0/L_0 = 0.143$). Box 10.1 Figure 4 is a graph showing the depth limits of these two kinds of waves for various periods and having steepnesses of less than 0.039.

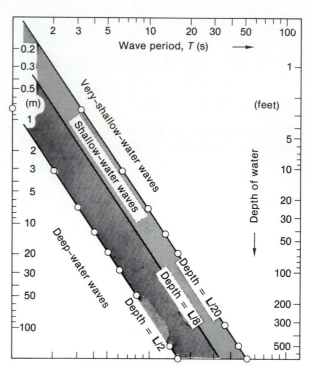

BOX 10.1 FIGURE 4.   Graph of depth of water (origin at top) vs. periods of waves (origin at left) showing depth limits for two kinds of shoaling waves having $H_0/L_0 < 0.039$. Wavelengths have been converted to wave periods using Figure 8-27. The lower curve (depth = L/2) defines a spectrum of depths to wave base. The upper curve (depth = L/20) defines the spectrum of depths at which shallow-water waves become very-shallow-water waves. The middle curve (depth = L/8) defines the spectrum of depths to transformation base landward of which the deep-water formulas no longer apply. (J. E. Sanders in G. M. Friedman and J. E. Sanders, 1978, fig. A-14, p. 473.)

flatten progressively landward until the ellipses disappear altogether and the motion in the bottom-boundary layer becomes only planar back-and-forth oscillations parallel to the bottom. This statement applies only to bottoms composed of bedrock or of compact, cohesive clay. In such material it is impossible for water "particles" to move either downward into the bottom or upward away from the bottom. The only possible motion is a planar oscillation. Where the bottom consists of permeable cohesionless sediments, such as sand, however, the upper part of the interstitial water in the bottom sediments *does* become involved in the orbital motions within the bottom-boundary layer above. Accordingly, at the bottom, the water "particles" describe flattened ellipses that do not quite close—at the end of the orbit, a water "particle" is landward of the point where its orbit began. (See Box 10.1 Figure 1, B.) As the "particles" travel along the

long sides of these ellipses, they move back and forth in paths that are nearly parallel with the bottom. In so doing, they apply significant shearing force to the bottom sediments.

The two sets of small-scale ripples just described characterize the shoaling-wave zone for shallow-water waves having steepnesses of > 0.039. (See Eq. 10-4.) Where low waves having steepnesses of < 0.039 are present, the shoaling-wave zone includes an inner part where shallow-water waves become very-shallow-water waves. The bed forms beneath shoaling very-shallow-water waves become more diverse toward the breaker zone than those beneath shallow-water waves (Figure 10-8). In the zone immediately landward of transformation base, where the shoaling waves are shallow-water waves, a belt of small-scale asymmetric ripples forms as described above. Farther land-

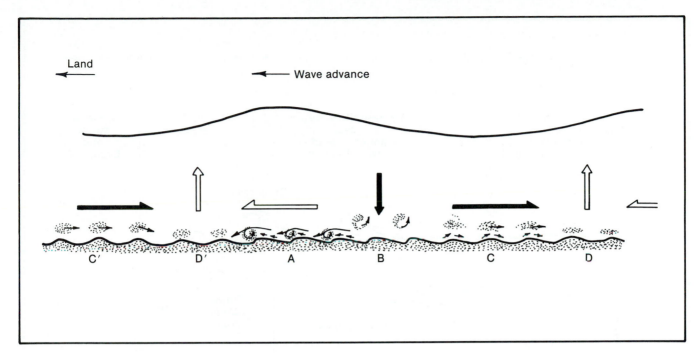

FIGURE 10-7. Orbital motion of water in bottom-boundary layer and origin of oscillation ripples on sand bottom shown by schematic vertical profile, at right angle to shore, through shoaling shallow-water waves and bottom. (Sizes of ripples on bottom not drawn to same scale as waves on water surface; both are schematic.) The four phases of the wave cycle, lettered A, B, C, D, are shown as they exist simultaneously under different parts of the wave profiles. As the waves travel shoreward, these four phases are repeated in sequence at each spot on the bottom.

Under wave crests (only one of which is shown, at A), waves induce motion in water within the bottom-boundary layer, which applies shearing force directed landward (shown by horizontal open arrow with one barb), in direction of wave advance. Bottom responds to shearing by forming asymmetrical ripples (as in small-scale "current ripples"); separation eddies (See Figure 5-21, D.), each containing abundant sand in suspension (shown by dots in water), form on "downshear" (landward) sides of each ripple. Such eddies tend to rise through the water, to enlarge, and to dissipate.

After wave crest has passed, water in bottom-boundary layer travels vertically downward (large vertical black arrow), carrying sediment back toward bottom (B).

Beneath wave trough (C and C'), water in bottom-boundary layer oscillates away from shore (large horizontal black arrows with one barb), applying shearing force directed away from shore. Bottom responds to shearing by forming ripples and sediment-laden eddies on both sides of ripple crests, which are now symmetrical or nearly symmetrical.

After trough of wave has passed (D and D'), water in bottom-boundary layer moves vertically upward (large vertical open arrows), lifting sediment-laden water. (Based on O. Kolp, 1958, figs. 4 and 5, p. 175.)

ward, however, at a depth of $L/20$, where some shoaling shallow-water waves become shoaling very-shallow-water waves, no small-scale ripples are present. Instead, one finds a belt of larger bed forms that we classify as small 3D dunes (See Table 7-2.), whose heights may be as much as 0.3 to 1 m, and wavelengths, 1 to 4 m. Landward migration of these small 3D dunes gives rise to trough cross strata that dip landward.

We consider that this zone of small 3D dunes ("lunate megaripples" of Clifton, Hunter, and Phillips, 1971) is diagnostic of the effects of shoaling very-shallow-water waves. Accordingly, where a shoreface shaped by shoaling

very-shallow-water waves has prograded seaward, the existence of large-scale trough cross strata between asymmetric wave-generated ripples below and beach strata above indicates that the conditions of waves and water depth enabled a zone of very-shallow-water waves to exist just seaward of the breakers. Using the conditions on the Pacific coast of North America as a basis, we infer that the shelf would be narrow (width not more than a few kilometers) and that deep water existed close to shore. By contrast, the absence of such trough cross strata implies that conditions precluded the formation of the near-breaker zone of very-shallow-water waves. The

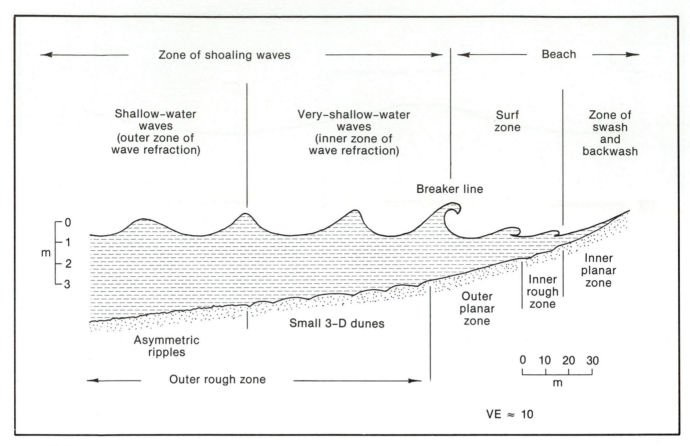

FIGURE 10-8.   Schematic profile at right angle to shore, Whalehead Cove, Oregon, showing relationships among zones of shoaling waves, breakers, surf, swash and backwash, and kinds of bed forms on sea floor. In outer part of outer rough zone (at left), the ripples are small and asymmetric; their steeper sides are toward shore. Inner part of outer rough zone features small 3D dunes (lunate megaripples). Along coasts lacking a zone of very-shallow-water waves, the small 3D dunes are not present. (After H. E. Clifton, R. E. Hunter, and R. L. Phillips, 1971, fig. 15, p. 661.)

inferred bathymetric setting was a wide shelf, with deep water several tens of kilometers away from the shoreface.

SOURCES: Bagnold, 1940, 1946, 1947, 1963; Clifton, R. E. Hunter, and R. L. Phillips, 1971; Friedman and Sanders, 1978; Middleton and Southard, 1984.

Effects of Large Waves.   These large waves include long-period swells, storm waves, and tsunami.

LONG-PERIOD SWELLS.   The periods of ordinary swells range from 6 to 14 s. In contrast are long-period swells, whose periods range from 14 to 22 s. Such swells are long, low waves that shift the boundary between inner- and outer shoreface seaward to depths where shorter-period waves do not stir the bottom.

Long-period swells are thought to be capable of causing short-term scouring of a sandy bottom to depths of a meter or more and of suspending and otherwise dispersing great volumes of sand.

Several published reports relate stories told by sea captains of finding sand on deck after storm waves had broken over the bows of their ships, which had been caught in storms near the shore. These reports imply that the storm waves stirred substantial quantities of sand into suspension. The veracity of stories of sand on the decks of ships during storms can be judged by considering the following true-life adventure story of some New Jersey clams.

A remarkable accumulation on the beach at Ocean City, New Jersey, of countless thousands of still-living pelecypods (*Mercenaria mercenaria*) took place on 27 and 28 February 1961 (Figure 10-9). Such organisms normally live in burrows on the inner shoreface. How could they possibly have been taken from their burrows and tossed on shore? Ordinarily the clams do not care for this sort of adventure. Quite the contrary; in order to remain part of the infauna, they are capable of deepening their burrows very rapidly.

Thus, if motions in the bottom-boundary layer cause the bottom to be lowered, the normal thing is for the clams immediately to dig in deeper. However, during the winter the temperature of the bottom water off New

FIGURE 10-9.   Bulldozer clearing thick layer of large shells of *Mercenaria mercenaria* that inundated the beach at Ocean City, New Jersey, on 27 and 28 February 1961. (Wide World Photos.)

FIGURE 10-10.   Sediments from shoreface off Fire Island barrier, Long Island, New York; relief peels made from median sections of cores. (CERC Core 68, depth 89 feet, lat.40°38′N; long. 72°55′W; N. Kumar and J. E. Sanders, 1975, fig. 6, p. 1554; CERC Core 72, depth 32 feet, lat.40°42′N; long.72°54′W, N. Kumar and J. E. Sanders, 1976, fig. 2, p. 148; cores courtesy of U.S. Army Corps of Engineers, Coastal Engineering Research Center.)

Jersey drops to 1 to 4°C. This is not cold enough to kill infaunal organisms, but it chills them to the point where they cannot react very quickly. Thus, if a storm takes place while cold bottom water has immobilized the infauna, the waves may lower the level of the bottom and the animals are helpless to do anything about it. The result is that the burrowing organisms are washed out of their burrows and become playthings for the waves.

If the foregoing explanation is correct, then on the days concerned, the waves off the New Jersey coast must have lowered the sandy bottom by at least 30 cm (depth of normal *Mercenaria* burrows). The active waters evidently held the sand in suspension and deposited only the chilled clams on the beach. The result was a gigantic accumulation of "clams in the whole shell," which, as the unwilling adventurers died, quickly became one vast stinking mess. What became of the sand? We do not know for sure, but can guess that as the storm waves died down most of it probably was deposited back on the shoreface once again. The sand definitely was not deposited on the beach along with the clams.

A hint about the fate of the New Jersey sand comes from long cores of modern shoreface sediments off Long Island, New York, which consist of thick (up to 2 m) sediment couplets. The basal parts of these couplets consist of structureless basal gravel, up to several tens of centimeters thick, and containing well-rounded pebbles of rock fragments up to 4 cm in diameter and a few large broken shells. This coarse basal zone is overlain by slightly micaceous, very well-laminated fine sands up to 2 m thick (Figure 10-10). No skeletal remains of any kind have been found in these well-laminated shoreface sands.

We infer that these sediment couplets resulted from the effects of severe storms. We surmise that at the height of the storm all the sand was kept in suspen-

sion and the gravel formed a widespread lag pavement on the shoreface. All the skeletal remains were separated from the terrigenous sediment and deposited on the beach or else broken into pieces too small to identify. As the storm waned, the sand was deposited from suspension. This must have happened under conditions of bottom shear where the stable bed configuration was a plane surface. We can infer this from the plane-parallel laminae. Very likely, then, conditions analogous to those in the transition from the lower-flow regime to the upper-flow regime of alluvial channels were involved (Chapter 7; but with possible modifications related to oscillating flow). The storm couplets are graded and exhibit a vertical sequence from base to top of massive graded sand to laminated sand (upper plane-bed deposition) to wave-ripple cross-laminated sand (fairweather sediment reworking).

If the waves could place into suspension enough sand to deposit 2 m of laminated sand after a storm, then the water column could very well have contained sand in suspension all the way to the surface as suggested by the sea-captains' stories.

Long-period swells unaccompanied by local winds nearly always generate a bottom current that flows landward. The situation may be comparable to the results of some wave-tank experiments in which the regular waves, generated in the deepest part of the tank, were propagated across a shelf break and a gently sloping shelf to a shoreface slope at the far end of the tank. In these experiments, some waves established a large-scale water circulation. These waves not only generated the usual orbital motions, but also established a pulsating current that flowed along the bottom toward the line of breakers. At the breakers, this bottom water flowed upward and returned outward at the surface.

In these same tank experiments, large bed forms consisting of linear, rounded ridges and troughs having long axes parallel to shore were formed (Figure 10-11). Such linear ridges of cohesionless sediment that are always submerged are one of several varieties of *longshore bars*; they are separated by linear depressions known as *longshore troughs*. Longshore bars may remain stable for long periods, or they may migrate. In some cases, the bars migrate onshore and disappear by adding their sand to the bottom just outside the outermost breakers.

SOURCES: Bagnold, 1946, 1947; Figuerido, Sanders, and Swift, 1982; Friedman and Sanders, 1978; Kumar and Sanders, 1976; M. C. Powers and B. Kinsman, 1953; F. P. Shepard, 1950a.

STORM WAVES.   Long-period swells may immediately precede large storm waves. As a result, it may not be easy to determine at what point the effects of the long-period swells end and those of the storm waves begin. Storm waves may be accompanied by changes in water level

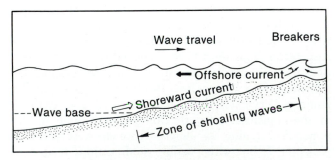

FIGURE 10-11.   Schematic profile at right angle to shore through zone of shoaling waves where bottom consists of sand. Currents in water and large bed forms on bottom based on wave-tank experiments conducted by R. A. Bagnold (1947). Further explanation in text. (J. E. Sanders in G. M. Friedman and J. E. Sanders, 1978, fig. A-20, p. 479.)

that create bottom currents. These currents may flow away from the land.

A feature common in sedimentary strata that have been inferred to be ancient shoreface deposits and thought to be a product of storms is *hummocky cross strata* (HCS). (See Figure 5-19, D.) HCS are irregularly curved strata that were deposited on a bottom characterized by hummock-and-swale bed forms lacking any preferred orientation. The laminae of HCS are even; the dip angles of both laminae and truncation surfaces are less than 15°. Individual sets of hummocky beds may be a few centimeters to about 6 m thick; the "wavelengths" between hummocks are from tens of centimeters to several meters. Sets of HCS are characterized by sharp bases that are presumed to be products of erosion. Some HCS display coarse basal lags, and wave-ripple-cross-laminated- or mud-draped tops that may be bioturbated.

The features described as laminae in cores of shoreface sands and -silts may actually be HCS. On the scale of small-diameter cores, the low-angle strata of HCS are not distinguishable from horizontal- or gently inclined strata.

Hummocks are inferred to form on the outer part of a shoreface, below everyday wave base, by interactions between oscillatory motions in the bottom-boundary layer set up by shoaling storm waves and unidirectional currents. The kind(s) of unidirectional currents are not known. Possibilities include those related to perturbations of the water surface resulting from air–sea interactions, tidal currents, and tsunami. Whether motions of the bottom-boundary layer associated with tsunami are considered as "oscillatory" or "unidirectional" is relative. The periods of oscillatory flow in the bottom-boundary layer associated with shoaling tsunami are 12 to 15 min. Compared to the periods of oscillatory flow of storm waves (say about 10 s), a flow lasting 12 min is "unidirectional." Compared with the periods of a tidal current (say 6 h), 12 min could be considered as "oscillatory."

Except in size, HCS resemble wavy beds that form on intertidal flats and the tiny "domes" associated with some convoluted laminae. (See Figure 7-29.) Hummocky cross strata probably form under conditions when the stable bed configuration is transitional to a planar surface and therefore slightly less energetic than for plane-parallel laminae. The precise mechanism of formation and the conditions under which HCS form are not yet known. HCS may occur within a vertical sequence inferred to result from storm activity like that described previously for graded storm layers: coarse lag to HCS to plane laminae, capped by fair-weather wave-ripple cross laminae.

HCS form primarily in silt and fine sand, whereas in coarse sand or pebbly sand, coarse-textured ripples up to 35 cm in height commonly form instead.

A recent discovery that HCS are actively forming in the surf zone of a storm-wave-dominated beach demonstrates that, as with so many other sedimentary structures, HCS can form under multiple sets of conditions. HCS formed in a surf zone should be expected to consist of smaller HCS sets because of shallow water depths, and for the same reason, sequences of HCS formed in the surf zone should be thin. Other examples of inferred HCS from environments other than marine shorefaces have also been found, for example from the shorefaces of lakes. HCS should no longer be regarded as diagnostic of marine shorefaces.

Another sedimentary structure found in marine rocks and believed to have been formed by storm-related processes is swaley cross strata (SCS). SCS resemble HCS, but instead of being dominated by convex-up hummocks, SCS are dominated by concave-up swales. Their origin is not yet understood.

After fair weather has returned following a storm, small waves create a rippled bottom on the inner shoreface and burrowing organisms recolonize and bioturbate the sediments on the outer shoreface. Once this has happened, ripple marks and wave-ripple laminae appear in the inner-shoreface sediments and the top 30 cm or so of the outer-shoreface sediments display the effects of bioturbation.

The shoreface is a complex zone between the shelf and the beach. As mentioned, the inner shoreface is dominated by sand that fair-weather waves shape into a rippled surface. On the outer shoreface, silt or clay may be present. During quiet times, these become thoroughly bioturbated. However, infrequent large storms can rework the outer-shoreface sediments to considerable depths (several meters or more) and deposit distinctive autocyclic sequences that may be graded and that commonly include, from base to top, coarse lag debris-HCS-plane laminae. The interaction of storms and bioturbating organisms creates interbedded sediments of contrasting characteristics.

A final word about the shoreface: It can be affected by the marine grass *Thalassia*. Luxuriant *Thalassia* meadows slow any currents and reduce the effects of waves on the water in the bottom-boundary layer. As a result, even on an exposed shoreface, fine sediments like those typically found in lagoons can be deposited. Because remains of *Thalassia* are not likely to be preserved in the geologic record, *Thalassia*-influenced shoreface fines may be misinterpreted as lagoon deposits. If the shoreface prograded, the relationships between underlying shelf sediments and overlying beach sediments may demonstrate the correct interpretation of these *Thalassia*-influenced shoreface deposits.

sources: Arnott and Southard, 1990; Dott and Bourgeois, 1982; Duke, 1985; Greenwood and Sherman, 1986; R. E. Hunter and Clifton, 1982; Kumar and J. E. Sanders, 1976; Mooers, 1967; Southard, Lambie, Federico, Pile, and Weidman, 1990.

TSUNAMI. Because of their ultralong periods, 12 to 15 min, and resulting enormous wavelengths, ca. 800 to 1200 km, tsunami are very-shallow-water waves even in the deep-sea basins, where the average water depth is 4 km. Thus the celerity of tsunami is governed by the square root of water depth. (See Box 10.1 Eq. 2.) As a matter of fact, using this equation and travel-time data on the celerity of tsunami, the correct average depth of the Pacific Ocean was calculated even before complete sounding data became available.

The slope of the bottom near shore affects the shoaling behavior of tsunami. Suitably sloping bottoms, as surround volcanic islands such as those in the Hawaiian chain, cause crests of tsunami to heap up to heights of 15 to 30 m or more.

Wherever they pass, tsunami must apply great shearing forces to the bottom that result in great shearing stresses. Such stresses increase where the tsunami shoal and attain maximum values where they break. What these stresses create in the way of sedimentary structures is not known.

sources: Bourgeois, Hansen, Wiberg, and Kaufman, 1988; Friedman and Sanders, 1978.

Effects of Water-Level Changes. When air–sea interactions perturb the surface level of the water, various currents are caused to flow in a direction so as to reestablish the normal surface. The directions of flow of such currents may be parallel to shore, toward shore, or away from shore. The duration of the water-surface perturbations that drive these currents depends on atmospheric conditions. The perturbation may result from wind shear, from extremely low atmospheric pressure in the eye of a hurricane, or from changes in the seasonal wind pattern.

The winds from a cyclonic storm can pile water against the shore, raising the water level (known as *wind setup*) and thus generating a compensatory flow along the bottom away from shore. (See Figure 8-7, C.) Currents resulting from surface perturbations associated with the passage of a cyclonic storm typically disappear in 12 to 18 h or less. Under exceptional conditions, they might persist for several days.

The effect of atmosphere-driven currents on shelf sediments could be considerable. In any case, the result will be influenced by the interaction of these currents with the tidal current(s) and wave-generated currents.

sources: Adams and Weatherly, 1981; Friedman and Sanders, 1978; Swift, Han, and Vincent, 1986; Weggel, 1972.

Effects of Perturbations of Pycnocline. Lurking beneath the surface of many shelf water bodies is a pycnocline that is subject to being perturbed by the passage along it of *internal waves* and *internal tides*. Whereas the relief of perturbations on the air–water interface may be only a few meters, that of perturbations on a pycnocline may be tens of meters (Figure 10-12). Such perturbations of a pycnocline can induce motions in the water within the bottom-boundary layer and thus can be the driving force behind sediment-transport mechanisms.

Compared with the vast amounts of knowledge that have been acquired from the study of surface waves, these perturbations of the pycnocline are little known. They are not easy to study. One of the most-detailed investigations was carried out by Otto Pettersson in the early twentieth century. (See Figure 8-36.)

Tidal Currents

The effects of tidal currents on shelf sediments range from insignificant to very important. In the southern part of the North Sea, tidal currents have created various bed forms as functions of celerity and sand supply (Figure 10-13). Most notable of these are the *longitudinal sand ridges* up to 65 km long that form where the current

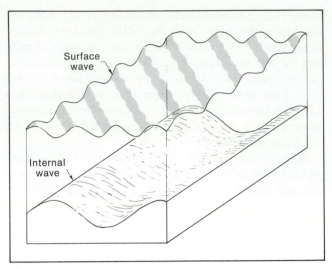

FIGURE 10-12.   Large internal wave on pycnocline beneath smaller surface waves; schematic. (After M. G. Gross, 1972, fig. 9-19, p. 261.)

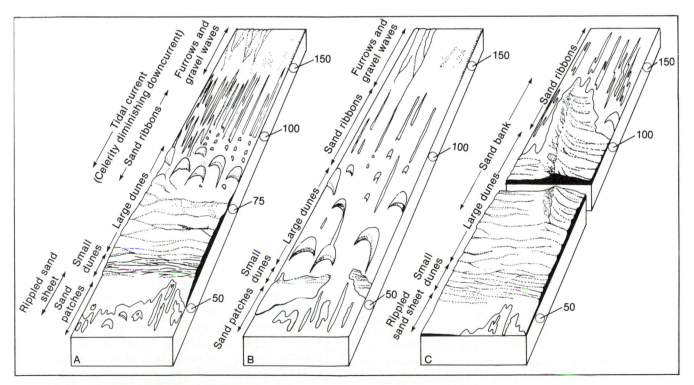

FIGURE 10-13.   Relationships among speeds of tidal currents, sand supply, and bed forms, based on observations in the North Sea. Tidal current flows from upper right to lower left with celerities of near-surface currents (cm/s) shown by circled numbers on right-hand sides of each schematic block diagram.
   A. Generalized distribution of bed forms in downcurrent direction.
   B. Bed-form pattern where sand supply is low.
   C. Bed-form pattern, featuring large longitudinal ridge, where sand is abundant. (After H. D. Johnson and C. T. Baldwin, 1986, p. 239.)

is swift and sand supply abundant. These ridges are comparable to seif dunes. (See Figure 14-24.) Measurements of tidal currents on the open shelf off the northeastern United States showed values of 10 cm/s or less. Such slow currents would not move sand at the bottom. (See Figure 7-19.) On shelves where the celerities of tidal currents are greater than about 50 cm/s, sand at the sea bottom does move. In the southern North Sea, the moving sand has built sand ridges. Depending on conditions, several kinds of sand ribbons, or large dunes that form fields covering thousands of square kilometers may form instead. In localities where the speeds of the tidal currents are slower than in the fields of linear sand ridges, sheetlike sand bodies covered with small-scale bed forms develop.

SOURCES: Howarth, 1982; Johnson and Baldwin, 1986; Stride, 1970, 1982.

Ocean Currents

Some shelves extend far enough seaward to intersect major ocean currents. These currents, which may be colder or warmer than the inshore waters, flow parallel to the outer parts of the shelf. They may exert several influences. In the western Gulf of Mexico, south of the Mississippi delta, the semipermanent currents coming from the Caribbean via the Yucatan Channel divide in such a way as to form two major circulation cells, an eastern and a western. In the western cell, the surface water flows counterclockwise off the Louisiana-East Texas sector of the shore. The northern branch of a second minor cell centered off Mexico flows clockwise off South Texas. At about lat. 27 to 28°N, waters from these two oppositely directed surface currents meet and coalesce to form a current directed offshore. The offshore-directed current entrains fine suspended sediment and transports it seaward.

Other examples of shelves affected by ocean currents are off northeastern South America, where the surface-ocean currents flow northeastward, and off eastern South Africa, where the surface-ocean currents flow southwestward. Off South Africa, the effect of the oceanic current is to sweep the outer shelf clean of fine terrigenous sediments and deposit pelagic carbonate sediments.

SOURCES: Curray, 1960; Flemming, 1971; Kuehl, Nittrouer, and DeMaster, 1982.

Biofacies Relationships

Organisms are integral parts of shallow-marine deposits, not only as contributors of skeletal remains to the intrabasinal sediments but also as makers of various tracks and trails and as the active agents in bioturbation. In shallow marine sediments, the effects of bioturbation are so pervasive that their absence requires some special explanation.

The factors involved in the distribution of organisms include not only the two we have discussed as products of air–sea interactions, temperature and water motions, but also numerous others, including depth, salinity- and transparency of the water, supplies of nutrients, and kinds of bottom. The kinds of organisms found in ancient sedimentary strata provide convincing evidence for interpreting the environment of deposition. The distribution of organisms tends to be zonal. The patterns of their remains in sediments permit recognition of various biofacies that reflect depth and distance from shore.

We illustrate some of these relationships by considering the distribution of marine microinvertebrates, the distribution of macroinvertebrates in level-bottom communities; rock-destroying organisms; and bioherms, reefs, and carbonate buildups.

MARINE MICROINVERTEBRATES

The chief microinvertebrates are the Foraminifera. Of the Foraminifera, the *planktonic* species represent a mobile population that floats mostly near the water surface and serves as an index to purely oceanic water masses. The distance between the innermost body of purely oceanic water and the shoreline controls the distribution of the planktonic species in the sediments.

By contrast, the *benthic* Foraminifera are affected by conditions on the bottom. On most shelves, populations of Foraminifera in shoreface sediments are small and consist of large, thick-walled species. Planktonic species are rare to absent. The ratio of planktonic- and benthic Foraminifera changes with depth and distance from shore (Figure 10-14). This results from the relatively greater abundance of planktonic species as distance from shore increases. On the inner parts of most shelves (shoreface to depths of 50 to 70 m), the planktonic/benthic ratio may be less than 0.1. On the outer parts of shelves, this ratio may range from 0.1 to 1.0.

Among benthic Foraminifera, the kinds that construct tests by gluing together sand particles (*agglutinating foraminifers*) are commonest in the middle parts of some shelves. In some outer-shelf areas, the variety of benthic Foraminifera increases, whereas the proportion of benthic kinds and of agglutinated types decreases.

SOURCES: Blatt, Berry, and Brande, 1991; Gevirtz, Park, and Friedman, 1971; Poag, 1981; Upshaw, Creath, and Brooks, 1966.

MARINE MACROINVERTEBRATES ON LEVEL BOTTOMS

The marine macroinvertebrates living on level bottoms form two life-style assemblages: (1) the epifauna and (2) the infauna. Among these two broad subdivisions various organisms form groups that are associated with specific

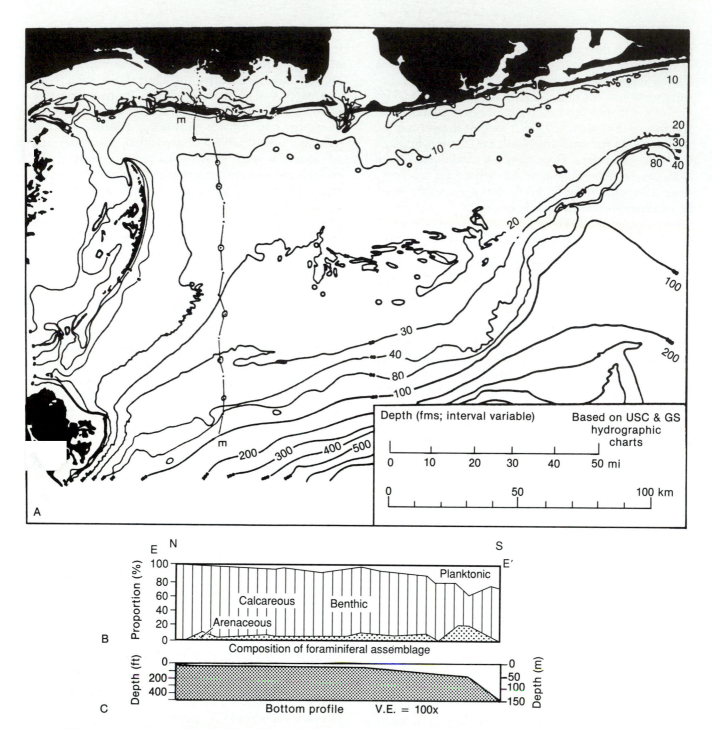

FIGURE 10-14.   Variations among Foraminifera with depth and distance from shore in traverse across shelf off Mississippi east of Mississippi delta. (C. F. Upshaw, W. B. Creath, and F. L. Brooks, 1966, fig. 2, p. 11; fig. 17, p. 36, and fig. 39, p. 67.)

kinds of bottom sediments, such as ooids, muddy sand, and mud. In addition, other organisms float near the surface or swim in the water (Figure 10-15). The skeletal remains of the organisms typically add calcium carbonate to the sediment.

In the sediments of nontropical shelves, mollusks dominate the epifauna. Toward the tropics, skeleton-

secreting algae and -corals increase. The term **biostrome** has been proposed for *sheetlike accumulations of skeletal debris, such as shell beds, crinoid beds, or coral beds*. Biostromes are not limited in composition to skeletal debris; planar stromatolites are also biostromes.

The infauna of a shallow (few tens of meters) sandy bottom includes a few genera of starfish, sea urchins,

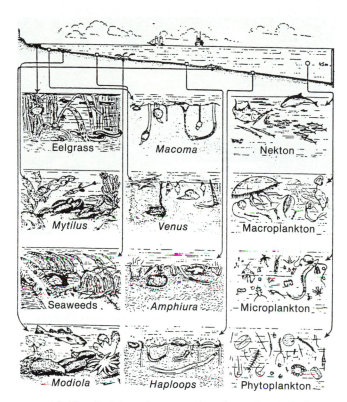

FIGURE 10-15. Sketches of communities of organisms affecting the bottom sediments in the Danish Sound from the coast of Sealand toward Kullen. (Gunnar Thorson and P. H. Winther, in J. W. Hedgpeth, 1957, fig. 1, p. 31.)

shrimps, pelecypods, gastropods, and pseudobranchs (worms). Representatives of these groups have colonized suitable bottoms in all climatic zones. However, species associations in different geographic zones are not identical. Instead, as one moves into a different geographic zone, one usually finds that species have substituted for one another.

Because the sediments of modern shelves preserve biofacies patterns that reflect depth and distance from shore, the rate of entry into the sediments of the skeletal remains must exceed the rate of transport of the sediments in a shelf-transverse direction.

SOURCES: Briggs, 1987; Ginsburg and N. P. James, 1974; Thorson, 1957.

ROCK-DESTROYING ORGANISMS

The rock-destroying organisms of the shallow tropical seas include species of algae, sponges, gastropods, pelecypods, chitons (segmented amphineurid mollusks) (Figure 10-16), and echinoids. These organisms remove rock from coastal cliffs. Such *bioerosion* leaves behind distinctive notches whose locations are closely controlled by sea level.

SOURCES: Acker and Risk, 1985; A. C. Neumann, 1966.

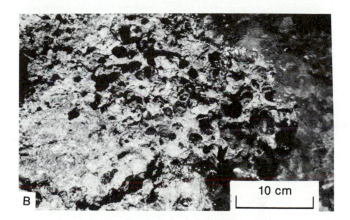

FIGURE 10-16. Irregular surfaces with notches, pits, and holes resulting from bioerosion of Pleistocene limestones.

A. Pitted surfaces, island in Mediterranean Sea off Nahariya, Israel. (Y. Nir.)

B. Close view at low tide of intertidal zone showing holes, each of which houses a chiton or a gastropod. Joulter's Cay, Bahamas. (Authors.)

C. A chiton removed from its rocky housing (center of photograph); view at low tide. Joulter's Cay, Bahamas. (Authors.)

BIOHERMS, REEFS, AND CARBONATE BUILDUPS

Many kinds of attached colonial invertebrates have developed a special mode of life peculiar to aquatic environments, especially to certain marine environments. Such invertebrates may create **bioherms**, *mound-shaped structures built by the in-situ growth of sedentary organisms*. Most bioherms are constructed by organisms that secrete mineralized skeletons of some kind. *Rigid, wave-resistant bioherms* are specially designated as **reefs**. This technical definition of a reef contrasts with the seaman's usage in which "reef" is applied to any solid object that forms a hazard to navigation (of course, reef-type bioherms, our reefs, are indeed hazards to navigation). (See Figure 7-32.)

We shall employ the term **non-reef bioherm** to designate *bioherms built by organisms that do not possess the ecologic potential to secrete a rigid wave-resistant structure*. Examples of non-reef bioherms are mound-like accumulations of phylloid algae, crinoids, ahermatypic corals, and oysters. (See Figure 13-3.) Some geologists employ the term "bank" for what we refer to as non-reef bioherms. We reject such a definition and shall use the term **bank** only in its oceanographic sense for *a large shoal, such as the Grand Banks or Georges Bank off northeastern North America*. In the modern oceans, many such shoals are underlain by carbonate sediments, for example, the Great Bahama Bank.

A useful noncommittal term for those who prefer not to use the words bioherm or reef is **carbonate buildup**, which designates *any carbonate structure that (1) differs in nature to some degree from laterally equivalent deposits and from surrounding- and overlying strata, (2) is typically thicker than laterally equivalent carbonate deposits, and (3) at some time(s) during its depositional history, probably stood topographically higher than the surrounding sediment*.

Among modern reef-building organisms, zooxanthellate corals predominate. Coralline red algae form a close second. These corals are confined to tropical waters. (See Figure 8-2.) Zooxanthellate reef-building corals survive only within a narrow temperature range from about 18 to about 30°C.

Reefs can generally be classified into two large groups: (1) elongate reefs and (2) isolated reefs (Figure 10-17). Elongate reefs include (1a) **fringing reefs**, which are *reefs contiguous to a landmass (which may be an island)*; and (1b) **barrier reefs**, which are *reefs trending parallel to shore, but separated from it by a lagoon*.

Isolated reefs generally are roughly circular. Two kinds are recognized: (2a) **pinnacle reefs**, which are *isolated, spikelike reef; towering several meters or more above the sea floor*; and (2b) **patch reefs**, which are *small isolated reefs*

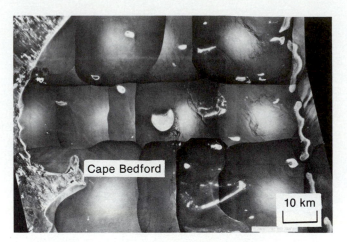

FIGURE 10-17.   View from air looking vertically down on northern part of the Great Barrier Reef tract, off Queensland coast, Australia, as seen in a mosaic of aerial photographs. Linear shelf-edge barrier reefs are separated by tidal channels (right). Open Pacific Ocean is at extreme right; Queensland coast at left. Light gray areas in lagoon are pinnacle reefs having diverse shapes. Figure 10-19 shows location of area photographed.

whose diameter measures only a few meters or tens of meters. Patch reefs are distinctive features on many continuous seismic-reflection profiles (Figure 10-18). *Clusters of corals and other skeleton-secreting sessile organisms form rigid structures that resemble patch reefs but that are considerably smaller than typical patch reefs (as small as a few coral heads and associated biota)*. These have been called **microatolls** or *microreefs*; they may represent the earliest stages in the origin of patch reefs.

In the modern world, the Great Barrier Reef, Australia, contains many examples of these various reefs.

Great Barrier Reef, Australia.   Along the outer edge of the continental shelf off the tropical Queensland coast of northeastern Australia, coral reefs stretch for a distance of nearly 2000 km, from lat.9°S to 24°S (Figure 10-19). The reef tract ends on the north where the Fly River of Papua-New Guinea discharges its turbid, silt-laden waters into the Gulf of Papua. To the south, the 18°C seawater isotherm limits the reef tract.

Despite the huge amounts of energy involved in the breaking of waves against the reef, the corals always grow most robustly toward the windward side. Such growth is stimulated by the surging waves and their oxygen- and nutrient-enriched waters. Flourishing reefs withstand the battering of all but the fiercest storms. However, major hurricanes can utterly destroy the living portions of reefs that have existed for thousands of years. Modern reefs include not only species that build massive colonies, as one might expect, but also delicate branching corals, which

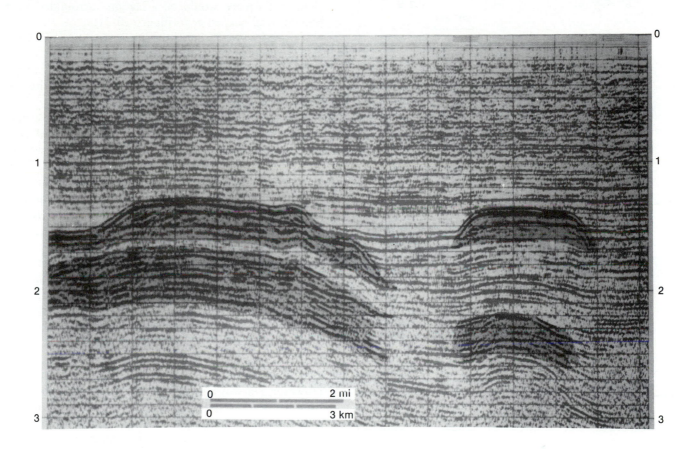

FIGURE 10-18.   Continuous seismic-reflection profile (multichannel seismic line) showing two different kinds of carbonate patch reefs. Within the younger reef at left a few reflecting interfaces dip steeply to the left. The reflecting interfaces beneath all four reefs are plane surfaces, but on the seismic record are convex upward. This results from the effect of the overlying reef within which seismic waves travel faster than outside the reef. This "instrumental" curvature of the reflections from a plane interface is known as "velocity (*sic*) pull up." (Shell Oil Company, 1987, fig. 20, p. 42.)

allow the force of the water to be dissipated through them.

The term "Great Barrier Reef" implies that a single continuous barrier reef intervenes between the open Pacific Ocean and a protected lagoon that stretches from reef to mainland. In reality, this is not the case. Rather, the reef tract consists of literally thousands of reefs that exhibit an enormous diversity of shapes and sizes. (See Figure 10-17.)

The reef province consists essentially of two parts, a northern and a southern. In the northern part, the reefs of the so-called outer barrier lie along the edge of the continental shelf and are separated from one another by channels through which tidal currents surge. Most of these shelf-edge reefs are small. Their lengths measure hundreds of meters to several kilometers. *Exceptionally long outer reefs* are known as **ribbon reefs**; the lengths

of these range up to 30 km. The tidal range is on the order of 3 m and more. Such tidal ranges create tidal currents having speeds of several knots (± 1 m/s) that flow through the channels between the reefs of the outer barrier. The shelf lagoon, which is about 40 to 50 m deep, is subject to vigorous wave action; it is floored by intrabasinal-skeletal- or coralgal debris mixed with extrabasinal sediment that decreases in abundance away from the coast. Because the northern outer reefs are inaccessible, and weather stormy for much of the year, few of these reefs have been studied intensively.

Numerous pinnacle reefs dot the floor of the shelf lagoon. Seaward of the shelf-edge reefs, the slope falls off precipitously to depths of 2000 m.

To the south, few shelf-edge reefs are present. Instead, pinnacle reefs are common there (Figure 10-20). Here,

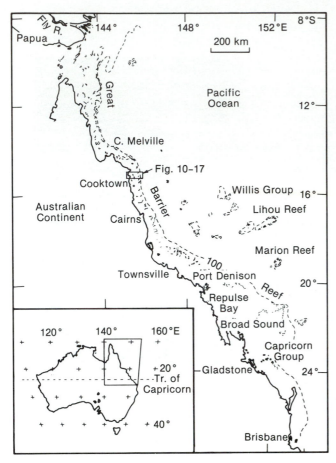

FIGURE 10-19.   Map of Great Barrier Reef Province showing distribution of reefs. (G. M. Friedman and J. E. Sanders, 1978, fig. 12-23, p. 368.)

storms deposit coarse reef rubble that builds above sea level around pinnacle reefs.

SOURCES: Belperio and Searle, 1988; Done, P. K. Dayton, A. E. Dayton, and Steger, 1991; P. G. Flood and Orme, 1988; Furnas, A. W. Mitchell, Gilmartin, and Revelante, 1990; J. F. Marshall, 1983; Orme, P. G. Flood, and Sargent, 1978; Orme, J. P. Webb, Kelland, and Sargent, 1978; Stoddart, 1978.

Hypotheses of Origin of Coral Atolls.   Since 1835, when Charles Darwin published his hypothesis of subsidence, the origin of modern tropical coral reefs and especially atolls has been much discussed. Darwin began with the premise that coral reefs flourish only in shallow, tropical seas. He postulated that under conditions of gradual submergence, the corals fringing an island or a mainland shore could grow upward as fast as the bottom subsided. Thus, according to Darwin, an originally conical volcanic island encircled by a reef and being submerged, for example, would become progressively smaller, and the lagoon between reef and island would progressively widen.

Ultimately, the island might altogether vanish from view. In its place would be a lagoon surrounded by a ring- or oval-shaped *atoll* (Figure 10-21).

If the initial reef fringed the linear shoreline of a large land mass, then submergence would create an elongate barrier reef. The barrier reef would trend parallel to the shoreline and an elongate lagoon would separate the reef from shore. (See Figure 10-21, B.)

Although since 1835 various aspects of the origin of reefs have been hotly debated, the factor of subsidence has been confirmed. Drilling through various mid-Pacific atolls has shown that reefs and associated carbonate sediments extend continuously from the surface to underlying basalt (Figure 10-22). Borings have proved that the thickness of reef- and reef-associated materials at Midway Atoll is 330 m; at Mururoa Atoll, 440 m; and in boreholes at Enewetak Atoll, 1267 and 1405 m.

However, the drilling and other studies made since Darwin's time have not disclosed any examples of an atoll built around the shores of an initial conical island. Thus Darwin's idea that the three kinds of reefs–fringing reefs, barrier reefs, and atolls—form a genetic sequence has not been supported directly. He envisaged that, where a conical island sank and the reefs grew upward, these three kinds of reefs would form a progressive sequence. However, Darwin also suggested that atolls could grow from a subsiding flat surface and thus not pass through the stages of fringing reef and barrier reef. All the atolls studied in detail have grown on circular platforms that were initially flat (or gently inclined), not conical. However, fossil reefs have been discovered in the subsurface forming at least partial rings around ancient buried volcanoes, providing circumstantial support for the conical-island hypothesis of atoll formation.

In the debate on the origin of reefs, the chief competition to Darwin's hypothesis of the drowning of conical islands was Daly's glacial-control concept. Daly argued that emergence accompanying glaciation enabled waves to plane off an exposed fringing reef. As sea level subsequently rose across a partially planed reef, a barrier reef could be formed. If the emerged reef had been completely planed, an atoll would result. Like Darwin's subsidence, so Daly's rise- and fall of sea level have now been confirmed as important factors in the evolution of modern reefs. But the confirmation has not been as Daly envisioned it. The episodes of planation should have produced cliffed headlands behind barrier reefs. The absence of such cliffed headlands is a strong point in the argument against the idea that in the origin of modern reefs, planation of fringing reefs was an important process.

SOURCES: Cullis, 1904; Goter and Friedman, 1987; Ladd, Tracey, and Gross, 1970; Schlanger, 1963; Thom, Orme, and Polach, 1978.

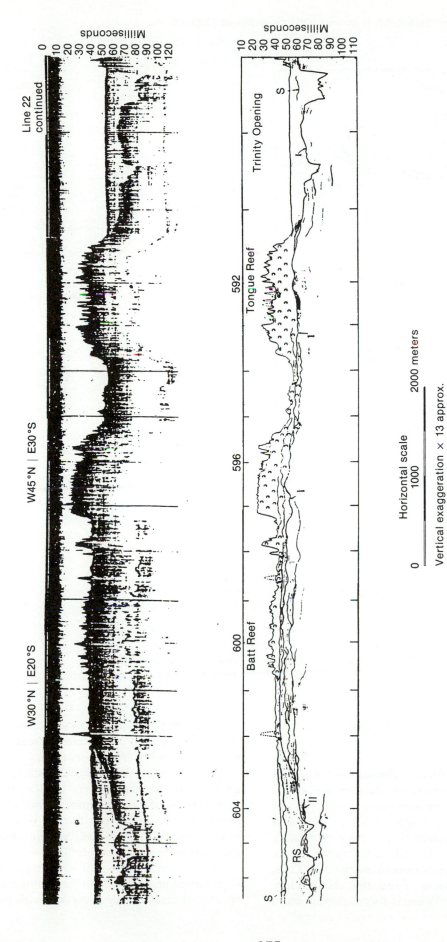

FIGURE 10-20. High-resolution continuous seismic-reflection profile along channel between two pinnacle reefs, Tongue Reef on the NE and Batt Reef on the SW, in the southern Great Barrier Reef lagoon, about 50 km N of Cairns, Queensland, Australia. (G. R. Orme, J. P. Webb, Kelland, and Sargent, 1978, fig. 2.)

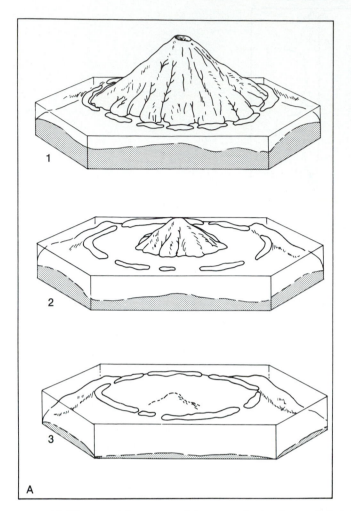

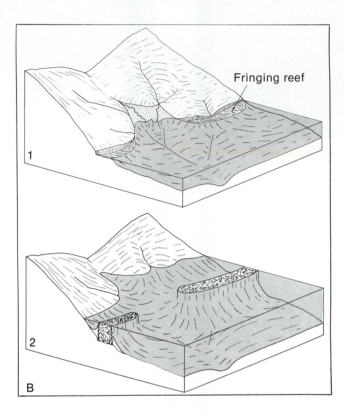

FIGURE 10-21.   Darwin's concept of the stages in development of atolls and barrier reefs.
    A. Atoll formed by submergence of conical volcanic island. (1) **Fringing reef**, *a reef attached to the shore of a landmass*, grows around perimeter of volcanic cone. (2) As cone subsides, reef grows upward, forming a barrier reef. (3) Eventually the cone becomes completely submerged; by growing upward, the reef has created an atoll.
    B. Reefs near mainland shore. (1) Fringing reef grows on flat surface cut by waves. (2) After submergence and vertical growth by the reef, the fringing reef has become a barrier reef. (A, after F. P. Shepard, 1963, fig. 175, p. 365, and C. R. Longwell, R. F. Flint, and J. E. Sanders, 1969, fig. 22-20, A, p. 565; B, after A. N. Strahler and A. H. Strahler, 1973, figs. 15.50 and 15.51, p. 409.)

Effects of Quaternary Oscillations of Climate and Sea Level

During the Quaternary Period, climate and sea level oscillated enormously. Graphs of temperature versus time clearly indicate that these changes have been both periodic and asymmetrical (Figure 10-23). To a first approximation we are certain that during cold times (Earth's ice-house modes), the water to form the greatly expanded continental glaciers came from the sea. Thus, glaciers grew at the expense of seawater. Accordingly, world sea level dropped. What is more, during warm times (Earth's greenhouse modes), continental glaciers shrank, the volume of seawater increased, and a worldwide high stand of the sea resulted.

After either of these two climatic modes has been established, the fluctuations of sea level closely reflect changes of world climate. However, as mentioned in Chapter 1, during the Earth's boiler-room climate modes, climate and sea level are out of phase. Thus at the times of highest temperatures, sea level was lowest.

The basic tenet of seismic stratigraphy is that the geologic record made by a high stand of the sea differs from the record made during a low stand. Moreover, from the

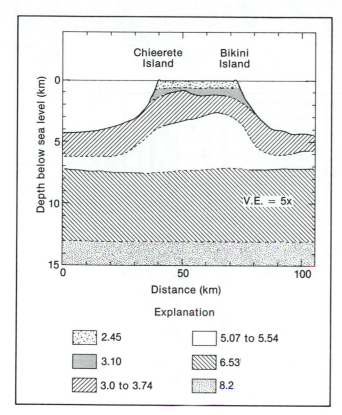

FIGURE 10-22. Geophysical profile across Bikini Atoll in the Pacific Ocean showing that reefs and associated carbonate sediments extend continuously from the surface to the underlying basalt. The numbers give the celerity of sound in kilometers per second, as follows: 2.45, reef-framework builders and unconsolidated- to partly consolidated carbonate sediment; 3.10, reef rock and lithified carbonate sediment; 3.0–3.74, vesicular- or fractured basalt; 5.07–5.54, basalt; 6.53, olivine basalt; 8.2 is an average celerity found below the Mohorovicic discontinuity (top of Earth's mantle). (Modified from R. W. Raitt, 1954, fig. 157, p. 522.)

continuous seismic-reflection-profile records one should be able to interpret the geologic history in light of these differences.

Before our very eyes today is a high-standing sea. What is it doing? How does what it is doing now differ from what happened during the recent low stand? In many places, the shelf seas are not leaving much of a record. Instead, the sea has simply submerged **relict sediments**, defined as *sediments deposited in an environment that existed during a low stand of the sea (or at some level lower than that of a high stand) and have not been covered by any products of the high-stand depositional environments.* In parts of the continental shelf, *relict sediments that are being reworked by physical- and/or biological processes at a high stand of the sea* are designated **palimpsest sediments.**

Although in parts of the shelf, the high-standing sea is not doing much depositing, a stratigraphic record involv-

ing shelf sediments does accumulate where a delta, for example, progrades across the shelf. In the process, the delta buries shelf sediments. Thus, a distinctive parasequence is formed: shelf strata below, delta strata above. Similarly, off desert coasts, wind blows sand into the water and progrades the shoreline seaward. Here, shelf sediments are capped by dune sands.

The effects of emergence on carbonate shelves are especially dramatic in reefs but also bring about wholesale changes in the metastable carbonate minerals. On carbonate shelves, the kinds of mesosequence boundaries (Chapter 6) can be determined readily.

During Pleistocene episodes of emergence, reefs were exposed subaerially and subjected to dissolution by slightly acid rain waters. (See Figure 3-5.) Drowned dissolutional features, such as sinkholes (known as blueholes after they have been submerged), have been recorded from most reefs and range in diameter up to 1/2 km and in depth down to 120 m. Using scuba gear in the Bahamas, one explorer has studied 54 such blueholes. (This is an extremely dangerous enterprise because of the complexity and tortuosity of bluehole cave systems.) Surfaces of unconformity resulting from the burial by marine carbonate sediments of older marine carbonates, now lithified and containing many dissolution cavities as a result of the action of rain water, have been recorded in numerous exposures and in boreholes of many localities, including the Florida Keys (Figure 10-24, A), the Great Barrier Reef Province, the central Pacific, and the Caribbean.

Coral reefs that were situated deeper than about 30 to 40 m below sea level at the beginning of the Holocene submergence were drowned when the rate of sea-level rise reached about 20 mm/yr.

At Mururoa Atoll in the Pacific, a major surface of unconformity is present 6 to 10 m below the surface of the reef. Above the surface of unconformity the ages of the reef range from 5400 to 10,000 yr B.P. Below the surface of unconformity, the minimum age is 80,000 yr in one hole and 120,000 yr in two others. Like other reefs, this atoll was emergent for nearly 100,000 yr and has been drowned only for the past several thousand years. Hence we postulate that modern reef-building organisms have reoccupied the weathered- and partially dissolved tops of ancient, formerly emerged reefs. Therefore the reef corals of Holocene age form but a thin veneer over an older karst substrate underlain by Pleistocene limestone, which may or may not be entirely of reef origin.

The two kinds of mesosequence boundaries are recognized on the basis of the extent to which the shelf emerged during a low stand. A wholly emergent shelf is the basis for a type-1 mesosequence boundary; a partially emergent shelf, for a type-2 mesosequence boundary.

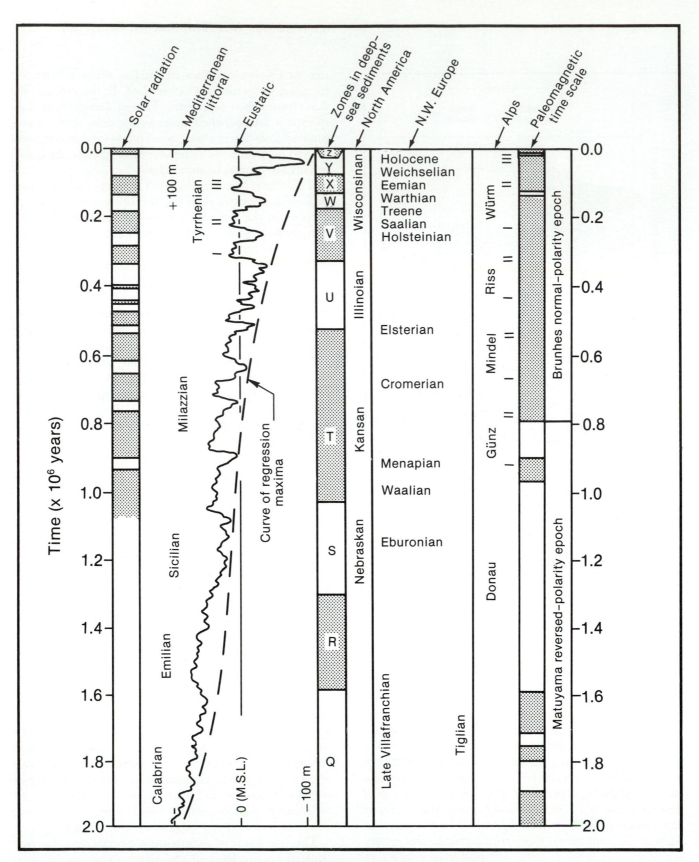

FIGURE 10-23. Relationship between oscillations of climate and sea level during the Quaternary Period. These graphs correlate the chronology of the Quaternary Period with changes in sea level, glacial- and interglacial cycles, paleomagnetic reversals, the Milankovitch cycles of solar radiation, and absolute ages determined by radiometric dating. The glacial ages for North America, northwestern Europe, the European Alps, and the Mediterranean coast are correlated with the zones of cold-warm cycles determined in cores of deep-sea sediments. (After R. W. Fairbridge, 1972, fig. 1, p. 104.)

FIGURE 10-24.   Effects of dissolution from subaerial exposure of reefs and related carbonate sediments.

A. Solution cavity in porous backreef facies of Key Largo Limestone (Pleistocene) infilled by heterogeneous debris. Windley Key, Florida. (W. Van Wie.)

B. Slabbed face of core of skeletal limestone from 70-m depth in boring on Midway Atoll, Hawaiian Islands. The upper, coarse skeletal limestone consists mostly of aragonite and the lower, fine skeletal limestone is composed mostly of calcite. Sharp contact marks surface of unconformity caused by dissolution. Large cavity in lower limestone is partly filled with micrite. (H. S. Ladd, J. I. Tracey, and M. G. Gross, 1967, fig. 18, p. A17; courtesy H. S. Ladd and J. I. Tracey.)

Because of the ease with which fresh water affects carbonate sediment, the evidence for identifying these mesosequence boundaries on a carbonate shelf is easy to see. When the whole carbonate shelf emerged, the effects of dissolution of metastable carbonate minerals extend all the way to the shelf margin.

The substrate of modern reefs in an atoll may be a dissolution rim around a central dissolution basin. For a barrier reef it may be a dissolution rim marginal to a karstic plain (Figure 10-25). The new reef overlies a type-1 mesosequence boundary.

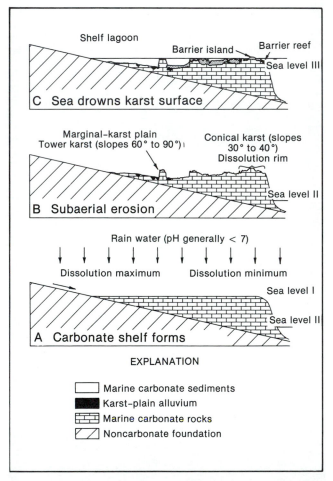

FIGURE 10-25.   Sequential stages in the development of an atoll as a result of changes of sea level and formation of karst landscape by dissolution.

A. After carbonate sediments have built a basin-marginal terrace whose top surface forms a carbonate shelf, sea level drops and the carbonates become subaerially exposed. Infiltrating fresh water leaches carbonates.

B. Differential dissolution results in conical karst surface featuring raised rims at the margin of the carbonate shelf. The result is a type-1 mesosequence boundary.

C. Rising sea drowns conical karst on carbonate shelf; differential deposition accentuates relief. (Modified from E. G. Purdy, 1974b, fig. 43, p. 71.)

If only part of the shelf emerged, then the distinctive dissolution effects will be limited to the parts that did emerge, presumably the proximal regions. Under such conditions, a type-2 mesosequence boundary forms.

SOURCES: Droxler, Morse, and Kornicker, 1988; Gradstein, Ludden, and others, 1990; Grigg and Epp, 1989.

In modern sediments along the Atlantic coast of the eastern United States, the present-day boundary between extrabasinal- and intrabasinal sediments is located south of Miami, Florida (lat.24°N). Although the location of such a boundary is subject to several factors, such as terrigenous influx, nevertheless, the climate exerts a major influence. During at least one boiler-room climatic episode during the Quaternary Period, this boundary shifted northward by several hundred kilometers. Submerged carbonate deposits (including ooids, beachrock, and coralline algae that have been dated by radiocarbon at 10,000 to 14,000 yr B.P.) have been found from Cape Hatteras, North Carolina (lat.35°N), southward and in water depths ranging from 50 to 110 m. Beachrocks are intertidal, ooids form at or near the intertidal level, and coralline algae thrive in very shallow warm water through which light must penetrate. Hence, 10,000 to 14,000 yr B.P., when these sediments were being deposited, the temperatures were higher than at present, but the sea level was much lower.

Clearly, these shelf-edge carbonate deposits are relict and antedate the Holocene submergence. From its original position near the edge of the shelf to its present level, the sea has risen 50 to 110 m. In so doing, it submerged these warm-climate intertidal-, near-intertidal-, and shallow-water carbonate deposits, but after submerging them, has not covered them with Holocene sediments. Between the time of the boiler-room-mode low stand of the sea/very warm climate and the present greenhouse-mode high stand, the mean temperature has decreased. The result is that carbonates on the seaward side of a shelf pass laterally into terrigenous sands near the present coast.

A similar arrangement exists along the shelf off western Florida, in the northeastern Gulf of Mexico. A belt of relict nearshore carbonate sediments has not been covered by the sea that is now depositing quartzose sediments off the modern coast.

In many parts of the Caribbean region, shelf-edge reefs are present where no reefs grow along the modern shores. Because the rate of upward growth by corals exceeds that of submergence during the Holocene transgression (Figure 10-26), simple drowning by rapid sea-level rise is not an acceptable explanation. The probable explanation involves changes in the water. Most investi-

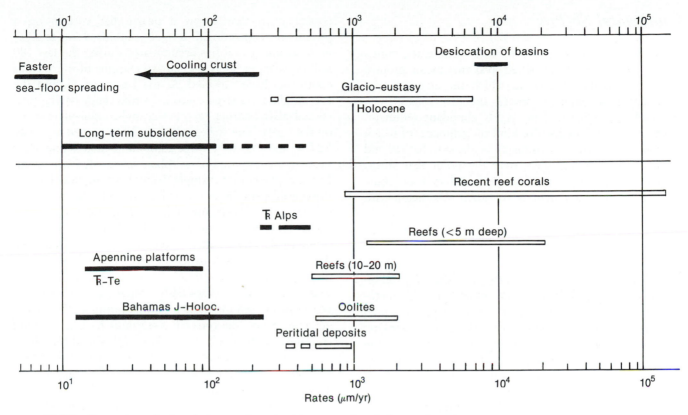

FIGURE 10-26.   Comparison of rates of change of sea level (upper part) and of upward coral growth and nearshore-sediment deposition (lower part). Open lines, Holocene; thick black lines, rates from geologic record. Notice that despite their overlapping ranges, the maximum known rate of coral growth is an order of magnitude faster than the fastest-known rate of Holocene submergence. (W. Schlager, 1981, fig. 5, p. 201.)

gators who have discussed this topic have not been aware of the climatic change leading up to and back again from a peak of warmth in a boiler-room episode that we think coincided with low stands of glacial-age sea level. Who knows what temperature changes accompanied the rapid rise of sea level? A change of temperature that could make the water unfit for corals (by either cooling or warming) eliminates the otherwise-mysterious paradox of how shelf-marginal reefs can be submerged.

SOURCES: Ginsburg and James, 1974; Lighty, MacIntyre, and Stuckenrath, 1978; MacIntyre, 1972; Sanders and Friedman, 1969; Schlager, 1981.

Tectonic Activities

Tectonic activities establish the background within which shelf sediments accumulate. Tectonic activities control the rate and extent of subsidence (See Figure 10-26.) and may create various gentle folds whose axes trend in many directions, from perpendicular to parallel with the strike of the shelf, or faults. Tectonic elevation of the land may boost the sediment yields and convert an autochthonous shelf to an allochthonous shelf. Finally, a fundamental change from a passive margin to a convergent margin may destroy the shelf and convert it into a foreland basin as part of an orogeny (Chapter 17). We elaborate further here only on the topic of rate of subsidence as compared with other rates.

The bar in the upper left of Figure 10-26 shows the rate of subsidence of a passive margin, which may be taken as typical for most shelf areas. Presumably, the maximum rate applies to the edge of the shelf and becomes progressively less toward the land, eventually reaching zero. As the shelf is tilted downward toward the deep-sea basin, it interacts with the sea, which is capable of moving up or down at rates that can be one or two orders of magnitude faster than the rate of subsidence. When a shelf is submerged, room for new sediments to accumulate, known as *accommodation space*, appears.

Continental Shelves

As mentioned, on the basis of their sediments, continental shelves can be classified into two major groups: (1) autochthonous shelves, in which the adjacent land yields negligible sediment; and (2) allochthonous shelves, in which the adjacent land yields abundant sediment. Further subdivisions involve kind of sediment, of which we concentrate on the two major kinds, extrabasinal and intrabasinal. We shall gloss over the point that volcanoes can be present on any shelf and that on some shelves, volcanic sediment is abundant. We begin with autochthonous shelves.

Autochthonous Shelves

The characteristics of an autochthonous shelf depend on (1) the kind of morphologic features and geologic materials that have been submerged, (2) the depth to which they have been submerged, (3) the kinds of reworking of materials that took place during submergence, (4) the ways in which the sea at its present level is reworking the bottom sediments, and (5) the effects of climate (which influences many of the foregoing factors).

Two major categories can be recognized: autochthonous shelves composed of extrabasinal sediment and autochthonous shelves composed of intrabasinal sediment. As a first approximation, these two categories reflect climate. The extrabasinal sediment lies mostly outside the tropics and the intrabasinal sediment, within the tropics.

AUTOCHTHONOUS SHELVES COMPOSED OF EXTRABASINAL SEDIMENTS

Two large-scale varieties of extrabasinal sediments forming autochthonous shelves are glacial sediments and coastal-plain sediments.

Submerged Glacial Sediments. In many parts of the Northern Hemisphere, Pleistocene continental glaciers flowed across the shelf areas. (See Box 6.1 Figure 1.) Accordingly, the Holocene transgression has drowned glacial sediments. As the sea submerged glacial sediments containing boulders, it winnowed away the associated fines and left bouldery and gravelly shoals on the sea floor. Low spots on the irregular sea floor are tending to be filled with mud.

SOURCES: C. W. Ellis, 1962; J. J. Fisher and J. R. Jones, 1982; C. A. Kaye, 1967.

Submerged Coastal-Plain Sediments. Around the margins of the North Atlantic Ocean, the shelf is underlain by a coastal plain that forms a nearly flat surface on a prominent depositional terrace composed of terrigenous sediment that has accumulated during the last 180 million years or so as the passive margins of the various continents have subsided. At the landward edge of the coastal plain is a stream-eroded strike valley having pre-coastal-plain bedrock on one side and coastal-plain strata on the other. The strike valley has been named the *inner lowland* and the strike ridge adjoining it that is underlain by seaward-dipping coastal-plain strata, an *inner cuesta*. In many places (for example, from Texas to New Jersey), the coastal-plain inner cuesta lies on land. Elsewhere (for example, from New York City to Labrador), this inner cuesta and its adjacent strike-valley lowland have been submerged. The result is a deep inner- or central-shelf depression bordered on its seaward side by a shallower marginal bank.

In other areas, especially where the underlying coastal-plain sediments are sandy, during each glacial-age emergence, the retreating sea must have left behind countless sandy beach ridges. Any of this sand that dried out would have been blown by the wind, which doubtless fashioned the sand into fields of dunes. During the still-stands at maximum emergence, new beaches and barriers were established at or close to the edge of the shelf.

During each emergence, major rivers were rejuvenated; they extended and deepened their valleys. They discharged their loads of fluvial sediments directly into the deep-sea basins. Although at the present high stand of the sea, many modern rivers enter the ocean in the coastal-plain region, few of them have built deltas. Instead, these rivers discharge into estuaries, which, until they have been filled completely, tend to serve as sediment traps.

As we have seen in a previous section of this chapter, one of the chief effects of shoaling waves is to drive sand-size- and coarser sediment toward shore and then to shift it parallel to shore. Such sand moves parallel to shore until it comes to some stopping place, or "sink." One kind of "sink" is a low-lying axis of a drowned shelf-transverse valley, now the mouth of an estuary or a bay. Another kind of "sink" is a locality, not necessarily a low-lying area, where two belts of longshore-drifted sediments moving in opposite directions converge.

During the Holocene transgression, such sand "sinks" remained active as the shoreline shifted across the shelf. Accordingly, great bodies of well-sorted fine sand named *shoal-retreat massifs*, have been deposited (Figure 10-27). In effect, then, the sand forming a shoal-retreat massif was prograded landward up the slope of the shelf-transverse valley or of the shelf itself.

Because most beaches are retreating, the sea must be eroding sediment from the land. But the place(s) of deposition of such sediment eroded from beaches are not well

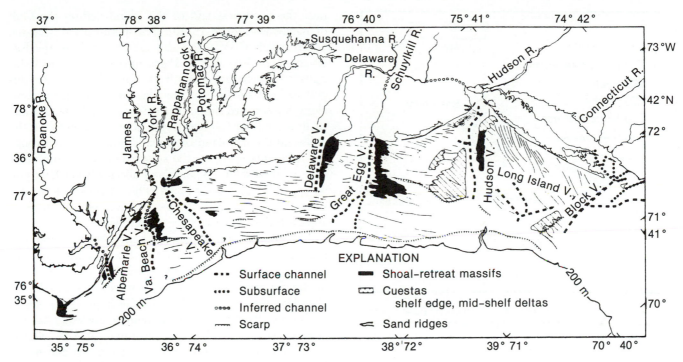

FIGURE 10-27. Map of shelf off eastern United States from Cape Hatteras to the eastern tip of Long Island, New York, showing distribution of shoal-retreat massifs, shelf-transverse channels, and sand ridges. (D. J. P. Swift, 1974, fig. 3, p. 120.)

known. Even where no delta-building rivers are present, fine sediment, eroded either from the shoreface or from the shelf bottom, is accumulating not on most parts of the shelf, but in deeper closed depressions in the shelf surface or in deep water seaward of the shelf break.

Near large rivers, the typical terrigenous shelf sediment is silty clay. Here the concentrations and supply of fine particles are such that, despite the tendencies to remove them, they are being deposited.

Elsewhere, the commonest terrigenous shelf sediment is quartz sand. Nearly all of this sand is being reworked by the high-standing sea, and some of it clearly is being dispersed by tidal currents and by other currents in the bottom-boundary layer. How the sand should be apportioned among sand that belongs totally in the modern high-stand cycle, palimpsest sand being reworked by the high-standing sea, and relict sand is being actively investigated.

SOURCES: M. E. Field, 1980; Swift, 1974, 1975a; Swift, Thorne, and Oertel, 1986.

AUTOCHTHONOUS SHELVES COMPOSED OF INTRABASINAL SEDIMENT (CARBONATE SHELVES)

As we noted at the beginning of this chapter, autochthonous shelves composed of intrabasinal sediment

are known as carbonate shelves. Three categories are important: (1) carbonate ramps, (2) rimmed shelves, and (3) carbonate platforms.

Carbonate Ramps. On the modern continental shelves, as one approaches the Equator and its warm waters, carbonate sediments become increasingly abundant. Near the Equator, modern carbonate sediments accumulate more rapidly than in high latitudes.

Despite this peak of carbonate abundance near the Equator, shelves away from the Equator, even those underlying cold water, may include carbonate sediment. The abundance of these nonequatorial carbonates is greatest where terrigenous sediment is not present.

Intrabasinal carbonate-shelf sediments consist of the skeletal debris of planktonic- and benthic Foraminifera (planktonic species are relatively more abundant seaward); of mollusks; and, near knolls, of fragments of algae, corals, and Foraminifera that are now living (or formerly lived at lower sea level).

Modern shelf carbonate sediments are accumulating on ramps on the Campeche Bank off the Yucatan Peninsula, Mexico; off the west coast of Florida; and on the Sahul shelf off northwestern Australia.

Campeche Bank, a shallow carbonate shelf lying seaward of the Yucatan Peninsula, Mexico (Figure 10-28), is an example of a modern carbonate ramp. Here, along the shoreface and beaches, where waves are actively

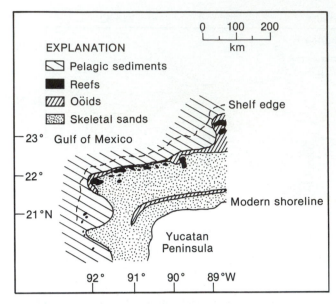

FIGURE 10-28.   Map of sediment facies, Campeche Bank, a shallow carbonate ramp lying off the Yucatan Peninsula, Mexico. (After B. W. Logan, J. L. Harding, W. M. Ahr, J. D. Williams, and R. G. Snead, 1969, fig. 37, p. 121.)

shoaling and breaking and currents are swift, the sediments are of sand size, are well sorted, and consist of ooids or skeletal debris and pellets. Seaward, where the waters are quiet, the sediment consists of pellet-lime mud containing planktonic Foraminifera.

Continuous seismic-reflection profiles across the margins of carbonate ramps show only low-dipping reflectors (Figure 10-29).

Examples of ancient limestones deposited on open carbonate shelves include those that in a landward direction pass from fine-textured foraminiferal muds or chalks through various skeletal-pellet muds and into oolitic- and/or skeletal calcarenites. Either such shelves were ramplike, or the depth of water at the shelf break was too great for reef growth. A carbonate shelf can be submerged and reefs not grow if a change in climate or rate of submergence (or both) create conditions that exceed the narrow tolerance limits of hermatypic corals and other reef-building organisms. Carbonate sediments deposited on an open shelf may pass shoreward into noncarbonate sediments, into carbonate sands (usually including ooids), or into reefs.

Rimmed Shelves: Gulf of Batabano, Cuba. Rimmed shelves are characterized by shelf-edge reefs that face open water and adjoin a shallow lagoon of variable width. In an earlier section of this chapter we mentioned the Great Barrier Reef, a particularly large example of a shelf-edge reef. On a smaller scale is the rimmed shelf south of Cuba. The curvature of the south shore of Cuba

near the northwest end of the island defines the Gulf of Batabano (Figure 10-30). The Peninsula de Zapata, which trends E-W, forms the southern side of the Ensenada de la Broa (Bay of Pigs), into which the Rio Hatiguanico discharges much fresh water. We consider this gulf to be a small-scale shelf lagoon; reefs are present along the seaward side of the eastern shelf edge, and the shallow sea floor (average depth about 7 m) between the reefs and the mainland of Cuba is underlain by carbonate sediments. Also, on the seaward side near the outer edge of the "shelf," is an island about 60 km long and 60 km wide, the Isla de Piños. This island provides considerable shelter from open-sea waves whose angle of approach lies between ESE and WSW.

The tropical climate, general bathymetric setting, and coastal configuration are such that one might expect to find hypersaline water at the north end of the Gulf or in the Ensenada de la Broa. However, because of the abundant rainfall on Cuba and the large discharge of fresh water from the Rio Hatiguanico, the water salinity *decreases* northward from a maximum of 37‰ at the south end to a minimum of 28‰ at the east end of the Ensenada de la Broa. Most of the Cuban shoreline of the gulf is fringed with mangrove swamps. In the southern part of the gulf the tidal amplitude is 60 to 75 cm, but it decreases to only 9 cm in the north. Despite the relief of the island of Cuba and Isla de Piños, carbonate sediments predominate. (See Figure 10-30, B.) A little quartz has been eroded from the Isla de Piños, but most of the sediments consist either of intrabasinal carbonate sediments (skeletal sands, ooids, pellets, and lime mud) or of extrabasinal particles of lithified limestone (weathered from the Upper Tertiary bedrock exposed along the coast).

The chief skeleton-secreting organisms include corals, various algae (*Halimeda, Penicillus, Acetabularia*), pelecypods, gastropods, foraminifers, echinoids, bryozoans, and ostracodes. The marine grass *Thalassia* carpets most of the northern- and central parts of the Gulf.

Skeletal sands, dominated by fragments of corals and algae (named *coralgal facies* because the dominant constituents are degradation products of corals and individual skeletons or small colonies of coralline algae), form a well-sorted lag concentrate along the southern margin of the gulf, both near the reefs (an area not mapped in detail; because of the political situation, the geologists were not able to finish their study) and along the southwestern edge. In the turbulent waters near the reefs, oolitic sands form local pockets. The most-widespread sediment in the gulf is a skeletal-pellet lime mud, in which the chief particles other than pellets are mollusk fragments.

Carbonate Platforms: The Bahama Banks.   Today, the only carbonate platform of any appreciable size is the Ba-

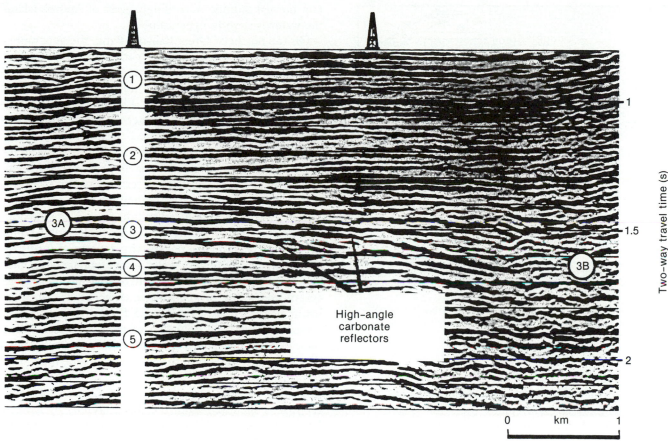

FIGURE 10-29.   Continuous seismic-reflection profile record from multichannel seismic survey off N. Africa (precise location proprietary) in Miocene-age strata showing prominent oblique traces from seaward-dipping (to R) margins of a carbonate shelf. The general concordance of the dipping traces with the others indicates that the shelf edge functioned in a "keep-up" mode. (1) Sandstone; (2) shale with limestone layers; (3) limestone; (3A) shelf deposits; (3B) pelagic shale and limestone; (4) shale; and (5) intercalated shale and limestone. (J. M. Fontaine, R. Cussey, J. Lacaze, R. Lanaud, and L. Yapaudjian, 1987, fig. 9, p. 288.)

hamas. We shall describe the Bahamas in detail because for interpreting carbonate sediments, this area has become a classic.

The Bahamas are an extensive series of low-lying islands and surrounding areas of shallow water lying east and southeast of the Florida peninsula, which form the irregular top surface of the Bahamas Platform (Figure 10-31). The Bahamas Platform lies within the belt of northeast Trade Winds and thus is bathed by warm ocean currents. Islands of Pleistocene bedrock, known as *cays* (pronounced "keys"), are exposed along the edges of the platform.

A thin veneer of modern carbonate sediment, generally < 3.5 m thick, overlies a thick (±6000 m) pile of carbonate bedrock ranging in age from Pleistocene to Cretaceous. This veneer of modern carbonate sediment, our concern here, has formed during the last 4000 years of the Holocene submergence.

The Bahamas Platform consists of two large banks, Little Bahama Bank and Great Bahama Bank. (See Figure 10-31.) The northern part of the Great Bahama Bank (Figure 10-32) has been studied in most detail, and hence our discussion focuses on it. The Great Bahama Bank lies between the Tongue of the Ocean, a deep (1300 to 2500 m) and wide (25 to 50 km) channel, and the Strait of Florida, a much-shallower (< 850 m) but wider (> 100 km) channel. The width of the northern Great Bahama Bank varies between 100 and 200 km; the effects of astronomic tidal oscillations are felt all across the bank.

More than 80% of the Great Bahama Bank is covered by water < 10 m deep; throughout more than half of the bank, water depth is < 6 m. As a result of tidal fluctuations (approximate range 1 m), water moves rapidly across the margins of the Bank. Andros Island (See Figure 10-31 and Box 12.3 Figure 1.) acts as a barrier to the water motions created by the northeast Trade Winds,

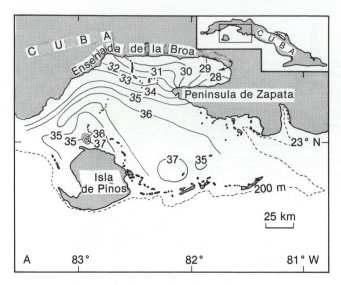

FIGURE 10-30.   Environments and sediments, Gulf of Batabano, Cuba.

A. Map showing bottom configuration and salinity (‰) of bottom water for August 1958. (After C. W. Hoskins, 1964, figs. 1, p. 1681, and 7, p. 1686.)

tain normal salinity. Accordingly, west of Andros Island the water is slightly hypersaline.

On the northern Great Bahama Bank, the geographic pattern of the five facies of carbonate sediments is related to wave- and current energy and to salinity. These facies are (1) pellet-lime mud, (2) grapestone, (3) ooid, (4) skeletal, and (5) reef (Figure 10-33; see also Figure 10-32). In detail, the pattern is much more complex than indicated in the figures. Although these facies are based on physical characteristics, the biological aspect of the sediment substrate is also distinct. Both energy level and biological characteristics of the substrate control the kinds of organic communities that develop.

West of Andros Island, pellet-lime mud floors the bottom of the Great Bahama Bank. Before deposit-feeding organisms make it into pellets, this lime mud forms by the accumulation of tiny skeletal components (aragonite needles) secreted within the cells of organisms such as certain codiacean- or green algae, especially the shaving-brush alga *Penicillus*. (See Box 2.5 Figures 1 and 2.) The *Penicillus* plant lives only two to three months. When the organism dies and its filaments disintegrate, its needles are added rapidly to the sediment as lime mud. The concentration of lime mud is a function of (1) rapid production of aragonite needles, and (2) the quiet water that is sheltered from the Trade Winds by Andros Island. Other important sources of lime mud are the

protecting the part of the Great Bahama Bank lying to the west (leeward side of the island) from the open ocean. Here, as evaporation takes place, the water is not sufficiently replenished by influx of marine waters to main-

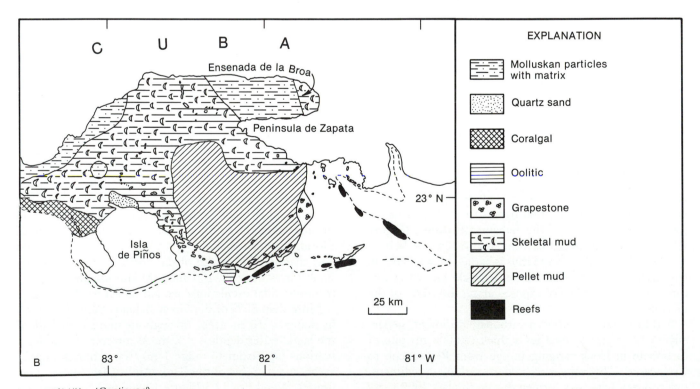

FIGURE 10-30.   (*Continued*)

B. Map of sediment facies. (After G. M. Friedman, 1964, fig. 49, p. 805, based on C. C. Daetwyler and A. L. Kidwell, 1959, and unpublished suggestions by E. G. Purdy.)

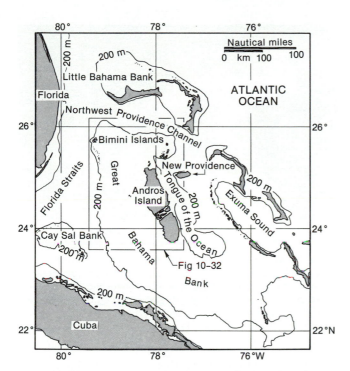

FIGURE 10-31. Index map of Great Bahama Bank and contiguous area. (G. M. Friedman and J. E. Sanders, 1978, fig. 12-25, p. 371.)

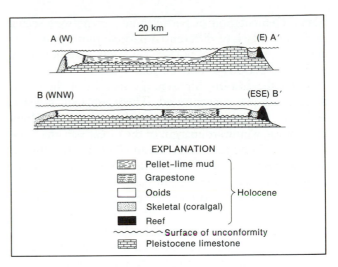

FIGURE 10-33. Idealized profiles and sections through Great Bahama Bank showing sediment facies and their relationships to water motion. Vertical scale much exaggerated; relationship to underlying Pleistocene limestone schematic; horizontal scale as in Figure 10-32. A-A', Pellet-lime mud floors protected area on leeward side of Andros Island. No grapestone facies. B-B', Grapestone floors interior of open parts of Bank; no pellet-lime mud facies. Further explanation in text. (G. M. Friedman and J. E. Sanders, 1978, fig. 12-27, p. 373.)

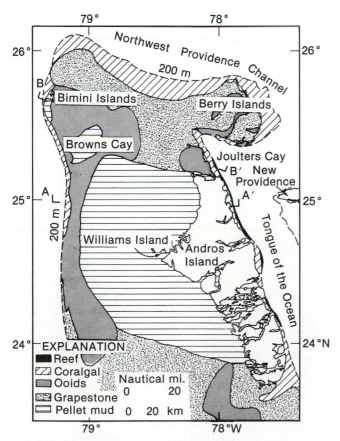

FIGURE 10-32. Map of sediment facies, northern part of Great Bahama Bank. (Modified from N. D. Newell, J. Imbrie, E. G. Purdy, and D. T. Thurber, 1959, fig. 8, p. 199, and E. G. Purdy, 1963b, fig. 1, p. 473.)

tiny *epibionts*, animals and plants that live on other living things, especially the ubiquitous seagrass *Thalassia*. It should not be supposed that all important biological sources of lime mud have been identified; however, clearly the various mud-producing species that are so numerous on the Great Bahama Bank are more than capable of generating the amount of mud found there. (This was discussed at greater length in Chapter 2.) The pellets in this facies, which are mostly of sand size, are so ubiquitous that some geologists refer to this facies as an *interior-platform sand*. Bioturbation proceeds actively and the conical mounds of *Callianassa* abound. (See Figure 7-46, A.)

Within the geographic area of the pellet-lime-mud facies, *whitings* (or whitenings), elongate patches of milky-white water tens to hundreds of meters long, are commonly seen from the air. These whitings are dense suspensions of aragonite needles. The most-likely explanation proposed for these whitings is that aragonite is being precipitated chemically from the water. In the Bahamas, some whitings have been observed when schools of fish or the propellers of boats traveling across the platform stirred up the soft bottom, but most whitings are not associated with these phenomena and also exhibit numerous characteristics suggesting that they involve precipitation of aragonite in the water column. (See Chapter 7.)

The grapestone facies occupies the interior parts of the Great Bahama Bank that are not sheltered by Andros

Island. Cementation of particles into grapestone seems to require a lack of differential particle movement. Evidently, short periods of bottom agitation are followed by prolonged periods of bottom stability.

Along the eastern, high-energy margin of the Bank, three facies are concentrated. These are reef, skeletal, and ooid. The reef facies is well developed only along the eastern, windward margin of the Great Bahama Bank facing the Tongue of the Ocean. (See Figure 10-32.) The northeast Trade Winds bring in nutrients and clear oxygen-enriched waters of normal marine salinity (35 to 36‰). By contrast, no reefs are present along the western margin of the bank. Water that has crossed the Bank is not suitable for reef growth. Such water is low in nutrients and turbid with suspended lime mud; its salinity is higher than normal. What is more, in winter, its temperature may be too cold for corals to survive. Still, we do not understand all the factors that determine why reef communities prosper only on the windward side of the bank.

As in the Gulf of Batabano, a coralgal facies forms a well-sorted lag concentrate in the turbulent waters along the margins of the Bank. This facies is present whether reefs are alongside or not. Although lime mud also forms constantly by mechanical- and biological degradation, the turbulent water winnows and removes this fine material.

The oolitic facies likewise forms a belt in shallow turbulent waters. Interestingly, near Joulter's Cay, reefs face the Tongue of the Ocean; skeletal debris underlies the shallow lagoon behind the reef, and then to the leeward, a very pronounced and sharp boundary separates the skeletal facies from the adjoining ooid facies. (See Figure 10-32.) This lateral boundary is so abrupt that it would almost seem possible to stand with one foot in the skeletal facies and the other in the ooid facies (Figure 10-34). The skeletal facies is confined to the subtidal lagoon, and the ooid facies develops where the sea bottom rises into the littoral zone. The benthic organic communities shun the littoral zone. Here, shoaling and breaking waves move the sediments. On these agitated particles, concentric layers of aragonite are precipitated. The turbulent waters polish and abrade the ooids; any skeletal particles present among the ooids are likewise highly polished and abraded. Although ooids form most prolifically in the littoral zone, tidal currents can move them elsewhere across areas underlain by skeletal-, grapestone-, and pellet-lime-mud facies, where the bottom is stable and where *Thalassia* grasses flourish. Spillover-delta lobes of the ooid facies, whose active surface is characterized by small dunes, migrate bankward and spread across other facies. Viewed from the air, the light-colored active ooid shoals stand out in sharp contrast to the dark-colored, plant-

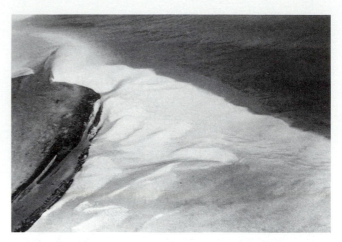

FIGURE 10-34.   View eastward from above, of ooid- (lower part of photograph, white with rippled surface) and coralgal (upper right) facies in sharp lateral contact; rock island (cay) and tidal channel at left of oblique aerial photograph. Distance on ground spanned by width of photograph is approximately 1.5 km. Joulter's Cay, Great Bahama Bank, 09 June 1970. (Authors.)

carpeted, stabilized areas that the ooids bury (Figure 10-35).

> SOURCES: Beach and Ginsburg, 1980; Cook, Hine, and Mullins, 1983; Crevello, J. L. Wilson, Sarg, and Read, 1989; Hine, Wilber, Bane, Neumann, and Lorenson, 1981; Hine, Wilber, and Neumann, 1981; Read, 1982.

Allochthonous Shelves

Allochthonous shelves are those receiving extrabasinal sediment from the adjacent land. Major categories are

FIGURE 10-35.   View from above of spillover lobe of rippled ooid facies (light gray, at right), extending across pellet-mud facies (darker gray, upper left). Black patches in area of pellet mud denote *Thalassia* grass growing on bottom. Pontoon of float plane is visible in upper right. Width of aerial view is approximately 200 m. West edge of Great Bahama Bank, 5 km south of Browns Cay. (Authors.)

shelves near deltas, shelves near sea-marginal fans, and shelves off deserts.

SHELVES NEAR DELTAS

The typical modern sediment on shelves near deltas is mud. Much depends on how the delta is situated with respect to the rest of the shelf. Off the Irrawaddy River, Burma, silts from the river's suspended load are spreading across the shelf to bury relict outer-shelf sands. On the Texas-Louisiana shelf, in the northwest Gulf of Mexico, the effect of the Mississippi River has now become somewhat local because this large delta has prograded completely across the shelf and is discharging into deep water. On this shelf, the depth at the shelf break usually is about 120 m. At the Mexican-Texas border the shelf width is 130 km but at the Texas-Louisiana border it expands to its maximum width of 210 km. Sand is present along the modern shoreface down to a depth of 9 m. Off Texas and Louisiana, modern shelf sediments display a continuous size decrease offshore, from the shoreface sand through silty sand and sand-silt-clay, to a large offshore area of silty clay. The proportions of Foraminifera and of echinoid remains are reliable keys to depth of water. Wood fibers indicate proximity to rivers. Except for the plume of modern shelf silty clay that extends about 32 km offshore to water depths of 27 m and the zone of water flowing offshore from the coalesced semipermanent ocean currents, mentioned in a previous section, the modern sediment remains close to shore. Cores from nearly all localities show bioturbated sediment.

Relict sediments include sands and some shelf-edge carbonate bioherms having relief of less than 4 m.

The Mediterranean shelf off southern Spain near the Ebro delta is 25 to 75 km wide and the depth to the shelf break is 160 m. This is a generally low-energy shelf with small waves and insignificant tidal currents. Air–sea interactions related to changes in wind direction cause the motions in the bottom-boundary layer to oscillate seasonally. During the SE winds of summer, the water flows NE. During the NE winds from the Alps (named the levants), especially during the stormy season from November to March, water motions reach maximum values.

The Ebro River discharges about 2.5 million tonnes/yr of sediment. Sand and gravel form the bottom out to depths of 20 m; elsewhere, a sandy mud consisting of 60% extrabasinal sediment and 40% intrabasinal carbonate forms homogeneous strata covering the shelf. Skeletal components in the 40%-carbonate fraction consist predominantly of mollusks on the inner shelf, of benthic foraminiferans on the middle shelf, and of planktonic foraminiferans on the outer shelf. Dating methods based on excess ^{210}Pb and ^{137}Cs indicate that modern

sediments are reaching the shelf, but the slow rate of upbuilding implies much is bypassed.

On the continental shelf off Washington in northwestern United States, the bottom on the inner shelf to depths of 40 to 60 m consists of modern sand. To seaward, on the mid- and outer shelf, in depths greater than 120 m, silt from the Columbia River is abundant.

On the inner shelf of the Canadian Beaufort Sea, seaward of the sandy shoreface, major storms deposit massive- or graded silt layers up to 20 cm thick. These silt layers are analogous to the thicker and coarser couplets (graded storm layers) described from the Long Island shelf.

Where large waves are not present, the depth to the sand-silt transition between inner- and outer shoreface may reflect time-averaged wave base. If this sand-silt boundary moves back and forth, the result may be the interbedding of layers of sand with layers of silt.

SOURCES: C. E. Adams, Jr., Swift, and J. M. Coleman, 1984; Curray, 1960; Kuehl, Nittrouer, and DeMaster, 1982; Nittrouer, Bergenback, DeMaster, and Kuehl, 1988; Rodolfo, 1975.

SHELVES NEAR SEA-MARGINAL FANS

Modern shelves offshore from sea-marginal fans are not numerous nor conspicuous, but they form a category that was important in the geologic past. The very narrow shelf in the Gulf of Aqaba is an example seaward of fans that lie alongside coral reefs. (See Figure 13-37.)

SOURCES: Friedman, 1988; Friedman and Sanders, 1978.

SHELVES OFF DESERTS

Off two modern deserts, the northwestern African shelf off the Sahara, and the westcentral shelf of the Persian Gulf off the Qatar Peninsula, the prevailing winds are blowing sand into the ocean, thus causing the shore to be prograded with sand.

Off Qatar, cores show a parasequence about 22 m thick, consisting of a basal transgressive lagoonal sandy mud, 2 m thick, which is overlain by 20 m of quartz sand. The lower 10 m of the quartz sand lacks bedding, and contains a few scattered mollusks and echinoids and associated bioturbation structures. This can be interpreted as a nearshore shelf deposit formed after the sea more or less reached its present high-stand level. Above the nonbedded sand is 4 m of fine sand displaying subhorizontal layers. The upper 6 m of the parasequence consists of sand displaying seaward-dipping planar cross strata deposited as foreset beds on the accreting slope of the shoreface. On shore are dunes that are being driven into the sea by the strong NW "shamal" winds.

In the northwestern part of the Persian Gulf, winds blowing off the desert create dust storms from which quartz silt falls out of the atmosphere into the water and is contributed to the shelf sediments.

SOURCES: Kukal and Saadallah, 1973; Purser, 1985; Shinn, 1973.

Mixing of Extrabasinal- and Intrabasinal Sediments

We have so far discussed extrabasinal- and intrabasinal shelf sediments as if they formed mutually exclusive categories. Although on many shelves one or the other of these predominates, in some places these contrasting kinds of sediments are mixing in various ways. This mixing ranges from the simple addition of carbonate skeletal debris to extrabasinal sandy muds, as on the southern Spanish Mediterranean shelf off the Ebro delta, to selective transport of one kind of sediment into the domain of the other. As a result, extrabasinal sediment and intrabasinal sediment form discrete layers that become interbedded as conditions oscillate. Recall the example of the New Jersey clams that were swept from their burrows and deposited on the adjacent beach, where the usual sediment is quartz sand. Had the local Department of Public Works not removed the clams, they would have become a carbonate shell bed with quartz sand above and below. On the east coast of Florida, extrabasinal quartz is being transported southward into the domain of intrabasinal tropical carbonate sediments. On the west coast of Florida, just the opposite is taking place: carbonate sediments are being transported into a province of extrabasinal sediments.

Other kinds of mixing result from changes of drainage or of coupled sea-level/climate changes.

CHANGES OF DRAINAGE

As we indicated earlier, modern tropical reefs do not grow in areas where the water is influenced by large rivers. An example is the north end of the Great Barrier Reef, where the Fly River enters the Gulf of Papua. Coral reefs and a large entering river can coexist if the river is situated downcurrent from the reef (Figure 10-36). In such a setting, the reef is exposed only to warm, saline water; the fresh-water plume from the river is carried alongshore away from the reef. However, if the point of discharge of the river shifts, then terrigenous sediment can bury the reef. (See Figure 10-36, B.) Under the conditions shown in Figure 10-36, B no reefs would colonize the submerged delta lobe for two reasons: (1) the substrate may not be suitable, and (2) the area is downcurrent from the delta and thus the fresh water from the river renders the salinity (and possibly the temperature) unfit for corals.

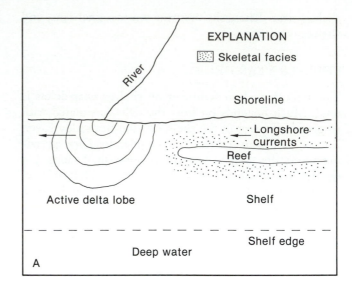

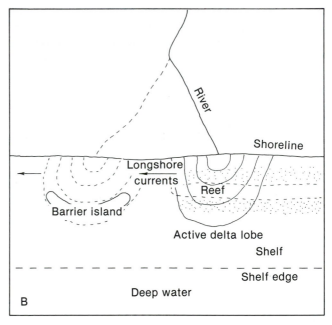

FIGURE 10-36.   Relationships between reefs and a delta.
   A. Active delta lobe, downcurrent from reef, deposits terrigenous sediment next to reef and reef-derived carbonate materials.
   B. River channel is relocated inland; old delta lobe is submerged and reworked into barrier island (Chapter 11). Reef is killed and buried by terrigenous deltaic sediment, mostly fine textured.
(G. M. Friedman and J. E. Sanders, 1978, fig. 12-40, p. 383.)

COUPLED SEA-LEVEL/CLIMATE CHANGES

The mixing of extrabasinal- and intrabasinal sediments associated with coupled changes of sea level and climate result from the corresponding changes in the position of the regional boundary between climate-controlled domains of these two kinds of sediment. In a previous section we reviewed the evidence from modern shelves in which widespread low-stand shelf-edge carbonates were supplanted at high stand by widespread mainland-marginal noncarbonates. Numerous repetitions of this kind of coupled changes would build up a stratigraphic record in which carbonates and noncarbonates, each forming their own domains, could be mixed.

In eastern Australia today this low-stand pattern prevails but at a high stand of the sea. Shelf-edge reefs are yielding intrabasinal carbonate sediments to the outer part of the shelf and the Australian continent is supplying extrabasinal quartz-rich sediments to form a mainland-marginal fringe on the inner shelf.

SOURCES: Belperio and Searle, 1988; Doyle and Roberts, 1988; Friedman and Sanders, 1978; Kendall and Schlager, 1981; Kopaska-Merkel, 1987a; Maxwell and Swinchatt, 1970; F. O. Meyer, 1989; Mount, 1984; K. R. Walker, Shanmugam, and Ruppel 1983.

Epeiric Seas

We shall discuss the general relationships of epeiric seas and then consider some inferred examples of their ancient sedimentary deposits.

General Relationships; Absence in the Modern World

The most-striking thing about epeiric seas is that they are nowhere present in the modern world. The closest modern counterpart to an ancient epeiric sea is the Bahamas Platform. Despite many similarities, the Bahamas Platform is puny by comparison with some of the epeiric seas that are inferred to have spread across Eurasia and North America during the geologic past. Other suggested modern counterparts to ancient epeiric seas include the Baltic Sea, Hudson's Bay, the Yellow Sea, and the Bering Sea, but all these are also much smaller than typical ancient epeiric seas.

The bottom slopes of epeiric seas have been estimated at 1:1,000 to 1:10,000 (1 to 10 cm/km). Paleowater depths in epeiric seas are not easy to reconstruct. When widespread sheets of carbonates were being deposited, most workers favor the concept that paleowater depth was 30 m or less.

The enormous areal extents of some ancient epeiric seas probably meant that they were subjected to wind-driven surface currents, such as those today moving in the upper 100 m or so of the open oceans. Despite their great expanses of open water, epeiric seas are thought to have been too shallow to permit extensive wave action. Long-period swells propagated from a deep-sea basin would have been damped at the shallow margins of an epeiric sea. In water 30 m deep, the largest possible wavelengths of deep-water waves would be 60 m and their periods, 6 s. (See Figure 8-27.) This is about the size of the largest sea waves blown up by ordinary storm winds in today's oceans. By contrast, the periods of long swells may be 20 s or more, and their wavelengths measure hundreds of meters (Chapter 8).

Presumably, winds would mix the shallow waters throughout an epeiric sea. Winds would also create changes of level, and these would cause unidirectional currents to flow. However, some deposits of epeiric seas contain phosphate nodules in black shales. These are interpreted as having resulted from upwelling and the development of anoxic conditions at and near the sea floor. Water depths of 100 m or more have been suggested for epeiric seas when black shales containing phosphorites were being deposited. Epeiric seas characterized by waters deeper than 100 m may have differed oceanographically from epeiric seas in which water depth was only a few tens of meters or less.

Over most of today's oceans, evaporation exceeds precipitation. Thus, except where large rivers entered, epeiric seas located within about 40° of the Equator probably tended to be hypersaline. To what extent any hypersaline waters circulated or were stratified as a result of differences in level or in density is not known.

Two examples of inferred epeiric seas include those that spread across most of North America during the Ordovician Period (Figure 10-37) and across the North American midcontinent during the Cretaceous Period.

Inferred Examples of Deposits from Ancient Epeiric Seas

The following discussion of epeiric seas deals with inferred examples of ancient strata deposited in them. We include strata composed of extrabasinal sediments and strata composed of intrabasinal sediments.

STRATA COMPOSED OF EXTRABASINAL SEDIMENTS

Inferred examples of extrabasinal sediments deposited in ancient epeiric seas include various shales and sandstones. The deposition of a mud (now a shale) in a shallow sea requires that the concentration of suspended sediment be large (probably > 150 mg/l) so that more fines are brought in than the waves and currents can take away. The gray marine shales associated with the Pennsylvanian and Permian coal measures in the midcontinent area of the United States are presumed examples.

Other examples of widespread marine shales that pass laterally into shoreface- and/or beach deposits composed of extrabasinal sediments include the Upper Cretaceous Pierre-Mancos-Lewis complex of the Great Plains-Rocky Mountains region of the United States.

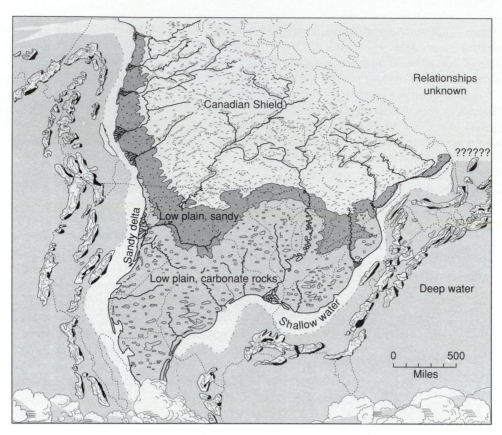

FIGURE 10-37.   Paleogeographic map of North America during Early Middle Ordovician Period showing extent of submergence beneath a shallow epeiric sea. (After G. M. Kay, 1951, pl. 1; C. R. Longwell, R. F. Flint, and J. E. Sanders, 1969, fig. 21-10, p. 526.)

Some geologists think that the depth of water in which these Upper Cretaceous shales were deposited amounted to a few hundreds of meters rather than tens of meters.

A widespread black shale of Late Devonian-Early Mississippian age, which occupies millions of square kilometers on the North American continent, the identical area occupied by an undoubtedly shallow carbonate-depositing epeiric sea earlier in the Paleozoic Era, is the Ohio-Chattanooga-New Albany complex. The bathymetric interpretation of this black shale has been and continues to be hotly disputed. Depth interpretations have ranged from "shallow" to "deep" (meaning at least enough depth to permit a long-term density stratification to be established). If this black mud (now shale) was indeed deposited in a density-stratified water body, then its vast areal extent implies that the depth was probably hundreds or even thousands of meters rather than tens of meters. Whether or not this is true, it should be pointed out that a bathymetric interpretation appropriate for a carbonate- or an evaporite deposit need not apply to an overlying- or underlying terrigenous shale, and vice versa. Almost certainly, the epeiric seas in which widespread strata of carbonates and evaporites were de-

posited were both shallow and tropical. Thus it seems reasonable to suppose, in contrast, that a mud (now a shale) deposited in an epeiric sea means either (1) the water was shallow and the climate cool or (2) if the climate was tropical, the water was deep.

From what we have seen in previous sections, it should be clear that ancient shallow-marine sandstones are likely to fit into one of the following categories: (1) inner-shoreface deposits (pure sands wave-ripple laminated throughout); (2) outer-shoreface deposits (pebbly sands, probably graded and associated with HCS sequences, inferred storm deposits, possibly interlayered with bio-turbated muds, products of quiet conditions between storms); (3) shoal-retreat massifs (pure sands, probably dominated by shoreface deposits, both inner and outer); (4) sand ridges of various kinds related to strong tidal currents or to storm conditions (probably displaying flat bottoms and convex-up tops and featuring large-scale cross strata). If one is trying to interpret a marine sheet sandstone, then an important point to establish is how the sand was dispersed. Was the sand dispersed under low-stand nonmarine conditions (by the wind, for example) and then covered and reworked in part by the sea during a following high stand? Or was the sand dispersed during submergence or during conditions prevailing during a high stand? Was the dispersal a product of a single set of conditions or of several sets of conditions that may have alternated?

In the Pleistocene high-stand sediments in the southeastern Virginia coastal plain are examples of inferred lower-shoreface pebbly sands that display horizontal laminae (Figure 10-38). When pebbles are numerous they inhibit attempts to prepare smooth, plane faces that display the sedimentary structures. Accordingly, we do not know if these sediments contain HCS.

Other more-ancient shelf sandstones include the Cape Sebastian Sandstone (Upper Cretaceous of southwestern Oregon), the Shannon Sandstone and Mossby Sandstone (Cretaceous, Rocky Mountains, western United States), and the Viking Formation (Lower Cretaceous) and Cardium Formation (Upper Cretaceous), Canadian Rockies, (Alberta).

SOURCES: E. A. Beaumont, 1984; Bourgeois, 1980; Ettensohn, 1985; Kumar and Sanders, 1976; Rice, 1984; Swift and Rice, 1984; Tillman and Martinsen, 1984;

Tillman and Siemers, 1984; R. G. Walker and C. H. Eyles, 1991.

STRATA COMPOSED OF INTRABASINAL SEDIMENTS

Many examples of inferred deposits of epeiric seas composed of intrabasinal sediment are known from the geologic record. Our examples include reefs from rimmed shelves and various other carbonate deposits.

The modern reef builders became prominent during the Paleocene Epoch (about 60 million years ago). Earlier in geologic history, other organisms built extensive reefs. Among these were the rudistids (extinct oyster-like, robust bivalve mollusks), whose heyday of reef building was during the Cretaceous Period (Figure 10-39).

Other significant framework builders of ancient (and in some cases, modern) reefs and bioherms have included calcareous algae, calcareous sponges, lime-secreting annelids (such as serpulids), "hydrocorallines," bryozoans, certain kinds of Permian brachiopods, various Paleozoic tabulate corals, tetracorals, and stromatoporoids, and the earliest skeletal framebuilders known, the Early Cambrian archeocyathids.

Most known modern stromatolites form primarily by trapping and binding of sediment particles in cyanobacterial mats, and, being nonrigid, are not reefs. However, some Proterozoic stromatolites may have formed mainly by biologically mediated calcium-carbonate precipitation within cyanobacterial mats. Such structures would constitute rigid and resistant frameworks like the skeletal frameworks just mentioned. They would be reefs. Nonrigid, mound-shaped stromatolites are non-reef bioherms.

Viewed broadly, a shelf-edge reef (and, for that matter, any steep shelf edge, even where no reefs are present) and the associated deep water on the exposed side and shallow water on the protected side can be divided into three energy belts: (1) the calm, low-energy, deep water on the exposed (basin) side; (2) the high-energy shelf-edge belt of vigorous water motions where the waves shoal and break; and (3) the shallow, low-energy, calm water of the protected (possibly backreef) side. These energy belts have been given letter designations as shown in Figure 10-40.

In energy-belt X, the bottom lies far below wave base for wind-generated waves. Therefore the bottom sediments are moved only by extremely long-period waves, by internal waves, by various gravity-powered flows of sediment along the bottom, and by various deep currents.

Energy-belt Y includes the inner part of the shoreface and adjacent areas where the waves break. Here, wind-

FIGURE 10-38.   Interbedded sands and gravels, inferred examples of ancient shoreface deposits of Pleistocene age, Benns Church, southeastern Virginia. (N. Kumar and J. E. Sanders, 1976, fig. 8, p. 157.)

E   W

FIGURE 10-39.   Molluscan reefs of Cretaceous age.

A. Views of core of massive Cretaceous shelf-edge reef, built by rudistids (extinct reef-building bivalve mollusks having a geologic range from Late Jurassic to Late Cretaceous, possibly into the Paleocene Epoch). Massive nonstratified rock, of which about 50 m is exposed, consists mostly of rudistids, many of which are still in growth positions. Dark vertical linear features are fractures; dark circular areas at lower right are caves that were occupied by people during Paleolithic- and later times. Nahal Hame'arot Reef, Mount Carmel, Israel, lat.32°40′N; long.34°55′W. (A. Bein.)

generated waves undergo shoaling transformations and become shallow-water waves, very-shallow-water waves, and eventually, breakers. The bottom water surges relentlessly back and forth every few seconds.

Energy-belt Z is so shallow that its waters are not affected appreciably by wind-generated waves.

Carbonate (as indeed do nearly all) sediments vary systematically according to their situations with respect to these energy belts. Hence, carbonate rocks formed from such sediments can be interpreted without much hesitation. For example, Y-belt limestones are derived from carbonate materials of one of three kinds: (1) reefs, (2) skeletal sands, and (3) oolitic sands. Because all these materials were deposited in shallow, well-aerated waters, the color of Y-belt limestones is light gray or tan. Because a

reef core consists of rigid framework-building organisms, the reef limestone lacks bedding. The original particles of skeletal- and oolitic sands were constantly in motion and not subject to extensive burrowing. Therefore their original strata, which usually were horizontal or included various cross strata formed by the migration of ripples and flow-transverse dunes, are preserved in the resulting limestones. Sorting of particles is usually excellent; interparticle matrix tends to be absent. Initial porosity is high. The pores may be filled with migrating petroleum hydrocarbons or with mineral cement. Accordingly, Y-belt materials are likely sites of stratigraphic-type oil traps.

Because the bottom in the X belt lies below wave base for wind-generated surface waves, fine carbonate particles accumulate. In the bottom sediments, organic matter may be abundant. If so, the resulting rocks tend to emit a petroliferous odor when freshly broken. Within the interstitial waters of the sediments, sulfate-reducing bacteria may actively create iron sulfides (See Eq. 3-3.), which impart a black color to the sediment (and thus to carbonate rocks), and the poisonous gas, hydrogen sulfide. If H_2S poisons the bottom waters, it snuffs out the burrowing and scavenging benthic organisms. In their absence, dead planktonic- or nektonic organisms that drop to the bottom from above will be preserved and stratification will not be disrupted. A typical limestone formed from carbonate sediments deposited in the X belt is a black, well-bedded micrite. Sediments from the X belt are thought by many to be source facies for petroleum hydrocarbons.

FIGURE 10-39.   (*Continued*)

B. Close view of both cross sections and longitudinal sections of reef-building rudistids in growth positions in a patch reef. Upper Cretaceous, Galilee, Israel. (Authors.)

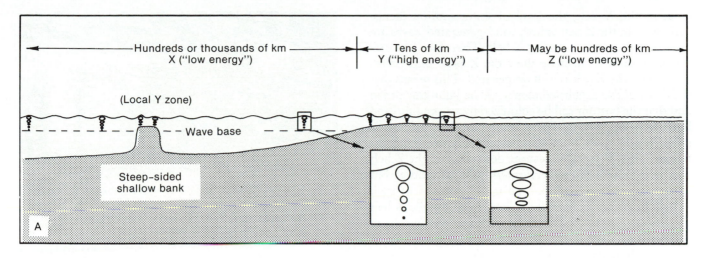

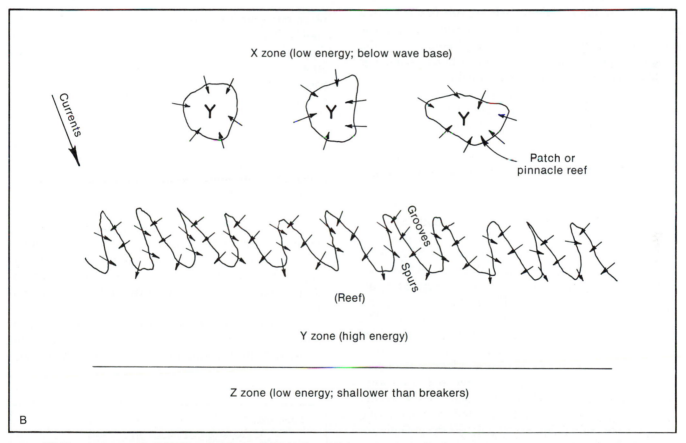

FIGURE 10-40.   Concept of three schematic energy belts, X, Y, and Z, based on intensity of water motions at margin of rimmed shelf or -platform.

A. Schematic profile showing relationships of energy belts to waves. In X zone, bottom lies below wave base; deep-water waves cause water "particles" to move in circular orbits. In Y zone, bottom lies above wave base; seaward of breaker zone, shallow-water (or very-shallow-water) waves cause water particles to move in elliptical orbits. Further explanation in text. (Modified from M. L. Irwin, 1965, fig. 3, p. 450.)

B. Sketch map showing relationship of reefs to energy belts X, Y, and Z. Reefs are located in the Y zone; three pinnacle reefs are shown lying on the windward side of barrier reef, which displays grooves and spurs on its open-sea side (toward top). Arrows indicate the direction of possible migration of hydrocarbons into the reefs following burial. (Modified from M. L. Irwin, 1965, fig. 4, p. 451.)

As in the X belt, the level of water motion (hence "energy") in the Z belt is low; wind-generated waves are dissipated in the intervening Y belt. Therefore fine carbonate sediments likewise floor the Z belt. But, unlike the X belt, the Z belt is well oxygenated. This means that the color of the Z-belt sediments will be light gray or tan and that the bottom will be colonized by scavenging- and burrowing organisms. These benthic organisms destroy stratification; they homogenize and pelletize the sediments. Thus a typical limestone derived from the Z belt is a light gray or tan pelletal micrite that is mottled but rarely displays lamination. The initial porosity of such Z-belt limestones generally is low. Accordingly, Z-belt limestones may form seals or cap rocks to the petroleum trapped in Y-belt reservoirs, and possibly generated in X-belt sediments.

Where the floor of the Z belt extends upward into a peritidal complex, sea-marginal evaporite minerals may form interstitially (as in the Holocene sediments at Abu Dhabi; Chapter 12) and the calcium-carbonate sediments may be dolomitized to form supratidal crusts.

Although most ancient examples in which the existence of reefs at the edge of a shelf has been inferred are clear enough on the seaward side, the paleogeographic setting landward of the reef (or shelf edge) may not be so clear. The possibilities include: (1) a shelf lagoon bounded by a continental landmass; (2) carbonate platforms of various sizes; and (3) an epeiric sea. Moreover, even where the paleogeographic setting is known, we find that one of these may pass into another.

The generalized relationships displayed by nearly all of these shallow areas that were bounded by steep slopes can be shown on an idealized profile and section. (See Figure 9-7.) The reef's rigid framework secreted by organisms forms the *reef core*. The material forming the reef core is devoid of bedding; it appears massive. The reef core is surrounded by its own skeletal debris, which is swept both landward and seaward. Storms break loose coarse particles, most of which accumulate on the open-sea (forereef) side. The coarsest forereef sediments form a *reef talus*: sand- to cobble-size fragments of skeletal debris derived from a reef and deposited in strata having an appreciable initial dip. Skeletal debris swept landward forms the backreef facies.

Because they are such prolific producers of calcium-carbonate particles, reefs tend to be buried in their own debris. (As a matter of fact, the continued survival of the reef depends on operation of mechanisms to remove this debris.) As a result, strata composed of skeletal debris overlie and dip away from the reef core. This same dipping relationship prevails even if other strata, not related to the reef (such as shale, for example), bury the reef cores. Toward the reef core, the thickness of covering strata progressively decreases; some strata may even

FIGURE 10-41. Bioherm, approximately 15 m thick at right margin of photograph, built by phylloid alga *Ivanovia*. Draping strata of skeletal debris appear and thicken to left. Pennsylvanian, San Juan Canyon, southeast Utah. (Authors.)

pinch out against the solid reef core (Figures 10-41 and 10-42). This thinning and pinching out of strata and dipping away from the reef constitutes a relationship known as *drape*. (See Figure 10-41.) As a result of compaction after burial, draping of shale around a reef is accentuated. Because seismic waves do not travel through the reef core and the draping shale with the same speeds, a reef that has been buried by shale shows clearly on seismic records. (See Figure 10-18.)

Many ancient shelf-edge reefs as well as reefs and bioherms of all kinds and locations are associated with evaporite minerals. Some reef-associated evaporites were deposited interstitially. (We discuss peritidal examples of interstitial evaporites in Chapter 12 and nonmarine examples in Chapter 14.) Other reef-associated evaporites are thought to have been deposited in open hypersaline waters whose depths may have ranged from "shallow" to "deep" (and generally are not easy to determine unambiguously).

Examples of ancient shelf-edge reefs include the subsurface Tertiary of Louisiana; the Triassic of the southern Alps, Italy; the Jurassic of the French Alps south

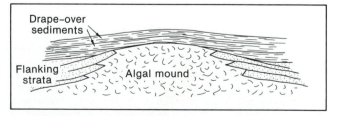

FIGURE 10-42. Schematic profile through small algal reef illustrates different relationships among reef core (labeled algal mound), flanking strata (which interfinger with debris eroded from reef core), and draped-over strata (which are entirely younger than the reef and its laterally equivalent flanking strata). Compare Figure 10-41. (G. M. Friedman and J. E. Sanders, 1978, fig. 12-38, p. 382.)

of Geneva, Switzerland, the Middle East; and the Permian of the Delaware Basin, West Texas and New Mexico, United States. One of the oldest-known possible examples is represented by early Proterozoic (1.89 Ga) stromatolitic boundstones that formed an accretionary rim on a shelf edge in the Rocknest Formation, Wopmay Orogen, Northwest Territories, Canada.

sources: Crevello, Wilson, Sarg, and Read, 1989;
Friedman, 1968; Friedman and Kopaska-Merkel, 1991;
Grotzinger, 1989; Head, 1987; Shaver, 1991;
Toomey, 1981.

During the Jurassic Period in the Middle East, a broad epeiric sea spread from the ancient Tethys Sea across the passive margin of the ancient Nubian continent in an area between what is the modern Persian Gulf and the modern Mediterranean Sea (Figure 10-43). To the west, near the present-day Mediterranean coast, the epeiric sea became a shelf sea. Close to the modern coast of Israel stood a hingeline marked by a precipitous slope between the shelf edge and the deep sea. The sediments east of the shelf edge were peritidal and included pellet-lime mud and lime mud with skeletal fragments. The sediments that formed at the shelf edge were high-energy deposits and included reefs, ooid shoals, and skeletal and pelletal lag sands.

The noteworthy formation of the Upper Jurassic of Saudi Arabia is the Arab Formation, which consists

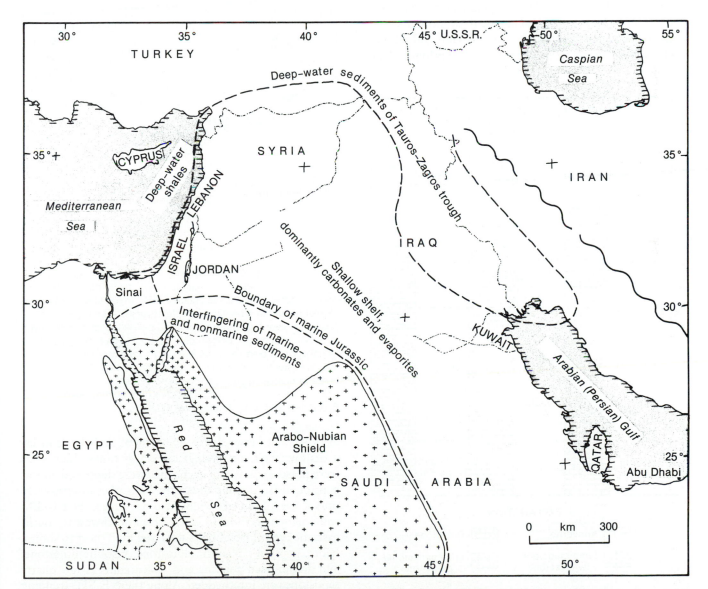

FIGURE 10-43. Paleogeographic map of Middle East showing inferred depositional environments that prevailed during most of the Jurassic Period, and the dominant sediments deposited in each environment. (After M. Goldberg and G. M. Friedman, 1974, fig. 1, p. 2.)

of interbedded carbonate rocks and anhydrite (Figure 10-44). The basal member, the Arab D, is the reservoir of some of the world's largest oil fields.

SOURCES: Alsharhan and Kendall, 1986; Ayres, Bilal, R. W. Jones, Slentz, Tartir, and A. O. Wilson, 1982; Cherven, 1986; Friedman and Sanders, 1978; R. W. Powers, 1962.

A classic area involving reefs, carbonate rocks, evaporites, and terrigenous strata is the Permian basin, of west Texas and southeastern New Mexico. Late in the Paleozoic, upward growth of the Central Basin Platform split the northern part of the overall basin into two smaller basins, the Delaware basin on the west (Figure 10-45), and the Midland basin on the east. The regional slope was from northwest (site of a low-lying landmass) toward the southeast (into a deep embayment connected at the south to the open sea). In this embayment, the salinity of the water was normal at some times and hypersaline

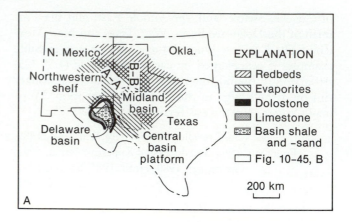

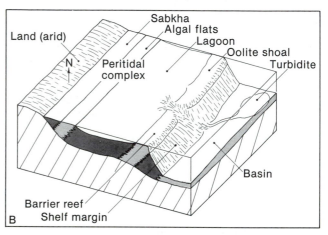

FIGURE 10-45.   Relationships in Delaware basin, southwestern United States, during Late Guadalupian (Permian) time.
A. Map of facies belts. (After W. W. Tyrrell, Jr., 1969, fig. 1, p. 81.)
B. Schematic block diagram showing inferred environmental conditions along NW basin margin. Progradation during submergence creates stratigraphic succession in which strata deposited closest to land in a peritidal complex overlie Z-belt lagoonal deposits. On their basinward side, the Z-belt lagoonal deposits overlie Y-belt shelf-edge-, barrier-reef-, and ooid-shoal deposits. Gravity transport of ooids and skeletal debris from reef down indentations in steep shelf margin (slope approximately 30°) spreads layers of coarse sediment (some as turbidites) across basin floor where otherwise only X-belt black lime mud accumulated. (After B. A. Silver and R. G. Todd, 1969, fig. 4, p. 2227.)

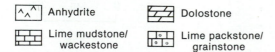

FIGURE 10-44.   Stratigraphic columns of Upper Jurassic rocks in Middle East. (A. S. Alsharhan and C. G. St. C. Kendall, 1986, fig. 11, p. 995.)

at others. In between the embayment and the land lay a shallow shelf sea (a shelf lagoon at times when reefs were present) on which evaporites were deposited almost continuously. At some times, the belts of evaporites were only a few kilometers wide. At other times, their widths were many tens of kilometers. From northwest to southeast, the typical facies belts were red mud (next to shore), halite, gypsum, aphanitic dolostone, pisolitic limestone, and dolomitized coquina and -calcarenite. A few quartz sandstones are interbedded. Along the NE-SW-trending Guadalupe Mountain front, reefs are exposed. Debris from the reefs and other shelf-edge sources was propelled by gravity down the submarine slopes and spread across the basin. Interbedded with the finely laminated terrigenous fine sandstones of the Delaware Basin are well-developed graded layers, some containing pebbles at their bases. Syndepositional slump structures, including overturned folds (See Figure 9-12.) and thrusts, are further indications that submarine gravity-powered mass movements took place.

Characteristic breccias are widespread. These are collapse breccias, which formed where soluble evaporites were uplifted from the deep-lying zone of salty ground water to the nearer-surface zone of fresh ground water. In the fresh-water zones, evaporite minerals were dissolved. Such breccias are not present in the subsurface where the evaporites are still intact. The collapse breccias indicate that at times, strata of evaporites (including gypsum and halite) extended continuously from the basin, across the basin margin (where reefs grew at other times), and all the way to the shoreline. The continuity of some of the thick gypsum strata is inferred to have been broken not only by subsurface dissolution and re-

sulting collapse of overlying strata but also by subsurface bacterial degradation and the formation of calcite. (See Eq. 3-3.)

A comparable stratigraphic pattern resulted from a combination of progradation and submergence in the Middle Permian rocks along the margins of the Midland basin, situated east of the Central Basin Platform. A restored stratigraphic diagram (Figure 10-46) has been prepared from data derived from borings. The noteworthy feature displayed by this diagram is the differentiation of the intrabasinal strata into X, Y, and Z energy belts and the basinward progradation by 20 km of the highstand Clear Fork strata during a net submergence of 1 km. The amount of associated extrabasinal sediment is small, perhaps a product of low rates of subaerial erosion in the arid climate of this region during Permian time.

A continuous seismic-reflection-profile record from a north-south-trending line nearby displays distinctive internal reflectors within the three basin-marginal parts of each member of the Clear Fork Formation (Figure 10-47).

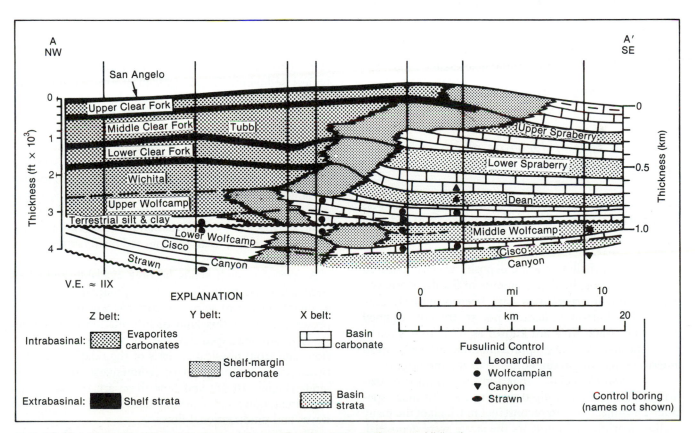

FIGURE 10-46.   Restored stratigraphic diagram of Lower Permian strata, northwestern Midland basin, Texas, showing Y-belt carbonates that prograded basinward (to right) by 20 km during a net submergence of 1 km. Compare with continuous seismic-reflection profile record of Figure 10-47. Surface of angular unconformity (wavy line) separates Upper Wolfcamp and younger strata from Middle Wolfcampian and older strata. (B. A. Silver and R. G. Todd, 1969, fig. 9, p. 2232.)

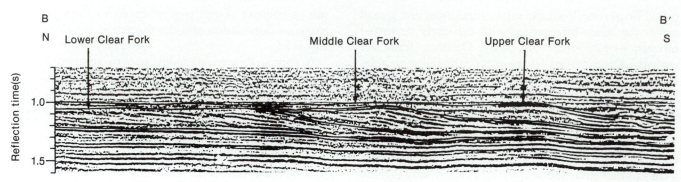

FIGURE 10-47. Multichannel continuous seismic-reflection profile from northern part of Midland basin, West Texas, along line B-B′ in Figure 10-45, A, showing three carbonate depositional sequences in the lower, middle, and upper Clear Fork Limestone (Middle Permian). All three consist of high-stand deposits displaying low-angle sigmoidal- to oblique reflectors indicative of *in-situ* upbuilding of shelf-marginal carbonates within a setting where the shelf margin was prograding basinward during net submergence. This pattern within shelf-edge carbonates has been named "catch-up" morphology. Any associated low-stand- or transgressive deposits are not seismically resolvable. (J. F. Sarg, 1988, fig. 3, B, between p. 168 and 169.)

The Permian limestones and -sandstones have yielded much petroleum, and the search for additional fields containing petroleum continues. Abundant subsurface information from oil-test borings and from continuous seismic-reflection profiles is available to complement the nearly continuous but commonly inaccessible exposures.

SOURCES: Sarg, 1988; Ward, Kendall, and Harris, 1986.

Characteristics of Deposits of Shelf Seas and Epeiric Seas

In the stratigraphic record the products of shelf seas and epeiric seas look alike, and on the scale of ordinary exposures, cannot be distinguished. Both were deposited in shallow seas.

As a group, shallow-marine deposits are characterized first and foremost by their contents of fossils. Moreover, many shallow-marine strata have been extensively burrowed (but burrow structures per se are not confined to marine strata; they are known as well from nonmarine deposits). As a first approximation, any widespread limestone may be presumed to be of marine origin.

In addition to deciding that a given deposit is of marine origin, one generally desires to be a bit more specific. The objective is to reconstruct in which of the many shallow marine environments the strata were deposited. Again, the faunal evidence commonly is diagnostic.

Ancient marine shelf deposits pass landward into and are likely to be interbedded with marginal-marine strata, such as those deposited on deltas, on fans, on barriers, and so forth. Carbonate skeletal sands consist of shallow-

water benthic organisms, such as various cyanobacteria. Reef-building organisms in growth position identify their own environment. Terrigenous strata vary widely, depending on the amounts of sediment in suspension; this factor usually varies according to location with respect to major deltas. Away from deltas, sand is the typical sediment. Coarse graded storm deposits associated with HCS and horizontal laminae should be common. Cross laminae from large, wave-generated ripples are present in many shallow-marine sands. Other kinds of cross strata may be present. A common variety in tidal-current ridges resembles the cross strata in seif dunes. Various patterns of parasequences are formed by shifting environments during submergences, emergences, transgressions, and regressions related to tectonic activity, sea-level fluctuation (possibly coupled with climatic changes), and varying sediment supply from land.

As cores of shelf sands become available and relief peels are made from their cut longitudinal faces, more evidence related to sequence patterns will be found. At present, more of this kind of information comes from study of surface exposures and cores of ancient formations and from physical analysis of the relationships between currents and primary sedimentary structures in sands (Figure 10-48) and how these are related to the combined flows from orbital motions of short-period waves and unidirectional flows.

The effects of seasonal air–sea interactions have been recognized in the Jurassic mudstones deposited in the northwest European epeiric basin.

SOURCES: Friedman and Sanders, 1978; Myrow and Southard, 1991; Oschmann, 1990.

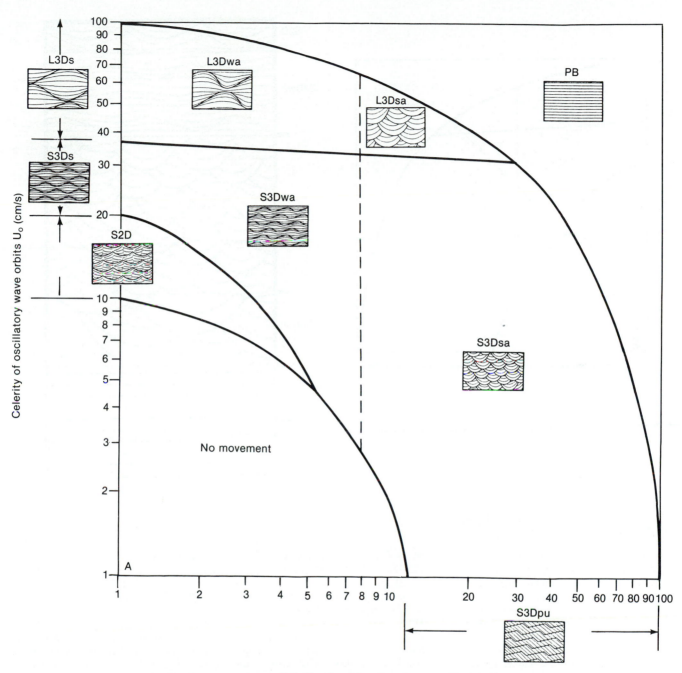

EXPLANATION

S2D	Small 2–D ripples
S3Dpu	Small 3–D ripples, purely unidirectional flow
S3Dwa	Small 3–D ripples, weakly asymmetric
S3Dsa	Small 3–D ripples, strongly asymmetric
S3Ds	Small 3–D ripples, symmetrical
L3Dwa	Large 3–D ripples, weakly asymmetric
L3Dsa	Large 3–D ripples, strongly asymmetric
L3Ds	Large 3–D ripples, symmetrical
PB	Plane bed

FIGURE 10-48.   Relationships among combined oscillatory flow from orbits of shallow-water waves and unidirectional currents and various bed configurations in fine sand.
   A. Bed-configuration stability diagram.

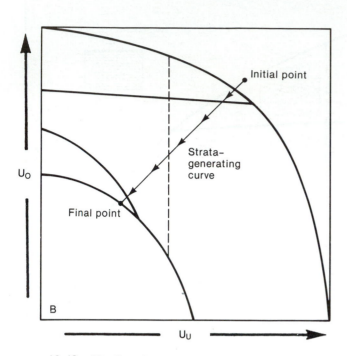

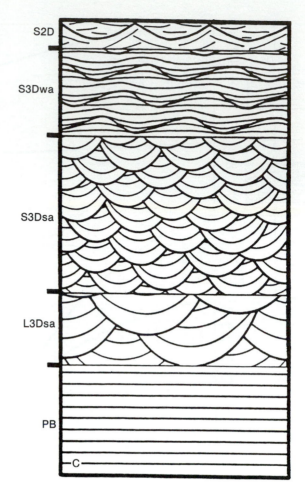

FIGURE 10-48. (*Continued*)

B. Schematic representation of bed-configuration stability diagram on which is plotted a line (strata-generating curve) extending from an initial point across various subfields of bed forms.

C. Schematic succession of strata formed by strata-generating curve. (P. M. Myrow and J. B. Southard, 1991, figs. 1, p. 204; 2 and 3, p. 205.)

Suggestions for Further Reading

AHR, W. M., 1973, The carbonate ramp: an alternative to the shelf model: Gulf Coast Association of Geological Societies Transactions, v. 23, p. 221–225.

AIGNER, T., 1982, Calcareous tempestites: storm-dominated stratigraphy in Upper Muschelkalk limestones (Middle Triassic, S.W. Germany), p. 248–261 *in* Einsele, G.; and Seilacher, A., eds., Cyclic and event stratification: New York, Springer-Verlag, 536 p.

BELPERIO, A. P., and SEARLE, D. E., 1988, Terrigenous and carbonate sedimentation in the Great Reef Province, p. 143–174 *in* Doyle, L. J., and Roberts, H. H., eds., Carbonate-clastic transitions: Developments in Sedimentology 42: New York, Elsevier, 304 p.

BOUMA, A. H., Berryhill, H. L., BRENNER, R. L., and Knebel, H. J., 1982, Continen-
tal shelf and epicontinental seaways, p. 281–327 *in* Scholle, P. A.; and Spearing, D., eds., Sandstone depositional environments: Tulsa, OK, American Association of Petroleum Geologists Memoir 31, 410 p.

BRENNER, R. L., 1980, Construction of process-response models for ancient epicontinental seaway depositional systems using partial analogs: American Association of Petroleum Geologists Bulletin, v. 64, p. 1223–1243.

CRAM, J. M., 1979, The influence of continental shelf (*sic*) width on tidal range: Paleooceanographic implication: Journal of Geology, v. 87, p. 175–228.

HILL, P. R., and NADEAU, O. C., 1989, Storm-dominated sedimentation on the inner shelf of the Canadian Beaufort Sea: Journal of Sedimentary Petrology, v. 59,
p. 455–468.

READING, H. G., ed., 1986, Sedimentary environments and facies, 2nd ed.: Oxford, Blackwell, 615 p. (Chapters 9 and 10: 229–341.)

REINECK, H. -E., and SINGH, I. B., 1980, Depositional sedimentary environments, 2nd ed.: New York, Springer-Verlag, 549 p.

SCHOLLE, P. A., BEBOUT, D. G., and MOORE, C. H., eds., 1983, Carbonate depositional environments: Tulsa, OK, American Association of Petroleum Geologists Memoir 33, 708 p.

TILLMAN, R. W., and SIEMERS, C. T., eds., 1984, Siliciclastic shelf sediments: Tulsa, OK, Society of Economic Paleontologists and Mineralogists Special Publication No. 34, 268 p.

CHAPTER 11

Beaches and Barriers

Modern coastal areas composed of sediment subjected to the action of waves and tides, and not directly influenced by large rivers, can be subdivided into two groups: (1) *mainland beaches* and *barrier-backbarrier complexes* and (2) *peritidal complexes*. We discuss the first group in this chapter and the second in Chapter 12.

Beaches and barriers are important to us for several reasons. Because they form at the edges of the sea or a lake, they are popular places for people to visit for recreation and to build places to live. Beaches and barriers consist of sediments deposited close to base level. Accordingly, their presence in the geologic record serves as a close approximation of ancient sea level. In addition, the sediments of beaches and barriers respond in characteristic ways to rises in water level. Thus, their study can yield important information about the relationship between the stratigraphic record and changes of sea level. The sediments deposited in many ancient barrier complexes are reservoirs filled with oil and gas. Accordingly, extensive investigations into the modern sediments from these complexes have been undertaken by the research staffs of major oil companies. Beaches are also sites where gold, diamonds, and other heavy minerals accumulate to form placer deposits.

Our objectives in this chapter are to become acquainted with the processes and sediments of modern mainland beaches and barrier-backbarrier complexes, to try to formulate criteria for recognizing the sedimentary products of such environments, and to study a few examples from the geologic record that have been interpreted as products of ancient beaches or ancient barrier complexes. As we shall see, the sedimentary record of barrier complexes differs drastically depending on whether it is created during a submergence or during progradation of the shore (at a stationary sea level or during slow submergence).

We begin with some fundamental definitions. After that we summarize the morphodynamics of beaches, describe beach sediments, summarize the various compo-

nents of barrier complexes (including beaches), review how autocyclic barrier successions are formed during progradation and submergence, and then mention inferred ancient examples.

Definitions

Because beaches are such popular places to visit for recreation, most of us tend to take it for granted that everyone knows what a beach is and hence, that the definition of a *beach* requires no elaboration. However, the scientific literature about beaches is replete with conflicting usages and contains many statements that tend to obscure important aspects of beaches. In the following discussion, we attempt to clarify the definition of what is a beach.

Everyone agrees that beaches are features found along the margins of bodies of water, and that beaches are related to waves. Three important factors in the scientific definition of a beach are (1) the kind of material; (2) the inner- or landward limit; and (3) the outer- or seaward (lakeward) limit.

"Beach" clearly is not a synonym for any and all kinds of shorelines. Rocky headlands and mudflats are not beaches. The term beach is appropriately applied only to a body of cohesionless sediment. This usually means sand-size particles, but can include gravel. In the definition of a beach, the composition of the cohesionless particles is not a factor. The sediment can be extrabasinal, intrabasinal, or volcanic. Typically, quartz particles predominate, but other beach material includes carbonate skeletal debris, carbonate nonskeletal particles, volcanic materials, and even tin cans, medical waste, and other human trash. Some beaches are active sites of formation of nonskeletal carbonate particles, such as ooids. In the tropics, carbonate beaches predominate. However, such beaches may be present in high latitudes under special conditions where extrabasinal sediment is not abundant and the calcareous skeletal particles are not swept off the beach.

In the scientific literature about beaches, no serious disagreement exists over where to draw the inner-, or landward, boundary. Along a sandy coast, the landward limit of a beach is placed at the upper limit of wave action. On other kinds of coasts, the landward limit of the beach may not be at the upper limit of wave action. In-stead, if the cohesionless sediment ends within the limit of wave action, the landward limit of a beach is placed at the contact between the cohesionless sediment and other distinctly contrasting material.

Where to place the outer- or water-side boundary of a beach remains a topic of much disagreement. At least four schools of thought have been expressed. (1) The low-water line (Figure 11-1, A) is the boundary that was

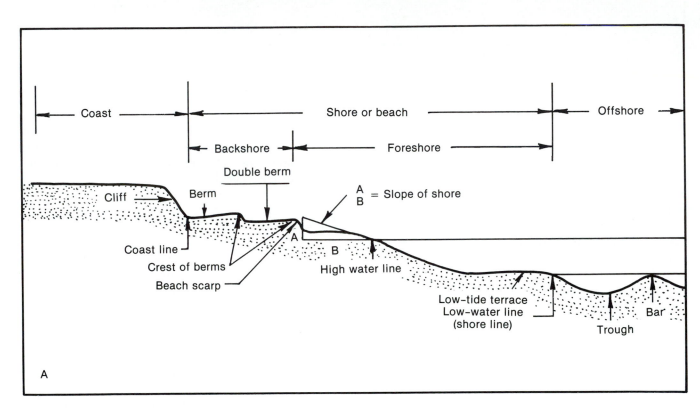

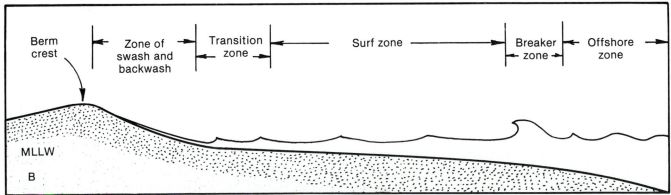

FIGURE 11-1.   Profiles at right angles to shore illustrating various ways of defining beaches and of naming their morphodynamic subdivisions. (Text defines terms.)

A. "Standard" definition of a beach formulated in 1933 by the U.S. Army Corps of Engineers Beach Erosion Board (BEB). By placing the seaward boundary of a beach at the low-water line, the BEB definition excludes the breaker zone, here assigned to the "offshore." (F. P. Shepard, 1948, fig. 33, p. 82.)

B. Slightly expanded definition of beach extended seaward to include the breaker zone and em-phasizing the relationships between surf zone and zone of swash and backwash. Area immediately seaward of the breakers, considered by us to be part of the inner shoreface, is assigned to the "offshore zone" (following BEB usage). (J. C. Ingle, Jr., 1966, fig. 11, p. 181.)

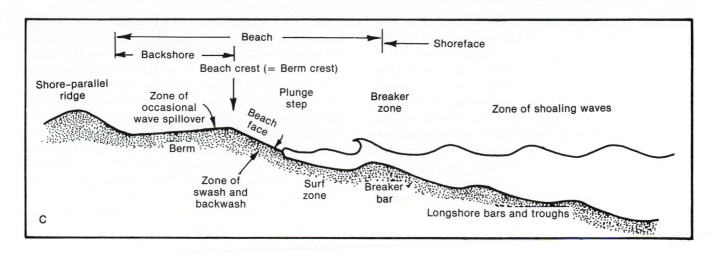

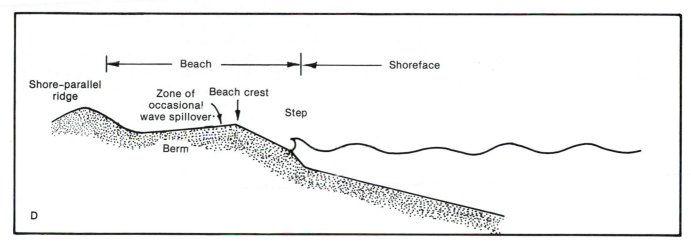

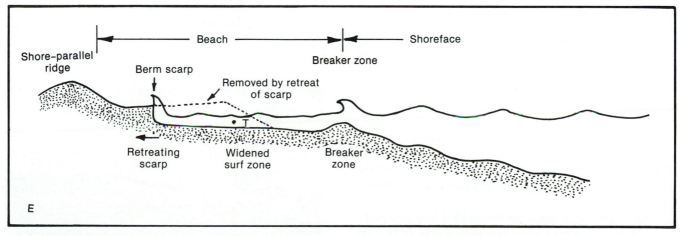

FIGURE 11-1. (*Continued*)

C. Beach inshore from gently sloping bottom exhibits longshore bars on inner shoreface beneath zone of shoaling waves, breaker bar, surf zone and trough, small plunge step, plane beach face, and berm. (Authors.)

D. Stepped beach inshore from steeply sloping bottom lacks breaker bar; waves come almost to water's edge before collapsing over a prominent step. (Based on R. L. Miller and J. M. Zeigler, 1953, fig. 1, p. 418.)

E. Profile of beach as in C, but one in which berm is being rapidly removed. Powerful, wide longshore current in surf zone directed along shore (axis of current represented by dot with letter T to show flow out of plane of profile and toward viewer) has destroyed beach face and created vertical beach scarp. (Authors.)

adopted in 1933 by the Beach Erosion Board of the U.S. Army Corps of Engineers and has been accepted and followed by many other investigators. This definition is convenient to apply, but it excludes the surf zone and the breaker zone. Many students of shore processes have argued that despite the practical convenience of doing so, a beach should not be delineated by the low-water line. But if one includes as parts of a beach features located under water, how far out is it appropriate to go? The other proposed outer boundaries are (2) as far out as the waves influence the bottom sediments; (3) at a depth of about 10 m on an open-sea coast (everyday wave base); and (4) at the outermost line of breakers. (See Figure 11-1, B.) Of these, proposed outer boundary (2) is clearly too broad; it considers as beach the entire zone of shoaling waves, which at times can extend all the way across a continental shelf. Moreover, if tsunami are taken into account, a beach thus defined extends to the bottom of the deep sea. Such a definition would give new meaning to a vacation at the beach. Proposed outer boundary (3) is not so broad as (2), but would consider as beach most of the zone of shoaling waves. Proposed outer boundary (4), preferred by us, provides a basis for defining beaches as features formed by the effects of *breaking*, as contrasted with those of *shoaling*, waves. Three of the four proposed outer boundaries share the problem that they would be difficult or impossible to recognize in the stratigraphic record. Potentially, the products of boundary (2) could be recognized more easily than those of the other boundaries, but it is defined too broadly to be useful.

In summary, then, as used in this book, a **beach** is defined as *a body of cohesionless sediment, along a coast, that is subject to the effects of breaking waves. The landward boundary is the upper limit of the action of water from breaking waves, or the contact between the cohesionless sediment and some other material, whichever is closer to the water. The boundary on the water side is the outermost line of breakers, as determined at low tide if the body of water is subject to tidal fluctuations.*

Lake beaches differ from marine beaches in two important ways: (1) the effects of the tide need not be considered, and (2) the scale is reduced because of the absence of the large waves that can form on the surface of the sea.

Seen in profile, a beach can be divided into two parts: (1) a *backshore*, whose top surface dips gently away from the water, and (2) a *foreshore*, a surface that generally dips toward the water. (See Figure 11-1, A.) *The wide part of the backshore that dips gently away from the water* is known as a **berm**. *The boundary between the backshore and the foreshore where the slope reverses* is usually a distinct feature, the **berm crest**. *The gently sloping part of the foreshore that is usually under water but that may be exposed at low tide* is a **low-tide terrace**. The landward, steeper part of the foreshore is a narrow, variable zone that displays at least three contrasting appearances: (1) *A surface that dips at a low angle (5° to 8°) from the top of the berm toward the water* known as a **beach face**. During many days of the year, this slope may be more or less planar (Figure 11-2). (2) A beach face that is ruffled or corrugated; it consists of rhythmically spaced series of curving headlands and bays, collectively designated as *beach cusps* (Figure 11-3). The horizontal spacing or wavelengths of cusps ranges from a few meters to many tens of meters and their local relief, from a few centimeters to several meters. Several kinds of these rhythmic beach forms are known. The most-symmetrical variety forms as a result of differential seaward progradation when waves are approaching the beach straight on. (3) A vertical- or even locally overhanging slope known as a *beach scarp* (Figure 11-4). Beach scarps form when the water rapidly undercuts and erodes the berm.

As a first approximation, the general behavior of the water is that the breaking waves form surges of water that flow up the beach face as *swash* and down the beach face as *backwash*. Farther along, we shall take up the complicating relationships among breakers, swash, backwash, and the shape of the beach.

From a geomorphologic point of view, beaches can be organized into three major categories: (1) *mainland beaches*, (2) *strand plains*, and (3) *barrier beaches*, which are parts of *barrier islands*. We define the first two briefly and examine barrier islands and associated environments in detail.

A **mainland beach** is *a beach adjoining a mainland*. In terms of environments, a mainland beach is an environment that lies between terrestrial environments and the

FIGURE 11-2.   View westward of plane beach face seaward of low beach scarp, Robert Moses State Park, Fire Island, Long Island, New York, on 26 September 1983. Light-colored beach-face strata, dipping toward ocean at left, abruptly truncate well-laminated berm-top strata consisting of alternating layers of light- and dark-colored sand. (Audrey Massa.)

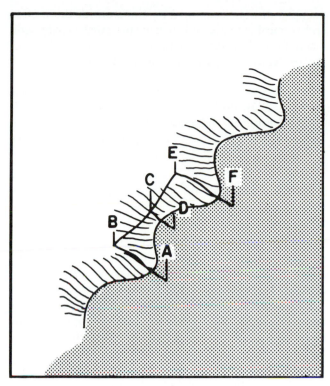

FIGURE 11-3.   Regular, symmetrical beach cusps, sketched in oblique view from above. Letters and lines indicate positions of profile-sections shown in Figure 11-18. Text defines cusps. (J. E. Sanders in G. M. Friedman and J. E. Sanders, 1978, fig. A-28, A, p. 487.)

FIGURE 11-4.   Initial stages in cutting of berm scarp. Onrushing very oblique swash moving toward beach face and away from observer has curled back after striking notch cut at base of small scarp (narrow dark area in center of view); at lower right, some swash water has flowed over the top of the small scarp. Robert Moses State Park, Fire Island, Long Island, New York, view W on 18 September 1975. (I. Baumgaertner.)

FIGURE 11-5.   Small triangular strand plain at SW end of Fire Island barrier, Long Island, New York, that prograded seaward on the east side of a long jetty built in 1939–1940, viewed obliquely from low-flying airplane on 12 October 1976 after dredge had removed all sand west of the jetty. Atlantic Ocean at right; Great South Bay and Long Island (not visible) at left. (B. Caplan.)

sea. Only small intertidal marshes are associated with mainland beaches.

A **strand plain** is *a prograded shore built seaward by waves and currents, and continuous for some distance along a coast.* A strand plain is characterized by parallel beach ridges with intervening swales (Figure 11-5).

A **barrier beach** is *a beach on a barrier island.* In turn, **barrier islands** (or barriers) are *elongate coastal islands, possibly built by large waves, and composed chiefly of cohesionless sediment (sand, gravel, shell debris), and separated from a mainland (or a very large island) by depressions that typically are filled with water (lagoons or bays)* (Figure 11-6). Barrier islands are separated by roughly shore-perpendicular channels, called *tidal inlets,* that are active sites of sediment movement.

Barriers consist of an exposed- or seaward side subjected to vigorous wave action and a protected- or landward or backbarrier side that may be influenced more by tides and by organisms than by waves. A **barrier complex** consists of *all the depositional environments and sediments associated with barrier islands (or barrier spits). All the depositional environments and sediments situated between a barrier island and a mainland (or a very large island)* are collectively designated as a **backbarrier complex.**

On the basis of their tidal ranges, coastal regions have been subdivided into *microtidal* (tidal range < 2 m), *mesotidal* (tidal range 2 to 4 m), and *macrotidal* (tidal range > 4 m) (Figure 11-7).

Microtidal coasts are characterized by long barrier islands and few inlets. These are wave-dominated coasts. On mesotidal coasts may be found barrier islands or peritidal complexes without barriers. Barrier islands of mesotidal coasts are shorter and their inlets more numerous

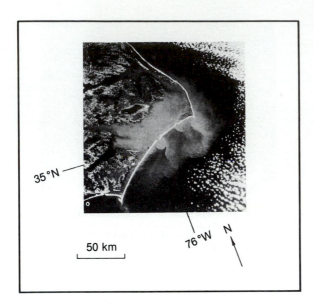

FIGURE 11-6.   Barrier islands and backbarrier complex, Outer Banks and Pamlico Sound, North Carolina, imaged from a satellite. Narrow barrier islands composed of sand appear as thin white strips. Two tidal inlets (marked by semicircular gray areas) break continuity of middle sector of barrier. On landward side, backbarrier sediment is insignificant. Light gray areas in center of view are sandy shoals (on seaward side) and suspended sediment in Pamlico Sound and in two tongues at seaward ends of inlets. The suspended sediment of Pamlico Sound may have been brought landward by tidal action. Asymmetry of sand lobes in shoals seaward of the barrier indicates transport parallel to the coast and toward the southwest. Irregular patterns of white spots at upper- and lower right are clouds over the ocean. (NASA.)

than are barrier complexes of microtidal coasts. Mesotidal coasts are strongly affected by both waves and tides. Macrotidal coasts generally lack barrier islands, because the high tidal range prevents coarse cohesionless sediment from accumulating at a single position (with reference to a line normal to shore). Macrotidal coasts are characterized by peritidal complexes.

The marginal regions of lakes very closely resemble those of oceans. Differences result largely from contrasts of scale. Although many sedimentary features found at the margins of the sea resemble those found along the shores of large lakes (Chapter 14), the features formed in the sea are distinctive in many respects. (1) The scale of marine beaches is much larger than that of lake beaches. (2) Wave action on the open sea never relents, as it does in lakes when the wind dies down. (3) Although the levels of lakes fluctuate, lakes generally lack the well-developed features formed by the ceaseless operation of the ocean's tides. (4) Even the largest lakes are confined to single climatic zones, whereas the shores of the sea stretch across all the world's climate zones. Hence it is possible to compare and contrast marine beaches or intertidal flats, for

example, of a moist-temperate climate with those of an arid-tropical climate, and to trace marginal-marine sediments formed in one climate zone into comparable sediments deposited in an adjacent climate zone.

SOURCES: Barwis and M. O. Hayes, 1979; J. L. Davies, 1964, 1972, 1980; Dolan, Hayden, and Lins, 1980; Friedman and Sanders, 1978; C. A. M. King, 1972; Komar, 1976; Massa, 1981.

Morphodynamics of Marine Beaches

The term *morphodynamics* refers to the interaction between the dynamic activities of some process and the morphologic features built by these dynamic activities. According to our definition of a beach, the basic dynamic activities on beaches are the result of breaking waves.

Based on their profile relationships (as seen along a line normal to the shoreline), kinds of breakers, and interaction between bores from the broken waves and backwash, three beach states have been established: *dissipative*, *transitional*, and *reflective* (Figure 11-8).

In a dissipative beach (also known as a *surf beach*), which is a typical product of low waves shoaling over a low-dipping foreshore, *spilling breakers* build a *breaker bar*. (See Figure 11-1, C.) After a wave has broken initially and the depth of water landward of the outer breaker bar is great enough, the water from the breakers is reorganized into smaller waves of oscillation. Such smaller waves of oscillation continue toward shore until they, too, break and form bores that move landward across the *surf zone*. As shown in Figure 11-1, B, the waves in the surf zone react with the outflow from the *zone of swash and backwash* in a *transition zone*.

A *reflective beach* (also known as a *surge beach*) forms where steep waves shoaling over a steep shoreface end their careers as *plunging breakers*, which generally form a *plunge step* rather than a breaker bar. (See Figure 11-1, D.) A plunge step, or simply a step, is defined as *a steep slope at the outer edge of the beach face*. The steep foreshore reflects the incident wave energy. Reflective beaches do not develop surf zones.

On some beaches, at high tide no surf zone is present; the waves break on a step. (See Figure 11-1, C.) However, at low tide, the waves may break above a submerged breaker bar and a surf zone is present. These are transitional beaches.

SOURCES: Bryant, 1982; Guza and Inman, 1975; Short, 1979; Sonu, 1973; Wright, Chappell, Thom, Bradshaw, and Cowell, 1979; Wright, Nielson, Short, and Green, 1982; Wright and Short, 1984; Wright, Short, and Green, 1985a.

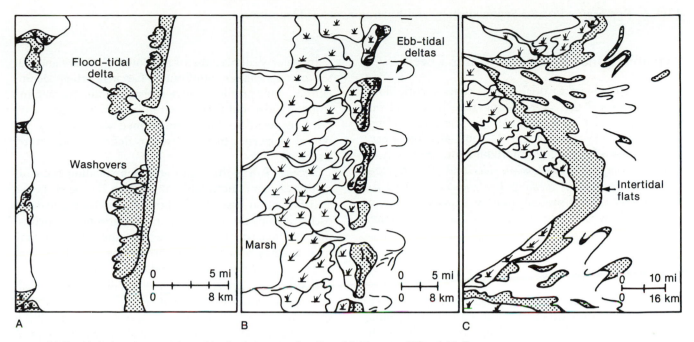

FIGURE 11-7. Variations in coastal sand-body shape as a function of tidal range. (After J. H. Barwis and M. O. Hayes, 1979.)

A. Microtidal coast features long, narrow barriers and few inlets.

B. Mesotidal coast is typified by short barriers and numerous inlets.

C. Macrotidal coast displays characteristic estuaries and prominent linear, tidal-current ridges in nearshore zone; barriers are not present.

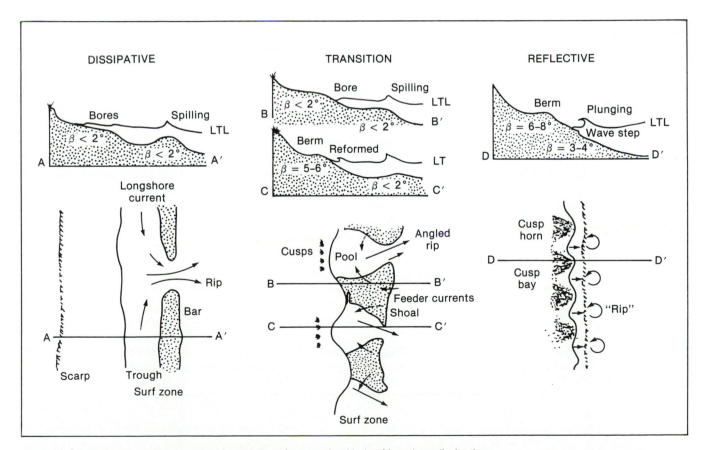

FIGURE 11-8. Profiles and sketch maps of segments of contrasting kinds of beaches: dissipative (compare Figure 11-1, C) and reflective (compare Figure 11-1, D) and an intermediate transition beach (compare Figure 11-1, B). (Edward Bryant, 1982, fig. 1, p. 433.)

Dynamic Zones Related to Water Levels

On a beach subject to tidal action, characteristic flows of water toward the shore, along the shore, and out to sea again take place at three dynamic levels related to the tides. These levels are: (1) subtidal zone, *a zone that is always submerged*, (2) intertidal zone, *the zone lying within the range of the astronomic tides*, and (3) supratidal zone, *the zone that is submerged only during spring tides*. Where tides are not present, only two zones are important: (1) below the water level and (2) above the water level.

SUBTIDAL ZONE

In the subtidal zone, three subzones are important: (1) the subzone within which the waves undergo shoaling transformations and are refracted; (2) the subzone in which the waves break; and (3) the subzone where breaking waves reform and/or are converted to sheets of water (surf) that move toward and/or along shore. Although we analyzed the zone of shoaling transformations on the inner shoreface in Chapter 10, we review important points here as they are related to beaches.

The following paragraphs sketch the changes that waves undergo as they approach the shore, break, and reform and/or are transformed into surf.

Zone of Shoaling Transformations. Between the zone where the waves do not affect the bottom (and thus are named deep-water waves) and the breaker zone, waves interact with the bottom in various ways. These interactions are collectively designated as shoaling transformations.

Between depth $L/2$ and $L/8$, changes in the waves are relatively insignificant. The orbits are still circular, wave profiles closely resemble those of deep-water waves, and wave celerities are about the same as they are in deep water.

In water shallower than $L/8$, however, waves and water motions change significantly. Circular orbits become ellipses, wave profiles change (crests become steeper and troughs wider and flatter than in deep water), and wave celerities decrease. Such transformed waves display the typical properties that are used to define shallow-water waves.

Because $L/8$ is the depth at which shoaling transformations become noticeable, this depth has been named transformation base. At the bottom, the back-and-forth oscillations of the water as it moves around in "orbits" that are much-flattened ellipses become strong enough to form small-scale ripples where the water–sediment interface is underlain by fine sand.

Before they break, some waves (all the low swells but almost none of the steep sea waves) enter a zone shallower than 1/20th of their wavelengths ($L/20$), at which point they become very-shallow-water waves. A fundamental principle that applies to all very-shallow-water waves is that their speeds are no longer controlled by their wavelengths but only by the depth of water. The mathematical function is that wave celerity equals the square root of the product of the Earth's gravitational acceleration times the depth of water. Table 11-1 tabulates some data related to these important changes.

Two implications of the numbers in this table are: (1) that at the depth where they are converted to very-shallow-water waves, the celerities of long-period swells are much reduced from their values in deep water (for 8-s waves, 7 m/s compared with 12 m/s) and that as water depth decreases to the other values shown in the table, celerity of such waves decreases even further (now being independent of the sizes of the waves and controlled solely by water depth); and (2) short-period waves will break before they become very-shallow-water waves. As a result, they never enter the inner zone of wave refraction within which swells are refracted so prominently outside the breaker zone.

TABLE 11-1. Some properties of waves compared

Period of waves (s)	Speed of wave in deep water (m/s)	Depth of water at wave base (L/2) (m)	Depth where waves become very-shallow-water waves (L/20) (m)	Speed of very-shallow-water waves at depth of conversion (m/s)
8	12	50	5.0	7
5	9	19.5	2.0	5
4	6.6	12.5	1.5	4
3	3.4	7	1.0	3.1

The fundamental paper on wave refraction by Munk and Traylor (1947) illustrates the behavior of long-period swells crossing a narrow shelf and refracting notably just outside the breaker zone as they become very-shallow-water waves and pass through the inner zone of wave refraction and then break. On the Atlantic coast, as they travel across the wide, shallow continental shelf, many swells are refracted in the outer zone of wave refraction. By the time such refracted swells become very-shallow-water waves and thus enter the inner zone of wave re-fraction, their crests have already become nearly parallel to the breaker zone.

SOURCES: Friedman and Sanders, 1978; Munk and Traylor, 1947; Seymour, 1989.

Breaker Zone. The key feature within the subtidal zone that we use to define the outer limit of the beach is the low-tide breaker zone. The typical morphologic expression of the breaker zone is a ridge (known as a bar) composed of coarse sediment (sand or even gravel) built by the surges of water associated with each line of breakers (toward the land on the sea side and toward the sea on the land side).

Surf Zone.   *Reconstituted waves occurring in the zone immediately shoreward of the breakers* form the surf. Where surf is present, we can define the surf zone; it is *the zone of vigorously moving water lying landward of the breaker zone and on the seaward side of the zone of swash and backwash* (Figure 11-9, A). Although in the surf zone the water at the surface flows generally landward, the water along the bottom flows away from the land. Oblique approach of the waves creates a *longshore current* in the surf zone. (See Figure 11-9, B.)

Apart from the currents in the surf zone, waves are present. These may move landward, seaward, and along the shore. The landward-moving waves are those that have reformed from the collapse of the water in the breaker zone. The seaward-moving waves are those that form when the swash water "piles up" against the water in the surf zone to form a wave known as a *soliton*. The waves that move along the shore are various *edge waves*.

What has been referred to as the surf zone does not fit neatly into the three morphodynamic zones being discussed. Strictly speaking, the surf zone belongs with the subtidal morphodynamic zone; it would not exist without the water. However, the part of a beach that is a surf zone at high tide can become dry at low tide. Therefore, this part of the beach belongs with the intertidal zone. Accordingly, important parts of the activities within the surf zone are discussed in a following section entitled intertidal zone. And, as is explained farther along, what is a surf zone at low tide can become part of the zone of wave transformations at high tide.

Water Circulation.   Oblique approach of waves drives water into the subtidal morphodynamic zone at an angle. As a result, the water tends to "pile up"; it adjusts by flowing parallel to shore. The "piling up" is most pronounced within the troughs landward of the breaker bar(s). At intervals, this water flowing parallel to shore within these troughs, particularly within the surf zone, cuts gaps through the breaker bar(s) and returns seaward as narrow currents named *rip currents*.

The collective name for *coast-parallel transport of sediment in any morphodynamic zone* is longshore transport (also known as *longshore drift*). The shore-parallel transport of sediment in the surf zone contributes significantly to the total longshore drift. Such longshore transport of sediment in the surf zone takes place independently of what may be happening in the other morphodynamic zones. In other words, if one uses a mathematical formula based on wave parameters to predict the amount of longshore sediment transport in the surf zone, then one has a tool that is valid only for predicting longshore sediment transport in the surf zone—period. This statement may sound ridiculous, but it has been made deliberately. Too many attempts have been made, for example, which purport to relate the longshore transport in the surf zone to erosion of the beach face. As we shall see, such erosion takes place in the intertidal zone or in the supratidal zone; it is not closely coupled to the transport in the subtidal zone.

Depending on the range of the tides and slope of the nearshore bottom, the kinds of water circulation just described may be active only at low tide. At high tide, this circulation pattern may not be active; instead, the sub-zone may become a part of the zone of shoaling waves. Strictly speaking, at high tide, that part of this subzone should be excluded from the beach as we have defined beach.

SOURCES: Friedman and Sanders, 1978; Reimnitz, Toimil, Shepard, and Gutierrez-Estrada, 1976; Seymour, 1989.

INTERTIDAL ZONE

As a first approximation, one may expect to find within the intertidal- and supratidal zones some of the same general patterns of water circulation just described for the landward part of the subtidal zone. These are sheets of water from breaking waves flowing landward, possibly obliquely, over ridges of sediment and into lower-lying areas to landward; these form currents of water that flow parallel to shore; and at selected points the water cuts channels through the ridge and within these channels, swift currents flow seaward.

Within the intertidal zone, the key points are how the surf behaves and how it relates to the swash and backwash that flow up and down the beach face, respectively.

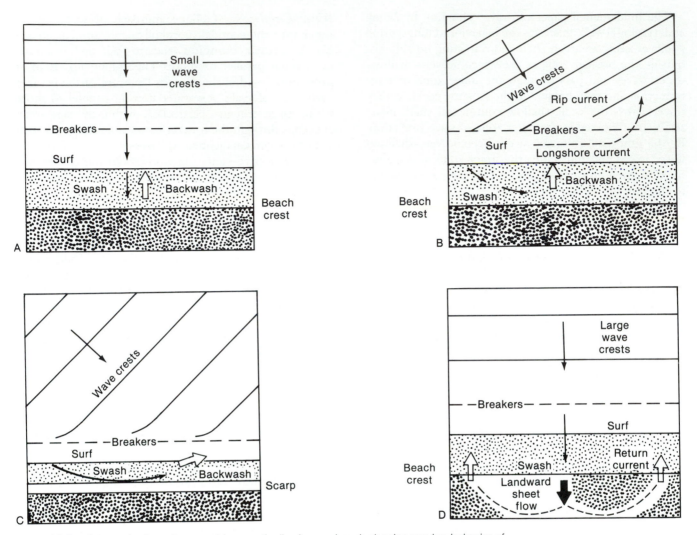

FIGURE 11-9.   Schematic views downward from vertically above a beach showing varying behavior of water that surges landward from breakers created by small waves (A and B) and large waves (C and D).
   A. When waves approach the shore head on, both swash and backwash flow straight up and straight down the beach face.
   B. Oblique approach of waves sends swash diagonally up beach face, but backwash returns straight down beach-face slope. Water from longshore current flowing parallel to shore returns seaward as rip currents through gaps in breaker bar.
   C. Extremely oblique approach of storm waves creates swash that grazes beach, eroding sloping beach face into vertical scarp. Backwash flows nearly parallel with shore. Riverlike flow of alongshore water creates flat-bottomed trough.
   D. Large waves coming straight in create vigorous swash, much of which does not become backwash but flows over berm crest and becomes landward sheet flow (spillover, thick black arrow). Eventually, spillover water creates pools, which drain by cutting backflow channels (thick open arrows) through berm-crest ridge. (Based on observations at Fire Island, Long Island, New York, by I. Baumgaertner and J. E. Sanders.) (J. E. Sanders in G. M. Friedman and J. E. Sanders, fig. A-29, p. 489.)

Waves that have not been fully refracted approach the shore obliquely. Such oblique waves develop swashes having directions of flow that are along- as well as up the beach face. Such oblique swashes travel diagonally up the beach face, but after becoming backwashes, tend to flow straight down the beach-face slope. Particles pushed along by diagonal swashes move not only up and down the beach face, but along it as well. *Movement of particles*

parallel to shore on the beach face is **beach drifting**. (See Figure 11-9, B.)

Water Circulation. The kind of circulation in the intertidal zone depends on the state of the tide. At high tide, the circulation is as just described for the subtidal zone. Longshore currents form inside the line of breakers and at intervals rip currents cut through the breaker bars to

FIGURE 11-10. Intertidal zone showing well-developed ridge and runnel viewed from vertically above at low tide, with west at top and north at right. When tide was high, water flowed parallel to shore in runnel, creating small flow-transverse dunes having long axes nearly perpendicular to shore. Flow seaward through gaps in ridge (top, center) generated small flow-transverse dunes having long axes parallel to shore. Linear features at right are traces of scarps that were eroded at various times during preceding four months. Small waves approaching obliquely from SW (upper left) break before they are refracted. Other, less-obvious waves are tiny sets of wind-generated waves traveling from W to E, nearly perpendicular to shore, and larger swells approaching from S, nearly parallel to shore. Robert Moses State Park, Fire Island, Long Island, New York, 09 April 1975. (B. Caplan.)

panies it on its landward side (the floor of the high-tide surf zone), which become dry at low water, have been named *ridge* and *runnel* (for the trough). On a ridge-and-runnel beach at low tide it is possible to see currents of water flowing parallel to shore as the tide goes out. These currents return seaward through exposed gaps in the ridge. (These are the same gaps through which the rip currents flow at high tide.)

The intertidal morphodynamic zone of a ridge-and-runnel beach on a microtidal coast displays two beach-face zones where swash and backwash are active: the inner beach face forming the seaward edge of the berm, active at high tide, and an outer beach face, on the seaward side of the ridge, active at low tide. When the tide is out, the zone of swash and backwash may be transferred to the outer, temporary beach face, and the floor of the runnel may become dry. Now, the inner beach face also is completely dry and is not affected by the water. (See Figure 11-10.) Along the base of the inner now-dry beach face, ground water from within the berm seeps out, forming a line of miniature springs.

When the tide returns, the low-tide zone of swash and backwash migrates up to the crest of the now-submerging ridge. Water floods into the runnel. Water from large swashes may spill over the crest of the ridge and pour into the runnel. These sheets of water do not become backwash; they flow landward and may deposit their entrained sand as tiny spillover delta lobes in the quiet water they encounter in the runnel (Figure 11-11). As the tide rises higher and higher, a point is reached where the zone of swash and backwash shifts abruptly

FIGURE 11-11. Spillover-delta lobes formed in intertidal morphodynamic zone, Robert Moses State Park, Fire Island, New York, viewed looking east at low tide on 27 February 1983. Each lobe is at landward end of a shallow, flat-bottom, steep-walled channel through which, at high tide, swash water that overtopped the crest of the ridge fed into the water filling the runnel. Dark area at left of center is silt that settled from suspension on the floor of the runnel partially covering the small ripples formed when the runnel was submerged. Also present are scattered small pools of water that did not drain away with the outgoing tide. (Audrey massa.)

return water seaward (Figure 11-10). Also at high tide, the inner edge of the intertidal circulation zone is the zone of swash and backwash on the beach face. As long as this remarkable zone of swash and backwash is operating, it serves as a buffer along the landward side of the longshore current in the high-tide surf zone. This flow of water parallel to shore in the high-tide surf zone tends to create a flat bottom. At low tide, this flat surface may be exposed. If so, then most workers would probably refer to it as a "low-tide terrace." This name is slightly misleading; its distinctive slope is determined by what happens at high tide.

At low tide, the breakers may shift seaward exposing the sand ridge that had been functioning as the innermost breaker bar. This ridge and the trough that accom-

from the outer, temporary beach face to the inner beach face. We think that this abrupt twice-daily change of the zone of swash and backwash from the outer- to the inner beach face exemplifies what might happen on a larger scale during submergence. (See discussion in a following section.)

SOURCES: Dabrio, 1982; Friedman and Sanders, 1978; Greenwood and Davidson-Arnott, 1979; Greenwood and Mittler, 1985; Howard and Reineck, 1981.

Formation of Berm Scarps. At certain times, and for reasons that are not fully understood, the behavior of water in the high-tide surf zone as just described is supplanted by a totally different pattern. One of the keys to this change is what might be referred to as "reverse refraction." Within the deeper water filling the trough landward of the breaker bar, waves that have broken re-form into new waves. The crests of these newly re-formed waves swing around to become perpendicular to shore rather than parallel to shore. (In normal refraction, the trends of wave crests tend to become parallel to shore. They are thus edge waves. In this "reverse refraction," just the opposite happens—the crests tend to become perpendicular to shore.) Under certain conditions, water motions in the surf zone cause the zone of swash and backwash to disappear. When this happens, the sheets of water undercut the sediment of the inner beach face and cause it to collapse and thus transform it into a vertical scarp. (See Figure 11-4.) Sheets of water flowing more parallel to the shore than toward it undermine the base of this scarp causing more and more sand to collapse along vertical fractures. (This behavior of the sand is related to the pore pressure of the interstitial capillary water; the surface tension of the water acts as a "cement" to hold the sand particles together.) If the undercutting mode is established on a rising tide, then the entire berm can be eroded away in a few hours. If the tide rises higher than normal, the landward-retreating scarp can "eat" into any higher ridges of sand that may be present along the landward edge of the berm ("shore-parallel ridges" as a general, non-genetic term; "dunes" of most popular usage). Such retreating berm scarps are the chief cause of erosion of the subaerial parts of beaches and adjacent sand bodies.

To summarize, the dynamic activities in the intertidal morphodynamic zone range from those of the surf zone at high tide to the results of wave spillover and associated return currents as the tide rises.

SOURCES: Baumgaertner, 1975; Friedman and Sanders, 1978; Sherman and Nordstrom, 1985.

SUPRATIDAL ZONE

In the supratidal zone, water appears only rarely, but when it does, incredibly swift changes can take place. At higher-than-normal spring tides (during a perigee/syzygy coincidence, for example) or when large swells are breaking, or both of these conditions prevail, sheets of swash water flow up to the crest of the inner beach face, building it upward. Within a few hours, the crest of the beach (berm crest) may stand well above the general level of the rest of the berm. How high depends on the tidal height and sizes of the waves; it could be as much as a meter or more. The swashes that wash completely across the berm crest continue to flow landward down the back slope that leads away from the crest. If the berm-top sand is dry, as it typically is during dry weather, then many of the first-arriving sheets of spilled-over swash disappear by sinking into the dry sand. Eventually, however, the berm-top sand may become saturated. After the sand has become saturated, the water from the spillover sheets begins to accumulate; it may create spillover pools of varying sizes. These spillover pools may continue to enlarge until their ever-rising water level finds low places to serve as outlets. Such outlets may be low swales along the crest-line of the berm, on the seaward side, or gaps in the crest lines of the shore-parallel ridge(s) ["dune(s)"] on the landward side. At Robert Moses State Park, Fire Island, Long Island, New York, some of these berm-top spillover pools have attained impressive dimensions. Their widths perpendicular to shore have reached many tens of meters; their lengths parallel to shore, a half a kilometer or so (Figure 11-12); and their maximum water depths, 2 meters. Water draining seaward from such substantial spillover pools has eroded gaps in the beach crest that were 15 meters wide and 2 meters deep. During one storm at Robert Moses State Park, Fire Island, the skipper of a small power boat in distress guided his vessel to safety from the waves by navigating through one of these return channels and beaching his craft on the landward side in the quiet water of a spillover pool. Two days later, he returned with a low-loader trailer and hauled his boat away overland.

Within these berm-top spillover pools, strong currents flow parallel to shore; in the outlet channels, powerful currents flow away from shore. The amounts of sand that can be shifted by the swift currents flowing in the intertidal circulation zone at high tide and along the top of the berm in the supratidal circulation zone can be tremendous.

SOURCES: Barwis and M. O. Hayes, 1979; Friedman and Sanders, 1978; M. O. Hayes, 1980; Rosalsky, 1949, 1964.

FIGURE 11-12.   Supratidal- and intertidal zones of an ocean beach in oblique view looking northward from small airplane at low tide on 04 November 1970, Robert Moses State Park, Fire Island, Long Island, New York. Dark area in foreground is water of Atlantic Ocean. White area in middle is sand, from an intertidal ridge and supratidal berm. Thin dark strip in middle is water draining seaward from a runnel via a gap in the ridge (lower right). Gray area crossing center of view is water from a long, narrow, shallow bermtop spillover pool containing water trapped between the sand of berm crest on the ocean side and that of the shore-parallel ridges ("dunes" of popular usage) within the triangular strand plain on the opposite side. (See Figure 11-5.) (B. Caplan.)

Responses to Changing Conditions

Beaches are extremely responsive to changes in their formative processes. We illustrate this point by considering effects of changes of winds and waves and of water level. In the water-level category we include only short-term changes here; the effects of longer-term changes in water level are discussed in a following section.

EFFECTS OF CHANGING WINDS AND WAVES

Although in the popular mind, waves are considered to be destructive to beaches, in reality the place where most waves are really destructive is the inner zone of the shoreface. Most shoaling waves incessantly "steal" sand from this zone of the shoreface. Much of the "stolen" sand goes to the adjacent beach and some goes farther offshore. Only now and then, when particular wave conditions are established during a rising tide, however, do waves "steal" sand from the beach. What becomes of all the sand eroded from the beach is not surely known; much of it moves along the shore to various sinks. How much goes back out to the shoreface is still a mystery. What this means is that not all "wave energy" is destructive to beaches. Quite the contrary, tremendous amounts of "wave energy" are expended in building beaches.

This contrasting effect of the waves is aptly illustrated by the well-known relationship that episodes of destructive erosion (removal of berm and formation of beach scarp) are followed almost immediately by a period of post-storm recovery. The sand to restore the eroded berms comes from the inner zone of the shoreface; it is delivered to the berms by onshore migration of various bars and/or by fluxing through the breaker bars.

Detailed analyses of beach response to changing atmospheric conditions (expressed by changes in barometric pressure, wind speed, and wind direction) and water conditions (expressed as breaker height and velocities of longshore currents) have been made on the east shore of Lake Michigan, on the Texas Gulf coast, and on the Pacific coast of Oregon (Figure 11-13). In the Oregon example, the typical summertime atmospheric pattern is established by the East Pacific high-pressure cell. Clockwise flow of air around this high-pressure cell creates longshore winds from the north. During the winter, this East Pacific high shifts southward and in its place is a low-pressure cell. Counterclockwise flow around this offshore low-pressure center creates longshore winds from the south. Summer storms are associated with low-pressure cells that move onshore. These reverse the direction of the longshore north winds, which are associated with high barometric pressure.

The time-series graphs show a generally close inverse match between barometric pressure and breaker height. That is, high breakers closely match times of low barometric pressure. Much less of a match exists between alongshore wind speed and breaker height. These relationships indicate that breakers reflect offshore atmospheric conditions more closely than they do the longshore wind. The curve of the velocities (speed plus direction) of the longshore current generally matches that of the velocities of longshore wind but with a lag of several days. Because no directional data for breakers are shown, the controls on the longshore current remain ambiguous.

The major mode of sand transport is via intertidal sand bars that form in water depths of 1 to 2 m and shift both toward the shore and along the shore. Shoreward rates ranged from 1 to 5 m/day; alongshore rates, from 10 to 15 m/day. Because of the rapidly shifting positions of the gaps in the line of bars, the shoreline remained straight.

The time-series analyses of the Lake Michigan beaches showed a close connection among barometric pressure, breaker height, longshore wind speed, and longshore current. The beach showed a cyclic series of responses to a change from fair-weather southwest winds to strong northwest winds as a storm center passed through the region. During a stretch of fair weather and gentle southwest winds, the small longshore current flowed

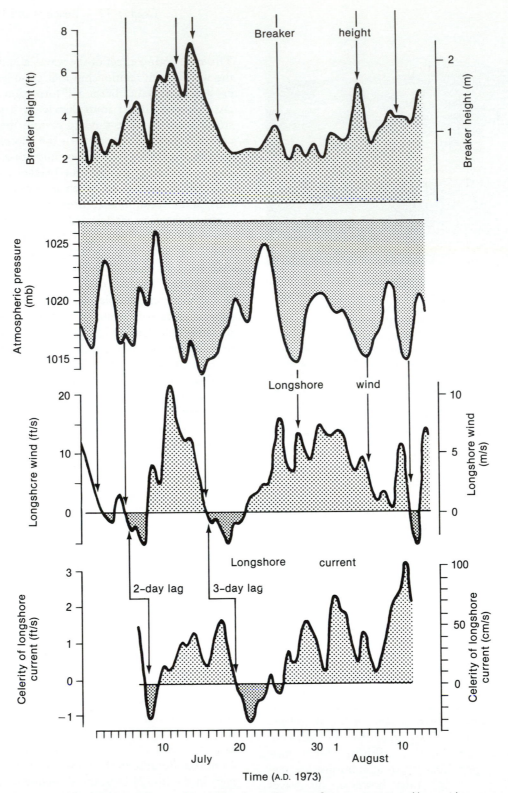

FIGURE 11-13.   Smoothed time-series graphs of conditions on Oregon coast near Newport in summer of 1973. Changes in barometric pressure (recorded hourly) are associated with passage of weak weather fronts that control the direction and speed of winds (north winds above X axis, wind speed increasing upward; south winds below X axis, wind speed increasing downward) and of longshore currents (toward south above X axis, celerity increasing upward; toward north below X axis, celerity increasing downward and indicated by negative numbers). (After W. T. Fox, Jr., and R. A. Davis, 1976, figs. 13, p. 15 and 14, p. 17.)

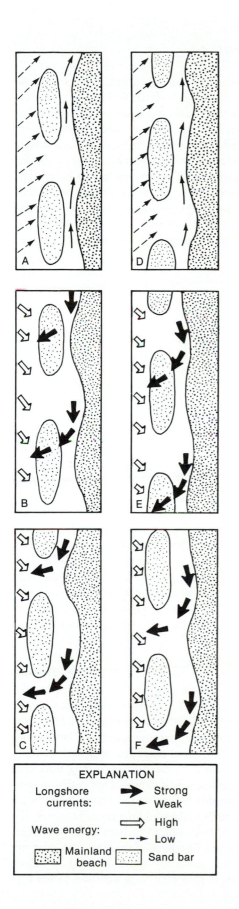

northward, and the mainland part of the beach prograded lakeward, more opposite the longshore bars than opposite gaps in the bar crest. The result was an embayed shoreline, with bays opposite gaps in the bar (Figure 11-14, A, D, and G). When strong northwest winds created storm waves from the northwest, the longshore current reversed and the headlands (opposite the bars) diverted its flow lakeward. As a result, the diverted current cut new gaps across the bars; the former gaps were filled. (See Figure 11-14, B and E.) The effect of one storm and the reestablishment of fair-weather conditions was to displace the gaps (and subsequently formed shoreline embayments). The displacement resulting from a second storm and subsequent reestablishment of fair-weather conditions restored the situation that existed prior to the first storm.

Where the wave regimen follows a seasonal pattern, the beaches may display a seasonal cyclicity. For exam-

FIGURE 11-14.   Schematic maps showing responses of beach on eastern shore of Lake Michigan to changing wave conditions. (After W. T. Fox, Jr., and R. A. Davis, 1976, fig. 7, p. 8.)

A. Typical fair-weather conditions of summer. Small waves from SW (dashed-line arrows) establish north-flowing longshore current (thin, solid-line arrows) between bars (at left) and mainland part of beach (at right). Notice embayments in mainland part of beach opposite gaps in line of bars and corresponding "headlands" opposite bars. This relationship has resulted from differential progradation westward of the mainland part of the beach as sand was added from the inner part of the shoreface (not shown; at left). More sand came ashore by moving through the bars than through the gaps.

B. Large waves from the NW (open arrows) associated with the passage of a cyclonic storm establish vigorous south-flowing longshore currents (thick black arrows). The projecting "headland" parts of the shoreline of the mainland segment of the beach divert these currents and cause them to flow offshore, where they erode gaps through the bars.

C. New configuration of bars, gaps, and shoreline of mainland segment of beach resulting from erosion by offshore-directed longshore currents. Projecting "headlands" are opposite bars; embayments are opposite gaps.

D. Conditions resulting from post-storm resumption of small waves from the SW and associated fair-weather differential westward progradation of the mainland part of beach. Notice that the bar in the middle of the map occupies the position of the gap between bars in A. Correspondingly, a "headland" in the shoreline corresponds with the location of the former embayment of map A.

E. Repetition of storm waves from NW (as in B); "headlands" divert currents offshore as in B.

F. End product of erosion of new gaps by strong offshore-diverted longshore currents. Situation comparable to that shown in map C, but with "headland" and new gap in F corresponding to former position of embayment and bar in C.

G. Conditions resulting from post-storm resumption of small waves from SW and associated fair-weather differential westward progradation of mainland part of beach. Compare with A.

ple, during the winter months, beaches in southern California typically are cut back. During the summer these beaches prograde.

SOURCES: Baumgaertner, 1975; W. T. Fox and R. A. Davis, Jr., 1976; Friedman and Sanders, 1978; M. O. Hayes and Boothroyd, 1987; Shepard, 1950.

EFFECTS OF SHORT-TERM CHANGES IN WATER LEVEL

We illustrate some effects of short-term changes in water level by examining the behavior of the two beach faces in the intertidal morphodynamic zone of a ridge-and-runnel beach on a microtidal coast during each rising tide and the upward accretion of berm-top strata as related to the lunar perigee/syzygy cycle of the tides. Because depth of water influences the kinds and locations of breakers, changing water levels may trigger important feedback mechanisms with respect to the waves.

As mentioned in a previous section, at low tide on a ridge-and-runnel beach the zone of swash and backwash is located at the outer beach face, on the seaward side of the ridge. As the tide rises, however, this zone creeps slowly up to the crest of the ridge. Meanwhile, the level of the water in the runnel is rising. As the crest of the ridge is flooded, the zone of swash and backwash disappears from the ridge. Until water over the ridge deepens to a meter or so, the ridge will dampen the arriving large waves. Eventually, the zone of swash and backwash shifts back to its high-tide position on the inner beach face. During each rising of the tide, therefore, the zone of swash and backwash shifts from the outer beach face to the inner beach face. In so doing, it jumps across the deeper water of the runnel. Comparable shifts of the zone of swash and backwash may take place when the sea submerges a barrier, as discussed in a later section. During submergence, the shifts may be from the barrier being submerged to a new barrier that forms when waves can pass over the former barrier, now submerged, without breaking.

During the three months preceding a perigee/syzygy maximum in the heights of spring tides, the maximum water levels attain progressively greater heights. Correspondingly, and given sufficiently high waves, new berm-top strata may be added.

The changing tide levels can affect the kinds of breakers formed by a given set of waves. Thus, at low tide, waves might form plunging breakers, whereas at high tide, they would form spilling breakers, or vice versa. Because the kinds of breaker exert such profound influences on beach morphodynamics, any change in kind of breaker is important.

Beach Sediments

According to our expanded definition of a beach and the discussion of the morphodynamics of the three active zones of an ocean beach, the kinds of sediments that form in a beach include much more variety than is found on the mid-tide level of the beach face. We emphasize this point because many investigators have been careful to sample the mid-tide level of a beach face as being representative of "beach sands." We review sandy beach sediments under the headings of planar strata, cross strata, cusps, and channel fills. We then show how these are related to the morphodynamic zones and how these respond to changing water levels and/or depth zones. We close the section on sandy beaches by showing how seaward prograding of these zones deposits a distinctive autocyclic succession of beach sands. Finally, we mention gravel beaches.

Planar Strata in Beach Sands

As mentioned, the beach-face sands, generally regarded as being representative of an entire sandy beach, are deposited by the activity of the swash and backwash. Beach-face strata dip toward the water at angles determined by the amount of slope of the beach face itself. (See Figure 11-2.) Other planar strata in beach sands are products of swash spillover onto the top of the adjacent berm. As a result, the attitude of these berm-top sandy strata conforms to the dip of the backshore. Seen in strike section (parallel to shore), berm-top strata are nearly horizontal. In a dip section (perpendicular to shore), they dip at a low angle away from the water. (See Figure 11-2.)

The effect of the swash on the beach-face sands has been found experimentally to differ from that of the backwash (Figure 11-15). The swash comes in as a thin sheet that flows up the beach face against friction and the tangential component of gravity (g_t). (See Figure 7-1.) Some of the water soaks into the sediment. The swash slows, stops, and begins to flow back down the slope. The important point seems to be that when the swash has stopped moving, some water is still present on the surface.

By contrast, the backwash flows down the beach face, in the same direction as g_t, gaining speed as it overcomes friction. The only thing that stops the backwash is a loss of water or a collision with the next-incoming bore that will become the next sheet of swash. This collision usually takes place just downslope of the beach face. As far as the beach-face sediments are concerned, however, the sequence of events in an outgoing backwash is increasing speed, decreasing depth, and then no water at all. When the water has gone, the action stops. Just before the water vanishes, it may form tiny antidunes (See Fig.

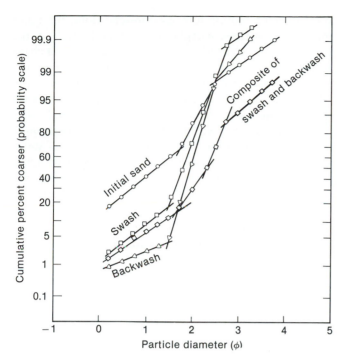

FIGURE 11-15. Cumulative frequency-distribution curves, probability scale, of sands from experiments in small wave tank (2 m long, 30 cm wide) designed to study effects of swash and backwash of waves having periods of about 3 s. Samples were collected using greased board after a beach face had been established (at upper left, initial sand). Swash curve is from sample of surface layer collected after passage of a swash but before the backwash. Backwash samples were collected before the next swash. Composite sample was collected after wave action stopped. In particle-size range between 1.5 and 3.0ϕ, curve of composite sample shows two straight-line segments joined by change of slope at 50th percentile, 2.3ϕ. This change of slope has been named the "saltation break." By contrast, in size range between 1.5 and about 2.4ϕ, curve of swash sample and curve of backwash sample show one straight-line segment. (Replotted from J. R. Kolmer, 1973; initial sample, fig. 3, p. 201; swash sample, fig. 6, p. 202, backwash sample, fig. 7, p. 203, composite sample, fig. 4, p. 202.)

7-22.) and small kinematic waves, known as roll waves, that are thought to be small-scale features comparable to the flood bulges in a river (Chapter 14). These waves exist for only 1 s or less. After the passing and vanishing of the backwash, the beach-face sediments are distinctively sorted. (See backwash curve in Figure 11-15.)

The swash adds layer upon layer of new sand to the beach face and thin sheets of spillover water build up the top of the berm. Eventually, the crest of the berm may stand as much as 2 m above mean sea level.

During some storms (when waves come straight in) the berm can accrete upward very rapidly. In August 1971, waves from Hurricane Doria added about 30 cm of new sand to the top of the berm at Fire Island, New

York. The limit to how high the berm can grow is set by the waves. Presumably, a berm can grow no higher than the height of the runup water from the breaking waves; runup height is controlled by wave height.

Berm-top sediments deposited by spillover sheets are of three kinds: (1) well-sorted sands that are almost pure heavy-mineral concentrates; (2) well-sorted sands almost totally devoid of heavy minerals and of shell debris; and (3) poorly sorted mixtures of coarse sand, fine pebbles, and miscellaneous shell debris. All three kinds can form nearly horizontal, plane laminae whose gentle inclination landward matches that of the surface of the berm top.

Cross Strata in Beach Sands

Although many investigators believe that beach sands lack cross strata, observations in many modern beaches prove otherwise. Cross strata in beach sands result from the unidirectional flow of thin sheets of water and resulting formation and migration of small dunes and ripples, as in flash-flood deposits (See Figure 5-23.), or from the growth into a standing body of water of sand that falls down underwater avalanche faces, as on a Gilbert-type delta (Chapter 14; Figure 11-16). Therefore, beach sands share many similarities with sediments deposited in shallow streams. However, the sediments deposited by shallow streams are dominated by products of a unidirectional current whose flow is governed by the downvalley gradient. In contrast, beach sediments are deposited by shallow sheets of water flowing unidirectionally at any

FIGURE 11-16. Cross strata in beach sands, Robert Moses State Park, Fire Island barrier, S shore of Long Island, New York. (Authors.)
A. Planar cross strata that dip steeply landward (toward upper right) deposited at high tide by waves shifting sediment across the top of the ridge and depositing it on the steep, angle-of-repose slope forming the landward face of the ridge, viewed in vertical side of small trench (wall about 0.5 m high and 8 m long) dug perpendicular to shore on inner side of ridge and exposed at low tide.

FIGURE 11-16. (*Continued*)
B. The same area 2 weeks later after vertical accretion of 25 cm of nearly horizontal strata.

one place and time but with directions of flow governed by many factors. The result is that cross strata in beach sands dip toward shore, along the shore in both directions, and away from shore.

Channels in Beach Sands

Channels are a second example of features many geologists assume are not present in beach sands. However, as with cross strata, channels of contrasting sizes are numerous in some beach sands. Large channels result where water washes through the crest of a berm (Figure 11-17). Smaller channels are cut by water draining sea-

FIGURE 11-17.   East side of washover channel cut through berm-top sediments of barrier island, Westhampton, S shore of Long Island, New York, photographed in July 1970. Crudely stratified sediment deposited in washover channel contrasts markedly with evenly laminated berm-top strata containing many layers that are heavy-mineral concentrates (lower left). After wash-through channel had been filled, the surface of the beach was built upward about 1.5 m by deposition of berm-top strata lacking concentrations of heavy minerals. Machete at lower right is 60 cm long. (Authors and D. M. Caldwell.)

ward through the sandy ridges present along the outer margins of all three morphodynamic levels. Indeed, the characteristics of the fillings of some of these small channels in beach sands may be diagnostic of marine beaches.

When spillover pools form on the top of the berm in the supratidal zone, a typical result is that tiny *spillover deltas* are built into them. Eventually, the ponded water will drain seaward, usually by cutting channels through the crest of the berm. (See Figure 5-9.) Flow of water along the top of the berm creates trough cross strata resembling those formed by flash floods in ephemeral streams. (See Figure 5-23.) The dip of such berm-top cross strata generally is parallel to the shoreline.

The fillings of the channels that cut through the berm-crest ridge may be diagnostic of marine beaches. If, before the tide has risen to its peak level, swash from large waves has overtopped the berm-crest ridge and has formed spillover pools on the top of the berm, and if the escaping water has already cut one or more back-flow channels through the berm-crest ridge, then as the tide continues to rise, both the crest of the berm and the floor of the channel will accrete upward. This will take place even while all the other water motions are in progress. The result is a distinctive kind of channel-fill stratification. (See Figure 5-9.)

Sediments in Beach Cusps

Sediments in beach cusps build upward from a nearly planar surface to form a surface having distinctive, rhythmically repeated relief features. (See Figure 11-3.) In this respect, cusp strata are somewhat analogous to sets of HCS, which are characterized by flat bases and convex-up tops (Chapter 5).

The thickness (and possibly also the content of heavy minerals) of each new layer that is added varies systematically. Layers that are only one particle diameter thick and that contain mostly heavy minerals in the "bays" thicken progressively in each direction toward the "headlands." As the layers thicken, their proportions of heavy minerals decrease. At the apex of the "headlands" the thickness of the layer may be several centimeters or more. In one example on Fire Island, New York, the thickness was 25 mm. This represents a factor of 100 increase from the thickness of the same layer in the adjacent "bay" (only 0.25 mm). Each lenticular layer adds on in the same way, thin over thin and thick over thick (Figure 11-18). Before long, the surface of the beach displays great relief. When waves strike the beach obliquely, they destroy the cusps. But if straight-in waves reappear, new cusps are deposited. The distinctive stratification of beach cusps is a diagnostic feature of beach sands. As mentioned, the only known analogous feature is hummocky cross stratification.

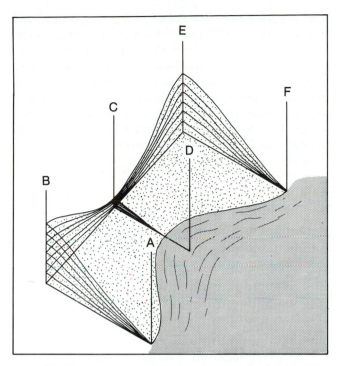

FIGURE 11-18. "Fence diagram" through sediments forming beach cusps showing variation in the thicknesses of individual layers of sand as seen in cutaway views of three trench faces normal to shore [AB and EF through cusp "headlands" and CD through cusp "bay" and one trench parallel to shore (BCE); locations of lines with respect to cusps shown in Figure 11-3]. Sketch is a composite of trenches dug at Great South Beach, Martha's Vineyard, Massachusetts, and at Robert Moses State Park, Fire Island, Long Island, New York. Thin layers of dark-colored heavy minerals at C (and extending toward D) and at A and F contrast with thicker layers composed of light-colored light minerals. (J. E. Sanders, D. Fornari, and W. Wilcox, 1976.)

Relationships to Dynamic Zones

The fundamental point to be emphasized about the relationship between the dynamic zones and their sediments derives from the close connection between motions of water and features in sand-size sediments. But, in beach sands, the result is somewhat paradoxical. The paradox is that waves and beaches form an inseparable combination, yet the diagnostic features generated by waves in sands, namely oscillation ripples, are rare in beach sands. Such wave-generated ripples are confined to the troughs in the inner parts of the subtidal zone and in the upper part of the intertidal zone (where they can form at high tide). Instead, the characteristic features of beach sands result from the distinctive flows of sheets of water in each morphodynamic zone. As mentioned previously, this distinctive pattern of water motions can be generalized in terms of sheets of water surging landward over a ridge and into a trough, flowing parallel to shore in the trough, and then cutting a channel through the con-

straining ridge and flowing back out to sea again. On a dissipative beach, waves collapse over a breaker bar. Smaller waves cross the surf zone or runnel and break on the inner beach face.

"Beach" sands classically are considered to be extremely well sorted. This conclusion is based on extensive studies of beach-face sands. However, as we have emphasized repeatedly, the beach face is only a small part of the total beach as herein used. Other parts that are important sites of sediment deposition are the top of the berm in the supratidal zone and the surf zone (intertidal zone). In these other two zones the sediments may be well sorted, as on the beach face, or (as mentioned) they may not be so well sorted. If sand bodies formed in high-energy environments are overlain by fine sediments deposited in a low-energy environment, the well-sorted sands may be modified; their open pores may be filled by fine particles that trickled down from above. In the absence of such trickled-down fines, beach sands can serve as excellent petroleum reservoirs. They also may serve as host rocks for uranium minerals.

SOURCE: Friedman and Sanders, 1978.

Autocyclic Succession

The tripartite division of marine beaches into subtidal-, intertidal-, and supratidal-zones permits deposition of autocyclic successions. Where a beach progrades, the sediments from these three zones accumulate one above the other, with the subtidal-zone sediments at the bottom and the supratidal-zone sediments at the top. Beach progradation forms an upward-shoaling parasequence. Given that beach sediment is usually coarser than the sediment on the inner zone of the adjacent foreshore, coastal progradation yields an upward-coarsening succession. The thickness of the beach-sediment part of a coastal parasequence is the vertical distance from the breaker zone to the top of the berm, usually not more than 5 meters.

If beach sediments are preserved during a submergence, then the succession will be beach sediments below and offshore sediments above. We discuss these topics further in a following section on barriers.

Gravel Beaches

The preceding discussion of beaches is derived primarily from research on sandy beaches. Gravel beaches share many characteristics with sandy beaches. Differences include (1) shape sorting of pebbles on the basis of differing tendencies to roll; (2) active sieve-type deposition as small particles move around within the pores among larger particles; (3) the inability of the wind to move pebbles; and (4) association with nearby sources of coarse sediment, such as fans or till. In the case of gravel beaches whose supply of sediment is till, waves typically remove all the fine fractions and leave a residue of boulders to form the beach. Because entrainment of gravel requires much more energy than entrainment of sand, the particles on some gravel beaches move only during intense storms. Given sufficiently high wave energy, however, even everyday waves move the gravel. In general, the sorting of gravel in the breaker zone is better, and the particles are both more spherical and coarser than in the swash zone. Shoreface gravels are less well sorted, less well rounded, and less spherical than are gravels found in more-landward positions. Disk-shaped particles accumulate preferentially on the top of the berm.

Active gravel beaches are noisy; the sound of particles banging together can approach that of a great roar. In the zone of swash and backwash on a gravel beach consisting of small pebbles, some individual stones are driven into other stones with such force that they fly right out of the water.

SOURCES: Bluck, 1967, 1969; Bourgeois and Leithold, 1984; R. W. G. Carter and Orford, 1984; Clifton, 1973; Dobkins and Folk, 1970; Fitzgerald and P. S. Rosen, 1987; Kirk, 1980; Leithold and Bourgeois, 1983; Maejima, 1982; Masari and Parea, 1988; L. H. Nielsen, P. N. Johannessen, and F. Surlyk, 1988; Orford, 1975, 1977, 1978 ms.; Postma and Nemec, 1990.

Barriers

Barriers can be divided into two contrasting sides, the open-sea (or exposed) side, and the backbarrier (or sheltered) side. In order from the land into the water, one usually finds shore-parallel ridges and/or coastal dunes, beach, tidal inlets and flood-tidal deltas, and spit platforms and spits.

Features of Open-Sea Sides

The highest part ("backbone") of a barrier consists of one or more shore-parallel ridges and/or coastal dunes. Whatever the origin of a ridge, if it consists of sand, the sand typically dries out and is blown by the wind. As a result, most people call these barrier ridges "dunes" (even if the sand has been heaped up by a bulldozer!). From the point of view of ridge origin and coastal management, it is important to establish whether the ridges are true eolian dunes or the windblown sand is only a thin veneer masking sand that was not heaped up by the wind. Well-demonstrated examples of both situations are known.

Shore-parallel ridges not composed entirely of wind-blown sand probably represent higher-than-normal berm crests built by sheets of extreme runup water. Sand in true coastal dunes can be blown off the beach to migrate inland, or can arrive at the beach by migrating toward the sea from an inland source of sand. The features of the strata should enable one to distinguish true coastal dune sand from dune-decorated ridges having cores of non-eolian sand (Figure 11-19). In the relief peel shown in Figure 5-23, the boundary between windblown sand and water-laid sand is readily apparent.

SOURCES: Massa, 1981; Tanner and Stapor, 1972.

BEACHES

The beaches on barriers display the same features as the mainland beaches discussed in detail in previous sections. Barrier beaches differ only in their association with sediments of inlets and spits on the one hand and with back-barrier sediments on the other.

TIDAL INLETS AND EBB-TIDAL DELTAS

Separating the individual islands of a barrier chain are **tidal inlets**, *deep, narrow channels through which the tide enters and leaves backbarrier depressions.* Inlets are somewhat analogous to meandering flood-plain rivers in the sense that both kinds of channels shift laterally and in so doing, leave behind patterned autocyclic successions of strata. Apart from the differences in the kind of water involved (and thus in organisms present), tidal inlets differ from flood-plain rivers in at least three important respects.

1. The current in a tidal inlet flows in one direction, then stops; flows in the opposite direction, stops; and repeats the sequence each time the tide rises and falls. The flow of water in an inlet is not always in the same direction throughout the cross section of the channel. The incoming tidal current starts to flow landward in one part of the channel while the outflowing ebb current is still active in another part. The same bi-directional flow prevails during the ebb tide. As a result, one part of the channel is characterized by ebb-dominant flow and another part by flood-dominant flow. This asymmetry of water flow is reflected in the directions of asymmetry of the bed forms in channel-floor sediments and in dominant dip directions of cross strata.

2. The shallow water at the margins of an inlet is subjected to wave action. Therefore, instead of finding finer sediments in the progressively shoaling water (as in a river), one finds that, because they are subjected to wave action, the sediments coarsen toward the top. Thus the autocyclic tidal-inlet succession combines the two kinds of shoaling-water successions. The sediments of one of these (a normal channel) become finer upward, and those of the other (from wave action), coarsen up-

FIGURE 11-19.   Coastal dune sands of various thicknesses.

A. View seaward of backbarrier side of barrier beach east of Moriches Inlet, Westhampton, S shore of Long Island, New York, showing thin veneer of windblown sand overlying horizontal strata deposited on top of berm that was built up to a height of about 3 m above mean sea level. Much of the area, whose surface is typical of that of coastal dunes, is underlain by sands deposited by water. In many places, the dune sands are not more than 0.5 m thick. (Authors.)

B. Exposure, about 1.5 m thick, of coastal dune sand, displaying cross strata having two contrasting dips (below) and nearly horizontal strata (above). By analogy with inland dunes described from Wyoming (J. R. Steidtmann, 1974, p. 1839), we presume that the low-dipping cross strata formed low on the downwind slope where the sand was moist. By contrast, the steep, angle-of-repose sets formed near the dune crest where the dry sand avalanched down the steep slip face after being blown over the crest. The position of the water table may have influenced the deposition of the nearly horizontal strata. During a severe storm, the wind may have blown away all the dry sand, leaving only a flat surface underlain by water-saturated sand. The wind would transport sand in horizontal sheets across such flat surfaces, and not rebuild dunes until a plentiful new sand supply had accumulated. North Stradbroke Island, off the coast of Queensland, Australia. (Authors.)

ward. Two such autocyclic successions also are present in deltas (Chapter 13), but with the upward-coarsening succession (related to wave action) below and the upward-fining succession (related to a distributary channel) at the top. In a tidal inlet the channel-dominated part of the succession is at the bottom and the wave-dominated part is at the top.

3. Although the amount of water flowing through a tidal inlet is subject to variations related to changing tidal heights, the flow in an inlet never undergoes the kind of extreme fluctuations that accompany a great river flood.

The sand brought to the seaward end of an inlet builds a shoal known as an *ebb-tidal delta*. In areas of low wave energy ebb-tidal deltas build up semicircular bodies of sand that influence the pattern of wave action. Storm waves tend to destroy ebb-tidal deltas, thus reducing the probability that ebb-tidal deltas will survive to become parts of the geologic record.

The autocyclic succession of strata deposited as a tidal inlet shifts was first determined at Fire Island Inlet, which is located in about the middle of the south shore of Long Island, New York (Figure 11-20).

Fire Island Inlet is remarkable because during the period from 1825 to 1940 it migrated about 8 km to the WSW at a rate that ranged between 5.4 and 251.4 m/yr but averaged 63.6 m/yr, just a little more than a meter per week. (See Figure 11-20.) All of the Fire Island barrier lying WSW of the lighthouse is less than 150 years old and exists where formerly the ocean stood. The remarkable point about the western end of Fire Island is that it appears to be no different from all the rest of the island. Thus, without the dated maps it would require an extremely detailed investigation to determine the manner in which the western end of this barrier island had grown.

The sequence of strata deposited by the lateral migration of Fire Island Inlet (Figure 11-21) is about 10 m thick (depth of water in the inlet) and consists of two major parts: (1) channel below and (2) spit (including spit platform) above. The channel can be subdivided further into three units: (1) *channel-floor lag*, (2) *deep-channel* (below −4.5 m) *sediments*, and (3) *shallow-channel sands* (−4.5 to −3.75 m depth). The spit platform consists of bottomsets, foresets, and topsets, as in a Gilbert-type delta. At Fire Island Inlet the thickness of spit-platform strata is 3.15 m, of which 0.75 m is bottomset, about 2 m is foreset, and the upper 0.4 m is topset. The spit sediments are about 3.6 m thick; they extend from about −0.6 m below mean low water to about 1 m above mean high water (the mean tidal range is about 2 m).

SPIT PLATFORMS AND SPITS

Spits consist of two parts: (1) a lower-level spit-platform and (2) an upper-level subaerial spit proper. Spits lengthen by first extending the submerged platform and then building the exposed part.

The spit platform is fed by longshore currents flowing laterally in the subtidal zone (and at high tide in a runnel if one is present in the intertidal zone). This longshore subaqueous accumulation of wave-driven sand goes on whenever waves approach the coast from the appropriate oblique direction. (If waves approach from the other oblique quadrant they erode the spit platform.)

The spit platform builds out as if it were the lobe of a Gilbert-type delta (Chapter 13). Foreset beds inclined at the angle of repose for fine sand are prominent. The waves and tidal currents sweeping over the spit platform flatten its surface at a level lying about half a meter or so below mean low water. Spit-platform sediments are integral parts of the succession deposited by a migrating tidal inlet. (See Figure 11-21.)

Before the barrier can grow by adding sediment to build shore-parallel ridges and/or coastal dunes, a spit platform with its flat upper surface must be available. Thus the inlet-filling succession may be overlain by beach sand or dune sand or both.

Given a spit platform, the subaerial part familiarly known as a spit can grow (Figure 11-22). The sand to build the upper level (the spit proper) may be transported along shore, either in the upper part of the intertidal zone when vigorous currents in the widened, flat-bottomed surf zone erode a beach scarp or in the coast-parallel flow of supratidal spilled-over water on the berm top, whose return path is along the top of the berm.

If the supply of sand or gravelly sand is plentiful, a new spit can extend itself rapidly forward to the distal edge of the spit platform. The tip of the spit may grow hundreds of meters during a single high tide until it reaches the edge of the spit platform. Thereafter, its further extension in its former direction of rapid growth practically stops and the tip recurves (Figure 11-23).

A subaerial spit that has established itself on the spit platform becomes engaged in a contest for survival. Powerful forces tend to strip the sand off the top of the spit platform. Only continued longshore supply from the adjacent beach guarantees the continued existence of the spit.

The rate of lateral growth of some spits has been phenomenal. The large spit at the mouth of the River Senegal in western Mauritania, Africa, has grown southward 17 km in 100 yr. This is an average rate of 170 m/yr or more than 3 m/wk (Figure 11-24)!

Nearly all the features found on marine beaches can be found on spits as well. Spits may exhibit superimposed dunes and wave ripples. The main differences between a spit and an ocean-side beach of a barrier arise from the contrasting ways in which spillover water can flow. On a beach that is backed by ridges (dune ridges? beach ridges?), spillover water becomes ponded. On spits, such

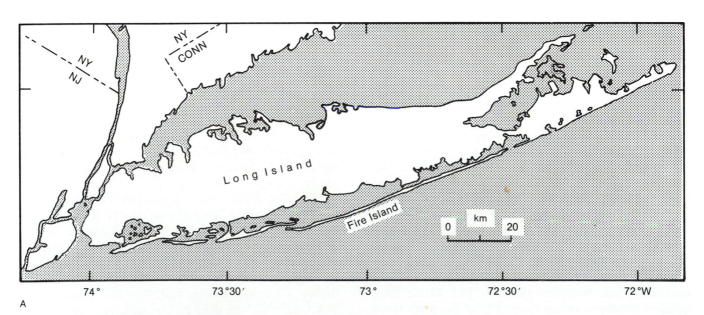

FIGURE 11-20. Alongshore migration of tidal inlet.
 A. Index map of Long Island, New York, showing location of Fire Island barrier.

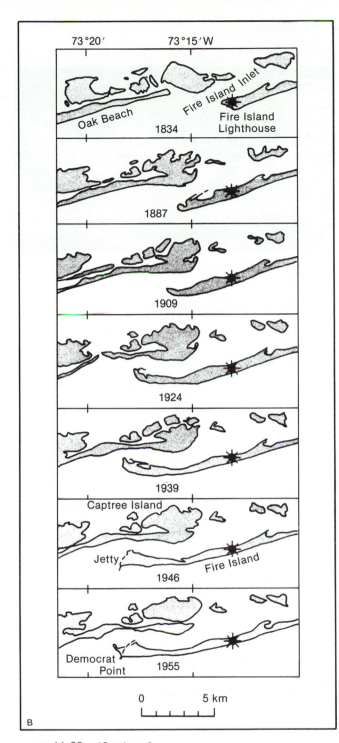

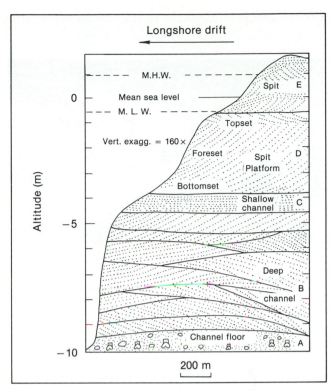

FIGURE 11-21. Kinds of strata in autocyclic inlet succession, schematic profile and section through east bank of Fire Island Inlet at Democrat Point spit. This inlet succession was deposited by lateral migration of an inlet channel 10 m deep. Direction of dip of cross strata in deep channel, topsets of spit platform, and spit are not related to plane of sketch, but have been drawn to show their distinctive arrangements. (After N. Kumar and J. E. Sanders, 1974, fig. 6, p. 506.)

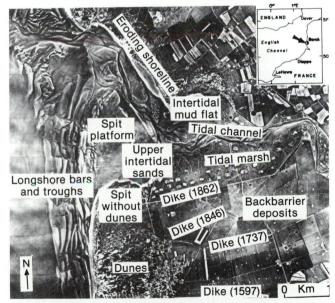

FIGURE 11-22. Authie Bay, France, viewed vertically from airplane at low tide showing two parts of spit: (1) a lower-level spit platform and (2) an upper-level subaerial spit proper. The northern part of the subaerial spit, near its submerged platform, is devoid of dunes. Note longshore bars and troughs on seaward side of spit. As this long spit grows northward, it develops into a barrier and dunes encroach over the barrier; backbarrier sediments likewise advance northward. Dikes built by farmers to reclaim land for agriculture at dates shown mark successive stages of northward progradation of backbarrier sediments. (J. LeFournier and G. M. Friedman, 1974, fig. 1, p. 497.)

FIGURE 11-20. (Continued)

B. Migration of Fire Island Inlet by westward growth of Fire Island barrier. Fire Island Lighthouse, built on shore of inlet in 1834, now stands 8 km away from the inlet. (Redrafted from N. Kumar, 1973, fig. 3, p. 130; also N. Kumar and J. E. Sanders, 1974, fig 3, p. 499, based on data from U.S. Army Corps of Engineers and S. Gofseyeff, 1953, and F. L. Panuzio, 1968.)

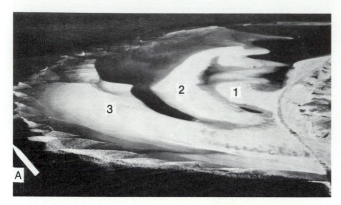

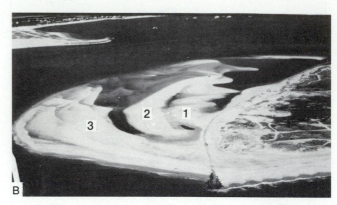

FIGURE 11-23. Spit complex and spit platform W of jetty at Democrat Point, Fire Island, Long Island, New York, viewed obliquely from above looking NE in April 1983. (Audrey Massa and Agi Nadi.)

A. On 01 April 1983 a new spit (No. 3, foreground), which had formed during a storm in March 1983, had grown from the jetty nearly to the edge of the spit platform edge outlined by breaking waves (white areas in otherwise black area of water). Two older, modified spits are marked by 1 and 2.

B. By 13 April 1983, spit No. 3 has grown to the edge of the spit platform and its tip has recurved to the NE.

ponding is possible only locally. Instead, spillover water crosses the spits and, where it enters the water on the sheltered side, deposits steep cross strata. These dip away from the open-sea side of the spit; they are foreset beds of spillover delta lobes (Figure 11-25). In some cases the spillover water erodes channels through the spit-berm crest. The waves quickly fill the channel and the non-filled parts of these eroded depressions become bays. By growth of successive recurved tips, spits likewise form bays (Figure 11-26). In spit bays, fine dark-colored sediments accumulate. These present striking contrasts to the coarse well-sorted spit sediments that enclose them.

If a spit acquires a high "backbone ridge" (a high wave-berm-crest ridge or coastal dune), then the frequent inundations of the spit sediments by sheets of spillover water cease. Because of the important influences on the sediments created by these spillovers, it is useful to use the presence of such a "backbone ridge" to distinguish spits from barrier spits.

In summary, spit sediments include the following four chief kinds: (1) spit beach-face sands, well laminated, with planar layers, gently inclined toward the sea; (2) spit berm-top sands, well laminated with laminae nearly horizontal and built up to several meters above mean high water; (3) steeply dipping foreset sands of spillover-delta lobes; and (4) dark gray clayey- and sandy silts that accumulate in the spit bays. Because waves readily drive spits toward the mainland, a typical vertical succession in spit sands (from base upward) is: steeply dipping spillover-delta foresets; nearly horizontal spit berm-crest strata; low-angle, seaward-dipping spit beach-face strata (Figure 11-27). Enclosed within such coarse sediments may be thick (several meters) localized bodies of dark gray silt. Other sedimentary structures, such as slumps with contorted laminae in the spit platform and ripple

cross laminae or trough cross laminae in the upper parts of spits, may also be found. If spit deposits are correctly identified in the rock record, then they permit precise

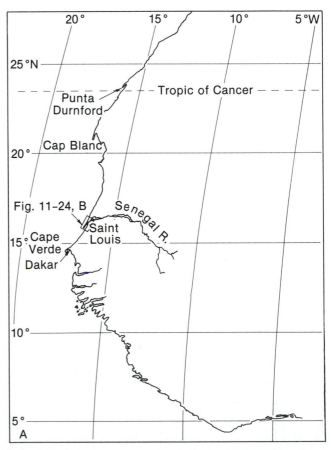

FIGURE 11-24. Rapid growth of spit near mouth of Senegal River, West Africa.

A. Index map.

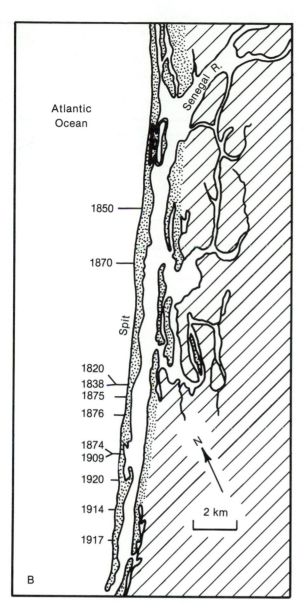

FIGURE 11-24. (*Continued*)
B. Map showing successive southward migration of large spit at mouth of Senegal River. The growth of this spit has been phenomenal: 170 m/yr, which is more than 3 m/wk. (After V. P. Zenkovich, 1967, fig. 273, p. 567.)

reconstruction of water depth and of sea level. However, unless careful attention is paid to the characteristics of associated strata, spit deposits can be confused with deposits formed in other settings that include giant cross beds such as point bars, giant dunes, and Gilbert-type deltas.

Features of Backbarrier Areas

The backbarrier (or sheltered) sides of barriers include many environments in which distinctive processes act upon sediments. We include lagoons/bays, flood-tidal deltas, washover fans, intertidal flats, and intertidal

FIGURE 11-25. Spillover delta built into bay on protected side of spit at Democrat Point, SW end of Fire Island barrier, S shore of Long Island, New York, low tide, 03 October 1970.
A. Side view of large lobe of spillover delta, which has prograded across dark-colored fine sediments deposited on floor of spit bay.
B. East wall of N-S trench dug in spillover lobe of A showing cross strata dipping steeply toward spit bay (upper left). As tide drops, flows of spillover water become very thin and thus concentrate heavy minerals. Compare with Figure 11-16. (Authors.)

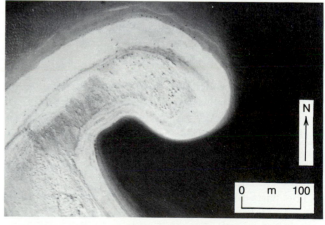

FIGURE 11-26. Recovered tip of spit at Democrat Point, SW end of Fire Island barrier, S shore of Long Island, New York, viewed vertically from a low-flying airplane on 07 December 1970. Spit has grown completely across spit platform (light gray, shallow-water area at upper left). By continued growth of the tip of the spit toward lower right, the spit bay that has just formed will lengthen. Flat spit berm top shows pronounced effects of strong northwest winds. (Bruce Caplan and authors.)

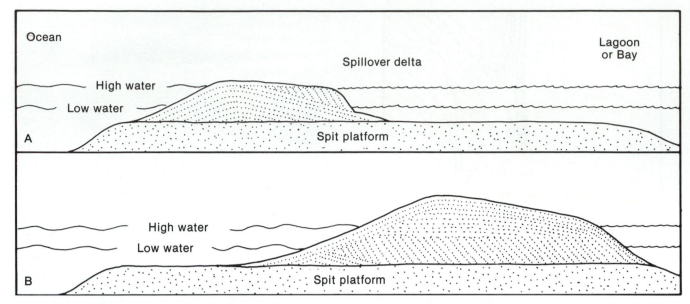

FIGURE 11-27.   Three major kinds of sands in a spit and changes in their proportions that result from migration of spit away from open ocean.

A. Initial condition. Bulk of spit consists of spit-beachface sands that dip seaward and of landward-dipping spit berm-top sands. Steeply dipping sands of spillover-delta lobe constitute only a minor proportion of the total and are confined to upper part of spit on the side away from the open sea.

B. After migration through a distance equal to its initial width and rearrangement of the sand, the basal part of the spit consists almost entirely of steeply dipping sands from a spillover-delta lobe. These are overlain by nearly horizontal spit berm-top sands that were built upward by sheets of spillover water from times of higher-than-normal waves or -tides. Top of spit consists of a thin ve-neer of spit-beachface sands that dip toward open sea (at left) and of spit berm-top sands that are inclined gently away from the open sea. Not shown is fine-textured sediment filling of spit bay. If present, such a deposit would occupy area where spillover-delta sands are shown. (J. E. Sanders in G. M. Friedman and J. E. Sanders, 1978, fig. 11-10, p. 315.)

marshes. The last two are examples of peritidal complexes that we discuss further in Chapter 12.

LAGOONS/BAYS

Coastal lagoons and many coastal bays (depending on local usage) are shallow bodies of water enclosed between a mainland shore and the sheltered side of a barrier island. In these water bodies, wave action is comparatively insignificant and the chief influences on the sediment are the rise and fall of the tides, the activities of organisms, and the climate.

The parts of lagoons that are not invaded by various bodies of sand or coarser sediment from the barrier island are filled by fine sediment that settles out of suspension. This material is commonly rich in first-cycle organic matter and thoroughly bioturbated. Eventually, most lagoonal fine sediments tend to be covered by coarser sediments derived from the various environments next discussed.

FLOOD-TIDAL DELTAS

At the landward ends of tidal inlets, the sediment transported by the flood-tidal currents accumulates as **flood-tidal deltas**, defined as *tidal deltas whose groundplan shape is convex away from the landward ends of tidal channels* (Figure 11-28). In contrast with ebb-tidal deltas, which waves tend to destroy, flood-tidal deltas built into backbarrier areas tend to be preserved in the geologic record and are of interest here.

The initial situation in an active flood-tidal delta is for a thin lobe of tidal-delta sediment (usually coarse and oxidized, hence light colored) to grow across the sediments of a lagoon floor or of an intertidal flat (Figure 11-29) (which usually are fine and reduced, hence dark colored).

Vigorous tidal currents extensively rework the sandy sediments; shifting bed forms give rise to various cross strata. Both planar-tabular cross laminae of 2D dunes and trough cross laminae of ripples and 3D dunes, cre-

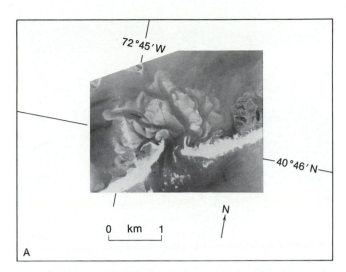

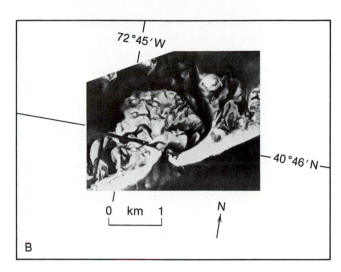

FIGURE 11-28.   Moriches flood-tidal delta, Westhampton, Atlantic coast of Long Island, New York.
   A. Photograph taken in June 1938, 7 years after an inlet formed when the barrier was breached during a storm, and 3 months before the record-breaking hurricane of September 1938.
   B. The same area photographed in 1947. The general area of the flood-tidal delta has not changed significantly despite the fact that the inlet was severely modified by the 1938 hurricane (it created a second inlet), later closed by waves, and then reopened. In 1947, the position of the seaward end of the inlet was about 0.2 km SW of its position in June 1939 and the main channel had curved abruptly to the west so that tidal currents were eroding the lagoon side of the barrier. New areas of tidal marsh have grown just E of the channel (and W of "peninsula" of marsh at right center). (Both photos U.S. Dept. of Agriculture.)

ated by both ebb- and flood-tidal currents, are common in flood-tidal deltas.

   Because the shallow distributary channels of a flood-tidal delta shift repeatedly, an active lobe can become inactive overnight. On an inactive lobe, swift currents are no longer active; in the sediment, the activities of organisms dominate. Simple rise and fall of the tide may cover the lobe with tidal silt. Burrowing organisms move in. Eventually, an intertidal marsh will cover it. After such

FIGURE 11-29.   View at low tide of thin lobe of flood-tidal-delta sand (light gray, at right) prograding across sediments of intertidal flat (dark gray, at left). Moriches flood-tidal delta, Westhampton, New York. (D. M. Caldwell, 1971 ms., fig. 9, p. 58.)

covering, the lobe of flood-tidal-delta sand has become part of the local stratigraphic record.

   Because the sediments of an active flood-tidal-delta lobe are so extensively cross stratified, we would normally expect to find cross strata in a buried lobe of flood-tidal-delta sediment. However, the burrowing organisms are active destroyers of strata. Buried inactive lobes of the Moriches flood-tidal delta scarcely resemble sands from the active lobes. Instead of consisting of cross-stratified sand, these buried flood-tidal-delta lobes have become nothing more than irregular pockets of coarse brown sand in a burrow-mottled matrix of fine gray sand. (See Figure 5-16, B.)

   This example from Moriches Bay is very significant. It illustrates the point we have been emphasizing repeatedly—namely, the sediments that become parts of the geologic record may not look anything like the accessible, easily studied, active, modern sediments.

   The flood-tidal delta built at the landward end of San Luis Pass, south of Galveston Island, Texas is forming along a microtidal coast. These deposits on a microtidal coast differ from those of tidal deltas in regions with greater tidal ranges in that they lack large-scale and high-angle sedimentary structures, may be more intensely bioturbated even before they are abandoned, contain washover shell deposits, and coarsen upward. In most respects, the buried deposits of the flood-tidal delta in San Luis Pass are very similar to those in Moriches Bay, where the tidal range is greater. Conversely, the upper parts of buried Holocene microtidal (tidal range = 0.9 m) flood-tidal deltas in Back Sound, North Carolina, behind the barrier island Shackleford Banks, include laminated- and cross-laminated sands. Thus, in some flood-tidal deltas organisms do not obliterate all current-caused laminae.

In summation, flood-tidal deltas consist of sheets of current-driven sand dominated by migration of flow-transverse, 2D dunes that deposit planar-tabular cross strata and of 3D dunes that form trough cross strata; various kinds of ripples are associated with these dunes. The distribution of the different kinds of bed forms is controlled by position within the flood-tidal delta and the differing paths taken by flood- and ebb-tidal currents. However, in the rock record, these primarily cross-laminated deposits are represented mostly by bioturbated sand.

SOURCES: Boothroyd, Friedrich, and McGuinn, 1985; Fraser, 1989; Howard and Frey, 1985; Hunter, Clifton, and Phillips, 1979; Israel, Ethridge, and Estes, 1987; Kayan and Kraft, 1979; Reading, 1986.

WASHOVER FANS

In our discussion of beaches we mentioned spillover water from sheets of swash that are more vigorous than normal. We need to contrast these thin sheets of spillover water with the thicker sheets of water that pour through gaps in the barrier ridge that are eroded during great storms.

Storm waves may erode or deposit sediments. During the same storm they may do both. A great storm in the spring of 1966 opened a large gap through the Outer Banks (North Carolina) barrier that was 145 m wide. Large ridges composed of cross-stratified fine gravel (Figure 11-30) were spread landward from the gap. During the late stages of the same storm the waves healed the breach and deposited a dikelike ridge of gravel whose top stood about 5 m above normal water level. Eleven years later, dune sand had not buried this gravel.

Numerous examples are known of sediments that were spread landward from gaps eroded in barriers (Figure 11-31). These fanlike bodies of sediment transported landward by water pouring through a temporary gap in a coastal barrier have been designated as *washover fans*. Two essential differences between a washover fan and a flood-tidal delta are: (1) A washover fan results from a one-time-only kind of activity whereas a tidal delta is the product of the repeated discharges of the flooding tide. (2) A washover fan is built subaerially whereas a flood-tidal delta grows mostly under water. In addition,

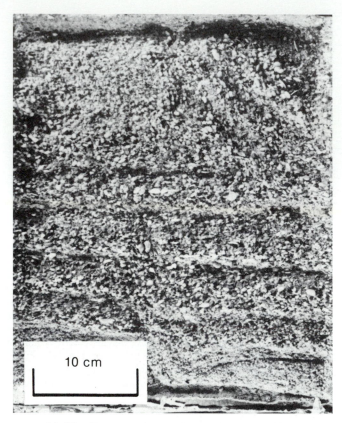

FIGURE 11-30.   Epoxy relief-peel of box core collected in longitudinal washover ridge composed of gravel. Beneath 30 cm of gravel is about 5 cm of laminated coarse sand. Outer Banks, North Carolina, near Caffey's Inlet. (J. A. Burger and authors.)

a washover gap generally is not eroded below sea level. (See Figure 11-17.) If it were, it would become a new inlet.

Washover fans can transport considerable amounts of sand landward of the barrier, and can contribute materially to landward barrier migration. Washover fans are most important in microtidal regimes. They form largely by the deposition of (1) upper-flow-regime plane-bed planar laminae on the surface of an intertidal marsh, and (2) deltalike foresets that prograde into the lagoon; very rapid flows of water during washover formation can even result in formation of antidunes.

Post-washover wind transport can remove much of the washover sand deposited on the intertidal marsh. Wind will remove dry sand only above the water table. On the backbarrier side of Nauset Spit, Cape Cod, Massachusetts, a winter storm on 06–07 February 1978 deposited about 400 m^3 of washover sand per meter of dune breach. Later, offshore winds blew about half of this sediment back onto the beach. Vegetation is the most-important stabilizer of washover deposits.

On coarse-textured poorly sorted beaches, washover fans may contain layers displaying distinct inverse grad-

FIGURE 11-31.   Small washover fans on part of barrier, Outer Banks, North Carolina, seen in vertical aerial photograph taken in March 1962 after severe Ash Wednesday storm. During the early part of the storm, a tremendous longshore current that flowed southeastward cut a prominent scarp and destroyed the berm. Later, large waves coming straight into the beach created the washover fans and buried most of the scarp. (NOAA.)

ing formed as the celerity of tidal currents increased toward high tide during individual washover events.

SOURCES: Fraser, 1989; Howard and Frey, 1985; Hunter, Clifton, and Phillips, 1979; Noe-Nygaard and Surlyk, 1988; Reading, 1986; Reineck and Singh, 1980.

INTERTIDAL FLATS

All along the landward sides of barriers subject to tidal action, intertidal flats are exposed at low tide. To complete the list of the parts of a barrier complex we simply

mention these flats here; we describe the details in Chapter 12.

INTERTIDAL MARSHES

If the landward side of a barrier accumulates enough sediment to shoal the lagoon to the depth appropriate for salt-tolerant grasses (Figure 11-32), intertidal marshes become established. These may grow on former lagoon-floor silts, on sands deposited on flood-tidal deltas, or on washover fans.

The roots of the marsh grasses stabilize the surface, whereas the blades "comb" the tidal waters and baffle out suspended silt and -sand, and floating debris. By this mechanism and continued growth of the marsh grasses, the nearly horizontal surface of an established intertidal marsh is maintained. During submergence, an intertidal marsh can accrete upward. On the Connecticut coast of Long Island Sound, marsh grasses were able to accrete upward during a rate of submergence equal to 1 mm/yr, but were not able to keep up with a submergence rate of 1.8 mm/yr. Upward accretion at the rate of 1 mm/yr enables 1 km² of active marsh to "store" 2000 tons/yr of tidal silt.

Intertidal-marsh sediments are dominated by the network of roots and accumulated debris of the other parts of the marsh grasses. Where they accumulated around tufts of grass, the laminae of silt and fine sand display a wavy aspect. Gastropods, mussels, fiddler crabs (genus *Uca*), and other crustaceans inhabit the marsh surface and their remains may be present in marsh sediments. Other skeletal debris may come from various organisms that live elsewhere but whose remains are spread across the marsh by high storm tides.

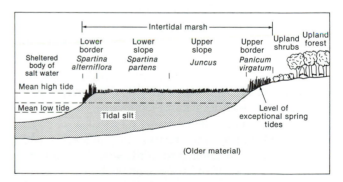

FIGURE 11-32.   Sketch profile through intertidal marsh soon after marsh turf has roofed over an intertidal mudflat. Thickness of salt-marsh peat equals half of mean tidal range. If submergence occurs at a rate that does not exceed rate of upward marsh growth, then peat can become much thicker than half the tidal range and, as it transgresses landward, will overlie whatever materials compose the shore, including solid bedrock. (W. R. Miller and F. E. Egler, 1950, fig. 6, p. 55.)

Careful studies of intertidal marshes have shown that many more environments exist than we have implied. Several distinct successions have been recognized, as have features such as levees along the creeks and salt ponds on the surface.

An important point about intertidal marshes on the landward sides of barriers is that all marshes look alike but from place to place, their ages may differ greatly. In other words, by mere inspection, one cannot tell the age of a marsh. To determine the age of a marsh, one must collect cores, establish the thicknesses of salt-marsh peat, and obtain radiocarbon dates from the peat. By doing so, one can chart the migrations of salt marshes and thus learn the sequence in which the different parts of a sedimentary basin were flooded during a submergence.

Autocyclic Patterns in Barrier Complexes

In barrier complexes, various situations exist for depositing autocyclic sediments. We discuss the characteristics of these autocyclic sediments and then compare their behavior on a prograding coast with that on a submerging coast.

Characteristics

Now that we have examined all the major kinds of sediments composing a barrier complex, let us consider the question of how these sediments form successions of strata that could be incorporated into the geologic record. On barriers we can identify two chief depositional slopes that can prograde and thus deposit autocyclic parasequences.

1. The ridge(s)–beach–shoreface surface, striking generally parallel to the coast, whose relief varies from about 5 m or so above sea level to a depth of 5 to 20 m or so below sea level (Figure 11-33). The vertical distance between two horizontal lines drawn through the top of the shore-parallel ridge and the outer toe of the shoreface thus ranges between 10 and 25 m. This distance equals the thickness of any succession of strata (barrier succession) that can form by lateral sedimentation along such an inclined depositional slope.

Figure 11-34 shows the relationships in a vertical succession between sediments from the shoreface and those of the adjacent shelf off a segment of the coast of Italy. Sediments from the outer part of the shoreface display the effects of extensive bioturbation.

2. The ridge(s)–spit–spit platform–tidal inlet slope, striking generally perpendicular to the coast, whose relief varies from about 5 m or so above sea level (as in the ridge(s)–beach–shoreface slope) to from 10 to 30 m below sea level (deeper because of the great depth of scour in the narrow part of most inlets). Therefore the thick-

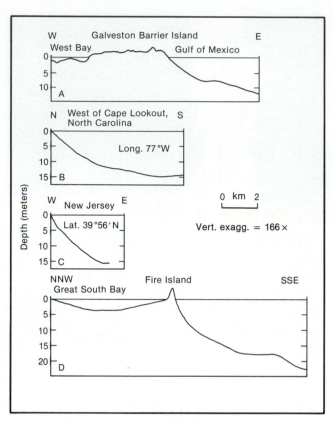

FIGURE 11-33. Profiles, drawn at right angles to shore, displaying the shorefaces of various modern barriers, all drawn to same scale as profile and section through Galveston Island barrier.
   A. Galveston Island, Texas.
   B. West of Cape Lookout, North Carolina.
   C. Central New Jersey.
   D. Fire Island, New York. Depths to lower limits of shoreface range from 8 to 18 m (A, after H. A. Bernard and R. J. LeBlanc, 1965, fig. 20, C, p. 158; B and C replotted from F. P. Shepard, 1960, fig. 9, p. 205.)

ness of any succession of strata (inlet succession) formed by lateral sedimentation along this inclined depositional slope ranges from 15 to 35 m.

The stratigraphic record of a barrier clearly will depend on the rates and extents of lateral sedimentation on the two depositional slopes just described. If the coast-parallel slope (No. 1, including the shoreface) accumulates sediment very slowly and the coast-perpendicular depositional slope (No. 2, including the inlet) shifts very rapidly, then beneath the visible parts of the barrier a sort of invisible prism-shaped "root" will form (Figure 11-35). The sediments composing this root will be those of the inlet succession. The part lying below sea level will be flat on top and concave up at the base. The width at sea level may be about 0.5 km or so; the thickness, 10 to 30 m; and the length, many hundreds of meters or even a few tens of kilometers.

In order for an inlet succession to expand into such a "barrier-root" prism, a tidal inlet must migrate ac-

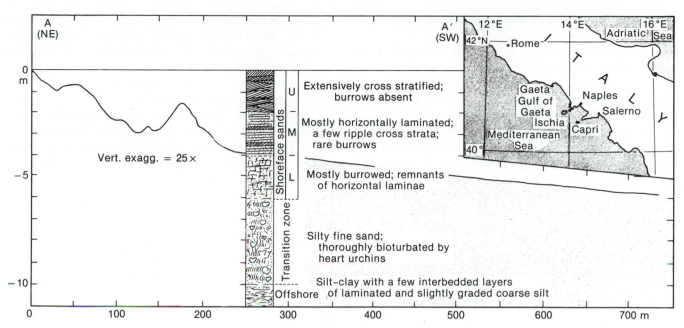

FIGURE 11-34. Vertical section through modern shoreface- and shelf deposits showing relationship between sedimentary structures and bathymetric profile. Gulf of Gaeta, Italy (see inset map). (After H. E. Reineck and I. B. Singh, 1973, fig. 462, p. 316.)

tively along the coast. For this to happen, one of the requirements is that the backbarrier area not be obliterated by sediment infilling but continue to exist as a depression into and out of which large quantities of water flow during each part of the tidal cycle. If the backbarrier depressions fill with sediment, then inlets will cease to exist. This survival relationship between inlets and backbarrier depressions restricts inlet successions to coasts that are submerging or have recently been submerged. Inlet successions are not likely to be found on prograding

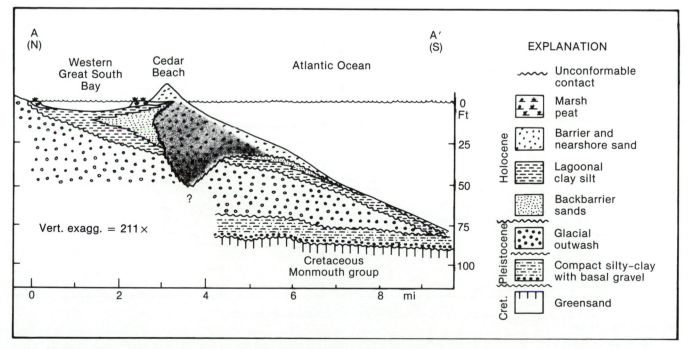

FIGURE 11-35. Profile and section through barrier island on S coast of Long Island, New York, shows "root" thought to have been built by migration of a tidal inlet parallel to shore. Locations of control borings not shown. (M. R. Rampino and J. E. Sanders, 1980, fig. 2, p. 1067.)

coasts where backbarrier depressions have been completely filled.

Some tidal inlets do not migrate, or if they do, they may not shift very far. Therefore, despite the widespread occurrences of tidal inlets themselves, the distribution of tidal-inlet sediments may be only local. We do not yet know how widespread inlet successions may be.

The only way a barrier succession can form is for the open-sea side of the barrier to advance seaward by adding sediment. A prograding barrier succession begins at the base with shoreface sediments and ends at the top with shore-ridge sediments. In the middle are the beach sediments. An example of such a succession based on numerous borings from Galveston Island, Texas, is shown in Figure 11-36. Notice how the increase in size of particles is expressed on the spontaneous-potential (SP) log of a boring through the succession.

With the basic relationships of these two kinds of successions from a barrier complex in mind, let us now see what happens to them under different situations at the shore. We look first at prograding coasts and then take up submerging coasts.

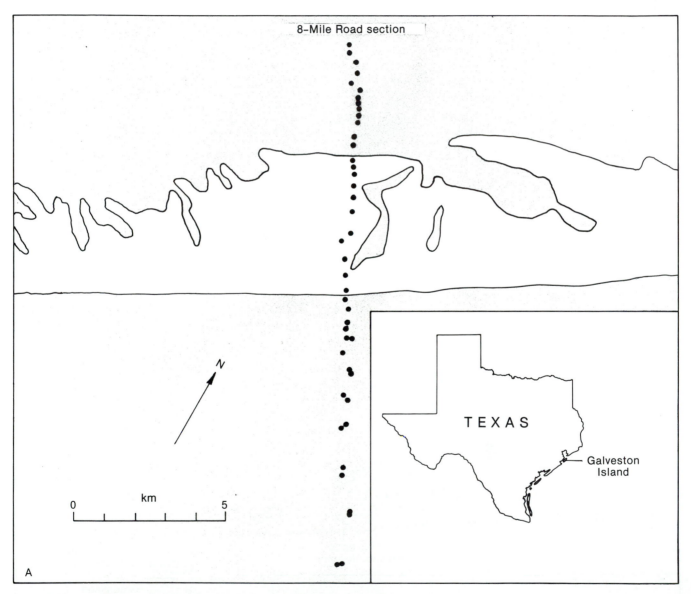

FIGURE 11-36. Summary of results of detailed study of surface- and subsurface sediments of Galveston Island, Texas.

A. Index map of Texas showing location of Galveston Island and map of the central part of the island (local index map on part C) showing locations of borings along 8-Mile Road section from which the subsurface data have been derived.

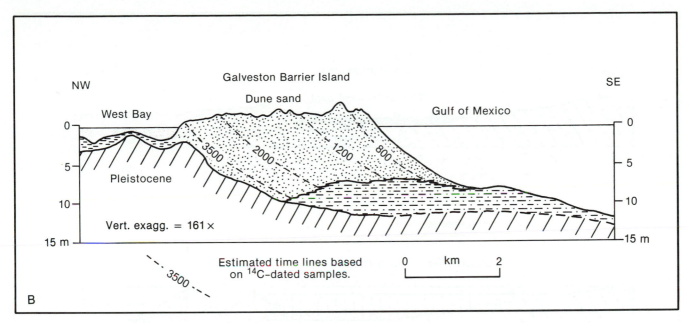

FIGURE 11-36. (*Continued*)
  B. Profile and section at right angles to shore along line of borings (8-Mile Road section; part A) through Galveston barrier island and adjacent areas. Position of top of underlying Pleistocene materials is based on 12 borings. Detailed subdivision of various kinds of sand within barrier (shoreface, beach, dune) not shown.

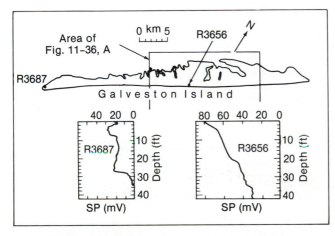

FIGURE 11-36. (*Continued*)
  C. Spontaneous-potential (SP) logs of two borings through contrasting kinds of barrier sediments. Boring R3656 shows typical upward-coarsening succession deposited by seaward progradation of shoreface. (Upward coarsening is indicated by the regular increase upward in the SP curve.) Boring R3687 displays sharp base and upward fining (in lower two-thirds)-upward coarsening (toward top) succession typical of sediments deposited by coast-parallel migration of tidal inlet. (Redrawn from H. A. Bernard and R. J. LeBlanc, Sr., 1965, A, fig. 18a, p. 155; and B, fig. 20C, p. 158; also H. A. Bernard; C. F. Major, Jr.; B. S. Parrott; and R. J. LeBlanc, Sr., 1970, A, fig. 50; B, fig. 51; C, fig. 45.)

Preservation on a Prograding Coast

The example of Galveston Island, just mentioned, illustrates a typical example of preservation of barrier sediments on a prograding coast. Under such conditions, every part of the ridge(s)–beach–shoreface slope adds new layers of sediment. In this way, every depositional environment on the open-sea side of the barrier starts accreting its own particular stratum of sediment, each possessing its own diagnostic characteristics, thickness, and fixed sequential relationships to the products of adjacent environments which form units lying above or below. Under these circumstances, even ebb-tidal deltas may be buried by prograding barrier complexes and therefore preserved. The result is an autocyclic parasequence.

Such prograding is the only way to guarantee the preservation in the geologic record of beach- and coastal-dune sands. It also requires huge amounts of sediment. Finally, such prograding may take place while sea level remains relatively fixed or while the sea slowly submerges the land. Under ideal circumstances, the sedimentary record contains clues about how sea level behaved. If the sea remained stable relative to the land while the barrier shoreface prograded, then, at the top of the succession, backbarrier sediments will not be present (Figure 11-37, A). By contrast, if the coast is being prograded during submergence, then on top of any coastal-dune sand may be backbarrier sediments or thick nonmarine

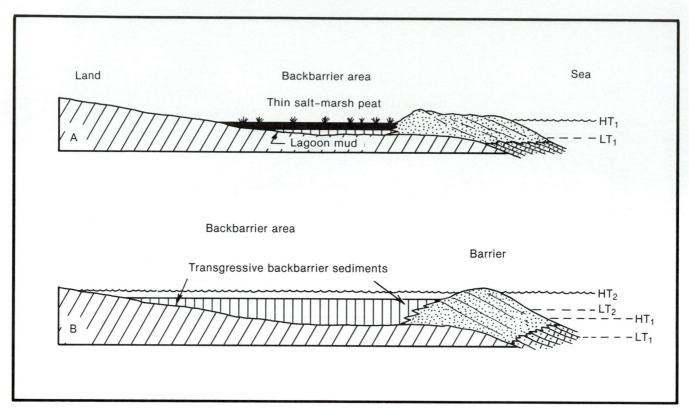

FIGURE 11-37.   Deposits formed when barrier coasts prograde under various conditions of relative sea level (schematic profiles and sections drawn perpendicular to shore and omitting inlet deposits). (J. E. Sanders in G. M. Friedman and J. E. Sanders, 1978, fig. 11-22, p. 324.)

A. Barrier progrades while sea remains at constant level. Backbarrier area, initially a lagoon, fills with lagoon sediment and then becomes an intertidal salt marsh. Thickness of salt-marsh peat equals half the mean tidal range. Contact between sediments deposited on lower part of beach and those deposited on upper part of shoreface is horizontal. Salt-marsh peat may interfinger with barrier sands but does not overlie barrier sands through any extensive area.

B. Barrier progrades during submergence (caused by subsiding coast, by rising sea, or by both). Ocean side of barrier advances seaward while lagoon widens in both landward- and seaward directions. In backbarrier area, many different conditions of sedimentation are possible, of which only one, a persistent lagoon, is shown here. Because the lagoon persists, lagoonal sediments thicken upward and transgress over mainland sediments (left) and over barrier sediments (right). The widening sheet of barrier sand "climbs" seaward, forming a three-part sequence consisting of shoreface sediments (below), barrier sands (in middle), and backbarrier sediments (at top). Contact between sediments deposited on lower part of beach and those deposited on upper part of shoreface is not horizontal but is inclined landward.

strata. However, during coastal progradation, backbarrier environments may rework the sediments previously deposited on the open-sea sides of barriers. Backbarrier tidal creeks on Kiawah Island, South Carolina, a microtidal coast, rework much of the underlying dune-, foreshore-, and shoreface sands into channel-lag and point-bar deposits. During progradation, the depth of tidal channels and rate of channel migration strongly affect the preservation of open-sea-side barrier deposits.

SOURCES: Barwis and M. O. Hayes, 1979; Friedman and Sanders, 1978; M. O. Hayes, 1980.

Preservation on a Submerging Coast

As we have just seen, some coasts may prograde conspicuously even while the coast is being submerged. In such cases it is easy to understand how a geologic record is accumulated. But what happens if the barriers do not prograde conspicuously during coastal submergence? Do any barrier sediments stand a chance of being preserved if the sea transgresses the land?

In order to answer this question we need to find out what kind of "retreat tracks" are left behind as various parts of a coast are submerged. We need to see how various parts of a barrier coast respond to rising sea level. Figures 11-38 and 11-39 illustrate two ways in which coastal sediments can respond to coastal submergence.

Figure 11-38 illustrates how a transgression along a high-energy coast destroys all record of the backbarrier marsh- and lagoonal sediments and deposits a thin layer of open-marine sediments on a wave-eroded surface. Wave action on the shoreface erodes preexisting coastal facies and forms what is called a ravinement surface, *a time-transgressive erosion surface eroded on the shoreface by wave action.* This is a surface of unconformity cut into preexisting coastal sediments, formed on the shoreface during a transgression (shoreline retreat). The ravinement surface is a product of landward migration of an erosional shoreface. Some sediment eroded from the shoreface during formation of the ravinement surface is redeposited farther offshore. This process destroys almost every trace of the varied and extensive coastal sedimentary complex. The only sediments likely to be preserved are the roots of tidal inlets, trunk tidal channels, or shoal-retreat massifs or other sediments that accumulated in former stream valleys, and any sediments that filled in preexisting depressions in the bedrock surface, such as sinkholes. These sediments may be topographically lower than the depth of incision by the ravinement surface and therefore may be buried by younger shoreface sediments and preserved. During the late Pleistocene-Holocene transgression on the Atlantic coast of the United States, this process of ravinement formation and coastal-facies destruction has been widespread. Backbarrier marsh deposits locally are exposed on beaches on the coasts of Long Island, Delaware, Virginia, and Texas. These exhumed backbarrier sediments are being eroded by the ravinement surface where it intersects the transgressing shoreline.

Figure 11-39 shows an alternative mode of transgression in which, rather than by moving continuously landward as in Figure 11-38, barriers retreat by discrete jumps. In the discrete-jumps model, coastal sediments are more likely to be completely preserved, because the surf zone and inner shoreface, where most of the cutting of the ravinement surface takes place, do not move across the entire barrier complex. Instead, the entire barrier complex, including the beach and inner shoreface, jumps from a seaward location to a more-landward one, leaving the lagoonal-, marsh-, and barrier sediments in the intervening area relatively undisturbed. Some parts of the Atlantic coast of the United States preserve a record of barrier jumping during the late Pleistocene–Holocene transgression.

It is also possible to preserve substantial portions of barrier-complex successions during coastal submergence without barrier jumping. At a high rate of submergence,

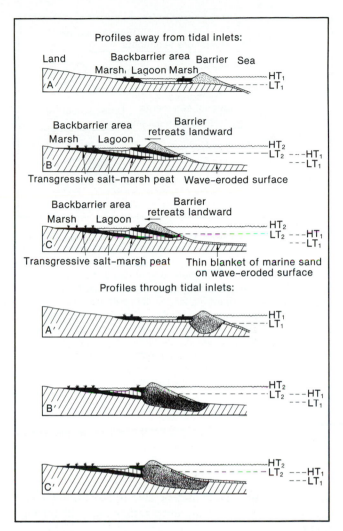

FIGURE 11-38.   Possible effects of barrier retreat landward during submergence shown by schematic profiles and sections drawn perpendicular to coast. (J. E. Sanders in G. M. Friedman and J. E. Sanders, 1978, fig. 11-23, p. 325.)

A. Initial conditions. Marshes have become established on both sides of lagoon.

A'. The same, but profile through a tidal inlet.

B. Barrier retreats landward and initial arrangements in backbarrier area persist (that is, widths of lagoon and marshes remain the same, salt-marsh peats thicken, and all backbarrier sediments advance landward). Basal deposit of transgression is salt-marsh peat (left). No sediment is left on sea floor by retreat of barrier. Waves erode surface on older materials but do not deposit sediment on this surface.

B'. The same as in B, but with profile drawn through tidal inlet. Body of inlet-filling sand widens landward. Waves erode top of inlet filling and older materials but deposit no sediment on this surface.

C. Barrier retreats landward, as in B, but leaves behind a thin blanket (about 1 to 3 m thick) of marine sand on wave-eroded surface cut into older materials.

C'. The same as in C, but with profile drawn through tidal inlet. Body of inlet-filling sand widens landward. Waves deposit thin blanket of marine sand on inlet-filling sand. (J. E. Sanders.)

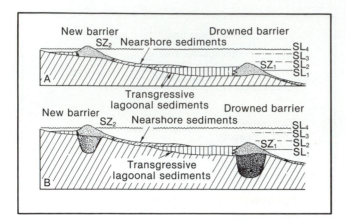

FIGURE 11-39.   Intermittent landward advance of barrier during submergence illustrated by schematic profile and section perpendicular to coast.

A. Profile away from tidal inlets. Barrier forms at initial sea level (SL1, at right). During submergence, this barrier persists and possibly thickens upward. As long as this barrier is present, the lagoon behind it will widen and deepen; lagoonal sediments transgress landward. Eventually, this first barrier is submerged (SL2); still later (SL3), a new barrier forms and with it a new lagoon. After new barrier has become established, nearshore marine sediments overlie submerged lagoonal sediments along a sharp contact. Surf zone of initial barrier (SZ1) skips landward to location on new barrier (SZ2) and does not disturb sediments lying between the two barriers. (J. E. Sanders and N. Kumar, 1975a, fig. 2, C and D, p. 66.)

B. The same, but with profile drawn through prism of inlet-filling sands. (J. E. Sanders in G. M. Friedman and J. E. Sanders, 1978, fig. 11-24, B, p. 326)

the ravinement surface may be able to rise rapidly above the barrier-complex sediments, rather than cut through them.

The shoreface may be eroded on nonbarrier coasts as well. Even under conditions of low wave energy, as on portions of the Gulf Coast of Florida, berm ridges retreat.

The controls on the mode of coastal retreat, whether continuous, with formation of a ravinement surface, discontinuous with preservation of coastal sediments, or by some combination of these processes, are now matters of hot debate. For example, Fire Island, New York, has demonstrably undergone both stepwise jumping and continuous retreat. The importance of knowing the circumstances governing each kind of barrier behavior becomes obvious when one considers the possible fate, with continued submergence, of the billions of dollars worth of roads and residential- and commercial buildings that have been constructed on the barrier islands and -spits that rim much of the world's sandy coasts. In the long run, the foundations of these roads and buildings are no more secure than that of a child's sand castle. The efforts of humans to control and stabilize barrier-island shape, size, and position have exerted little long-term

FIGURE 11-40.   Damage from Hurricane Hugo, which struck coastal South Carolina during the night of 21/22 September 1989, viewed obliquely from low-flying airplane. The storm created a new inlet (closed by an emergency work crew by the time of this photo) and caused a house to collapse into it. Two formerly neighboring houses are nowhere to be seen, having been destroyed or moved by the storm out of the area visible in the photograph. (Peter J. R. Buttner, 1989, fig. 40, p. 167.)

effect on the landward movement of barrier-island sediment that is driven by relative sea-level rise or on the alongshore movement of barrier-island sediment driven by longshore currents. The shoreline of the barrier-island system of Long Island, New York, is retreating at an average rate of about 1 m/yr. In addition to the daily processes of sediment movement, infrequent high-energy (catastrophic) events are significant factors. The effects on Galveston Island of Hurricane Alicia in 1985, and on the coast of South Carolina of Hurricane Hugo on 21/22 September 1989 (Figure 11-40) are but two grim reminders of the large-scale movements of coastal sediment that are a part of this dynamic environment.

SOURCES: Belknap and Kraft, 1981, 1985; R. A. Davis, 1985; Donselaar, 1989; Fraser, 1989; Kraft and John, 1979; Kraft, Chrzastowski, Belknap, Toscano, and Fletcher, 1987.

Inferred Examples of Ancient Barrier Complexes

What we regard as ancient examples of barrier complexes are of two kinds: (1) offlap successions, formed while the shoreline prograded; and (2) transgressive successions, formed during submergence. It should be remembered, as was previously mentioned, that preserved sedimentary successions in the rock record are disproportionately composed of the deposits of rare, high-energy events, when compared to modern sediments. This is especially true in coastal settings affected by relatively infrequent major storms (tropical hurricanes or winter storms).

Ancient Offlap Sequences

The best-exposed and most carefully studied offlap successions preserving ancient barrier sediments probably are those of Cretaceous age in the Rocky Mountain region in the states of New Mexico, Utah, Colorado, Wyoming, and Montana. Even in the absence of any detailed comparative materials from modern coastal sediments to go by, the early geologists who studied the Cretaceous strata in this vast area were able to make correct interpretations. These results have been substantiated and refined but not significantly modified by observers well acquainted with the details of modern coastal sediments.

A restored stratigraphic section based on the continuous exposures in the Book Cliffs, central-eastern Utah and western Colorado (Figure 11-41), illustrates the relationships among the various sediments that accumulated along the western margin of the late Cretaceous seaway. In the Book Cliffs region, the shoreline trended NE-SW. Abundant sediment supplied by streams from rising highlands to the NW prograded the shoreline southeastward through a distance of 217 km (135 miles) during a submergence of 610 m (2000 ft). After the shoreline had been pushed seaward, it remained southeast of the Book Cliffs, while subsidence and accumulation of an additional 610 m (2000 ft) of nonmarine strata took place. (See Figure 11-41, B.)

The episode of progradation was not continuous. On numerous occasions, rapid submergence enabled the sea to transgress northwestward and to deposit marine shale above the beach sands and other sediments that had accumulated in the coastal lowlands landward of the beaches. During each individual period of coastal progradation, however, a continuous sheet of sandstone, about 10 to 20 m thick, was deposited above marine shale. The bases of these sandstones are gradational, and their tops are sharply truncated (Figure 11-42).

The interpretation that the sands were deposited on ancient beaches is well supported by locally abundant heavy minerals and successions of kinds of bedding. Scattered exposures of fine sediments containing oyster shells are interpreted as ancient lagoonal deposits. However, the close association of extensive coals directly overlying the ancient beach sands does not suggest deposits formed in or near a salt-water lagoon (although the coals have been interpreted as part of the "lagoonal" deposits). Rather, the coals imply extensive fresh-water swamps, such as are found today on the fluvial plains associated with marine deltas or associated with broad channeled tidal flats (for example, swamps in south Louisiana and in South Carolina in which widespread peats are forming). Possibly the beach sands were deposited on numerous beach ridges that were parts of a delta complex

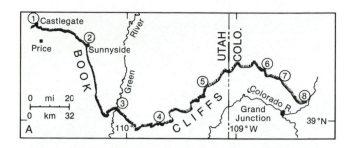

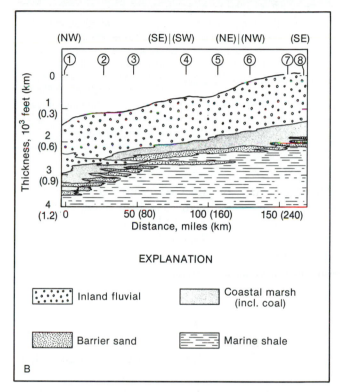

FIGURE 11-41.   Upper Cretaceous marine- and nonmarine strata exposed in Book Cliffs, Utah and Colorado.
  A. Index map showing locations of Book Cliffs and measured sections (numbers in circles) used as the basis for constructing the stratigraphic diagram.
  B. Stratigraphic diagram showing relationships among four principal kinds of sediments. The exposures trending NW-SE perpendicular to the Cretaceous shoreline (1 to 4 and 6 to 8), show pronounced lateral changes, whereas those trending NE-SW parallel to the shoreline (4 to 6) show few lateral changes. (After R. G. Young, 1955; A, fig. 1, p. 178; B, fig. 3, p. 194.)

resembling that of the modern Rhône delta in the Mediterranean. (See Figure 13-29.) If the sediments were delivered to the sea by many medium-size rivers, then possibly the waves redeposited most of the sands on barrier beaches. Detailed studies of the sandstones (difficult to make on the vertical cliffs) will be required to determine if any true deltaic sediments are present among the barrier-beach sands.

Other examples are found in the lower Mesaverde Formation (Point Lookout Sandstone and overlying

FIGURE 11-42.   View northward of sandstone, about 20 m thick, thought to be the deposit of an ancient barrier that prograded eastward into the Late Cretaceous seaway. Lens of fine-textured materials at left is interpreted as backbarrier deposit; to the west, the backbarrier deposits thicken and contain coal. At its base, the massive sandstone is interstratified with nearshore marine thin sandstones, -siltstones, and -shales. Eastward, the massive sandstone gives way to similar fine-textured marine strata. Castlegate Formation (Upper Cretaceous), Coal Creek Canyon, near Price, Utah. (Authors.)

Menefee strata), of late Cretaceous age; some of the gas-producing sands in the San Juan basin, New Mexico; and in the Eocene and Miocene strata of the coast ranges in California (Figure 11-43). These ancient examples perhaps are more closely allied to the barriers on the topset parts of deltas (as on the Rhône delta in the Mediterranean; see Figure 13-29) or near fans (as along the coast of the Red Sea), than to barriers that do not overlie deltaic deposits, such as at Galveston Island, Texas. (See Figure 11-36, B.)

SOURCES: C. V. Campbell, 1971; Masters, 1967.

Ancient Transgressive Sequences

Examples of ancient transgressive barrier (or -beach) deposits are not so easily identified as are ancient offlap successions. This is partly because, in many cases, these deposits have been reworked onto the marine shelf by transgressing marine waters. Under circumstances of rapid subsidence and rapid sediment deposition during transgression, such successions may be preserved. Examples of inferred ancient transgressive barrier (or beach) deposits include discontinuous sand bodies found at the base of a widespread marine shale that have been described from the upper part of the subsurface Mesaverde Formation (the Cliff House Sandstone of Cretaceous age in the San Juan basin, New Mexico). Discontinuous thick sandstones ("benches") are associated with abrupt changes in the level of the basal contact with respect to a reference surface (the "steps"). Another example that resulted from a transgression is the Pleistocene sands and gravels exposed in a sand pit at Benns Church, Virginia. (See Figure 10-38.)

FIGURE 11-43.   Examples of inferred nearshore sands deposited on ancient barrier of Miocene age (Branch Canyon Sandstone), southeastern Caliente Range, California. (N. Kumar.)
A. Pebbles in even, horizontal strata, probably deposited by storms on former foreshore seaward of breaker zone.
B. Lenses of pebbles and cross-stratified sandstones, probably deposited near shore close to zone of breaking waves.

Perhaps also belonging in the category of transgressive sands are many of the subsurface "shoestring sands" of Pennsylvanian age in Oklahoma and Kansas. An example of a "shoestring sand" of Holocene age, inferred to consist entirely of inlet-filling sands, has been reported from the continental shelf off Long Island, New York. This lens of Holocene shelf sand is inferred to have formed at a still stand of the sea at −24 m (local datum), about 8500 to 9000 yr B.P. After the former barrier had been submerged, the exposed portion is thought to have washed away, leaving only the deeper, lens-shaped "root" consisting of inlet-filling sands.

SOURCES: Friedman and Sanders, 1978;
Sanders and Kumer, 1975; Sabins, 1963.

Suggestions for Further Reading

BALL, M. M., 1967, Carbonate sand bodies of Florida and the Bahamas: Journal of Sedimentary Petrology, v. 37, p. 556–591.

BRYANT, EDWARD, 1982, Behavior of grain size (*sic*) changes on reflective (*sic*) and dissipative foreshores, Broken Bay, Australia: Journal of Sedimentary Petrology, v. 52, p. 431–450.

FOX, W. T., and DAVIS, R. A., Jr., 1976, Weather patterns and coastal processes, p. 1–23 in Davis, R. A., Jr., and Ethington, R. L., eds., Beach (*sic*) and nearshore sedimentation: Tulsa OK, Society of Economic Paleontologists and Mineralogists Special Publication 24, 187 p.

KASSNER, J., and BLACK, J. A., 1982, Efforts to stabilize a coastal inlet: a case study of Moriches Inlet, New York: Shore and Beach, v. 50, no. 2, p. 21–29.

KUMAR, NARESH; and SANDERS, J. E., 1974, Inlet sequence: a vertical succession of sedimentary structures and textures (*sic*) created by the lateral migration of tidal inlets: Sedimentology, v. 21, p. 491–532.

LEITHOLD, E. L.; and BOURGEOIS, JOANNE, 1983, Characteristics of coarse-grained sequences deposited in nearshore, wave-dominated environments–examples from the Miocene of south-west Oregon: Sedimentology, v. 31, p. 749–775.

MORTON, R. A., 1988, Nearshore responses to great storms, p. 7–22 *in* Clifton, H. E., ed., Sedimentologic consequences of convulsive geologic events: Boulder, CO, Geological Society of America Special Paper 229, 157 p.

POSTMA, GEORGE; and NEMEC, WOJCIECHK, 1990, Regressive (*sic*) and transgressive sequences in a raised Holocene gravelly beach, southwestern Crete: Sedimentology, v. 37, p. 907–920.

RAMPINO, M. R., and SANDERS, J. E., 1980, Holocene transgression in south-central Long Island, New York: Journal of Sedimentary Petrology, v. 50 p. 1063–1080.

RAMPINO, M. R., and SANDERS, J. E., 1981, Evolution of the barrier islands of southern Long Island, New York: Sedimentology, v. 28, p. 37–47.

SHIPP, R. C., 1984, Bedforms (*sic*) and depositional sedimentary structures of a barred nearshore system, eastern Long Island, New York: Marine Geology, v. 60, p. 235–359.

WEIMER, R. J., HOWARD, J. D., and LINDSAY, D. R., 1982, Tidal (*sic*) flats and associated tidal channels, p. 191–245 *in* Scholle, P. A., and Spearing, D., eds., Sandstone depositional environments: Tulsa, OK, American Association of Petroleum Geologists Memoir 31, 410 p.

WOLFF, M. G., 1982, Evidence for onshore sand transfer along the south shore of Long Island, New York and its implication against the "Bruun Rule": Northeastern Geology, v. 4, p. 10–16.

Marginal Flats: Peritidal Environments

General Setting

Modern coastal areas composed of sediment subjected to the action of waves and tides, and not directly influenced by large rivers, are subdivided into two groups: (1) shoreface-beach-barrier complexes and (2) peritidal complexes. In Chapter 11 we described the first group. In this chapter we take up the subject of peritidal complexes.

Most peritidal complexes occur in protected settings, commonly behind barrier islands. Where the coastline adjoins the flat-lying mainland and where no barrier islands or lagoons are present, peritidal complexes become especially important.

In Chapter 11 we noted how coastal features are systematically distributed with respect to tidal range. Further insights result from plotting wave height against tidal range (Figure 12-1). As noted in Chapter 11, long barriers characterize wave-dominated coasts. At the other extreme are macrotidal coasts, which are characterized by peritidal complexes with very broad and well-developed *intertidal flats* and *intertidal marshes* (Figure 12-2).

Our objectives in this chapter are to become acquainted with the processes and sediments in modern peritidal complexes, to try to formulate criteria for recognizing the sedimentary products of such complexes, and to look at a few examples from the geologic record that are inferred to be deposits of ancient peritidal complexes. It is easy enough to accomplish part of this task—for openers, all one has to do is go down to the sea shore, watch what is happening, and collect a few samples to take into the laboratory for study. But having done this, what does one really know? Are the samples collected during fair weather from easily accessible parts of these deposits the kinds of sediments that will go into the geologic record? It is becoming increasingly apparent that vast quantities of data have been collected about modern coastal sediments of kinds that may almost never enter the geologic record. In our discussion we shall concentrate on features that we think will go into the geologic record. In addition we shall try to explain the mechanisms by which these coastal sediments become parts of the geologic record.

Peritidal sediments are important to us for several reasons. Many deposits formed in ancient peritidal complexes are reservoirs filled with oil and gas. Accordingly, extensive investigations into the modern sediments from these complexes have been undertaken by the research staffs of major oil companies. Peritidal complexes in tropical climates have provided important and in some cases revolutionary data on the origins of dolomite, of certain microbial structures, and of some kinds of evaporites.

Although we have mentioned tidal action and various aspects of tidal sediments (in Chapter 11 and elsewhere), we attach such great significance to sediments that are deposited chiefly under the influence of the tides that we shall discuss these sediments in their own separate section. Under the heading of peritidal complexes we include *backbarrier complexes* in both nontropical (including polar) and tropical climate zones and various settings in the tropics where tidal action influences carbonate sediments alone or in association with evaporites.

Peritidal environments are extremely diverse and dynamic places; in them many processes operate. This diversity is expressed in the wide range of features found in peritidal sediments. Few other depositional settings are characterized by the intimate juxtaposition of "high-energy" and "low-energy" conditions; of marine- and nonmarine materials; of vertical- and lateral sedimentary processes involving waves, currents in channels, and currents as sheet flows; and of the distinctive effects of organisms on the sediments. Perhaps in no other accessible depositional setting is it possible to find as many features as in a peritidal complex.

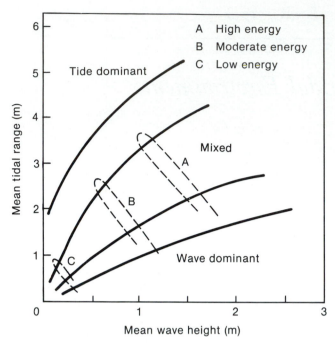

Mean wave height versus mean tidal range. Domains of tide-dominant-, wave-dominant-, and mixed coasts are indicated, as are high-energy-, moderate-energy-, and low-energy coasts. On wave-dominated coasts, barriers are well developed. On tide-dominated coasts, barriers commonly are not present. (After M. O. Hayes, 1979.)

Peritidal complexes not associated with barrier complexes are most characteristic of macrotidal coasts (tidal ranges > 4 m) where wave energy is relatively low. On coasts where tidal ranges are high, a common association is peritidal complexes and offshore **tidal sand ridges,** *sand ridges formed at some distance from shore that are*

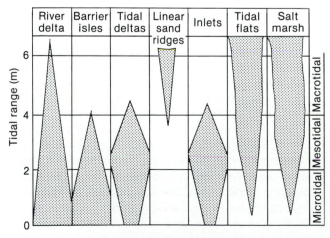

FIGURE 12-2. Variation of coastal morphology as a function of tidal range. (Modified from M. O. Hayes, 1975.)

(1) shaped mainly by tidal action, (2) submerged, and (3) oriented at an oblique angle to the direction of flow of tidal currents.

The marginal flats of nonmarine basins share many features with marine-marginal flats. However, nonmarine basins generally lack tides and intertidal zones. In addition, in many nonmarine basins, waves are very small. Thus physical conditions are different, but the effects of organisms and the kinds of sediments found on the marginal flats of nonmarine basins resemble more closely those of marine marginal flats. Nevertheless, this chapter is of necessity mostly concerned with the marginal flats of basins that are connected to the sea.

SOURCES: Boothroyd, Friedrich, and McGuinn, 1985; Fraser, 1989; Friedman, 1985; Friedman and Krumbein, 1985; Harms, Southard, and Walker, 1982; Howard and Frey, 1985; McCave, 1985; Penland and Suter, 1985; Reading, 1986; Reineck and Singh, 1980; Scholle and Spearing, 1982; D. G. Smith, Reinson, Zaitlin, and Rahman, 1991.

Zonation of Peritidal Areas

Based on their relationship to the fluctuating levels of the sea, peritidal areas fall naturally into three categories: (1) *subtidal,* (2) *intertidal,* and (3) *supratidal* (Chapter 11).

Subtidal Zone

The subtidal zone is lower than the low-water level of even the greatest of spring tides, and their cover of seawater is usually shallow. Given normal salinity in these shallow-water areas, a normal marine fauna thrives and the bottom may be subject to gentle wave action.

Intertidal Zone

The intertidal zone includes areas lying between the levels of normal low- and high tide. The intertidal zone is synonymous with the *littoral zone* as defined by ecologists. The intertidal zone is subject to regular flooding and exposure by astronomic tides on a daily (daily tide) or twice-daily basis (semidaily tide).

Supratidal Zone

The supratidal zone is the peritidal area lying above, but close to, the mean high-tide level and not higher than the level of the highest storm tide. Although these zones are easy to define and a logical basis exists for each of them, the reference levels are not fixed planes. The amplitude of the "normal" astronomical tide varies daily and the levels of storm surges are not related to the spring-tidal levels. Notice, also, that the definition of the zones does not mention the tidal range. Obviously, the appearance of a coastal area and peritidal sediments will be influenced by the magnitude of the normal tidal fluctuation. The level of the highest storm tide may not be known. Such tides may be so infrequent that, within historic time when accurate records have been kept, they may not have happened. Evidence of prehistoric high storm tides may not exist, or, as a result of subsequent sea-level changes, may not be properly interpreted. The reference levels undergo long-term secular change as a result of isostatic glacial rebound, tectonic uplift or subsidence, eustatic sea-level change, and other factors that affect local sea level. Despite these difficulties, peritidal sediments may be used to infer the ranges of ancient tides. At present, this subject is being actively investigated. For example, tidal ranges have been estimated in Cambrian cyclic peritidal carbonates of the Virginia Appalachians. Thickness of cycle tops composed of dolomitized laminites (thinly laminated strata) formed by microbial mats, coupled with reasonable estimates for subsidence rates and sedimentation rates, yield estimates of about 2-m tidal ranges. Many important practical- and theoretical problems could be cleared away if it were possible to determine ancient tidal ranges for many localities quantitatively and systematically.

In the ensuing discussion we shall first summarize some general relationships involving the conditions in peritidal complexes (physical conditions, relationships among the organisms, and kinds of sediment), including criteria for their recognition. Then we shall describe some modern examples of various peritidal complexes and close with some inferred examples of ancient peritidal complexes.

SOURCES: J.R.L. Allen, 1985; Fraser, 1989; Frey, Howard, Han, and Park, 1989; Friedman and Krumbein, 1985; Grant, 1988; Kopaska-Merkel, 1987a; Kopaska-Merkel and Grannis, 1990; Mazzullo and Birdwell, 1989; Shinn, 1983; D. G. Smith, Reinson, Zaitlin, and Rahmani, 1991; Wanless and others, 1988; Warren and Kendall, 1985.

Characteristic Features of Peritidal Sediments

As we have defined the term, peritidal designates various areas ranging from the polar zones to the equatorial region, along all kinds of coasts. Hence we shall be involved with a wide variety of physical conditions, or-

ganisms, and sediments. Certain characteristic features are typical of peritidal sediments. No one of them alone can be considered unique to peritidal sediments. As with most kinds of sediments, the origin is inferred on the basis of a community of characteristics and of the relationships to adjoining deposits.

The characteristics of peritidal sediments result from a geographic setting at the edge of the sea; from the flow of tidal currents and resultant shifting of *tidal channels*; from the effects of small shallow-water waves at high tide; from the draining away of the waters at low tide, thus exposing the sediments of the intertidal zone to the atmosphere; from the coincidence of highest water levels with maximum wave action during storms; from the distribution and activities of organisms; and from various physical- and chemical processes that affect the sediments.

In the modern world, a geographic setting at the edge of the sea is self-evident. In the geologic record, the appropriate paleogeographic setting would have to be reconstructed from the lateral (facies) relationships of a given stratigraphic unit or from the changes in a vertical succession. Ideally, ancient peritidal sediments should be sandwiched between nonmarine- and marine strata. The surest distinctions between such strata are made through fossils (Chapter 16).

Physical- and Chemical Conditions

Important physical conditions in peritidal complexes include agitation of the water, amount of suspended sediment, and the climate (as it affects the salinity and temperature of the water and the behavior of water in the atmosphere).

Because our concern will be chiefly for the environments where fine-textured sediments accumulate, we shall be concentrating on areas not subject to vigorous wave action. Although we exclude such areas from our discussion, we need to point out that if the concentration of suspended sediment in the water becomes greater than 100 mg/l (which approaches the levels found in delta-building rivers, Table 13-1), then fine-textured sediments can accumulate even in high-energy coastal areas. Another way in which fine-textured sediments can accumulate in areas intermittently subject to high-energy conditions is by deposition of **marine snow**: *very large flakes (up to 1 cm across) that are aggregates of silt- and clay-size particles.* Peritidal complexes in estuaries and near large marine deltas are discussed in Chapter 13.

The peritidal areas of our concern here are sheltered from the waves of the open sea by a *barrier island* or *barrier spit*, by a coastal indentation, or by a wide expanse of very shallow water.

Salinity values of the coastal seawater will range from brackish to hypersaline (60 to 65‰); the water temperature from near zero to +30°C; and the climate from cold and moist to tropical, both wet and dry.

Physical effects on the sediments of peritidal environments include such things as the alternate deposition at a given spot of sand and mud, the building of patterned sequences by lateral- and vertical sedimentation, the formation of various void spaces in the sediment by the escape of gas bubbles and by desiccation, and the upheaval of the surface of the sediment by the growth of crystals precipitated from the *interstitial waters*. The alternate deposition of sand and mud creates interbedded successions of sandstone and mudstone. (See Figure 5-22.) Layers of sand and mud alternate in many other kinds of environments as well.

The chief chemical effects in some peritidal sediments include precipitation of evaporite minerals. The waters evaporated may be the hypersaline waters of a lagoon or bay, hypersaline interstitial waters, or both.

The flow of tidal currents creates channels (Figure 12-3). Typically tidal channels are of two sizes: (1) large trunk channels in which water flows, even at low tide, and (2) smaller channels that may be nearly dry at low tide. Some tidal channels are sinuous and may migrate laterally. Others, whether sinuous or not, do not migrate laterally. Channels that migrate laterally engage in *lateral sedimentation*. Such activity creates patterned autocyclic parasequences whose sediment particles become finer upward (exception: inlets, where the upward particle-size gradation may be coarse to finer and back to coarse). The cross-sectional shapes of the numerous tidal channels may be preserved only where inactive channels have become filled with other kinds of sediment. In such fillings, the sizes of the sediment particles are controlled by processes that are not related to the flow of water in

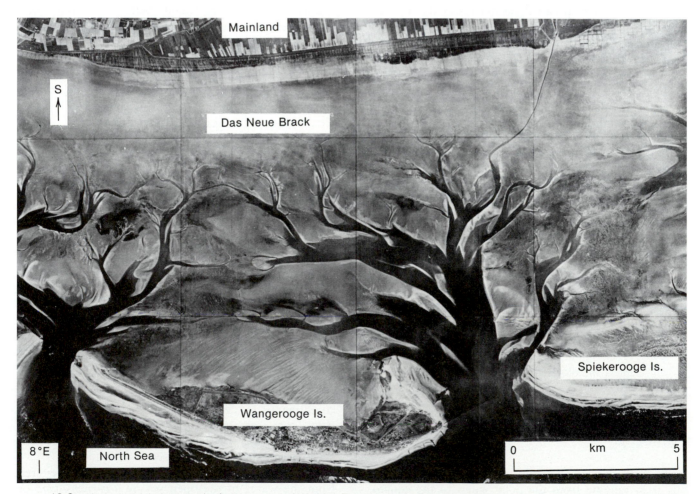

FIGURE 12-3.   Vertical aerial photograph of eastern end of Wadden Zee at low tide (location on Figure 12-10) shows dendritic pattern of tidal channels and extensive intertidal flat. Most of the intertidal flat is underlain by silty clay (appears on photo as medium-gray tone). Exposed sand (white) is confined to barriers (at bottom) and to natural levees bordering tidal channels. (H.-E. Reineck.)

the channel. Systematic particle-size gradations in such fillings are entirely fortuitous. As important as channels and channel sequences are for recognizing peritidal sediments, channels alone do not constitute proof of peritidal origin. Other sediments featuring channels are deposited by floodplain rivers and on deltas, to name two examples.

Another effect of tidal currents is to deposit **herring-bone cross laminae**, which are *sets of cross laminae in which the dip directions of successive sets of cross laminae are nearly opposite*. This is caused by the reversal of tidal currents with rising- and falling tide. Herringbone cross laminae strongly suggest that the deposit was tidally influenced. However, herringbone cross laminae are not restricted to tidal deposits; they have been found in river deposits and in longshore troughs of large lakes.

The frequent reversal of tidal currents on tidal flats produces other structures characteristic of this setting. One such structure is a **reactivation surface**, *a curved surface that truncates normal cross strata and that formed because changes in the current created a new profile on the bed form which differed from the previous profile*. This commonly occurs when bed forms that accreted during one half of the tidal cycle are moderately eroded by tidal currents flowing in the opposite direction.

Other tidal structures are distinctive cross strata named **tidal bundles**, *sets of cross laminae formed by the stronger set of currents in an asymmetric tidal cycle*. Because of this asymmetry, herringbone cross laminae are not preserved. The cross laminae within a tidal bundle may display subtle changes from base to top in response to the changes in the current during each tidal cycle. Tidal bundles are set off from one another by reactivation surfaces. Tidal bundles may superficially resemble portions of fluvial dune cross-stratal sets, which may also contain reactivation surfaces, but differ in details and in association with other tidal features. Because tides vary in amplitude on a monthly cycle (*neap-spring tidal cycle*, Chapter 8), the thickness of tidal bundles may vary on a monthly cycle. The passage of months is thus recorded in some ancient tidal deposits.

The effects of small shallow-water waves at high tide create various ripples in sandy sediments. Therefore, peritidal complexes in which the upper parts of the former intertidal zone were sandy display many varieties of ripples whose long axes are diversely oriented. Here again, the ripples alone are not diagnostic. Ripples are found in various nontidal settings, such as lakes, rivers, and numerous marine environments, including the deepsea floor. Larger-scale bed forms, subaqueous dunes, are also abundant on sandy intertidal flats.

The late-stage drainage of thin sheets of water from intertidal flats modifies ripples and other sedimentary features (Figures 12-4 and 12-5). As typical as is such

FIGURE 12-4.   Small ripples shaped in very fine sand by reversing tidal currents. Dominant flood current toward left created asymmetrical current ripples having steep sides toward left. Later, during the ebb tide, sand was shifted toward the right, thus creating the narrow, peaked profiles, including those having lines of closed depressions along their crests (as in center). Pencil is 16 cm long. Intertidal flats north of Five Islands, Nova Scotia. (Authors.)

draining away of water, it is not unique to intertidal environments. Thin sheets of water likewise can drain from sediments on parts of river flood plains and of lake bottoms. Similarly, the exposure to the atmosphere of the surface of sediments in the intertidal zone affords the possibility of preserving footprints of birds and of other creatures that do not inhabit the sea, and for allowing certain rare subaerial sedimentary structures to form and be preserved. These may include raindrop-impact craters (See Figure 14-15.) (known in rocks as old as 2.7 billion

FIGURE 12-5.   Intertidal wavy-laminated pellet lime grainstone interlaminated with massive heavily bioturbated pellet lime grainstone. The ripple forms containing wavy laminae have been truncated, possibly by ebb-tidal currents. Weathered vertical outcrop, Whirlwind Formation, middle Middle Cambrian, lower Middle Limestone unit, southern Drum Mountains, westcentral Utah. (D. C. Kopaska-Merkel, 1987a, fig. 4-3, p. 137.)

years), adhesion ripples, seafoam impressions, mud microwashovers, *runzelmarken* (wrinklemarks), desiccation polygons, and fenestral fabric. Intermittent subaerial exposure of sediments deposited under water is not confined to intertidal environments. Intermittent drying of wet sediments is a normal part of the regimen of playas and of flood plains.

A distinctive feature of modern peritidal complexes, and presumably, therefore, one that also existed in the past, is the association of highest water levels with greatest wave activity. This association is the work of great storms, in which *wind "tides" (storm surges)* accompany storm waves. These high-energy storm events may result in deposition of layers of sediment a few millimeters to a few centimeters thick. Because of the effects of wind on confined bodies of coastal water however, highest water levels in one area may be accompanied by lowest water levels not far away. In such places, maximum wave activity takes place at lowest water levels.

SOURCES: J. R. L. Allen, 1985; DeBoer, van Gelder, and Nio, 1988; Fitzgerald and Penland, 1987; Fraser, 1989; Frey, Howard, Han, and Park, 1989; Friedman, 1985; Friedman and Krumbein, 1985; Gilbert and Aitken, 1989; Hine, Belknap, Hutton, Osking, and Evans, 1988; Howard and Frey, 1985; Kopaska-Merkel, 1987a; Kopaska-Merkel and Grannis, 1990; Mazzullo and Birdwell, 1989; Nio and Young, 1991; Noe-Nygaard and Surlyk, 1988; Rosen and others, 1988; Scholle and Spearing, 1982; D. G. Smith, Reinson, Zaitlin, and Rahmani, 1991; Thorbjarnarson and others, 1985; Wanless and others, 1988; Warren, 1988.

Organisms

In peritidal sediments, the effects of organisms, including both marine invertebrates and various plants, are numerous and varied. Effects include burrowing, sorting by selective feeding, building of mats, trapping of sediments, and the creation of pellets. None of these activities is confined to peritidal environments. We emphasize two relationships: organisms and climate and organisms and sediments.

Climate exerts a major effect on the distribution of organisms in peritidal regions. Two conspicuous relationships between organisms and climate are the change from salt-water grasses (the marsh builders) of the temperate regions to mangroves of the tropics, and the preponderance of the skeletons of carbonate-secreting organisms (both plant and animal) among nearshore sediments of most tropical regions that are not influenced by terrigenous sediment delivered to the sea by large rivers.

Also, in many parts of the tropics the coast is fringed with coral reefs (Chapter 10).

Of particular importance for peritidal sediments are the cyanobacteria, the dominant organisms in most microbial mats. (See Figures 7-40 and 7-41.)

MICROBIAL MATS

Many organisms, notably cerithid gastropods, graze on *microbial mats*; these gastropods concentrate in "herds" that can advance across a fresh mat leaving behind only pellets on a mat-free surface (Figure 12-6). Cyanobacteria exhibit a wide tolerance for variations in salinity and exposure to the atmosphere. The salinity tolerance of cerithid gastropods is less than that of the mat-forming cyanobacteria. Therefore microbial mats can flourish where the salinity (or other factors) keeps away the cerithids. The distribution of these two kinds of organisms is mutually exclusive. In waters having near-normal salinity, cerithid gastropods (and other grazing invertebrates, whose fossil record extends back into the Early Cambrian) hold sway and microbial mats are absent. In waters of elevated salinity, subtidal, intertidal, or supratidal, cerithid gastropods cannot survive. In such salty surroundings, cyanobacteria and other mat-forming organisms such as diatoms prosper. In the hypersaline brines of Hamelin Basin Pool, Shark Bay, Australia, where cerithids are not present, flat-laminated stromatolites typify the upper intertidal zone, and domal- and columnar stromatolites the lower intertidal zone. Algal **thrombolites** *(cryptalgal structures related to stromatolites*

FIGURE 12-6. A herd of *Cerithium* gastropods grazing on an algal mat. Gastropods have not yet eaten some of the algal mat (light gray at top), but where *Cerithium* are numerous, no algal mat remains (at bottom). Note sharp edge of algal mats where *Cerithium* ceased "nibbling." Light gray curved lines at left are *Cerithium* trails. Intertidal flats, western Andros Island, Bahamas. (G. M. Friedman, A. J. Amiel, M. Braun, and D. S. Miller, 1973; fig. 6, p. 547.)

but lacking lamination and characterized by a macroscopic clotted fabric) in the Early Devonian Manlius Formation have been interpreted as subtidal. They are associated with skeletal remains from a very restricted fauna consisting of 1 species of brachiopod, 1 species of gastropod, and 1 species of ostracode. The environment is inferred to have been characterized by variable- or elevated salinities that excluded algal predators as in the modern Hamelin Pool. During much of Precambrian and early Paleozoic time, mat-forming organisms flourished in the absence of mat-eating organisms (which did not yet exist). During this period, mat-formers probably spread across all parts of the peritidal territory including subtidal settings. The abundance, large size, wide environmental distribution, and morphologic diversity of Proterozoic stromatolites (some containing the fossils of mat-forming microorganisms) attest to how successful the microbial-mat communities were in the absence of predators.

PLANTS

In temperate regions, very soon after shoaling has progressed to the point where sediment, usually mud, is exposed to the atmosphere for more than 6 hours, species of the salt-tolerant grass *Spartina* move in and make a closely grown marsh turf over the mudflat. These salt-marsh grasses establish and perpetuate a nearly horizontal upper surface whose level coincides closely with the mean high-tide mark. (See Figure 11-32.)

The landward fringes of the peritidal zones in tropical climates are closely grown with mangroves (Figure 12-7). The mangroves create a nearly impenetrable barrier to human passage. Because plant debris is so plentiful in both salt marshes and mangrove thickets, peat forms.

ANIMALS

The sediments in a few peritidal environments, including mounds of sediment in some tidal channels, are stabilized by colonies of the blue mussel, *Mytilus edulis* (Figure 12-8). Each mussel attaches itself firmly to the bottom by secreting a thin, tough thread, known as a byssus. If enough mussels take hold on a gravel bar, they are able to prevent the gravel from being shifted by all but the most-violent currents. On intertidal marshes, mussels fulfill a different role. Here they may produce a substantial proportion of marsh sediment in the form of fecal pellets composed of suspended sediment trapped by the mussels during suspension feeding.

As mentioned in the section on tidal deltas in Chapter 11, one of the most-significant and pervasive influences of animals on intertidal flats is burrowing. Most intertidal-flat sediments have been extensively bioturbated. Especially characteristic are vertical dwelling burrows and surficial feeding trails. Commonly the kinds of burrows exhibit a shore-parallel zonation controlled by sediment type and by degree of subaerial exposure. Some burrowing organisms form mounds that protrude above the sediment surface. Mound-forming tube dwellers that tend to occur together in aggregations or colonies, such as the polychaete worm *Lanice conchilega*, may affect patterns of sediment movement and deposition and form bioherms that may be preserved in the rock record.

FIGURE 12-7.   Mangrove thicket along shores of Ten Thousand Islands, Florida. (Florida News Bureau.)

FIGURE 12-8.   Mussels attached to pebbles on floor of shallow tidal channel, peritidal area, Moriches Bay, Long Island, New York. (C. Olsen.)

ORGANISMS AS ENVIRONMENTAL INDICATORS

The organisms that are adapted to life in the upper parts of the intertidal zone, or in the supratidal zone, help establish a firm case for recognizing an ancient peritidal complex. However, not all such organisms are confined to these zones. Reliable indicator organisms are mangroves and salt-water grasses. Unreliable organisms include mat-forming cyanobacteria and algae, which can exist in saline lakes far from the sea and can also thrive in subtidal zones where high salinity or other conditions keep out mat-eating organisms and burrowers. Even so-called reliable organisms may not be precise indicators of the intertidal zone. Red mangroves in the Florida Everglades grow along waterways as much as 10 miles inland. These trees grew from seedlings brought in several decades ago on floodwaters from a major hurricane. In addition, thriving colonies of mangroves may be found at an elevation of about 1 m above sea level in cypress domes in the Everglades at least 20 miles from the present shoreline (Figure 12-9). These colonies grew from seedlings brought into the area before construction of the Pa-hay-okee Road, which now prevents floodwaters bearing mangrove seedlings from reaching the cypress domes.

SOURCES: Aitken, Risk, and Howard, 1988; Breyer and McCabe, 1986; Browne and Demicco, 1987; Carey, 1987; Frey, Howard, Han, and Park, 1989; Frey, Howard, and Hong, 1987; Frey and Pemberton, 1985; Friedman and Krumbein, 1985; Grant, 1988; Scholl, Bebout, and Moore, 1983; Shinn, 1982; Smith and Frey, 1985.

Kinds of Sediments

The sediments in peritidal complexes span the complete spectrum of sedimentary particles (Chapter 2). Intrabasinal carbonate particles are dominant in tropical areas, whereas extrabasinal particles dominate in most nontropical peritidal complexes, but may also be common in tropical peritidal settings.

Backbarrier-Peritidal Complexes in Nontropical Climates

In discussing the sediments of peritidal complexes, we again need to keep in mind the differences between lateral- and vertical sedimentation. Lateral sedimentation requires some kind of sloping depositional surface that can move as sediment is added to it. Vertical sedimentation requires a generally horizontal surface that accretes upward.

FIGURE 12-9. Mangroves growing in a cypress dome, Everglades National Park, Florida, between Pa-hay-okee road and Rock Reef Pass, 20 miles from the coastline. Further discussion in text. (R. S. Merkel.)
  A. Proproots of mature red mangroves in cypress dome.
  B. Close view of red mangrove seedling in cypress dome.

Backbarrier-peritidal complexes in high latitudes (e.g., north of the Arctic circle) differ but little in sedimentary processes, sedimentary structures, and sediment/organism interactions from temperate backbarrier-peritidal complexes. However, on high-latitude tidal flats, ice-driven processes of sediment movement and of deformation of strata may be important, hence sediment particles of gravel- to boulder size may be moved about in these regions. We shall not discuss these processes further.

In the backbarrier regions under consideration, two depositional slopes are particularly prominent. These are (1) the surface leading from the top of a tidal marsh across the high parts of the intertidal flat, past the low parts of the intertidal flat, and ending at the bottom of a central tidal channel, and (2) the surface leading from the low parts of the intertidal flat into the channels of meandering watercourses (designated as "creeks") that cross the low parts of the intertidal flat.

The variations present along profile (1) are intertidal marsh–high intertidal flat–low intertidal flat–tidal channel. They depend on the concentration of suspended sediment in the water and on the relative effectiveness of wave- and tidal action. Where the concentration of suspended sediment is low (< 100 mg/l), wave-dominated coastal water bodies develop marginal belts of sand and central deposits of silt and clay. Where tides dominate, the silt and clay coat the shores and the sand is confined to the deeper parts where tidal currents flow swiftly. (See Figure 13-7.) If, however, the concentration of suspended sediments in the coastal waters exceeds 100 mg/l, then silt and clay may be deposited everywhere. We shall describe an example of a peritidal complex adjacent to a coastal water body having variable width in which wave action is appreciable at one end and less important at the other end and in which suspended sediment is less than 100 mg/l.

At high tide between the individual members of the chain of Frisian Islands that are situated between the southern waters of the North Sea and the European mainland of The Netherlands and Germany is a large lagoon known as the Wadden Zee (or Wadden Sea in English; Figure 12-10). At low tide much of the area between the Frisian Islands and the mainland becomes exposed. (See Figure 12-3.) The tidal range in this area is between 2.4 and 1 m. At the western end of the island chain, south of Ameland, The Netherlands, the width of the Wadden Sea is greatest. Because of wave action at high tide, fine sand forms the higher, marginal parts of the intertidal flats, and silt and clay the lower, central parts. At the eastern end of the Frisian Islands the width of the Wadden Sea is least. Here wave action at high tide is a minor factor. Thus fine sand is confined to the lower parts of the intertidal flats and is concentrated in

the tidal channels and on the low natural levees flanking these channels. (See Figure 12-3.)

Because intertidal flats are places where mud and sand accumulate in various proportions, many varieties of bedding are formed. In some areas (e.g., where waves dominate), only sand may be deposited; all the mud is kept in suspension and comes to rest elsewhere. In other areas (where tides dominate), the sediments may consist entirely of mud. In still other areas, sand may be deposited in one part of the tidal cycle (during flood- or ebb tide, for example) and mud in a different part of the cycle (at slack water). The alternating deposition of sand and mud creates various kinds of layers; a chief factor is the proportion of sand to mud. Where the proportions of sand and mud preserved in the sedimentary record are about equal the result is an interbedding of continuous layers of sand and mud. Ordinarily the upper surfaces of the sand layers will be rippled; the thickness of the resulting layers thus varies laterally (greatest thicknesses coincide with ripple crests and least thicknesses with ripple troughs). (See Figure 5-22, A.)

Where sand is the more-abundant constituent the mud may never form continuous layers but simply drape over the sand ripples as wavy thin layers or as isolated meniscus-shaped lenses. (See Figure 5-22, B.) These *wavy layers and lenses of mud* are known by the German term **flasers**, which means vein, as in wood or rock. *Stratification in successions consisting of alternating rippled*

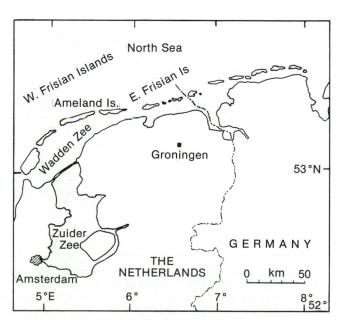

FIGURE 12-10. Southern coast of North Sea, The Netherlands and Germany, showing chain of barriers, the Frisian Islands; associated lagoon, the Wadden Zee (Sea in English); and adjacent mainland.

sand and mud drapes in which mud is the minor constituent and mud layers are discontinuous is **flaser bedding.** (See Figure 5-22, B.)

Where mud is the dominant constituent the sand is confined to isolated ripples, a few of which may be joined together. (See Figure 5-22, C.)

At high tide the fine sand on the wave-dominated high tidal flats of The Netherlands becomes extensively rippled. After the water recedes, the effects of burrowing organisms become apparent. Sediment-feeding worms, such as *Arenicola marina* (Figure 12-11, A) and others, various pelecypods, and crustaceans, live in burrows in

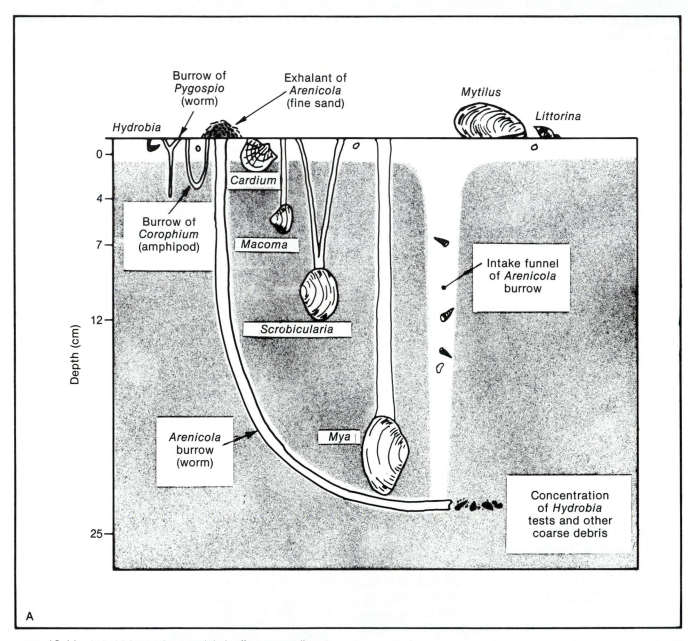

FIGURE 12-11.   Intertidal organisms and their effects on sediment.
    A. Selected members of the intertidal epifauna (top) and infauna in living positions; schematic profile through fine sandy intertidal flat along shores of North Sea. White parts of diagram indicate oxidized sediment that appears brownish in nature; gray tone designates reduced sediment whose natural color is dark gray to black. Although the gastropod *Hydrobia* lives on the surface of the intertidal flat, most of the *Hydrobia* skeletal material is preserved at a depth of about 25 cm; the shells fall down the *Arenicola* intake funnels and are not eaten by the worms. Further explanation in text. (After K. O. Emery and R. W. Stevenson, 1957, fig. 27, p. 723.)

FIGURE 12-11. (*Continued*)

B. Thousands of mounds, each about 2 cm high, of fine sand squeezed up by burrowing worms and completely covering rippled surface of lower intertidal flat. View toward lagoon (Pamlico Sound), near Oregon Inlet, North Carolina. (J. E. Sanders and J. A. Burger, May 1966.)

FIGURE 12-12.   Sandy intertidal flat exposed at low tide, Jade Bay, north German coast, near Wilhelmshaven. As meandering intertidal water courses (*Prielen*, of local usage) shift laterally, they rework the sediments by eroding on the outside of the meander bends and by depositing on the insides. (F. Wunderlich and H.-E. Reineck.)

the fine sand. *Arenicola* has been found to sort the sediments extensively, chiefly by concentrating the tests of the small gastropod *Hydrobia* at the depth of the worm's feeding. Debris of various sizes falls down the worms' intake funnels, but only fine sand appears in the mounds that surround the exhalant openings. (See Figure 12-11, B.) The material coarser than fine sand forms a definite thin layer lying at a depth of about 25 cm beneath the surface of the sandy intertidal flat. (See Figure 12-11, A.)

Mud accumulates on the lower parts of the Dutch tidal flats. The chief agents for reworking these muddy sediments are the small meandering watercourses (named creeks) that cross the low flats (Figure 12-12). As the creek channels shift by undercutting their steep concave banks, they spread out a channel-floor lag composed chiefly of the disarticulated shells of large pelecypods. The level of the shell pavement coincides with the depth of the creek channel. Because the discharge of these creeks is steady (unlike a stream subject to periodic floods), the observed everyday relationships between channel-floor lag and discharge are the only factors that influence the sedimentary record. Accordingly, the thickness of the tidal-creek channel succession closely approximates the depth of water of a normal discharge (rather than the depth of flood discharge, as in a stream in an alluvial valley).

The rates of lateral migration of creeks on sandy tidal flats have been found to range from about 25 to 100 m/yr; on muddy flats the rate is slower (25 m/yr). Study of dated maps and aerial photographs of the tidal flats in the lee of barrier island Wangerooge, Germany

(See Figure 12-3.) has shown that 58% of the area of the tidal flats was reworked by the creeks in 68 years.

The shells forming the channel-floor lag are eroded from the sediments on the outsides of the meander bends. The shifting channel releases shells of dead clams from their burrows. These shells fall to the floor of the channel, where they tend to accumulate in convex-up positions. As the tidal channels advance, they bury the channel-floor lags and create autocyclic tidal-channel parasequences (Figure 12-13).

The autocyclic parasequences deposited by tidal-channel migration are only a few meters thick. The channel-floor lag at the base consists largely of shell debris, but may include chunks of peat or of mud. The lower part of the succession typically contains alternating distinct layers of sand and mud, including flasers; bioturbation structures are rather uncommon. Most creeks crossing intertidal flats are too small to create large-scale dunes. Therefore, in this tidal-channel succession, large-scale cross strata are not present. The upper part of the parasequence is finer than the lower part; typically, the upper sediments have been extensively modified by burrowing organisms.

Because the wandering creeks rework the intertidal sediments so completely, the preserved record of the intertidal sediments probably contains more skeletal debris in lenticular beds (channel-floor lag deposits) than shells in growth position in burrows.

Various microbial mats and mussel colonies spread across the surfaces of the intertidal flats. Both these kinds of organisms stabilize the sediment below; their binding action tends to keep the sediment from being moved by currents. The effects of microbial mats on the sedi-

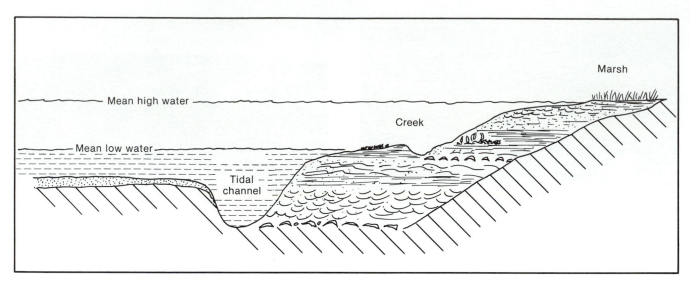

FIGURE 12-13. Results of progradation of a peritidal complex during stillstand of sea; schematic profile and section based on relationships in Dutch Wadden Sea. Depositional sequence contains channel-fill successions of two sizes. Further explanation in the text. (J. E. Sanders in G. M. Friedman and J. E. Sanders, l978, fig. E-20, p. 552.)

ments of nontropical intertidal flats composed of silicate minerals have not been studied in as much detail as have the effects of microbial mats in tropical intertidal flats composed of carbonate sediments (discussed in following sections). Clusters of mussel colonies generally are surrounded by accumulations of fecal pellets which may represent deposition of substantial amounts of sus-

pended sediment trapped by the filter-feeding mollusks. At low tide various animals walk and crawl across the surface of the sediments, leaving imprints and trails (Figure 12-14).

Channel-floor sediments of the trunk tidal channels which connect the tidal creeks to the inlets include debris of mussel shells and other invertebrates that in-

FIGURE 12-14. Footprints in soft modern dolomite sediments and in ancient dolostone.

A. Emu tracks in glistening white sediment flooring first lagoon-marginal ephemeral lake east of Salt Creek, Coorong Lagoon area, South Australia. This freshly deposited primary precipitate of dolomite crystals is underlain by hydromagnesite. (Authors.)

B. Dinosaur footprints (of coelosaurian genus *Elaphrosaurus*) preserved on bedding surface in Lower Cenomanian (Cretaceous) fine-textured dolostone, village of Beit Zait, Israel, near Jerusalem. Before lithification, these carbonate sediments (now dolostone rock) were firm enough to retain these footprints. (G. M. Friedman and J. E. Sanders, 1967, pl. II, B, p. 318.)

habit backbarrier areas. The water in these channels is deep enough and usually flows swiftly enough to transport sand and to create large-scale dunes and various other bed forms. The larger tidal channels are bordered by natural levees built of fine sand dropped when the rising tide spills out of the channel. As the trunk tidal channels shift laterally they deposit patterned autocyclic parasequences of strata (See Figure 12-13.), rather comparable to point-bar sequences in rivers (Chapter 14). Some intertidal flats, for example, those in high-latitude regions that are severely affected by formation of sea ice, lack tidal channels. These intertidal flats produce different parasequences than do intertidal flats characterized by networks of laterally migrating tidal channels.

Progradation of a peritidal complex (See Figure 12-13.) creates a parasequence whose basic thickness is a close approximation to the mean tidal amplitude plus the mean depth of the trunk tidal channels. Rarely will the thicknesses of successions deposited by the progradation of individual peritidal complexes exceed 5 m. Such a complete peritidal parasequence typically contains two hierarchies of upward-fining channel successions: (1) the thicker trunk-channel deposits, at the base; and, next above, (2) the thinner creek deposits. (See Figure 12-13.) These two sets of upward-fining successions are overlain by the sediments deposited on the high intertidal flats, which, depending on dominance by waves or by tides, are sandy or muddy, respectively. At the top may be a salt-marsh peat, whose normal initial thickness equals about half the mean tidal range. If, after an autocyclic peritidal succession has prograded, progressive gradual submergence takes place at a rate that can be matched by the rate of vertical accretion of the tidal marshes (a maximum of about 3 mm/yr), then the lower part becomes buried and is incorporated into the geologic record with its initial thickness essentially unchanged, and the intertidal-marsh peat at the top progressively thickens in response to the allocyclic sea-level rise. After the landward parts of the newly deepened lagoon have shoaled appropriately, such a marsh-capped peritidal complex may extend itself laterally. However, as noted, the thicknesses of any parts of this complex that are deposited by processes of lateral sedimentation do not increase. By contrast, the thicknesses of the two vertically sedimented units, the salt-marsh peat on the landward side and the lagoon-bottom sediments on the opposite side, steadily increase.

In a wave-dominated lagoon, where the higher parts of the intertidal zone are sandy, a totally different situation prevails. Whether the high intertidal-zone sands would thicken during submergence (as does the salt-marsh peat) remains to be determined.

SOURCES: Aitken, Risk, and Howard, 1988; Carey, 1987; Fitzgerald and Penland, 1987; Fraser, 1989; Frey, Howard, Han, and Park, 1989; Frey, Howard, and Hong, 1987; Howard and Frey, 1985; Nio and Young, 1991; Reading, 1986; Reineck and Singh, 1980; D. G. Smith, Reinson, Zaitlin, and Rahmani, 1991; Thorbjarnarson, Nittrouer, DeMaster, and McKinney, 1985.

Backbarrier-Peritidal Complexes in Hot, Arid Climates

The backbarrier-peritidal complexes of the high-latitude, temperate climate zones are characterized by large astronomical tidal ranges (up to 5 or 10 m, perhaps more), by waters whose salinity generally is less than that of normal seawater, by marsh-building grasses, and by a predominance of particles of silicate minerals. By contrast, in hot, arid climates the salinity of the lagoon waters exceeds that of normal seawater, the marsh-forming grasses of the temperate zones are absent, and carbonate particles commonly predominate. Carbonates, sulfates, and even halite may be precipitated as primary minerals. Moreover, in many low-latitude zones the amplitudes of the astronomical tides are small (only a few meters or even less); in fact, in some low-latitude regions the amplitude of the lunar tide becomes zero. Box 12.1 contains an example of backbarrier sediments in hot, arid climate zones: the Coorong Lagoon, South Australia.

Peritidal Complexes Marginal to Tropical Bays

The peritidal complexes examined so far owe their protection from the sea to a barrier island or barrier spit. Other peritidal complexes are protected from vigorous wave action by an indented coastal outline; they lie along the shores of bays. Two Australian bays, Shark Bay and Broad Sound, illustrate some of the kinds of sediments that are being deposited along the shores of indented coasts (Box 12.2).

Peritidal Complexes in Domains of Carbonate Sediments

In peritidal regions dominated by carbonate sediments, the sediment may consist almost exclusively of carbonate material, or the carbonates may be mingled with

BOX 12.1

Coorong Lagoon, South Australia

The Coorong Lagoon, 100 km long and about 3 km wide, has become separated from the open water of Encounter Bay (a part of the Great Australian Bight of the Indian Ocean) by the northwestward growth of carbonate skeletal sands forming the spectacular barrier spit known as Younghusband Peninsula (Box 12.1 Figure 1). At its northwest extremity the depth of the lagoon water is about 10 m; elsewhere

the average depth is only 2 or 3 m. The salinity increases progressively toward the southeast, away from the connection to the open sea. Maximum values as high as 60‰ have been reported. The landward shore of the Coorong Lagoon consists of an emerged Pleistocene beach ridge.

The sediments of the Coorong Lagoon consist chiefly of pellet lime muds containing much

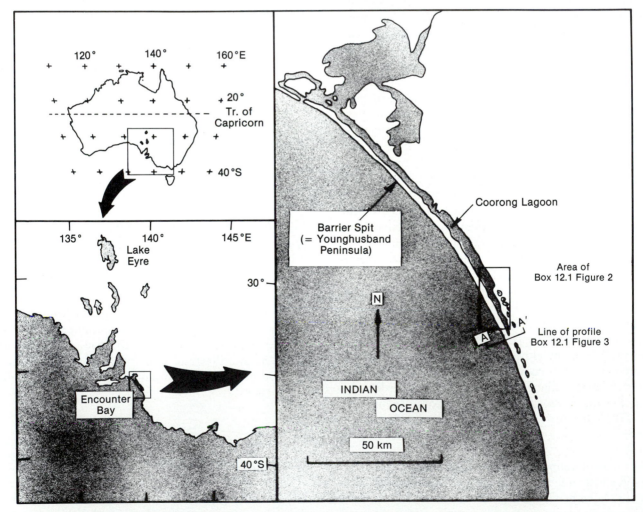

BOX 12.1 FIGURE 1.   Location maps of Coorong Lagoon, south coast of Australia. (Map at right from C. C. von der Borch, 1976, fig. 1, p. 414.)

organic matter and composed of aragonite and high-magnesian calcite. The muds are about 1 m thick. These are underlain by skeletal sands containing abundant mollusk debris.

Adjacent to the Coorong Lagoon are ephemeral lakes (Box 12.1 Figure 2) in which dolomite is a primary precipitate. In addition to dolomite, gypsum and the carbonate minerals aragonite, calcite, high-magnesian calcite, magnesite, and hydromagnesite occur as primary precipitates in the upward-shallowing successions of the Coorong lakes. The levels of the bottoms of these lakes are about 1 m above the level of the water in the Coorong Lagoon (Box 12.1 Figure 3). The lakes are situated between Pleistocene beach-dune ridges; some of the lakes were once connected to the Coorong Lagoon, whereas others have never been so connected. The formation of the beach-dune ridges, and therefore of the Coorong lakes, was controlled by Pleistocene sea-level changes. The alternating Pleistocene beach ridges and swales with saline lakes landward of the Coorong Lagoon resemble a strandplain complex, with the notable difference that a well-developed barrier- and lagoon complex is present at its seaward margin. Two of these lakes contain stromatolites.

The microbial mats in the "northern stromatolite lake" are only about 2 or 3 cm thick; they are underlain by 2.5 m of pellet muds created by cerithid gastropods. These pellets doubtless constitute the only remains of former microbial mats that were devoured by the voracious cerithids. After the salinity reached a level higher than that which the cerithids tolerate (evidently only a short while ago) these gastropods disappeared from the lake and the mats expanded.

When the lake water evaporates completely, as it does annually, dolomite, magnesite, gypsum, and other minerals are precipitated. The dolomite crystals form a soft sediment in which the three-toed tracks of large birds (*Emus*) are preserved. (See Figure 12-14.)

SOURCES: Lock, 1982; Rosen, Miser, and Warren, 1988; von der Borch, 1976; von der Borch and Lock, 1979; Warren, 1982, 1988; Warren and Kendall, 1985.

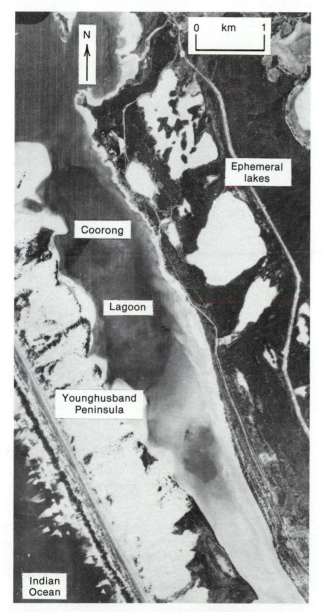

BOX 12.1 FIGURE 2. View looking vertically downward from an airplane on part of Coorong Lagoon, South Australia, and adjacent areas (location on Box 12.1 Figure 1). Further explanation in text. (Australian Government photograph, courtesy D. E. Lock, Flinders University of South Australia.)

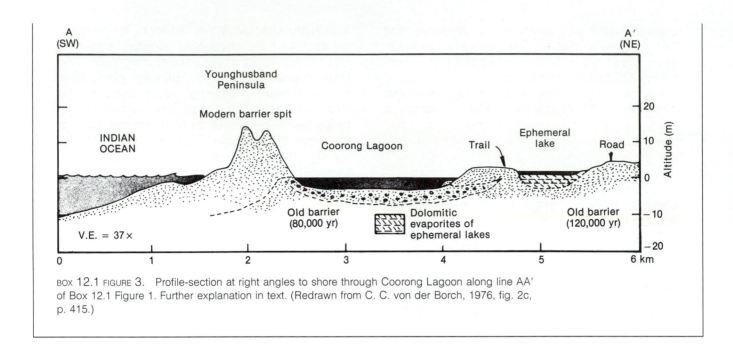

BOX 12.1 FIGURE 3.   Profile-section at right angles to shore through Coorong Lagoon along line AA' of Box 12.1 Figure 1. Further explanation in text. (Redrawn from C. C. von der Borch, 1976, fig. 2c, p. 415.)

extrabasinal- and/or evaporitic material. In Box 12.3 we consider an example of the former, and in Box 12.4 an example of the latter.

Inferred Examples of Ancient Peritidal Complexes

Numerous examples of deposits that are inferred to have been made in ancient peritidal complexes have been reported in the geologic literature since 1960.

The characteristics of peritidal environments overlap those of other environments. However, features useful in recognizing ancient peritidal sediments include the following:

1. Stratigraphic association with deposits made by a shifting shoreline. (Peritidal sediments overlie marine strata where the coast prograded and overlie nonmarine strata or an old erosion surface where the sea transgressed.)
2. Upward-fining successions of two distinct thicknesses may be found. The thicker successions normally underlie the thinner. The basal parts of such successions may consist of abundant skeletal debris (coquina or shell hash).
3. Sediments (not deposited in deep channels) consisting of repeatedly alternating thin layers of sand and mud.
4. Presence of skeletal debris of organisms that inhabit lagoonal waters (either brackish or hypersaline);

may be mixed locally with skeletal debris washed in from adjacent marine environments.
5. Presence of salt-water grasses or abundant mangroves (and/or related peats, which can become coals).
6. Fenestral fabrics formed by desiccation of sediments, especially microbial stromatolites.
7. Desiccation polygons and locally abundant chips formed by disruption of desiccated lime mud (flat-pebble conglomerate).
8. Juxtaposition of marine indicators such as marine fossils, with indicators of penecontemporaneous subaerial exposure such as adhesion ripples, seafoam marks, and runzelmarken (wrinklemarks).
9. Nodular anhydrite (or gypsum).
10. Basal transgressive peats or coals (formed from decay of mangroves or salt-water grasses) or basal transgressive dolostones (formed from supratidal lithified crusts whose bedding surfaces never display footprints).

As with our examples of modern peritidal complexes, we shall organize our discussion under three headings: (1) terrigenous rocks, (2) carbonate rocks devoid of evaporites, and (3) carbonate rocks associated with evaporites.

In Terrigenous Rocks

Examples of inferred ancient peritidal complexes range in age from Pleistocene to Precambrian.

Examples of a succession of thin-bedded, inferred peritidal sediments of Pleistocene age are exposed along the

BOX 12.2

Peritidal Complexes of Two Tropical Bays

Shark Bay, Western Australia

Shark Bay, Western Australia, is an elongate, complex embayment that trends NW-SE and faces the Indian Ocean at coordinates approximately lat.25°S and long.113°E. (See Figure 10-4.) The climate of the region is semiarid. Rainfall averages 230 mm/yr whereas evaporation averages 2200 mm/yr.

The range of the mixed-diurnal tides decreases along the axis of Shark Bay from 0.61 to 1.70 m in the northwest to 0.45–0.61 to 0.9 m at the southeast end.

Many of the shores of Shark Bay are steep cliffs a meter or so high, composed of Pleistocene limestone. On the northeast, however, the Bay is bordered by a low alluvial plain lying between two delta complexes (now largely inactive). Here the sea-marginal landscape is low and flat; hence even with the small tidal amplitude, the width and areal extent of the peritidal complex are substantial. The prevailing southerly winds of summer change the water level enough to expose some areas, normally flooded, for periods of as long as a week.

The sediments of the peritidal complex consist of extrabasinal terrigenous quartz from the deltas and carbonates derived from the marine waters. Carbonates include skeletal debris, ooids, pellets, and algal- and cyanobacterial stromatolites.

The peritidal complex has been subdivided into five zones: (1) upper supratidal, (2) lower supratidal, (3) upper intertidal, (4) middle intertidal, and (5) lower intertidal.

The upper supratidal zone is almost never flooded by modern marine waters, but the lower supratidal zone is flooded by marine waters once every 2 to 3 years. The surface generally is a dry crust of aragonite-cemented pellet mud. Root casts (rhizoliths) are common. Gypsum, halite, and possibly dolomite are precipitated interstitially.

The upper intertidal zone is flooded daily or twice daily by marine waters for a period of 4 or 5 days, during spring tides, then is dry for about 10 days until the next interval of spring tides. The surface of the upper intertidal zone is covered in part by microbial mats and in part by crusts of aragonitic carbonates. The middle intertidal zone is regularly flooded by marine waters once or twice a day. The lower intertidal zone is exposed during the low parts of the tidal cycle for 4 to 5 days, and then flooded for 10 days. The lower intertidal zone contains domal- and columnar stromatolites that have been reported from only one other region in the modern world (the Bahamas, where they are subtidal). A well-defined latitudinal zonation of growth form results from variations in subaerial exposure. The domal stromatolites are elongated perpendicular to shore, probably in response to the movements of tidal water. Ancient tidal stromatolites have also been found in which a comparable orientation perpendicular to shore has been inferred (Box 12.2 Figure 1).

A southward change from mangroves to microbial mats coincides with an increase in salinity from oceanic to hypersaline. In cores, the flat-laminated stromatolites form a layer up to 30 cm thick over an aragonitic pellet mud, which in turn is underlain by foraminiferal-skeletal sand of the lower intertidal zone. Such a parasequence demonstrates that the complex prograded seaward.

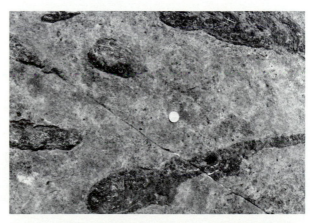

BOX 12.2 FIGURE 1. Stromatolites formed in a setting comparable to that of Shark Bay. Wide ends of stromatolites face into the deeper water of an embayment. Bedding-plane exposure with original relief of stromatolites planed off by modern weathering. Pierson Cove Formation (upper Middle Cambrian), southern Drum Mountains, westcentral Utah. (Authors.)

Broad Sound, Queensland, Australia

Broad Sound, Australia, is situated on the east coast, at long. 150°E and lat. ≈ 22°30'S (Box 12.2 Figure 2). The climate is subhumid and subtropical. The temperature ranges from a mean daily maximum of 23°C in July to 31°C in January. The mean annual rainfall, which is seasonal and arrives mostly in the summer (December to March), is about 100 cm. Mean annual evaporation at Rockhampton, 100 km to the south, is 170 cm.

The salinity of the surface water of Broad Sound is 35.5 to 36.5‰. Because Broad Sound receives little river water for much of the year, the salinity increases at the southeastern (landward) end. In the tidal creeks draining the mangrove thickets, salinity may reach 45‰.

Wave activity is vigorous and the tidal range is up to 10 m, thus creating an extensive peritidal complex.

The sediments in the lower part of the intertidal zone consist of quartz- and calcareous sands, and in the upper part, of black muds, which form a narrow belt adjacent to a zone of mangroves that mark the inner edge of the intertidal zone. Beyond the mangroves is a supratidal area up to 5 km wide consisting of shallow channels and intervening mudflats. Coastal grasslands border the landward edge of the supratidal zone. (See Box 12.2 Figure 2.) The surface of the supratidal muds is encrusted with salt. The topmost layer has broken into desiccation polygons. Algae grow only where the supratidal mudflats consist of 2% sand and 98% silt and clay.

The salinities of interstitial waters from the supratidal sediments are about three times that of normal seawater; the Mg/Ca and K/Ca ratios both increase with salinity. This is inferred to be a function of loss of Ca^{2+} by precipitation of $CaCO_3$ and $CaSO_4$.

Near Charon Point dolomitic concretions have been found 10 to 20 cm beneath the surface of

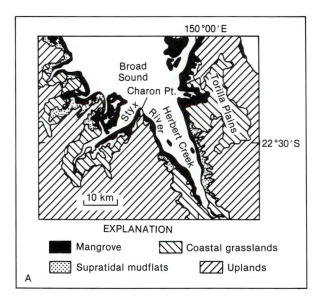

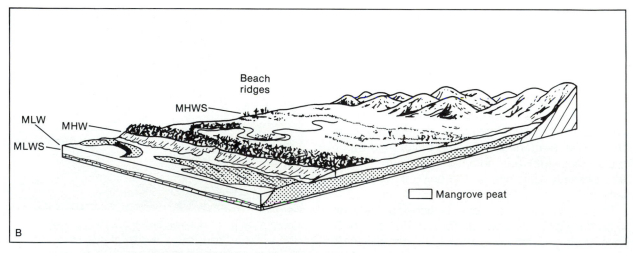

BOX 12.2 FIGURE 2.   Peritidal complex, Broad Sound, Queensland, Australia.
  A. Map of various peritidal depositional environments.
  B. Schematic representation of depositional environments in the Charon Point area of Broad Sound. (Redrafted after P. J. Cook, 1973, figs. 1 and 2, p. 999.)

the supratidal muds. (See Box 12.2 Figure 2.) These concretions are noteworthy because their host sediments are not carbonates but silicates (mean content of $CaCO_3$, 4.3%).

SOURCES: Cook, 1973; Dravis, 1983; Logan, Davies, Read, and Cebulski, 1970; Mayo, 1976; Playford and Cockbain, 1976; Riding, Awramik, Winsborough, Griffin, and Dill, 1991; Shinn, 1983.

south bank of the St. Mary's River, near Chester, Florida (Figure 12-15, A) and in Virginia southeast of Norfolk (Figure 12-15, B). The strata in both of these examples are only a few meters thick, extend through small areas, and lie just landward of barrier sands. Neither photograph shows any autocyclic successions formed by lateral sedimentation in shifting channels.

Examples from sedimentary bedrock include the J sandstone of the Dakota Group (lying above the Skull Creek Shale and below the Graneros Shale, both marine), west of Denver, Colorado; the lower Cretaceous Fall River Formation of the northern Black Hills, north-eastern Wyoming; the strata near the top of the Lower Coal (Jurassic), Bornholm, Denmark (overlies fluvial strata, overlain by marine strata); parts of the Jurassic of Israel and the Sinai (Figure 12-16); and parts of the Bald Eagle Formation (Upper Ordovician) of eastern Pennsylvania (Reedsville marine strata below; Bald Eagle non-marine strata above).

Other examples of inferred ancient peritidal deposits include the Thonkman Quartzite (Lower Ordovician) of northeastern British Columbia, and the Eureka Quartzite (Ordovician); the Zabriskie Quartzite (Cambrian); and the middle member of the Wood Canyon Formation (Upper Precambrian) of eastern California

FIGURE 12-15. Inferred examples of terrigenous peritidal sediments of Pleistocene age.
  A. View facing S of thin-bedded succession, S shore of St. Mary's River, NE of Chester, Florida. Pencil at right gives scale. (Authors.)
  B. Close view of mud-draped sand ripple, SE of Norfolk, Virginia. (Authors.)

FIGURE 12-16. Slabbed side of core through ancient terrigenous strata, inferred to be of peritidal origin, Daya Formation (lower middle Jurassic), core 18 Daya 1 well, Israel (lat.31.1°N, long.35.05°E). Layers appearing dark gray to black are silty claystones; light gray areas, very fine sandstone. Irregularity of bedding has resulted from initial interlayering of sand and mud. Present are flasers (indicated by thin black streaks in the sandy layers) and thin lenses- and wavy layers of sand in mud (indicated by thin light areas in the black layers). (See Figure 5-22.) Other irregularity results from the effects of burrowing organisms (irregular oval patches made by organisms tunneling parallel to the layers), and from compaction. (M. Goldberg and G. M. Friedman, 1974, pl. 5, fig. 2.)

BOX 12.3

Carbonates Only: Andros Island, Bahamas

The Bahamas, as we saw in Chapter 10, especially the area known as the Great Bahama Bank or Andros Platform, have been classically studied as the most-comparable available known counterpart to the epeiric seas that deposited large parts of the rock record on the continents. On the eastern margin of the Bank is a large island known as Andros Island (Box 12.3 Figure 1). This island acts as a physical barrier to the trade winds blowing from the northeast. Along the leeward western margin of the island, the sea-marginal flats occupy a belt up to 15 km wide and about 12 km long. In cross section, the deposits of the intertidal flats form a wedge that pinches out against the exposed Pleistocene rock of Andros Island to

the east. If the Great Bahama Bank were buried by sediment, this contact would form a classical onlap. We shall examine the peritidal complex starting with the marine side and progressing eastward toward the island.

Subtidal Zones

The subtidal zones include (1) the nearshore marine environment (known as the adjacent marine subtidal zone), (2) tidal channels, and (3) tidal ponds.

The *adjacent marine subtidal zone* forms a belt that parallels the peritidal complex and extends seaward some 15 to 30 km. The sediments of the adjacent marine subtidal zone consist of fecal pellets. Because burrowers rework the sediments so completely, neither beds nor laminae can be seen. Much of the reworking is performed by the shrimp *Callianassa*. In the subtidal zone, any sediment that has not been completely homogenized by burrowers displays mottled structures.

Tidal channels, which are similar to those from noncarbonate terrains, form meandering and twisting tributaries and distributaries. The main channels range in width from 30 to 100 m; their depths are as great as 4 m. These watercourses characterize a *channeled belt* (Box 12.3 Figure 2), 1.5 to 3 km wide, that parallels the western margin of Andros Island. Current speeds within channel systems vary widely and result in carbonate sediments that range in size from mud to pebbles. The coarse particles consist predominantly of shells, shell fragments, and chips of dried-out lime mud (intraclasts). (See Box 2.6 Figure 3.) In some channels, undercutting and caving of the cut banks cause large blocks of lime mud to collapse into the channel. As in fluvial channels, bars accumulate on the slip-off slope of meander bends. On the west side of Andros Island, the sediment of these bars consists of a mixture of tests of organisms and semilithified chips of lime mud. The sediment deposited in these bars may be either nonbedded or cross stratified. Cerithid gastropods live in profusion on the intertidal flats but do not live in the channels. However, as a result of the undercutting and erosion of the banks, tremendous numbers of such gastropod shells accumulate in channel bars. These deposits show the diagnostic

BOX 12.3 FIGURE 1.   Andros peritidal complex, Bahamas.
   A. Sketch of part of Great Bahama Bank (Bahama Platform) showing location of Andros Island and adjacent peritidal complex.
   B. Profile and section of Bahama Platform along line A-A' of map showing peritidal sediments lapping across western edge of Andros Island. (E. A. Shinn, R. M. Lloyd, and R. N. Ginsburg, 1969, fig. 1, p. 1203.)

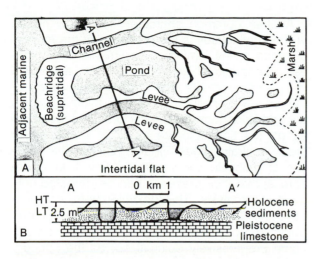

BOX 12.3 FIGURE 2. Detail of part of Andros peritidal complex.
A. Schematic map at mean low water.
B. Idealized profile and section along line A-A'. Intertidal zone occupies only narrow areas. (Ht, normal high tide; Lt, normal low tide.) (E. A. Shinn, R. M. Lloyd, and R. N. Ginsburg, 1969, fig. 8, p. 1207.)
C. Aerial photograph of "three creeks" area, peritidal flat, northwestern Andros Island. Tidal channels, levees, and ponds are all clearly visible. (Photograph courtesy of R. N. Ginsburg.)

upward-fining autocyclic succession common to all channels.

The channels are flanked by low natural levees. (See Box 12.3 Figure 2.) The relief of the natural levees allows them to serve as dams to hold water in the low spaces between the channels. The *levee-dammed bodies of tidal water* are **intertidal ponds**. The sediments of the ponds consist of lime mud. Because it has been homogenized by burrowing organisms, this mud generally does not display any primary sedimentary layer structures. As a result of intermittent drying-out of the ponds, these sediments may exhibit deep desiccation cracks (up to 30 cm deep).

Intertidal Zone

On the western margin of Andros Island the intertidal zone is narrow and restricted to the gently sloping levees and bars on the inner banks of channel meanders and to the margins of the ponds. (See Box 12.3 Figure 2.) The sediments of the intertidal zone consist mostly of pellets; bioturbation has generally obliterated all sedimentary structures. Cerithid gastropods are abundant and roots and root hairs of the red mangrove penetrate the sediment.

Supratidal Zone

Three geomorphic areas compose the supratidal zone (See Box 12.3 Figure 2.): (1) beach ridges, which form the boundary between the adjacent marine subtidal zone and the remainder of the tidal complex, (2) the tops of the natural levees bordering the tidal channels, and (3) marshes.

Beach ridges are of two kinds: (1) high (1 to 2 m) and (2) low (<60 cm). High ridges are typical storm deposits (Box 12.3 Figure 3). Some storm deposits consist of festoon-cross-bedded concentrations of gastropod shells of coarse-sand- to pebble size. Such concentrations resemble deposits accumulating on high-energy intertidal beaches on which waves break daily. By contrast, low ridges are composed of pellets and fine skeletal debris that accumulated in thin laminae as stromatolites. The laminated sediments of these low beach ridges contrast strikingly with the homogenized- and mottled sediments of the subtidal- and intertidal zones only a few meters away.

The thinly laminated deposits of natural levees closely resemble those of low beach ridges.

BOX 12.3 FIGURE 3. Holocene storm deposit (date of storm not known) of carbonate skeletal debris, mostly molluskan, on bank of Crane Key, Florida. Comparable storm deposits are known in the Bahamas. (Authors.)

The chief particles composing natural levees are pellets. Laminae are thin and discontinuous. Both low beach ridges and natural levees display *small vugs (or voids) between the laminae of the sediment*. These vugs, known as **birdseye structures**, are of two kinds: (1) planar isolated vugs 1 to 3 mm high by several millimeters wide, and (2) isolated more-or-less-bubblelike vugs 1 to 3 mm in diameter. Planar vugs form by shrinkage resulting from desiccation of exposed sediment; the bubblelike vugs are caused by gas bubbles. In ancient limestones such vugs have been filled with calcite or anhydrite but likewise are known as birdseye structures or *fenestral fabric* (Box 12.3 Figure 4). Birdseye structures are abundant in supratidal sediments, sporadic in intertidal sediments, and true birdseyes never occur in subtidal sediments. However, voids which strongly resemble birdseye structures *can* occur in subtidal sediments. In some cases, it may be possible to distinguish the two only by reference to the associated sediments. Therefore, birdseye structures cannot be considered diagnostic of supratidal- and intertidal sediments, even though the subtidal "pseudobirds-eyes" appear to be relatively rare. Another problem of recognizing birdseye structures is the tendency for muddy sediments that contain a great many birdseyes to resemble grainstones. These have been termed *diagenetic grainstones*.

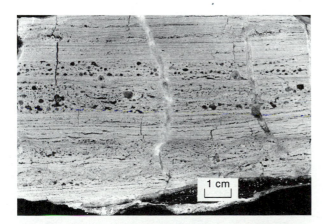

BOX 12.3 FIGURE 4.   Polished slab of ancient limestone having birdseye structure interpreted as stromatolitic intertidal deposit showing both flat- and bubble-shaped void spaces (vugs; they appear black on the picture). Horizontal planar vugs probably resulted from the rotting away of cyanobacterial cells. Circular vugs probably represent gas bubbles. Vertical stepped cracks are desiccation cracks. Pillara Limestone (Devonian), Laidlow Range, Canning Basin, Western Australia. (P. E. Playford.)

Another common characteristic of low ridges and natural levees is the presence of the mineral dolomite in the laminated sediments. The highest proportions of dolomite (to 80%) are within hard surface crusts that form by lithification of the underlying pellet mud. These dolomitic crusts are firm underfoot; they never preserve footprints as do the soft dolomite muds of the Coorong area, Australia. (See Figure 12-14.) The crystals of dolomite are so small (only 1 to 2 μm across) that their presence can be detected only by X-ray analysis. This dolomite does not possess the fully ordered crystal structure of true dolomite. Both skeletal fragments and pellets become progressively dolomitized; as the concentration of dolomite increases, the outlines of pellets are obscured and ultimately cease to be visible. Evaporative processes on exposed intertidal flats have been invoked for explaining the formation of dolomite in this setting.

Dolomite not associated with lithified crusts may also be dispersed in much lower concentrations within intertidal-flat sediments. Dolomite is more abundant in association with low-salinity ground water.

Additional characteristic features of low beach ridges and natural levees are desiccation cracks (or mudcracks) (See Box 2.6 Figure 3.) and the presence of abundant *flat, dried, and eroded chips of lime mud.* (See Box 2.6 Figure 3.) These flat chips are *invariably laminated.* Such flat chips, *once cemented, form deposits* known in the rock record as **flat-pebble conglomerate**, *intraformational conglomerate,* or *edgewise conglomerate.* Mudcracks may form in grainstones but are most common in lime muds. The weathered sediment that infills cracks and covers polygons is more porous and permeable than the polygons themselves. These permeable infilling sediments, because of their greater fluid transmissibility, may be selectively dolomitized.

Marsh deposits occupy the largest area in the peritidal complex. On the most-landward side of the flats they form a broad belt as much as 3 km wide. As with so many sediments of other parts of the peritidal complex, the sediments of the marsh consist of pellet mud. And as in the low ridges and levees, the marsh sediment is finely laminated. Marsh surfaces are commonly covered with mats of the fresh-water alga *Scytonema*, which is patchily coated with high-magnesian calcite to form *algal tufa*. As a result of the greater admixture of organic matter, the marsh laminae are accentuated compared to those in ridges and levees. Laminae of pellet mud, light tan, white, or gray in color, alternate with dark-brown laminae rich in organic matter. Between the laminae, birdseye vugs are abundant. Cores of marsh sediment reveal abundant marsh grass and the roots of both black- and red mangroves. Commonly, the relatively coarse and thick layers deposited by hurricanes are distinctly visible intercalated between laminated intervals. These coarse layers can be traced laterally for up to tens of meters, much farther than the thinner storm layers found in the deposits of levees and low beach ridges.

The Bahamian peritidal complexes and their carbonate deposits are located in a humid- or semiarid climatic zone in which evaporite minerals are not being precipitated. In Box 12.4 we describe the carbonate deposits of a peritidal complex in an arid climate in which sulfates, such as gypsum and anhydrite, are being precipitated interstitially in the sediment.

SOURCES: Gebelein, Steinen, Garrett, Hoffman, Queen, and Plummer, 1980; Mazzullo and Birdwell, 1989; Reading, 1986; Reineck and Singh, 1980; Scholle, Bebout, and Moore, 1983; Shinn, 1968a, c; 1982, 1983; Shinn, Lloyd, and Ginsburg, 1969.

and western Nevada. All of these examples from the North American Cordillera are hundreds of meters thick and occupy areas ranging into the tens of thousands of square kilometers. Such great thicknesses and areal extents do not seem compatible with the scales of peritidal deposits discussed so far. Examples of ancient inferred peritidal deposits that are not as thick nor as widespread as these Cordilleran examples include the Lower Devonian near Koblenz, Germany; the Upper Devonian of eastern New York State; the Clinton Formation (Silurian) in eastern Pennsylvania; the Potsdam Sandstone (Upper Cambrian) in northeastern New York; the Eriboll Sandstone (Cambrian) of northwestern Scotland; and the lower fine-textured quartzite (middle Dalradian; Precambrian) of Islay, Scotland. Although the Devonian of New York State is predominantly a terrigenous deposit, locally it contains abundant skeletal debris of peritidal origin that forms limestones (Figure 12-17).

In Carbonate Rocks Devoid of Evaporites

In ancient peritidal complexes involving carbonate sediments, not only the sedimentary structures but also the kind of carbonate rock are closely controlled by the various subenvironments. Carbonate peritidal complexes always lie next to true marine deposits, which typically consist of skeletal particles or of homogeneous pelletal limestones. In addition to these adjacent-marine subtidal sediments, the subtidal deposits include sediments deposited in tidal channels. The bases of the channel deposits sharply truncate underlying strata. The channel-fill strata contain well-sorted skeletal debris or small- to large-size chunks composed of channel-marginal fine-textured materials that were undercut and thus collapsed into the channel, where they became surrounded by a matrix of sand-size debris. Finally, they may display autocyclic upward-fining *parasequences* whose thicknesses approximate the depths of the former channel. These upward-fining parasequences and the materials composing them may form bundles of strata having parallel, horizontal contacts (parasequence surfaces) above and below. The margins of former channels may not be preserved.

Carbonate materials deposited within intertidal zones may show effects of desiccation: production of flakes or chips (See Box 2.6 Figure 3.), which typically become intraformational conglomerates, and production of some (but not all) fenestral fabrics. In the intertidal zone, sand and mud may be complexly interstratified forming flaser bedding, lenticular bedding, or wavy bedding. (See Figures 5-22 and 12-16.) If the salinity of the tidal waters becomes high enough to keep out the gastropods, then microbial mats may flourish, producing variously shaped stromatolites in the rock record.

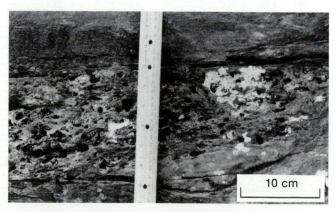

FIGURE 12-17. Ancient shell hash that has been deposited at base of tidal channel (margin of channel at lower right); most fossils (here represented by molds and thus appearing as irregular black areas) are of brachiopods. Middle- or Upper Devonian, Catskill Mountains, New York State. (K. G. Johnson and G. M. Friedman, 1969, fig. 22, p. 475.)

BOX 12.4

Peritidal Complexes with Carbonates and Noncarbonate Evaporites: Sea-Marginal Sabkhas of the Arabian or Persian Gulf

Here we need to introduce the term **sabkha**, *a surface of deflation formed by the removal of dry, loose particles down to the ground-water table or to the zone of capillary water.* Sabkhas are arid supratidal regions characterized by condi-

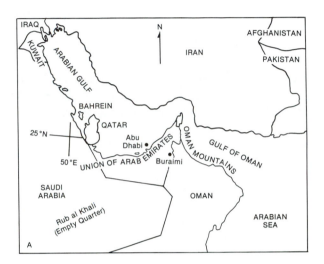

tions under which evaporite minerals are precipitated within the sediment. Sabkhas are of two kinds, continental and sea marginal. Continental sabkhas are described in Chapter 14; we discuss sea-marginal sabkhas here. The most-studied sea-marginal sabkhas are those of the Arabian- or Persian Gulf (the gulf is known by both names), specifically along the coast of Abu Dhabi (Box 12.4 Figure 1).

The Arabian- or Persian Gulf lies in a subtropical, arid area where the average annual rainfall is less than 5 cm. The average annual evaporation of waters from the Gulf is 124 cm. The salinity of the surface water of the Gulf ranges from 36.5‰ at the southeast end to ±40.5‰ at the northwest (landward) end. Temperature of surface water decreases from 24°C at the southeastern end to ±19°C at the northwestern end.

The normal astronomical tidal range at Abu Dhabi is about 2.5 m in the open Gulf, but on the landward side of the lagoon decreases to about 1 m.

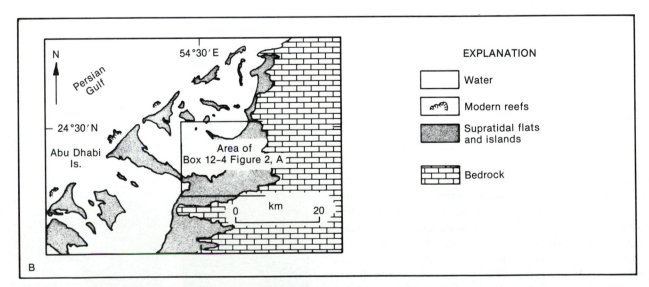

BOX 12.4 FIGURE 1.   Sketch maps of Arabian- or Persian Gulf (labeled Arabian Gulf in figure) and coast of Abu Dhabi.

A. Arabian- or Persian Gulf area. (After R. S. Patterson and D. J. J. Kinsman, 1981, fig. 1, p. 1458.)

B. Major geologic features of coast near Abu Dhabi. (After B. H. Purser and G. Evans, 1973, fig. 8, p. 220.)

At Abu Dhabi, the Holocene submergence inundated a landscape carved in Pleistocene and Tertiary carbonate rocks and brown quartzose-carbonate dune sands. Radiocarbon dates from numerous borings through the Holocene sediments indicate that the present coastal area began to take shape when the sea transgressed rapidly to a position about 3 m relatively higher than at present. An open-water shoreline stood about 10 km landward of the inner margins of the present Abu Dhabi lagoon (Box 12.4 Figure 2). The highest Holocene shoreline is approximately 4000 years old. Quartz-carbonate beach sands grade seaward into gray, muddy, carbonate sand containing abundant tests of cerithids and other invertebrates. The top of this has been lithified to a crust several centimeters thick.

Peritidal conditions began about 3750 years B.P., when an emergence of about 1 m took place. Conditions comparable to those along the modern Abu Dhabi lagoon were established and the lagoonal shoreline prograded seaward. During 2700 years the lagoon shore has advanced about 7 km. This advance was accompanied by an additional emergence of about 0.6 m. (See Box 12.4 Figure 2.)

Perhaps starting about 4000 years B.P., barrier islands consisting largely of oolitic sand were built around a few aligned high knobs of the bedrock. The combined effects of waves and tides have built "tails" of sediment landward from each of the barrier islands.

Inlets between the barriers connect to a network of tidal channels. At the seaward ends of these inlets are ebb-tidal deltas composed of oolitic sand. Lying seaward of the barriers at positions between the tidal deltas are small reefs.

The salinity of the waters of the main part of the Abu Dhabi lagoon ranges from 53.6 to 66.9‰; still-higher salinities have been found in some iso-

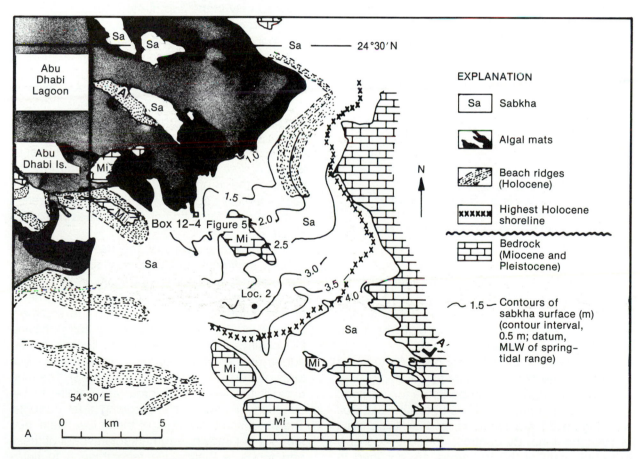

BOX 12.4 FIGURE 2.   Sabkha east of Abu Dhabi.
A. Detailed map of surficial materials based on plane-table surveys. Marks at A and A′ show locations of each end of profile and section, of part B. (After D. J. J. Kinsman, 1970, figs. 3, p. 8, and 5, p. 10.)

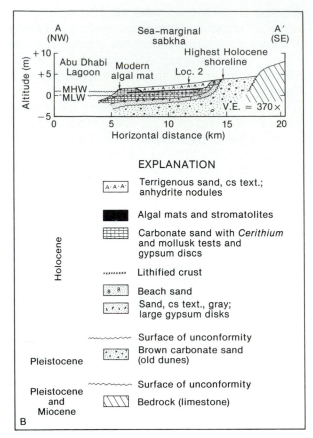

BOX 12.4 FIGURE 3. Surface of desiccated microbial mat showing polygons having upturned edges along which new mats (black) are growing. Black bars on rod equal 10 cm. Uppermost intertidal zone, Abu Dhabi peritidal complex. (Authors.)

BOX 12.4 FIGURE 2. (*Continued*)

B. Profile and section along AA' of part A. Late in December 1971 and in January 1972, the water table at Loc. 2 (location projected onto line of section) stood at depth of 1.11 m beneath the surface of the sabkha (altitude + 1.64 m above mean low water in Abu Dhabi lagoon). (Subsurface stratigraphic relationships modified from D. J. J. Kinsman, 1970, fig. 10, p. 15; G. Evans, V. Schmidt, P. Bush, and H. Nelson, 1969, figs. 4, p. 148, and 5, p. 149; P. Bush, 1973, fig. 10, p. 405; and K. J. Hsü and J. Schneider, 1973, fig. 4, p. 420.)

lated pools on the inner edge of the intertidal flats, where microbial mats are growing. The temperature of the lagoon water ranges from a daily minimum of ±12°C in the winter to a daily maximum of ±46°C in the summer.

The floor of Abu Dhabi lagoon is covered by carbonate mud containing chiefly pellets, tests of cerithid gastropods, and imperforate foraminifers.

The inner margins of the intertidal flats (See Box 12.4 Figure 2.) are characterized by thick mats of cyanobacteria. By contrast, the less-extensive microbial mats of the Bahamas are confined to the supratidal zone. As noted previously, microbial mats can grow only where the high salinity (or other conditions) protect them from the predatory cerithid gastropods. At Abu Dhabi the waters of

the inner part of the lagoon serve this function. In the Bahamas the protective levels of salinity are reached only in the supratidal areas.

Desiccation of microbial mats creates shrinkage polygons (Box 12.4 Figure 3). The edges of the polygons commonly curl upward and in the cleft between the polygons vigorous algal or cyanobacterial growth continues. As we have discussed in Chapter 7, fine suspended matter, especially lime mud, that is washed across the mucilaginous mats becomes glued to them. After the microbes have trapped and bound a sheet or lamina of lime mud, they secrete another layer of mat over the trapped mud particles. The result of such alternate trapping of sediment and secretion of mats is the formation of *stromatolites* (Box 12.4 Figure 4).

The uppermost part of the intertidal zone is a bare sediment surface underlain by carbonate material containing abundant pellets. Flow of thin sheets of water across these sediments on the outgoing tide has sculpted the sediment surface into a series of linear bed forms having low positive relief. These include linear rounded ridges and shorter, broader tongue-shaped features named *setulfs* (Box 12.4 Figure 5). The long axes of the linear bed forms are parallel to the direction of flow of the outgoing tidal currents. The mechanics of origin of these low linear bed forms is not known. A remarkable aspect of these bed forms is their close resemblance to flutes (Chapter 5).

The supratidal zone, between the microbial mats and the Cenozoic bedrock, consists of broad flats, up to 6 km wide, whose surface slopes la-

BOX 12.4 FIGURE 4. Section cut at a right angle to stromatolitic layers in deposits of modern microbial mat, uppermost intertidal zone, Abu Dhabi peritidal complex. Vertical cracks separate adjacent desiccation polygons. Growth of gypsum-crystal mush created the mottled area at upper left. (Authors.)

BOX 12.4 FIGURE 6. Surface of sabkha near Abu Dhabi, showing thin crust of halite (white area in distance beyond Land Rover). In foreground, mottled surface has resulted from comprehensive burrowing, probably by crabs. Gray sediment from beneath white crust has been spread around the numerous burrow entrances. (Authors.)

goonward at 1:2500 (40 cm/km). These broad flats are *sea-marginal sabkhas*. The sabkha surface lies above the level of normal spring tides, but on occasion is inundated by a wind tide. This surface water vanishes, partly by seeping into the sediments and partly by evaporation. Final drying involves precipitation of a crust of halite (Box 12.4 Figure 6), which may blow away, be covered by the sediment brought up from below by burrowing organisms, be dissolved by the next sheet of

water that floods the sabkha, or be covered by windblown sand.

Cores and pits dug into the supratidal sabkha sediments indicate that a continuous deposit of microbial mats, about 0.3 m thick, underlies the area that is now a sabkha. Beneath this mat is carbonate sand containing cerithid gastropods. Both were deposited starting about 3750 years B.P. and continuing to 1000 years B.P. while the inner shore of the lagoon was prograding seaward and the sea stood 1.6 to 1.0 m higher than at present, as mentioned previously. Carbonate mud or windblown sand, both containing various evaporite minerals, the modern sabkha deposit, cap the sequence. (See Box 12.4 Figures 2, B and 7.) The modern mats began to grow only rather recently.

In the sediments of the sabkha, eight or nine different authigenic minerals have been precipitated. Although halite, aragonite, gypsum, anhydrite, dolomite, celestite, magnesite, huntite [$Mg_3Ca(CO_3)_4$], and perhaps calcite form in the subsurface environment of the sabkha, anhydrite and dolomite are of particular interest. In pits dug at various sites across the sabkha, anhydrite stands out most prominently. This mineral forms white nodules, less than 1 mm to more than 15 cm in diameter and varying in shape from spherical to strongly flattened, that contrast decidedly with the tan-to-brown host sediment. As anhydrite grows, the mechanical force of its crystal growth displaces the surrounding sediment and lifts the surface of the sabkha between 30 cm and 1 m. Parallel-layered stromatolites of

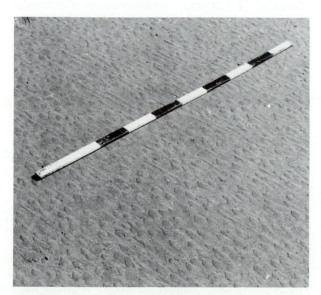

BOX 12.4 FIGURE 5. Linear, positive-relief bed forms on sediment underlying uppermost intertidal sediments at Abu Dhabi include straight, rounded ridges and spatulate setulfs. (G. M. Friedman and J. E. Sanders, 1974, fig. 3, p. 182.)

BOX 12.4 FIGURE 7. Wall of pit, dug down from sabkha surface near Abu Dhabi. Top 25 cm consist of sabkha sediments (homogeneous, enclosing white nodular anhydrite; divisions on scale, 10 cm). Below are laminated stromatolites of former uppermost intertidal zone, which became covered when the sabkha prograded seaward. (G. M. Friedman, 1972, fig. 19, p. 706; M. Goldberg and G. M. Friedman, 1974, pl. 1, fig. 7.)

the host sediment become physically displaced and bent around the nodules. The elevation of the sabkha surface is ultimately controlled by the permanent capillary zone, in which damp sediment resists erosion. Winds deflate the surface down to this level, truncating structures such as contorted anhydrite layers. The position of the capillary zone varies seasonally and over longer time periods. The nodular anhydrite is the signature of a sea-marginal sabkha in an arid climate. Anhydrite nodules or their diagenetic alteration products (e.g., megaquartz nodules with anhydrite inclusions) have been used to identify rock units as former sabkha deposits in rocks as old as 2.4 billion years (Gordon Lake Formation, Ontario).

In the sabkhas of the Arabian- or Persian Gulf, gypsum crystals commonly form below the sediment surface in the lower supratidal zone,

and, following progradation of the sabkha, are dehydrated to anhydrite in the middle supratidal zone. The anhydrite occurs as nodules that continue to grow displacively after replacing the gypsum crystals, as mentioned above. However, in the middle supratidal zone of Kuwait, laminated gypsum is forming under a few centimeters of water in an abandoned tidal channel. This is caused by the drop in elevation between the upper-supratidal vegetated sabkha, which has been uplifted by the formation of anhydrite nodules, and the barren middle-supratidal sabkha, which lacks anhydrite nodules. At this topographic discontinuity the ground-water table intersects the sediment surface, and in a distance of 50 m, the salinity increases from 100 to 140‰ in the subsurface to 240‰ in the shallow pool where the gypsum is forming. The gypsum crystals resemble those that typically form subaqueously in deep evaporite basins, yet they are forming in the supratidal zone. This suggests that some "basinal" evaporites in the rock record could have formed when the basins were dried out and sabkhaized. In any case, this example shows that gypsum-crystal form should be used with caution as an indicator of environment of formation.

The precipitation of aragonite, gypsum, and anhydrite increases the molar ratio of magnesium to calcium in the pore fluids. Whereas this ratio is 5:1 in seawater, measurements in the interstitial brines of the sabkha have yielded values of 12:1. Under these conditions dolomite forms; the lime mud underlying the sabkha becomes dolomitized. The dolomite crystals are only 1 to 2 μm across. Dolomitization releases calcium ions, which are then available for further precipitation of anhydrite. Because the organic matter of stromatolites is a Mg-organic complex, the presence of buried stromatolites also enhances the precipitation of dolomite. This sabkha dolomite is readily recognizable in the rock record because it (1) consists of small crystals, (2) is found at the tops of upward-shoaling parasequences, and (3) is enriched in the heavy isotope of oxygen, [18]O. Sabkha dolomites are indirectly significant to hydrocarbon exploration. Although sabkha dolomites are not usually permeable, they are commonly situated below impermeable strata containing evaporites, and above nearshore-marine carbonate strata that may be extremely porous and permeable. Thus sabkha dolomites are like fingers, pointing to possible stratigraphic traps.

The hydrologic behavior of the waters responsible for the interstitial precipitation of the evaporite minerals is not fully understood. What is definitely known is that evaporation at the sabkha surface causes the interstitial fluids to be concentrated brines having salinities as great as 10 times that of normal seawater. The interstitial water lost by evaporation must be replaced by subsurface flow of other water that is less saline and less dense than the water from which the evaporite minerals are precipitated. In a modern sabkha on the south shore of the Arabian or Persian Gulf SE of Abu Dhabi (Box 12.4 Figure 8), the "replacement" water is continental ground water, supplemented by marine waters that periodically flood the seaward portions of the sabkha as a result of the action of strong onshore winds. This water infiltrates the sabkha surface, adding a component of downward flow to the general seaward ground-water flow regime, and refluxing beneath the lagoon. Farther inland, beyond the limit of marine floods, the vertical component of ground-water flow is directed upward as a result of evaporative losses at the surface. What is not known is exactly by what mechanism the water rises as it evaporates. (Is it entirely by capillary flow? Entirely by hydrostatic flow?

Or by some combination of these mechanisms?) Salinity data from interstitial waters at Um Said, Qatar, where the sabkha sediments are quartz sands, suggest that the water in the upper parts of the sabkha flows landward from the sea and that as evaporation increases its density, the water that is not evaporated refluxes seaward at depth.

The hydrologic regimes of different sabkhas are not all the same. The broad sabkha at Abu Dhabi, beneath which ground water flows seaward, is actively aggrading. Its preservation potential is excellent. By contrast, the narrow sabkha at Um Said, which lacks connection to a seaward ground-water flow regime, is being deflated at its landward end even as it prograding seaward. Its potential for preservation is slight. Beneath a small sabkha on the Capricorn Coast of Queensland, Australia, that is underlain by extrabasinal sediments, subsurface flow of interstitial water is landward. Such flow responds to two factors: (1) evaporative pumping during the dry winter and (2) the existence of a permeable buried tidal channel that serves as a conduit for marine waters to enter the sediment below the sabkha.

We emphasize that not all marine sabkhas are identical to the one at Abu Dhabi. Other modern sabkhas that have been studied, including

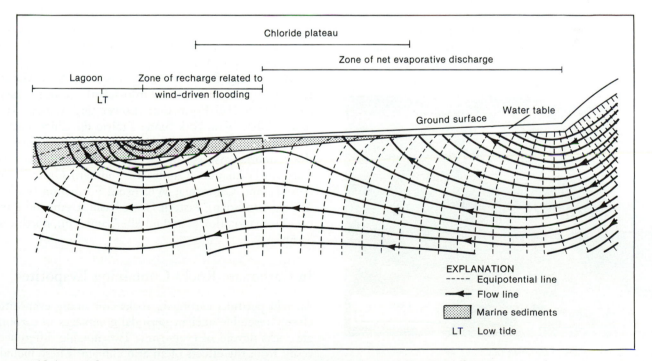

BOX 12.4 FIGURE 8.   Schematic flow net for section through sabkha southeast of Abu Dhabi. Section oriented perpendicular to coastline. Marine sediments are assumed to be much less permeable than underlying materials, which are not differentiated. (R. S. Patterson and D. J. J. Kinsman, 1981, fig. 13, p. 1472.)

the Gavish Sabkha in the Gulf of Elat (Aqaba), Red Sea; various North African sabkhas; and sabkhas at Ras Gharib, Gulf of Suez; and on Bonaire Island in the Caribbean Sea, present an array of hydrologic regimes, sedimentary successions, diagenetic minerals and -fabrics, and geomorphic expressions. For example, unlike the Abu Dhabi sabkha, many modern sabkhas are situated landward of beach ridges. Saline water is recharged mainly by ground-water flow and by percolation through the beach-ridge sediments. Saline springs are present. Also, behind many of these barriers are saline ponds or -lagoons. At the Gavish Sabkha the salinity of surface brines reaches 350‰. This sabkha is underlain by beds of gypsum nearly 1 m thick. Other sabkha-associated pools are underlain by much-

more-extensive deposits of subaqueous evaporites, including halite. The sediments underlying the Gavish Sabkha contain little dolomite or anhydrite, whereas at Abu Dhabi these minerals are abundant. Knowledge of the diversity of modern sabkhas is growing rapidly, yet synthetic models that explain the differences and similarities among all kinds of sabkhas are still to be devised.

SOURCES: Epstein and Friedman, 1983; Friedman and Krumbein, 1985; Gavish, 1980; Gavish, Krumbein, and Tamir, 1978; Gunatilaka and Shearman, 1988; Larsen, 1980; McKenzie, 1980; McKenzie, Hsü, and Schneider, 1980; Patterson and Kinsman, 1981; Purser, 1985; Schreiber, 1986; Sneh and Friedman, 1985.

Rocks formed in the supratidal zone may consist entirely of flat-laminated limestone or -dolostone or of interlaminated limestone and dolostone (See Figure 7-41.), fenestral fabrics, and desiccation cracks. (See Box 12.4 Figure 3.)

Burrows and burrow-mottled sediments are present from the subtidal zone to the supratidal zone. Supratidal burrow-mottled sediments may include intricate mixtures of limestone and dolostone (Figure 12-18). Where interstitial waters of supratidal sediments were saline enough to form dolomite, the supratidal zone may be

underlain by a crust of dolostone. If a gradual submergence takes place, such a dolomite crust becomes the basal deposit of the transgressive sequence.

Inferred examples of carbonate rocks deposited in ancient peritidal complexes, where the climate was not arid enough to cause precipitation of evaporites, have been described from all parts of the geologic record. Examples include the Dachstein Limestone (Triassic), northern Alps; Comblanchien Formation (mid-Jurassic), near Digon, France; Calcare Massiccio (late Mesozoic) of central Apennines, Italy; Manlius Formation (Lower Devonian), New York State; Pillara Formation (Devonian), Canning Basin, Western Australia; New Market Limestone (Lower Middle Ordovician), western Maryland; Tribes Hill Formation (Lower Ordovician), Mohawk Valley, New York State; Prairie du Chien Group (Lower Ordovician), upper Mississippi Valley; parts of the Cambrian System of the Grand Canyon, Arizona; Carrarra Formation (Cambrian), southern Great Basin, Nevada, and California; Whirlwind Formation (middle Middle Cambrian), eastern Great Basin, Utah; and Lancara Formation (lower and middle Cambrian), northwest Spain.

In Carbonate Rocks Containing Evaporites

Ancient peritidal carbonate rocks containing evaporites closely resemble ancient peritidal complexes in carbonate rocks devoid of evaporites. Two notable differences result from the effects of an arid climate. These include a greater abundance of dolomite and the presence of gypsum, anhydrite, halite, and possibly other evaporite minerals. In addition, stromatolites display a great variety of shapes, such as domed and columnar.

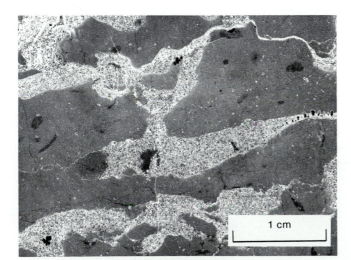

FIGURE 12-18.   View of dolomite patches (lighter mottles) filling burrows in micrite (darker gray shade), as displayed in print made by using as a negative an acetate peel made from etched and polished face of slab of mottled dolomitic micrite. Tribes Hill Formation (Lower Ordovician), New York State. (M. Braun and G. M. Friedman, 1969, fig. 13, p. 124.)

FIGURE 12-19.   Sawed face of core from bore hole cut through ancient limestone (dark gray) containing nodules of anhydrite (white). Evaporite Member of Ardon Formation (Jurassic), Core No. 2, Makhtesh Katan 2 well, Israel (lat.30.95°N; long.35.25°E). (M. Goldberg and G. M. Friedman, 1974, pl. 1, fig. 6.)

The typical expression of evaporite minerals in peritidal carbonates is nodular anhydrite (Figure 12-19). In some carbonate host sediments isolated euhedral crystals of evaporite minerals have grown. If the evaporites dissolve, then cavities form having shapes of the dissolved crystals. Evaporite minerals may replace the carbonate. For example, in the subsurface Silurian of the Michigan basin, halite has replaced an algal stromatolite. (See Figure 4-12.)

Examples of ancient peritidal carbonates associated with evaporites include the Miocene of Sicily and other parts of the Mediterranean area; Ardon Formation (Jurassic) of Israel; Macumber Formation of Windsor group (Mississippian), Maritime Provinces, Canada; Alapah Limestone (Mississippian) of the Brooks Range, Alaska; Arroyo Penasco Formation (Mississippian), northcentral New Mexico; Upper Devonian of Alberta, Canada; St. George Group (Ordovician) of western Newfoundland; and the Ordovician filling of the Brent meteor crater, Ontario.

SOURCES: Aigner, 1982; J.R.L. Allen, 1981; Bose, Ghosh, Shome, and Bardhan, 1988; Breyer and McCabe, 1986; Browne and Demicco, 1987; Buick, Dunlop, and Groves, 1981; Chandler, 1988; Cudzil and Driese, 1987; Epstein and Friedman, 1983; Fraser, 1989; Galloway and Hobday, 1983; Harms, Southard, and Walker, 1982; Hine and Snyder, 1985; Koerschner and Read, 1989; Kopaska-Merkel, 1987a; Lock, 1982; McCrory and Walker, 1986; Noe-Nygaard and Surlyk, 1988; Nummedal, Pilkey, and Howard, 1987; Pratt and James, 1986; Reading, 1986; Strasser, 1988; Uhlir, Akers, and Vondra, 1988; van der Westhuizen, Grobler, Loock, and Tordiffe, 1989; Warren and Kendall, 1985.

Suggestions for Further Reading

ALLEN, J. R. L., 1982, Sedimentary structures: their characteristics and physical basis: Amsterdam, Elsevier, v. 1, 593 p., v. 2, 663 p.

BALL, M. M., 1967, Carbonate sand bodies of Florida and the Bahamas: Journal of Sedimentary Petrology, v. 37, p. 556–591.

DE BOER, P. L., VAN GELDER, A. and NIO, S. D., eds., 1988, Tide-Influenced sedimentary environments and facies: Dordrecht, D. Reidel, 530 p.

FRIEDMAN, G. M., and KRUMBEIN, W. E., eds., 1985, Hypersaline ecosystems, the Gavish Sabkha: Berlin-Heidelberg-New York, Springer-Verlag, 484 p.

GINSBURG, R. N., ed., 1975, Tidal deposits: New York, Springer-Verlag, 428 p.

REINECK, H.-E., and SINGH, I.B., 1980, Depositional sedimentary environments, 2nd ed.: New York, Springer-Verlag, 549 p.

SCHOLLE, P. A., BEBOUT, D. G., and MOORE, C. H., eds., 1983, Carbonate depositional environments: Tulsa, OK, American Association of Petroleum Geologists Memoir 33, 708 p.

WEIMER, R. J., HOWARD, J. D., and LINDSAY, D. R., 1982, Tidal flats and associated tidal channels, p 191–245 *in* Scholle, P. A.; and Spearing, D., eds., Sandstone depositional environments: Tulsa, OK, American Association of Petroleum Geologists Memoir 31, 410 p.

The Mouths of Rivers: Estuaries, Deltas, and Fans at the Sea Shore

Chapter 13 is the last of three chapters that deal with the marginal areas of the sea. Chapters 11 and 12 summarize what happens in sea-marginal areas not associated with the mouths of major rivers. In Chapter 13 the emphasis is on settings near the mouths of rivers, primarily *estuaries* and *deltas*. In such transitional environments, marine- and nonmarine processes interact. Marine processes are controlled by the salinity and circulation of seawater and by waves, currents, and tides. The chief nonmarine processes are controlled by the characteristics of the fresh-water flows of rivers.

Transitional Environments Associated with the Mouths of Large Rivers

The mouths of large rivers are "battlegrounds," so to speak, on which many geologic processes are actively engaged. Some of these processes augment one another; others cancel out. Whatever their combinations, the outcome is to create a series of coastal configurations and to deposit various bodies of sediment. The physical forces of the sea, chiefly waves and tidal currents, may be powerful or subdued. Large rivers bring in fluctuating amounts of water and sediment. We shall consider two situations associated with the mouths of large rivers: (1) *estuaries* and (2) *deltas*.

General Relationships

Before we begin the discussion of these features, it will be useful to summarize some of their broader-scale relationships. An estuary exists at the mouth of a river because no delta is being built. Many factors affect this situation. Estuaries form where the sediment loads of rivers are small, where subsidence is taking place faster than the buildup of whatever quantities of sediments are being delivered to the mouth of the river, or where the

tidal range is large and tidal currents are vigorous. Because of the effects of the Flandrian submergence, the latest sea-level rise, estuaries are probably more common today than they have been during much of Earth's history.

Some of these relationships are apparent from the data shown in Table 13-1, which lists the major rivers of the world in order of increasing content of suspended sediment. The mouths of nearly all rivers having concentrations of suspended sediment of less than 160 mg/l are estuaries. For example, the mouth of the world's largest river, the Amazon, is an estuary (Figure 13-1). The mean concentration of suspended sediment in the Amazon is 156.2 mg/l. By contrast, deltas have been built at the mouths of nearly all rivers in which the concentration of suspended sediment exceeds 225 mg/l.

Rivers having mean concentrations of suspended sediment lying between 190 and 230 mg/l form a transitional group. The Orinoco River (195.7 mg/l) has built a Holocene delta into a somewhat protected part of the western Atlantic Ocean. Where it enters the Indian Ocean, the mouth of the Zambesi River (200.0 mg/l) is neither an estuary nor a delta. The Rio de la Plata (215.0 mg/l) enters a large estuary, but a Holocene delta is growing into its apex.

Exceptions to this threefold grouping are as follows:

1. The low mean concentration of suspended sediment of the Mackenzie River is 34.1 mg/l. This should place it well within the estuary class. However, the Mackenzie River has built a small Holocene delta into the Arctic Ocean. The Mackenzie delta probably exists because the floating ice pack suppresses wave action and thus even small amounts of sediment can accumulate to form a delta.

2. The mouth of the Orange River, whose 1681.3 mg/l place it high in the delta-building group, is neither an estuary nor a delta. Despite the high concentration of suspended sediment in the Orange River, the total

TABLE 13-1. The major rivers of the world arranged according to increasing concentration of sediment in suspension and showing water discharge and characteristics of their mouths

River (location)	Yearly discharge of water (km^3)	Mean concentration of suspended sediment (mg/l)	Characteristics of mouth of river
Rhine (Netherlands)	68.5	6.6	Estuary[a]
St. Lawrence (Canada)	304	11.8	Estuary
Columbia (USA)	174	20.7	Estuary
Yenisey (USSR)	610	21.6	Estuary
Lena (USSR)	511	30.1	Estuary
Mackenzie (Canada)	15	34.1	Small delta
Ob (USSR)	400	39.5	Estuary
Congo (The Congo)	1,350	47.9	Estuary
Amur (USSR)	350	71.1	[b]
Amazon (Brazil)	3,187.5	156.2	Estuary
Orinoco (Venezuela)	442	195.7	Delta
Zambesi (Zaire)	500	200.0	[b]
Rio de la Plata (Uruguay-Argentina)	600	215.0	Delta in estuary
Niger (Nigeria)	293	228.7	Delta
Danube (Romania)	199	339.2	Delta
Po (Italy)	48.7	369.6	Delta
Mekong (Vietnam)	387	438.2	Delta
Yukon (Alaska, USA)	185	475.7	Delta
Tigris-Euphrates (Iran-Iraq)	210	500.0	Delta
Rhône (France)	52.7	597.7	Delta
Irrawaddy (Burma)	428	698.6	Delta
Mississippi (Louisiana, USA)	600	833.3	Delta
Nile (Egypt)	70	1,578.5	Delta
Orange (Southwest Africa)	91	1,681.3	[b]
Ganges-Brahmaputra (India-Bangladesh)	1,210	1,799.3	Estuary
Indus (India-Pakistan)	175	2,488.0	Delta
Yangtze (China)	690	2,734.6	Delta
Colorado (California, USA)	20.3	6,666.1	Delta
Yellow (China)	126	14,975.4	Delta

[a] The mouth of the Rhine in the North Sea is now an estuary, but during the Pleistocene Epoch and earlier, the Rhine built a delta. Much of the sediment from the Alps, which formerly was transported to the Rhine delta in the North Sea, is now trapped in the Rhine's lacustrine delta in the Bodensee.
[b] Neither a delta nor an estuary.

SOURCE: A. P. Lisitzin, 1972.

yearly discharge of water is so small (91 km^3) that the waves of the open Atlantic Ocean can dissipate all the river's load.

3. The mouth of the Ganges-Brahmaputra River, whose mean sediment concentration of 1799.3 mg/l ranks it fifth among the world's major rivers in terms of suspended load, and whose 1210 km^3/yr of water are exceeded only by the Amazon and the Congo, is an estuary. In this case, strong tidal currents transport the large quantities of sediment to the head of a major submarine canyon that indents the narrow shelf and from there gravity propels them into the deep sea (Chapter 9).

Active subsidence may be another contributing factor in the absence of a delta at the mouth of this mighty river system.

Both estuaries and fjords are coastal indentations at the mouths of rivers. The main differences between these two classes of coastal features are in bottom configuration and its control on water circulation. Typically, an estuary widens and deepens seaward, whereas a fjord is a deep basin formed by glacial modification of a river valley. Between the deep basin and the open sea is a shallow threshold and also, commonly, a rather narrow entrance. Fjords are not described here because fjord sedimenta-

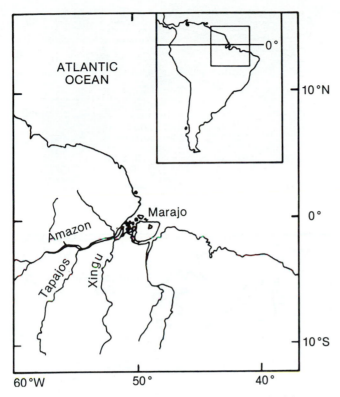

FIGURE 13-1.   Sketch map of the estuary at the mouth of the Amazon River, northeastern South America. The sediment load of the world's largest river is not great enough to overwhelm the effects of the waves, tides, and subsidence; therefore, no delta is being built.

tion does not produce a volumetrically significant contribution to the rock record.

We now turn to the characteristics of estuaries. We shall summarize some examples of modern sediments deposited in them and then review a few inferred ancient examples from the geologic record. In this discussion we speak of estuaries in the geomorphic sense. From the point of view of hydrodynamics, any river entering the sea may be considered an estuary if marine waters can move up the river channel as a salt-water wedge, such as in the Mississippi River during low flow.

Estuaries

As mentioned, in the geomorphic sense an estuary is the drowned mouth of a river. An important characteristic of estuaries is the pattern of water circulation. Many oceanographers define estuaries as "places where estuarine circulation prevails." This may be a useful functional definition to the "initiated," but it does not help newcomers. What is "estuarine circulation?" For our purposes, we shall define an **estuary** as *a coastal indentation, at the mouth of a river, in which seawater can circulate freely*. In estuaries the fresh water and salt water usually flow in opposite directions at least part of the time.

In many estuaries a situation obtains that may seem totally paradoxical. The rivers flowing into some estuaries demonstrably discharge negligible sediment from the land, but nevertheless, these estuaries are filling rapidly with silt. Where does this silt come from? It must be from the sea, where waves can place fine sediment in suspension and the tide can bring it landward.

Provenance studies of the sediment have provided geologic proof that the silt in some estuaries comes from the sea and not from the rivers. For example, in Charleston Harbor, South Carolina, the estuary of the Cooper River is filling rapidly with silt that contains abundant hornblende, a mineral not found in the sediment discharged by the Cooper River. The estuaries along the southern margin of the North Sea are being filled chiefly with sediment eroded from the Quaternary glacial deposits now forming the floor of the North Sea. This material is of Scandinavian provenance; it differs markedly from the sediment discharged from the Rhine River, which comes from the Alps and central-western Europe. In some estuaries, such as that of the James River, in Virginia, the concentration of sea-derived material in the sediment decreases landward to the limit of intrusion by seawater into the estuary, whereas the distribution of land-derived material is complementary.

Most examples of estuaries that are being filled chiefly with sediment derived from the sea floor are located in humid-temperate climates, where, because of the complete plant cover, rivers transport negligible suspended sediment. Estuaries in semiarid climates receive abundant terrigenous sediment from their rivers. Such estuaries may be filling with sediment derived from both the land and the sea.

The sediments of estuaries contain skeletal debris of estuarine organisms and usually show marked effects of bioturbation. The patterns of sediment distribution result from the interaction of tides, waves, and density currents.

ESTUARINE ORGANISMS

The waters of estuaries and comparable coastal indentations contain abundant nutrients washed in from the land. Despite their changeable conditions of salinity, temperature, and depth, estuaries are the habitats of many organisms. In addition, numerous migratory fish spawn in estuaries. Typical estuarine skeleton-secreting invertebrates are oysters and mussels that grow attached to the bottom; various gastropods that roam the surface of the sediments; and many organisms, chiefly arthropods, pelecypods, and worms, that live in burrows within the sediment. Algal mats coat parts of the bottoms of many estuaries.

Ever since Cretaceous times, oysters have built calcareous shell pavements, -mounds, and -reefs in estuaries

FIGURE 13-2.   Close view of cluster of oysters in growth position. Small barnacles have attached themselves to oyster shells at top. (Gene Tobler and authors.)

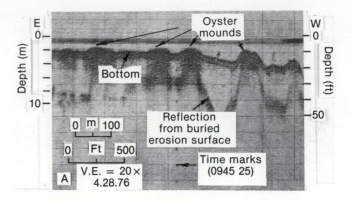

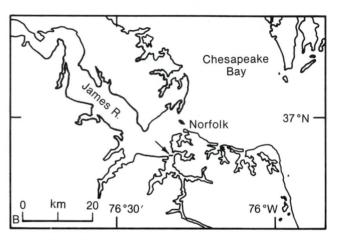

FIGURE 13-3.   Estuarine oyster bioherms.

A. Sonoprobe (continuous seismic reflection profile) record through seven mounds created by the common oyster, *Crassostrea virginica*, on traverse across Elizabeth River, opposite Craney Island, Norfolk Harbor, Virginia, along lat. 36°53.6'N made on 5 June 1964 (location on B). Six of the mounds began to grow on high parts of a buried erosion surface cut on Miocene marine strata. As sediment accumulated around them, the oysters bioherms grew upward, some as much as 5 m.

B. Location map. (J. E. Sanders in G. M. Friedman and J. E. Sanders, 1978; A, fig. 10-9, p. 278; B, fig. 10-1, p. 271.)

and other coastal water bodies. The individual oysters cement themselves to the bottom and to one another (Figure 13-2). In this way they can construct mounds that project above the surrounding sediment and that are capable of growing upward as sediment accumulates (Figure 13-3).

ESTUARINE SEDIMENTS

An estuary is a local morphologic depression that tends to be filled with sediment derived from the sea, from the land, or from both. Tidal action concentrates silt in estuaries. Rivers that discharge abundant sediment build deltas in the estuaries. Almost certainly, all modern marine deltas began by first building a delta in and then by filling an estuary, which formed when the rapidly rising water of the Flandrian submergence drowned the river valleys that had been deepened during the preceding Quaternary low stands. Among modern estuaries we can trace all stages. In some estuaries, small deltas have been built at the landward ends (Figure 13-4). Other estuaries have been completely filled and an alluvial plain has been extended across them (Figure 13-5). In the eastern Mediterranean coastal region, filling of estuaries by deltas during historic time has transformed once-thriving seaports into landlocked towns. Where deltas have filled

estuaries, the estuarine sediments are overlain by deltaic sediments, and the deltaic sediments, in turn, by sediments deposited on an alluvial plain. These deltaic- and alluvial sediments form autocyclic deposits.

As long as open water exists, both the tides and waves will be active. Depending on their relative intensities, one or the other may predominate. If tides predominate, then fine sediment coats the shore and coarser sediment occupies the center of the estuary (Figure 13-6), commonly forming linear sand ridges oriented roughly parallel to the axis of the estuary. In a wave-dominated estuary, the reverse is true. Coarser sediments become concentrated near the shore and fine sediments are shifted offshore. In either case, peritidal sediments (Chapter 12) will be formed along the landward fringes

FIGURE 13-4.   Map of part of Atlantic coast of South America showing small delta at the head of the Rio de la Plata estuary. Delta at mouth of the Paraña River contrasts with lack of delta at mouth of the Uruguay River. Inset shows regional location.

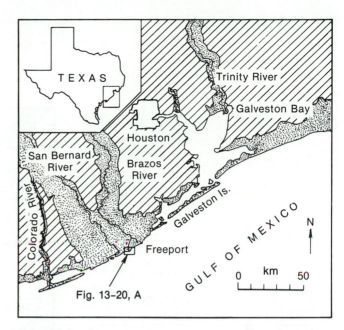

FIGURE 13-5.   Contrasting quantities of alluvial fills in some modern estuaries, central Texas coast. San Bernard and Brazos rivers have completely filled former estuaries and now discharge directly into Gulf of Mexico. Trinity River has extended its alluvial plain only a short distance into Galveston Bay. Stippled areas, Holocene sediments; diagonal lines, areas underlain by Pleistocene sediments. (After R. J. LeBlanc, Sr., and W. D. Hodgson, 1959b, fig 6, p. 202–203.)

of the estuary. Estuaries that are protected by barriers will contain flood-tidal-delta and washover deposits.

If the estuarine fill is deposited under the control of waves and tides, then no matter where the sediment came from, the pattern of the deposits will be established by marine factors. Such an estuarine fill may begin with a basal tidal-channel sand. Above this may be a tidal silt, and at the top, marsh deposits (Figure 13-7). The higher the tidal- or wave energy in the estuary, the better mixed are the fresh- and marine waters. In microtidal regimes, when river discharge is low and if wave action is limited, then a well-defined stratification of waters may form, with the less-dense fresh water overriding the denser seawater. If the river is in flood, if the estuary is not protected from the waves, or if the regime is macrotidal, then fresh- and seawater are well mixed in the estuary. The degree of mixing, as well as temporal variations in mixing efficiency, profoundly affect both the organisms that live in estuaries and estuarine sedimentation. For example, estuaries that experience seasonal- or intermittent mixing are repeatedly colonized in their seaward portions by organisms that cannot tolerate fresh water. These organisms are killed during mixing events that lower the salinity of the bottom water. Subsequently, when a stable stratification is reestablished, the fresh-water-intolerant organisms return.

The Hudson estuary, New York, is an example of an estuary in which tides predominate over waves. Studies have been made of the distribution of the radioactive isotopes ^{137}Cs, which comes from atmospheric fallout generated by nuclear-weapons testing since 1959 and from the Indian Point nuclear power-generating station, and ^{134}Cs and ^{60}Co, which come only from low-level releases from the power plant. In many cores collected since 1973, the top 12 to 15 cm contain abundant ^{137}Cs, which is virtually absent deeper in the cores. The net rate of accumulation in the areas sampled by these cores is 1 mm/yr, the regional rate of submergence. Elsewhere, especially in dredged channels, ^{137}Cs is abundant throughout the top 2 m of cores. In these areas the rate of accumulation has been from 5 to as much as 20 or 30 cm/yr, a rate based on the quantity supplied.

INFERRED EXAMPLES OF THE DEPOSITS OF ANCIENT ESTUARIES

Ancient estuarine deposits do not appear to form widespread stratigraphic units among the exposed bedrock strata of the continents. By definition, estuarine

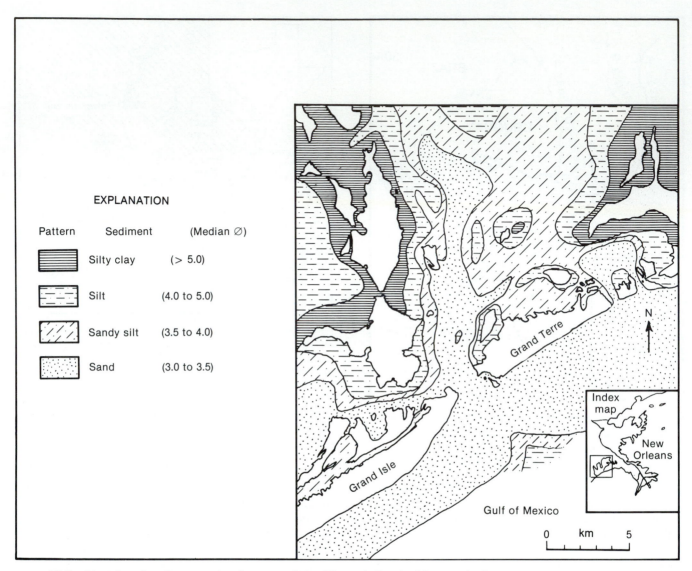

FIGURE 13-6.   Map of median diameters of sediments on floor of Barataria Bay, Louisiana, and adjacent Gulf of Mexico portrays contrasting pattern of tide-dominated- and wave-dominated coastal water bodies. In tide-dominated Barataria Bay, sand is restricted to tidal channels, the deepest parts of the area, and gives way shoreward to belts of silt and silty clay. The only exception is at lower left, where the wind blows sand into the bays from the coastal dunes. By contrast, in the wave-dominated Gulf of Mexico, sand forms the belt nearest shore and gives way to silt in the deeper water offshore. (After W. C. Krumbein, 1939, p. 186, fig. 4; sample points omitted here.)

deposits are confined to former river valleys. Moreover, the existence of deep valleys along a coast implies that a previous episode of great emergence took place during which the valleys were cut by rejuvenated rivers. Afterward, an episode of submergence was responsible for creating the estuaries. Despite their somewhat limited lateral extents, however, estuarine sediments can be economically important. The basal sand of a tide-dominated estuary is usually covered by an impermeable silty clay, a combination that has been found to create petroleum reservoirs. In addition, because they are coastal deposits,

estuarine sediments are valuable indicators of the locations of ancient shorelines and of past sea-level fluctuations.

Without much doubt, the lower reaches of every large modern river contain estuarine sediments that have been deposited during the Flandrian submergence. "Tidal silt" is a common deposit in many of these places. In the Hudson estuary, New York, the tidal silt, which contains scattered skeletal debris of oysters, mussels, and other invertebrates, is as much as 35 m thick. This silt has been accumulating continuously for about 11,000 yr, with only

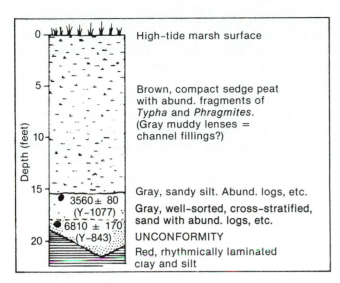

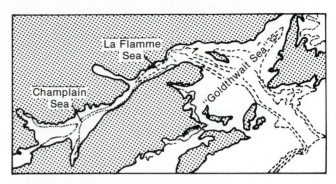

FIGURE 13-8. Maximum extent of Late Quaternary estuary in St. Lawrence lowland, Canada, and Champlain Valley lowland, United States, restored from distribution of estuarine sediments, some of which are now exposed at altitudes ranging up to many tens of meters above modern sea level. (After J. A. Elson, 1969b, fig. 3, p. 252.)

FIGURE 13-7. Columnar section through estuarine sediments of Late Quaternary age, southern Quinnipiac Valley, Connecticut. Varved sediments from proglacial lake, at base, are overlain unconformably by coarse pebbly sand and cross-stratified sand containing abundant plant debris. Numbers show radiocarbon ages for wood samples. Silty clay that overlies sand grades upward into modern tidal-marsh deposits. Sand is interpreted here as the deposit of tidal currents. The silt was deposited from suspension along the shallow margins of the estuary, which ultimately became mudflats and were colonized by salt-tolerant grasses. (After A. L. Bloom and C. W. Ellis, 1965, fig. 2, p. 2.)

slight indications of salinity changes suggested by the kinds of foraminifers present.

During the retreat of the last glacier to cover northeastern North America, about 12,500 yr ago, the sea invaded the low parts of the St. Lawrence and Champlain valleys. Although this former body of water has been named the Champlain Sea, it really was an estuary (Figure 13-8). The sea was able to flood this lowland because the weight of the glacier had depressed the surface of the Earth's lithosphere. When the glacier melted, the lithosphere rebounded slowly. While the surface of the lithosphere was still low, and after the glacier had retreated far enough north to uncover the valleys, the sea entered and remained for about 1000 yr. Then the surface of the lithosphere was elevated, the land emerged, and the former floor of the estuary became land. The chief sedimentary product of this Late Pleistocene estuary is gray silt of the kind deposited by tidal action.

Other estuarine deposits of Pleistocene age have been found in the shallow subsurface beneath coastal-plain rivers and elsewhere. Examples are known from southeastern Virginia and Georgia. In southeastern Virginia, the Great Bridge Formation is inferred to be an example of an estuarine silt. This former estuary was completely filled and its position does not coincide with that of any large modern river.

Sandstones inferred to have been deposited by tidal currents in ancient estuaries have been described from the Fall River Formation (Upper Cretaceous) exposed northeast of the Black Hills in the northeastern corner of Wyoming, United States. In this area the trend of the Cretaceous shoreline was east-west. Land lay to the south and the open sea to the north. The inferred estuarine sand bodies occupy north-south-trending channels that are incised into the inferred intertidal facies of the Fall River Formation.

The supposed estuarine sandstones range in thickness from 10 to about 30 m, in width from 400 to 1600 m, and in length from 1200 to 5600 m. The deepest and narrowest ends of these channels are at their southern (landward) ends.

Other inferred examples include the subsurface sediments beneath the Niger delta, in western Africa (Miocene to Quaternary), and parts of the Jurassic of Yorkshire, England (the "Estuarine Series").

CHARACTERISTICS OF ESTUARINE DEPOSITS

The main proof that an ancient sea-marginal sediment was deposited in an estuary consists of its gross shape, that of a valley filling. In small exposures, the deposits of an estuary are not distinguishable from sediments deposited in bays, in lagoons, on the intertidal flats and -marshes in backbarrier areas (Chapter 12), or in many of the environments fringing the subaerial part of a large marine delta, which we discuss in upcoming sections of this chapter. However, provenance information may reveal the input of fluvial sediment that is characteristic of estuaries and less characteristic of bays, lagoons, and intertidal flats not associated with estuaries.

Estuaries that form in areas affected by hurricanes, such as the Gulf Coast of Florida, contain distinctive storm deposits that strongly resemble the marine-shelf storm layers discussed in detail in Chapter 10. Estuarine storm layers may be graded units consisting of sandy shell gravel at the base to slightly shelly quartz sand at the top, or thin, homogeneous shelly sands and gravels. These sediment layers, if not reworked by waves or organisms, will be interstratified with lower-energy deposits in the rock record of such an estuary.

SOURCES: Barnes, 1980; Boothroyd, 1978; Clifton, 1982; Emery, Stevenson, and Hedgpeth, 1957; Fairbridge, 1980; Fraser, 1989; Nichols and Biggs, 1985; Peterson, Scheidegger, Komar, and Niem, 1984; Redfield, 1965; H. J. Simpson, Williams, Olsen, and Hammond, 1976.

Deltas

Deltas are so dynamic that even casual observers have been impressed by the scope and speed of the changes on deltas. Some lakes have disappeared because deltaic sediment filled their basins. Sediment discharged from rivers has filled estuaries and bays, some of which formerly served as harbors. Many deltas have built new land along the open sea coast. The land areas formed by growth of deltas are some of the world's most densely populated regions.

Deltaic sediments form a significant proportion of the geologic record. Exactly what proportion is not known, because the diverse characteristics of deltaic sediments have only recently been fully appreciated. Deltas consist primarily of autocyclic deposits. Deltas are important factors in the localization of deposits of fossil fuels. Many millions of tons of the world's reserves of coal were deposited as parts of ancient deltas. Moreover, much petroleum has been discovered in ancient deltaic sediments.

The following paragraphs discuss the definition of a delta, the interaction of variables, and the stratigraphic patterns that result from the growth of deltas. This material is applicable to all deltas. This section is followed by a discussion of delta-marginal plains, some examples of marine deltas, modern and ancient, and a section on deltas and cyclothems.

Definition

The term *delta* was first applied by the early Greek geographer Herodotus to the triangular area of land built by the Nile River at its mouth in the Mediterranean Sea. He chose this term because the shape of the triangular area of land resembles the shape of the Greek capital letter delta.

As a geologic term, the word *delta* includes more than triangular-shaped land areas at the mouths of rivers entering the sea. The outlines of the land built by many deltas are not triangular. Moreover, a few deltas have not built any land (yet); they consist of bodies of sediment that have been deposited close to, but entirely beneath, sea level. Still other deltas are bodies of sediment that were deposited in lakes. Therefore, as far as a geologic definition is concerned, neither the shapes of river-built land areas, nor even the existence of river-built land, are the critical factors, nor is the sea. In a general sense, we can define the individual unit of a **delta** as *a lobate body consisting of sediment that has been transported to the end of a channel by a current of water, and deposited mostly or entirely subaqueously, at the margin of the standing water into which the channel discharged or is still discharging.* Actually, some deltaic sediment is deposited before reaching the end of the channel; for example, **crevasse splays**, *fan-shaped bodies of sediment spread away from a stream channel by water flowing in a crevasse.* An important point to emphasize about this definition of deltas is that it embraces only the basic unit deposited by each channel. Such a unit is known as a *lobe*. In a large delta, many individual distributaries exist. All the lobes they deposit coalesce to form a *deltaic complex.* Deltas are commonly loci of maximum sediment deposition and form greatly thickened successions relative to time-equivalent nondeltaic strata.

A logical question to ask about deltas is why they are distinguished from fans (Chapters 9 and 14). Both fans and deltas are lobate accumulations of sediment built at the ends of channels. The main difference is that fans are deposited away from the influences of a free surface of a body of water, such as that in a lake or the sea. Thus fans are not built by nearshore processes. Most deltas, unlike fans, consist of a complex of associated nearshore-, subaqueous-, and terrestrial sedimentary deposits. By contrast, fans are products of processes that operate entirely in the subaerial realm or entirely in the subaqueous realm well below the depths of influence of nearshore processes. One consequence of this is that fans are unlikely to contain coal deposits, although, like deltas, fan sediments may contain important hydrocarbon reservoirs. Despite these differences, however, fans and deltas may be closely related. A delta may build a land area across which a fan may prograde. Sediment initially deposited as a delta is likely to be transported downslope and redeposited on a subaqueous fan in deeper water.

Interacting Variables

Although deltas share a few common features, they are influenced by many variable factors. Most of these variables can be organized under the following headings:

(1) the channel and its discharges, (2) the water in the basin, (3) climate, (4) tectonic movements, and (5) non-tectonic deleveling.

CHANNEL: CHANNEL SIZE AND DISCHARGES OF WATER AND OF SEDIMENT

One of the essential factors in a delta is the channel; in fact, no channel means no delta. Typically, the channel is that of a river. Channel is used in the singular to emphasize the unit situation. Actually, a large river divides into many channels known as *distributaries* (Figure 13-9). All large deltas consist of sediments discharged from a river. However, the underlying principles apply not only to rivers of all sizes but also to tidal channels and even to small spillover channels on a beach (Figure 13-10) or a spit.

The size of the channel can range widely. Widths may be a few tens of centimeters to several kilometers, and depths from a few centimeters to many tens of meters.

The channel discharges water and sediment, which includes two contrasting loads. The suspended load is dispersed within the water that is conveyed by the channel. The bed load is impelled along the bottom of the channel. The proportions of sediment transported in these two loads exert an important control on the deltaic complex. As we shall see, a delta built chiefly of sediment transported as bed load looks totally unlike one that consists largely of sediment from the suspended load. In either case, a delta is composed of sediment that is

FIGURE 13-10. View landward at low tide in March 1972 from crest of ridge across water-filled runnel, Robert Moses State Park, Fire Island barrier, S shore of Long Island, New York. Small wave-generated ripples are visible through thin cover of water that remains on the floor of runnel. Landward side of ridge is crossed by numerous evenly spaced, vertical-sided, shallow channels, each of which ends in a tiny spillover delta lobe. Thin heavy-mineral concentrates are present at seaward ends of channels where sheets of washover water coalesced upon entering the channels. (Authors.)

deposited in such a way that the channel can be extended. Therefore, the channel flows over the sediment discharged from it.

Climate and tectonic movements are major factors in the discharges of both water and sediments. Accordingly, we include further discussion under those headings.

WATER IN THE BASIN: PROPERTIES, MOVEMENT, DEPTH, ORGANISMS

The water in the basin influences a delta in many ways. The chief influence results from the density, and whether the density of the water in the basin is the same as or differs substantially from that discharged from the channel. The density of water is a function of its temperature and salinity and the load of suspended sediment. The density of seawater, about 1.025 g/cm^3, exceeds that of nearly all rivers. The density of lake water is extremely variable; it may range from less than 1.000 g/cm^3 to more than 1.025 g/cm^3. Therefore the density of the water of a lake may be greater than, less than, or equal to the density of the water discharged from a river (Figure 13-11).

The density situation determines what happens to the water and its sediment loads when they reach the end of the channel. If the density of the water-and-suspended-sediment mixture discharged from the channel is the same as that of the water in the basin, then the chief effects are the spreading, slowing down, and diffusing of the entering water. Typically the flow

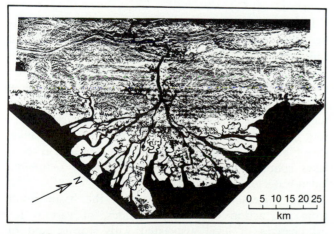

FIGURE 13-9. Branching distributaries of Mahakam Delta, which discharges into Makassar Strait (water appears black on this side-looking radar image), eastern Kalimantan (formerly Borneo), Indonesia, about long.117°E, lat.1°S. Comparatively straight distributaries of river contrast with meandering tidal channels. Dominance of tidal activity over waves causes ends of channels to flare. Curvilinear features on land at top of image are resistant edges of folded strata. (Ph. Magnier; T. Oki; and L. Witoelar Kartaadiputra, 1975, fig. 4, p. 9, courtesy Ph. Magnier.)

0 5 10 15 20 25
km

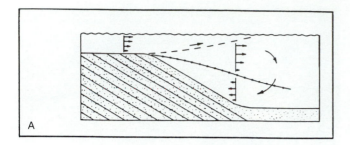

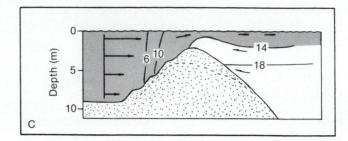

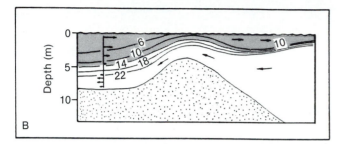

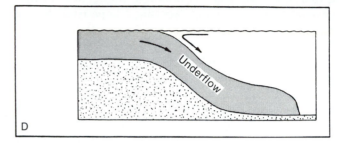

FIGURE 13-11.   Schematic profiles showing combinations of water densities at mouths of rivers. Density values in parts per thousand ([‰]). (Suggested by concepts in C. C. Bates, 1953.)

A. Density of river water same as water in basin. (After A. V. Jopling, 1965a, fig. 4, p. 56.)

B. Density of river water less than that of basin water; low-stage discharge with salt-water wedge and low distributary-mouth bar. (After L. D. Wright and J. M. Coleman, 1974, fig. 7, p. 762.)

C. Density of river water less than that of basin water; flood-stage discharge with salt-water wedge flushed out and high distributary-mouth bar. (After L. D. Wright and J. M. Coleman, 1974, fig. 12, p. 762.)

D. Density of river water exceeds that of basin water.

separates. At the surface, the water flows away from the channel. At the bottom, the flow is toward the channel. (See Figure 13-11, A.) The bed load may accumulate on a steep slope at the angle of repose of the particles. This is a typical situation where a river flows into a fresh-water lake or in at the end of a tidal inlet.

If the density of the water discharged from the channel is significantly less than that in the basin, then, near the end of the channel, a density stratification will be established. A wedge of the dense, usually saline, water enters the lower reaches of the channel and this forces the water flowing in the channel out of contact with the floor of the channel. Along the sloping interface between the two bodies of water, mixing takes place. A surface current flows out of the channel, but a bottom current flows into the channel. The bed load stops at the tip of the salt-water wedge; silt and clay accumulate on the channel floor beneath the salt water. (See Figure 13-11, B.)

As the tide rises and falls, and as the quantity of water discharged from the channel fluctuates, the wedge shifts. Despite these shifts, however, as long as it remains in the channel, the wedge acts as a dam for the bed load. During floods, when the high water piles up a thicker-than-normal column of fresh water, the wedge retreats from the channel.

The extra thickness of the piled-up water exerts more hydrostatic pressure at the bottom and this new pressure forces out the salt water. *Only at such times is the bed-load sediment discharged out of the channel.* (See Figure 13-11, C.)

If the density of the water discharged from the channel exceeds that of the basin, then the river discharge sinks and flows along the bottom as a *density underflow.* The suspended load, and, depending on the speed of the flow, possibly the bed load as well, flow along the bottom as *a continuous turbidity current* (Chapter 9). While an underflow is active, a surface current flows toward the point where the underflow sinks. (See Figure 13-11, D.) Underflows are common in some lakes containing water less dense than the sediment-laden water of the rivers that feed them.

The movements of the water in the basin result from the action of waves, of tides, of various local changes of level associated with storms, and of the regional oceanographic setting. The interactions among the factors of channel discharges, waves, and tides largely determine the shapes of marine deltas, that is, *deltas built into the sea.* Because these factors control delta morphology, they form the basis of a classification of deltas into *sediment-discharge-dominated, wave-dominated,* and *tide-dominated* (Figure 13-12).

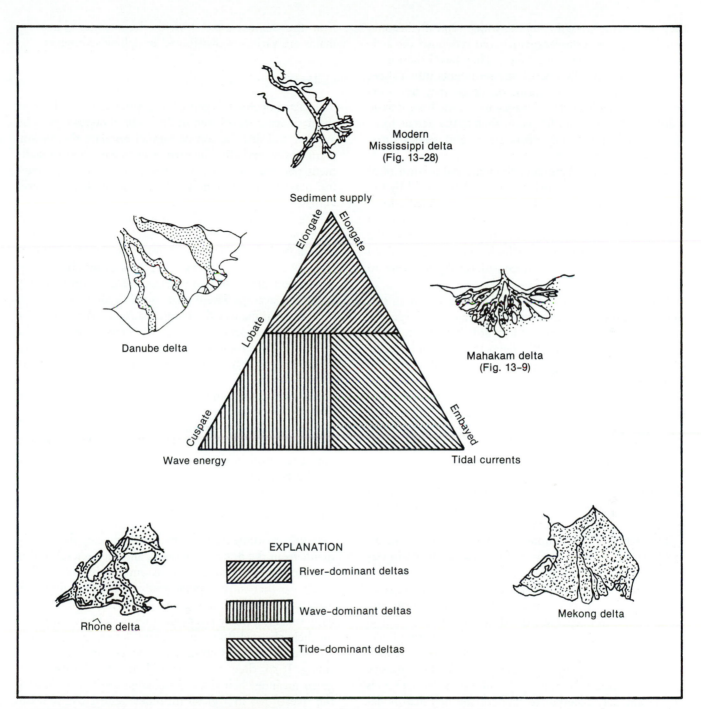

FIGURE 13-12.   Triangular diagram showing relationships among supply of sediment, waves, and tidal currents and morphology of resultant deltas, with several examples of marine deltas. (After W. E. Galloway, 1975, fig. 3, p. 92.)

If sediment discharge is much greater than the power of waves and tides to remove sediment, then the coastline progrades rapidly, and a delta like the modern Mississippi delta is formed.

If wave action dominates over sediment discharge, then a wave-dominated delta, like the marine Rhône delta, described below, is formed. If the power of the waves is too great compared to the quantity of sediment discharged from the channel, then no delta is deposited. Instead, the waves redistribute the sediment along the shore and this may cause long stretches to prograde. If the waves prograde the shore with sand, they build a wide *beach plain* (or strandplain). If the waves prograde the shore with silt and clay, they build wide coastal *mudflats*, which typically become marshes. Waves soon reorganize and redeposit the sediments from inactive delta lobes. They bevel the top of the lobe and typically rework the sediments into a sheet of sand. The effects of storms on deltas may be to increase the effectiveness of waves to redistribute deltaic sediment. If storms are frequent, then in the deltaic foreset deposits, hummocky cross strata and swaley cross strata may be common.

If tidal currents dominate, then the delta consists of *linear sand ridges* whose orientation is determined by the directions of movement of tidal currents. Tidally dominated coastal areas where rivers enter the sea tend to form estuaries rather than deltas (e.g., the mouth of the Ganges-Brahmaputra River, which some have classified as a tide-dominated delta, is considered by us to be an estuary).

The depth of water in the basin affects a delta in two ways. First, other things being equal, depth is a large factor in determining the rate of forward growth. The same amount of sediment builds a delta lobe forward rapidly in "shallow" water but less rapidly in "deep" water. We have put quotation marks around these terms because we apply them only in a relative sense. What is referred to here as "deep," as far as a delta is concerned, may be only a few hundred meters, as contrasted with a few tens of meters for "shallow." On a scale that includes the deepsea basins, both obviously are "shallow." The second way in which water depth affects a delta has to do with the stratigraphic pattern of the delta. In this regard, a critical relationship is the ratio of channel depth to depth in the basin at the point of discharge.

The depth of the channel may be about the same as, much less than, or much more than that in the basin water at the point of discharge. If a long underwater slope is available, various slope failures may take place; sediments from the delta are displaced along the bottom and come to rest in deeper water. Moreover, if the bed load from a large channel is deposited on thick, **hydroplastic sediments** (*water-saturated, and possibly gas-charged, fine-textured deposits having negligible strength*), then massive amounts of sand may sink down and thus thick sand pods or -prisms may accumulate. Where the bed load is deposited on a firm substrate, this does not happen. Viewed in the short term, fine-textured sediments from a single lobe can be only as thick as the depth of the water into which they are discharged. Therefore, the hydroplastic deltaic sediments do not become thick enough for large-scale foundering of the bed load unless the delta lobe builds forward into "deep" water.

In the growth of a delta, the organisms living in the basin are more passive spectators than active participants. Bottom-dwelling organisms are affected by the influx of sediment. Organisms that do not live on the bottom are forced to adjust according to the ways in which the water

from the channel mixes with the basin water. Their remains are variously distributed in deltaic sediments.

CLIMATE

The factor of climate controls the water regimen, both in the drainage network and in the basin of deposition. The amount and distribution of rainfall regulate the natural plant cover, and this, in turn, affects the yield of sediment per unit area. In fluvial networks, the areas having maximum sediment yields are situated in semiarid climates.

The amount and distribution of rainfall determine both the base flow and flood discharges of rivers. On a delta built by a large river, water is abundant; therefore, even in a desert region, plants grow abundantly. The topset plains of many deltas are the sites of lushly vegetated swamps where peat forms in abundance.

Climate influences the temperature of the water in the basin. The temperature, hence the density, of the water in both the river and the basin may fluctuate seasonally. These seasonal changes can promote or prevent density underflows.

Deltas may form in lakes or at the margin of the sea in glacially influenced regions. In such cases, the sediment-deforming effects of grounded ice may leave their mark on the shallow-water deltaic sediments. More importantly, sea ice can dramatically reduce wave energy and its effects on the reworking of deltaic sediments.

TECTONIC MOVEMENTS

Tectonic movements can affect both the supply of sediments from the drainage network and the general setting in which a delta grows. It is axiomatic that rivers flow into areas that subside persistently. Therefore, deltas always form in subsiding regions. As the land surface gradually subsides, many rivers are able to deposit sediments on their channel floors and natural levees and thus may be able to aggrade upward as fast as the land is subsiding. If so, then despite subsidence, the location of a given meander belt does not change greatly. As long as the river flows in a given meander belt, it continues to deliver sediment to its own particular delta lobe and the lobe grows forward.

Naturally, a channel on a subsiding alluvial plain cannot grow upward forever. Eventually, probably during a flood, a crevasse will open across a natural levee and part, or even all of the discharge may pour through it. The result is flow in a new channel. If the outcome is *an abrupt abandonment of a segment of the river channel*, then the name **avulsion** is applied. Diversion is not always total; sometimes the river bifurcates and its discharge is shared equally by two distributaries for hundreds of years (Figure 13-13). When large amounts of deltaic

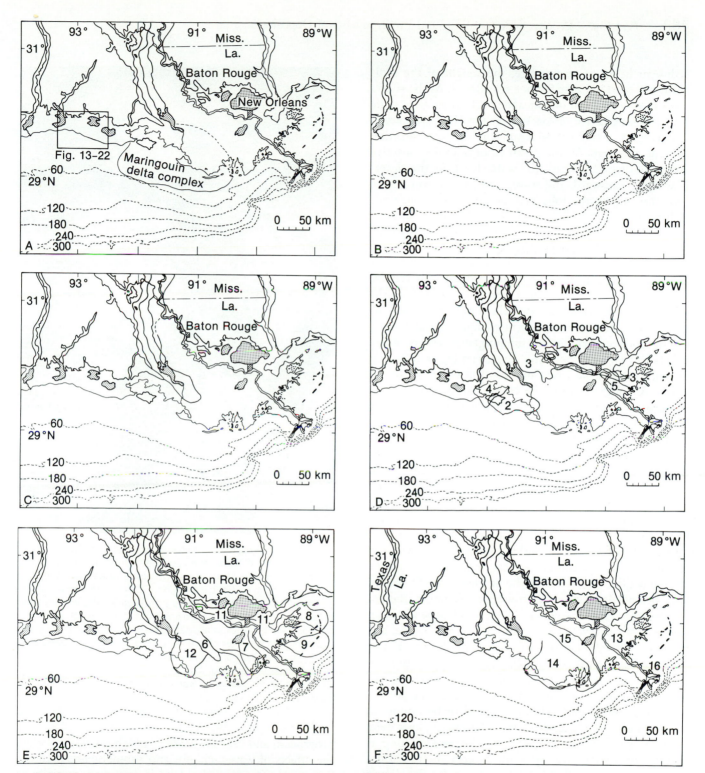

FIGURE 13-13.   Successive stages in development of lobes of Mississippi delta complex during last stages of Holocene (Flandrian) submergence. (Redrawn by J. E. Sanders from data in D. E. Frazier, 1967, figs. 11 and 12, p. 307 and 308.)

A. Maringouin delta complex builds during stillstand at about −15 m (local datum).

B. Rapid submergence drowns Maringouin delta; creates broad estuary in lower end of Mississippi's alluvial valley. River flowing in Teche meander belt.

C. Oldest of three Teche delta lobes fills SW side of estuary.

D. Division of flow into present course (eastern meander belt) begins; river builds delta that fills remainder of estuary and extends well beyond New Orleans (lobes 3 and 5). Water from Teche meander belt builds two more delta lobes (lobes 2 and 4).

E. St. Bernard lobes (lobes 7, 8, 9, and 11) and Lafourche lobes (lobes 6, 10, and 12).

F. After St. Bernard delta complex became inactive, the Mississippi River built the Plaquemines (lobe 13) and modern deltas (lobe 16). Lafourche channel builds lobes 14 and 15.

487

sediment begin to accumulate, the weight of the sediment will tend to cause further subsidence. As the river's discharge shifts from one meander belt to another, delta lobes become active or inactive (Figure 13-14). Inactive deltaic lobes may become significant sites of deposition of peat and clay. As the basin floor subsides, inactive lobes are carried beneath the water level. As the shoreline shifts landward, the waves rework the top of the lobe. Eventually the lobe may be buried by nondeltaic sediments.

NONTECTONIC DELEVELING

Subsidence is a major factor in changing the relative position of water level on a delta, but it is not the only factor. Other factors include compaction and various short-term- and long-term changes of water level. Compaction causes further submergence. Severe storm winds can change the water level, causing it to go up in some cases (*wind setup*) and down in others (*wind setdown*). Along the coast of the Gulf of Mexico, hurricanes can create storm surges on the order of 4 or 5 m. During these surges, marine water is driven far inland.

Long-term deleveling not related to subsidence is brought about by eustatic changes of sea level. Such changes have been major events of the last 10 million years or so. When sea level drops, deltas emerge, and the rivers incise valleys into them. When sea level rises, the deepened valleys become estuaries and, after these have been filled, deltas begin to grow on the more-open parts of the coast. Although the rise in sea level is an allocyclic phenomenon, the bulk of the deposits that form during and after the sea-level rise (estuarine and deltaic deposits) are autocyclic. Naturally, the response of a delta to changing sea level depends upon the rate of sea-level change, the rate of subsidence, and the rate of sediment supply. For example, the Fraser River delta, British Columbia, Canada, has continued to prograde throughout the Holocene sea-level rise because the rate of sediment supply exceeded the sum of subsidence plus sea-level rise. Instead of becoming an estuary during peak sea-level rise, the Fraser River delta merely experienced a decrease in the rate of progradation from about 6.5 m/yr to about 1 m/yr. Formation of the Fraser River delta is driven by the deltaic engine (autocyclic), but slightly modified by the (allocyclic) sea-level rise.

Stratigraphic Patterns

Deltaic deposits form stratigraphic units in response to three first-order processes: (1) *prograding*, or forward advance; (2) *aggrading*, or upbuilding, which is possible only above those parts of a delta that have previously

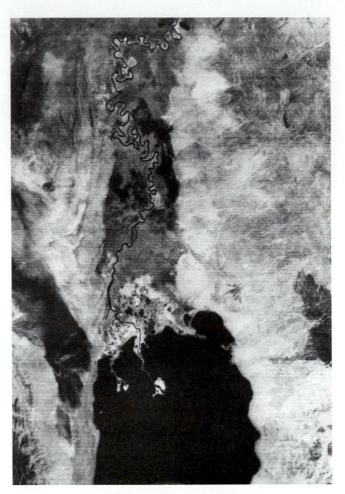

FIGURE 13-14.   Relationship between channel diversion and delta lobes as a result of subsidence seen in high-altitude image from ERT satellite of three delta lobes (two active, one inactive) of Omo River at the north end of Lake Rudolf, Ethiopia and Kenya. Avulsion at a point in the center of the alluvial valley about 30 km upstream from the lake caused easternmost lobe to be abandoned. The east side of the Omo alluvial valley is receiving abundant sediment from two rows of evenly spaced fans (white triangular areas); the west side of this valley is receiving little or no sediment. Dark areas on either side of the north end of Lake Rudolf are playas having little or no water when this image was made. (NASA.)

prograded; and (3) *transgressing*, which takes place when the effects of waves, chiefly, and submergence predominate, and the top of the delta is reworked. Actually, in a single large marine delta, all three processes may be going on simultaneously in different places. In general, the foresets (delta front) and channel deposits prograde; the topsets (delta plain), bottomsets (prodelta sediments), and the channel deposits can aggrade; and all sediments can be transgressed.

Although the shapes of deltas vary enormously, deltas form only a comparatively small number of stratigraphic patterns. Delta morphology is related to sand/mud ratio and to kind and orientation of sand bodies. Thus the kinds and amounts of sediment input exert powerful effects on delta morphology (constructional control). The destructional processes of waves and tides are also important factors in determining delta morphology.

We discuss three main kinds of deltas: (1) deltas composed largely of the channel's bed-load sediments; (2) deltas composed largely of the channel's suspended-load sediments, but including prisms of bed load deposits, not overthickened, that were dropped on a firm substratum; and (3) deltas composed largely of the channel's suspended-load sediments, but including overthickened pods of bed-load deposits that collapsed into a hydroplastic substratum.

DELTAS CONSISTING LARGELY OF BED-LOAD SEDIMENTS (GILBERT-TYPE DELTAS)

The first analysis of deltas that attained wide distribution, and which formed the basis of geologic ideas for several generations, was made by G. K. Gilbert (1843–1918) and published in 1885. In his studies of the shoreline features of ancient Lake Bonneville, Gilbert encountered natural sections through deltas composed largely of coarse sand and gravel that had been built by shallow channels into the "deep" waters of Lake Bonneville. He subdivided the sediments of these deltas into foresets, bottomsets, and topsets (Figure 13-15).

Foresets.   The **foresets** of the deltas described by Gilbert are *large-scale cross strata deposited at the angle of repose on the steep subaqueous front of the delta.* The difference between the cross strata deposited by a migrating sand wave, for example, and delta foresets inclined at the angle of repose is that in simple cases the sand wave migrates because part of it is eroded. By contrast, in a growing delta lobe new sediment is continually being added.

Experiments conducted since Gilbert's time have been the basis for important interpretations about the conditions under which the foresets are deposited. Slow

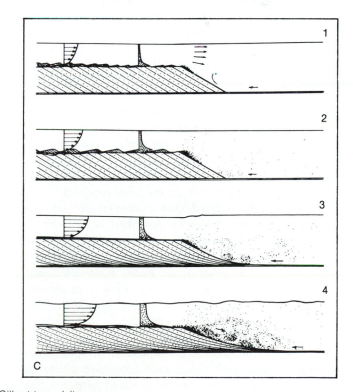

FIGURE 13-15.   Idealized sketch profile and section through coarse Gilbert-type deltas.

A. Profile-section at right angles to shore through entire delta.

B. Enlarged section through delta showing vertical succession that resulted from forward growth. (Redrawn from G. K. Gilbert, 1890, figs. 14 and 15, p. 68 and 70.)

C. Effect on delta foreset slope of increasing current speed, shown by schematic profiles parallel to current in laboratory tank. Stippled profiles above sediment show quantity of suspended sediment versus height above bed. Graph with arrows relates speed of current to height above bed (schematic). (1) Planar foresets make sharp-angled contact with bottomsets, slow current. (2) Foreset slope still planar, ripples in channel slightly larger, and current speed somewhat faster than in (1). (3) Tangential lower contact of concave-up foresets, faster current. (4) Low-angle, concave-up foresets with tangential lower contact, fastest current. (A. V. Jopling, 1965b, fig. 3, p. 779.)

currents having thin traction carpets deposit planar fore-sets that are truncated at both top and bottom. (See Figure 13-15, C, 1.) At slightly greater speeds, the backflow eddy, part of which travels along the bottom toward the delta, begins to transport fine sediment. (See Figure 13-15 C, 2.) This countercurrent along the bottom may even flow up the foresets, forming small-scale current ripples with cross laminae dipping in a direction that is opposite to the dip of the foresets. (Compare Figures 7-1, C, and 7-20, A.)

At still-greater current speeds, when a thick bed load travels as a liquefied cohesionless-particle flow down the foreset slope, the angle of the foresets decreases, and the distinctness of the break in slope at the bases of foresets becomes blurred. (See Figure 13-15, C, 3 and 4; compare Figure 7-21, type-D dune.)

Bottomsets.   The **bottomsets** are *the fine-textured strata deposited as a fringe adjacent to, but beyond, the foreset slope.* At appropriate speeds of the entering current, a special variety, named **toesets**, appears. These are *fine sandy sed-iments deposited at the toe of the foreset slope of a delta by the countercurrent flowing along the bottom toward the delta.* As mentioned, the kind of contact between bottomsets (including possibly toesets) and the foresets varies and is governed by the speed of the current in the channel.

Topsets.   The **topsets** described by Gilbert consist of two kinds of strata: (1) *nearly horizontal, coarse sediments spread by the stream channel across the top of the foresets* (See Figure 13-15, B, top.) and (2) fan sediments that prograded across the nearly horizontal top of the delta. (See Figure 13-15, A, at tip of arrow from the label topsets.)

Gilbert clearly realized that *a stratigraphic succession through the coarse deltas he described begins with the finest particles in the horizontal bottomsets at the base; includes as the dominant part the inclined foresets, composed of sand and pebbles; and is capped by the nearly horizontal topsets, which consist of the coarsest sediment.* This is an *upward-coarsening* succession. (See Figure 13-15, B.) Such deltas have been named **Gilbert-type deltas.**

Many examples of Gilbert-type deltas have been described from the sediments deposited in Quaternary meltwater lakes. Such deltas may constitute a valuable source of sand and gravel for construction purposes (Figure 13-16). The distinctive steep foresets of Gilbert-type deltas have been displayed on high-resolution, shallow-penetration, continuous-reflection seismic profiles (Figure 13-17).

Most Gilbert-type deltas were probably built by shallow channels that deposited their bed-load sediment in water significantly deeper than the channel. Common examples of such channels are those of braided streams and of washovers on beaches and spits. When a shallow channel prograles, it does not erode the foresets to any

FIGURE 13-16.   Sectional view of Gilbert-type delta built into Late Quaternary Lake Hitchcock, now exposed in sand-and-gravel pit, Sunderland, Massachusetts. Height of exposure, which is nearly equal to the depth of water in the former lake, is about 20 m. Sandy foresets dip steeply westward (to left); horizontal topsets (at top) consist of gravel and coarse sand. Steep slope to right (in center of picture) is subaerial angle of repose made by particles falling from temporarily inactive working face to floor of excavation. (Authors.)

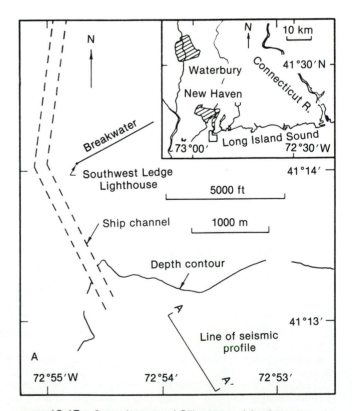

FIGURE 13-17.   Steep foresets of Gilbert-type delta that prograded southward into a fresh water lake of Late Quaternary age that formerly occupied the depression now filled by seawater of Long Island Sound.

A. Index map, showing location of seismic-reflection profile. Inset shows regional location.

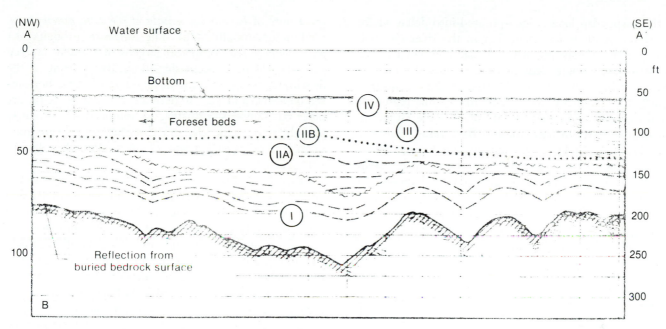

FIGURE 13-17. (*Continued*)

B. Continuous seismic-reflection profile through sediments underlying the approach to New Haven harbor, Connecticut. (Location on A). Prominent diagonal reflectors within Unit III are interpreted as foreset beds. Delta prograded into water that was 25 m deep at northwest end of profile but deepened to 30 m at southeast end of profile. Water surface stood about 25 m lower than present level of Long Island Sound. Units I and II, underlying the delta foresets, probably consist of fine-textured strata. Layers of Unit I drape over the bedrock surface; they may be the offshore deposits of an older lake, probably also of Quaternary age. Unit IV, above foresets, consist of fine-textured Holocene sediments deposited in Long Island Sound. (Bolt, Beranek, and Newman, Inc., courtesy J. E. Leonard.)

great depth, hence nearly all parts of the foresets tend to be preserved. *Coarse Gilbert-type deltas, rich in gravel, most or all of which are fed by braided streams,* have been named **braid deltas.** Braid deltas represent a coarse end member of the broader category of Gilbert-type deltas.

A special kind of Gilbert-type delta is formed when relatively coarse foreset strata are remobilized as debris flows and turbidity currents, transporting relatively poorly sorted and coarse-textured sediment into the bottomset environment. In such a delta, conglomerates exhibiting stratal features indicating deposition by a variety of gravity-driven sedimentary processes are interstratified with fine-textured typical bottomset strata.

DELTAS COMPOSED LARGELY OF SUSPENDED-LOAD SEDIMENTS

In total contrast to Gilbert-type deltas are those deltas that have been built predominantly of fine-textured sediment from the river's suspended load. Such deltas are products of rivers having large deep channels that are flanked by natural levees. Such rivers may flow into lakes or the sea. The natural levees extend out into the water and, in part, form the shorelines of interdistributary bays. Local lobes of fine sand may

be deposited during floods as crevasse splays (See Figure 14-42.) that poured into the bay water. Gravity-powered sediment-transport processes such as creep, slumping, and turbidity currents are common on the fronts of fine-textured deltas and modify delta-front slopes in response to sediment supply.

Based on the thickness of the bed-load sand, it is possible to recognize two chief varieties: (1) deltas in which the prisms of bed-load sand are of normal thickness, meaning that the thicknesses of the channel-floor sand bodies are determined by the flood discharge of the river; and (2) deltas in which the bed-load sand forms pods that are unusually thick, because the bed-load sand foundered into the thick, fine-textured hydroplastic delta deposits formed where suspended-load sediments were deposited in "deep" water.

"Shoal-Water" Deltas: Bed-Load Sand Prisms of Normal Thickness. As mentioned previously, water depth determines the thickness of the *foresets,* or *delta-front* sediments, in the usage of those who feel that the term foresets should be applied only to steep, angle-of-repose, coarse-textured strata of Gilbert-type deltas. Because during flood stages the depths of water in rivers may be several tens of meters or more, much

of the suspended load is incorporated into deltas where the water depth is less than that in the river channel. Much of the Mississippi River deltaic complex consists of shoal-water deltaic sediments. This includes numerous lacustrine deltas on the Mississippi delta plain that rapidly fill ephemeral lakes that form in this region. A small example of a "shoal-water" delta is the Holocene Guadalupe Delta in San Antonio Bay, Texas. This bay is 29 km long, 11 km wide, and, on the average, about 2.1 m deep (Figure 13-18). The *topset parts* (delta plain) *of* this *delta* include (1) distributary channels and their flanking natural levees and (2) interchannel areas, such as marshes, lakes, and shallow bays. The foreset (delta-front) deposits, consisting of well-laminated and ripple- cross-laminated sands and silts, are localized at the fronts of *bars at the mouths of the delta distributaries* (distributary-mouth bars). The *bottomset* (prodelta) *deposits*, which are composed largely of vaguely layered silty clay, form crescent-shaped sheets in front of the distributaries. Beyond the distal edges of these bottomset deposits, and also underlying the entire delta, are older bay deposits consisting of homogeneous mixtures of light gray sand, silt, and clay that have been thoroughly mottled as a result of burrowing. (See Figure 5-16, A.)

At the base of the deltaic sequence, which overlies the poorly sorted bay sediments that also are generally fine textured and locally as thick as 24 m, is about 0.5 m of bottomset silty clay. Above are the well-laminated and cross-laminated foreset sands and -silts, about 2 m thick. At the top are the various members of the topset complex. These range from thin sheets of silty clays that accumulated in small interdistributary bays and marshes, to incised channel-floor prisms, and lenticular natural-levee deposits, chiefly root-disrupted laminated silty clays. Other silty clay fills the parts of abandoned channels where water formerly flowed. Some prisms of channel-floor sands have cut down completely through the foreset sediments and are in contact with the bottomsets;

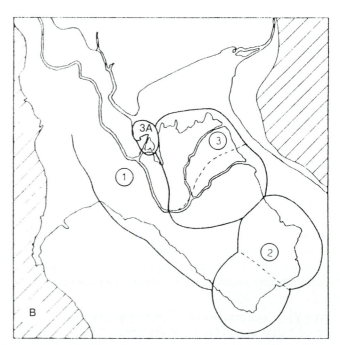

FIGURE 13-18.   Views of small delta built into "shallow" water and having bed-load sand prisms of normal thickness, Guadalupe Delta, San Antonio Bay, Texas.

A. Location map, with inset for regional setting, and schematic profile along line A-A' paralleling axis of San Antonio Bay. On map, stippled areas indicate Holocene sediments; diagonal ruled lines, areas of Pleistocene sediments. On section, stippled areas designate barrier sediments; closely spaced vertical lines, sediments of Guadalupe delta; and short horizontal lines, sediments of San Antonio Bay.

B. Sketch maps of successive lobes of Guadalupe Delta, numbered from oldest (no. 1) to youngest (nos. 3 and 3A). (After C. A. Donaldson, R. H. Martin, and W. H. Kanes, 1970: A, figs. 2 and 4, p. 110; B, fig. 4, p. 112.)

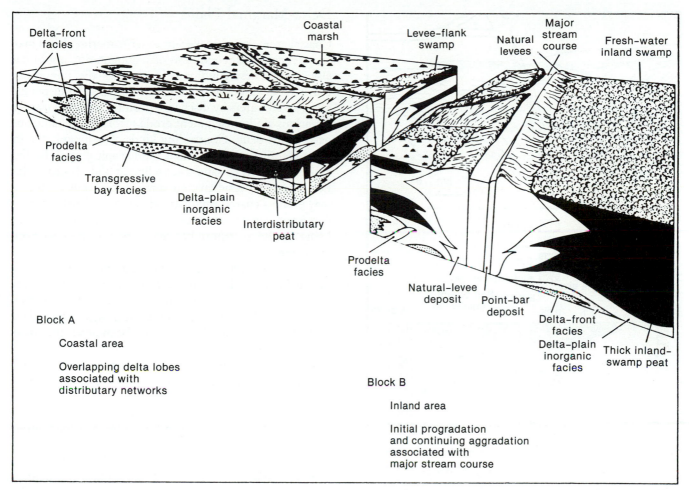

FIGURE 13-19.   Facies relationships on a delta of a large river having natural levees at margins of distributaries; schematic block diagram. (D. E. Frazier, 1967, fig. 3, p. 291.)

others cut completely through the sediments of that particular delta lobe (Figure 13-19). The channel-floor prisms were emplaced during floods. The average thickness of the sheetlike topset deposits is about 1 to 1.5 m, whereas that of the incised channel-floor prisms is as much as 5 m.

Away from a channel-floor sand prism, sheetlike topset silty clays directly overlie the foresets. At such localities the particle size changes from fine (silty clay) at the base to slightly coarse (silt) in the middle and back to fine (silty clay) at the top. In sequences through a channel-floor sand prism, the overall size gradation is from fine (silty clay) at the base to coarse (fine sand) at the top (Figure 13-20). However, within the channel-floor prism, the gradation typically is from coarser sand at the base to finer sand or silt at the top.

SOURCES: Bernard, Major, Jr., Parrott, and LeBlanc, Sr., 1970; Gilbert, 1890; Postma and Roep, 1985; Stanley and Surdam, 1978.

"Bird-Foot" Deltas: Bed-Load Sand in Greatly Overthickened Pods.   The term "bird-foot" delta has been applied in two ways: (*1*) To *the many deltas having the same general planimetric outlines as that of the modern Mississippi delta,* and (*2*) To *the few deltas having shapes resembling that of the modern Mississippi delta, but in addition also possessing overthickened pods of bed-load sand* (Figure 13-21). The author of the term, H. N. Fisk (1908–1964), restricted "bird-foot" delta to those deltas containing overthickened bodies of bed-load sand. In this book we shall follow Fisk's usage.

When he proposed the concept of a "bird-foot" delta, Fisk inferred that the bed-load sand bodies were continuous prisms which had formed (1) by prograding of the seaward sides of the *bars at the mouths of the main distributaries* and (2) by foundering of these bar sands into the underlying hydroplastic sediments. Accordingly, Fisk termed these *"bar-finger sands."*

Information based on additional borings indicates that these thick sand bodies are not prisms, but pods.

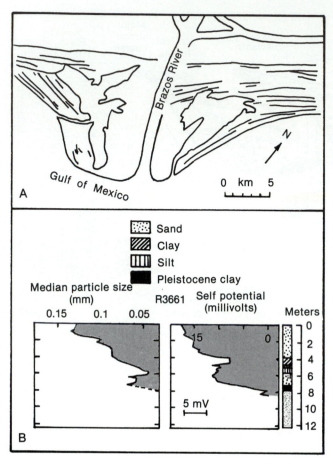

FIGURE 13-20.   Upward-coarsening sequence through small shoal-water marine delta illustrated by lithologic log and self-potential log in boring R3661, through new Brazos delta, S of Freeport, Texas. For location see Figure 13-5.

A. Location map (dated 1954) of boring R3661 on new Brazos delta, which began in 1929 after U.S. Army Corps of Engineers dredged a new channel west of Freeport.

B. Self-potential (SP) curve (edge of shaded area) alongside lithologic log of boring R3661. Increase in particle size upward is well shown by upward increase in self potential. Despite the overall upward coarsening, however, the lithologic log does not represent a simple clay-to-silt-to-sand transition. (After H. A. Bernard, C. F. Major, Jr., B. S. Parrott, and R. J. LeBlanc, 1970, fig. 41.)

(See Figure 13-21.) The origin of these sand bodies is still somewhat enigmatic. Without doubt they result more from rapid progradation of large quantities of the channel-floor bed load during floods than from day-to-day deposition on the seaward face of the distributary-mouth bar. Because they definitely are not fingers and may not have been deposited on the distributary-mouth bars, we have not used the term "bar-finger sands."

SOURCES: Coleman, 1988; Coleman, Suhayada, Whelan, and Wright, 1974; Curtis, 1970; Frazier, 1967, 1974; Holle, 1952; Kosters, 1989; H. H. Roberts, Adams, and Cunningham, 1980; Tye and Kosters, 1986.

Delta-Marginal Plains

Coastal areas that are adjacent to large marine deltas may experience special conditions of sedimentation not found elsewhere. These conditions result from the alongshore transport of suspended sediment that has been dispersed into coastal waters by discharge from the distributaries or of various kinds of sediment eroded from the subaqueous parts of the delta. Contrasting kinds of sedimentation in areas marginal to large deltas are (1) spit growth by the alongshore drift of sand-size sediment and (2) progradation of coastal mudflats. In some circumstances, waves may erode a coastal mudflat and construct small beaches consisting chiefly of skeletal debris from invertebrates living on the nearshore bottom or within the nearshore sediments.

In coastal southwestern Louisiana and adjacent parts of northeastern Texas, *a nearly flat plain has been constructed by seaward growth of the coast as a result of addition of sediments derived mainly from the Mississippi delta and the nearshore gulf bottom.* This area has been given the name of "chenier plain."

The word chenier is from the French, meaning oak tree. Chenier was applied because oak trees, which grow on the low beach ridges that extend across the area (Figure 13-22), are the most-conspicuous components of the vegetation. After comparable features lacking abundant oak trees were found near other large modern marine deltas, a decision was made to change the name used as a general term to **delta-marginal plain**, defined as *a low-lying coastal region, marginal to a large marine delta, in which many parallel- or subparallel beach ridges,*

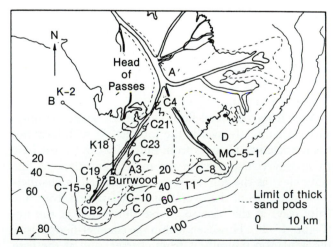

FIGURE 13-21.   Overthickened pods of channel-floor sands. Mississippi's "bird-foot" delta.

A. Outline map of "bird-foot" delta showing locations of borings and lines of profiles and sections. (Map and limits of thick sands from J. M. Coleman, J. N. Suhayada, T. Whelan, and L. D. Wright, 1974, figs. 3 and 6, p. 52 and 53.)

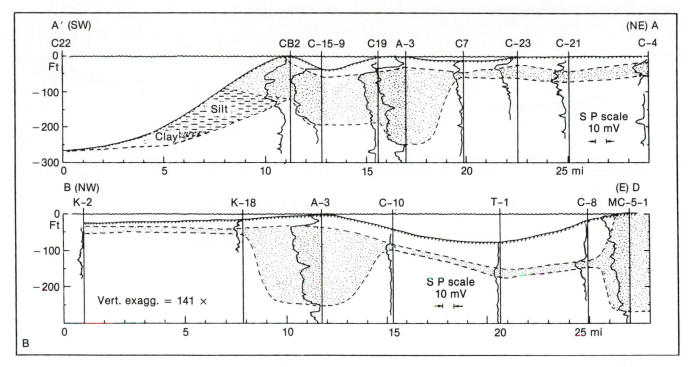

FIGURE 13-21. (*Continued*)

B. Profiles and sections based on samples and SP logs of borings. The SP logs show characteristic upward-coarsening gradation in particle size. Compare patterns of SP logs penetrating thick sand bodies with SP logs penetrating thin sand bodies. The tops of the sand bodies extend upward above sea level. The bases extend downward, locally forming isolated pods. (Redrawn from data provided from Louisiana State University, Coastal Studies Institute, courtesy J. M. Coleman.)

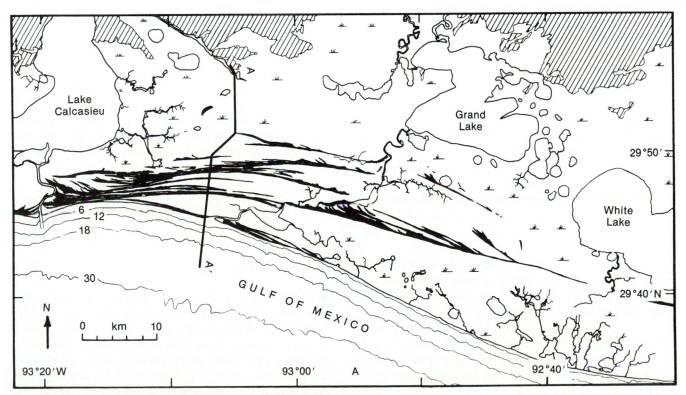

FIGURE 13-22. Map of portion of delta-marginal ("chenier") plain, southwestern Louisiana, and stratigraphic section based on line of closely spaced borings. For location, see Figure 13-13.

A. Map showing generalized distribution of surface sediments. Closely spaced diagonal parallel lines, Prairie Formation (Pleistocene); black, beach ridges of Holocene age; swamp pattern, coastal mudflats and marshes of Holocene age. Line A-A′ locates stratigraphic section of B. Depth contours in feet.

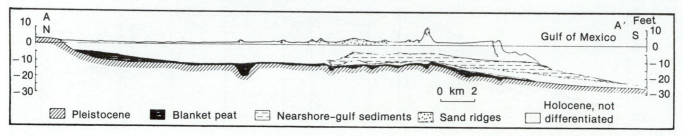

FIGURE 13-22. (*Continued*)

B. Profile and section along line A-A' of A, based on numerous closely spaced borings to depth of eroded Prairie Formation and careful paleoecologic analysis of Holocene sediments. The initial deposit made by the advancing sea is the so-called blanket peat (black). It is overlain by open-gulf silt, indicating that the breaker zone shifted abruptly from a location south of the basal peat to the northern end of the profile. Thereafter the breaker zone shifted southward, as spits grew along the coast and the low areas between spits were filled with silt, clay, and marsh deposits. (After J. V. Byrne, D. O. LeRoy, and C. M. Riley, 1959, pls. 1 and 2.)

commonly including abundant skeletal debris of nearshore marine invertebrates, alternate with intervening low areas underlain by fine-textured sediments, chiefly silts and silty clays.

The origin of the delta-marginal plain of southwestern Louisiana and adjacent parts of Texas has been disputed. One school contends that the conditions of deposition in the delta-marginal plain have been closely controlled by the direction of discharge of the Mississippi distributaries. According to this view, which is the dominant one, when the Mississippi discharges southward, abundant mud progrades the delta-marginal plain and builds mudflats. By contrast, when the Mississippi discharges eastward, fine-textured sediments become scarce along the delta-marginal plain and waves can erode the bottom. The waves build beach ridges with this eroded sediment.

A contrasting view relates the variation in behavior of the delta-marginal plain to the kind of sediment discharged from the Mississippi River. Adherents of this view argue that the important variable is not direction of discharge, but rather what is discharged. When sand appears, during floods, the coast progrades with sand. Most of the time, however, only mud is available, and the coast progrades with mud.

Although students of the area have acknowledged that spits are present, no one seems to have appreciated the significance of spits. If, indeed, the beach ridges grew chiefly by spit elongation, as is suggested by their plan views (See Figure 13-22.), then open water must have been present to enable spits to grow. Spit growth would also account for the subsurface stratigraphic relationships shown by a line of closely spaced borings. (See Figure 13-22.)

A coastline resembling the "chenier plain" of southwestern Louisiana occurs in Suriname, northeastern South America. The Suriname shelf is characterized by very high suspended-sediment concentrations (upwards of 1000 mg/l) and by huge mud banks, 10 by 20 km in size, which are oriented at a large angle to the coast, and which migrate westward alongshore in response to winds and waves. The interbank areas experience coastal erosion and winnowing with the result that coast-parallel ridges of very-fine sand are constructed, which are separated by swales underlain by clay and silt. The Amazon River is the source of the fine-textured sediment that composes the subaqueous mud banks and the coastal plain, which therefore is a delta-marginal plain. However, the construction of this delta-marginal plain differs from that inferred for the delta-marginal plain of southwestern Louisiana. In the case of Suriname, the sand ridges are inferred to be beach ridges formed in interbank areas by the winnowing action of waves and currents on the foreshore. As the mud banks migrate alongshore, so too do the zones of active beach-ridge construction, with the result that parallel elongate ridges of very fine sand are formed (Figure 13-23).

SOURCES: Augustinius, 1980; LeBlanc, 1972; Rine and Ginsburg, 1985; J. P. Wells and J. M. Coleman, 1981a, b.

Marine Deltas

In this section we examine a few examples of modern marine deltas. Many studies of modern deltas are available from which to select examples to analyze and compare. We have chosen three that display widely contrasting characteristics. These are (1) Atchafalaya's pre-1972 submerged marine delta, (2) Mississippi delta, and (3) Rhône delta in the Mediterranean. We also briefly review the post-1972 development of the Atchafalaya's marine delta. Doubtless many geologists believe that the Atchafalaya's pre-1972 submerged marine delta is not even a delta at all. Therefore they would surely look upon it as a

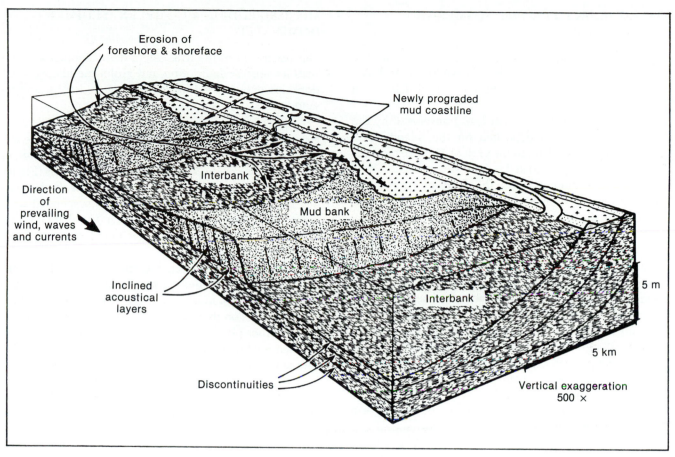

FIGURE 13-23.   Idealized block diagram showing how sediment is being added by two mechanisms: (1) additions to the shoreface (concave-up lines on front panel of block,) and (2) migration
of mud banks parallel to shore (within which are inclined layers that appear on seismic-reflection
profiles (labeled inclined acoustical layers)). Waves build thin bodies of beach sand parallel to
shore. (J. M. Rine and R. N. Ginsburg, 1985, fig. 13, p. 650.)

surprising choice; they and other readers may wonder why we have even mentioned it. We have included this strange and atypical deposit because we think it conveys an important message: a river can deposit a subaqueous delta bulge that consists of fine-textured sediment lacking sand, contains no subaerial topsets, and has not been modified by the advance of any distributary channels. In addition, this odd delta is destined to become the precursor of a future lobe of the giant Mississippi delta. We have included the Mississippi delta for the obvious reasons of its location; because it has been studied for many years and in great detail, both on the surface and by means of borings; because it is the deposit of a major river, which has contributed a mixture of suspended- and bed-load sediments to a relatively protected body of water; because the deposits of its distributaries display how growth into "shallow" water (<20 m) contrasts with growth into "deep" water (>20 m); and because, on the foreset slope of the modern "deep"-water delta, individual layers of sand are not being transported into deep water but instead are forming thick pods by foundering into

the thick fine-textured sediments forming the foresets and bottomsets. This is an example of a delta whose form is dominated by sediment input. We discuss the Rhône delta in the Mediterranean, a wave-dominated delta, because of its high proportion of sand; its seaward growth pattern, which includes many beach ridges; and its laterally persistent sand layers among the foresets, each such layer being the product of a flood. No example of a tide-dominated delta is described, though some of the world's important rivers enter the sea in tide-dominated coastal regions. In coastal areas dominated by very high tidal ranges, even rivers carrying large volumes of sediment, such as the Ganges-Brahmaputra River, tend to produce estuaries rather than deltas. However, the open funnel-shaped estuaries characterized by tidal sand ridges, that form at the mouths of rivers dominated by tidal currents, are classified by some as tide-dominated deltas. Examination of continuous-seismic-reflection profiles and nautical charts at river mouths of many estuaries reveals prominent progradational bulges below water forming subaqueous deltas.

ATCHAFALAYA'S PRE-1972 SUBMERGED MARINE DELTA

The Atchafalaya River has built two deltas. One, the inland delta, is located in Grand Lake-Six Mile Lake. (See Figure 14-41.) The second, the coastal delta, is in the Gulf of Mexico.

Until 1972 all the sandy bed load transported by the Atchafalaya River was deposited on the inland (lake) delta. From the lake delta to the Gulf of Mexico, the river sediments formerly consisted only of silt and clay. These fine-textured sediments built the floor of Atchafalaya Bay up to an equilibrium level. The fine-textured sediment from the river was being delivered continuously to the bay, but all of it formerly passed through the bay and built an unusual delta in the Gulf of Mexico (Figure 13-24).

A remarkable feature of the Atchafalaya's pre-1972 submerged marine delta was its lack of subaerial topsets. Just as the waves did in Atchafalaya Bay, the waves of the Gulf of Mexico built an equilibrium surface across the prograding mound of sandless deltaic sediment. The Atchafalaya's pre-1972 submerged marine delta consisted entirely of clay-size sediment. Seen in the rock record, the lithified equivalent of such a delta would probably not even be considered a delta; it would be a shale or a claystone about 3 m thick.

In 1972 the Atchafalaya's inland delta had built completely across the two lakes and began to discharge sand into Atchafalaya Bay. This change was initiated when unusually high flood discharges in 1972–1975 caused large volumes of fine sand to be carried in suspension into Atchafalaya Bay. Subaerial sandy topsets of the Atchafalaya marine delta have grown rapidly, and the bay may be completely filled by about the turn of the century. The Atchafalaya marine delta now strongly resembles a young lobe of the Mississippi River delta complex, described next.

SOURCES: H. H. Roberts, Adams, and Cunningham, 1980; L. M. Smith, Dunbar, and Britsch, 1986; W. C. Thompson, 1955; Tye, 1987; Van Harden, 1983.

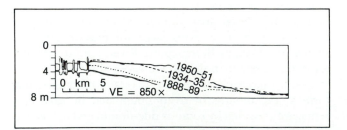

FIGURE 13-24.   Stages of seaward growth of Atchafalaya submerged marine delta, consisting of silty clay, shown by profile-section at right angles to shore. Further explanation in text. (W. C. Thompson, 1955, fig. 6, p. 61.)

MISSISSIPPI DELTA COMPLEX (SEDIMENT DOMINATED)

Our discussion of the Mississippi delta complex emphasizes its general description and Holocene history, relationships between sediments and water depths, and its sand bodies.

General Description and Holocene History.   The delta complex of the Mississippi is being built into the Gulf of Mexico by one of the world's major river systems (Figure 13-25). The Mississippi's annual fresh-water discharge is 600 km^3 and its mean concentration of suspended sediment is 833.3 mg/l. (See Table 13-1.) The delta consists of lobes, one or two active at a given time, covering about 30,000 km^2 apiece.

One of the prominent characteristics of the Mississippi delta is that the channel-marginal fine sandy natural levees extend continuously from the lower reaches of the meander belts on the alluvial plain onto the distributaries of the topset plain of the delta complex (Figure 13-26). The natural levees flanking an active distributary channel extend down to the shore; as the channel lengthens, the levees grow right on out into the sea. Because the natural levees are more or less the same everywhere, they serve as useful sandy ribbons tying together the interchannel

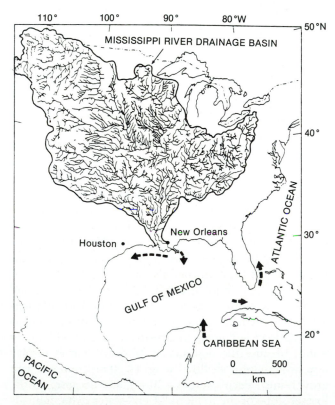

FIGURE 13-25.   Map showing drainage network of the Mississippi River. Dashed arrows signify currents. (J. M. Coleman, 1988, fig. 1, p. 1000.)

fresh-water overbank environments of the alluvial flood-plain with the interdistributary salt-water environments of the marginal- and shallow parts of the sea. Although the change from nonmarine to marginal marine involves some distinct differences, in general, no sharp boundary separates the deposits of the valley-floor overbank fresh-water environments from those of the coastal marshes and shallow marginal-marine bays.

The entire Gulf coastal region through which the natural-levee ridges wind their way is a vast morass of swamps, shallow lakes and channels (*bayous* in Louisiana language), coastal marshes, and bays. In these swamps and marshes, plant life is abundant. The dead plant material accumulates to form widespread peat layers. Deltaic peats may, with burial, become important coal deposits or hydrocarbon source rocks. During great storm surges, the rise of sea level can inundate vast areas of the topset plain of the delta as well as the lower parts of the Mis-

sissippi alluvial plain. In this way, a thin layer of marine organisms and marine sediment can be spread out in areas where otherwise such organisms and sediment would not be found.

Oceanographic measurements made east of the Mississippi delta have shown that the water of the Gulf of Mexico contrasts sharply with that discharged from the Mississippi River. Away from the delta, or near it at depths of greater than 15 m, open-gulf water is transparent, typically containing only trifling amounts of suspended sediment ($\pm$ 50 mg/l), and its salinity is 35‰ (dark-gray and black areas in Figure 13-26). The suspended sediment of inshore Gulf water depends on wave activity. During storms such water may contain 600 to 700 mg/l of suspended sediment, more than is present in the river water during most times of the year. The salinity of the river water is essentially zero and its load of suspended sediment varies from as little as 25 mg/l at low stages to about 400 mg/l or more during high water. An important problem not yet solved concerns

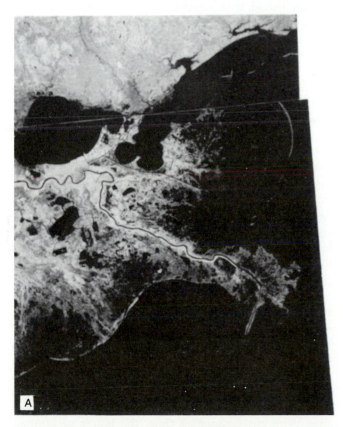

FIGURE 13-26.   Mississippi delta, northern Gulf of Mexico, viewed from ERT satellite (whose name has since been changed to LANDSAT) on two different days, 16 January 1973 (B) and 08 October 1974 (A). Image was made from computer-enhanced signals generated by closely spaced linear sweeps of multispectral scanner (MSS). (NASA and U.S. Geological Survey.)

A. Natural levees (narrow white strips along river channel) are emphasized in image in which water shows as solid black. This image also displays two currently inactive delta lobes (at left and at top), which are being submerged and reworked by the sea. (Compare Figure 13-13, E.) MSS band 7.

B. Suspended sediment (light gray areas in water), discharged from distributaries, spreads along front and sides of delta. Two plumes are being transported alongshore toward the northwest. MSS band 5.

the mechanism by which the sediment suspended in the outflowing river water reaches the bottom to enlarge the delta. Presumably this happens during storms or floods. At low stages of the river and during fair weather, nearly opaque fresh water and its load of suspended sediment are underlain by nearly transparent Gulf water having a minuscule content of suspended sediment.

Surveys made during the flood of February 1973 showed that complex processes operate during high water. The height of the distributary-mouth bar increased by 2.5 m above its low-water height for a length along the channel of about 1 km. Using 160 m for the width of the bar, the calculated volume of sediment that was added to the crest of the bar is 200,000 m^3. Intense flow of water was directed toward the bar crest from both the seaward- and landward sides. Such dual flow toward a bar is analogous to the conditions in a breaker bar (Chapter 11). After the flood crest had moved out of the lower reaches of the river, waves and currents eroded sand from the flood-stage bar and transferred the sand to the seaward slope of the bar. Thus, during the flood, the crest of the bar grew upward, but as soon as discharge returned to normal, the seaward face of the bar was prograded rapidly seaward.

The topset parts of the Mississippi delta complex are continuous with the alluvial deposits at the south end of the Mississippi's alluvial valley. This valley is a relatively flat-bottomed and steep-sided feature, about 80 km wide, that extends from the latitude of Baton Rouge, Louisiana (30°25′N), northward for many hundreds of kilometers. On the floor of the alluvial valley are two prominent meander belts, each about 10 km wide. These are (1) the belt in which the present Mississippi River flows, located along the east side of the valley floor, and (2) the Teche belt, now inactive, situated along the western side of the valley floor. In between these two meander belts is the low-lying Atchafalaya basin, which is about 50 km wide and 200 km long. (See Figure 14-41.) A future meander belt of the Mississippi River will form here.

The history of the Mississippi delta complex during the past 10,000 years is known in detail from study of hundreds of borings, from numerous ^{14}C dates on peats, and from stratigraphic analysis based on recognition of the three fundamental kinds of deposits: (1) progradational, (2) aggradational, and (3) marine (including marginal marine). Early in the Flandrian submergence of what is now south Louisiana, the sea paused at a relative depth of about −15 m with respect to present sea level. During the interval that extended from about 10,000 years B.P. to 6500 B.P., the river, flowing in the Teche meander belt, deposited the Maringouin delta complex. (See Figure 13-13, A.) About 6200 B.P., rapid submergence resumed and the sea formed a broad estuary that extended northward at least to Baton Rouge.

(See Figure 13-13, B.) Since 6200 B.P., no other estuaries have formed; the history has been one of delta growth. Subsequently, the river has constructed 16 delta lobes. The earliest delta was built along the west side of the estuary by a river flowing in the Teche meander belt. This lobe prograded rapidly southward into water about 12 m deep; the lobe was soon capped with a deltaic plain. The first of the modern Teche open-gulf delta lobes was built near the present coast. (See Figure 13-13, C.) From about 5650 to 4650 B.P., a gradual diversion took place at a location situated about 80 km northwest of Baton Rouge. The discharge into the eastern-, or present, meander belt increased steadily. As a result, a delta prograded down the east side of the estuary. This delta pushed the shoreline southward and extended as a long, narrow lobe eastward past New Orleans. Approximately 3900 B.P., the last of the delta lobes that was fed from a river flowing in the Teche meander belt became inactive. (See Figure 13-13, D.)

After it had abandoned the Teche meander belt, the Mississippi bifurcated again into two main distributaries. This time, the point of diversion was located near Donaldsonville, about 100 km WNW of New Orleans. Each main distributary began building a series of delta lobes. The distributary that flowed eastward past New Orleans built the St. Bernard series of delta lobes; the distributary that flowed southward built the Lafourche series. (See Figure 13-13, E.) At about 3500 B.P., the St. Bernard distributary bifurcated near New Orleans. The new southern distributary built a narrow delta lobe southward in a location lying just west of the modern delta. This lobe became inactive about 2000 years ago (lobe 7, 3400 to 2000 B.P.). About 800 years ago, when the entire discharge of the Mississippi broke through the natural levees near New Orleans and built the broad Plaquemines lobe (the modern lobe, lobe 13), the St. Bernard delta complex became inactive. About 150 years B.P., the modern delta grew southeastward completely across the shelf. (See Figure 13-13, F.) Flow in the Lafourche channel ceased when the north end was dammed off in 1904.

Since about 6200 B.P., the behavior of the Mississippi River can be summarized as consisting of episodes when the entire discharge was concentrated in a single trunk channel and intervening episodes when the discharge has been subdivided into two distributaries of nearly equal size. With the passage of time, the points of bifurcation have migrated southeastward. At the present time, a new bifurcation is developing at a point not far from the most inland of the three previous bifurcations. The Atchafalaya River is slowly but surely gaining discharge at the expense of the Mississippi.

The ultimate effect of the increased flow into the Atchafalaya basin will be to construct a new meander belt down the central part of the alluvial valley. As mentioned previously, a new delta lobe has started to grow in the Gulf of Mexico as a result of the discharge of sand from the Atchafalaya River. As a matter of fact, were it not for the efforts of the U.S. Army Corps of Engineers to prevent this very thing from happening, the primary locus of sediment deposition might have already switched from the present Mississippi delta to the nascent Atchafalaya delta.

Relationships Between Sediments and Water Depths.   As in many other deltas built chiefly of suspended-load sediments, the bulk of the Mississippi delta consists of silt and clay. In view of the average low-stage suspended load, which consists about equally of silt (48%) and clay (50%) and a minor amount of fine sand (2%), this is not surprising.

The sediments accumulating on the delta vary with depth of water. This variation is evident both in the particle-size distribution (Figure 13-27) and in primary sedimentary structures (Figure 13-28).

The bottomsets consist of clay and the foresets of silt and fine sand. (See Figure 13-27.) As the delta lobe advances seaward, the partial halo of bottom-

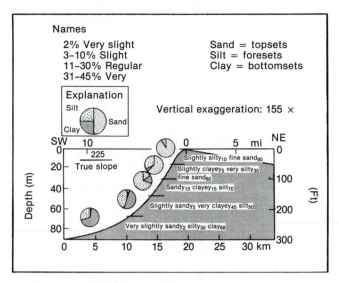

FIGURE 13-27.   Schematic profile, at right angles to shore, extending seaward from Southwest Pass, Mississippi delta, showing progressive decrease in particle sizes of bottom sediment with depth and distance from the distributary-mouth bar. (Based on A. C. Trowbridge, 1930: profile, fig. 12, p. 892; size analyses, figs. 10 and 11, p. 890 and 891.)

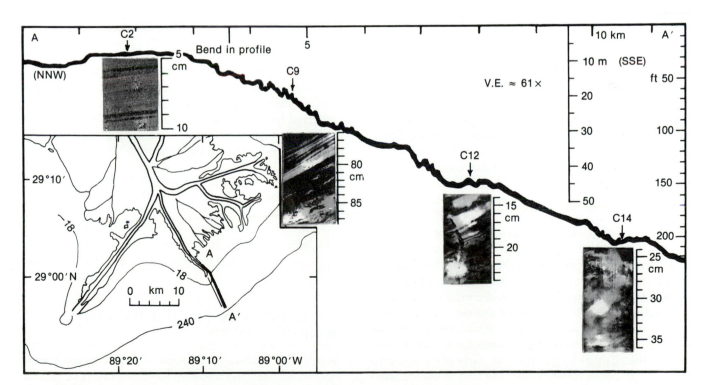

FIGURE 13-28.   Profile of bottom and primary sedimentary structures in cores of bottom sediment off South Pass, Mississippi delta, along A-A' of inset map. Core numbers shown above profile; depths beneath sediment/water interface of core segments illustrated shown by numbers at right of cores. Core C2 is a photograph of core sediment; cores C9, C12, and C14 are X-radiographs of cores. (Prepared from material furnished by J. M. Coleman, Coastal Studies Institute of Louisiana State University.)

set clays spreads over the adjacent sea-floor sediments. In many places the nondeltaic sea-floor sediments are coarse sands, but locally, algal limestones are found. Similarly, the silty lower foresets spread over the inner edges of the bottomset clays already deposited. In this way, a fixed upward-coarsening sequence of strata, consisting of clay grading upward into silt, is deposited. (See Figure 13-27.)

The primary sedimentary structures in the bottomset clays consist of features made by organisms; nearly all vestiges of any stratification have been destroyed (core C14, at right in Figure 13-28). Stratification appears in the silty foresets and becomes progressively more apparent upward, as illustrated by cores C12 and C9 in the middle of Figure 13-28. The inclination of the layers in cores C12 and C9 probably resulted from the effects of slumping. Gravitational rearrangement of sediments on the front of the Mississippi delta and other deltas composed of fine-textured sediment has been found to be a major process. Creep, slumping, and turbidity currents are all common bulk-sediment-transport processes on the fronts of fine-textured deltas. The irregularities of the bottom profile shown by the echogram (See Figure 13-28.) are thought to be the result of slumping. The topset sands of the crest of the distributary-mouth bar (core C2 at left in Figure 13-28) display prominent lamination.

Sand Bodies.   In the Mississippi delta complex, sand is deposited in two contrasting ways. These are (1) as elongate, thick ($>$10 m) prisms or lines of pods that radiate outward from the coast and that are products of seaward growth of distributary channels and their flanking natural levees, and (2) as thin ($<$5 m) sheets that may contain local thicker parts in the form of elongate bodies parallel to shore and that are products of the effects on inactive delta lobes of shoaling- and breaking waves.

The channel floor in the lower reaches of the main channel of the Mississippi River consists of a series of giant dunes, some having relief of 30 m or so (about 100 ft). These dunes imply that bed-load sediment is being driven by traction along the floor of the channel. What is not known is the condition of these dunes during floods and the depth to which the sand is mobilized by flood flows. Such large dunes are not present from mile 29 to the ends of the distributaries. Presumably the seaward end of the field of large dunes will shift downriver as the channel lengthens. The parts of the channel having large dunes coincide with the area of older shoal-water delta lobes.

Because the thickness of such shoal-water delta lobes is only about 20 m, the channel of the Mississippi River between mile 100 and mile 20 cuts completely through the deltaic deposits across which it is flowing and has also incised itself at least another 40 m into whatever

sediment lies below. In these localities the thickness of the channel sand bodies bears no relationship whatever to the thickness or the sequence of deltaic sediments. Moreover, the advancing channel has destroyed some of the deltaic sediments and in their places has deposited an elongate body of sand.

From the Head of Passes seaward, the Mississippi has deposited "deep-water" deltas. Presumably, the weight of the sand piled up on the distributary-mouth bars during floods is so great that the sand founders into the thick underlying hydroplastic silts and -clays, thus building pods. (See Figure 13-21.) Because of the presumed foundering of the sand that piles up on the distributary-mouth bars during floods, not much, if any, sand flows as sheets down the front of the delta.

The effects of waves that accompany the reworking of inactive delta lobes are well shown on Figure 13-26, A, by the Chandeleur Islands (thin, arcuate strips of sand north of the Head of Passes) and by the barrier islands west of the Head of Passes. As the breaker zone migrates across an inactive delta lobe from the combined effects of marine erosion and subsidence, the waves disperse the fine-textured sediments and leave a sheet of sand as a lag deposit, commonly forming barrier islands. If any actual barrier islands or substantial remnants of barrier islands become preserved, they would form elongate sand ridges that trend parallel to the shore. Ultimately, during submergence, the reworked barrier complex forms a sand shoal. This shoal is further reworked by marine processes until it is buried or until submergence removes it from the realm of active sand movement. Barrier islands, like the Chandeleur Islands (See Figure 13-26.), form and are preserved in areas of abandoned deltaic lobes through a combination of reworking (a horizontal movement) and submergence (a vertical movement).

SOURCES: J. M. Coleman, 1988; J. M. Coleman, Suhayada, Whelan, and L. D. Wright, 1974; Curtis, 1970; Farrell, 1987; Frazier, 1967, 1974; Holle, 1952; Kosters, 1989; Penland, Boyd, and Suter, 1988; Roberts, Suhayada, and J. M. Coleman, 1980; Trowbridge, 1930; Tye and Kosters, 1986; L. D. Wright and J. M. Coleman, 1974.

RHÔNE DELTA IN THE MEDITERRANEAN (WAVE DOMINATED)

As with the Mississippi delta, we shall discuss the Rhône delta in the Mediterranean under the headings of general description and Holocene history, relationships between sediments and water depths, and sand bodies.

General Description and Holocene History.   The Rhône River is building a marine delta into the Mediterranean Sea (Figure 13-29, A) where wave action is moderate

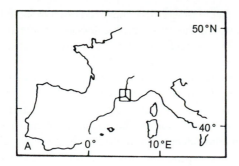

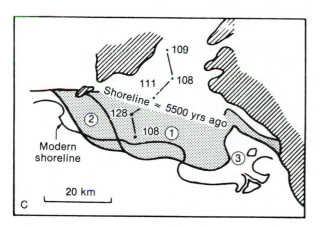

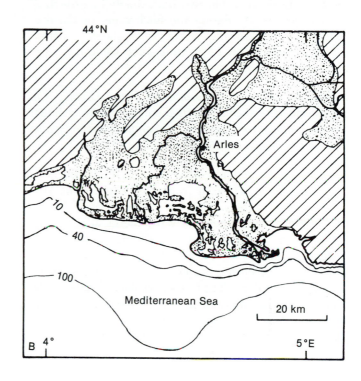

FIGURE 13-29.   Rhône delta in the Mediterranean.

A. Regional index map, showing location of map shown in B.

B. Generalized map of lower Rhône River and adjacent Mediterranean Sea. Isobaths in meters. Holocene sediments stippled; areas of bedrock, diagonal parallel lines.

C. Progressive southward growth of Rhône delta lobes during the past 5500 years. Line segments and numbers of borings locate subsurface section shown in Figure 5-40. Numbers 1, 2, and 3 signify successive lobes. (P. C. Scruton, 1960; B, fig. 11, p. 905; C, fig. 16, p. 101.)

and the tidal range negligible. The slope of the bottom near the delta is such that depths of 40 m are found at distances of 4 to 6 km out from shore. (See Figure 13-29, B.) The mean annual fresh-water discharge of the Rhône River is 52.7 km³/yr and the mean concentration of suspended sediment is 597.7 mg/l. (See Table 13-1.)

As did the Mississippi during the Holocene Epoch, the Rhône has constructed a complex consisting of more than one delta lobe. Since about 5500 years ago the Rhône has built three lobes. (See Figure 13-29, C.) The generalized Holocene history of the Rhône delta complex can be illustrated by means of a subsurface stratigraphic section compiled from borings. (See Figure 5-40.)

In the northern half of the diagram, the strata are inferred to be entirely of nonmarine origin. The only suggestion of changes in regimen are the four buried soils. By contrast, in the southern half, deltaic strata predominate and it is possible to recognize four depositional events. From oldest to youngest, the parasequences developed as follows: (1) Deposition of coarse valley-fill al-

luvium, controlled by a level of the Mediterranean that was at least 60 m lower than at present. (2) Deposition of fine-textured fluvial sediments by flood-plain rivers having gradients much reduced by the rising sea level. (3) Rapid submergence and resulting delta progradation during slight submergence. The rapid submergence of 24 m, about 5500 years B.P., shifted the shoreline 18 km inland to the vicinity of boring 111. This allocyclic shift began a new parasequence. During delta growth, the shore was prograded about 20 km southward into water about 35 m deep. As the deltas grew, a gradual submergence of about 4 m took place.

After the water depth had decreased to only a few meters, additional prograding seems to have been by means of coastal barrier islands or mainland beaches. Because of the net submergence of 4 m during progradation (from −10 to −6 m with respect to modern sea level), coastal backbarrier fine-textured sediments were able to bury the coastal-barrier sands. (4) A second rapid submergence, of only 4 m this time, shifted the shoreline about 13 km northward and started a new parasequence. Again, the coast prograded, but this time chiefly by growth of coastal barriers. As during the final stages of the preceding cycle, a gradual submergence of about 2 m (from −2 m to present sea level) took place while the shore prograded by about 13 km.

Relationships Between Sediments and Water Depths. On the Rhône delta, the fair-weather conditions are such that sand is confined to water depths of about 4 m and less. At greater depths the sediments consist of silt and clay.

Sand Bodies. The sand bodies found in the Rhône delta complex differ significantly from those found in the Mississippi delta complex. This is a consequence of the relatively greater importance of wave energy affecting the Rhône River delta. Because the Rhône delta complex prograades significantly by the outgrowth of beaches with formation of beach ridges and barrier bars, the constructional phase of the delta's history results in coast-parallel sand sheets. In addition, in the Rhône delta, nothing resembling the Mississippi's massive channel-sand pods has been found. Submergence of inactive lobes and reworking by the waves gives the same result in both delta complexes: deposition of sheets of lag sand containing marine fossils.

The foresets of the Rhône delta in the Mediterranean and the Mississippi delta differ strikingly. Among the foresets of the Rhône delta are found numerous persistent layers of fine sand; each such sand can be correlated with a flood-stage discharge of the Rhône River. By contrast, sand layers found in closely spaced cores among the foresets of the Mississippi delta cannot be matched one with another. They may have never been spread out widely or perhaps were formerly in sheets but these were broken by subaqueous slumping. The deposits of the Rhône River delta are less extensive than those of the Mississippi River delta, because the Rhône River contributes less fine-textured sediment and because wave reworking removes much fine material from the deposits of the Rhône River delta and distributes it widely throughout the basin.

SOURCES: Oomkens, 1970; Scruton, 1960; van Straaten, 1960.

CHARACTERISTICS OF DELTAIC SEDIMENTS

Delta deposits differ drastically depending upon the abundance and distribution of sand. The coarse Gilbert-type deltas, with their distinctive steeply dipping foresets, are considered to be so characteristic as to require no further comment. The problem is how to recognize fine-textured deltas in which the initial dip of foreset beds is negligible.

As a general rule, deltaic deposits overlie marine strata and pass upward into nonmarine strata. After many delta lobes have been prograded, have coalesced, and have been submerged, the stratigraphic succession consists of many complexly interfingered units of marine- and nonmarine strata.

Within the delta deposits themselves, the greatest uniformity prevails in the bottomsets and lower foresets and, to some extent, in the upper foresets. Enormous variability is typical among the topsets; much depends on whether the succession includes distributary channels. The most-typical deposit of a fine-textured delta is evenly laminated silt of the upper foresets (core C9, Figure 13-28). Any unit of silt-size sediment that is underlain by clay that grades up into silty clay that is more than a few meters thick and that is poorly stratified in its lower part and well stratified in its upper part should be suspected of being a deltaic deposit. Such upward coarsening from clay to silt typifies the lower parts of deltaic sequences deposited on fine-textured deltas.

The upper parts of deltaic sequences depend on the locations of the sequences with respect to distributary channels. Away from a distributary, the sediments may be fine textured; they are the products of deposition in bays, marshes, lakes, and fresh-water swamps. On many deltas the typical topset deposit is peat, which may become coal in the geologic record. Where the organic plant material that becomes peat is diluted by fine-textured terrigenous material, organic-rich clays or silts may form instead.

Successions that include distributaries are characterized by sand. The thickness and lateral relationships of the distributary sands depend on the size of the channel and on the depth of water into which the delta prograded. Large distributaries can incise themselves deeply into and even entirely through the deltaic deposits. In "deep"-water "bird-foot" deltas, the distributary-channel sands are deposited as linear groups of thick pods. The distributary-channel sands contain upward-fining successions, as in the deposits of floodplain rivers (Chapter 14), and these sands form linear prisms or lines of pods that branch in a seaward direction.

Once a delta has built a flat topset surface, almost any kind of sediment can accumulate on it. Commonly, alluvial deposits of various kinds are deposited, either in streams or by fans (Chapter 14). Once streams have become established on a topset plain, they may be able to persist during subsidence. If they do, then thick nonmarine sediments overlie the deltaic sediments. In such cases it may not be easy to decide where the delta sediments stop and the other kinds begin.

EXAMPLES OF INFERRED DEPOSITS OF ANCIENT MARINE DELTAS

Armed with the foregoing information about modern marine deltas, let us now turn to the stratigraphic record to describe a few ancient deposits that are considered to be the work of former marine deltas. We include the Cenozoic of the northwestern Gulf of Mexico, United States; the Pennsylvanian of the interior of the United States; parts of the Carboniferous in northern England; and parts of the Middle and Upper Devonian in eastern New York State, United States.

Cenozoic, Northwestern Gulf of Mexico, United States. A belt of fine-textured terrigenous sediments of Cenozoic age can be traced from southern Mississippi to the east side of the Mississippi's alluvial valley and from the west side of this alluvial valley southwestward into Texas. The strata dip toward the Gulf of Mexico; depositional strikes curve parallel with the present Gulf shoreline. Downdip, in the subsurface, the strata thicken greatly. From the information gained by the drilling of thousands of borings made in the search for petroleum, reconstructions indicate that these Cenozoic strata for the most part are best interpreted as the deposits of many deltaic complexes. In northeastern Texas the inferred Cenozoic deltas are concentrated beneath the modern major rivers (Figure 13-30).

In south Louisiana deltaic parasequences evolved starting with a series of overlapping delta lobes that underlie the south end of the Mississippi Valley embayment (See Figure 13-30, B.) and become progressively younger toward the south and east. A marine shale underlies each lobe. That shale is overlain by various silty shales and sands that are inferred to be the delta deposits. These, in turn, are overlain by other sands and shales

that are interpreted as the nonmarine deposits of alluvial plains. After several such lobes had been built side by side, they were submerged rapidly. The sea transgressed northward across the alluvial deposits and left a thin bed of shells and sand at the base of a marine shale. This marine shale passes upward into another thick autocyclic succession that begins with silty shale and coarsens upward, eventually giving way to non-marine strata.

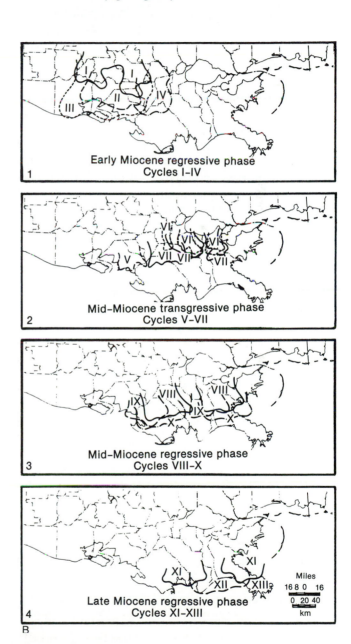

FIGURE 13-30.   Subsurface indications of deltas in Cenozoic strata of northwestern Gulf of Mexico.
A. Rockdale system of deltas in Wilcox Group (Eocene) underlying Texas coast. (After W. L. Fisher and J. H. McGowen, 1967, fig. 6, p. 115.)
B. Inferred succession of delta lobes in subsurface Miocene strata, south Louisiana, reconstructed from data provided by petroleum-exploration borings. The sites of the deltas generally shifted southward with time, and these Miocene subsurface deltas underlie the general area of the modern Mississippi deltaic complex. (D. M. Curtis, 1970, fig. 6, p. 302.)

Many of the sands, particularly those of Miocene age, are prolific reservoirs of petroleum. Many oil pools have formed where the Miocene and other sands have been deformed because of diapiric emplacement of salt plugs.

SOURCES: W. L. Fisher and McGowen, 1967; LeBlanc and Hodgson, 1959b.

Pennsylvanian, Interior United States.   The Pennsylvanian strata of the interior parts of the United States contain interbedded marine- and nonmarine strata. The surface exposures have been studied intensively in connection with the mining of vast deposits of coal. Subsurface study by means of geophysical logs, samples, and cores has been concentrated on the search for petroleum in the sandstones. Beginning in the nineteenth century, numerous investigators have emphasized the regularly repeated patterns of strata that are associated with the coal. Each layer of coal overlies a distinctive clay or shale that occurs at the top of a nonmarine sandstone. Generally, the coal is overlain by a pyrite-bearing black shale that grades up into marine shale and/or marine limestone. These patterned successions were named *cyclothems* and much

study was devoted to the question of their origin and significance. We discuss cyclothems further in a following section.

Detailed mapping of individual Pennsylvanian stratigraphic units in terms of their inferred environments of deposition has indicated that deltas were numerous (Figure 13-31). A series of major rivers drained southward and southwestward and emptied into a seaway that spread across Kansas, Missouri, and Oklahoma. Recent subsurface studies of the lower Pennsylvanian strata of northeastern Oklahoma have concentrated on comparisons between the prisms of sandstone, so well shown on the SP logs from petroleum test holes, and the pods and lenses of sands on the modern Mississippi "bird-foot" delta (Figure 13-32).

SOURCES: L. F. Brown Jr., 1979; Busch, 1974; Kasino and Davies, 1979; R. C. Moore, 1949; Wanless, J. R. Baroffio, Gamble, Horne, Orlopp, Rocha-Campos, Souter, Trescott, Vail, and Wright, 1970; Wengerd, 1962.

Carboniferous, Northern England.   In a significant stratigraphic study made early in the nineteenth century, John Phillips analyzed the patterned successions within the Yoredale Series, of Carboniferous age, in Yorkshire. He

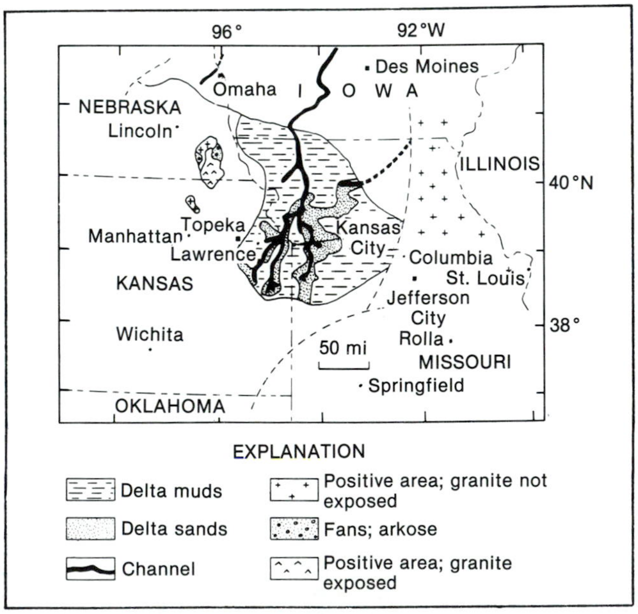

FIGURE 13-31.   Environmental map of Walter Johnson Sandstone Member (and related strata of the same age), Marmaton Group (Upper Desmoinesian), middle Pennsylvanian, shows delta lobe that prograded southward. This delta was built by a river that flowed south across Iowa and emptied into the northern side of the Pennsylvanian seaway. This delta may have been deposited by the same major drainage system that built the earlier B Booch delta in central-eastern Oklahoma. (See Figure 13-32.) The northward shifting of the locations of the deltas from this drainage network reflects the progressive submergence of the interior of the North American continent. (After H. R. Wanless and others, 1970, fig. 20, p. 236.)

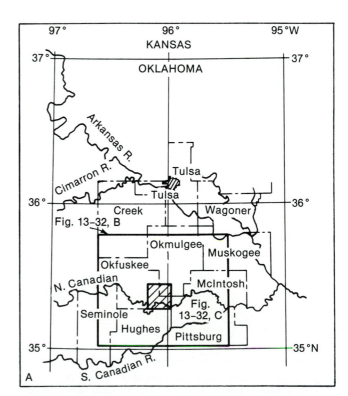

FIGURE 13-32.   Thickness maps and electric-log cross section of Booch (a Dutch name, hence pronounced "Boke") sandstone (in McAlester Formation, Lower Pennsylvanian) in central-eastern Oklahoma (Creek, Okmulgee, Muskogee, McIntosh, Okfuskee, Seminole, and Hughes counties).

A. Index map.

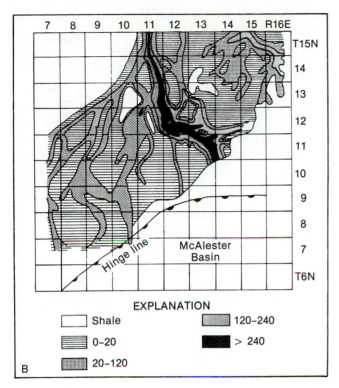

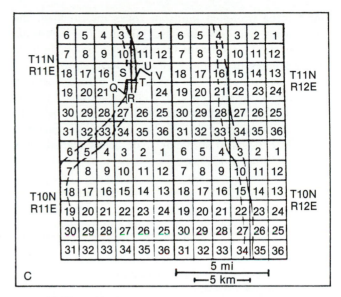

FIGURE 13-32. (*Continued*)

C. Booch sandstone channel prisms in four townships located in southcentral part of B. Inside shaded area, thickness of sandstone in lower part of Booch interval is 20 to 120 ft. In areas not shaded, thickness of sandstone in this interval is <20 ft. At least some Booch sandstone is present in all wells studied throughout these four townships. (Based on subsurface correlations and -mapping by J. E. Sanders.)

FIGURE 13-32. (*Continued*)

B. Regional thickness map of sandstone in Booch interval. Elongate, narrow prisms of sandstone are interpreted as being "bar-finger" sand bodies of the Booch "bird-foot" delta complex that prograded southward into the deep water of the McAlester Basin. Thickness classes of sandstone in feet. (After D. A. Busch, 1974, fig. 102, p. 130.)

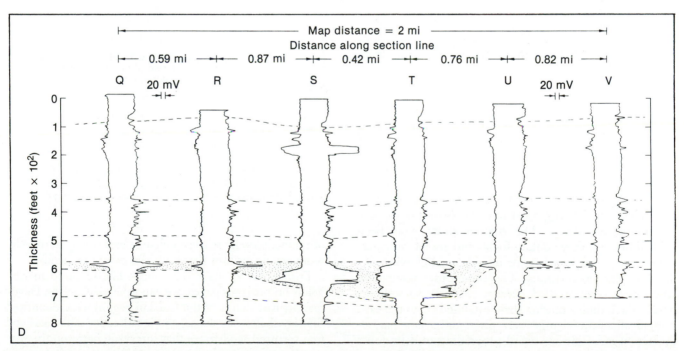

FIGURE 13-32. (*Continued*)

D. Electric-log cross section through Booch sandstone based on six exploratory holes drilled at locations shown on C. Compare Figure 13-21. (Correlations by J. E. Sanders.)

found three principal kinds of rocks interbedded, the usual order being (from base upward) (1) limestone, (2) shale, and (3) sandstone (named gritstone by Phillips). Closer studies made more recently have shown that a typical sandstone may be overlain by a three-member sequence, in upward order (a) a clay penetrated by fossil roots (rootlet bed), (b) a coal, and (c) various shales. In addition, the lower shale, no. 2 in the Phillips sequence, generally consists of silty-clay shale below and siltstone higher up (Figure 13-33).

These cyclic successions have now been interpreted as deltaic deposits. In making a deltaic interpretation, British geologists have emphasized the upward change in particle size from clay to silt to sand and the distribution of the sandstones as well as the coals that overlie the sandstones. Because bar-finger sands are not present, it has been inferred that the Carboniferous deltas were of the shoal-water variety.

SOURCES: Belt, 1975; Collinson, 1969; Fielding, 1986; Haszeldine, 1984; J. Phillips, 1835.

Parts of Middle- and Upper Devonian, New York, United States. Since the classic work of Joseph Barrell early in the twentieth century, the thick complex of nonmarine "Catskill" redbeds overlying and interfingering with marine shales and sandstones has been recognized as the work of ancient deltas. Modern sedimentologic studies have shown that the Devonian of the Catskill area includes sediments that were deposited in point-bar sequences and associated overbank deposits of floodplain rivers (Chapter 14), on fans, in braided streams, on intertidal flats, in lagoons, on barrier beaches, in various parts of marine deltas, and in shallow seas away from shore.

Previous work has outlined and supported the concept that in the Catskill region during the Devonian Period, deltas were present. Nevertheless, the actual strata that were deposited on marine deltas themselves have scarcely been mentioned. Rather, the chief emphasis has been placed on the marine strata upon which the marine deltas must have prograded, and on the nonmarine alluvial deposits of floodplain rivers, of braided streams, and of fans, which aggraded upward above the topsets, as supratopset strata. The marine strata have been studied because they contain fossils and are laterally persistent; they are the basis for making stratigraphic subdivisions and correlations. The supratopset nonmarine strata have been studied because they are so well exposed. Each time a marine shale extended far to the east, a rapid allocyclic change took place and a new parasequence began. The westward prograding of delta lobes built autocyclic upward-coarsening successions that were capped by coarse alluvial sediment. It was the interplay

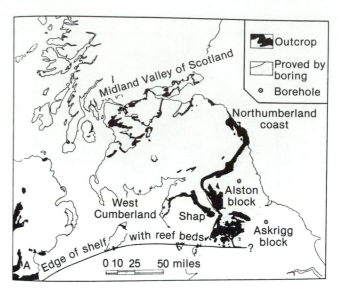

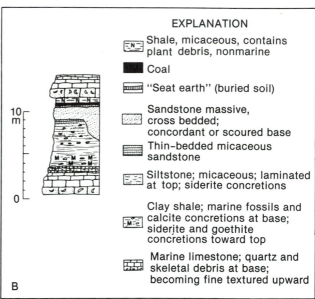

FIGURE 13-33.   Ancient deltas in the Yoredale Series, Lower Carboniferous, Great Britain.
    A. Index map showing location of Yoredale strata.
    B. Idealized profile of Yoredale cyclothem subdivided using marine limestones. Thickness of noncalcareous parts may be as much as 60 m. (D. Moore, 1959; A, fig. 3, p. 525; B, sketch by J. E. Sanders based on Moore's descriptions, p. 532–533.)

of rapid allocyclic sea-level change and steady autocyclic sediment deposition that built the parasequences.

From what we have seen about fine-textured marine deltas on previous pages, we can infer that on the Devonian marine deltas of the Catskill region were deposited the clay shales, silt shales, and interbedded silt shales and fine-textured sandstones that intervene between the marine limestones or dark-colored marine shales below and the thick nonmarine strata above.

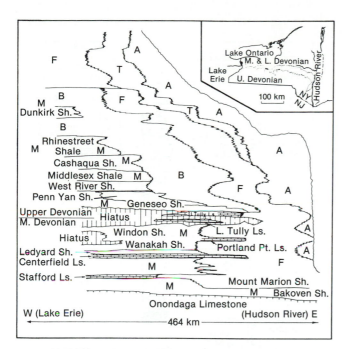

FIGURE 13-34. Restored schematic stratigraphic diagram along outcrop belt that trends east-west across central New York State (location on inset index map), showing Catskill tectonic fan-delta complex. Simultaneous subsidence, influxes of great quantities of terrigenous sediment, and changes of sea level resulted in westward progradation of the shoreline during submergence and accumulation of thick alluvial strata, probable fan deposits (A) above the deltaic strata (T, inferred topsets; F, inferred foresets). Bottomset shales (B) interfinger to the west with marine shales (M) and -limestones. Further explanation in text. (Modified from L. V. Rickard, 1975.)

If this concept is correct, then from the generalized restored stratigraphic section (Figure 13-34), we can infer that at least four episodes of westward prograding by marine deltas must have taken place; each formed part of a parasequence. Because the thickness of the inferred marine-deltaic strata is such a small proportion of the total of the Middle- and Upper Devonian succession, we can conclude that the marine deltas were numerous, that they prograded seaward very rapidly with respect to the rate of subsidence, and that most of the subsidence took place after alluvial plains had become established on the topset parts of the marine deltas. Hummocky cross-stratified storm-influenced Upper Devonian marine sandstones and -siltstones interbedded with shaly strata on the southern New York state line form upward-coarsening sequences 15 to 30 m thick whose origin may be related to delta-lobe switching of the Catskill marine deltas.

Because the entire succession contains such thick supratopset strata, we can surmise further that the marine deltas must have built westward to the point where the water deepened. This follows from the fact that the rate of westward growth of the marine-delta lobes was checked while the supratopset strata thickened so conspicuously. A further indication that the Devonian marine deltas of the Catskill region were of the shoal-water variety is the absence of bar-finger sand prisms and -pods. If any such sands are present, they should be located at the western edge of the former shallow platform, an area now covered by younger strata.

Deltaic patterned sequences comparable to those from the Late Paleozoic-age cyclothems have not been recognized in New York State. Possibly this situation has resulted from the poor exposures of the supposed deltaic strata.

To emphasize its distinctive aspects, the Devonian succession in the Catskill region has been named a "tectonic delta complex." If the foregoing remarks about the Devonian marine-delta deposits are correct, then the thick, well-exposed supratopset beds should be separated from the true deltaic strata. Both were products of progradation and submergence. The deltas formed first, and afterward, as submergence continued, fans aggraded. In order to emphasize the inferred presence of both marine deltas and supratopset fans, we suggest that the term "tectonic delta complex" be abandoned and in its place the term *tectonic fan-delta complex* be substituted.

SOURCES: Barrell, 1913, 1914; Craft and Bridge, 1987; Ettensohn, 1985a, b; G.M. Friedman, 1988; Glaeser, 1979; Halperin, 1987; Rickard, 1975; Van Tassell, 1987; Woodrow, 1985; Wright and Sevon, 1985.

DELTAS AND CYCLOTHEMS

The term *cyclothem* originated from study of the Pennsylvanian strata of the central United States, as explained previously. A columnar section of a typical cyclothem of Pennsylvanian age is shown in Figure 13-35. Various students of cyclothems have been impressed with one or more of the following characteristics: (1) the surfaces of truncation at the bottoms of the nonmarine sandstones, (2) the coal beds, and (3) the marine limestones. Cyclothems are classical examples of parasequences.

In cross section, some of the sandstones form spectacular channels that cut tens of meters or so downward into the underlying strata. Before the modern concepts of marine deltaic sedimentation had been developed, such surfaces of truncation were thought to be ancient landscapes that were eroded subaerially when the interior of the continent emerged during a drop in sea level. In the days when stratigraphers were constantly on the lookout for "natural" breaks subdividing the successions they studied, it was to be expected that they would use these presumed subaerial-erosion

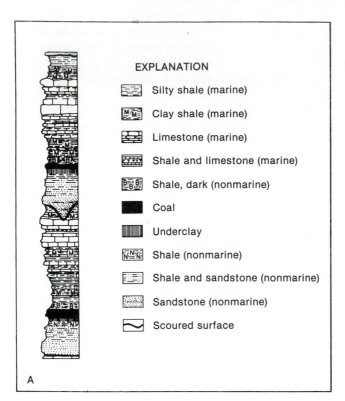

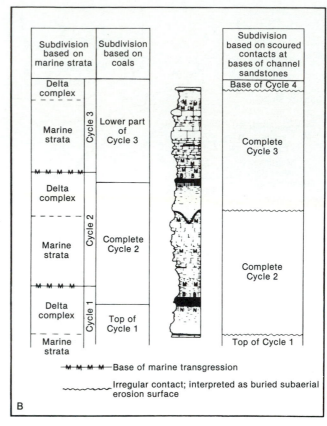

FIGURE 13-35.   Pennsylvanian cyclothem and ways of subdividing cyclothems.

A. Schematic columnar section of two cyclothems of the kind that are common in the Pennsylvanian and Permian strata of Kansas. Typical thickness of such cyclothems is about 30 m. (Modified from R. C. Moore, 1964, fig. 1, p. 290; also 1949, fig. 154, p. 212.)

B. Ways of subdividing cyclothems. At left, subdivision is based on marine strata; base of cyclothem placed at deposit made by a marine transgression. After submergence and accumulation of marine strata, forming lower part of the cycle, a delta complex progrades. The water shoals, the shoreline advances seaward, and coastal swamps form in which coal can accumulate. Depth of water into which delta lobes prograded determines thickness of the deltaic part of the cyclothem.

In center, subdivision is based on coal beds.

At right, subdivision is based on scoured contacts at bases of channel sandstones. Further explanation in text. (Ideas compiled from various sources.)

surfaces for the zero reference points in numbering the strata of the cyclothems. (See Figure 13-35, B, right.)

Another alternative would be to subdivide the cyclic successions by means of the coals. (See Figure 13-35, B, center.) This procedure takes advantage of the wide lateral extents of the coals and is encouraged by their economic significance.

The third possibility, subdividing on the basis of the marine limestones, compares with the second, the use of coals, in taking advantage of units having wide lateral persistence. In addition, the emphasis on limestones coincides with the effects of marine submergences. Because the British Carboniferous cyclothems were subdivided on the basis of the marine limestones (See Figure 13-35, B, at left.), their deltaic origin became apparent. The top of a marine limestone is the surface across which a marine delta lobe was prograded (autocyclic).

The next-higher marine limestone is significant because it represents the effects of a submergence (allocyclic) and termination of that particular delta lobe.

Within each delta lobe will probably be found a surface of truncation that marks the base of each distributary-channel sand. Subdivision into units bounded by these truncation surfaces would, and in the case of the Pennsylvanian cyclothems in the United States, did indeed, mask the fundamental deltaic sequences. If the boundaries of the typical Pennsylvanian cyclothems in the United States are selected using the marine units or the coals, then the deltaic origin becomes obvious.

SOURCES: Klein and Willard, 1989; D. G. Moore, 1959; R. C. Moore, 1964; Peterson and Hite, 1969; Rahmani and Flores, 1984.

Fans at the Sea Shore (Fan Deltas)

In a few places in the modern world, the circumstances of high relief and abundant sediment discharge are such that fans have been built at the sea shore. These bodies of sediment have become known as fan deltas. Use of this term may confuse beginners. No fan can grow at the sea shore until a delta topset plain has formed. A better term is *fan-delta complex*.

Modern Examples

A few examples of modern fans at the sea shore include the Gulf of Aqaba, Red Sea (Figure 13-36), the south coast of Puerto Rico, and the shores of the Gulf of California. In all of them, high mountains come down to the sea. Other fan-delta complexes have been built at the sea shore from the meltwater of glaciers. (See Figure 14-55.)

A fan-delta complex may persist despite the tendency for the waves to destroy it. Under certain conditions a fan-delta complex may be protected by a barrier reef along its distal margin (Figure 13-37). We discussed reefs in Chapter 10; suffice it to say here that reefs do not grow where terrigenous sediment is being supplied copiously from the land. Hence the fan-delta complex shown probably was inactive for a long time, and in the absence of terrigenous sediment, the reef colony established itself.

In other settings the fan-delta complexes serve merely as conduits for the transport of terrigenous sediment to the sea. The waves sweep the sediment away from the distal margins of such fan-delta complexes and build out the shore as a series of beach ridges. (Figure 13-38; fuller discussion in Chapter 11.) Where wave- and current energies are weak, as along the shores of the Dead Sea, coarse- and fine-textured sediments may be interlayered at the distal ends of sea-marginal- or lake-marginal fans, in response to episodic changes (such as floods resulting from storms) of volumes and particle-size distributions of fan-derived sediment.

The preservation of significant quantities of fan-delta-complex sediments in a marginal-marine setting requires a persistent supply of sediment to the fan and rapid subsidence. The result is a wall of coarse debris, probably bounded on one side by a steep surface (a fault scarp

A

FIGURE 13-36. View vertically downward from an airplane of fans of several sizes and ages at the shore of the Red Sea. Bedrock (at top) consists of Precambrian gneisses composing Arabo-Nubian Shield. Seaward side of large fan underlain by light-colored sediment is fringed by coral reef (scarcely visible in this view). The dry bed of the incised straight channel (about 2 km long), which crosses fan from apex to periphery, displays a well-developed braided-stream pattern. Drowning of seaward end of this dry channel has created a small rectangular embayment in the periphery of the fan. Radiating network of faintly incised channels covers surface of large fan. Compare Figure 14-8. Several small fans (upper left and lower right) have grown across parts of the large fan. Waves approaching from lower right have built a pointed spit on the left side of the large fan.

B

FIGURE 13-37. Fans that have built across reefs, Ras Abu Gallum, west side of Gulf of Aqaba, viewed obliquely from the air.

A. Distant view showing lobe of fan in foreground that has prograded completely across the reef.

B. Closer view of smaller lobe of fan (at top of A) that has been modified by the waves, which have built a series of beach ridges (lower left). (Authors.)

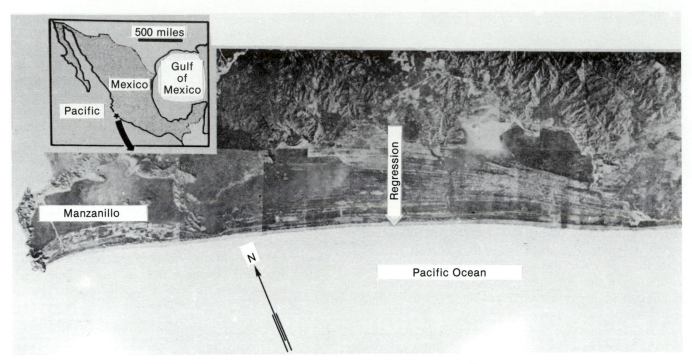

FIGURE 13-38.   View from airplane of beach plain that has prograded seaward on west coast of Mexico using sediment eroded from fans at the shore. Width of beach plain at arrow is approximately 7 km. (P. C. Scruton.)

or steep mountain front; see Figure 14-7.) and passing laterally and abruptly into marine strata.

> SOURCES: Bellaiche and others, 1981; Ethridge and Wescott, 1984; Fraser and Suttner, 1986; Hayward, 1985; McPherson, Shanmugam, and Moiola, 1987; Nilsen, 1985; Postma, 1984; Sneh, 1979; Wescott and Ethridge, 1980.

Examples of Inferred Ancient Fan-Delta Complexes

Possible examples of ancient fan-delta-complex deposits built at the margin of a sea include the San Onofre Breccia, of Oligocene-Miocene age in the Coast Ranges of California; the coarse parts of the Paradox and Eagle Valley formations (Pennsylvanian) of southwestern Colorado; and parts of the Devonian Catskill complex, as previously described.

The San Onofre Breccia accumulated in southern California along the margins of the deep marine basin in the center of which the Monterey Formation was deposited. Some parts of this breccia contain coarse debris that was eroded from former lands lying west of the present

outcrops. These lands subsequently vanished by subsiding beneath the Pacific Ocean or by being accreted onto the western margin of North America.

The coarse parts of the Paradox Formation in Colorado, the inferred sea-marginal fan deposits, interfinger with salt that was deposited in the Paradox Basin (Figure 13-39). Correspondingly, the coarse-textured parts of the Eagle Valley Formation interfinger with the sulfate evaporites (containing local salt) of the Eagle Basin. The relationships within these basins of Pennsylvanian age in Colorado are known only from subsurface data obtained from petroleum test borings. The highland blocks that flanked these basins were elevated actively during both Pennsylvanian and Permian times.

During the Mesozoic Era, the whole region subsided more or less as a single unit and was buried by younger strata.

> SOURCES: Amajor, 1986; Bergh and Torske, 1986; Chough, Hwang, and Choe, 1990; Gloppen and Steel, 1981; Massari, 1984; Porebski, 1981; Ricci Lucchi, Colella, Ori, Ogliani, and Colalongo, 1981; Wescott and Ethridge, 1983.

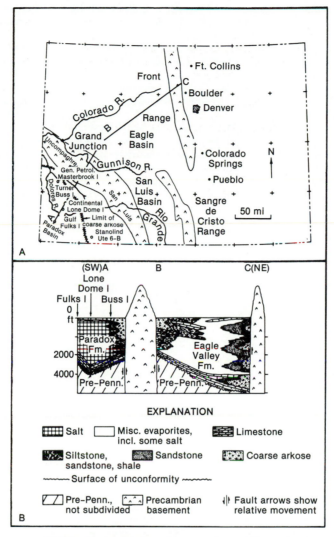

FIGURE 13-39.   Reconstruction of fans at shore of Pennsylvanian sea in southwestern Colorado.

A. Sketch map showing part of Pennsylvanian sea and locations of elongate highland blocks of basement rock that were rapidly elevated to form mountainous islands about which fans formed.

B. Schematic profile-section with approximate tops of Paradox and Eagle Valley formations restored to horizontal datum at former sea level. Fan sediments consisting of debris eroded from Uncompaghre highland block interfinger with salt deposited in Paradox Basin. Compare with Figure 14-7. (Redrawn by J. E. Sanders from S. A. Wengerd, 1962, sec. 9, fig. 20, p. 297; and J. A. Peterson and R. J. Hite, 1969, fig. 4, p. 888.)

Suggestions for Further Reading

BOOTHROYD, J. C., 1978, Mesotidal inlets and estuaries (*sic*), p. 287–360 *in* Davis, R. A., Jr., ed., Coastal sedimentary environments: New York, Springer-Verlag, 420 p.

CAMPBELL, C. V., and OAKS, R. Q., JR., 1973, Estuarine sandstone filling tidal scours, Lower Cretaceous Fall River Formation, Wyoming: Journal of Sedimentary Petrology, v. 43, p. 765–768.

CHOUGH, S. K., HWANG, I. G., and CHOE, M. Y., 1990, The Miocene Doumsan fan-delta, southeast Korea: A composite fan-delta system in back-arc margin: Journal of Sedimentary Petrology, v. 60, p. 445–455.

COLEMAN, J. M., 1988, Dynamic changes and processes in the Mississippi River delta: Geological Society of America Bulletin, v. 100, p. 999–1015.

ELLIOTT, T., 1986, Deltas, p. 113–154 *in* Reading, H. G., ed., Sedimentary environments and facies (*sic*), 2nd ed.,: Oxford, England, Blackwell Scientific Publications, 615 p.

EMERY, K. O., STEVENSON, R. E., and HEDGPETH, J. W., 1957, Estuaries and lagoons, p. 673–750 *in* Hedgpeth, J. W., ed., Treatise on marine ecology and paleoecology: Geological Society of America Memoir 67, v. 1, Ecology, 1296 p.

GALLOWAY, W. E., 1989, Genetic stratigraphic sequences in basin analysis II: Application to northwest Gulf of Mexico Cenozoic Basin: American Association of Petroleum Geologists Bulletin, v. 73, p. 143–154.

MORGAN, J. P., 1970, Deltas, a résumé: Journal of Geological Education, v. 18, p. 107–117.

PENLAND, SHEA; BOYD, RON; and SUTER, J. R., 1988, Transgressive depositional systems of the Mississippi delta plain: A model for barrier shoreline- and shelf sand development: Journal of Sedimentary Petrology, v. 58, p. 932–949.

REDFIELD, A. C., 1965, Ontogeny of a salt marsh (*sic*) estuary: Science, v. 147, p. 50–55.

ROBERTS, H. H., ADAMS, R. D., and CUNNINGHAM, R. W., 1980, Evolution of sand-dominant subaerial phase, Atchafalaya delta, Louisiana: American Association of Petroleum Geologists Bulletin, v. 64, p. 264–279.

WANLESS, H. R., TUBB, J. B., GEDNETZ, D. E., and WEINER, J. L., 1963, Mapping sedimentary environments of Pennsylvanian cycles: Geological Society of America Bulletin, v. 74, p. 437–486.

WATELEY, M. K. G., and PICKERING, K. T., 1989, Deltas: sites and traps for fossil fuels: Geological Society of London Special Publication, No. 41, 360 p.

WRIGHT, L. D., 1977, Sediment transport and deposition (*sic*) at river mouths: A synthesis: Geological Society of America Bulletin, v. 88, p. 837–868.

WRIGHT, L. D., and COLEMAN, J. M., 1973, Variations in morphology of river deltas as functions of ocean wave (*sic*) and river discharge (*sic*) regimes: American Association of Petroleum Geologists Bulletin, v. 57, p. 370–398.

Nonmarine Environments:
Deserts, Rivers, Lakes, and Glaciers

Nonmarine environments include deserts and semiarid regions, rivers, swamps, lakes, glaciers, jungles, rocky highlands, mountainous regions, grasslands, and forests. These are environments of great diversity and ecological variability. Paradoxically, although nonmarine environments are the ones best known to humans and the most easily studied, the sedimentary products of nonmarine environments in the geologic record have taken a back seat in geologic research to those of marine environments. Exploratory drilling for petroleum in basins worldwide has indicated that many of them, notably in China, have been filled by nonmarine sediments.

By definition, most nonmarine environments are located above the base level of erosion and their deposits are easily eroded. Unless the basement beneath the nonmarine environments subsides greatly, they may be only temporary sites of deposition. Indeed, most nonmarine environments supply sediments that are later deposited in marine environments. Only when nonmarine sediments subside below sea level more rapidly than they can be eroded will nonmarine strata become part of the geologic record. In various places in the world, nonmarine successions have been preserved far from the sea in grabens or in great crustal downwarps (Figure 14-1). A parent deposit or -deposits supplying vast amounts of sediment to fill the graben or downwarp is essential for building a thick sequence of nonmarine strata.

Nonmarine strata are also preserved as sea-marginal deposits where fans or rivers feed nonmarine sediments that push back the sea and prograde across marine sediments. During a later submergence, the nonmarine sediments are buried beneath marine strata.

Another reason why nonmarine rocks are less well represented in the rock record than are marine rocks is that during much of their geologic history, large parts of present continents were submerged beneath shallow seas and thus accumulated only marine deposits. The present situation of high-standing continents is unusual. Even so, the geologic record preserves enough examples of nonmarine rocks to enable us to infer that at times in the past nonmarine environments were widespread.

Deserts and Semiarid Environments: Modern- and Ancient Deposits

The word *desert* means literally a deserted, nonoccupied, or noncultivated area and has become synonymous with arid land, whether deserted or not. Low precipitation and a high ratio of evaporation to precipitation characterize a desert.

In general, desert areas receive no more than 25 cm of precipitation per year. In Early Paleozoic times, before land plants had evolved, even humid areas must have displayed the plantless aspect of modern dry deserts. Prominent desert environments include (1) *bare rock surfaces*, (2) *pediments*, (3) *fans*, (4) *intermittent streams*, (5) *dunes*, (6) *sabkhas*, and (7) *playas* (Figure 14-2). Deserts are commonly thought of as confined to warm climatic settings. However, deserts exist even in the Arctic.

Bare Rock Surfaces

Weathering breaks down the rocks of a desert. *Wind removes any dry, loose particles* (a process known as **deflation**), thereby lowering the surface of the desert. *Particles picked up by the wind sandblast the surfaces of the rock*; this process is called **abrasion**. Such action loosens additional particles, which the wind then carries off. Blasting by wind-carried particles smooths the bare rock surfaces. Such *wind-smoothed pavement forming a rocky desert* or **hamada** may be swept clear of sand.

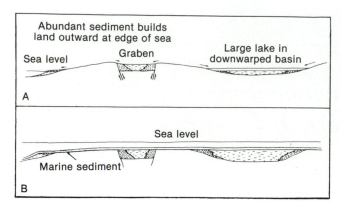

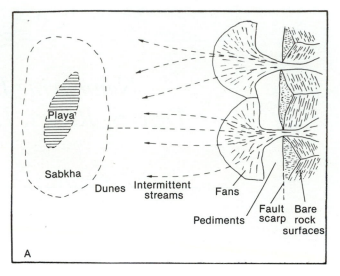

FIGURE 14-1.   Schematic cross sections showing how nonmarine strata can be preserved in the rock record.

A. Nonmarine sediments accumulate in graben, in downwarped basin, and at the margin of the sea.

B. Nonmarine sediments are preserved beneath marine strata deposited by a transgressing sea. (Authors.)

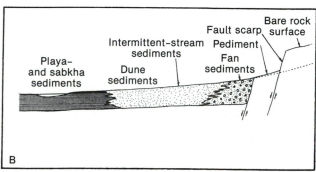

FIGURE 14-2.   Sedimentary environments in a desert basin adjoining a steep mountain block.

A. Schematic map view.

B. Profile-section. (Authors.)

Residual gravel (a **deflation lag**) may form *a veneer on bare rock surfaces* (Figure 14-3). Bare rock surfaces of deserts have not been recognized in the rock record.

Pediments

Pediments (Figures 14-4 and 14-5) are *sloping surfaces that are adjacent to highlands and that were cut across bedrock by periodic floods that formed sheets of water* known as *sheet-wash*. The surfaces of pediments slope gently toward basins and away from highlands. Bare rock may compose the surfaces or a thin veneer of loose sediment may cover them; commonly dry channels cut by *intermittent streams*, known as **arroyos** (or **wadis**) dissect them. Pediments may be preserved in the rock record as surfaces of unconformity. Preservation is especially likely if sulfates, such as gypsum or anhydrite, have impregnated the loose sediments veneering a pediment.

Fans

Fans, sometimes known as **alluvial fans**, are *deposits having surfaces that are segments of cones and that radiate downslope from the points where, along a fault scarp or precipitous slope, streams emerge from rocky highlands.* Sediments are transported to the fan by a single trunk stream. In plan view, the outline of a single fan is approximately triangular and the apex of the triangle points upstream (Figure 14-6). Surfaces of fans are concave upward and may slope as much as 25° at the head to as little as 1° at the toe. Most fan surfaces slope from 5 to 10°. Mean particle size and the surface slope of a fan are positively correlated. Both decrease exponentially from head to toe of the fan. At the edges of highlands, *adjacent fans may grow and coalesce to form a merged complex, in which the distinctive*

forms of individual fans have been lost, and which is known by the French term **piedmont**, literally, foot of the mountain. The Spanish word for such coalesced fan complexes is *bajada*.

Thick fan deposits that accumulate adjacent to fault scarps or adjacent to escarpments of erosion may be extremely coarse (Figures 14-7 and 14-8). Uplift on the faults may create a precipitous landscape and a parent deposit that provides a persistent supply of coarse debris and increases the competency of streams. The rate and magnitude of uplift of the adjacent highlands, the composition of the bedrock, and the frequency and amount of precipitation in the drainage basin of a fan control the site, rate, and magnitude of deposition on the fan. Fans may form at the foot of elevated regions in both humid- and arid climates. Submarine fans that head at the mouths of submarine canyons (Chapter 9) share many features with subaerial fans.

Three intergradational transporting mechanisms are recognized as important in fans: (1) stream flow, (2) debris flow, and (3) mudflow.

FIGURE 14-3.   Exposed bedrock in desert of Gebel Maghara, Sinai Peninsula. Alternating strata of terrigenous- and carbonate rocks (Jurassic) underlie slopes. Carbonate strata stand out as ledges. Residual gravel forms a veneer on rock surface in foreground. (Authors.)

Stream flow may deposit sediment when sediment-laden waters surge from the end of the stream channel and spread out on the fan. These surges deposit sheets of silt, sand, and gravel, with little visible clay; such sediments are distributed by the network of streams. When the water flowing across a fan mixes with sediment, the density and viscosity of the mixture increase. *A water and sediment mixture that behaves more like a plastic mass than a Newtonian fluid* is a **debris flow**. Deposits of debris flows are poorly sorted or nonsorted, are coarse textured, and include cobbles and boulders, some of which weigh tons. These large particles are set in a fine-textured matrix of mud, which, however, rainwash may remove. *A debris flow in which the entrained particles are mostly sand size and finer and in which mud is dominant* is known as a **mudflow** (Figure 14-9). The proportion of stream-flow-, debris-flow-, and mudflow deposits in fans varies according to the amount, frequency, and intensity of rainfall. In general, in fans of more-arid regions, debris- and mudflow sediments predominate. In humid regions, stream-flow deposits form the bulk of fans. The stream-flow deposits

typically contain 5% or less of clay-size particles; mudflow deposits contain generally 25% or more of clay-size particles. A striking characteristic of mudflows is their capacity to carry large blocks, a meter or more across. (See Figure 14-9.) This is made possible by the high effective density, viscosity, and plastic nature of the mud-water mixture, especially in the core of the flow.

From apex (proximal part) to periphery (distal part), many large fans can be divided into three segments (Figure 14-10). The names of these subdivisions (with their typical sediments mentioned in parentheses) are (1) proximal fan (mostly conglomerate), (2) mid fan (mixed conglomerate and sandstone), and (3) distal fan (coarse- to very coarse-textured sandstone). *Rudites deposited in fans* are known as **fanglomerates**. In many areas of the world, Quaternary fan sediments are important reservoirs for ground water.

In the rock record, fan deposits typically display the following diagnostic features that may aid in their recognition. (1) In sectional view, they are lenticular or wedge shaped. (2) Water-laid deposits are coarse-textured sand-

FIGURE 14-4.   Talus cones at bottom of fault scarp at edge of mountains (in background) disintegrate *in situ* into particles that are removed by water and accumulate on fan (surface sloping gently to left, in right foreground). Homogeneous, nonlayered lower part of mountain is Precambrian basement; overlying bedded strata are Lower Paleozoic deposits of intermittent streams. The contact between Precambrian and Lower Paleozoic rocks is a flat pediment surface where Paleozoic streams cut across bedrock of the basement. As a result of faulting, the pediment surface appears in three segments at different elevations. Sinai Peninsula, northwestern Gulf of Aqaba. (Authors.)

FIGURE 14-5.   Dissected pediment underlain by Cretaceous strata slopes toward foreground from fault scarp at edge of highlands (in distance). Foreground shows floor of flood plain of Arava Valley, between Dead Sea and Gulf of Aqaba, Red Sea. (Authors.)

stones and conglomerates, well sorted or moderately well sorted; in debris-flow deposits, sand particles, cobbles, and boulders are embedded in a poorly sorted shaly and silty matrix; in mudflow deposits, shale is dominant and large particles are minor. (3) Stratification ranges from

FIGURE 14-6.   Upper reach of Slims River, Yukon (61°55'N, 138°38'W), showing marked contrast between tributary fans and braided trunk river, viewed from vertically above in aerial photograph. This arid region of the Canadian Arctic is an example of a *cold desert*. Tributary fans constrict the trunk river. Dark areas on fans are vegetated. Width of view about 7.5 km, north toward top of photograph. Trunk river flows from right to left. (Air photo (A15517-20), copyright Her Majesty the Queen in Right of Canada, reproduced from the collection of the National Air Photo Library with permission of Energy, Mines and Resources Canada. B. R. Rust and E. H. Koster, 1984, fig. 5, p. 55.)

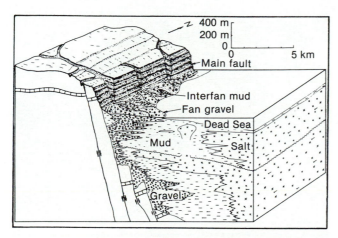

FIGURE 14-7.   Schematic block diagram between Judaean Mountains and Dead Sea, Israel, showing partial view of sediment filling of graben. Proximal fans consist of gravel, and distal fans of interbedded gravel and muds. Center of basin is occupied by lacustrine muds and -salt. (Modified from Y. Langozky, in Y. Langozky and A. Sneh, 1966, fig. 2, p. 9.)

FIGURE 14-9.   Section through mudflow deposit of Quaternary age shows boulders embedded in fine-textured matrix. Fan at fault scarp between Judaean Mountains and Dead Sea, Israel. Boulder in center of photograph is 1 m in diameter. (Authors.)

good in water-laid deposits to poor in debris-flow deposits to nearly nonrecognizable in mudflow deposits; water-laid deposits may be cross stratified or plane bedded. (4) Water-laid deposits from streams that were confined to channels appear as channel fills truncated at their bases; those from sheetfloods form extensive sheetlike beds. (5) First-cycle organic material is absent. (6) The colors of the rocks are yellow or brown or red, indicating that the iron is in an oxidized state. (7) Typically, fan deposits interfinger with other products of desert environments, such as those of playas.

SOURCES: Blair, 1987a, b; G. S. Fraser, 1989; Fraser and Suttner, 1986; Frostick and Reid, 1987; Koster and Steel, 1984; Miall, 1990.

Intermittent Rivers

Some features of deserts are best described simply as dry rivers. Such dry rivers are not parts of fans, yet are closely allied to fans. In each major desert these *dry rivers* have received local names: **arroyo** (southwestern United States), **wadi** (Arabia), **oued** (Sahara), and **omibiri** (Kala-

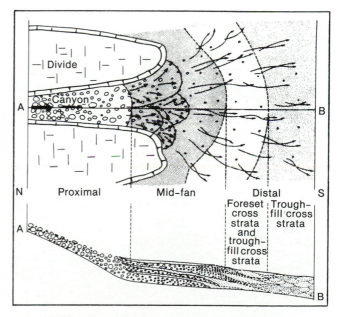

FIGURE 14-10.   Plan view and cross section of Van Horn fan, West Texas. Widths of arrows, in plan view, indicate relative competency of stream flow. Profile and section show downfan decreases in slope and size of particles and the kinds of stratification: (1) proximal fan, massive conglomerate; (2) mid fan, interbedded conglomerate and cross-stratified sandstone; and (3) distal fan, cross-stratified sandstone. (J. H. McGowen and C. G. Groat, 1971, fig. 31, p. 39.)

FIGURE 14-8.   Trunk stream emerging from edge of mountains and spreading out into fan. Network of braided channels laces fan. Ras Abu Gallum, Sinai Peninsula. (Authors.)

hari). Although these obvious watercourses are dry most of the time, they do discharge water now and then, notably during floods. During periods of flow, water may become a raging torrent; it then erodes, transports, and deposits huge amounts of sediment. Whereas fan deposits show characteristic shapes and occur at abrupt changes of slope at the foot of a highland, the deposits of dried-up beds of intermittent rivers may be hundreds of kilometers long and several kilometers wide. Several such dried-up river beds of the Sahara are more than 800 km long.

As seen in satellite photographs, the most-conspicuous features of modern deserts, despite their aridity, are the beds of dry rivers, the work of streams (Figure 14-11). Flash floods occur only occasionally in any limited area; once a year or less. Yet their energy both in cutting through sediment and bedrock and in transporting and depositing sediment is so enormous that intermittent streams leave a marked imprint on the desert landscape.

When infrequent cloudbursts take place in highland areas near or within the desert, water may flow intensely in these intermittent rivers for several days. When such flows diminish, they deposit all particle sizes in layers of variable thickness up to several meters (Figure 14-12). Stratification ranges from excellent to poor. Even-

FIGURE 14-12.   Poorly sorted conglomerate (of Quaternary age) sandwiched between coarse sandstone in deposit of intermittent stream (wadi) of North African desert (Libya). Eyeglass case in center of photograph is 15 cm long. (Authors.)

tually, as the supply of water diminishes, the erstwhile river vanishes because its water sinks below the surface of the river bed.

An example of deposits of intermittent streams in the rock record includes the so-called Nubian Sandstone facies, which is a sequence of fluvial sandstones that ranges in age from Cambrian to Holocene. This sandstone is exposed on both sides of the Red Sea in positions marginal to the Arabo-Nubian Shield. tually, as the supply of water diminishes, the erstwhile During early Paleozoic time, sheetfloods spread sands over the distal parts of the shield and cut pediments across the underlying basement rocks of Precambrian age. (See Figure 14-4.) Nubian Sandstone facies of Jurassic age from cores in boreholes consists of coarse-textured, cross-stratified sandstones having anhydrite mottles that confirm an origin under arid conditions.

To distinguish products of subaerial fans in the rock record from those of intermittent streams may be difficult if the shapes of the deposits are not known. After all, many fans change progressively downslope to intermittent rivers. Deposits of both generally consist largely of conglomerates or of coarse-textured sandstones. Products of debris flows are more common in fan deposits than in those of intermittent rivers; steep slopes and nearness to parent deposit favor deposition of debris-flow sediment in fans. In deposits of intermittent rivers, fine-textured sandstones, siltstones, and shales tend to be uncommon; the sorting of the coarse debris may range from good in stream-laid deposits to poor in those of debris-flow origin. The presence of sulfates, as gypsum in surface exposures or as anhydrite in the subsurface, or of other evaporite minerals as a cement

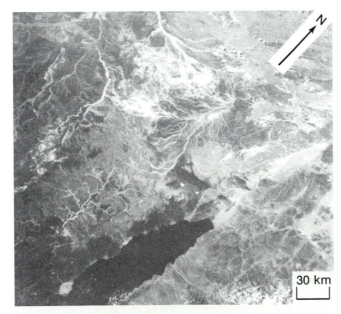

FIGURE 14-11.   Satellite photograph of northern Sinai Peninsula, southwestern Jordan, and Negev Desert of southern Israel. Landscape in Negev and Sinai, west of water body (Gulf of Aqaba), shows the dominant network of various tributaries of Wadi El-Arish and Wadi Watir, which drain into the Gulf. Arava Valley, a graben (See Figures 14-20 and 14-21.), extends from head of Gulf to the north. (NASA.)

30 km

between the particles, confirms an origin under arid conditions. Intertonguing of deposits of fans or intermittent streams with evaporites of playa- or sabkha origin likewise confirms that the depositional environment was arid. However, such an origin would not distinguish between fan- and intermittent-stream deposits. The presence of various bed forms of both upper- and lower-flow regimes would preclude a debris-flow origin for such deposits. But both stream flow and debris flow are active agents both on fans and in intermittent streams.

In the section on fluvial environments in this chapter we discuss how to distinguish the products of fans and intermittent streams from deposits of meandering rivers.

SOURCES: Ethridge and Flores, 1981; Langford, 1989; Sneh, 1983.

Wind Deposits

During transport of sediment, wind selects particles that span fewer size grades than does water. Size discrimination by the wind generally shows up in three areally segregated kinds of deposits: (1) **loess**, *windblown deposits composed chiefly of silt but including also fine sand;* (2) **sand seas** or **ergs**, *areally extensive wind-affected deposits of sand-size particles characterized by large bed forms of various kinds (dunes);* and (3) **deflation lags** known by several names (reg, serir, desert armor, or desert pavement), *lag deposits consisting of particles generally larger than 0.5 mm as well as those of very-fine-sand- and silt sizes not yet carried away because (a) the finer particles are shielded by larger particles, (b) boundary-layer turbulence is insufficient to waft them into suspension, or (c) they form a surface armor that is difficult for the wind to move.* Particles composing the loess move by suspension, sediments composing dunes move chiefly by traction, and deflation lags, when moved, travel as a traction carpet.

To demonstrate conclusively that the particles of a body of rock accumulated as a subaerial dune, one should seek several lines of evidence. Such converging lines of evidence have been gathered for the Permian Lyons Sandstone of the Rocky Mountains. The Lyons Sandstone displays large-scale, high-angle cross strata with steep dips, commonly exceeding 25°. Large-scale, high-angle cross strata, beveled by a horizontal or slightly dipping bounding plane, are characteristic of dune deposits (Figures 14-13 and 14-14). In the Lyons Sandstone, individual sets of these cross strata range in thickness from 3 to 14 m. Commonly these sets can be traced for hundreds of meters. The sets of cross strata are bounded by horizontal- or slightly dipping erosional surfaces formed by the migration of large bed forms. (See Figure 5-36.) By analogy with modern dunes and Pleistocene eolianites, evidence

FIGURE 14-13.   Exposure of Coconino Sandstone (Permian) of Arizona showing cross strata of the kind commonly interpreted as eolian. (Authors.)

of this kind has been used for routine identification of dune deposits. The following characteristics have helped in upgrading the odds in interpreting the Lyons Sandstone. (1) Abundant pits with raised rims strikingly similar to impressions of raindrops dot the upper surfaces of many cross strata (Figure 14-15). In modern deserts, such as that in which the photograph of Figure 14-15, B, was taken, the imprints of raindrops are sometimes preserved for months. (2) Tracks of animals have been preserved in remarkable detail. These tracks are of two kinds: (a) those of scorpions and (b) those variously interpreted as amphibian, reptile, or mammal-like reptile (Figure 14-16). Experimental studies have shown that tracks showing these preservational characteristics can be formed only subaerially by animals walking on dry sand. (3) Ripple marks on steep avalanche faces compare with those on modern dunes; ripple crests are commonly

FIGURE 14-14.   Lyons Sandstone (Permian) of Colorado Front Range viewed in quarry wall shows large-scale, high-angle cross strata and erosional surface (arrows) between two inferred superimposed dune deposits. (Authors.)

FIGURE 14-15.   Impressions of raindrops.
A. Pits with raised rims interpreted as impressions of raindrops on avalanche face of inferred upper dune deposits in Figure 14-14. (Authors.)
B. Impressions of raindrops on modern desert floor, Negev Desert, Israel. (Authors.)

(although not always) parallel to the dip of the face (Figure 14-17).

Since the time when the Lyons Sandstone was first interpreted as an eolian deposit, the products of additional eolian processes to those listed above have been

FIGURE 14-16.   Tracks of *Limnopus coloradoensis*, variously interpreted as an amphibian, a reptile, or a mammal-like reptile, in inferred ancient dune deposit, Lyons Sandstone. (Collection of University of Colorado; courtesy of T. R. Walker.)

FIGURE 14-17.   Views of ripple axes that trend in the dip direction of avalanche slopes on dunes.
A. Ripple marks on avalanche face of inferred dune deposit below erosional surface (arrows) in Figure 14-14. (Authors.)
B. Ripple marks on avalanche face of modern dune deposit, Sinai Peninsula. (Authors.)

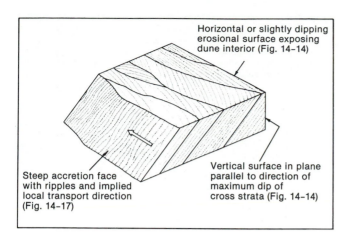

FIGURE 14-18.   Block diagram showing bed forms of Lyons Sandstone. (T. R. Walker and J. C. Harms, 1972, fig. 7, p. 285.)

FIGURE 14-19.   Sabkha, in foreground, veneered by gravelly deflation lag. Hummocks of blown sand surround sparse vegetation. Fault scarp and highlands are in background. Shurat El Rarkana, Sinai Peninsula. (Authors.)

recognized. Migration of wind ripples over the surfaces of eolian dunes or flats, where accompanied by deposition, commonly produces thin inversely graded laminae that are very regular in thickness. These structures are particularly useful for identifying eolian sands because they are extremely common, and they can be recognized even in small exposures or where interbedded with subaqueous deposits.

Taken together, these lines of evidence indicate that the cross-stratified beds of the Lyons Sandstone, as reconstructed in a block diagram (Figure 14-18), originated in subaerial dunes.

In detailed studies, recognizing deposits as eolian is merely the first step in reconstructing the eolian environment. Where eolian sandstones are exceptionally well exposed, as at some locations on the Colorado Plateau, it is possible to identify products of the depositional processes that operated on different parts of individual dunes, to make estimates of dune size, and to use cyclic cross strata to determine dune migration speeds or to determine the shape, trend, and migration direction of small dunes that were superimposed on larger dunes.

SOURCES: Ahlbrandt and Fryberger, 1982; Bagnold, 1941; Brookfield, 1984; Clemmenson, Olson, and Blakey, 1989; Goudie, 1989; Halsey, Cato, and Rutter, 1990; Kocurek and Nielson, 1986; Lancaster, 1989; Lancaster and Balkema, 1989; E. D. McKee, 1982; Pye and Tsoar, 1990; Rubin and Hunter, 1985, 1987; Rubin and Ikeda, 1990; Tsoar, 1983.

Sabkhas

When intense cloudbursts dump their moisture, intermittent streams flow violently. The flood water sweeps mud and coarser sediments across the pediments and fans into the stream channels. Part of the derived bed load accumulates on the sabkha (Figures 14-19 and 14-20;

see Chapter 12). Within the framework of the sedimentary particles underlying a sabkha, evaporite minerals, especially gypsum or anhydrite, and even halite, can be precipitated. If sufficient moisture falls, it soaks into the sediment, but soon afterward evaporation draws it up again. As this water vanishes at the surface, it leaves behind additional evaporite minerals.

Sabkhas are of two kinds: (1) *interior continental* and (2) *sea marginal*. We discussed sea-marginal sabkhas in Chapter 12. Here we present interior continental sabkhas.

As mentioned, sabkhas are equilibrium deflation-sedimentation surfaces or *deflation-sedimentation windows* through to the local water table. The capillary fringe above the water table marks the base level of wind deflation. The wind tends to blow away sediment above this

FIGURE 14-20.   Sabkha, in foreground, being encroached upon by migrating dune sand. Fault scarp and highlands are in background. Southern Arava Valley, north of Gulf of Aqaba, Red Sea. (Authors.)

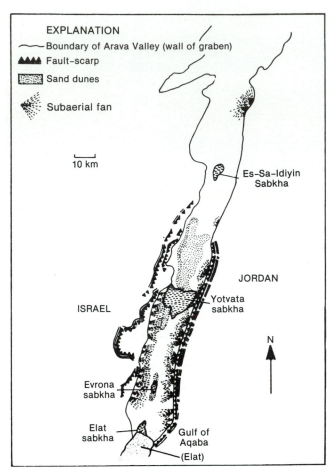

FIGURE 14-21.   Sketch map of Arava Valley, north of Gulf of Aqaba, Red Sea, showing depositional environments of desert: fans, sand dunes, sabkhas, and flood plain of intermittent stream (Wadi Arava; white area). Fan deposits line wall of gravel. (E. Orni and E. Efrat; modified by A. J. Amiel and G. M. Friedman, 1971, fig. 2, p. 583.)

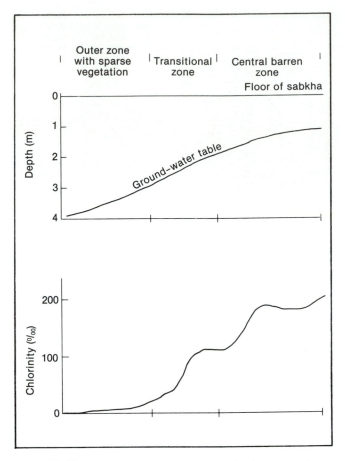

FIGURE 14-22.   Profiles showing changes in depth to ground-water table and chlorinity across Yotvata Sabkha in Arava Valley between Dead Sea and Red Sea. (A. J. Amiel and G. M. Friedman, 1971, fig. 10, p. 587.)

capillary fringe. Left behind is a surface controlled by the level of the ground-water table.

An example of a continental sabkha is the Yotvata Sabkha in the southern Arava Valley between the Dead Sea and the Red Sea (Figure 14-21). Within the valley, short intermittent streams do not form a continuous network. Some of them flow southward toward the Red Sea, but most of them do not reach it. Fans block these watercourses. When water is present in these streams, it tends to seep into the sediment. The Yotvata Sabkha lies between two dune fields, an extensive one to the north and a narrow one to the south. In places, dunes have encroached over part of the sabkha.

The sediments underlying the sabkha consist of sand, silt, and clay brought in by the intermittent streams from the slopes of the graben wall and by wind from the north. The sabkha surface slopes away from the graben wall; in the center of the valley it is nearly horizontal. Depth

to ground water, composition of ground water, and distribution of vegetation form the basis for subdividing the sabkha into three zones: (1) a central zone, barren of vegetation, (2) a transitional zone, and (3) an outer zone with sparse vegetation (including tamarisks, rushes, and palms). Traced from the outer- to the central zone, the ground-water table rises toward the sabkha floor. In this direction, also, chlorinity increases (Figure 14-22).

Within the framework of sedimentary particles in the outer zone and in part of the transitional zone, gypsum is precipitated. In part of the transitional zone and in the central zone, halite is precipitated (Figure 14-23). Commonly in deserts, wind deflates the interdune areas down to the water table, hence forming local sabkhas (Figure 14-24).

Sabkhas may be adjacent to playa lakes. Subsurface Permian deposits underlying northwestern Europe have been interpreted as products of a continental sabkha that was next to a playa lake (Figure 14-25). The inferred sabkha deposits, known from samples collected from boreholes, consist of interbedded sandstones and shales.

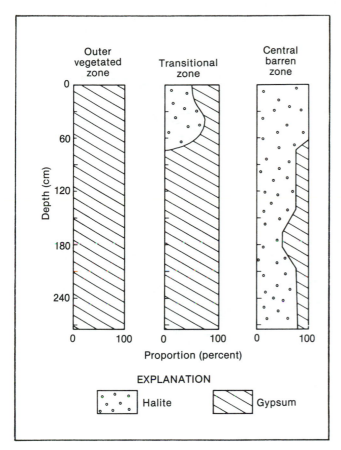

FIGURE 14-23.   Distribution of gypsum and halite in sediments of Yotvata Sabkha, Arava Valley. Percentages refer to authigenic evaporite minerals only. (A. J. Amiel and G. M. Friedman, 1971, fig. 10, p. 587.)

These are analogous to the sands, silts, and clays of the Yotvata Sabkha. In the Permian example, the calcium-sulfate mineral is anhydrite rather than gypsum. Under subsurface conditions, anhydrite is the stable form. The deposits of the inferred Permian sabkha interfinger with sediments that have been regarded as products of dunes, playa lakes, and intermittent streams.

SOURCES: Glennie, 1983; Handford, Loucks, and
G. R. Davies, 1982; Warren, 1989.

Playas

The word **playa** is Spanish, and means a shore, strand, or bank of a body of water. English-speaking geologists commonly apply the term to the *dry bed of a playa lake.* Playas occupy *broad, shallow depressions in desert regions.* On occasion, these depressions are covered with thin sheets of water. (See Figure 14-2.)

The closed depressions in which playas form are usually the lowest places locally available; they generally lack external drainage. The playa water drains away by in-

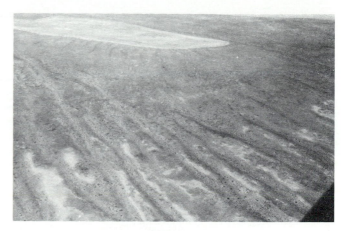

FIGURE 14-24.   Oblique aerial view of linear dunes (foreground) that end in the middle part of the photo. Beyond the ends of the dunes is a sabkha (dark gray) and beyond it, a playa (light gray), South Australia. (Authors.)

filtrating into the sediments and by evaporation. These waters dissolve and remove minerals. Other losses result from deflation.

Infrequent floods bring sediments to playas. In some areas the floods are annual and the playa is covered by an ephemeral lake. In other areas, playas are flooded by infrequent storms that drop moisture in the surrounding mountains. Such storms may happen at any time of the year. In either case, the water brings fine sediments to the playa. Materials coarser than sand typically are left behind on fans, on pediments, and on the floors of the intermittent rivers. Between floods, the water of the playa leaves behind evaporite minerals and the suspended fines. As a result, playa sediments consist of interbedded thin layers of sand, silt, clay, and evaporite minerals. Because playas are dry most of the time, conditions in the surface sediment remain oxidizing. The colors of surface sediments are red and brown. Below a thin oxidized layer, reducing conditions may prevail and the fine sediments are gray or black.

The alternate flooding and drying tends to leave behind cyclic sediments that consist of *couplets* or even of *triplets.* The couplets include a coarse sandy- or silty layer that grades upward into a much finer-textured clay. A triplet is formed if the clay layer is overlain by or contains a thin film of evaporite minerals, such as calcite, anhydrite, dolomite, gypsum, halite, or the sodium carbonates trona and shortite. Such couplets and triplets form when the coarser sediments settle out of the flood waters and are further distributed by wind-driven waves and -currents. The finest sediments may be kept in suspension until the water disappears. The next flood may locally rework some of the upper sediments, redissolve some soluble minerals, and deposit another couplet. The thickness of playa cycles ranges from a few centimeters

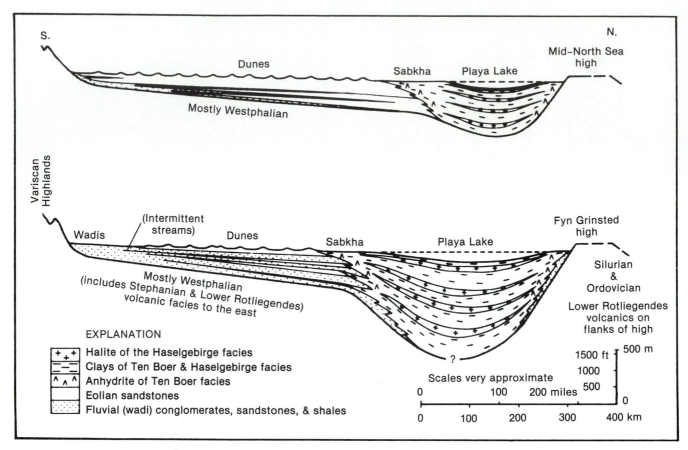

FIGURE 14-25.   Restored stratigraphic diagram through the Permian Rotliegendes Basin in southern North Sea and eastern Netherlands just prior to the Zechstein (Permian) transgression. Products of several depositional environments are interstratified. (W. Glennie, 1972, fig. 17, p. 1067.)

to a few meters. Cyclic playa deposits are similar to those deposited in large lakes, described in a following section of this chapter.

The floor of Salt Flat Graben, an intermontane basin in West Texas, is a typical playa (Figure 14-26). The av-

FIGURE 14-26.   View of old playa lake bottom and sediments underlying it, Salt Flat graben, West Texas. Dark-gray area in distance is the west-facing escarpment of the Guadalupe Mountains. (Authors.)

erage width of this graben is approximately 17 km. Wells have been drilled to depths of between 500 and 600 m without penetrating the entire succession of Quaternary basin-fill sediments. The sediments of this fill are partly extrabasinal and partly intrabasinal. The color of the upper 0.5 m of the sediments is brown, indicating oxidizing conditions. Below this depth the color for the most part is gray, which indicates reducing conditions. The evaporite deposits are finely laminated (Figure 14-27). The predominant minerals are gypsum and halite, but calcite, aragonite, and dolomite locally are also present. Some layers have been blackened by iron sulfide; in them are discontinuous lenses of sulfur. If a rod is punched through the dolomite that locally forms a surface hardpan, abundant hydrogen sulfide is released. This association of hydrogen sulfide, iron sulfide, and native sulfur suggests that gypsum has been destroyed by bacterial processes. (See Eq. 3-10.)

An example in which evaporites have accumulated in the center of a structural basin is the region of the Dead Sea (See Figure 14-7.), where, during the Pleistocene Epoch, a salt sequence at least 4000 m thick was deposited under nonmarine conditions. Because the salt

FIGURE 14-27. View of wall of a trench in playa-lake sediments that consist of finely laminated layers of differing minerals. Dolomite interbeds, blackened by iron sulfides, are present at the bottom and a short distance above the bottom. (G. M. Friedman, 1966, fig. 2, p. 264.)

interfingers with fan sediments, one cannot escape the conclusion that tectonic subsidence amounted to 4000 m. The average rate of subsidence was 4 m/1000 yr.

Another example is Lake Eyre, central Australia. The area of Lake Eyre is 9200 km². It occupies a graben depression in the driest part of Australia and its drainage area is about 1,300,000 km². As with the modern Dead Sea, the surface of Lake Eyre lies below sea level.

An example of a structural basin of Permian age in which nonmarine-playa evaporites accumulated and interfinger with the products of other desert depositional environments is the Rotliegendes Basin in the southern North Sea and eastern Netherlands. (See Figure 14-25.) The playa lake lay at the center of the structural basin and thus it and the sediments from associated desert environments became thicker than the sediments deposited in other settings. In these Permian deposits, halite interfingers with clays.

SOURCES: Handford, Loucks, and G. R. Davies, 1982; Southgate, Lambert, Donnelly, Etminan, and Weste, 1989; Ward, 1988; Warren, 1989.

Fluvial Environments: Modern- and Ancient Deposits

Rivers fall into two major categories: (1) braided and (2) meandering- or flood-plain rivers. Braided streams are *streams characterized by their relatively low sinuosity (the ratio length of channel : length of valley) and by their multiple unstable channels separated by bars that are largely immobile except at flood stage.* By contrast, meandering streams are *highly sinuous streams that generally possess a single channel*

with stable cohesive banks. Meanders migrate downstream in a regular fashion, producing characteristic vertical successions (described in a later section) unlike any produced by braided streams. In addition, braided streams tend to carry coarser sediment and possess higher gradients than do meandering streams, although the two kinds overlap considerably.

Because of changing water levels, fluvial sediments may be exposed to the atmosphere for longer time periods than they are covered by water. Accordingly, fluvial sediments are subjected to the effects of the wind. Box 14.1 explores some interactions between fluvial- and eolian deposits.

Braided Streams

The depositional processes of braided streams are less well known than those of the more sedate meandering streams for three reasons. (1) Braided streams are generally distributed in remote- or sparsely settled areas, such as contiguous to active glaciers or in far-distant deserts. (2) Braided streams flow swiftly and can incorporate large particles into large bed forms. As a result, field studies are dangerous and flume studies impractical. (3) Drilling through sand and gravel, which compose most deposits of braided streams, is difficult and expensive. Thus, in comparison with meandering rivers, our knowledge of the stratigraphy of braided-stream deposits is less complete. No general model for braided alluvium has yet been derived, but models have been proposed for certain kinds of braided streams. In the following paragraphs we illustrate some models that may be broadly applicable.

The flow in braided streams fluctuates. In arid regions, braided streams flow only seasonally, and then only for a few days or even only hours. Most of the time the streams are dry. Braided streams form on surfaces of moderate- to high slope and at abrupt changes in slope. Commonly the water is shallow (a few meters or less). They are characterized by multiple unstable channels. Within their channels braided streams deposit longitudinal bars (Figure 14-28). Transverse bars are known, but are less common than longitudinal bars. As a longitudinal bar builds vertically above the stream channel, the channel bifurcates. Bars build up rapidly, and newly formed channels cut earlier bars. Many braided streams are almost choked in their own sediment. The particle sizes of bar sediments decrease downstream, especially where particles are very large (maximum particle size, tens to hundreds of centimeters). In our discussion of fans we briefly noted the work of braided streams. Braided alluvium on fans is characterized by larger particle sizes and steeper alongstream particle-size gradients. In addition, a fan deposit forms a segment of a cone characterized by divergent current directions, whereas braided-stream deposits form linear shoestring-like sand- or gravel bodies

BOX 14.1

Fluvial-Eolian Interactions

Where arid regions meet perennial streams, fluvial- and eolian depositional processes interact. These interactions may take several forms and result in characteristic sedimentary deposits. Several combinations are possible: (1) rivers bring water and sediment into a setting where the sediment can be dried out and blown by the wind, (2) running water may erode and rework eolian deposits, or (3) the wind may erode and rework fluvial deposits. Because water is a far-more-competent agent of erosion than is wind, rivers are more likely to erode eolian deposits than vice versa. However, eolian erosion of fluvial deposits may be locally important on bare rock surfaces in deserts, on the unconsolidated deposits of intermittent streams, or where a climatic change toward increasing aridity causes an eolian sand sea to prograde across and to rework a fluvial deposit starved by the progressive loss of stream flow. Water from a stream may flow into a playa and dissolve evaporite minerals. However,

the major fluvial effects on eolian deposits are erosional and depositional. Rivers may erode eolian dunes (Box 14.1 Figure 1). When they overflow, rivers may carry sediment into eolian depositional environments. This can happen during flood stages or when, by avulsion, river channels are diverted into new territory where eolian deposits may have been accumulating. Flood waters may both erode dunes and deposit sediment in interdune areas in a single event. Repetitive flooding of an eolian environment by a nearby river will result in an alternating sequence of upward-fining fluvial sediments (from base to top: basal scour, coarse pebbly lag deposits, normally graded sand- and silt sequences with upwardly decreasing thickness of individual cross-bed sets, and clay drape) and eolian dune- and interdune deposits.

SOURCES: Langford, 1989; Langford and Chan, 1989.

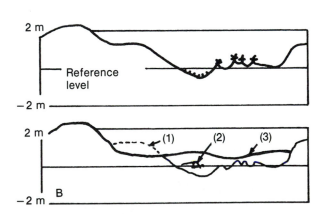

BOX 14.1 FIGURE 1.   Modification of eolian sediment by rivers.
   A. Photograph showing eolian sand dunes that were eroded by Medano Creek, Colorado, during the 1985 flood. The dune on the right, 4.5 m high, was completely removed. (R. P. Langford, 1989, fig. 8, p. 1030 and figs. 4 A and 4 B, p. 1027.)
   B. Profiles of interdunes showing the effects of fluvial/eolian interactions. The upper sketch is a measured profile of a nonflooded eolian interdune surface at the Mojave River Wash, California. The lower sketch shows the changes to the upper sketch that would result from overbank flooding, including (1) erosion and reworking of the interdune surface, (2) rapid deposition of fluvial sediment, and (3) deposition of a clay drape over interdune and flanking dunes. (After R. P. Langford, 1989, fig. 4, p. 1027.)

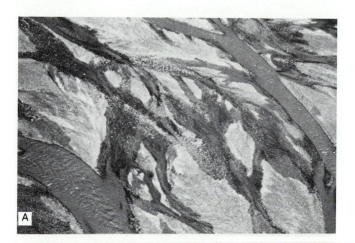

FIGURE 14-28.   Bar- and channel morphology of a gravelly fan.

A. Aerial view of coarse gravel longitudinal bars on the upper Scott fan. Downstream is to the upper left; bar lengths range from 30 to 50 m.

B. Ground view looking downstream in the same general vicinity as the aerial view of A. The vehicle is parked on a thin, sheet-like longitudinal bar. (J. C. Boothroyd and D. Nummedal, 1978, fig. 9, p. 652.)

porting elsewhere. The channel sediments in transit during times of vigorous flow consist of sand and gravel that accumulates under conditions of the upper-flow regime giving type J and type K dunes. (See Table 7-2.) As the current wanes and the water disappears, fine sediment settles out of suspension to make a fine-textured top of the sequence. Ideally then, an upward-fining succession, or normally graded sequence, would result. At the base of such a vertical succession would be an irregular surface scoured by the channel. On it would be a coarse basal lag. Next above are variously cross-stratified sands, coarser below and finer above. At the top, completing the graded succession, may be silt or clay. However, finer-textured tops of such upward-fining successions are rarely preserved. The next flood usually removes them.

The deposits of braided streams consist mostly of sand and gravel; muds are subordinate or absent. This paucity of muds or shales in the rock record distinguishes the deposits of braided streams from those of meandering streams. The bulk of deposits of sandy braided streams has been reworked by traction into cross-laminated- and horizontally laminated sands. The proportion of these two kinds of sands depends on hydrodynamic- and geomorphic settings.

Braided streams may be classified as gravelly or sandy. Gravelly- and sandy braided streams form a continuum controlled by proximity to a source of coarse sediment. In addition to proximity to the parent deposit(s) and the correlated systematic variation in slope, braided streams are also strongly affected by water depth. The effects of this variation are better known for sandy braided streams than for gravelly braided streams.

SOURCES: Cant, 1982; Miall, 1978; Rundle, 1985; Rust and Gibling, 1990.

SEDIMENTS OF GRAVELLY BRAIDED STREAMS

Gravelly braided streams are found in high-relief areas. The areal extents of such streams are limited. Downstream they grade into sandy braided streams. In gravelly braided streams, particle sizes decrease rapidly downstream (although much less rapidly than on coarse fans). Vertical sequences formed by proximal gravelly braided streams include the deposits of debris flows and horizontally bedded- and imbricated gravels formed by the migration of longitudinal gravel bars (Figure 14-29). Trough- and tabular cross strata characterize the deposits of distal gravelly braided streams.

Paleozoic gravelly alluvial sequences include gravelly braidplains (rare in the modern world) formed in humid regions before terrestrial vegetation had colonized upland areas. In contrast to the semiarid- and paraglacial

with parallel current directions. (See Figure 14-6.) *A surface underlain by braided alluvium not deposited on fans but not confined by valleys is a* braidplain. Braidplain deposits resemble those of braided streams in that current directions are relatively uniform. However, the vertical sequences underlying braidplains differ from those formed by braided channels, because on braidplains, lateral confinement of channels is less important than in braided channels.

Vertical sequences of modern braided streams are much harder to establish than are those of meandering streams. Nevertheless, some attempts have been made to formulate general models.

A process basic to all braided streams is erosional scour. Stream beds are paved with a lag concentrate of coarse debris that the stream has not succeeded in trans-

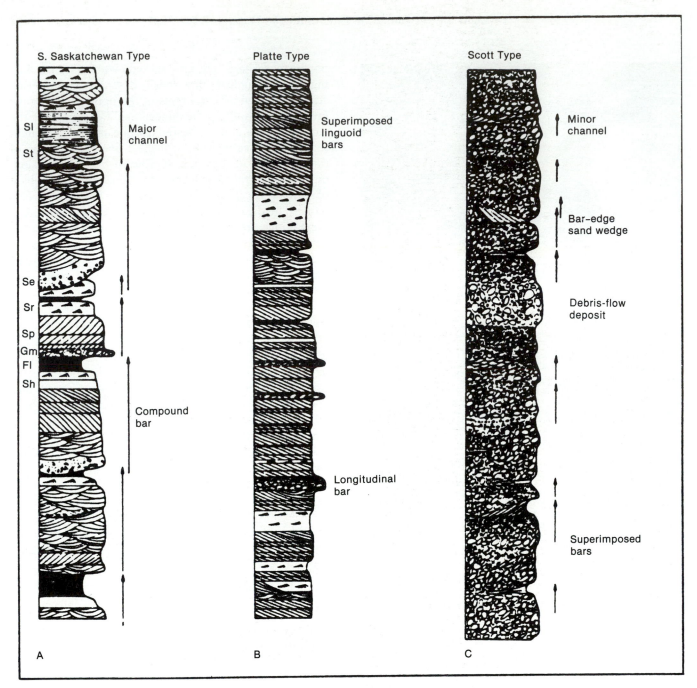

FIGURE 14-29. Vertical facies models for braided-stream deposits. Arrows show small-scale cyclic sequences. Conglomerate clasts are not shown to scale. SL = sand, fine; St = sand, medium to very coarse; Se = erosional scours; Sr = sand, very fine to coarse; Sp = sand, medium to very coarse, may be pebbly, Gm = massive or crudely bedded gravel; Fl = sand, silt, mud; Sh = sand, very fine to very coarse; Gt = gravel, stratified.

A. South Saskatchewan River model. The South Saskatchewan River is a deep sandy braided stream, and contains abundant trough cross strata in the lower parts of upward-fining sequences.

B. Platte River model. The Platte River is a shallow sandy braided stream and lacks trough cross strata.

C. Scott fan model. The Scott fan is composed of gravelly braided alluvium, and its vertical facies sequences resemble those of proximal gravelly braided rivers. The vertical sequence of the Scott fan is dominated by massive gravelly deposits of longitudinal bars and of debris flows. These facies sequences are models, synthesized from numerous observations of the deposits of the South Saskatchewan and Platte rivers, and of the Scott fan, as well as of other braided rivers and fans, and do not constitute actual observed sequences. Further discussion in text. (A. D. Miall, 1978, fig. 1, p. 600–601.)

settings within which modern gravelly braided streams are confined, gravelly Paleozoic braidplains are thought to have formed in regions that experienced deep- and frequent floods. Consequently, planar-tabular cross strata formed by modification of longitudinal bars during waning-stage flow are found in sandstones that constitute an important though subordinate facies of these Paleozoic deposits.

SOURCES: Ashmore, 1991; Koster and Steel, 1984; Miall, 1978; Steel and Thompson, 1983.

SEDIMENTS OF SANDY BRAIDED STREAMS

Sandy braided streams differ from meandering streams (described below) in several respects. Sandy braided streams flow on steeper slopes, transport greater loads of sediment that is too coarse to be carried in suspension under normal conditions, and possess relatively unstable banks. Because the banks of sandy braided streams are unstable, massive sand bodies in the deposits of such streams are attributed to failure and slumping of bank material. Sandy braided streams may be classified as shallow or deep, with the boundary between the two kinds probably falling between 2 and 3 m at bankfull stage. Vertical sequences of shallow sandy braided streams are dominated by upward-shallowing successions consisting of planar-tabular sets of cross strata (formed by the migration of longitudinal rhomboid- or linguoid bars) overlain by ripple-cross-laminated sand and capped by muddy overbank deposits. (See Figure 14-29, A.) By contrast, deep sandy braided streams contain both planar-tabular and trough cross strata. The trough cross strata are formed by the migration of cuspate dunes in the deeper portions of the sandy braided stream. (See Figure 14-29, B.) Both kinds of sandy braided streams produce vertical sequences in which pebble lags overlie a basal erosion surface. These give way upward into cross-stratified sands, which may be capped by overbank- or stagnant-pool muds. Commonly, sandy braided streams grade downstream into meandering (flood-plain) rivers.

An example of an inferred deep-water sandy braided-stream deposit from the rock record is the Devonian Battery Point Sandstone of Canada. In this formation the idealized vertical sequence starts with a truncation surface (eroded at the former channel floor). Above this the sequence displays alternating trough- and planar-tabular cross-stratified sandstone capped by small sets of planar-tabular cross strata, approximately 7.5 m thick, and overlying alternating rippled sandstone, shale, and some low-angle cross-stratified sandstone, 1.5 to 2 m thick (Figure 14-30). The lower part of the sequence has been interpreted as being formed by shifting channels that are overlain by bar-top deposits. The top part

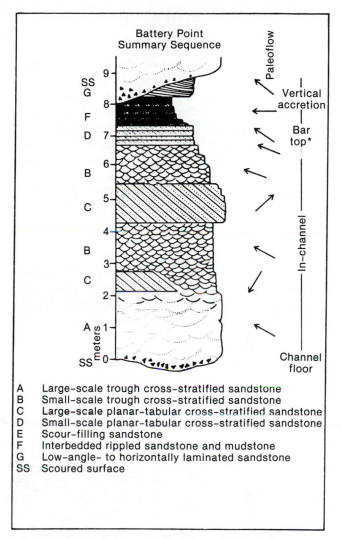

A   Large-scale trough cross-stratified sandstone
B   Small-scale trough cross-stratified sandstone
C   Large-scale planar-tabular cross-stratified sandstone
D   Small-scale planar-tabular cross-stratified sandstone
E   Scour-filling sandstone
F   Interbedded rippled sandstone and mudstone
G   Low-angle- to horizontally laminated sandstone
SS Scoured surface

FIGURE 14-30.   Idealized vertical facies sequence, Devonian Battery Point Sandstone. This facies sequence is a local model, synthesized from a measured section composed of many actual sequences, and does not constitute an actual observed sequence. This vertical-sequence model is similar to that of the South Saskatchewan River. Arrows show inferred paleoflow directions; read each as if it were a map view of a compass needle with North at the top. (D. J. Cant and R. G. Walker, 1976, fig. 16, p. 114.)

of the sequence consists of products inferred to have formed during the waning stages of stream floods. In places, some progressive upward decrease in particle size has been recognized, but upward fining is not as common as in meandering rivers. The vertical sequence just described resembles that of the South Saskatchewan River (See Figure 14-29, A.), which may be a good model for deep sandy braided streams.

SOURCES: J. R. L. Allen, 1983; Blodgett and Stanley, 1980; Cant, 1978; R. G. Walker and Cant, 1984.

Meandering Streams: Modern and Ancient Deposits

In our discussion of fans and braided streams we have already touched upon a few aspects of fluvial sedimentation. Fans and braided streams commonly occupy arid environments or mountainous terrain or are confined to valleys having steep profiles in the interiors of continents at some distance from the sea. However, any stream incapable of transporting its bed load can be a braided stream. In this section we concern ourselves with meandering streams that spread out in low-relief valleys in the interiors of continents or on low-relief coastal plains.

Under the heading of fluvial environments we include the sites of sediment deposition associated with rivers having a substantial **base flow**, that is, *the day-in, day-out quantity of water contributed to a stream by groundwater sources*. The stream of water itself will be our chief concern. Thanks to the base flow, such streams contain enough water to keep the channel substantially full. We

focus on sinuous stream channels associated with flood plains, and on associated environments located on **alluvial plains**, *surfaces having low relief that are underlain by sediments deposited mostly by rivers.* We shall discuss the everyday discharge levels and the effects of floods on flood-plain rivers.

SOURCES: Baker, 1978; Carson, 1984; Collinson and Lewin, 1983; G. M. Friedman and Sanders, 1978; Miall, 1978; R. M. H. Smith, 1987; R. G. Walker and Cant, 1984.

SEDIMENTS OF MEANDERING RIVERS: BASE-FLOW CONDITIONS

Meandering rivers are active streams that flow in definite channels. The channels are generally under water most of the time; *the deepest part of a stream channel* is called the **thalweg**. Typically, such water-filled channels meander and are bordered on each side by *low, rounded ridges of very fine sand and coarse silt* known as **natural levees**. *A*

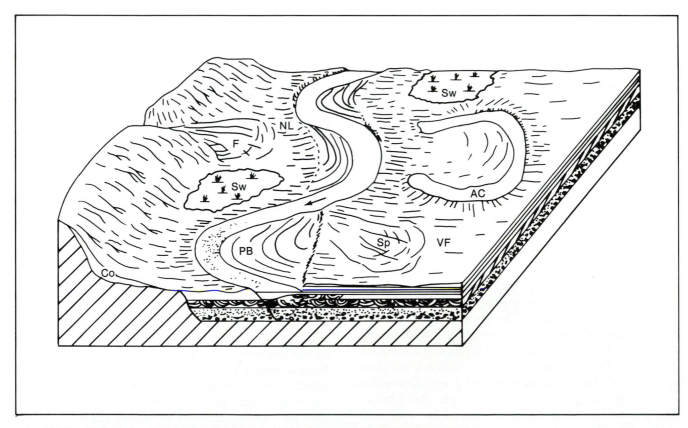

FIGURE 14-31.   Oblique view from above of schematic block diagram of one edge of an alluvial plain having a master stream with well-developed meanders. Slope wash from valley wall at left contributes colluvium (Co). Tributary stream at left has built fan (F) onto alluvial plain. NL = natural levee; PB = point bar; Sp = splay deposit; Sw = swamp; AC = abandoned channel (now an ox-bow lake and filling with fine sediments); VF = valley floor. (Suggested by drawings in S. C. Happ; G. Rittenhouse; and S. C. Dobson, 1940, fig. 6, p. 27; J. R. L. Allen, 1964, fig. 4, p. 168; R. J. LeBlanc, 1972, fig. 15, p. 151; and G. S. Visher, 1972, fig. 11, p. 94.)

meandering channel and its flanking discontinuous natural levees constitute a **meander belt**.

Extending from the meander belts to the lateral margins of the valley-floor lowland are **flood plains**: *low, flat valley-bottom areas that are covered by water only during a flood. Closed depressions within a flood plain that may be under water for long periods* are **flood basins** or **backswamps** (Figure 14-31).

The deposits of meandering rivers consist of three contrasting suites: (1) the channel deposits, (2) the channel-marginal deposits (natural levees and associated sediments), and (3) the **overbank deposits** (*the collective term for all sediments deposited by river water on a valley floor outside the stream channel*). As we shall see in the following discussion, the channel deposits grow chiefly by lateral sedimentation, whereas most of the overbank deposits accumulate by vertical sedimentation.

Stream Channel. We have already noted that water flowing in a definite channel tends to form meanders (Figure 14-32). Meanders result from a large-scale spiral secondary flow pattern within the turbulent water. At the surface, the current flows toward the outside of the meander bend. As a result, along the outside of each bend, the water piles up; in technical terms, its surface becomes superelevated. This *outer, steep side of the channel in a meander* is also known as the **cutbank**. The heaped-up water creates a pressure gradient. The combination of pressure gradient and cross-stream variation in bed shear as a result of the change in angular acceleration causes helical flow. A component of this helical flow continues as a countercurrent along the bottom directed toward the convex side of the channel (as seen in map view; Figure 14-33). The spiral-flow pattern deepens the channel in

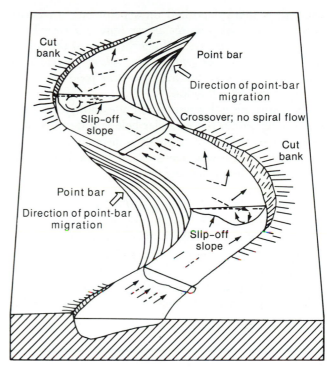

FIGURE 14-33. Schematic view from above looking downstream at two meander bends and intervening crossover. Heaping up of water against outside shores of bends creates spiral flow that deepens channel in bends. In between meander bends (crossover reaches), water becomes shallow and lacks spiral flow. Solid arrows indicate directions of current at surface; dashed arrows, directions of current along bottom; circular arrows having only one barb, directions of spiral flow in plane of shaded transverse sections. Further explanation in text. (J. E. Sanders in G. M. Friedman and J. E. Sanders, 1978, fig. 8-39, p. 221; based on J. F. Friedkin, 1945.)

the bend. Some eroded sand is deposited on the opposite bank; much more is deposited on the convex bend of the next-downstream meander. Immediately downstream from each bend is *a short straight reach between meanders* known as a **crossover**, where the water becomes shallower. Farther downstream, the surface current flows toward the bank and heaps up water against the outside bank of the next bend. This causes another spiral flow in the opposite sense from the previous spiral, a deepened channel, and so forth. Thus, using reference directions, right and left as one looks downstream in a meandering channel, the spiral flow is clockwise in a meander bend that curves to the left and counterclockwise in a meander bend that curves to the right.

These flow spirals undermine the sediment composing the concave banks (cutbanks); the undermined sediment collapses into the current. Some of the sediment that thus enters the stream is carried across to the convex bank directly opposite, but most of it is carried downstream to the next convex bank and, unless it is very fine

FIGURE 14-32. View from airplane of small meandering stream in western Kansas, United States. White areas next to the stream are sands deposited on point bars. (Authors.)

textured, is there deposited on a **point bar**, *a crescent-shaped bar built out from the convex bank of a meandering stream channel* (white areas along the sides of the stream in Figure 14-32). *The gently sloping side of the channel of a meandering stream that leads from the point bar to the thalweg* is the **slip-off slope**.

The entire meander system migrates slowly downvalley, keeping within the meander belt as it does so. This migration of the channel builds a distinctive sequence composed of crossover deposits and the sediments described in the following section.

Point-Bar Sequence. *The sequence of sediments deposited by the lateral migration of a sinuous stream channel* has been named a **point-bar sequence**. The characteristics of the sequence vary according to the size and depth of the channel. We shall examine the point-bar sequence as established by detailed studies of the Brazos River meander belt southwest of Houston, Texas (Figure 14-34).

The relationships beneath the water of the low stages of the Brazos River are not well known, but the low-stage depth in a meander bend is about 5 m, which qualifies it as a "deep" channel. We refer to it as "deep" because its depth exceeds 4.5 m (about 15 ft), the minimum depth required for the formation of large-scale dunes (heights of 2 m or so and wave lengths of > 50 m). In the following analysis we illustrate how a sequence can form by the lateral shifting of a channel at base-flow discharge.

At the very bottom of a channel, no matter what its depth, one finds a **channel-floor lag**, *a deposit of the coarsest material available that accumulates on the floor of an alluvial channel*. The channel-floor lag can contain pebbles, mud chips, water-saturated logs and -twigs, bones, refrigerators, and just about any other kind of coarse debris. The minimum thickness of the channel-floor lag equals the thickness of a *single layer* of coarse particles. Associated with and overlying this lag may be a layer of coarse particles that are moved only under extreme conditions.

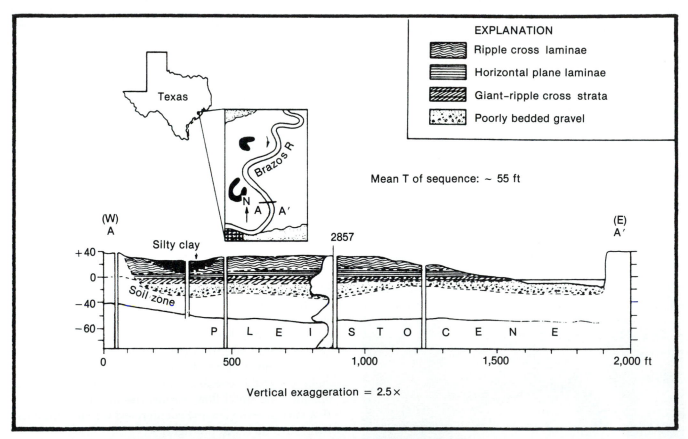

FIGURE 14-34.   Profile and section through point-bar sequence deposited by lateral shifting of Brazos River near Richmond, Texas. Contrast the mean thickness (about 55 ft or 16.7 m) of the point-bar sequence with the depth of water in the channel at a low stage (15 feet or 4.5 m). To left of boring R2857 is sketched an idealized SP log that might be expected from a hole drilled through a point-bar sequence. For explanation of SP log, see Chapter 16. Boring R2857 was not actually logged. Compare this idealized expected SP trace with the actual trace of an SP log recorded in boring R2857-A, only 30 ft (9 m) away from boring R2857. (See Figure 14-37.) (Modified from H. A. Bernard, C. F. Major, Jr., B. S. Parrott, and R. J. LeBlanc, Sr., 1970, fig. 6, facing p. 7.)

These associated coarse particles may occupy interstices between the particles of the channel-floor lag, and also may form a coarse layer overlying the channel-floor lag. In a large, deep channel, the thickness of this associated layer of coarse particles may be several meters or more.

The downstream migration without change of profiles of large-scale dunes creates large-scale cross strata. Assuming that migration occurs under conditions of the lower-flow regime and that fine sand is present, then small-scale cross laminae from migrating backflow ripples will underlie the large-scale cross strata. (See Figure 7-20, A.)

In the parts of the channel that are shallower than 4.5 m, no large-scale dunes can form, so the scale of cross strata is much smaller than in the deep-channel sediments; such smaller cross strata form by the migration of large- and small ripples. The small ripples are present only in fine sand and very fine sand. Rippled sand and plane-bedded laminated sand are found not only in the shallow parts of the channel, but also on the exposed surface of point bars, where they form during the falling stages of flows that covered the point bars.

The point-bar sequence spreads in the direction in which the channel migrates. The extent of channel migration is thought to be controlled by sinuosity. A stream having low sinuosity lacks well-developed meanders. Such a stream can migrate completely across the valley floor and thus the point-bar sequence may form a blanket extending across the entire alluvial plain.

It is clear from detailed studies of numerous cores from the meander-belt point-bar sediments of the Brazos River that the sedimentary structures and bedding characteristics created in the channel at low-water stages almost certainly will not become a part of the stratigraphic record. At low stages, the channel of the Brazos River is only about 5 m deep. By contrast, borings indicate that the point-bar succession is about 15.5 m thick. The base of the point-bar sequence extends downward 4 to 5 m below the level of the bottom of the low-water channel and the top of the sequence extends upward 5 to 6 m above the level of the water in the channel at low-stage discharges. Therefore, if we are to understand the story of the point-bar succession, we must find out what happens during floods.

SOURCES: Baker, 1978; G. M. Friedman and Sanders, 1978; Langford and Bracken, 1987; Miall, 1978; Walker and Cant, 1984.

SEDIMENTS OF MEANDERING RIVERS: FLOOD CONDITIONS

When a stream floods, many new environmental conditions come into play. These conditions differ according to whether the *flood bulge* is on its way into a given reach

(rising stage) or whether the crest of the bulge has passed through a given reach (falling stage). The large influx of water may come from rainfall, from melting snow or ice, or from sudden bursting of a dam. Increased runoff into the tributaries and from the tributaries into the master stream can cause a bulge of water to pile up faster than it can flow away down the main channel. A **flood bulge** itself is *a long, low wave of water that moves along with, yet independently of, the base flow.* A simplified illustration of the two motions of the flood bulge and the base flow is provided by a person (compares to flood wave) walking on and in the same direction as a continuously moving conveyor belt (compares to base flow), such as are found in certain large airports.

Such a bulge of water is known as a *kinematic wave* (Figure 14-35). In engineering terms, a kinematic wave gives rise to a condition known as unsteady (or, as we would write, nonsteady) flow. One of the characteristic features of a kinematic wave is that it enables the time of high water to become decoupled from the time of maximum concentration of suspended sediment. The suspended sediment is locked into the base flow, so to speak. But the crest of the kinematic wave moves along the top of the base flow.

During the passage of the flood wave, the water is deeper and flows faster than it does during times of base flow. The base flow existed before the flood wave arrived and will still go on after the flood bulge has passed down the channel. Because of this before-and-after similarity of conditions within the channel, the flood may effectively cover its tracks; that is, after the flood, the sediments in the channel may appear to be very much as they were before the flood arrived. Did the flood really do anything in the channel? A closer look is required to discover the effects of a flood on the channel. By contrast, flood effects outside the channel are easy to see.

SOURCES: Baker, 1978; Coleman, 1969; G. M. Friedman and Sanders, 1978; Lane and Borland, 1953; Langford and Bracken, 1987; Miall, 1978; R. G. Walker and Cant, 1984.

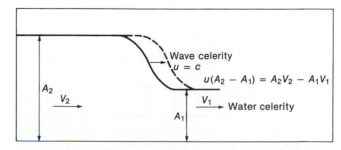

FIGURE 14-35.   Celerity of flood bulge in a river as a kinematic wave. *A* = cross-sectional area; *V* = speed of water. Further discussion in text. (P. B. Bedient and W. C. Haber, 1988, fig. 4.8, a, p. 240.)

Relationships of Point-Bar Sequence to Floods. During a flood, as everyone knows, the absolute altitude of the water surface of the channel increases. Down below, at the bottom of the channel, the situation during a flood is not so obvious; the spectrum of possibilities ranges rather widely. One extreme condition is deep scour; that is, the flood waters mobilize and disperse some of the sediments that normally lie motionless beneath the channel floor. The absolute altitude of the channel floor is lowered. Thus the channel is deepened both by adding more water at the top and by taking away sediment from the bottom (Figure 14-36). It is not known exactly what happens to these dispersed sediments nor if all bottom sediments behave in the same way. Surely the sediments become *dilated* (See Figure 7-2, B.), and some of them may even go into turbulent suspension. How far they travel downvalley and what kind of deposits they form are intriguing questions to which we have no answers. The other extreme is represented by deposition of sediment in the channel. The circumstances governing channel scour or -deposition during floods are little known. In a stream having well-developed meanders, a flood tends to scour in the meander bends and to fill in the crossovers. When such a flood subsides, it tends to fill in the meander bends and to erode the crossovers.

Although the details of how the various units were deposited have not been observed during floods, the numerous borings in the Brazos River point bars have established a definite autocyclic point-bar succession that

resembles a parasequence (Figure 14-37). At the base is a poorly bedded gravel about 7.5 m thick. This is overlain by a unit of medium- to fine-textured sand that is distinguished by giant-ripple cross strata. Above it comes a zone of fine- to very fine-textured sand, 4 m thick. At the top is a unit of ripple-laminated, very fine sand to coarse silt that is 4 m thick.

The upper part of a point-bar sequence varies according to local circumstances. A generalized point-bar-sequence model resembles the Brazos River point-bar sequence but is capped by fine-textured vertical-accretion deposits (Figure 14-38). Where lobes of sediment spill over high areas and prograde into local depressions, they form medium-scale planar-tabular cross strata (Figure 14-39). In the example shown in Figure 14-39, such planar-tabular cross strata are overlain by horizontal plane laminae, which cap the point-bar sequence.

The Brazos River point-bar sequence just described is coarse at the base and becomes fine at the top. The upward decrease in particle size may be accompanied by a decrease in the scale of sedimentary structures.

The change of particle size is reflected on the self-potential (SP) curve of geophysical logs made in borings through a point-bar sequence. The SP curve measures the amount of current that flows between the drilling mud in the hole and the material surrounding the mud. In coarse, porous materials a relatively large current flows, and this causes the trace of the curve on the log to move to the left of the zero line (taken arbitrarily at the right). In fine-textured materials having low porosity and permeability, the amount of current that flows between the drilling mud and the material walling the hole is low. Thus the SP curve remains close to the zero line. An example of an SP curve through the recent point-bar sediments of the Brazos River is shown in Figure 14-37. In Chapter 16 we present further details on SP curves and other geophysical logs.

One should expect to find such upward-fining sequences wherever any sinuous meandering river channel migrates laterally. Upward-fining point-bar sequences, then, are the distinctive products into which meandering rivers fashion their bed loads.

If the depth of the channel during a flood is not so greatly changed as it is in the Brazos River, then the channel-floor sediments, subjected only to lower-flow-regime conditions during base flow, may be subjected to upper-flow-regime conditions. If this happens, then any large-scale dunes that were present during conditions of the lower-flow regime will be changed progressively into washed-out Type-I dunes (See Figure 7-21.) or the channel may attain a plane-bed condition with vigorous sediment transport. (See Figure 7-23.) The resulting features in the sediment are low-angle cross strata or plane laminae lacking cross strata. If

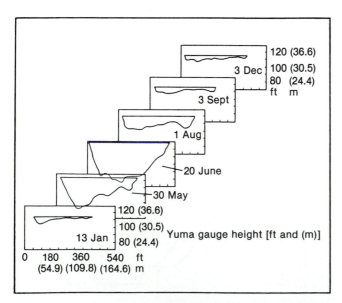

FIGURE 14-36. Profiles through Colorado River, at Yuma, Arizona, showing flow cross sections on various dates during 1912. (J. E. Sanders in G. M. Friedman and J. E. Sanders, 1978, fig. 8-42, p. 224; modified from E. W. Lane and W. M. Borland, 1953, fig. 4, p. 1076.)

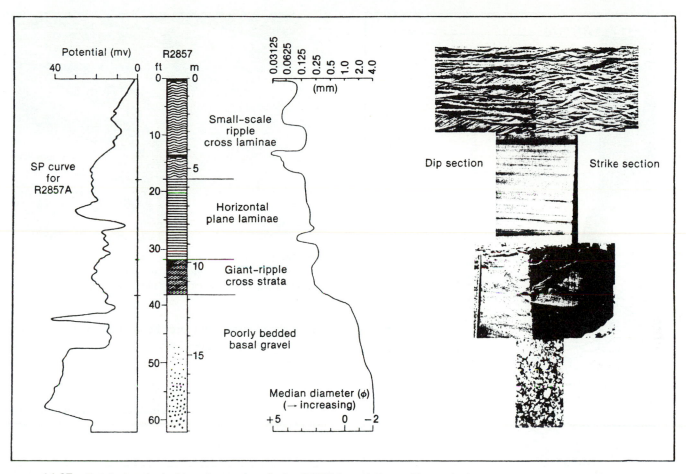

FIGURE 14-37. Detailed geological log of cores from boring R2857 through Brazos River point-bar sequence near Richmond, Texas (See Fig. 14-34.) and SP log recorded from boring R2857A, 9 m away from boring R2857. Explanation of symbols on geological log is shown in Figure 14-34. (Geologic log of boring R2857 modified from H. A. Bernard, C. F. Major, Jr., B. S. Parrott, and R. J. LeBlanc, Sr., 1970, fig. 23 on back of p. 8; SP log of boring R2857A from Shell Development Co., courtesy R. J. LeBlanc, Sr.)

dunes persist but their profiles change by alternating back and forth between the profiles of ordinary dunes of the lower-flow regime and the profiles of washed-out dunes of the transition to the upper-flow regime, then the resulting cross strata in the sediment may be distinctive. Such cross strata will consist of two alternating sets: (1) sets with about 30° dips formed by the migration of the steep avalanche faces of the lower-flow-regime profile and (2) sets with about 10° dips formed by migration of the low profiles of washed-out dunes (Figure 14-40). This alternation of two different sets of cross strata illustrates an important point, which needs to be emphasized: namely, the internal morphology of a dune cannot always be predicted simply from inspection of its external shape. In other words, one cannot predict with assurance the kind of cross strata that will be present by assuming that observed bed forms will simply migrate without ever changing their shapes. Always apply the shovel test: trenches do not deceive,

you fail to cut two trenches at right angles to each other.

Another effect of most floods is to submerge the point bars. Water flowing across a point bar may erode a new channel across the bar. This happens because the stable radius of curvature of meander bends is partly a function of discharge. Hence the most-stable channel under conditions of base flow generally is not the same as the most-stable channel during a flood. If a point bar is cut off during flood stage, then its meander is abandoned and the river assumes a straighter course in that reach.

All things considered, a flood may affect the point-bar sequence in four ways: (1) by scouring the channel deeper, it causes the sequence to attain its maximum thickness; (2) it may deposit sediments having bedding characteristics of the upper-flow regime; (3) it may cut a channel through a previously deposited point-bar succession and then promptly fill this channel with sediment

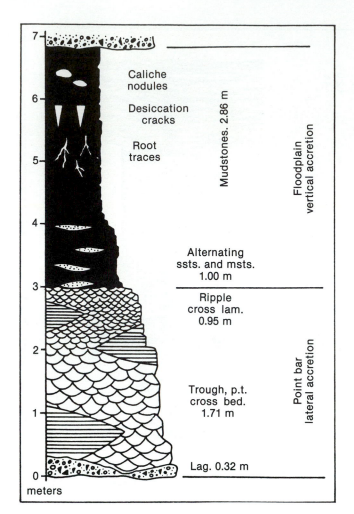

FIGURE 14-38.   Schematic vertical sequence of point-bar- and overlying vertical-accretion deposits. Data summarized from Devonian Old Red Sandstone of Great Britain and Devonian Catskill rocks of the eastern United States. (J. R. L. Allen, 1970.) Note that parallel laminae can replace trough cross strata or ripple cross laminae, or both. Thicknesses of sedimentary facies based upon averages in the data of J. R. L. Allen (1970). (R. G. Walker and D. J. Cant, 1984, fig. 2, p. 73.)

FIGURE 14-39.   Flat-bedded, laminated sand overlying large-scale planar-tabular cross strata in excavation made in point bar, Arkansas River, Tulsa, Oklahoma. (Authors.)

gins of the channel, and later beyond the margins of the channel, wherever it goes on the flood plain. (See Figure 5-11, B.)

Deposits at Margins of Channels.   Two kinds of deposits form at the margins of the channel: (1) natural levees, and (2) crevasse splays.

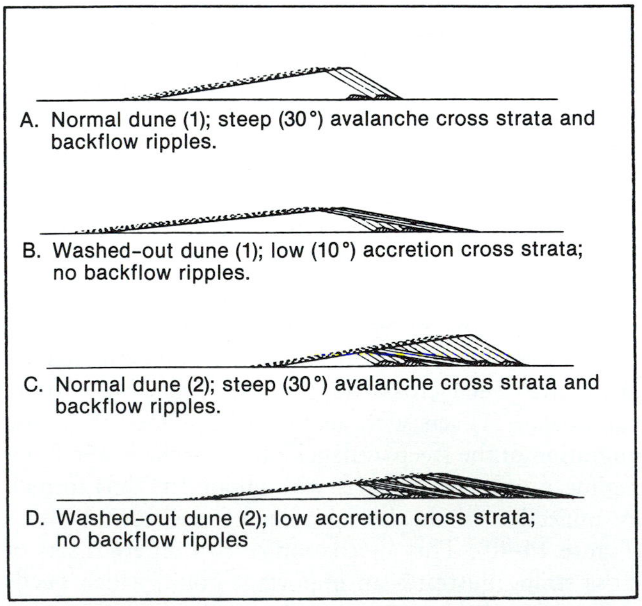

A. Normal dune (1); steep (30°) avalanche cross strata and backflow ripples.

B. Washed-out dune (1); low (10°) accretion cross strata; no backflow ripples.

C. Normal dune (2); steep (30°) avalanche cross strata and backflow ripples.

D. Washed-out dune (2); low accretion cross strata; no backflow ripples

FIGURE 14-40.   Downstream migration of sand wave in a deep (> 4.5 m) stream channel having discharge that alternates between lower-flow regime and the transition to the upper-flow regime, and a bed load consisting of fine sand. Resulting sets of steep-angle avalanche cross strata are truncated diagonally by downcurrent-dipping surfaces that are parallel to and form set boundaries of low-angle accretion cross strata. (J. E. Sanders in G. M. Friedman and J. E. Sanders, 1978, fig. 8-45, p. 226.)

of the kind moving down the channel during the flood stage; and (4) it acts to deposit large-scale cross strata in the upper part of a point-bar sequence.

So far we have been discussing what happens to the bed load under various circumstances. We now consider the suspended load. Under conditions of base flow, most of the suspended load moves continuously down the channel with the water and is not deposited anywhere in the alluvial valley. During a flood, however, the water level may exceed the bankfull stage and thus flood water may spread beyond the channel and leave overbank deposits on parts or all of the flood plain. Overbank deposits result from *vertical accretion* whereas channel deposits grow by *lateral accretion*. As the water spreads, it begins to deposit its suspended load, first at the mar-

Natural levees build upward more or less uniformly where the flood waters spill out of the channel as broad, shallow sheets. The suspended fine sand and -silt build up laminated sheets. The levees form only along channels carrying sand and silt in suspension. Until recently, along the lower reaches of the southward-flowing Atchafalaya River, Louisiana, south of the combined Grand Lake and Six Mile Lake and Atchafalaya Bay, the Atchafalaya River lacked natural levees (Figure 14-41). Upstream from the north end of Grand Lake, the Atchafalaya deposited its sand and silt on a delta at the north end of the lake.

If the level of the water in the channel is higher than the level outside the channel, then thin sheets of water

FIGURE 14-42.   Aerial view of lobate crevasse-splay deposit on natural levee of Brahmaputra River, Bangladesh. (J. M. Coleman, 1969, fig. 11, B, p. 158.)

flow down the gently inclined sides of the levees that slope away from the channel. The levee sediment is easily eroded, however, so the sheet flows tend to be concentrated into channels and to erode gaps through the levees. Such gaps have been named *crevasses*. The water pouring through a crevasse deposits *a lobate body of sediment*, a **crevasse splay**, *which forms an arc extending away from the levee* (Figure 14-42).

Overbank Deposits.   Water flowing through crevasses, from overflowing the natural levees, and from tributary drainage combines to inundate the flood plain. Commonly, such waters will contain much sediment in suspension. The flood waters may flow slowly downvalley or may become fully ponded. The slowly moving water deposits sheets of silt; the ponded water forms swamp deposits.

The flood waters that inundate a flood plain flow slowly downvalley but at a speed that is less than that of the water still hemmed in by the sides of the channel. Hence suspended sediment, typically silt, but ranging from fine sand to silty clay, drops out of the flood waters and forms a widespread, thin blanket as overbank sediments on the flood plain.

After the flood has passed down the channel, much of the water will disappear from the flood plain; however, small lakes or swamps may remain in which clays accumulate. If vegetation is abundant, these clays may contain much organic matter. Such swamp deposits form thin pockets or lenses within the overbank silts.

In contrast to the chaotic assemblages of coarse- and fine-textured sediments dropped by mudflows, the sediments deposited by a stream in flood are well sorted, are evenly laminated (See Figure 5-11, A.), and may be predominantly fine textured. However, the original lamination of flood-plain sediments may be obliterated by the

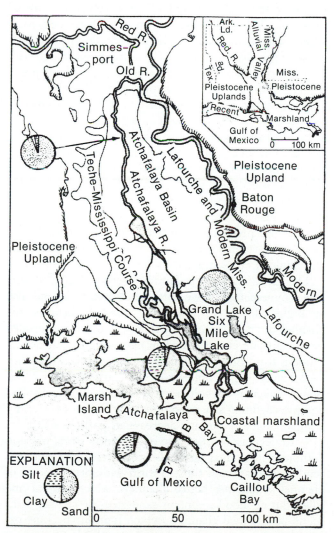

FIGURE 14-41.   Map of Atchafalaya River, Louisiana, indicating particle-size distribution of bed-load sediment (shown by circular diagrams displaying proportions of sand, silt, and clay). Prior to 1970 in the lower reaches of Atchafalaya River, no sand was present in the bed load; hence natural levees were absent. (Map modified from W. C. Thompson, 1955, fig. 1, p. 53; circular diagrams from data in fig. 4, p. 58.)

effects of burrowing organisms and by the disturbing effects of plant roots, except at the base of the succession where the rate of aggradation may have been sufficient to outpace bioturbation.

All the overbank deposits of the flood plain are subject to subaerial exposure and drying; therefore the silts and clays form desiccation cracks. Other consequences of subaerial exposure include formation of tiny impact craters from falling raindrops, the indenting of the surface by the feet and crawling tracks of animals, and chemical weathering. Such weathering can oxidize iron compounds and create nodules and crusts of caliche. If a soil forms, rooted plants can grow. The roots of plants tend to destroy sedimentary features.

BIOLOGICAL ACTIVITIES IN THE DEPOSITS OF MEANDERING RIVERS

The deposits of meandering rivers are the habitats of various nonmarine invertebrates, chiefly pelecypods and gastropods, which leave trails on sediment surfaces and dig burrows within the sediment. As an example, nonmarine *Unio*-type pelecypods plow distinctive trails in the surface sediment and burrow into the sediment to keep below the ground-water table. Their trails on the parts of point bars exposed at low water lead directly toward the low-water channel. These linear trails end either in the water or in a burrow.

Because of abundant water, dense plant populations can grow along meandering rivers. As a result, in flood-plain sediments, leaf litter is abundant.

THREE-DIMENSIONAL SEDIMENTATION PATTERN BUILT BY A MEANDERING RIVER

As previously mentioned, the laterally accreted deposits of a meandering-stream channel and the vertically accreted deposits spread across the flood plain by major floods are volumetrically the most-important part of an alluvial deposit formed by a meandering stream. Thus the rock body formed by a meandering stream is dominated by the channel- and overbank suites. A meander belt tends to be confined laterally by the erosion-resistant clayey deposits that fill abandoned meanders (*ox-bow lakes*). As a consequence, the entire meander belt accretes vertically until the river is significantly elevated above the surrounding floodplain. This metastable situation persists until during a flood, the river's natural levee is catastrophically ruptured. *The entire river system below the levee break switches to a lower part of the flood plain, abandoning the downstream part of the old meander belt.* This process is called avulsion. Once again, the river builds a zone of *clay-plugged abandoned meanders* on each side of the active channel, which accretes verti-

cally until the next avulsion. Over a long period of time, a three-dimensional body of sediment is constructed in which *elongate meander belts far thicker than the depth of the active river* (multistory sand bodies) are embedded in fine-textured flood-plain deposits. Some examples of deposits inferred to have been formed by ancient meandering rivers are described in Box 14.2.

CHARACTERISTICS FOR RECOGNIZING THE DEPOSITS OF ANCIENT MEANDERING RIVERS

The fundamental basis for recognizing the deposits of ancient river flood plains is the distinct separation into two suites of materials: (1) relatively coarse bed-load deposits of the channels and (2) relatively fine-textured overbank deposits, mostly from the suspended load, which were spread across the ancient flood plain. Because of the wide spectrum of conditions associated with both the channels and flood plains of rivers, the deposits of each of these suites can display great diversity.

For example, the channels may have always been shallow (depth < 4.5 m), always deep (depth > 4.5 m), or shallow during base-flow discharge and deep during floods. The bed load of sediment in the channel can range from fine- to coarse sand. The factors of depth of water, speed of flow, and sizes of sediment available combine to create many kinds of sedimentary structures. In particular, the interactions among these three factors control the presence or absence of bed forms. If bed forms are present, these three factors control the scale and kinds of cross strata.

The channel deposits always overlie a scoured surface. Immediately above this scoured surface (and also at higher levels in multistory sand bodies) is the channel-floor lag, commonly featuring pebbles of eroded bedrock; shale clasts; abundant carbonized plant debris, ranging from bits of twigs to large fossil logs (See Box 14.2 Figure 1.); or bones and teeth of terrestrial vertebrates. Taken as a whole, each channel sequence displays an upward-fining size gradation. (See Figure 14-37.) Accompanying this upward decrease in particle size is an upward decrease in the scale of any cross strata present. Although individual channels are narrow, the channels tend to migrate and thus to form widespread sheets of channel sediments. Active channels, which are cut and filled during floods, contain sand-size- or coarser sediment. Abandoned channels (which typically become ox-bow lakes) tend to fill with fine clay. Such fine clay is an exception to the generalization that the channel suite includes only coarse sediment.

The overbank sediments include everything from ancient soils to coals. Ancient flood plains included such diverse sedimentary settings as well-drained flats that

BOX 14.2

Some Examples of Deposits of Ancient Meandering Rivers

Modern fluvial sediments range all the way from the coarse deposits of fans to the fine clays of deltas. Rivers in large lowlands can deposit sheets of sand or mixed channel sands and overbank sheets of silt and clay. Many such deposits of alluvial plains extend down to the seashore and form the top parts of large marine deltas. Most of these delta deposits are closely related to marine deposits that they may overlie, underlie, or pass laterally into. Because of this association with marine deposits, we discussed deltas separately in Chapter 13. In this section, we concentrate on inferred ancient examples of fluvial deposits consisting of related suites of channel- (Box 14.2 Figure 1) and overbank materials that are not immediately interbedded with marine formations. Well-described examples include the White River beds (Oligocene) of the Great Plains, United States; parts of the North Horn Formation (Paleocene and Cretaceous) of central- and eastern Utah, United States; Jurassic of Sinai Peninsula (See Box 14.2 Figure 1.); parts of the Newark Group (Upper Triassic-Lower Jurassic) of northeastern

North America; Permo-Carboniferous redbeds of southwestern United States; Carboniferous nonmarine strata of the northern Appalachians, eastern Canada; Pennsylvanian strata in Oklahoma and elsewhere in the United States; parts of the "Catskill redbeds" (Devonian), United States; and the Old Red Sandstone (Devonian) of northern Europe.

We shall add some details from two examples of those ancient fluvial deposits that have been described by investigators who were particularly concerned with the cyclic sequences deposited by the lateral shifting of meandering channels. We include the Catskill (Devonian) strata of New York State and the Old Red Sandstone (Devonian) of northern Europe.

Catskill (Devonian), New York State

The strata named the "Catskill redbeds" include various nonmarine-, chiefly fluvial, conglomerates and coarse sandstones, which probably were deposited as fans and by braided streams, and cyclically interbedded sandstones and red- and

BOX 14.2 FIGURE 1.   Remains of coalified fossil logs in channel sediments of ancient flood-plain river, lower Member, Inmar Formation (Jurassic), Wadi Rajabia, Gebel Maghara Mountains, Sinai Peninsula. (M. Goldberg and G. M. Friedman, 1974, pl. 3, figs. 1 and 2.)

A. Overall view of section through channel deposits. Dark holes about 40 cm in diameter (center and left center) are cavities left by removal of fragments of fossil logs. Shadow at base of sandstone face marks recessed area in cliff where underlying overbank deposits (siltstone and shale) have been eroded.

B. Close view of coalified fragment of fossil tree trunk.

green siltstones, which probably were deposited on flood plains of meandering rivers. These fluvial deposits are part of a vast wedge of sediment that was spread out along the southeast margin of the Appalachian seaway after its shore had been prograded westward by several hundred kilometers by the deposition of many deltas. A wide alluvial plain came to occupy part of the former seaway. (See Figure 13-34.)

A characteristic of many parts of the "Catskill" strata is the diagnostic segregation of sandstones showing upward-fining cyclic patterns, former channel deposits; and siltstones of various colors (red, green, and dark gray here), former overbank deposits. A representative section exposing 100 m of fluvial strata is located 2.4 km east of Haines Falls, New York. About half the succession consists of upward-fining channel cycles and half of overbank red siltstones. The lowest channel sandstone is a two-storied unit 12 m thick; it consists of two channel sequences, each 6 m thick (Box 14.2 Figure 2). The lower of these sequences displays a section from base upward: a channel-floor lag deposit, fine- to medium-textured sandstone with large-scale cross strata, medium-textured flat-bedded sandstone, and medium-textured small-scale cross-laminated sandstone. Because the large-scale cross strata extend all the way to the bottom of the channel sequence, we can infer that this channel succession was formed by a flood discharge in which only lower-flow-regime conditions prevailed, even in the deepest parts.

The flat-bedded, medium-textured sandstone in the middle of the sequence interrupts the continuity of the upward-fining size progression. When this upper unit was deposited, the channel, even in flood, may have been too shallow to create large dunes. The cross strata in this succession have not been examined with a view toward determining whether they originated by lower-flow-regime migration of ordinary dunes having avalanche faces on their downstream sides and backflow ripples in their troughs (See Figure 7-20.), or they formed by migration of washed-out dunes having accretion-type surfaces on their downstream sides and no backflow ripples in their troughs. (See Figure 7-21.)

The red siltstones of the inferred overbank strata are reasonably typical of a great many ancient flood-plain deposits. The red color of such materials results from interstitial hematite, which forms a coating on the framework particles and a matrix among these particles. We discuss the origin of hematitic pigment in sediments in Box 14.3.

Old Red Sandstone (Devonian), Northern Europe

The Old Red Sandstone complex of northern Europe, long renowned for its early fossil fishes, has attracted detailed study because of its variety of upward-fining cycles. The Old Red displays many parallels with the Catskill complex, which is of roughly the same age and was deposited at about the same paleolatitude. Near its northern highland source areas, the Old Red strata are interpreted as fanglomerates. To the south, in southern England (Devon and Cornwall), Belgium, and Germany, the equivalent Devonian strata are marine deposits. Recognition by Adam Sedgwick (1785–1873) and R. I. Murchison (1792–1871) of this change southward from nonmarine to marine was one of the earliest demonstrations of a lateral change of facies in the stratigraphic record. Geographically intermediate between the fan deposits and the marine strata are the cyclic deposits of river flood plains. These are exposed in the Welsh borderland. We have selected two cyclothems that are distinctive because in each of them large-scale cross strata are lacking. The cyclothem at Tugford is 9.3 m thick, and about 8.5 m of it consists of various channel deposits. (See Box 14.2 Figure 2.) The only cross strata in this channel sequence are ripple cross laminae.

We interpret the absence of large-scale cross strata to mean that no conditions prevailed for the preservation of deposits formed in the lower-flow regime. During base flow, large-scale dunes may have existed in the channel. If so, they were completely destroyed during a flood. During the maximum flood, the depth of water in the channel must have been at least 6.2 m, equal to the thickness of the channel deposits.

The channel deposits exposed at Mitcheldean include only flat-bedded and ripple cross-laminated strata. (See Box 14.2 Figure 2.) These are interpreted as the products of channels having water depths ranging from 0.5 to 2.0 m too shallow, even during flood discharges, for the development of large-scale dunes.

The red siltstones interbedded with both kinds of channel sands display burrows, calcareous concretions, and desiccation cracks.

SOURCES: J. R. L. Allen, 1964; J. R. L. Allen and Friend, 1968; Goldberg and G. M. Friedman, 1974; Woodrow and Sevon, 1985.

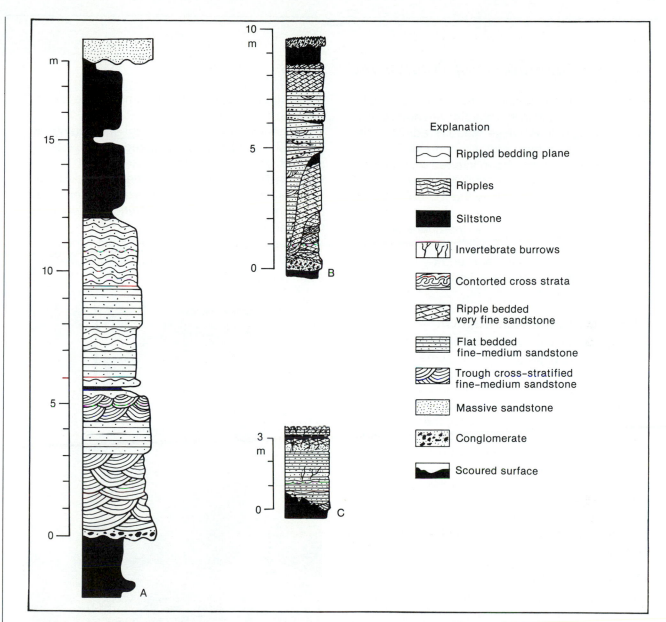

BOX 14.2 FIGURE 2.   Columnar sections through various examples of inferred deposits of ancient flood-plain rivers of Devonian age.

A. Cyclothem A1, north side of N.Y. Route 23 A, about 1.5 miles east of Haines Falls, New York. (After J. R. L. Allen and P. F. Friend, 1965, fig. 9, p. 51.)

B. Cyclothem of early Devonian age (Dittonian) exposed in stream at Tugford, Shropshire (SO566873). (After J. R. L. Allen, 1964, fig. 10, p. 181.)

C. Lowermost cyclothem of early Devonian age (Breconian), exposed in wilderness quarry (SO672185), 2/3 mile east of Mitcheldean, Gloucestershire. (After J. R. L. Allen, 1964, fig. 12, p.188.)

were subaerially exposed most of the time to swamps that were always under water. Therefore, the color of flood-plain sediments can range from red through various shades of green or gray to black. Because they contain hematite pigment, many ancient overbank sediments are red. The nature and environmental significance of redbeds are discussed in Box 14.3. Bedding surfaces of overbank deposits may display stumps of trees in growth position, footprints of land-dwelling vertebrates, the tracks and trails of insects, fresh-water pelecypods and gastropods, raindrop-impact craters, and desiccation cracks.

BOX 14.3

Climatic Significance of Nonmarine Redbeds

Many ancient nonmarine formations, especially the kinds discussed in this chapter, are noteworthy for their red color. Because modern reddish soils are confined to a belt lying within about 30°N or S of the Equator, it has been inferred that ancient *redbeds* indicate warm climates. However, a lively debate has ranged over how the red color came to be and whether the ancient climate was dry or wet. We shall first describe the characteristics of the red pigment of redbeds, then discuss the origin of the red pigment, and end by summarizing the relationships between iron-oxide pigments and modern climate.

Characteristics of Red Pigment

The significant pigment of redbeds is finely divided hematite (Fe_2O_3), which in some cases has been shown to occur as tiny hexagonal crystals having diameters ranging from 20 to 0.03 μm. In some fine-textured redbeds, about the only iron present is in the pigment. The pigment forms an optically opaque matrix among the quartz particles of siltstone (Box 14.3 Figure 1) or as a coating on the quartz particles. Where some quartz particles have been pressed against other quartz particles, the hematite coating is absent. The pigment of a few redbeds includes the mineral maghemite (magnetic ferric oxide, γ-Fe_2O_3) as an accessory constituent.

Origin of Red Pigment

The following five mechanisms have been proposed as ways for creating the red pigment of redbeds:

1. The hematite formed in a lateritic soil in a humid-tropical climate by oxidation of the ferrous iron derived from rock-making ferromagnesian silicate minerals (chiefly pyroxene, hornblende, biotite, and chlorite) and from magnetite of the bedrock. The hematite was physically transported to the site of deposition, and after deposition was not subject to any major changes.

2. The hematite formed by *in-situ* alteration of hydrated brownish iron oxides derived from oxidation of the ferrous iron in iron-bearing minerals (ferromagnesian silicates and magnetite, chiefly) in a hot, arid climate. The fresh minerals containing ferrous iron were deposited as initially nonred alluvium, altered *in situ* to brownish hy-

drated iron oxides, and with time, these oxides became hematite.

3. The hematite formed by the aging of yellowish- and brownish hydrated ferric oxides in a moist, tropical climate; the oxides got into alluvial sediments either (a) as finely divided soil-weathering products transported and deposited by rivers or (b) as *in-situ* alteration products of fresh ferromagnesian minerals.

4. The hematite was inherited from a preexisting redbed unit and was redeposited as second-cycle particles.

BOX 14.3 FIGURE 1.   Enlarged view of red siltstone from lower Newark Group (Upper Triassic), Haverstraw, New York. Quartz particles (white), mostly sharply angular (but a few are subrounded to well rounded), are set in an opaque cryptocrystalline matrix of hematite (black, below). At top, the light-colored material surrounding the quartz particles is calcite, which has replaced the hematite (or its precursory hydrated iron-oxide minerals); photomicrograph in plane-polarized light of thin section. (J. E. Sanders in G. M. Friedman and J. E. Sanders, 1978, fig. 8-56, p. 235.)

5. The hematite was precipitated directly out of seawater during the initial stages of evaporation.

All of these mechanisms can and have produced hematite, which colors redbeds red. The problem is to decide which mechanism operated to form a given red deposit. In the nonmarine redbeds under discussion, we can eliminate mechanism 5, which operates only under marine conditions. Mechanism 4 applies in some cases, notably the Pleistocene redbeds derived from the recycling of the Triassic-Jurassic redbeds of northeastern North America, but does not shed any light on the general problem of the origin of the red pigment. Hematite as films on particles is clearly a chemical precipitate.

Mechanisms 1, 2, and 3 differ in their climatic implications (moist for 1 and 3, dry for 2) and in the times and sites of origin of the pigment. In mechanism 1, hematite is created in the upland soils and is transported as such to its site of deposition; in mechanism 2, no pigment of any kind is transported with the sediment but all pigment forms by *in-situ* alteration; in mechanism 3, no pigment is transported as hematite, but all pigment starts out as yellowish- and brownish hydrated iron oxides. Some such pigment is transported and some forms *in situ*, but afterward all converts to hematite.

Relationships of Iron-Oxide Pigments to Modern Climates

Modern red soils are confined to tropical regions. The results of recent research in tropical Mexico, Colombia, and Puerto Rico indicate that, although some soils are red, the pigment in alluvium derived from them is brown. With time and in a suitable chemical environment (See Figure 7-55.), this brown pigment ultimately becomes hematite, and only then does the color of the alluvium match that of ancient nonmarine redbeds. The chief reactions involved are (1) creation of brown pigment from minerals containing ferrous iron, and (2) conversion of brown pigment to red pigment. These changes can take place in alluvium that is subjected to either a hot dry or a tropical moist climate.

The brown-to-red conversion evidently requires times ranging from hundreds of thousands of years to a few million years. Presumably, both the alteration of the ferrous iron of minerals to brown pigment and the conversion of this brown pigment to red hematite could take place in a soil profile as well as intrastratally within alluvium. However, the requirement of a long conversion time tends to work against the creation of abundant hematitic pigment in soils and to favor its generation in alluvium. We can conclude from all the foregoing that red color can form in either a moist or a dry climate, and that most of the pigment probably formed *in situ* but some of it may have been transported with the other mineral particles. The precise climatic significance of redbeds must be assessed from the clues contained in faunal- and floral remains, and in associated sedimentary deposits.

SOURCE: T. R. Walker, 1974.

Interbedded with these flood-plain deposits may be sheets of fine sand, which prograded away from the channels as crevasse splays, natural levees, or flood surges from a trunk stream into a ponded tributary. Such sheets of sand may be graded; their basal bedding surfaces may display counterparts of various current marks; their internal layers may be deformed as a result of the interplay among the moving current, the substrate over which the current moved, and the sediment being deposited; and their top surfaces may display animal tracks and trails and/or raindrop-impact craters. All kinds of fluvial sediments may display structures made by the burrowing of nonmarine invertebrates.

Absent from the deposits of meandering rivers are abundant glauconite and fossil remains of marine organisms, except as debris eroded from older marine formations.

Lake Environments: Modern- and Ancient Deposits

A **lake** is defined as *a landlocked body of water occupying a morphologic basin.* Most lakes are products of the blockage of a throughflowing drainage system or of local internal drainage. The important point is that the lake persists because inflowing water from all sources exceeds losses. The inflow comes from rainfall directly or indirectly, as surface drainage (stream runoff or sheetflow), as subsurface seepage (from the local reservoirs of ground water), or as meltwater (from snow or ice). The losses include outflow (if the lake is part of a blocked throughgoing network) and evaporation (affects all lakes). Because the water of most lakes is derived from rainfall, the salinity of most lake water is low; the water is "fresh." However, some lake waters have always been saline because the lakes are former parts of the sea that became landlocked as a result of crustal movements. The Caspian "Sea," areally the world's largest lake, was created by isolation from its former connection with the Mediterranean Sea. Still-other lakes become salty by prolonged evaporation of their waters. For example, the Salton Sea, in southern California, became a fresh-water lake in 1905 when floods broke through the irrigation works and the entire Colorado River poured into the Salton depression. After two years of hard work, engineers managed to get the river to flow back into the Gulf of California. Because it receives water from a vast network of irrigation ditches, the Salton Sea has become progressively more saline. By 1960 its salinity had reached 33.68‰ (nearly normal marine salinity). Lake waters may also become saline because they are fed by waters that have dissolved older layers of salt.

Lakes may be classified as *hydrologically open* or *hydrologically closed*. Hydrologically closed lakes lack a permanent outlet for their water (an example is the Great Salt Lake in Utah). These lakes lose water only by evaporation and seepage. In contrast to hydrologically closed lakes, hydrologically open lakes, from which permanent rivers flow, rarely become very saline.

The *scientific study of lakes* is limnology. The science of limnology is to lakes what oceanography is to oceans. Limnology includes the biology of aquatic organisms, study of pollen in lake sediments, chemistry of lake waters, analyses of organic matter from organisms and bottom sediments, circulation of lake waters and its effects on dissolved oxygen and other gases, and the relationships between lake waters and physiography, weather, and climate. Many geologists are interested in lake sediments, the depositional processes in lakes, shoreline features of lakes, and the interpretation of ancient lake sediments in the stratigraphic record. Lake sediments that record the changing seasons can be used to establish a chronology in which duration of time can be found by counting layers, as with tree rings. Lake sediments also form depositional sequences, many of which are cyclic, that record changing climatic- and other environmental parameters that affected lake-water levels and the chemical composition of the water. Depositional episodes produce characteristic mesosequences that are similar to marine mesosequences.

Nearly all modern lakes have been affected by the activities of our high-energy, industrial civilization and by increasingly large numbers of people. Lake-bottom sediments are valuable archives that preserve records of the impact of people on lakes and allow the preindustrial conditions to be determined. Lakes which may *not* have been affected by modern civilization include lakes buried under 3 to 4 km of glacial ice in Antarctica, and those

such as Lake Anguissaq, in Greenland, which bears a permanent cover of ice at least 2 m thick!

In the discussion that follows we shall take up the subjects of lake basins and the interaction between lake waters and lake sediments. Examples of ancient lake deposits are discussed in Box 14.4.

SOURCES: P. A. Allen and Collinson, 1986; Fouch and Dean, 1988; Wetzel, 1983.

Lake Basins

Many large modern lakes occupy depressions that have been created by faulting or by crustal warping. Modern lakes in fault-bounded depressions include the Dead Sea in Israel and Jordan (Figure 14-43); Lake Baikal in Siberia, the world's largest (in volume), deepest, and oldest lake; Lakes Tanganyika, Edward, and Mobutu Sese Seko (formerly Lake Albert) in eastern Africa; and Lake George in New York State, United States. Modern lakes whose basins have been created by crustal warping include the Caspian and Aral Seas in the Soviet Union, Lake Biwa in Japan, Lake Victoria in eastern Africa, Lake Chad in central Africa, and Lake Maracaibo in Venezuela. Many of these lakes have existed continuously since the Miocene or Pliocene epochs. Lake Baikal is 50 to 75 million years old; it may have originated in the Mesozoic Era.

Vast numbers of modern lakes are direct effects of the Quaternary continental glaciers that spread over the northern parts of North America and of Eurasia. The glaciers deepened many lowlands and created closed depressions that became lakes. Retreating glaciers left behind marginal ridges of debris. These ridges (moraines) blocked many valleys whose former river discharge was thus impounded to form lakes. Glacier ice itself blocked a few large rivers. Whenever the margin of the glacier retreated, the lakes that were bounded on one side by a glacier drained catastrophically. Because modern lakes and some former Quaternary lakes are so closely related, we shall discuss them together.

Glacial Lake Missoula, in the northwestern part of the coterminous United States, is an example of a former glacier-marginal lake that drained catastrophically on many occasions (Figure 14-44). The Lake Missoula floods illustrate an important point about lakes that are situated in the upper reaches of drainage basins. The water of such high-altitude lakes is a potential hazard to lower-lying parts of the drainage network. Any steep-sided valley in a mountainous region can become a lake suddenly, and just as quickly can lose all its waters in a catastrophic flood. Temporary dams are created by glaciers, by avalanches of snow or debris, or by crustal movements. Several kinds of dams commonly appear during earthquakes.

BOX 14.4

Inferred Examples of Ancient Lake Deposits

The best places to make the transition from recent- to ancient lake deposits are in tectonically active regions such as Japan, East Africa, or the Middle East, where around the borders of large modern lakes, older lake sediments are exposed. We shall use as a first example the Oligocene-Quaternary deposits of the Dead Sea graben, Israel and Jordan. Following this, we describe two selected examples of ancient lake sediments that form significant parts of the bedrock: (1) the deposits of the long-lived lake in the Uinta Basin of Utah, Colorado, and Wyoming, particularly the older Flagstaff Limestone (Paleocene), and the younger Green River Formation (Eocene) and (2) parts of the Portland Formation (Upper Triassic-Jurassic) of southern Connecticut.

Oligocene-Pleistocene Deposits of Dead Sea Graben, Israel and Jordan

Conditions analogous to those prevailing today have persisted in the Dead Sea graben since Oligocene times. (See Figure 14-43.) A tectonic lowland having steep walls has served as a collecting basin for waters and sediments from the surrounding highlands. More than 4000 m of strata have accumulated. Along the steep margins of the basin, fanglomerates predominate. In the center of the basin the strata consist of lake deposits, chiefly fine-textured terrigenous sediments and various evaporites, including halite rock. (See Figure 14-7.)

Stratigraphic names have been given to the exposed upper part of the succession, which is of Pleistocene age. The lower part of the succession is known from a few borings; the relationships between the exposed units and those encountered only in bore holes have not yet been established. The exposed strata of Pleistocene age have been subdivided into the Samra Formation (below) and Lisan Formation (above). The Samra Formation consists of oolitic limestone of nonmarine origin, calcareous sandstones and -siltstones, gypsum, and conglomerates. The Lisan Formation includes terrigenous sediments, gypsum, and rhythmically interlaminated light-colored aragonite-rich layers and dark-colored calcitic layers. The light-colored aragonitic layers compare closely with the white

aragonitic sediments that are precipitated onto the floor of the modern Dead Sea during periods of most-intense evaporation. The dark-colored calcitic layers are interpreted as the products of bacterial degradation of what was originally evaporitic gypsum.

SOURCE: Garber, Levy, and G. M. Friedman, 1987.

Ancient Lake Deposits in Uinta Basin, Rocky Mountains, Western United States

During the Paleocene, Eocene, and Oligocene epochs, structural basins in Wyoming, Utah, and Colorado in the Rocky Mountains of the western United States contained various large lakes. One of these basins, the Uinta Basin, was periodically filled with water to form a large lake (Box 14.4 Figure 1). At other times, the lakes were much smaller. Lake sediments and related fluvial deposits and fanglomerates up to 4500 m thick were deposited. As in the Dead Sea graben, fans surrounded the lake. Accordingly, fanglomerates and braided-stream deposits form a continuous marginal zone between the lake sediments and the bedrock of the circumferential highlands. Deep drilling has recently shown that two lacustrine units, previously considered to have been deposited in separate lakes, in fact accumulated in a single large lake. The older of these lacustrine formations is the Flagstaff Limestone; the younger is the Green River Formation (Box 14.4 Figure 2).

The outlines of Lake Flagstaff shown in Box 14.4 Figure 1 are based on the inferred extent of the Flagstaff Formation, which includes one of the thickest nonmarine limestones known (50 to 200 m). This limestone forms the caprock of the Wasatch Plateau in central Utah (indicated by largest and irregularly shaped black area on Box 14.4 Figure 1). The minimum area occupied by Lake Flagstaff is 71,680 km^2 (28,000 mi^2). By comparison, the area of Lake Superior, by far the largest of the Great Lakes, is 82,900 km^2 (32,008 mi^2).

During the early part of the Eocene Epoch, the influx of lake-marginal sediments, already large, increased as a result of uplift in the parent areas. This increased sedimentation and crowded

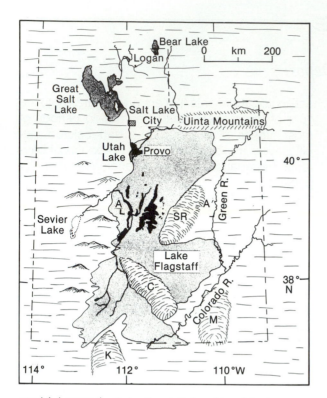

BOX 14.4 FIGURE 1.   Lake Flagstaff, the older of the large Early Tertiary lakes in the Uinta Basin, Utah, and adjacent areas, restored boundaries shown on map of Utah. Displayed are principal modern drainage features, areas of outcrop of Flagstaff Formation (black), inferred maximum extent of Lake Flagstaff (in late Paleocene time), and names of surrounding elevated areas (SR = San Rafael Swell; C = Circle Cliffs upwarp; M = Monument upwarp; K = Kaibab upwarp). Compare Box 14.4 Figure 3, A. (J. E. Sanders in G. M. Friedman and J. E. Sanders, 1978, fig. 9-24, p. 256; outcrop areas from A. La Rocque, 1960, fig. 2, p. 10; extent of Lake Flagstaff and adjacent highlands after C. B. Hunt, 1956, fig. 55, p. 76.)

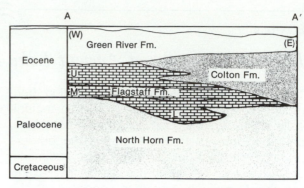

BOX 14.4 FIGURE 2.   Schematic restored stratigraphic section of Flagstaff Formation and contiguous strata, central Utah. North Horn and Colton formations are inferred to be deposits of flood-plain rivers; Flagstaff and Green River formations, ancient lake deposits. (A. La Rocque, 1960, fig. 1, p. 8.)

back the shoreline, reducing the size of the lake somewhat. With further subsidence and plentiful water, however, the lake expanded once more. During this second expanded phase, the Green River Formation was deposited. Green River strata were deposited in two connected lakes: Lake Uinta in the Uinta Basin and Lake Gosiute in the adjacent Green River Basin north of the Uinta Mountains (Box 14.4 Figure 3).

In the center of the Uinta Basin, open-water lake deposits of the Green River Formation include many varieties that have been subdivided into three major units. We shall not use the formal names, but simply refer to the major units as lower, middle, and upper.

The lower unit includes shell marls; algal reefs, as individuals or groups of reefs; algal-pebble beds; and ostracode-bearing and oolitic lime-stones. Algae flourished in the lake and built reefs that expanded broadly over the smooth lake floor. Fish, mollusks, crustaceans, and aquatic insect larvae were abundant in the waters of the lake; turtles, crocodiles, birds, and small camels, as well as myriads of winged insects frequented its shores.

The middle part of the Green River Formation contains "oil shales," which are more accurately described as kerogen-rich dolomitic marls. They are siliceous carbonates whose essential minerals are calcite, dolomite, fine-textured quartz, and clay minerals. Other minerals include large quantities of halite, trona, and shortite ($Na_2CO_3 \cdot 2CaCO_3$), and lesser amounts of other saline- and unusual authigenic minerals, such as silicates, borosilicates, fluorides, phosphates, and complex carbonates. Sandstones are more abundant than in the lower part; those in contact with the kerogen-bearing strata are saturated with gilsonite, a distinctive solid hydrocarbon. During this time the lake contracted to a small size, the outlet ceased to exist, and the waters became more saline and chemically stratified. Bottom water became extremely toxic; hence no bottom-dwelling fauna existed. Upwelling of hydrogen sulfide would have been instantly fatal for near-surface organisms that otherwise flourished in the oxygenated upper water. Beautiful specimens of fossil fish, some killed in large numbers by upwelling or mixing events, litter bedding planes in the Green River Formation. (See Figure 14-49.) Many of these specimens are exhibited in museums and private collections worldwide.

SOURCES: Grande, 1980; Smoot, 1983; Surdam and Stanley, 1980.

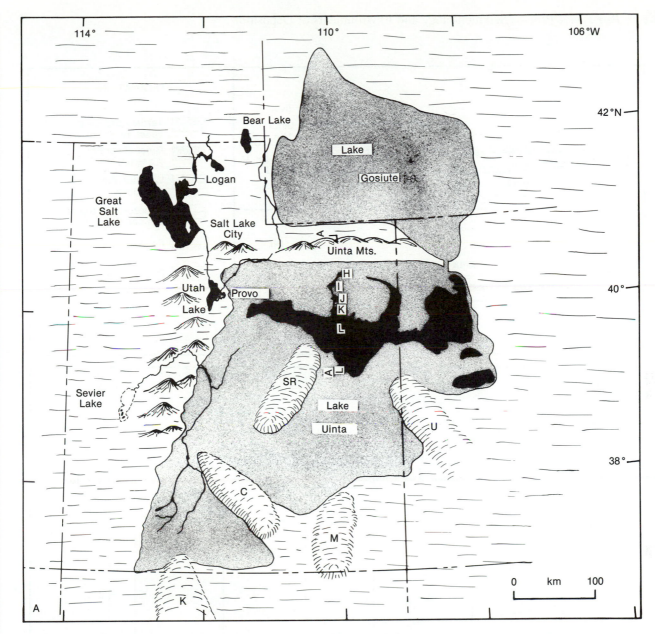

BOX 14.4 FIGURE 3.   Geographic- and geologic setting of Green River Formation, deposited in younger of large Early Tertiary lakes in the Uinta Basin, Utah, and environs.

A. Map of Utah and adjacent parts of Wyoming and Colorado showing major modern drainage features, outcrops of Green River Formation in Uinta Basin, inferred maximum extents of Lakes Gosiute and Uinta, and surrounding high areas (SR = San Rafael Swell; U = Uncompaghre upwarp; C = Circle Cliffs upwarp; M = Monument upwarp). (J. E. Sanders in G. M. Friedman and J. E. Sanders, 1978, fig. 9-26, p. 258–259; Lake Gosiute and outcrops of Green River Formation in Colorado from W. H. Bradley, 1929, fig. 13, p. 89; extent of Lake Uinta and locations of highlands after C. B. Hunt, 1956, fig. 56, p. 78; outcrops of Green River Formation and locations of control wells in northeastern Utah after M. D. Picard, 1955, fig. 1, p. 77.)

B. (page 550, top) Profile and section across Uinta Basin (long. 109°55'W) along A-A' of A, showing present-day configuration of strata. (Redrawn by J. E. Sanders from A. J. Eardley, 1951, fig. 237, p. 400.)

C. (page 550, top) Restored profile and section through Lake Uinta at close of deposition of Green River Formation. Lake level is shown as lying at altitude +1000 ft (about 305 m). The northern shore of Lake Uinta stretched along the south flank of the Uinta Mountains, which were being elevated during the Eocene Epoch. At times, the southern shore of Lake Uinta lay just north of the present outcrop edge along the south side of the Tavaputs Plateau (left-hand side of section). At other times, however, the south shore of the lake shifted many kilometers to the south of the present outcrop edge.

D. (page 550, bottom) Subsurface stratigraphic section of Green River Formation from near axis of Uinta Basin to outcrop edge of Green River Formation along line of wells shown in A. Deltaic sediments pass southward into fan deposits. Laterally equivalent strata consist of fine-textured lake deposits in Sinclair's Juniper Hills No. 1 and northward. In Roosevelt No. 2 well, the Green River Formation was not subdivided. At northern edge of Uinta Basin, all lake sediments probably grade into deltas and/or fans, but this transition is not exposed and wells shown did not penetrate it. (Replotted by J. E. Sanders from M. D. Picard, 1955, fig. 3, p. 86.)

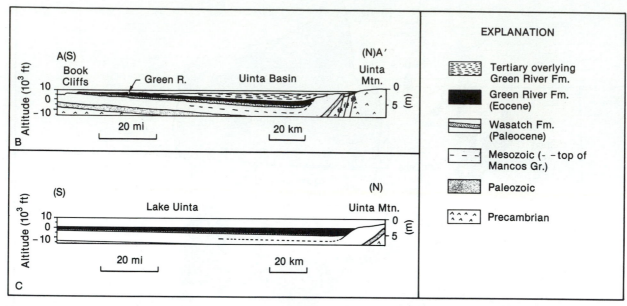

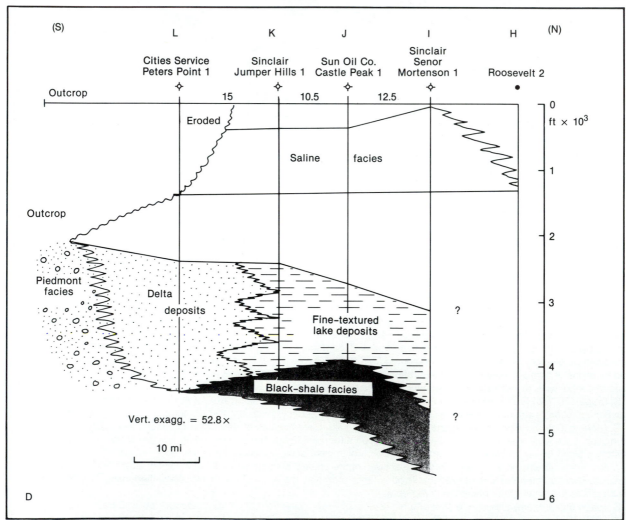

BOX 14.4 FIGURE 3.   *(Continued)*

FIGURE 14-43.   Dead Sea and adjacent desert areas of Israel (left) and Jordan (right); oblique view from satellite. Lisan Peninsula divides Dead Sea into northern- and southern basins. A-A' indicates location of profile, Figure 14-53. Where water reaches bedrock along faults at margin of graben, the shore is nearly straight. Light-colored areas south of Dead Sea are flat surfaces at top of wall-to-wall sediment fill of graben floor. (NASA).

The hazards posed by lakes in mountainous areas were familiar to Sir Charles Lyell (1797–1875). In Volume 1 of the first edition of his famous book, *Principles of Geology*, published in 1830, Lyell described numerous examples of the catastrophic draining of lakes in mountain regions. He also emphasized how important these floods are to

the erosion of the valleys. Late in the nineteenth century and early in the twentieth century a school of thought, which we shall refer to as "gradualism," became established in the United States. This school included staunch uniformitarians who opposed any idea that hinted of catastrophic activities. Such ideas were stigmatized under the heading of "catastrophism." In this connection, the "Great Spokane Flood" is an instructive object lesson.

The idea of the "Great Spokane Flood," proposed in 1925 by J. H. Bretz (1882–1982), became synonymous with Bretz's career. During the 1920s Bretz published six papers on various aspects of his idea that several great floods had crossed the part of eastern Washington known as the Channeled Scabland (Figure 14-45). Bretz was unable to convince many of his skeptical colleagues that the supposed flood(s) had occurred. He was not able to cite an acceptable source for the water, and at that time did not have the advantage of distant views that had been made from far-enough away to appreciate the gigantic sizes of the flood-derived structures. All this changed beginning in 1956 when air photos were examined. Still later, images from satellites added many other striking views.

In a complete reversal of the previous situation, most modern geologists now accept Bretz's idea that catastrophic floods flowed across northeastern Washington State when the waters of Glacial Lake Missoula drained suddenly.

Glacial processes are responsible for the formation of most modern lakes, whereas during much of Earth's history, glaciers were of limited extent and tectonic forces were the primary creators of lakes. At times when the Earth was not, and had not recently been, glaciated the numbers- and distributions of lakes must have been much different than today. Lakes are also formed by volcanoes, rivers, meteorite impacts, and coastal processes. Lakes formed by volcanoes either fill dormant craters or calderas, or are formed when lava flows, ash falls, or volcanoes themselves block valleys. Meandering rivers form ox-bow lakes by the process of channel avulsion and meander abandonment as a natural part of their evolution. Flood-plain lakes form when rivers overflow their banks. Many of these are shallow and variable in extent. During the flood season, the area of Grand Lac, on the flood plain of the Mekong River, varies from 2500 to 11,000 km^2 (965 to 4247 $miles^2$). Meteorites form craters similar to volcanic craters, which may be filled with water and become lakes; the Ries crater in southern Germany is a famous example. Lakes may also form by coastal impoundments behind marginal marine barriers.

Viewed from the perspective of geologic time, lakes are temporary features that are fated to disappear. Some lakes, such as playas or salinas, retain their waters for only a few hours or days after a flood (Figure 14-46). The

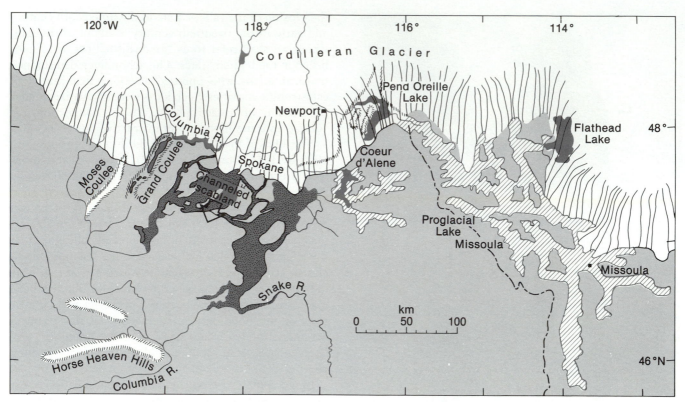

FIGURE 14-44.   Parts of northwestern Montana, northern Idaho, and northeastern Washington showing proglacial Lake Missoula. This former lake was dammed against the southern margin of the Cordilleran glacier. Waters from Lake Missoula flowed southwestward many times in the form of gigantic sheetfloods. Map shows possible flow routes to the Channeled Scabland in northeastern Washington. See also Figure 14-45. (J. E. Sanders in G. M. Friedman and J. E. Sanders, 1978, fig. 9-2, p. 239; replotted from J. H. Bretz, H. T. U. Smith, and G. E. Neff, 1956, figs. 23 and 24, p. 1036 and 1038.)

lives of lakes created by glacial action are tied to the existence of the glaciers themselves. However, the existence of the Great Lakes, tied to Pleistocene glacial events, no longer depends on glaciers. Large lakes in subsiding areas have persisted for up to 50 million years or longer. Nevertheless, a strong natural tendency exists for lake basins to be filled with sediment or, in the cases of lakes situated above sea level, for their waters to drain away and for a flow to the sea to be established. Lakes occupying depressions whose levels lie at or below sea level ultimately will cease to exist because seawater will invade them. For example, during the Pleistocene Epoch, each time sea level dropped during a glacial age, the Black Sea became a lake. Borings indicate that the sediments in lake basins near the sea, such as Lake Maracaibo, Venezuela, and Lake Biwa, Japan, contain interbedded layers of marine sediment. These indicate that the basins periodically have been invaded by seawater but later became lakes again.

SOURCES: V. R. Baker, 1973, 1978; Degens and D. A. Ross, 1971; D. A. Ross, 1971; Sarmiento and Kirby, 1962.

Lake Waters and Lake Sediments

The relationships between the characteristics of the water in a lake and the bottom sediments are very close and are delicately adjusted to the climate. In particular, the circulation of the water, the salinity, and the temperature affect the fate of first-cycle organic matter and control chemical precipitation. In addition, climate affects vegetation, and this affects sediments in three ways: (1) the kinds and abundances of plants are reflected in the variety and amounts of plant debris that accumulate in the lake, (2) the continuity of the plant cover affects the yields of terrigenous sediment from the drainage basin, and (3) tropical lakes may be largely or entirely covered with aquatic vegetation, which can cause deoxygenation of the lake water and concomitant changes in lake sediments (Figure 14-47).

A lake itself can affect plant life in two ways: (1) around a perennial lake, water is always abundant and the water promotes the growth of trees, shrubs, and grasses; and (2) the lake water is the habitat for many aquatic plants. These include *unicellular floating plants*, which compose

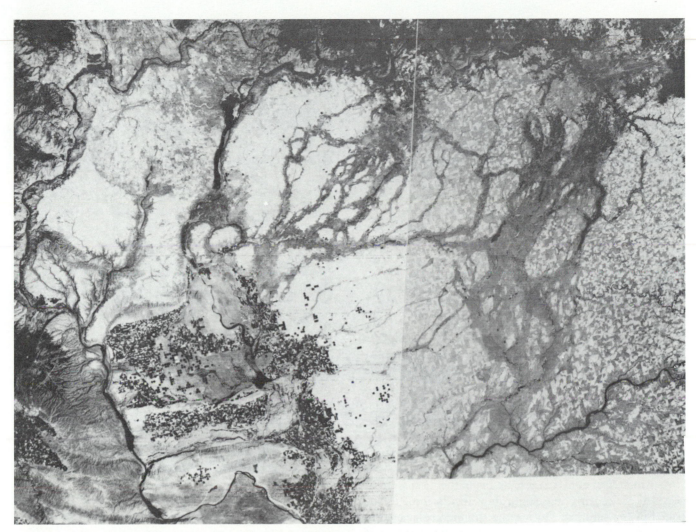

FIGURE 14-45.   Channeled Scabland, Washington, site of the "Great Spokane Floods," viewed from LANDSAT satellite on 27 August 1973 (left frame) and on 8 August 1973 (right frame); north at top. Compare with Figure 14-44. Columbia River enters view in northeast corner of left frame and flows in a broad arc along the north-, west-, and south sides of the western part of the Columbia Plateau, leaving the field of view near Hanford. Snake River crosses lower part of right frame. Two kinds of channels, now inactive, are present: (1) steep gorges (Moses Coulee and Grand Coulee), at center of left frame, which are retreat tracks of waterfalls on a former river about the size of the present Columbia River, and (2) shallow braided, rock-floored channels (dark gray areas in right half of combined frames), which discharged the great floods. One complex of these braided channels extends northeast-southwest and joins the Columbia River near the south end of Grand Coulee. The other complex extends more nearly north-south from Spokane to the Snake River near the town of Hooper. The small, light-gray squares are wheat fields. Multispectral scan, band 5. (NASA).

the **phytoplankton,** and various other aquatic plants, including algae, aquatic grasses, reeds, and water lilies, which grow along the margins of lakes or float on their surfaces.

The abundance of nutrients from phytoplankton serves as a basis for classifying lake waters into what are called *trophic levels.* Thus *lake waters that are deficient in plant nutrients* are **oligotrophic,** *lake waters that contain moderate supplies of plant nutrients* are **mesotrophic,** and *lake waters that contain an abundance of plant nutrients* are **eutrophic.**

The organic matter from the plants is the food for many aquatic animals. Much of the first-cycle organic matter that is not actively involved in the food chain tends to be oxidized; the end products are carbon dioxide and water.

EFFECTS OF CLIMATE ON LAKES

We shall illustrate the close relationship between climate and the behavior of lakes by discussing first the seasonal cycle of lakes in temperate climates and then proglacial

FIGURE 14-46.   Gettel Playa, Tule Valley, western Utah, filled with fresh water after a rare heavy rain. In foreground, dirt road, which is usually completely dry, passes directly into the ephemeral lake. Mountains in distance consist of dark brown Fish Haven Dolomite above white Ordovician Eureka Quartzite. (Authors.)

lakes and lakes in tropical climates. As we shall see, climate influences not only the sediments but also the levels of some lakes.

Seasonal Cycles in Lakes of Temperate-Climate Zones.   The bases for the seasonal cycles in lakes are the effects exerted by temperature, density, and salinity of the water on circulation and dissolved gases.

As a matter of convenience, we begin the discussion of seasonal cycles with the relationships that become established in the summer. During the summer, solar energy warms the air and water. Accordingly, the temperature of the water in the upper meter or two of the lake increases and its density correspondingly decreases. Anyone who has ever plunged below the surface waters of a pond or small lake while swimming or diving is well aware that the water is stratified (Figure 14-48). The warm water at the surface contrasts decidedly with the cooler water below; a pronounced *thermocline* separates the two water layers. Because it stays in contact with the atmosphere, *the thin layer of warm water at the surface* (**epilimnion**) remains well aerated. However, *the colder, denser water below the thermocline* (**hypolimnion**) is cut off from the atmosphere. Therefore its dissolved oxygen is no longer renewed and is subject to depletion by respiration and by combining with first-cycle organic matter. (See Figure 14-48, A.) The extent of oxygen loss will depend on the abundance of aquatic organisms and on the amounts of organic matter that drop into this lower layer of water. In some lakes the oxygen content may decline to the point where anaerobic bacteria become active. One of the by-products of the metabolism of the kinds of anaerobic bacteria that reduce sulfates is H_2S, a poisonous gas (Eq. 3-10).

FIGURE 14-47.   View of the floating fern *Salvinia*, which reproduces rapidly enough to choke the light and life out of ponds and lakes in the tropics. When the plant cover has become complete, as shown here, the water quickly becomes anoxic and this utterly changes the characteristics of sediments deposited. (M. J. Burgis and P. Morris, 1987, nonnumbered fig. preceding p. 139.)

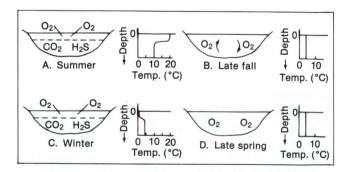

FIGURE 14-48.   Schematic profiles through small lake in temperate climate zone with temperature vs. depth graphs for water at various seasons of the year. Relationships for late fall (B) shown after annual overturn and upwelling has occurred; water is isothermal at 3.98°C. Further explanation in text. (After W. H. Bradley, 1948, fig. 2, p. 637.)

The midwater stratification of summer lasts until late in the fall. As the days grow shorter and cooler and the nights colder and longer, the temperature of the surface waters decreases, and the density correspondingly increases. Eventually the temperature of the surface waters approaches +3.98°C, the point of maximum density. (See Figure 8-35.) Now the density contrast between the surface water and deep water becomes insignificant, and surface water may even become denser than the deep water. The surface water sinks and displaces the oxygen-deficient bottom water upward. (See Figure 14-48.)

As the oxygen-deficient water rises to the surface, it brings up a rich supply of nutrient elements, particularly phosphorus and potassium, that were not removed from the lower water. The upwelling water may also contain dissolved silica, that enables diatoms to flourish. Therefore, at the time of the autumn overturn, the phytoplankton proliferate or *bloom*, and the effect is to discolor the water temporarily.

If the bottom waters became toxic during the period of summer density stratification, then the fall upwelling may cause mass mortality of the organisms in the lake. The carcasses of the instantly killed fish and other organisms settle to the bottom and, if bottom-scavenging animals are not present, specimens may be preserved in their entirety (Figure 14-49). Eventually, however, dissolved oxygen reappears in the upwelled water through photosynthesis and diffusion from the air.

In winter, with further cooling, a new stratification is established. Because the density of fresh water decreases slightly as the temperature drops below +3.98°C, stratification develops between less-dense, almost-freezing water at the surface and the underlying denser water at near +3.98°C. (See Figure 14-48, C.) During the winter the denser bottom water is cut off from the atmosphere and its dissolved oxygen is not renewed. Oxygen depletion begins, but now the supply of first-cycle organic matter is less than in the summer, and respiratory oxygen consumption is less, because at low temperatures the pace of vital processes slows. If the surface of the lake freezes over, then the oxygen supply of even the surface water is cut off, except for a small amount that may be produced by plants below the ice.

During the spring the surface waters warm to the point of maximum density, and for a brief period the density stratification in the lake disappears. Some ventilation of the lower water is now possible. (See Figure 14-48, D.) With further warming, the summer midwater stratification is reestablished and the cycle is complete.

This well-developed seasonal cycle may or may not be expressed in the bottom sediments. If the rate of sedimentation is slow and the bottom sediments are disturbed by deposit-feeding organisms and by burrowers, the sedimentary record may become blurred. Within uniform-appearing lake-bottom sediments, seasonal variations in abundance of pollen, other organic materials, or organic carbon or calcium carbonate may be recorded.

Seasonal Cycles in Proglacial Lakes. A **proglacial lake** is *a lake fed by meltwater from a glacier.* A proglacial lake experiences only two seasons: (1) "summer," when the water is open, and (2) "winter," when the lake, and possibly all the streams connected to it, are frozen. The length of the "winter" season may equal or exceed that of the "summer" season.

During the summer thaw, the surface waters renew their supplies of oxygen and the streams coming from melting glaciers contribute terrigenous sediment. Because the streams carry abundant suspended sediment, and discharge cold water from the melting glacier, their densities may exceed that of the lake water. Hence the stream water becomes an underflow, generating a steady turbidity current of low density, or an *interflow*, flowing along the thermocline (Figure 14-50). The steady, low-density turbidity currents transport silt and very fine sand along the bottom, building a laminated or even a micro-cross-laminated deposit. The finer sizes remain suspended in the lake water and do not settle immediately. In this way a fairly persistent supply of silt and very fine sand is spread across parts of the bottom. Because of occasional storms, slumping, or other sediment-moving events, spasmodic turbidity currents of higher density than the underflows may be generated. Spasmodic turbidity currents generally deposit their suspended fines immediately after depositing their silt or very-fine sand. Therefore the deposit of each spasmodic turbidity current may be graded and/or ripple cross laminated. If several spasmodic turbidity currents are triggered during

FIGURE 14-49. Carbonized remains of small fish preserved on bedding surface of fine-textured calcareous ancient lake-bottom sediment. Fish probably was killed suddenly when a rapid overturn disrupted a previously stratified water mass whose lower layer had become charged with H_2S. Preservation of minute details of soft parts of fish indicates that no bottom-scavenging organisms were present. Green River Formation (Eocene), Wyoming, United States. Personal collection of G. M. Friedman. (Gene Tobler.)

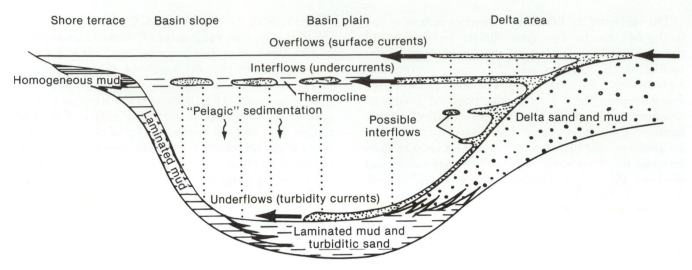

Shore terrace   Basin slope   Basin plain   Delta area

Overflows (surface currents)

Interflows (undercurrents)

Homogeneous mud

Thermocline

"Pelagic" sedimentation

Possible
interflows

Delta sand and mud

Laminated mud

Underflows (turbidity currents)

Laminated mud and
turbiditic sand

FIGURE 14-50.   Distribution mechanisms and kinds of sediment that may be present in oligotrophic lakes having an annual cycle of thermal stratification shown in schematic profile-section. The top of the basin-marginal terrace lies above the thermocline. No scale. (M. Sturm and A. Matter, 1978, fig. 10, p. 162.)

a given ice-free season, then the summer deposit may be many centimeters thick and may consist of several graded- or more-complex layers.

During the winter, the lake freezes over, and even the water in the streams may freeze. The ice cover shuts off any contact with the atmosphere. Dissolved oxygen is slowly depleted and, at the same time, the concentration of CO_2 gradually increases. The fine particles still suspended in the water column from the previous summer's turbidity currents and other sediment-dispersing processes slowly sink to the lake floor. They reach bottom according to their settling speeds, which are functions of their sizes, shapes, and densities. The vertically sedimented winter deposit is thus also a graded layer. Typically, the gradation is from silty clay at the base to fine clay at the top. The winter layer usually is dark colored, lacks laminae, and is a few millimeters to 1 cm or so thick.

The vertically sedimented winter layer overlies the laterally accumulated summer layer gradationally. In contrast, the winter layer ends upward along a clearcut contact. (See Figure 5-11, C.) This is a result of the sudden influx of coarser sediment brought to the bottom by the first of the summer's turbidity currents.

Because small icebergs are common in proglacial lakes (Figure 14-51), large rock fragments can be rafted out from the lake shore and dropped into the fine-textured sediments. Such *ice-rafted fragments* are called **dropstones** (Figure 14-52).

Varves.   The abundant sediment brought in along the bottom during the summer ice-free period and the sparse

FIGURE 14-51.   Debris-laden iceberg, approximately 30 × 15 m, floating in Malaspina Lake, near southeast margin of Malaspina Glacier, Alaska. (T. C. Gustavson, courtesy Harlequin Productions, M. O. Hayes and J. H. Hartshorn.)

FIGURE 14-52.   Well-rounded dropstone, long diameter approximately 20 cm, enclosed within laminated sediment of proglacial Malaspina Lake, Alaska, exposed when the lake level dropped. (T. C. Gustavson, courtesy Harlequin Productions, M. O. Hayes and J. H. Hartshorn.)

sediment deposited out of suspension during the long frozen period constitute a distinctive seasonal deposit. This creates one kind of *varve*. The term varve was introduced in 1912 by Baron Gerhard deGeer, whose classic studies in the chronology of the retreat of the Scandinavian ice sheet were instrumental in establishing a firm association between varves and the deposits of proglacial lakes and, in particular, in conveying the impression that a varved deposit is characterized by graded sediment couplets, in this case formed by the contrasting winter- and summer deposits. Since 1912, however, the term varve gradually has been applied to yearly deposits of *all* kinds, marine as well as nonmarine, including not only couplets but quadruplets and other patterns, some graded and some not. Furthermore, even some varved proglacial lake sediments have been shown to be more complex than simple successions of couplets.

In making counts of varved proglacial lake deposits, one must be careful to distinguish the winter clay layers from any clays deposited during the summer by spasmodic turbidity currents. Because of the effects of spasmodic turbidity currents, *each graded layer* in proglacial lake sediments does not *necessarily equate with one year*.

The years can be determined accurately by counting only the winter clay layers, and then only in sections where no winter layers have been removed by erosion by turbidity flows or other currents.

SOURCES: Antevs, 1957; deGeer, 1912; Sauramo, 1923.

Lakes in Tropical-Climate Zones. In tropical regions, large, deep lakes may become permanently stratified. The warm surface layer never cools to the point at which it can displace the subsurface water beneath the thermocline. As a result, the bottom waters may become permanently anoxic. An example of a permanently stratified large, deep lake in a hot, dry tropical climate is the Dead Sea. The extreme range of recorded temperatures for the water in the upper 50 m of the Dead Sea is 10 to 45°C. The mean yearly range is 19 to 36°C (Figure 14-53, right). The salinity of the Dead Sea ranges between 285 and 330‰, making it one of the modern world's most-hypersaline bodies of water.

The surface waters of the Dead Sea contain a small amount of dissolved oxygen. However, water below a depth of 50 m is charged with hydrogen sulfide (Figure 14-53, left).

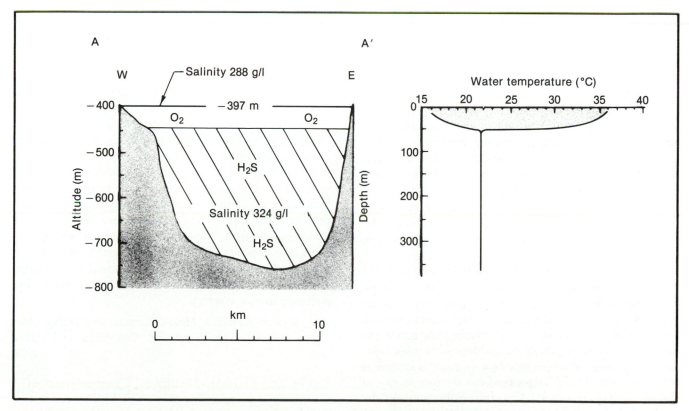

FIGURE 14-53. Stratified water of the Dead Sea, Israel and Jordan, a tropical lake, shown by topographic profile and hydrographic section along line A-A' of Figure 14-43. Dissolved oxygen in surface water is about 1.5 ml/l; lower water mass contains 10 to 11 ml/l of H₂S; graph at right shows variation in temperature of surface water and constant temperature below a depth of 50 m. (Data from D. Neev and K. O. Emery, 1967, map facing p. 10 and fig. 33, p. 55.)

Because of its high salinity, the Dead Sea precipitates evaporite minerals. These range from aragonite to halite, but the chief product is gypsum. In the surface waters and on the bottom where the water is shallow and where dissolved oxygen is present, gypsum is ubiquitous. Aragonite is precipitated abundantly only when the surface water temperature exceeds 35°C, which happens only in the summer about every fifth year. At such times the waters become white with tiny aragonite crystals.

The foregoing description of the Dead Sea is accurate up to 1979, at which time the hydrologic regime of the Dead Sea underwent a major change. Its stable midwater stratification was destroyed; surface waters are now mixing with much of the deeper, fossil water mass. This mixing event appears to have resulted from an increase in salinity of surface waters following a diversion for irrigation of fresh water from the Jordan River. However, such mixing events may occur naturally in the Dead Sea. A similar event may have taken place 270 years ago.

On 21 August 1986, catastrophic overturn occurred in another deep tropical lake: Lake Nyos, in Cameroon, West Africa. The sudden release of more than 1.0 km^3 of CO_2 from the lake killed about 1700 people and more than 3000 cattle (concentrations of CO_2 greater than 10% are lethal to humans). The gas release was violent enough to inundate an 80-m-high rock promontory on the southwestern shore of the lake.

ORGANIC MATTER IN LAKE WATERS AND LAKE SEDIMENTS; EUTROPHICATION

Because of the abundant plant populations around their shores and the phytoplankton in their near-surface waters, many lakes are charged with organic matter that becomes incorporated into their sediments. For example, the content of organic carbon in the silty-clay sediments in the central parts of Lake George, New York, constitutes 9 to 17% of the dry weight of the sediment. The organic matter comes from the phytoplankton within the lake and from the debris of vegetation washed in from the surrounding watershed.

The natural tendency of many lakes is to be crowded out of existence by the growth of plants. Through time, therefore, the waters of the lake may become more and more overwhelmed by organic matter, so that, despite ordinary circulation, the dissolved oxygen is exhausted by the demands of the oxidizable organic matter. *Loss of dissolved oxygen in a body of water* is known as eutrophication. In the tropics, where floating ferns can almost completely cover lakes, drastically reducing the rate of photosynthetic oxygen production in the water column by phytoplankton while at the same time increasing the input of organic material, eutrophication may set in rapidly. In temperate climates, rooted aquatic vegetation or floating sedges may encroach on lakes centripetally, greatly reducing oxygen levels in sediments and waters of the shallow, nearshore zone.

Added to the natural tendency for lakes to undergo progressive eutrophication are the effects on the oxygen supply of certain materials dumped into lakes by people. Two common elements that are important plant foods are nitrogen and phosphorus. In fact, the growth of algae and other plants in most lakes is limited by the supply of one or both of these elements, usually phosphorus. Human feces contain nitrogen, detergents are rich in phosphorus, and fertilizers contain both elements. When these materials enter a lake they enable the populations of algae and other plants to increase. *Explosive population increases of phytoplankton*, called blooms, may result. When this happens, the amount of dissolved oxygen decreases. The eutrophication of Lake Erie has brought this word to the public's attention. Eutrophication causes drastic changes in lake ecology, which may include the retention of huge volumes of organic material in lake sediments. If oxygen levels in bottom waters remain low, anaerobic bacteria may attack this organic material, producing toxic gases, such as methane, as a result. Even before this happens, however, the fall in oxygen levels in lake waters causes massive kills of indigenous fish and their replacement by other species tolerant of low levels of dissolved oxygen. These species are generally considered less valuable for both commercial- and sport fishing. If oxygen levels continue to fall, even these undesirable species may be eliminated and the lake waters may be rendered virtually barren of life except for certain bacteria. The water becomes an embalming fluid.

Much of the organic matter deposited in the sediments of highly productive or eutrophic lakes is preserved and buried, rather than oxidized or metabolized by anaerobes. After burial and exposure to geothermal heating, some of the organic matter is altered to the long-chain hydrocarbon polymers known as kerogens. If their compositions are appropriate, kerogens are the sources of hydrocarbons from "oil-shales," many of which are ancient lake deposits. The People's Republic of China is particularly rich in hydrocarbons formed in lacustrine sediments and these deposits are of major economic importance to that country.

SOURCES: Battarbee, Mason, Renberg, and Talling, 1990; Fraser, 1989; Füchtbauer, 1988; Matter and Tucker, 1978.

Lakes and Human-Produced Contamination

A side effect of many modern industrial-age activities is the release of heavy metals into the atmosphere and into the waters of rivers, lakes, and the sea. Most of these heavy metals are present in lake sediments only in trace quantities, but some are toxic even in trace amounts. Examples include mercury, lead, cadmium, copper, chromium, zinc, and plutonium. Analyses of the trace-metal content of cores of lake sediments have been made in many localities. These typically display a large increase toward the top of the core—an effect of our industrial age.

To return to the example of Lake George, analyses have been made for copper, chromium, and zinc. Cores from the southern basin show an upward increase in trace-metal content, starting at a subbottom depth of 13 cm. Similar distributions of these and other heavy metals have been reported from many lakes in North America and in Europe.

Another pollutant, asbestos, although not a heavy metal, has been found to be a hazard to human health. Asbestos has been identified in the ground-up waste products dumped into Lake Superior by iron-ore processing plants on the shores. Before asbestos was discovered in them, these wastes were thought to be harmless.

Possibly even more harmful to human health than either heavy metals or asbestos are a wide variety of complex organic compounds introduced to lake sediments by human activity. These include pesticides like DDT and its breakdown products, PCBs (polychlorinated biphenyls), herbicides, and the by-products of manufacturing processes that involve organic materials. Some of these compounds are highly toxic at the lowest concentrations measurable; others are powerful carcinogens for which the minimum safe level may be as small as a few molecules. Siskiwit Lake, a tiny lake on Isle Royale, an uninhabited island in northern Lake Superior, has become a natural laboratory for the measurement of global atmospheric transport and deposition of toxic chemicals. Siskiwit Lake and Isle Royale have never been occupied by people, and the lake-within-a-lake has never had any direct connection to Lake Superior, so the only source of contaminants is the atmosphere. Cores of sediments in Siskiwit Lake show essentially no organic contaminants traceable to human activity before the end of World War II. This marked the onset of the modern chemical age, and the youngest sediments in this isolated lake show a tremendous variety of toxic chemicals including furans, dioxins, toxaphene (a poison at least as toxic as DDT whose nearest known source is the cotton fields of the southeastern United States), and PCBs and DDE (a breakdown product of DDT) in concentrations two to ten times those in Lake Superior. The DDT in the youngest sediments in Siskiwit Lake probably did not come from the United States, where it has been banned for more than 15 years, but from Central America, Asia, or the USSR, where it is still used extensively.

SOURCES: Battarbee, Mason, Renberg, and Talling, 1990; Sanders, 1989.

Clues to Recognizing Lake Deposits

Nearly all lake deposits follow a similar general pattern. From the center of the lake to the shore one can usually recognize a central-lake suite, typically composed of fine materials, and a marginal suite that may be either coarse or fine. Coarse marginal deposits are the products of deltas, fans, fluvial plains, and beaches. Because many or even most large modern lakes are situated in fault troughs, and because many ancient large lakes were similarly situated, the typical coarse marginal deposits are fan sediments. Fine marginal facies are the products of marshes or carbonate-evaporite flats. Where the marginal sediments have built appreciable underwater slopes, gravity flows may shift sediment outward to the central-lake area. Where a single point source feeds a density underflow into a lake, a coarse central-lake facies may develop, a fine-textured marginal-slope facies accumulates where suspension deposition dominates, and a coarse marginal facies where wave activity operates.

This general pattern likewise prevails along the margins of the sea. The chief differences between the pattern in lake deposits and that in marine deposits result from differences in scale. Lake deposits pass laterally into marginal materials in all directions; within an area of comparable size, marine deposits would appear to be one-sided and to display open-water deposits for indefinite distances.

Stratigraphic units composed of lake deposits and their associated marginal materials are built in response to three dynamic factors. These are (1) subsidence of the basin, (2) supply of sediment, and (3) fluctuations of water level. Exactly the same can be said for materials deposited at the edge of the sea. Because depositional patterns and dynamic factors involved in forming stratigraphic units are similar one would expect lake sediments and marine sediments to be much alike, and indeed they are. In some cases it is not easy to distinguish ancient lake sediments from ancient marine deposits. Few hard-and-fast generalizations have withstood close scrutiny.

The best chances for making a distinction between lacustrine and marine sediments exist when one alternative is the sediment deposited in a fresh-water lake. The contrasts in the organisms that inhabit fresh water and those that inhabit the sea are numerous and striking. Insect larvae, for example, are common in fresh-water lake deposits but rare in marine sediments.

The value of fossils decreases when saline lake waters are involved. If the lake started as a cut-off arm of the sea, then an additional complication may be introduced. Initially, the organisms are of ordinary marine kinds. If the change from salt water to fresh water takes place gradually, however, some of these organisms may be able to adapt to the changing salinity. Thus their offspring may ultimately survive in lake water having a salinity quite different from that of seawater. As a result, the faunal contrast between lake- and marine sediments may not be very great.

The most-useful indication of an ancient lake deposit comes from regional stratigraphic relationships. The lake-marginal deposits completely encircle the open-water lake deposits.

The contrast in size between large lakes and the sea is expressed in the sediments deposited in the zone of shoaling waves and on beaches. Although wind waves of comparable size may be generated by storms on both lakes and the sea, nevertheless lakes do not experience swells or appreciable lunar tides. Therefore the effects of long-period swells and of tidal fluctuations are found only in the sea.

Because the size of the waves affects the thickness and characteristics of sequences deposited on prograded beaches and zones of shoaling waves, it should be possible to separate lake-marginal from sea-marginal wave-influenced sediments. Thicknesses of such successions deposited in lakes range downward from 5 m or so (Figure 14-54); those deposited in sea-marginal settings typically range upward from 10 m, but on protected coasts can be equal to those of lake-marginal successions. Because of this overlap, it is not possible to cite a specific limiting thickness that can always be used to distinguish sediments deposited on lake beaches and zones of shoaling waves from those deposited in comparable marine settings.

Box 14.4 describes examples of inferred ancient lake deposits.

SOURCES: R. A. Davies, 1983; Fraser, 1989; Matter and Tucker, 1978.

Glacial Environments: Modern- and Ancient Deposits

Ten percent of the Earth's surface is covered by glacial ice, which constitutes 75% of the Earth's fresh water. However, during the Quaternary glaciations up to 30% of the Earth's surface was covered by the vast continental ice sheets and lesser mountain glaciers.

Our purposes in discussing glacial environments and glacial sediments are very limited. We do not intend to delve as deeply into glacial subjects as we have into our discussions of other environments. We defend this lack of balance because of the widespread practice of dealing with glacial deposits in courses entitled glacial geology, Pleistocene geology, or geomorphology. Because these courses exist, it is the usual practice to omit glacial topics from sedimentology courses. Despite this practice,

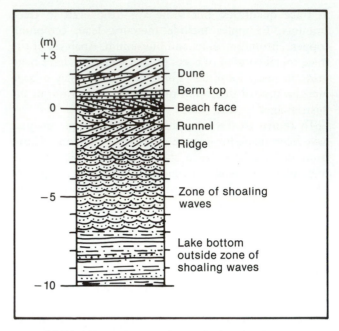

FIGURE 14-54.   Upward-coarsening vertical succession (schematic) formed by progradation of a beach and related sediments on the shores of a large lake. Succession begins at the base with silty clay deposited lakeward of zone of shoaling waves, and grades upward through ripple-marked sands of the zone of shoaling waves and through sediments of the various parts of a beach to dune sand at the top. The thickness from the base of the sand deposited in the zone of shoaling waves to the base of the beach-face deposits, 7 m in this sketch, is a basis for inferring the time-averaged depth to the outer margin of the depth of shoaling waves. This depth, in turn, reflects the size of the lake and the size of the waves on its water surface. The sketched profile is about as thick as such lake sediments ever become. The comparable deposits in most lakes would be thinner, possibly only 2 or 3 m. (J. E. Sanders, in G. M. Friedman and J. E. Sanders, 1978, fig. 9-19, p. 253.)

however, we think that it is important to introduce at least a short summary of glaciers and glacial deposits.

Modern Glaciers and Glacial Deposits

These [subglacial] streams of the glaciers are remarkable for the whitish-blue colour of their waters, which they preserve for a distance of several miles. This colour is ascribed to the numerous particles of rocky matter which the torrents bring down in a state of the greatest comminution effected by attrition. (Wittich, 1845).

According to the glossary of geology prepared by the American Geological Institute (Bates and Jackson, 1987) a **glacier** is *"a large mass of ice formed, at least in part, on land by the compaction and recrystallization of snow, moving slowly by creep downslope or outward in all directions due to the stress of its own weight, and surviving from year to year."*

From the earliest days of geology, glaciers have been recognized as powerful agents for erosion and for transport of sediment of all sizes, including huge boulders. However, the concepts of the Pleistocene glacial- and interglacial ages, which are now so widely accepted, began to be developed in Europe only in the 1830s and 1840s. Even earlier, in 1795, the Scottish geologist James Hutton (1726–1797) was one of the first to recognize that glaciers once were more extensive than they are now. The Swiss-American geologist Louis Agassiz (1807–1873) (1842) was most responsible for disseminating the idea of continental glaciations in the middle part of the nineteenth century.

Glaciers accumulate new snow directly from the atmosphere, and, if they are near steep slopes, from snow avalanches. With atmospheric snowfalls come dust particles. From the slopes may come rocky debris, falling as individual fragments or coming all at once as a layer deposited by a rock avalanche. (See Figure 7-3.)

A key area for understanding the effects of a glacier on sediments is the margin, or terminus (Figure 14-55). Many important activities take place beneath the ice and well back from the margin, but these are hidden from direct observation. By contrast, at the margin, particularly of a retreating glacier, one can see many of the important- and distinctive sedimentologic processes in action

FIGURE 14-55.   Snout of wasting Woodworth Glacier, Alaska (upper right), viewed obliquely on 12 August 1938 from airplane looking toward the northeast. Parallel ridges and grooves at right center were made by the flow of the glacier over till. By 1970 the grooved till had become covered with trees. Esker (center), which shows both sinuous- and braided pattern, ends in large fan at lower left. At upper left is a hummocky area underlain by debris-covered stagnant ice. Braided streams carry abundant sand-size sediment away from glacier. (Bradford Washburn; remarks on situation in 1970 by J. H. Hartshorn.)

FIGURE 14
Barnard (
logical Su

that hav
medial r
The fl
deposits
and may

GRINDI

Particles
posited w
come par
and may
along cres
If a pe
lower par
abraded b
ice just al
grooved (l

Tl
of
po
lal

ge
ha
wi
tio
the

rel
by
or
firs
By
ref
pos
or

an
eitl
equ
do
bal
bal
me
gla
a te
rap
stat
was

SN

As 1
mu
talli
mat
ever
Bec
and
and
as i
per
and
cal
Ant
kno
and
of fe

ities afforded by deposition against or surrounding large masses of ice and by a supply of freshly ground sediments of nearly all possible sizes. Much outwash is deposited by braided streams on fans (See Figure 14-55, lower left.) and on fluvial plains. (See Figure 14-55, lower right.)

The distinctively glacial aspects of outwash are found in those bodies of sediment that were deposited against one or more walls of ice. For example, *the channel deposits of rivers that flow in ice tunnels* can build upward; the sediment is supported by the ice. After the ice has melted, these deposits form ridges, known as **eskers** (Figure 14-63; see Figure 14-55, center).

SOURCE: M. Edwards, 1986.

Ancient Glacial Deposits

Ancient glacial deposits can be divided into two categories: (1) those of Quaternary- and late Tertiary ages, which form large parts of the regolith and modern landscape in many northern temperate- and arctic areas, and (2) those of pre-Mesozoic age, which form parts of the bedrock in numerous parts of the world (Figure 14-64). The late Cenozoic tills have not been deeply buried and generally have not been lithified. The pre-Mesozoic glacial deposits consist of well-compacted sediment or of hard rock.

SOURCE: Hambrey and Harland, 1981.

Suggestions for Further Reading

AHLBRANDT, T. S., and FRYBERGER, S. G., 1982, Introduction to eolian deposits, p. 11–47 *in* Scholle, P. A.; and Spearing, D., eds., Sandstone depositional environments: Tulsa, OK, American Association of Petroleum Geologists Memoir 31, 410 p.

ALLEN, P. A., and COLLINSON, J. D., 1986, Lakes, p. 63–94 *in* Reading, H. G., ed., Sedimentary environments and facies, 2nd ed.: Oxford, Blackwell Scientific Publishers, 615 p.

BURGIS, M. J.; and MORRIS, PAT, 1987, The Natural history of lakes: Cambridge, England, Cambridge University Press, 218 p.

FROSTICK, L. E.; and REID, E., 1987, Desert sediments: ancient and modern: Geological Society of London Special Publication 35, 491 p.

KOCUREK, G.; and NIELSON, J., 1986, Conditions favorable for the formation of warm-climate sand seas: Sedimentology, v. 33, p. 795–816.

LANGFORD, R. P.; and BRACKEN, BRYAN, 1987, Medano Creek, Colorado, a model for upper-flow-regime fluvial deposition: Journal of Sedimentary Petrology, v. 57, p. 863–870.

RUBIN, D. M., and HUNTER, R. E., 1985, Why deposits of longitudinal dunes are rarely recognized in the geological record: Sedimentology, v. 32, p. 147–157.

RUBIN, D. M., and HUNTER, R. E., 1987, Bedform alignment in directionally varying flows: Science, v. 237, p. 276–278.

WETZEL, R. G., 1983, Limnology, 2nd ed.,: Philadelphia-London-Toronto, W. B. Saunders Co., 767 p.

PART V

LARGE-SCALE PATTERNS OF SEDIMENTARY DEPOSITS

In this, the final part of the book, we get back to the global perspective with which we began. The common theme uniting the three chapters of this part is the emphasis on study of sedimentary strata at large scales, ranging from large parts of single basins to the entire world.

In Chapter 15 we discuss extraterrestrial influences on sedimentation and on patterns of sedimentary layers. We specifically include planetary orbital effects, tidal phenomena, Milankovitch climatic variation, and the possible influence of the Sun's orbit.

In recent years it has become very clear that extraterrestrial factors subtly but pervasively affect terrestrial sedimentation. In particular, cyclic sequences at a variety of scales show evidence of extraterrestrial influence. One of the main purposes of this chapter is to acquaint the reader with some of the extraterrestrial factors that could be responsible for cyclicity in terrestrial sedimentary sequences.

In Chapter 16 we return to the subject of stratigraphy. Here we consider some of the complexities of correlation, as well as some specifics about field-study methods, definition and recognition of stratigraphic units, more on unconformity surfaces (their recognition and significance), and paleogeographic reconstruction. This chapter also includes a discussion of the different kinds of stratigraphic units: litho-, bio-, magneto-, chrono-, and polarity-chronostratigraphic units. The goal of much of this material is to explain how and why one establishes formal, mappable stratigraphic units of all kinds, and might be called "classical stratigraphy." In addition, we briefly review biostratigraphy, biogeography, and magnetostratigraphy with an emphasis on their practical contributions to stratigraphic studies.

We begin Chapter 17 by looking at sedimentary basins from a historical perspective. Then follows a section in which different kinds of basins (defined using plate-tectonic concepts) are described in terms of their modes of formation, morphology, distribution, and content of sedimentary strata. Thereafter the concept of the geosyncline is contrasted with the plate-tectonic interpretations. The chapter concludes with a consideration of where the study of sedimentary deposits is today, and where it is going.

CHAPTER 15

Extraterrestrial Forcing Functions

For many years geologists tended to believe that everything necessary for understanding the Earth's behavior was confined to processes operating right here on the Earth. They recognized, of course, that solar energy comes from the outside and that the oceanic tides are controlled by changing relationships in the alignment of the Earth, the Moon, and the Sun. But these two familiar phenomena were generally taken for granted and did not form the basis for general inquiries into extraterrestrial factors that might be significant influences on the Earth. A few exceptions should be noted. For example, some of the earliest geologists knew about the geologic implications of studies of the Solar System by astronomers. One of John Playfair's (1748–1819) arguments in favor of the validity of Hutton's novel ideas about the Earth was their consistency with astronomy. In recent years many geologists have investigated the possible effects on the Earth of impacts by extraterrestrial objects, such as large meteorites or even asteroids, or of the Earth's periodic passage through parts of the galaxy where the quantities of matter in outer space are greater than elsewhere. We illustrate how James Hutton was aware of astronomic factors by quoting part of a paragraph from Playfair's book (Playfair, 1802, p. 437):

Note XX. Par. 118. Inequalities in the Planetary Motion.

The orbits of the planets change not only in their position, but even their magnitude and their form: the longer axis of each has a slow angular motion; and, though its length remains fixed, the shorter axis increases and diminishes, so that the form of the orbit approaches to that of a circle, and recedes from it by turns. In the same manner, the obliquity of the ecliptic, and the inclination of the planetary orbits, are subject to change; but the changes are small, and, being first in one direction, and then the opposite...

As mentioned in Chapter 1, in 1864 James Croll (1821–1890) proposed the hypothesis that the variations in the Earth's orbital elements through time were sufficient causes of the climatic changes during the Pleistocene Period in which glacial ages alternated with nonglacial ages. After many downs and ups, a version of Croll's hypothesis, modified by Milankovitch, is enjoying great popularity.

The extinction of species on Earth has long been an object of popular fascination and of serious scientific research. Almost all species that have ever lived are now extinct. Currently, the mass extinctions of large numbers of species, especially at the end of the Cretaceous, have been attributed to extraterrestrial forcing through the impact of a large asteroid. Although the question is by no means settled, many geologists advocate a causal relationship between impact and extinction.

What we propose to do in this chapter is to examine the astronomic situation within the Solar System with respect to possible effects on the Earth's environmental conditions. We begin with some fundamental definitions of planetary orbits, review the Earth–Moon–Sun relationships that control the astronomic tides, and then examine the mechanics of the entire Solar System, within which all climate mechanisms function.

Elements of Planetary Orbits

Under this heading we explore some relationships that have been established among planetary orbits. We begin with the fundamental definitions.

Definitions

The astronomic factors that we need to consider are (1) properties of elliptical orbits, (2) the progression of the longitude of perihelion, (3) axial inclination, and (4) the angles between critical axes and planes.

PROPERTIES OF ELLIPTICAL ORBITS

Figure 15-1 shows an ellipse and the fundamental geometric relationships. As a first approximation of the arrangements in the Solar System, one can suppose, as did Johann Kepler (1571–1630), that the center of the Sun occupies one of the foci of a planetary ellipse. As we shall see in a following section, although this approximation has served many useful purposes, it is not strictly correct. Indeed, a literal application of it causes one to overlook what may be profoundly significant dynamic relationships.

The locations of the two foci determine key aspects of a planet's elliptical orbit. The ellipticity is the ratio of $\overline{OF}$ to a, the semimajor axis. Ellipticity ranges between zero for a circle and 1.0 for a straight line. An important climatic implication of ellipticity is based on the surface-to-surface separation distance between the Sun and a planet.

If such a thing as a circular orbit of one planetary body around the center of a sun could exist, then the surface-to-surface separation between the two bodies would always be the same. And because of the inverse-spreading effects of Snell's law, by which intensity of radiated energy decreases as the square of the center-to-center distances between a sun and a planet, the greater the distance of such a hypothetical planet away from such a hypothetical sun, the less intense would be the energy received at the surface of the planet. Even granting a constant source of solar energy (a much-debated point to be discussed further in a following section), then any factor that causes the distance from sun to planet to change will affect the intensity of solar energy received.

The length of the semimajor axis of an orbital ellipse is the mean distance of that planet from the Sun. Astronomers argue that this distance is a constant but that the length of

the semiminor axis can change. If the length of the semiminor axis increases, the orbit becomes more circular. As an orbit becomes more circular, the distance to the Sun becomes more constant in all positions occupied by the planet. As the orbit becomes more elliptical, the distance to the Sun becomes more variable in all parts of the orbit; that is, in some parts of the orbit the planet is closer to the Sun than in other parts (Figure 15-2).

Because of the changing positions of the other planets in the Solar System, the ellipticity of the Earth's orbit changes from nearly circular to an ellipse having an eccentricity (ellipticity) of about 5%. The length of time required for this change in ellipticity to take place is 93,408 yr.

PROGRESSION OF THE LONGITUDE OF PERIHELION

On a planetary orbit, the point where the planet is closest to the Sun is where its path crosses the semimajor axis on the side of the focus occupied by the Sun. This *point of closest approach of a planet (or any body orbiting the Sun) to the Sun* is known as **perihelion**. Because of the changing gravitational attraction of all the other plan-

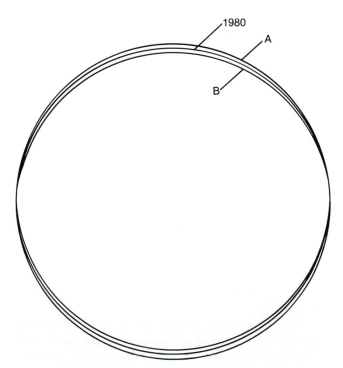

FIGURE 15-2.   Changing ellipticity of Earth's orbit.

A. Nearly circular mode; distance from the Sun, hence intensity of solar radiation received at the top of the Earth's atmosphere, is virtually uniform at all points along the orbital path.

B. Elliptical mode, with maximum ellipticity resulting from decrease in length of semiminor axis. In this mode the Earth is closer to the Sun as it occupies the sides of the ellipse. At such times the intensity of solar energy received attains maximum values. (J. E. Sanders, 1981, fig. 1.3, p. 20.)

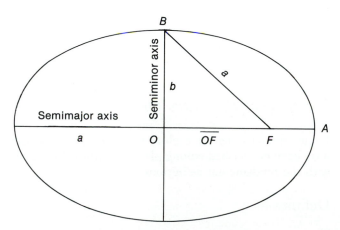

FIGURE 15-1.   An ellipse showing fundamental geometric relationships. The Sun approximately occupies one of the foci of a planetary ellipse. (J. E. Sanders, 1981, fig. D.1, p. 562.)

ets on the orbits of all planets, the semimajor axis pivots around within the plane of the planet's orbit. This means that through time, the orientation of the perihelion position will change. This motion is known technically as *the progression of the longitude of perihelion* (Figure 15-3).

The direction of the major-axis progression is counterclockwise. The time required for the Earth's semimajor axis to progress through a complete circle in the plane of the Earth's orbit is 93,408 yr. Notice that this is the same amount of time required for the change in ellipticity. Although these two periods are equal, they are not locked into the same phase relationships.

AXIAL INCLINATION

The tilt of the Earth's polar axis is the angle between this axis and the Ecliptic Pole, a line normal to the plane of the ecliptic. This angle varies between 22°29'36" and 23°50'30" on a cycle of 17,280 yr.

The changing angle of tilt operates simultaneously with a slow clockwise change in the direction of inclination of the Earth's pole of rotation. This motion is a kind of "wobble" comparable to the gyrations of a spinning toy top (Figure 15-4). As noted, the Earth's pole gyrates clockwise. This is just the opposite of most of the other orbital motions, which are counterclockwise (when viewed looking down from above the plane of the ecliptic). Movement of the pole is proved by the fact that the so-called pole star keeps changing with time. Nowadays, the pole star is Polaris, but in the past, other stars have occupied this important position (Figure 15-5). The time required for the Earth's polar axis to make a complete circle (seen from above) is 25,920 yr. The name applied to the major "wobble" of the Earth's polar axis of rotation is *precession*.

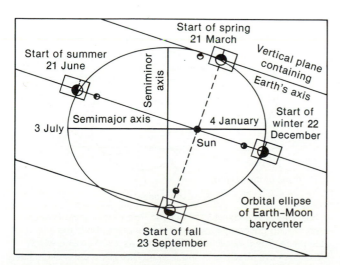

FIGURE 15-3. Counterclockwise motion of semimajor axis of orbital ellipse causes the longitude of the perihelion position to progress, that is, to shift through time.

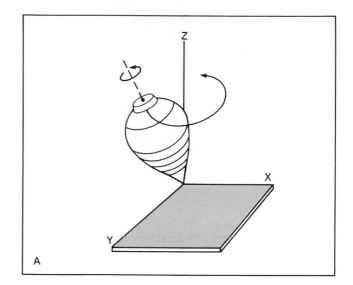

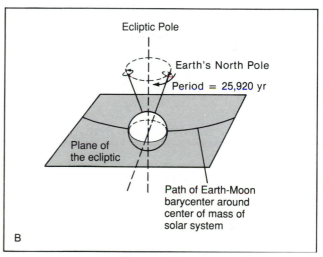

FIGURE 15-4. Precession of a pole of rotation, comparing a child's toy top and the Earth's pole.

A. As its rate of rotation decreases, the rotation axis of a child's toy top spinning on the *XY* plane begins to wobble (precess) around a pole (*Z*) normal to this plane. Notice that the direction of spinning and of wobble are the same.

B. Schematic view from obliquely above the plane of the ecliptic showing a cone that represents the loci of points generated by the precession of the Earth's North Pole at a constant angle with respect to the Ecliptic Pole. Notice that the direction of the Earth's spinning on its polar axis is counterclockwise when viewed from above the North Pole, just the opposite sense from the clockwise direction of polar-axis precession. (After J. A. Hynek, 1989, figs. 2, p. 238, and 3, p. 239.)

The relationships between the changing orientation of the Earth's polar axis undergoing its axial precession and the changing angle of tilt can be expressed by a diagram of the position of the pole as seen from above the Ecliptic Pole (shown as a dot on Figure 15-6). Two reference circles are generated by the loci of key angles: (1) an inner circle defined by the minimum tilt angle and (2) an outer circle defined by the maximum tilt angle. The

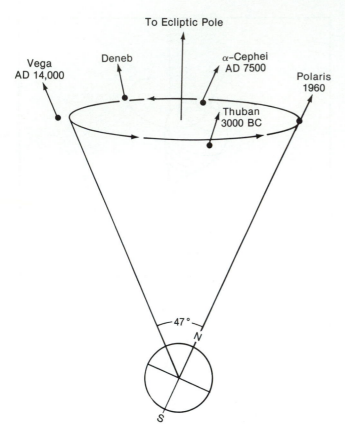

FIGURE 15-5.   Changing pole stars through time is proof that the orientation of the Earth's polar axis does not remain constant. This diagram has been drawn on the assumption that the only polar motion is a gyration at a constant angle with the Ecliptic Pole. In reality, the Earth's pole of rotation not only gyrates, but the angle it makes with the Ecliptic Pole also varies. (A. N. Strahler.)

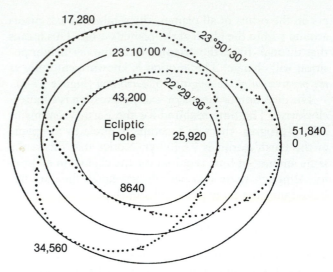

FIGURE 15-6.   Changing tilt and polar-axis precession portrayed as viewed looking down from above the Ecliptic Pole (dot in center of circle). Outer circle is loci of points generated by the maximum tilt angle; inner circle, by minimum tilt angle; and middle circle, by mean tilt angle. The dots show the path of the Earth's North Pole, starting at 0 yr (at right), a time of maximum tilt, and proceeding clockwise. Half-cycle positions of 8640 (minimum), 17,280 (maximum), 25,920 (minimum), 34,560 (maximum), 43,200 (minimum), and 51,240 (maximum) are labeled.

Notice that after one full gyration in 25,920 yr, the North Pole has returned to the same orientation as at 0 when the cycle began, but now the tilt angle is at a minimum. A second cycle of 25,920 yr is required to return the North Pole to its original alignment of both orientation and maximum tilt angle, as at 0. (Authors.)

position of the Earth's pole lies between these two circles. The two factors interact in such a way that after one full axial-precessional gyration the tilt angle has changed from its maximum value to its minimum value. In order to generate a full cycle of maximum tilt to minimum tilt and back to maximum tilt in the same orientation position, a second gyration is necessary, thus making a complete tilt + axial precession cycle of 25,920 × 2 or 51,840 yr.

ANGLES BETWEEN CRITICAL AXES AND PLANES

Another plane that needs to be defined in this discussion is the *mean plane of all the planetary orbits,* known in astronomy as the **invariable plane.** In astronomical terms, *the line formed by the intersection of two orbit planes* is named a **node.** *The plane of the Earth's orbit* is known as the **plane of the ecliptic.** The plane of the ecliptic makes an angle with the invariable plane of 1°36'. The *Earth's node,* the line of intersection of these two planes, rotates around in the invariable plane in a period of 362,880 yr.

As a result of the Earth's nodal cycle, the Ecliptic Pole (pole normal to the plane of the ecliptic) generates a circle of twice 1°36', or 3°12', around the **Invariable Pole** (*pole normal to the invariable plane;* Figure 15-7).

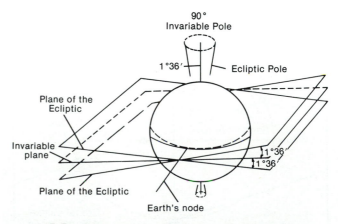

FIGURE 15-7.   Relationship of Ecliptic Pole to Invariable Pole as seen looking obliquely downward from a point above the invariable plane. The Invariable Pole appears as a dot around which the successive positions of the Ecliptic Pole generate a circle having an angular radius of 1°36'. This motion of the Ecliptic Pole around the Invariable Pole results from the motion of the Earth's node. (J. E. Sanders, 1981, fig. 13.31b, p. 328.)

The motions of the three poles we have been discussing interact. These motions and poles are the larger-angle and shorter-term gyration of the Earth's pole of rotation (1) about the Ecliptic Pole (2) and the smaller-angle and longer-term gyration of the Ecliptic Pole around the Invariable Pole (3). Their resultant angular relationships determine the latitudes of the departure from the Equator of the subsolar point. (See Figure 8-1.)

Having now defined the major elements of the Earth's orbit, we are in a position to add in the elements of the Moon's orbit to understand tidal variations, and afterward, to understand the so-called Milankovitch factors as related to climatic cycles.

Oceanic Tides: Earth–Moon and Sun

In Chapter 8 we presented the general relationships between the Earth–Moon pair and the Sun that are responsible for the astronomic tides on the surface of the water of the Earth's oceans. Here we explore lunar-orbital variables further, discuss how these variables interact, list some additional lunar cycles not mentioned in Chapter 8, and discuss how these cycles affect the tides and possibly other environmental aspects of the Earth.

Lunar Variables

The three chief variables that can be noticed easily on a monthly time scale are phase, Earth–Moon separation distance, and declination. Other lunar variables operate on longer-than-monthly time scales.

PHASES OF THE MOON

The phases of the Moon result from the ever-changing alignments of the Sun, the Earth, and the Moon. Twice during each orbit the Moon experiences what are known as syzygy phases; that is, *times when the centers of the Sun, the Earth, and the Moon lie along a straight line.* At the syzygy of New Moon, the Moon lies between the Earth and the Sun. (See Figure 8-32, A.) At the syzygy of Full Moon, the Earth lies between the Sun and the Moon. (See Figure 8-32, C.) At the quarter phases, a line from the Earth to the Moon makes a right angle with a line from the Earth to the Sun. (See Figure 8-32, B and D.)

EARTH–MOON SEPARATION DISTANCE

A major factor affecting the amplitudes of astronomic tides is the Earth–Moon separation distance. This distance changes cyclically as a result of both short-term and long-term factors. The short-term change results from the ellipticity of the Moon's orbit. The long-term change results from planetary factors that change the dimensions of the Moon's orbit.

Each month the Moon's orbit takes it through a *perigee* position (closest to the Earth) and an *apogee* position (farthest from the Earth).

Because of planetary-position effects on the size of the Moon's orbit, the Earth–Moon separation distance undergoes long-term oscillations. At the present time we are in that part of the cycle in which the Moon and the Earth are moving farther apart.

LUNAR DECLINATION

Lunar declination changes regularly because the Moon's orbit lies in a plane that makes an angle of 5°09′ with the plane of the Earth's orbit. (See Figure 8-30.) Therefore, in each lunar orbit, the Moon will reach its maximum north declination and maximum south declination once, and will pass above the Earth's Equator (point of zero declination) twice, once during each passage from the points of maximum declination.

OTHER LUNAR VARIABLES

Other lunar variables include rotation of *the long axis of the Moon's orbit* (named by astronomers the lunar apse) in the plane of the Moon's orbit, the rotation of the lunar node (*line formed by the intersection of the Moon's orbital plane and the Earth's orbital plane*, the plane of the ecliptic), and changes in ellipticity of the Moon's orbit.

INTERACTIONS AMONG LUNAR VARIABLES

The three relationships (1) lunar phase, (2) lunar apogee and -perigee, and (3) lunar declination shift rapidly and regularly. As a result, their various coincidences generate many short-term lunar cycles. The factors of lunar phase and lunar declination always coincide in the same way with the key positions of the Earth's declination with respect to the Sun. This creates an annual (seasonal) phase-declination lunar cycle. (See Chapter 8.) The lunar apse and lunar node move in opposite directions. Therefore, in addition to their own individual cycles, their times of coincidence give rise to yet-another cycle, the apse-node cycle.

Some Lunar Cycles

The lunar cycles discussed here include perigee-syzygy cycle, lunar apse cycle, lunar nodal cycle, and apse-node cycle.

PERIGEE-SYZYGY CYCLE

Because the position of the lunar perigee keeps shifting, the perigee position and lunar phase form an important cycle of about 14 months. The cycle is expressed as the time required for lunar perigee to coincide first with one of the syzygy phases, then to shift to coincide with the other syzygy phase, and finally to shift back to a coincidence with the initial syzygy phase.

The changing factor of distance between the Earth and the Moon, as displayed in the perigee-syzygy cycle, also affects the amount of the centrifugal force around the Earth–Moon barycenter and causes it to vary cyclically. The variation is connected with the rate of rotation of the pair of orbiting bodies; this rate is faster when the Moon is at perigee and slower when it is at apogee.

LUNAR-APSE CYCLE

The lunar apse cycle comes about because the long axis of the Moon's orbit (the apse) rotates clockwise through a full circle in the plane of the Moon's orbit. The time required for this rotation is 8.849 yr.

LUNAR-NODAL CYCLE

The lunar-nodal cycle of 18.6134 yr results from the counterclockwise rotation of the Moon's node (when viewed from above). Figure 15-8 shows the relationships of the longitude of the Moon's node and mean annual tidal ranges for Boston, Massachusetts; New York City; and Charleston, South Carolina. The longitude positions of zero and 180° are times when the node points directly at the Sun. These are times when full eclipses of the Sun take place during each lunar orbit.

Figure 15-9, a computer-prepared time series of the lunar tide-generating force, clearly displays the lunar-nodal cycle. The graph shows maxima for 1908, 1926, 1945, and 1963. Times of computed minima are 1898.9, 1917.9, 1931.1, 1954.7, and (not shown on the graph) 1973.3 and 1991.9.

> SOURCES: Fairbridge and Sanders, 1987; Kaye and Stuckey, 1973.

APSE-NODE CYCLE

The lunar apse-node cycle consists of coincidences between the lunar apse and the lunar node. Because the apse moves clockwise and the node counterclockwise,

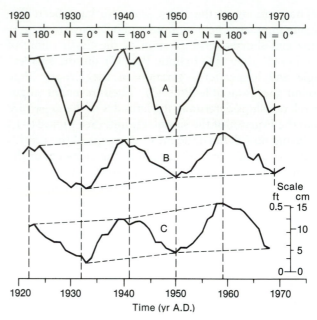

FIGURE 15-8.   Curves of annual mean tidal range for (A) Boston, (B) New York, and (C) Charleston, South Carolina. Dashed vertical lines mark times when the orientation of the lunar node (N) was pointed at the Sun, longitude 0 and 180°. Dotted lines show long-term trends of rising sea level. (C. A. Kaye and G. W. Stuckey, 1973, fig. 2, p. 143.)

the period of their coincidences is not the same as that of either. Instead, the period is 2.998 yr.

OTHER CYCLES

When the lunar node and apse at a perigee-syzygy coincidence are pointing directly at the Sun when the Earth–Moon pair are at perihelion, the tide-generating force reaches a maximum. The last time this alignment was attained was on 08 January 1340. An approximation of this alignment takes place at various seasons every 62.013 years, most recently on 16 January 1961. This is the apse-perihelion cycle. A coincidence of the lunar node and perihelion defines a cycle of 93 yr.

Although the apse-node cycle is 2.998 years, a larger-scale repetition takes place every 186 years (10 nodal cycles and 21 apse cycles).

Table 15-1 summarizes some short-term lunar cycles.

> SOURCE: Fairbridge and Sanders, 1987.

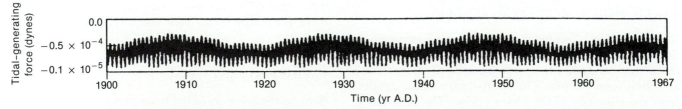

FIGURE 15-9.   Graph of tide-generating force, 1900 to 1967, computed using all tidal components and machine plotted. (Courtesy J. T. Kuo, Columbia University.)

TABLE 15-1.   Some short-period lunar cycles

Cycle	Period
Earth's rotation/Moon's orbit	
Twice daily	11 h, 55 min
Daily	23 h, 50 min
Monthly (cycle of lunar phases)	29.531 days (synodic period)
Seasonal (coincidence of syzygy phases and lunar declination with solar declination)	Quarterly
Perigee-syzygy cycle (the time required for lunar perigee at one syzygy phase to return to this same phase again after coinciding with the other syzygy phase)	14 months
Lunar node-apse cycle (the time required for the apse and node, which move in opposite senses, to coincide)	2.998 yr
Lunar apse cycle (the time required for a complete progression of the lunar apse, i.e., the long axis of its orbit, within the plane of the Moon's orbit)	8.849 yr
Lunar nodal cycle (the time required for the Moon's node, i.e., the line formed by the intersection of the Moon's orbital plane and the plane of the Earth's orbit, to rotate 360°)	18.6134 yr

SOURCE: Sanders, 1989, table 4, p. 12; compiled from Pettersson, 1912, 1914b, 1930.

TIDAL CYCLES

The cyclic behavior of the astronomic tides is a function of the interaction of the numerous lunar cyclic factors. The effects of these cyclic factors are well known; they are entered into the harmonic analyses that are combined to yield the predictions for tidal heights at various coastal localities. The first-order variables are lunar phase, Earth–Moon separation distance, and lunar declination.

Lunar phase yields the well-known bimonthly cycle of spring tides and neap tides. The Earth–Moon separation distance affects the size of the tidal bulge; the bulge is largest at perigee positions. Lunar declination affects the position of the tidal water bulge and this determines the relative amplitudes of the morning tide and the evening tide at localities where the tide rises twice each day. If the heights of the morning tide and evening tide differ, the condition is referred to as *diurnal inequality*.

The monthly spring tides reach maximum values when the perigee position coincides with one of the syzygy phases. (On Figure 8-33, the syzygy phases are New Moon on the 11th and Full Moon on the 25th. Notice that the amplitude of the spring tides associated with the Full Moon that coincides with perigee exceeds that of the spring tides associated with the New Moon that coincides with apogee. Notice, also, how both syzygy phases coincide with the Moon's position above the Earth's Equator and thus with zero diurnal inequality.) Diurnal inequality disappears at zero lunar declination no matter what part of the phase cycle coincides with the position of the Moon over the Earth's Equator.

The effect of the lunar perigee-syzygy cycle on tidal amplitudes is illustrated by Figures 15-10 and 15-11, which are plots of the predicted tidal heights at Fire Island Inlet, New York, for 1987 and 1988. The two yearly graphs clearly show how the maximum tidal amplitudes shift back and forth between peaks at New-Moon perigee and Full-Moon perigee. Late in December 1986, New Moon and perigee coincided just a few days before perihelion. During the first three months of 1987, the amplitude peak coinciding with the New-Moon phases decreased, whereas that coinciding with the Full-Moon phases increased. By April they were equal. Thereafter the amplitude peaks of the Full-Moon phases became the larger ones, reaching a maximum in early July, and thus completing half of a perigee-syzygy cycle. During the rest of 1987, the Full-Moon

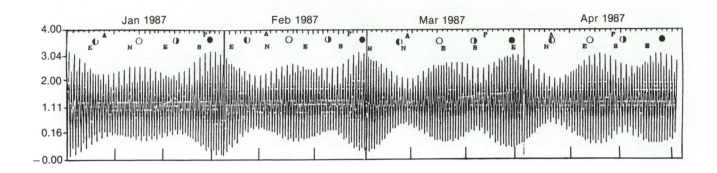

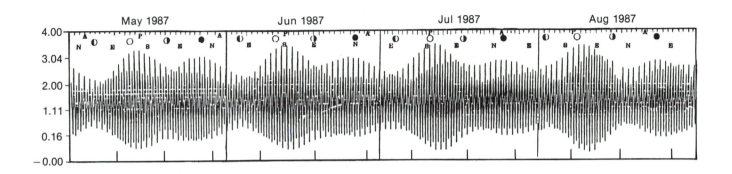

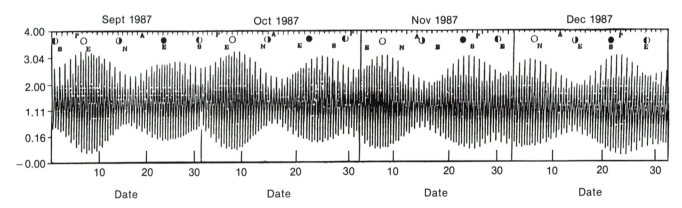

FIGURE 15-10.   Predicted heights of tides for Fire Island Inlet, New York, 1987. Astronomic data (from the NOS tide-table publication) added at top are apogee (A), perigee (P), phases of the Moon, and declination of the Moon (N = farthest north, E = above the Earth's Equator, and S = farthest south). (Authors.)

amplitude peaks decreased, whereas the New-Moon peaks increased. Equality was reached in November 1987, after which time the New-Moon peaks exceeded the Full-Moon peaks. The New-Moon maxima of January and February 1988 completed a full cycle. Amplitude equality was reached in late May-early June 1988, and the Full-Moon amplitude maximum appears in late September.

The importance of the perigee-syzygy cycles in coastal activities has been emphasized by Fergus J. Wood (1976)

in his monograph entitled "The strategic role of perigean spring tides in nautical history and North American coastal flooding, 1635–1976."

In coastal areas the times of higher-than-normal astronomic tides represent what might be termed *windows of vulnerability*. During intense coastal storms, the higher-than-normal wind-generated waves typically are accompanied by storm surges. Potential for severe damage varies according to the relationships between times of maximum waves and/or storm surge and times of

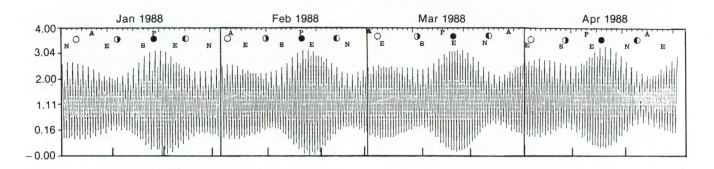

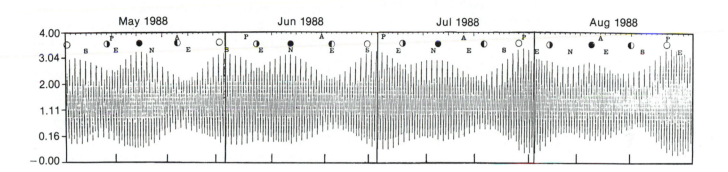

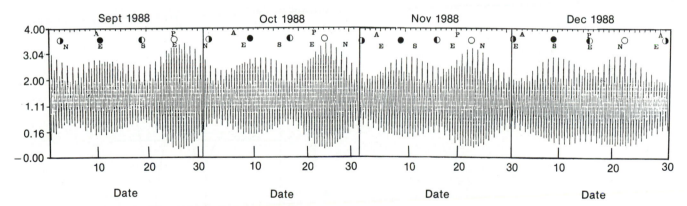

FIGURE 15-11.   Predicted heights of tides for Fire Island Inlet, New York, 1988. Astronomic data (from the NOS tide-table publication) added at top are apogee (A), perigee (P), phases of the Moon, and declination of the Moon (N = farthest north, E = above the Earth's Equator, and S = farthest south). (Authors.)

maximum astronomic tides. For example, the last hurricane to graze the New York metropolitan area made landfall a few hours after high tide. As a result, coastal flooding was not serious. By contrast, in March 1962, at a time when New Moon and perigee coincided (on the 6th) and the Moon was over the Earth's Equator (on the 8th), a severe coastal storm raged for several days. This storm was one of the most destructive in modern times along a segment of the Atlantic coast reaching from Long Island (New York)

to North Carolina. High wind-generated waves and wind setup persisted through five consecutive high tides that were not only higher than normal because of the perigee-syzygy coincidence but also of equal height because of the Moon's equatorial position.

Study of the strata deposited when sand builds up beaches or spits suggests that most activity takes place during a few days when successive tidal high-water levels increase as the amplitude maxima of a perigee-syzygy coincidence are approaching.

ATMOSPHERIC CYCLES HAVING LUNAR-CYCLE PERIODS

Time-series graphs of temperature using signal-processing- and statistical techniques to extract the amplitude of an 18.6-yr wave from the data clearly indicate that the variations registered vary with the lunar-nodal period (Figure 15-12). Only the time interval 1910 to 1965 is represented on both graphs. Notice that the curves from the two regions, Africa and India, are out of phase. For example, the 1918, 1936, and 1955 troughs on the African curve match peaks on the Indian curve. Clearly, the temperatures in each locality are cyclic. However, if the data from the two regions were averaged, the result would not be cyclic. The two out-of-phase cycles would cancel each other.

Similar graphs have been prepared for variations in air pressure (Figure 15-13). As with the two temperature curves of Figure 15-12, the cyclic air-pressure curves for

Kimberley, in the interior of South Africa, and for coastal sites in southern Africa are 180° out of phase.

Finally, graphs of amplitude vs. time display the 18.6-yr cycle of precipitation, as registered by drought maxima in northeastern China and by a flood-area index in India (Figure 15-14).

Clearly, three of the fundamental aspects of climate, temperature, air pressure, and precipitation, display an 18.6-yr cycle whose peaks and/or troughs coincide with the 18.6-yr peaks and/or troughs on the graph of the lunar tide-generating curve. (See Figure 15-9.) What this coincidence means in terms of physical cause-and-effect relationships remains to be determined. But the coincidences are so definite that one cannot escape the conclusion that some important messages lurk within the data.

A final feature of many such graphs is displayed in Figure 15-14, A, in the span from 1920 to 1940. A phase shift takes place in this 20-yr interval. In the language of signal processing, this is known as a *bistable flip-flop*.

SEDIMENTARY CYCLES DISPLAYING LUNAR PERIODS

The variation in thickness of rhythmically laminated sediments has been determined for some formations and the thickness values indicate tidal periodicities. One example is from the Precambrian of Australia (Figure 15-15).

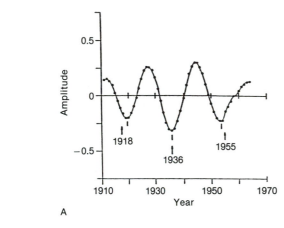

A

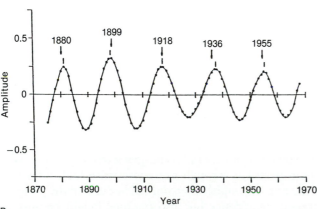

B

FIGURE 15-12. Time-series graphs of amplitude functions of air temperatures extracted from 18.6-yr cycle and plotted on identical horizontal scales. Notice that the phase relationships of the two curves differ by 180°. Further explanation in text. (R. G. Currie, 1987; A, fig. 22-14; B, fig. 22-15, p. 391.)
   A. Sites in Africa.
   B. Sites in India.

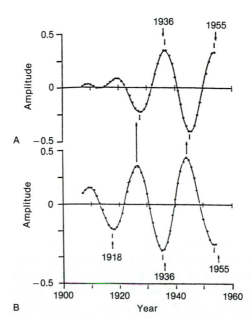

B

FIGURE 15-13. Time-series graphs of amplitude functions of air pressures extracted from an 18.6-yr cycle. Notice that the two curves are out of phase with each other. (R. G. Currie, 1987, fig. 22-13, p. 391.)
   A. Kimberley, located in the interior of South Africa.
   B. Coastal sites, South Africa.

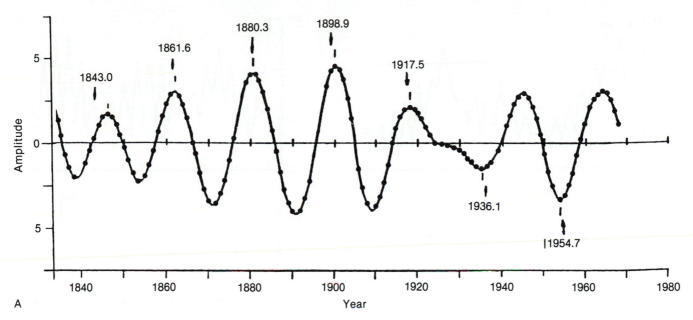

FIGURE 15-14.   Time-series graphs of amplitude functions of precipitation extracted from an 18.6-yr cycle and plotted to the same horizontal scale. (R. G. Currie, 1987; A, fig. 22-2, p. 380; B, fig. 22-17, p. 393.)
A. Drought maxima, northeastern China.

Other patterns of variation in thickness of laminae in the Mansfield Formation (Pennsylvanian) of Indiana involve siltstones and claystones (Figure 15-16).

SOURCES: Currie, 1987; Kvale, Archer, and H. R. Johnson, 1989; G. E. Williams, 1989.

Whatever is behind these relationships, the influence of some kind of extraterrestrial forcing function seems virtually certain. An extraterrestrial interpretation for these phenomena on such short-, or calendar-, time periods as a few decades contrasts with the kind of extraterrestrial forcing functions connected with the longer-term Milankovitch climatic factors, to which we now turn.

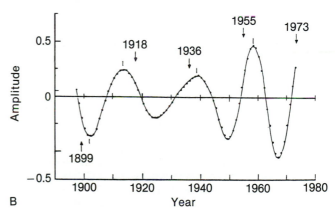

FIGURE 15-14.   (*Continued*)
B. Flood-area index, India.

Milankovitch Climatic Factors

As we indicated in Chapter 1, Milankovitch climatic cycles have been cited as driving factors in the Earth's climatic oscillations. In Chapter 1 we presented a Milankovitch curve but did not explain the fundamental factors on which it is based. Here we discuss the astronomic variables that give rise to five cycles in the so-called Milankovitch bandwidth, that is, cycles having periods from about 20,000 to about 400,000 yr. These are two cycles of "precession" of about 19 ka (ka = kiloyear, 1000 yr) and 23 ka; "obliquity," of about 41 ka; and two cycles of ellipticity (or eccentricity), one of about 100 ka, and the other of about 413 ka. In the following paragraphs we review each of these.

The astronomic factors on which Milankovitch based his curve of the secular march of insolation (See Figure 1-6.) are those related to (1) the Earth–Sun separation, (2) the maximum latitude attained by the annual migration of the subsolar point (tilt angle and Earth's node), and (3) the relationship between the ecliptic and the longitude of perihelion. We discuss changing ellipticity first.

Changing Orbital Ellipticity

The factor of changing ellipticity of the Earth's orbit affects the Earth–Sun separation distance and thus the intensity of solar energy received at the top of the Earth's atmosphere. This is the "ca.-100 ka" factor assigned to the Milankovitch "ellipticity" cycle. This cycle is

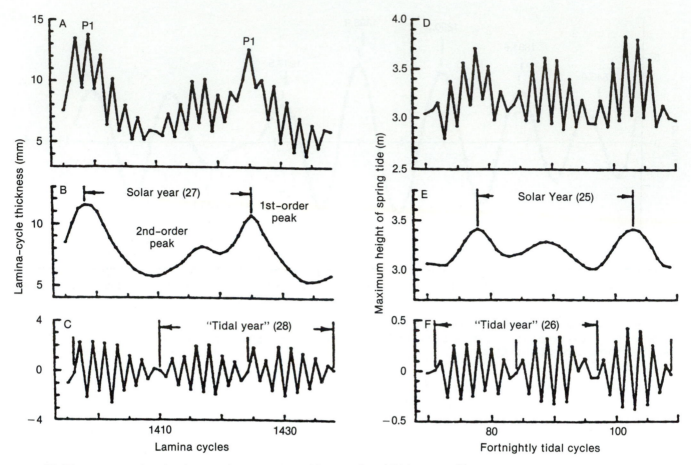

FIGURE 15-15.   Evidence of ancient lunar cycles reconstructed from studies of thicknesses of laminated Precambrian sedimentary rocks, Australia. At left, thicknesses of the laminae; at right, tidal patterns for Townsville, Queensland.
  A. Nonsmoothed curve. First-order peaks (P1) define yearly maxima.
  B. Smoothed curves, with 27 cycles between yearly maxima.
  C. Residual curve (A minus B). Vertical lines mark changes of phase.
  D. Maximum height, fortnightly tidal cycles 19 October 1968 to 03 June 1970.
  E. Smoothed curve of D, showing 25 fortnightly cycles between yearly maxima.
  F. Residual curve (D minus E). (G. E. Williams, 1989, fig. 7, p. 167.)

based on the cycle of changing ellipticity, which is 93,408 yr. The time period is the same as that required for a complete cycle of the progression of the longitude of perihelion, but these two cycles are not linked. Both are governed by what has been called the *All-Planets Restart Cycle,* the time required for all the planets in the Solar System to resume a given set of initial positions.

According to astronomers, the orbital factor of mean distance (*a* in Figure 15-1, the length of the semimajor axis of the Earth's elliptical orbit) does not change. If this conclusion is correct, then the variable factor is *b*, the length of the semiminor axis. In its most-elliptical mode, *b* attains its minimum values; at such times the Earth is closest to the Sun and thus the intensity of incoming solar radiation is greatest. Times of maximum elliptical orbit, therefore, should coincide with whatever climate regime is triggered by maximum solar energy (Figure 15-17).

Maximum Latitude Attained by the Annual Migration of the Subsolar Point

The inclination of the Earth's polar axis of rotation with respect to the Earth's orbit plane, and the sensibly constant position maintained by the Earth's polar axis of rotation on a time scale of a single orbit (Figure 15-18) explain the annual migration of the subsolar point. (See Figure 8-1.)

Precession of the Equinoxes

Although we can consider the position of the Earth's pole as being virtually constant on the scale of a single orbit, when one examines the pole on a time scale of several thousand years its position is definitely not constant. The Earth's pole undergoes motions connected with its clockwise precession and with changes of tilt angle. Moreover, the clockwise polar-axis precession does not act in isolation; its climatic effect is the result of interaction with another factor, the changing orientation of the long axis of the Earth's orbit, namely, the counterclockwise *progression of the longitude of perihelion*. The result of this interaction is that the position of the ecliptic migrates along the Earth's orbital path, a movement named *precession of the equinoxes* (Figure 15-19). To a nonastronomer, this dual usage of the word "precession" seems to invite confusion. However, the usage is

deeply embedded in the language of orbital variations and is not likely to be changed. Therefore, whenever the word "precession" appears, readers should be on the alert to discern which meaning is intended. In this book we shall avoid using "precession" alone, but always specify either "polar-axis precession" or "precession of the equinoxes."

The cycle known as precession of the equinoxes results from an interaction between the clockwise change in direction of orientation of the Earth's polar axis (following its polar-axis precession cycle of 25,920 yr) and the counterclockwise change in orientation of the long axis of the Earth's orbital ellipse (the progression of the longitude of perihelion, having a cycle of 93,408 yr). The result is that the equinox migrates along the orbit and makes a complete circuit in 20,293 yr. In terms of climate, the critical factor seems to be the changing relationship between winter solstice in the Northern Hemisphere and perihelion. In A.D. 1325 these two coincided (Figure 15-19, A), but ever since, they have been slowly drifting apart. (Notice that in these figures showing conditions in the past, the polar-axis precession shifts

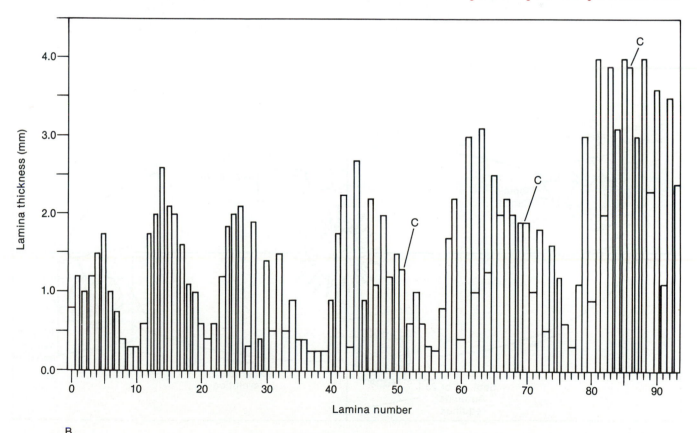

FIGURE 15-16.   Characteristics and thicknesses of laminae in Mansfield Formation (Pennsylvanian), Indiana. (E. P. Kvale, A. W. Archer, and H. R. Johnson, 1989, A, fig. 3, p. 367; B, fig. 4, p. 367.)
   A. Idealized sketch showing couplet consisting of two siltstones of unequal thickness separated by thin claystones.
   B. Bar graph of thicknesses of siltstone laminae. Points where the thicknesses of the two siltstones of the couplets become equal are indicated by the letter C.

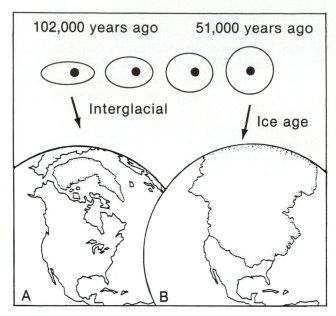

FIGURE 15-17.   Earth's changing ellipticity and climate.

A. Interglacial climate of 102,000 yr ago during last time of maximum orbital ellipticity (much exaggerated in sketch at top, left).

B. Glacial climate of 51,000 yr ago during time of last minimum orbital eccentricity; ice sheet covered much of North America. The current ellipticity is near the mean value of 2%, but is decreasing. (J. E. Sanders, 1981, fig. 13.30, p. 327.)

counterclockwise and the progression of the longitude of perihelion, clockwise, just the reverse of their current motions. Their directions of motion have not changed. What has happened is an apparent change caused by these attempts to reconstruct conditions in the past.) As shown in Figure 15-19, C, the opposite condition, perihelion and summer solstice in the Northern Hemisphere, prevailed in 8824 B.C. The previous coincidence between winter solstice in the Northern Hemisphere and perihelion was in 18,968 B.C. (Figure 15-19, E).

Obliquity of the Ecliptic

The diagrams of Figure 15-19 do not include the factor of tilt of the Earth's polar axis with respect to the Ecliptic Pole. Recall from Figure 15-6 that two polar-axis precession cycles are required to include a full coincidence of tilt + orientation. Thus two cycles of the precession of the equinoxes are required to play out a single cycle of a full range of tilt changes with respect to the orbit (*obliquity of the ecliptic*). The cycle of $2 \times 20,293 = 40,586$ is what is meant by the Milankovitch "obliquity" cycle. The tilt cycle involving only the Earth's polar axis and the Ecliptic Pole is 17,280 yr (which seems to

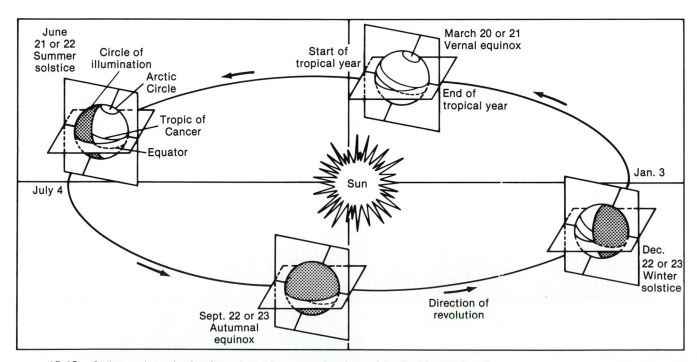

FIGURE 15-18.   Oblique schematic view from above down onto the plane of the Earth's orbit (small squares of which have been drawn around the Earth in each of the four positions shown) showing the virtual constancy of the orientation of the Earth's pole of rotation (contained in the vertical plane drawn through the Earth in each of the four positions shown) during each orbit. Notice that the equinox- and solstice positions shown do not coincide with the axes of the orbital ellipse. Ellipticity of Earth's orbit much exaggerated. Sizes of Earth and Sun not to scale. (C. R. Longwell, R. F. Flint, and J. E. Sanders, 1969, fig. 5-17, p. 116.)

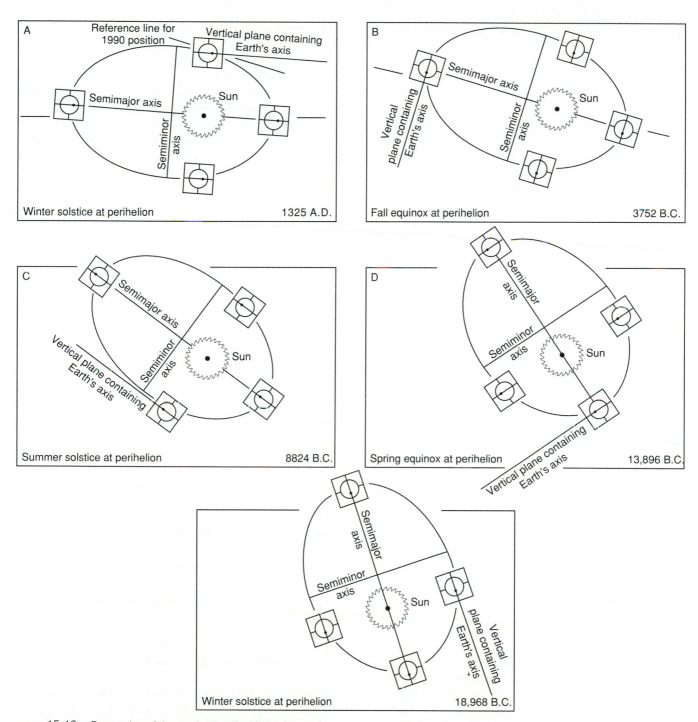

FIGURE 15-19. Precession of the equinoxes showing a full cycle.

A. In A.D. 1325, perihelion and winter solstice in the Northern Hemisphere coincided. Horizontal line segments in the middle of the diagram mark position of vertical plane containing the Earth's axis at winter solstice in 1989.

B. In 3752 B.C., the fall equinox was at perihelion.

C. In 8824 B.C., the summer solstice was at perihelion.

D. In 13,896 B.C., the spring equinox was at perihelion.

E. In 18,968 B.C., the winter solstice was again at perihelion, as in A.

In their forward-running mode, the effects of progression are clockwise and those of polar-axis precession counterclockwise. Because the diagrams show conditions going backward in time, the sense of motion of these two factors is just the opposite of their forward-running senses. Not shown here are the changes in the angle of the Earth's polar axis of rotation. (Authors, corrected from J. E. Sanders, 1981, fig. 13.32, p. 329.)

be the second Milankovitch "precession" cycle of about 19 ka).

The **equinox** is defined by astronomers as *the position where the path of the line between the center of the Earth and the center of the Sun crosses the plane of the ecliptic.* (See Figure 15-18.)

As mentioned in a previous section, the full range of angular motion displayed by the Earth's axis of rotation involves not only the angle of the pole with respect to the plane of the Earth's orbit, but also with respect to the invariable plane of the solar system, with which the Earth's orbital plane makes a slight angle. Therefore the full migration of the subsolar point includes a factor controlled by the Earth's nodal cycle.

Earth's Nodal Cycle

The Milankovitch cycle of "about 413 ka" has been defined only vaguely as some kind of "second eccentricity." We think it more likely that this long-term cycle is not related to orbital eccentricity, but to the angular relationships brought about by the Earth's nodal cycle of 362,880 yr. (See Figure 15-7.)

The foregoing astronomic factors are the mechanisms by which Milankovitch (and others) have expressed how one of the variables in the Earth's climate equation varied through time. Recall from Chapter 1 that Milankovitch recognized, but did not attempt to deal with, variable composition of the Earth's atmosphere, changes in the Earth's albedo (reflectivity), and changes in the Sun. We showed in Chapter 1 that the proportion of carbon dioxide in the Earth's atmosphere undergoes cycles having periods in the Milankovitch bandwidth. Next, we take up the Sun, a climatic factor of great importance.

The Sun's Orbit

Although the astronomic variations in the Earth's orbital elements have been brought into discussions of the Earth's climate changes since the middle of the nineteenth century, these variations have been seriously considered as forcing functions for the Earth's climate only in the past few decades. One particular connection that seems to have been established to the satisfaction of all is that the Earth's orbital elements vary cyclically and predictably and the reason they do so is that they are forced to by the effects of the changing orbits of all the planets in the Solar System. In short, everyone agrees that the changing orbital situations of the planets cause changes in the orbital elements of the Earth. A convenient fiction taken for granted is that the chang-

ing orbital elements of the Earth can be regarded as the chief variables. In other words, the Sun can be ignored. Recent results based on a computer-generated ephemeris of the Solar System prove that assuming the Sun is not affected by planetary motions perpetuates a serious error. In the following section we present evidence proving that the Sun is not at rest at one focus of the planetary orbital ellipses, as is widely but erroneously believed, but rather is itself making an orbit around the center of mass of the Solar System, which is the true focus of planetary orbital ellipses. Moreover, it has now been demonstrated that the motions of the planets force the Sun to orbit and that the Sun's orbit is cyclic and predictable. Finally, if planetary motions cause the Sun to undergo short-period variations of a few tens to a few hundreds of years, then we can expect that the planets will also force the Sun to be involved in longer-term variations, say of the so-called Milankovitch periods. Let us continue with some fundamental definitions.

Definitions

Ever since the time of the great Egyptian astronomer Ptolemy, astronomers have made great progress using a geocentric basis for their telescopic measurements. Even after astronomers accepted the idea that the Earth orbits the Sun and not vice versa, they have continued to use geocentric coordinates and to consider that the center of the Sun can be taken as a fixed point of reference. But, as Isaac Newton (1642–1727) so clearly realized, the Sun cannot be considered as fixed; it is obliged to orbit around the center of mass of the Solar System. If the Sun could be viewed from a point in space against a background of so-called "fixed" stars, then it would be seen to oscillate. In reality, in making its orbit, the Sun shifts by about one solar diameter. The Sun's oscillation would be taken as proof that it is part of an orbiting system with nonluminous (thus invisible at great distances) planets. Indeed, astronomers have attempted to find out if other Sun-like stars are being orbited by planetary bodies. Their chief method is to look for small oscillations of the star of interest against the background of other so-called "fixed" stars. The point is that, as a matter of principle, students of celestial mechanics are well aware that the Sun must be orbiting the center of mass of the Solar System. But, as a matter of everyday practical convenience, they find it easier to ignore the Sun's orbit. The days of such ignoring should cease.

Figure 15-20 is a map of the position of the center of the Sun in its current orbit around the center of mass (barycenter) of the Solar System. The most-distant point in the orbit from the barycenter is the *apobac* and the

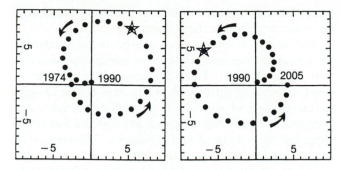

FIGURE 15-20.   Just-completed solar orbit and current orbit represented by "maps" looking down on the plane of the Earth's Equator extended into space. Dots (sizes schematic) show positions of the center of the Sun at 200-day intervals in its orbit around the center of mass of the Solar System (intersection of lines in center of squares). (Sides calibrated in units of 10^{-3} Astronomic Units, or about 150,000 km. Thus the lengths of the sides of the squares are about 3 million km.) Stars mark apobac positions. Arrows show direction of Sun's motion.

A. Previous orbit started with the peribac of 19 December 1974 and ended with the peribac of 25 April 1990.

B. Current orbit began on 25 April 1990 and will end in the year 2005. (After R. W. Fairbridge and J. E. Sanders, 1987, fig. 26-2, p. 449.)

least-distant point, the *peribac*. The Sun moves toward its apobac when Jupiter and Saturn line up on the same side of the Sun (as on 23 August 1981).

The Sun shifts toward its peribac when Jupiter and Saturn line up on opposite sides of the Sun (as in October 1991). The next Jupiter-Saturn alignment on the same side of the Sun will take place in December 2001. The orbital period of Jupiter is about 12 yr and that of Saturn about 30 yr. If one assumes, for purposes of easy calculation, that their orbits are circles and their periods exactly 12 and 30 yr, respectively, then Jupiter advances $360/12 = 30°/\text{yr}$, whereas Saturn advances $360/30 = 12°/\text{yr}$ (the degrees advanced per year of one equals the orbital period of the other). Thus in 5 yr Jupiter advances $5 \times 30 = 150°$ while Saturn advances only $5 \times 12 = 60°$. By comparing the two, it is evident that every 5 yr Jupiter "gains" $90°$ on Saturn and thus in 20 yr will gain $360°$, or in effect, will, in the terminology of racing-car drivers or of long-distance runners, "gain a lap" on Saturn (Figure 15-21). This Saturn-Jupiter lap cycle (abbreviated SJL) is the primary driving force of the Sun's orbit. When the exact orbital periods of the two are used for the calculations, the period of the SJL is 19.8 yr.

Solar Cycles

We have mentioned briefly the subject of the "solar constant." According to the nuclear physicists who formu-

lated the now-accepted hypothesis of the origin of the Sun's energy, statistical constancy is what one should expect. So firmly was this idea of constancy established among the scientific teams assembled by the U.S. National Aeronautics and Space Administration (NASA) to devise experiments to be carried out and geophysical instruments to be left on the surface of the Moon by the U.S. astronauts in the late 1960s, that a meter to measure solar radiation was not even included. This decision may eventually rank with one of the all-time great missed opportunities in the history of science.

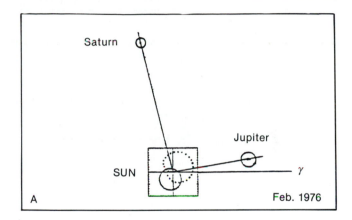

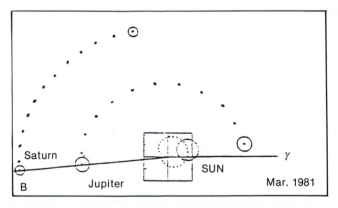

FIGURE 15-21.   Relationships of Jupiter, Saturn, and the Sun during the just-completed solar orbit shown by schematic "maps" of positions in the X–Y plane (Earth's equatorial plane extended outward in all directions). Intersection of axes that bisect the small reference squares (each measuring about 3 million km on a side) marks center of mass of Solar System. Positions of centers of Sun, Jupiter, and Saturn iterated at intervals of 200 Julian days (dots); data from NASA JPL computerized ephemeris, computations from JPL tapes made at NASA's Goddard Institute of Space Studies, New York, hand plotted. The direction of the positive X axis is the equinox of 1950. γ is the ecliptic of 1950.0.

A. In February 1976, Jupiter and Saturn were situated about $90°$ apart. The center of mass was within the Sun, but close to the Sun's surface.

B. In March 1981, Jupiter and Saturn were aligned. The Sun had moved far enough away from the center of mass on the opposite side from the Jupiter-Saturn alignment so that the center of mass lay completely outside the body of the Sun.

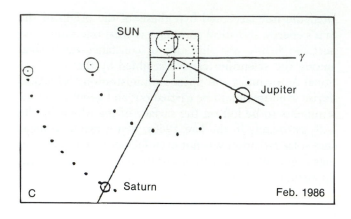

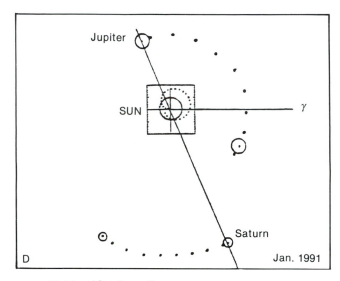

FIGURE 15-21. (*Continued*)

C. In February 1986, Jupiter and Saturn were 90° apart and the Sun occupied an opposing position. Notice that the center of mass still lay completely outside the body of the Sun.

D. In January 1991, Jupiter and Saturn were aligned, but on opposite sides of the Sun, which was very close to the center of mass of the solar system. (J. E. Sanders, 1989, fig. 8, p. 13.)

SUNSPOT CYCLE

One feature of the Sun that has been noticed for centuries is a series of small dark areas known as sunspots. The numbers of sunspots and their positions on the face of the Sun vary. They have been seen and counted and various methods have been devised for expressing their abundances. The so-called Wolf number, proposed in 1858, represents the average number of sunspots in a given year. Over time, the Wolf number varies cyclically (Figure 15-22). Although the cyclic variation has been referred to as "the solar cycle," no one has yet figured out how to predict sunspots nor proposed a mechanism for their origin that has been generally adopted. Many studies have been made in which numbers of sunspots have been linked to variations in the Earth's environment-

al conditions, such as rainfall, temperature, barometric pressure, or geomagnetism (Figure 15-23).

The study of sunspots has proved to be something of a scientific medusa; as soon as one group of scientists thinks they have killed the subject by cutting off its head, it pops up again with many new heads. Despite its current lack of scientific respectability, the subject of sunspots is acknowledged to indicate that the Sun is undergoing some kind of cyclic behavior. Moreover, evidence that has been recently reinterpreted as a product of solar cyclicity has been found in Precambrian rocks in Australia (Figure 15-24).

Although the new evidence about the orbit of the Sun does not explain sunspots, it does demonstrate a hitherto-ignored basis for cyclic behavior of the Sun. Even in its current status as a subject new to science, the planetary–Sun interactions imply that a fundamental cyclicity pervades the Earth's environmental conditions. What is more, these solar variables are driven by the dynamics of the Solar System, and therefore are subject to the same kinds of long-term variables that are responsible for the Milankovitch cycles.

Sun's Orbit and Environmental Variables on Earth

Considerable research as well as extensive debate has centered on the importance of extraterrestrial influences on the Earth's climate. Most of the discussions have centered on two contrasting time scales, long term (so-called Milankovitch cycles with periods ranging from about 20,000 to about 400,000 yr) and short term (solar cycles with periods of a few years to a few tens of years).

As mentioned, Milankovitch cyclic variations in the Earth's astronomic elements are brought about by cyclic changes in the elements in the orbits of the other planets. Some attempts have been made to connect the sunspot cycle with planetary influences on the Sun, but these have generated more controversy than scientific progress. A totally new set of possibilities follows from the recognition of planetary effects on the Sun's orbit. As mentioned, a fundamental solar-orbital cycle results from the Saturn-Jupiter Lap (SJL) of 19.8 yr. When longer series of solar-orbit maps are prepared (Figure 15-25), they show another significant cycle of 178 yr. Still-other cycles are suggested by the small shifts in orientations of each orbit.

Another predictable solar factor that varies cyclically is the distance between the center of the Sun and the Solar-System barycenter (Figure 15-26). This distance graph clearly displays a cyclicity of 178 yr. Analyses of varves, of Greenland ice, and of tree rings have shown periods of nearly 180 yr that have been attributed to solar effects.

A value of twice this cycle, or 356 yr, has been reported for episodes of cold climate in northern Europe. The last such episode, the so-called Little Ice Age, spanned the interval of 1645–1715 (the reign of King Louis XIV in France; but the name Little Ice Age has also been given to a cold period that took place some 500 yr before the one discussed here). This is also the time span of the Maunder Minimum of sunspots. Using this cycle period, one would expect another cold interval to begin 356 yr after 1645, or in the year A.D. 2001.

Oscillations having periods ranging from about 200 to 100,000 yr have been found in the cores that provide a continuous record, 1000 m thick, representing a duration of at least 200,000 yr of the Castile and Bell Canyon formations (Permian) of the Delaware Basin, southeastern New Mexico and southwestern Texas (Figure 15-27). These oscillations could be connected with solar variation.

If it can be shown that the Sun's responses to planetary effects are sufficient to generate variable activity that can affect the Earth's climate, then a whole new category of factors will have to be considered in paleoclimatic interpretations. Variations of many kinds may exist in the Milankovitch bandwidth. What is more, given the demonstrated natural cyclic variability of carbon dioxide and the possible variability of the Sun, the most-important factors affecting the Earth's climate may be those that Milankovitch did not include. And the factors he did include may prove to be of secondary, even trivial, importance.

What is needed is a series of detailed investigations of the sedimentary record made to answer questions about the existence of and durations of cycles that may have been caused by the effects of solar variation. These investigations will have to be made on levels of sophistication not previously attempted.

However, as matters now stand, geologists attribute at least some sedimentary cycles, such as Neogene third-order cycles (Table 1-1), to the growth and decay of continental glaciers resulting from the effects of extraterrestrial forcing through Milankovitch cycles. Milankovitch astronomic factors introduce

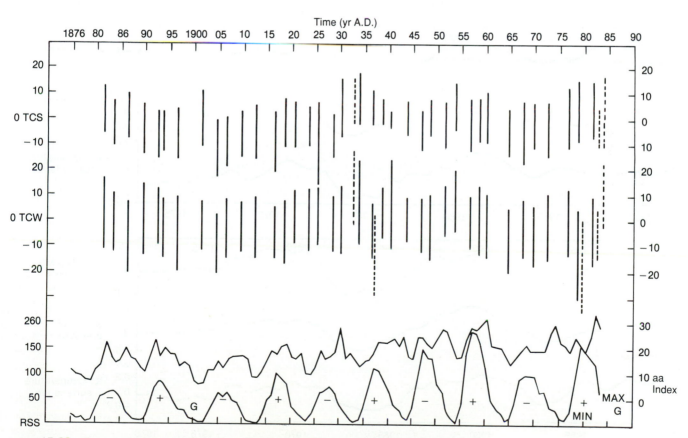

FIGURE 15-22. Sunspot cycle, 1875 to 1983, shown as a plot of sunspot numbers (Zurich Relative Sunspot Numbers and its recent successor, International Sunspot Numbers), all plotted against a positive Y axis, but with the signs of magnetic polarities indicated. Also shown are a curve of geomagnetic disturbance (aa index) and seasonal departures of temperatures in the conterminous United States, for winter (TCW)- and summer (TCS) seasons. (H. D. Willett, 1987, fig. 23-2, p. 408.)

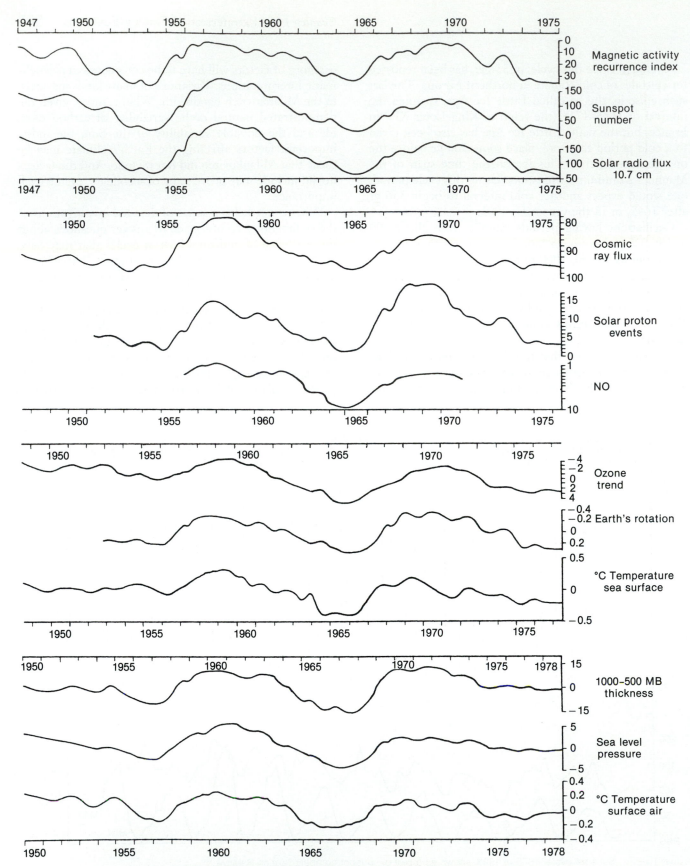

FIGURE 15-23. Graphic comparisons of numerous geophysical variables and sunspot number, presented in groups of three curves, the time axes of some being slightly shifted as indicated. MB = millibars, NO = nitrous oxide. (G. Wollin; J. E. Sanders; and D. B. Ericson, 1987, fig. 13-5, p. 252.)

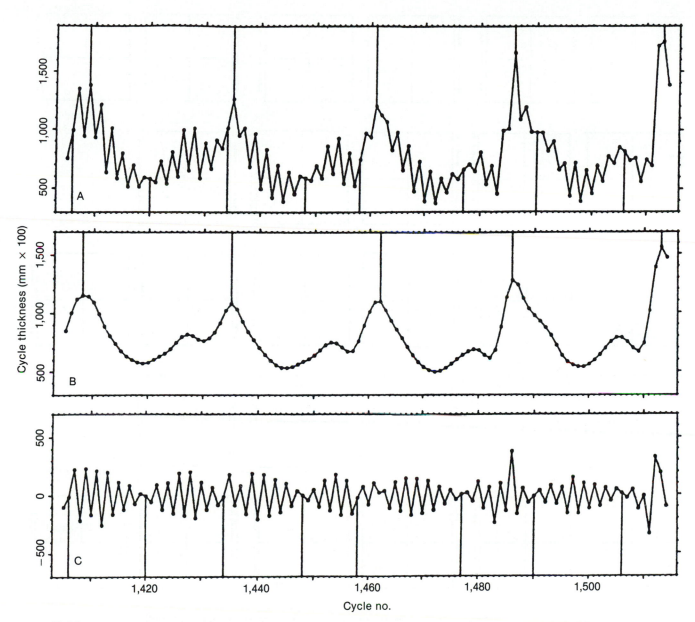

FIGURE 15-24.   Evidence of solar cycles in laminated rhythmites of Elatina Formation, Precambrian, Australia. (G. E. Williams and C. P. Sonett, 1986, fig. 3, p. 525.)

A. Thickness of 110 cycles of about 12-yr VT (varve thickness); cycles display distinct maxima every 25 to 27 cycles.

B. Smoothed curve of data shown in A. Vertical lines above curves A and B mark boundaries of lamina cycles.

C. Resultant curve formed by subtracting B from A. Vertical lines below curves mark changes of phase.

important cyclic variations into the Earth's climate, which affect not only sea level, but also such factors as humidity and aridity, which in turn control specific patterns of facies. Parasequences and mesosequences and hence seismic stratigraphy ultimately relate to extraterrestrial forcing factors such as Milankovitch climatic cycles.

Whatever is decided about these matters, one other significant relationship needs to be included here: that between sediment yield and rainfall.

SOURCES: R. Y. Anderson, 1982; Fairbridge and Sanders, 1987; Landscheidt, 1987; Wollin, Sanders, and Ericson, 1987.

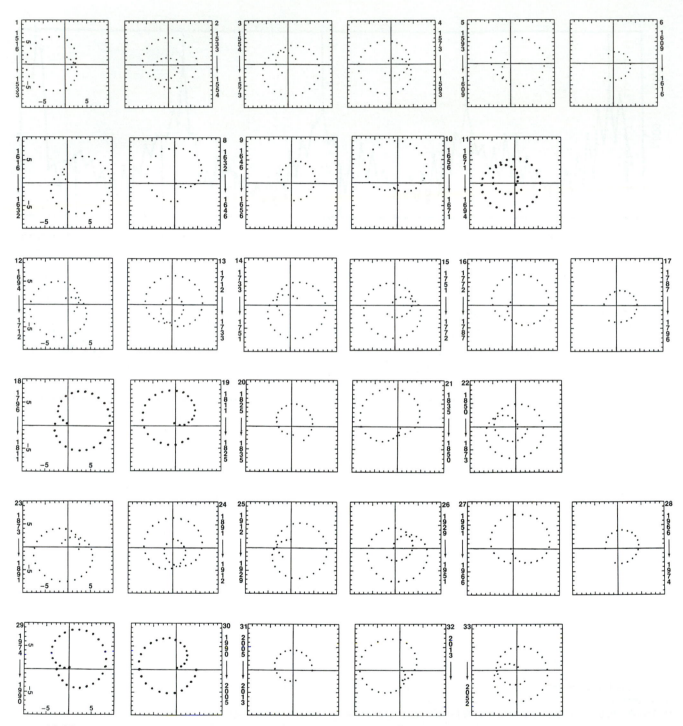

FIGURE 15-25. Solar-orbit maps for 3 of the fundamental 178-yr cycles of 11 orbits each, starting on 24 Jan 1516 and extending to 13 Jul 2050. Figure 15-20 shows orbits 29 and 30 in the numbering scheme shown here. (Authors.)

Sediment Production and Sediment Yield

Definitions

The term **sediment** usually designates *regolith that has been transported*, and **regolith** designates the *nonconsolidata materials that overlie the bedrock*. The production of

sediment then becomes a matter of production of regolith. Much regolith has resulted from the breakdown of the bedrock. This breakdown can be a comparatively *in-situ* process in which weathering converts bedrock into residual regolith. Or it may result from some transport process in which breakage and grinding are significant factors, as at the base of a glacier, the results of gigantic

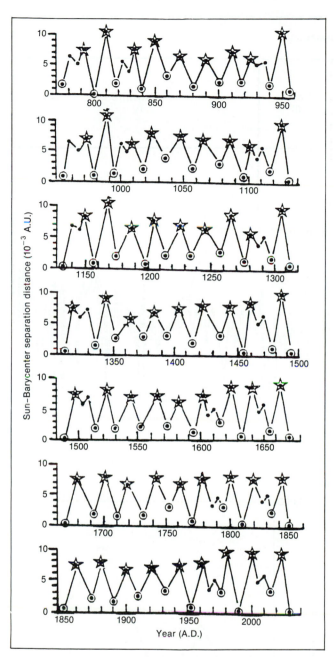

FIGURE 15-26. Sun–barycenter separation distance through time, with rows of points separated into 178-yr cycles. The right-hand end points on each row are repeated as the left-hand end points on the row next below. Stars, apobacs; circles, peribacs of major orbits of the Sun. A.U. = astronomic units; 10^{-3} A.U. = approx. 150,000 km. (After R. W. Fairbridge and J. E. Sanders, 1987, fig. 26-4, p. 453.)

may be exported. The rate of export is usually what is meant by sediment yield. Sediment yield is expressed as mass per unit area (such as tonnes/km^2). It is usually determined by measuring the quantity of sediment in a stream and by relating the water in the stream to the discharge area.

Relationship to Hydrologic Cycle

The *hydrologic cycle* represents the circulation of water into and out of an area. The hydrologic cycle is intimately related to the circulation of the atmosphere. The amount of rainfall and its distribution affect the plant cover of an area and also the quantity of runoff. Sediment yield has been found to reach maximum values where annual rainfall is not great enough to support a full plant cover but does suffice to supply intermittent streams with occasional large loads of water (Figure 15-28).

In the United States, the U.S. Geological Survey maintains a network of stream-gauging stations at which quantities of water, of sediment, and of other variables are determined daily or at some other interval. Because of the interest in understanding the large-scale pollution of the upper Hudson River by PCBs (polychlorobiphenyl or polychlorinated biphenyls, a group of synthetic-organic compounds), the New York State Department of Environmental Conservation and the U.S. Geological Survey entered into a cooperative

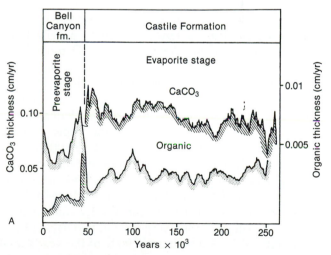

FIGURE 15-27. Thickness data for components of Bell Canyon and Castile formations, Permian, Delaware Basin, southeastern New Mexico and southwestern Texas, based on 1000 m of core representing more than 200,000 yr. (R. Y. Anderson, 1982; A, fig. 1, p. 7286; B, fig. 3, p. 7288.)

A. Smoothed curve of thicknesses of layers of calcium carbonate and of organic matter for entire cored interval; smoothing interval, 10,000 yr. When evaporitic conditions became established, the thickness of calcium-carbonate layers increased and that of layers of organic matter decreased.

slope failures that become rock avalanches, or impacts by extraterrestrial objects (meteorites, asteroids). An important exception to the breakdown-of-bedrock variety of regolith is the "instant" regolith created by volcanic explosions.

Whatever its origin, the regolith/sediment blanket may accumulate where the new regolith is forming or

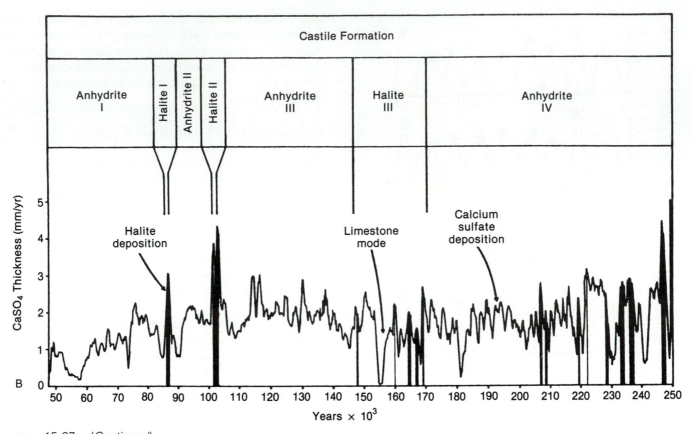

FIGURE 15-27.   (*Continued*)

B. Smoothed curve of thicknesses of calcium sulfate; smoothing interval 1000 yr. Major episodes of halite deposition (shown in black) coincide with peaks of calcium sulfate.

agreement to expand the network of gauging stations and to increase the frequency of suspended-sediment determinations from a now-and-then to a daily basis. Starting in the Water Year 1976 (01 October 1976 through 30 September 1977), suspended sediment was measured daily in the upper Hudson River at Waterford and in the Mohawk River at Cohoes (Figure 15-29). The results showed that the total quantity of suspended sediment transported in the upper Hudson during the Water Year 1977 was about 390,000 tons, a quantity nearly 8 times the 50,000 tons that had been calculated previously using measurements made on widely scattered days. The combined upper Hudson-Mohawk discharge of suspended sediment was 1.2 million tons, and nearly half of it was transported in two days (14 and 15 March 1977). Approximately 90% of the year's total suspended sediment was transported in the three months of October 1976, March 1977, and April 1977 (Figure 15-30).

Although the drainage areas and quantity of water from each subbasin were about the same for the upper Hudson and the Mohawk, the sediment yield from the Mohawk was 83.4 tonnes/km² whereas that from the upper Hudson was 29.6 tonnes/km². The mean for the

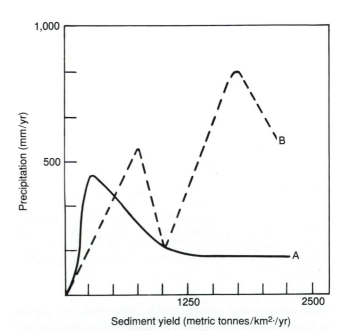

FIGURE 15-28.   Graph of rainfall vs. sediment yields. Solid curve from Langbein and Schumm, 1958, using ca. 265 basins in the United States. Dashed curve from Lee Wilson, 1969, using 1500 basins from throughout the world. Data adjusted to basin area of 259 km² (100 mi²). (Lee Wilson, 1973, fig. 1, p. 336.)

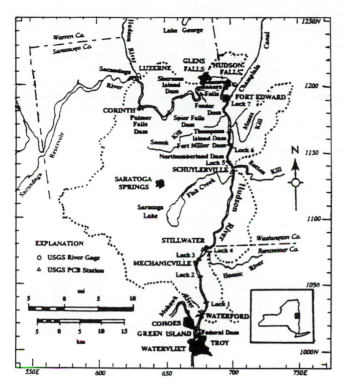

FIGURE 15-29.   Location map of U.S. Geological Survey gauging stations, upper Hudson and Mohawk rivers, New York State, with inset map of New York State. (J. E. Sanders, 1989, fig. 3, p. 8.)

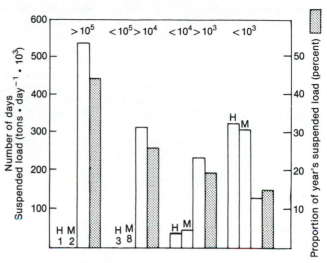

FIGURE 15-30.   Bar graphs of suspended sediment, upper Hudson and Mohawk Rivers, Water Year 1977, organized by order-of-magnitude classes. Categories at top indicate range of daily suspended load (short tons). In each daily-tonnage category are shown 4 rectangles (bars). The left-hand pair of each group shows the number of days when discharge fell into the indicated category (scale at left), labeled H for upper Hudson and M for Mohawk. Of the right-hand pair in each group, the plain rectangle indicates the total amount of suspended load transported by the combined rivers (scale at left) and the stippled rectangle the proportion of the year's total represented by the discharge falling within that class (scale at right). For example, on one day the upper Hudson and on 2 days the Mohawk discharged more than 100,000 tons. The aggregate total discharge for those days was about 540,000 tons, which was 43% of the year's total. (Authors.)

conterminous United States is 64.7 tonnes/km^2 (W. F. Curtis and others, 1973, p. 10). The fact that so much of the year's total sediment was transported on only a few days during heavy rains and floods illustrates a general principle that has been demonstrated countless times in many areas; namely, that the truly large-scale results are associated with intense rainfall and floods. Moreover, such atmospheric events may be influenced by extraterrestrial forcing functions.

SOURCES: W. F. Curtis, Culbertson, and Chase, 1973; Sanders, 1982, 1989; Wilson, 1973.

Suggestions for Further Reading

BROECKER, W. S., and DENTON, G. H., 1990, What drives glacial cycles?: Scientific American, v. 262, no. 1, p. 49–56.

DENTON, G. H., and HUGHES, T. J., 1983, Milankovitch theory (*sic*) of ice ages; hypothesis of ice-sheet linkage between regional insolation and global climate: Quaternary Research, v. 20, p. 125–144.

EDDY, J. A., 1976, The Maunder Minimum: Science, v. 192, p. 1189–1202.

FISCHER, A. G., 1986, Climatic rhythms recorded in strata, p. 351–376 *in* Wetherill, G. W., Albee, A. L., and Stehli, F. G., eds., Annual Review of Earth (*sic*) and Planetary Sciences, v. 14: 1986: Palo Alto, California, Annual Reviews Inc., 593 p.

HALFMAN, J. D., and JOHNSON, T. C., 1988, High-resolution record of cyclic climatic change during the past 4 ka from Lake Turkana, Kenya: Geology, v. 16, p. 496–500.

HAMEED, SULTAN, 1984, Fourier analysis of Nile flood levels: Geophysical Research Letters, v. 1, p. 843–845.

HAYS, J.; IMBRIE, JOHN; and SHACKLETON, N. J., 1976, Variations in the Earth's orbit; pacemaker of the ice ages: Science, v. 194, p. 1121–1132.

HAYS, J. E.; and PITMAN, WALTER, III, 1973, Lithospheric plate motion, sea level (*sic*) changes, and climatic and ecological consequences: Nature, v. 246, p. 18–22.

HOUSE, M. R., 1985, A new approach to an absolute time scale from measurements of orbital cycles and sedimentary microrhythms: Nature, v. 316, p. 721–725.

IMBRIE, JOHN; and IMBRIE, J. Z., 1980, Modeling the climatic response to orbital variations: Science, v. 207, p. 943–953.

KAYE, C. A., and STUCKEY, G. W., 1973, Nodal tidal cycle of 18.6 yr.: Geology, v. 1, p. 141–144.

KEMPE, STEPHAN; and DEGENS, E. G., 1979, Varves in the Black Sea and in Lake Van (Turkey), p. 309–318 *in* Schluchter, C., ed., Moraines and varves: Rotterdam, The Netherlands, A. A. Balkema, 441 p.

KVALE, E. P., ARCHER, A. W., and JOHNSON, H. R., 1989, Daily, monthly and yearly tidal cycles within laminated siltstones of the Mansfield Formation (Pennsylvanian) of Indiana: Geology, v. 17, p. 365–368.

MITCHELL, J. M., Jr., 1976, An overview of climatic variability and its causes: Quaternary Research, v. 6, p. 481–493.

OLSEN, P. E., 1986, A 40-million year (*sic*) lake record of orbital climatic forcing: Science, v. 234, p. 842–847.

RODHE, H.; and VIRJI, H., 1976, Trends and periodicities in east African rainfall data: Monthly Weather Review, v. 104, p. 307–315.

VAN HOUTEN, F. B., 1986, Search for Milankovitch patterns among oolitic ironstones: Paleooceanography, v. 1, p. 459–466.

WANLESS, H. R., and SHEPARD, F. P., 1936, Sea level (*sic*) and climatic changes related to late (*sic*) Paleozoic cycles: Geological Society of America Bulletin, v. 47, p. 1177–1206.

WILLIAMS, G. E., 1989, Tidal rhythmites: geochronometers for the ancient Earth–Moon system: Episodes, v. 12, p. 162–171.

WILLIAMS, G. E., and SONETT, C. P., 1986, Solar signature in sedimentary cycles from the late (*sic*) Precambrian Elatina Formation, Australia: Nature, v. 318, p. 523–527.

WILSON, LEE, 1973, Variations in mean annual sediment yield as a function of mean annual precipitation: American Journal of Science, v. 273, p. 335–349.

Principles of Stratigraphy

Now that we have summarized the chief modern environments of deposition and their sedimentary products and have examined extraterrestrial forcing functions that affect sediment production and -deposition, we are ready to take on the task of synthesizing all the information that can be obtained from the study of strata, or the subdiscipline of *stratigraphy*. We approach stratigraphy by first summarizing some aspects of the thickness of strata. We next turn to exposures and borings, including presentations about the field study of exposed strata and about subsurface strata. Then we review various branches of stratigraphy and their special kinds of stratigraphic units. Thereafter come the topics of establishing lateral relationships among strata and criteria for time equivalence. We close with two major topics that are fundamental to stratigraphic syntheses: interpreting breaks in the stratigraphic record and paleogeographic reconstructions.

We begin our discussion of analyzing ancient strata with the important topic of thickness.

Thickness of Strata

The thickness of strata is not only an obvious property, but also one that involves many aspects of the origin and history of the deposit. We shall organize our discussion under two headings: significance and measurement.

Significance

The significance of the thickness of strata includes the volume of the deposit, problems of preservation in the geologic record, and indicators of crustal subsidence.

VOLUME OF DEPOSIT

The volume of a sedimentary deposit is one of its fundamental attributes, but it is not always easily determined.

Ideally, the preserved volume is determined by multiplying the thickness by the areal extent. We shall examine the thickness aspects here and consider areal extent in a following section.

The thickness of a sedimentary deposit at a given locality is some function of the amount of sediment supplied to the basin of deposition and of the amount of crustal subsidence. Obviously, if no sediment is supplied, no strata will accumulate. But some of the sediment that is supplied to a given spot may not accumulate as strata there. Instead, the sediment may move elsewhere. The term **bypassing** designates *the movement of sediment particles of a particular size range beyond a potential site of deposition to some other place*. **Total passing** designates *a condition in which all sediment supplied to a potential site of accumulation moves elsewhere*. For example, from about 1890 to 1970, all the fine-textured sediment that was being delivered by the Atchafalaya River to Atchafalaya Bay was passing out of the bay and coming to rest in the Gulf of Mexico. (See Figures 14-41 and 13-24.)

Crustal subsidence greatly influences bypassing and total passing. To return to the example of Atchafalaya Bay, had the floor of this bay been lowered by crustal movements during the time when total passing was going on, then the water would have been deepened. Accordingly, sediments would have been added until the bottom once again had shoaled to the depth appropriate for total passing. The thickness of sediment added would have been a function only of the amount of subsidence.

The amount of sediment supplied to a basin depends on several factors. These include size of the area being eroded and rate at which the area yields sediments. The sediment yield from normal weathering and erosion by water is a function of climate; rock composition and fabric; and steepness, plant cover, and other surficial characteristics of slope. (See Figure 15-25.)

If the strata can be assigned numerical ages, then the thickness of sediment can be employed to compute the rates of significant geologic processes. These processes include rate of sedimentation (applicable in deep-water areas where all sediment supplied can accumulate) or rate of subsidence (where the only sediment that accumulates results from subsidence).

PROBLEMS OF PRESERVATION

The volume of sedimentary strata actually preserved in the geologic record rarely is a simple result of volume of sediment accumulated. To show the range of possible complications we shall discuss several examples, from simple to complex.

For a simple example, consider the volume of a sheet of extrusive igneous rock. This sheet is a product of the flow of a certain thickness of lava, which spread over a given area. Assuming that the volcanic rock was not extensively eroded before it became covered with other strata (either sedimentary or volcanic), then the volume of the sheet of extrusive igneous rock preserved in the geologic record is practically identical to the volume of lava that was spread out over the Earth's surface.

As mentioned previously, strata of halite and of other easily soluble evaporites can be completely dissolved. This happens when the strata are uplifted from positions of deep, salty formation waters and the evaporites are brought into contact with near-surface fresh ground water. Layers of salt at depth can be greatly thinned as a result of diapiric upward flow of salt plugs, as is common in the U.S. Gulf Coast.

CRUSTAL SUBSIDENCE

The importance of strata as indicators of crustal subsidence will be discussed in Chapter 17. In passing, we simply note two points here. (1) The thicknesses of strata are controlled by subsidence. (2) The kinds of strata deposited reflect the relationship and timing of rates of subsidence and rate of supply of sediments (Figure 16-1).

Exposures and Borings

Many of the aspects of exposures and borings are self evident and require no further elaboration. Nevertheless, a few points merit special emphasis. To begin with, we reemphasize that the basic information on ancient strata almost always comes from sectional views. Such views are derived from exposures (natural or artificial), from borings, and from seismic-reflection data. Distinctive aspects of exposures include the possibilities for large lateral view, the weathered expression of the strata, availability of data from bedding surfaces, relief and distinctness of contacts, and the possibility of collecting large

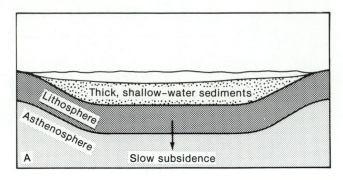

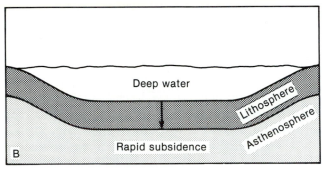

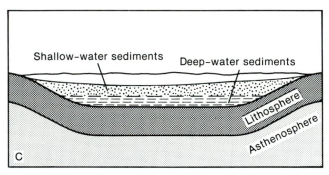

FIGURE 16-1.   Interaction of subsidence and sediment supply, schematic profile-sections.

A. Basin subsides slowly, sediment supply keeps pace; resulting thick sediments display shallow-water characteristics throughout. Total thickness of sediment results from subsidence; is not related to depth of water.

B. Basin subsides rapidly, sediment supply negligible; result is deep-water basin.

C. Basin fills with sediment (additional subsidence negligible). Sediment fill consists of deep-water sediment in lower part, shallow-water sediment in upper part. Total thickness of sediment equals depth of water. (Authors.)

specimens of the strata and their fossils. Obvious characteristics of borings are data from strata that are not exposed or that are poorly exposed, lack of lateral view, the opportunity to collect nonweathered material, and the possibility of finding distinctive signatures on instrumental logs.

Exposures and many borings, especially those deeper than about 300 m, differ in two fundamental respects. First, many borings, notably those in the fillings of sedimentary basins, encounter strata that are thicker than exposed strata of the same ages, and that may be very different from exposed same-age strata near the margins of basins. Also, borings may encounter strata that are not seen anywhere at the surface. Second, deep borings that penetrate below the level of fresh ground water encounter salty formation waters, whose salinity may be 10 times as great as that of seawater. Accordingly, such deep borings penetrate the regions where halite and other evaporite minerals that are soluble in fresh water are preserved. For example, halite, a widespread constituent of deeply buried marine strata, is almost never seen at the surface. Instead of the halite that is present in the subsurface realm, one finds only *collapse breccias* (See Figure 3-26.) at the surface. The fact that halite is not preserved in surficial exposures is only one of the many deficits of information in exposed strata. It emphasizes the point that full understanding requires full information. More-complete information is available when both surface- and subsurface data (including seismic data) are combined.

Field Study of Exposed Strata

Field study of exposed strata is one of the most-fundamental operations in working out the geologic relationships in any area. It is the beginning step in many kinds of sedimentologic analyses, and forms the basis for geologic mapping, for interpreting geologic structure, for synthesizing the geologic history, for the intelligent extraction of resources, for engineering projects, and for rational land-use planning. Therefore, we need to cultivate a systematic approach to the field study of strata.

Criteria for recognizing strata were introduced in Chapter 5 in order to provide a basis for discussion of layers. Here we start our discussion of exposed strata with some ways to measure thickness.

MEASURING THICKNESS

The thickness of strata can be measured in many ways. These range from use of some convenient scale, folding rule, or a tape on the exposed edges of strata, to measurements made from true-scale profiles drawn from large-scale geologic maps. The method employed necessarily depends on the ingenuity of the investigator, characteristics of the strata, purpose of the measurements, conditions of exposure, time available, and standard of accuracy and scale of topographic maps or of aerial photographs of the area. We shall note the measurement of thicknesses of horizontal strata, vertical strata, and inclined strata.

Horizontal Strata. The thicknesses of horizontal strata are most easily measured directly with a rule or by de-

termining differences of altitude (using hand-level sightings, surveying barometer, surveying instruments, or points taken from an accurate topographic map).

Vertical Strata. The thicknesses of vertical strata are easily measured using such direct methods as a rule or tape. Alternative methods include pacing the distance perpendicular to strike or locating contacts accurately on an accurate topographic map and using the map scale to find the distance.

Inclined Strata. The thickness of accessible inclined strata can be measured directly with a rule or a tape. Another method is to measure the strike and dip, plot the contacts on a large-scale, accurate topographic map, draw a true-scale section, and measure the thickness desired from the section. A useful, rapid, yet accurate method involves traversing in the field and plotting the results on a scale large enough to permit measurements to be made from the plotted traverse. We shall refer to this as the method of traversing and describe it in detail.

The method of traversing is applicable to dipping strata exposed along valley sides, in highway cuts or railway cuts, and in ditches or tunnels. Traversing is especially valuable for measuring the thicknesses of large units and of covered intervals or of inaccessible strata. Ideally, the route of traverse should be reasonably level and the strike and dip should be uniform throughout. The simple graphic solution to be described is based on a level traverse and uniform strike and dip. Appropriate modifications can be made to adjust for nonlevel traverse routes and for changes in strike and dip.

The compass bearing of each segment of the traverse route is measured and distances along each segment determined, either by laying out a tape or by pacing. A sample of the kind of notes to be recorded in the field is shown in Figure 16-2.

The graphic plot of the field data is shown in Figure 16-3. Select a suitable scale and lay out the bearing and distances recorded on the traverse. Draw strike lines through Sta. 0 and the last station (Sta. 4 in example). Construct a line perpendicular to the strike, and draw between strike lines, and extend beyond J to H. This line perpendicular to strike (XJH) is the edge view of the vertical plane in which the dip angle is measured. (Recall that dip is defined as the angle down from the horizontal measured in the vertical plane perpendicular to the strike direction.) By rotation of 90° around this line, we bring this vertical plane into the horizontal plane of the plot of the traverse (below and to the right of line XJH on Figure 16-3). Having made this rotation, we can now lay off the dip angle from this line at two points X and J. We next draw the lines XZ, which is the edge view of the lowest stratum exposed, and JK, which

Sta.	Bearing	Distance (m)	Thickness (m)	Remarks
O				Base of gray siltstone; N30°W, 30°NE dip.
	N15°E	0+23		Sta. 1
1	N30°E	1+19.5	17	Top of gray siltstone; base of gray limestone.
		1+24		Sta. 2
2	N40°E	2+21.5		Sta. 3
3	N70°E	3+10	17.5	Top of white limestone; base of tan dolostone.
		3+31		Sta. 4, end of exposure;
			11.5	top of tan dolostone.

FIGURE 16-2. Example of form for recording field data. (Sta. = station.) (J. E. Sanders, in G. M. Friedman and J. E. Sanders, 1978, fig. 13-7, p. 410.)

is the edge view of the highest stratum exposed. A line drawn perpendicular to lines XZ and JK represents the total thickness of the interval measured. This thickness can be scaled directly from the sketch map of the traverse. The thickness of various parts of the total interval can be determined by drawing one set of lines parallel to strike from the points on the traverse to line XJH, and another set of lines parallel to XZ and JK. The distances along the line perpendicular to XZ and JK are the thicknesses of each subinterval. They can be scaled off directly.

GENERAL COMMENTS

The way strata appear to the eye in exposures is strongly influenced by bedding-surface partings, which in turn may be influenced by extent of cementation or of contrasts in particle sizes and in minerals. When one is measuring the thickness of strata it is useful to record carefully in one's field notebook the exact basis for deciding where a given stratum ends and the next begins. Until more attention has been paid to the relationships between bedding-surface partings and other expressions of stratification, it will not be possible to know to what extent our ideas about stratification may be in error because of a natural tendency to emphasize units defined by partings.

Various attempts have been made to measure the thicknesses of strata and then to treat the thickness values statistically. The stratification index, (number of beds × 100)/thickness, appears to be a useful device for the detailed study of strata.

The number of sandy layers per 10-m length of core of deep-sea sediment has been used as a basis for distinguishing turbidites from contourites. Other valuable information about strata might be obtained by determining the ratio of laminated strata to nonlaminated strata. Presumably, all laminated strata originated as a result of the interaction of a moving current with silt- or sand-size sediments on the bottom. By contrast, nonlaminated sediment reflects various other circumstances of deposition, including rapid dumping from suspension, sediment finer than silt size, slumped fine-textured sediment, debris flows, and thorough bioturbation.

SOURCES: Friedman and Sanders, 1978; Prothero, 1990.

Information About Subsurface Strata

Information about strata in the subsurface comes from various kinds of borings and from seismic exploration. We shall summarize this information and then discuss how it is used for stratigraphic purposes.

Kinds of Borings

Borings are made for various purposes and into different kinds of materials. So-called soil-test borings are drilled to learn about the regolith for various engineering or agricultural projects. Most soil-test borings are made by pounding a sampler into the ground. Hundreds or thousands of exploratory borings, some to depths in excess of 30,000 ft (9230 m), are made each year in search of oil and gas.

Nearly all holes bored in search of petroleum are drilled using a rotary rig and circulating dense drilling fluid ("mud"). A rotary drilling rig consists of four main components: (1) mast, or tower, (2) drill pipe, (3) power unit for the *rotary table* and the *draw works*, and (4) mud pump. The mast needs to be strong enough to support the weight of the column ("string") of steel drill pipe and tall enough to enable 60-ft sections of pipe to be stacked against it. The power unit provides the torque for rotating the drill pipe via the rotary table; for running the draw works, which raises and lowers the steel pipe; for operating the mud pump; and for sundry other small devices.

After the location of the hole has been "staked" (as a result of a careful survey of position and of the altitude of the ground at the drill site), the rotary rig moves on location. After the machinery has been set

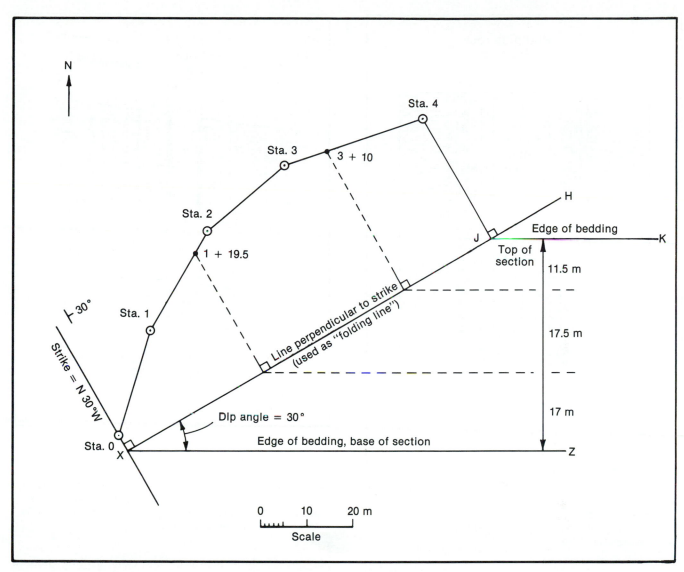

FIGURE 16-3.   Example of graphic plot of field data. Explanation in text. (J. E. Sanders, in G. M. Friedman and J. E. Sanders, 1978, fig. 13-8, p. 411.)

up, a large-diameter hole is drilled to receive the *surface casing pipe*. The size and depth of the surface casing vary with projected depth of the exploratory boring, with local geologic conditions, and with governmental regulations. The surface casing must perform three functions: (1) stay in place against any fluid pressures encountered during the drilling; (2) protect all fresh-water aquifers from contamination by drilling mud, by salt water, or by petroleum; and (3) hold open the top of the hole so that the drill pipe can enter and leave the hole without complications.

The drill bit fits on the lower end of the *drill collars*, which attach to the string of drill pipe. The bit cuts a hole whose diameter is slightly larger than that of the pipe. The *drilling mud* is circulated under pressure down the inside of the pipe. This mud passes through the bit

and returns to the surface in the annular space between the pipe and the subsurface formations (Figure 16-4, A). Because it is under great pressure, the drilling mud penetrates the formations and creates a *"mud cake,"* composed of its particulate and high-viscosity components, around the hole (Figure 16-4, B).

After the large-diameter hole for the surface casing has been drilled to the designated depth, the slightly smaller steel casing pipe is cemented in place. This is done by mixing quick-setting cement and pumping it down the hole to fill the space between the hole and the pipe using the mud-pumping system.

After the cement has set, drilling operations resume and the hole is continued through the cement inside the casing pipe. Below that, the bit encounters rock, and a stream of samples of small chips of the subsurface

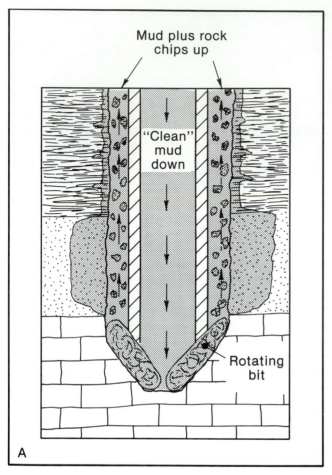

FIGURE 16-4.   Hole drilled by rotary tools using mud as drilling fluid.

A. Schematic profile at lower end of column of drilling pipe. (Details of bit and its coupling to lower end of pipe not shown.) Further explanation in text.

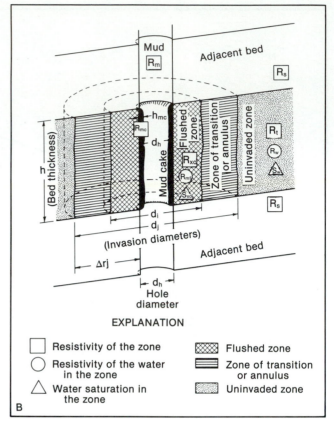

FIGURE 16-4.   *(Continued)*

B. Schematic sketch of borehole drilled with rotary tools showing mud cake lining the wall of the hole (black) and surrounding zones grading outward into the parts of the strata that are not affected by the drilling mud. Symbols are for standardized terms used in the interpretation of electric logs. (Schlumberger-Doll Research Center, courtesy C. Clavier.)

formations penetrated (the "cuttings") comes to the surface with the drilling mud. The *cuttings* are separated from the drilling mud on a device known as the *shale shaker*, and the mud is recirculated down the hole.

Cores can be collected with a rotary rig. However, because the process of collecting cores involves much expensive rig time over and above that required for regular rotary drilling, cores are not collected routinely.

Some rotary drilling is done with compressed air or gas as a drilling fluid. A few holes are drilled with a drop-the-bit pounding method using equipment known as a *cable tool*. The heavy bit, held at the end of a steel cable, is raised and dropped repeatedly. The impact of the bit breaks the subsurface rock into small pieces, which are cleaned out periodically using an open-ended cylinder known as a *bailer*. In some petroleum test holes, cable-tool methods are used to drill the last few feet into a sand. The object of this procedure is to avoid contaminating the potential reservoir with drilling mud forced into the

pores under high pressure. The cable-drilling method was invented before the rotary method; it was once used for all drilling.

Core drilling is done for the purpose of collecting cylindrical samples 1 to 3 in. (or more) in diameter for geologic study and/or chemical assay. This is the only kind of drilling whose prime objective is to collect geologic samples. A cylindrical diamond-studded drill bit is used, somewhat as in rotary drilling. The difference is that a core-drilling bit does not grind up all the rock encountered, but cuts only a circular annulus surrounding a central cylinder of rock (the core). The core is collected in a core barrel, which is brought to the surface when it is full (every 10 ft or more).

Samples

The various drilling methods described yield many kinds of samples. These range from collections of particles of sands and clays penetrated by a soil-test rig to small chips of rock sent up during rotary drilling, to cylindrical samples of rock yielded by core drilling.

For most engineering purposes, the samples from a soil-test boring can be placed into small jars. If geologic samples are desired from a soil-test boring rig, it is generally necessary for the geologist who wants the samples to be present during the drilling and to take personal charge of collecting the samples. With a soil-test boring rig, one can collect cores routinely from silts and clays, but almost never from water-saturated sands and gravels that are thicker than about 0.5 m. Such sands and gravels generally flow out the bottom when the sampler is raised to the surface.

When a rotary rig is drilling a hole to explore for petroleum, the sampling program is planned by the geologist who monitors the drilling (this is known as "sitting on the hole" or "well sitting"). Samples of the chips are collected at specified intervals. These are marked according to depth by the amount of drill pipe in the hole. Because of the lag time involved in mud circulation and also because the samples may sink in the mud column, the depths marked on the sample bags may not correspond to the true depths from which the chips originated, hence the geologist must calculate the drill-time lag. Nowadays, drill-time lag is commonly calculated right away, and the correct depths are marked on the bags to start with, but samples from older wells may show nonlagged depths. An additional problem presented by cuttings from rotary drilling is that the rising mud column or the rotating drill pipe may spall off pieces of the rock section already drilled through ("cavings"). The cuttings are a mixture of cavings and fragments of rock from the current bottom of the hole. Despite these complications, a skilled geologist can prepare an excellent log from the cuttings, which yields valuable data on the kinds of rocks penetrated. Cuttings should be examined carefully using a binocular stereomicroscope.

Cuttings from cable-tool drilling are likewise examined with a binocular stereomicroscope. The depth ranges of cable-tool cuttings are known with more accuracy than are those of rotary cuttings.

Cores of rock are stored in special cardboard or compartmented wooden boxes which are carefully labeled with well, location, and depth information. Later, the cores may be split or sawed longitudinally to expose fresh surfaces for study.

Instrumental Logs

Many kinds of instrumental methods are available for supplementing the information yielded by samples from boring. Most of these methods have been invented to increase the information gained from holes drilled by the rotary method, in which the main objective is to complete the hole using as little expensive rig time as possible. Some common kinds of logs are (1) drilling time, (2) self-potential and electrical resistivity, (3) hole-diameter, (4) radioactivity, (5) sonic, (6) dipmeter, (7) deviation, and (8) temperature.

A *drilling-time log* records the rate at which the drill string advances downward. Such a log gives an accurate representation of the depths to contacts between shales and hard limestones or sandstones, or other hard-to-drill rocks. Drilling-time logs are recorded during actual drilling; they require no extra rig time.

The *self-potential* (SP) and *electrical-resistivity* logs are the so-called E-logs (E for electric), which can be used only (a) where drilling mud has been used and (b) before production casing pipe has been set. Usually electric logs are recorded after the hole has been drilled to (or close to) the projected final depth. In very deep wells, drilling may be halted at several points (e.g., at 5000-ft, or 1524-m, intervals) in order for E-logs to be run. The hole is conditioned for logging by circulating the mud without further rotation of the bit. After the string of drill pipe has been removed from the hole, the logging tool is adjusted and lowered to the bottom of the hole. The log is made while the tool is raised at a constant rate. One trace (the one on the left side of the printed log) records the spontaneous potential created by reaction of the drilling mud with the wall rock. The other traces (on the right side of the printed log) record the electrical resistance to the passage of a current between electrodes of known spacing on the logging tool. Modern versions of the E-log are called *induction logs*.

Where only sands and shales are present, the interpretation of E-logs is relatively straightforward. The deflections toward the left on the SP curve mark sands. (See Figures 14-34, A; 14-37; 13-20; and 13-21, B.) The behavior of the resistivity curve indicates the kinds of fluids. Low resistivity indicates that an electrically conductive material, such as salt water or pyrite, is present; high resistivity indicates oil or gas (Figure 16-5).

Carbonate-cemented sands and limestones complicate the interpretation of E-logs. The E-log trace of a limestone commonly resembles that of an oil-saturated sand.

A log of the diameter of the hole is made by means of a spring *caliper* that is attached to other logging devices lowered into a mud-filled hole. The caliper records a profile of the side of the bore hole. In firmly cemented strata, the side of the hole is vertical and hole diameter equals bit diameter. In shales or noncemented sands (including many sands that have been saturated with petroleum), the wall collapses and hence it is no longer vertical and its diameter is larger than that of the bit. The variations recorded on the caliper log closely correspond to variations in many kinds of strata (Figure 16-6, A and B). In shales with high proportions of swelling

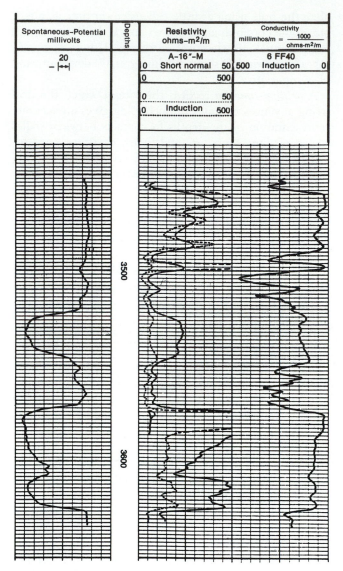

FIGURE 16-5. Portion of a modern induction-electric log, between depths of 3450 and 3630 ft (1052 to 1106 m), for a hole drilled into Cenozoic strata in the U.S. Gulf Coast. Unit that is 30 ft (9 m) thick and having top at a depth of 3572 ft (1089 m) appears to be a sand containing oil or gas in its upper 8 ft (2.4 m), but salt water in its lower parts. Sand, 14 ft (4.3 m) thick, with top at 3608 ft (1100 m), contains salt water throughout. (Schlumberger-Doll Research Center, courtesy C. Clavier and V. Hepp.)

tivity logs can also be made from inside the production-casing pipe. This is valuable because it permits modern logs to be made on old wells that may not have been logged mechanically before the pipe was set. One part of a radioactivity log measures the gamma rays emitted by natural radioactive components of the strata. The other part is based on the reaction of the strata to bombardment with gamma rays or with neutrons. (See Figure 16-6, A.) The response of the strata to the bombardment yields data on density of the material and on fluid contents. The most-commonly used radioactive log is the *compensated neutron-formation density log* (CNL-FDC).

Sonic logs are subject to the same conditions as electric logs. The sonic-logging tool measures and records the speed of sound in the strata penetrated. On modern sonic logs, corrections are made for variations in the diameter of the hole, as indicated by the caliper log. Such corrected logs are labeled with the letters BHC, which stand for *bore-hole compensated* logs. Sonic logs yield data on the speed of sound (the so-called velocity) in the strata. Such information assists in displaying stratification, in assessing the amounts of porosity and the fluid contents of porous layers (See Figure 16-6, A and B.), in analyzing seismic-reflection data, and in constructing seismic models of the rock sequence. An increase in travel time on the sonic log means a decrease in the speed of sound in the strata; such a decrease commonly corresponds to an increase in porosity (or a change in mineral composition).

Dipmeter logs are another kind of in-the-mud survey made in a newly drilled hole. A dipmeter consists of four resistivity tools, mounted in articulated pairs on arms at right angles. On each arm is a caliper tool for measuring the diameter of the hole. A compass on tool no. 1 measures and records the azimuth of the tool from magnetic north (Figure 16-7, A, left). A computer is required to match the four resistivity curves (Figure 16-7, A, right) with changes in orientation of the tool as it is pulled up the hole and to compare each curve with the others. By these comparisons the computer determines the attitudes of surfaces where the resistivity changes (Figure 16-7, B). The resistivity changes at stratification surfaces and at various other kinds of interfaces created by diagenetic- or organic processes.

Two modern logs that visually display the same kind of information acquired by a dipmeter log are the *Borehole Televiewer*, which carries a rotating TV camera, and the *Formation MicroScanner*, which bears arrays of microelectrodes that measure resistivity. The data are computer analyzed and displayed as an oriented grayscale- or color image. Thus both the Borehole Televiewer and the Formation MicroScanner provide images that resemble core photographs, without the necessity of coring.

clays (clays such as montmorillonite, which possess an affinity for fresh water and absorb it into their crystal lattice, expanding greatly in the process), the hole diameter may actually be smaller than the bit diameter, and the drill pipe may become stuck. The caliper log is used in interpreting the traces of logs (such as the sonic log, described below), whose responses are affected by hole diameter.

As with electric logs, *radioactivity logs* can be made in a mud-filled hole. However, unlike electric logs, radioac-

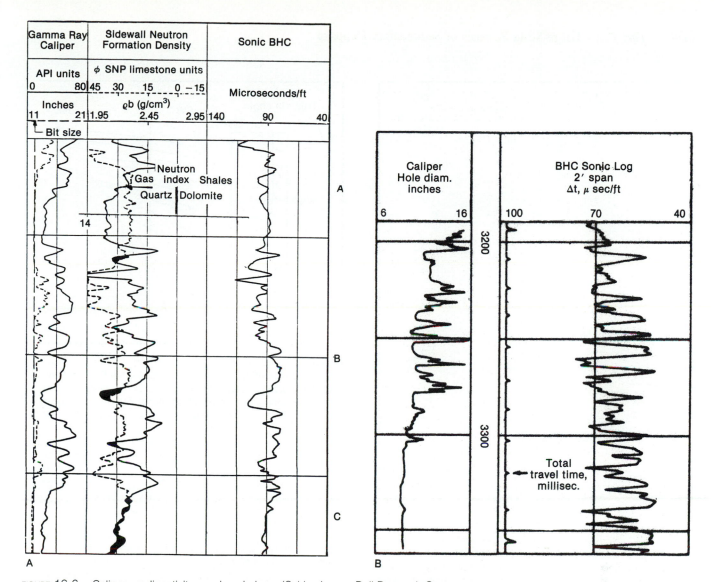

Gamma Ray Caliper	Sidewall Neutron Formation Density	Sonic BHC
API units	φ SNP limestone units	
0　　　　80	45　30　　15　　0　−15	Microseconds/ft
Inches	ρb (g/cm³)	
11　　　　21	1.95　　2.45　　2.95	140　　90　　40
Bit size		

Neutron
Gas index Shales
Quartz | Dolomite

14

A

B

C

Caliper
Hole diam.
inches

6　　　　16

3200

3300

BHC Sonic Log
2′ span
Δt, μ sec/ft

100　　70　　40

Total
travel time,
millisec.

A

B

FIGURE 16-6.　Caliper-, radioactivity-, and sonic logs. (Schlumberger-Doll Research Center, courtesy C. Clavier and V. Hepp.)

A. Displays of parts of three porosity logs from Khafji sand-shale succession (Cretaceous), northern Saudi Arabia (depths not available). Dashed line at extreme left is caliper log (referenced to scale at top log marked inches). Solid curve next to caliper log records gamma rays emitted by the natural radioactivity of the strata penetrated (measured in standardized API units). Two curves in center are based on radioactive bombardment of the strata. Neutron bombardment generates the sidewall-neutron curve, a dashed line referenced to an effective porosity scale calibrated in limestone sidewall neutron porosity (SNP). A SNP limestone unit is referenced to limestone with pores filled with pure water. Corrections must be made for actual kind of rock and the composition of the fluid in the pores. Bombardment with gamma rays gives solid curve referenced to density (g/cm³). Gas sand coincides with zone A (at top), where dashed curve lies to right of solid curve by more than three porosity units. Zone B contains many layers of shale. Many shales are naturally highly radioactive and for these the solid curve lies to the right of the dashed curve. Zone C (at bottom) is an oil-bearing sand (irregular black area where dashed curve lies to right of solid curve). The sonic log (at right) displays speed of sound as a function of the time required for the sound pulses to travel one foot (in microseconds per foot). Notice that travel time decreases toward the right (speed of sound increases toward the right) and that a decrease in travel time corresponds to a decrease in apparent porosity.

B. Caliper- and sonic logs. On the caliper log (at left) the variations resulting from collapse of the walls of the hole are pronounced from a depth of 3200 to 3300 ft (975 to 1006 m). Below 3300 ft (1006 m) the hole maintains a reasonably uniform diameter of about 8 in. (20 cm), the bit size used. For every pronounced indentation in the wall of the hole (that is, an abrupt increase in the diameter of the hole), the speed of sound decreases. Below 3300 ft (1006 m), numerous changes with depth in the speed of sound take place, but these are not accompanied by corresponding changes in the diameter of the hole.

603

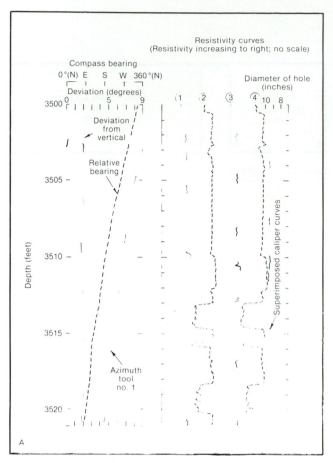

FIGURE 16-7. Dipmeter logs and their interpretation.

A. Displays of deviation-azimuth survey (at left) and curves of four resistivity tools (paired 1, 3 and 2, 4 on arms at right angles) plus superimposed caliper curves (at right) for hole drilled in Gulf Coast through channel-type sandstones and interbedded shales. Sampling rate: 60 readings per foot (197 readings per meter).

Deviation-azimuth log shows that hole deviates 2° from perpendicular toward the east (based on separation of 70 to 80° between dashed-line relative-bearing curve and solid-line curve for the azimuth of tool no. 1, referenced to scale 0 to 360° at top). In being raised from depth of 3521 to 3500 ft (1073 to 1067 m), the logging tool rotated clockwise 215°, or about 10°/ft (33°/m).

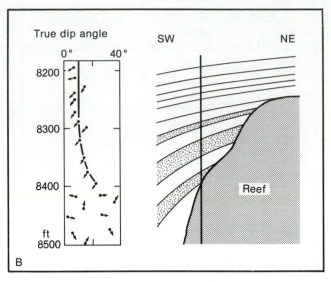

FIGURE 16-7. (*Continued*)

B. Plot of dips versus depth (at left) and schematic section showing an interpretation of the dips. Horizontal placement of dots shows amount of dip (0° at left, 40° at right). Arrows point to direction of dip (using dot as center of a circle with N at the top and E at right).

The log is from a hole drilled into the Abo reef (Middle Permian) in the Delaware Basin, SE New Mexico. The dipmeter found a group of dips toward the SW whose steepness increases downward from about 8° at 8200 ft (2499 m) to 22° at 8400 ft (2560 m). This is taken as an indication of drape over a reef, with maximum drape close to the reef. (Compare Figures 10-41, photo, edge of bioherm, and 10-42, schematic profile through reef.) Below 8400 ft (2560 m), dips show great diversity. By themselves, such diversity of dips could mean several things: (1) diversely oriented low-angle and steep-angle cross strata deposited by currents, (2) organic- or diagenetic boundaries in reef rock, or (3) artifacts of a now-discarded technique of correlation of the measured data. (J. A. Gilreath and J. J. Maricelli, 1964, fig. 14, p. 1908, courtesy Schlumberger-Doll Research Center.)

indicates not only the geothermal gradient, but also the progress of setting of the cement.

SOURCES: Friedman and Sanders, 1978; Helander, 1983; Schlumberger, 1986; Serra, 1985; Streltsova, 1988; Theys, 1991.

Geologic Uses of Subsurface Information

In addition to the kinds of stratigraphic studies that can be made from continuous seismic-reflection profiles (Chapter 6), subsurface information is the basis for preparing maps and cross sections of the subsurface structure at various depths and for various stratigraphic studies and -maps.

STRUCTURE MAPS

The data to make a structure map are taken from one of the instrumental logs, typically the E-log or the density

Deviation surveys indicate whether or not the hole is vertical (boreholes are rarely vertical). *Deviation logs* record the amount and direction that the hole deviates from the vertical. (See Figure 16-7, A, left.) In some cases the magnitude of deviation is sufficient to increase by substantial amounts the apparent stratigraphic thickness of units penetrated in a borehole. If noncorrected thicknesses are used to construct geological cross sections, faulty interpretations are likely to result.

Temperature logs are sometimes recorded in the mud, but always are made inside the production casing to check the cement. As the calcite in the cement crystallizes, the temperature rises. Thus a temperature survey

log of vertically drilled holes. The starting point is some uniform geologic surface that can be regarded as having been initially a horizontal plane. Such planes include thin limestones, layers of tephra, coal beds, or evaporites. Maps of the tops of sands have to be made for reservoir studies, but the tops of sands are not the best kinds of surfaces for regional structural mapping for two reasons: (1) the original top of the sand may not have been horizontal, and (2) the sand may not be laterally persistent.

Once the depth information to the mapping surface has been determined from the logs, it must be reduced to a standard datum surface, usually sea level. This requires accurate altitude data of the drill site. Such data are usually expressed in three ways: as *ground level* (G.L.), *derrick floor* (D.F.), and *drive bushing* (D.B.), or *Kelly bushing* (K.B.—the level of the top of the rotary table, which is the standard zero point in most logs). One assumes that the hole is vertical or deviates only by 2 or 3° from the vertical. As mentioned previously, the deviation of the hole can be measured; this should be done if it is likely that natural conditions during drilling could deflect the bit. In order to obtain structural data, complicated corrections need to be made to the logs of directionally drilled, nonvertical holes.

The ground-level altitudes of many wells have never been surveyed (or else were surveyed and were not recorded on the logs). Instrumental logs exist for many such holes, but unless the altitude of the drill site can be established, these logs cannot be employed for structure mapping. Modern topographic maps on a scale of 1/24,000 are accurate enough to enable such altitudes to be determined from the map in cases where the map location of the well can be established accurately. Unfortunately, accurate map locations of some wells are not available.

STRATIGRAPHIC STUDIES AND -MAPS

Subsurface data form the basis of many kinds of stratigraphic studies and -maps. A few examples include thicknesses of particular stratigraphic intervals (*isopach* maps; Figure 16-8, A) or of sandstones or other rock types, proportions of various kinds of materials within a given interval (e.g., percentages of sandstone and shale), porosity, resistivity, directions of dip from dipmeter surveys, or about anything at all that can be expressed as a number (Figure 16-8, B). Interpretive maps based on numerical data can be drawn showing the distribution of environments of deposition or of diagenetic features. These maps can also be drawn as three-dimensional relief contoured plots for ease of interpretation (Figure 16-8, C). These 3D plots are particularly useful when geologic data have to be presented to nongeologists, who com-

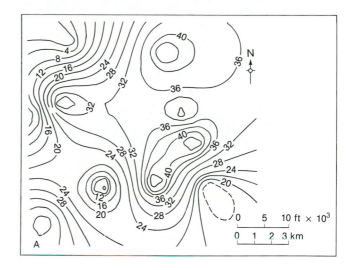

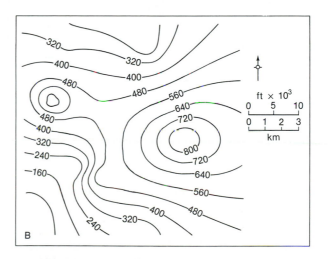

FIGURE 16-8. Subsurface maps.
A. Isopach map (contour interval = 4 ft) of oolitic grainstone of Permian San Andres Formation, Mabee Field, Texas. (S. K. Ghosh and G. M. Friedman, 1989, fig. 5-3, A, p. 104.)
B. Contour map (contour interval = 80 porosity-feet) of porosity-feet of oolitic grainstone; location as in A. The parameter porosity-feet is calculated by multiplying the thickness of a stratigraphic interval by its average porosity. (S. K. Ghosh and G. M. Friedman, 1989, fig. 5-10, B, p. 107.)

monly lack experience interpreting conventional contour maps.

Subsurface data may also be presented as cross sections, which provide an interpretive view of a vertical slice through the rock strata. Modern computer graphics allow plan- and cross-sectional views to be combined in color plots that are truly three dimensional.

In addition, subsurface studies may be used to help delineate depositional sequences (Chapter 6), to establish lateral correlation of strata, and for event stratigraphy.

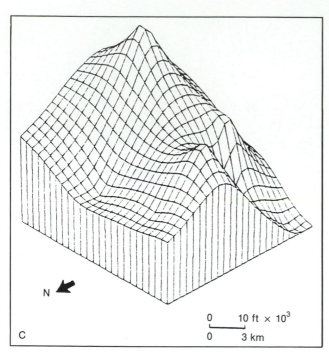

FIGURE 16-8. *(Continued)*
 C. Three-dimensional projection of the contour map in B. This kind of display makes the variation in the parameter mapped (in this case, porosity-feet) more obvious to those not accustomed to interpreting contour maps. Such 3-D projections are sometimes used to convey geological information to juries. (S. K. Ghosh and G. M. Friedman, 1989, fig. 5-10, A, p. 107.)

Both of these topics are discussed in a following section of this chapter.

SOURCES: Blatt, Berry, and Brande, 1991; Dresser Atlas, 1982; Ghosh and Friedman, 1989; Lemon, 1990; Nelson and others, 1987; Pirson, 1977; Prothero, 1990; Rider and Laurier, 1979; Schlumberger, 1981; Schlumberger Educational Services, 1987; Serra, 1985; Swanson, 1981.

Subdivisions of Stratigraphy and Their Distinctive Kinds of Stratigraphic Units

In this section we examine the various subdivisions of stratigraphy and explore the distinctive kinds of stratigraphic units associated with each. In lithostratigraphy, strata are subdivided on the basis of their physical characteristics that do not involve fossils. In biostratigraphy, stratal subdivisions are made on the basis of their contents of fossils. In chronostratigraphy, units are defined by their time relationships. In magnetostratigraphy, units are established using their magnetic characteristics. In each section, we also summarize the formal procedures for naming and describing these units as prescribed in the North American Stratigraphic Code (1983), prepared by the North American Commission on Stratigraphic Nomenclature.

Lithostratigraphy and Lithostratigraphic Units

Lithostratigraphy is the study of the physical characteristics of strata. Despite the connotation of rock implied by the prefix litho, this branch of stratigraphy includes strata composed of sediments that have not been lithified and are not properly designated as rocks. Therefore, the prefix litho includes sediments as well as rocks.

A **lithostratigraphic unit** is *a defined body of sedimentary-, extrusive igneous-, metasedimentary-, or metavolcanic strata that is distinguished and delimited on the basis of lithic characteristics and stratigraphic position* (North American Stratigraphic Code, 1983; all quotations in this section are from the Code).

The features used to define lithostratigraphic units can be as varied as the strata themselves. Boundaries separating lithostratigraphic units may be placed at sharp contacts or at arbitrary levels within zones of gradation.

One of the great problems with drawing boundaries in strata is that, in terms of lateral extent, all strata are limited, but some strata are more limited than others. *Strata having wide lateral persistence* are designated as **key beds**. According to the North American Stratigraphic Code (1983, Art. 23), *Key beds may be used as boundaries for a formal lithostratigraphic unit.* However, a recommended limitation on this practice is that such usage be restricted to areas within which "the internal lithic characteristics of the unit remain relatively constant." The Code adds further: "Where the rock between key beds becomes drastically different ... a new unit should be applied, even though the key beds are continuous."

The purpose of selecting such units is to carry out some geologic objective, such as geologic mapping, a sedimentologic study, assessment of a mineral resource, study of the distribution of fossils, or collection of specimens for some laboratory measurements (such as paleomagnetism).

If areal geologic mapping is the objective, then a major factor that enters into the recognition of units is what is known as *mappability*. Many considerations affect the question of mappability. These include the size of the area to be mapped, its relief, scale of base maps (1/24,000 is regarded as the standard), purpose of the project and time assigned for completing the mapping, kind and number of exposures of the strata, the experience and skill of the mapper(s), and extent of the previous geologic study and mapping of surrounding areas. Units that may serve one geologist's purposes may be totally inadequate for another geologist's different purposes.

The ultimate objective of subdividing the strata is to find and identify all of them and then to assemble a framework of nonoverlapping units for designating them.

FORMAL NAMES

The cornerstone of formal lithostratigraphic units is the **Formation**: *the fundamental formal unit of lithostratigraphic classification. The formation is of intermediate rank in the hierarchy of lithostratigraphic units and is the only formal unit used for completely subdividing the entire stratigraphic column all over the world into named units on the basis of lithostratigraphic characteristics.*

Formations may be subdivided into **members** (*named or nonnamed lithologic entities within a formation*) or **beds** (*named or nonnamed distinctive individual layers*). A member "is always part of some formation," but "a formation need not be divided into members unless a useful purpose is served by doing so." Some formations may be completely divided into members; others may contain only certain parts designated as members; still others may lack members. A member may extend laterally from one formation to another.

Formations can be assembled into **groups** (*two or more formations*). *Several associated groups* may be constituted into a **supergroup**. Or, on occasion, *a group may be subdivided* into **subgroups**. A particular unit may be named a formation in one area, and a group or member in another, without change in name (except if the unit rank forms part of the name).

According to the International Stratigraphic Guide (Hedberg, 1976), the formal name of a lithostratigraphic unit "should be formed from the name of an appropriate local geographic feature, combined with the appropriate term for its rank (group, formation, member, bed), or with the dominant rock type of which the unit is composed, or with both; e.g., Gafsa Formation, Fortuna Sandstone, Taylor Coal Member. Descriptive adjectives need not be included." Initial letters of all words in formal stratigraphic names are capitalized.

SOURCES: Blatt, Berry, and Brande, 1991; Cohee, 1974; Forgotson, 1957; Hedberg, 1976; International Subcommission on Stratigraphic Classification, 1987; Lemon, 1990; North American Commission on Stratigraphic Nomenclature, 1983; Prothero, 1990; Whittaker et al., 1991.

Biostratigraphy and Biostratigraphic Units

The science of biostratigraphy is based upon the law of faunal succession, first formulated by William Smith (1769–1839) in the late eighteenth century. In essence, this law states that on the bases of their fossil contents, sedimentary rocks formed during a particular time interval can be distinguished from strata deposited during other time intervals. In order to appreciate the use of fossils in biostratigraphy, we must review the classification of organisms, the factors that affect organisms in time at a given locality, and those that affect the distribution of organisms on the Earth's surface.

CLASSIFICATION OF ORGANISMS

The basis of the classification of organisms was established by the Swedish naturalist Carl von Linnaeus (1707–1778); his classification is known as the Linnaean classification. It organizes living creatures into hierarchies or ranks, and the modern version starts at the top with kingdoms, ending at the bottom with subspecies as the lowest-ranked formal unit (Table 16-1).

The term **taxon** (plural taxa) is a general term used to refer to *any one of the taxonomic groups in the Linnaean hierarchy*. For the purposes of biostratigraphy, the species is the fundamental taxonomic unit. Biologists define **species** on the basis of *groups of individuals (populations) that share a common gene pool and that are interbreeding (or are potentially interbreeding)*. Paleontologists usually deal only with the morphologic aspects of preserved skeletal remains or hard parts. Therefore, fossil species are referred to as *morphospecies*.

In the following discussion of organisms we shall assume the best of all possible worlds and therefore assume that all workers who have published scientific papers about fossils have reliably identified all their specimens and that all conflicts between "lumpers" and "splitters" have been resolved. Accordingly, we shall ignore the devastating effect that "erroneous" fossil classifications exert on the time significance of fossils. Anyone having detailed acquaintance with any group of fossil organisms will realize that our "best-of-all-possible-worlds" approach amounts to a quantum jump over some very thorny territory.

ORGANISMS THROUGH TIME

The essence of Charles Darwin's (1809–1882) great contributions to the study of organisms is contained in his phrase "descent with modification." The implication of

TABLE 16-1. Biological hierarchical classification of organisms, with most-inclusive categories at the top

Kingdom
    Phylum
        Class
            Order
                Family
                    Genus
                        Species
                          Subspecies

NOTE: Intermediate hierarchical levels are used as needed by adding the prefixes super- for larger groups, sub- for smaller groups, and infra- for yet-smaller groups. Of these in-between levels, only the subspecies is shown, because it is of cardinal importance to biostratigraphers.

this concept is that one can obtain from the fossil record a series of fossils representing successive generations of a given stock (i.e., a *lineage*). In such a lineage one can expect to find that systematic morphologic changes have taken place. The shapes of some of the older fossils will not be the same as those of younger fossils. Moreover, the same progression of changes of shape will be found in all places where such fossils are present. These changes in the shapes of fossil remains are the basis for defining fossil species (morphospecies).

Darwin's insights have been given the title of "evolution." His interpretation was that the driving force for the changes is the combined effects of "environment" (viewed in the broadest sense to include not only the geographical setting but also other organisms), population pressures, and variation brought about by sexual reproduction. The interaction of these three factors resulted in natural selection, which has been described by the phrase "survival of the fittest."

The term "evolution," whether it operates as Darwin thought or by other mechanisms, involves the repeated origin of new species and extinction of existing species. As a result, each fossil species survived for only a particular time span that may be long or short. Moreover, the fossil remains of a given species are restricted to rocks deposited during a corresponding long- or short time interval.

In the 1970s, considerable intellectual foment began among paleontologists over the subject of how new species develop and how this affects the distribution of species in the fossil record. Debate continues between partisans of two contrasting positions: phyletic gradualism on the one hand, and punctuated equilibrium on the other. Without entering into the details of this continuing debate, we note that, according to either viewpoint, a new species is thought to appear when individuals possessing one or more new or altered heritable attributes are born in isolation from their closest relatives. If the new species is to prosper, then individuals belonging to it must spread from wherever the speciation occurred to other parts of the world, and there interbreed and pass along the new genetic code so that succeeding generations will display the new trait(s).

The effect of recent biological discoveries about the ways in which some cells are replicated following "the master instructions" contained in DNA (deoxyribonucleic acid, an essential component found in the chromosomes of nearly all living organisms; DNA transmits the hereditary pattern) has brought many new insights into the field of genetics. The genetic code, as it has been designated, governs two fundamental biologic processes. (1) The code enables some kinds of cells to be replicated innumerable times with almost perfect fidelity to the master template. (2) The genetic code of the indi-

viduals in a new generation is subject to modification when DNA from male and female parents is combined as a result of sexual reproduction. The code contains instructions not only for individual cells, but also for the growth path of individual organisms. The fertilized egg of a brachiopod does not grow up to become a elephant.

What is not known, but what would explain some otherwise-puzzling features of the paleontologic record, is if the genetic code itself is "encoded"; that is, if a controlling template directs lineages in a way analogous to the template which controls the growth of individuals. If such a lineage template exists, it would explain how, within a given lineage, new morphologic attributes appear more or less simultaneously (in a geologic sense) among widely separated and genetically isolated populations. It would eliminate the need to postulate that a new morphologic feature begins at a single point and that, for this feature to become widespread, individuals carrying the genetic code for this new feature must migrate. It could also add a new dimension to the subject of extinction. (In this connection, an idea close to this one was proposed in the 1930s under the title of "racial senescence." However, "racial senescence" was soundly rejected because it seemed to be only another version of the much-condemned concept of "straight-line evolution," or orthogenesis, a word about equal to "sin" among right-thinking evolutionary biologists and -paleobiologists.)

We close this brief peek at an enormously complex field of scholarship with the reminder that the **law of faunal succession**, which states that *distinctive assemblages of fossils occur in the same stratigraphic order in different areas*, works, and that it was proposed by William Smith (1769–1839) about half a century before Darwin's profound book on the origin of species was published.

SOURCES: Barigozzi, 1983; Blatt, Berry, and Brande, 1991; Cloud, 1948, Coyne and Barton, 1988; Eldredge and Gould, 1972, 1977; Gingerich, 1985; Gould, 1980, 1982; Gould and Eldredge 1977, 1986; Hallam, 1978; D. E. Kellogg, 1983; Lemon, 1990; Levinton, 1988; Levinton and Simon, 1980; P. R. Sheldon, 1987; S. M. Stanley, 1976, 1979, 1985; Vbra, 1980; West, 1979; Williamson, 1981.

DISTRIBUTION OF ORGANISMS IN THE MODERN WORLD; BIOGEOGRAPHY

Few taxa, living or fossil, are ubiquitous throughout the globe. Most species and higher taxa survive and reproduce under a certain set of environmental conditions and perish without issue under others. *The study of the dis-*

tribution of organisms in the modern world and the factors that influence this distribution is known as **biogeography**. An important conclusion of biogeographic studies is that organisms are distributed throughout regions known as *faunal provinces*. Moreover, some kind of barrier, visible or invisible to the casual observer, separates these provinces.

By applying the principles of plate tectonics, biogeographers have brought about a renaissance of their discipline. For example, the separation of continents provides a mechanism for establishing new faunal provinces. In this case, land animals could become segregated onto different continents and no unusual land bridges nor migration schemes need be invoked as an explanation.

We have already discussed (Chapter 10) biogeography as related to reefs. Here we add a few remarks about barriers to migration.

A consequence of the "must-migrate" view of the appearance of new species is that biologists have studied extensively the ways in which organisms can migrate and the factors preventing them from migrating. Even sessile benthic organisms pass through larval stages that are long enough to enable the progeny to move away from their parents. The consensus view is that even if an organism moves "at a snail's pace" it can spread widely fast enough so that its first-appearance datum surface can be treated geologically as synchronous.

A contrasting result of biogeographic- and ecologic research is that certain marine invertebrates, for example, are restricted to kinds of environments in which a particular kind of sediment is deposited. *Organisms that are tied to a particular environment* are known as **facies fossils**. They migrate to a new area only when conditions, such as a submergence or an emergence, cause the boundaries of their particular habitats to shift.

SOURCES: Blatt, Berry, and Brande, 1991; Campbell and Valentine, 1977; Lemon, 1990; Levinton, 1982; Pielou, 1979; Scheltema, 1977; Valentine, 1977; Vermeij, 1978.

A **biostratigraphic unit** is *a body of rock defined or characterized by its fossil content.* Relatively young biostratigraphic units may be defined or characterized by the remains of organisms that are technically not yet fossilized, but for the purpose of defining biostratigraphic units, these are treated no differently from true fossils.

The fundamental proposition underlying biostratigraphic units is that new species appear (as a result of *speciation*, however that works) and disappear (because they become extinct).

The application of the restriction of fossil species to particular time intervals, and to the rocks deposited within them, is through the definition of **zones**, *the fundamental units of biostratigraphy.*

ZONES

The first use of zones was by Carl Albert von Oppel (1831–1865), in 1856. Since that time, numerous kinds of zones have been defined. The more-common kinds in use today are described in the following sections.

Biostratigraphic units are of three general kinds (*interval zones, assemblage zones,* and *abundance zones*). In all cases, biostratigraphic units are defined on the basis of the spatiotemporal distribution of one or more taxa (taxonomic groups, such as species).

Interval Zones. *The body of strata between two specific, well-established lowest- and/or highest occurrences of single taxa constitutes an* **interval zone.** Three kinds of interval zones are recognized. The interval between the lowest- and highest occurrences of a single taxon is a *taxon-range zone.* The interval between the observed lowest occurrence of one taxon and the observed highest occurrence of another taxon is named a *concurrent-range zone* if the two taxa overlap stratigraphically, and a *partial-range zone* if they do not. The interval between observed successive lowest occurrences or successive highest occurrences of two taxa is another kind of interval zone. If the two taxa are inferred to be successive species within a single evolutionary lineage, and if the boundaries of the zone are both lowest occurrences, then according to the International Stratigraphic Guide, the zone is a *lineage zone.* However, for most taxonomic groups, lineages have not been well established or are disputed, and it is usually best to refrain from defining or employing so-called lineage zones.

Assemblage Zones. An **assemblage zone** is *a biostratigraphic unit characterized by the co-occurrence of three or more taxa.* If an assemblage zone's boundaries are based on two or more observed first- and/or last occurrences of the included characterizing taxa, then according to the North American Stratigraphic Code, it is called an *Oppel zone.*

Abundance Zones (Acme Zones). An **abundance zone** is *a biostratigraphic unit characterized by quantitatively distinctive maxima of relative abundance of one or more taxa.* Abundance zones are subjectively defined and should not be employed except in those rare cases in which a taxon was substantially more abundant during one brief period of time than it was at any other time during its existence.

Because, as mentioned above, no taxa are found throughout the entire globe, no zonation scheme is applicable to every locality. Each zone exists throughout a geographic region, which may be very large, in which, theoretically, its characterizing taxa can be found. Moreover, a complementary region exists in which the characterizing taxa are absent.

In practice, for a given part of the world and a given interval of time, characterized by relatively constant environmental conditions, one or more zonation schemes may coexist, each of which is based upon a particular fossil taxon. For example, many zonation schemes are based upon planktonic marine organisms. Continental sedimentary rocks deposited during the time corresponding to a marine fossil zone cannot be dated using that zone. In some cases, separate zonal schemes, utilizing different taxonomic groups, can be developed for a single period of time for normal marine-, brackish-, and fresh waters. In such a case, correlations among the three zonal schemes can be established in transitional environments. This allows subsequent correlation of marine- and lacustrine strata, based upon fossil zones, even though the lacustrine fossils may nowhere co-exist with the marine fossils.

Zones are based on *index fossils*. The ideal index fossil is one that spread widely (such as a planktonic marine invertebrate), contains skeletal material that can be preserved in the geologic record (forget jellyfish), is so abundant that its fossil remains literally pop out of every outcrop, is easily recognized by simple inspection in the field (thus eliminating the need to make a statistically significant collection of specimens and submit them to a specialist for possibly interminable study in the laboratory), and that did not endure for a very long time (ideally, less than a million years).

Biostratigraphic units are fundamentally distinct from other kinds of stratigraphic units, and their definitions are independent of the definitions or distribution of other kinds of stratigraphic units. An important corollary is attached to the "must-migrate" concept of speciation, which states that a species begins when the requisite number of individuals appear in an isolated area and requires them and/or their direct progeny to migrate. The corollary states that a certain amount of time must elapse between the birth of the original individual(s) of the new species and the first appearance of fossil remains of this species in other localities. If the species appeared everywhere at the same time, then the first appearance would mark a synchronous surface in the geologic record, one that is parallel to a time plane. If migration time is factored into consideration, then the first appearance marks a *diachronous surface*, that is, it crosses time planes. The start-here-migrate-there approach holds that *the boundaries of most biostratigraphic units . . . are both characteristically and conceptually diachronous*. Such diachroneity distinguishes biostratigraphic units from chronostratigraphic units, whose boundaries are synchronous. Nevertheless, throughout large geographic areas, the boundaries of many biostratigraphic units are approximately synchronous; all biostratigraphic units are ultimately limited by time (Figure 16-9). Therefore, biostratigraphic units are

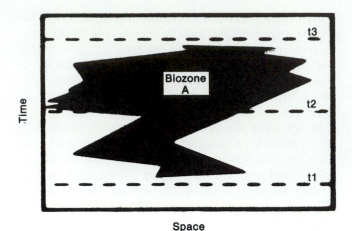

FIGURE 16-9.   Schematic space-time diagram illustrating the diachronous nature of the boundaries of biostratigraphic units. Biozone A which lies between the two timelines t1 and t3, displays diachronous boundaries. Note that before time t2, Biozone A was highly provincial, whereas after time t2, it was relatively cosmopolitan. Although the boundaries of Biozone A are not yet synchronous and it cannot be recognized worldwide, the finding of this biozone in a rock sequence demonstrates that the rocks in question are younger than time t1 and older than time t3.

commonly used for time correlation and "are effective for interpreting chronostratigraphic relations (*sic*)." To repeat what we stated previously, biostratigraphy works.

So far, we have concentrated on the lower boundaries of biostratigraphic units, defined by the first appearances in the geologic record of the fossil remains of new organisms. The upper boundaries of biostratigraphic units are defined by other first appearances or by extinctions, a topic upon which much recent effort has been concentrated. True **extinction** means *the dying out of all members of a given lineage*. Extinction contrasts with **extermination**, which is *the disappearance of particular kinds of individuals from a local area*. (The person who comes to rid one's house of pests, such as fleas, roaches, or mice, therefore, is an exterminator, not an extinctor.)

Two kinds of extinctions have been defined: background extinctions and mass extinctions. Great debates are in progress over the causes of these kinds of extinctions. Do lineages experience something comparable to the aging process in individuals? Are terrestrial changes operating unassisted, such as climate changes or rises and falls of sea level, sufficient? To what extent are impacts of extraterrestrial objects involved?

SOURCES: L. W. Alvarez, W. Alvarez, Asaro, and Michel, 1980; Blatt, Berry, and Brande, 1991; Bower, 1984, 1985; D. K. Elliott, 1986; Hancock, 1977; Jablonski, 1980, 1984, 1986; Lemon, 1990; Lewin, 1986; Prothero, 1990; Raup, 1986; Raup and Boyajian, 1988; Raup and Sepkoski, 1984, 1986, 1988; Schock, 1989; Silver and Schultz, 1982; Whittaker et al., 1991.

Chronostratigraphy and Chronostratigraphic Units

Chronostratigraphy is the study of stratigraphic units based on their time relationships, however these can be established. The objective is to establish surfaces that are everywhere the same age; that is, they are synchronous. A stratigraphic surface that is not everywhere the same age is said to be diachronous.

A chronostratigraphic unit is *a body of rock having synchronous boundaries and established to serve as the material reference for all rocks formed during the same span of time.* Each of its boundaries is synchronous. Thus a chronostratigraphic unit consists of *all rocks, and only those rocks, formed during [a specified] time span.*

"Boundaries of chronostratigraphic units should be defined...on the basis of observable paleontological- or physical features of the rocks."

Each chronostratigraphic unit-term (interval of rock strata) is matched by a corresponding geochronologic unit-term (interval of geologic time). Geochronologic units lack material referents, but are *divisions of time, each of which corresponds to the time span of an established chronostratigraphic unit.*

If one is to make chronostratigraphic subdivisions, it must be presupposed that one possesses the capability of recognizing time markers in the strata. Granting such a capability, then the number and kinds of local strata that belong in any particular chronostratigraphic unit may vary enormously from place to place. The only basis for including the strata is that they fall within the designated time span.

SOURCES: Blatt, Berry, and Brande, 1991; Lemon, 1990; Prothero, 1990; Schock, 1989; Watson, 1983; Whittaker et al., 1991.

Magnetostratigraphy and Magnetostratigraphic Units

The science of magnetostratigraphy is based upon the evidence that supports the following inference: in the geologic past the Earth's magnetic field has reversed its polarity numerous times. This inference may be made because the minute particles of ferrous minerals that are found in molten igneous material and in fine-textured sedimentary rocks orient themselves to the Earth's magnetic field. Once lava or magma has solidified, and once sediments have become lithified, the contained ferrous particles are not free to rotate, and they retain their telltale magnetic polarity unless they are deeply weathered or heated. Measurement of the magnetic polarity of particles within a sedimentary- or igneous rock establishes the orientation of the Earth's magnetic field when that rock formed. If such measurements are made through

a substantial stratigraphic interval, one finds that within certain subintervals the polarities of the rocks are approximately parallel to the Earth's current magnetic field, and within others, the polarities are opposite.

The interpretations of the behavior of the Earth's magnetic field are based on comparisons with spectrographic measurements of magnetic fields on the Sun. On the Sun, magnetic fields having one polarity disappear and are replaced by fields having the opposite polarity. As applied to the Earth, the implication is that the Earth's field can change by reversing itself and that a reversal involves a temporary loss of the magnetic field and its reappearance with polarity reversed. In a geologic sense, magnetic-polarity reversals are thought to have been virtually instantaneous. If this is true, then all sedimentary rocks deposited at a particular time, if they were capable of acquiring a magnetic orientation, acquired a common orientation parallel to the Earth's magnetic field at that time. The record of paleomagnetic reversals forms a means of fingerprinting sequences of sedimentary strata, by comparing the sequence and relative thicknesses of successive normal- and reversed polarity zones. Sedimentary sequences that lack major stratigraphic breaks and that represent sufficient time to display a unique sequence of paleomagnetic zones may be precisely and accurately correlated to other sequences of the same age.

A magnetostratigraphic unit is *a body of rock unified by specified remanent-magnetic properties and is distinct from underlying- and overlying such units having different magnetic properties.*

Diagenetic- or metamorphic processes may alter primary magnetic polarity signatures. Therefore, *although transitions between polarity reversals are of global extent, a magnetopolarity unit does not contain within itself evidence that the polarity is primary, or criteria that permit its unequivocal recognition in...other areas.* Other criteria, such as fossil content, radiometric ages, or physical tracing of beds must be used to establish the correlation of magnetostratigraphic units. In practice, distinctive sequences of magnetopolarity units are often correlated using pattern-matching techniques, especially where approximate ages are known.

POLARITY-CHRONOSTRATIGRAPHIC UNITS

In cases where magnetic polarity is inferred to be primary, then polarity-chronostratigraphic units, analogous to chronostratigraphic units, may be defined. A polarity-chronostratigraphic unit is *a body of rock that contains the primary magnetic-polarity record imposed when the rock was deposited, or crystallized, during a specific interval of geologic time.... Each polarity-chronostratigraphic unit is the record of the time during which the rock formed and the Earth's magnetic field had a designated polarity.* Polarity-chronostratigraphic units are analogous to chronostratigraphic units in that they are based on material standards (e.g., measurements on individual rock units) and encompass all rocks formed during a certain span of time. Polarity-chronostratigraphic units differ from chronostratigraphic units in that their fundamental units, polarity zones, form a continuous nonoverlapping series of units whose synchronous boundaries are based on the polarity reversals of the Earth's magnetic field.

Polarity-chronologic units are divisions of time, analogous to geochronologic units, but corresponding to the time spans of established polarity-chronostratigraphic units.

SOURCES: W. Alvarez, Arthur, Fischer, Lowrie, Napoleone, Premoli-Silva, and Roggenthen, 1977; Blatt, Berry, and Brande, 1991; Lowrie and W. Alvarez, 1981; Schock, 1989.

Other Kinds of Stratigraphic Units

Other kinds of stratigraphic units not described in detail here include **pedostratigraphic units** (consisting of *intervals defined by pedologic horizons, or soils*), **allostratigraphic units** (consisting of *lithologically similar stratigraphic units separated by bounding discontinuities*), **lithodemic units** (*stratigraphic units consisting of intrusive- highly deformed-, and/or highly metamorphosed rock*), **geochronologic units** (*divisions of time and therefore not, strictly speaking, stratigraphic units*), and **diachronic units** (*the unequal spans of time represented either by a specific lithostratigraphic-, allostratigraphic-, biostratigraphic-, or pedostratigraphic unit, or by an assemblage of such units*). The depositional sequences of sequence stratigraphy are not formally recognized by the North American Stratigraphic Code.

Naming Stratigraphic Units; The Stratigraphic Code

In this section we discuss the formal procedures that should be followed in publications that deal with stratigraphic units. These procedures should be followed not only for describing new stratigraphic units but also for the redefinition of any already existing units. For lithostratigraphic units, emphasis should be on lithologic characteristics and for biostratigraphic units, on paleontologic characteristics. For chronostratigraphic units, emphasis should be on characteristics relating to age and time correlation.

Descriptions of a stratigraphic unit should include (1) its name, the derivation of its name, and its type locality (this locality may be a place or a subsurface borehole); (2) rank of unit (such as Group, Formation,

Member, Bed); (3) historical background (original references, priorities, assurances against unnecessary duplication of already existing units); (4) geologic- and geographic identification (maps, columnar sections, structural sections, photographs, and diagnostic taxa, if biostratigraphic) of **stratotypes**, *the original, or subsequently designated, type representative of a named stratigraphic unit or of a stratigraphic boundary, identified as a specific interval or as a specific point in a specific sequence of rock strata, and constituting the standard for the definition and recognition of that stratigraphic unit or boundary* (*International Stratigraphic Guide*, 1976, p. 14; appears in References as Hedberg, 1976); (5) description of unit at type locality, such as thickness, lithologic- and biostratigraphic characteristics, structural attitude, discontinuities, nature of boundaries of unit (whether, for example, these boundaries are sharp, transitional, or unconformable), and any distinguishing or identifying features; (6) regional aspect, such as geographic extent, variations in thickness and lithologic characteristics, stratigraphic relationships with other stratigraphic units or with marker zones; (7) origin of rocks of unit; (8) correlation with other units; (9) geologic age; and (10) references.

In recent years, the International Commission on Stratigraphy of the International Union of Geological Scientists has embarked on a program to systematize global chronostratigraphic units. This involves the naming of formal boundary stratotypes for chronostratigraphic units that are recognized globally.

The *International Stratigraphic Guide* (Hedberg, 1976), designed as a guide to stratigraphic classification, terminology, and procedure, should be used as supplementary information by those contemplating formal descriptions of stratigraphic units.

SOURCES: Hedberg, 1976; International Subcommission on Stratigraphic Classification, 1987; North American Stratigraphic Code, 1983; Schock, 1989; Whittaker et al., 1991.

Establishing Lateral Relationships of Strata

So far we have been concentrating on the vertical succession of strata as seen in individual exposures or as recorded on logs of borings. After the vertical order of strata at a single point has been determined, one has established a firm basis for extending the investigation laterally. This is done in several ways, including direct visual tracing, matching of successions, and matching of marker beds.

Direct Visual Tracing

One of the most-satisfying methods for establishing the lateral continuity of a stratigraphic unit is direct visual tracing. This can be done in the field by walking on the unit as far as it is exposed. Other kinds of visual tracing include study of satellite- or aerial photographs or of other kinds of images recorded by instruments that were flown overhead or by making aerial observations from a helicopter or an airplane. If actual bedrock is not exposed throughout, the unit's presence may be indicated by distinctive morphologic characteristics of the areas underlain by it, by weathering residues, or by the plant communities that grow selectively upon the soil formed from the unit.

Matching of Successions

Where direct lateral continuity is broken or, whatever the conditions of exposures, where rapid checking is sufficient for inferring the lateral extent of stratigraphic units, the units can be extended laterally by comparing successions at various localities. This method is useful but it is not infallible. It works best only where one is familiar with the entire stratigraphic succession and where field time is available to check the continuity of a few units by the method of direct tracing. Errors can arise wherever several sequences that look alike are present, and the observer is either unaware of the existence of lookalike sequences, or, alternatively, if the observer knows how many lookalikes there are, but nonetheless is unable to distinguish between (or among) them.

Various computerized statistical pattern-matching routines have been developed for stratigraphic analysis. These may be of great help in establishing the correct matching of sequences that contain lookalikes.

Matching of Marker Beds

In many kinds of deposits, the original lateral extents of the strata vary enormously. Some strata give way to others within short distances. For example, a body of sandstone deposited in a former stream channel may be only a few tens or perhaps hundreds of meters wide and may be entirely surrounded by shale. Some strata, such as storm deposits, may extend for several kilometers, or may be discontinuous. Still other strata, such as coals, marine limestones, tephra, or sheets of extrusive igneous rock, may extend rather widely. In such cases one tries to match the widely persistent marker beds; indeed, they may be the only strata that one is able to match.

Marker beds may be characterized by any identifiable or measurable parameter, including particle size, color, mineral composition, fossil content, electric-log response, paleomagnetic polarity, chemical composition, or stable-isotope ratios of selected components. In some cases, no one bed serves as a reliable marker, but relatively thin successions of strata may exhibit a unique set of characteristics, such as a unique electric-log pattern.

After one has made the vertical subdivisions and has determined their lateral continuity, one comes to the subject of correlation, or establishing the time relationships among the strata. We examine some aspects of time next.

Criteria of Time Equivalence

The time relationships of strata provide the means for assembling numerous local successions into a coherent, worldwide stratigraphic framework of nonoverlapping units. Two objectives need to be realized; these can be met concurrently or successively. The first is establishing time equivalence among units within the area of their preserved local extents. The second is deciding upon the positions of the units in the standard geologic time scale.

Time Equivalence Within Local Areas

Time equivalence on a local scale can be established by the intertonguing of various strata, by tracing of marker beds that were deposited more or less synchronously throughout their extents (e.g., ash-fall layers or tektite horizons), and by tracing continuous seismic reflections on continuous seismic profiles.

INTERTONGUING

Where strata having homogeneous main bodies pass laterally into strata composed of different kinds of material, a transition zone may be present. Within this transition zone, tongues from one body of material may extend outward toward the main body of a second material. The result may be an interstratification of the two kinds of material. (See Figure 11-37.) Although at any one point, a tongue composed of one material is either older or younger than interbedded tongues composed of the other kind of material, the mutual intertonguing demonstrates that parts (or all) of the main bodies of contrasting materials are of the same age, thus are time equivalents.

The method of intertonguing can be used in two ways. As just explained, it works in a straightforward manner for the intertongued strata. In addition, by extension and use of a third unit, the time equivalence of unlike strata that do not intertongue with each other can be established. For example, the time equivalence of a pelagic chalk with a nonmarine conglomerate might be established by the method of intertonguing even though the chalk and the conglomerate are nowhere in direct contact. The chalk might intertongue with the silts and sands of a marine delta, whereas the topset sands, -silts, and -coals of this same delta might interfinger with the nonmarine conglomerate. This interfingering of the pairs via the common intermediate unit establishes the quasi-time equivalence of all three unlike units.

CONTINUITY OF SYNCHRONOUSLY DEPOSITED MARKER BEDS

Certain marker beds were deposited simultaneously (or virtually simultaneously) throughout their entire extents. Examples include vertically sedimented strata such as tephra or pelagic sediments; products of extremely rapid lateral sedimentation, such as individual turbidites or sheets of extrusive igneous rock; or products of growth within a widespread established environment, such as a coal swamp. (Not all coals are synchronous deposits; if the edges of the coal swamp migrated slowly in one direction through a long time interval, a continuous but nonsynchronous layer of coal would be deposited.)

Correct identification of a synchronously deposited key bed establishes its time equivalence. Similarly, the ages of strata enclosed between two such key beds are established as falling within the interval of time that elapsed between deposition of the key beds. For example, the time equivalence of fine-textured lake strata and coarse fanglomerates in the Triassic-Jurassic rocks underlying the Hartford basin in central Connecticut and Massachusetts has been established by lateral tracing of key "beds" consisting of extrusive basalt.

CONTINUITY OF TRACES ON CONTINUOUS SEISMIC-REFLECTION PROFILES

Establishing time equivalence from the continuity of a given trace on a continuous seismic-reflection profile record is based on the concept that the impedance contrast within the strata from which the seismic energy was reflected coincides with a time plane. Where such a time plane can be shown to match a depositional interface, the continuous traces on the seismic-reflection record may be treated as marker beds and traced directly. Some continuous traces on seismic-reflection records may be expressions of energy reflected from impedance contrasts that resulted from diagenetic reactions. Thus, for example, equal depth of burial and thus exposure of the strata to equal conditions of temperature and pressure may establish the conditions for precipitation of a given cement that is more or less the same age throughout. The continuous trace of a diagenetic interface can serve as a time surface.

In some cases, seismic data can be tied to subsurface lithologic- or well-log data at two or more points, espe-cially involving sonic logs. If this can be done, then the time equivalence of strata observed in different boreholes may be determined.

Position with Respect to Standard Geologic Time Scale

The position of a stratigraphic unit with respect to the standard geologic time scale can be established in three ways. These are (1) by establishing the time equivalence of local units with the established stratotype, (2) by use of fossils, and (3) by methods not involving fossils, such as radiometric-age determinations, stable-isotope ratios, or geomagnetic-polarity reversals.

TIME EQUIVALENCE OF LOCAL UNITS WITH ESTABLISHED STRATOTYPE

Once a stratotype for a part of the standard geologic time scale has been established, it is possible to employ the methods of time equivalence within local areas, and thus to determine the geologic age of strata near the stratotype. This procedure requires no additional comment.

USE OF FOSSILS

If two conditions are met, one can employ fossils for determining the position with respect to the standard geologic time scale of strata exposed anywhere on Earth. First, the appropriate stratotypes must be designated. Whether the stratotypes that have already been designated are altogether appropriate or not is subject to debate. Nonetheless, by the middle of the nineteenth century, the standard stratotypes for the abundantly fossiliferous parts of the geologic record were more or less completely established and the names that are still in use today had been proposed (Table 16-2). Second, one applies the principle that strata enclosing like fossils are of like geologic ages. Once it has been ascertained that strata at two or more locations contain fossils of a common age, then the time-equivalence of these strata is established.

METHODS NOT INVOLVING FOSSILS

Correlation with the standard geologic time scale may be made using radiometric dating. In this method, the ratio of an unstable isotope to its stable daughter isotope gives an absolute age. Pairs of isotopes used for radiometric dating include $^{14}C/^{14}N$, $^{40}K/^{40}Ar$, and $^{235,238}U/^{206,207}Pb$. ^{14}C decays relatively quickly to ^{14}N, hence the ratio of these two isotopes yields absolute-age determinations in specimens no older than about 80,000 yr. ^{40}K and $^{235,238}U$ decay much more slowly than does ^{14}C, and the ratios of these isotopes to their stable daughters permit

TABLE 16-2. Names of parts of the Geologic Time Scale (arranged in order of date proposed)

Name (Source)	Unit (Locality)	Date	Author
Tertiary	Strata in a special kind of mountain range, N Italy	1760	Giovanni Arduino
Jurassic System (Jura Mtns.)	Thick limestones (Jura Mtns., Switzerland)	1799	A. von Humboldt
Carboniferous (containing carbon)	Coal-bearing strata (N England)	1822	Rev. W. D. Conybeare and W. Phillips
Cretaceous (L., chalk)	Chalk (NW Europe)	1822	J. J. d'Omalius d'Halloy
Quaternary		1829	J. Desnoyers
Newer Pliocene	Subdivisions of Tertiary (W and S Europe)	1833	Charles Lyell
Older Pliocene			
Miocene			
Eocene (Pliocene; Gr., "major recent"); Micene; Gr. "minor recent"); Eocene; Gr., "dawn of the recent")			
Recent and "post-Pliocene" ("Recent" = "actual period;" "time when Earth has been tenanted by Man")	Subdivisions of "post-Tertiary"	1833	Charles Lyell
Triassic (originally Trias, for the three formations)	A group of three formations in S Germany, from top downward: Keuper marls, Muschelkalk, and Buntsandstein	1834	von Alberti
Silurian System (Silures, Welsh tribe of freedom fighters who repelled the Roman invaders)	Marine strata beneath Old Red Sandstone (Welsh borderland); "the world's oldest fossiliferous system" (Murchison)	1835	R. I. Murchison
Cambrian System (Cambria, Latin name for what is now Wales)	Lower part of the "killas and slate" of William Smith's geologic map of 1815; "the world's thickest system" (Sedgwick)	1835	Rev. Adam Sedgwick
Devonian System (Devonshire, county, SW England)	Marine equivalent of Old Red Sandstone	1838	R. I. Murchison and Rev. Adam Sedgwick
Permian System (Perm, a province in Russia)	Marine- and nonmarine strata overlying Russian units correlative with the British Carboniferous (Ural Mountains)	1841	R. I. Murchison (Permian supplanted the earlier terms Dyas and Penéen as used by J. J. d'Omalius d'Halloy)
Pleistocene	Substitute for the unit originally named the "Newer Pliocene"		Charles Lyell
Pleistocene	Post-Tertiary glacial deposits	1846	E. Forbes
Oligocene	Strata between Lyell's original Miocene and original Eocene	1854	E. Beyrich
Paleocene	Strata containing the oldest of the 5 assemblages of Tertiary flora; older than Lyell's original Eocene	1874	W. P. Shimper
Ordovician System (Ordovicii, a second tribe of Welsh freedom fighters who fought the Romans valiantly)	Part of the "killas and slate" of the W. Smith 1815 map; the disputed strata at the top of Sedgwick's Cambrian and base of Murchison's Silurian	1879	C. Lapworth
Holocene	Time elapsed since retreat of last continental glaciers		

SOURCES: Lyell, Charles, 1830–1833, Principles of geology: London, John Murray, 3 volumes.

Sedgwick, Rev. Adam; and Murchison, R. I., 1839, On the classification of the older stratified rocks of Devonshire and Cornwall: Geological Society of London, Quarterly Journal, v. 3, p. 121–123.

Wilmarth, M. G., 1925, The geologic time classification of the U.S. Geological Survey compared with other classifications, accompanied by the original definitions of Era, Period, and Epoch terms: U.S. Geological Survey Bulletin 769, 138 p.

determination of the absolute ages of specimens that are millions or even billions of years old.

The ratio of ^{18}O to ^{16}O can be used to make stratigraphic correlations. The $^{18}O/^{16}O$ ratio in seawater is strongly affected by temperature and by sequestering of ^{16}O in glacial ice. Thus, during glacial- and interglacial periods, the $^{18}O/^{16}O$ ratio in marine fossils changes systematically. The patterns of change in the isotope ratio in two or more stratigraphic sections can be used to correlate marine strata.

In a comparable way, the stable isotopes of carbon, ^{13}C and ^{12}C, can be used for stratigraphic correlation.

Strontium-isotope stratigraphy works differently. The $^{87}Sr/^{86}Sr$ ratio in seawater has changed continuously and roughly linearly through geologic time (at least since the Eocene). Therefore, if the $^{87}Sr/^{86}Sr$ ratio of calcite in a marine fossil of Eocene age or younger is known precisely, a single sample can give an absolute date by its position on the $^{87}Sr/^{86}Sr$ decay curve.

Elemental composition of strata may be used to determine the absolute age of strata in yet-another way. Some occurrences of rare elements are so unusual that they can be used as geochemical marker beds. One of these is the iridium anomaly at the Cretaceous-Tertiary boundary. This is a global marker of known age and permits precise location of the Cretaceous-Tertiary boundary.

Geomagnetic-polarity reversals may be used to correlate strata to the standard geologic time scale as follows. The alternating bands of normal-polarity strata and reversed-polarity strata form recognizable patterns because width of these bands is proportional to the span of time they represent. Thus polarity-reversal patterns underlying different parts of the ocean floor can be correlated. Because the absolute ages of particular polarity intervals are known, identification of particular sequences of polarity bands permits correlation to the geologic time scale.

SOURCES: Arthur and others, 1983; Hedberg, 1976; Lemon, 1990; Mahaney, 1984; Odin, 1982; Schock, 1989; Sheriff, 1989.

Stratigraphic Synthesis

After one has determined the lateral extents of stratigraphic units and has assigned them geologic ages, then the way is clear for making broader-scale stratigraphic syntheses. Items to be discussed in this category include relationship of strata to changes of sea level, identification and interpretation of breaks in the stratigraphic record, and reconstruction of paleogeography. Chapter 17 is devoted to the broad subject of basin analysis.

Relationships of Strata to Changes of Sea Level

As emphasized in Chapter 6, changes of sea level exert profound effects on the deposition of strata. Renewed emphasis is being placed on study of the features in strata that result from changes of sea level. Recall from Figure 6-7 that the position of the sea can rise or fall in the vertical direction and that of the shoreline can shift landward or seaward. The unqualified term "regression" means different things to different groups of geologists. To geologists studying Pleistocene changes of sea level associated with the waxing and waning of continental glaciers, "regression" means emergence. To geologists working in the Tertiary of the Gulf Coast of the United States or in the Cretaceous of the Rocky Mountains, for example, "regression" implies progradation (usually during submergence!). One proposal is to use depositional regression for the effects of progradation. We prefer to avoid the term "regression" and instead to use an unambiguous substitute, such as progradation or emergence, as the case may be.

SOURCES: J. R. Curray, 1964.

EVENT STRATIGRAPHY

The concept of *event stratigraphy* encompasses several contrasting approaches to the study of sedimentary deposits. These several approaches hold in common the idea that attention should be focused on *events* and what they reveal about stratigraphy and sedimentation. Event stratigraphy explicitly argues that the principle of *uniformitarianism* does not completely describe sedimentary processes. Event stratigraphy can be conveniently classified according to the scale of events. Relatively frequent events (still rare at the scale of a human lifetime), such as major storms and major turbidity currents, are studied at the lowest level of analysis. The second level is concerned with events that are less common and whose effects are more profound. These include shifts in the rate and/or magnitude of sea-level change. At the third level are extremely infrequent catastrophic events whose effects are felt globally: the impacts of giant meteors, very large volcanic eruptions, and the like.

The three levels of analysis correspond to different kinds of events that may be used in different ways to facilitate stratigraphic analysis.

The deposits laid down by storms or by turbidity currents, although they represent only a minute portion of the Earth's history, stand out from "background" sedimentation both physically and in their implications. Storm deposits may form resistant ledges in outcrop, may contain abundant well-preserved fossils, and may interrupt and complicate sedimentary cycles. Storms may remove the deposits of "background" sedimentary processes and may leave in their stead highly condensed sections consisting primarily of the coarsest particles. Turbidites transport large volumes of sediment from shallow- to deep water, in the process rearranging the sedimentary particles into the distinctive patterns of Bouma sequences. Turbidites may disrupt the regular cyclic sediment patterns of fondothems, or may generate new cyclic patterns resulting from the interaction of lateral- and vertical sedimentation. Turbidites may cause metastable carbonate particles (composed of aragonite or high-magnesian calcite) to be preserved at depths far below their compensation depths. Thus, sedimentary successions dominated by storms or by turbidity currents can only be properly understood if the infrequent high-energy events are accounted for.

The second level of event stratigraphy refers to a refinement of the ideas about depositional cycles and relative sea-level changes discussed in Chapter 6. These concepts are based on the idea that depositional cycles controlled by eustatic changes ought to be correlatable globally. Thus, if one can recognize the products deposited at the point of maximum submergence of the sea in a series of boreholes or of outcrops and show that they were created during the same eustatically controlled facies cycle, then these points define a time plane (which appears as a line on a stratigraphic diagram). This inference is possible because the point of maximum transgression (maximum landward shift of facies) is a unique point that is readily recognized in a stratigraphic succession. The same is true of the point of maximum emergence.

The power of this approach to stratigraphy is that time planes may be established within strata that lack diagnostic fossils and that cannot be dated radiometrically or by other means. The keys in this second-level event stratigraphy are (1) identifying the points of maximum transgression and of maximum emergence and (2) making sure that a set of points of maximum transgression or -emergence do in fact belong to the same eustatically controlled cycle.

The highest level of event stratigraphy is concerned with very rare catastrophic events. Such an event is the giant-meteor impact that is inferred to have occurred at the end of the Cretaceous, causing a mass-extinction event that wiped out the dinosaurs as well as many other taxa. Whether or not this interpretation is correct, it serves to illustrate the principle that catastrophic effects (the mass extinction) may be best explained with reference to catastrophic events that are outside our modern experience and therefore outside the purview of uniformitarianism (the giant meteor impact).

SOURCES: Einsele and Seilacher, 1982; Füchtbauer, 1988; Griffiths, Kopaska-Merkel, and Schott, in press; Kopaska-Merkel, 1987a; Prothero, 1990; Van Wagoner and others, 1990.

Breaks in the Stratigraphic Record

As noted in several preceding sections, sedimentary strata record the passage of time. However, some strata record more time than others. Not all sediments accumulate at a regular rate, as for example, does the well-sorted sand that passes through the neck of an hourglass. Many sedimentary processes operate only intermittently. In addition, parts of the stratigraphic record, whether they accumulated intermittently or not, have been eroded and thus still other gaps have been created. The stratigraphic record may be considered as a series of deposits formed by discrete events. We now examine the discontinuous aspects of sedimentation. Following this we consider the nature and significance of gaps in the record, represented by surfaces of unconformity, a subject already broached in Chapters 5 and 6.

DISCONTINUOUS ASPECTS OF SEDIMENTATION

Here we take up again a subject introduced in Chapter 5. It is probably correct to infer that only the thinnest sedimentary laminae approach the state of true continuous sedimentation. Any sedimentary stratigraphic unit of mappable scale is at best an incomplete record of time. From the previous chapters that described the operation of sedimentary environments, recall that many sediments accumulate only now and then. Some sediments are delivered only during storms. Others depend on the season of the year. Sediment, no matter how intermittently supplied, is totally passed from some parts of the sea floor and comes to rest at other sites. In these and other examples, the average rate of sedimentation may be very slow, yet the layers that do accumulate are deposited individually within short time intervals. The situation is somewhat akin to a desert area having a mean annual rainfall of 5 cm/yr but that receives a cloudburst of 60 cm of rainfall every dozen years and no rain in between. *An interval of time that elapses while strata are not actively accumulating within an environment of deposition* is a hiatus or *diastem*. Use of the term diastem is appropriate when the durations of the nonrepresented time intervals are short (in a geologic sense) and where no significant thicknesses of sediment accumulated close by. Many geologists believe that every bedding-surface parting coincides with a diastem.

Two phenomena associated with the discontinuous nature of sedimentation relate to the scale of observation. These are the relationship between the frequency and magnitude of events that may deposit or erode sediments and the relationship between the completeness of the stratigraphic record and the amount of elapsed time studied.

As a general rule, sedimentary processes that are episodic vary in intensity in a predictable way: the intensity and frequency are inversely related. This is the principle behind the calculation of the so-called hundred-year flood for river basins. In the case of storm deposition, for example, minor storms are frequent, more-severe storms occur less often, and extremely violent storms are rare. Thus sedimentation in many environments is characterized not only by intermittent deposition but by a hierarchy of relatively frequent minor depositional- or erosional events and relatively infrequent major depositional- or erosional events. Observations of the modern world cover a time span of only a few centuries, so events that occur significantly less often than once per century are virtually unknown. These relatively infrequent events may have affected deposition or erosion in sedimentary environments, which commonly persist for millennia or longer.

CONCEPT OF UNCONFORMITY

As long as a given sedimentary environment accumulates strata (even if its times of strata building are liberally laced with diastems), the succession that is deposited is described as being *conformable*. The continuity of the strata can be broken in several ways. In a simple case, a subsiding area that has been collecting sediments is uplifted, some of the sediments are removed, and the area afterwards subsides and accumulates sediment once again. This series of events creates *a relationship* described as one *of unconformity*. *The surface cut on the older materials during removal of some older materials prior to deposition of younger covering strata* is a **surface of unconformity**. This subject was introduced in Chapter 5, but we explore it in greater detail here.

The relationship of unconformity can involve many geologic factors. Moreover, surfaces of unconformity are sites where valuable natural resources can be concentrated. Finally, surfaces of unconformity provide important information for deciphering the geologic history of a region.

Factors Involved. Some of the factors connected with a surface of unconformity can be determined by direct observations of tangible geologic evidence. Such evidence includes the kinds of materials present below and above the surface of unconformity, the attitudes of strata, and the configuration of the surface itself. Other factors involve indirect geologic aspects that have to be inferred. These include the length of time that elapsed between the time represented by the materials below and above the surface of unconformity and changes in the major environmental realm (atmosphere or hydrosphere).

The materials found next to a surface of unconformity can include anything found at the surface of the Earth. Where sedimentary strata lie beneath the surface of unconformity, it is important to establish whether, before their subsequent burial, these older strata were or were not involved in any geologic structural events. The following kinds of questions need to be answered. Was the regional attitude of the older strata disturbed in any way? If not, then the surface is named a *disconformity* (or, if physical evidence of unconformity is not present, the surface is named a *paraconformity*). If the regional attitude of the older strata was disturbed, then a lack of parallelism will be apparent between the older, disturbed strata and the younger covering strata. This lack of parallelism has prompted the designation of such examples as *relationships of angular unconformity* or of *discordance* (Figure 16-10). Do any features such as faults or dikes cut the older strata and are they themselves truncated by the surface of unconformity? If the answer is yes, then obviously such features will not extend upward beyond the surface of unconformity, and they indicate major geologic events that occurred after deposition of the strata below the surface of unconformity and before formation of the unconformity surface. Radiometric age dates or paleotemperature indicators (such as color of conodont elements or even present-day geothermal gradient) may show discontinuity at unconformity surfaces.

Where uplift and erosion have been extremely extensive, all the strata of sedimentary rock may have been removed and the underlying basement consisting of metamorphic- or igneous rocks may become exposed at the Earth's surface. Later subsidence and covering with sedimentary strata of a surface of unconformity cut across basement rocks results in a *relationship of nonconformity*.

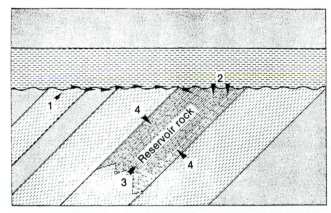

FIGURE 16-10.   Schematic cross section of strata above and below a major surface of angular unconformity. The reservoir rock, which is sealed by the unconformity (wavy horizontal line), now dips to the left at a high angle. Four possible sources of hydrocarbons for this reservoir (numbers on the diagram) are explained in text.

INTERPRETING STRATIGRAPHIC BREAKS

As mentioned previously, the analysis of a stratigraphic break involves several kinds of information. First, one seeks to determine how much time elapsed while the local record was being interrupted. Rarely is it possible to formulate an answer in terms of years. More commonly, one tries to compare the strata above and below with a succession elsewhere that seems not to have been interrupted. By doing this, one can make judgments based on the differences between the two successions.

Second, it is important to establish whether the interruption in the record of marine strata resulted from conditions on the sea floor or whether it involved an interval of uplift, subaerial exposure, and resubmergence. Although in the past, the belief was widely held that an ancient sea would always deposit sediment, we know from modern geologic observations of the present sea floor and sediments beneath it, that in many areas, no sediment is accumulating. Currents may be so strong and so persistent in some areas that they keep sediment from accumulating on the bottom through broad areas and during long periods of time. When no new particles are being added to the bottom, the materials forming the sea floor may react chemically or biochemically with seawater. In this way, crusts of manganese oxides, glauconite, or phosphatic material may form. The accreting chemical- or biochemical deposit may lock into itself a few large skeletal fragments that fell from above and were not swept away. Examples include shark's teeth; fish bones, scales, and teeth; and (during the Cenozoic) earbones of whales. Under conditions not yet fully understood, carbonate cements may form at the water/sediment interface. (See Chapter 3.) Geologic structures may be formed in strata beneath the sea, and parts of these structures may be truncated at the sea floor. Tidal currents and various gravity-powered sediment flows can incise channels into the sea-floor sediments.

Sediments occurring below the modern sea floor exhibit numerous stratigraphic breaks, many in places that cannot ever have been subaerially exposed, but must have always existed well below sea level. Stratigraphic breaks spanning millions of years are found in sediments underlying the abyssal plains, in water depths of several kilometers. For example, all upper Eocene strata, representing a time span of about 10 million years, are missing from a borehole drilled beneath 6 km of water off northwest Australia by Ocean Drilling Program Leg 123 in 1988.

Subaerial exposure of marine sediments results in oxidation of pyrite, dissolution of aragonite and high-magnesian calcite, precipitation of calcite cement, formation of weathering residues (including such things as bauxite and laterite, and release of chert and silicified shells from limestones), creation of karst features (enlarged joints, caves, sinkholes, and so forth), and the sculpting of subaerial landscapes (but many subaerial morphologic features are not easily distinguished from such things as channels cut in a peritidal complex). The presence of basement rock other than oceanic basalt beneath marine strata implies uplift and erosion that were so great that subaerial exposure almost certainly was responsible.

The determination of the conditions of origin of a given stratigraphic break within marine strata may not be easy to make. However, this difficulty should not deter one from seeking critical evidence. Subaerial exposure may or may not have taken place; when it did happen, it usually left behind useful clues.

The other direct tangible aspect of a surface of unconformity is the configuration of the surface itself. One should ask the following questions. Is it a smooth plane? Or does it display relief such as that associated with buried hills, ridges, shore cliffs, and stream valleys? In eroded carbonate rocks, are any sink holes, solution pits, caves, or solution-enlarged joints present? (See Figure 3-26.)

The indirect aspects involved with surfaces of unconformity are more or less independent of the directly observable geologic characteristics. Two indirect points to be established concern (1) the length of time represented by the unconformable contact and (2) whether or not any major change of environmental realm was involved. This means, for example, finding out if marine strata were uplifted to become land, subaerially eroded, and later resubmerged. The time represented by the unconformable contact may be conceptually divided into two parts: time represented by sediment that had been deposited before formation of the surface of unconformity and that was eroded during unconformity-surface formation, and the subsequent time period during which the surface of unconformity formed and during which little or no sediment was deposited. Both the depth of erosion and duration of exposure of an unconformity surface may vary significantly from place to place. Careful mapping may allow estimation of the time of onset of unconformity-surface formation if the surface of unconformity formed roughly synchronously but depth of erosion varied substantially, as is commonly the case.

The significant factors pertaining to surfaces of unconformity in regional geologic studies are (1) the continentwide extent of many stratigraphic discontinuities and (2) the covering and concealing of geologic features that have been buried after the sea resubmerged a landscape. As explained in Chapters 5 and 6, continentwide surfaces of unconformity have been employed as the basis for defining Sequences. Beneath each sequence is an

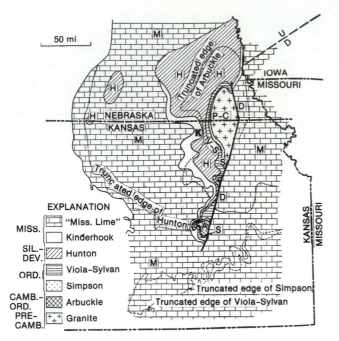

FIGURE 16-11. Paleogeologic map showing relationships at pre-Pennsylvanian land surface in eastern Kansas and southeastern Nebraska. The outcrop pattern proves that folding, faulting, and erosion took place after the Mississippian limestones had been deposited and prior to deposition of the Pennsylvanian covering strata. The Precambrian rocks, which stood in relief above the Paleozoic strata, are known as the Nemaha Ridge. (A. I. Levorsen, 1967, fig. 13-13, p. 606, from Geology of Petroleum, 2nd ed., copyright 1967 by W. H. Freeman and Co. Printed with permission.)

old land surface for which it may be possible to construct a paleogeologic map (Figure 16-11).

SOURCES: Blatt, Berry, and Brande, 1991; Lemon, 1990; Levorsen, 1960; Prothero, 1990.

ECONOMIC DEPOSITS RELATED TO SURFACES FORMED BY SUBAERIAL EXPOSURE

Surfaces of unconformity that resulted from subaerial exposure are commonly associated with important deposits of valuable natural resources. Two such examples are (1) petroleum traps in which the sealing unit discordantly overlies the truncated reservoir and (2) various residual accumulations and deposits of certain metals that were secondarily enriched as a result of weathering.

Petroleum Traps in Which a Sealing Unit Discordantly Overlies a Truncated Reservoir.   The reservoirs of many of the giant petroleum fields in North America are truncated abruptly upward by surfaces of unconformity. (In North America a giant petroleum field is generally defined as one having recoverable reserves greater than 100 million bbl of oil or greater than 1 trillion ft^3 of gas.) Examples include the continent's two largest, Prudhoe Bay, Alaska; and the East Texas field; as well as the Oklahoma City field, Oklahoma; the Hugoton-Panhandle

field, Texas, Oklahoma, and Kansas; and the Golden Lane fields, Veracruz Province, Mexico. In each of these fields, and in countless others similar to them, the tilted edges of a reservoir formation of marine origin have been eroded and discordantly and unconformably overlain by an impermeable sealing stratum, in many (but not necessarily all) cases, also a formation of marine origin. Examples of seals that have trapped petroleum are anhydrite and shale.

The history of such petroleum accumulations involves two episodes of subsidence and of deposition that were separated by a time of uplift and of subaerial erosion. The older succession of marine strata was tilted and subaerially eroded. In most of the examples cited, at the time the strata that now are the reservoirs were at the surface, no petroleum was present in them. In a few cases, however, petroleum was present, but when the reservoir was exposed at the former land surface, the petroleum leaked out. Along some such ancient natural seeps, leaking oil may have become such a sticky, tarry residue that it sealed the outcrop edge against further escape of petroleum.

In the carbonate-rock reservoirs of many giant fields, dissolution of the limestone by circulating fresh ground water created a vast network of open spaces that later were filled with petroleum to become the reservoir. Clearly, such carbonate rocks could not possibly have become petroleum reservoirs until after their pore space had been created during a time of uplift and chemical weathering.

Eventually the uplifted area subsided anew and the sea once more submerged the region. After continued subsidence and deposition of a thick, impermeable cover, the formerly open reservoir became closed. At this point it was able to stop leaking petroleum (if that is what it was doing unchecked while it was open to the surface), or to start trapping petroleum (if it had not done so previously).

How the petroleum (or the solid *kerogen*, a raw material capable of generating petroleum) got into these truncated reservoirs is not well known. One explanation is that the petroleum was generated in some nearby fine-textured "source bed" and then migrated into the reservoir via the channelway along the unconformable contact (No. 1 on Figure 16-10). For example, if such a "source bed" is identified as a unit lying below the surface of unconformity, then the idea is that the petroleum migrated laterally out of the source bed, moved up the dip until it came to the sealing layer, then traveled along the unconformable contact beneath the sealing layer, and finally entered the reservoir from its updip side. Another possibility is that the petroleum was generated in the unconformably overlying fine-textured impermeable sealing formation and, once formed, filled the subjacent reservoir (No. 2 on Figure 16-10). If this were the case, then the unconformably overlying unit would have had to perform two contrasting functions. Its double duty would be first, to generate the petroleum fluids and then to allow them to flow downward into the reservoir; and second, and by total contrast, after the petroleum fluids had moved downward into the reservoir, to prevent them from escaping upward out of the reservoir.

Still another possible explanation is that the makings of petroleum, but not the actual petroleum itself, were dispersed in the reservoir formation, but that these materials did not become mobile petroleum fluids and thus were not concentrated in the trap until after the truncated reservoir had been resealed and deeply buried. This alternative is possible only if the petroleum-making precursory materials were in a chemically stable, physically immobile, probably solid form. In such a form they could survive both exposure at the surface and the inevitable flushing with fresh ground water that would have taken place. Only one substance is known that could have survived under these conditions: kerogen. The method(s) by which the kerogen could have become dispersed in the reservoir formation are not fully understood. In a reservoir composed of terrigenous sediment, the kerogen may have accumulated along with the quartz particles as a result of having been recycled by erosion of an ancient kerogen-bearing shale. In a marine carbonate reservoir, the kerogen may have come from the burial diagenesis of raw organic matter of marine origin, which was initially deposited along with the marine carbonate sediments. It has been suggested that ooids contain sufficient quantities of organic material when they are formed to fill all of the primary interparticle pore space in an ooid grainstone with hydrocarbons after maturation (No. 3 on Figure 16-10). In a variant of this alternative, the petroleum-making precursory materials were dispersed in a rock unit stratigraphically adjacent to the reservoir formation, or even in an organic-rich lateral equivalent of the reservoir unit (No. 4 on Figure 16-10).

In either case, the idea is that during the first cycle of subsidence, the strata were not buried very deeply. The exact depth of their burial may not be known, but it is presumed to have been less than the depth required to convert the kerogen into petroleum (possibly 1.5 to 2 km). During the second episode of subsidence, however, the depth of burial was great enough to enable geothermal heat to convert the kerogen to petroleum. Such a delayed conversion to petroleum is possible via kerogen because kerogen is both a solid (hence stays put even in porous and permeable surroundings) and is chemically resistant (not easily oxidized).

SOURCES: Chenoweth, 1972; G. E. Dorsey, 1933; F. J. Gardner, 1940; H. P. Jones and Spears, 1976; Levorsen, 1927, 1931a, b, 1933, 1934, 1960; Morgridge and W. B. Smith, Jr., 1972; Sanders, 1982; Viniegra and Castillo-Tejero, 1970; Walters, 1946; Webb, 1976.

Mineral Deposits Enriched During Weathering. During an episode of subaerial exposure, numerous geologic processes can operate both at the Earth's solid surface and in the shallow subsurface as a result of changes in the position of the ground-water table. As a result, metallic elements in sulfide minerals, which are easily oxidized, may become concentrated as oxide minerals, and the silicate minerals in rocks may be converted into oxides such as bauxite. These changes do not involve sedimentary deposits, per se; hence, we merely mention them. A kind of change related to the weathering of silicate rocks that does affect sedimentary deposits is the release of valuable accessory minerals that might become concentrated as placers and the removal of magnetite and resulting concentration of ilmenite (Chapter 2).

SOURCE: Blatt, Berry, and Brande, 1991.

Paleogeographic Reconstructions

One of the important objectives of studying sedimentary strata is to reconstruct the geographic conditions that prevailed when the strata were deposited. This usually means figuring out the broad distributions of ancient lands and seas and then determining the relief and drainage of the land and the water depth and conditions within the ancient sea. Inferences about ancient climates and about ancient latitudes are especially important. Information about the Earth's magnetic field at the time of deposition enables one to compare inferred paleolatitude (geographic) with paleomagnetic latitude.

We illustrate some of the problems of paleogeography by considering three topics: (1) comparing the preserved edge of a unit with its original extent; (2) islands, real and imaginary; and (3) ancient equatorial zones.

PRESERVED LIMIT COMPARED WITH ORIGINAL EXTENT

After one has traced a stratigraphic unit to its limits in a given direction, one seeks the answer to two questions: (1) Does this limit coincide with the original depositional edge of the unit? (2) If the preserved edge and the original depositional edge do not coincide, is it possible to determine how far the preserved edge lies from the depositional edge?

Characteristics of Original Depositional Edges. Two common kinds of depositional edges are generally easy to recognize: (a) the margin of a nonmarine depositional basin having a border fault that was active during sedimentation, and (b) the edge of the sea (that is, an ancient shoreline). In each case, predictable patterns of sediment were arranged in approximately parallel belts that were controlled by an initial depositional slope away from the edge in question. Generally, the trend of this slope parallels the fault or the shoreline. Moreover, this depositional slope extended from the apex of fans to the bottom of a playa or a long-lived lake (in the case of a nonmarine basin ending at an active fault, see Figure 14-7) or from the shoreline out onto the continental shelf or from a reef into a deep basin at the edge of the sea. *The trend of the boundaries between slope-controlled belts of sediment* is defined as **depositional strike.** *The inclination of the depositional slope measured in a vertical plane at right angles to depositional strike* is **depositional dip.** In effect, then, depositional strike and -dip describe the attitude of the depositional slope, which was the shape of the land surface of the nonmarine basin or of the marginal parts of a sea floor.

The original edge of a nonmarine basin bounded by a fault that was active during deposition is characterized by coarse fan sediments and by the actual fault itself or, alternatively, by the steep basin-marginal land surface close to the fault. Former edges of the sea, of course, are marked by estuarine sediments, by peritidal deposits, by beach sediments, by marine deltas, by buried morphologic features carved by waves along a rocky shore (such as sea stacks, for example), or by reefs.

Distance from Preserved Edge to Depositional Edge. From the preserved sedimentary record it is not always possible to determine just how far the preserved edge lies from the original depositional edge. In examples involving marine strata, the sedimentary- and faunal data may support an approximate estimate of former depth and distance from the ancient shoreline. More commonly, however, if the unit being studied has been tilted at low angles, it is possible to gather facies information from various localities along the present tectonic strike. This information can then be evaluated to determine whether or not the facies boundaries, which coincide with original depositional strike, match present tectonic strike. If one can follow the present tectonic strike and stay within a single facies belt, then one can infer that the present tectonic strike is subparallel to the ancient depositional strike. One may not be able to reconstruct the distance to the former shoreline, but one can know with some assurance that the overall trend of the modern tectonic strike parallels the former shoreline (as inferred from ancient depositional strike). Moreover, if these two strike trends are not parallel, then one can infer the trend of one or more postdepositional structural features (Figure 16-12). If the preserved edge crosses a sufficient number of facies belts, then one can make accurate estimates of the location of the former shoreline.

Other clues to possible former extent come from provenance studies of the strata compared with distinctive rock types that may now be exposed nearby (Chapter 2).

Analysis of the significance of the edge of a truncated stratigraphic unit is routinely made in subsurface mapping. By contrast, only rarely has such an approach been applied to outcrop edges of exposed formations.

SOURCES: Chenoweth, 1967, 1972; Levorsen, 1943, 1960, 1967; A. B. Shaw, 1964; Walters, 1946.

ISLANDS: REAL AND IMAGINARY

The geologic record contains examples of islands that projected out of ancient seas. A few convincing examples of such islands are known from the Paleozoic rocks in the United States: the Cambrian Baraboo Island of southcentral Wisconsin; the Early Paleozoic Ozark Island, Missouri; the Pennsylvanian Wichita-Amarillo Island, Oklahoma and Texas; and the Pennsylvanian Uncompaghre-San Luis Island, southwestern Colorado.

The Baraboo Island, Wisconsin. The Baraboo Island (Figure 16-13) can be definitively reconstructed from exposures within and around the elongate Baraboo Range whose modern-day relief is about 200 ft (65 m) (ridge tops at an altitude of 1100 ft [365 m]; altitudes of adjacent lowlands, about 900 ft [300 m]). Upper Cambrian conglomerates and sandstones surround a linear synclinal ridge underlain by resistant Precambrian quartzite, which formed a group of islands about 25 miles (40 km) long and 11 miles (18 km) wide. Quartzite clasts range in size from 30 ft (9 m) to 0.08 inches (2 mm) in diameter. Of these, only those smaller than 4.5 ft (1.5 m) have been rounded and some of these have been found within beds of Cambrian sandstone as much as 1000 ft (300 m) from the ancient shore cliffs. Ubiquitous horizontal lamination compares closely with relief peels made from modern shoreface sediments inferred to be ancient storm deposits.

Evidence demonstrating the existence of the two examples of the Pennsylvanian-age islands is presented in figures located in other chapters. (See Figure 13-39 for the Uncompaghre-San Luis Island of Colorado, and Figure 17-8 for the Wichita-Amarillo Island emergent in the Anadarko Basin.)

SOURCES: Dalziel and Dott, 1970; Dott, 1974; Raasch, 1958.

Case of the Phantom Island. Improper use of stratigraphic data can result in the conclusion that an island emerged where, in fact, just the opposite, a submergence, is what actually happened (Figure 16-14). In this example, the original island was submerged by marine strata that graded from sand (later sandstone) through

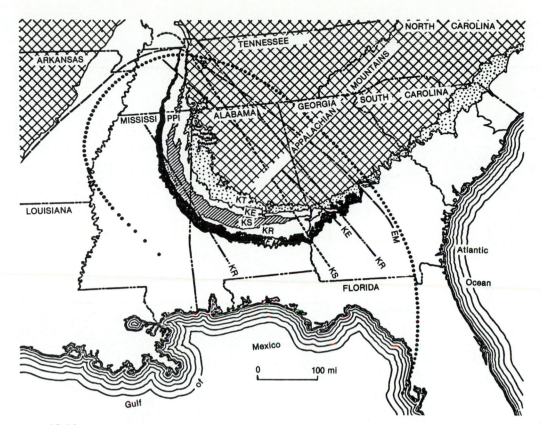

FIGURE 16-12. Geologic sketch map of parts of southeastern United States showing curving out-crop belts of Cretaceous formations at SW end of Appalachians and inferred depositional strikes of each major unit (defined by various line patterns connecting NW and SE points of like facies, one each within the units marked EM, KE, and KS; two from the KR; and none from the KT; limits of sand in the KT shown by large black dots). Cross-hatched pattern, pre-Cretaceous rocks, not differentiated. Cretaceous units are: KT, Tuscaloosa Formation; KE, Eutaw; KS, Selma; and KR, Ridley. Eocene unit, EM is Midway Formation. Many authors have inferred that the cresentic outcrop belts of these formations are parallel to the Cretaceous shorelines in the region. The fact that the depositional strike lines trend NW-SE, however, suggests that the trend of the Cretaceous shore-line was also NW-SE. The curving outcrop belts have resulted from post-Cretaceous doming of the Appalachians along an axis that strikes NE-SW and plunges gently SW and from post-doming downdip erosion. Thus, if one were to start from a point in NW Alabama and proceed along the outcrop of a given unit shown in a counterclockwise direction to the Alabama-Georgia border, one would begin in a facies deposited closest to the former shoreline and then encounter facies deposited progressively farther from the former shoreline. One would encounter the facies most distant from the shoreline in central western Alabama. From there eastward, one would encounter facies deposited progressively closer to the former shoreline. (A. I. Levorsen, 1943, fig. 15, p. 904.)

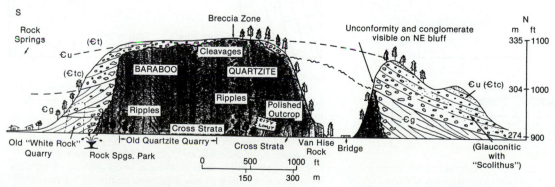

FIGURE 16-13. The Baraboo Island, Wisconsin.

A. North-south profile along west side of Upper Narrows of Baraboo River, Rock Springs, Wis-consin. The truncation of the upper Cambrian conglomerates and sandstones against the vertical Baraboo Quartzite (Precambrian) and the distribution and shapes of the gravel prove that the syn-clinal ridge of the Precambrian rock stood as an island in the Cambrian sea.

623

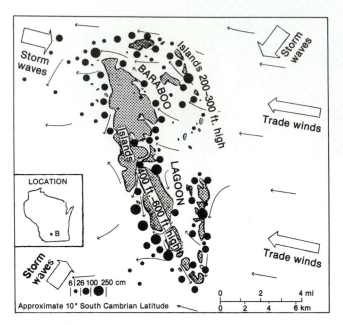

FIGURE 16-13.   (*Continued*)
   B. Reconstructed map of the Baraboo Island group showing sizes of gravel, inferred wind patterns, directions of storm waves, and paleolatitude. (R. H. Dott, Jr., 1974; A, fig. 2, p. 244; B, fig. 3, p. 244.)

silt (later shale) to carbonate sediment (later limestone). The boundaries separating these kinds of material are not time planes; rather, the time planes cut across these boundaries. (See Figure 16-14, A, B.)

By making the assumption that the lithologic boundaries are time surfaces, one geologist drew three paleogeographic maps for each time interval in which the

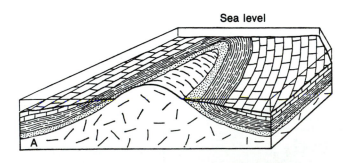

FIGURE 16-14.   Real and phantom islands.
   A. Island composed of Precambrian basement rocks being submerged by rising sea, schematic block showing sediment pattern on sea floor. Marine sediments consist of marginal sand, nearshore mud, and offshore carbonates.
   B. Island has been completely buried; schematic profile only.

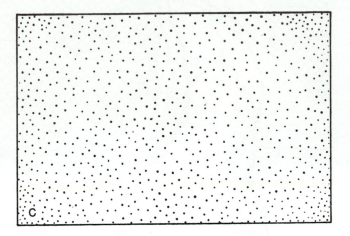

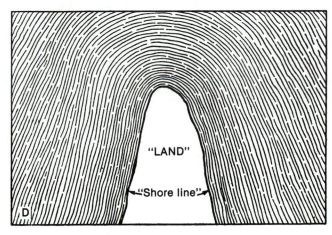

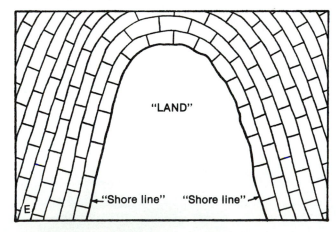

FIGURE 16-14.   (*Continued*)
   C through E, Paleogeographic maps drawn by geologist who supposed that the boundaries between the sediments are time surfaces.
   C. Paleogeographic map of sand sea.
   D. Paleogeographic map of situation when nearshore mud was deposited. Area underlain by sand is inferred to mark an island.
   E. Paleogeographic map of situation when offshore carbonates were deposited. "Island" has expanded to include the area underlain by the mud. Further explanation in text. (A. B. Shaw, 1964; A, fig. 9-2 (a); B, fig. 9-2 (b), p. 59; C, fig. 9-3, p. 60; D, fig. 9-4, p. 60; E, fig. 9-5, p. 61.)

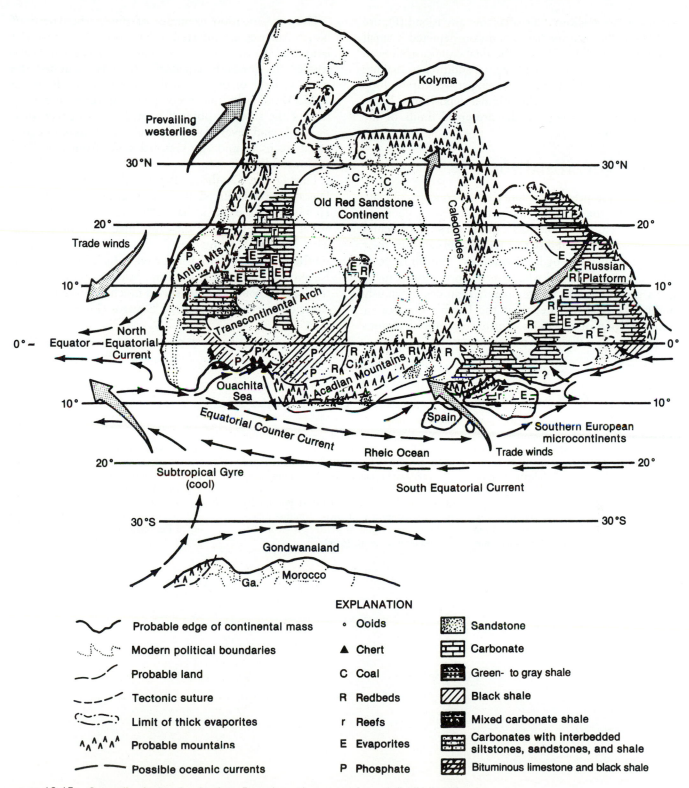

FIGURE 16-15.   Generalized map showing Late Devonian paleogeography and distribution of major lithofacies. (F. R. Ettensohn, 1985.)

initial condition showed a sea lacking any island (Figure 16-14, C), the intermediate condition depicted a small island (Figure 16-14, D), and the final condition showed a larger island (Figure 16-14, E). In addition to making an erroneous correlation, the geologist who drew these three maps inferred that the original depositional limits could be projected beyond the preserved limits of the lithologic units.

ANCIENT EQUATORIAL ZONES

Given the widespread acceptance of the concepts of plate tectonics and the reliability of the results from paleomagnetic research, few geologists express surprise or reluctance about the possibility that continents can shift their geographic positions by large amounts. Thus most geologists are not surprised when an interpretation is made that a region now in the temperate latitudes previously might have been situated near the Earth's Equator or even near one of the poles. Former equatorial regions are identified by their low paleomagnetic inclinations.

Here we mention examples of times when parts of North America are inferred to have occupied equatorial latitudes. We are not aware that geologists who have studied the formerly equatorial regions examined the strata in light of the seasonal reversals of winds in the Trade-Wind belts that accompany the seasonal migration of the subsolar point (Chapter 8). Moreover, such seasonal wind reversals would cause the waters in equatorial basins to undergo seasonal shifts from upwelling to downwelling.

During the Cambrian Period, the Baraboo Island, Wisconsin, is inferred to have been in the Southern Hemisphere, nearly at lat.10°S (See Figure 16-13.), about in the middle of the latitudinal zone of seasonal wind reversals. During the Silurian Period, the region of southeastern New York is inferred to have been near the Equator. A reconstruction of the paleogeography during the Late Devonian Period shows much of present North America and Eurasia near the Equator (Figure 16-15).

SOURCES: Bambach, Scotese, and Ziegler, 1980; Dott, 1974; Marsaglia and Klein, 1983.

Suggestions for Further Reading

CHENOWETH, P. A., 1967, Unconformity analysis: American Association of Petroleum Geologists Bulletin, v. 51, p. 4–27.

DOTT, R. H., Jr., 1974, Cambrian tropical storm waves in Wisconsin: Geology, v. 2, no. 5, p. 243–246.

HAMBLIN, W. K., 1961, Paleogeographic evolution of the Lake Superior region from late Keeweenawan to Late Cambrian time:

Geological Society of America Bulletin, v. 71, p. 1–18.

HEDBERG, H. D., ed., 1976, International stratigraphic guide: New York, John Wiley & Sons, 200 p.

KRUMBEIN, W. C., 1942, Criteria for subsurface recognition of unconformities: American Association of Petroleum Geologists Bulletin, v. 26, p. 36–62.

North American Commission on Stratigraphic Nomenclature, 1983, North American stratigraphic code: American Association of Petroleum Geologists Bulletin, v. 67, p. 841–875.

RAASCH, G. O., 1958, Baraboo monadnock and paleowind direction: Alberta Society of Petroleum Geologists Journal, v. 6, p. 183–187.

CHAPTER 17

Basin Analysis

Fundamentals of Basin Analysis

Basin analysis is *the study of when and how basins form, evolve, and especially, are filled with sediment, and are destroyed.* Basin evolution is controlled by the interaction of the tectonics of the lithosphere, sedimentation, sea-level change, and climate.

Subsidence creates new space below base level. Wherever the lithosphere subsides and sediment is available in the depositional environment, this new space below base level is filled. Because subsidence and sedimentation can interact in a variety of ways, the patterns of strata that fill basins may also vary. The interaction between these two factors is modulated by changes in sea level and in climate that are cyclic at several scales (Chapters 6 and 15). Examination of the patterns of sedimentary strata that constitute the basin fill therefore permits inferences about subsidence, sea-level change, sedimentation history, and vertical movements of the lithosphere.

Since the middle of the nineteenth century, geologists have sought to identify broad patterns in the distribution of various kinds of sedimentary deposits. They have investigated such patterns with the idea of formulating a unifying hypothesis to explain how the Earth behaves. In this quest they have been joined by geophysicists, who have measured subsurface temperatures, determined the local values of the Earth's gravity, measured and mapped the Earth's magnetic field, and studied the behavior of seismic waves. Many geophysical models of the interior of the Earth have been formulated from calculations made after the geophysicists assigned numbers to various parameters that they believed affect the measurements. Some such geophysical models have illuminated tectonic theory; others have conflicted with tectonic theory or have led tectonic philosophers into what later turned out to be false ideas.

Before delving into the main body of this chapter, the reader should if necessary review the basic principles of plate tectonics, described in Box 17.1. In this chapter we begin by considering where and how sequences of sedimentary rocks are deposited. This focus on sediments, rather than on the igneous rocks of the lithosphere beneath them, is appropriate in a course entitled sedimentology, stratigraphy, or historical geology. The sediment-oriented approach allows us to focus on the ways in which tectonic movements affect sedimentation.

Subsidence and Sedimentation

Sediments accumulate in low areas; the material deposited in one place is derived from parent deposit(s) that has (have) been elevated relative to the place where the strata are being deposited. *The low areas in which sedimentary strata accumulate* are called **basins**. Because basins contain most of the world's hydrocarbons and ground water, and many of its mineral- and metal deposits, they are the focus of considerable practical interest. Basins also contain thick sequences of sedimentary strata, whose characteristics reveal much about the subsidence that provided room for sediment accumulation. Only rarely does the geologic evidence enable us to locate the former elevated areas, which commonly have been entirely eroded away. Accordingly, we concentrate on the strata, the products of erosion.

It is axiomatic that sedimentary strata result from the interaction of the rate at which a basin subsides and the rate at which sediment is supplied. As stated, this proposition implies that these two major factors are independent. To a certain extent, that is true. This is shown by the extremes of **stuffed basins**, *tectonic basins that are full of sediment* and **starved basins**, *tectonic basins that lack thick sediment fills.* Lacking sediment, a starved basin is filled with either water or air. However, the two factors are related in at least three ways: (1) once an area begins to subside and to accumulate sediments, the weight of the sediments may create such a load on the lithosphere

BOX 17.1

Concept of Sea-Floor Spreading and Plate Tectonics

Since the eighteenth century, when physical concepts began to be applied by scientists in attempts to understand the Earth's history, the study of sedimentary deposits and developments in geophysics have been closely linked. This is so because large-scale motions of the outer parts of the Earth exert profound effects on sedimentation. In this Box we review the development of the revolutionary new understanding of the Earth engendered by the theory of plate tectonics.

The concept of sea-floor spreading and plate tectonics resulted from exploration of the deep-sea floor. The key feature in all the model making has been the mid-oceanic ridge system (Box 17.1 Figures 1 and 2). The existence of such a ridge in the South Atlantic Ocean was shown by the *Meteor* Expedition in the 1920s. However, not until the precision recording echo sounder had been developed in the 1950s, systematically used, and the results compiled, was it possible to determine the detailed morphology of this ridge. After morphology had been combined with seismology, a striking result appeared. The crest of the ridge system was found to be closely associated with a narrow belt of shallow-focus earthquakes. Moreover, it was shown to extend for 80,000 km

and to be part of a worldwide fracture system. In addition, the ridge displays such features as (1) a median rift valley and various linear scarps and depressions on its flanks that are inferred to be products of normal faulting resulting from regional tensional forces, (2) higher-than-normal heat flow, (3) volcanoes and volcanic islands localized along the ridge crest, and (4) unusual seismic-, gravitational-, and magnetic characteristics in the crust beneath the ridge.

In the 1960s, paleomagnetic evidence on polarity was systematically collected from continental volcanic rocks. Analysis of this evidence led to the remarkable conclusion that the polarity of the Earth's magnetic field had reversed repeatedly. The volcanic rocks involved were dated by radiometric methods and thus it became possible to construct a geomagnetic-reversal time scale. Magnetic measurements made at sea along traverses at right angles to mid-ocean ridges showed that the intensity of the Earth's magnetic field varies in a pattern that is astonishingly symmetrical. (See Box 17.1 Figure 2.) The cause of this pattern was explained as being the expression of opposite senses of polarity in the magnetite within the

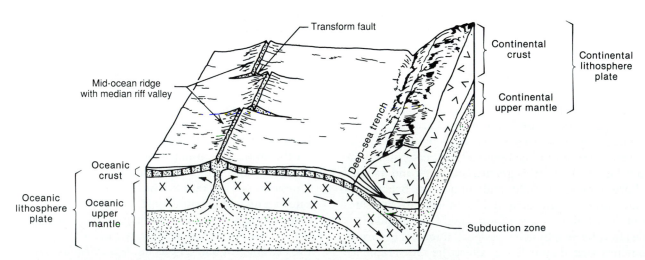

BOX 17.1 FIGURE 1.   Sea-floor spreading by divergent motion as plates move away from mid-ocean ridge. In converging motion at ocean trench (right), the plate is subducted; ocean-floor sediment may be scraped from top of plate and piled against trench wall. Plane of subduction beneath deep-sea trench that dips toward continent at an angle of about 45° and along which earthquake foci cluster is known as Benioff zone. Transform faults cut mid-ocean ridge. (E. R. Oxburgh, 1974, fig. 1, p. 301.)

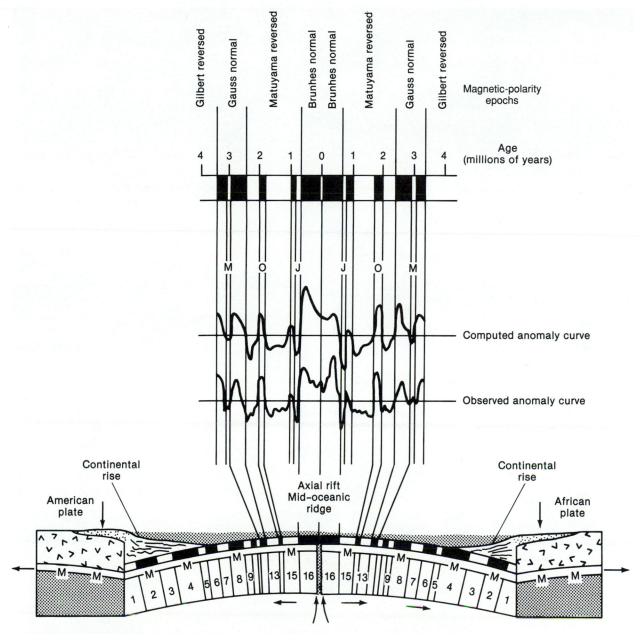

BOX 17.1 FIGURE 2. Stripes of magnetic highs and lows distributed symmetrically about spreading mid-ocean ridge. At top of diagram are shown the known intervals of normal (black) and reversed (white) polarity of the Earth's magnetic field, with their names. A time interval of 3.3 million years is represented by the width of the diagram; the width of the stripes is proportional to the durations of the time intervals for normal- and reversed polarity. (J. E. Sanders, 1981, fig. 18.20, p. 459.)

mafic igneous rocks forming vertical dikelike slabs, which extend from the sea floor downward to the depth where the geothermal gradient would cause the temperature to reach the Curie point of magnetite (578°C at a pressure of 1 atm). Deeper rocks are not magnetic and thus could not be affecting the anomalies. Positive anomalies were thought to match slabs having polarity that is the same as the Earth's present polarity (normal). Negative anomalies were thought to overlie slabs having polarity opposite to the Earth's present field (reversed). If this explanation is correct, then the symmetrical anomaly pattern could be neatly explained by assuming that the sea floor had been built by regular additions of new igneous rock through a huge dike located in the center of the mid-oceanic ridge.

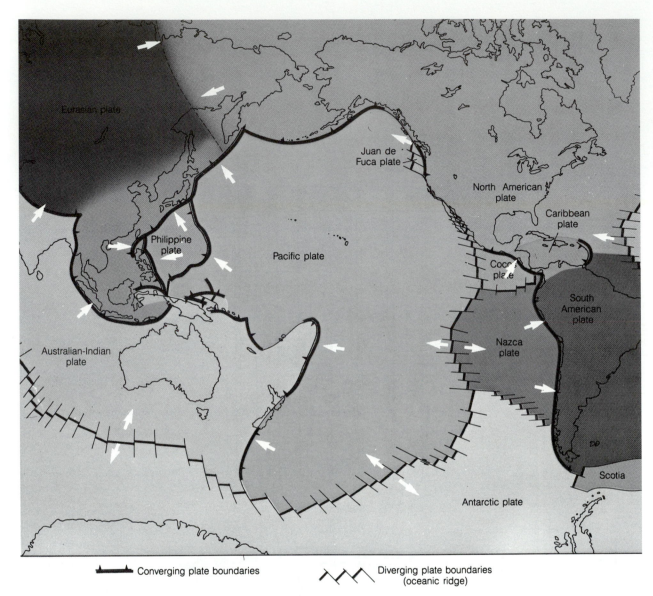

BOX 17.1 FIGURE 3.   Map of the Earth showing inferred plates (W. K. Hamblin, 1991, fig. 15.11, p. 267.) A. Pacific Ocean and surrounding area.

(See Box 17.1 Figure 2.) The idea is that before the magma in the center of the dike has time to solidify, the walls separate and new magma rises in the middle. (The movement of the sea floor away from the axis of the ridge probably contributes to the formation of new magma. Release of pressure on hot but solid rocks greatly compressed at depth is all that is necessary to cause the silicate materials to change state from solid to liquid. Hence, if some great force is capable of moving the deep-sea floor, then the same force will be responsible for generating magma beneath the place where the pressure is reduced.)

The successive slabs (appearing as stripes on magnetic maps) are thought to be related to normal- and reversed polarity intervals as determined from radiometrically dated sequences of volcanic rocks on land. If this is true, then it enables spreading rates of the deep-sea floor to be calculated. Computed rates range between 2 and 18 cm/yr. If these rates can be extended into the geologic past, then one must conclude that all the area of the present deep-sea floor in the Atlantic could have been created by sea-floor spreading in the last 180 million years, or since the Jurassic Period. This amounts to only the last 4% of geologic time.

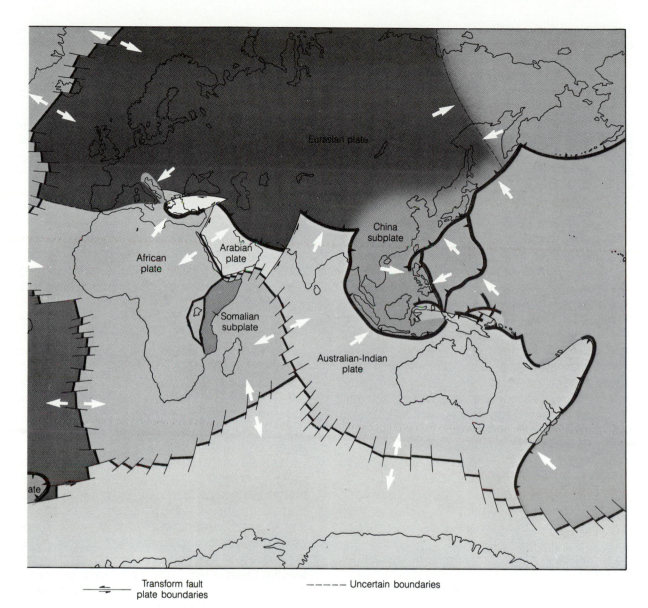

Transform fault
plate boundaries

----- Uncertain boundaries

BOX 17.1 FIGURE 3.   (*Continued*)
  B. Indian Ocean and surrounding area.

The evidence that the ocean basins have been created by the solidification of mafic magma (with various amounts of associated lava in the form of deep-sea extrusives) is now overwhelming. The magnetic-anomaly map of the Atlantic Ocean has been employed to predict correctly the ages of the basement rocks at various points. Continuous seismic-reflection profiles have shown that the thickness of deep-sea sediments increases progressively with distance from the center of the ridge. Moreover, results of the Deep-Sea Drilling Project (DSDP) and its successor, the Ocean Drilling Program (ODP), have confirmed that the ages of the oldest sediments agree with the predicted ages of the underlying basement rock.

The great deep-sea trenches and geographically associated Benioff (seismic) zones (planes, beneath the trenches, that dip towards the continents at an angle of about 45°, along which earthquake foci cluster), have been proposed as disposal sites for amounts of old lithospheric material equal to all the new material being created at the spreading ridges. (See Box 17.1 Figure 1.)

The deep-sea trenches are related to mid-ocean ridges in a grand scheme of global tectonics, or plate tectonics. (See Box 17.1 Figure 1.)

The basis for the plate-tectonic extension of the concept of sea-floor spreading is that a map showing the distribution of earthquakes can become the basis for drawing a map of the Earth's lithospheric plates (Box 17.1 Figure 3).

According to plate-tectonic concepts, the surface of the Earth is underlain by a mosaic of a dozen or so major plates (and many minor ones) that move with respect to one another, and may shift in any direction with respect to the solid parts of the upper mantle underlying the asthenosphere. Three kinds of motion along plate boundaries are possible: (1) diverging motion, (2) converging motion, and (3) transcurrent (or lateral) motion.

The effects of divergence can be illustrated by considering what would happen to a segment of continental lithosphere that is stretched to the point where its ductile lower part begins to thin and its brittle upper part to crack, to form a series of concave-up normal faults. This process typically forms what have been termed *rift basins.* If the divergence causes the lithosphere to be greatly thinned, its surface may subside beneath sea level. If the lithosphere is ruptured, then a new spreading ridge may form as the adjacent plates continue to move away from each other. At the spreading center, magma (plus lava) supplied from below creates new ocean floor and adds material to the trailing edges of the two separating plates. The new lithosphere, thin at the spreading center, thickens progressively as it cools and moves away. As its density increases during progressive cooling, its top surface subsides isostatically.

Where two plates move directly toward each other and one passes beneath the other, the result is an asymmetrical arrangement that has been named *subduction.*

Where two plates pass each other moving laterally, great faults are formed along which the motion may persist for millions of years. The names of such lateral zones are *strike-slip faults* (or *wrench faults*) on continents or wherever new material is not involved. The San Andreas fault, California, is an example of a major strike-slip fault. Some strike-slip faults, known as *transform faults,* end against the crests of spreading ridges.

These three kinds of motion create the three major kinds of regional forces that have been cited as the active mechanisms by which strata have been deformed. These are (1) extensional forces, which cause normal faults; (2) compressional forces, which cause folds, reverse faults, and thrusts; and

(3) shearing couples, which cause strike-slip (or transcurrent) faults and associated local diagonal belts of *en-echelon* normal faults and folds (Box 17.1 Figure 4).

Lithosphere plates can be characterized by the kind of material forming their capping layer. For the most part, all deep-ocean basins are underlain by lithosphere capped by thin (5 to 6 km) *sima* (rocks rich in silica and magnesium), which typically consists of an upper, pillowed basalt (1 to 2 km thick) and a lower stratiform body of ultramafic rock (3 to 4 km thick). These two layers, plus variable amounts of ultramafic rock from the underlying mantle, constitute a distinctive group of ultramafic rocks collectively designated as the *ophiolite suite.* In the present deep-ocean basins, no known rocks are older than Middle Jurassic.

By contrast, the upper parts of the lithosphere underlying continental masses consist of *sial* (rocks rich in silica and aluminum) 25 to 70 km thick, including rocks of varying ages ranging back about 4 billion years. Thus continental rocks have managed to accumulate faster than they have been destroyed by erosion. Does the Earth get rid of its old oceanic rocks by converting some (or all) of them into continental rocks?

Because of the lower density of its uppermost rocks, a plate topped by a continental mass is thought to be unable to vanish down a subduction zone. Therefore, at the margin of a continental plate, a subduction zone may be steadily consuming and disposing of ocean-floor plate materials, but if a plate containing another continental mass arrives, the downward disposal is thought to cease. Collision of two plates having continental toppings is thought to create mountain chains and, in the process, to fuse the two continental masses together (Box 17.1 Figure 5). The complex join between two such continental masses has been termed a *suture.* Ophiolites, inferred to have been squeezed upward, sideways, and downward at such collision boundaries, may mark past sutures.

Continental materials evidently can be destroyed only by erosion. If prolonged erosion could remove all the sial, so that only sima were left, isostasy would cause the surface to subside, possibly to oceanic depths. Another possible fate for a continental mass is to be split apart. This is thought to begin by domal elevation of the surface through a broad circular area to form a *prerift arch* or *dome* (Box 17.1 Figure 6). Such an arch or dome is thought by some to be the surface expression of a deep-lying hot spot in the mantle, the top of a vertical *mantle plume* by which heat from the Earth's core is conducted upward to the surface. An example of a place where such doming is thought to have occurred is the Arabo-Nubian Shield. (See Figure 10-43.) This dome began to form in

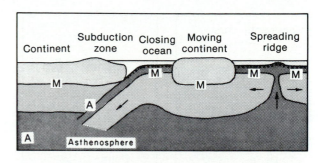

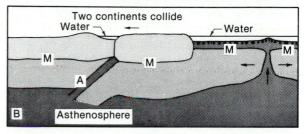

BOX 17.1 FIGURE 5.   Concept of disposal of lithosphere at a subduction zone (A); hypothetical schematic sections.

A. As new lithosphere is generated at a spreading ridge (right), old lithosphere disappears down a subduction zone (left). Short, dark, vertical lines at top of ocean crust indicate alternating zones of rock having opposite magnetic polarities. Subduction is thought to be possible only for oceanic crust. Former ocean closes as subduction progresses. M = M-discontinuity, an interface within the lithosphere characterized by a contrast in density.

B. Collision of two continents, a process held to be the cause of mountain building. Continental crust is considered to be unable to travel down a subduction zone. (After J. M. Bird and J. F. Dewey, 1970, fig. 9, p. 1047; J. F. Dewey, 1972, p. 64.)

the Cambrian Period. Other episodes of uplift took place during the Paleozoic and Mesozoic Eras. As the dome developed, it was extensively eroded, and thick bodies of fluvial sandstones, known as Nubian sandstones, were deposited around it.

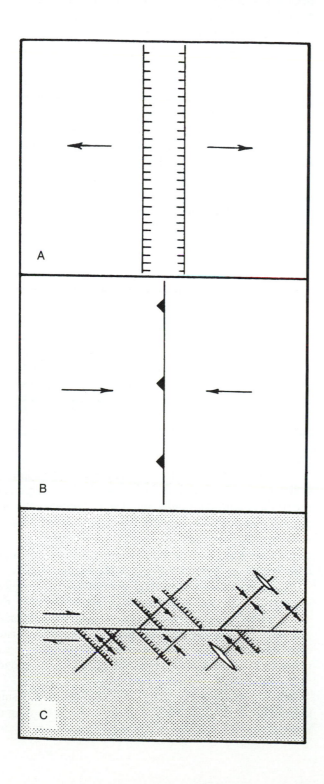

BOX 17.1 FIGURE 4.   Kinds of forces that deform strata as related to major kinds of inferred plate motions; schematic sketch maps.

A. Plates moving apart create tension; initial movement creates graben bounded by normal faults (hachured lines, hachures on downthrown side).

B. Converging plates create compression; result is thrusting, including both overthrusts and underthrusts. (Simple thrust fault shown by line with black triangles on upper plate.)

C. Two plates passing in translatory motion, creating major shearing couple. Right-lateral shearing couple shown causes vertical strike-slip fault (or transform fault where new material is added) and adjacent sets of folds oriented at 45° to strike-slip fault, and normal faults oriented on the other 45° diagonal and at right angles to the axes of the folds. (After J. E. Sanders, 1974b, fig. 9A, p. 31.)

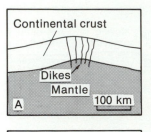
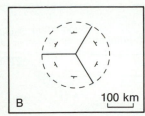
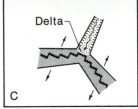
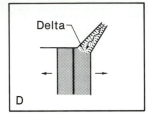

BOX 17.1 FIGURE 6.   Concept of fracture and spreading of continental crust after initial doming; schematic.

A. Profile and section through dome. Lithosphere bulged up from below.

B. Map view of dome, showing major fractures at 120°; dikes injected along fractures.

C. Spreading ridges form along two fractures; graben subsides along third direction. River flowing along length of graben deposits delta at one end.

D. Spreading along one fracture only. One of two other fractures functions as a transform fault; graben forms along third fracture system. (After K. C. A. Burke and J. F. Dewey, 1973, fig. 2, p. 408; W. R. Dickinson, 1974b, fig. 9, p. 17.)

The next stage in the sequence of events is tensional rupture along three linear rifts oriented at angles of 120°. (See Box 17.1 Figure 6, B.) This is the rift-valley stage. At first, the rifts are thought to be confined to the domed-up region, but even-

tually they are inferred to extend well beyond it. Spreading may become active along all three of these linear rifts. In that case, the end product will be three spreading ridges that meet at a single point, called a *triple junction*. (See Box 17.1 Figure 6, C.) Alternatively, spreading may proceed along only two of the arms and leave the third as an aborted feature that remains a graben. (See Box 17.1 Figure 6, D.)

According to plate-tectonic concepts, a continental mass may thus be rent asunder and a new ocean basin begin to form. This ocean may widen by spreading away from an active ridge. At a later time, the spreading away from this ridge may stop and a new pattern of plate motions may be established. In the new arrangement, the recently formed ocean may be consumed by subduction at one or both of its margins. Thus two continents formed from a single continental mass, after splitting apart, later may be joined. *A sequence of events that begins with a continent and involves the rupture of this continent and the opening and later closing of an ocean to restore a single continental mass* is known as a **Wilson cycle.**

SOURCES: Benioff, 1954; Burke and Dewey, 1972; Burke and Wilson, 1972; S. W. Carey, 1975; Dewey and Bird, 1971; Dewey and Burke, 1974; Dewey and Spall, 1975; Dott and Batten, 1988; Hayes and Pitman, 1973; Heezen, 1968; Heezen and Ewing, 1963; Heezen, Tharp, and Ewing, 1959; Hess, 1965; Miall, 1990; Pitman and Talwani, 1972; J. T. Wilson, 1968.

that this weight increases the tendency to subside; (2) as the sediments are compacted, their upper surface, where new sediments are being added, can subside at a greater rate than that of the underlying lithosphere; and (3) tectonic activity controls rates and sites of uplift (the elevated areas are the chief suppliers of sediment) and sites of volcanic activity. In addition, the fluctuation of eustatic sea level modulates the effect of both subsidence and sedimentation.

We can illustrate the effects of the nearly infinite numbers of combinations of the interplay between rates of subsidence and rates of supply of sediment by considering four combinations. These are (1) rapid subsidence and copious supply of sediment, (2) rapid subsidence and sparse supply of sediment, (3) slow subsidence and copious supply of sediment, and (4) slow subsidence and sparse supply of sediment.

1. Where rates of both subsidence and of supply of sediment are rapid, stratal thicknesses become large and the strata are buried rapidly and deeply. Under such conditions, sediments literally pour into the subsiding basin. A landmass having lofty relief must be undergoing rapid erosion to supply the vast bulk of sediments. Such a landmass implies pronounced uplift of a contiguous belt.

In the parent-bedrock area, the effects of chemical weathering will probably be negligible. If so, then the sediment will be compositionally immature (Chapter 2) and include minerals that are usually destroyed during chemical weathering. If feldspar-bearing bedrock is exposed in the parent area, then feldspar will be abundant in the sediments.

What happens in the basin of deposition that is subsiding rapidly is not at all well understood. According to some geologists, rapid subsidence and copious sediment supply can prevent environmental processes from "doing their job" of sorting and abrading the sediment. In fact, it has even been claimed (wrongly, we think) that rapid tectonic activity can override the effects of environment altogether. Some special results have been claimed when sediments are "poured in" to a rapidly subsiding basin. These ideas are so ingrained that we think they deserve some comment.

The only sediments known to us that have been deposited so rapidly that they have not been reworked by water or by wind are the catastrophic deposits of rock avalanches, debris flows, and mud flows. (See Figure 14-9.) We know of no case in which subsidence has been so rapid that sediments delivered to a shoreline are buried before waves can sort them. It takes only a half dozen or so waves to do the major part of the sorting that will be done. Because the periods of most water waves range from 5 to 10 s, the time required is less than a minute. Similarly, only a short distance of transport in a stream of water is necessary to sort out various sand populations. The results of flume experiments demonstrate that effective sorting can take place in the distance of only a few meters. Some geologists get so caught up in contemplating the huge spans of geologic time that they forget that some geological processes take place in what we might call "everyday time."

Gravity transport down steep subaqueous slopes can create jumbled deposits, for example, wildflysch (Figure 17-1). However, such deposits result from gravity-powered mass transport of sediment along the sea floor and the mixing of layers of sediment, each of which may have been well sorted earlier. Gravity-powered subaqueous mass-flow deposits do not indicate that tectonics have somehow overprinted the effects of environment.

2. Where the rate of subsidence greatly exceeds the rate of supply of sediment, the floor of the basin sinks to greater and greater depths. Such a basin will become starved.

3. Where subsidence is slow and supply of sediment is great, the basin becomes filled with sediment. The shore may prograde completely across a marine embayment and convert the former water area into land. Fans may prograde across this land surface. Under these circumstances, sediment bypasses the basin and does not remain there.

4. Where subsidence is slow and rate of sediment supply is small, shallow marine conditions, as in an epeiric sea, may persist for a long time. The waves and currents are able to rework the sediments extensively before the next layer arrives.

SOURCES: Allen and Allen, 1990; Bally and Snelson, 1980; Barton and Wood, 1984; Bond and Kominz, 1984; Miall, 1990.

Kinds of Basins

Basins may be classified in any number of ways. A current approach to basin classification is based upon the concepts of plate tectonics (Box 17.1). Using plate tectonics we can recognize three major conditions with respect to plates that control the formation of basins: (1) divergent, (2) convergent, and (3) transcurrent. These give rise to major categories of basins. Within those categories we can recognize subcategories based on other characteristics, such as size and position with respect to plate boundaries or continental margins.

Extensional Basins (Divergent Setting)

We take up the subject of extensional basins first. These are divided into small extensional basins, such as grabens, aulacogens, and rifts; and large extensional basins, such as the major ocean basins of the world.

SMALL EXTENSIONAL BASINS

Grabens. *A linear fault-bounded subsiding area* is a **graben**. Grabens are common and are straightforward kinds of structural features. They range in size up to lengths of tens of thousands of kilometers and widths of tens to hundreds of kilometers.

An example of a graben is the Triassic-Jurassic Hartford basin of the eastern United States for which Joseph

FIGURE 17-1. Two large blocks and various smaller blocks of shallow-water limestone and -sandstone set in a dark-colored, deep-water shale. The blocks in this wildflysch are interpreted as having moved down a steep slope from near a shelf edge to a basin or a basin margin. Upper boulder is nearly a perfect sphere. Rysedorph Conglomerate (Middle Ordovician), Rensselaer, New York. (Authors.)

Barrell (1869–1919) proposed the interpretation that is accepted today (Figure 17-2). Barrell showed that the Triassic-Jurassic strata occupy a fault-bounded trough.

More recently, grabens have been given other names, especially if they are interpreted as parts of a plate-spreading situation. Examples of these are aulacogens, discussed next, together with rifts.

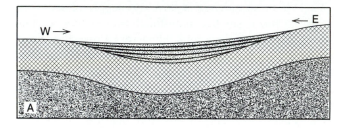

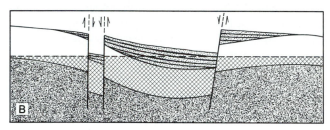

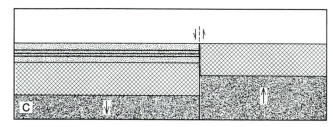

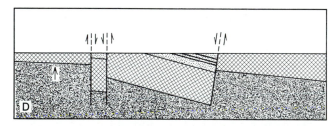

FIGURE 17-2. Two versions of the tectonic relationships that were responsible for the Triassic-Jurassic strata underlying the Hartford Basin in central Connecticut and Massachusetts. A and B. Pre-1915 version adopted by J. D. Dana (1813–1895).

In A, the strata are shown as filling a broad synclinal downwarp and extending laterally well beyond their present limits, both to the W and to the E. Three thick lines indicate sheets of extrusive mafic igneous rock that are interlayered with the sedimentary strata.

Present arrangement shown in B, after large movement on normal faults and erosion to present land surface (dashed line). Faulting is postdepositional.

C and D. Version showing basin as graben with postdepositional tilting and erosion. Crossed-diagonal pattern, crust; irregular stippled pattern, upper mantle. (After W. M. Davis, 1898; J. Barrell, 1915, figs. 4–6 facing p. 24, 28; J. E. Sanders, 1963b, fig. 3, p. 507.)

Aulacogens and Rifts. A continental mass may be ruptured, commonly along three linear rifts oriented at angles of 120°. (See Box 17.1 Figure 6, B.) Spreading may become active along all three of these linear rifts. In that case, the end product will be a **triple junction**, *a single point where three spreading ridges meet.* (See Box 17.1 Figure 6, C.) Alternatively, spreading may proceed along only two of the arms and leave the third as an aborted feature that remains a graben. *An arm of a triple junction that does not continue to spread* has been termed a **failed arm** of a triple junction. *A failed arm of a triple junction that becomes a narrow V-shaped depression with the apex of the V away from a rifted continental margin and which subsides and fills with sediment* is termed an **aulacogen**. (See Box 17.1 Figure 6, D.)

Most aulacogens begin as narrow, fault-bounded grabens down which major rivers commonly flow (Figure 17-3). At their mouths, these rivers may deposit major deltas. Examples of such deltas at the open ends of failed triple junctions are thought to include those built by the Mississippi River in North America and by the Benue-Niger, Zambesi, and Limpopo rivers in Africa. The Rhine River reached the North Sea through a failed triple junction when sea level stood lower. In India, the Godavari, Mahanadi, Narmada, and Ganges rivers all flow along what are thought to be failed arms. In South

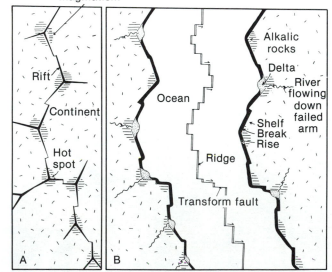

FIGURE 17-3. Two stages in formation of aulacogens.

A. Continental crust is thought to be domed over hot spots of alkaline magmatism; after doming, ruptures develop along linear rifts oriented at angles of 120°.

B. Ocean forms as divergent motion spreads new plates apart; rivers flow down failed arms of triple junctions to feed deltas at reentrants on continental margins. As failed arms of triple junction subside and fill with sediment, they become known as aulacogens. (J. F. Dewey and K. C. A. Burke, 1974, fig. 1, p. 58.)

America, the Amazon reaches the Atlantic Ocean along the failed arm of what has been inferred to have been a Late Paleozoic junction that did not spread at all.

The initial graben of an aulacogen may become a broader downwarp that may itself be ultimately broken by faults. Aulacogens stretch into the interiors of continents.

Rifts that are destined to become the spreading centers of ocean basins resemble aulacogens in the early stages of their development. Rifts ultimately become permanently connected to marine basins and are flooded by marine waters. This permits the formation of thick evaporitic sequences (Chapter 10), the deposition of strata composed primarily of the skeletons of marine organisms, and other effects caused by the circulation of marine waters or by the presence of marine organisms. However, aulacogens may be temporarily connected to marine basins and hence their contained strata may resemble early rift-fill strata.

An aulacogen still connected to an open ocean is the Gulf of Suez, which connects with the Red Sea and with a transform fault that underlies the Dead Sea-Gulf of Aqaba (Figure 17-4). In this aulacogen, Miocene- and younger evaporites, limestones, and terrigenous rocks attain a thickness of about 4 km.

If old oceans close by continental collision, an aulacogen may be the only feature left to indicate the former presence of a rifted continental margin. An example of a North American aulacogen is the Ottawa-Bonnechere

Graben of eastern Canada, which is about 500 km long and 40 km wide. This graben opened along preexisting Precambrian fractures. It is thought to be the failed arm of a three-armed system. The other two arms are presumed to have continued to open, forming the ocean floor of Iapetus (a former Paleozoic ocean that is thought to have occupied more or less the same site as the present Atlantic Ocean). The arms that opened to the scale of an ocean have been completely destroyed in later events. Exactly where the opening rifts lay is now impossible to determine. Cambrian and Ordovician sedimentary rocks, mostly limestones with some basal sandstones, about 1 km thick, accumulated in this aulacogen.

Another example is the southern Oklahoma aulacogen. The Oklahoma aulacogen is located where the Ouachita extension of the Appalachian orogen makes a reentrant into the North American craton (Figure 17-5). This aulacogen began as a graben underlain by Precambrian granitic rocks and was filled by as much as 5000 m of coarse immature terrigenous rocks; extrusive basalt, spilite, and rhyolite; and shallow sills of Early to Middle Cambrian age. From Late Cambrian to Late Ordovician time, the trough was a broad downwarp in which up to 3100 m of limestone accumulated. With the passage of time, the rate of subsidence decreased. For example, in this downwarp, the thickness of the Silurian and Devonian rocks is only 335 m, which is no more than their thickness elsewhere on the platform. During Late Paleozoic time, the downwarp was compressed and broken by faults. This faulting resulted in a complex pattern of a median uplift and marginal yoked basins that we discuss in a following section.

LARGE EXTENSIONAL BASINS (OCEAN BASINS)

Ocean basins are first-order features of the Earth's relief. As we indicated earlier, ideas about the ocean basins have ranged through great extremes. One dated concept is that they originated early in the Earth's history and have remained as permanent fixtures ever since. The modern concept is that the sea floor is mobile and can spread laterally away from mid-ocean ridges. We shall describe the ocean basins as they are now and then present the inferred sequence of events in their origin according to the concept of plate tectonics.

The bottom of an ocean such as the Atlantic stands highest along the median ridge crest. Laterally away from the ridge crest in both directions are progressively deeper areas known as the ridge flanks. Still farther away are deep-lying basin floors; locally, these are abyssal plains.

According to the concept of plate tectonics, a developing ocean basin goes through several stages. The se-

FIGURE 17-4.   Inferred triple junction consisting of three arms occupied by (1) Red Sea (foreground), (2) Gulf of Aqaba (right), and (3) Gulf of Suez (left) Gemini XII photograph (looking north). Although the Gulf of Suez is on trend with the northern Red Sea and is a site where extensive Miocene rifting and evaporite deposition took place, it is not on a plate boundary. A plate boundary coincides with the Red Sea and turns up the Gulf of Aqaba into the Dead-Sea fault zone. (See Figure 14-7.) The Gulf of Suez is a failed arm, an aulacogen; the Gulf of Aqaba is a linear lowland along a major strike-slip fault. (NASA.)

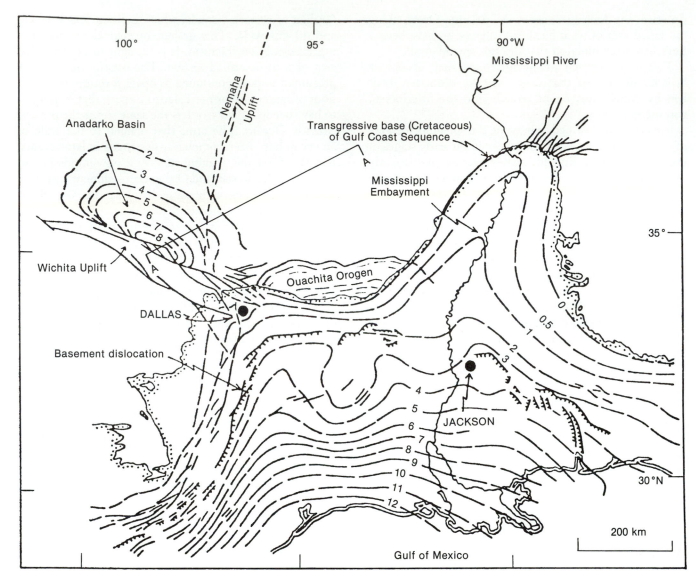

FIGURE 17-5.   Generalized tectonic map, Oklahoma aulacogen and its relationship to Mississippi Embayment and to Gulf of Mexico. This aulacogen, the Anadarko Basin, one of the deepest Paleozoic basins of the North American craton, is located where the Ouachita extension of the Appalachian orogen makes a reentrant into the North American craton. Contours in kilometers on base of Paleozoic sequence. (K. C. A. Burke and J. F. Dewey, 1973, fig. 10, p. 420.)

quence is thought to begin with a *prerift arch*. (See Box 17.1 Figure 6, A.) This is followed by the *rift-valley* stage. As continued crustal distention induces the rift valleys to subside further, waters from the ocean may flood the rifts to form the next stage, that of *proto-oceanic* gulfs. (See Box 17.1 Figure 6, C.) In these restricted gulfs, immense thicknesses of evaporites may accumulate. For example, beneath parts of the Red Sea, the thickness of evaporites is as much as 5 to 7.5 km. Further divergent motion of the two separated continental blocks may allow new oceanic crust and -lithosphere to form. This next stage is that of a narrow ocean. The final stage is that of the open ocean.

Along the subsiding *passive continental margin* of an open ocean, a continental terrace may be deposited. Along the top surface of this terrace are the continental shelf, continental slope, and continental rise that lead to the deep-sea floor. Commonly, the shelf deposits consist of carbonates and the deep-sea deposits of terrigenous muds.

A classic ancient example of the deposits of the passive margin of an open ocean is the Cambro-Ordovician sequence formed in the Appalachian region, northeastern United States. During the Cambrian and Ordovician periods, most of the North American continent was a shallow epeiric sea in which carbonate sediments were deposited. At the eastern edge of this sea, that is, at the eastern edge of the continent, a steep paleoslope existed that was probably an active hinge line between the continent to the west and the deep ocean to the east. (See Figure 9-10.) Slides, slumps, turbidity currents, mudflows, and sand flows moved down the steep slope to oceanic depths. There, the displaced shallow-water sediments were deposited at the margin of the deep-water basin. (See Figure 9-11.) Along the basin margin, the shallow-water deposits, mostly limestone debris, including brecciolas, interfinger with hemipelagic shale. The brecciolas formed along the original eastern edge of the carbonate shelf for hundreds of kilometers. Similarly, in the Jurassic Period, a hinge line formed a precipitous slope near the present-day Mediterranean coast of Israel between shelf edge and deep sea. Skeletal- and peloidal carbonate debris from a shallow-water carbonate shelf was displaced downslope by turbidity currents to make graded layers that interfinger with black- and green shales, the deep-water basin facies (See Figure 10-43.).

If an ocean basin forms by the mechanism of sea-floor spreading, then a distinctive sequence of rocks should form as the sea floor moves from ridge crest, to ridge flank, and to deep-basin floor. At spreading centers, the dominant lithology is that of the ophiolite suite. Progressively farther away, the basement acquires a cover of pelagic sediments. In tropical seas, this pelagic cover consists of calcareous pteropod- (See Figure 7-52.) and *Globigerina* oozes (See Figure 9-26.), and siliceous radiolarian ooze. Because the floors of the basins generally lie deeper than the carbonate-compensation depth, mostly brown clay, with lesser amounts of siliceous ooze, accumulates there. (See Figure 7-51.) On a section of the sea floor in the area between the ridge flank and the deep basin, delicate skeletal remains are selectively dissolved. The first to be dissolved, at shallower depths, are the pteropods. Next to go are the tests of the *Globigerinas;* they persist to greater depths. Where land masses are nearby, abundant sediment may be supplied that can be gravity propelled to the deep-sea floor. If any plate moves away from the Equator and toward the poles, the kind of pelagic sediments deposited may shift as the plate enters different latitude zones. The tropical-water assemblage of pteropods-*Globigerinas*-radiolarians may give way to diatoms. In cold polar waters, diatom ooze is the dominant pelagic sediment.

According to the tenets of plate tectonics, the life expectancy of ocean basins is limited. Ocean basins grow laterally and they are thought to vanish laterally. Of all basins, they are the largest, by an order of magnitude. Throughout most of their extents, ocean basins are invariably starved.

Modern deep oceans, such as the Atlantic and Pacific oceans, are examples of ocean basins. The Gulf of Mexico is a small ocean basin.

Example of a Divergent Setting: Eastern North America. The continental terrace and continental margin off eastern North America are underlain by a complex pile of strata, which has been accumulating under the present general passive-margin tectonic setting that has prevailed at least since Late Jurassic time. Sediment derived from the elevated Appalachians has been brought to the sea by many short rivers and forms the tapering body of continental-terrace strata (Figure 17-6). Repeated cycles of submergence and emergence have taken place during an interval of oceanward tilting of the margin of the continental basement rocks. The body of terrace sediments ends on the east at the continental slope. Seaward of the base of the slope lies a thick mass of sediment whose upper surface forms the continental rise. Much of the sediment underlying the continental rise consists of terrigenous contourites. The terrace sediment also includes substantial contributions from the sea in the form of skeletal materials and glauconite.

Underlying the Upper Cretaceous and younger strata of the continental terrace are elongate basin fillings of probable Late Jurassic to Early Cretaceous age. One of these basins, the Baltimore Canyon Trough (See Figure 17-6.), has been the center of extensive seismic research in connection with a series of dry holes drilled in search of petroleum. Other basins are of Late Triassic-Early Jurassic age. These are inferred to be buried parts of the same group of tectonic features to which the Hartford basin (See Figure 17-2.) and other related basins belong. These basins preserve remnants of formerly more-extensive filling strata. Their implied postdepositional history includes extensive regional uplift, normal faulting, folding, strike-slip faulting, and deep erosion. Probably erosion of the elevated central parts of these Triassic-Jurassic basins supplied much of the sediment that fills the Baltimore Canyon Trough.

SOURCES: Allen and Allen, 1990; Aydin and Nur, 1982; Badley and others, 1988; Baker and Morgan, 1981; Bechtel, Forsyth, and Swain, 1987; Epstein and Friedman, 1983; Ingersoll, 1988; Kingston, Dishroon, and Williams, 1983; Miall, 1990; Poag, 1985.

Compressional Basins (Convergent Setting)

ARC-TRENCH SYSTEMS

The great systems of *island arcs* and deep-sea trenches that are such prominent features of the Pacific Ocean have long been of interest to geologists. Four major

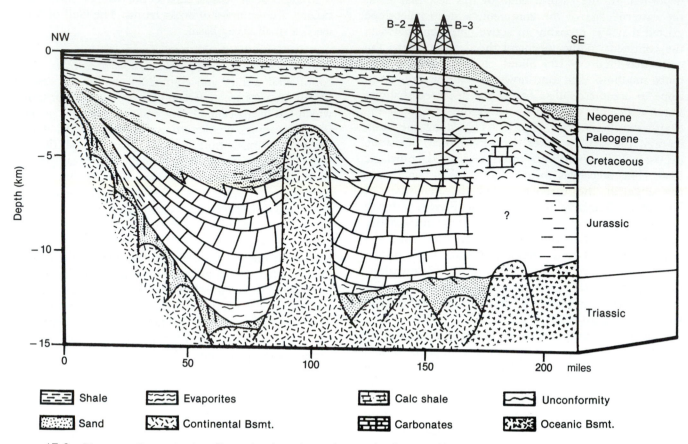

FIGURE 17-6. Diagrammatic structural profile-section through passive margin of eastern North America off New Jersey (Baltimore Canyon Trough), showing locations of test borings B-2 and B-3. (S. A. Epstein and G. M. Friedman, 1983, fig. 4, p. 955; modified from C. W. Poag, 1979, fig. 3, p. 1456.)

morpho-tectonic elements are found in most arc-trench systems. These are (1) the **deep-sea trench** itself, *a narrow, elongate, steep-sided, rock-walled depression that generally is deeper than the adjacent sea floor by 2000 m or more and that is floored by oceanic crust;* (2) the **forearc basin**, *a sedimentary basin, 50 to 250 km wide, lying between a deep-sea trench and its adjoining island arc;* (3) the island arc and associated *intra-arc basins;* and (4) the **backarc area**, which is *the outer area adjoining an island arc on the side away from a deep-sea trench* (Figure 17-7).

The initial studies were devoted to measuring gravity. Great linear belts of gravity anomalies were found to coincide with the arc-trench systems. Negative anomalies followed along the trenches; positive anomalies characterized the adjacent island-arc chain. Many island arcs are double; that is, they contain two sets of islands. The islands nearest the trench are nonvolcanic, whereas the other islands contain volcanoes.

The trenches possess rocky walls that are steeper on the side toward the island arc and less steep on the opposite side. A great geophysical discovery related to trenches is that earthquake foci are concentrated along

them and that the depths of these foci increase systematically away from the trench on the island-arc side. This zone of frequent earthquakes is called a *Benioff zone.* Not all trenches adjoin island arcs; for example, the Peru-Chile trench lies just off the continent of South America. Here the Benioff zone dips beneath the continent. (See Box 17.1 Figure 1.)

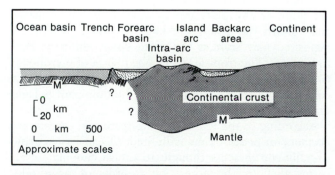

FIGURE 17-7. Relationship of basins near an island arc, schematic profile. Further explanation in text. (After W. R. Dickinson, 1974b, fig. 10b, p. 18.)

The amount of sediment fill in the trenches varies from almost none to complete filling. The amount of fill differs with the ages of the trenches and their access to sources of terrigenous sediment from continents.

The origin of deep-sea trenches has not been altogether settled. Contrasting ideas include (a) graben, formed by crustal extension or (b) great subduction zones, where, according to plate-tectonic theory, two crustal plates converge and the oceanic plate plunges beneath a continental plate at a *convergent plate margin.* According to the plate-tectonic interpretation, the steep inner wall of the trench is underlain by a deformed subduction complex of mixtures of various kinds of rocks, especially ophiolite scraps. The mass of this subduction complex is thought to increase as additional oceanic crustal material is scraped from the top of the lithosphere that descends beneath the subduction zone. The sediments deposited directly in the trench are largely turbidites. Transport by turbidity currents within a trench is mostly by longitudinal flow along the trench axis. In addition, according to plate-tectonic concepts, as the lithosphere descends into the subduction zone, the arriving oceanic plate carries with it sediments from outside the trench, including turbidites and pelagic sediments. These sediments are thought to have been deposited over extensive areas of the ocean floor that are wholly outside the arc-trench system. They are presumed to end up in the trenches by being scraped off the descending more-rigid oceanic plate that underlies them. Very little concrete evidence of such activity has been found in modern trenches. If the subduction idea is correct, then one might argue that the Pacific trenches are so young that they do not show the predicted effects. This position is untenable; some trenches contain strata of at least Mesozoic age. Alternatively subduction may take place only at deeper levels. The upper part of the mantle and overlying crust may not go downward, but upward to form great thrust sheets (*obducted slabs*). The relationship of the trenches to the postulate of massive subduction involving oceanic sediments has not yet been determined satisfactorily. We discuss this matter further in a subsequent section.

In modern arc-trench systems, the widths of forearc basins, located between trench and arc, vary from 50 to 250 km. In such basins the thicknesses of the sediment fills can become very great. The thicknesses of trough-filling turbidites and subsea fan complexes in some fore-arc basins reach 12 to 15 km. The foundation rocks of a forearc basin are plutonic- or volcanic. The forearc basins receive their sediments, mostly of volcanic- and plutonic rock debris, from the contiguous arc. Normal faults commonly bound the basins on the arc side.

The island arcs are the curvilinear belts of nonvolcanic- and volcanic islands that roughly parallel the trenches.

The arcs include island chains, along which parts of the volcanoes are submerged beneath the sea, and subaerial volcanoes built near continental margins. Granitic- and granodioritic plutons may be emplaced in the metamorphic roots of continental-margin arcs. The volcanism of the arcs is characterized by explosive eruptions to form stratified piles of tephra and interbedded subaerial extrusives. Most eruptions in continental-margin arcs are subaerial and commonly, ignimbrites predominate. By contrast, in intraoceanic arcs, pillowed extrusives and subaqueous tephra are more abundant than subaerial extrusives and airborne tephra.

Within the arc may be intra-arc basins. The submarine morphologic configuration of such basins determines how the intra-arc sediments are dispersed. Tephra and fragments of volcanic rocks within the intra-arc basins may be dispersed by fluvial-, deltaic-, nearshore-marine-, and deep-sea processes that act on submarine slopes. Examples of inferred ancient intra-arc sedimentary deposits include intermontane continental redbeds interstratified with Cenozoic volcanic rocks in Peru and Bolivia; Upper Cenozoic marine strata capped by lacustrine deposits on Honshu, Japan; and Upper Cenozoic marine strata in the Tofua Trough adjacent to the active volcanic chain of Tonga in the Pacific Ocean.

Within the backarc area may lie (1) an *interarc basin* floored by oceanic crust and (2) a *retroarc basin* floored by continental crust and separated from the arc by a group of folds and thrusts. Among sedimentary strata in modern interarc basins are wedges of turbidites composed of volcanic debris and/or pelagic materials. Deposits in retroarc basins include fluvial-, deltaic-, and marine strata as much as 5 km thick. The parent deposits of the sediments in retroarc basins may be the magmatic rocks of the arc or the uplifted strata in the contiguous folded- and thrust-faulted belt.

SOURCES: Allen and Allen, 1990; Burchfiel and Royden, 1982; Edwards and Santogrossi, 1990; Flemings and Jordan, 1989; Lash, 1985; Miall, 1990; Stockmal, Beaumont, and Boutilier, 1986; Thornburg and Kulm, 1987.

Elongate Basins Along Transcurrent Faults and Fracture Zones (Transform Setting)

Along many major strike-slip faults, and in depressions between higher basement blocks in the sea-floor fracture zones, are elongate basins that are bounded by transcurrent faults. (See Gulf of Aqaba in Figure 17-4.) In many respects, these fault-bounded basins resemble grabens.

Whatever their origins, linear troughs bounded by faults display characteristic sedimentary patterns. The fillings of such troughs exposed on land include various nonmarine limestone lake sediments that interfinger marginally with fanglomerates (See Figure 14-7.) and redbeds. Other filling materials may consist of fluvial- and deltaic sediments, evaporites, extrusive- and intrusive basalts, and tephra. The fillings of fault-bounded basins on the deep-sea floor, such as in the Gulf of Aqaba (See Figure 17-4.), consist of turbidites and pelagic oozes.

SOURCES: Allen and Allen, 1990; Allen and others, 1986; Beaumont and Tankard, 1987; Biddle and Christie-Blick, 1985; Coward, 1986; Dott, 1988; Edwards and Santogrossi, 1990; Friedman, 1985, 1988; Garfunkel, Zak, and Freund, 1981; Manspeizer, 1988; Zak and Freund, 1981.

Flexural Basins (Vertical-Tectonics Setting)

SYNCLINAL BASINS YOKED TO ANTICLINAL UPLIFTS OF BASEMENT ROCKS

Some basins that extend across cratons are generally synclinal but are closely tied (or yoked) to adjacent anticlinal- or fault-block uplifts that act as major sources of sediments. The classic example of such synclinal basins became active early in the Pennsylvanian Period; it extended from southeastern Oklahoma into Colorado, over the site of the older Oklahoma aulacogen. As mentioned previously, this aulacogen became inactive during the Silurian Period. It remained inactive until Early Pennsylvanian time. Active uplift along a median axis (Wichita-Amarillo uplift) accompanied vigorous synclinal downwarping on both sides, forming the Anadarko Basin (Figure 17-8) on the northeast and the Hollis Basin on the southwest. As a result, an elongate island grew upward out of the Pennsylvanian epeiric seaway and began to shed coarse sediment into the adjacent subsiding basins. At first, the debris came only from the covering strata of Paleozoic age, which consisted chiefly of the thick Cambro-Ordovician carbonate rocks. Eventually

the Precambrian basement rock was exposed. Altogether, as much as 7000 m of terrigenous debris of Pennsylvanian and Permian ages, much of it fanglomerate, was deposited during several pulses of uplift and faulting. The Anadarko Basin is one of the deepest Paleozoic basins on the North American craton. By Late Permian time the relief of the elevated tract had been reduced; eventually it was buried completely by marine deposits. A comparable area of uplifts and yoked basins of almost the same age as the feature just described stretched across southwestern Colorado and adjacent parts of Utah. (See Figure 13-39.) Both the Texas-Oklahoma and the southwest Colorado basins and uplifts have been extensively drilled and have produced large quantities of hydrocarbons.

SOURCES: Borak and Friedman, 1981; Brewer, Good, Oliver, Brown, and Kaufman, 1983; Friedman, Reeckmann, and Borak, 1981; Garner and Turcotte, 1984; Miall, 1990.

INTRACRATONIC BASINS

We have previously defined a craton as a broad, stable, central part of a continent. As originally defined, craton included the central parts of oceans as well. According to the concepts of plate tectonics, craton should be restricted to continents; its use for the sea floor is inappropriate.

The term **intracratonic basin** designates a *tectonic basin within a craton*. Most intracratonic basins were stuffed; that is, during crustal movement, great thicknesses of strata accumulated in them. Examples are the Michigan and Illinois basins in the United States and the Paris-London basin in Europe. The sedimentary strata filling these basins are in part intrabasinal, such as carbonates and evaporites. Some of the sediments were derived from distant cratonic sources. Examples include such terrigenous sediments as shales and sandstones. Positive areas,

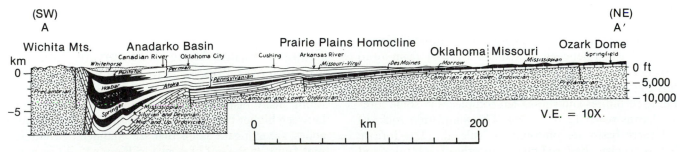

FIGURE 17-8.   Cross section of Anadarko Basin showing (1) synclinal downwarp on north flank of Wichita Uplift (See Figure 17-5.), (2) exposed Precambrian basement rock of the uplift, and (3) strata of accumulated terrigenous debris in downwarp adjacent to upwarp. Line of section indicated on Figure 17-5. (P. B. King, 1951, fig. 12C, p. 56.)

known as *epeirogenic uplifts*, adjoin the two basins mentioned in the United States. These positive areas are the Ozark Dome in Missouri, Arkansas, Oklahoma, and Kansas; and the Cincinnati Arch in Kentucky, Tennessee, Ohio, and Indiana. Intracratonic basins are formed by a process called epeirogeny, discussed further in a following section.

FORELAND BASINS

Foreland basins are *sedimentary basins located between the front of a mountain range and the adjacent craton, and related to overthrusting at the convergent plate margin.* Foreland basins may be formed by flexural subsidence resulting from loading of supracrustal rocks emplaced during thrusting in the mountain chain, a concept known as *thrust loading.* Thrusting is the active basin-forming mechanism. The mountain chain that thrusting creates sheds vast amounts of terrigenous debris into the foreland basin, ultimately filling it with thick prograding wedges of nonmarine-, transitional-, and marine strata, such as the Catskill fan-delta complex. (See Figure 13-34.) In the Himalaya Mountains, thrusting and uplift resulted in the formation of gravels that prograded more than 110 km into the foreland basin; these gravels serve as syntectonic indicators of thrusting.

EPEIROGENY

Broad vertical movements of the lithosphere unaccompanied by crumpling of strata are **epeirogenic movements.** Downward, or "negative," epeirogenic movements enable thick successions of sedimentary strata to accumulate. Upward, or "positive," epeirogenic movements form such features as the Colorado Plateau of the southwestern United States.

Epeirogenic uplifts and -downwarps, broad cratonic regions that have been bent either upward or downward but not otherwise significantly deformed, have been recognized for many years, as mentioned. Until recently it was generally agreed that the relatively nondeformed and nonmetamorphosed strata of such regions never experienced conditions other than gradual burial followed (in some cases) by gradual unroofing. The evidence cited in support of this contention was the fact that the sedimentary strata underlying epeirogenic uplifts have not been deformed, as have those of fold-mountain belts, and they do not appear to have moved laterally from the places where they formed. However, detailed petrographic- and geochemical study of strata exposed at the surface in nondeformed areas of epeirogenic basins and domes has revealed evidence that these

strata formerly were heated to high temperatures. Such intense heating implies a former great depth of burial, because generally no evidence supports the concept of a local supply of excessive heat. The inferred former great depth of burial of strata now exposed at the Earth's surface implies that the lithosphere has undergone large vertical motions. These vertical motions represent the effects of processes distinct from the lateral motions of tectonic plates.

The kinds of evidence used to infer former intense heating, and therefore former great depth of burial of nondeformed sedimentary strata now exposed at the Earth's surface, include the occurrence of saddle-dolomite cements with oxygen- and carbon-isotopic signatures characteristic of formation under conditions of deep burial, of fluid inclusions within crystals that resemble primary fluid inclusions formed at high temperatures, of conodonts (toothlike microfossils) that have been altered by heating, and of organic material that has been heated in place to a degree of organic maturity only possible when the strata are subjected to temperatures considerably greater than those associated with their present depths of burial. Exotic geochemical techniques, such as fission-track analysis and $^{40}Ar/^{39}Ar$ spectrum analysis, indicate former intense heating and great depth of burial, followed by large-scale vertical uplift of the lithosphere. These features that suggest former intense heating commonly occur without any associated evidence of former geothermal activity, which would permit intense heating under near-surface conditions.

Using these lines of evidence, Silurian strata of the northern Appalachian Basin are inferred to have been buried to a depth of 5 km, Devonian strata in the Catskill Mountains of New York State to a depth of about 6.5 km, and Lower Ordovician carbonates of the northern Appalachian Basin to a depth of more than 7 km. These depths are all much greater than those previously inferred on other criteria. If the petrographic- and geochemical evidence for former deep burial of nondeformed strata is correct, then new paleogeographic reconstructions must be made. The uplift of sedimentary strata from burial depths of 5 to 7 km to the surface of the Earth implies (1) deposition of 5 to 7 km of strata younger than those now at the surface, followed by (2) removal of these formerly overlying strata by erosion. The thicknesses of strata that were deposited in these basins and then eroded away appear to be much greater than previously thought. Where did these vanished sediments come from? Where did they go? Such questions must now be answered if earlier estimates of former depth of burial of sedimentary strata in epeirogenic uplifts and downwarps are too low.

In addition, intracratonic basins formed by epeirogenic downwarping, such as the Michigan Basin of eastern North America, are sites of huge hydrocarbon reserves. These hydrocarbons were generated within the basins from organic matter deposited with the sediments as the basins formed. The intense heating of the sediments that accompanied their deep burial by epeirogenic downwarping contributed to the conversion of organic matter (kerogen) to petroleum and natural gas. If epeirogenic downwarping and -uplift have been much larger than previously supposed, then potentially valuable reserves of hydrocarbons may lie, unsuspected, trapped close beneath the surface in rocks whose former depths of burial have been regarded as trivial.

SOURCES: Bhattacharyya and Friedman, 1979, 1983, 1984; Borak and Friedman, 1981; Cercone, 1984; Coward, Dewey, and Hancock, 1987; Crawford, 1981; Dickson and Coleman, 1980; Dott, 1988; England, 1983; Friedman, 1987a, b, c, 1990; Friedman and Kopaska-Merkel, 1990; Friedman, Reeckmann, and Borak, 1981; Friedman and Sanders, 1982; Grover and Read, 1983; Johnsson, 1986; Watts, Karner, and Steckler, 1982.

The Concept of the Geosyncline

Historical Perspective

The concepts of basin analysis relate to ideas about subsidence and basin filling in various tectonic settings. The 1970s and 1980s were a dynamic era in the science of basin analysis. Through plate tectonics (Box 17.1), seismic- and sequence stratigraphy, and advances in sedimentology, basin analysis has been revolutionized. Yet pre-plate-tectonic models that explain fundamental concepts of deep subsidence are comparable to modern ideas for some megascale basins. To explain these models we review concepts about geosynclines, the term used by some geologists for any surface that subsided.

The concept of a geosyncline was inspired by the geologic relationships that were worked out for the Appalachian Mountains, particularly for their western parts, a region known as the Valley and Ridge province. The generalizations made about this part of the Appalachians prompted similar studies in Europe, but with contrasting results. This contrast becomes apparent when one compares American views of the Appalachians with European views of the Alps.

VIEWS OF EARLY AMERICAN GEOLOGISTS

The originators of the geosynclinal concept were the American geologists James Hall (1811–1898) and James

Dwight Dana (1813–1895). Hall observed that, where the Paleozoic marine strata in the interior of North America are thin (thicknesses of only a few hundreds or a few thousands of meters), they are flat lying. By contrast, in the Appalachians, thicknesses of equivalent strata amount to tens of thousands of meters and the strata are not horizontal. Hall hypothesized that the subsidence of the strata within a trough, where they would be extra thick, provided the mechanism for folding them (Figure 17-9).

Shortly afterward, in 1873, James Dwight Dana argued that subsidence alone would not fold the strata. Instead, Dana proposed that the strata had become thick by sinking unmolested in a great synclinal trough (which he named a geosynclinal, later renamed *geosyncline*). According to Dana, only afterward were these thick, trough-filling strata folded (Figure 17-10). Dana suggested that the deformation of the crust beneath a geosyncline was the cause of subsidence and sedimentation and also the cause of the subsequent deformation of the sedimentary strata and concomitant mountain building.

Both Hall and Dana emphasized an important inference about the Appalachian area that had subsided. This inference stated that throughout the thousands of meters of vertical sinking, the depth of the marine waters had remained "shallow." In other words, subsidence had been more or less exactly matched by accumulation of sedi-

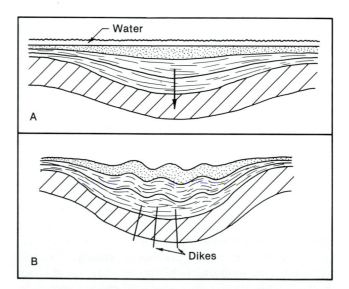

FIGURE 17-9.   Sketch of origin of geosynclinal sediments and -structures, according to the ideas of James Hall (1859a, p. 69–70).

A. Trough subsides; strata in center become thicker than on sides.

B. With great subsidence, crust beneath the center of the trough fails, cracks open on stretched part along base of trough, and dikes are intruded. Strata above are folded by squeezing together of limbs of trough.

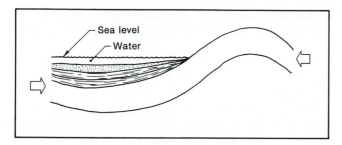

FIGURE 17-10. Sketch of origin of geosynclinal sediments and -structures, according to ideas of J. D. Dana. Dana visualized five steps, as follows: (1) Trough subsides and sediment accumulates; (2) downbending of crust requires the complementary upward flexure of a geanticline (stages 1 and 2 sketched above); (3) heat rising from below weakens the bottom of the geosyncline; (4) the weakened trough yields as a result of lateral pressure; (5) "the stratified rocks become, in the partial collapse, upturned or folded, and pressed into a narrower space than they occupied before; a mountain range exists as a result" (J. D. Dana, 1880, p. 820). (Sketch based on J. D. Dana, 1873, p. 170 and 1880, p. 819–821.)

ment. The original idea that a part of the sea floor might subside and yet sediment could accumulate fast enough to keep the water depth from changing very much had been published by Charles Darwin.

In the Appalachian Valley and Ridge province, this relationship between great subsidence and the existence of thick shallow-water sediments was so striking that it was made part of the "geosynclinal doctrine." According to most American geologists, the identifying characteristic of a geosyncline was that it is a place where subsidence proceeds at a rate exactly matched by the rate of sedimentation. To these American geologists, "geosyncline" became synonymous with continuously shallow water.

SOURCES: Dott and Batten, 1988; Friedman and Sanders, 1978; Kay, 1951.

VIEWS OF EARLY EUROPEAN GEOLOGISTS

Even before they had heard about geosynclines, the Swiss geologists studying the Alps had noticed a striking contrast between two suites of terrigenous sedimentary rocks. They named one of these suites the *flysch* and the other the *molasse*. The term flysch is a corruption of the German verb *fliessen*, which means *to flow*. This term was applied because the outcropping parts of the shaly flysch were especially prone to slope failures. The term molasse refers to the soft sandstone that was locally quarried for use in buildings.

Flysch. The strata named flysch (See Figure 5-6, A.) are *a thick succession of marine sedimentary strata consisting* of repetitively interbedded alternating and laterally persistent sands (and/or coarser sediments) and shales found in the interior of a fold-mountain chain. During the nineteenth- and first half of the twentieth centuries, the environment of deposition of the flysch was the subject of much controversy. Nearly every conceivable environment ranging from mangrove swamps to the deep sea was proposed, but no agreement could be reached. Although early Austrian authors favored a deep-sea origin for the flysch, this idea seemed to be untenable in light of the lack of coarse materials in the *Challenger* collection of modern deep-sea sediments. The flysch remained a sedimentologic enigma. Somewhat half heartedly, and on the basis of its coarse layers, ripple marks, and plant debris, it was assigned a shallow-water origin. However, nearly every such interpretation contained a comment that the flysch did not resemble other formations whose shallow-water origin had been firmly established.

Associated with "ordinary" flysch was a peculiar variety named the wildflysch. (See Figure 17-1.) This term was applied to *a spectacular deposit consisting of small- to enormous blocks of sedimentary-, igneous-, and metamorphic rocks set in a matrix of fine-textured, typically dark-colored marine shale, -siltstone, or -mudstone.* In modern terms, we would designate the wildflysch as one kind of *diamictite* (Chapter 4).

Because the flysch remained such a difficult sedimentologic enigma, most Alpine geologists were satisfied that the peculiar properties of flysch had nothing to do with its depositional environment, but rather were distinctively related to "tectonics." They applied to the flysch such words as *"preorogenic,"* or *"pre-paroxysmal."* This labeling was thought to solve the mystery of how the flysch had acquired its peculiar characteristics. Others thought that flysch was peculiar because it had been "the final filling" of the geosyncline. This idea of "finality" picked up a few adherents in Switzerland but died a quiet death in the Carpathians. In southern Poland, the Carpathian flysch ranges in age from Jurassic to early Cenozoic; it spans the entire time of the filling of the Carpathian geosyncline.

Molasse. By contrast, the molasse designated a *tectono-stratigraphic unit consisting of a wedge-shaped body of extrabasinal sediments typified by patterned successions of shallow-marine- and nonmarine strata, whose particles were derived from erosion of the older rocks, including flysch, that composed the rising mountain chain.* The molasse consists of autocyclic nearshore-marine sediments, fluvial- and deltaic strata, and fans. Copious sediment eroded from the rising Alpine chain filled the foreland basin that extended along the north side of the Alps. This shallow sea was filled by fine sediments deposited as parts of marine deltas. On the delta topset plains, coarse fan sediment accumulated, forming fan-delta complexes. Hence, even though the sedimentologic characteristics of the molasse were not in doubt, it acquired a pigeonhole in the tectonic vocabulary as a "*postorogenic*" deposit.

In our opinion, no single concept has caused more confusion in the study of sedimentary deposits than has this idea that the diagnostic properties of sediments were uniquely related to their tectonic setting with respect to the timing of "orogeny." Each of these supposedly distinctive "tectonic" suites of sediments can be considered as "*postorogenic*," "*synorogenic*," or "*preorogenic*," and the term used depends on how one defines "orogeny." Clearly, the Alpine flysch of Mesozoic and early Tertiary age was deposited before it was elevated, eroded, and recycled into the Miocene molasse of central Switzerland. In that sense, flysch is certainly "preorogenic." But what about the implications of tectonics during the deposition of the flysch? All that coarse sediment means that somewhere an extensive parent area must have been uplifted. Flysch, therefore, must be "post-" with respect to any "orogeny" involved in that uplift. And the great thickness of the flysch implies great subsidence. Whether one considers subsidence to be "orogeny" or not, one must admit that the origin of the flysch involved active subsidence. We infer that this subsidence was so rapid that it formed a deep-water basin. If so, then the flysch is "post" with respect to whatever one chooses to call that movement. Or, if one considers that subsidence accompanied sedimentation, then the flysch must be "syn-" with respect to that downward movement.

As for the Alpine Miocene molasse, it is the filling of a foreland basin whose subsidence was a consequence of the overthrusting and uplift of the flysch. Hence, as is generally supposed, the molasse is "post-" this deformation. Yet the thickness of this molasse is many thousands of meters. Moreover, because this molasse contains only shallow-water deposits, great subsidence must have accompanied deposition. Therefore the Swiss molasse is "syn-" with respect to this subsidence during its deposition. Finally, along its southern parts, the Miocene molasse has been overthrust by the Oligocene and older flysch. Therefore the molasse is "pre-" with respect to the final overthrusting.

In modern terms we would interpret the flysch as an extensive marine fondothem containing abundant graded layers and possibly associated with an adjacent marine clinothem. We think that the flysch is much more likely the deposit of extremely "deep" rather than "shallow" water. (However, we admit that bathymetric interpretations are difficult to prove.) We infer that the flysch did not begin to accumulate until after a deep basin had

formed and steep submarine slopes were available to enable gravity-powered, bottom-following sediment flows to move.

We interpret the molasse as a kind of clastic wedge filling a foreland basin associated with an elevated land area formed by overthrusting. The molasse can be regarded as a terrigenous undathem that accumulated when a vast body of sediment prograded away from a steep mountain chain. The Miocene molasse of Switzerland compares with the Devonian of the Old Red Continent of northern Europe and of the Catskills in New York State, with the Carboniferous Coal Measures of Europe and the Appalachians, with parts of the Cretaceous of the Rocky Mountain region, and with many similar deposits of various ages and localities. For those who do not like the term "molasse," the alternative of "*tectonic fan-delta complex*" is available.

These two contrasting kinds of sedimentary suites commonly are related and are synchronous. Today, in southern Europe, for example, a molasse of Holocene age is accumulating in the Po Valley. Nearby, Holocene flysch is being deposited in the deep parts of the Adriatic Sea. What seems to be an important tectonic milestone in the history of an orogenic belt is the final uplift and obliteration of the flysch basin(s). This event signals the end of the deep-water stage and implies that the ultimate conversion of a part of the sea floor to a land area will soon be complete (as far as that cycle is concerned). Some geologists have proposed a causal connection between deposition of a flysch sequence and a later orogenic event (e.g., Hall's model of geosyncline formation, previously discussed). Modern plate-tectonic concepts imply that no such connection exists. However, if a flysch basin persists for a long time, it is likely to be caught up in an orogenic event, be uplifted, and be deformed.

Shortly after the American ideas about geosynclines, based on the Appalachians, had been proposed, European geologists picked up the concepts and applied them to the Alps. The progression of ideas went something like this: (a) geosynclines are supposed to be the antecedents of fold-chain mountains; (b) the Alps are a fine example of fold-chain mountains; (c) therefore the Alps probably were developed from a precursor geosyncline. Accordingly, (d) the paleogeographic history inferred from study of the strata in the Alps should contribute useful information toward a generalized concept of what a geosyncline is and is not.

A final major point about geosynclines is their position with respect to a continental mass. All geosynclines were supposed to stretch along one side of a broad stable central part of a continental mass, which Wilhelm Hans Stille (1876–1966) named a *craton*. (As originally defined, the term craton also includes stable parts of ocean basins.)

SOURCES: Bamford, 1979; Kay, 1947, 1951, 1967; Stille, 1924, 1936a, b.

Where Do Sedimentology and Stratigraphy Stand Today?

At the end of a chapter in which we discuss the revolution in stratigraphy engendered by the development of modern ideas of basin analysis and of plate tectonics, it seems appropriate to pause and reflect on where sedimentology and stratigraphy stand today. Should we interpret all sedimentary strata according to the tenets of plate tectonics? Or should we keep trying to straighten out all the entangled ideas about geosynclines? What have sedimentology and stratigraphy to contribute to all the new tectonic hypotheses based on recently developed geophysical models derived from studies of the sea floor?

Before we attempt to discuss the first two of these questions, let us emphasize the key ways in which sedimentology and stratigraphy can contribute to geophysical models.

Sedimentology, Stratigraphy, and Geophysical Models

Sedimentology and stratigraphy are now starting to make their full impact felt on those who conceive geophysical models. Sedimentology and stratigraphy can yield data that are important for geophysical models in two subject areas: (1) amounts and rates of subsidence, and (2) implications for horizontal movements.

STRATA AND RATES OF SUBSIDENCE

Study of the thicknesses and characteristics of sedimentary strata can provide numerical information on crustal subsidence. Although sedimentology and stratigraphy may not provide the mechanisms and the reasons for such subsidence, they can provide numbers for the amounts and the rates. If these numbers have been properly determined, they represent a kind of reality that must be considered as a starting point for geophysical models. For example, for the Mediterranean basin, sedimentologic- and stratigraphic evidence seems to require rapid and catastrophic subsidence of nearly the entire basin to oceanic depths since the Pliocene Epoch, yet geophysicists consider this to be mechanically impossible.

The geologic record preserved in the Triassic and Jurassic strata of the Austrian Alps evidently contains strata whose interpretation likewise requires us to infer that an earlier, large and rapid depth change took place. *In-situ* Upper Triassic coral reefs are overlain by deep-sea pelagic deposits of Jurassic age whose inferred depth of deposition is 3000 to 4000 m. Such catastrophic subsidence is regarded likewise as mechanically impossible. We say to our geophysicist colleagues: "Look again at your hypotheses. They might need a few adjustments. Let the stratigraphic record set the limits on the operation of the models, not vice versa."

SEDIMENTARY STRATA AND HORIZONTAL MOVEMENTS

Although nearly all sedimentary strata yield some kind of information about vertical tectonic movements, only a few sedimentary deposits provide information critical for analyzing lateral movements of the kinds that are such integral parts of plate tectonics. We discuss three examples: (1) mélanges, (2) fillings of deep-sea trenches, and (3) sediments whose distribution patterns form belts that are parallel to lines of latitude and are controlled by an ocean-atmosphere circulation pattern comparable to today's.

Mélanges. A peculiar deposit resembling and commonly difficult to distinguish from wildflysch is a **mélange**. A mélange is defined as *a heterogeneous mixture of huge angular-, poorly sorted-, and exotic tectonic blocks of rocks of diverse kinds and provenances dispersed tectonically in a pervasively sheared fine-textured matrix of marine origin.* A mélange forms *a special kind of tectonic breccia that resulted from deformation of consolidated rocks under an overburden of consolidated rocks, so that blocks were detached from an underlying unit and incorporated within an upper unit that was moving as an overthrust or a gravity-sliding mass.* Most mélanges are thought to have resulted from the deformation of deep-sea trench fillings in a subduction zone along a convergent plate margin. Huge blocks in a fine-textured matrix characterize both mélanges and wildflysch. However, unlike simple wildflysch, a mélange is a true tectonite. A wildflysch that has been sheared may not be distinguishable from a true mélange.

Each mélange includes two kinds of blocks: (1) **exotic blocks**, defined as *tectonic inclusions detached from rock units whose compositions contrast with that of the main body of the mélange,* and (2) **native blocks**, which are *blocks derived from disrupted brittle layers that once were interbedded with the ductilely deformed matrix.*

A mélange surely represents a collection of blocks that imply lateral tectonic movements. In the plate-tectonic scheme of things, mélange is interpreted as scrapings from a lithosphere plate that is undergoing subduction

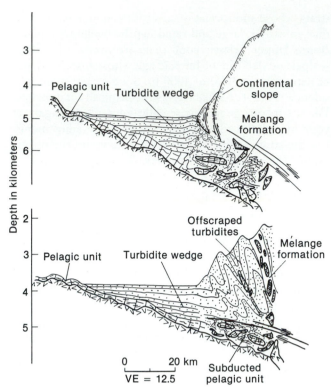

FIGURE 17-11. Hypothetical modes of formation of mélange as scrapings from a lithosphere plate that is undergoing subduction. In upper diagram, subduction leads to formation of mélange consisting of terrigenous- and pelagic blocks together with scraps of oceanic- and continental crust. Lower diagram shows selective offscraping of terrigenous deposits and subduction of pelagic strata. (D. W. Scholl and M. S. Marlow, 1974, fig. 5, p. 202.)

(Figure 17-11). If this idea is correct, then mélanges are key indicators for the presence of ancient subduction zones.

A well-described example of a mélange is found in the Franciscan Formation (Jurassic) of the California Coast Ranges. Other examples are known from elsewhere in the circum-Pacific belt of mountain chains, from the Alpine-Tethyan chains, and from the Hercynian chain of southern Ireland.

Fillings of Deep-Sea Trenches. According to plate-tectonic principles, ocean-floor crust and its covering of pelagic- and other sedimentary strata are transported down and disappear within subduction zones that dip away from deep-sea trenches toward the adjacent island arcs. If one grants that this concept may be correct, then using the rates of sea-floor spreading that have been inferred from magnetic-anomaly curves, one is faced with the problem of what became of all the pelagic sediment that the moving ocean floor brought to the trenches. Ac-

cording to some interpretations, the crust of the entire area of the Pacific Ocean has been manufactured completely since Late Jurassic time. If the rocky crust is lost by subduction and the sea-floor sediments are scraped off in the trenches, then colossal quantities of offscraped pelagic sediment should fill the trenches and any mountain chains derived from uplift of older trenches.

According to one estimate, in 100 million years, pelagic sediments 200 to 500 m thick going into a trench 1000 km long should form an offscraped deposit having a volume of 2.5×10^6 km^3. No such volume of pelagic sediments has been found, either in the trenches or in recently elevated fold belts. Perhaps this enigma may be explained by the very slow deposition of pelagic sediments; therefore the amount of such sediment accumulating is not as great as has been computed.

Subduction is such an appealing concept that the trenches have been suggested as sites for disposal of our garbage, toxic industrial wastes, and radioactive waste products. The evidence from pelagic strata strongly indicates that some major tinkering needs to be done to the concept of subduction. If trenches do not function like huge garbage disposals, then the use of trenches as garbage dumps could result in the poisoning of the deep ocean and the destruction of a vital part of the Earth's biosphere.

Sedimentary Strata Forming Latitude-Parallel Belts. Sedimentary strata forming latitude-parallel belts in the modern world, such as reefs and evaporites, are useful tools for analyzing the trends of ancient latitude lines. Information of this kind from the geologic record places severe constraints on the amount of continental displacement in a north-south direction.

SOURCES: Allen and Allen, 1990; Bally, 1982; Beaumont and Tankard, 1987; Cameron, 1990; Cloetingh, McQueen, and Lambeck, 1985; Ingersol, 1988; Miall, 1990.

Mesosequences and Basin Analysis

Through the 1970s, basins were analyzed in terms of the concept of the geosyncline. Classical stratigraphy, as practiced by most stratigraphers, placed little emphasis on the analysis of genetically related groups of strata that form what we call mesosequences. A revolution had already begun in the 1960s, and by the end of the 1970s all of stratigraphy had changed. Basin analysis focused, as it does today, on the analysis of basins with respect to their origin as inferred from the concepts of plate tectonics. The study of the sedimentary fill of basins, stratigraphy, had begun the shift that continues today, to an emphasis on a hierarchy of genetically related strata. The bottom of this hierarchy, laminae, formed a part of pre-1960 stratigraphy, but the new way of looking at how laminae are assembled into larger and larger units did not. Parasequences, facies tracts, mesosequences, and

Sequences define a continuum of scale that permits the sedimentary fill of a basin to be studied as an entity, the result of the interacting processes of subsidence, sedimentation, and sea-level change that both created the basin and filled it in.

Suggestions for Further Reading

ALLEN, P. A.; HOMEWOOD, P.; and WILLIAMS, G. D., 1986, Foreland basins: An introduction, p. 3–12 *in* Allen, P. A.; Homewood, P.; and Williams, G. D., eds., Foreland basins: International Association of Sedimentologists Special Publication 8, 453 p.

BAKER, P. H.; and MORGAN, P., 1981, Continental rifting: Progress and outlook: EOS, American Geophysical Union, Transactions, v. 62, p. 585–586.

BALLY, A. W., 1982, Musings over sedimentary basin (*sic*) evolution: Royal Society of London Philosophical Transactions, Series A, v. 305, p. 325–338.

CLOETINGH, S.; McQUEEN, H.; and LAMBECK, K., 1985, On a tectonic mechanism for regional sea level (*sic*) variations: Earth and Planetary Science Letters, v. 75, p. 157–166.

COWARD, M. P., 1986, Heterogeneous stretching, simple shear and basin development: Earth and Planetary Science Letters, v. 80, p. 325–336.

DOTT, R. H., 1988, Something old, something new, something borrowed, something blue—A hindsight and foresight of sedimentary geology: Journal of Sedimentary Petrology, v. 58, p. 358–364.

EDWARDS, J. D., and SANTOGROSSI, P. A., eds., 1990, Divergent/passive margin (*sic*) basins: Tulsa, OK, American Association of Petroleum Geologists Memoir 48, 256 p.

FRIEDMAN, G. M., 1985, Gulf of Elat (Aqaba). Geological and sedimentological framework, p. 39–71 *in* Friedman, G. M., and Krumbein, W. E., eds., Hypersaline ecosystems: Heidelberg-Berlin-New York, Springer-Verlag, 484 p.

FRIEDMAN, G. M., 1987, Vertical movements of the crust: Case histories from the northern Appalachian Basin: Geology, v. 15, p. 1130–1133.

INGERSOLL, R. V., 1988, Tectonics of sedimentary basins: Geological Society of America Bulletin, v. 100, p. 1704–1719.

KINGSTON, D. R., DISHROON, C. P., and WILLIAMS, P. A., 1983, Global basin classification: American Association of Petroleum Geologists Bulletin, v. 67, p. 2175–2193.

MIALL, A. D., 1990, Principles of sedimentary basin (*sic*) analysis, 2nd ed., Chapter 9, Sedimentation and plate tectonics, p. 499–615: New York, Springer-Verlag, 668 p.

MOLNAR, P., 1988, Continental tectonics in the aftermath of plate tectonics: Nature, v. 335, p. 131–137.

WATTS, A. B., and THORNE, J., 1984, Tectonics, global changes in sea level and their relationship to stratigraphic sequences at the US Atlantic continental margin: Marine Petroleum Geology, v. 1, p. 319–339.

Bibliography

ACKER, K. L., and RISK, M. J., 1985, Substrate destruction and sediment production by the boring sponge *Cliona caribbaea* on Grand Cayman Island: Journal of Sedimentary Petrology, v. 55, p. 705–711.

ADAMS, C. E., JR., and WEATHERLY, G. L., 1981, Suspended-sediment transport and benthic boundary-layer dynamics, p. 1–18 *in* Nittrouer, C. A., ed., Sedimentary dynamics of continental shelves: Developments in sedimentology, v. 32: Amsterdam, Elsevier Scientific Publishing Company, 449 p.

ADEY, W. H., 1978, Coral reef (*sic*) morphogenesis: a multidimensional model: Science, v. 202, p. 831–837.

ADEY, W. H.; and BURKE, R., 1976, Holocene bioherms algal ridges and bank-barrier reefs of eastern Caribbean: Geological Society of America Bulletin, v. 87, p. 95–109.

ADEY, W. H.; MACINTYRE, I. G.; STUCKENRATH, R.; and DILL, R. F., 1977, Relict barrier reef (*sic*) system off St. Croix: its implications with respect to late Cenozoic coral reef (*sic*) development in the Western Atlantic: International Coral Reef Symposium, 3rd, Proceedings: University of Miami, v. 1, p. 15–21.

AGER, D. V., 1981, Major marine cycles in the Mesozoic: Geological Society of London, Journal, v. 138, p. 159–166.

AGTERBERG, F. P., and GRADSTEIN, F. M., 1988, Recent developments in quantitative stratigraphy: Earth-Science Reviews, v. 25, p. 1–73.

AHLBRANDT, T. S., and FRYBERGER, S. G., 1982, Introduction to eolian deposits, p. 11–47 *in* Scholle, P. A.; and Spearing, D., eds., Sandstone depositional environments: Tulsa, OK, American Association of Petroleum Geologists Memoir 31, 410 p.

AHR, W. M., 1973, The carbonate ramp: an alternative to the shelf model: Gulf Coast Association of Geological Societies Transactions, v. 23, p. 221–225.

AHRENS, L. H., 1965, Distribution of the elements in our planet: New York, McGraw-Hill Book Company, 110 p.

AHRENS, L. H., ed., 1968, Origin and distribution of the elements: Oxford, Pergamon Press, 1178 p.

AIGNER, T., 1982, Calcareous tempestites: storm-dominated stratigraphy in Upper Muschelkalk limestones (Middle Triassic, S.W. Germany), p. 248–261 *in* Einsele, G.; and Seilacher, A., eds., Cyclic and event stratification: New York, Springer-Verlag, 536 p.

AISSAOUI, D. M., 1988, Magnesian calcite cements and their diagenesis: dissolution and dolomitization, Mururoa Atoll: Sedimentology, v. 35, p. 821–841.

AITKEN, A. E., RISK, M. J., and HOWARD, J. D., 1988, Animal-sediment relationships on a subarctic intertidal flat, Pangnirtung Fiord, Baffin Island, Canada: Journal of Sedimentary Petrology, v. 58, p. 769–778.

AITKEN, J. D., 1978, Revised models for depositional grand cycles, Cambrian of the southern Rocky Mountains, Canada: Canadian Petroleum Geology Bulletin, v. 26, p. 515–542.

AITKEN, J. D., and MCILREATH, I. A., 1981, Depositional environments of the Cathedral Escarpment near Field, British Columbia, p. 35–44 *in* Aitken, J. D., compiler, and Taylor, M. E., ed., The Cambrian system in the southern Canadian Rocky Mountains, Alberta and British Columbia: Second international symposium on the Cambrian system, Guidebook for Field Trip 2, 61 p.

AITKEN, J. D., and MCILREATH, I. A., 1984, The Cathedral Reef escarpment, a Cambrian great wall with humble origins: Geos, v. 13, p. 17–19.

ALEINIKOFF, J. N., MOENCH, R. H., and LYONS, J. B., 1985, Carboniferous U-Pb age of the Sebago batholith, southwestern Maine: metamorphic (*sic*) and tectonic implications: Geological Society of America Bulletin, v. 96, p. 990–996.

ALEXANDER, G. B.; HESTON, W. M., and ILER, R. K., 1954, The solubility of amorphous quartz in water: Journal of Physical Chemistry, v. 58, p. 453–459.

ALGEO, T.; and WILKINSON, B. H., 1988, Periodicity of mesoscale Phanerozoic sedimentary cycles and the role of Milankovitch orbital modulation: Journal of Geology, v. 96, p. 313–322.

ALLEN, J. R., and PSUTY, N. P., 1987, Morphodynamics of a single-barred beach with a rip channel, Fire Island, New York: Coastal Sediments '87: American Society of Civil Engineers, p. 1964–1975.

ALLEN, J. R. L., 1964, Studies in fluviatile sedimentation: six cyclothems from the Lower Old Red Sandstone, Anglo-Welsh Basin: Sedimentology, v. 3, p. 163–198.

ALLEN, J. R. L., 1970a, The avalanching of granular solids on dune (*sic*) and similar slopes: Journal of Geology, v. 78, p. 326–351.

ALLEN, J. R. L., 1970b, Physical processes of sedimentation: an introduction: Earth Science Series, no. 1: London, George Allen and Unwin, 248 p.

ALLEN, J. R. L., 1981, Lower Cretaceous tides revealed by cross-bedding with mud drapes: Nature, v. 289, p. 579–581.

ALLEN, J. R. L., 1982, Sedimentary structures: their characteristics and physical basis, v. 1: Amsterdam, Elsevier Scientific Publishing Company, 593 p.

ALLEN, J. R. L., 1983, Studies in fluviatile sedimentation: bars, bar complexes and sandstone sheets (low-sinuosity braided streams) in the Brownstones (L. Devonian), Welsh Borders: Sedimentary Geology, v. 33, p. 237–293.

ALLEN, J. R. L., 1985a, Mud micro-washovers: an intertidal sedimentary structure indicating atmospheric exposure: Journal of Sedimentary Petrology, v. 55, p. 240–242.

ALLEN, J. R. L., 1985b, Principles of physical sedimentology: London, George Allen and Unwin, 272 p.

ALLEN, J. R. L., and FRIEND, P. F., 1968, Deposition of the Catskill facies, Appalachian region; with notes on some other Old Red Sandstones basins, p. 2–74 *in* Klein, G. deV., ed., Late Paleozoic and Mesozoic continental sedimentation, northeastern North America, a symposium: Geological Society of America Special Paper 106, 309 p.

ALLEN, P. A., 1985, Hummocky cross-stratification is not produced purely under progressive gravity waves: Nature, v. 313, p. 562–564.

ALLEN, P. A., and ALLEN, J. R., 1990, Basin analysis. Principles and applications: Oxford, Blackwell Scientific Publications, 451 p.

ALLEN, P. A., and COLLINSON, J. D., 1986, Lakes, p. 63–94 *in* Reading, H. G., ed., Sedimentary environments and facies (2nd ed.): Oxford, Blackwell Scientific Publications, 615 p.

ALLEN, P. A.; HOMEWOOD, P.; and WILLIAMS, G. D., 1986, Foreland basins: an introduction, p. 3–12 *in* Allen, P. A.; and Homewood, P., eds., Foreland Basins: International Association of Sedimentologists Special Publication 8: Oxford, Blackwell Scientific Publications, 453 p.

ALLER, R. C., 1982, The effects of macrobenthos on chemical properties of marine sediments and overlying water, p. 53–102 *in* McCall, P. L., and Tevesz, M. J. S., eds., Animal-sediment relations (*sic*): The biogenic alteration of sediments: New York, Plenum Press, 336 p.

ALMASI, M. N.; HOSKIN, C. M.; REED, J. K.; and MILO, J., 1987, Effects of natural and artificial *Thalassia* on rates of sedimentation: Journal of Sedimentary Petrology, v. 57, p. 901–906.

ALTSCHULER, Z. S.; JAFFE, E. G.; and CUTTITTA, F., 1956, The aluminum phosphate (*sic*) zone in the Bone Valley Formation, Florida, and its uranium deposits: U.S. Geological Survey Professional Paper 300, p. 495–504.

ALVAREZ, L. W.; ALVAREZ, WALTER; ASARO, F.; and MICHEL, H. V., 1980, Extraterrestrial cause for the Cretaceous-Tertiary extinction: Science, v. 208, p. 1095–1108.

ALVAREZ, WALTER; ARTHUR, M. A.; FISCHER, A. G.; LOWRIE, W.; NAPOLEONE, G.; PREMOLI-SILVA, I.; and ROGGENTHEN, W. R., 1977, Upper Cretaceous-Paleocene magnetic stratigraphy at Gubbio, Italy: V. Type section for the Late Cretaceous-Paleocene geomagnetic reversal time scale: Geological Society of America Bulletin, v. 88, p. 383–389.

ALVAREZ, WALTER; COLACICCHI, ROBERTO; and MONTANARI, ALESSANDRO, 1985, Synsedimentary slides and bedding formation in Apennine pelagic limestones: Journal of Sedimentary Petrology, v. 55, p. 720–734.

AMAJOR, L. C., 1986, Alluvial fan (*sic*) facies in the Miocene-Pliocene coastal plain (*sic*) sands, Niger Delta, Nigeria: Sedimentary Geology, v. 49, p. 1–20.

AMIEL, A. J., and FRIEDMAN, G. M., 1971, Continental sabkha in Arava Valley between Dead Sea and Red Sea: significance for origin of evaporites: American Association of Petroleum Geologists Bulletin, v. 55, p. 581–592.

ANDERS, M. H., KRUEGER, S. W., and SADLER, P. M., 1987, A new look at sedimentation rates and the completeness of the stratigraphic record: Journal of Geology, v. 95, p. 1–14.

ANDERSON, R. S.; and HALLET, B., 1986, Sediment transport by wind: toward a general model: Geological Society of America Bulletin, v. 97, p. 523–535.

ANDERSON, R. Y., 1982, A long geoclimatic record from the Permian: Journal of Geophysical Research, v. 87, p. 7285–7294.

ANDRESEN, A.; and BJERRUM, L., 1967, Slides in subaqueous slopes in loose sand and silt, p. 221–242 *in* Richards, A. F., ed., Marine geotechnique, Internat. Res. Conf. Marine Geotechnology, 1966, Proceedings: Urbana, University of Illinois Press, 327 p.

ANDREWS, E. D., 1983, Entrainment of gravel from naturally sorted river-bed material: Geological Society of America Bulletin, v. 94, p. 1225–1231.

ANDREWS, E. D., 1984, Bed-material entrainment and hydraulic geometry of gravel-bed rivers in Colorado: Geological Society of America Bulletin, v. 95, p. 371–378.

ANDREWS, W. E., 1938 ms., Restoration and protection of Fire Island: Babylon, New York, Long Island State Park Commission.

ANGEVINE, C. L., and TURCOTTE, D. L., 1983, Porosity reduction by pressure solution: a theoretical model for quartz arenites: Geological Society of America Bulletin, v. 94, p. 1129–1134.

ANSTEY, N. A., 1982, Simple seismics: Boston, MA, International Human Resources Development Corp., 168 p.

ANTEVS, ERNST, 1922, The recession of the last ice sheet in New England [with a preface and contributions by J. W. Goldthwait]: New York, American Geographical Society, Research Series, No. 11, 120 p.

ANTEVS, ERNST, 1925, Retreat of the last ice-sheet (sic) in eastern Canada: Canada Geological Survey Memoir 146, no. 126, 142 p.

ANTEVS, ERNST, 1957, Geological tests of the varve (sic) and radiocarbon chronologies: Journal of Geology, v. 65, p. 129–148.

ARCHIE, G. E., 1950, Introduction to petrophysics of reservoir rocks: American Association of Petroleum Geologists Bulletin, v. 34, p. 943–961.

ARNOTT, R. W., and SOUTHARD, J. B., 1990, Exploratory flow-duct experiments on combined-flow bed configurations, and some implications for interpreting storm-event stratification: Journal of Sedimentary Petrology, v. 60, p. 211–219.

ARTHUR, M. A.; ANDERSON, T. F.; KAPLAN, I. R.; VEIZER, J.; and LAND, L. S., 1983, Stable isotopes in sedimentary geology: Society of Economic Paleontologists and Mineralogists Short Course 10, 435 p.

ARTHUR, M. A., and FISCHER, A. G., 1977, Upper Cretaceous-Paleocene magnetic stratigraphy at Gubbio, Italy, I. Lithostratigraphy and sedimentology: Geological Society of America Bulletin, v. 88, p. 367–371.

ARTHUR, M. A., and GARRISON, R. E., 1986, Cyclicity in the Milankovitch band through geologic time—an introduction: Paleoceanography, v. 1, p. 369–372.

ASHLEY, G. M.; SHAW, J.; and SMITH, N. D., 1985, Glacial sedimentary environments: Tulsa, OK, Society of Economic Paleontologists and Mineralogists Short Course 16, 246 p.

ASHLEY, G. M., and others, 1990, Classification of large-scale subaqueous bedforms: a new look at an old problem: Journal of Sedimentary Petrology, v. 60, p. 160–172.

ASHMORE, P. E., 1991, How do gravel-bed rivers braid?: Canadian Journal of Earth Sciences, v. 28, p. 326–341.

ASTON, S. R., 1980, Nutrients, dissolved gases and general biogeochemistry in estuaries, p. 233–262 in Olausson, E.; and Cato, I., eds., Chemistry and biogeochemistry of estuaries: New York, John Wiley & Sons, 452 p.

ATHEARN, W. D., 1965, Sediment cores from the Cariaco Trench, Venezuela: Woods Hole, MA, Woods Hole Oceanographic Institute, Technical Report 65-37, 20 p.

AUGUSTINIUS, P. G. E. F., 1980, Actual development of the chenier coast of Suriname (South America): Sedimentary Geology, v. 26, p. 91–114.

AUGUSTINIUS, P. G. E. F., 1989, Cheniers and chenier plains: a general introduction: Marine Geology, v. 90, p. 219–230.

AWRAMIK, S. M., 1984, Ancient stromatolites and microbial mats, p. 1–22 in Cohen, Y.; Castenholz, R. W.; and Halvorson, H. O., eds., Microbial mats: stromatolites: New York, Alan R. Liss, Inc., 498 p.

AYALA-CASTANARES, AUGUSTIN; and PHLEGER, F. B., eds., 1969, Lagunas costeras, un simposio: Simposio Internacional sobre lagunas costeras, Mexico City, Mexico, 28–30 Noviembre 1967, memoria: Mexico Universidad Nacional Autonoma (WMAS-UNESCO), 686 p.

AYDIN, A.; and NUR, A., 1982, Evolution of pull-apart basins and their scale independence: Tectonics, v. 1, p. 91–105.

BAAS-BECKING, L. G. M.; KAPLAN, I. R.; and MOORE, D., 1960, Limits of natural environment in terms of pH and oxidation-reduction potentials: Journal of Geology, v. 68, p. 243–284.

BABA, M.; THOMAS, K. V.; and KUMAR, M. P., 1982, Accretion of the Valiathura (Trivandrum) beach in the post-monsoon period. Techn. Rept. No. 17-1982, Centre for Earth Science Studies, Trivandrum, India, 21 p.

BADLEY, M. E.; PRICE, J. D.; RAMBECH DAHL, C.; and AGDESTEIN, T., 1988, The structural evolution of the northern Viking Graben and its bearing upon extensional modes of basin formation: Geological Society of London, Journal, v. 145, p. 455–472.

BAGNOLD, R. A., 1941, The physics of blown sand and desert dunes: London, Methuen and Company, 165 p. (Reprinted 1954.)

BAGNOLD, R. A., 1947, Sand movement by waves; some small-scale experiments with sand of very low density: Institute of Civil Engineers, Journal, v. 28, paper no. 5554, p. 447–469.

BAGNOLD, R. A., 1954, Experiments on a gravity-free dispersion of large solid spheres in a Newtonian fluid under shear: Royal Society (London) Proceedings, (A), v. 255, p. 49–63.

BAGNOLD, R. A., 1956, The flow of cohesionless grains in fluids: Royal Society of London, Philosophical Transactions, Ser. A, v. 249, p. 235–297.

BAGNOLD, R. A., 1963, Beach (sic) and nearshore processes. Part 1. Mechanics of marine sedimentation, p. 507–528 in Hill, M. N., gen. ed., The sea. Ideas and observations on progress in the study of the seas. Volume 3. The Earth beneath the sea. History: New York, Wiley-Interscience, 963 p.

BAGNOLD, R. A., 1968, Deposition in the process of hydraulic transport: Sedimentology, v. 10, p. 45–56.

BAGNOLD, R. A.; and Barndorff-Nielsen, O., 1980, The pattern of natural size distributions: Sedimentology, v. 27, p. 198–207.

BAILEY, S. W., 1980, Structure of layer silicates, p. 1–10 in Brindley, G. W.; and Brown, G., eds., Crystal structures of clay minerals and their X-ray identification: London, Mineralogical Society Monograph 5, 495 p.

BAJORUNAS, L.; and DUANE, D. B., 1967, Shifting offshore bars and harbor shoaling: Journal of Geophysical Research, v. 72, p. 6195–6205.

BAKER, P. H.; and MORGAN, P., 1981, Continental rifting: progress and outlook: Eos, American Geophysical Union Transactions, v. 62, p. 585–586.

BAKER, V. R., 1973, Paleohydrology and sedimentology of Lake Missoula flooding (sic) in eastern Washington: Geological Society of America Special Paper 144, 69 p.

BAKER, V. R., 1978, The Spokane flood controversy and the Martian outflow channels: Science, v. 202, p. 1249–1256.

BAKER, V. R., 1988, Geological fluvial geomorphology: Geological Society of America Bulletin, v. 100, p. 1157–1167.

BALDWIN, BREWSTER; and BUTLER, C. O., 1985, Compaction curves: American Association of Petroleum Geologists Bulletin, v. 69, p. 622–626.

BALLY, A. W., 1975, A geodynamic scenario for hydrocarbon occurrences, panel discussion: Global Tectonics and Petroleum Occurrence; World Petroleum Congress, 9th, Tokyo, Japan, Proceedings, v. 2 (Geology), p. 33–44.

BALLY, A. W., comp., 1981, Geology of passive continental margins: Tulsa, OK, American Associa-

tion of Petroleum Geologists Short Course Notes 19, variously paginated.

BALLY, A. W., 1982, Musings over sedimentary basin evolution: Royal Society (London), Philosophical Transactions, v. A305, p. 325–338.

BALLY, A. W., ed., 1987, Atlas of seismic stratigraphy: Tulsa, OK, American Association of Petroleum Geologists, Studies in Geology 27, v. 1, 125 p.

BALLY, A. W.; and SNELSON, S., 1980, Realms of subsidence, p. 9–75 in Miall, A. D., ed., Facts and principles of world petroleum occurrence: Canadian Society of Petroleum Geologists Memoir 6, 1003 p.

BAMFORD, D., 1979, Seismic constraints on the deep geology of the Caledonides of northern Britain, p. 93–96 in Harris, A. L., Holland, C. H., and Leake, B. E., eds., The Caledonides of the British Isles—reviewed: Geological Society (London) Special Publication 8, 768 p.

BARIGOZZI, C., ed., 1982, Mechanisms of speciation: New York, Alan R. Liss, 546 p.

BARNABY, R. J., and RIMSTIDT, J. D., 1989, Redox conditions of calcite cementation interpreted from Mn and Fe contents of authigenic calcites: Geological Society of America Bulletin, v. 101, p. 795–804.

BARNES, R. S. K., 1980, Coastal lagoons: Cambridge, Cambridge University Press, 106 p.

BARNOLA, J. M.; RAYNAUD, D.; KOROTKEVICH, V. S.; and LORIUS, C., 1987, Vostok ice core provides 160,000-year record of atmospheric CO_2: Nature, v. 329, p. 408–414.

BARRELL, JOSEPH, 1913, The Upper Devonian delta of the Appalachian geosyncline, Part I: The delta and its relations to the interior sea: American Journal of Science, 4th series, v. 36, p. 429–472.

BARRELL, JOSEPH, 1914a, The Upper Devonian delta of the Appalachian geosyncline, Part II: American Journal of Science, 4th series, v. 37, p. 87–104, 225–253.

BARRELL, JOSEPH, 1914b, The strength of the Earth's crust: Journal of Geology, v. 22, p. 28–48, 145–165, 209–236, 289–314, 441–468, 537–555, 655–683, 729–741; 1915, v. 23, p. 27–44, 425–443, 499–515.

BARRELL, JOSEPH, 1915, Central Connecticut in the geologic past: Connecticut Geological and Natural History Survey Bulletin 23, 44 p.

BARRON, E. J., 1989, Severe storms during Earth history: Geological Society of America Bulletin, v. 101, p. 601–612.

BARROWS, M. H., and CLUFF, R. M., 1984, New Albany Shale Group (Devonian–Mississippian) source rocks and hydrocarbon generation in the Illinois basin, p. 111–138 in Demaison, Gerard; and Murris, R. J., eds., Petroleum geochemistry and basin evaluation: Tulsa, OK, American Association of Petroleum Geologists Memoir 35, 426 p.

BARTHEL, K. W., 1981, Lithophaga obesa (Philippi) reef-dwelling and cementing pelecypod—a survey of its boring, p. 649–659 in Gomez, E. D., and others, eds., International Coral Reef (sic) Symposium, 4th, Proceedings, v. 2, 785 p.

BARTON, P.; and WOOD, R., 1984, Tectonic evolution of the North Sea basin: crustal stretching and subsidence: Royal Astronomical Society, Geophysical Journal, v. 79, p. 987–1022.

BARWIS, J. H., and HAYES, M. O., 1979, Regional patterns of modern barrier island (sic) and tidal inlet (sic) deposits as applied to paleoenvironmental studies, p. 472–498 in Ferm, J. C., Horne, J. C., and others, eds., Carboniferous depositional environments in the Appalachian region: University of South Carolina, Carolina Coal Group, 760 p.

BASAN, P. B., ed., 1978, Trace fossil (sic) concepts: Tulsa, OK, Society of Economic Paleontologists and Mineralogists Short Course 5, 181 p.

BASCOM, WILLARD, 1964, Waves and beaches; the dynamics of the ocean surface: Garden City, New York, Doubleday and Company, 267 p.

BASU, ABHIJIT; and MOLINAROLI, EMANUELA, 1989, Provenance characteristics of detrital opaque

Fe-Ti oxide (*sic*) minerals: Journal of Sedimentary Petrology, v. 59, p. 922–924.

BATES, C. C., 1953, Rational theory of delta formation: American Association of Petroleum Geologists Bulletin, v. 37, p. 2119–2162.

BATES, N. B.; and BRAND, UWE, 1990, Secular variation of calcium-carbonate mineralogy (*sic*); an evaluation of ooid and micrite chemistries (*sic*): Geologische Rundschau, v. 79, p. 27–46.

BATES, R. L., and JACKSON, J. A., 1987, Glossary of geology, 3rd edition: Alexandria, Virginia, American Geological Institute, 788 p.

BATHURST, R. G. C., 1975, Carbonate sediments and their diagenesis, 2nd ed.: Amsterdam, Elsevier Scientific Publishing Company, 658 p.

BATHURST, R. G. C., 1980, Deep crustal diagenesis in limestones: Revista Inst. Investigaciones Geol., Barcelona, v. 34, p. 89–100.

BATHURST, R. G. C., 1983, Early diagenesis of carbonate sediments, p. 349–377 *in* Parker, A.; and Sellwood, B. W., eds., Sediment diagenesis: Boston, D. Reidel Publishing Company, 427 p.

BATHURST, R. G. C., 1987, Diagenetically enhanced bedding in organic platform limestones: stratified cementation and selective compaction: Sedimentology, v. 34, p. 749–778.

BATTARBEE, R. W.; MASON, SIR J.; RENBERG, I.; and TALLING, J. F., eds., 1990, Paleolimnology and lake acidification: London, Royal Society of London, 219 p.

BAULD, J., 1981, Occurrence of benthic microbial mats in saline lakes: Hydrobiologia, v. 81, p. 87–111.

BAUMGAERTNER, IMRE VAN, 1975, A major beach erosional (*sic*) cycle at Robert Moses State Park, Fire Island, during the storm of 1–2 December 1974: the confirmation of "grazing-swash undercutting" as a major beach erosional (*sic*) mechanism, p. 259–278 *in* Wolff, M. P., ed., New York State Geological Association, Annual Meeting, 47th, Hofstra University, Hempstead, New York, 31 Oct.–02 Nov., 1975, Guidebook: Department of Geology, Hofstra University, Hempstead, New York, 327 p.

BAYER, U.; and SEILACHER, A., eds., 1985, Sedimentary (*sic*) and evolutionary cycles: New York, Springer-Verlag, 465 p.

BEACH, D. K., and GINSBURG, R. N., 1980, Facies succession of Pliocene–Pleistocene carbonates, northwestern Great Bahama Bank: American Association of Petroleum Geologists Bulletin, v. 64, p. 1634–1642.

BEADLE, L. C., 1974, The inland waters of tropical Africa: London, Longman, 365 p.

BEAUCHAMP, B.; KROUSE, H. R.; HARRISON, J. C.; NASSICHUK, W. W.; and ELIUK, L. S., 1989, Cretaceous cold-seep communities and methane-derived carbonates in the Canadian Arctic: Science, v. 244, p. 53–56.

BEAUDRY, DESIREE; and MOORE, G. F., 1985, Seismic stratigraphy and Cenozoic evolution of West Sumatra forearc basin: American Association of Petroleum Geologists Bulletin, v. 69, p. 742–759.

BEAUMONT, C.; QUINLAN, G.; and HAMILTON, J., 1988, Orogeny and stratigraphy: numerical models of the Palaeozoic in the eastern interior of North America: Tectonics, v. 7, 389–416.

BEAUMONT, C.; and TANKARD, A. J., eds., 1987, Sedimentary basins and basin-forming mechanisms: Canadian Society of Petroleum Geologists Memoir 12, 527 p.

BECHTEL, T.; FORSYTH, D.; and SWAIN, C., 1987, Mechanisms of isostatic compensation in the vicinity of the East African rift, Kenya: Royal Astronomical Society, Geophysical Journal, v. 90, 445–465.

BECKMANN, W. C.; ROBERTS, A. C.; and LUSKIN, BERNARD, 1959, Subbottom depth recorder: Geophysics, v. 24, p. 749–760.

BEDIENT, P. B., and HABER, W. C., 1988, Hydrology and floodplain analysis: Reading, MA, Addison-Wesley Publishing Company, 650 p.

BEERBOWER, J. R., 1964, Cyclothems and cyclic depositional mechanisms in alluvial plain (*sic*) sedimentation, p. 31–42 *in* Merriam, D. F., ed., Symposium on cyclic sedimentation: Kansas Geological Survey Bulletin 169, 2 vols. (v. 1, p. 1–380; v. 2, p. 381–636).

BEGET, T.; and HAWKINS, D. B., 1989, Influence of orbital parameters on Pleistocene loess deposition in central Alaska: Nature, v. 337, p. 151–153.

BEGET, T.; STONE, D. B.; and HAWKINS, D. B., 1990, Paleoclimatic forcing of magnetic susceptibility (*sic*) variations in Alaskan loess during the Late Quaternary: Geology, v. 18, p. 40–43.

BEHRENS, E. W., 1969, Hurricane effects on a hypersaline bay, p. 301–311 *in* Ayala-Castanares, Augustin; and Phleger, F. B., eds., Lagunas costeras, un simposio: Simposio Internacional sobre lagunas costeras, Mexico City, Mexico, 28–30 Noviembre 1967, memoria: Mexico Universidad Nacional Autonoma (WMAS-UNESCO), 686 p.

BELKNAP, D. F., KELLEY, J. T., and SHIPP, R. C., 1987, Quaternary stratigraphy of representative Maine estuaries: initial examination by high-resolution seismic reflection profiling, p. 178–207 *in* Fitzgerald, D. M., and Rosen, P. S., eds., Glaciated coasts: San Diego, Academic Press, 364 p.

BELKNAP, D. F., and KRAFT, J. C., 1981, Preservation potential of transgressive coastal lithosomes on the U.S. Atlantic coast: Marine Geology, v. 42, p. 424–442.

BELKNAP, D. F., and KRAFT, J. C., 1985, Influence of antecedent geology on stratigraphic preservation potential and evolution of Delaware's barrier systems: Marine Geology, v. 63, p. 235–262.

BELLAICHE, G.; DROZ, L.; ALOIS, J.-C.; and others, 1981, The Ebro and Rhône deep-sea fans: first comparative study: Marine Geology, v. 43, p. M75–M85.

BELLANCA, A.; and NERI, R., 1986, Evaporite carbonate cycles of the Messinian, Sicily, stable isotopes, mineralogy, textural features, and environmental implications: Journal of Sedimentary Petrology, v. 56, p. 614–621.

BELPERIO, A. P., and SEARLE, D. E., 1988, Terrigenous and carbonate sedimentation in the Great Barrier Reef Province, p. 143–174 *in* Doyle, L. J., and Roberts, H. H., eds., Carbonate-clastic transitions: Developments in Sedimentology No. 42: New York, Elsevier, 304 p.

BELT, E. S., 1975, Scottish Carboniferous cyclothem patterns and their paleoenvironmental significance, p. 427–449 *in* Broussard, M. L., ed., Deltas. Models for exploration: Houston, TX, Houston Geological Society, 555 p.

BENNETT, H. H., 1939, A permanent loss for New England: soil erosion resulting from the hurricane: Geographical Review, v. 29, p. 196–204.

BENNETT, P. C., and SIEGEL, D. I., 1987, Increased solubility of quartz in water due to complexing by inorganic compounds: Nature, v. 236, p. 684–686.

BENNETT, R. H., BRYANT, W. R., and HULBERT, M. H., 1991, Microstructure of fine-grained sediments: from mud to shale: New York, Springer-Verlag, 582 p.

BENSON, R. H., 1972, Ostracodes as indicators of threshold depth in the Mediterranean during the Pliocene, p. 63–75 *in* Stanley, D. J., ed., The Mediterranean Sea: a natural sedimentation laboratory: Stroudsburg, PA, Dowden, Hutchinson, and Ross, 765 p.

BENTOR, Y. K., ed., 1980, Marine phosphorites: Tulsa, OK, Society of Economic Paleontologists and Mineralogists Special Publication 29, 249 p.

BERELSON, W. M., and HERON, S. D., JR., 1985, Correlations between Holocene flood tidal (*sic*) delta and barrier island inlet (*sic*) fill sequences: Back Sound–Shackleford Banks, North Carolina: Sedimentology, v. 32, p. 215–222.

BERG, O. R., and WOLVERTON, D. B., eds., 1985, Seismic stratigraphy II, an integrated approach: Tulsa, OK, American Association of Petroleum Geologists Memoir 39, 276 p.

BERGER, A.; IMBRIE, J.; HAYS, J. D.; KUKLA, G.; and SALTZMAN, B., eds., 1984, Milankovitch and climate. Understanding the response to astronomic forcing, 2 vols.: Dordrecht, The Netherlands, D. Reidel Publishing Company, v. 1, 510 p.; v. 2, 384 p.

BERGER, W. H., 1967, Foraminiferal ooze: solution (*sic*) at depths: Science, v. 156, p. 383–385.

BERGER, W. H., 1968, Planktonic foraminifera: selective solution (*sic*) and paleoclimatic interpretation: Deep-Sea Research, v. 15, p. 31–43.

BERGER, W. H., 1972, Deep sea (*sic*) carbonates: dissolution facies and age-depth constancy: Nature, v. 236, p. 392–395.

BERGER, W. H., 1974, Deep-sea sedimentation, p. 213–241 *in* Burk, C. A., and Drake, C. L., eds., The geology of continental margins: New York, Springer-Verlag, 1009 p.

BERGER, W. H., 1975, Deep-sea carbonates: dissolution profiles from foraminiferal preservation, p. 82–86 *in* Sliter, W. V., Be, A. W. H., and Berger, W. H., eds., Dissolution of deep-sea carbonates: Cushman Foundation for Foraminiferal Research Special Publication 13, 159 p.

BERGER, W. H., and WINTERER, E. L., 1974, Plate stratigraphy and the fluctuating carbonate line, p. 11–48 *in* Hsü, K. J., and Jenkyns, H. C., eds., Pelagic sediments: on land and under the sea: International Association of Sedimentologists Special Publication 1, 447 p.

BERGGREN, W. A., KENT, D. V., FLYN, J. J., and VAN COUVERING, J. A., 1985, Cenozoic geochronology: Geological Society of America Bulletin, v. 96, p. 1407–1418.

BERGH, S.; and TORSKE, T., 1986, The Proterozoic Skoadduvarri Sandstone Formation, Alta, northern Norway: a tectonic fan-delta complex: Sedimentary Geology, v. 47, p. 1–25.

BERNARD, H. A., and LEBLANC, R. J., SR., 1965, Resume of the Quaternary geology of the northwestern Gulf of Mexico province, p. 137–185 *in* Wright, H. E., Jr., and Frey, D. G., eds., The Quaternary of the United States: Princeton, NJ, Princeton University Press, 922 p.

BERNARD, H. A., MAJOR, C. F., JR., PARROTT, B. S., and LEBLANC, R. J., SR., 1970, Recent sediments of Southeast Texas—a field guide to the Brazos alluvial (*sic*) and deltaic plains and the Galveston barrier island (*sic*) complex: Texas University at Austin, Bureau of Economic Geology, Guidebook 11, 16 p. + reprint appendices [83 p.].

BERNER, R. A., 1970, Pleistocene sea levels possibly indicated by buried black sediments in the Black Sea: Nature, v. 227, p. 700.

BERNER, R. A., 1980, Early diagenesis—a theoretical approach: Princeton, NJ, Princeton University Press, 241 p.

BERNER, R. A., 1989, Biogeochemical cycles of carbon and sulfur and their effect on atmospheric oxygen over Phanerozoic time: Global and Planetary Change, v. 1, p. 97–122.

BETHKE, C. M.; HARRISON, W. J.; UPSON, C.; and ALTANER, S. P., 1988, Supercomputer analysis of sedimentary basins: Science, v. 239, p. 261–267.

BHATIA, M. R., and CROOK, K. A. W., 1986, Trace element (*sic*) characteristics of graywackes and tectonic setting (*sic*) discrimination of sedimentary basins: Contributions to Mineralogy and Petrology, v. 92, p. 181–193.

BHATTACHARYYA, AJIT; and FRIEDMAN, G. M., 1979, Experimental compaction of ooids and lime mud and its implications for lithification during burial: Journal of Sedimentary Petrology, v. 49, p. 1279–1286.

BHATTACHARYYA, AJIT; and FRIEDMAN, G. M., eds., 1983a, Modern carbonate environments: Benchmark papers in geology, v. 74: Stroudsburg, PA, Dowden, Hutchinson, and Ross, 376 p.

BHATTACHARYYA, AJIT; and FRIEDMAN, G. M., 1983b, Mineralogical and paramorphic textural changes in modern ooids by heat and compaction: Geology, v. 11, p. 596–598.

BHATTACHARYYA, AJIT; and FRIEDMAN, G. M., 1984, Experimental compaction of ooids under deep-burial diagenetic temperatures and pressures: Journal of Sedimentary Petrology, v. 54, p. 362–372.

BIDDLE, K. T.; and CHRISTIE-BLICK, N., eds., 1985, Strike-slip deformation, basin formation and sedimentation: Society of Economic Paleontologists and Mineralogists Special Publication 37, 386 p.

BIGGS, R. B., 1978, Coastal bays, p. 69–99 *in* Davis, R. A., Jr., ed., Coastal sedimentary environments: New York, Springer-Verlag, 420 p.

BIRCH, FRANCIS; SCHAIRER, J. F.; and SPICER, H. C., 1942, Handbook of Physical Constants: Geological Society of America Special Paper 36, 325 p.

BIRD, E. C. F., 1969, Coasts. An introduction to systematic geomorphology: Cambridge, MA, Massachusetts Institute of Technology Press, 246 p.

BIRD, E. C. F., and SCHWARTZ, M. L., eds., 1985, The world's coastline: New York, Van Nostrand Reinhold Company, 1071 p.

BIRKENMEIER, W. A., 1985, Time scales of nearshore profile (*sic*) changes, p. 1507–1521 *in* International Conference on Coastal Engineering, 19th, Houston, Texas, Proceedings: New York, American Society of Civil Engineers.

BIRNBAUM, S. J., and WIREMAN, J. W., 1985, Sulfate-reducing bacteria and silica solubility: a possible mechanism for evaporite diagenesis and silica precipitation in banded iron formations: Canadian Journal of Earth Sciences, v. 22, p. 1904–1909.

BITSCHENE, P. R., and SCHMINCKE, H. U., 1990, Fallout tephra layers; composition and significance, p. 48–82 *in* Heling, Dietrich et al., eds., Sediments and environmental geochemistry; selected aspects and case history: Berlin, Springer-Verlag, 371 p.

BITTERLI, P., 1963, Aspects of the genesis of bituminous rock sequences: Geologie en Mijnbouw, N. S., v. 42, p. 183–201.

BLAIR, T. C., 1987a, Sedimentary processes, vertical stratification sequences, and geomorphology of the Roaring River alluvial fan, Rocky Mountain National Park, Colorado: Journal of Sedimentary Petrology, v. 57, p. 1–18.

BLAIR, T. C., 1987b, Tectonic (*sic*) and hydrologic controls on cyclic alluvial fan (*sic*), fluvial (*sic*) and lacustrine rift-basin sedimentation, Jurassic-Lowermost Cretaceous Todos Santos Formation, Chiapas, Mexico: Journal of Sedimentary Petrology, v. 57, p. 845–862.

BLATT, HARVEY, 1967a, Original characteristics of clastic quartz grains: Journal of Sedimentary Petrology, v. 37, p. 401–424.

BLATT, HARVEY, 1967b, Provenance determinations and recycling of sediments: Journal of Sedimentary Petrology, v. 47, p. 1031–1044.

BLATT, HARVEY, 1982, Sedimentary petrology: San Francisco, W. H. Freeman and Company, 564 p.

BLATT, HARVEY, 1985, Provenance studies and mudrocks: Journal of Sedimentary Petrology, v. 55, p. 69–75.

BLATT, HARVEY, 1989, Flux of siliciclastic grains in sediments: Journal of Geological Education, v. 37, p. 243–249.

BLATT, HARVEY; BERRY, W. B. N.; and BRANDE, SCOTT, 1991, Principles of stratigraphic analysis: Boston, Blackwell Scientific Publications, 512 p.

BLATT, HARVEY; MIDDLETON, G. V.; and MURRAY, R., 1972, Origin of sedimentary rocks: Englewood Cliffs, NJ, Prentice-Hall, 634 p. (2nd ed., 1980, 782 p.)

BLODGETT, R. H., and STANLEY, K. O., 1980, Stratification, bedforms and discharge relations of the Platte braided river system, Nebraska: Journal of Sedimentary Petrology, v. 50, p. 139–148.

BLOM, W. M., and ALSOP, D. B., 1988, Carbonate mud (*sic*) sedimentation on a temperate shelf: Bass Basin, southeastern Australia: Sedimentary Geology, v. 60, p. 269–280.

BLOOM, A. L., 1965, The explanatory description of coasts [with German and French abstracts]: Zeitschrift fur Geomorphologie, v. 9, p. 422–436.

BLOOM, A. L., 1977, Atlas of sea-level curves: Ithaca, NY, Cornell University, IGCP-200.

BLOOM, A. L., and ELLIS, C. W., 1965, Postglacial stratigraphy and morphology of coastal Connecticut: Connecticut Geological and Natural History Survey, Guidebook 1, 10 p.

BLUCK, B. J., 1967, Sedimentation of gravel beaches: examples from South Wales: Journal of Sedimentary Petrology, v. 37, p. 128–156.

BLUCK, B. J., 1969, Particle rounding in beach gravel: Geological Magazine, v. 106, p. 1–14.

BOBB, W. H., and BOLAND, R. A., JR., 1969, Channel improvement, Fire Island Inlet, New York: Vicksburg, Mississippi, U.S. Army Engineering Experiment Station, Technical Report H-69-16, 205 p.

BOERSMA, J. R., 1967, Remarkable types of mega cross-stratification in the fluviatile sequence of a sub-Recent distributary of the Rhine, Amerongen, The Netherlands: Geologie en Mijnbouw, v. 46, p. 217–235.

BOGEN, J., 1983, Morphology and sedimentology of deltas in fjords and fjord valley (*sic*) lakes: Sedimentary Geology, v. 36, p. 245–267.

BOGGS, SAM, JR., 1987, Principles of sedimentology and stratigraphy: Columbus, OH, Merrill Publishing Company, 784 p.

BOJE, R.; and TOMCZAK, M., eds., 1978, Upwelling ecosystems: New York, Springer-Verlag, 300 p.

BOKUNIEWICZ, H. J., 1981a, The seasonal beach at East Hampton, New York: Shore and Beach, v. 49, no. 3, p. 28–33.

BOKUNIEWICZ, H. J., 1981b, Monitoring seasonal beach responses: an educational and public service (*sic*) program: Journal of Geological Education, v. 29, p. 121–127.

BOKUNIEWICZ, H. J., 1985 ms., The condition of the beach at the Fire Island Pines, NY, 1984–1985: Stony Brook, NY, State University of New York, Marine Sciences Research Center, Report to the Fire Island National Seashore Advisory Board, 36 p. + App.

BOKUNIEWICZ, H. J., GORDON, R. B., and KASTENS, K. A., 1977, Form and migration of sand waves in a large estuary, Long Island Sound: Marine Geology, v. 24, p. 185–199.

BOLLI, H. M., and SILVA, I. P., 1973, Oligocene to Recent planktonic Foraminifera and stratigraphy (*sic*) of the leg 15 sites in the Caribbean Sea, p. 475–497 *in* Edgar, N. T., Kaneps, A. G., and Herring, J. R., eds., Initial Reports of the Deep-Sea Drilling Project, v. 15: Washington, DC, U.S. Government Printing Office, 1137 p.

BOLTON, B. R., and FRAKES, L. A., 1985, Geology and genesis of manganese oolite, Chiatura, Georgia, U.S.S.R.: Geological Society of America Bulletin, v. 96, p. 1398–1406.

BOND, G. C., and KOMINZ, M. A., 1984, Construction of tectonic subsidence curves for the early Paleozoic miogeocline, southern Canadian Rocky Mountains: implications for subsidence mechanisms, age of breakup, and crustal thinning: Geological Society of America Bulletin, v. 95, p. 155–173.

BOND, G. C., and KOMINZ, M. A., 1988, Evolution of thought on passive continental margins from the origin of geosynclinal theory (1860) to the present: Geological Society of America Bulletin, v. 100, p. 1909–1933.

BOOTHROYD, J. C., 1978, Mesotidal inlets and estuaries, p. 287–360 *in* Davis, R. A., Jr., ed., Coastal sedimentary environments: New York, Springer-Verlag, 420 p.

BOOTHROYD, J. C., FRIEDRICH, N. E., and McGUINN, S. R., 1985, Geology of microtidal coastal lagoons, Rhode Island: Marine Geology, v. 63, p. 35–76.

BOOTHROYD, J. C.; and NUMMEDAL, D., 1978, Proglacial braided outwash: a model for humid alluvial fan (*sic*) deposits, p. 641–668 *in* Miall, A. D., ed., Fluvial sedimentology: Calgary, Alberta, Canadian Society of Petroleum Geologists Memoir 5, 859 p.

BOPP, R. F.; SIMPSON, H. J.; OLSEN, C. R.; TRIER, R. M.; and KOSTYK, NADIA, 1981, Chlorinated hydrocarbons and radionuclide chronologies in sediments of the tidal Hudson River, New York: Environmental Science and Technology, v. 15, p. 210–216.

BORAK, B.; and FRIEDMAN, G. M., 1981, Textures of sandstones and carbonate rocks in the world's deepest wells: Anadarko Basin, Oklahoma: Sedimentary Geology, v. 29, p. 133–151.

BORCHERT, H., 1960, Genesis of marine sedimentary iron ore: Institute of Mining and Metallurgy Transactions, v. 69, p. 261–279.

BORNHOLD, B. D.; and GIRESSE, PIERRE, 1985, Glauconitic sediments on the continental shelf off Vancouver Island, B.C., Canada: Journal of Sedimentary Petrology, v. 55, p. 653–664.

BOSE, P. K.; GHOSH, GAUTAM; SHOME, SABYASACHI; and BARDHAN, SUBHENDU, 1988, Evidence of superimposition of storm waves on tidal currents in rocks from the Tithonian-Neocomian Unia Member, Kutch, India: Sedimentary Geology, v. 54, p. 321–329.

BOSELLINI, A., 1984, Progradation geometries (*sic*) of carbonate platforms: examples from the Triassic of the Dolomites, northern Italy: Sedimentology, v. 31, p. 1–24.

BOUMA, A. H., 1962, Sedimentology of some flysch deposits. A graphic approach to facies interpretation: Amsterdam, Elsevier Scientific Publishing Company, 168 p.

BOUMA, A. H., 1979, Continental slopes, p. 1–15 *in* Doyle, L. J., and Pilkey, O. H., eds., Geology of continental slopes: Tulsa, OK, Society of Economic Paleontologists and Mineralogists Special Publication 27, 374 p.

BOUMA, A. H., 1987, Megaturbidite; an acceptable term?: GeoMarine Letters, v. 7, p. 63–67.

BOUMA, A. H., BERRYHILL, H. L., BRENNER, R. L., and KNEBEL, H. J., 1982, Continental shelf and epicontinental seaways, p. 281–327 *in* Scholle, P. A.; and Spearing, D., eds., Sandstone depositional environments: Tulsa, OK, American Association of Petroleum Geologists Memoir 31, 410 p.

BOUMA, A. H., and BRYANT, W. R., 1969, Rapid delta growth in Matagorda Bay, Texas, p. 171–189 *in* Ayala-Castanares, Augustin; and Phleger, F. B., eds., Lagunas costeras, un simposio: Simposio Internacional sobre lagunas costeras, Mexico City, Mexico, 28–30 Noviembre 1967, memoria: Mexico Universidad Nacional Autonoma (WMAS-UNESCO), 686 p.

BOUMA, A. H., MOORE, G. T., and COLEMAN, J. M., eds., 1978, Framework, facies, and oil-trapping characteristics of the upper continental margin: Tulsa, OK, American Association of Petroleum Geologists, Studies in Geology 7, 326 p.

BOUMA, A. H., NORMARK, W. K., and BARNES, N. E., 1985, Deepsea fans and related turbidite sequences: New York, Springer-Verlag, 351 p.

BOURGEOIS, JOANNE; HANSEN, T. A.; WIBERG, P. L.; and KAUFFMAN, E. G., 1988, A tsunami deposit at the Cretaceous-Tertiary boundary in Texas: Science, v. 241, p. 567–570.

BOURGEOIS, J.; and LEITHOLD, E. L., 1984, Wave-worked conglomerates—depositional processes and criteria for recognition, p. 331–343 *in* Koster, E. H., and Steel, R. J., eds., Sedimentology of gravels and conglomerates: Canadian Society of Petroleum Geologists Memoir 10, 441 p.

BOWEN, A. J., and INMAN, D. L., 1969, Rip currents 2: Laboratory (*sic*) and field observations: Journal of Geophysical Research, v. 74, p. 5479–5490.

BOWEN, A. J., INMAN, D. L., and SIMMONS, V. P., 1968, Wave "set-down" and "set-up": Journal of Geophysical Research, v. 73, p. 2569–2577.

BOWER, J. C., 1984, The relative biostratigraphic values of fossils: Computers and Geosciences, v. 10, p. 111–132.

BOWER, J. C., 1985, The index fossil (*sic*) concept and its application to quantitative stratigraphy, p. 43–64 *in* Gradstein, F. M., and others, eds., Quantitative stratigraphy: Hingham, MA, D. Reidel Publishing Company, 598 p.

BRADLEY, W. H., 1929, The varves and climate of the Green River epoch: U.S. Geological Survey Professional Paper 140, p. 87–110.

BRADLEY, W. H., 1948, Limnology and the Eocene lakes of the Rocky Mountain region: Geological Society of America Bulletin, v. 59, p. 635–648.

BRAMBATI, ANTONIO, 1972, Sedimentology and pollution in the Mediterranean: a discussion, Part 12, Epilogue, p. 711–721 *in* Stanley, D. J., ed., The Mediterranean Sea: a natural sedimentation laboratory: Stroudsburg, PA, Dowden, Hutchinson, and Ross, 765 p.

BRAMLETTE, M. N., and BRADLEY, W. H., 1940, Geology and biology of North Atlantic deep-sea cores: U.S. Geological Survey Professional Paper 196-A, p. 1–34.

BRAUN, MOSHE; and FRIEDMAN, G. M., 1969, Carbonate lithofacies and environments of the Tribes Hill Formation (Lower Ordovician) of the Mohawk Valley, New York: Journal of Sedimentary Petrology, v. 39, p. 113–135.

BRAYSHAW, A. C., 1985, Bed microtopography and entrainment thresholds in gravel-bed rivers: Geological Society of America Bulletin, v. 96, p. 218–223.

BRENNER, R. L., 1980, Construction of process-response models for ancient epicontinental seaway depositional systems using partial analogs: American Association of Petroleum Geologists Bulletin, v. 64, p. 1223–1243.

BRENNER, R. L., SWIFT, D. J. P., and GAYNOR, G. C., 1985, Reevaluation of coquinoid sandstone (*sic*) depositional model, Upper Jurassic of central Wyoming and south-central Montana: Sedimentology, v. 32, p. 363–372.

BRENNINKMEYER, B. M., and NWANKWO, A. F., 1987, Source of pebbles at Mann Hill beach, Scituate, Massachusetts, p. 251–277 *in* Fitzgerald, D. M., and Rosen, P. J., eds., Glaciated Coasts: New York, Academic Press, 364 p.

BRETZ, J. H., SMITH, H. T. U., and NEFF, G. E., 1956, Channeled scabland of Washington—new data and interpretations: Geological Society of America Bulletin, v. 67, p. 957–1049.

BREWER, J. A.; GOOD, R.; OLIVER, J. E.; BROWN, L. D.; and KAUFMAN, S., 1983, COCORP profiling across the southern Oklahoma aulacogen: overthrusting across the Wichita Mountains and compression within the Anadarko Basin: Geology, v. 11, p. 109–114.

BREYER, J. A., and McCABE, P. J., 1986, Coals associated with tidal sediments in the Wilcox Group (Paleogene), South Texas: Journal of Sedimentary Petrology, v. 56, p. 510–519.

BRIDGE, J. R., 1981, Hydraulic interpretation of grain-size distributions using a physical model for bed load (*sic*) transport: Journal of Sedimentary Petrology, v. 51, p. 1109–1124.

BRIDGES, P. H., 1982, Ancient offshore tidal deposits, *in* Stride, A. H., ed., Offshore tidal sands: New York, Chapman and Hall.

BROCKERHOFF, F. G., and FRIEDMAN, G. M., 1987, Paleodepth of burial of Middle Ordovician Chazy Group carbonates in New York and Vermont: Northeastern Geology, v. 9, p. 51–58.

BROECKER, W. S., and DENTON, G. H., 1989, The role of ocean-atmosphere reorganizations in glacial cycles: Geochimica et Cosmochimica Acta, v. 53, p. 2465–2501.

BROECKER, W. S., and DENTON, G. H., 1990, What drives glacial cycles? Scientific American, v. 262, no. 1, p. 48–56.

BROECKER, W. S.; LI, Y-H.; and PENG, T-H., 1971, Carbon dioxide—Man's unseen artifact, Chapter 11, *in* Hood, D. W., ed., Impingement of man on the oceans: New York, Wiley-Interscience, 738 p.

BROMELY, R. G., and EKDALE, A. A., 1984, *Chondrites*: a trace fossil (*sic*) indicator of anoxia in sediments: Science, v. 224, p. 872–874.

BROOKFIELD, M. E., 1984, Eolian sands, p. 91–104 *in* Walker, R. G., ed., Facies models, 2nd ed.: Geoscience Canada Reprint Series 1, 317 p.

BROOKINGS, D. G., 1988, Eh-pH diagrams for geochemistry: Berlin, Springer-Verlag, 176 p.

BROWN, L.; ANDO, C.; KLEMPERER, S.; OLIVER, J.; KAUFMAN, S.; CZUCHRA, B.; WALSH, T.; and ISACHSEN, Y. W., 1983, Adirondack-Appalachian crustal structure: the COCORP northeast traverse: Geological Society of America Bulletin, v. 94, p. 1173–1184.

BROWN, L. F., JR., 1979, Deltaic sandstone facies of the midcontinent, p. 35–63 *in* Hyne, N. J., ed., Pennsylvanian sandstones of the mid-continent: Tulsa Geological Society, 360 p.

BROWN, L. F., JR., and FISHER, W. L., 1980, Seismic stratigraphic (*sic*) interpretation and petroleum exploration: Geophysical principles and techniques: Tulsa, OK, American Association of Petroleum Geologists, Continuing Education Course Notes 16, 56 p.

BROWN, R. G., 1969, The Coorong Lagoon, South Australia, p. 191–192 *in* Ayala-Castanares, Augustin; and Phleger, F. B., eds., Lagunas costeras, un simposio: Simposio Internacional sobre lagunas costeras, Mexico City, Mexico, 28–30 Noviembre 1967, memoria: Mexico Universidad Nacional Autonoma (WMAS-UNESCO), 686 p.

BROWNE, K. M., and DEMICCO, R. V., 1987, Thrombolites of the Lower Devonian Manlius Formation of central New York, Carbonates and Evaporites, v. 2, p. 149–155.

BRUNEAU, L.; JERLOV, N. G.; and KOCZY, F. F., 1953, Physical (*sic*) and chemical methods: Goteborg, Swedish Deep-Sea Expedition 1947–48, Reports, v. 3.

BRUNNER, C. A., and NORMARK, W. R., 1985, Biostratigraphic implications for turbidite depositional processes on the Monterey deep-sea fan, central California: Journal of Sedimentary Petrology, v. 55, p. 495–505.

BRUUN, A. F., 1957, Deep sea and abyssal depths, p. 641–672 *in* Hedgpeth, J. W., ed., Treatise on marine ecology and paleoecology: Geological Society of America Memoir 67, v. 1, Ecology, 1296 p.

BRUUN, P. M., 1969, Tidal inlets on alluvial shores, p. 349–365 *in* Ayala-Castanares, Augustin; and Phleger, F. B., eds., Lagunas costeras, un simposio: Simposio Internacional sobre lagunas costeras, Mexico City, Mexico, 28–30 Noviembre 1967, memoria: Mexico Universidad Nacional Autonoma (WMAS-UNESCO), 686 p.

BUCZYNSKI, Chris; and CHAFETZ, H. S., 1990, Habit of bacterially induced precipitates of calcium carbonate: examples for laboratory experiments and Recent sediments, p. 20 *in* Rezak, Richard; and Lavoie, Dawn, eds., Carbonate microfabrics symposium and workshop, Program: College Station, TX, Texas A&M University, 60 p.

BUDD, D. A., 1988, Aragonite-to-calcite transformation during fresh-water diagenesis of carbonates: insights from pore-water chemistry: Geological Society of America Bulletin, v. 100, p. 1260–1270.

BUFFINGTON, E. C., 1966, Sounding, p. 823–829 *in* Fairbridge, R. W., ed., The encyclopedia of oceanography: Encyclopedia of Earth Sciences, v. 1: New York, Reinhold Publishing Corporation, 1021 p.

BUICK, R.; DUNLOP, J. S. R.; and GROVES, D. I., 1981, Stromatolite recognition in ancient rocks: an appraisal of irregularly laminated structures in an Early (*sic*) Archaean chert-barite unit from North Pole, Western Australia: Alcheringa, v. 5, p. 161–181.

BURBANK, D. W.; BECK, R. A.; RAYNOLDS, R. G. H.; HOBBS, R.; and TAHIRKHELI, R. A. K., 1988, Thrusting and gravel progradation in foreland basins: a test of post-thrusting gravel dispersal: Geology, v. 16, p. 1143–1146.

BURCHFIEL, B. C.; and ROYDEN, L., 1982, Carpathian foreland fold and thrust belt and its relation to Pannonian and other basins: American Association of Petroleum Geologists Bulletin, v. 66, p. 1179–1195.

BURGIS, M. J.; and MORRIS, P., 1987, The natural history of lakes: Cambridge, Cambridge University Press, 218 p.

BURK, C. A., and DRAKE, C. L., 1974, The geology of continental margins: New York, Springer-Verlag, 1009 p.

BURKE, K. C. A., and DEWEY, J. F., 1972, Orogeny in Africa, p. 583–608 *in* Dessauvagie, T. F. J., and Whiteman, A. J., eds., African geology: Ibadan, Nigeria, University of Ibadan, 668 p.

BURKE, K. C. A., and DEWEY, J. F., 1973, Plume-generated triple junctions: key indicators in applying plate tectonics to old rocks: Journal of Geology, v. 81, p. 406–433.

BURKE, K. C. A., and WILSON, J. T., 1972, Is the African plate stationary? Nature, v. 239, p. 387–390.

BURNETT, W. C., and RIGGS, S. R., 1990, Phosphate deposits of the world. Volume 3, Neogene to modern phosphorites: Cambridge, Cambridge University Press, 464 p.

BURRUS, J., ed., 1986, Thermal modeling in sedimentary basins, First IFP Exploration Research Conference, Carcans, France, June 3–7, 1985. Paris, Editions Technip, 600 p.

BURST, J. F., 1958, "Glauconite" pellets: their mineral nature and applications to stratigraphic interpretations: American Association of Petroleum Geologists Bulletin, v. 42, p. 310–327.

BUSCH, D. A., 1974, Stratigraphic traps in sandstones: exploration techniques: Tulsa, OK, American Association of Petroleum Geologists Memoir 21, 174 p.

BUSCH, R. M., and ROLLINS, H. B., 1984, Correlation of Carboniferous strata using a hierarchy of transgressive-regressive units: Geology, v. 12, p. 471–474.

BUSH, P., 1973, Some aspects of the diagenetic history of the sabkha in Abu Dhabi, Persian Gulf, p. 395–407 *in* Purser, B. H., ed., The Persian Gulf. Holocene carbonate sedimentation and diagenesis in a shallow epicontinental sea: New York, Springer-Verlag, 471 p.

BUTLIN, K. R., 1953, The bacterial sulphur (*sic*) cycle: Research (London), v. 6, p. 184–191.

BUTTNER, P. J. R., 1987, The barrier system of the south shore of Long Island, New York, U.S.A.—response to the expected rise in sea level: Northeastern Environmental Science, v. 6, p. 1–22.

BUTTNER, P. J. R., 1988, An artificial intelligence (*sic*) framework for the analysis of coastal change, Long Island barrier system, New York: Phase 1, monitoring data base 1985–1989: Northeastern Environmental Science, v. 7, p. 71–82.

BUTTNER, P. J. R., 1989, Shoreline dynamics of eastern Jones Island, Long Island, New York: Phase I, Summary of monitoring data base 1985–1989: Northeastern Environmental Science, v. 8, p. 135–171.

BUTTON, A., 1976, Iron-formation (*sic*) as an end member in carbonate sedimentary cycles in the Transvaal Supergroup, South Africa: Economic Geology, v. 71, p. 193–201.

BUURMAN, P.; and VAN DER PLAS, L., 1968, The occurrence of halloysite and gibbsite in peneplain deposits of the Belgian Condroz: Geologie en Mijnbouw, v. 47, p. 345–348.

BYERS, C. W., 1974, Shale fissility: relation (*sic*) to bioturbation: Sedimentology, v. 21, p. 479–484.

BYERS, C. W., 1977, Biofacies patterns in euxinic basins: a general model, p. 5–17 *in* Cook, H. E., and Enos, P. E., eds., Deep-water carbonate environments: Tulsa, OK, Society of Economic Paleontologists and Mineralogists Special Publication 25, 336 p.

BYERS, C. W., 1979, Biogenic structures in black shale (*sic*) paleoenvironments: Postilla, no. 174, 43 p.

BYERS, C. W., 1982, Stratigraphy—the fall of continuity: Journal of Geological Education, v. 30, p. 215–221.

BYRNE, J. V., LEROY, D. O., and RILEY, C. M., 1959, The chenier plain and its stratigraphy, southwestern Louisiana: Gulf Coast Association of Geological Societies Transactions, v. 9, p. 237–260.

CALDWELL, D. M., 1971 ms., A sedimentological study of an active part of a modern tidal delta, Moriches Inlet, Long Island New York: New York, Columbia University, Department of Geology, Master's thesis, 70 p.

CALDWELL, J. M., 1966, Coastal processes and beach erosion: Boston Society of Civil Engineers, Journal, v. 53, p. 142–157.

CALDWELL, N. E., and WILLIAMS, A. T., 1985, The role of beach profile (*sic*) configuration in the discrimination between (*sic*) different depositional environments affecting coarse clastic beaches: Journal of Coastal Research, v. 1, p. 129–139.

CALVERT, S. E., 1974, Deposition and diagenesis of silica in marine sediments, p. 273–299 *in* Hsü, K. J., and Jenkyns, H. C., eds., Pelagic sediments: on land and under the sea: International Association of Sedimentologists Special Publication 1, 447 p.

CAMERON, GORDON, 1990, Discovery of Jurassic crust and sediment in the Pacific: The Resolution Report, v. 6, p. 24.

CAMPBELL, C. A., and VALENTINE, J. W., 1977, Comparability of modern (*sic*) and ancient marine faunal provinces: Paleobiology, v. 3, p. 49–57.

CAMPBELL, C. V., 1967, Lamina, lamina set, bed, and bedset: Sedimentology, v. 8, p. 7–26.

CAMPBELL, C. V., 1971, Depositional model—Upper Cretaceous Gallup beach shoreline, Ship Rock area, northwestern New Mexico: Journal of Sedimentary Petrology, v. 41, p. 395–409.

CAMPBELL, M. R., 1929, The coal fields of the United States: U.S. Geological Survey Professional Paper 100, p. 1–33.

CAMPBELL, M. R., 1930, Coal as a recorder of incipient rock metamorphism: Economic Geology, v. 25, p. 675–696.

CANT, D. J., 1978, Development of a facies model for sandy braided river sedimentation: comparison of the South Saskatchewan River and the Battery Point Formation, p. 627–639 *in* Miall, A. D., ed., Fluvial sedimentology: Calgary, Alberta, Canadian Society of Petroleum Geologists Memoir 5, 859 p.

CANT, D. J., 1982, Fluvial facies models and their application, p. 115–138 *in* Scholle, P. A.; and Spearing, D., eds., Sandstone depositional environments: American Association of Petroleum Geologists Memoir 31, 410 p.

CAPUTO, M. V., and CROWELL, J. C., 1985, Migration of glacial centers across Gondwana during Paleozoic Era: Geological Society of America Bulletin, v. 96, p. 1020–1036.

CARDOSO, J.; WATTS, C. D.; MAXWELL, J. R.; GOODFELLOW, R.; EGLINTON, G.; and GOLUBIC, S., 1978, A biogeochemical study of the Abu Dhabi algal mats: a simplified ecosystem: Chemical Geology, v. 23, p. 273–291.

CAREY, D. A., 1987, Sedimentological effects and palaeoecological implications of the tube-building

polychaete *Lanice conchilega* Pallas: Sedimentology, v. 34, p. 49–66.

CARLSON, P. R., 1978, Holocene slump on the continental margin off the Malaspina Glacier, Gulf of Alaska: American Association of Petroleum Geologists Bulletin, v. 62, p. 2412–2426.

CARMALT, S. W.; and ST. JOHN, B., 1986, Giant oil and gas fields, p. 11-53 *in* M. T. Halbouty, ed., Future petroleum provinces in the world: Tulsa, OK, American Association of Petroleum Geologists Memoir 40, 708 p.

CARNEY, R. S., 1981, Bioturbation and biodeposition, p. 357–399 *in* Boucot, A. J., ed., Principles of benthic marine paleoecology: New York, Academic Press, 463 p.

CAROZZI, A. V., 1965, Lavoisier's fundamental contribution to stratigraphy: Ohio Journal of Science, v. 65, p. 71–85.

CAROZZI, A. V., 1986, New eustatic model for the origin of carbonate cyclic sedimentation: Arch. Sci., Geneve, v. 39, p. 53–66.

CARPENTER, A. B., TROUT, M. L., and PICKETT, E. E., 1974, Preliminary report on the origin and chemical evolution of Pb-Zn-rich oil field (*sic*) brines in central Mississippi: Economic Geology, v. 69, p. 1191–1206.

CARPENTER, W. B., 1871, On the Gibraltar current, the Gulf Stream, and the general oceanic circulation: Geographical Society Proceedings, v. 15, p. 54–88.

CARPENTER, W. B., 1872a, On the temperature and animal life in the deep sea: Royal Institute of Great Britain Proceedings, v. 6 (1870–1872), p. 63–82.

CARPENTER, W. B., 1872b, On the latest scientific researches in the Mediterranean: Royal Institute of Great Britain Proceedings, v. 6 (1870–1872), p. 236–259.

CARPENTER, W. B., 1872c, On the temperature and other physical conditions of inland seas, in their relation (*sic*) to geological inquiry: Geological Magazine, v. 9, p. 545–551.

CARPENTER, W. B., 1872d, On the temperature and movements of the deep sea: Popular Science Review, v. 11, p. 121–135.

CARPENTER, W. B., 1872e, Report on scientific researches carried on during the months of August, September and October 1871 in H. M. Surveying-ship "Shearwater": Royal Society of London Proceedings, v. 20, p. 535–644.

CARPENTER, W. B., 1873, On the physical geography of the Caspian Sea, in its relation (*sic*) to geology: British Association for the Advancement of Science, Reports, v. 43, p. 165–167.

CARPENTER, W. B., 1874, On the physical cause of oceanic circulation: Philosophical Magazine, v. 47, p. 359–362.

CARPENTER, W. B., 1882, Land and sea considered in relation (*sic*) to geological time: Royal Institute of Great Britain Proceedings, v. 9 (1880–1882), p. 268–282.

CARR, A. P., 1969, Size grading along a pebble beach, Chesil Beach, England: Journal of Sedimentary Petrology, v. 39, p. 297–311.

CARR, A. P.; GLEASON, R.; and KING, A., 1970, Significance of pebble size and shape sorting by waves: Sedimentary Geology, v. 4, p. 89–101.

CARSON, M. A., 1984, The meandering-braided river threshold: a reappraisal: Journal of Hydrology, v. 73, p. 315–334.

CARTER, J. G., ed., 1990, Skeletal biomineralization. Patterns, processes and evolutionary trends, vol. 1: New York, Van Nostrand Reinhold, 827 p.

CARTER, R. W. G., 1988, Coastal environments. An introduction to the physical (*sic*), ecological (*sic*) and cultural systems: London, Academic Press, 617 p.

CARTER, R. W. G., and ORFORD, J. D., 1984, Coarse clastic barrier beaches: a discussion of the distinctive dynamic and morphosedimentary characteristics: Marine Geology, v. 60, p. 377–389.

CARTER, T. G., and others, 1972, A new bathymetric chart and physiography of the Mediterranean Sea, p. 19–23 *in* Stanley, D. T., ed., The Mediterranean Sea: a natural sedimentation laboratory: Stroudsburg, PA, Dowden, Hutchinson, and Ross, 765 p.

CASE, J. E., and HOLCOMBE, T. L., 1980, Geologic-tectonic map of the Caribbean region: U.S. Geological Survey Misc. Invest. Ser., map I-1100.

CASPERS, H., 1957, Black Sea and Sea of Azov, p. 801–889 *in* Hedgpeth, J. W., ed., Treatise on marine ecology and paleoecology: Geological Society of America Memoir 67, v. 1, Ecology, 1296 p.

CECIL, C. B., 1990, Paleoclimate controls on stratigraphic repetition of chemical (*sic*) and siliciclastic rocks: Geology, v. 18, p. 533–536.

CECILE, M. P., and CAMPBELL, F. H. A., 1978, Regressive stromatolite reefs and associated facies, Middle Goulburn Group (lower Proterozoic) in Kilohigok Basin, Northwest Territories: an example of environmental control of stromatolite form: Canadian Petroleum Geology Bulletin, v. 26, p. 237–267.

CERCONE, K. R., 1984, Thermal history of Michigan Basin: American Association of Petroleum Geologists Bulletin, v. 68, p. 130–136.

CHAFETZ, H. S., 1978, A trough cross-stratified glaucarenite: a Cambrian tidal inlet (*sic*) accumulation: Sedimentology, v. 25, p. 545–559.

CHAFETZ, H. S., 1979, Petrology of carbonate nodules from a Cambrian tidal inlet (*sic*) accumulation, central Texas: Journal of Sedimentary Petrology, v. 49, p. 215–222.

CHAFETZ, H. S., 1982, The Upper Cretaceous Beartooth Sandstone of southwestern New Mexico: a transgressive deltaic complex on silicified paleokarst: Journal of Sedimentary Petrology, v. 52, p. 157–169.

CHAFETZ, H. S., 1986, Marine peloids: a product of bacterially induced precipitation of calcite: Journal of Sedimentary Petrology, v. 56, p. 812–817.

CHAMBERLIN, R. T., 1914, Diastrophism and the formative processes; VII, Periodicity of Paleozoic orogenic movements: Journal of Geology, v. 22, p. 315–345.

CHAMBERLIN, T. C., 1898a, The ulterior basis of time divisions and the classification of geologic history: Journal of Geology, v. 6, p. 449–462.

CHAMBERLIN, T. C., 1898b, The influence of great epochs of limestone formation upon the constitution of the atmosphere: Journal of Geology, v. 6, p. 609–621.

CHAMBERLIN, T. C., 1898c, Continental shelf distinguished from sea shelf: Journal of Geology, v. 6, p. 524–526.

CHAMBERLIN, T. C., 1899, An attempt to frame a working hypothesis of the cause of glacial periods on an atmospheric basis: Journal of Geology, v. 7, p. 545–584, 667–685, 751–787.

CHAMBERLIN, T. C., 1906, On a possible reversal of deep-sea circulation and its influence on geologic climates: American Philosophical Society Proceedings, v. 45, p. 33–43.

CHAMBERLIN, T. C., 1909, Diastrophism as the ultimate basis of correlation: Journal of Geology, v. 17, p. 685–693.

CHAMBERLIN, T. C., 1913a, Map of North America during the great ice age. Scale, 104 miles = 1 inch. Chicago, IL.

CHAMBERLIN, T. C., 1913b, Diastrophism and the formative processes; I, introduction: Journal of Geology, v. 21, p. 517–522.

CHAMBERLIN, T. C., 1913c, Diastrophism and the formative processes; II, shelf seas and certain limitations of diastrophism: Journal of Geology, v. 21, p. 523–533.

CHAMBERLIN, T. C., 1913d, Diastrophism and the formative processes; III, the lateral stresses within the continental protuberances and their relations (*sic*) to continental creep and sea-transgression (*sic*): Journal of Geology, v. 21, p. 577–587.

CHAMBERLIN, T. C., 1913e, Diastrophism and the formative processes; IV, rejuvenation of the continents: Journal of Geology, v. 21, p. 673–682.

CHAMBERLIN, T. C., 1914a, Diastrophism and the formative processes; V, the testimony of the deep-sea deposits: Journal of Geology, v. 22, p. 131–144.

CHAMBERLIN, T. C., 1914b, Diastrophism and the formative processes; VI, foreset beds and slope deposits: Journal of Geology, v. 22, p. 266–274.

CHAMBERLIN, T. C., 1914c, Diastrophism and the formative processes; VIII, the quantitative element in circumcontinental growth: Journal of Geology, v. 22, p. 516–528.

CHAMBERLIN, T. C., 1918, Diastrophism and the formative processes; IX, a specific mode of self-promotion (*sic*) of periodic diastrophism: Journal of Geology, v. 26, p. 193–197.

CHAMLEY, HERVE, 1989, Clay sedimentology: New York, Springer-Verlag, 623 p.

CHANDLER, F. W., 1988, Diagenesis of sabkha-related, sulphate nodules in the Early Proterozoic Gordon Lake Formation, Ontario, Canada: Carbonates and Evaporites, v. 3, p. 75–94.

CHANG, S.-G., 1967, A new sandstone classification (*sic*) scheme: Geological Society of China Proceedings, no. 10, p. 107–114.

CHARLES, R. G.; and BLATT, HARVEY, 1978, Quartz, chert, and feldspars in modern fluvial muds and sands: Journal of Sedimentary Petrology, v. 48, p. 427–432.

CHARNOCK, H.; REES, A. I.; and HAMILTON, N., 1972, Sedimentation in the Tyrrhenian Sea, p. 615–629 *in* Stanley, D. J., The Mediterranean Sea: a natural sedimentation laboratory: Stroudsburg, PA, Dowden, Hutchinson, and Ross, 765 p.

CHAVE, K. E., 1964, Skeletal durability and preservation, p. 377–387 *in* Imbrie, John; and Newell, N. D., eds., Approaches to paleoecology: New York, John Wiley & Sons, 432 p.

CHOQUETTE, P. W., and PRAY, L. C., 1970, Geological nomenclature and classification of porosity in sedimentary carbonates: American Association of Petroleum Geologists Bulletin, v. 54, p. 207–250.

CHOUGH, S. K.; and HESSE, R., 1985, Contourites from Eirik Ridge, south of Greenland: Sedimentary Geology, v. 41, p. 185–199.

CHOUGH, S. K., HWANG, I. G., and CHOE, M. Y., 1990, The Miocene Doumsan fan-delta, southeast Korea: a composite fan-delta system in back-arc margin: Journal of Sedimentary Petrology, v. 60, p. 445–455.

CHOW, N.; and JAMES, N. P., 1987, Cambrian grand cycles: a northern Appalachian perspective: Geological Society of America Bulletin, v. 98, p. 418–429.

CHOW, T. J.; BRULAND, K. W.; BERTINE, K. K.; SOUTAR, A.; KOIDE, M.; and GOLDBERG, E. D., 1973, Lead pollution (*sic*) records in southern California coastal sediments: Science, v. 181, p. 551–552.

CHOWDHURI, K. R., and REINECK, H.-E., 1978, Primary sedimentary structures and their sequence in the shoreface of barrier island Wangerooge (North Sea): Senckenbergiana Maritima, v. 10, p. 15–29.

CHRISTIANSEN, C.; BLAESILD, P.; and DALSGAARD, K., 1984, Reinterpreting 'segmented' grain-size curves: Geological Magazine, v. 121, p. 47–51.

CHRISTIE, R. L., 1978, Sedimentary phosphate deposits—an interim review: Geological Survey of Canada, Paper 78-20, 9 p.

CHRISTIE-BLICK, N., and others, 1988, Chronology of fluctuating sea levels since the Triassic, comments and replies: Science, v. 241, p. 596–602.

CHURNET, H. G., MISRA, K. C., and WALKER, K. R., 1982, Deposition and dolomitization of Upper Knox carbonate sediments, Copper Ridge district, East Tennessee: Geological Society of America Bulletin, v. 93, p. 76–86.

CITA, M. B.; BEGHI, C.; CAMERLENGHI, A.; KASTENS, K. A.; MCCOY, F. W.; NOSETTO, A.; PARISI, E.; SCOLARIS, F.; and TOMADIN, L., 1984, Turbidites and megaturbidites from the Heredotus (*sic*) abyssal plain (eastern Mediterranean) unrelated to seismic events: Marine Geology, v. 55, p. 79–101.

CLARK, PETER; and KARROW, P. F., 1983, Till stratigraphy in the St. Lawrence Valley near Malone, New York: revised glacial history and stratigraphic nomenclature: Geological Society of America Bulletin, v. 94, p. 1308–1318.

CLEMENS, K. E., and KOMAR, P. D., 1989, Oregon beach-sand compositions produced by the mixing of sediments under a transgressing sea: Journal of Sedimentary Petrology, v. 58, p. 519–529.

CLEMMENSEN, L. B.; OLSEN, HENRIK; and BLAKEY, R. C., 1989, Erg-margin deposits in the Lower Jurassic Moenave Formation and Wingate Sandstone, Southern Utah: Geological Society of America Bulletin, v. 101, p. 759–773.

CLIFTON, H. E., 1973, Pebble segregation and bed lenticularity in wave-worked versus alluvial gravel: Sedimentology, v. 20, p. 173–187.

CLIFTON, H. E., 1981, Progradational sequence in Miocene shoreline deposits, southeastern Caliente Range, California: Journal of Sedimentary Petrology, v. 51, p. 165–184.

CLIFTON, H. E., 1982, Estuarine deposits, p. 179–189 *in* Scholle, P. A.; and Spearing, D., eds., Sandstone depositional environments: American Association of Petroleum Geologists Memoir 31, 410 p.

CLIFTON, H. E., ed., 1988, Sedimentologic consequences of convulsive geologic events: Boulder, CO, Geological Society of America Special Paper 229, 157 p.

CLIFTON, H. E., HUNTER, R. E., and GARDNER, J. V., 1988, Analysis of eustatic (*sic*) tectonic (*sic*), and sedimentologic influences on transgressive (*sic*) and regressive cycles in the late (*sic*) Cenozoic Merced Formation, San Francisco, California, p. 109–128 *in* Kleinspehn, K. L.; and Paola, C., eds., New perspectives in basin analysis: New York, Springer-Verlag.

CLIFTON, H. E., HUNTER, R. E., and PHILLIPS, R. L., 1970, Underwater observations in Oregon nearshore zone: U.S. Geological Survey Professional Paper 700-A, p. A100.

CLIFTON, H. E., HUNTER, R. E., and PHILLIPS, R. L., 1971, Depositional structures and processes in the non-barred high-energy nearshore: Journal of Sedimentary Petrology, v. 41, p. 651–670.

CLOETINGH, S.; MCQUEEN, H.; and LAMBECK, K., 1985, On a tectonic mechanism for regional sea level (*sic*) variations: Earth and Planetary Science Letters, v. 75, p. 157–166.

CLOUD, P. E., JR., 1948, Some problems and patterns of evolution exemplified by fossil invertebrates: Evolution, v. 2, p. 322–350.

COATES, D. R., and VITEK, J. D., eds., 1980, Thresholds in geomorphology: Symposium held 19–21 October 1978, Binghamton, NY: London, George Allen and Unwin, 485 p.

COHEN, Y.; CASTENHOLZ, R. W.; and HALVORSON, H. O., eds., 1984, Microbial mats: stromatolites. MBL lectures in biology 3: New York, Alan R. Liss, 498 p.

COLEMAN, J. M., 1969, Brahmaputra River: channel processes and sedimentation: Sedimentary Geology (special issue), v. 3, p. 122–239.

COLEMAN, J. M., 1981, Deltas—processes of deposition and models for exploration (2nd ed.): Minneapolis, MN, Burgess Publishing Company, 124 p.

COLEMAN, J. M., 1988, Dynamic changes and processes in the Mississippi River delta: Geological Society of America Bulletin, v. 100, p. 999–1015.

COLEMAN, J. M., and PRIOR, D. B., 1980, Deltaic sand bodies: American Association of Petroleum Geologists Short Course 15, 171 p.

COLEMAN, J. M., and PRIOR, D. B., 1982, Deltaic environments of deposition, p. 139–178 *in* Scholle, P. A.; and Spearing, D., eds., Sandstone depositional environments: American Association of Petroleum Geologists Memoir 31, 410 p.

COLEMAN, J. M., and PRIOR, D. B., 1983, Deltaic influences on shelf-edge instability processes, p. 121–127 *in* Stanley, D. J., and Moore, G. T., eds., The shelfbreak, critical interface on continental margins: Society of Economic Paleontologists and Mineralogists Special Publication 33, 467 p.

COLEMAN, J. M., and ROBERTS, H. H., 1988, Sedimentary development of the Louisiana continental shelf related to sea level (*sic*) cycles: Part II: Seismic response: Geo-Marine Letters, v. 8, p. 109–119.

COLEMAN, J. M.; SUHAYADA, J. N.; WHELAN, T.; and WRIGHT, L. D., 1974, Mass movements of Mississippi River delta sediments: Gulf Coast Association of Geological Societies Transactions, v. 24, p. 49–68.

COLEMAN, N. L., 1982, Velocity profiles with suspended sediment: Journal of Hydraulic Research, v. 19, p. 211–229.

COLLINS, A. G., 1970, Geochemistry of some petroleum-associated waters from Louisiana: U.S. Bureau of Mines, Report of Investigations 7326, 31 p.

COLLINS, A. G., 1975, Geochemistry of oilfield waters. Developments in petroleum science 1: Amsterdam, Elsevier Scientific Publishing Company, 496 p.

COLLINSON, J. D., 1969, The sedimentology of the Grindslow Shales and the Kinderscout Grit; a deltaic complex in the Namurian of northern England: Journal of Sedimentary Petrology, v. 39, p. 194–221.

COLLINSON, J. D.; and LEWIN, J., eds., 1983, Modern and ancient fluvial systems: International Association of Sedimentologists Special Publication 6, 575 p.

COLLINSON, J. D., and THOMPSON, D. B., 1982, Sedimentary structures: London, George Allen and Unwin, 194 p.

CONIGLIO, M., 1987, Biogenic chert in the Cow Head Group (Cambro-Ordovician), western Newfoundland: Sedimentology, v. 34, p. 813–823.

CONSTANTZ, B. R., 1986, The primary surface area of corals and variations in their susceptibility to diagenesis, p. 53–76 *in* Schroeder, J. H., and Purser, B. H., eds., Reef diagenesis: New York, Springer-Verlag, 455 p.

COOK, H. E., HINE, A. C., and MULLINS, H. T., eds., 1983, Platform margin (*sic*) and deep water carbonates: Tulsa, OK, Society of Economic Paleontologists and Mineralogists Short Course Lecture Notes 12, 573 p.

COOK, P. J., 1973, Supratidal environment and geochemistry of some Recent dolomite concretions, Broad Sound, Queensland, Australia: Journal of Sedimentary Petrology, v. 43, p. 998–1011.

COOK, P. J., and SHERGOLD, J. H., eds., 1979, Proterozoic-Cambrian phosphorites. Sponsored by International Geological Correlation Programme, Project 156: Canberra, Australia, Canberra Publishing and Printing Company, 106 p.

CORRENS, C. W., 1950, Zur Geochemie der Diagenese. I. Das Verhalten von $CaCO_3$ und SiO_2: Geochimica et Cosmochimica Acta, v. 1, p. 49–54.

COWARD, M. P., 1986, Heterogeneous stretching, simple shear and basin development: Earth and Planetary Science Letters, v. 80, p. 325–336.

COWARD, M. P., DEWEY, J. F., and HANCOCK, P. L., eds., 1987, Continental Extensional Tectonics: Geological Society (London) Special Publication 28, 637 p.

COYLE, W. G., III; and EVANS, K. R., 1987, Phylloid algal mounds in the Frisbie Limestone (Pennsylvanian), northeastern Kansas [abstract]: Society of Economic Paleontologists and Mineralogists, MidYear Meeting, 4th, Abstracts, Tulsa, p. 17 (only).

COYNE, J. A., and BARTON, N. H., 1988, What do we know about speciation?: Nature, v. 331, p. 485–486.

CRAFT, J. H., and BRIDGE, J. S., 1987, Shallow-marine sedimentary processes in the Late Devonian Catskill sea, New York State: Geological Society of America Bulletin, v. 98, p. 338–355.

CRAM, J. M., 1979, The influence of continental shelf (sic) width on tidal range: Paleo-oceanographic implication: Journal of Geology, v. 87, p. 175–228.

CRAWFORD, M. L., 1981, Phase equilibria in aqueous fluid inclusions, p. 75–99 in Hollister, L. S., and Crawford, M. L., eds., Short course in fluid inclusions: applications to petrology: Mineral Association of Canada, 304 p.

CRENSHAW, M. A., 1989, Biomineralization mechanisms, p. 1–9 in Carter, J. G., ed., Skeletal biomineralization: patterns, processes and evolutionary trends: American Geophysical Union, Short Course in Geology, v. 5, pt. 2.

CROWELL, J. C., 1978, Gondwanan glaciation, cyclothems, continental positioning, and climate change: American Journal of Science, v. 278, p. 1345–1372.

CROWLEY, K. D., 1983, Large-scale bed configurations (macroforms), Platte River Basin, Colorado and Nebraska: primary structures and formative processes: Geological Society of America Bulletin, v. 94, p. 117–133.

CROWLEY, K. D., 1984, Filtering of depositional events and the completeness of the sedimentary record: Journal of Sedimentary Petrology, v. 54, p. 127–136.

CUBITT, J. M., and REYMENT, R. A., eds., 1982, Quantitative stratigraphic correlation: New York, John Wiley & Sons, 301 p.

CUDZIL, M. R., and DRIESE, S. G., 1987, Fluvial, tidal and storm sedimentation in the Chilhowee Group (Lower Cambrian), northeastern Tennessee, USA: Sedimentology, v. 34, p. 861–883.

CUFFEY, R. J., 1985, Expanded reef-rock textural classification and the geologic history of bryozoan reefs: Geology, v. 13, p. 307–310.

CUOMO, M. C., and RHOADS, D. C., 1987, Biogenic sedimentary fabrics associated with pioneering polychaete assemblages: modern and ancient: Journal of Sedimentary Petrology, v. 57, p. 537–543.

CURRAY, J. R., 1965, Late Quaternary history, continental shelves of the United States, p. 723–735 in Wright, H. E., Jr., ed., The Quaternary of the United States: Princeton, NJ, Princeton University Press, 922 p.

CURRAY, J. R., EMMEL, F. J., and CRAMPTON, P. J. S., 1969, Holocene history of a strand plain, lagoonal coast, Nayarit, Mexico, p. 63–100 in Ayala-Castanares, Augustin; and Phleger, F. B., eds., Lagunas costeras, un simposio: Simposio Internacional sobre lagunas costeras, Mexico City, Mexico, 28–30 Noviembre 1967, memoria: Mexico Universidad Nacional Autonoma (WMAS-UNESCO), 686 p.

CURRIE, R. G., 1984, Periodic (18.6-year) and cyclic (11-year) induced drought and flood in western North America: Journal of Geophysical Research, v. 89 (D5), p. 7215–7230.

CURRIE, R. G., 1987, Examples and implications of 18.6- and 11-yr terms in world weather records, Chapter 22, p. 378–403 in Rampino, M. R., Sanders, J. E., Newman, W. S., and Konigsson, L.-K., eds., Climate. History, periodicity, and predictability: New York, Van Nostrand Reinhold Company, 588 p.

CURTIS, C. D., 1978, Possible links between sandstone diagenesis and depth related (sic) geochemical reactions occurring in enclosing mudstones: Geological Society of London Journal, v. 135, p. 107–117.

CURTIS, D. M., 1970, Miocene deltaic sedimentation, Louisiana Gulf Coast, p. 213–308 in Morgan, J. P., and Shaver, R. H., eds., Deltaic sedimentation modern and ancient: Tulsa, OK, Society of

Economic Paleontologists and Mineralogists Special Publication 15, 312 p.

CURTIS, W. F., CULBERTSON, J. K., and CHASE, E. B., 1973, Fluvial sediment (sic) discharge to the oceans from the conterminous United States: U.S. Geological Survey, Circular 67, 17 p.

DABRIO, C. J., 1982, Sedimentary structures generated on the foreshore by migrating ridge and runnel (sic) systems on microtidal and mesotidal coasts of South Spain: Sedimentary Geology, v. 32, p. 141–151.

DAETWYLER, C. C., and KIDWELL, A. L., 1959, The Gulf of Batabano, a modern carbonate basin: World Petroleum Congress, 5th, New York, Proceeding, Section 1, p. 1–21.

DALRYMPLE, R. A., and THOMPSON, W. W., 1977, Study of equilibrium beach profiles, p. 1277–1296 in International Conference on Coastal Engineering, 15th, Honolulu, Hawaii, Proceedings.

DALY, R. A., 1936, Origin of submarine "canyons": American Journal of Science, 5th series, v. 31, p. 401–420.

DALZIEL, I. W. D., and DOTT, R. H., JR., 1970, Geology of the Baraboo district, Wisconsin: Wisconsin Geological and Natural History Survey, Information Circular No. 14, 164 p.

DAMON, P. E., 1971, The relationship between late Cenozoic volcanism and tectonism and orogenic-epeirogenic periodicity, p. 15–36 in Turekian, K. K., ed., Late Cenozoic glacial ages: New Haven, CT, Yale University Press, 606 p.

DAMON, P. E., 1977, Solar induced (sic) variations of energetic particles at one A.U., p. 429–448 in White, O. R., ed., The solar output and its variations: Boulder, Colorado, Colorado Associated Universities Press, 526 p.

DAMON, P. E.; LEMAN, J. C.; and LONG, A., 1978, Temporal fluctuations of atmospheric ^{14}C: causal factors and implications: Annual Reviews of Earth and Planetary Sciences, v. 6, p. 457–494.

DAMON, P. E.; LONG, A.; and GREY, D. C., 1966, Fluctuations of atmospheric C^{14} during the last six millennia: Journal of Geophysical Research, v. 71, p. 1055–1071.

DAMUTH, J. E., 1975, Echo character (sic) of the western Equatorial Atlantic floor and its relationship to the dispersal and distribution of terrigenous sediments: Marine Geology, v. 18, p. 17–45.

DAMUTH, J. E., 1979, Migrating sediment waves created by turbidity currents in the northern South China Sea Basin: Geology, v. 7, p. 520–523.

DAMUTH, J. E., and EMBLEY, R. W., 1979, Upslope flow of turbidity currents on the southwest flank of the Ceara Rise, western equatorial Atlantic: Sedimentology, v. 26, p. 825–834.

DAMUTH, J. E., and EMBLEY, R. W., 1981, Mass-transport processes on Amazon Cone: western equatorial Atlantic: American Association of Petroleum Geologists Bulletin, v. 65, p. 629–643.

DANA, J. D., 1873, On some results of the Earth's contraction from cooling, including a discussion of the origin of mountains and the nature of the Earth's interior: American Journal of Science, 3rd series, v. 5, p. 423–443; v. 6, p. 6–14, 104–115, 161–172.

DANA, J. D., 1880, Manual of geology, 3rd ed.: New York, Ivison, Blakeman & Co., 911 p.

DARBY, D. A., 1984, Trace elements in ilmenite: a way to discriminate provenance or age in coastal sands: Geological Society of America Bulletin, v. 95, p. 1208–1218.

DARBY, D. A., and TSANG, Y. W., 1987, Variation in ilmenite element composition within and among drainage basins: implications for provenance: Journal of Sedimentary Petrology, v. 57, p. 831–838.

DARBYSHIRE, J., 1952, The generation of waves by wind: Royal Society of London Proceedings, series A, v. 215, p. 299–328.

DARWIN, CHARLES, 1846, Geological observations on South America: London, Smith, Elder, 279 p.

DARWIN, CHARLES, 1881, Formation of vegetable mould through the action of worms, with observations on their habit: London, John Murray, 326 p.

DARWIN, CHARLES, 1896, The structure and distribution of coral reefs, 3rd ed.: New York, D. Appleton and Company, 344 p.

DAVIES, H. G., 1965, Convolute lamination and other structures from the Lower Coal Measures of Yorkshire: Sedimentology, v. 5, p. 305–325.

DAVIES, J. L., 1964, A morphogenetic approach to world shorelines: Zeitschrift fur Geomorphologie, v. 8 (Sonderheft), p. 127–142.

DAVIES, J. L., 1977, Geographical variation in coastal development: New York, Longman, 204 p.

DAVIES, T. R. H., and TINKER, C. C., 1984, Fundamental characteristics of stream meanders: Geological Society of America Bulletin, v. 95, p. 505–512.

DAVIS, J. C., 1986, Statistics and data analysis in geology, 2nd ed.: New York, John Wiley & Sons, 646 p.

DAVIS, R. A., JR., 1965, Underwater study of ripples, southeastern Lake Michigan: Journal of Sedimentary Petrology, v. 35, p. 857–866.

DAVIS, R. A., JR., 1983, Depositional systems—a genetic approach to sedimentary geology: Englewood Cliffs, NJ, Prentice-Hall, 669 p.

DAVIS, R. A., JR., ed., 1985, Coastal sedimentary environments, 2nd ed.: New York, Springer-Verlag, 420 p.

DAVIS, R. A., JR., KNOWLES, S. C., and BLAND, M. J., 1989, Role of hurricanes in the Holocene stratigraphy of estuaries: examples from the Gulf coast of Florida: Journal of Sedimentary Petrology, v. 59, p. 1052–1061.

DAVIS, T. L., 1984, Seismic-stratigraphic facies model, p. 311–317 in Walker, R. G., ed., Facies models, 2nd ed.: Toronto, Ontario, Geoscience Canada Reprint Series 1, 317 p.

DAVIS, W. M., 1898, The Triassic Formation of Connecticut: U.S. Geological Survey, Annual Report, 18th, Part 2, p. 1192.

DE BOER, P. L.; VAN GELDER, A.; and NIO, S. D., eds., 1988, Tide-influenced sedimentary environments and facies: Dordrecht, The Netherlands, D. Reidel Publishing Company, 530 p.

DE GRACINSKY, P. C.; DEROO, G.; HERBIN, J. P.; JACQUIN, T.; MAGNIEZ, F.; MONTADERT, L.; MULLER, C.; PONSOT, C.; SCHAAF, A.; and SIGAL, J., 1986, Ocean-wide stagnation episodes in the Late Cretaceous: Geologische Rundschau, v. 75, p. 17–41.

DE MILLE, G.; SHOULDICE, J. R.; and NELSON, H. W., 1964, Collapse structures related to evaporites of Prairie Formation, Saskatchewan: Geological Society of America Bulletin, v. 75, p. 307–316.

DE SIEVEKING, G.; and HART, M. B., 1986, The scientific study of flint and chert: Cambridge, Cambridge University Press, 290 p.

DE SIMONE, D. J., and LA FLEUR, R. G., 1986, Glaciolacustrine phases in the northern Hudson Lowland and correlatives in western Vermont: Northeastern Geology, v. 8, p. 218–229.

DEACON, E. L., and WEBB, E. K., 1962, Interchange of properties between sea and air. 3. Small-scale interactions, p. 43–87 in Hill, M. N., ed., The sea. Ideas and observations on progress in the study of the seas. Volume 1, Physical oceanography: New York, Wiley-Interscience, 864 p.

DEAN, R. G., and MAURMEYER, E. M., 1983, Models of beach profile (sic) response, p. 151–165 in Komar, P. D., and Moore, J. R., eds., Handbook of coastal processes and erosion (sic): Boca Raton, FL, CRC Press, 305 p.

DEAN, W. E., and GARDNER, J. V., 1986, Milankovitch cycles in Neogene deep-sea sediment: Paleoceanography, v. 1, p. 539–553.

DEAN, W. E., and SCHREIBER, B. C., eds., 1978, Notes for a short course on marine evaporites: Tulsa, OK, Society of Economic Paleontologists and Mineralogists Short Course 4, 188 p.

DECIMA, ARVEDO; MCKENZIE, J. A.; and SCHREIBER, B. C., 1988, The origin of "evaporative" limestones: an example from the Messinian of Sicily (Italy): Journal of Sedimentary Petrology, v. 58, p. 256–272.

DEGEER, GERHARD, 1912, A geochronology of the last 12,000 years: International Geological Congress, 11th, Stockholm, 1910, Comptes Rendus, v. 1, p. 241–258.

DEGENS, E. T., and ROSS, D. A., eds., 1974, The Black Sea—geology, chemistry, and biology: Tulsa, OK, American Association of Petroleum Geologists Memoir 20, 633 p.

DEMAISON, G. J., and MOORE, G. T., 1980, Anoxic environments and oil source bed (*sic*) genesis: American Association of Petroleum Geologists Bulletin, v. 64, p. 1179–1209.

DENNISON, J. M., and HEAD, J. W., 1975, Sealevel variations interpreted from the Appalachian basin Silurian and Devonian: American Journal of Science, v. 275, p. 1089–1120.

DENTON, G. H., and HUGHES, T. J., 1983, Milankovitch theory of ice ages; hypothesis of ice-sheet linkage between regional insolation and global climate: Quaternary Research, v. 20, p. 125–144.

DEPARTMENT OF ENERGY, 1988, Site Characterization Plan Overview, Yucca Mountain Site, Nevada Research and Development Area, Nevada: Washington, DC, Department of Energy, 164 p.

DEUSER, W. G., 1974, Evolution of anoxic conditions in Black Sea during Holocene, p. 133–136 *in* Degens, E. T., and Ross, D. A., eds., The Black Sea—geology, chemistry, and biology: Tulsa, OK, American Association of Petroleum Geologists Memoir 20, 633 p.

DEUSER, W. G., 1975, Reducing environments, p. 1–37 *in* Riley, J. P.; and Skirrow, G., eds., Chemical Oceanography, vol. 3, 2nd ed.: London, Academic Press.

DEWEY, J. F., and BURKE, K. C. A., 1974, Hot spots and continental break-up (*sic*): implications for collisional orogeny: Geology, v. 2, p. 57–60.

DICKEY, P. A., 1966, Patterns of chemical composition in deep subsurface waters: American Association of Petroleum Geologists Bulletin, v. 50, p. 2472–2478.

DICKEY, P. A., 1969, Increasing concentration of subsurface brines with depth: Chemical Geology, v. 4, p. 361–370.

DICKINSON, W. R., 1971, Plate tectonics in geologic history, new global tectonic (*sic*) theory (*sic*) leads to revised concepts of geosynclinal deposition and orogenic deformation: Science, v. 174, p. 17–113.

DICKINSON, W. R., 1974, Plate tectonics and sedimentation, p. 1–27 *in* Dickinson, W. R., ed., Tectonics and sedimentation: Tulsa, OK, Society of Economic Paleontologists and Mineralogists Special Publication 22, 204 p.

DICKINSON, W. R., 1985, Interpreting provenance relations (*sic*) from detrital modes of sandstones, p. 333–361 *in* Zuffa, G. G., ed., Provenance of arenites: Dordrecht, The Netherlands, D. Reidel Publishing Company, 379 p.

DICKINSON, W. R., and SEELY, D. R., 1979, Structure and stratigraphy of forearc regions: American Association of Petroleum Geologists Bulletin, v. 63, p. 2–31.

DICKINSON, W. R., and SUCZEK, C. A., 1979, Plate tectonics and sandstone composition: American Association of Petroleum Geologists Bulletin, v. 63, p. 2164–2182.

DICKMAN, M.; and ARTUZ, I., 1979, Mass mortality of photosynthetic bacteria as a mechanism for dark lamina (*sic*) formation in sediments of the Black Sea: Nature, v. 275, p. 191–195.

DICKSON, J. A. D., and COLEMAN, M. L., 1980, Changes in carbon (*sic*) and oxygen isotope (*sic*) composition during limestone diagenesis: Sedimentology, v. 27, p. 1–12.

DIJKEMA, K. S.; REINECK, H.-E.; and WOLFF, W. J., eds., 1980, Geomorphology of the Wadden Sea Area: Leiden, Report 1, 135 p.

DILL, R. F.; SHINN, E. A.; JONES, A. T.; Kelly, K.; and Steinen, R. P., 1986, Giant subtidal stromatolites forming in normal salinity (*sic*) waters: Nature, v. 324, p. 55–58.

DIMROTH, E., 1975, Paleo-environment of iron-rich sedimentary rocks: Geologische Rundschau, v. 64, p. 751–767.

DINGLE, R. V., 1980, Large allochthonous sediment masses and their role in the construction of the continental slope and rise (*sic*) off southwestern Africa: Marine Geology, v. 37, p. 333–354.

DINGLER, J. R., and ANIMA, R. J., 1989, Subaqueous grain flows at the head of Carmel Submarine Canyon, California: Journal of Sedimentary Petrology, v. 59, p. 280–286.

DIXON, W. J., and MASSEY, F. J., JR., 1957, Introduction to statistical analysis, 2nd ed: New York, McGraw-Hill Book Company, 488 p.

DOBKINS, J. E., and FOLK, R. L., 1970, Shape development on Tahiti-Nui: Journal of Sedimentary Petrology, v. 40, p. 1167–1203.

DOLAN, J. F.; BECK, C.; and AGAWA, Y., 1989, Upslope deposition of extremely distal turbidites. An example from the Tiburon Rise, west-central Atlantic: Geology, v. 17, p. 990–994.

DOLAN, ROBERT; HAYDEN, BRUCE; and LINS, HARRY, 1980, Barrier islands: American Scientist, v. 68, p. 16–25.

DOMINGUEZ, J. M. L.,; and WANLESS, H., in press, Facies architecture of a falling sea level (*sic*) strand plain, Doce River coast, Brazil, *in* Swift, D. J. P.; Oertel, G. W.; and Tillman, R. W., eds., Shelf sand (*sic*) and sandstone (*sic*) bodies. Geometry (*sic*), facies, and distribution: International Association of Sedimentologists Special Publication 12.

DONALDSON, C. A., MARTIN, R. H., and KANES, W. H., 1970, Holocene Guadalupe delta of Texas Gulf coast, p. 107–137 *in* Morgan, J. P., and Shaver, R. H., eds., Deltaic sedimentation, modern and ancient: Tulsa, OK, Society of Economic Paleontologists and Mineralogists Special Publication 15, 312 p.

DONSELAAR, M. E., 1989, The Cliff House Sandstone, San Juan Basin, New Mexico: model for the stacking of "transgressive" barrier complexes: Journal of Sedimentary Petrology, v. 59, p. 13–27.

DOTT, R. H., JR., 1963, Dynamics of subaqueous gravity depositional (*sic*) processes: American Association of Petroleum Geologists Bulletin, v. 47, p. 104–128.

DOTT, R. H., JR., 1974, Cambrian tropical storm waves in Wisconsin: Geology, v. 2, p. 243–246.

DOTT, R. H., JR., 1978, Tectonics and sedimentation a century later: Earth-Science Reviews, v. 14, p. 1–34.

DOTT, R. H., JR., 1983, Episodic sedimentation—how normal is average? How rare is rare? Does it matter? Journal of Sedimentary Petrology, v. 53, p. 5–23.

DOTT, R. H., JR., 1988, Perspectives: Something old, something new, something borrowed, something blue—a hindsight and foresight of sedimentary geology: Journal of Sedimentary Petrology, v. 58, p. 358–364.

DOTT, R. H., JR., and BATTEN, R. L., 1981, Evolution of the Earth, 3rd ed.: New York, McGraw-Hill Book Company, 573 p.

DOTT, R. H., JR., and BATTEN, R. L., 1988, Evolution of the Earth, 4th ed.: New York, McGraw-Hill Book Company, 643 p.

DOTT, R. H., JR.; and BOURGEOIS, JOANNE, 1982, Hummocky stratification: significance of its variable bedding sequences: Geological Society of America Bulletin, v. 93, p. 663–680.

DOYLE, L. J., and PILKEY, O. H., eds., 1979, Geology of continental slopes: Tulsa, OK, Society of Economic Paleontologists and Mineralogists Special Publication 27, 374 p.

DRAKE, C. L.; EWING, M.; and SUTTON, G. H., 1959, Continental margins. Geosynclines; the east coast of North America north of Cape Hatteras,

p. 110–198 *in* Physics and chemistry of the earth, vol. 3, Elmsford, NY, Pergamon Press, 464 p.

DRAVIS, J. J., 1983, Hardened subtidal stromatolites, Bahamas: Science, v. 219, p. 385–386.

DRESSER-ATLAS, 1982, Interpretive methods for production well logs, 2nd ed.: Dresser-Atlas Industries, 159 p.

DROSER, M. L., and BOTTJER, D. J., 1986, A semiquantitative field classification of ichnofabric: Journal of Sedimentary Petrology, v. 56, p. 558–559.

DROXLER, A. W., MORSE, J. W., and KORNICKER, W. A., 1988, Controls on carbonate mineral (*sic*) accumulation in Bahamian basins and adjacent Atlantic Ocean sediments: Journal of Sedimentary Petrology, v. 58, p. 120–130.

DUC, A. W., and TYE, R. S., 1987, Evolution and stratigraphy of a regressive barrier/backbarrier complex: Kiawah Island: South Carolina: Sedimentology, v. 34, p. 237–251.

DUFF, P. MC. D.; HALLAM, A.; and WALTON, E. K., 1967, Cyclic sedimentation: Amsterdam, Elsevier Scientific Publishing Company, 280 p.

DUKE, W. L., 1985, Hummocky cross-stratification, tropical hurricanes, and intense winter storms: Sedimentology, v. 32, p. 167–194.

DULLIEN, F. A. L., 1979, Porous media. Fluid transport and structure: New York, Academic Press, 396 p.

DUNBAR, C. O.; and RODGERS, JOHN, 1957, Principles of stratigraphy: New York, John Wiley & Sons, 356 p.

DUNBAR, R. B., and BERGER, W. H., 1981, Fecal pellet (*sic*) flux to modern bottom sediment of Santa Barbara basin (California) based on sediment trapping: Geological Society of America Bulletin Part 1, v. 92, p. 212–218.

DUNHAM, R. J., 1962, Classification of carbonate rocks according to depositional texture, p. 108–121 *in* Ham, W. E., ed., Classification of carbonate rocks: Tulsa, OK, American Association of Petroleum Geologists Memoir 1, 279 p.

DUNNE, L. A., 1988, Discussion of fan-deltas (*sic*) and braid deltas: varieties of coarse-grained deltas: Geological Society of America Bulletin, v. 100, p. 1308–1310 (includes reply by McPherson, J. G.; Shanmugam, Ganapathy; and Moiola, R. J.).

DURAND, BERNARD, ed., 1980, Kerogen, insoluble organic matter from sedimentary rocks: Paris, Editions Technip, 499 p.

DUTCHER, R. R., HACQUEBARD, P. A., SCHOPF, J. M., and SIMON, J. A., eds., 1974, Carbonaceous materials as indicators of metamorphism: Geological Society of America Special Paper 153, 108 p.

DUTRO, J. T., JR., DIETRICH, R. V., and FOOSE, R. M., eds., 1989, AGI Data Sheets for Geology in the Field, Laboratory, and Office, 3rd ed.: Alexandria, Virginia, American Geological Institute, variously paginated.

DUTTON, C. E., 1871, The causes of regional elevations and subsidences: American Philosophical Society Proceedings, v. 12, p. 70–72.

DUTTON, C. E., 1889, On some of the greater problems of physical geology (with discussion by G. K. Gilbert and R. S. Woodward): Philosophical Society of Washington Bulletin, v. 11, p. 51–64.

DUTTON, SHIRLEY P., and LAND, L. S., 1988, Cementation and burial history of a low-permeability quartzarenite, Lower Cretaceous Travis Peak Formation, East Texas: Geological Society of America Bulletin, v. 100, p. 1271–1282.

DZULYNSKI, STANISLAW, 1965, New data on experimental production of sedimentary structures: Journal of Sedimentary Petrology, v. 35, p. 196–212.

DZULYNSKI, STANISLAW; and SANDERS, J. E., 1962, Current marks on firm mud bottoms: Connecticut Academy of Arts and Sciences Transactions, v. 42, p. 57–96.

EARDLEY, A. J., 1951, Structural geology of North America: New York, Harper and Brothers, 624 p.

EARDLEY, A. J.; SHUEY, R. I.; GVOSDETSKY, V.; and others, 1973, Lake cycles in the Bonneville Basin, Utah: Geological Society of America Bulletin, v. 84, p. 211–215.

EASTERBROOK, D. J., 1982, Characteristic features of glacial sediments, p. 1–10, *in* Scholle, P. A.; and Spearing, D., eds., Sandstone depositional environments: American Association of Petroleum Geologists Memoir 31, 410 p.

EBINGER, C. J., 1989, Tectonic development of the western branch of the East African rift system: Geological Society of America Bulletin, v. 101, p. 885–903.

EDDY, J. A., 1976, The Maunder Minimum: Science, v. 192, p. 1189–1202.

EDDY, J. A., 1978, Evidence for a changing Sun, p. 11–33 *in* Eddy, J. A., ed., The new solar physics: Boulder, CO, Westview Press.

EDGAR, T.; RYAN, W. B. F.; and HSÜ, K. J., 1973, Plans for future drilling in the Mediterranean and Black Sea [with discussion], p. 23–25 *in* Symposium sur la Geodynamique de la region Mediterranenne, Comm. Int. L'Explor. Sci. Mer Mediterr., v. 22, No. 2a.

EDWARDS, J. D., and SANTOGROSSI, P. A., eds., 1990, Divergent/passive margin (*sic*) basins: American Association of Petroleum Geologists Memoir 48, 252 p.

EDWARDS, L. E., 1978, Range charts and no-space graphs: Computers and Geosciences, v. 4, p. 247–255.

EDWARDS, M., 1986, Glacial environments, p. 445–470 *in* Reading, H. G., ed., Sedimentary environments and facies (*sic*): Oxford, Blackwell Scientific Publications, 615 p.

EHHALT, D. H., 1973, Methane in the atmosphere, p. 144–158 *in* Woodwell, G. M., and Pecan, E. V., eds., Carbon and the biosphere: Washington, DC, U.S. Atomic Energy Commission, Technical Information Center, 392 p.

EHHALT, D. H., 1974, The atmospheric cycle of methane: Tellus, v. 26, p. 58–70.

EHHALT, D. H., HEIDT, L. E., LUEB, R. H., and MARTELL, E. A., 1975, Concentrations of CH_4, CO, CO_2, H_2, H_2O, and N_2O in the upper atmosphere: Journal of the Atmospheric Sciences, v. 32, p. 163–169.

EINSELE, G.; and SEILACHER, ADOLF, eds., 1982, Cyclic (*sic*) and event stratigraphy: New York, Springer-Verlag, 536 p.

EKDAHL, C. A., and KEELING, C. D., 1973, Atmospheric carbon dioxide and radiocarbon in the natural carbon cycle. I. Quantitative deductions from records at Mauna Loa Observatory and at the South Pole, p. 51–85 *in* Woodwell, G. M., and Pecan, E. V., eds., Carbon and the biosphere: Washington, DC, U.S. Atomic Energy Commission, Technical Information Center, 392 p.

EKDALE, A. A., BROMLEY, R. G., and PEMBERTON, S. G., 1984, Ichnology: the use of trace fossils in sedimentology and stratigraphy: Tulsa, OK, Society of Economic Paleontologists and Mineralogists Short Course 15, 317 p.

EKDALE, A. A., and MASON, T. R., 1988, Characteristic trace-fossil associations in oxygen-poor sedimentary environments: Geology, v. 16, p. 720–723.

ELDER, O. L., and FOWLER, S. W., 1977, Polychlorinated biphenyls: penetration into the deep ocean by zooplankton fecal pellet (*sic*) transport: Science, v. 1997, p. 459–461.

ELDER, W. P., 1988, Geometry of Upper Cretaceous bentonite beds: implications about volcanic source areas and paleowind patterns, Western Interior, United States: Geology, v. 16, p. 835–838.

ELDREDGE, NILES, 1971, The allopatric model of phylogeny in Paleozoic invertebrates: Evolution, v. 25, p. 156–167.

ELDREDGE, NILES, 1972, Systematics and evolution of *Phacops rana* (Green, 1832) and *Phacops iowensis* Delo, 1935 (Trilobita) from the Middle Devonian of North America: American Museum of Natural History Bulletin, v. 147, no. 2, p. 45–114.

ELDREDGE, NILES, 1973, Systematics of Lower (*sic*) and Lower Middle Devonian species of the trilobite *Phacops* Emmrich in North America: American Museum of Natural History Bulletin, v. 151, no. 4, p. 285–337.

ELDREDGE, NILES; and GOULD, S. J., 1972, Punctuated equilibria: an alternative to phyletic gradualism, p. 82–115 *in* Schopf, T. J. M., ed., Models in paleobiology: San Francisco, Freeman, Cooper, and Company, 250 p.

ELDREDGE, NILES; and GOULD, S. J., 1977, Evolutionary models and biostratigraphic strategies, p. 25–40 *in* Kauffman, E. G., and Hazel, J. E., eds., Concepts and methods in biostratigraphy: Stroudsburg, PA, Dowden, Hutchinson, and Ross, 658 p.

ELEY, B. E., and JULL, R. K., 1982, Chert in the Middle Silurian Fossil Hill Formation of Manitoulin Island, Ontario: Bulletin of Canadian Petroleum Geology, v. 30, p. 208–215.

ELLIOT, T., 1986, Deltas, p. 113–154 *in* Reading, H. G., ed., Sedimentary environments and facies, 2nd ed.: New York, Elsevier, 615 p.

ELLIOTT, D. K., ed., 1986, Dynamics of extinction: New York, John Wiley & Sons, 294 p.

ELLIS, J. P., and MILLIMAN, J. D., 1985, Calcium carbonate suspended in Arabian Gulf and Red Sea waters: biogenic and detrital, not "chemogenic": Journal of Sedimentary Petrology, v. 55, p. 805–808.

ELLWOOD, B. B.; CHRZANOWSKI, T. H.; HROUDA, FRANTISEK; LONG, G. J.; and BUHL, M. L., 1988, Siderite formation in anoxic deep-sea sediments: a synergetic bacterially controlled process with important implications in paleomagnetism: Geology, v. 16, p. 980–982.

ELMORE, R. D., 1984, The Copper Harbor Conglomerate: A late (*sic*) Precambrian fining-upward alluvial fan (*sic*) sequence in northern Michigan: Geological Society of America Bulletin, v. 95, p. 610–617.

ELMORE, R. D., PILKEY, O. H., CLEARY, W. J., and CURRAN, H. A., 1979, Black Shell turbidite, Hatteras abyssal plain, western Atlantic Ocean: Geological Society of America Bulletin, v. 90, p. 1165–1176.

ELSON, J. A., 1969, Late Quaternary marine submergence of Quebec: Rev. Geogr. Montreal, v. 23, p. 247–258.

ELSTON, D. P., and MCKEE, E. H., 1982, Age and correlation of the late Proterozoic Grand Canyon disturbance, northern Arizona: Geological Society of America Bulletin, v. 93, p. 681–699.

EMBLEY, R. W., 1975 ms., Studies of deep-sea sedimentation processes using high-frequency seismic data: New York, Columbia University, Department of Geological Sciences, Ph.D. Dissertation, 334 p.

EMBLEY, R. W., 1980, The role of mass transport in the distribution and character (*sic*) of deep-ocean sediments in the Atlantic: Marine Geology, v. 38, p. 23–50.

EMBRY, A. F., and KLOVAN, J. E., 1971, A Late Devonian (*sic*) reef tract on Northeastern Banks Island, N.W.T.: Canadian Petroleum Geology Bulletin, v. 19, p. 730–781.

EMERY, K. O., 1960a, Basin plains and aprons (*sic*) off southern California: Journal of Geology, v. 68, p. 464–479.

EMERY, K. O., 1960b, The sea off southern California. A modern habitat of petroleum: New York, John Wiley & Sons, 366 p.

EMERY, K. O., HEEZEN, B. C., and ALLEN, T. D., 1966, Bathymetry of the eastern Mediterranean Sea: Deep-Sea Research, v. 13, p. 173–192.

EMERY, K. O.; and HULSEMANN, JOBST, 1962, The relationship of sediments, life and water in a marine basin: Deep-Sea Research, v. 8, p. 165–180.

EMERY, K. O., and HUNT, J. M., 1974, Summary of Black Sea investigations, p. 575–590 *in* Degens, E. T., and Ross, D. A., eds., The Black Sea—geology, chemistry, and biology: Tulsa, OK, American Association of Petroleum Geologists Memoir 20, 633 p.

EMERY, K. O., and STEVENSON, R. W., 1957, Estuaries and lagoons, p. 673–750 *in* Hedgpeth, J. W., ed., Treatise on marine ecology and paleoecology: Geological Society of America Memoir 67, v. 1, Ecology, 1296 p.

EMERY, K. O.; and UCHUPI, ELAZAR, 1972, Western North Atlantic Ocean: topography, rocks, structure, water, life, and sediments: Tulsa, OK, American Association of Petroleum Geologists Memoir 17, 532 p.

EMERY, K. O.; UCHUPI, ELAZAR; PHILLIPS, J. D.; BOWIN, C. O.; BUNCE, E. T.; and KNOTT, S. T., 1970, Continental rise of eastern North America: American Association of Petroleum Geologists Bulletin, v. 54, p. 44–108.

ENGLAND, P. C., 1983, Constraints on extension of continental lithosphere: Journal of Geophysical Research, v. 88, p. 1145–1152.

EPSTEIN, S. A., and FRIEDMAN, G. M., 1982, Processes controlling precipitation of carbonate cement and dissolution of silica in reef (*sic*) and near-reef settings: Sedimentary Geology, v. 33, p. 157–172.

EPSTEIN, S. A., and FRIEDMAN, G. M., 1983, Depositional and diagenetic relationships between Gulf of Elat (Aqaba) and Mesozoic of U.S. east coast offshore: American Association of Petroleum Geologists Bulletin, v. 67, p. 953–962.

EREZ, JONATHAN, 1978, Vital effect on stable-isotope composition seen in foraminifera and coral skeletons: Nature, v. 273, p. 199–202.

EREZ, JONATHAN, 1979, Modification of the oxygen-isotope record in deep-sea cores by Pleistocene dissolution cycles: Nature, v. 281, p. 535–538.

EREZ, JONATHAN, 1983, Calcification rates, photosynthesis and light in planktonic foraminifera: p. 307–312 *in* Westbroek, P.; and de Jong, E. W., eds., Biomineralization and biological metal accumulation: Dordrecht, The Netherlands, D. Reidel Publishing Company, 546 p.

ESCHNER, T. B.; and KOCUREK, GARY, 1986, Marine destruction of eolian sand seas: origin of mass flows: Journal of Sedimentary Petrology, v. 56, p. 401–411.

ETHRIDGE, F. G., and FLORES, R. M., eds., 1981, Recent and ancient nonmarine depositional environments: models for exploration: Tulsa, OK, Society of Economic Paleontologists and Mineralogists Special Publication 31, 349 p.

ETHRIDGE, F. G., and WESCOTT, W. A., 1984, Tectonic setting, recognition and hydrocarbon reservoir potential of fan-delta deposits, p. 217–235 *in* Koster, E. H., and Steel, R. J., eds., Sedimentology of gravels and conglomerates: Canadian Society of Petroleum Geologists Memoir 10, 441 p.

ETTENSOHN, F. R., 1985a, The Catskill Delta complex and the Acadian orogeny, p. 39–49 *in* Woodrow, D. L., and Sevon, W. D., eds., The Catskill Delta: Boulder, CO, Geological Society of America Special Paper 201, 246 p.

ETTENSOHN, F. R., 1985b, Controls on development of Catskill Delta complex basin-facies (*sic*), p. 65–77 *in* Woodrow, D. L. and Sevon, W. D., eds., The Catskill Delta: Boulder, CO, Geological Society of America Special Paper 201, 246 p.

EUGSTER, H. P., and HARDIE, L. A., 1978, Saline lakes, p. 237–294 *in* Lerman, Abraham, ed., Lakes: chemistry, geology, physics: New York, Springer-Verlag, 363 p.

EUGSTER, H. P., and KELTS, K., 1983, Lacustrine chemical sediments, p. 321–368 *in* Goudie, A. S.; and Pye, K., eds., Chemical sediments and geomorphology: London, Academic Press.

EVANS, G.; and BUSH, P., 1969, Some oceanographical (*sic*) and sedimentological observations on a Persian Gulf lagoon, p. 155–169 *in* Ayala-Castanares, Augustin; and Phleger, F. B., eds., Lagunas costeras, un simposio: Simposio Internacional sobre lagunas costeras, Mexico City, Mexico, 28–30 Noviembre 1967, memoria: Mexico Universidad Nacional Autonoma (WMAS-UNESCO), 686 p.

EVANS, G.; SCHMIDT, V.; BUSH, P.; and NELSON, H., 1969, Stratigraphy and geologic history of the sabkha, Abu Dhabi, Persian Gulf: Sedimentology, v. 12, p. 145–159.

EVERTS, C. H., 1987, Continental shelf (*sic*) evolution in response to a rise in sea level, p. 49–58 *in* Nummedal, Dag; Pilkey, O. H.; and Howard, J. D., eds., Sea-level fluctuation and coastal evolution: Tulsa, OK, Society of Economic Paleontologists and Mineralogists Special Publication 41, 267 p.

EWING, J. I., and HOLLISTER, C. D., 1972, Regional aspects of Deep Sea Drilling in the Western North Atlantic, p. 951–976 *in* Kaneps, A. G., ed., Initial reports of the Deep Sea Drilling Project, v. XI: Washington, DC, U.S. Government Printing Office, 1077 p.

EWING, M.; WOOLLARD, G. P.; and VINE, A. C., 1938, Geophysical investigations in emerged and submerged Atlantic coastal plain. Part III: Barnegat Bay, N.J., section: Geological Society of America Bulletin, v. 50, p. 257–296.

EWING, M.; WOOLLARD, G. P.; and VINE, A. C., 1940, Geophysical investigations in the emerged and submerged Atlantic coastal plain. Part IV: Cape May, New Jersey, section: Geological Society of America Bulletin, v. 51, p. 1821–1840.

EXUM, F. A., and HARMS, J. C., 1968, Comparison of marine-bar with valley-fill stratigraphic traps, western Nebraska: American Association of Petroleum Geologists Bulletin, v. 52, p. 1851–1868.

EYLES, C. H., 1987, Glacially influenced submarine-channel sedimentation in the Yakataga Formation, Middleton Island, Alaska: Journal of Sedimentary Petrology, v. 57, p. 1004–1017.

EYLES, NICHOLAS; and CLARK, B. M., 1988, Storm-influenced deltas and ice scouring in a Late Pleistocene glacial lake: Geological Society of America Bulletin, v. 100, p. 793–809.

EYLES, NICHOLAS; EYLES, C. H.; and MIALL, A. D., 1983, Lithofacies types and vertical profile models; an alternative approach to the description and environmental interpretation of glacial diamict sequences: Sedimentology, v. 30, p. 393–410.

FAGERSTROM, J. A., 1987, The evolution of reef communities: New York, John Wiley & Sons, 600 p.

FAIRBRIDGE, R. W., 1961, Eustatic changes of sea level, p. 99–185 *in* Ahrens, L. H., and others, eds., Physics and chemistry of the Earth, v. 4: London, Pergamon Press.

FAIRBRIDGE, R. W., 1972, Quaternary sedimentation in the Mediterranean region controlled by tectonics, paleoclimates and sea level, p. 99–113 *in* Stanley, D. J., ed., The Mediterranean Sea: a natural sedimentation laboratory: Stroudsburg, PA, Dowden, Hutchinson, and Ross, 765 p.

FAIRBRIDGE, R. W., 1980, The estuary: its definition and geodynamic cycle, p. 1–35 *in* Olausson, E.; and Cato, I., eds., Chemistry and biochemistry of estuaries: New York, John Wiley & Sons, 452 p.

FAIRBRIDGE, R. W., 1987, The spectra of sea level in a Holocene time frame, p. 127–142 *in* Rampino, M. R., Sanders, J. E., Newman, W. S., and Konigsson, L. K., eds., 1987, Climate: history, periodicity, and predictability: New York, Van Nostrand Reinhold Company, 588 p.

FAIRBRIDGE, R. W., ed., 19??, The encyclopedia of atmospheric sciences and astrogeology: The Encyclopedia of the Earth Sciences, v. 2: New York, Reinhold Publishing Corporation, 1200 p.

FAIRBRIDGE, R. W.; and BOURGEOIS, JOANNE, eds., 1978, The Encyclopedia of Sedimentology: Stroudsburg, PA, Dowden, Hutchinson, and Ross, 901 p.

FAIRBRIDGE, R. W., and SANDERS, J. E., 1987, The Sun's orbit, A.D. 750–2050: basis for new perspectives on planetary dynamics and Earth-Moon linkage, p. 446–471, *in* Rampino, M. R., Sanders, J. E., Newman, W. S., and Konigsson, L. K., eds., 1987, Climate: history, periodicity, and predictability: New York, Van Nostrand Reinhold Company, 588 p.

FALVEY, D. A., 1974, The development of continental margins in plate tectonic (*sic*) theory: Australian Petroleum Exploration Association, Journal, p. 95–106.

FARRELL, K. M., 1987, Sedimentology and facies architecture of overbank deposits of the Mississippi River, False River Region, Louisiana, p. 111–

120 *in* Ethridge, F. G., Flores, R. M., and Harvey, M. D., eds., Recent developments in fluvial sedimentology: Tulsa, OK, Society of Economic Paleontologists and Mineralogists Special Publication 9, 389 p.

FARROW, G. E., and FYFE, J. A., 1988, Bioerosion and carbonate mud (*sic*) production on high-latitude shelves: Sedimentary Geology, v. 60, p. 281–297.

FAUGERES, J. C.; STOW, D. A. V.; and GOUTHIER, E., 1984, Contourite drift moulded by deep Mediterranean outflow: Geology, v. 12, p. 296–300.

FAURE, G., 1977, Principles of isotope geology: New York, John Wiley & Sons, 464 p.

FENSCHEL, T. M., and RIEDL, R. J., 1970, The sulfide system: a new biotic community underneath the oxidized layer of marine sand bottoms: Marine Biology, v. l7, p. 255–268.

FERREL, WILLIAM, 1863, On the causes of the annual inundation of the Nile: American Journal of Science, 2nd series, v. 35, p. 62–64.

FERTL, W. H., and TIMKO, D. J., 1970, How abnormal-pressure detection (*sic*) techniques are applied: Oil and Gas Journal, January 12, p. 62–71.

FETH, J. H., 1964, Review and annotated bibliography of ancient lake deposits (Precambrian to Pleistocene) in the western states: U.S. Geological Survey, Bulletin 1080, 119 p.

FEUILLET, J.-P.; and FLEISCHER, PETER, 1980, Estuarine circulation: controlling factor of clay mineral (*sic*) distribution in James River estuary, Virginia: Journal of Sedimentary Petrology, v. 50, p. 267–279.

FIELD, M. E., 1980, Sand bodies on coastal plain (*sic*) shelves: Holocene record of the U.S. Atlantic inner shelf off Maryland: Journal of Sedimentary Petrology, v. 50, p. 505–528.

FIELD, M. E., 1981, Sediment mass transport in basins: controls and patterns, p. 61–84 *in* Douglas, R. G., Colburn, I. P., and Gorsline, D. S., eds., Depositional systems of active continental margin (*sic*) basins: Short Course Notes: Pacific Section, Society of Economic Paleontologists and Mineralogists, 165 p.

FIELD, M. E., MEISBURGER, E. P., STANLEY, E. A., and WILLIAMS, S. J., 1979, Upper Quaternary peat deposits on the Atlantic inner shelf of the United States: Geological Society of America Bulletin, v. 90, p. 618–628.

FIELDING, C. R., 1986, Fluvial channel and overbank deposits from the Westphalian of the Durham coalfield, NE England: Sedimentology, v. 33, p. 119–140.

FINKEL, E. A., and WILKINSON, B. H., 1990, Stylolitization as a source of cement in Mississippian Salem Limestone, west-central Indiana: Geological Society of America Bulletin, v. 74, p. 174–186.

FISCHER, A. G., 1982, Long term (*sic*) climatic oscillations recorded in stratigraphy, p. 97–104 *in* Berger, W. H., and Crowell, J. C., eds., Climate in Earth history: Washington, DC, U.S. National Academy Press, 335 p.

FISCHER, A. G., 1986, Climatic rhythms recorded in strata: Annual Reviews of Earth and Planetary Science, v. 14, p. 351–376.

FISHER, D. W., 1987, Lower Devonian limestones, Helderberg Escarpment, New York, p. 119–132 *in* Roy, D. C., ed., Geological Society of America, Centennial Field Guidebook, v. 5, Northeastern Section, 481 p.

FISHER, J. H., ed., 1977, Reefs and evaporites—concepts and depositional models: Tulsa, OK, American Association of Petroleum Geologists, Studies in Geology 5, 196 p.

FISHER, R. V., 1961, Proposed classification of volcaniclastic sediments and rocks: Geological Society of America Bulletin, v. 72, p. 1409–1414.

FISHER, R. V., 1963, Bubble-wall texture and its significance: Journal of Sedimentary Petrology, v. 33, p. 224–227.

FISHER, R. V., 1966a, Mechanism for deposition from pyroclastic flows: American Journal of Science, v. 264, p. 350–363.

FISHER, R. V., 1966b, Rocks composed of volcanic fragments and their classification: Earth-Science Reviews, v. 1, p. 287–298.

FISHER, R. V.; and SCHMINKE, H.-U., 1984, Pyroclastic rocks: New York, Springer-Verlag, 528 p.

FISHER, W. L., and McGOWEN, J. H., 1967, Depositional systems in the Wilcox Group of Texas and their relationship to occurrence of oil and gas: Gulf Coast Association of Geological Societies Transactions, v. 17, p. 105–125.

FITZGERALD, D. M.; and PENLAND, S., 1987, Backbarrier dynamics of the East Friesian Islands: Journal of Sedimentary Petrology, v. 57, p. 746–754.

FITZGERALD, D. M., and ROSEN, P. J., eds., 1987, Glaciated Coasts: New York, Academic Press, 364 p.

FLEISCHER, PETER, 1972, Mineralogy (*sic*) and sedimentation history, Santa Barbara Basin, California: Journal of Sedimentary Petrology, v. 42, p. 49–58.

FLEISS, J. L., 1973, Statistical methods for rates and proportions: New York, John Wiley & Sons, 223 p.

FLEMING, R. H., 1957, General features of the oceans, p. 87–108 *in* Hedgpeth, J. W., ed., Treatise on marine ecology and paleoecology: Geological Society of America Memoir 67, v. 1, Ecology, 1296 p.

FLEMING, R. H.; and REVELLE, R., 1939, Physical processes in the ocean, p. 48–141 *in* Trask, P. D., ed., Recent marine sediments—a symposium: Tulsa, OK, American Association of Petroleum Geologists, 936 p. (Reprinted, revised and enlarged ed., 1955, Society of Economic Paleontologists and Mineralogists Special Publication 4; also in paperback, 1968, New York, Dover.)

FLEMINGS, P. B., and JORDAN, T. E., 1989, A synthetic stratigraphic model of foreland basin (*sic*) development: Journal of Geophysical Research, v. 94, p. 3851–3866.

FLEMMING, B. W., 1988, Pseudo-tidal sedimentation in a non-tidal shelf environment (southeastern African margin), p. 167–180 *in* Deboer, P. L.; van Gelder, A.; and Nio, S.-D., eds., Tide-influenced sedimentary environments and facies (*sic*): Dordrecht, The Netherlands, D. Reidel Publishing Company, 530 p.

FLINT, JEAN-JACQUES; and LOLCAMA, J., 1986, Buried ancestral drainage between Lakes Erie and Ontario: Geological Society of America Bulletin, v. 97, p. 75–84.

FLINT, R. F., 1943, Growth of North American ice sheet during the Wisconsin age: Geological Society of America Bulletin, v. 54, p. 325-362.

FLINT, R. F., 1947, Glacial geology and the Pleistocene Epoch: New York, John Wiley & Sons, 589 p.

FLINT, R. F., 1957, Glacial and Pleistocene geology: New York, John Wiley & Sons, 533 p.

FLINT, R. F., 1971, Glacial and Quaternary geology: New York, John Wiley & Sons, 892 p.

FLINT, R. F.; SANDERS, J. E.; and RODGERS, JOHN, 1960a, Symmictite: a name for nonsorted terrigenous sedimentary rocks that contain a wide range of particle sizes: Geological Society of America Bulletin, v. 71, p. 507–510.

FLINT, R. F.; SANDERS, J. E.; and RODGERS, JOHN, 1960b, Diamictite, a substitute term for symmictite: Geological Society of America Bulletin, v. 71, p. 1809–1810.

FLINT, S.; CLEMMEY, H.; and TURNER, P., 1986, The Lower Cretaceous Way Group of northern Chile: an alluvial fan-fan delta complex: Sedimentary Geology, v. 46, p. 1–22.

FLOOD, P. G., and ORME, G. R., 1988, Mixed siliciclastic/carbonate sediments of the northern Great Barrier Reef Province, Australia, p. 175–205 *in* Doyle, L. J., and Roberts, H. H., eds., Carbonate-clastic transitions: Developments in sedimentology 42: New York, Elsevier, 304 p.

FLOOD, P. G., and WALBRAN, P. D., 1986, A siliciclastic coastal sabkha, Capricorn Coast, Queensland, Australia: Sedimentary Geology, v. 48, p. 169–181.

FLOOD, R. D., 1983, Classification of sedimentary furrows and a model for furrow initiation and evolution: Geological Society of America Bulletin, v. 94, p. 630–639.

FLORES, R. M., BLANCHARD, L. F., SANCHEZ, J. D., MARLEY, W. E., and MULDOON, W. J., 1984, Paleogeographic controls of coal accumulation, Cretaceous Blackhawk Formation and Star Point Sandstone, Wasatch Plateau, Utah: Geological Society of America Bulletin, v. 95, p. 540–550.

FLORKE, O. W., JONES, J. B., and SEGNIT, E. R., 1975, Opal-CT crystals: Neues Jahrb. Mineral. Monatsh., Jahrgang 1975, H. 8, p. 369–377.

FOLK, R. L., 1956, The role of texture and composition in sandstone classification: Journal of Sedimentary Petrology, v. 26, p. 166–171.

FOLK, R. L., 1959, Practical petrographic classification of limestones: American Association of Petroleum Geologists Bulletin, v. 43, p. 1–38.

FOLK, R. L., 1965, Some aspects of recrystallization in ancient limestones, p. 14–48 *in* Pray, L. C., and Murray, R. C., eds., Dolomitization and limestone diagenesis, a symposium: Tulsa, OK, Society of Economic Paleontologists and Mineralogists Special Publication 13, 180 p.

FOLK, R. L., 1966, A review of grain-size parameters: Sedimentology, v. 6, p. 73–93.

FOLK, R. L., 1975, Glacial deposits identified by chattermark trails in detrital garnets: Geology, v. 3, p. 473–475.

FÖLLMI, K. B., 1989, Evolution of the Mid-Cretaceous triad, Lecture notes in earth sciences 23: Berlin, Springer-Verlag, 153 p.

FORBES, D. L.; and BOYD, R., 1987, Gravel ripples on the inner Scotian shelf: Journal of Sedimentary Petrology, v. 57, p. 46–54.

FORCE, E. R., 1976, Metamorphic source rocks of titanium placer deposits—a geochemical cycle: U.S. Geological Survey Professional Paper 959-B, 16 p.

FORCE, E. R., 1984, A relation (*sic*) among geomagnetic reversals, seafloor spreading (*sic*) rate, paleoclimate, and black shales: EOS, American Geophysical Union Transactions, v. 65, p. 18–19.

FOREL, F. A., 1885, Les ravins sous-lacoustre des fleuves glaciaires: Academie des Sciences, Paris, Comptes Rendus, v. 101, p. 725–728.

FOUCH, T. D., and DEAN, W. E., 1982, Lacustrine deposits, p. 87–114, *in* Scholle, P. A.; and Spearing, D., eds., Sandstone depositional environments: Tulsa, OK, American Association of Petroleum Geologists Memoir 31, 411 p.

FRASER, G. S., 1989, Clastic depositional sequences. Processes of evolution and principles of interpretation: Englewood Cliffs, NJ, Prentice-Hall, 459 p.

FRASER, G. S., and HESTER, N. C., 1977, Sediments and sedimentary structures of a beach ridge (*sic*) complex, southwestern shore of Lake Michigan: Journal of Sedimentary Petrology, v. 47, p. 1187–1200.

FRASER, G. S.; and SUTTNER, L., 1986, Alluvial fans and fan deltas: a guide to exploration for oil and gas: Boston, International Human Resource Development Corporation, 199 p.

FRASER, H. J., 1935, Experimental study of porosity and permeability of clastic sediments: Journal of Geology, v. 43, p. 910–1010.

FRASSETTO, ROBERTO, 1960, A preliminary survey of thermal microstructure in the Strait of Gibraltar: Deep-Sea Research, v. 7, p. 152–162.

FRAZIER, D. E., 1967, Recent deltaic deposits of the Mississippi River: their development and chronology, p. 287–315 *in* Symposium on the geological history of the Gulf of Mexico, Antillean-Caribbean region: Gulf Coast Association of Geological Societies Transactions, v. 17, 532 p.

FRAZIER, D. E., 1974, Depositional-episodes (*sic*): their relationship to the Quaternary stratigraphic framework in the northwestern portion of the Gulf basin: Austin, TX, University of Texas, Bureau of Economic Geology, Geological Circular 44-1, 28 p.

FREDSOE, J., 1982, Shape and dimensions of stationary dunes in rivers: American Society of Civil Engineers, Hydraulics Division, Journal, v. 108, p. 932–947.

FREEMAN-LYNDE, R. P., WHITLEY, K. F., and LOHMANN, K. C., 1986, Deep-marine origin of equant spar cements in Bahama Escarpment limestones: Journal of Sedimentary Petrology, v. 56, p. 799–811.

FREEZE, R. A., and CHERRY, J. A., 1979, Groundwater: Englewood Cliffs, NJ, Prentice-Hall, 604 p.

FREY, R. W.; HOWARD, J. D.; HAN, S.-J.; and PARK, B.-K., 1989, Sediments and sedimentary sequences on a modern macrotidal flat, Inchon, Korea: Journal of Sedimentary Petrology, v. 59, p. 28–44.

FREY R. W.; HOWARD, J. D.; and HONG, J.-S., 1987, Prevalent lebensspuren on a modern macrotidal flat, Inchon, Korea: ethological and environmental significance: Palaios, v. 2, p. 571–593.

FREY R. W., and PEMBERTON, S. G., 1985, Biogenic structures in outcrops and cores. I. Approaches to Ichnology: Canadian Petroleum Geology Bulletin, v. 33, p. 72–115.

FREY R. W., PEMBERTON, S. G., and SANDERS, T. D. A., 1990, Ichnofacies and bathymetry: a passive relationship: Journal of Paleontology, v. 64, p. 155–158.

FREY R. W., and WHEATCROFT, R. A., 1989, Organism-substrate relations (*sic*) and their impact on sedimentary petrology: Journal of Geological Education, v. 37, p. 261–279.

FRIEDKIN, J. F., 1945, A laboratory study of the meandering of alluvial rivers, Vicksburg, Mississippi, U.S. Army Corps of Engineers, Mississippi River Comm., Waterways Experiment Station, 40 p. (Reprinted 1972, p. 237–281 *in* Schumm, S. A., ed., River morphology: Benchmark papers in geology: Stroudsburg, PA, Dowden, Hutchinson, and Ross, 429 p.)

FRIEDMAN, G. M., 1959, The Samreid Lake sulfide deposit, Ontario, an example of a pyrrhotite-pyrite iron formation: Economic Geology, v. 54, p. 268–284.

FRIEDMAN, G. M., 1961, Distinction between (*sic*) dune, beach and river sands from their textural characteristics: Journal of Sedimentary Petrology, v. 31, p. 514–529.

FRIEDMAN, G. M., 1962, On sorting, sorting coefficients and the lognormality of the grain-size distribution of sandstones: Journal of Geology, v. 70, p. 737–753.

FRIEDMAN, G. M., 1964, Early diagenesis and lithification in carbonate sediments: Journal of Sedimentary Petrology, v. 34, p. 777–813.

FRIEDMAN, G. M., 1965a, On the origin of aragonite in the Dead Sea: Israel Journal of Earth Sciences, v. 14, p. 79–85.

FRIEDMAN, G. M., 1965b, Terminology of crystallization textures and fabrics in sedimentary rocks: Journal of Sedimentary Petrology, v. 35, p. 643–655.

FRIEDMAN, G. M., 1965c, Occurrence and stability relationships of aragonite, high-magnesian calcite, and low-magnesian calcite under deep-sea conditions: Geological Society of America Bulletin, v. 76, p. 1191–1192.

FRIEDMAN, G. M., 1966, Occurrence and origin of Quaternary dolomite of Salt Flat, West Texas: Journal of Sedimentary Petrology, v. 36, p. 263–267.

FRIEDMAN, G. M., 1967, Dynamic processes and statistical parameters compared for size frequency (*sic*) distribution of beach (*sic*) and river sands: Journal of Sedimentary Petrology, v. 37, p. 327–354.

FRIEDMAN, G. M., 1968, The fabric of carbonate cement and matrix and its dependence on the salinity of water, p. 11–20 *in* Müller, Germann; and Friedman, G. M., eds., Recent developments in carbonate sedimentology in central Europe: New York, Springer-Verlag, 255 p.

FRIEDMAN, G. M., 1972a, "Sedimentary facies": products of sedimentary environments in Catskill

Mountains, Mohawk Valley, and Taconic sequence, eastern New York state: Society of Economic Paleontologists and Mineralogists, Eastern Section, Guidebook, 48 p.

FRIEDMAN, G. M., 1972b, Thin-section petrography of the Mediterranean evaporites: Petrographic data and comments on the depositional environment of the Miocene sulfates and dolomites at sites 124, 132, and 134, western Mediterranean Sea, p. 695–708 *in* Ryan, W. B. F., Hsü, K. J., and others, eds., Initial reports of the Deep Sea Drilling Project, v. 13, Washington, DC, U.S. Government Printing Office, 1447 p.

FRIEDMAN, G. M., 1975a, The making and unmaking of limestones or the downs and ups of porosity: Journal of Sedimentary Petrology, v. 45, p. 379–398.

FRIEDMAN, G. M., 1975b, Can reefs and evaporites form in stratigraphic contact during one depositional cycle? Answer from Quaternary example: American Association of Petroleum Geologists Bulletin, v. 59, p. 1736.

FRIEDMAN, G. M., 1978, Solar Lake: a sea-marginal pond of the Red Sea (Gulf of Aqaba or Elat) in which algal mats generate carbonate particles and laminites, p. 227–235 *in* Krumbein, W. E., ed., Environmental biogeochemistry and geomicrobiology. The aquatic environment, vol. 1: Ann Arbor, MI, Ann Arbor Sci.

FRIEDMAN, G. M., 1979, Differences in size distributions of populations of particles among sands of various origins: Sedimentology, v. 26, p. 3–23.

FRIEDMAN, G. M., 1985a, Gulf of Elat (Aqaba). Geological and sedimentological framework, p. 39–71 *in* Friedman, G. M., and Krumbein, W. E., eds., Hypersaline ecosystems. The Gavish Sabkha. Ecological Studies 53: New York, Springer-Verlag, 484 p.

FRIEDMAN, G. M., 1985b, The problem of submarine cement in classifying reefrock: an experience in frustration, p. 117–121 *in* Schneidermann, N.; and Harris, P. M., eds., Carbonate cements: Tulsa, OK, Society of Economic Paleontologists and Mineralogists Special Publication 36, 379 p.

FRIEDMAN, G. M., 1987a, Deep-burial diagenesis: its implications for vertical movements of the crust, uplift of the lithosphere and isostatic unroofing, a review: Sedimentary Geology, v. 50, p. 67–94.

FRIEDMAN, G. M., 1987b, Addendum: deep-burial diagenesis: its implications for vertical movements of the crust, uplift of the lithosphere and isostatic unroofing, a review: Sedimentary Geology, v. 54, p. 165–167.

FRIEDMAN, G. M., 1987c, Vertical movements of the crust: case histories from the northern Appalachian Basin: Geology, v. 15, p. 1130–1133.

FRIEDMAN, G. M., 1988a, Methane-derived authigenic carbonates formed by subduction-induced pore-water expulsion along the Oregon/Washington margin: discussion: Geological Society of America Bulletin, v. 100, p. 622.

FRIEDMAN, G. M., 1988b, Slides and slumps: Earth Science, fall, p. 21–23.

FRIEDMAN, G. M., 1988c, The Catskill tectonic fan-delta complex: northern Appalachian Basin: Northeastern Geology, v. 10, p. 254–257.

FRIEDMAN, G. M., 1990a, Vertical parasequences of Lower Devonian limestones, Helderberg Escarpment: the Indian Ladder Trail at the John Boyd Thacher State Park near Albany, New York: Northeastern Geology, v. 12, nos. 1/2, p. 14–18.

FRIEDMAN, G. M., 1990b, Anthracite and concentration of alkaline feldspar (microcline) in flatlying undeformed (*sic*) Paleozoic strata: a key to large-scale vertical crustal uplift, p. 16–28 *in* Heling, D.; Rothe, P.; Förstner, U.; and Stoffers, P., eds., Sediments and environmental geochemistry: Berlin, Springer-Verlag, 371 p.

FRIEDMAN, G. M., 1991, Methane-generated lithified dolostone of Holocene age: Eastern Mediterranean: Journal of Sedimentary Petrology, v. 61, p. 188–194.

FRIEDMAN, G. M.; AMIEL, A. J.; BRAUN, MOSHE; and MILLER, D. S., 1973, Generation of carbonate particles and laminites (*sic*) in algal mats—example from sea-marginal hypersaline pool, Gulf of Aqaba, Red Sea: American Association of Petroleum Geologists Bulletin, v. 57, p. 541–557.

FRIEDMAN, G. M.; AMIEL, A. J.; and SCHNEIDERMANN, N., 1974, Submarine cementation in reefs: example from the Red Sea: Journal of Sedimentary Petrology, v. 44, p. 816–825.

FRIEDMAN, G. M., FABRICAND, B. P., IMBIMBO, E. S., BREY, M. E., and SANDERS, J. E., 1968, Chemical changes in interstitial waters from continental shelf (*sic*) sediments: Journal of Sedimentary Petrology, v. 38, p. 1313–1319.

FRIEDMAN, G. M., and FONER, H. A., 1982, pH and Eh changes in sea-marginal algal pools of the Red Sea and their effects on carbonate precipitation: Journal of Sedimentary Petrology, v. 52, p. 41–46.

FRIEDMAN, G. M., and JOHNSON, K. G., 1982, Exercises in Sedimentology: New York, John Wiley & Sons, 208 p.

FRIEDMAN, G. M., and KEITH, B. D., 1983, Economic evaluation of the Taconic Region—New York, Vermont: Northeastern Geology, v. 5, p. 132–136.

FRIEDMAN, G. M., and KOPASKA-MERKEL, D. C., 1990, Late Silurian pinnacle reefs of the Michigan Basin, p. 89–100, *in* Catacosinos, P. A., and Daniels, P. A., Jr., eds., Early sedimentary evolution of the Michigan Basin, Geological Society of America Special Paper 256, 248 p.

FRIEDMAN, G. M., and KRUMBEIN, W. E., 1985, eds., Hypersaline ecosystems. The Gavish Sabkha. Ecological Studies 53: New York, Springer-Verlag, 484 p.

FRIEDMAN, G. M.; REECKMANN, S. A.; and BORAK, B., 1981, Carbonate deformation mechanisms in the world's deepest wells (9 km)? Tectonophysics, v. 74, p. 715–719.

FRIEDMAN, G. M., and SANDERS, J. E., 1967, Origin and occurrence of dolostones, p. 267–348 *in* Chilingar, G. V., Bissell, H. J., and Fairbridge, R. W., eds., Carbonate rocks, origin, occurrence, and classification: Amsterdam, Elsevier Scientific Publishing Company, 471 p.

FRIEDMAN, G. M., and SANDERS, J. E., 1970, Coincidence of high sea level with cold climate and low sea level with warm climate: evidence from carbonate rocks: Geological Society of America Bulletin, v. 81, p. 2457–2458.

FRIEDMAN, G. M., and SANDERS, J. E., 1974, Positive-relief bedforms on modern tidal flat that resemble molds of flutes and grooves: implications for geopetal criteria and for origin and classification of bedforms: Journal of Sedimentary Petrology, v. 44, p. 181–189.

FRIEDMAN, G. M., and SANDERS, J. E., 1978, Principles of sedimentology: New York, John Wiley & Sons, 792 p.

FRIEDMAN, G. M., and SANDERS, J. E., 1982: Time-temperature-burial significance of Devonian anthracite, implies former great (6.5 km) depth of burial of Catskill Mountains, New York: Geology, v. 10, p. 93–96.

FRIEDMAN, G. M., SANDERS, J. E., and MARTINI, I. P., 1982, Excursion 17A, sedimentary facies: products of sedimentary environments in a cross section of the classic Appalachian Mountains and adjoining Appalachian Basin in New York and Ontario, Field excursion guide book: International Association of Sedimentologists, variously paginated.

FRIEDMAN, G. M.; SNEH, A.; and OWEN, R. W., 1985, The Ras Muhammad Pool: implications for the Gavish Sabkha, p. 218–237 *in* Friedman, G. M., and Krumbein, W. E., eds., 1985, Hypersaline ecosystems. The Gavish Sabkha. Ecological Studies 53: Berlin, Springer-Verlag, 484 p.

FROSTICK, L. E.; and REID, E., 1987, Desert Sediments: ancient and modern: Geological Society (London) Special Publication 35, 401 p.

FÜCHTBAUER, HANS, 1959, Zur Nomenklatur der Sedimentgesteine: Erdol und Kohle: v. 12, p. 605–613.

FÜCHTBAUER, HANS, ed., 1988, Sediment und Sedimentgesteine. Sediment-Petrologie Part II: Stuttgart, E. Schweizerbart'sche Verlagsbuchhandlung, 1141 p.

FÜCHTBAUER, HANS; and LEGGEWIE, R., 1984, Korngrossenbeziehungen zwischen Silt- und Sandsteinen: Neues Jahrbuch Geol. Palaont. Abh., v. 167, p. 133–161.

FÜCHTBAUER, HANS; and RICHTER, D. K., 1988, Karbonatgestiene, p. 233–434 *in* Füchtbauer, Hans, ed., Sediment und Sedimentgesteine. Sediment-Petrologie Part II: Stuttgart, E. Schweizerbart'sche Verlagsbuchhandlung, 1141 p.

FÜCHTBAUER, HANS; and VALETON, IDA, 1988, Kieselgesteine, p. 501–542 *in* Füchtbauer, Hans, ed., Sediment und Sedimentgesteine. Sediment-Petrologie Part II: Stuttgart, E. Schweizerbart'sche Verlagsbuchhandlung, 1141 p.

FYFE, W. S., PRICE, N. J., and THOMPSON, A. B., 1978, Fluids in the Earth's crust. Their significance in metamorphic, (*sic*) tectonic, (*sic*) and chemical transport processes: Amsterdam, Elsevier Scientific Publishing Company, 376 p.

GALLOWAY, W. E., 1975, Process framework for describing the morphologic and stratigraphic evolution of deltaic depositional systems, p. 87–98 *in* Broussard, M. L., ed., Deltas. Models for exploration: Houston, TX, Houston Geological Society, 555 p.

GALLOWAY, W. E., 1987, Depositional (*sic*) and structural architecture of prograding clastic continental margins: tectonic influence on patterns of basin filling: Norsk Geologisk Tidsskrift, v. 67, p. 237–251.

GALLOWAY, W. E., 1989a, Genetic stratigraphic sequences in basin analysis I: architecture and genesis of flooding-surface bounded (*sic*) depositional units: American Association of Petroleum Geologists Bulletin, v. 73, p. 124–142.

GALLOWAY, W. E., 1989b, Genetic stratigraphic sequences in basin analysis II: application to northwest Gulf of Mexico Cenozoic Basin: American Association of Petroleum Geologists Bulletin, v. 73, p. 143–154.

GALLOWAY, W. E., 1989c, Clastic facies models, depositional systems, sequences and correlation: a sedimentologist's view of the dimensional (*sic*) and temporal resolution of lithostratigraphy, p. 459–477 *in* Cross, T. A., ed., Quantitative dynamic stratigraphy: Englewood Cliffs, NJ, Prentice-Hall, 432 p.

GALLOWAY, W. E., and HOBDAY, D. K., 1983, Terrigenous clastic depositional systems: New York, Springer-Verlag, 423 p.

GALLOWAY, W. E., and MORTON, R. A., 1989, Geometry (*sic*), genesis and reservoir characteristics of shelf sandstone facies, Frio Formation (Oligocene), Texas coastal plain, p. 89–115, *in* Morton, R. A.; and Nummedal, Dag, eds., Shelf sedimentation, shelf sequences, and related hydrocarbon accumulation: Society of Economic Paleontologists and Mineralogists, Gulf Coast Section, Annual Research Conference, 7th, 1 April 1989, Proceedings, 211 p.

GARBER, R. A.; LEVY, YITZHAK; and FRIEDMAN, G. M., 1987, The sedimentology of the Dead Sea: Carbonates and Evaporites, v. 2, p. 43–57.

GARDNER, J. V., and KIDD, R. B., 1987, Sedimentary processes on the northwestern Iberian continental margin viewed by long-range side-scan sonar and seismic data: Journal of Sedimentary Petrology, v. 57, p. 397–407.

GARFUNKEL, Z., 1984, Large-scale submarine rotational slumps and growth faults in the eastern Mediterranean: Marine Geology, v. 55, p. 305–324.

GARFUNKEL, Z.; ZAK, I.; and FREUND, R., 1981, Active faulting in the Dead Sea Rift: Tectonophysics, v. 80, p. 1–26.

GARNER, D. L., and TURCOTTE, D. L., 1984, The thermal and mechanical evolution of the Anadarko Basin: Tectonophysics, v. 107, p. 1–24.

GARRISON, R. E., and FISCHER, A. G., 1969, Deep-water limestones and radiolarites of the Alpine Jurassic, p. 20–56 *in* Friedman, G. M., ed., Depositional environments in carbonate rocks—a symposium: Tulsa, OK, Society of Economic Paleontologists and Mineralogists Special Publication 14, 209 p.

GAT, J. R., 1984, Stable isotope (*sic*) composition of Dead Sea waters: Earth and Planetary Science Letters, v. 71, p. 361–376.

GAUTIER, D. L., 1982, Siderite concretions: indicators of early diagenesis in the Gammon Shale (Cretaceous): Journal of Sedimentary Petrology, v. 52, p. 859–871.

GAVISH, E., 1980, Recent sabkhas marginal to the southern coasts of Sinai, Red Sea, p. 233–251 *in* Nissenbaum, A., ed., Hypersaline brines and evaporite environments: Amsterdam, Elsevier Scientific Publishing Company, 270 p.

GAVISH, E.; KRUMBEIN, W. E.; and TAMIR, N., 1978, Recent clastic (carbonate) sediments and sabkhas marginal to the gulfs of Eilat and Suez. Field excursion guidebook IAS 10th International Congress, Jerusalem II, p. 309–332.

GAYNOR, G. C., and SWIFT, D. J. P., 1988, Shannon Sandstone depositional model: sand ridge (*sic*) formation on the Campanian western interior shelf: Journal of Sedimentary Petrology, v. 58, p. 868–880.

GEBELEIN, C. D.; STEINEN, R. P.; GARRETT, PETER; HOFFMAN, E. J.; QUEEN, J. M.; and PLUMMER, L. N., 1980, Subsurface dolomitization beneath the tidal flats of central west Andros Island, Bahamas, p. 31–49 *in* Zenger, D. H., Dunham, J. B., and Ethington, R. L., eds., Concepts and models of dolomitization, Society of Economic Paleontologists and Mineralogists Special Publication 28, 320 p.

GEESLIN, J. H., and CHAFETZ, H. S., 1982, Ordovician Aleman ribbon cherts: an example of silicification prior to carbonate lithification: Journal of Sedimentary Petrology, v. 52, p. 1283–1293.

GERMANN, K.; BOCK, W. D.; and SCHROTER, T., 1984, Facies development of Upper Cretaceous phosphorites in Egypt: sedimentological and geochemical aspects, p. 354–362 *in* Klitzsch, E.; Said, Rushti; and Schrank, E., eds., SFB 69: Results of the special research project Arid Areas, Period 1981–1984: Berliner Geowiss., Abh. A, v. 50, 457 p.

GEVIRTZ, J. L.; PARK, R. A., and FRIEDMAN, G. M., 1971, Paraecology of benthonic (*sic*) Foraminifera and associated micro-organisms of the continental shelf off Long Island, New York: Journal of Paleontology, v. 45, p. 153–177.

GHOSH, S. K., and FRIEDMAN, G. M., 1989, Petrophysics of a dolostone reservoir: San Andres Formation (Permian), West Texas: Carbonates and Evaporites, v. 4, p. 45–119.

GIBBONS, M. J., 1978, The geochemistry of sabkha and related deposits: Newcastle, United Kingdom, unpubl. Ph.D. dissertation.

GIBBS, R. J., 1985, Settling velocity, diameter, and density for flocs of illite, kaolinite, and montmorillonite: Journal of Sedimentary Petrology, v. 55, p. 65–68.

GIBBS, R. J., MATTHEWS, M. D., and LINK, D. A., 1971, The relationship between sphere size and settling velocity: Journal of Sedimentary Petrology, v. 41, p. 7–18.

GIERMANN, G., 1961, Erlauterungen zur bathymetrischen Karte der Strasse von Gibraltar: Inst. Oceanogr. Monaco Bulletin 1218, p. 1–28.

GILBERT, C. M., 1954, Sedimentary rocks, p. 251–384 *in* Williams, Howel; Turner, F. J.; and Gilbert, C. M., Petrography: San Francisco, W. H. Freeman and Company, 406 p.

GILBERT, G. K., 1890, Lake Bonneville: U.S. Geological Survey, Monograph, v. 1, 438 p.

GILBERT, G. K., 1895, Sedimentary measurement of Cretaceous time: Journal of Geology, v. 3, p. 121–127.

GILBERT, R., 1983, Sedimentary processes of Canadian Arctic fjords: Sedimentary Geology, v. 36, p. 147–175.

GILBERT, R.; and AITKEN, A., 1981, The role of sea ice in biophysical processes on intertidal flats at Pangnirtung (Baffin Island), N.W.T.: Proceedings, workshop on ice action on shores: National Research Council of Canada, Assoc. Comm. for Research on Shoreline Erosion and Sedimentation (ACROSES), Publication No. 3, p. 89–103.

GILREATH, J. A., and MARICELLI, J. J., 1964, Detailed stratigraphic control through dip computations: American Association of Petroleum Geologists Bulletin, v. 48, p. 1902–1910.

GINGERICH, P. D., 1985, Species in the fossil record: concepts, trends, and transitions: Paleobiology, v. 11, p. 27–41.

GLAESER, J. W., 1979, Catskill delta slope (*sic*) sediments in central Appalachians basin: source and reservoir deposits: p. 343–358 *in* Doyle, L. J., and Pilkey, O. H., Jr., eds., Geology of continental slopes: Society of Economic Paleontologists and Mineralogists Special Paper 27, 374 p.

GLASS, B. P., and CROSBIE, J. R., 1982, Age of Eocene/Oligocene boundary based on extrapolation from North American microtektite layer: American Association of Petroleum Geologists Bulletin, v. 66, p. 471–476.

GLENN, S. M., 1983 ms., A continental shelf (*sic*) bottom boundary layer (*sic*) model. The effect of waves, currents and a movable bed: Woods Hole, Massachusetts, Woods Hole Oceanographic Institution, Ph.D. dissertation (WHOI-83-6), 336 p.

GLENNIE, K. W., 1972, Permian Rotliegendes of northwest Europe interpreted in light of modern desert sedimentation studies: American Association of Petroleum Geologists Bulletin, v. 56, p. 1048–1071.

GLENNIE, K. W., 1983, Lower Permian Rotliegend desert sedimentation in the North Sea area, p. 521–524 *in* Brookfield, M. E., and Ahlbrandt, T. S., eds., Eolian sediments and processes, Developments in sedimentology 38: Amsterdam, Elsevier Scientific Publishing Company, 660 p.

GLOPPEN, T. G., and STEEL, R. J., 1981, The deposits, internal structure, and geometry in six alluvial fan-fan delta bodies (Devonian-Norway): a study in the significance of bedding sequence in conglomerates, p. 49–69 *in* Ethridge, F. G., and Flores, R. M., eds., Recent and ancient nonmarine depositional environments: models for exploration: Society of Economic Paleontologists and Mineralogists Special Publication 31, 349 p.

GOFSEYEFF, S., 1953, A case history of Fire Island Inlet, New York, p. 272–305 *in* Johnson, J. W., ed., Conference on Coastal Engineering, 3rd, Cambridge, MA, Proceedings, 398 p.

GOLDBERG, M.; and FRIEDMAN, G.M., 1974, Paleoenvironments and paleogeographic evolution of the Jurassic System in southern Israel: Geological Survey of Israel Bulletin 61, 44 p.

GOLDHAMMER, R. K., DUNN, P. A., and HARDIE, L. A., 1987, High frequency (*sic*) glacio-eustatic sea-level oscillations with Milankovitch characteristics recorded in Middle Triassic platform carbonates in northern Italy: American Journal of Science, v. 287, p. 853–892.

GOLDICH, S. S., 1938, A study of rock weathering: Journal of Geology, v. 46, p. 17–58.

GOLDRING, ROLAND; and AIGNER, T., 1982, Scour and fill: the significance of event separation, p. 354–362 *in* Einsele, G.; and Seilacher, A., eds., Cyclic (*sic*) and event stratification: Berlin, Springer-Verlag, 536 p.

GOLDSTEIN, AUGUST, JR., 1942, Sedimentary petrologic provinces in the northern Gulf of Mexico: Journal of Sedimentary Petrology, v. 12, p. 77–84.

GOLE, M. J.; and KLEIN, C., 1981, Banded iron-formations through much of Precambrian time: Journal of Geology, v. 89, p. 169–183.

GOODWIN, P. W., and ANDERSON, E. J., 1985, Punctuated aggradational cycles: a general hypothesis of episodic stratigraphic accumulation: Journal of Geology, v. 93, p. 515–533.

GOODWIN, P. W., ANDERSON, E. J., and GOODMAN, W. M., 1986, Punctuated aggradational cycles: implications for stratigraphic analysis: Paleoceanography, v. 1, p. 417–429.

GOODWIN, R. H., and PRIOR, D. B., 1989, Geometry (*sic*) and depositional sequences of the Mississippi Canyon, Gulf of Mexico: Journal of Sedimentary Petrology, v. 59, p. 318–329.

GORSLINE, D. S., 1981, Fine sediment (*sic*) transport and deposition in active margin basins: p. 39–60 *in* Douglas, R. G., Colburn, I. P., and Gorsline, D. S., eds., Depositional systems of active continental margin basins: Short Course Notes: Pacific Section, Society of Economic Paleontologists and Mineralogists, 165 p.

GOTER, E. R., and FRIEDMAN, G. M., 1987, Deposition and diagenesis of the Windward Reef of Enewetak Atoll: Carbonates and Evaporites, v. 2, p. 157–170.

GOUDIE, S. A., 1989, Wind erosion in deserts: Geologists' Association Proceedings, v. 100, pt. 1, p. 83–92.

GOULD, H. R., 1951, Some quantitative aspects of Lake Mead turbidity currents, p. 34–52 *in* Hough, J. L., ed., Turbidity currents and the transportation (*sic*) of coarse sediments to deep water: Tulsa, OK, Society of Economic Paleontologists and Mineralogists Special Publication 2, 107 p.

GOULD, H. R., 1960a, Character (*sic*) of the accumulated sediment, part N, p. 149–186 *in* Comprehensive survey of sedimentation in Lake Mead, 1948–49: U.S. Geological Survey Professional Paper 295, 254 p.

GOULD, H. R., 1960b, Turbidity currents, part Q, p. 201–207 *in* Comprehensive survey of sedimentation in Lake Mead, 1948–49: U.S. Geological Survey Professional Paper 295, 254 p.

GOULD, S. J., 1980, Is a new (*sic*) and general theory of evolution emerging?: Paleobiology, v. 6, p. 119–130.

GOULD, S. J., 1982, Chapter 5, the meaning of punctuated equilibrium and its role in validating a hierarchical approach to macroevolution, p. 83–104, *in* Milkman, R., ed., Perspectives on evolution: Sunderland, MA, Sinauer Associates, 241 p.

GOULD, S. J., 1985, The paradox of the first tier: an agenda for biology: Paleobiology, v. 11, p. 2–12.

GOULD, S. J.; and ELDREDGE, NILES, 1977, Punctuated equilibria: the tempo and mode of evolution reconsidered: Paleobiology, v. 3, p. 115–151.

GOULD, S. J.; and ELDREDGE, NILES, 1986, Punctuated equilibrium at the third stage: Systematic Zoology, v. 35, p. 143–148.

GRABAU, A. W., 1913, Principles of stratigraphy: New York, A. G. Seiler and Company, 1185 p.

GRABAU, A. W., 1920, Geology of the nonmetallic mineral deposits other than silicates. vol. 1, Principles of salt deposition: New York, McGraw-Hill Book Company, 435 p.

GRACIANSKY, P. C.; DEROO, G.; HERBIN, J. P., MONTADERT, L.; MULLER, C.; SCHAAF, A.; and SIGAL, J., 1984, Ocean wide (*sic*) stagnation episode in the Late Cretaceous: Nature, v. 308, p. 346–349.

GRADSTEIN, FELIX; and LUDDEN, JOHN, 1989, The birth of the Indian Ocean: Nature, v. 337, p. 506–507.

GRADSTEIN, FELIX; LUDDEN, JOHN; and others, 1990, Initial Reports of the Ocean Drilling Program, v. 123: College Station, TX, Texas A&M University Press, 716 p.

GRADSTEIN, F. M., and others, eds., 1985, Quantitative stratigraphy: Hingham, MA, D. Reidel Publishing Company, 598 p.

GRAF, W. H., 1984, Hydraulics of sediment transport: Littleton, CO, Water Resources Publications, 513 p.

GRANDE, LANCE, 1980, Paleontology of the Green River Formation, with a review of the fish fauna: Geological Survey of Wyoming Bulletin, vol. 63, 333 p.

GRANT, J., 1988, Intertidal bedforms, sediment transport, and stabilization by benthic microalgae, p. 499–510 *in* De Boer, P. L.; van Gelder, A.; and Nio, S. D., eds., Tide-influenced sedimentary environments and facies: Dordrecht, The Netherlands, D. Reidel Publishing Company, 530 p.

GRASSHOFF, K., 1975, The hydrochemistry of landlocked basins and fjords, p. 455–597 *in* Riley, J. P.; and Skirrow, G., eds., Chemical oceanography, vol. 2, 2nd ed.: London, Academic Press.

GRAVENOR, C. P., 1982, Chattermarked garnets in Pleistocene sediments: Geological Society of America Bulletin, v. 93, p. 751–758.

GREENWOOD, BRIAN; and DAVIDSON-ARNOTT, R. G. D., 1979, Sedimentation and equilibrium in wave-formed bars: a review and case study: Canadian Journal of Earth Science, v. 16, p. 312–332.

GREENWOOD, BRIAN; and DAVIS, R. A., JR., eds., 1984, Hydrodynamics and sedimentation in wave-dominated coastal environments: Marine Geology, v. 60 (special issue).

GREENWOOD, BRIAN; and MITTLER, P. R., 1985, Vertical sequence and lateral transitions in the facies of a barred nearshore environment: Journal of Sedimentary Petrology, v. 55, p. 366–373.

GREENWOOD, BRIAN; and SHERMAN, D. J., 1986, Hummocky cross-stratification in the surf zone: flow parameters and bedding genesis: Sedimentology, v. 33, p. 33–45.

GRIFFITHS, CEDRIC; KOPASKA-MERKEL, D. C.; and SCHOTT, MICHEL, in press, Sedimentary sequences in a deep ocean basin: Argo Abyssal Plain, Northwest Australia, *in* Gradstein, Felix; Ludden, John; and others, Scientific Results of the Ocean Drilling Program, v. 123: College Station, TX, Texas A&M University Press.

GRIFFITHS, J. C., 1967, Scientific method in analysis of sediments: New York, McGraw-Hill Book Company, 508 p.

GRIGG, R. W.; and EPP, DAVID, 1989, Critical depth for the survival of coral islands: effects on the Hawaiian Archipelago: Science, v. 243, p. 638–641.

GRIM, R. E., 1968, Clay mineralogy, 2nd ed.: New York, McGraw-Hill Book Company, 596 p.

GROEN, P., 1969, Physical hydrology of coastal lagoons, p. 275–289 *in* Ayala-Castanares, Augustin; and Phleger, F. B., eds., Lagunas costeras, un simposio: Simposio Internacional sobre lagunas costeras, Mexico City, Mexico, 28–30 Noviembre 1967, memoria: Mexico Universidad Nacional Autonoma (WMAS-UNESCO), 686 p.

GROSS, G. A., 1980, A classification of iron formations based on depositional environments: Canadian Mineralogist, v. 18, p. 215–222.

GROSS, M. G., 1972, Oceanography. A view of the Earth: Englewood Cliffs, NJ, Prentice-Hall, 581 p. Chapters 11 (p. 295–325) and 12 (p. 327–355).

GROTZINGER, J. P., 1986a, Evolution of Early Proterozoic passive-margin carbonate platform, Rocknest Formation, Wopmay Orogen: North West Territories, Canada: Journal of Sedimentary Petrology, v. 56, p. 831–847.

GROTZINGER, J. P., 1986b, Upward-shallowing platform cycles: a response to 2.2 billion years of low-amplitude, high-frequency (Milankovitch band) sea level (*sic*) oscillations: Paleoceanography, v. 1, p. 403–416.

GROVER, G., JR.; and READ, J. F., 1983, Paleoaquifer and deep-burial related cements defined by regional cathodoluminescent patterns, Middle Ordovician carbonates, Virginia: American Association of Petroleum Geologists Bulletin, v. 67, p. 1275–1303.

GROVER, N. C., and HOWARD, C. S., 1938, The passage of turbid water through Lake Mead: American Society of Civil Engineers Transactions, v. 103, p. 720–790.

GRUNER, J. W., 1932, The crystal structure of kaolinite: Zeitschrift fur Krystallographie, v. 83, p. 75–88.

GUIDISH, T. M.; LERCHE, I.; KENDALL, G. G. ST. C.; and O'BRIEN, J. J., 1984, Relationship between eustatic sea level (sic) changes and basement subsidence: American Association of Petroleum Geologists Bulletin, v. 68, p. 164–177.

GUILBERT, J. M., and PARK, C. F., JR., 1986, The geology of ore deposits: New York, W. H. Freeman and Company, 985 p.

GUILCHER, ANDRE, 1965, Continental shelf and slope (continental margin), Chapter 13, p. 281–311 in Hill, M. N., gen. ed., The sea. Ideas and observations on progress in the study of the seas. Volume 3, The Earth beneath the sea. History: New York, Wiley-Interscience, 963 p.

GUMBEL, E. J., 1958, Statistics of extremes: New York, Columbia University Press, 375 p.

GUNATILAKA, H. A., and SHEARMAN, D. J., 1988, Gypsum-carbonate laminites in a Recent sabkha, Kuwait: Carbonates and Evaporites, v. 3, p. 67–73.

GUO, BAIYING, and FRIEDMAN, G. M., 1990, Petrophysical characteristics of Holocene beach rock: Carbonates and Evaporites, v. 5, p. 223–243.

GURNEY, G. G., and FRIEDMAN, G. M., 1987, Burial history of the Cherry Valley carbonate sequence, Cherry Valley, New York: Northeastern Geology, v. 9, p. 1–11.

GUZA, R. T., and BOWEN, A. J., 1975, The resonant instabilities of long waves obliquely incident on a beach: Journal of Geophysical Research, v. 80, p. 4529–4534.

GVIRTZMAN, G.; and FRIEDMAN, G. M., 1977, Sequence of progressive diagenesis in Quaternary reefs, Red Sea: American Association of Petroleum Geologists, Studies in Geology 4, p. 357–380.

GYGI, R. A., 1981, Oolitic iron formations: marine or not marine? Eclogae Geologicae Helvetiae, v. 74, p. 233–254.

HAECKEL, ERNST, 1925, Kristallseelen: Leipzig, Alfred Kroener Verlag, 168 p.

HALL, JAMES, 1859, Paleontology: vol. III, containing descriptions and figures of the organic remains of the Lower Helderberg Group and the Oriskany Sandstone: New York Geological Survey, Natural History of New York, part 6, 532 p.

HALLAM, ANTHONY, 1976, Stratigraphic distribution and ecology of European Jurassic bivalves: Lethaia, v. 9, p. 245–259.

HALLAM, ANTHONY, ed., 1977, Patterns of evolution: Amsterdam, Elsevier Scientific Publishing Company, 591 p.

HALLAM, ANTHONY, 1978, How rare is phyletic gradualism and what is its evolutionary significance? Evidence from Jurassic bivalves: Paleobiology, v. 4, p. 16–25.

HALLEMEIER, R. J., 1977, Calculating a yearly limit depth to the active beach profile: Fort Belvoir, Virginia, U.S. Army Corps of Engineers Coastal Engineering Research Center, Technical Paper 77-9.

HALLEMEIER, R. J., 1980, American Society of Civil Engineers, Waterways Division, Journal, v. 105, p. 299–318.

HALLEMEIER, R. J., 1981, Coastal Engineering, v. 4, p. 253–277.

HALLEMEIER, R. J., 1982, Continental Shelf (sic) Research, v. 1, p. 159–190.

HALLEMEIER, R. J., 1984, American Society of Civil Engineers, Waterways, Ports, Coast and Ocean Engineering Division, Journal, v. 110, p. 34–49.

HALLEY, R. B., and HARRIS, P. M., 1979, Freshwater cementation of a 1000-year-old oolite: Journal of Sedimentary Petrology, v. 49, p. 969–988.

HALPERIN, A., 1987 ms., A marine/nonmarine transition in the Upper Devonian West Falls Group of south-central New York [unpubl. M.S. thesis]: Binghamton, NY, State University of New York, 72 p.

HALSEY, L. A., CATO, N. R., and RUTTER, N. W., 1990, Sedimentology and development of parabolic dunes, Grande Prairie dune field, Alberta: Canadian Journal of Earth Sciences, v. 27, p. 1762–1772.

HALSEY, S. D., 1979, Nexus: new model of barrier island development, p. 185–210 in Leatherman, S. P., ed., 1979, Barrier islands from the Gulf of St. Lawrence to the Gulf of Mexico: New York, Academic Press, 325 p.

HAMBLIN, W. K., 1961, Paleogeographic evolution of the Lake Superior region from late Keeweenawan to Late Cambrian time: Geological Society of America Bulletin, v. 71, p. 1–18.

HAMBREY, M. J., and HARLAND, W. B., eds., 1981, Earth's pre-Pleistocene glacial record: London, Cambridge University Press, 1004 p.

HANCOCK, J. M., 1977, The historic development of concepts of biostratigraphic correlation, p. 3–22 in Kauffman, E. G., and Hazel, J. E., eds., Concepts and methods in biostratigraphy: Stroudsburg, PA, Dowden, Hutchinson, and Ross, 658 p.

HANDFORD, C. R., LOUCKS, R. G., and DAVIES, G. R., eds., 1982, Depositional and diagenetic spectra of evaporites—a core workshop: Society of Economic Paleontologists and Mineralogists Core Workshop 3: Calgary, Canada, 395 p.

HANDS, E. B., 1980, Prediction of shore retreat and nearshore profile (sic) adjustments to rising water levels in the Great Lakes: U.S. Army Corps of Engineers Coastal Engineering Research Center, Technical Paper 80-7.

HANDS, E. B., 1983, The Great Lakes as a test model for profile responses to sea level (sic) changes, p. 167–190 in Komar, P. D., ed., CRC handbook of coastal processes and erosion (sic): Boca Raton, FL, CRC Press, 305 p.

HAPP, S. C.; RITTENHOUSE, GORDON; and DOBSON, S. C., 1940, Some principles of accelerated stream (sic) and valley sedimentation: U.S. Department of Agriculture, Technical Bulletin 695, 134 p. (p. 22–31 reprinted 1972 as p. 336–346 in Schumm, S. A., ed., River morphology: Benchmark papers in geology: Stroudsburg, PA, Dowden, Hutchinson, and Ross, 429 p.)

HAQ, B. U., 1984, Paleoceanography: a synoptic overview of 200 million years of ocean history, p. 201–231 in Haq, B. U., and Milliman, J. D., eds, Marine geology and oceanography of Arabian Sea and coastal Pakistan: New York, Van Nostrand Reinhold Company, 382 p.

HAQ, B. U.; HARDENBOL, JAN; and VAIL, P. R., 1987, Chronology of fluctuating sea levels since the Triassic: Science, v. 235, p. 1156–1167.

HARDIE, L. A., SMOOT, J. P., and EUGSTER, H. P., 1978, Saline lakes and their deposits: a sedimentological approach, p. 7–42 in Matter, Albert; and Tucker, M. E., eds., Modern and Ancient Lake Sediments: Oxford, Blackwell Scientific Publications (International Association of Sedimentologists, Special Publication 2), 290 p.

HARMS, J. C., 1975a, Stratification produced by migrating bed forms, Chapter 3, p. 45–61 in Harms, J. C., Southard, J. B., Spearing, D. R., and Walker, R. G., eds., Depositional environments as interpreted from primary sedimentary structures and stratification sequences: Tulsa, OK, Society of Economic Paleontologists and Mineralogists, Dallas 1975 Short Course No. 2, Lecture Notes, 161 p.

HARMS, J. C., 1975b, Stratification and sequence in prograding shoreline deposits, Chapter 5, p. 81–102 in Harms, J. C., Southard, J. B., Spearing, D. R., and Walker, R. G., eds., Depositional environments as interpreted from primary sedimentary structures and stratification sequences: Tulsa, OK, Society of Economic Paleontologists and Mineralogists, Dallas 1975 Short Course No. 2, Lecture Notes, 161 p.

HARMS, J. C., 1979, Primary sedimentary structures: Annual Reviews of Earth and Planetary Sciences, v. 7, p. 227–248.

HARMS, J. C., and FAHNESTOCK, R. K., 1965, Stratification, bed forms, and flow phenomena (with an example from the Rio Grande), p. 84–115 in Middleton, G. V., ed., Primary sedimentary structures and their hydrodynamic interpretation—a symposium: Tulsa, OK, Society of Economic Paleontologists and Mineralogists Special Publication 12, 265 p.

HARMS, J. C., SOUTHARD, J. B., SPEARING, D. R., and WALKER, R. G., 1975, Depositional environments as interpreted from primary sedimentary structures and stratification sequences: Tulsa, OK, Society of Economic Paleontologists and Mineralogists, Dallas 1975 Short Course 2, Lecture Notes, 161 p.

HARMS, J. C., SOUTHARD, J. B., and WALKER, R. G., 1982, Structures and sequences in clastic rocks: Tulsa, OK, Society of Economic Paleontologists and Mineralogists, Calgary 1982 Short Course 2, Lecture Notes, not consecutively paginated.

HARRIS, R. J., 1975, A primer of multivariate statistics: New York, Academic Press, 332 p.

HARVEY, J. G., 1976, Atmosphere and ocean: our fluid environments: London, Artemis Press, 143 p.

HASSAN, F. A., and STUCKI, B. R., 1987, Nile floods and climatic change, p. 37–46 in Rampino, M. R., Sanders, J. E., Newman, W. S., and Konigsson, L. K., eds., 1987, Climate: history, periodicity, and predictability: New York, Van Nostrand Reinhold Company, 588 p.

HASZELDINE, R. S., 1984, Muddy deltas in freshwater lakes, and tectonism in the Upper Carboniferous coalfield of NE England: Sedimentology, v. 31, p. 811–822.

HATCH, F. H., WELLS, A. K., and WELLS, M. K., 1949, The petrology of the igneous rocks, 10th ed.: Textbook of petrology, v. 1: London, Thomas Murby and Company, 469 p.

HAURWITZ, B., 1941, Dynamic meteorology: New York, McGraw-Hill Book Company, 365 p.

HAY, W. W., 1977, Modulation of marine sedimentation by the continental shelves, p. 569–605 in Andersen, N. R.; and Malahoff, A., eds., The fate of fossil (sic) CO_2 in the oceans: New York, Plenum Press, 749 p.

HAY, W. W., 1988, Paleoceanography: a review for the GSA centennial: Geological Society of America Bulletin, v. 100, p. 1934–1956.

HAYDEN, BRUCE; DOLAN, ROBERT; and FELDER, W., 1979, Spatial (sic) and temporal analysis of shoreline variations: Coastal Engineering, v. 2, p. 351–361.

HAYES, M. O., 1975, Morphology of sand accumulations in estuaries, p. 3–22 in Cronin, L. E., ed., Estuarine Research, vol. 2: Geology and engineering: New York, Academic Press, 587 p.

HAYES, M. O., 1979, Barrier island (sic) morphology as a function of tidal (sic) and wave regime, p. 1–27 in Leatherman, S. P., ed., Barrier Islands from the Gulf of St. Lawrence to the Gulf of Mexico: New York, Academic Press, 325 p.

HAYES, M. O., 1980, General morphology and sediment patterns in tidal inlets: Sedimentary Geology, v. 26, p. 139–156.

HAYS, J. D.; IMBRIE, J.; and SHACKLETON, N. J., 1976, Variations in the Earth's orbit: pacemaker of the ice ages: Science, v. 194, p. 1121–1132.

HAYWARD, A. B., 1985, Coastal alluvial fans (fan deltas) of the Gulf of Aqaba (Gulf of Eilat), Red Sea: Sedimentary Geology, v. 43, p. 241–260.

HEATHERSTRAW, A. D., 1981, Comparisons of measured (sic) and predicted sediment transport (sic) rates in tidal currents, p. 75–103 in Nittrouer, C. A., ed., Sedimentary dynamics of continental shelves: Developments in sedimentology, v. 32. Amsterdam, Elsevier Scientific Publishing Company, 449 p.

HECKEL, P. H., 1972, Recognition of ancient shallow marine (sic) environments, p. 226–286 in Rigby, J. K., and Hamblin, W. K., eds., Recognition of ancient sedimentary environments: Tulsa, OK, Society of Economic Paleontologists and Mineralogists Special Publication 16, 340 p.

HEDBERG, H. D., ed., 1972, An international guide to stratigraphic classification, terminology, and usage. Introduction and summary: Lethaia, v. 5, p. 283–295 (Introduction), p. 297–323 (Summary).

HEDBERG, H. D., ed., 1976, International stratigraphic guide: New York, John Wiley & Sons, 200 p.

HEDGPETH, J. W., ed., 1957a, Treatise on marine ecology and paleoecology: Geological Society of America Memoir 67, v. 1, Ecology, 1296 p.

HEDGPETH, J. W., 1957b, Classification of marine environments, p. 17–27 *in* Hedgpeth, J. W., ed., Treatise on marine ecology and paleoecology: Geological Society of America Memoir 67, v. 1, Ecology, 1296 p.

HEEZEN, B. C., 1965, Turbidity currents, Chapter 27, p. 742–775 *in* Hill, M. N., gen. ed., The sea. Ideas and observations on progress in the study of the seas. Volume 3, the Earth beneath the sea. History: New York, Wiley-Interscience, 963 p.

HEEZEN, B. C., 1968, The Atlantic continental margin: University of Missouri Rolla, Journal, v. 1, p. 5–25.

HEEZEN, B. C., and DRAKE, C. L., 1964, Grand Banks slump: American Association of Petroleum Geologists Bulletin, v. 48, p. 221–225.

HEEZEN, B. C.; GRAY, C.; SEGRE, A. G.; and ZARUDSKI, E. F. K., 1971, Evidence of foundered continental crust beneath the central Tyrrhenian Sea: Nature, v. 229, p. 327–329.

HEEZEN, B. C., HOLLISTER, C. D., and RUDDIMAN, W. F., 1966, Shaping of the continental rise by deep geostrophic contour currents: Science, v. 152, p. 502–508.

HEEZEN, B. C., and JOHNSON, G. L., III, 1969, Mediterranean undercurrent and microphysiography west of Gibraltar: Inst. Oceanogr. Monaco Bulletin, v. 69, p. 185–197.

HEEZEN, B. C., and LAUGHTON, A. S., 1965, Abyssal plains, Chapter 14, p. 312–364 *in* Hill, M. N., gen. ed., The sea. Ideas and observations on progress in the study of the seas. Volume 3, The Earth beneath the sea. History: New York, Wiley-Interscience, 963 p.

HEEZEN, B. C., and McGREGOR, I. D., 1973, eds., Initial Reports of the Deep Sea Drilling Project, v. 20: Washington, DC, U.S. Government Printing Office, 958 p.

HEEZEN, B. C., and MENARD, H. W., 1965, Topography of the deep-sea floor, Chapter 12, p. 233–290, *in* Hill, M. N., gen. ed., The sea. Ideas and observations on progress in the study of the seas. Volume 3, The Earth beneath the sea. History: New York, Wiley-Interscience, 963 p.

HEEZEN, B. C.; THARP, MARIE; and EWING, MAURICE, 1959, The floors of the oceans: Geological Society of America Special Paper 65, 122 p.

HEIM, A., 1882, Der Bergsturz von Elm: Deutsch. Geol. Gesellsch., Zeitschr., Band 34, Heft 1, p. 74–115.

HEIM, A., 1932, Bergsturz und Menschenleben: Zürich, Fretz and Wasmuth Verlag, 218 p.

HEIN, F. J., 1982, Depositional mechanisms of deep-sea coarse clastic sediments, Cap Enrage Formation, Quebec: Canadian Journal of Earth Science, v. 19, p. 267–287.

HEIN, F. J., ROBB, G. A., WOLBERG, A. C., and LONGSTAFFE, F. J., 1991, Facies descriptions and associations in ancient reworked (?transgressive) shelf sandstones: Cambrian and Cretaceous examples: Sedimentology, v. 38, p. 405–431.

HELANDER, D. P., 1983, Fundamentals of formation evaluation: Tulsa, OK, Oil and Gas Consultants International Publications, 332 p.

HELING, DIETRICH, 1988, Ton und Siltsteine, p. 185–232 *in* Füchtbauer, Hans, ed., Sedimente und Sedimentgesteine: Stuttgart, E. Schweitzerbart, 1141 p.

HELMOLD, K. P., and KAMP, P. C. van de, 1984, Diagenetic mineralogy and controls on albitization and laumontite formation in Paleogene arkoses, Santa Inez Mountains, California, p. 239–276 *in* McDonald, D. A., and Surdam, R. C., eds., Elastic diagenesis, American Association of Petroleum Geologists Memoir 37, 434 p.

HERBIN, J. P.; MONTADERT, L.; MULLER, C.; GOMEZ, R.; THUROW, J.; and WIEDMANN, J., 1986, Organic-rich sedimentation at the Cenomanian-Turonian boundary in oceanic (*sic*) and coastal basins in the North Atlantic and Tethys, p. 389–422 *in* Summerhayes, C. P., and Shackleton, N. J., eds., North Atlantic paleoceanography: Geological Society of London Special Publication 21, 473 p.

HERMAN, Y.; and ROSENBERG, P. E., 1969, Mineralogy (*sic*) and micropaleontology (*sic*) of a goethite-bearing Red Sea core, p. 448–459 *in* Degens, E. T., and Ross, D. A., eds., Hot brines and recent heavy metal (*sic*) deposits in the Red Sea. A geochemical and geophysical account: New York, Springer-Verlag, 600 p.

HERSEY, J. B., 1965, Continuous reflection profiling, Chapter 4, p. 47–72 *in* Hill, M. N., gen. ed., The sea. Ideas and observations on progress in the study of the seas. Volume 3, The Earth beneath the sea. History: New York, Wiley-Interscience, 963 p.

HERSEY, J. B.; EDGERTON, H. E.; RAYMOND, S. O.; and HAYWARD, G., 1961, Pingers and thumpers advance deep-sea exploration: Instrument Society of America, Journal, v. 8, p. 72–77.

HESS, H. H., 1946, Drowned ancient islands of the Pacific basin: American Journal of Science, v. 244, p. 772–791.

HESSE, REINHARD, 1987, Selective and reversible carbonate-silica replacements in Lower Cretaceous carbonate bearing (*sic*) turbidites of the eastern Alps: Sedimentology, v. 34, p. 1055–1077.

HESSE, REINHARD, 1988, Diagenesis #13. Origin of chert: diagenesis of biogenic siliceous sediments: Geoscience Canada, v. 15, p. 171–192.

HEWARD, A. P., 1981, A review of wave-dominated clastic shoreline deposits: Earth-Science Reviews, v. 17, p. 223–276.

HEYDEMANN, ANNEROSE, 1966, Ueber die chemische Verwitterung von Tonmineralen (experimentalle Untersuchungen): Geochimica et Cosmochimica Acta, v. 30, p. 995–1035.

HIECKE, W., 1984, A thick Holocene homogenite from the Ionian Abyssal Plain (eastern Mediterranean): Marine Geology, v. 55, p. 63–78.

HILDRETH, WES; and MAHOOD, GAIL, 1985, Correlation of ash-flow tuffs: Geological Society of America Bulletin, v. 96, p. 968–974.

HILL, B. T., and JONES, S. J., 1990, The Newfoundland ice extent and the solar cycle from 1860 to 1988 (89JC03120): Journal of Geophysical Research, v. 95, no. C4, p. 5384–5394.

HILL, P. R., and NADEAU, O. C., 1989, Storm-dominated sedimentation on the inner shelf of the Canadian Beaufort Sea: Journal of Sedimentary Petrology, v. 59, p. 455–468.

HINE, A. C., 1979, Mechanisms of berm development and resulting beach growth along a barrier spit complex: Sedimentology, v. 26, p. 333–351.

HINE, A. C., BELKNAP, D. F., HUTTON, J. G., OSKING, E. B., and EVANS, M. W., 1988, Recent geological history and modern sedimentary processes along an incipient, low-energy, epicontinent-sea coastline: northwest Florida: Journal of Sedimentary Petrology, v. 58, p. 567–579.

HINE, A. C., EVANS, M. W., DAVIS, R. A., JR., and BELKNAP, D. F., 1987, Depositional response to seagrass mortality along a low-energy barrier-island coast: west-central Florida: Journal of Sedimentary Petrology, v. 57, p. 431–439.

HINE, A. C., and SNYDER, S. W., 1985, Coastal lithosome preservation: evidence from the shoreface and inner continental shelf off Bogue Banks, North Carolina: Marine Geology, v. 63, p. 307–330.

HINE, A. C., WILBER, R. J., and NEUMANN, A. C., 1981, Carbonate sand bodies along contrasting shallow bank margins facing open seaways: northern Bahamas: American Association of Petroleum Geologists Bulletin, v. 65, p. 261–290.

HINO, M., 1975, Theory on formation of rip current and cuspidal coast, p. 901–919 *in* Conference on Coastal Engineering, 14th, Copenhagen, Proceedings.

HIRSCH, P., 1978, Microbial mats in a hypersaline Solar Lake: types, composition and distribution, p. 189–201 *in* Krumbein, W. E., ed., Environmental biogeochemistry and geomicrobiology: Ann Arbor, MI, Ann Arbor Sci., 396 p.

HISCOTT, R. N., 1980, Depositional framework of sandy mid-fan complexes of Tourelle Formation, Ordovician, Quebec: American Association of Petroleum Geologists Bulletin, v. 64, p. 1052–1077.

HISCOTT, R. N., 1981, Deep sea fan (*sic*) deposits in the Macigno Formation (Middle-Upper Oligocene) of the Gordana Valley, northern Appennines, Italy: Discussion: Journal of Sedimentary Petrology, v. 51, p. 1015–1021.

HISCOTT, R. N., and JAMES, N. P., 1985, Carbonate debris flows, Cow Head Group, western Newfoundland: Journal of Sedimentary Petrology, v. 55, p. 735–745.

HISCOTT, R. N., and MIDDLETON, G. V., 1980, Fabric of coarse deep-water sandstones, Tourelle Formation, Quebec, Canada: Journal of Sedimentary Petrology, v. 50, p. 703–722.

HIXON, M. A., and BROSTOFF, W. N., 1982, Fish grazing and community structure of Hawaiian reef algae, p. 507–514 *in* Gomez, E. D., and others, eds., International Coral Reef (*sic*) Symposium, 4th, Proceedings, v. 2, 785 p.

HJULSTRÖM, F., 1935, Studies of the morphological activities of rivers as illustrated by the River Fyris: University of Uppsala, Geol. Inst., Bull., v. 25, p. 221–527.

HJULSTRÖM, F., 1939, Transportation of detritus by moving water, p. 5–31, part 1, Transportation, *in* Trask, P. D., ed., Recent marine sediments; a symposium: Tulsa, OK, American Association of Petroleum Geologists, 736 p. (London, Thomas Murby; reprinted, 1968, New York, Dover.)

HOBBS, W. H., 1899, The diamond field of the Great Lakes: Journal of Geology, v. 7, p. 375–388.

HOBDAY, D. K., and BANKS, N. L., 1971, A coarse-grained pocket beach complex, Tanafjord (Norway): Sedimentology, v. 16, p. 129–134.

HODELL, D. A., ELMSTROM, K. M., and KENNETT, J. P., 1986, Latest Miocene benthic $\delta^{18}O$ changes, global ice volume, sea level and the "Messinian salinity crisis": Nature, v. 320, p. 411–414.

HODGINS, D. O., SAYNO, L. J., KINSELLA, E. D., and MORGAN, P. W., 1986, Nearshore sediment dynamics—Beaufort Sea. The 1986 monitoring program: Environmental Studies Revolving Funds Report 054: Ottawa, Canada, 195 p.

HOFMANN, H. J., and SNYDER, G. L., 1985, Archean stromatolites from the Hartfille Uplift, eastern Wyoming: Geological Society of America Bulletin, v. 96, p. 842–849.

HOGG, N., 1980, Effects of bottom topography on ocean currents, p. 167–205 *in* GARP Publication 23, Orographic effects in planetary flows: Geneva, World Meteorological Organization.

HOLDAWAY, H. K., and CLAYTON, C. J., 1982, Preservation of shell microstructure in silicified brachiopods from the Upper Cretaceous Wilmington Sands of Devon: Geological Magazine, v. 119, p. 371–382.

HOLGREN, D. A., MOODY, J. D., and EMMERICH, H. H., 1975, The structural settings for giant oil and gas fields: World Petroleum Congress, 9th, Tokyo, Proceedings, v. 2, p. 45–54.

HOLLAND, H. D., 1967, Gangue minerals in hydrothermal deposits, Chapter 9, p. 382–436 *in* Barnes, H. L., ed., Geochemistry of hydrothermal ore deposits: New York, Holt, Rinehart, and Winston, 670 p. (2nd ed., 1979, New York, Wiley-Interscience, 798 p.)

HOLLAND, H. D., 1968, The abundance of CO_2 in the Earth's atmosphere through geologic time, p. 949–954 *in* Ahrens, L. H., Cameron, A. G. W., and others, eds., Origin and distribution of the elements: New York, Pergamon Press, 1178 p.

HOLLAND, H. D., 1978, The chemistry of the atmosphere and oceans: New York, Wiley-Interscience, 351 p.

HOLLE, C. G., 1952, Sedimentation at the mouth of the Mississippi River; p. 111–129 *in* Johnson, J. W., ed., Conference on Coastal Engineering, 2nd, Houston, TX, Nov. 1951, Proceedings: Berkeley, CA, University of California, Council on Wave Research, 393 p.

HOLMES, ARTHUR, 1928, The nomenclature of petrology: New York, Hafner Publishing Company, 284 p.

HOLMES, C. W., 1982, Geochemical indices of fine sediment (*sic*) transport, northwest Gulf of Mexico: Journal of Sedimentary Petrology, v. 52, p. 307–321.

HOLMES, R. W., 1957, Solar radiation, submarine daylight, and photosynthesis, p. 109–128 *in* Hedgpeth, J.W., ed., Treatise on marine ecology and paleoecology: Geological Society of America Memoir 67, v. 1, Ecology, 1296 p.

HOSKINS, C. W., 1964, Molluscan biofacies in calcareous sediments, Gulf of Batabano, Cuba: American Association of Petroleum Geologists Bulletin, v. 48, p. 1680–1704.

HOTCHKISS, F. S.; and WUNSCH, C., 1982, Internal waves in Hudson Canyon with possible geological implications: Deep-Sea Research, v. 29, p. 415–442.

HOUBOLT, J. J. H. C., and JONKER, J. B. M., 1968, Recent sediments in the eastern part of the Lake of Geneva (Lac Léman): Geologie en Mijnbouw, v. 47, p. 131–148.

HOUSE, M. R., 1985, A new approach to an absolute time scale from measurements of orbital cycles and sedimentary microrhythms: Nature, v. 316, p. 721–725.

HOUSEKNECHT, D. W., 1988, Intergranular pressure solution (*sic*) in four quartzose sandstones: Journal of Sedimentary Petrology, v. 58, p. 228–246.

HOWARD, J. D., and FREY R. W., 1985, Physical and biogenic aspects of backbarrier sediment systems, Georgia coast, USA: Marine Geology, v. 63, p. 77–127.

HOWARD, J. D., and REINECK, H.-E., 1981, Depositional facies of high-energy beach-to-offshore sequence: comparison with low-energy sequence: American Association of Petroleum Geologists Bulletin, v. 65, p. 807–830.

HOWARD, P. F., 1979, Phosphate: Economic Geology, v. 74, p. 192–194.

HOWARTH, M. J., 1982, Tidal currents of the continental shelf, p. 10–26, *in* Stride, A. H., ed., Offshore tidal sands: New York, Chapman and Hall, 239 p.

HOWELL, D. G., and NORMARK, W. R., 1982, Sedimentology of submarine fans, p. 365–404 *in* Scholle, P. A.; and Spearing, D., eds., Sandstone depositional environments: American Association of Petroleum Geologists Memoir 31, 410 p.

HSÜ, K. J., 1986, Sedimentary petrology and biologic evolution: Journal of Sedimentary Petrology, v. 56, p. 729–732.

HSÜ, K. J., 1989, Physical principles of sedimentology: Berlin, Springer-Verlag, 233 p.

HSÜ, K. J.; and SCHNEIDER, J., 1973, Progress report on dolomitization-hydrology of Abu Dhabi sabkhas, Arabian Gulf, p. 409–422 *in* Purser, B. H., ed., The Persian Gulf: Holocene carbonate sedimentation and diagenesis in a shallow epicontinental sea: New York, Springer-Verlag, 471 p.

HSÜ, K. J., and WEISSERT, H. J., eds., 1985, South Atlantic paleoceanography: Cambridge, Cambridge University Press, 350 p.

HUBBARD, D. K., and BARWIS, J. H., 1979, Discussion of tidal inlet sand deposits: examples from the South Carolina coast, p. II-128–II-142 *in* Hayes, M. O., and Kana, T. W., eds., Terrigenous clastic depositional environments, some examples: a field course sponsored by the American Association of Petroleum Geologists: Technical Report No. 11-CRD, Department of Geology, University of South Carolina, 131 and 184 p. (numbered in two parts).

HUBBARD, D. K.; MILLER, A. I.; and SCATURO, DAVID, 1990, Production and cycling of calcium carbonate in a shelf-edge reef system (St. Croix, U.S. Virgin Islands): applications to the nature of reef systems in the fossil record: Journal of Sedimentary Petrology, v. 60, p. 335–360.

HUBERT, J. F., 1960, Petrology of the Fountain and Lyons formations, Front Range, Colorado: Colorado School of Mines, Quarterly, v. 55, 242 p.

HULSEMANN, JOBST; and EMERY, K. O., 1961, Stratification in recent sediments of Santa Barbara Basin controlled by organisms and water character: Journal of Geology, v. 69, p. 279–290.

HUNT, J. M., 1973, Unsolved problems concerning origin and migration of petroleum: American Association of Petroleum Geologists Bulletin, v. 57, p. 785.

HUNT, J. M., 1979, Petroleum geochemistry and geology (*sic*): San Francisco, W. H. Freeman and Company, 617 p.

HUNTER, R. E., 1985, Subaqueous sand-flow cross strata: Journal of Sedimentary Petrology, v. 55, p. 886–894.

HUNTER, R. E., and CLIFTON, H. E., 1982, Cyclic deposits and hummocky cross-stratification of probable storm origin in Upper Cretaceous rocks of the Cape Sebastian Areas, south-western Oregon: Journal of Sedimentary Petrology, v. 52, p. 127–143.

HUNTER, R. E., CLIFTON, H. E., and PHILLIPS, R. L., 1979, Depositional processes, sedimentary structures and predicted vertical sequences in barred nearshore systems, southern Oregon coast: Journal of Sedimentary Petrology, v. 49, p. 411–726.

HUNTER, R. E., and KOCUREK, G., 1986, An experimental study of subaqueous slipface deposition: Journal of Sedimentary Petrology, v. 56, p. 387–394.

HUNTLEY, D. A., 1976, Long period (*sic*) waves on a natural beach: Journal of Geophysical Research, v. 81, p. 6441–6449.

HUNTLEY, D. A., and BOWEN, A. J., 1975, Comparison of the hydrodynamics of steep (*sic*) and shallow beaches, p. 69–109 *in* Hails, J.; and Carr, A., eds., Nearshore sediment dynamics and sedimentation (*sic*): New York, John Wiley & Sons.

HURST, J. M.; and SURLYK, F., 1983, Depositional environments along a carbonate ramp to slope (*sic*) transition in the Silurian of Washington Land, North Greenland: Canadian Journal of Earth Science, v. 20, p. 473–449.

HUTCHEON, IAN, 1983, Aspects of the diagenesis of coarse-grained siliciclastic rocks: Geoscience Canada, v. 10, p. 4–14.

HUTCHINSON, D. R., and GROW, J. A., 1985, New York Bight fault: Geological Society of America Bulletin, v. 96, p. 975–989.

HUTCHINSON, D. R., KLITGORD, K. D., and DETRICK, R. S., 1986, Rift basins of the Long Island platform: Geological Society of America Bulletin, v. 97, p. 688–702.

HUTHNANCE, J. M., 1982, On one mechanism forming linear sand banks: Estuar. Coast., Shelf. Sci., v. 14, p. 79–99.

HYNE, N. J., COOPER, W. A., and DICKEY, P. A., 1979, Stratigraphy of intermontane, lacustrine delta, Catatumbo River, Lake Maracaibo, Venezuela: American Association of Petroleum Geologists Bulletin, v. 63, p. 2042–2057.

ILER, R. K., 1979, Chemistry of silica: New York, Wiley-Interscience, 866 p.

ILLING, L. V., l954, Bahamian calcareous sands: American Association of Petroleum Geologists Bulletin, v. 38, p. 1–95.

ILLINOIS GEOLOGIC SURVEY, 1931, Bulletin 60 (a special volume devoted to cyclical Pennsylvanian strata in many states).

IMBRIE, JOHN, 1985, A theoretical framework for the Pleistocene ice ages: Geological Society of London, Journal, v. 142, p. 417–432.

IMBRIE, J.; and BUCHANAN, H., 1965, Sedimentary structures in modern carbonate sands of the Bahamas, p. 149–172 *in* Middleton, G. V., ed., Primary sedimentary structures and their hydrodynamic interpretation: Tulsa, OK, Society of Economic Paleontologists and Mineralogists Special Publication 12, 265 p.

IMBRIE, JOHN; and IMBRIE, J. Z., 1980, Modeling the climatic response to orbital variations: Science, v. 207, p. 943–953.

IMPERATO, D. P., SEXTON, W. J., and HAYES, M. O., 1988, Stratification and sediment characteristics of a mesotidal ebb-tidal delta, North Edisto Inlet, South Carolina: Journal of Sedimentary Petrology, v. 58, p. 950–958.

INGERSOLL, R. V., 1988, Tectonics of sedimentary basins: Geological Society of America Bulletin, v. 100, p. 1704–1719.

INGERSOLL, R. V., and ERNST, W. G., eds., 1987, Cenozoic basin development of coastal California, Rubey Volume VI: Englewood Cliffs, NJ, Prentice-Hall, 496 p.

INGLE, J. C., JR., 1966, The movement of beach sand—an analysis using fluorescent grains: Developments in sedimentology, v. 5: Amsterdam: Elsevier Scientific Publishing Company, 221 p.

INMAN, D. L., 1953, Areal (*sic*) and seasonal variations in beach (*sic*) and nearshore sediments at La Jolla, California: U.S. Army Corps of Engineers, Beach Erosion Board, Technical Memo 34, 82 p.

INMAN, D. L., 1963, Ocean waves and associated currents, Chapter 3, p. 49–81 *in* Shepard, F. P., Submarine geology, 2nd ed.: New York, Harper & Row, 557 p.

INMAN, D. L., and BAGNOLD, R. A., 1963, Beach (*sic*) and nearshore processes. Part II. Littoral processes, p. 529–553 *in* Hill, M. N., gen. ed., The sea. Ideas and observations on progress in the study of the seas. Volume 3, The Earth beneath the sea. History: New York, Wiley-Interscience, 963 p.

INMAN, D. L., and GUZA, R. T., 1982, The origin of swash cusps on beaches: Marine Geology, v. 49, p. 133–148.

INTERNATIONAL GEOL. CORRELATION PROGRAMME, 1982, Phosphorites: Project 156, Newsletters No. 10 (54 p.), No. 11 (45 p.), and No. 12 (43 p.).

INTERNATIONAL SUBCOMMISSION ON STRATIGRAPHIC CLASSIFICATION, 1987, Unconformity-bounded stratigraphic units: Geological Society of America Bulletin, v. 98, p. 232–237.

IRWIN, H., 1980, Early diagenetic carbonate precipitation and pore fluid (*sic*) migration in the Kimmeridge Clay of Dorset, England: Sedimentology, v. 27, p. 577–591.

IRWIN, M. L., 1965, General theory of epeiric clear water (*sic*) sedimentation: American Association of Petroleum Geologists Bulletin, v. 49, p. 445–459.

ISAACS, C. M., 1984, Disseminated dolomite in the Monterey Formation, Santa Maria and Santa Barbara areas, California, p. 155–170 *in* Garrison, R. E.; Kastner, M.; and Zenger, D. H., eds., Dolomites of the Monterey Formation and other organic-rich units: Society of Economic Paleontologists and Mineralogists, Pacific Section, Symposium volume.

ISRAEL, A. M., ETHRIDGE, F. G., and ESTES, E. L., 1987, A sedimentologic description of a peritidal, flood-tidal delta, San Luis Pass, Texas: Journal of Sedimentary Petrology, v. 57, p. 288–300.

JABLONSKI, DAVID, 1980a, Apparent (*sic*) versus real biotic effects of transgressions and regressions: Paleobiology, v. 6, p. 397–407.

JABLONSKI, DAVID, 1980b, Keeping time with mass extinctions: Paleobiology, v. 10, p. 139–145.

JABLONSKI, DAVID, 1986, Background (*sic*) and mass extinctions: the alternative to macroevolutionary regimes: Science, v. 231, p. 129–133.

JABLONSKI, DAVID; and FLESSA, K. W., 1986, The taxonomic structure of shallow-water marine faunas: implication for Phanerozoic extinctions: Malacologia, v. 27, p. 43–66.

JACHENS, R. C., and HOLZER, T. L., 1982, Differential compaction (*sic*) mechanism for earth fissures near Casa Grande, Arizona: Geological Society of America Bulletin, v. 93, p. 998–1012.

JACKSON, M. P. A., and TALBOT, C. J., 1986, External shapes, strain rates, and dynamics of salt structure: Geological Society of America Bulletin, v. 97, p. 305–323.

JAMES, H. L., 1966, Chemistry of the iron-rich sedimentary rocks, Chapter W in Data of geochemistry, 6th ed.: U.S. Geological Survey Professional Paper 440, 61 p.

JAMES, N. P., 1981, Megablocks of calcified algae in the Cow Head Breccia, western Newfoundland: vestiges of a Cambro-Ordovician platform margin: Geological Society of America Bulletin, Part 1, v. 92, p. 799–811.

JAMES, N. P., and CHOQUETTE, P. W., eds., 1988, Paleokarst: New York, Springer-Verlag, 416 p.

JAMES, N. P., and GINSBURG, R. N., 1979, The deep seaward margin of Belize barrier and atoll reefs: International Association of Sedimentologists Special Publication 3, 201 p.

JENSEN, A., 1977, Character (*sic*) and provenance of the opaque minerals in the Nexo sandstone, Bornholm: Geological Society of Denmark Bulletin, v. 26, p. 69–76.

JETT, G. A., and HELLER, P. L., 1988, Tectonic significance of polymodal compositions in melange sandstones, western melange belt, North Cascade Range, Washington: Journal of Sedimentary Petrology, v. 58, p. 52–61.

JIPA, D. C., 1980, Orogenesis and flysch sedimentology–critical remarks on the Alpine model: Sedimentary Geology, v. 27, p. 229–239.

JOHNSON, D. A., 1982a, Abyssal teleconnections: interactive dynamics of the deep ocean (*sic*) circulation: Palaeogeography, Palaeoclimatology, and Palaeoecology, v. 38, p. 93–128.

JOHNSON, D. A., 1982b, Abyssal teleconnections II. Initiation of Antarctic bottom water flow in the southwestern Atlantic, p. 243–281 *in* Hsü, K. J., and Weissert, H. J., eds., South Atlantic paleoceanography: Cambridge, Cambridge University Press, 350 p.

JOHNSON, D. A., and DAMUTH, J. E., 1979, Deep thermohaline flow and current-controlled sedimentation in the Amirante Passage: western Indian Ocean: Marine Geology, v. 33, p. 1–44.

JOHNSON, D. P., 1982, Sedimentary facies of an arid zone (*sic*) delta: Gascoyne Delta, Western Australia: Journal of Sedimentary Petrology, v. 52, p. 547–563.

JOHNSON, D. W., 1931, Stream sculpture on the Atlantic slope, a study in the evolution of Appalachian Rivers: New York, Columbia University Press, 142 p.

JOHNSON, D. W., 1939, The origin of submarine canyons. A critical review of hypotheses: New York, Columbia University Press, 126 p. (Reprinted 1967, New York, Hafner Publishing Company.)

JOHNSON, H. D., and BALDWIN, C. T., 1986, Shallow siliciclastic seas, Chapter 9, p. 229–282 *in* Reading, H. G., ed., Sedimentary Environments and Facies (*sic*), 2nd ed: Oxford, Blackwell Scientific Publications, 615 p.

JOHNSON, J. G., 1990, Method of multiple working hypotheses: a chimera: Geology, v. 18, p. 44–45.

JOHNSON, K. G., and FRIEDMAN, G. M., 1969, The Tully clastic correlatives (Upper Devonian) of New York State: a model for recognition of alluvial, dune(?), tidal, nearshore (bar and lagoon), and offshore sedimentary environments in a tectonic delta complex: Journal of Sedimentary Petrology, v. 39, p. 451–485.

JOHNSON, MARKES E.; RONG, JIA-YU; and YANG, XUE-CHANG, 1985, Intercontinental correlation by sea-level events in the Early (*sic*) Silurian of North America and China (Yangtze Platform): Geological Society of America Bulletin, v. 96, p. 1384–1397.

JOHNSON, T. C.; HALFMAN, J. D.; BUSCH, W. H.; and FLOOD, R. D., 1984, Effects of bottom currents and fish on sedimentation in a deep-water, lacus-

trine environment: Geological Society of America Bulletin, v. 95, p. 1425–1436.

JOHNSSON, M., 1986, Distribution of maximum burial temperatures across the northern Appalachian Basin and implications for Carboniferous sedimentation patterns: Geology, v. 14, p. 383–387.

JOHNSSON, M. J., 1990, Overlooked sedimentary particles from tropical weathering environments: Geology, v. 18, p. 107–110.

JONES, B. F., EUGSTER, H. P., and RETTIG, S. L., 1977, Hydrochemistry of the Lake Magadi Basin, Kenya: Geochimica et Cosmochimica Acta, v. 41, p. 53–72.

JONES, BRIAN; and MACDONALD, R. W., 1989, Micro-organisms and crystal fabrics in cave pisoliths from Grand Cayman, British West Indies: Journal of Sedimentary Petrology, v. 59, p. 387–396.

JONES, BRIAN; and NG, K.-C., 1988, The structure and diagenesis of rhizoliths from Cayman Brac, British West Indies: Journal of Sedimentary Petrology, v. 58, p. 457–467.

JONES, BRIAN; and PEMBERTON, S. G., 1987, Experimental formation of spiky calcite through organically mediated dissolution: Journal of Sedimentary Petrology, v. 57, p. 687–694.

JONES, D. L., and KNAUTH, L. P., 1979, Oxygen isotopic (*sic*) and petrographic evidence relevant to the origin of the Arkansas Novaculite: Journal of Sedimentary Petrology, v. 49, p. 581–597.

JONES, J. B., and SEGNIT, E. R., 1971, The nature of opal 1. Nomenclature and constituent phases: Geological Society of Australia, Journal, v. 18, p. 57–68.

JONES, K. P. N., MCCAVE, I. N., and PATEL, P. D., 1988, A computer-interfaced sedigraph for modal size analysis of fine-grained sediment: Sedimentology, v. 35, p. 163–172.

JOPLING, A. V., 1965a, Laboratory study of the distribution of grain sizes in cross-bedded deposits, p. 53–65 *in* Middleton, G. V., ed., Primary sedimentary structures and their hydrodynamic interpretation: Tulsa, OK, Society of Economic Paleontologists and Mineralogists Special Publication 12, 265 p.

JOPLING, A. V., 1965b, Hydraulic factors and the shape of laminae: Journal of Sedimentary Petrology, v. 35, p. 777–791.

JUNGE, C. E., 1960, Sulfur in the atmosphere: Journal of Geophysical Research, v. 65, p. 227–237.

KAPLAN, I. B., and RITTENBERG, S. C., 1965, Basin sedimentation and diagenesis, Chapter 23, p. 583–619 *in* Hill, M. N., gen. ed., The sea. Ideas and observations on progress in the study of the seas. Volume 3. The Earth beneath the sea. History: New York, Wiley-Interscience, 963 p.

KARL, H. A., CACCHIONE, D. A., and CARLSON, P. R., 1986, Internal-wave currents as a mechanism to account for large sand waves in Navarinsky Canyon head: Bering Sea: Journal of Sedimentary Petrology, v. 56, p. 706–714.

KARSON, J. A., ELTHON, D. L., and DE LONG, S. E., 1983, Ultramafic intrusions (*sic*) in the Lewis Hills Massif, Bay of Islands complex, Newfoundland: implications for igneous processes at oceanic fracture (*sic*) zones: Geological Society of America Bulletin, v. 94, p. 15–29.

KASINO, R. E., and DAVIES, D. K., 1979, Environments and diagenesis, Morrow sands, Cimmaron County (OK), and significance to regional exploration, production and well completion (*sic*) practices, p. 169–194 *in* Hyne, N. E., ed., Pennsylvanian sandstones of the mid-continent: Tulsa Geological Society Special Publication 1, 360 p.

KASPER, D. C., LARUE, D. K., and MEEKS, Y. J., 1987, Fine-grained Paleogene terrigenous turbidites in Barbados: Journal of Sedimentary Petrology, v. 57, p. 440–448.

KASTNER, M., 1981, Authigenic silicates in deep-sea sediments: formation and diagenesis, p. 915–980 *in* Emiliani, Cesare, ed., The Sea, v. 7: New York, Wiley-Interscience, 1728 p.

KASTNER, M.; and KEENE, J. B., 1975, Diagenesis of pelagic siliceous oozes: International Congress of Sedimentology, 9th, Nice, theme 7, p. 89–98.

KAUFFMAN, E. G., 1977, Evolutionary rates and biostratigraphy, p. 109–141 *in* Kauffman, E. G., and Hazel, J. E., eds., 1977, Concepts and methods in biostratigraphy: Stroudsburg, PA, Dowden, Hutchinson, and Ross, 658 p.

KAUFFMAN, E. G., 1988, Concepts and methods of high-resolution event stratigraphy: Annual Reviews of Earth and Planetary Sciences, v. 16, p. 605–654.

KAUFFMAN, E. G., and HAZEL, J. E., eds., 1977, Concepts and methods in biostratigraphy: Stroudsburg, PA, Dowden, Hutchinson, and Ross, 658 p.

KAY, G. M., 1951, North American Geosynclines: Geological Society of America Memoir 48, 143 p.

KAY, G. M., 1967, On geosynclinal nomenclature: Geological Magazine, v. 104, p. 311–316.

KAYAN, I.; and KRAFT, J. C., 1979, Holocene geomorphic evolution of a barrier salt marsh system, southwest Delaware Bay: Southeastern Geology, v. 20, p. 79–99.

KAYE, C. A., and STUCKEY, G. W., 1973, Nodal tidal cycle of 18.6 yr.: Geology, v. 1, p. 141–144.

KEHEW, A. E., 1982, Catastrophic flood (*sic*) hypothesis for the origin of the Souris spillway, Saskatchewan and North Dakota: Geological Society of America Bulletin, v. 93, p. 1051–1058.

KEIGWIN, L. D., JR., 1982, Isotopic paleooceanography of the Caribbean and East Pacific: role of Panama uplift in Late Neogene time: Science, v. 217, p. 350–353.

KEITH, B. D., and FRIEDMAN, G. M., 1973, Recognition of a paleoslope environment (abstract): Geological Society of America, Abstracts with Programs, v. 5, no. 7, p. 688–689.

KEITH, B. D., and FRIEDMAN, G. M., 1977, A slope-fan-basin-plain model, Taconic Sequence, New York and Vermont: Journal of Sedimentary Petrology, v. 47, p. 1220–1241.

KELLER, G.; and BARRON, J. A., 1983, Paleoceanographic implications of Miocene deep sea (*sic*) hiatuses: Geological Society of America Bulletin, v. 94, p. 590–613.

KELLER, G.; and BARRON, J. A., 1987, Paleodepth distribution of Neogene deep-sea hiatuses: Paleoceanography, v. 2, p. 697–714.

KELLER, G. H., and LAMBERT, D. N., 1972, Geotechnical properties of submarine sediments, p. 401–415 *in* Stanley, D. J., ed., The Mediterranean Sea: a natural sedimentation laboratory: Stroudsburg, PA, Dowden, Hutchinson, and Ross, 765 p.

KELLER, W. D., STONE, C. G., and HOERSCH, A. L., 1985, Textures of Paleozoic chert and novaculite in the Ouachita Mountains of Arkansas and Oklahoma and their geological significance: Geological Society of America Bulletin, v. 96, p. 1353–1363.

KELLING, GILBERT; and STANLEY, D. J., 1972, Sedimentation in the vicinity of the Strait of Gibraltar, p. 489–519 *in* Stanley, D. J., ed., The Mediterranean Sea: a natural sedimentation laboratory: Stroudsburg, PA, Dowden, Hutchinson, and Ross, 765 p.

KELLOGG, D. E., 1983, Phenology of morphologic change in radiolarian lineages from deep-sea cores: implications for macroevolution: Paleobiology, v. 9, p. 355–362.

KELLOGG, T. B., 1987, Glacial-interglacial changes in global deepwater circulation: Paleoceanography, v. 2, p. 259–271.

KELTS, K.; and ARTHUR, M. A., 1981, Turbidites after ten years of deep-sea drilling—wringing out the mop?, p. 91–127 *in* Warme, J. E., Douglas, R. G., and Winterer, E. L., eds., The deep sea (*sic*) drilling project: a decade of progress: Tulsa, OK, Society of Economic Paleontologists and Mineralogists Special Publication 32, 564 p.

KENDALL, A. C., 1979a, Continental (*sic*) and supratidal (sabkha) evaporites, p. 145–158 *in* Walker, R. G., ed., Facies models: Geoscience Canada Reprint Series 1, 211 p.

KENDALL, A. C., 1979b, Subaqueous evaporites, p. 159–174 *in* Walker, R. G., ed., Facies Models: Geological Association of Canada, Reprint Series No. 1, 211 p.

KENDALL, C. G. St. C., and WARREN, J. K., 1987, A review of the origin and setting of tepees and their associated fabrics: Sedimentology, v. 34, p. 1007–1027.

KENNEDY, J. F., 1980, Bed forms in alluvial streams: some views on current understanding and identification of unsolved problems, p. 6a.1–6a.13 *in* Shen, H. W.; and Kikkawa, H., eds., Application of stochastic processes in sediment transport: Fort Collins, CO, Water Resources Publications, variously paginated.

KENNEDY, W. J., and SELLWOOD, B. W., 1970, *Ophiomorpha nodosa* Lundgren, a marine indicator from the Sparnacian of southeast England: Geologists' Association (London) Proceedings, v. 81, p. 99–110.

KENNETT, J. P., 1982, Marine Geology: Englewood Cliffs, NJ, Prentice-Hall, 812 p.

KENNETT, J. P., ed., 1985, The Miocene ocean—paleoceanography and biogeography: Geological Society of America Memoir 163, 337 p.

KENNETT, J. P., and SHACKLETON, N. J., 1976, Oxygen isotopic (*sic*) evidence for the development of the psychrosphere 38 m.y. ago: Nature, v. 260, p. 513–515.

KENT, D. V., and GRADSTEIN, F. M., 1985, A Cretaceous and Jurassic geochronology: Geological Society of America Bulletin, v. 96, p. 1419–1427.

KENYON, N. H., BELDERSON, R. H., and STRIDE, A. H., 1978, Channels, canyons and slump folds on the continental slope between southwest Ireland and Spain: Oceanologica Acta (Paris), v. 1, p. 369–380.

KENYON, P. M., and TURCOTT, D. L., 1985, Morphology of a delta prograding by bulk sediment (*sic*) transport: Geological Society of America Bulletin, v. 96, p. 1457–1465.

KERANS, CHARLES, 1988, Karst-controlled reservoir heterogeneity in Ellenburger Group carbonates of west Texas: American Association of Petroleum Geologists Bulletin, v. 72, p. 1160–1183.

KERANS, CHARLES; and LUCIA, F. J., 1989, Recognition of second (*sic*), third (*sic*), and fourth/fifth (*sic*) order scales of cyclicity in the El Paso Group and their relation (*sic*) to genesis and architecture of Ellenburger reservoirs, p. 105–110 *in* Cunningham, B. K., and Cromwell, D. W., eds., The Lower Paleozoic of West Texas and southern New Mexico—modern exploration concepts: Midland, TX, Society of Economic Paleontologists and Mineralogists, Permian Basin Section, Publication No. 89-31, 223 p.

KERR, R. A., 1987, Geophysics smorgasbord was spread in Baltimore: Science, v. 236, p. 1425.

KESSLER, L. G., II; and MOORHOUSE, K., 1984, Depositional processes and fluid mechanics of Upper Jurassic conglomerate accumulations, British North Sea, p. 383–397 *in* Koster, E. H., and Steel, R. J., eds., Sedimentology of gravels and conglomerates: Calgary, Alberta, Canadian Society of Petroleum Geologists Memoir 10, 441 p.

KETCHUM, B. H., REDFIELD, A. C., and AYERS, J. C., 1951, The oceanography of New York Bight: Cambridge, MA, Massachusetts Institute of Technology and Woods Hole Oceanographic Institution, Papers in Physical Oceanography and Meteorology, v. 12, 46 p.

KHALAF, F. I., 1988, Petrography and diagenesis of silcrete from Kuwait, Arabian Gulf: Journal of Sedimentary Petrology, v. 58, p. 1014–1022.

KHALAF, FIKRY, 1989, Textural characteristics and genesis of the aeolian sediments in the Kuwaiti desert: Sedimentology, v. 36, p. 253–271.

KIDDER, D. L., 1985, Petrology and origin of phosphate nodules from the midcontinent Pennsylvanian epicontinental sea: Journal of Sedimentary Petrology, v. 55, p. 809–816.

KIDWELL, S. M., 1984, Outcrop features and origin of basin margin (*sic*) unconformities in the lower Chesapeake Group (Miocene, Atlantic coastal

plain), p. 37–58 *in* Schlee, J. S., ed., Interregional unconformities and hydrocarbon accumulation: Tulsa, OK, American Association of Petroleum Geologists Memoir 36, 184 p.

KIDWELL, S. M., 1986, Models of fossil concentrations: paleobiologic implications: Paleobiology, v. 12, p. 6–24.

KIDWELL, S. M., 1989, Stratigraphic condensation of marine transgressive records: origin of major shell deposits in the Miocene of Maryland: Journal of Geology, v. 97, p. 1–24.

KIDWELL, S. M., and HOLLAND, S. M., 1991, Field description of coarse bioclastic fabrics: Palaios, v. 6, p. 426–434.

KING, C. A. M., 1959, Beaches and coasts: London, Edward Arnold Ltd., 403 p.

KING, C. A. M., 1972, Beaches and coasts, 2nd ed.: New York, St. Martin's Press, 570 p.

KING, D. T., JR., 1986, Waulsortian-type buildups and resedimented (carbonate-turbidite) facies, Early Mississippian Burlington Shelf, central Missouri: Journal of Sedimentary Petrology, v. 56, p. 471–479.

KING, D. T., JR., 1990, Upper Cretaceous marl-limestone sequences of Alabama: possible products of sea-level change, not climate forcing: Geology, v. 16, p. 19–26.

KING, P. B., 1951, The tectonics of middle America—Middle North America east of the Cordilleran system: Princeton, NJ, Princeton University Press, 203 p.

KINGSLEY, R. J., 1989, Calcium carbonate (*sic*) spicules in the invertebrates, p. 27–33 *in* Carter, J. G., ed., Skeletal biomineralization: patterns, processes and evolutionary trends: American Geophysical Union Short Course in Geology, v. 5, pt. 2.

KINGSTON, D. R., DISHROON, C. P., and WILLIAMS, P. A., 1983, Global basin classification: American Association of Petroleum Geologists Bulletin, v. 67, p. 2175–2193.

KINSMAN, D. J. J., 1970, Early diagenesis of carbonate sediments in a supratidal setting: Princeton, NJ, Princeton University, Department of Geology, A.P.I. Research Project, Semi-Annual Progress Report, 3rd, 42 p.

KIRK, R. M., 1975, Aspects of surf (*sic*) and runup processes on mixed sand (*sic*) and gravel beaches: Geografiska Annaler, v. 57, p. 117–133.

KIRK, R. M., 1980, Mixed sand and gravel beaches: morphology, processes and sediments: Progress in Physical Geography, v. 4, p. 189–210.

KIRWAN, A. D., JR.; DOYLE, L. J.; BOWLES, W. D.; and BROOKS, G. R., 1986, Time-dependent hydrodynamic models of turbidity currents analysed with data from the Grand Banks and Orleansville events: Journal of Sedimentary Petrology, v. 56, p. 379–386.

KITTLEMAN, L. R., 1979, Tephra: Scientific American, v. 251, no. 12, p. 160–177.

KLEIN, G. DEV., 1963, Analysis and review of sandstone classifications in the North American geological literature 1940–1960: Geological Society of America Bulletin, v. 74, p. 555–576.

KLEIN, G. DEV., 1977, Tidal circulation (*sic*) model for deposition of clastic sediment in epeiric (*sic*) and mioclinal shelf seas: Sedimentary Geology, v. 18, p. 1–12.

KLEIN, G. DEV., 1982, Probable sequential arrangement of depositional systems on cratons: Geology, v. 10, p. 17–22.

KLEIN, G. DEV.; MARSAGLIA, K. M.; SWIFT, D. P.; NUMMEDAL, DAG; and DUKE, W. L., 1987, Hummocky cross-stratification, tropical hurricanes, and intense winter storms (discussions and reply): Sedimentology, v. 34, p. 333–359.

KLEIN, G. DEV., and others, 1982, Sedimentology of a subtidal, tide-dominated sand body in the Yellow Sea, southwest Korea: Marine Geology, v. 50, p. 221–240.

KLEIN, G. DEV., and WILLARD, D. A., 1989, Origin of the Pennsylvanian coal-bearing cyclothems of North America: Geology, v. 17, p. 152–155.

KLING, G. W., CLARK, M. A., COMPTON, H. R., DEVINE, J. D., EVANS, W. C., HUMPHREY,

A. M., KOENIGSBERG, E. J., LOCKWOOD, J. P., TUTTLE, M. L., and WAGNER, G. N., 1987, The 1986 Lake Nyos gas disaster in Cameroon, West Africa: Science, v. 236, p. 169–175.

KNAUTH, L. P., 1979, A model for the origin of chert in limestone: Geology, v. 7, p. 274–277.

KNEBEL, H. J., 1981, Processes controlling the characteristics of the surficial sand sheet, U. S., Atlantic outer continental shelf, p. 349–368 *in* Nittrouer, C. A., ed., 1981 Sedimentary dynamics of continental shelves: Developments in sedimentology, v. 32: Amsterdam, Elsevier Scientific Publishing Company, 449 p.

KOBLUK, D. R.; and NOOR, IQBAL, 1990, Coral microatolls and a probable Middle Ordovician example: Journal of Paleontology, v. 64, p. 39–43.

KOCH, G. S., JR., and LINK, R. F., 1970–1971, Statistical analysis of geological data (vols. 1 & 2 bound as one): New York, Dover, 375 p. and 438 p.

KOCUREK, G.; and NIELSON, J., 1986, Conditions favorable for the formation of warm-climate sand seas: Sedimentology, v. 33, p. 795–816.

KOERSCHNER, W. F., III, and READ, J. F., 1989, Field and modelling studies of Cambrian carbonate cycles, Virginia Appalachians: Journal of Sedimentary Petrology, v. 59, p. 654–687.

KOLMER, J. R., 1973, A wave tank (*sic*) analysis of the beach foreshore grain size (*sic*) distribution: Journal of Sedimentary Petrology, v. 43, p. 200–204.

KOLODNY, YEHOSHUA, 1981, Phosphorites, p. 981–1023 *in* Emiliani, Cesare, ed., The sea, vol. 7: New York, Wiley-Interscience, 1728 p.

KOLP, O., 1958, Sedimentsortierung und -Umlagerung am Meeresboden durch Wellenwirkung: Petermanns Geographische Mitteilungen Jahrgang., v. 102, p. 173–178.

KOMAR, P. D., 1976, Beach processes and sedimentation (*sic*): Englewood Cliffs, NJ, Prentice-Hall, 429 p.

KOMAR, P. D., 1983, Nearshore currents and sand transport on beaches, p. 76–109 *in* Johns, B., ed., Physical oceanography of coastal (*sic*) and shelf seas: Amsterdam, Elsevier Scientific Publishing Company, 234 p.

KOMAR, P. D., 1985, The hydraulic interpretation of turbidites from their grain sizes and sedimentary structures: Sedimentology, v. 32, p. 393–407.

KOMAR, P. D., 1988, Discussion of sedimentology and paleohydrology of glacial-lake outburst deposits in southeastern Saskatchewan and northwestern North Dakota: Geological Society of America Bulletin, v. 100, p. 1311–1312 (includes reply by Lord, M. L., and Kahew, A. E.).

KOMAR, P. D., and LI, ZHENLIN, 1986, pivoting analyses of the selective entrainment of sediments by shape and size with application to gravel threshold: Sedimentology, v. 33, p. 425–436.

KOPASKA-MERKEL, D. C., 1987a, Depositional environments and stratigraphy of a mixed carbonate/terrigenous platform deposit: Cambrian of west-central Utah: Carbonates and Evaporites, v. 2, p. 133–148.

KOPASKA-MERKEL, D. C., 1987b, Microporosity in ooids: Mesozoic and Paleozoic of Texas: Carbonates and Evaporites, v. 2, p. 125–132.

KOPASKA-MERKEL, D. C., 1989, Eustasy, not eustacy: Northeastern Environmental Science, v. 8, p. 87.

KOPASKA-MERKEL, D. C., and FRIEDMAN, G. M., 1989, Petrofacies analysis of carbonate rocks: example from the Lower Paleozoic Hunton Group of Oklahoma and Texas: American Association of Petroleum Geologists Bulletin, v. 73, p. 1289–1306.

KOPASKA-MERKEL, D. C., and GRANNIS, JONATHAN, 1990, Detailed structure of wrinkle marks: Journal of the Alabama Academy of Science, v. 61, p. 236–243.

KOPASKA-MERKEL, D. C., and MANN, S. D., 1990 ms. (in press), Classification of lithified carbonates using ternary plots of pore facies: example from the Jurassic Smackover Formation, *in* Rezak, Richard; and Lavoie, Dawn, eds., Carbonate microfabrics: Frontiers in sedimentary geology series: New York, Springer-Verlag.

KOSTASCHUK, R. A., 1985, River mouth (*sic*) processes in a fjord-delta (*sic*), British Columbia, Canada: Marine Geology, v. 69, p. 1–23.

KOSTECKI, J. A.; VEN KATARATHNAM, K.; and RAY, P. K., 1980, Lithology and structures of Quaternary sediments of Indus Fan: American Association of Petroleum Geologists Bulletin, v. 64, p. 734–735.

KOSTER, E. H., and STEEL, R. J., eds., 1984, The sedimentology of gravels and conglomerates: Calgary, Alberta, Canadian Society of Petroleum Geologists Memoir 10, 441 p.

KOSTERS, E. C., 1989, Organic-clastic facies relationships and chronostratigraphy of the Barataria interlobe basin, Mississippi Delta Plain: Journal of Sedimentary Petrology, v. 59, p. 98–113.

KRAFT, J. C., ALLEN, E. A., and MAURMEYER, E. M., 1978, The geological and paleogeomorphological evolution of a spit system and its associated coastal environments. Cape Henlopen spit, Delaware: Journal of Sedimentary Petrology, v. 48, p. 211–226.

KRAFT, J. C., CHRZASTOWSKI, M. J., BELKNAP, D. F., TOSCANO, M. A. and FLETCHER, C. H., III, 1987, The transgressive barrier-lagoon coast of Delaware: morphostratigraphy, sedimentary sequences and responses to relative rise in sea level, p. 129–143 *in* Nummedal, Dag; Pilkey, O. H.; and Howard, J. D., eds., Sea-level fluctuation and coastal evolution, Society of Economic Paleontologists and Mineralogists Special Publication 41, 267 p.

KRAFT, J. C., and JOHN, C. J., 1979, Lateral and vertical facies relationships of transgressive barriers: American Association of Petroleum Geologists Bulletin, v. 63, p. 2145–2163.

KRAUSE, F. F., and OLDERSHAW, A. E., 1979, Submarine carbonate breccia beds—a depositional model for two-layer, sediment gravity flows from the Sekwi Formation (Lower Cambrian), Mackenzie Mountains, N.W. Territories: Canadian Journal of Earth Sciences, v. 16, p. 189–199.

KRAUSKOPF, K. B., 1956, Dissolution and precipitation of silica at low temperatures: Geochimica et Cosmochimica Acta, v. 10, p. 1–26.

KREISA, R. D., and MOIOLA, R. J., 1986, Sigmoidal tidal bundles and other tide-generated sedimentary structures of the Curtis Formation, Utah: Geological Society of America Bulletin, v. 97, p. 381–387.

KRINSLEY, D. H., and SMALLEY, I. J., 1972, Sand: American Scientist, v. 60, p. 286–291.

KRUMBEIN, W. C., 1933, Textural and lithological variations in glacial (*sic*) till: Journal of Geology, v. 41, p. 382–408.

KRUMBEIN, W. C., 1936, Application of logarithmic moments to size-frequency distributions of sediments: Journal of Sedimentary Petrology, v. 6, p. 35–47.

KRUMBEIN, W. C., 1939, Tidal lagoon (*sic*) sediments on the Mississippi delta, p. 178–194 *in* Trask, P. D., ed., Recent marine sediments. A symposium: Tulsa, OK, American Association of Petroleum Geologists, 736 p.

KRUMBEIN, W. C., 1940, Flood gravels of San Gabriel Canyon, California: Geological Society of America Bulletin: v. 51, p. 639–676.

KRUMBEIN, W. C., and GARRELS, R. M., 1952, Origin and classification of chemical sediments in terms of pH and oxidation-reduction potentials: Journal of Geology, v. 60, p. 1–33.

KRUMBEIN, W. C., and PETTIJOHN, F. J., 1938, Manual of sedimentary petrography: New York, Appleton-Century-Crofts, 549 p.

KRUMBEIN, W. C., and SLOSS, L. L., 1951, Stratigraphy and sedimentation: San Francisco, W. H. Freeman and Co., 497 p. (2nd ed. 1963, 660 p.; Chapter 2, The stratigraphic column, p. 8–52; Chapter 3, Stratigraphic procedures, p. 53–92; Chapter 12, Stratigraphic maps, p. 432–500).

KRYNINE, P. D., 1948, The megascopic study and field classification of sedimentary rocks: Journal of Geology, v. 56, p. 130–165.

KU, T.-L., 1969, Uranium series (*sic*) isotopes in sediments from the Red Sea hot-brine area, p. 512–524 *in* Degens, E. T., and Ross, D. A., eds., Hot brines and Recent heavy metal (*sic*) deposits in the Red Sea: New York, Springer-Verlag, 600 p.

KUEHL, S. A., HARIU, T. M., and MOORE, W. S., 1989, Shelf sedimentation off the Ganges-Brahmaputra river system: evidence for sediment bypassing to the Bengal fan: Geology, v. 17, p. 1132–1135.

KUEHL, S. A., NITTROUER, C. A., and DEMASTER, D. J., 1982, Modern sediment accumulation and strata formation on the Amazon continental shelf: Marine Geology, v. 49, p. 279–300.

KUENEN, PH. H., 1935, Geological interpretation of the bathymetrical results: The Snellius Expedition, v. 55, part 1: Leyden, E. J. Brill, 124 p.

KUENEN, PH. H., 1937, Experiments in connection with Daly's hypothesis on the formation of submarine canyons: Leidsche Geol. Meded., v. 8, p. 327–335.

KUENEN, PH. H., 1938, Density currents in connection with the problem of submarine canyons: Geological Magazine, v. 75, p. 241–249.

KUENEN, PH. H., 1939, Sediments of the East Indian archipelago, p. 348–355 *in* Trask, P. D., ed., Recent marine sediments. A symposium: Tulsa, OK, American Association of Petroleum Geologists; London, Thomas Murby, 936 p. (Reprinted, revised, and enlarged edition, 1955, Tulsa, OK, Society of Economic Paleontologists and Mineralogists Special Publication 4; also in paperback, 1968, New York, Dover.)

KUENEN, PH. H., 1948, Turbidity currents of high density: International Geological Congress, 18th, Great Britain, Reports, part 8, p. 44–52.

KUENEN, PH. H., 1950, Marine geology: New York, John Wiley & Sons, 568 p.

KUENEN, PH. H., and MIGLIORINI, C. I., 1950, Turbidity currents as a cause of graded bedding (*sic*): Journal of Geology, v. 58, p. 91–127.

KUENEN, PH. H., and SANDERS, J. E., 1956, Sedimentation phenomena in Kulm and Flozleeres greywackes, Sauerland and Oberharz, Germany: American Journal of Science, v. 254, p. 649–671.

KUKLA, GEORGE, 1987, Loess stratigraphy in central China: Quaternary Science Reviews, v. 6, p. 191–219.

KUKLA, GEORGE; HELLER, F.; LIEU, X-M.; XU, T-L.; LIU, T-S.; and AN, Z-S., 1988, Pleistocene climates in China dated by magnetic susceptibility: Geology, v. 16, p. 811–814.

KULHMANN, H. H., 1989, Korallenforschung im Museum fur Naturkunde: Humboldt-Universitat zu Berlin, Wissenschaftlich Zeitschrift, R. Math./Nat. wiss., v. 38, p. 407–414.

KULM, L. D., ROUSH, R. C., HARLETT, J. C., NEUDECK, R. H., CHAMBERS, D. M., and RUNGE, E. J., 1975, Oregon continental shelf (*sic*) sedimentation: interrelationships of facies distribution and sedimentary processes: Journal of Geology, v. 83, p. 145–176.

KUMAR, N., 1973, Modern (*sic*) and ancient barrier sediments: new interpretation based on stratal sequence in inlet-filling sands and on recognition of nearshore storm deposits: New York Academy of Science, Annals, v. 220, art. 5, p. 245–340.

KUMAR, N.; and SANDERS, J. E., 1974, Characteristics of near-shore storm deposits: examples from modern (*sic*) and ancient sediments (abstract): American Association of Petroleum Geologists, Annual Meetings Abstracts, San Antonio, TX, v. 1, p. 55.

KUMAR, N.; and SANDERS, J. E., 1976, Characteristics of shoreface storm deposits: modern and ancient examples: Journal of Sedimentary Petrology, v. 46, p. 145–162.

KVALE, E. P., ARCHER, A. W., and JOHNSON, H. R., 1989, Daily (*sic*), monthly (*sic*), and yearly tidal cycles within laminated siltstones of the Mansfield Formation (Pennsylvanian) of Indiana: Geology, v. 17, p. 365–368.

LA ROCQUE, A., 1960, Molluscan faunas of the Flagstaff Formation of central Utah: Geological Society of America Memoir 78, 100 p.

LADD, H. S., ed., 1957, Treatise on marine ecology and paleoecology: Geological Society of America Memoir 67, v. 2, Paleoecology, 1077 p.

LADD, H. S., TRACEY, J. I., and GROSS, M. G., 1967, Drilling on Midway Atoll, Hawaii: Science, v. 118, p. 1088–1094.

LAINE, E. P., and HOLLISTER, C. D., 1981, Geological effects of the Gulf Stream system on the western Bermuda Rise: Marine Geology, v. 39, p. 277–310.

LAJOIE, JEAN, 1979, Facies models 15. Volcaniclastic rocks: Geoscience Canada, v. 6, p. 129–139.

LAKATOS, S.; and MILLER, D. S., 1983, Fission-track analysis of apatite and zircon defines a burial depth of 4.5 to 7 km for lowermost Upper Devonian, Catskill, New York: Geology, v. 11, p. 103–104.

LAL, D., 1977, The oceanic microcosm of particles: Science, v. 198, p. 997–1010.

LAMB, H. H., 1972, Climate. Present, past and future. vol. 1: Fundamentals and climate now: London, Methuen and Company, New York, Barnes and Noble Books, 613 p.

LANCASTER, N., 1989, The dynamics of star-dunes: an example from the Gran Desierto, Mexico: Sedimentology, v. 36, p. 273–289.

LANCASTER, N.; and BALKEMA, A. A., 1989, The Namib Sand Sea: dune forms, processes and sediments: Rotterdam, 180 p.

LANDSCHEIDT, THEODOR, 1987, Long-range forecasts of solar cycles and climate change, p. 421–445 *in* Rampino, M. R., Sanders, J. E., Newman, W. S., and Konigsson, L. K., eds., 1987, Climate: history, periodicity, and predictability. New York, Van Nostrand Reinhold Company, 588 p.

LANE, E. W., and BORLAND, W. M., 1953, River-bed scour during floods: American Society of Civil Engineers Proceedings, v. 79, p. 254–1–254-14.

LANGFELDER, J.; STAFFORD, D.; and AMEIN, M., 1968, A reconnaissance of coastal erosion in North Carolina: Raleigh, North Carolina State University, Department of Civil Engineering, 127 p.

LANGFORD, R. P., 1989, Fluvial-aeolian interactions: Part I, modern systems: Sedimentology, v. 36, p. 1023–1036.

LANGFORD, R. P.; and BRACKEN, Bryan, 1987, Medano Creek, Colorado, a model for upper-flow-regime fluvial deposition: Journal of Sedimentary Petrology, v. 57, p. 863–870.

LANGFORD, R. P., and CHAN, M. A., 1989, Fluvial-aeolian interactions: Part II, ancient systems: Sedimentology, v. 36, p. 1037–1051.

LANGMUIR, DONALD, 1971, Eh-pH determinations, p. 597–635, *in* Carver, R. E., ed., Procedures in sedimentary petrology: New York, Wiley-Interscience, 653 p.

LANGOZKY, Y.; and SNEH, A., 1966, The Dead Sea—Arava Rift Valley project: Israel Inst. Petroleum Research and Geophysics, Report 1018, p. 5–10.

LAPORTE, L. F., ed., 1974, Reefs in time and space (selected examples from the Recent and Ancient): Tulsa, OK, Society of Economic Paleontologists and Mineralogists Special Publication 18, 256 p.

LARSEN, H., 1980, Ecology of hypersaline environments, p. 23–29 *in* Nissenbaum, A., ed., Hypersaline brines and evaporitic environments: Amsterdam, Elsevier Scientific Publishing Company, 270 p.

LARUE, D. K., 1985, Quartzose turbidites of the accretionary complex of Barbados, II. Variations in bedding styles, facies and sequences: Sedimentary Geology, v. 42, p. 217–253.

LASCHET, C., 1984, On the origin of cherts: Facies, v. 10, p. 257–290.

LASH, G. G., 1985, Recognition of trench-fill (*sic*) in orogenic flysch sequences: Geology, v. 13, p. 867–870.

LASH, G. G., 1986, Sedimentology of channelized turbidite deposits in an ancient (early Paleozoic) subduction complex, central Appalachians: Geological Society of America Bulletin, v. 97, p. 703–710.

LAUBSCHER, H. P., 1985, Large-scale, thin-skinned thrusting in the southern Alps: kinematic models: Geological Society of America Bulletin, v. 96, p. 710–718.

LAUNDER, B. E., 1985, Progress and prospects in phenomenological turbulence models, *in* Dwoyer, D. L., Hussaini, M. Y., and Voigt, R. G., eds., Theoretical approaches to turbulence: New York, Springer-Verlag.

LAVOISIER, A. L., 1789, Observations generales sur les couches horizontales, qui ont été deposées par la mer, et sur les consequences qu'on peut tirer de leurs dispositions, relativement a l'ancienneté du globe terrestre: France, Academie Royale des Sciences Memoir. (For translation into English, see Carozzi, A. V., 1965, Lavoisier's fundamental contribution to stratigraphy: Ohio Journal of Science, v. 65, no. 2, p. 71–85.)

LAWSON, J. D., 1979, Fossils and lithostratigraphy: Lethaia, v. 12, p. 189–191.

LEADBEATER, B. S. C.; and RIDING, R., eds., 1986, Biomineralization of lower plants and animals (*sic*): Symposium held at University of Birmingham, April 1985, Proceedings: Systematics Association, Special Volume 30: Oxford, Clarendon Press; New York, Oxford University Press, 401 p.

LEATHERMAN, S. P., 1979a, Beach and dune interactions during storm conditions: Quarterly Journal of Engineering Geology, Ireland, v. 12, p. 281–290.

LEATHERMAN, S. P., ed., 1979b, Barrier islands from the Gulf of St. Lawrence to the Gulf of Mexico: New York, Academic Press, 325 p.

LEATHERMAN, S. P., 1983, Barrier island evolution in response to sea-level rise—discussion: Journal of Sedimentary Petrology, v. 53, p. 1026–1031.

LEATHERMAN, S. P., and ZAREMBA, R. E., 1986, Dynamics of a northern barrier beach: Nauset Spit, Cape Cod, Massachusetts: Geological Society of America Bulletin, v. 97, p. 116–124.

LEATHERMAN, S. P., and ZAREMBA, R. E., 1987, Overwash and aeolian processes on a United States northeast coast barrier: Sedimentary Geology, v. 52, p. 183–206.

LEBLANC, R. J., SR., 1972, Geometry of sandstone reservoir bodies, p. 133–189 *in* Cook, T. D., ed., Underground waste management and environmental implications: Tulsa, OK, American Association of Petroleum Geologists Memoir 18, 412 p.

LEBLANC, R. J., SR., and HODGSON, W. D., 1959, Origin and development of the Texas shoreline: Gulf Coast Association of Geological Societies Transactions, v. 9, p. 197–220.

LECKIE, DALE, 1988, Wave-formed, coarse-grained (*sic*) ripples and their relationship to hummocky cross-stratification: Journal of Sedimentary Petrology, v. 58, p. 607–622.

LECKIE, DALE; and KRYSTINIK, L. F., 1989, Is there evidence for geostrophic currents preserved in the sedimentary record of inner (*sic*) to middle shelf deposits?: Journal of Sedimentary Petrology, v. 59, p. 862–870.

LECKIE, DALE; and WALKER, R. G., 1982, Storm (*sic*) and tide-dominated shoreline in Cretaceous Moosebar-Lower Gates interval—outcrop equivalents of Deep Basin gas trap in western Canada: American Association of Petroleum Geologists Bulletin, v. 66, p. 138–157.

LECKIE, R. M., and WEBB, P. N., 1983, Late Oligocene-early Miocene glacial record of the Ross Sea, Antarctica: evidence from DSDP Site 270: Geology, v. 11, p. 578–582.

LEEDER, M. R., 1982, Sedimentology: process and product: London, George Allen and Unwin, 344 p.

LEFOURNIER, J.; and FRIEDMAN, G. M., 1974, Rate of lateral migration of adjoining sea-marginal sedimentary environments shown by historical records, Authie Bay, France: Geology, v. 2, p. 497–498.

LEG 123 SHIPBOARD SCIENTIFIC PARTY, 1988, Sedimentology of the Argo and Gascoyne abyssal plains, NW Australia: report on Ocean Drilling Program Leg 123 (Sept. 1–Nov. 1, 1988): Carbonates and Evaporites, v. 3, p. 201–212.

LEIBNITZ, G. G., 1749, Protogaea: Göttingen, Bibliopolae University, 86 p.

LEINFELDER, R. R.; and HARTKOPF-FRÖDER, CHRISTOPH, 1990, In situ (*sic*) accretion mechanism of concavo-convex lacustrine oncoids ("swallow nests") from the Oligocene of the Mainz Basin, Rhineland, FRG: Sedimentology, v. 37, p. 287–301.

LEITHOLD, E. L.; and BOURGEOIS, J., 1984, Characteristics of coarse-grained (*sic*) sequences deposited in nearshore, wave-dominated environments—examples from the Miocene of south-west Oregon: Sedimentology, v. 31, p. 749–775.

LEMON, R. R., 1990, Principles of stratigraphy: Columbus, OH, Merrill Publishing Company, 559 p.

LEONARD, J. E.; and CAMERON, BARRY, 1979, Origin of a high latitude (*sic*) carbonate beach: Mount Desert Island, Maine: Northeastern Geology, v. 1, p. 133–145.

LEROY, S. D., 1981, Grain-size and moment measures: a new look at Karl Pearson's ideas on distributions: Journal of Sedimentary Petrology, v. 51, p. 625–630.

LEVINTON, J. S., 1982, Marine ecology: Englewood Cliffs, NJ, Prentice-Hall, 526 p.

LEVINTON, J. S., 1988, Genetics, paleontology, and macroevolution: Cambridge, Cambridge University Press, 637 p.

LEVINTON, J. S., and SIMON, C. M., 1980, A critique of the punctuated equilibria (*sic*) model and implications for the detection of speciation in the fossil record: Systematic Zoology, v. 29, p. 130–142.

LEVORSEN, A. I., 1943, Discovery thinking: American Association of Petroleum Geologists Bulletin, v. 27, p. 887–928.

LEVORSEN, A. I., 1967, Geology of petroleum, 2nd ed.: San Francisco, W. H. Freeman and Company, 724 p.

LEWIN, R., 1980, Evolutionary theory under fire: Science, v. 210, p. 883–887.

LEWIN, R., 1986, Mass extinctions select different victims: Science, v. 231, p. 219–220.

LEWIN, R., 1988, A lopsided look at evolution: Science, v. 241, p. 291–293.

LEWIS, D. W., 1984, Practical sedimentology: Stroudsburg, PA, Dowden, Hutchinson, and Ross, 229 p.

LIGHTY, R. G., 1977, Relict shelf-edge Holocene coral reef: southeast coast of Florida: International Coral Reef Symposium, 3rd, Proceedings, University of Miami, v. 2, p. 215–221.

LIGHTY, R. G.; MACINTYRE, I. G.; and STUCKENRATH, R., 1978, Submerged early Holocene barrier reef southeast Florida shelf: Nature, v. 275, p. 59–60.

LINDSAY, D. W., 1982, Punctuated equilibria and punctuated environments: Nature, v. 296, p. 611–612.

LLOYD, R. M., PERKINS, R. D., and KERR, S. D., 1987, Beach (*sic*) and shoreface ooid deposition on shallow interior banks, Turks and Caicos Islands, British West Indies: Journal of Sedimentary Petrology, v. 57, p. 976–982.

LOCK, D. E., 1982, Groundwater controls on dolomite formation in the Coorong region of South Australia and its ancient analogues: The Flinders University of South Australia, unpublished Ph.D. dissertation, 275 p.

LOGAN, B. W., and CEBULSKI, D. E., 1970, Sedimentary environments of Shark Bay, Western Australia, p. 1–37 *in* Logan, B. W., Davies, G. R., Read, J. F., and Cebulski, D. E., Carbonate sedimentation and environments, Shark Bay, Western Australia: Tulsa, OK, American Association of Petroleum Geologists Memoir 13, 223 p.

LOGAN, B. W., HARDING, J. L., AHR, W. M., WILLIAMS, J. D., and SNEAD, R. G., 1969, Late (*sic*) Quaternary sediments of Yucatan Shelf, Mexico, p. 1–128 *in* Logan, B. W., and others, Carbonate sediments and reefs, Yucatan Shelf, Mexico: Tulsa, OK, American Association of Petroleum Geologists Memoir 11, 355 p.

LOGVINENKO, N. V., 1982, Origin of glauconite in the Recent bottom sediments of the oceans: Sedimentary Geology, v. 31, p. 43–48.

LONGINELLI, A., 1979/1980, Isotope geochemistry of some Messinian evaporites: paleoenvironmental implications: Palaeogeography, Palaeoclimatology, and Palaeoecology, v. 29, p. 95–123.

LONGMAN, M. W., 1980, Carbonate diagenetic textures from near surface (*sic*) diagenetic environments: American Association of Petroleum Geologists Bulletin, v. 64, p. 461–487.

LONGWELL, C. R., FLINT, R. F., and SANDERS, J. E., 1969, Physical geology: New York, John Wiley & Sons, 685 p.

LONSDALE, P. F., 1982, Sediment drifts of the northeast Atlantic and their relationship to the observed abyssal currents: Bull. Inst. Geol. Bassin d'Aquitaine, Bordeaux, v. 31, p. 141–149.

LORIUS, C.; JOUZEL, J.; RITZ, D.; MERLIVAT, L.; BARKOV, N. I.; KOROTKEVICH, Y. S.; and KOTLYAKOV, V. M., 1985, A 150,000-year climatic record from Antarctic ice: Nature, v. 316, p. 591–596.

LOUGHNAN, F. C., 1969, Chemical weathering of the silicate-minerals (*sic*): New York, American Elsevier Publishing Company, 154 p.

LOUTIT, T. S., and KENNETT, J. P., 1981, Australian Cenozoic sedimentary cycles, global sea level (*sic*) changes and deep sea (*sic*) sedimentary record: Oceanologica Acta Special Volume, p. 45–63.

LOVELL, J. B. P., and STOW, D. A. V., 1981, Identification of ancient sandy contourites: Geology, v. 9, p. 347–349.

LOWE, D. R., 1982, Sediment gravity flows: II. Depositional models with special reference to the deposits of high-density turbidity currents: Journal of Sedimentary Petrology, v. 52, p. 279–297.

LOWE, D. R., 1988, Suspended load fall-out rate as an independent variable in the analysis of current structures: Sedimentology, v. 35, p. 765–776.

LOWENSTAM, H. A., 1981, Minerals formed by organisms: Science, v. 211, p. 1126–1131.

LOWENSTAM, H. A.; and WEINER, S., 1989, On biomineralization: New York, Oxford University Press, 324 p.

LOWRIE, W.; and ALVAREZ, W., 1981, One hundred million years of geomagnetic polarity (*sic*) history: Geology, v. 9, p. 392–397.

LOWRIE, W.; and HELLER, F., 1982, Magnetic properties of marine limestones: Reviews of Geophysics and Space Physics, v. 20, p. 171–192.

LUDVIGSEN, ROLF, 1989, The Burgess Shale: not in the shadow of the Cathedral Escarpment: Geoscience Canada, v. 16, p. 51–59.

LUEPKE, G., 1980, Opaque minerals as aids in distinguishing between source (*sic*) and sorting effects on beach-sand mineralogy (*sic*) in southwestern Oregon: Journal of Sedimentary Petrology, v. 50, p. 489–496.

LUMSDEN, D. N., 1988, Characteristics of deep-marine dolomite: Journal of Sedimentary Petrology, v. 58, p. 1023–1031.

LUSKIN, BERNARD; HEEZEN, B. C.; EWING, MAURICE; and LANDISMAN, MARK, 1954, Precision measurement of ocean depth: Deep-Sea Research, v. 1, p. 131–140.

LUTERNAUER, J. L., and LIAM FINN, W. D., 1983, Stability of the Fraser River delta front: Canadian Geotechnical Journal, v. 20, p. 603–616.

LYON-CAEN, H.; and MOLNAR, P., 1989, Constraints on the deep structure and dynamic processes beneath the Alps and adjacent regions from an analysis of gravity anomalies: Geophys. Jour. Int., v. 99, p. 19–32.

MACARTHUR, G. G., 1916, Solubility of oxygen in salt solutions and the hydrates of these salts: Journal of Physical Chemistry, v. 20, p. 495–502.

MACCLINTOCK, PAUL, 1940, Weathering of the Jerseyan till: Geological Society of America Bulletin, v. 51, p. 103–116.

MacGinitie, G. E., 1934, The natural history of *Callianassa californiensis* Dana: American Midland Naturalist, v. 15, p. 166–177.

MacIntyre, I. G., 1970, Sediments off the west coast of Barbados: diversity of origins: Marine Geology, v. 9, p. 5–23.

MacIntyre, I. G., 1972, Submerged reefs of eastern Caribbean: American Association of Petroleum Geologists Bulletin, v. 56, p. 720–738.

MacIntyre, I. G., 1985, Submarine cements—the peloidal question, p. 109–116 *in* Schneidermann, Nahum; and Harris, P. M., eds., Carbonate cements: Tulsa, OK, Society of Economic Paleontologists and Mineralogists Special Publication 36, 379 p.

MacIntyre, I. G., and Glynn, P. W., 1976, Evolution of modern Caribbean fringing reef, Galeta Point, Panama: American Association of Petroleum Geologists Bulletin, v. 60, p. 1052–1072.

Maejima, W., 1982, Texture and stratification of gravelly beach sediments, Enju beach, Kii peninsula, Japan: Journal of Geoscience, Osaka City University, v. 25, p. 35–51.

Magnier, Ph.; Oki, T.; and Witoelar Kartaadiputra, L., 1975, The Mahakam Delta, Kalimantan, Indonesia: World Petroleum Congress, 9th, Tokyo, Panel discussion 4, Deltaic deposits and petroleum. Preprints, 13 p.

Mahaney, W. C., 1984, Superposed Neoglacial and late Pinedale (Wisconsinan) tills, Titcomb Basin, Wind River Mountains, western Wyoming: Palaeogeography, Palaeoclimatology, and Palaeoecology, v. 45, p. 149–163.

Malde, H. E., 1968, The catastrophic late Pleistocene Bonneville flood in the Snake River plain, Idaho: U.S. Geological Survey Professional Paper 596, 52 p.

Maliva, R. G.; and Siever, Raymond, 1988, Pre-Cenozoic nodular cherts: evidence for opal-CT precursors and direct quartz replacement: American Journal of Science, v. 288, p. 798–809.

Maliva, R. G.; and Siever, Raymond, 1989, Nodular chert formation in carbonate rocks: Journal of Geology, v. 97, p. 421–433.

Maloney, N. J., 1966, Univ. Oriente, Inst. Oceanogr, Bol., v. 1, p. 396–473.

Mann, P.; Hempton, M. R.; Bradley, D. C.; and Burke, K., 1983, Development of pull-apart basins: Journal of Geology, v. 91, p. 529–554.

Mann, S. D., 1988, Subaqueous evaporites of the Buckner Member, Haynesville Formation, Northeastern Mobile Co., Alabama: Gulf Coast Association of Geological Societies Transactions, v. 38, p. 187–196.

Manohar, Madhav, 1955, Mechanics of bottom sediment (*sic*) movement due to wave action: U.S. Army Corps of Engineers Beach Erosion Board Technical Memorandum 75.

Maracus, W. A., 1989, Lag-time routing of suspended sediment (*sic*) concentrations during unsteady flow: Geological Society of America Bulletin, v. 101, p. 644–651.

Marsaglia, K. M., and Klein, G. deV., 1983, The paleogeography of Paleozoic (*sic*) and Mesozoic storm depositional systems: Journal of Geology, v. 91, p. 117–142.

Marshak, Stephen, 1986, Structure and tectonics of the Hudson Valley fold-thrust belt, eastern New York State: Geological Society of America Bulletin, v. 97, p. 354–368.

Marshak, Stephen; and Tabor, John, 1989, Structure of the Kingston oroclines in the Appalachian fold-thrust belt, New York: Geological Society of America Bulletin, v. 101, p. 683–701.

Mason, C.; Sallenger, A. H.; Holman, R. A.; and Birkemeier, W. A., 1985, Duck 82—a coastal storm processes (*sic*) experiment, p. 1913–1928 *in* International Conference on Coastal Engineering, 19th, Houston, Texas: New York, American Society of Civil Engineers.

Massa, A. A., 1981, Genesis of shore ridges at the western end of Fire Island, New York: Northeastern Geology, v. 3, p. 235–242.

Massari, F., 1984, Resedimented conglomerates of a Miocene fan-delta complex, southern Alps, Italy, p. 259–278 *in* Koster, E. H., and Steel, R. J., eds., Sedimentology of gravels and conglomerates: Canadian Society of Petroleum Geology Memoir 10, 441 p.

Massari, F.; and Parea, G. C., 1988, Progradational gravel beach sequences in a moderate- to high-energy, microtidal marine environment: Sedimentology, v. 35, p. 881–913.

Masters, C. D. 1967, Use of sedimentary structures in determination of depositional environments, Mesaverde Formation: Williams Fork Mountains, Colorado: American Association of Petroleum Geologists Bulletin, v. 51, p. 2033–2043.

Matlack, K. S., Houseknecht, D. W., and Applin, K. R., 1989, Emplacement of clay into sand by infiltration: Journal of Sedimentary Petrology, v. 59, p. 77–87.

Matter, Albert; and Tucker, M. E., ed., 1978, Modern (*sic*) and ancient lake deposits: International Association of Sedimentologists Special Publication 2, 290 p.

Matthews, E. R., 1980, Observations of beach gravel transport, Wellington Harbor entrance, New Zealand: New Zealand Journal of Geology and Geophysics, v. 23, p. 209–222.

Matthews, R. K.; and Frohlich, C., in press, Toward convergence of dynamic stratigraphy and seismic sequence stratigraphy: orbital forcing of low-frequency glacio-eustacy (*sic*): Journal of Geophysical Research.

Mattick, R. E., Weaver, N. L., Foote, R. Q., and Ruppel, B. D., 1973, A preliminary report on U.S. Geological Survey geophysical studies of the Atlantic outer continental shelf (abstract): Technical Program, East Coast Offshore Symposium—Baffin Bay to the Bahamas, Eastern Section, American Association of Petroleum Geologists, Atlantic City, p. 9 (only).

Maude, A. D., and Whitmore, R. J., 1958, A generalized theory of sedimentation: British Journal of Applied Physics, v. 9, p. 477–482.

Maynard, J. B., 1983, Geochemistry of sedimentary ore deposits: New York, Springer-Verlag, 305 p.

Mazzullo, Jim, 1986, Sources and provinces of late (*sic*) Quaternary sand on the East Texas-Louisiana continental shelf: Geological Society of America Bulletin, v. 97, p. 638–647.

Mazzullo, S. J., and Birdwell, B. A., 1989, Syngenetic formation of grainstones and pisolites from fenestral carbonates in peritidal settings: Journal of Sedimentary Petrology, v. 59, p. 605–611.

McBride, E. F., 1962, Flysch and associated beds of the Martinsburg Formation (Ordovician) central Appalachians: Journal of Sedimentary Petrology, v. 32, p. 32–91.

McCall, P. L., and Tevesz, M. J. S., eds., 1982, Animal-sediment relations (*sic*), the biogenic alteration of sediments: New York, Plenum Press, 336 p.

McCammon, R. B., 1970, On estimating the relative biostratigraphic value of fossils: Uppsala University, Geol. Inst. Bulletin, new series, v. 2, p. 49–57.

McCann, S. B., ed., 1980, The coastline of Canada: littoral processes and shore morphology: Geological Survey of Canada, Paper 80-10, 439 p.

McCann, S. B., and Kostaschuk, R. A., 1987, Fjord sedimentation in northern British Columbia, p. 33–49 *in* Fitzgerald, D. M., and Rosen, P. S., eds., Glaciated coasts: San Diego, Academic Press, 364 p.

McCave, I. N., 1975, Vertical flux of particles in the ocean: Deep-Sea Research, v. 22, p. 491–502.

McCave, I. N., 1984, Erosion, transport and deposition of fine-grained marine sediments, p. 35–69 *in* Stow, D. A. V., and Piper, D. J. W., eds., Fine-grained sediments: deep-water processes and facies (*sic*): Geological Society of London Special Publication 15 (Oxford, Blackwell Scientific Publications), 659 p.

McCave, I. N., 1985, Recent shelf clastic sediments, p. 49–65 *in* Brenchley, P. J., and Williams,

B. P. J., eds., Sedimentology: recent developments and applied aspects: Oxford, Blackwell Scientific Publications, 338 p.

McCave, I. N., Lonsdale, P. F., Hollister, C. D., and Gardner, W. D., 1980, Sediment transport over the Hatton and Gardar contourite drifts: Journal of Sedimentary Petrology, v. 50, p. 1049–1062.

McCrory, V. L. C., and Walker, R. G., 1986, A storm (*sic*) and tidally-influenced prograding shoreline—Upper Cretaceous Milk River Formation of southern Alberta, Canada: Sedimentology, v. 33, p. 47–60.

McDonald, D. A., and Surdam, R. C., eds., 1984, Clastic diagenesis: Tulsa, OK, American Association of Petroleum Geologists, Memoir 37, 434 p.

McDougall, Ian, 1985, K-Ar and ^{40}Ar/^{39}Ar dating of the hominid-bearing Pliocene-Pleistocene sequence at Koobi Fora, Lake Turkana, northern Kenya: Geological Society of America Bulletin, v. 96, p. 159–175.

McGowen, J. H., Granata, G. E., and Serri, S. J., 1979, Depositional framework of the Lower Dockum Group (Triassic), Texas Panhandle: Texas Bureau of Economic Geology, Report of Investigations 97, 60 p.

McGowen, J. H., and Groat, C. G., 1971, Van Horn Sandstone, West Texas; an alluvial fan (*sic*) model for mineral exploration: Texas Bureau of Economic Geology, Report of Investigations 72, 57 p.

McGregor, B.; and Bennett, R. H., 1979, Mass movement of sediment on the continental slope and rise seaward of the Baltimore Canyon Trough: Marine Geology, v. 33, p. 163–174.

McIlreath, I. A., and James, N. P., 1978, Facies models 13. Carbonate slopes: Geoscience Canada, v. 5, p. 189–199.

McIntire, W. G.; and Ho, C., 1969, Development of barrier island (*sic*) lagoons, p. 49–61 *in* Ayala-Castanares, Augustin; and Phleger, F. B., eds., Lagunas costeras, un simposio: Simposio Internacional sobre lagunas costeras, Mexico City, Mexico, 28–30 Noviembre 1967, memoria: Mexico Universidad Nacional Autonoma (WMAS-UNESCO), 686 p.

McKee, B. A., 1983, Concepts of sediment deposition and accumulation (*sic*) applied to continental shelf near the mouth of the Yangtze River: Geology, v. 11, p. 631–633.

McKee, E. D., 1982, Sedimentary structures in dunes of the Namib desert, Southwest Africa: Geological Society of America Special Paper 108, 64 p.

McKee, E. D., and Weir, G. W., 1953, Terminology for stratification and cross-stratification (*sic*) in sedimentary rocks: Geological Society of America Bulletin, v. 64, p. 381–389.

McKenzie, J. A., 1980, Holocene dolomitization of calcium carbonate (*sic*) sediments from the coastal sabkhas of Abu Dhabi UAE: a stable isotope study: Journal of Geology, v. 89, p. 185–198.

McKenzie, J. A., Hsü, K. J., and Schneider, J. F., 1980, Movement of surface waters under the sabkha, Abu Dhabi UAE, and its relation to evaporative dolomite genesis, p. 11–30 *in* Zenger, D. H.; Dunham, R. J.; and Ethington, R., eds., Concepts and Models of Dolomitization: Tulsa, OK, Society of Economic Paleontologists and Mineralogists Special Publication 28, 320 p.

McKenzie, J. A.; and Oberhansli, H., 1985, Paleoceanographic expressions of the Messinian salinity crisis, p. 99–123 *in* Hsü, K. J., and Weissert, H. J., eds., South Atlantic paleoceanography: Cambridge, Cambridge University Press, 350 p.

McKinney, T. F., and Friedman, G. M., 1970, Continental shelf (*sic*) sediments of Long Island, New York: Journal of Sedimentary Petrology, v. 40, p. 213–248.

McLaren, D. J., 1959, The role of fossils in defining rock units with examples from the Devonian of western (*sic*) and arctic Canada: American Journal of Science, v. 257, p. 734–751.

McLaren, Patrick; and Bowles, Donald, 1985, The effects of sediment transport on grain-size distributions: Journal of Sedimentary Petrology, v. 55, p. 457–470.

McMURCHY, R. C., 1934, Crystal structure of chlorites: Zeitschrift fur Krystallographie, v. 88, p. 420–430.

McNEAL, R. P., 1959, Lithologic analysis of sedimentary rocks: American Association of Petroleum Geologists Bulletin, v. 43, p. 854–879.

McPHERSON, J. G.; SHANMUGAM, GANAPATHY; and MOIOLA, R. J., 1987, Fandeltas (*sic*) and braid deltas: varieties of coarse-grained deltas: Geological Society of America Bulletin, v. 99, p. 331–340.

MEADE, R. H., 1966, Factors influencing the early stages of the compaction of clays and sands—review: Journal of Sedimentary Petrology, v. 36, p. 1085–1101.

MEHTA, A. J., HAYTER, E. J., PARKER, W. R., KRONE, R. B., and TEETER, A. M., 1989, Cohesive sediment transport; I, process description: Journal of Hydraulic Engineering, v. 115, p. 1076–1093.

MELVIN, JOHN, 1986, Upper Carboniferous fine-grained turbiditic sandstones from southwest England: a model for growth in an ancient, delta-fed subsea fan: Journal of Sedimentary Petrology, v. 56, p. 19–34.

MENARD, H. W., and LADD, H. S., 1965, Oceanic islands, seamounts, guyots and atolls, p. 365–387 *in* Hill, M. N., gen. ed., The sea. Ideas and observations on progress in the study of the seas. Volume 3, The Earth beneath the sea. History: New York, Wiley-Interscience, 963 p.

MERRILL, G. K., 1986, Map location (*sic*) literacy—How well does Johnny Geologist read: Geological Society of America Bulletin, v. 97, p. 404–409.

MESOLELLA, K. J., ROBINSON, J. D., McCORMICK, L. M., and ORMISTON, A. R., 1974, Cyclic deposition of Silurian carbonates and evaporites in Michigan Basin: American Association of Petroleum Geologists Bulletin, v. 58, p. 34–62.

MEZZADRI, GIOVANNI; and SACCANI, EMILIO, 1989, Heavy mineral (*sic*) distribution in Late Quaternary sediments of the southern Aegean Sea: implications for provenance and sediment dispersal in sedimentary basins at active margins: Journal of Sedimentary Petrology, v. 59, p. 412–422.

MIALL, A. D., ed., 1978, Fluvial sedimentology: Calgary, Alberta, Canadian Society of Petroleum Geologists Memoir 5, 859 p.

MIALL, A. D., ed., 1980, Facts and principles of world petroleum occurrence: Canadian Society of Petroleum Geologists, Memoir 6, 1003 p.

MIALL, A. D., 1984, Deltas, p. 105–118 *in* Walker, R. G., ed., Facies models, 2nd ed.: Geoscience Canada Reprint Series 1, 317 p.

MIALL, A. D., 1990, Principles of sedimentary basin (*sic*) analysis, 2nd ed.: New York, Springer-Verlag, 668 p.

MIDDLEMISS, F. A., 1962, Vermiform burrows and rate of sedimentation in the Lower Greensand: Geological Magazine, v. 99, p. 33–40.

MIDDLETON, G. V., 1973, Johannes Walther's law of the correlation of facies: Geological Society of America Bulletin, v. 84, p. 979–988.

MIDDLETON, G. V., 1978, Facies, p. 323–325 *in* Fairbridge, R. W.; and Bourgeois, Joanne, eds., The encyclopedia of sedimentology: Encyclopedia of the earth sciences, vol. 6: Stroudsburg, PA, Dowden, Hutchinson, and Ross, 901 p.

MIDDLETON, G. V., and NEAL, W. J., 1989, Experiments on the thickness of beds deposited by turbidity currents: Journal of Sedimentary Petrology, v. 59, p. 297–307.

MILANKOVITCH, M., 1941, Kanon der Erdbestrahlung und seine Anwendung auf das Eiszeitproblem: Belgrade, Acad. Royal Serbe, édns. spéc. 133, 633 p.

MILKMAN, R., ed., 1982, Perspectives on evolution: Sunderland, MA, Sinauer Associates, 241 p.

MILLER, M. C., McCAVE, I. N., and KOMAR, P. D., 1977, Threshold of sediment motion in unidirectional currents: Sedimentology, v. 24, p. 507–528.

MILLER, R. L., and ZEIGLER, J. M., 1958, A model relating dynamics and sediment pattern in equilibrium in the region of shoaling waves, breaker zone, and foreshore: Journal of Geology, v. 66, p. 417–441.

MILLER, W. R., and EGLER, F. E., 1950, Vegetation of the Wequetequock-Pawcatuck tidal-marshes (*sic*), Connecticut: Ecological Monographs, v. 20, p. 141–172.

MILLIMAN, J. D., and EMERY, K. O., 1968, Sea levels during the past 35,000 years: Science, v. 162, p. 1121–1123.

MILNER, H. B., 1962, Sedimentary petrography, 2nd ed.: vol. 1, Methods in sedimentary petrography; vol. 2, Principles and applications: New York, Macmillan Publishing Company, v. 1, 643 p.; v. 2, 715 p.

MILNES, A. R., and TWIDALE, C. R., 1983, An overview of silicification in Cainozoic landscapes of arid central and southern Australia: Australian Journal of Soil Research, v. 21, p. 387–410.

MITCHUM, R. M., JR., 1985, Seismic stratigraphic (*sic*) expression of submarine fans, p. 117–136 *in* Berg, O. R., and Woolverton, D. G., eds., Seismic stratigraphy II: an integrated approach: Tulsa, OK, American Association of Petroleum Geologists Memoir 39, 276 p.

MITCHUM, R. M., JR.; VAIL, P. R.; and SANGREE, J. B., 1977, Stratigraphic interpretation of seismic reflection (*sic*) patterns in depositional sequences, p. 117–133 *in* Payton, C. E., ed., Seismic stratigraphy—applications to hydrocarbon exploration: Tulsa, OK, American Association of Petroleum Geologists Memoir 26, 516 p.

MITCHUM, R. M., JR.; VAIL, P. R.; and THOMPSON, S., III, 1977, Seismic stratigraphy and global changes of sea level, part 2: the depositional sequence as a basic unit for stratigraphic analysis, p. 53–62 *in* Payton, C. E., ed., Seismic stratigraphy–applications to hydrocarbon exploration: Tulsa, OK, American Association of Petroleum Geologists Memoir 26, 516 p.

MITTERER, R. M.; and CUNNINGHAM, ROBERT, JR., 1985, The interaction of natural organic matter with grain surfaces: implications for calcium carbonate (*sic*) precipitation, p. 17–31 *in* Schneidermann, Nahum; and Harris, P. M., eds., Carbonate cements: Tulsa, OK, Society of Economic Paleontologists and Mineralogists Special Publication 36, 379 p.

MOLDVAY, L., 1961, On the laws governing sedimentation from eolian suspensions: Univ. Szeged, Acta Mineralogica-Petrographica, v. 14, p. 75–109.

MOLNAR, PETER, 1988, Continental tectonics in the aftermath of plate tectonics: Nature, v. 335, p. 131–137.

MONAGHAN, G. W., and LARSON, G. J., 1986, Late Wisconsinan drift stratigraphy of the Saginaw Ice Lobe in south-central Michigan: Geological Society of America Bulletin, v. 97, p. 324–328.

MONAGHAN, G. W., LARSON, G. J., and GEPHART, G. D., 1986, Late Wisconsinan drift stratigraphy of the Lake Michigan Lobe in southwestern Michigan: Geological Society of America Bulletin, v. 97, p. 329–334.

MONTY, C. L. V., 1968, D'Orbigny's concepts of stage and zone: Journal of Paleontology, v. 42, p. 689–701.

MOOERS, C. N. K., 1976, Introduction to the physical oceanography (*sic*) and fluid dynamics of continental margins, p. 7–21 *in* Stanley, D. J., and Swift, D. J. P., eds., Marine sediment transport and environmental management: New York, John Wiley & Sons, 602 p.

MOORE, C. H., 1989, Carbonate diagenesis and porosity: Developments in sedimentology 46: Amsterdam, Elsevier Scientific Publishing Company, 338 p.

MOORE, D. G., 1959, Role of deltas in the formation of some British Lower Carboniferous cyclothems: Journal of Geology, v. 67, p. 522–539.

MOORE, D. G., 1960, Acoustic-reflection studies of the continental shelf and slope off Southern Cali-

fornia: Geological Society of America Bulletin, v. 71, p. 1121–1136.

MOORE, D. G., and SCRUTON, P. C., 1957, Minor internal structures of some recent unconsolidated sediments: American Association of Petroleum Geologists Bulletin, v. 41, p. 2723–2751.

MOORE, R. C., 1949, Introduction to historical geology, lst ed.: New York, McGraw-Hill Book Company, 582 p.

MOORE, R. C., 1964, Paleoecological aspects of Kansas Pennsylvanian and Permian cyclothems, p. 287–380 *in* Merriam, D. F., ed., Symposium on cyclic sedimentation: Kansas Geological Survey Bulletin 169, 2 vols., v. l, p. 1–380; v. 2, p. 381–636.

MORETTI, I.; and TURCOTTE, D. L., 1985, A model for erosion, sedimentation, and flexure with application to New Caledonia: Journal of Geodynamics, v. 3, p. 155–168.

MORISON, J. R., and CROOKE, R. C., 1953, The mechanics of deep water (*sic*), shallow water (*sic*), and breaking waves: U.S. Army, Corps of Engineers, Beach Erosion Board, Technical Memo. 40, 14 p.

MORLEY, J. J.; PSIAS, N. G.; and LEINEN, M., 1987, Late Pleistocene time series of atmospheric (*sic*) and oceanic variables recorded in sediments from the Subarctic Pacific: Paleoceanography, v. 2, p. 21–48.

MOROD, SADOON; and AL DAHAN, A. A., 1986, Alteration of detrital Fe-Ti oxides in sedimentary rocks: Geological Society of America Bulletin, v. 97, p. 567–578.

MORRIS, K. A., 1980, A comparison of major sequences of organic-rich shales in the British Jurassic: Geological Society of London Journal, v. 137, p. 157–170.

MORSE, J. W., and MACKENZIE, F. T., eds., 1990, Geochemistry of sedimentary carbonates: Developments in sedimentology 48: Amsterdam, Elsevier Scientific Publishing Company, 707 p.

MORTON, A. C., 1985, Heavy minerals in provenance studies, p. 249–277 *in* Zuffa, G. G., ed., Provenance of arenites: Dordrecht, The Netherlands, D. Reidel Publishing Company.

MOSLOW, T. F., 1980, Stratigraphy of mesotidal barrier islands: Unpublished Ph.D. dissertation, University of South Carolina, Columbia, 187 p.

MOSS, A. J., 1972, Bed-load sediments: Sedimentology, v. 18, p. 159–219.

MOUNT, J. F., 1984, The mixing of siliciclastic (*sic*) and carbonate sediments in shallow shelf environments: Geology, v. 12, p. 432–435.

MOUNT, J. F., 1985, Mixed siliciclastic (*sic*) and carbonate sediments: a proposed first-order textural and compositional classification: Sedimentology, v. 32, p. 435–442.

MOZLEY, P. S., 1989, Relation (*sic*) between depositional environment and the elemental composition of early diagenetic siderite: Geology, v. 17, p. 704–706.

MÜLLER, GERMAN, 1988, Salzgesteine (Evaporites), p. 435–500 *in* Füchtbauer, Hans, ed., Sedimente und Sedimentgesteine: Stuttgart, E. Schweizerbart, 1141 p.

MULLINS, H. T., HEATH, K. C., VAN BUREN, H. M., and NEWTON, C. R., 1984, Anatomy of a modern open-ocean carbonate slope: northern Little Bahama Bank: Sedimentology, v. 31, p. 141–168.

MULLINS, H. T.; THOMPSON, J. B.; McDOUGALL, K.; and VERCOUTERE, T. L., 1985, Oxygen-minimum zone (*sic*) edge effects: evidence from the central California coastal upwelling system: Geology, v. 13, p. 491–494.

MURCHISON, D.; and WESTOLL, T. S., eds., 1968, Coal and coal-bearing strata: New York, Elsevier, 418 p.

MURRAY, S. P., 1970, Settling velocities (*sic*) and vertical diffusion of particles in turbulent water: Journal of Geophysical Research, v. 75, p. 1647–1654.

MUSCATINE, L.; and CERNICHIARI, E., 1969, Assimilation of photosynthetic products of zooxanthellae by a reef coral: Biological Bulletin, v. 137, p. 506–523.

MUSSMAN, W. J., and READ, J. F., 1986, Sedimentology and development of a passive- to convergent-margin unconformity: Middle Ordovician Knox unconformity, Virginia Appalachians: Geological Society of America Bulletin, v. 97, p. 282–295.

MUTTI, EMILIO, 1979, Turbidités et cones sous-marins profonds, p. 353–419 *in* Homewood, P., ed., Sedimentation detritique (fluviatile, littorale et marine): Institut de Geologie de l'Université de Fribourg, Short Course 1979.

MUTTI, EMILIO, 1985, Turbidite systems and their relations (*sic*) to depositional sequences, p. 65–93 *in* Zuffa, G. G., ed., Provenance of arenites: Boston, D. Reidel Publishing Company, 379 p.

MUTTI, EMILIO; and JOHNS, D. R., 1978, The role of sediment bypassing in the genesis of fan fringe (*sic*) and basin plain turbidites in the Hecho Group system (south-central Pyrenees, Spain): Mem. Soc. Geol. Italia, v. 18, p. 15–22.

MUTTI, EMILIO; and RICCI LUCCHI, F., 1972, Le torbiditi dell'Appenino settentrionale, introduzione all'analisi di facies: Soc. Geol. Italiana, Mem., v. 11, p. 161–199. (Translated in 1975 and published in International Geological Reviews, v. 20, p. 125–166.)

NAGLE, J. S., 1967, Wave (*sic*) and current orientation of shells: Journal of Sedimentary Petrology, v. 37, p. 1124–1138.

NAKATO, TATSUAKI, 1990, Tests of selected sediment-transport formulas: American Society of Civil Engineers, Journal of Hydraulic Engineering, v. 116, no. 3, p. 362–379.

NANSON, G. C., 1986, Episodes of vertical accretion and catastrophic stripping: a model of disequilibrium flood-plain development: Geological Society of America Bulletin, v. 97, p. 1467–1475.

NANSON, G. C., and HICKIN, E. J., 1986, A statistical analysis of bank erosion and channel migration in western Canada: Geological Society of America Bulletin, v. 97, p. 497–504.

NATIONAL RESEARCH COUNCIL, INTERDIVISIONAL COMMISSION ON DENSITY CURRENTS, SUBCOMMITTEE ON LAKE MEAD, 1949, Lake Mead density currents investigations, 1937–40, v. 1, 2: 1940–46, v. 3: Washington, DC, U.S. Bureau of Reclamation, 3 vols., 904 p.

NATLAND, M. L., and KUENEN, PH. H., 1951, Sedimentary history of the Ventura Basin, California, and the action of turbidity currents, p. 76–107 *in* Hough, J. L., ed., Turbidity currents and the transportation of coarse sediments to deep water: Tulsa, OK, Society of Economic Paleontologists and Mineralogists Special Publication 2, 107 p.

NAYLOR, M. A., 1980, Origin of inverse grading in muddy debris flow (*sic*) deposits—a review: Journal of Sedimentary Petrology, v. 50, p. 1111–1116.

NEDERLOFF, M. H., 1959, Structure and sedimentology of the Upper Carboniferous of the upper Pisuerga valleys, Cantabrian Mountains, Spain: Leidse Geologisch Meded., v. 24, p. 603–703.

NEEV, DAVID; and EMERY, K. O., 1967, The Dead Sea—depositional processes and environments of evaporites: Israel Geological Survey Bulletin 41, 147 p.

NELSEN, J. E., JR.; and GINSBURG, R. N., 1986, Calcium carbonate (*sic*) production by epibionts on *Thalassia* in Florida Bay: Journal of Sedimentary Petrology, v. 56, p. 622–628.

NELSON, C. H., and NILSEN, T. H., 1984, Modern (*sic*) and ancient deep-sea fan sediments: Tulsa, OK, Society of Economic Paleontologists and Mineralogists Short Course Notes 14, 404 p.

NELSON, C. M., 1985, Facies in stratigraphy: from "terrains" to "terranes": Journal of Geological Education, v. 33, p. 175–187.

NELSON, C. S., 1988, Non-tropical shelf carbonates—modern and ancient: Sedimentary Geology, v. 60 (special issue).

NEWELL, N. D., 1967, Paraconformities, p. 349–367 *in* Teichert, Curt; and Yochelson, E. L., eds., Essays in paleontology and stratigraphy: Lawrence, KS, University of Kansas Press, 626 p.

NEWELL, N. D.; IMBRIE, J.; PURDY, E. G.; and Thurber, D. T., 1959, Organism communities and bottom facies, Great Bahama Bank: American Museum of Natural History Bulletin, v. 117, p. 117–228.

NEWTON, R. S., 1968, Internal structure of wave-formed ripple marks in the nearshore zone: Sedimentology, v. 11, p. 275–292.

NEWTON, R. S.; SEIBOLD, EUGEN; and WERNER, F., 1973, Facies distribution (*sic*) patterns on the Spanish Sahara continental shelf mapped with side-scan sonar: Meteor. Forsch. Engl., C15, p. 55–77.

NICHOLS, M. M.; and ALLEN, G., 1981, Sedimentary processes in coastal lagoons, p. 27–80 *in* Coastal lagoon research, present and future: UNESCO Technical Papers in Marine Science 33.

NICHOLS, M. M., and BIGGS, R. B., 1985, Estuaries, p. 77–186 *in* Davis, R. A., Jr., ed., Coastal sedimentary environments, 2nd ed., New York, Springer-Verlag, 420 p.

NIEDORODA, A. W., SWIFT, D. J. P., and THORNE, J. A., 1989, Modeling shelf storm beds: controls of bed thickness and bedding sequence, p. 15–39 *in* Morton, R. A.; and Nummedal, Dag, eds., Shelf sedimentation, shelf sequences, and related hydrocarbon accumulation: Society of Economic Paleontologists and Mineralogists, Gulf Coast Section, Annual Research Conference, 7th, 1 April 1989, Proceedings, 211 p.

NIELSEN, L. H.; JOHANNESSEN, P. N.; and SURLYK, FINN, 1988, A late Pleistocene coarse-grained (*sic*) spit-platform sequence in northern Jylland, Denmark: Sedimentology, v. 35, p. 915–937.

NIILER, P. P., 1975, A report on the continental shelf (*sic*) circulation and coastal upwelling: Reviews of Geophysics and Space Physics, v. 13, p. 609–614.

NILSEN, T. H., 1980, Modern (*sic*) and ancient submarine fans: discussion of papers by R. G. Walker and W. R. Normark: American Association of Petroleum Geologists Bulletin, v. 64, p. 1094–1101.

NILSEN, T. H., 1982, Alluvial fan (*sic*) deposits, p. 49–86 *in* Scholle, P. A.; and Spearing, D., eds., Sandstone depositional environments: American Association of Petroleum Geologists Memoir 31, 410 p.

NILSEN, T. H., 1985, Modern and ancient alluvial fan (*sic*) deposits: New York, Van Nostrand Reinhold Company, 372 p.

NINKOVICH, DRAGISLAV; and HEEZEN, B. C., 1965, Santorini tephra, p. 413–452 *in* Whittard, W. F.; and Bradshaw, R., eds., Submarine geology and geophysics; Colston Research Society Symposium, 17th, University of Bristol, 5–9 April 1965, Proceedings: London, Butterworths, 464 p.

NIO, S. D., and NELSON, C. H., 1982, The North Sea and northeastern Bering Sea: a comparative study of the occurrence and geometry of sand bodies of two shallow epicontinental shelves, p. 105–113 *in* Nelson, C. H., and others, eds., The northeastern Bering Shelf: new perspectives of epicontinental shelf processes and depositional products: Geologie en Mijnbouw, v. 61 (special issue).

NIO, S. D., and YANG, C. S., 1991, Sea-level fluctuations and the geometric variability of tide-dominated sandbodies: Sedimentary Geology, v. 70, p. 161–193.

NISSENBAUM, A., ed., 1980, Hypersaline brines and evaporitic environments: Amsterdam, Elsevier Scientific Publishing Company, 270 p.

NITTROUER, C. A., ed., 1981, Sedimentary dynamics of continental shelves: Developments in sedimentology 32: Amsterdam, Elsevier Scientific Publishing Company, 449 p.

NITTROUER, C. A., KUEHL, S. A., DE MASTER, D. J., and KOWSMANN, R. O., 1986, The deltaic nature of Amazon shelf sedimentation: Geological Society of America Bulletin, v. 97, p. 444–458.

NITTROUER, C. A., and STERNBERG, R. W., 1981, The formation of sedimentary strata in an allochthonous shelf environment: the Washington continental shelf, p. 201–232 *in* Nittrouer, C. A., ed., Sedimentary dynamics of continental shelves: Developments in sedimentology 32: Amsterdam, Elsevier Scientific Publishing Company, 449 p.

NOBLE, J. P. A., and VAN STEMPVOORT, D. R., 1989, Early burial quartz authigenesis in Silurian platform carbonates, New Brunswick, Canada: Journal of Sedimentary Petrology, v. 59, p. 65–76.

NOE-NYGAARD, NANNA; and SURLYK, FINN, 1988, Washover fan and brackish bay sedimentation in the Berriasian-Valanginian of Bornholm, Denmark: Sedimentology, v. 35, p. 197–217.

NORMARK, W. R., 1978, Fan-valleys, channels and depositional lobes on modern submarine fans: characteristics for the recognition of sandy turbidite environments: American Association of Petroleum Geologists Bulletin, v. 61, p. 912–931.

NORMARK, W. R., 1989, Observed parameters for turbidity-current flow in channels, Reserve Fan, Lake Superior: Journal of Sedimentary Petrology, v. 59, p. 423–431.

NORTH AMERICAN COMMISSION ON STRATIGRAPHIC NOMENCLATURE, 1983, North American Stratigraphic Code: American Association of Petroleum Geologists Bulletin, v. 67, p. 841–875.

NOTHOLT, A. J. G., 1980, Economic phosphatic sediments: mode of occurrence and stratigraphical distribution: Geological Society of London, Journal, v. 137, p. 793–805.

NOTHOLT, A. J. G.; and JARVIS, I., 1989, A decade of phosphorite research and development: Agid News, no. 57, p. 21–23.

NOTHOLT, A. J. G.; and JARVIS, I., eds., 1990, Phosphorite research and development: Geological Society (London) Special Publication 52, 330 p.

NOTHOLT, A. J. G., SHELDON, R. P., and DAVIDSON, D. F., 1989, Phosphate deposits of the world, vol. 2, Phosphate rock resources: International Geological Correlation Programme Project 156: Phosphorites: Cambridge, Cambridge University Press, 566 p.

NUMMEDAL, DAG; and FISCHER, I., 1978, Process-response models for depositional shorelines: the German and Georgia bights: American Society of Civil Engineers Proceedings, 16th Coastal Engineering Conference, p. 1215–1231.

NUMMEDAL, DAG; PILKEY, O. H.; and HOWARD, J. D., eds., 1987, Sea-level fluctuation and coastal evolution: Tulsa, OK, Society of Economic Paleontologists and Mineralogists Special Publication 41, 267 p.

NUMMEDAL, DAG; and SWIFT, D. J. P., 1987, Transgressive stratigraphy at sequence bounding (*sic*) unconformities—some principles derived from Holocene (*sic*) and Cretaceous examples, p. 241–260 *in* Nummedal, Dag; Pilkey, O. H.; and Howard, J. D., eds., Sea-level fluctuation and coastal evolution: Tulsa, OK, Society of Economic Paleontologists and Mineralogists Special Publication 41, 267 p.

O'BRIEN, M. P., 1969, Dynamics of tidal inlets, p. 397–406 *in* Ayala-Castanares, Augustin; and Phleger, F. B., eds., Lagunas costeras, un simposio: Simposio Internacional sobre lagunas costeras, Mexico City, Mexico, 28–30 Noviembre 1967, memoria: Mexico Universidad Nacional Autonoma (WMAS-UNESCO), 686 p.

O'BRIEN, N. R., 1987, The effects of bioturbation on the fabric of shale: Journal of Sedimentary Petrology, v. 57, p. 449–455.

O'BRIEN, N. R., and SLATT, R. M., 1990, Argillaceous rock atlas: New York, Springer-Verlag, 141 p.

ODIN, G. S., 1982, Numerical dating in stratigraphy. Part I: New York, John Wiley & Sons, 630 p.

OELE, E.; SCHUTTENHELM, R. T. E.; and WIGGERS, A. J., eds., 1979, The Quaternary history of the North Sea: Uppsala, Sweden, Almquist and Wilksell.

OERTEL, G. F., 1979, Barrier island (*sic*) development during the Holocene recession, southeast United States, p. 273–290 *in* Leatherman, S. P., ed., 1979, Barrier islands from the Gulf of St. Lawrence to the Gulf of Mexico: New York, Academic Press, 325 p.

OERTEL, G. F., and LEATHERMAN, S. P., eds., 1985, Barrier islands: Marine Geology, v. 63 (special issue).

OESCHGER, H.; and LANGWAY, C. C., JR., eds, 1989, The environmental record in glaciers and ice sheets. Report of the Dahlem workshop on the environmental record in glaciers and ice sheets: Berlin, 13–18 March 1988: New York, Wiley-Interscience, 401 p.

OGREN, D. E., and WAAG, C. J., 1986, Orientation of cobble (*sic*) and boulder beach clasts: Sedimentary Geology, v. 47, p. 69–76.

OKAMOTO, G.; TAKESHI, O.; and KATSUMI, G., 1957, Properties of silica in water: Geochimica et Cosmochimica Acta, v. 12, p. 123–132.

OLAUSSON, E., 1980, The carbon dioxide-calcium carbonate system in estuaries, p. 298–305 *in* Olausson, E.; and Cato, I., eds., Chemistry and biogeochemistry of estuaries: New York, John Wiley & Sons, 452 p.

OLSEN, C. R.; SIMPSON, H. J.; PENG, T.-H.; BOPP, R. F.; and TRIER, R. M., 1981, Sediment mixing and accumulation rate (*sic*) effects on radionuclide depth profiles in Hudson estuary sediments: Journal of Geophysical Research, v. 86, p. 11020–11028.

OLSEN, P. E., 1986, A 40-million year (*sic*) lake record of orbital climatic forcing: Science, v. 234, p. 842–847.

OLSEN, P. E., 1990, Researchers invited to join drilling project: Geotimes, v. 35, no. 4, p. 7.

OOMKENS, E., 1970, Depositional sequences and sand distribution in the postglacial Rhône delta complex, p. 198–212 *in* Morgan, J. P., and Shaver, R. H., eds., Deltaic sedimentation modern and ancient: Tulsa, OK, Society of Economic Paleontologists and Mineralogists Special Publication 15, 312 p.

ORFORD, J. D., 1975, Discrimination of particle zonation on a pebble beach: Sedimentology, v. 22, p. 441–463.

ORFORD, J. D., 1977, A proposed mechanism for beach sedimentation: Earth Surface Processes and Landforms, v. 2, p. 381–400.

ORFORD, J. D., 1978 ms., Methods of identifying and interpreting the dynamics of littoral zone (*sic*) facies using particle size and form (*sic*), with special reference to beach gravel (*sic*) sedimentation: Reading, University of Reading, unpublished Ph.D. dissertation, 422 p.

ORFORD, J. D., and CARTER, R. W. G., 1982, Crestal overtop and washover sedimentation on a fringing sandy gravel barrier coast, Carnsore Point, southwest Ireland: Journal of Sedimentary Petrology, v. 52, p. 265–278.

ORNI, E.; and EFRAT, E., 1964, Geography of Israel: Jerusalem, Israel Program for Scientific Translations, 335 p.

OSBORN, N. I., CIESIELSKI, P. F., and LEDBETTER, M. T., 1983, Disconformities and paleoceanography in the southeast Indian Ocean during the past 5.4 million years: Geological Society of America Bulletin, v. 94, p. 1345–1358.

O'SULLIVAN, P. E., 1983, Annually laminated lake sediments and the study of Quaternary environmental changes—a review: Quaternary Science Reviews, v. 1, p. 245–313.

OTVOS, E. G., 1981, Barrier island (*sic*) formation through nearshore aggradation—stratigraphic (*sic*) and field evidence: Marine Geology, v. 43, p. 195–243.

OWEN, M. R., and CAROZZI, A. V., 1986, Southern provenance of upper Jackfork Sandstone, southern Ouachita Mountains: cathodoluminescence petrology: Geological Society of America Bulletin, v. 97, p. 110–115.

OWENS, E. H., 1977, Temporal variations in beach (*sic*) and nearshore dynamics: Journal of Sedimentary Petrology, v. 47, p. 168–190.

OWENS, E. H., and FROBEL, D. H., 1977, Ridge and runnel (*sic*) in the Magdalen Islands, Quebec: Journal of Sedimentary Petrology, v. 47, p. 191–198.

OXBURGH, E. R., 1974, The plain man's guide to plate tectonics: Geologists Association (London) Proceedings, v. 85, p. 299–357.

PALMER, A. A., 1986, Cenozoic radiolarians as indicators of neritic (*sic*) versus oceanic conditions in continental-margin deposits: U.S. mid-Atlantic coastal plain: Palaios, v. 1, p. 122–132.

PALMER, A. R., 1963, Biomere—a new kind of biostratigraphic unit: Journal of Paleontology, v. 39, p. 149–153.

PALMER, A. R., 1982, Biomere boundaries: a possible test for extraterrestrial perturbation of the biosphere, p. 469–475 SILVER, L. T., and SCHULTZ, P. H., eds., Geological implications of impacts of large asteroids and comets on the Earth: *in* Geological Society of America Special Paper 190, 528 p.

PALUSKA, A.; and DEGENS, E. T., 1979, Climatic (*sic*) and tectonic events controlling the Quaternary in the Black Sea region: Geologische Rundschau, v. 68, p. 284–301.

PANAGEOTOU, WILLIAM; and LEATHERMAN, S. P., 1986, Holocene-Pleistocene stratigraphy of the inner shelf off Fire Island, New York: implications for barrier-island migration: Journal of Sedimentary Petrology, v. 56, p. 528–537.

PANUZIO, F. L., 1968, The Atlantic coast of Long Island, p. 1222–1241 *in* Conference on Coastal Engineering, 11th, London, September 1968, Proceedings, v. 2, p. 745–1585.

PARCHURE, T. M., and MEHTA, A. J., 1985, Erosion of soft cohesive sediment deposits: Journal of Hydraulic Engineering, v. 111, p. 1308–1326.

PARKER, G., 1982, Conditions for the ignition of catastrophically erosive turbidity currents: Marine Geology, v. 46, p. 307–327.

PARKER, G.; LANFREDI, N.; and SWIFT, D. J. P., 1982, Substrate response to flow in a southern hemisphere (*sic*) ridge field, Argentine inner shelf: Sedimentary Geology, v. 33, p. 195–216.

PARKINSON, N.; and SUMMERHAYES, C., 1985, Synchronous global sequence boundaries: American Association of Petroleum Geologists Bulletin, v. 69, p. 685–687.

PARRISH, J. T., 1982, Upwelling and petroleum source rocks, with reference to the Paleozoic: American Association of Petroleum Geologists Bulletin, v. 66, p. 750–774.

PARRISH, J. T., and CURTIS, R. L., 1982, Atmospheric circulation, upwelling, and organic-rich rocks in the Mesozoic and Cenozoic Eras: Palaeogeography, Palaeoclimatology, and Palaeoecology, v. 40, p. 31–66.

PARRISH, J. T., GAYNOR, G. C., and SWIFT, D. J. P., 1984, Circulation in the Cretaceous Western Interior seaway of North America, a review, p. 221–231 *in* Stott, D. F., and Glass, D. J., eds., The Mesozoic of middle North America: Canadian Society of Petroleum Geologists Memoir 9, 573 p.

PARRISH, J. T., ZIEGLER, A. M., and SCOTESE, C. R., 1982, Rainfall patterns and distribution of coals and evaporites in the Mesozoic and Cenozoic: Palaeogeography, Palaeoclimatology, and Palaeoecology, v. 40, p. 67–101.

PARSONS, B., 1982, Causes and consequences of the relation (*sic*) between area and age of the ocean floor: Journal of Geophysical Research, v. 87, p. 289–302.

PATTERSON, R. S., and KINSMAN, D. J. J., 1981, Hydrologic framework of a sabkha along Arabian Gulf: American Association of Petroleum Geologists Bulletin, v. 65, p. 1457–1475.

PAVICH, M. J., and OBERMEIER, S. F., 1985, Saprolite formation beneath Coastal Plain (*sic*) sediments near Washington, DC: Geological Society of America Bulletin, v. 96, p. 886–900.

PAYTON, C. E., ed., 1977, Seismic stratigraphy—applications to hydrocarbon exploration: American Association of Petroleum Geologists Memoir 26, 516 p.

PEARSON, M. J., 1985, Some chemical aspects of diagenetic concretions from the Westphalian of Yorkshire, England: Chemical Geology, v. 31, p. 225–244.

PENLAND, SHEA; BOYD, R.; and SUTER, J. R., 1988, Transgressive depositional systems of the Mississippi Delta plain: a model for barrier shoreline and shelf sand development: Journal of Sedimentary Petrology, v. 58, p. 932–949.

PENLAND, SHEA; and SUTER, J. R., 1985, Low profile (*sic*) barrier islands. Overwash and breaching in the Gulf of Mexico: Proceedings of the 19th Coastal Engineering Conference, v. III, Ch. 157, p. 2339–2345.

PENLAND, SHEA; SUTER, J. R.; and BOYD, R., 1985, Barrier island (*sic*) arcs along abandoned Mississippi River deltas: Marine Geology, v. 63, p. 197–233.

PENTECOST, A.; and RIDING, R., 1986, Calcification in cyanobacteria, p. 73–90 *in* Leadbeater, B. S. C.; and Riding, R., eds., Biomineralization of lower plants and animals (*sic*): symposium held at University of Birmingham, April 1985, Proceedings: Systematics Association, Special Volume 30: Oxford, Clarendon Press; New York, Oxford University Press, 401 p.

PERKINS, B. F., ed., 1971, Trace fossils, a field guide to selected localities in Pennsylvanian, Permian, Cretaceous, and Tertiary rocks of Texas and related papers: Society of Economic Paleontologists and Mineralogists, Field Trip: Baton Rouge, Louisiana State University, Publication 71-1, 148 p.

PESTANA, HAROLD, 1985, Carbonate sediment (*sic*) production by *Sargassum* epibionts: Journal of Sedimentary Petrology, v. 55, p. 184–186.

PETERSON, CURT; SCHEIDEGGER, KENNETH; KOMAR, PAUL; and NIEM, WENDY, 1984, Sediment composition and hydrography in 6 high-gradient estuaries of the northwestern United States: Journal of Sedimentary Petrology, v. 54, p. 86–97.

PETERSON, J. A., and HITE, R. J., 1969, Pennsylvanian evaporate-carbonate cycles and their relation to petroleum occurrence, Southern Rocky Mountains: American Association of Petroleum Geologists Bulletin, v. 53, p. 884–908.

PETHICK, J. S., 1980, Velocity surges and asymmetry in tidal channels: Estuarine and Coastal Marine Science, v. 11, p. 331–345.

PETHICK, J. S., 1981, Long-term accretion rates on tidal salt marshes: Journal of Sedimentary Petrology, v. 51, p. 571–579.

PETHICK, J. S., 1984, An introduction to coastal geomorphology: London, Edward Arnold, Ltd.

PETTERSSON, OTTO, 1912, The connection between hydrographical and meteorological phenomena: Royal Meteorological Society Quarterly Journal, v. 38, p. 173–191.

PETTERSSON, OTTO, 1914a, Climatic variations in historic and prehistoric time: Svenska Hydrogr. Biol. Komm., Skriften, No. 5, 26 p.

PETTERSSON, OTTO, 1914b, On the occurrence of lunar periods in solar activity and the climate of the earth. A study in geophysics and cosmic physics: Svenska Hydrogr. Biol. Komm., Skriften.

PETTERSSON, OTTO, 1915, Long periodical (*sic*) variations of the tide-generating force: Conseil Permanente International pour l'Exploration de la Mer (Copenhagen), Pub. Circ. No. 65, p. 2–23.

PETTERSSON, OTTO, 1930, The tidal force. A study in geophysics: Geografiska Annaler, v. 18, p. 261–322.

PETTIJOHN, F. J., 1960, The term graywacke: Journal of Sedimentary Petrology, v. 30, p. 627.

PETTIJOHN, F. J., 1975, Sedimentary rocks, 3rd ed.: New York, Harper & Row, 628 p.

PETTIJOHN, F. J.; POTTER, P. E.; and SIEVER, RAYMOND, 1972, Sand and sandstone: New York, Springer-Verlag, 618 p.

PETTIJOHN, F. J.; POTTER, P. E.; and SIEVER, RAYMOND, 1987, Sand and sandstone, 2nd ed.: New York, Springer-Verlag, 553 p.

PEVEAR, D. R., and MUMPTON, F. A., 1989, Quantitative mineral analysis of clay: Clay Minerals Society Workshop Lectures, vol. 1, 171 p.

PHILLIPS, JOHN, 1835, Illustrations of the geology of Yorkshire, II. The Mountain limestone district: London, John Murray, 253 p.

PHILLIPS, S. E., and SELF, P. G., 1987, Morphology, crystallography, and origin of needle-fibre calcite in Quaternary pedogenic calcretes of South Australia: Australian Journal of Soil Research, v. 25, p. 249–264.

PHILLIPS, SANDRA, 1987 ms., Shelf sedimentation and depositional sequence (*sic*) stratigraphy of the Upper Cretaceous Woodbine-Eagle Ford groups, east Texas: Ithaca, NY, Cornell University, Department of Geological Sciences, Ph.D. dissertation, 232 p.

PHLEGER, F. B., 1960a, Ecology and distribution of Recent Foraminifera: Baltimore, MD, The Johns Hopkins University Press, 297 p.

PHLEGER, F. B., 1960b, Sedimentary patterns of microfaunas in northern Gulf of Mexico, p. 267–301 *in* Shepard, F. P., Phleger, F. B., and van Andel, Tj. H., eds., Recent sediments, northwest Gulf of Mexico. A symposium summarizing the results of work carried out in Project 51 of the American Petroleum Institute 1951–1958: Tulsa, OK, American Association of Petroleum Geologists, 394 p.

PHLEGER, F. B., 1969, Some general features of coastal lagoons, p. 5–25 *in* Ayala-Castanares, Augustin; and Phleger, F. B., eds., Lagunas costeras, un simposio: Simposio Internacional sobre lagunas costeras, Mexico City, Mexico, 28–30 Noviembre 1967, memoria: Mexico Universidad Nacional Autonoma (WMAS-UNESCO), 686 p.

PHLEGER, F. B.; and AYALA-CASTANARES, AUGUSTIN, 1969, Marine geology of Topolobampo lagoon, Sinaloa, Mexico, p. 101–136 *in* Ayala-Castanares, Augustin; and Phleger, F. B., eds., Lagunas costeras, un simposio: Simposio Internacional sobre lagunas costeras, Mexico City, Mexico, 28–30 Noviembre 1967, memoria: Mexico Universidad Nacional Autonoma (WMAS-UNESCO), 686 p.

PICARD, M. D., 1955, Subsurface stratigraphy and lithology of Green River formation in Uinta Basin, Utah: American Association of Petroleum Geologists Bulletin, v. 39, p. 75–102.

PICKARD, G. L., 1956, Physical features of British Columbia inlets: Royal Society of Canada Transactions, series 3, v. 50, p. 47–58.

PICKERING, K. T., 1979, Possible retrogressive flow slide deposits from the Kongsfjord Formation: a Precambrian submarine fan, Finnmark, N. Norway: Sedimentology, v. 26, p. 295–305.

PICKERING, K. T., 1982, The shape of deep-water siliciclastic systems—a discussion: Geomarine Letters, v. 2, p. 41–46.

PICKERING, K. T., and HISCOTT, R. N., 1985, Contained (reflected) turbidity currents from the Middle Ordovician Cloridorme Formation, Quebec, Canada: an alternative to the antidune hypothesis: Sedimentology, v. 32, p. 373–394.

PIELOU, E. C., 1979, Biogeography: New York, John Wiley & Sons, 351 p.

PIERSON, T. C., 1981, Dominant particle support (*sic*) mechanisms in debris flows at Mt. Thomas, New Zealand, and implications for flow mobility: Sedimentology, v. 28, p. 49–60.

PIERSON, T. C., 1985, Initiation and flow behavior of the 1980 Pine Creek and Muddy River lahars, Mount St. Helens, Washington: Geological Society of America Bulletin, v. 96, p. 1056–1069.

PILKEY, O. H., LOCKER, S. D., and CLEARY, W. J., 1980, Comparison of sand-layer geometry on flat floors of ten modern depositional basins: American Association of Petroleum Geologists Bulletin, v. 64, p. 841–856.

PIPER, D. J. W., 1978, Turbidite muds and silts on deep sea (*sic*) fans and abyssal plains, p. 164–175 *in* Stanley, D. J.; and Kelling, G., eds., Sedimentation in submarine canyons, fans and trenches: Stroudsburg, PA, Dowden, Hutchinson, and Ross, 395 p.

PIPER, D. J. W., and NORMARK, W. R., 1983, Turbidite depositional patterns and flow characteristics, Navy submarine fan, California borderlands: Sedimentology, v. 30, p. 681–694.

PIRAZZOLI, P. A.; MONTAGGIONI, L. F.; VERGNAUD-GRAZZINI, C.; and SALIEGE, J. F., 1987, Late Holocene sea levels and coral reef (*sic*) development in Vahitahi Atoll, eastern Tuamotu Islands, Pacific Ocean: Marine Geology, v. 76, p. 105–116.

PIRSON, S. J., 1977, Geologic well log (*sic*) analysis, 2nd ed., Houston, TX, Gulf Publishing Company, 475 p. (3rd ed., 1983.)

PITTMAN, W. C., III, 1978, Relationship between eustacy (*sic*) and stratigraphic sequences of passive margins: Geological Society of America Bulletin, v. 89, p. 1389–1403.

PITTMAN, E. D., 1970, Plagioclase feldspar as an indicator of provenance in sedimentary rocks: Journal of Sedimentary Petrology, v. 40, p. 591–598.

PITTMAN, E. D., 1974, Porosity (*sic*) and permeability changes during diagenesis of Pleistocene corals, Barbados, West Indies: Geological Society of America Bulletin, v. 85, p. 1811–1820.

PITTMAN, E. D., 1982, Effect of fault-related granulation on porosity and permeability of quartz sandstones, Simpson Group (Ordovician), Oklahoma: American Association of Petroleum Geologists Bulletin, v. 65, p. 2381–2387.

PITTMAN, E. D., 1988, Diagenesis of Terry Sandstone (Upper Cretaceous), Spindle Field, Colorado: Journal of Sedimentary Petrology, v. 58, p. 785–800.

POAG, C. W., 1979, Stratigraphy and depositional environments of Baltimore Canyon Trough: American Association of Petroleum Geologists Bulletin, v. 63, p. 1452–1467.

POAG, C. W., ed., 1985a, Geological evolution of the United States Atlantic margin: New York, Van Nostrand Reinhold Company, 383 p.

POAG, C. W., 1985b, Depositional history and stratigraphic reference section for central Baltimore Canyon Trough, p. 217–264 *in* Poag, C. W., ed., Geologic evolution of the United States Atlantic margin: New York, Van Nostrand Reinhold Company, 383 p.

POREBSKI, S. J., 1981, Swiebodzice Succession (Upper Devonian-Lower Carboniferous; western Sudeten): a prograding, mass flow dominated (*sic*) fan-delta complex: Geologia Sudetica, v. 16, p. 102–192.

POSAMENTIER, H. W., and VAIL, P. R., 1988, Eustatic controls on clastic deposition 2—sequence and systems tract (*sic*) models, p. 125–154 *in* Wilgus, C. K., Hastings, B. S., Kendall, C. G. St. C., Posamentier, H. W., Ross, C. A., and Van Wagoner, J. C., eds., Sea-level changes: an integrated approach: Tulsa, OK, Society of Economic Paleontologists and Mineralogists Special Publication 42, 407 p.

POSAMENTIER, H. W., JERVEY, M. T., and VAIL, P. R., 1988, Eustatic controls on clastic deposition 1—conceptual framework, p. 109–124 *in* Wilgus, C. K., Hastings, B. S., Kendall, C. G. St. C., Posamentier, H. W., Ross, C. A., and Van Wagoner, J. C., eds., Sea-level changes: an integrated approach: Tulsa, OK, Society of Economic Paleontologists and Mineralogists Special Publication 42, 407 p.

POSTMA, GEORGE, 1983, Water escape (*sic*) structures in the context of a depositional model of a mass flow-dominated conglomeratic fan-delta (Abrioja Formation, Pliocene, Almeria Basin, SE Spain): Sedimentology, v. 30, p. 91–103.

POSTMA, GEORGE, 1984, Slumps and their deposits in fan delta front (*sic*) and slope: Geology, v. 12, p. 27–30.

POSTMA, GEORGE; and NEMEC, WOJCIECH, 1990, Regressive (*sic*) and transgressive sequences in a raised Holocene gravelly beach, southwestern Crete: Sedimentology, v. 37, p. 907–920.

POSTMA, GEORGE; and ROEP, T. B., 1985, Resedimented conglomerates in the bottomsets of Gilbert-type gravel deltas: Journal of Sedimentary Petrology, v. 55, p. 874–885.

POTTER, P. E., MAYNARD, J. B., and PRYOR, W. A., 1980, Sedimentology of shale: New York, Springer-Verlag, 306 p.

POTTER, P. E., and PETTIJOHN, F. J., 1963, Paleocurrents and basin analysis: Berlin, Springer-Verlag, 296 p.

POWERS, M. C., 1953, A new roundness scale for sedimentary particles: Journal of Sedimentary Petrology, v. 23, p. 117–119.

PRATT, B. R., 1982, Stromatolite decline—a reconsideration: Geology, v. 10, p. 512–515.

PRATT, B. R., and JAMES, N. P., 1986, The St George Group (Lower Ordovician) of western Newfoundland: tidal flat (*sic*) island model for carbonate sedimentation in shallow epeiric seas: Sedimentology, v. 33, p. 313–343.

PRATT, L. M., KAUFFMAN, E. G., and ZELT, F. B., eds., 1985, Fine-grained deposits and biofacies of the Cretaceous Western Interior Seaway: evidence of cyclic sedimentary processes: Tulsa, OK, Society of Economic Paleontologists and Mineralogists, Field Trip Guidebook 4, 249 p.

PRAY, L. C., and WRAY, J. L., 1963, Porous algal facies (Pennsylvanian), Honaker Trail San Juan Canyon, Utah, p. 204–234 *in* Bass, R. O., and Sharps, S. L., eds., Shelf Carbonates of the Paradox Basin, a Symposium: Four Corners Geological Society, Field Conference, Fourth, 12–16 June 1963: Durango, CO, Petroleum Information, 273 p.

PRIOR, D. B., and COLEMAN, J. M., 1978, Disintegrating retrogressive landslides on very-low-angle subaqueous slopes, Mississippi Delta: Marine Geotechnology, v. 3, p. 37–60.

PRIOR, D. B., and COLEMAN, J. M., 1982, Active slides and flows in underconsolidated marine sediments on the slopes of the Mississippi delta, p. 121–149 *in* Saxov, S.; and Nieuwenhuis, J. K., eds., Marine slides and other mass movements: New York, Plenum Press, 353 p.

PROTHERO, D. R., 1990, Interpreting the stratigraphic record: New York, W. H. Freeman and Company, 410 p.

PULLEN, S. E.; HUNTYER, J. A.; and GILBERT, R., 1983, A shallow seismic survey on the intertidal flats at Pangnirtung, Baffin Island, N.W.T.: Geological Survey of Canada, Paper 83-1B, p. 273–277.

PURDY, E. G., 1963a, Recent calcium carbonate (*sic*) facies of the Great Bahama Bank. 1. Petrography and reaction groups: Journal of Geology, v. 71, p. 334–355.

PURDY, E. G., 1963b, Recent calcium carbonate (*sic*) facies of the Great Bahama Bank. 2. Sedimentary facies: Journal of Geology, v. 71, p. 472–497.

PURDY, E. G., 1974, Reef configurations: cause and effect, p. 9–76 *in* Laporte, L. F., ed., Reefs in time and space (selected examples from the Recent and Ancient): Tulsa, OK, Society of Economic Paleontologists and Mineralogists Special Publication 18, 256 p.

PURSER, B. H., 1985, Coastal evaporite systems, p. 72–102 *in* Friedman, G. M., and Krumbein, W. E., eds., Hypersaline ecosystems. The Gavish Sabkha. Ecological Studies 53: New York, Springer-Verlag, 484 p.

PURSER, B. H.; and EVANS, G., 1973, Regional sedimentation along the Trucial Coast, SE Persian Gulf, p. 211–231 *in* Purser, B. H., ed., The Persian Gulf, Holocene carbonate sedimentation and diagenesis in a shallow epicontinental sea: New York, Springer-Verlag, 471 p.

PYE, KENNETH, 1990, Aeolian sand and sand dunes: Harper-Collins Academic, 384 p.

QUINLAN, G. M.; and BEAUMONT, C., 1984, Appalachian overthrusting, lithospheric flexure, and the Paleozoic stratigraphy (*sic*) of the eastern interior of North America: Canadian Journal of Earth Sciences, v. 21, p. 973–996.

RAASCH, G. O., 1958, Baraboo monadnock and paleowind direction: Alberta Society of Petroleum Geologists Journal, v. 6, p. 183–187.

RAHMANI, R. A., and FLORES, R. M., eds., 1984, Sedimentology of coal and coal-bearing sequences: International Association of Sedimentologists Special Publication 7: Oxford, Blackwell Scientific Publications, 412 p.

RAISWELL, R., 1988, Chemical model for the origin of minor limestone-shale cycles by anaerobic methane oxidation: Geology, v. 16, p. 641–644.

RAITT, R. W., 1954, Seismic-refraction studies of Bikini and Kwajalein Atolls: U.S. Geological Survey Professional Paper 260-K, p. 506–527.

RAMPINO, M. R., and SANDERS, J. E., 1980, Holocene transgression in south-central Long Island, New York: Journal of Sedimentary Petrology, v. 50, p. 1063–1080.

RAMPINO, M. R., and SANDERS, J. E., 1981, Evolution of the barrier islands of southern Long Island, New York: Sedimentology, v. 28, p. 37–47.

RAMPINO, M. R., and SANDERS, J. E., 1982, Holocene transgression in south-central Long Island, New York—reply: Journal of Sedimentary Petrology, v. 52, p. 1020–1025.

RAMPINO, M. R., and SANDERS, J. E., 1983, Barrier island evolution in response to sea level rise—reply: Journal of Sedimentary Petrology, v. 53, p. 1031–1033.

RAMPINO, M. R., SANDERS, J. E., NEWMAN, W. S., and KONIGSSON, L. K., eds., 1987, Climate: history, periodicity, and predictability: New York, Van Nostrand Reinhold Company, 588 p.

RAUP, D. M., 1986, Biological extinctions in Earth history: Science, v. 231, p. 1528–1533.

RAUP, D. M., and BOYAJIAN, G. E., 1988, Generic extinction in the fossil record: Paleobiology, v. 14, p. 116.

RAUP, D. M., and SEPKOSKI, J. J., JR., 1984, Periodicity of extinction in the geologic past: U.S. National Academy of Sciences Proceedings, v. 81, p. 801–805.

RAUP, D. M., and SEPKOSKI, J. J., JR., 1986, Periodic extinction of families and genera: Science, v. 231, p. 833–836.

RAUP, D. M., SEPKOSKI, J. J., JR., STIGLER, S. M., and WAGNER, M. J., 1988, Testing for periodicity of extinction, discussion and reply: Science, v. 241, p. 94–99.

RAUP, O. B., 1970, Brine mixing: an additional mechanism for formation of basin evaporites: American Association of Petroleum Geologists Bulletin, v. 54, p. 2246–2259.

REA, D. K.; LEINEN, M.; and JANECEK, T. R., 1985, Geologic approach to the long-term history of atmospheric circulation: Science, v. 227, p. 721–725.

READ, J. F., 1982, Carbonate platforms of passive (extensional) continental margins: types, character (sic) and evolution: Tectonophysics, v. 81, p. 195–212.

READ, J. F., and GOLDHAMMER, R. K., 1988, Use of Fischer plots to define third-order sea-level curves in Ordovician peritidal carbonates: Geology, v. 16, p. 895–899.

READING, H. G., ed., 1986, Sedimentary environments and facies, 2nd ed.: Oxford, Blackwell Scientific Publications, 615 p.

READING, H. G., 1991, The classification of deep-sea depositional systems by sediment calibre and feeder system: Journal of the Geological Society, London, vol. 148, p. 427–430.

REDFIELD, A. C., 1958a, Preludes to the entrapment of organic matter in the sediments of Lake Maracaibo, p. 968–981 in Weeks, L. G., ed., Habitat of oil. A symposium: Tulsa, OK, American Association of Petroleum Geologists, 1384 p.

REDFIELD, A. C., 1958b, The influence of the continental shelf on the tides of the Atlantic coast of the United States: Journal of Marine Research, v. 17, p. 432–448.

REDFIELD, A. C., 1965, Ontogeny of a salt marsh (sic) estuary: Science, v. 147, p. 50–55.

REDFIELD, A. C., KETCHUM, B. H., and RICHARDS, F. A., 1963, The influence of organisms on the composition of sea-water (sic), Chapter 2, p. 26–77 in Hill, M. N., gen. ed., The sea. Ideas and observations on progress in the study of the sea. Volume 2. The composition of sea-water (sic). Comparative and descriptive oceanography: New York, Wiley-Interscience, 554 p.

REID, J. L., 1979, On the contribution of the Mediterranean Sea outflow to the Norwegian-Greenland Sea: Deep-Sea Research, v. 26, p. 1199–1223.

REINECK, H.-E., 1963, Sedimentgefuge im Bereich der sudlichen Nordsee: Senckenbergische natur-

forschende Gesellschaft, Abhandlungen, No. 505, p. 1–138.

REINECK, H.-E., 1970, Marine sandkorper, rezent und fossil: Geologische Rundschau, v. 60, p. 302–321.

REINECK, H.-E., 1984, Aktuogeologie klastischer Sedimente: Frankfurt, Waldemar Klein, 348 p.

REINECK, H.-E.; GUTTMANN, W. F.; and HERTWECK, G., 1967, Das Schlickgebiet sudlich Helgoland als Beispiel rezenter Schelfablagerungen: Senckenbergiana Lethaea, v. 48, p. 219–275.

REINECK, H.-E.; and SINGH, I. B., 1973, Depositional sedimentary environments: New York, Springer-Verlag, 439 p. (2nd ed., 1980, 549 p.)

REINECK, H.-E.; and WUNDERLICH, F., 1968, Classification and origin of flaser (sic) and lenticular bedding: Sedimentology, v. 11, p. 99–104.

REINSON, G. E., 1984, Barrier-island and associated strand-plain systems, p. 119–140 in Walker, R. G., ed., Facies models, 2nd ed.: Geoscience Canada Reprint Series 1, 317 p.

RETALLACK, G. J., 1981, Fossil soils: indicators of ancient terrestrial environments, p. 55–102 in Niklas, K. J., ed., Paleobotany, paleoecology, and evolution, vol. 1: New York, Praeger Publishers, 297 p.

RETALLACK, G. J., 1984, Completeness of the rock (sic) and fossil record: some estimates using fossil soils: Paleobiology, v. 10, p. 59–78.

REYMENT, R. A., 1980, Paleo-oceanology and paleobiogeography of the Cretaceous south Atlantic Ocean: Oceanologica Acta, v. 3, p. 127–133.

RHOADS, D. C., and MORSE, J. W., 1971, Evolutionary (sic) and ecologic significance of oxygen-deficient marine basins: Lethaia, v. 4, p. 413–428.

RHODES, E. G., 1982, Depositional model for a chenier plain, Gulf of Carpenteria, Australia: Sedimentology, v. 29, p. 201–221.

RICCI LUCCHI, F., 1978, Turbidite dispersal in a Miocene deep-sea plain: the Marnoso-Arenacea of the northern Apennines: Geologie en Mijnbouw, v. 57, p. 559–576.

RICCI LUCCHI, F.; COLELLA, A.; ORI, G. G.; OGLIANI, F.; and COLALONGO, M. L., 1981, Pliocene fan deltas of the Intra-Apenninic basin, Bologna, p. 81–162 in Ricci Lucchi, F., ed., International Association of Sedimentologists, 2nd European Regional Meeting, Bologna, Italy, Excursion Guidebook, 342 p.

RICCHI LUCCHI, F.; and VALMORI, E., 1980, Basin-wide turbidites in a Miocene, over-supplied deep-sea plain: a geometrical analysis: Sedimentology, v. 27, p. 241–270.

RICH, J. L., 1950, Flow markings, groovings, and intrastratal crumplings as criteria for recognition of slope deposits with illustrations from Silurian rocks of Wales: American Association of Petroleum Geologists Bulletin, v. 34, p. 717–741.

RICH, J. L., 1951, Three critical environments of deposition and criteria for recognition of rocks deposited in each of them: Geological Society of America Bulletin, v. 62, p. 1–20.

RICHARDS, F. A., 1957, Oxygen in the ocean, p. 185–238 in Hedgpeth, J. W., ed., Treatise on marine ecology and paleoecology: Geological Society of America Memoir 67, v. 1, Ecology, 1296 p.

RICHARDS, F. A., 1975, The Cariaco basin (trench), p. 11–67 in Barnes, Harold, ed., Oceanography and marine biology, an annual review, v. 13: New York, Hafner Press, 465 p.

RICKARD, L. V., 1975, Correlation of the Silurian and Devonian rocks of New York State: New York State Museum and Science Service, Map and Chart Series, no. 24, 16 p.

RICKARD, L. V., and FISHER, D. W., 1973, Middle Ordovician Normanskill Formation, eastern New York: age, stratigraphic (sic), and structural position: American Journal of Science, v. 273, p. 580–590.

RICKARDS, R. B., 1977, Patterns of evolution in the graptolites, p. 333–358 in Hallam, Anthony, ed., Patterns of evolution: Amsterdam, Elsevier Scientific Publishing Company, 591 p.

RICKEN, WERNER, 1986, Diagenetic bedding. Lecture Notes in Earth Sciences: Berlin, Springer-Verlag, 210 p.

RIDER, M. H.; and LAURIER, D., 1979, Sedimentology using a computer treatment of well logs: Transactions, Society of Professional Well Log Analysts, 6th European Symposium, 12 p.

RIDING, ROBERT, 1991, Calcareous algae and stromatolites: Berlin, Springer-Verlag, 571 p.

RIDING, ROBERT; AWRAMIK, S. M.; WINSBOROUGH, B. M.; GRIFFIN, K. M.; and DILL, R. F., 1991, Bahamian giant stromatolites: microbial composition of surface mats: Geological Magazine, v. 128, p. 227–234.

RIECH, V.; and VON RAD, ULRICH, 1979, Silica diagenesis in the Atlantic Ocean: diagenetic potential and transformations, p. 315–340 in Talwani, Manik; Hay, W.; and Ryan, W. B. F., eds., Deep drilling results in the Atlantic margins and paleoenvironment: American Geophysical Union, Maurice Ewing Series 3, 437 p.

RIECK, R. L., and WINTERS, H. A., 1982, Low-altitude organic deposits in Michigan: evidence for pre-Woodfordian Great Lakes and paleosurfaces: Geological Society of America Bulletin, v. 93, p. 725–734.

RIGGS, S. R., 1986, Proterozoic and Cambrian phosphorites—specialist studies: phosphogenesis and its relationship to exploration for Proterozoic and Cambrian phosphorites, in Cook, P. J., and Shergold, J. H., eds., Proterozoic and Cambrian phosphorites: Cambridge, Cambridge University Press, 386 p.

RINDSBERG, A. K., 1990, Ichnological consequences of the 1985 International Code of Zoological Nomenclature: Ichnos, v. 1, p. 59–63.

RINE, J. M., and GINSBURG, R. N., 1985, Depositional facies of a mud shoreface in Suriname, South America—a mud analogue to sandy, shallow-marine deposits: Journal of Sedimentary Petrology, v. 55, p. 633–652.

ROBERTS, W. L.; RAPP, G. R., Jr.; and WEBER, Julius, 1974, Encyclopedia of minerals: New York, Van Nostrand Reinhold Co., 693 p.

ROBERTS, H. H., ADAMS, R. D., and CUNNINGHAM, R. W., 1980, Evolution of sand-dominant subaerial phase, Atchafalaya delta, Louisiana: American Association of Petroleum Geologists Bulletin, v. 64, p. 264–279.

ROBERTS, H. H., and COLEMAN, J. M., 1988, Sedimentation styles and accumulation rates on the Louisiana shelf and slope: stacked condensed (sic) and expanded sections: Gulf Coast Association of Geological Societies Transactions, v. 38, p. 584–591.

ROBERTS, H. H., SUHAYADA, J. N., and COLEMAN, J. M., 1980, Sediment deformation and transport (sic) on low-angle slopes: Mississippi River Delta, p. 131–167 in Coates, D. R., and Vitek, J. D., eds., Thresholds in geomorphology: London, George Allen and Unwin, 498 p.

ROCHA-CAMPOS, A. C., 1967, The Tubarâo Group in the Brazilian portion of the Paraná Basin (1), p. 27–122, Pls. 3–35, in Bigarella, J. J., Becker, R. D., and Pinto, I. D., eds., Problems in Brazilian Gondwana geology: Curitiba-Paraná, Brazil, Brazilian contribution to the International Symposium on Gondwana Stratigraphy and Paleontology, 1st, 344 p.

RODGERS, JOHN, 1950, The nomenclature and classification of sedimentary rocks: American Journal of Science, v. 248, p. 297–311.

RODHE, H.; and VIRJI, H., 1976, Trends and periodicities in east African rainfall data: Monthly Weather Review, v. 104, p. 307–315.

ROEP, TH. B.; and LINTHOUT, K., 1989, Precambrian storm wave-base deposits of Early Proterozoic age (1.9 Ga), preserved in andalusite-cordierite rich granofels and quartzite (Ramsberg area, Varmland, Sweden): Sedimentary Geology, v. 61, p. 239–251.

ROLL, H. U., 1945, Wassernahes windprofil und wellen auf dem Wattenmeer: Ann. Meteorol., v. 1, p. 139–151.

RONA, P. A., 1970, Submarine canyon (*sic*) origin on upper continental slope off Cape Hatteras: Journal of Geology, v. 78, p. 141–152.

RONA, P. A., DENLINGER, R. P., FISK, M. R., HOWARD, K. J., TAGHON, G. L., KLITGORD, K. D., McCLAIN, J. S., McMURRAY, G. R., and WILTSHIRE, J. C., 1990, Major off-axis hydrothermal activity on the northern Gorda Ridge: Geology, v. 18, p. 493–496.

RONA, P. A., and WISE, D. U., 1974, Symposium: global sea level and plate tectonics through time: Geology, v. 2, p. 133–134.

ROSEN, M. R., MISER, D. E., and WARREN, J. K., 1988, Sedimentology, mineralogy, and isotopic analysis of Pellet Lake, Coorong region, South Australia: Sedimentology, v. 35, p. 105–122.

ROSEN, P., 1979, Boulder barricades in central Labrador: Journal of Sedimentary Petrology, v. 49, p. 1113–1123.

ROSEN, P. S.; and LEACH, KENNETH, 1987, Sediment accumulation forms, Thompson Island, Boston Harbor, Massachusetts, p. 234–250 *in* Fitzgerald, D. M., and Rosen, P. S., eds., 1987, Glaciated coasts: New York, Academic Press, 364 p.

ROSENBUSCH, H.; and IDDINGS, J. P., 1905, Microscopical physiography of the rock-making minerals. An aid to the microscopical study of rocks, 4th ed.: New York, John Wiley & Sons, 367 p.

ROSS, C. A., 1986, Paleozoic evolution of southern margin of Permian basin: Geological Society of America Bulletin, v. 97, p. 536–554.

ROSS, D. A., and DEGENS, E. T., 1974, Recent sediments of Black Sea, p. 183–199 *in* Degens, E. T., and Ross, D. A., eds., 1974, The Black Sea—geology, chemistry, and biology: Tulsa, OK, American Association of Petroleum Geologists Memoir 20, 633 p.

ROSS, D. A.; UCHUPI, E.; PRADA, K. E.; and MacILVAINE, J. C., 1974, Bathymetry and microtopography of Black Sea, p. 1–10 *in* Degens, E. T., and Ross, D. A., eds., The Black Sea—geology, chemistry, and biology: Tulsa, OK, American Association of Petroleum Geologists Memoir 20, 633 p.

ROWLEY, P. D., MAC LEOD, N. S., KUNTZ, M. A., and KAPLAN, A. M., 1985, Proximal bedded deposits related to pyroclastic flows of May 18, 1980, Mount St. Helens, Washington: Geological Society of America Bulletin, v. 96, p. 1371–1383.

ROY, P. S., THOM, B. G., and WRIGHT, L. D., 1980, Holocene sequences on an embayed high-energy coast: an evolutionary model: Sedimentary Geology, v. 21, p. 1–19.

RUBIN, D. M., 1987, Cross-bedding (*sic*), bedforms (*sic*), and paleocurrents: Tulsa, OK, Society of Economic Paleontologists and Mineralogists, Concepts in sedimentology and paleontology, v. 1, 185 p.

RUBIN, D. M., and HUNTER, R. E., 1985, Why deposits of longitudinal dunes are rarely recognized in the geological record: Sedimentology, v. 32, p. 147–157.

RUBIN, D. M., and HUNTER, R. E., 1987, Bedform alignment in directionally varying flows: Science, v. 237, p. 276–278.

RUBIN, D. M.; and IKEDA, HIROSHI, 1990, Flume experiments on the alignment of transverse, oblique, and longitudinal dunes in directionally varying flows: Sedimentology, v. 37, p. 673–684.

RUPPEL, S. C., and WALKER, K. R., 1984, Petrology and depositional history of a Middle Ordovician carbonate platform: Chackagauga Group, northeastern Tennessee: Geological Society of America Bulletin, v. 95, p. 568–583.

RUST, B. R., and GIBLING, M. R., 1990, Braidplain evolution in the Pennsylvanian South Bar Formation, Sydney Basin, Nova Scotia, Canada: Journal of Sedimentary Petrology, v. 60, p. 59–72.

RÜTZLER, K.; and REIGER, G., 1973, Sponge burrowing: fine structure of *Cliona lampa* penetrating calcareous substrate: Marine Biology, v. 21, p. 144–162.

RYAN, W. B. F., 1972, Stratigraphy of Late (*sic*) Quaternary sediments in the eastern Mediterranean,
p. 149–169 *in* Stanley, D. J., ed., The Mediterranean Sea: a natural sedimentation laboratory: Stroudsburg, PA, Dowden, Hutchinson, and Ross, 765 p.

RYAN, W. B. F., and CITA, M. B., 1977, Ignorance concerning episodes of oceanwide stagnation: Marine Geology, v. 23, p. 197–215.

RYAN, W. B. F., STANLEY, D. J., HERSEY, J. B., FAHLQUIST, D. A., and ALLAN, T. D., 1970, The tectonics and geology of the Mediterranean Sea, Chapter 12, p. 387–392 *in* Maxwell, A. E., gen. ed., The sea. Ideas and observations on progress in the study of the seas. Volume 4, New concepts of sea floor (*sic*) evolution. Part II. Regional observations. Concepts: New York, Wiley-Interscience, 664 p.

RYAN, W. B. F.; WORKUM, F.; and HERSEY, J. B., 1965, Sediments of the Tyrrhenian abyssal plain: Geological Society of America Bulletin, v. 76, p. 1261–1282.

RYER, T. A., 1981, Deltaic coals of Ferron sandstone member of Mancos shale: predictive model for Cretaceous coal-bearing strata of Western Interior: American Association of Petroleum Geologists Bulletin, v. 65, p. 2323–2340.

SABINE, P. A., 1974, How should rocks be named? (essay review): Geological Magazine, v. 111, p. 165–176.

SABINS, F. F., JR., 1963, Anatomy of stratigraphic trap, Bisti Field, New Mexico: American Association of Petroleum Geologists Bulletin, v. 47, p. 193–228.

SACKETT, W. M., POAG, C. W., and EDIE, B. J., 1974, Kerogen recycling in the Ross Sea, Antarctica: Science, v. 185, p. 1045–1047.

SADLER, P. M., 1981, Sediment accumulation (*sic*) rates and the completeness of stratigraphic sections: Journal of Geology, v. 89, p. 569–584.

SALLENGER, A. H., JR., 1979, Inverse grading and hydraulic equivalence in grain-flow deposits: Journal of Sedimentary Petrology, v. 49, p. 553–562.

SALLENGER, A. H., HOLMAN, R. A., and BIRKEMEIER, W. A., 1985, Storm-induced response of a nearshore-bar system: Marine Geology, v. 64, p. 237–257.

SANDBERG, P. A., 1983, An oscillating trend in Phanerozoic non-skeletal carbonate mineralogy: Nature, v. 305, p. 19–22.

SANDERS, J. E., 1963, Late Triassic tectonic history of the northeastern United States: American Journal of Science, v. 261, p. 501–524.

SANDERS, J. E., 1965, Primary sedimentary structures formed by turbidity currents and related resedimentation mechanisms, p. 192–219 *in* Middleton, G. V., ed., Primary sedimentary structures and their hydrodynamic interpretation: Tulsa, OK, Society of Economic Paleontologists and Mineralogists Special Publication 12, 165 p.

SANDERS, J. E., 1968, Stratigraphy and primary sedimentary structures of fine-grained, well-bedded strata, inferred lake deposits, Upper Triassic, central and southern Connecticut, p. 265–305 *in* Klein, G. deV., ed., Late Paleozoic and Mesozoic continental sedimentation, northeastern North America. A symposium: Geological Society of America Special Paper 106, 308 p.

SANDERS, J. E., 1978, Graywacke, p. 389–391 *in* Fairbridge, R. W.; and Bourgeois, Joanne, eds., The encyclopedia of sedimentology: Encyclopedia of earth sciences, Volume VI: Stroudsburg, PA, Dowden, Hutchinson, and Ross, 901 p.

SANDERS, J. E., 1981, Principles of physical geology: New York, John Wiley & Sons, 624 p.

SANDERS, J. E., 1982a, Petroleum in unconformity traps: from kerogen in truncated permeable wedges?, p. 77–91 *in* Halbouty, M. T., ed., The deliberate search for the subtle trap: Tulsa, OK, American Association of Petroleum Geologists Memoir 32, 351 p.

SANDERS, J. E., 1982b, The PCB-pollution problem in the upper Hudson River from the perspective of the Hudson River PCB Settlement Advisory Committee: Northeastern Environmental Science, v. 1, p. 7–18.
SANDERS, J. E., 1989, PCB pollution in the upper Hudson River: from environmental disaster to "environmental gridlock": Northeastern Environmental Science, v. 8, p. 1–86.

SANDERS, J. E., and FRIEDMAN, G. M., 1967, Origin and occurrence of limestones, p. 169–265 *in* Chilingar, G. V., Bissell, H. J., and Fairbridge, R. W., eds., Carbonate rocks: Amsterdam, Elsevier Scientific Publishing Company, 471 p.

SANDERS, J. E., and FRIEDMAN, G. M., 1969, Position of regional carbonate/noncarbonate boundary in nearshore sediments along a coast: possible climatic indicator: Geological Society of America Bulletin, v. 80, p. 1789–1796.

SANDERS, J. E.; and KUMAR, N., 1975, Evidence of shoreface retreat and in-place drowning during Holocene submergence of barriers, shelf off Fire Island, New York: Geological Society of America Bulletin, v. 86, p. 65–76.

SANDERSON, D. J., and DONOVAN, R. N., 1974, The vertical packing of shells and stones on some recent beaches: Journal of Sedimentary Petrology, v. 44, p. 680–688.

SANSONE, F. J., TRIBBLE, G. W., BUDDEMEIER, R. W., and ANDREWS, C. C., 1988, Time (*sic*) and space scales of anaerobic diagenesis within a coral reef (*sic*) framework: Proceedings of the 6th International Coral Reef (*sic*) Symposium, Australia, v. 3, p. 367–372.

SARMIENTO, R.; and KIRBY, R. A., 1962, Recent sediments of Lake Maracaibo: Journal of Sedimentary Petrology, v. 32, p. 698–724.

SARNTHEIN, M.; TETZLAFF, G.; KOOPMANN, B.; WOLTER, K.; and PFLAUMANN, U., 1981, Glacial (*sic*) and interglacial wind regimes over the eastern subtropical Atlantic and NW Africa: Nature, v. 293, p. 193–196.

SARNTHEIN, M.; THIEDE, J.; PLAUMANN, U.; ERLENKEUSER, K.; FUETTERER, D.; KOOPMANN, B.; LANGE, H.; and SEIBOLD, E., 1982, Atmospheric (*sic*) and oceanic circulation patterns off northwest Africa during the past 25 million years, p. 545–604 *in* von Rad, Ulrich; Hinz, K.; and others, eds., Geology of the northwest African continental margin: Berlin, Springer-Verlag, 712 p.

SAURAMO, MATTI, 1923, Studies on the Quaternary varve sediments in southern Finland: Finland Comm. Geol., Bull., no. 60, 124 p.

SAVAGE, S. B., 1984, The mechanics of rapid granular flows: Advances in Applied Mechanics, v. 24, p. 288–366.

SAVARD, MARTINE; and BOURQUE, P.-A., 1989, Diagenetic evolution of a Late (*sic*) Silurian platform, Gaspe Basin, Quebec, based on cathodoluminescence petrography: Canadian Journal of Earth Sciences, v. 26, p. 791–806.

SAVRDA, C. E., and BOTTJER, D. J., 1988, Limestone concretion (*sic*) growth documented by trace-fossil relations (*sic*): Geology, v. 16, p. 908–911.

SAXOV, S.; and NIEUWENHUIS, J. K., eds., 1982, Marine slides and other mass movements: NATO Conference Series IV: New York, Plenum Press, 353 p.

SCHELTEMA, R. S., 1977, Dispersal of marine invertebrate (*sic*) organisms: paleobiogeographic (*sic*) and biostratigraphic implications, p. 73–108 *in* Kauffman, E. G., and Hazel, J. E., eds., 1977, Concepts and methods in biostratigraphy: Stroudsburg, PA, Dowden, Hutchinson, and Ross, 658 p.

SCHINDEL, D. E., 1980, Microstratigraphic sampling and the limits of paleontologic resolution: Paleobiology, v. 6, p. 408–426.

SCHLAGER, WOLFGANG, 1981, The paradox of drowned reefs and carbonate platforms: Geological Society of America Bulletin, Part I, v. 92, p. 197–211.

SCHLANGER, S. O., ARTHUR, M. A., JENKYNS, H. C., and SCHOLLE, P. A., 1986, The Cenomanian Turonian oceanic anoxic event. I. Stratigraphy and distribution of organic carbon rich (*sic*) beds and the marine $\delta^{13}C$ excursion, p. 371–399 *in* Brooks, J.; and Fleet, A., eds, Marine petroleum source rocks: Geological Society of London Special Publication 26, 444 p.

SCHLANGER, S. O., and JENKYNS, H. C., 1976, Cretaceous ocean anoxic events, causes and consequences: Geologie en Mijnbouw, v. 55, p. 179–184.

SCHLUMBERGER, 1981, Dipmeter interpretation. Volume I, fundamentals: New York, Schlumberger, Ltd., 61 p.

SCHLUMBERGER EDUCATIONAL SERVICES, 1987, Log interpretation. Principles/applications: Houston, TX, Schlumberger Educational Services, 198 p.

SCHMALZ, R. F., 1969, Deep-water evaporite deposition: a genetic model: American Association of Petroleum Geologists Bulletin, v. 53, p. 798–823.

SCHMALZ, R. F., 1991, The Mediterranean salinity crisis: alternative hypotheses: Carbonates and Evaporites, v. 6, p. 121–126.

SCHMID, R., 1981, Descriptive nomenclature and classification of pyroclastic deposits and fragments: Recommendations of the IUGS Subcommission on the Systematics of Igneous Rocks: Geology, v. 9, p. 41–43.

SCHMINKE, HANS-ULRICH, 1988, Pyroklastische Gesteine, p. 731–778 in Füchtbauer, Hans, ed., 1988, Sediment und Sedimentgesteine. Sediment-Petrologie Part II: Stuttgart, E. Schweizerbart, 1141 p.

SCHMOKER, J. W., and HALLEY, R. B., 1982, Carbonate porosity versus depth: a predictable relation (sic) for South Florida: American Association of Petroleum Geologists Bulletin, v. 66, p. 2561–2570.

SCHNEIDER, E. D., 1970 ms., Downslope (sic) and across-slope sedimentation as observed in the westernmost North Atlantic: New York, Columbia University, Department of Geology, Ph.D. dissertation, 301 p. (2 vols.)

SCHNEIDERMANN, NAHUM; and HARRIS, P. M., eds., 1985, Carbonate cements: Tulsa, OK, Society of Economic Paleontologists and Mineralogists Special Publication 36, 379 p.

SCHOCK, R. M., 1989, Stratigraphy. Principles and methods. New York: Van Nostrand Reinhold, 375 p.

SCHOELL, M., 1988, Multiple origins of methane in the Earth: Chemical Geology, v. 71, p. 1–10.

SCHOETTLE, MANFRED; and FRIEDMAN, G. M., 1971, Fresh water (sic) iron-manganese nodules in Lake George, New York: Geological Society of America Bulletin, v. 82, p. 101–110.

SCHOLL, D. W., and MARLOW, M. S., 1974, Sedimentary sequence (sic) in modern Pacific trenches and the deformed circum-Pacific eugeosynclines, p. 193–211 in Dott, R. H., Jr., and Shaver, R. H., eds., Modern (sic) and ancient geosynclinal sedimentation: Tulsa, OK, Society of Economic Paleontologists and Mineralogists Special Publication 19, 380 p.

SCHOLLE, P. A., BEBOUT, D. G., and MOORE, C. H., eds., 1983, Carbonate depositional environments: Tulsa, OK, American Association of Petroleum Geologists Memoir 33, 708 p.

SCHOLLE, P. A., and HALLEY, R. B., 1985, Burial diagenesis: out of sight, out of mind!, p. 309–334 in Schneidermann, N.; and Harris, P. M., eds., Carbonate cements: Tulsa, OK, Society of Economic Paleontologists and Mineralogists Special Publication 36, 379 p.

SCHOLLE, P. A.; and SPEARING, D., eds., 1982, Sandstone depositional environments: Tulsa, OK, American Association of Petroleum Geologists Memoir 31, 410 p.

SCHOLTEN, ROBERT, 1974, Role of the Bosporus in Black Sea chemistry and sedimentation, p. 115–126 in Degens, E. T., and Ross, D. A., eds., The Black Sea—geology, chemistry, and biology: Tulsa, OK, American Association of Petroleum Geologists Memoir 20, 633 p.

SCHOPF, T. J. M., ed., 1972, Models in paleobiology: San Francisco, Freeman, Cooper and Company, 250 p.

SCHOPF, T. J. M., 1980, Paleoceanography: Cambridge, MA, Harvard University Press, 341 p.

SCHOPF, T. J. M.; and HOFFMANN, A., 1983, Punctuated equilibrium and the fossil record: Science, v. 219, p. 438–439.

SCHOVE, D. J., 1987, Sunspot cycles and weather history, Chapter 21, p. 355–377 in Rampino, M. R., Sanders, J. E., Newman, W. S., and Konigsson, L.-K., eds., Climate. History, periodicity, and predictability: New York, Van Nostrand Reinhold Company, 588 p.

SCHREIBER, B. C., 1986, Arid shorelines and evaporites, p. 189–228, in Reading, H. G., ed., Sedimentary environments and facies, 2nd ed., Oxford, Blackwell Scientific Publications, 615 p.

SCHREIBER, B. C., and FRIEDMAN, G. M., 1974, Upper Miocene evaporite deposits of Sicily (abstract): Geological Society of America, Abstracts with Programs, v. 5, no. 1, p. 69 (only).

SCHREIBER, B. C.; FRIEDMAN, G. M.; DECIMA, ARVEDO; and SCHREIBER, E., 1977, The depositional environments of the Upper Miocene (Messinian) evaporite deposits of the Sicilian Basin: Sedimentology, v. 23, p. 729–760.

SCHROEDER, J. H., and PURSER, B. H., eds., 1986, Reef diagenesis: New York, Springer-Verlag, 455 p.

SCHUCHERT, C., 1923, Sites and natures of the North American geosynclines: Geological Society of America Bulletin, v. 34, p. 151–260.

SCHUCHERT, C., 1935, Historical geology of the Antillean-Caribbean region: New York, John Wiley & Sons, 811 p.

SCHUHMACHER, H.; and ZIBROWINS, H., 1985, What is hermatypic? A redefinition of ecological groups in corals and other organisms: Coral Reefs, v. 4, p. 1–7.

SCHUMM, S. A., and STEVENS, M. A., 1973, Abrasion in place: a mechanism for rounding and size reduction of coarse sediments in rivers: Geology, v. 1, p. 37–40.

SCHWAB, F. L., 1975, Framework mineralogy (sic) and chemical composition of continental margin-type sandstone: Geology, v. 3, p. 487–490.

SCHWAB, W. C., and LEE, H. J., 1988, Causes of two slope-failure types in continental-shelf sediment, northeastern Gulf of Alaska: Journal of Sedimentary Petrology, v. 58, p. 1–11.

SCHWALBACH, J. R., and GORSLINE, D. S., 1985, Holocene sediment budgets for the basins of the California continental borderland: Journal of Sedimentary Petrology, v. 55, p. 829–842.

SCHWARZACHER, WERNER, 1954, Die Grossrhythmik des Dachstein Kalkes von Lofer: Tshermaks Mineral. Petrog. Mitt, v. 4, p. 44–54.

SCHWEBEL, D. A., 1983, Quaternary stratigraphy of the south-east of South Australia: Unpublished Ph.D. dissertation, the Flinders University of South Australia.

SCHWELLER, W. J., and KULM, L. D., 1978, Depositional patterns and channelized sedimentation in active eastern Pacific trenches, p. 311–324 in Stanley, D. J.; and Kelling, G., eds., Sedimentation in submarine canyons, fans and trenches: Stroudsburg, PA, Dowden, Hutchinson, and Ross, 382 p.

SCOFFIN, T. P., 1987, An introduction to carbonate sediments and rocks (sic): Glasgow, Blackie, 274 p.

SCOFFIN, T. P., 1988, The environments of production and deposition of calcareous sediments on the shelf west of Scotland: Sedimentary Geology, v. 60, p. 107–124.

SCOFFIN, T. P.; STEARN, C. W.; BOUCHER, D.; FRYDL, P.; HAWKINS, C. M.; HUNTER, I. G.; and McGEACHY, J. K., 1980, Calcium carbonate (sic) budget of a fringing reef on the west coast of Barbados, Pt. II: erosion, sediments, and internal structure: Bulletin of Marine Science, v. 302, p. 457–508.

SCOTT, A. J., HOOVER, R. A., and McGOWEN, J. H., 1969, Effects of Hurricane "Beulah," 1967 on Texas coastal lagoons and barriers, p. 221–236 in Ayala-Castanares, Augustin; and Phleger, F. B., eds., Lagunas costeras, un simposio: Simposio Internacional sobre lagunas costeras, Mexico City, Mex-

ico, 28–30 Noviembre 1967, memoria: Mexico Universidad Nacional Autonoma (WMAS- UNESCO), 686 p.

SCOTT, G. H., 1985, Homotaxy and biostratigraphical theory: Paleontology, v. 28, p. 717–782.

SCOTT, P. J. B., and RISK, M. J., 1988, The effect of Lithophaga (Bivalvia: Mytilidae) boreholes on the strength of the coral Porites lobata: Coral Reefs, v. 7, p. 145–151.

SCRUTON, P. C., 1953, Deposition of evaporites: American Association of Petroleum Geologists Bulletin, v. 37, p. 2498–2512.

SCRUTON, P. C., 1960, Delta building and the deltaic sequence, p. 82–102 in Shepard, P. F., Phleger, F. B., and van Andel, Tj. H., eds., Recent sediments, northwest Gulf of Mexico: Tulsa, OK, American Association of Petroleum Geologists, 394 p.

SEGERSTRALE, S. G., 1957, Baltic Sea, p. 751–802 in Hedgpeth, J. W., ed., Treatise on marine ecology and paleoecology: Geological Society of America Memoir 67, v. 1, Ecology, 1296 p.

SEIBOLD, E.; and BERGER, W. H., 1982, The sea floor: Berlin, Springer-Verlag, 288 p.

SEIDEMANN, D. E., MASTERSON, W. D., DOWLING, M. P., and TUREKIAN, K. K., 1984, K-Ar dates and $^{40}Ar/^{39}Ar$ age spectra for Mesozoic basalt flows of the Hartford Basin, Connecticut, and the Newark Basin, New Jersey: Geological Society of America Bulletin, v. 95, p. 594–598.

SEMENIUK, V.; and JOHNSON, D. P., 1982, Recent and Pleistocene beach/dune sequences, Western Australia: Sedimentary Geology, v. 32, p. 301–328.

SERRA, O., 1985, Sedimentary environments from wireline logs: Houston, TX, Schlumberger Well Services, 211 p.

SHACKLETON, N. J.; HALL, M. A.; LINE, J.; and SHUXI, C., 1983, Carbon isotope (sic) data in core V19-30 confirm reduced carbon dioxide (sic) concentration in the ice age (sic) atmosphere: Nature, v. 306, p. 319–322.

SHANMUGAM, G.; and MOIOLA, R. J., 1990, Submarine fan "lobe" models: implications for reservoir properties: Gulf Coast Association of Geological Societies Transactions, v. 40, p. 777–791.

SHARP, R. P., and MALIN, M. C., 1984, Surface geology from Viking landers on Mars: a second look: Geological Society of America Bulletin, v. 95, p. 1398–1412.

SHEARMAN, D. J., 1966, Origin of marine evaporites by diagenesis: Institute of Mining and Metallurgy Transactions, v. 75, section B Bulletin 717, p. 208–215 (discussion, p. B82–B86).

SHELDON, P. R., 1987, Parallel gradualistic evolution of Ordovician trilobites: Nature, v. 330, p. 561–563.

SHELDON, R. P., 1967, Long-distance migration of oil in Wyoming: Mountain Geologist, v. 4, p. 53–65.

SHELDON, R. P., 1981, Ancient marine phosphorites: Annual Reviews of Earth and Planetary Sciences, v. 9, p. 251–284.

SHELDON, R. P., 1987, Association of phosphorites, organic-rich shales, chert and carbonate rocks: Carbonates and Evaporites, v. 2, p. 7–14.

SHELL OIL Co., 1987, Atlas of seismic stratigraphy, p. 15–71 in Bally, A. W., ed., 1987, Atlas of seismic stratigraphy: Tulsa, OK, The American Association of Petroleum Geologists Studies in Geology 27, v. 1, 125 p.

SHEPARD, F. P., 1936, The underlying causes of submarine canyons: U.S. National Academy of Sciences Proceedings, v. 22, no. 8, p. 496–512.

SHEPARD, F. P., 1948, Submarine geology: New York, Harper and Brothers, 348 p.

SHEPARD, F. P., 1950a, Longshore-bars and longshore-troughs (sic), p. 121–156 in U.S. Army, Corps of Engineers, Beach erosion board, Technical Memo. 15.

SHEPARD, F. P., 1950b, Beach cycles in southern California: U.S. Army Corps of Engineers Beach Erosion Board Technical Memorandum 20, 26 p.

SHEPARD, F. P., 1960, Gulf coast barriers, p. 197–220 *in* Shepard, F. P., Phleger, F. B., and van Andel, Tj. H., eds., Recent sediments, northwest Gulf of Mexico: Tulsa, OK, American Association of Petroleum Geologists, 394 p.

SHEPARD, F. P., 1963a, Thirty-five thousand years of sea level, p. 1–10 *in* Clements, T., ed., Essays in marine geology: Los Angeles, CA, University of Southern California Press.

SHEPARD, F. P., 1963b, Submarine geology, 2nd ed.: New York, Harper & Row, 557 p.

SHEPARD, F. P., 1965, Submarine canyons, Chapter 20, p. 480–506 *in* Hill, M. N., gen. ed., The sea. Ideas and observations on progress in the study of the seas. Volume 3, The Earth beneath the sea. History: New York, Wiley-Interscience, 963 p.

SHEPARD, F. P., 1981, Submarine canyons: multiple causes and long-time persistence: American Association of Petroleum Geologists Bulletin, v. 65, p. 1062–1077.

SHEPARD, F. P., MARSHALL, N. F., MCLAUGHLIN, P. A., and SULLIVAN, G. G., 1979, Currents in submarine canyons and other sea valleys: Tulsa, OK, American Association of Petroleum Geologists Studies in Geology 8, 173 p.

SHERIDAN, R. E., 1974, Atlantic continental margin of North America, p. 391–407 *in* Burk, C. A., and Drake, C. L., 1974, The geology of continental margins: New York, Springer-Verlag, 1009 p.

SHERIFF, R. E., 1980, Seismic stratigraphy: Boston, MA, International Human Resources Development Corporation, 227 p.

SHERIFF, R. E., 1989, Geophysical methods: Englewood Cliffs, NJ, Prentice-Hall, 605 p.

SHERMAN, D. J., and NORDSTROM, K. F., 1985, Beach scarps: Zeitschrift fur Geomorphologie, Neue Folge, v. 29, no. 2, p. 139–152.

SHI, N. C., and LARSEN, L. H., 1984, Reverse sediment transport induced by amplitude modulated (*sic*) waves: Marine Geology, v. 54, p. 181–200.

SHIDELER, G. L., 1984, Suspended sediment (*sic*) responses in a wind-dominated estuary of the Texas Gulf Coast: Journal of Sedimentary Petrology, v. 54, p. 731–745.

SHIMKUS, K. M., and TRIMONIS, E. S., 1974, Modern sedimentation in Black Sea, p. 249–278 *in* Degens, E. T., and Ross, D. A., eds., The Black Sea—geology, chemistry, and biology: Tulsa, OK, American Association of Petroleum Geologists Memoir 20, 633 p.

SHINN, E. A., 1971, Holocene submarine cementation in the Persian Gulf, p. 63–67 *in* Bricker, O. P., ed., Carbonate cements: Baltimore, The Johns Hopkins University Press, 376 p.

SHINN, E. A., 1982, Birdseyes, fenestrae, shrinkage pores, and loferites–a reevaluation, Journal of Sedimentary Petrology, v. 53, p. 619–628.

SHINN, E. A., 1983, Tidal flat environment, p. 171–210 *in* Scholle, P. A., Bebout, D. G., and Moore, C. H., eds., Carbonate depositional environments, American Association of Petroleum Geologists Memoir 33, 708 p.

SHINN, E. A., LLOYD, R. M., and GINSBURG, R. N., 1969, Anatomy of a modern carbonate tidal-flat (*sic*), Andros Island, Bahamas: Journal of Sedimentary Petrology, v. 39, p. 1202–1228.

SHINN, E. A., STEINEN, R. P., LIDZ, B. H., and SWART, P. K., 1989, Whitings, a sedimentologic dilemma: Journal of Sedimentary Petrology, v. 59, p. 147–161.

SHOKES, R. F., TRABANT, P. K., PRESLEY, B. J., and REID, D. F., 1977, Anoxic hypersaline basin in the northern Gulf of Mexico: Science, v. 196, p. 1443–1446.

SHORT, A. D., 1975, Multiple offshore bars and standing waves: Journal of Geophysical Research, v. 80, p. 3838–3840.

SHORT, A. D., 1979, Three dimensional (*sic*) beach stage model: Journal of Geology, v. 87, p. 553–571.

SHORT, A. D., 1984, Beach (*sic*) and nearshore facies: southeast Australia: Marine Geology, v. 60, p. 261–281.

SHREVE, R. L., 1966, Sherman landslide, Alaska: Science, v. 154, p. 1639–1643.

SHROCK, R. R., 1948, Sequence in layered rocks. A study of features and structures useful for determining top and bottom or order of succession in bedded (*sic*) and tabular rock bodies: New York, McGraw-Hill Book Company, 507 p. (p. 9–17; 34–40; 43–56; 92–113; 141–144; 156–161; 162–166; 177–181; 188–208; 242–253.)

SIBLEY, D. F., and GREGG, J. M., 1987, Classification of dolomite rock textures: Journal of Sedimentary Petrology, v. 57, p. 967–975.

SIEGENTHALER, U.; and OESCHGER, H., 1978, Predicting future atmospheric carbon dioxide (*sic*) levels: Science, v. 199, p. 388–395.

SIEMERS, C. T., TILLMAN, R. W., and WILLIAMSON, C. R., 1981, Deep-water clastic sediments: a core workshop: Tulsa, OK, Society of Economic Paleontologists and Mineralogists Core Workshop 2, 416 p.

SIEVEKING, G. deG., and HART, M. B., eds., 1986, The scientific study of flint and chert: Cambridge, Cambridge University Press.

SIGURDSSON, H.; SPARKS, R. S. J.; CAREY, S. N.; and HUANG, T. C., 1980, Volcanogenic sedimentation in the Lesser Antilles arc: Journal of Geology, v. 88, p. 523–540.

SILVER, B. A., and TODD, R. G., 1969, Permian cyclic strata, northern Midland and Delaware basins, West Texas and southeastern New Mexico: American Association of Petroleum Geologists Bulletin, v. 53, p. 2223–2251.

SILVER, E. A., and BEUTNER, E. C., 1980, Penrose conference report: Melanges: Geology, v. 8, p. 32–34.

SILVER, L. T., and SCHULTZ, P. H., eds., 1982, Geological implications of impacts of large asteroids and comets on the Earth: Boulder, Colorado, Geological Society of America Special Paper 190, 528 p.

SILVESTER, R., 1955, Practical application of Darbyshire's method of hindcasting ocean waves: Australian Journal of Applied Science, v. 6, p. 261–266.

SIMKISS, K., 1964, Possible effects of zooxanthellae on coral growth: Experientia, v. 20, p. 140.

SIMONS, D. B., RICHARDSON, E. V., and NORDIN, C. F., JR., 1965, Sedimentary structures generated by flow in alluvial channels, p. 34–52 *in* Middleton, G. V., ed., Primary sedimentary structures and their hydrodynamic interpretation: Tulsa, OK, Society of Economic Paleontologists and Mineralogists Special Publication 12, 165 p.

SIMONSON, B. M., 1985, Sedimentological constraints on the origins of Precambrian iron-formations (*sic*): Geological Society of America Bulletin, v. 96, p. 244–252.

SIMPSON, CAROL; and SCHMID, S. M., 1983, An evaluation of criteria to deduce (*sic*) the sense of movement in sheared rocks: Geological Society of America Bulletin, v. 94, p. 1281–1288.

SIMPSON, G. G., 1952, Probability of dispersal in geologic time: American Museum of Natural History Bulletin, v. 99, p. 163–176.

SIMPSON, H. J., OLSEN, C. R., WILLIAMS, S. C., and TRIER, R. M., 1976, Man-made radionuclides and sedimentation in the Hudson River estuary: Science, v. 194, p. 179–183.

SKYRING, G. W.; CHAMBERS, L. A.; and BAULD, J., 1983, Sulfate reduction in sediments colonized by cyanobacteria, Spencer Gulf, South Australia: Australian Journal of Marine and Freshwater Research, v. 34, p. 359–374.

SLATER, R. A., 1984, A numerical model of tides in the Cretaceous seaway of North America: Journal of Geology, v. 93, p. 333–345.

SLATT, R. M., 1972, Texture and composition of till derived from parent rocks of contrasting textures, southeastern Newfoundland: Sedimentary Geology, v. 7, p. 283–290.

SLEEP, N. H., 1971, Thermal effects of the formation of Atlantic continental margins by continental breakup: Royal Astronomical Society, Geophysical Journal, v. 24, p. 325–350.

SLEEP, N. H., 1976, Platform subsidence (*sic*) mechanisms and eustatic sea-level changes: Tectonophysics, v. 36, p. 45–56.

SLINGERLAND, R., 1986, Numerical computation of co-oscillating paleotides in the Catskill epeiric sea of eastern North America: Sedimentology, v. 33, p. 487–497.

SLY, P. G., 1978, Sedimentary processes in lakes, p. 65–89 *in* Lerman, Abraham, ed., Lakes: chemistry, geology, physics: Berlin, Springer-Verlag, 363 p.

SMIRNOW, L. P., 1958, Black Sea basin. Its position in the Alpine structure and its richly organic Quaternary sediments, p. 982–994 *in* Weeks, L. G., ed., Habitat of oil. A symposium: Tulsa, OK, American Association of Petroleum Geologists, 1384 p.

SMITH, D. G.; REINSON, J. G.; ZAITLIN, B.; and RAHMANI, R., eds., 1991, Clastic tidal sedimentology: Calgary, Alberta, Canadian Society of Petroleum Geologists Memoir 16, 390 p.

SMITH, G. A., 1986, Coarse-grained nonmarine volcaniclastic sediment: terminology and depositional process: Geological Society of America Bulletin, v. 97, p. 1–10.

SMITH, J. M., and FREY R. W., 1985, Biodeposition by the ribbed mussel *Geukensia demissa* in a salt marsh, Sapelo Island, Georgia: Journal of Sedimentary Petrology, v. 55, p. 817–828.

SMITH, L. M., DUNBAR, J. B., and BRITSCH, L. D., 1986, Geomorphological investigation of the Atchafalaya Basin, Area West, Atchafalaya Delta, and Terrebonne Marsh, vol. 1: U.S. Army Corps of Engineers, Waterways Expt. Sta., Tech. Rpt. GL-86-3, 262 p.

SMITH, R. M. H., 1987, Morphology and depositional history of exhumed Permian point bars in the southwestern Karoo, South Africa: Journal of Sedimentary Petrology, v. 57, p. 19–29.

SMITH, W. O., 1958, Recent underwater surveys using low-frequency sound to locate shallow bedrock: Geological Society of America Bulletin, v. 69, p. 69–98.

SMOOT, J. P., 1983, Depositional subenvironments in an arid closed basin; the Wilkins Peak Member of the Green River Formation (Eocene), Wyoming, U.S.A.: Sedimentology, v. 30, p. 801–828.

SNEDDEN, J. W.; and NUMMEDAL, DAG, in press, Origin and geometry (*sic*) of storm-deposited storm beds in modern sediments of the Texas continental shelf, *in* Swift, D. J. P., Oertel, G. W.; and Tillman, R. W., eds., Shelf sand (*sic*) and sandstone (*sic*) bodies. Geometry (*sic*), facies, and distribution: International Association of Sedimentologists Special Publication 12.

SNEH, AMIHAI, 1979, Late Pleistocene fan-deltas (*sic*) along the Dead Sea (*sic*) rift: Journal of Sedimentary Petrology, v. 49, p. 541–552.

SNEH, AMIHAI, 1983, Desert stream sequences in the Sinai Peninsula: Journal of Sedimentary Petrology, v. 53, p. 1271–1279.

SNEH, AMIHAI; and FRIEDMAN, G. M., 1985, Hypersaline sea-marginal flats of the Gulfs of Elat and Suez, p. 103–135 *in* Friedman, G. M., and Krumbein, W. E., eds., Hypersaline ecosystems. The Gavish Sabkha. Ecological Studies 53: New York, Springer-Verlag, 484 p.

SNELLING, N. J., ed., 1983, The chronology of the geological record: Geological Society of London Memoir 10, 343 p.

SOUCIE, G., 1973, Where beaches have been going: into the ocean: Smithsonian, v. 4, no. 3, p. 55–61.

SOULSBY, R. L., 1981, Measurements of the Reynolds stress components close to a marine sand bank, p. 35–47 *in* Nittrouer, C. A., ed., Sedimentary dynamics of continental shelves: Developments in sedimentology, 32: Amsterdam, Elsevier Scientific Publishing Company, 449 p.

SOUTHAM, J. R., PETERSON, W. H., and BRASS, G. W., 1982, Dynamics of anoxia: Palaeogeography,

Palaeoclimatology, and Palaeoecology, v. 40, p. 183–198.

SOUTHARD, J. B., LAMBIE, J. M, FEDERICO, D. C., PILE, H. T., and WEIDMAN, C. R., 1990, Experiments on bed configurations in fine sands under bidirectional purely oscillatory flow, and the origin of hummocky cross-stratification: Journal of Sedimentary Petrology, v. 60, p. 1–17.

SOUTHGATE, P. N.; LAMBERT, I. B.; DONNELLY, T. H.; HENRY, R.; ETMINAN, H.; and WESTE, G., 1989, Depositional environments and diagenesis in Lake Parakeelya: a Cambrian alkaline playa from the Officer Basin, South Australia: Sedimentology, v. 36, p. 1091–1112.

SPICER, H. C., 1942, Observed temperatures in the earth's crust, p. 279–291 *in* Birch, Francis; Schairer, J. F., and Spicer, H. C., eds., Handbook of physical constants, Geological Society of America Special Paper 36, 325 p.

STACH, E.; MACKOWSKY, M.-Th.; TEICHMULLER, M.; TAYLOR, G. H.; CHANDRA, D.; and TEICHMULLER, R., 1982, Stach's textbook of coal petrology, 3rd ed.: Berlin, Gebruder Borntraeger, 535 p.

STAFF, G. M.; STANTON, R. J., JR.; POWELL, E. N.; and CUMMINS, HAYS, 1986, Time-averaging (sic), taphonomy, and their impact on paleocommunity reconstruction: death assemblages in Texas bays: Geological Society of America Bulletin, v. 97, p. 428–443.

STANLEY, D. J., ed., 1972, The Mediterranean Sea: a natural sedimentation laboratory: Stroudsburg, PA, Dowden, Hutchinson, and Ross, 765 p.

STANLEY, D. J.; and KELLING, G., eds., 1978, Sedimentation in submarine canyons, fans and trenches: Stroudsburg, PA, Dowden, Hutchinson, and Ross, 395 p.

STANLEY, D. J., and MOORE, G. T., eds., 1983, The shelfbreak: critical interface on continental margins: Tulsa, OK, Society of Economic Paleontologists and Mineralogists Special Publication 33, 467 p.

STANLEY, K. O., and SURDAM, R. C., 1978, Sedimentation on the front of Eocene Gilbert-type deltas, Washakie Basin, Wyoming: Journal of Sedimentary Petrology, v. 48, p. 557–573.

STANLEY, S. M., 1976, Stability of species in geologic time: Science, v. 192, p. 267–269.

STANLEY, S. M., 1979, Macroevolution: pattern and process: San Francisco, W. H. Freeman and Company, 332 p.

STANLEY, S. M., 1985, Rates of evolution: Paleobiology, v. 11, p. 13–26.

STEEL, R. J., and THOMPSON, D. B, 1983, Structures and textures in Triassic braided stream conglomerates ('Bunter' Pebble Beds) in the Sherwood Sandstone Group, North Staffordshire, England: Sedimentology, v. 30, p. 341–367.

STEIDTMANN, J. R., 1974, Evidence for eolian origin of cross-stratification in sandstone of the Casper Formation, southernmost Laramie Basin, Wyoming: Geological Society of America Bulletin, v. 85, p. 1835–1842.

STEINEN, R. P., 1974, Phreatic (sic) and vadose diagenetic modifications of Pleistocene limestone: petrographic observations from subsurface of Barbados, West Indies: American Association of Petroleum Geologists Bulletin, v. 58, p. 1008–1024.

STEINEN, R. P., and MATTHEWS, R. K., 1973, Phreatic (sic) vs. vadose diagenesis: stratigraphy and mineralogy of a cored borehole on Barbados, W. I.: Journal of Sedimentary Petrology, v. 43, p. 1012–1020.

STEINHORN, I., 1985, The disappearance of the long term (sic) meromictis stratification of the Dead Sea: Limnology and Oceanography, v. 30, p. 451–472.

STEINHORN, I.; ASSAF, D.; GAT, J. R.; NISHRY, A.; NISSENBAUM, A.; STILLER, M.; BEYTH, M.; NEEV, D.; GARBER, R.; FRIEDMAN, G. M.; and WEISS, W., 1979, The Dead Sea: deepening of the mixolimnion signifies the overture to overturn of the water column: Science, v. 206, p. 55–57.

STENO, NICOLAUS (STENONIS, NICOLAI), 1667, Elementorum myologiae specimen, se musculi de-

scripti geometrica: Florence, Italy, Stellae, 123 p.

STEVENS, C. H., 1986, Evolution of the Ordovician through Middle Pennsylvanian carbonate shelf in east-central California: Geological Society of America Bulletin, v. 97, p. 11–25.

STILLE, HANS, 1924, Grundfragen der vergleichenden Tektonik: Berlin, Borntraeger, 443 p.

STILLE, HANS, 1936a, The present tectonic state of the Earth: American Association of Petroleum Geologists Bulletin, v. 20, p. 848–880.

STILLE, HANS, 1936b, Wege und Ergebnisse der geologischtectonischen Forschung: Wissenschaftlich Forhandl. Gesellsch 25 Jahr Kaiser Wilhelm, Bd. 2, p. 84–85.

STOCKMAL, G. S.; BEAUMONT, C.; and BOUTILIER, R., 1986, Geodynamic models of convergent tectonics: the transition from rifted margin to overthrust belt and consequences for foreland basin (sic) development: American Association of Petroleum Geologists Bulletin, v. 70, p. 181–190.

STOCKMAN, K. W., GINSBURG, R. N., and SHINN, E. A., 1967, The production of lime mud by algae in South Florida: Journal of Sedimentary Petrology, v. 37, p. 633–648.

STODDART, D. R., 1969, Ecology (sic) and morphology of recent coral reefs: Biological Review, v. 44, p. 433–498.

STOKES, W. L., 1968, Multiple parallel-truncation bedding planes—a feature of wind-deposited sandstone formations: Journal of Sedimentary Petrology, v. 38, p. 510–515.

STOW, D. A. V., 1979, Distinguishing between fine-grained turbidites and contourites on the Nova Scotian deep water (sic) margin: Sedimentology, v. 26, p. 371–387.

STOW, D. A. V., 1981, Laurentian fan: morphology, sediments, processes, and growth pattern: American Association of Petroleum Geologists Bulletin, v. 65, p. 375–393.

STOW, D. A. V., 1985, Deep-sea clastics: where are we and where are we going?, p. 67–93 *in* Brenchley, P. J., and Williams, B. P. J., eds., Sedimentology: recent developments and applied aspects: Oxford, Blackwell Scientific Publications, 342 p.

STOW, D. A. V., and BOWEN, A. J., 1980, A physical model for the transport and sorting of fine-grained sediment by turbidity currents: Sedimentology, v. 27, p. 31–46.

STOW, D. A. V., and LOVELL, J. P. B., 1979, Contourites: their recognition in modern (sic) and ancient sediments: Earth-Science Reviews, v. 14, p. 251–291.

STOW, D. A. V., and PIPER, D. J. W., eds., 1984, Fine-grained sediments: deep-water processes and facies: Geological Society of London Special Publication 15: Oxford, Blackwell Scientific Publications, 659 p.

STOW, D. A. V.; and SHANMUGAM, G., 1980, Sequence of structures in fine-grained turbidites: comparison of Recent deep-sea and ancient flysch sediments: Sedimentary Geology, v. 25, p. 23–42.

STRASSER, ANDRE, 1988, Shallowing-upward sequences in Purbeckian peritidal carbonates (lowermost Cretaceous, Swiss and French Jura Mountains): Sedimentology, v. 35, p. 369–383.

STRIDE, A. H., ed., 1982, Offshore tidal sands: processes and deposits: London, Chapman and Hall, 222 p.

STRØM, K. M., 1937, Land-locked waters. Hydrography and bottom deposits in badly-ventilated (sic) Norwegian Fjords, with remarks upon sedimentation under anaerobic conditions: Norske Vid. Ak., Oslo, Mth. naturw. Kl., Skr., v. 1, 85 p.

STRØM, K. M., 1939, Land-locked waters and the deposition of black muds, p. 356–372 *in* Trask, P. D., ed., Recent marine sediments. A symposium: Tulsa, OK, American Association of Petroleum Geologists; London, Thomas Murby, 936 p. (Reprinted, revised, and enlarged edition, 1955, Tulsa, OK, Society of Economic Paleontologists and Mineralogists Special Publication No. 4; also in paperback, 1968, New York, Dover.)

STRONG, P. G., and WALKER, R. G., 1981, Deposition of the Cambrian continental rise: the St. Roch Formation near St. Jean-Port-Joli, Quebec: Canadian Journal of Earth Sciences, v. 18, p. 1320–1335.

STUMPF, R. P., 1983, The process of sedimentation on the surface of a salt marsh: Estuarine and Coastal Shelf Science, v. 17, p. 495–508.

STURM, M.; and MATTER, A., 1978, Turbidites and varves in Lake Brienz (Switzerland): deposition of clastic detritus by density currents, p. 145–166 *in* Matter, A.; and Tucker, M. E., eds., Modern and Ancient Lake Sediments: International Association of Sedimentologists Special Publication 2, 290 p.

SUESS, E.; KULM, L. D.; and KILLINGLEY, J. S., 1982, Mechanism of dolomitization of Peru convergent margin (sic) sediments; isotope (sic) and mineral record (abstract): EOS, American Geophysical Union Transactions, v. 64, no. 45, p. 1000 (only).

SUESS, E.; and THIEDE, J., eds., 1983, Coastal upwelling—its sediment record, Part A: Responses of the sedimentary regime to present coastal upwelling: New York, Plenum Press, 604 p.

SUESS, H. E., 1965, Secular variations of the cosmic-ray-produced carbon 14 in the atmosphere and their interpretations: Journal of Geophysical Research, v. 70, p. 5937–5952.

SUESS, H. E., 1970a, Bristlecone-pine calibration of the radiocarbon time scale 5200 B.C. to the present, p. 303–309 *in* Olsson, I. U., ed., Radiocarbon variations and absolute chronology: Nobel Symposium, Twelfth, Proceedings: New York, John Wiley & Sons, 652 p.

SUESS, H. E., 1970b, The three causes of the ^{14}C fluctuations, their amplitudes and time constant, p. 595–604 *in* Olsson, I. U., ed., Radiocarbon variations and absolute chronology: Nobel Symposium, Twelfth, Proceedings: New York, John Wiley & Sons, 652 p.

SUESS, H. E., 1974, Natural radiocarbon evidence bearing on climatic changes, p. 311–317 *in* Labeyrie, J., ed., Les methodes quantitatives d'etude des variations du climat au cours de Pleistocene: Paris, Editions du Centre National de la Recherche Scientifique, 317 p.

SUESS, H. E., 1978, LaJolla measurements of radiocarbon in tree-ring dated (sic) wood: Radiocarbon, v. 10, p. 1–18.

SUESS, H. E., 1980a, The radiocarbon record in tree rings of the last 8,000 years: Radiocarbon, v. 22, p. 200–209.

SUESS, H. E., 1980b, Radiocarbon geophysics: Endeavour (new series), v. 4, p. 113–117.

SUESS, H. E., 1981, Solar activity, cosmic-ray produced carbon 14, and the terrestrial climate, p. 307–310 *in* Sun and climate: Conference on Sun and climate, Toulouse, Proceedings: Paris, Centre National d'Etudes Spatiale.

SUMMERFIELD, M. A., 1983, Petrography and diagenesis of silcrete from the Kalahari Basin and Cape Coastal Zone, southern Africa: Journal of Sedimentary Petrology, v. 53, p. 895–910.

SUNAMURA, T., 1984, Quantitative predictions of beach-face slopes: Geological Society of America Bulletin, v. 95, p. 242–245.

SURDAM, R. C., and STANLEY, K. O., 1980, Effects of changes in drainage-basin boundaries on sedimentation in Eocene Lakes Gosiute and Uinta of Wyoming, Utah and Colorado: Geology, v. 8, p. 135–139.

SURLYK, F., 1978, Submarine fan sedimentation along fault scarps on tilted fault blocks (Jurassic/Cretaceous boundary, East Greenland): Bull. Gron. Geol. Unders., v. 128, 108 p.

SUTER, J. R., and BERRYHILL, H. L., 1985, Late (sic) Quaternary shelf margin (sic) deltas, northwest Gulf of Mexico: American Association of Petroleum Geologists Bulletin, v. 69, p. 77–91.

SUTER, J. R.; BERRYHILL, H. L., JR.; and PENLAND, S., 1987, Late Quaternary sea level (sic) fluctuations and depositional sequences, southwest Louisiana continental shelf, p. 199–219 *in* Nummedal, Dag; Pilkey, O. H.; and Howard, J. D., eds., Sea-level fluctuation and coastal evolution: Tulsa, OK, Society of Economic Paleontologists and Mineralogists Special Publication 41, 267 p.

SUTER, J. R.; and PENLAND, S., 1987, Evolution of Cat Island Pass, Louisiana: Coastal Sediments '87, American Society of Civil Engineers, 16 p.

SUTTNER, L. J., 1989, Recent advances in study of the detrital mineralogy of sand and sandstone: implications for teaching: Journal of Geological Education, v. 37, p. 235–240.

SVERDRUP, H. U., JOHNSON, M. W., and FLEMING, R. H., 1942, The oceans: Englewood Cliffs, NJ, Prentice-Hall, 1087 p.

SWIFT, D. J. P., and FIELD, M. E., 1981, Evolution of a classic sand ridge (*sic*) field: Maryland sector North American inner shelf: Sedimentology, v. 28, p. 461–482.

SWIFT, D. J. P., FIQUEIREDO, A. G., JR., FREELAND, G. L., and OERTEL, G. F., 1983, Hummocky cross-stratification and megaripples: a geological double standard: Journal of Sedimentary Petrology, v. 53, p. 1295–1317.

SWIFT, D. J. P.; HAN, GREGORY; and VINCENT, C. E., 1986, Fluid processes and sea-floor response on a modern storm-dominated shelf: Middle Atlantic shelf of North America. Part I: The storm-current regime, p. 99–119 *in* Knight, R. J., and McLean, J. R., eds., Shelf sands and sandstone reservoirs: Calgary, Alberta, Canadian Society of Petroleum Geologists Memoir 11, 554 p.

SWIFT, D. J. P.; HUDELSON, P. M.; BRENNER, R. L.; and THOMPSON, PETER, 1987, Shelf construction in a foreland basin: storm beds, shelf sand bodies and shelf-slope sequences in the Upper Cretaceous Mesaverde Group, Book Cliffs, Utah: Sedimentology, v. 34, p. 423–457.

SWIFT, D. J. P., and LYALL, A. K., 1967, Bay of Fundy: reconnaissance by sub-bottom profiles: Maritime Sediments, v. 3, p. 67–70.

SWIFT, D. J. P., and MASLOW, T. F., 1982, Holocene transgression in south central Long Island, New York—discussion: Journal of Sedimentary Petrology, v. 52, p. 1014–1019.

SWIFT, D. J. P., and NIEDORODA, A. W., 1985, Fluid and sediment dynamics on continental shelves, p. 47–134 *in* Tillman, R. W., and others, eds., Shelf sands and sandstone reservoirs: Tulsa, OK, Society of Economic Paleontologists and Mineralogists Short Course Notes 13, 708 p.

SWIFT, D. J. P., NIEDORODA, A. W., VINCENT, C. E., and HOPKINS, T. S., 1985, Barrier island (*sic*) evolution, middle Atlantic shelf, USA. Part I: Shoreface dynamics: Marine Geology, v. 63, p. 331–361.

SWIFT, D. J. P.; PHILLIPS, SANDRA; and THORNE, J. A., in press, Sedimentation on continental margins, Part V: Parasequences, *in* Swift, D. J. P., Oertel, G. W., and Tillman, R. W., eds., Shelf sand (*sic*) and sandstone (*sic*) bodies. Geometry (*sic*), facies, and distribution: International Association of Sedimentologists Special Publication 12.

SWIFT, D. J. P., SEARS, P. C.,; BOHLKE, B.; and HUNT, R., 1978, Evolution of a shoal retreat (*sic*) massif, North Carolina shelf; inferences from areal geology: Marine Geology, v. 27, p. 19–42.

SWIFT, D. J. P., and THORNE, J. A., in press, Sedimentation on continental margins. Part I: A general model for shelf sedimentation, *in* Swift, D. J. P., Oertel, G. W., and Tillman, R. W., eds., Shelf sand (*sic*) and sandstone (*sic*) bodies. Geometry (*sic*), facies, and distribution: International Association of Sedimentologists Special Publication 12.

SWIFT, D. J. P., THORNE, J. A., and OERTEL, G. F., 1986, Fluid processes and sea-floor response on a storm-dominated shelf: Middle Atlantic shelf of North America. Part II: Response of the shelf floor, p. 191–211 *in* Knight, R. J., and McLean, J. R., eds., Shelf sands and sandstone reservoirs: Calgary, Alberta, Canadian Society of Petroleum Geologists Memoir 11, 554 p.

SWIFT, J. H.; AAGAARD, K.; and MALMBERG, S.-A., 1980, The contribution of the Denmark Strait overflow to the deep North Atlantic: Deep-Sea Research, v. 27A, p. 29–42.

SWINEFORD, ADA; and FRYE, J. C., 1945, A mechanical analysis of wind-blown dust compared with analyses of loess: American Journal of Science, v. 243, p. 249–255.

SWIRYDCZUK, K.; WILKINSON, B. H.; and SMITH, G. R., 1980, The Pliocene Glenns Ferry Oolite—II: Sedimentology of oolitic lacustrine terrace deposits: Journal of Sedimentary Petrology, v. 50, p. 1237–1248.

SYLVESTER-BRADLEY, P. C., 1977, Biostratigraphical tests of evolutionary theory, p. 41–63 *in* Kauffman, E. G., and Hazel, J. E., eds., Concepts and methods of biostratigraphy: Stroudsburg, PA, Dowden, Hutchinson, and Ross, 658 p.

SYVITSKI, J. P. M., and FARROW, G. E., 1983, Structures and processes in bayhead deltas: Knight and Bute Inlets, British Columbia: Sedimentary Geology, v. 36, p. 217–244.

SZALAY, A.; and KONCZ, I., 1980, Prozesse der kohlenwasserstoffbildung und der migration in den Neogen-Senken des Pannon-Beckens: 26th International Geological Congress, Abstracts, p. 1070.

TADA, R.; and IIJIMA, A., 1983, Petrology and diagenetic changes of Neogene siliceous rocks in northern Japan: Journal of Sedimentary Petrology, v. 53, p. 911–930.

TANDON, S. K., and FRIEND, P. F., 1989, Near-surface shrinkage (*sic*) and carbonate replacement (*sic*) processes, Arran Cornstone Formation, Scotland: Sedimentology, v. 36, p. 1113–1126.

TANKARD, A. J., and BARWIS, J. H., 1982, Wave-dominated deltaic sedimentation in the Devonian Bokkeveld Basin of South Africa: Journal of Sedimentary Petrology, v. 52, p. 959–974.

TANNER, W. F., 1969, The particle size (*sic*) scale: Journal of Sedimentary Petrology, v. 39, p. 809–812.

TARLING, D. H., 1983, Paleomagnetism: principles and applications in geology, geophysics and archaeology: New York, Chapman and Hall.

TAYLOR, J. M., 1950, Pore-space reduction in sandstones: American Association of Petroleum Geologists Bulletin, v. 34, p. 701–716.

TAYLOR, J. M., 1980, Origin of the Werraanhydrit in the United Kingdom southern North Sea—a reappraisal, p. 91–113 *in* Füchtbauer, Hans; and Peryt, T. M., eds., The Zechstein Basin, with emphasis on carbonate sequences: Contributions to Sedimentology, v. 9, 328 p.

TEICHMULLER, M.; and TEICHMULLER, R., 1981, The significance of coalification studies to geology—a review: Centres de Recherches Exploration-Production Elf-Aquitaine Bulletin, v. 5 (2), p. 491–534.

TERS, MIREILLE, 1987, Variations in Holocene sea level on the French Atlantic coast and their climatic significance: chapter 12, p. 204–237 *in* Rampino, M. R., and others, eds., Climate. History, periodicity, and predictability: New York, Van Nostrand-Reinhold Company, 588 p.

TERUGGI, M. E., 1957, The nature and origin of Argentine loess: Journal of Sedimentary Petrology, v. 27, p. 322–332.

TERWINDT, J. H. J., and BROUWER, M. J. H. M., 1986, The behavior of intertidal sandwaves during neap-spring tide cycles, and the relevance for paleoflow reconstructions: Sedimentology, v. 33, p. 1–31.

TERZAGHI, K. C., and PECK, R. B., 1967, Soil mechanics in engineering practice, 2nd ed.: New York, John Wiley & Sons, 729 p.

TESSON, M.; GENSOUS, B.; ALLEN, G. P.; and RAVENNE, C., 1990, Late (*sic*) Quaternary lowstand wedges on the Rhône continental shelf, France: Marine Geology, v. 91, p. 325–332.

THIEDE, J.; and SUESS, E., eds., 1983, Coastal upwelling—its sediment record, Part B: sedimentary records of ancient coastal upwelling: New York, Plenum Press, 610 p.

THIEDE, J.; and VAN ANDEL, Tj. H., 1977, The paleoenvironment of anaerobic sediments of the late Mesozoic south Atlantic Ocean: Earth and Planetary Science Letters, v. 33, p. 301–309.

THIERSTEIN, H. R., and BERGER, W. H., 1978, Injection events in ocean history: Nature, v. 276, p. 461–466.

THOMAS, K. V.; and BABA, M., 1986, Berm development on a monsoon-influenced microtidal beach: Sedimentology, v. 33, p. 537–546.

THOMAS, W. A., and BACK, G. H., 1982, Paleogeographic relationship of a Mississippian barrier-island (*sic*) and shelf-bar system (Hartselle Sandstone) in Alabama to the Appalachian-Ouachita orogenic belt: Geological Society of America Bulletin, v. 93, p. 6–19.

THOMPSON, T. L., 1976, Plate tectonics in oil and gas exploration of continental margins: American Association of Petroleum Geologists Bulletin, v. 60, p. 1463–1501.

THOMPSON, W. C., 1955, Sandless coastal terrain of the Atchafalaya Bay area, Louisiana, p. 52–97 *in* Hough, J. L., and Menard, H. W., eds., Finding ancient shorelines, a symposium: Tulsa, OK, Society of Economic Paleontologists and Mineralogists Special Publication 3, 129 p.

THOMSEN, ERIK, 1989, Seasonal variability in the production of Lower Cretaceous calcareous nannoplankton: Geology, v. 17, p. 715–717.

THORARINSSON, SIGURDUR, 1954, The tephra-fall (*sic*) from Hekla on March 29th 1947: Reykjavik Soc. Sci. Icelandica, v. 2, no. 3, 68 p.

THORBJARNARSON, K. W., NITTROUER, C. A., DeMaster, D. J., and McKINNEY, R. B., 1985, Sediment accumulation in a back-barrier lagoon, Great Sound, New Jersey: Journal of Sedimentary Petrology, v. 55, p. 856–883.

THORNBURG, T. M., and KULM, L. D., 1987, Sedimentation in the Chile Trench: depositional morphologies, lithofacies and stratigraphy: Geological Society of America Bulletin, v. 98, p. 33–52.

THORNE, J. A.; GRACE, EUGENIA; SWIFT, D. J. P.; and NIEDORODA, A. W., in press, Sedimentation on continental margins, Part III. The depositional fabric: an analytical approach to stratification and facies, *in* Swift, D. J. P., Oertel, G. W.; and Tillman, R. W., eds., Shelf sand (*sic*) and sandstone (*sic*) bodies. Geometry (*sic*), facies, and distribution: International Association of Sedimentologists Special Publication 12.

THORNE, J. A., and SWIFT, D. J. P., in press, Sedimentation on continental margins, Part II: application of the regime concept, *in* Swift, D. J. P., Oertel, G. W.; and Tillman, R. W., eds., Shelf sand (*sic*) and sandstone (*sic*) bodies. Geometry (*sic*), facies, and distribution: International Association of Sedimentologists Special Publication 12.

THORNE, J. A., and WATTS, A. B., 1984, Seismic reflectors and unconformities at passive continental margins: Nature, v. 311, p. 365–368.

THORSON, GUNNAR, 1957, Bottom communities (sublittoral or shallow shelf), p. 461–535 *in* Hedgpeth, J. W., ed., Treatise on marine ecology and paleoecology (*sic*): Geological Society of America Memoir 67, v. 1, Ecology, 1296 p.

THORSON, R. M., and DIXON, E. J., Jr., 1983, Alluvial history of the Porcupine River, Alaska: role of glacial-lake overflow from northwest Canada: Geological Society of America Bulletin, v. 94, p. 576–589.

THUNNELL, R. C., and WILLIAMS, D. F., 1983, Paleotemperature (*sic*) and paleosalinity history of the eastern Mediterranean during the late Quaternary: Palaeogeography, Palaeoclimatology, and Palaeoecology, v. 44, p. 23–29.

THUNNELL, R. C., WILLIAMS, D. F., and BELYEA, P. R., 1984, Anoxic events in the Mediterranean Sea in relation to the evolution of late Neogene climates: Marine Geology, v. 59, p. 105–134.

THUNNELL, R. C.; WILLIAMS, D. F.; and HOWELL, M., 1987, Atlantic-Mediterranean water exchange during the late Neogene: Paleoceanography, v. 2, p. 661–678.

TILLMAN, R. W., and SIEMERS, C. T., eds., 1984, Siliciclastic shelf sediments: Tulsa, OK, Society of Economic Paleontologists and Mineralogists Special Publication 34, 268 p.

TILLMAN, R. W., SWIFT, D. J. P., and WALKER, R. G., eds., 1984, Shelf sands and sandstone reservoirs: San Antonio, TX, Society of Economic Paleontologists and Mineralogists Short Course 13, 708 p.

TISSOT, B. P., 1987, Migration of hydrocarbons in sedimentary basins, a geological (*sic*), geochemical (*sic*), and historical perspective, p. 1–19 *in* Doligezz, B., ed., Migration of hydrocarbons in sedimentary basins: Paris, Editions Technip.

TISSOT, B.; DEROO, G.; and HERBIN, J. P., 1979, Organic matter in Cretaceous sediments of the North Atlantic: Contribution to sedimentology and paleogeography, p. 362–374 *in* Talwani, Manik; Hay, W.; and Ryan, W. B. F., eds., Deep drilling (*sic*) results in the Atlantic Ocean: continental margins and paleoenvironment: American Geophysical Union, Maurice Ewing Series, v. 3, 437 p.

TISSOT, B. P., and WELTE, D. H., 1978, Petroleum formation and occurrence—a new approach to oil and gas exploration: Berlin, Springer-Verlag, 538 p.

TORRESAN, M. E., and SCHWAB, W. C., 1987, Fabric and its relation (*sic*) to sedimentologic (*sic*) and physical properties of near-surface sediment, Shelikof Strait and Alsek prodelta, Alaska: Journal of Sedimentary Petrology, v. 57, p. 408–418.

TRASK, P. D., 1953, Chemical studies of sediments of the western Gulf of Mexico: Massachusetts Institute of Technology-Woods Hole Oceanographic Institution, Papers in Physical Oceanography and Meteorology, v. 12, p. 47–120.

TRIBBLE, G. W., SANSONE, F. J., and SMITH, S. V., 1990, Stoichiometric modeling of carbon diagenesis within a coral reef (*sic*) framework: Geochimica et Cosmochimica Acta, v. 54, p. 2439–2449.

TRIPLEHORN, D. M., 1966, Morphology, internal structure, and origin of glauconite pellets: Sedimentology, v. 6, p. 247–266.

TROWBRIDGE, A. C., 1930, Building of Mississippi delta: American Association of Petroleum Geologists Bulletin, v. 14, p. 867–901.

TRUSWELL, J. F., 1972, Sandstone sheets and related intrusions (*sic*) from Coffee Bay, Transkei, South Africa: Journal of Sedimentary Petrology, v. 42, p. 578–583.

TSOAR, H., 1983, Dynamic processes acting on a longitudinal (seif) sand dune: Sedimentology, v. 30, p. 567–578.

TUCKER, M. E., and WRIGHT, V. P., 1990, Carbonate sedimentology: Oxford, Blackwell Scientific Publications, 482 p.

TUDHOPE, A. W., and RISK, M. J., 1985, Rate of dissolution of carbonate sediments by microboring organisms, Davies Reef, Australia: Journal of Sedimentary Petrology, v. 55, p. 440–447.

TWENHOFEL, W. H., and SHROCK, R. R., 1935, Invertebrate paleontology: New York, McGraw-Hill Book Company, 511 p.

TWICHELL, D. C., GRIMES, C. B., JONES, R. S., and ABLE, K. W., 1985, The role of erosion by fish in shaping topography around Hudson submarine canyon: Journal of Sedimentary Petrology, v. 55, p. 712–719.

TYE, R. S., 1987 ms., Nonmarine Atchafalaya deltas: Unpublished Ph.D. dissertation, Louisiana State University, 224 p.

TYE, R. S., and COLEMAN, J. M., 1989, Depositional processes and stratigraphy of fluvially dominated lacustrine deltas: Mississippi delta plain: Journal of Sedimentary Petrology, v. 59, p. 973–996.

TYE, R. S., and KOSTERS, E. C., 1986, Styles of interdistributary basin sedimentation, Mississippi Delta Plain, Louisiana: Gulf Coast Association of Geological Societies Transactions, v. 36, p. 575–588.

TYRRELL, W. W., JR., 1969, Criteria useful in interpreting environments of unlike but time-equivalent carbonate units (Tansill-Capitan-Lamar), Capitan reef complex, West Texas and New Mexico: p. 80–97 *in* Friedman, G. M., ed., Depositional environments in carbonate rocks: Tulsa, OK, Society of Economic Paleontologists and Mineralogists Special Publication 14, 209 p.

UCHUPI, E.; and AUSTIN, J. A., 1979, The stratigraphy and structure of the Laurentian Cone region: Canadian Journal of Earth Sciences, v. 16, p. 1726–1752.

UDDEN, J. A., 1914, Mechanical composition of clastic sediments: Geological Society of America Bulletin, v. 25, p. 655–744.

UHLIR, D. M.; AKERS, ARTHUR; and VONDRA, C. F., 1988, Tidal flat (*sic*) sequence, Sundance Formation (Upper Jurassic), north-central Wyoming: Sedimentology, v. 35, p. 739–752.

UMBROVE, J. H. F., 1942, The pulse of the Earth: The Hague, Nijohff, 179 p.

UNDERWOOD, M. B., 1986, Transverse infilling of the central Aleutian Trench by unconfined turbidity currents: Geo-Marine Letters, v. 6, p. 7–13.

UNITED STATES DEPARTMENT OF ENERGY, 1988, What is tuff? Office of Civilian Radioactive Waste (*sic*) Management Nevada Nuclear Waste (*sic*) Storage Investigations, 2 p.

USIGLIO, J., 1849, Analyse de l'eau de la Mediterranée sur le Cotes de France: Ann. des Chem. Phys., 3rd series, v. 27, p. 92–107.

VACHER, H. L., BENGTSSON, T. O., and PLUMMER, L. N., 1990, Hydrology of meteoric diagenesis: residence time of meteoric ground water in island freshwater lenses with application to aragonite-calcite stabilization rate in Bermuda: Geological Society of America Bulletin, v. 102, p. 223–232.

VAIL, P. R., 1987, Seismic stratigraphic (*sic*) interpretation procedure, p. 1–10 *in* Bally, A. W., ed., Atlas of seismic stratigraphy: Tulsa, OK, The American Association of Petroleum Geologists, Studies in Geology 27, v. 1, 125 p.

VAIL, P. R.; MITCHUM, R. M., JR.; and THOMPSON, S., III, 1977, Seismic stratigraphy and global changes of sea level, Part 4: Global cycles and relative changes of sea level, p. 83–97 *in* Payton, C. E., ed., Seismic stratigraphy; applications to hydrocarbon exploration: Tulsa, OK, American Association of Petroleum Geologists Memoir 26, 516 p.

VALENTIN, HELMUT, 1952, Die Kusten der Erde. Beitrage zur allgemeinen und regionalen Kustengeomorphologie: Berlin, Justs Perthes Gotha, 118 p.

VALENTINE, J. W., 1973, Evolutionary paleoecology of the marine biosphere: Englewood Cliffs, NJ, Prentice-Hall, 511 p.

VALENTINE, J. W., 1977, Biogeography and biostratigraphy, p. 143–162 *in* Kauffman, E. G., and Hazel, J. E., eds., 1977, Concepts and methods in biostratigraphy: Stroudsburg, PA, Dowden, Hutchinson, and Ross, 658 p.

VALENTINE, P. C., COOPER, R. A., and UZMANN, J. R., 1984, Submarine sand dunes and sedimentary environments in Oceanographer Canyon: Journal of Sedimentary Petrology, v. 54, p. 704–715.

VAN ANDEL, TJ. H., 1968, Deep-sea drilling for scientific purposes: a decade of dreams: Science, v. 160, p. 1419–1424.

VAN ANDEL, TJ. H., 1981, Consider the incompleteness of the geological record: Nature, v. 294, p. 397–398.

VAN ANDEL, TJ. H., and VEEVERS, J. J., 1967, Morphology and sediments of the Timor Sea: Australia Department of National Development, Bureau of Mineral Resources, Geology and Geophysics Bulletin 83, 173 p.

VAN ARSDALE, ROY, 1982, Influence of calcrete on the geometry (*sic*) of arroyos near Buckeye, Arizona: Geological Society of America Bulletin, v. 93, p. 20–26.

VAN DER PLAS, L., 1966, The identification of detrital feldspars: Amsterdam, Elsevier Scientific Publishing Company, 305 p.

VAN DER WESTHUIZEN, W. A., GROBLER, N. J., LOOCK, J. C., and TORDIFFE, E. A. W., 1989, Raindrop imprints in the Late Archaean-Early Proterozoic Ventersdorp Supergroup, South Africa: Sedimentary Geology, v. 61, p. 303–309.

VAN HARDEN, I. L., 1983, Deltaic sedimentation in eastern Atchafalaya Bay, Louisiana: Baton Rouge, Louisiana Sea Grant College Program, Center for Wetland Resources, Louisiana State University, 117 p.

VAN HINTE, J. E., CITA, M. B., and VAN DER WEIJDEN, C. H., eds., 1987, Extant (*sic*) and ancient anoxic basin (*sic*) conditions in the eastern Mediterranean: Marine Geology, v. 75 (Special Issue), 281 p.

VAN HOUTEN, F. B., 1990, Paleozoic oolitic ironstones on North American craton (abstract): Geological Society of America, Northeastern Section, Abstracts With Programs, v. 22, no. 2, p. 76 (only).

VAN HOUTEN, F. B., and BHATTACHARYYA, D. B., 1982, Phanerozoic oolitic ironstones, geologic record and facies model: Annual Reviews of Earth and Planetary Sciences, v. 10, p. 441–457.

VAN HOUTEN, F. B., and PURUCKER, M. E., 1985, On the origin of glauconite and chamositic granules: Geo-Marine Letters, v. 5, p. 47–49.

VAN RIJN, L. C., 1984a, Sediment transport, part I: Bed load (*sic*) transport: American Society of Civil Engineers, Journal of Hydraulic Engineering, v. 110, no. 10, p. 1431–1456.

VAN RIJN, L. C., 1984b, Sediment transport, part II: Suspended load (*sic*) transport: American Society of Civil Engineers, Journal of Hydraulic Engineering, v. 110, no. 10, p. 1613–1641.

VAN RIJN, L. C., 1984c, Sediment transport, part III: Bed forms and alluvial roughness: American Society of Civil Engineers, Hydraulic Engineering, v. 110, no. 12, p. 1733–1754.

VAN STRAATEN, L. M. J. U., 1960, Some recent advances in the study of deltaic sedimentation: Liverpool & Manchester Geological Journal, v. 2, p. 411–442.

VAN STRAATEN, L. M. J. U., 1970, Holocene and late-Pleistocene sedimentation in the Adriatic Sea: Geologische Rundschau, v. 60, p. 106–131.

VAN STRAATEN, L. M. J. U., 1972, Holocene stages of oxygen depletion in deep waters of the Adriatic Sea, p. 631–643 *in* Stanley, D. J., ed., The Mediterranean Sea: a natural sedimentation laboratory: Stroudsburg, PA, Dowden, Hutchinson, and Ross, 765 p.

VAN TASSEL, J., 1981, Silver abyssal plain (*sic*) carbonate turbidity: flow characteristics: Journal of Geology, v. 89, p. 317–333.

VAN TASSELL, J., 1987, Upper Devonian Catskill delta margin cyclic sedimentation: Brallier, Scherr, and Foreknobs Formations of Virginia and West Virginia: Geological Society of America Bulletin, v. 99, p. 414–426.

VAN WAGONER, J. C., MITCHUM, R. M., CAMPION, K. M., and RAHMANIAN, V. D., 1990, Siliciclastic sequence stratigraphy in well logs, cores, and outcrops: Tulsa, OK, American Association of Petroleum Geologists, Methods in Exploration Series 7, 55 p.

VAN WAGONER, J. C.; POSAMENTIER, H. W.; MITCHUM, R. M., JR.; VAIL, P. R.; SARG, J. F.; LOUTIT, T. S.; and HARDENBOL, J., 1988, An overview of the fundamentals of sequence stratigraphy and key definitions, p. 39–45 *in* Wilgus, C. K.; Hastings, B. S.; Kendall, C. G. St. C.; Posamentier, H. W.; Ross, C. A.; and Van Wagoner, J. C., eds., Sea-level changes: an integrated approach: Tulsa, OK, Society of Economic Paleontologists and Mineralogists Special Publication 42, 407 p.

VAN WATERSCHOOT VAN DER GRACHT, W. A. J. M., 1929, Sind jetzt Muttergesteine kunstige Erdollagerstatten in Bildung begriffen: Petrol. Zeitschrift, v. 25, p. 183–191.

VAUGHAN, T. W., 1910a, Geology of the Keys, the marine bottom deposits, and the recent corals of southern Florida: Carnegie Institution of Washington, Yearbook 8, p. 140–144.

VAUGHAN, T. W., 1910b, A contribution to the geologic history of the Floridian Plateau: Carnegie Institution of Washington Publication 133, p. 99–185.

VAUGHAN, T. W., 1915, The geologic significance of the growth-rate (sic) of the Floridian and Bahaman shoal-water corals: Washington Academy of Sciences Journal, v. 5, p. 591–600.

VAUGHAN, T. W., 1916a, The present status of the investigation of the origin of barrier coral reefs: American Journal of Science, 4th series, v. 41, p. 131–135.

VAUGHAN, T. W., 1916b, The results of investigations of the ecology of the Floridian and Bahaman shoal-water corals: U.S. National Academy of Sciences Proceedings, v. 2, p. 95–100.

VAUGHAN, T. W., 1916c, Some littoral (sic) and sublittoral physiographic features of the Virgin and northern Leeward Islands and their bearing on the coral reef (sic) problem: Washington Academy of Sciences Journal, v. 6, p. 53–66.

VAUGHAN, T. W., and SHAW, E. W., 1916, Geologic investigations of the Florida coral reef (sic) tract: Carnegie Institution of Washington Yearbook 14 (1915), p. 232–238.

VEATCH, A. C., and SMITH, P. A., 1939, Atlantic submarine valleys of the United States and the Congo Submarine Valley: Boulder, CO, Geological Society of America Special Paper 7, 101 p.

VEEVERS, J. J., and POWELL, C. MCA., 1987, Late Paleozoic glacial episodes in Gondwanaland reflected in transgressive-regressive depositional sequences in Euramerica: Geological Society of America Bulletin, v. 98, p. 475–496.

VENEC-PEYRE, M.-T., 1987, Boring foraminifera in French Polynesian coral reefs: Coral Reefs, v. 5, p. 205–212.

VENKATARATHNAM, KOLLA; BISCAYE, P. E.; and RYAN, W. B. F., 1972, Origin and dispersal of Holocene sediments in the eastern Mediterranean Sea, p. 455–469 in Stanley, D. J., ed., The Mediterranean Sea: a natural sedimentation laboratory: Stroudsburg, PA, Dowden, Hutchinson, and Ross, 765 p.

VERCOUTERE, T. L.; MULLINS, H. T.; MCDOUGALL, KRISTIN; and THOMPSON, J. B., 1987, Sedimentation across the central California oxygen minimum zone: an alternate coastal upwelling sequence: Journal of Sedimentary Petrology, v. 57, p. 709–722.

VERMEIJ, G. J., 1978, Biogeography and adaptation: patterns of marine life: Cambridge, MA, Harvard University Press, 332 p.

VERMEIJ, G. J., and PETUCH, E. J., 1986, Differential extinction in tropical American molluscs: endemism, architecture, and the Panama land bridge: Malacologia, v. 27, p. 29–41.

VINCENT, C. E., 1986, Processes affecting sand transport on a storm-dominated shelf, p. 121–132 in Knight, R. J., and McLean, J. R., eds., Shelf sands and sandstone reservoirs: Calgary, Alberta, Canadian Society of Petroleum Geologists Memoir 11, 554 p.

VINCENT, C. E.; SWIFT, D. J. P.; and HILLARD, BRUCE, 1981, Sediment transport in the New York Bight, North American Atlantic shelf, p. 369–398 in Nittrouer, C. A., ed., Sedimentary dynamics of continental shelves: Developments in sedimentology 32: Amsterdam, Elsevier Scientific Publishing Company, 449 p. (Same paper in Marine Geology, v. 42, p. 369–398.)

VINCENT, P. J., 1986, Differentiation of modern beach and coastal dune sands—a logistic regression approach using the parameters of the hyperbolic function: Sedimentary Geology, v. 49, p. 167–176.

VISHER, G. S., 1965, Fluvial processes as interpreted from ancient (sic) and recent fluvial deposits,

p. 116–132 in Middleton, G. V., ed., Primary sedimentary structures and their hydrodynamic interpretation: Tulsa, OK, Society of Economic Paleontologists and Mineralogists Special Publication 12, 265 p.

VISHER, G. S., 1969, Grain size (sic) distributions and depositional processes: Journal of Sedimentary Petrology, v. 39, p. 1074–1106.

VISHER, G. S., 1972, Physical characteristics of fluvial deposits, p. 84–97 in Rigby, J. K., and Hamblin, W. K., eds., Recognition of ancient sedimentary environments: Tulsa, OK, Society of Economic Paleontologists and Mineralogists Special Publication 16, 340 p.

VISSER, M. J., 1980, Neap-spring cycles reflected in Holocene subtidal large-scale bedform deposits: a preliminary note: Geology, v. 8, p. 543–546.

VOLLBRECHT, R., 1990, Marine and meteoric diagenesis of submarine Pleistocene carbonates from the Bermuda carbonate platform: Carbonates and Evaporites, v. 5, p. 13–95.

VON DER BORCH, C. C., 1976, Stratigraphy of stromatolite occurrences in carbonate lakes of the Coorong Lagoon area, South Australia, p. 413–420 in Walter, M. R., ed., Stromatolites. New York, Elsevier, 790 p.

VON DER BORCH, C. C. and LOCK, D. L., 1979, Geological significance of Coorong dolomites: Sedimentology, v. 26, p. 813–824.

VRBA, E., 1980, Evolution, species and fossils: how does life evolve? South African Journal of Science, v. 76, p. 61–84.

WALKER, J. C. G., and ZAHNLE, K. J., 1986, Lunar nodal tide and distance to the Moon during the Precambrian: Nature, v. 320, p. 600–602.

WALKER, K. R., JERNIGAN, D. G., and WEBER, L. J., 1990, Petrographic criteria for the recognition of marine, syntaxial overgrowths, and their distribution in geologic time: Carbonates and Evaporites, v. 5, p. 141–152.

WALKER, R. G., 1976, Facies models 2. Turbidites and associated coarse clastic deposits: Geoscience Canada, v. 3, p. 25–36.

WALKER, R. G., 1978, Deep water (sic) sandstone facies and ancient submarine fans: models for exploration for stratigraphic traps: American Association of Petroleum Geologists Bulletin, v. 62, p. 932–966.

WALKER, R. G., ed., 1984, Facies models, 2nd ed.: Geoscience Canada Reprint Ser. 1, 317 p.

WALKER, R. G.; and CANT, D. J., 1984, Sandy fluvial systems, p. 71–89 in Walker, R. G., ed., Facies Models, 2nd ed., Geoscience Canada Reprint Ser. 1, 317 p.

WALKER, T. R., and HARMS, J. C., 1972, Eolian origin of flagstone beds, Lyons Sandstone (Permian), type area, Boulder County, Colorado, p. 279–288 in Environments of sandstone, carbonate, and evaporite deposition: The Mountain Geologist, v. 9, p. 279–299.

WALTER, L. M., and BURTON, E. A., 1990, Dissolution of Recent platform carbonate sediments in marine pore fluids: American Journal of Science, v. 288, p. 601–643.

WALTON, W. R., and SMITH, W. T., 1969, Ancient coastal lagoons, p. 237–248 in Ayala-Castanares, Augustin; and Phleger, F. B., eds., Lagunas costeras, un simposio: Simposio Internacional sobre lagunas costeras, Mexico City, Mexico, 28–30 Noviembre 1967, memoria: Mexico Universidad Nacional Autonoma (WMAS-UNESCO), 686 p.

WANLESS, H. R., 1979, Limestone response to stress: pressure solution and dolomitization: Journal of Sedimentary Petrology, v. 49, p. 437–462.

WANLESS, H. R., and WELLER, J. M., 1932, Correlation and extent of Pennsylvanian cyclothems: Geological Society of America Bulletin, v. 43, p. 1003–1016.

WANLESS, H. R.; BAROFFIO, J. R.; GAMBLE, J. C.; HORNE, J. C.; ORLOPP, D. R.; ROCHA-CAMPOS, A.; SOUTER, J. E.; TRESCOTT, P. C.; VAIL, R. S.; and

WRIGHT, C. R., 1970, Late (sic) Paleozoic deltas in the central and eastern United States, p. 215–245 in Morgan, J. P., ed., Deltaic sedimentation, modern and ancient: Tulsa, OK, Society of Economic Paleontologists and Mineralogists Special Publication 15, 312 p.

WANLESS, H. R., TYRRELL, K. M., TEDESCO, L. P., and DRAVIS, J. J., 1988, Tidal-flat sedimentation from Hurricane Kate, Caicos Platform, British West Indies: Journal of Sedimentary Petrology, v. 58, p. 724–738.

WARD, J. D., 1988, Aeolian (sic), fluvial (sic), and pan (playa) facies of the Tertiary Tsondab Sandstone in the central Namib Desert, Namibia: Sedimentary Geology, v. 55, p. 143–162.

WARDLAW, N. C., 1976, Pore geometry (sic) of carbonate rocks as revealed by pore casts and capillary pressure: American Association of Petroleum Geologists Bulletin, v. 60, p. 245–257.

WARESBACK, D. B., and TURBEVILLE, B. N., 1990, Evolution of a Plio-Pleistocene volcanogenic-alluvial fan: the Puye Formation, Jemes Mountains, New Mexico: Geological Society of America Bulletin, v. 102, p. 298–314.

WARME, J. E., 1969, Mugu Lagoon, coastal southern California: origin, sediments and productivity, p. 137–154 in Ayala-Castanares, Augustin; and Phleger, F. B., eds., Lagunas costeras, un simposio: Simposio Internacional sobre lagunas costeras, Mexico City, Mexico, 28–30 Noviembre 1967, memoria: Mexico Universidad Nacional Autonoma (WMAS-UNESCO), 686 p.

WARME, J. E., DOUGLAS, R. G., and WINTERER, E. L., 1981, The Deep Sea Drilling Project: a decade of progress: Tulsa, OK, Society of Economic Paleontologists and Mineralogists Special Publication 32, 564 p.

WARREN, J. K., 1982, The hydrological setting, occurrence and significance of gypsum in late Quaternary salt lakes in South Australia: Sedimentology, v. 29, p. 609–637.

WARREN, J. K., 1983, Tepees, modern (southern Australia) and ancient (Permian–Texas and New Mexico)—a comparison: Sedimentary Geology, v. 34, p. 1–19.

WARREN, J. K., 1986, Source rock (sic) potential of shallow water evaporitic settings: Journal of Sedimentary Petrology, v. 56, p. 442–454.

WARREN, J. K., 1988, Sedimentology of Coorong dolomite in the Salt Creek region, South Australia: Carbonates and Evaporites, v. 3, p. 175–199.

WARREN, J. K., 1989, Evaporite sedimentology: Englewood Cliffs, NJ, Prentice Hall, 285 p.

WARREN, J. K., and KENDALL, C. G. St. C., 1985, Comparison of sequences formed in marine sabkha (subaerial) and salina (subaqueous) settings—modern and ancient, American Association of Petroleum Geologists Bulletin, v. 69, p. 1013–1023.

WARSHAW, C. M.; and ROY, RUSTUM, 1961, Classification and a scheme for the identification of layer silicates: Geological Society of America Bulletin, v. 72, p. 1455–1492.

WATABE, NORIMITSU, 1989, Calcium phosphate (sic) structures in invertebrates and protozoans, p. 35–44 in Carter, J. G., ed., Skeletal biomineralization: patterns, processes and evolutionary trends: American Geophysical Union Short Course in Geology, v. 5, pt. 2.

WATSON, R. A., 1983, A critique of chronostratigraphy: American Journal of Science, v. 283, p. 173–177.

WATTS, A. B., KARNER, G. D., and STECKLER, M. S., 1982, Lithospheric flexure and the evolution of sedimentary basins: Philosophical Transactions, Royal Society, London, v. A305, p. 249–281.

WATTS, A. B.; and THORNE, J., 1984, Tectonics, global changes in sea level and their relationship to stratigraphic sequences at the US Atlantic continental margin: Marine Petroleum Geology, v. 1, p. 319–339.

WAVRA, C. S., ISAACSON, P.E., and HALL, W. E., 1986, Studies of the Idaho black shale (*sic*) belt: stratigraphy, depositional environment, and economic geology of the Permian Dollarhide Formation: Geological Society of America Bulletin, v. 97, p. 1504–1511.

WAY, J. H., JR., 1968, Bed thickness (*sic*) analysis of some Carboniferous fluvial sedimentary rocks near Joggins, Nova Scotia: Journal of Sedimentary Petrology, v. 38, p. 424–433.

WEAVER, C. E., 1967, Potassium, illite and the ocean: Geochimica Cosmochimica Acta, v. 31, p. 2181–2196.

WEEKS, L. G., 1953, Factors of sedimentary basin development that control oil occurrence: New York Academy of Sciences Transactions, series II, v. 15, no. 7, p. 228–234.

WEEKS, L. G., 1957, Origin of carbonate concretions in shales, Magdalena Valley, Colombia: Geological Society of America Bulletin, v. 68, p. 95–102.

WEEKS, L. G., ed., 1958, Habitat of oil, a symposium: Tulsa, OK, American Association of Petroleum Geologists, 1384 p.

WEGENER, ALFRED, 1922, Die Entstehung der Kontinente und Ozeane, 3rd ed: Braunschweig, Germany, Friedr. Vieweg & Sohn Akt. Ges., 144 p.

WEIMER, R. J., HOWARD, J. D., and LINDSAY, D. R., 1982, Tidal flats and associated tidal channels, p. 191–245 *in* Scholle, P. A.; and Spearing, D., eds., Sandstone depositional environments: American Association of Petroleum Geologists Memoir 31, 410 p.

WEINER, JONATHAN, 1989, Glacier bubbles are telling us what was in ice age (*sic*) air: Smithsonian, v. 20, no. 2, p. 78–84, 86–87.

WEINSCHENK, E. H., 1916, The fundamental principles of petrology, translated by Albert Johannsen: New York, McGraw-Hill Book Company, 214 p.

WEISE, B. R., 1980, Wave-dominated delta systems of the Upper Cretaceous San Miguel Formation, Maverick Basin, South Texas: Texas Bureau of Economic Geology, Report of Investigations 107, 39 p.

WEISS, C. P. and WILKINSON, B. H., 1988, Holocene cementation along the central Texas coast: Journal of Sedimentary Petrology, v. 58, p. 468–478.

WELLS, J. P., and COLEMAN, J. M., 1978, Longshore transport of mud by waves: northeastern coast of South America: Geologie en Mijnbouw, v. 57, p. 353–359.

WELLS, J. P., and COLEMAN, J. M., 1981a, Physical processes and fine-grained sediment dynamics, coast of Surinam, South America: Journal of Sedimentary Petrology, v. 51, p. 1053–1068.

WELLS, J. P., and COLEMAN, J. M., 1981b, Periodic mudflat progradation, northeastern coast of South America: a hypothesis: Journal of Sedimentary Petrology, v. 51, p. 1069–1075.

WELLS, J. T., 1987, Effects of sea level (*sic*) rise on deltaic sedimentation (*sic*) in south central (*sic*) Louisiana, p. 157–166 *in* Nummedal, Dag; Pilkey, O. H.; and Howard, J. D., eds., Sea-level fluctuation and coastal evolution: Tulsa, OK, Society of Economic Paleontologists and Mineralogists Special Publication 41, 267 p.

WELLS, J. T., and ROBERTS, H. H., 1980, Fluid mud (*sic*) dynamics and shoreline stabilization Louisiana Chenier plain, p. 1382–1400 *in*, International Conference on Coastal Engineering, 17th, Proceedings: New York, American Society of Civil Engineers.

WENGERD, S. A., 1962, Pennsylvanian sedimentation in Paradox Basin, Four Corners region, p. 264–330 *in* Branson, C. C., ed., Pennsylvanian System in the United States: Tulsa, OK, American Association of Petroleum Geologists, 508 p.

WENTWORTH, C. K., 1922, A scale of grade and class terms for clastic sediments: Journal of Geology, v. 30, p. 377–392.

WESCOTT, W. A., and ETHRIDGE, F. C., 1980, Fan-delta sedimentology and tectonic setting, Yal-lahs fan delta: American Association of Petroleum Geologists Bulletin, v. 64, p. 374–399.

WESCOTT, W. A., and ETHRIDGE, F. C., 1983, Eocene fan delta-submarine fan deposition in the Wagwater Trough, east-central Jamaica: Sedimentology, v. 30, p. 235–247.

WEST, R. M., 1979, Apparent prolonged evolutionary stasis in the primitive Eocene hoofed mammal *Hypsodus*: Paleobiology, v. 5, p. 252–260.

WESTGATE, J., 1982, Discovery of a large-magnitude, late Pleistocene volcanic eruption (*sic*) in Alaska: Science, v. 218, p. 798–790.

WETZEL, R. G., 1983, Limnology, 2nd ed.: Philadelphia, W. B. Saunders, 767 p.

WEYL, P. K., 1968, The role of the oceans in climatic change: a theory (*sic*) of the ice ages: Meteorological Monographs, v. 8, p. 37–62.

WHITE, N.; and MCKENZIE, D. P., 1988, Formation of the steer's head geometry (*sic*) of sedimentary basins by differential stretching of the crust and mantle: Geology, v. 16, p. 250–253.

WHITE, R. M., 1990, The great climate debate: Scientific American, v. 263, no. 1, p. 36–43.

WHITEHURST, JOHN, 1778, An inquiry into the original state and formation of the Earth deduced (*sic*) from facts and the laws of nature. To which is added an appendix, containing some general observations on the strata in Derbyshire. With sections of them, representing their arrangement, affinities, and the mutations they have suffered at different periods of time. Intended to illustrate the preceding inquiries, and as a specimen of subterraneous geography: London, printed for the author, by J. Cooper in Drury-Lane, 199 p., 8 pls.

WHITTAKER, A.; COPE, J. C. W.; COWIE, J. W.; GIBBONS, W.; HAILWOOD, E. A.; HOUSE, M. R.; JENKINS, D. G.; Rawson, P. F.; RUSHTON, A. W. A.; SMITH, D. G.; THOMAS, A. T.; and WIMBLEDON, W. A., 1991, A guide to stratigraphical procedure: Journal of the Geological Society, London, v. 148, p. 813–824.

WHITTAKER, R. H., and LIKENS, G. E., 1973, Carbon in the biota, p. 281–302 *in* Woodwell, G. M., and Pecan, E. V., eds., Carbon and the biosphere: Washington, DC, U.S. Atomic Energy Commission, Technical Information center, 392 p.

WHITTINGTON, H. B., 1985, The Burgess Shale: New Haven, CT, Yale University Press, 151 p.

WILCOXEN, P. J., 1986, Coastal erosion and sea level rise—implications for ocean beaches and San Francisco's Westside transport project: Coastal Zone Management Journal, v. 4, p. 173–192.

WILDE, P.; NORMARK, W. R.; and CHASE, T. E., 1978, Channel sands and petroleum potential of Monterey deep-sea fan, California: American Association of Petroleum Geologists Bulletin, v. 62, p. 967–983.

WILGUS, C. K.; HASTINGS, B. S.; KENDALL, C. G. St. C.; POSAMENTIER, H. W.; ROSS, C. A.; and VAN WAGONER, J. C., eds., 1988, Sea-level changes: an integrated approach: Tulsa, OK, Society of Economic Paleontologists and Mineralogists Special Publication 42, 407 p.

WILKERSON, F. P.; KOBAYASHI, D.; and MUSCATINE, L., 1988, Mitotic index and size of symbiotic algae in Caribbean reef corals: Coral Reefs, v. 7, p. 29–36.

WILKINSON, B. H., 1979, Biomineralization, paleooceanography, and the evolution of calcareous marine organisms: Geology, v. 7, p. 524–527.

WILKINSON, B. H., 1982, Cyclic cratonic carbonates and phanerozoic (*sic*) calcite seas: Journal of Geological Education, v. 30, p. 189–203.

WILLETT, H. C., 1987, Climatic responses to variable solar activity—past, present, and predicted, Chapter 23, p. 404–414 *in* Rampino, M. R., Sanders, J. E., Newman, W. S., and Konigsson, L.-K., eds., Climate. History, periodicity, and predictability: New York, Van Nostrand Reinhold Company, 588 p.

WILLIAMS, ALWYN, 1989, Biomineralization in the lophophorates, p. 67–82, *in* Carter, J. G., ed., Skeletal biomineralization: patterns, processes and evolutionary trends: American Geophysical Union Short Course in Geology, v. 5, pt. 2.

WILLIAMS, A. T., and SCOTT, R. G., 1985, Discriminant analyses of some sediment statistical parameters obtained from the Long Island barrier chain, New York, U.S.A.: Northeastern Geology, v. 7, p. 37–42.

WILLIAMS, G. E., 1971, Flood deposits of the sand-bed ephemeral streams of central Australia: Sedimentology, v. 17, p. 1–40.

WILLIAMS, G. E., 1989, Tidal rhythmites: geochronometers for the ancient Earth-Moon system: Episodes, v. 12, p. 162–171.

WILLIAMS, G. E., and SONETT, C. P., 1986, Solar signature in sedimentary cycles from the late (*sic*) Precambrian Elatina Formation, Australia: Nature, v. 318, p. 523–527.

WILLIAMS, HOWELL; TURNER, F. J.; and GILBERT, C. M., 1954, Petrography: San Francisco, W. H. Freeman and Company, 406 p.

WILLIAMS, H. F. L., and ROBERTS, M. C., 1989, Holocene sea-level change and fan growth: Fraser River delta, British Columbia: Canadian Journal of Earth Science, v. 26, p. 1657–1666.

WILLIAMS, L. A., and CREAR, D. A., 1985, Silica diagenesis, II, general mechanisms: Journal of Sedimentary Petrology, v. 55, p. 312–321.

WILLIAMS, L. A., PARKS, G. A., and CREAR, D. A., 1985, Silica diagenesis, I, solubility controls: Journal of Sedimentary Petrology, v. 55, p. 301–311.

WILLIAMSON, P. G., 1980 ms., Evolutionary implication of late (*sic*) Cenozoic freshwater molluscs from the Turkana Basin, North Kenya: Bristol, England, University of Bristol, unpublished Ph.D. Dissertation.

WILLIAMSON, P. G., 1981, Paleontological documentation of speciation in Cenozoic molluscs from Turkana Basin: Nature, v. 292, p. 439–443.

WILSON, J. L., 1975, Carbonate facies in geologic history: New York, Springer-Verlag, 471 p.

WILSON, LEE, 1973, Variations in mean annual sediment yield as a function of mean annual precipitation: American Journal of Science, v. 273, p. 335–349.

WINKER, C. D., and BUFFLER, R. T., 1983, Evolution and seismic expression of Mesozoic and Cenozoic shelf margins, Gulf of Mexico and vicinity: American Association of Petroleum Geologists Bulletin, v. 67, p. 570.

WINKLER, C. D., 1982, Cenozoic shelf margins, northwest Gulf of Mexico basin: Gulf Coast Association of Geological Societies Transactions, v. 32, p. 427–448.

WITTICH, W., 1845, Curiosities of physical geography: London, Charles Knight & Co., ser. I, 225 p.; ser. II, 190 p.

WOFSY, S. C., 1976, Interactions of CH_4 and CO in the Earth's atmosphere: Annual Review of Earth and Planetary Sciences, v. 4, p. 441–469.

WOLCOTT, JOHN, 1988, Nonfluvial control of bimodal grain-size distributions in river-bed gravels: Journal of Sedimentary Petrology, v. 58, p. 979–984.

WOLFE, D. A., ed., 1986, Estuarine variability: New York, Academic Press, 509 p.

WOLFF, MONIKA, 1988, Torf und Kohle, p. 683–730 *in* Füchtbauer, Hans, ed., Sedimente und Sedimentgesteine: Stuttgart, E. Schweizerbart, 1141 p.

WOLFF, M. P., 1982, Evidence for onshore sand transfer along the south shore of Long Island, New York—and its implications against the "Bruun Rule": Northeastern Geology, v. 4, p. 10–16.

WOLLIN, GOESTA; SANDERS, J. E., and ERICSON, D. B., 1987, Abrupt geomagnetic variations—predictive signals for temperature changes 3–7 yr in advance, Chapter 13, p. 241–255 *in* Rampino, M. R., Sanders, J. E., Newman, W. S., and Konigsson, L.-K., eds., Climate. History, periodicity, and predictability: New York, Van Nostrand Reinhold Company, 588 p.

WOODROW, D. L., 1985, Paleogeography, paleoclimatology, and sedimentary processes of the Late Devonian Catskill Sea (Devonian), New York and Pennsylvania: Geological Society of America Bulletin, v. 96, p. 459–470.

WOODROW, D. L., and SEVON, W. D., eds., 1985, The Catskill delta: Boulder, CO, Geological Society of America Special Paper 201, 246 p.

WOOLNOUGH, W. G., 1937, Sedimentation in barred basins, and source rocks of oil: American Association of Petroleum Geologists Bulletin, v. 21, p. 1101–1157.

WORSLEY, T. R., NANCE, R. D., and MOODY, J. B., 1986, Tectonic cycles and the history of the Earth's biogeochemical (*sic*) and paleoceanographic record: Paleoceanography, v. 1, p. 233–263.

WRATHER, W. E., and LAHEE, F. H., eds., Problems of petroleum geology. A symposium: Tulsa, OK, American Association of Petroleum Geologists, Sidney Powers Memorial Volume, 1073 p. (London, Thomas Murby and Company).

WRIGHT, E. K., 1987, Stratification and paleocirculation of the Late Cretaceous Western Interior seaway of North America: Geological Society of America Bulletin, v. 99, p. 480–490.

WRIGHT, L. D., 1982, Field observations of long-period, surf zone (*sic*) standing waves in relation to contrasting beach morphologies: Australian Journal of Marine and Freshwater Research, v. 33, p. 181–201.

WRIGHT, L. D., 1989, Dispersal and deposition of river sediments in coastal seas: models from Asia and the tropics: Netherlands Journal of Sea Research, v. 23, no. 4, p. 493–500.

WRIGHT, L. D.; BOON, J. D., III; KIM, S. C.; and LIST, J. H., 1991, Modes of cross-shore sediment transport on the shoreface of the Middle Atlantic Bight: Marine Geology, v. 96, p. 19–51.

WRIGHT, L. D., and COLEMAN, J. M., 1974, Mississippi River mouth processes: effluent dynamics and morphologic development: Journal of Geology, v. 82, p. 751–778.

WRIGHT, L. D.; MAY, S. K., SHORT, A. D., and GREEN, M. O., 1985, Beach (*sic*) and surf zone (*sic*) equilibria and response times, p. 2150–2164 *in* International Conference on Coastal Engineering, 19th, Houston Texas, Proceedings: New York, American Society of Civil Engineers.

WRIGHT, L. D.; NIELSON, P.; SHI, N. C.; and LIST, J. H., 1986, Morphodynamics of a bar-trough surf zone: Marine Geology, v. 70, p. 251–286.

WRIGHT, L. D.; NIELSON, P.; SHORT, A. D.; and GREEN, M. O., 1982, Morphodynamics of a macrotidal beach: Marine Geology, v. 50, p. 97–128.

WRIGHT, L. D., and SEVON, W. D., 1985, The Catskill Delta: Geological Society of America Special Paper 201, 246 p.

WRIGHT, L. D., and SHORT, A. D., 1984, Morphodynamic variability of surf zones and beaches: a synthesis: Marine Geology, v. 56, p. 93–118.

WRIGHT, L. D.; SHORT, A. D.; BOON, J. D., III; HAYDEN, B.; KIMBALL, S.; and LIST, J. H., 1987, The morphodynamic effects of incident wave (*sic*) groupiness and tide range on an energetic beach: Marine Geology, v. 74, p. 1–20.

WRIGHT, L. D., SHORT, A. D., and GREEN, M. O., 1985, Short-term changes in the morphodynamic states of beaches and surf zones: an empirical predictive model: Marine Geology, v. 62, p. 339–364.

WRIGHT, L. D., and THOM, G. B., 1977, Coastal morphodynamics: Progress in Physical Geography, v. 1, p. 412–459.

WRIGHT, L. D.; WISEMAN, W. J.; BORNHOLD, B. D.; PRIOR, D. B.; SUHAYADA, J. N.; KELLER, G. H.; YANG, Z.-S.; and FANG, Y. B., 1988, Marine dispersal and deposition (*sic*) of Yellow River silts by gravity driven (*sic*) underflows: Nature, v. 332, p. 629–632.

WRIGHT, L. D.; WISEMAN, W. J. JR.; YANG, Z.-S.; BORNHOLD, B. D., KELLER, G. H.; PRIOR, D. B.; and SUHAYADA, J. N., 1990, Processes of marine dispersal and deposition (*sic*) of suspended silts off the modern mouth of the Huanghe (Yellow River): Continental Shelf (*sic*) Research, v. 10, no. 1, p. 1–40.

WRIGHT, ROBYN, 1986, Cycle stratigraphy as a paleogeographic tool: Point Lookout Sandstone, southeastern San Juan basin, New Mexico: Geological Society of America Bulletin, v. 96, p. 661–673.

WRIGHT, ROBYN; and ANDERSON, J. B., 1982, The importance of sediment gravity flow to sediment transport and sorting (*sic*) in a glacial marine (*sic*) environment: Eastern Weddell Sea, Antarctica: Geological Society of America Bulletin, v. 93, p. 951–963.

WRIGHT, P., 1984, Facies development on a barred (ridge and runnel) coastline: the case of south-west Lancashire (Merseyside), United Kingdom, p. 105–118 *in* Clark, M., ed., Coastal Research: United Kingdom Perspectives: Norwich, England, Geobooks, 131 p.

WÜNSCH, G., 1978, Quantitative bestimmungs-und trennungsmethoden, Band 6: Elemente der sechsten nebengruppe; wolfram: Handbuch der Analytischen Chemie: Berlin, Springer-Verlag, v. 3, 286 p.

WÜST, G.; BROGMUS, W.; and NOODT, E., 1954, Die Zonale Verteilung von Salzgehalt, Niederschlag, Verdunstung, Temperatur und Dichte an der Oberflache der Ozeane: Kieler Meeresforschungen, v. 10, p. 137–161.

WYRTKI, K., 1962, The oxygen minima in relation to oceanic circulation: Deep-Sea Research, v. 9, p. 11–23.

YEO, R. K., and RISK, M. J., 1981, The sedimentology, stratigraphy and preservation potential of intertidal deposits in the Minas Basin system, Bay of Fundy: Journal of Sedimentary Petrology, v. 51, p. 245–260.

YOUNG, H. R., and NELSON, C. S., 1988, Endolithic biodegradation of cool-water skeletal carbonates on Scott shelf, northwestern Vancouver Island, Canada: Sedimentary Geology, v. 60, p. 251–267.

YOUNG, R. G, 1955, Sedimentary facies and intertonguing in the Upper Cretaceous of the Book Cliffs, Utah-Colorado: Geological Society of America Bulletin, v. 66, p. 177–202.

YOUNG, S. W., 1976, Petrographic textures of detrital polycrystalline quartz as an aid to interpreting crystalline source rocks: Journal of Sedimentary Petrology, v. 46, p. 593–603.

YOUNG, T. P., and TAYLOR, W. E. G., eds., 1989, Phanerozoic ironstones: London, Geological Society of London, 251 p.

ZAHN, R.; and SARNTHEIN, M., 1987, Benthic isotope (*sic*) evidence for changes of the Mediterranean outflow during the late Quaternary: Paleoceanography, v. 2, p. 543–559.

ZAK, I., and FREUND, R., 1981, Asymmetry and basin migration in the Dead Sea rift: Tectonophysics, v. 80, p. 27–38.

ZARTMAN, R. E., 1988, Three decades of geochronologic studies in the New England Appalachians: Geological Society of America Bulletin, v. 100, p. 1168–1180.

ZARTMAN, R. E., and NAYLOR, R. S., 1984, Structural implications of some radiometric ages of igneous rocks in southeastern New England: Geological Society of America Bulletin, v. 95, p. 522–539

ZELLER, E. J., 1964, Cycles and psychology, p. 631–636 *in* Merriam, D. F., ed., Symposium on cyclic sedimentation: Kansas Geological Survey Bulletin 169, 2 volumes (v. 1, p. 1–380; v. 2, p. 381–636).

ZENKOVICH, V. P., 1967, Processes of coastal development: New York, Wiley-Interscience, 738 p. (Ed. by J. A. Steers, transl. by D. G. Fry.)

ZENKOVICH, V. P., 1969, Origin of barrier beaches and lagoon coast, p. 27–37 *in* Ayala-Castanares, Augustin; and Phleger, F. B., eds., Lagunas costeras, un simposio: Simposio Internacional sobre lagunas costeras, Mexico City, Mexico, 28–30 Noviembre 1967, memoria: Mexico Universidad Nacional Autonoma (WMAS-UNESCO), 686 p.

ZHENXIA, LIU; YICHANG, HUANG; and QINIAN, ZHANG, 1989, Tidal current ridges in the southwestern Yellow Sea: Journal of Sedimentary Petrology, v. 59, p. 432–437.

ZHONG, S., and MUCCI, A., 1989, Calcite (*sic*) and aragonite precipitation from seawater solution of various salinities: precipitation rate and overgrowth compositions: Chemical Geology, v. 78, p. 283–299.

ZEIGLER, J. M., 1969, Some observations and measurements of wind driven (*sic*) circulation in a shallow coastal lagoon, p. 335–339 *in* Ayala-Castanares, Augustin; and Phleger, F. B., eds., Lagunas costeras, un simposio: Simposio Internacional sobre lagunas costeras, Mexico City, Mexico, 28–30 Noviembre 1967, memoria: Mexico Universidad Nacional Autonoma (WMAS-UNESCO), 686 p.

ZUFFA, G. G., 1980, Hybrid arenites: their composition and classification: Journal of Sedimentary Petrology, v. 50, p. 21–29.

ZUFFA, G. G., ed., 1985, Provenance of arenites: NATO Advanced Scientific Institute. Dordrecht, The Netherlands, D. Reidel Publishing Co

Index

T = table; D = defined; * = figure; *C = figure caption; E = equation; F = formula; R = reference; S = source note.

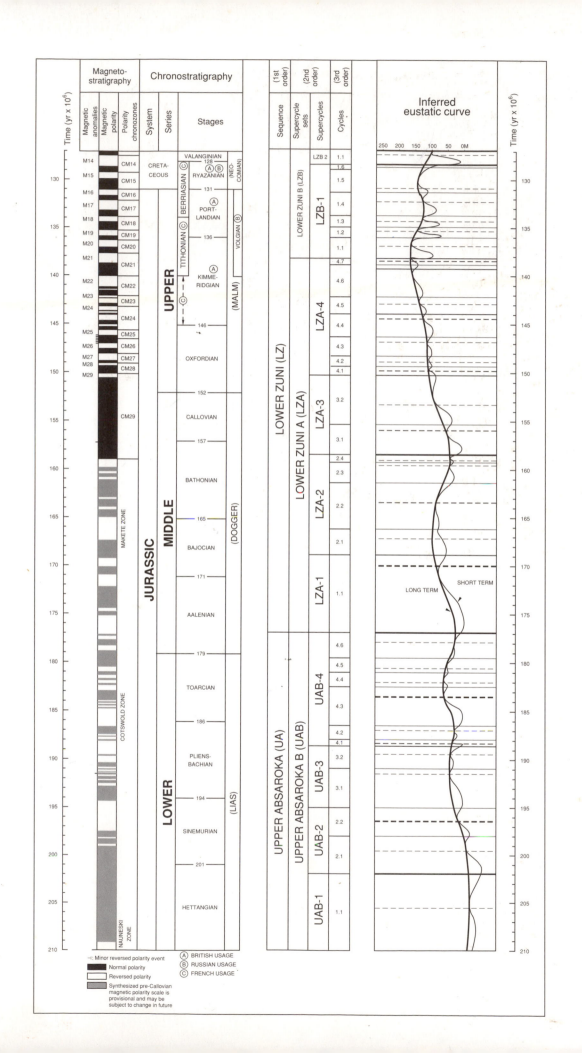